Infinite Dimension

A Hitchhiker's Guide

3rd Edition

Charalambos D. Aliprantis
Kim C. Border

Infinite Dimensional Analysis

A Hitchhiker's Guide

Third Edition

With 38 Figures
and 1 Table

 Springer

Professor Charalambos D. Aliprantis
Department of Economics
Krannert School of Management
Rawls Hall, Room 4003
Purdue University
100 S. Grant Street
West Lafayette IN 47907-2076
USA
E-mail: aliprantis@mgmt.purdue.edu

Professor Kim C. Border
California Institute of Technology
Division of the Humanities and Social Sciences
228-77
1200 E. California Boulevard
Pasadena CA 91125
USA
E-mail: kcborder@caltech.edu

Cataloging-in-Publication Data
Library of Congress Control Number: 2006921177

ISBN-10 3-540-29586-0 3rd ed. Springer Berlin Heidelberg New York
ISBN-13 978-3-540-29586-0 3rd ed. Springer Berlin Heidelberg New York
ISBN 3-540-65854-8 2nd ed. Springer Berlin Heidelberg New York

Springer is a part of Springer Science+Business Media
springeronline.com

© Springer-Verlag Berlin Heidelberg 1999, 2006

Cover design: Erich Kirchner
Production: Helmut Petri
Printing: Strauss Offsetdruck

SPIN 11572817   42/3153 – 5 4 3 2 1 0

In memoriam

Yuri Abramovich
Jeffrey Banks
Taesung Kim
Richard McKelvey

... colleagues, collaborators, friends.

Preface to the third edition

This new edition of *The Hitchhiker's Guide* has benefitted from the comments of many individuals, which have resulted in the addition of some new material, and the reorganization of some of the rest.

The most obvious change is the creation of a separate Chapter 7 on convex analysis. Parts of this chapter appeared in elsewhere in the second edition, but much of it is new to the third edition. In particular, there is an expanded discussion of support points of convex sets, and a new section on subgradients of convex functions. There is much more material on the special properties of convex sets and functions in finite dimensional spaces.

There are improvements and additions in almost every chapter. There is more new material than might seem at first glance, thanks to a change in font that reduced the page count about five percent.

We owe a huge debt to Valentina Galvani, Daniela Puzzello, and Francesco Rusticci, who were participants in a graduate seminar at Purdue University and whose suggestions led to many improvements, especially in chapters five through eight. We particularly thank Daniela Puzzello for catching uncountably many errors throughout the second edition, and simplifying the statements of several theorems and proofs. In another graduate seminar at Caltech, many improvements and corrections were suggested by Joel Grus, PJ Healy, Kevin Roust, Maggie Penn, and Bryan Rogers.

We also thank Gabriele Camera, Chris Chambers, John Duggan, Federico Echenique, Monique Florenzano, Paolo Ghirardato, Dionysius Glycopantis, Aviad Heifetz, John Ledyard, Fabio Maccheroni, Massimo Marinacci, Efe Ok, Uzi Segal, Rabee Tourky, and Nicholas Yannelis for their corrections and questions, their encouragement, and their (not always heeded) advice.

Finally, we acknowledge our intellectual debt to our mentor Wim Luxemburg, and the constant support of the late Yuri Abramovich.

Roko Aliprantis
KC Border
November 2005

Preface to the second edition

In the nearly five years since the publication of what we refer to as *The Hitch-hiker's Guide*, we have been the recipients of much advice and many complaints. That, combined with the economics of the publishing industry, convinced us that the world would be a better place if we published a second edition of our book, and made it available in paperback at a more modest price.

The most obvious difference between the second and the original edition is the reorganization of material that resulted in three new chapters. Chapter 4 collects many of the purely set-theoretical results about measurable structures such as semirings and σ-algebras. The material in this chapter is quite independent from notions of measure and integration, and is easily accessible, so we thought it should come sooner. We also divided the chapter on correspondences into two separate chapters, one dealing with continuity, the other with measurability. The material on measurable correspondences is more detailed and, we hope, better written. We also put many of the representation theorems into their own Chapter 14. This arrangement has the side effect of forcing the renumbering of almost every result in the text, thus rendering the original version obsolete. We feel bad about that, but like Humpty Dumpty, we doubt we could put it back the way it was.

The second most noticeable change is the addition of approximately seventy pages of new material. In particular, there is now an extended treatment of analytic sets in Polish spaces, which is divided among Sections 3.14, 12.5, and 12.6. There is also new material on Borel functions between Polish spaces in Section 4.11, a discussion of Lusin's Theorem 12.8, and a more general treatment of the Kolmogorov Extension Theorem in Section 15.6. There are many other additions through out the text, including a handful of additional figures. The truly neurotic reader may have noticed that by an almost unimaginable stroke of luck every chapter begins on a recto page.

We revised the exposition of numerous proofs, especially those we could no longer follow. We also took the opportunity to expunge dozens of minor errors and misprints, as well as a few moderate errors. We hope that in the process we did not introduce too many new ones. If there are any major errors, neither we nor our students could find them, so they remain.

We thank Victoria Mason at Caltech and Werner Müller, our editor at Springer–Verlag, for their support and assistance.

In addition to all those we thanked in the original edition, we are grateful for conversations (or email) with Jeffrey Banks, Paolo Battigalli, Owen Burkinshaw, John Duggan, Mark Fey, Paolo Ghirardato, Serena Guarnaschelli, Alekos Kechris, Antony Kwasnica, Michel Le Breton, John Ledyard, Massimo Marinacci, Jim

Moore, Frank Page, Ioannis Polyrakis, Nikolaos Sofronidis, Rabee Tourky, Nick Yannelis, ... and especially Yuri Abramovich for his constant encouragement and advice.

Roko Aliprantis
KC Border
May 1999

Preface to the first edition

This text was born out of an advanced mathematical economics seminar at Caltech in 1989–90. We realized that the typical graduate student in mathematical economics has to be familiar with a vast amount of material that spans several traditional fields in mathematics. Much of the material appears only in esoteric research monographs that are designed for specialists, not for the sort of generalist that our students need be. We hope that in a small way this text will make the material here accessible to a much broader audience. While our motivation is to present and organize the analytical foundations underlying modern economics and finance, this is a book of mathematics, not of economics. We mention applications to economics but present very few of them. They are there to convince economists that the material has some relevance and to let mathematicians know that there are areas of application for these results. We feel that this text could be used for a course in analysis that would benefit mathematicians, engineers, and scientists. Most of the material we present is available elsewhere, but is scattered throughout a variety of sources and occasionally buried in obscurity. Some of our results are original (or more likely, independent rediscoveries).

We have included some material that we cannot honestly say is necessary to understand modern economic theory, but may yet prove useful in future research. On the other hand, we wished to finish this work in our children's lifetimes, so we have not presented everything we know, or everything we think that you should learn. You should not conclude that we feel that omitted topics are unimportant. For instance, we make no mention of differentiability, although it is extremely important. We would like to promise a second volume that would address the shortcomings of this one, but the track record of authors making such promises is not impressive, so we shall not bother.

Our choice of material is a bit eccentric and reflects the interaction of our tastes. With apologies to D. Adams [4] we have compiled what we like to describe as a hitchhiker's guide, or low budget touring guide, to analysis. Some of the areas of analysis we explore leisurely on foot (others might say in a pedestrian fashion), other areas we pass by quickly, and still other times we merely point out the road signs that point to interesting destinations we bypass. As with any good hitchhiking adventure, there are detours and probably wrong turns.

We have tried to write this book so that it will be useful as both a reference and a textbook. We do not feel that these goals are antithetical. This means that we sometimes repeat ourselves for the benefit of those who start in the middle, or even at the end. We have also tried to cross-reference our results as much as possible so that it is easy to find the prerequisites. While there are no formal exercises, many of the proofs have gaps indicated by the appearance of the words "How" and "Why." These should be viewed as exercises for you to carry out.

We seize this opportunity to thank Mike Maxwell for his extremely conscientious job of reading the early drafts of this manuscript. He caught many errors and obscurities, and substantially contributed to improving the readability of this text. Unfortunately, his untimely graduation cut short his contributions. We thank Victoria Mason for her valuable support and her catering to our eccentricities. We give special thanks to Don Brown for his moral support, and to Richard Boylan for nagging us to finish. We also thank Wim Luxemburg for his enlightening conversations on difficult issues, and for sharing his grasp of history. We acknowledge beneficial conversations with Yuri Abramovich, Owen Burkinshaw, Alexander Kechris, Taesung Kim, and Nick Yannelis. We thank the participants in the seminar at Caltech: Richard Boylan, Mahmoud El-Gamal, Richard McKelvey, and Jeff Strnad. We also express our gratitude to the following for working through parts of the manuscript and pointing out errors and suggesting improvements: Kay-yut Chen, Yan Chen, John Duggan, Mark Fey, Julian Jamison, John Ledyard, Katya Sherstyuk. Michel Le Breton and Lionel McKenzie prompted us to include some of the material that is here. We thank Werner Müller, our editor at Springer–Verlag, for his efficiency and support. We typed and typeset this text ourselves, so we truly are responsible for *all* errors—mathematical or not.

Don't Panic

Roko Aliprantis
KC Border
May 1994

Contents

Preface to the third edition   **vii**

A foreword to the practical   **xix**

1 Odds and ends   **1**

    1.1   Numbers 1
    1.2   Sets 2
    1.3   Relations, correspondences, and functions 4
    1.4   A bestiary of relations 5
    1.5   Equivalence relations 7
    1.6   Orders and such 7
    1.7   Real functions 8
    1.8   Duality of evaluation 9
    1.9   Infinityies 10
    1.10 The Diagonal Theorem and Russell's Paradox 12
    1.11 The axiom of choice and axiomatic set theory 13
    1.12 Zorn's Lemma 15
    1.13 Ordinals 18

2 Topology   **21**

    2.1   Topological spaces 23
    2.2   Neighborhoods and closures 26
    2.3   Dense subsets 28
    2.4   Nets 29
    2.5   Filters 32
    2.6   Nets and Filters 35
    2.7   Continuous functions 36
    2.8   Compactness 38
    2.9   Nets vs. sequences 41
    2.10 Semicontinuous functions 43
    2.11 Separation properties 44
    2.12 Comparing topologies 47
    2.13 Weak topologies 47
    2.14 The product topology 50
    2.15 Pointwise and uniform convergence 53

2.16   Locally compact spaces 55
2.17   The Stone–Čech compactification 58
2.18   Stone–Čech compactification of a discrete set 63
2.19   Paracompact spaces and partitions of unity 65

3   Metrizable spaces   69

3.1   Metric spaces 70
3.2   Completeness 73
3.3   Uniformly continuous functions 76
3.4   Semicontinuous functions on metric spaces 79
3.5   Distance functions 80
3.6   Embeddings and completions 84
3.7   Compactness and completeness 85
3.8   Countable products of metric spaces 89
3.9   The Hilbert cube and metrization 90
3.10   Locally compact metrizable spaces 92
3.11   The Baire Category Theorem 93
3.12   Contraction mappings 95
3.13   The Cantor set 98
3.14   The Baire space $\mathbb{N}^\mathbb{N}$ 101
3.15   Uniformities 108
3.16   The Hausdorff distance 109
3.17   The Hausdorff metric topology 113
3.18   Topologies for spaces of subsets 119
3.19   The space $C(X, Y)$ 123

4   Measurability   127

4.1   Algebras of sets 129
4.2   Rings and semirings of sets 131
4.3   Dynkin's lemma 135
4.4   The Borel σ-algebra 137
4.5   Measurable functions 139
4.6   The space of measurable functions 141
4.7   Simple functions 144
4.8   The σ-algebra induced by a function 147
4.9   Product structures 148
4.10   Carathéodory functions 153
4.11   Borel functions and continuity 156
4.12   The Baire σ-algebra 158

5   Topological vector spaces   163

5.1   Linear topologies 166
5.2   Absorbing and circled sets 168
5.3   Metrizable topological vector spaces 172
5.4   The Open Mapping and Closed Graph Theorems 175
5.5   Finite dimensional topological vector spaces 177

5.6	Convex sets	181
5.7	Convex and concave functions	186
5.8	Sublinear functions and gauges	190
5.9	The Hahn–Banach Extension Theorem	195
5.10	Separating hyperplane theorems	197
5.11	Separation by continuous functionals	201
5.12	Locally convex spaces and seminorms	204
5.13	Separation in locally convex spaces	207
5.14	Dual pairs	211
5.15	Topologies consistent with a given dual	213
5.16	Polars	215
5.17	$\mathfrak{S}$-topologies	220
5.18	The Mackey topology	223
5.19	The strong topology	223

6 Normed spaces **225**

6.1	Normed and Banach spaces	227
6.2	Linear operators on normed spaces	229
6.3	The norm dual of a normed space	230
6.4	The uniform boundedness principle	232
6.5	Weak topologies on normed spaces	235
6.6	Metrizability of weak topologies	237
6.7	Continuity of the evaluation	241
6.8	Adjoint operators	243
6.9	Projections and the fixed space of an operator	244
6.10	Hilbert spaces	246

7 Convexity **251**

7.1	Extended-valued convex functions	254
7.2	Lower semicontinuous convex functions	255
7.3	Support points	258
7.4	Subgradients	264
7.5	Supporting hyperplanes and cones	268
7.6	Convex functions on finite dimensional spaces	271
7.7	Separation and support in finite dimensional spaces	275
7.8	Supporting convex subsets of Hilbert spaces	280
7.9	The Bishop–Phelps Theorem	281
7.10	Support functionals	288
7.11	Support functionals and the Hausdorff metric	292
7.12	Extreme points of convex sets	294
7.13	Quasiconvexity	299
7.14	Polytopes and weak neighborhoods	300
7.15	Exposed points of convex sets	305

8 Riesz spaces 311

8.1 Orders, lattices, and cones 312
8.2 Riesz spaces 313
8.3 Order bounded sets 315
8.4 Order and lattice properties 316
8.5 The Riesz decomposition property 319
8.6 Disjointness 320
8.7 Riesz subspaces and ideals 321
8.8 Order convergence and order continuity 322
8.9 Bands 324
8.10 Positive functionals 325
8.11 Extending positive functionals 330
8.12 Positive operators 332
8.13 Topological Riesz spaces 334
8.14 The band generated by E' 339
8.15 Riesz pairs 340
8.16 Symmetric Riesz pairs 342

9 Banach lattices 347

9.1 Fréchet and Banach lattices 348
9.2 The Stone–Weierstrass Theorem 352
9.3 Lattice homomorphisms and isometries 353
9.4 Order continuous norms 355
9.5 AM- and AL-spaces 357
9.6 The interior of the positive cone 362
9.7 Positive projections 364
9.8 The curious AL-space BV_0 365

10 Charges and measures 371

10.1 Set functions 374
10.2 Limits of sequences of measures 379
10.3 Outer measures and measurable sets 379
10.4 The Carathéodory extension of a measure 381
10.5 Measure spaces 387
10.6 Lebesgue measure 389
10.7 Product measures 391
10.8 Measures on $\mathbb{R}^n$ 392
10.9 Atoms 395
10.10 The AL-space of charges 396
10.11 The AL-space of measures 399
10.12 Absolute continuity 401

11 Integrals   403

11.1 The integral of a step function 404
11.2 Finitely additive integration of bounded functions 406
11.3 The Lebesgue integral 408
11.4 Continuity properties of the Lebesgue integral 413
11.5 The extended Lebesgue integral 416
11.6 Iterated integrals 418
11.7 The Riemann integral 419
11.8 The Bochner integral 422
11.9 The Gelfand integral 428
11.10 The Dunford and Pettis integrals 431

12 Measures and topology   433

12.1 Borel measures and regularity 434
12.2 Regular Borel measures 438
12.3 The support of a measure 441
12.4 Nonatomic Borel measures 443
12.5 Analytic sets 446
12.6 The Choquet Capacity Theorem 456

13 L_p-spaces   461

13.1 L_p-norms 462
13.2 Inequalities of Hölder and Minkowski 463
13.3 Dense subspaces of L_p-spaces 466
13.4 Sublattices of L_p-spaces 467
13.5 Separable L_1-spaces and measures 468
13.6 The Radon–Nikodym Theorem 469
13.7 Equivalent measures 471
13.8 Duals of L_p-spaces 473
13.9 Lyapunov's Convexity Theorem 475
13.10 Convergence in measure 479
13.11 Convergence in measure in L_p-spaces 481
13.12 Change of variables 483

14 Riesz Representation Theorems   487

14.1 The AM-space $B_b(\Sigma)$ and its dual 488
14.2 The dual of $C_b(X)$ for normal spaces 491
14.3 The dual of $C_c(X)$ for locally compact spaces 496
14.4 Baire vs. Borel measures 498
14.5 Homomorphisms between $C(X)$-spaces 500

15 Probability measures   **505**

15.1   The weak∗ topology on $\mathcal{P}(X)$ 506
15.2   Embedding X in $\mathcal{P}(X)$ 512
15.3   Properties of $\mathcal{P}(X)$ 513
15.4   The many faces of $\mathcal{P}(X)$ 517
15.5   Compactness in $\mathcal{P}(X)$ 518
15.6   The Kolmogorov Extension Theorem 519

16 Spaces of sequences   **525**

16.1   The basic sequence spaces 526
16.2   The sequence spaces $\mathbb{R}^{\mathbb{N}}$ and φ 527
16.3   The sequence space c_0 529
16.4   The sequence space c 531
16.5   The ℓ_p-spaces 533
16.6 ℓ_1 and the symmetric Riesz pair $\langle \ell_\infty, \ell_1 \rangle$ 537
16.7   The sequence space ℓ_∞ 538
16.8   More on $\ell_\infty' = ba(\mathbb{N})$ 543
16.9   Embedding sequence spaces 546
16.10 Banach–Mazur limits and invariant measures 550
16.11 Sequences of vector spaces 552

17 Correspondences   **555**

17.1   Basic definitions 556
17.2   Continuity of correspondences 558
17.3   Hemicontinuity and nets 563
17.4   Operations on correspondences 566
17.5   The Maximum Theorem 569
17.6   Vector-valued correspondences 571
17.7   Demicontinuous correspondences 574
17.8   Knaster–Kuratowski–Mazurkiewicz mappings 577
17.9   Fixed point theorems 581
17.10 Contraction correspondences 585
17.11 Continuous selectors 587

18 Measurable correspondences   **591**

18.1   Measurability notions 592
18.2   Compact-valued correspondences as functions 597
18.3   Measurable selectors 600
18.4   Correspondences with measurable graph 606
18.5   Correspondences with compact convex values 609
18.6   Integration of correspondences 614

19 Markov transitions   **621**

    19.1   Markov and stochastic operators 623

    19.2   Markov transitions and kernels 625

    19.3   Continuous Markov transitions 631

    19.4   Invariant measures 631

    19.5   Ergodic measures 636

    19.6   Markov transition correspondences 638

    19.7   Random functions 641

    19.8   Dilations 645

    19.9   More on Markov operators 650

    19.10 A note on dynamical systems 652

20 Ergodicity   **655**

    20.1   Measure-preserving transformations and ergodicity 656

    20.2   Birkhoff's Ergodic Theorem 659

    20.3   Ergodic operators 661

References   **667**

Index   **681**

A foreword to the practical

Why use infinite dimensional analysis?

Why should practical people, such as engineers and economists, learn about infinite dimensional spaces? Isn't the world finite dimensional? How can infinite dimensional analysis possibly help to understand the workings of real economies?

Infinite dimensional models have become prominent in economics and finance because they capture natural aspects of the world that cannot be examined in finite dimensional models. It has become clear in the last couple of decades that economic models capable of addressing real policy questions must be both stochastic and dynamic. There are fundamental aspects of the economy that static models cannot capture. Deterministic models, even chaotically deterministic models, seem unable to explain our observations of the world.

Dynamic models require infinite dimensional spaces. If time is modeled as continuous, then time series of economic data reside in infinite dimensional function spaces. Even if time is modeled as being discrete, there is no natural terminal period. Furthermore, models including fiat money with a terminal period lead to conclusions that are not tenable. If we are to make realistic models of money or growth, we are forced to use infinite dimensional models.

Another feature of the world that arguably requires infinite dimensional modeling is uncertainty. The future is uncertain, and infinitely many resolutions of this uncertainty are conceivable. The study of financial markets requires models that are both stochastic and dynamic, so there is a double imperative for infinite dimensional models.

There are other natural contexts in which infinite dimensional models are natural. A prominent example is commodity differentiation. While there are only finitely many types of commodities actually traded and manufactured, there are conceivably infinitely many that are not. Any theory that hopes to explain which commodities are manufactured and marketed and which are not must employ infinite dimensional analysis. A special case of commodity differentiation is the division of land. There are infinitely many ways to subdivide a parcel of land, and each subdivision can be regarded as a separate commodity.

Let us take a little time to briefly introduce some infinite dimensional spaces commonly used in economics. We do not go into any detail on their properties here—indeed we may not even define all our terms. We introduce these spaces now as a source of examples. In their own way each of these spaces can be thought of as an infinite dimensional generalization of the finite dimensional Euclidean space $\mathbb{R}^n$, and each of them captures some salient aspects of $\mathbb{R}^n$.

Spaces of sequences

When time is modeled as a sequence of discrete dates, then economic time series are sequences of real numbers. A particularly important family of sequence spaces is the family of ℓ_p-spaces. For $1 \leqslant p < \infty$, ℓ_p is defined to be the set of all sequences $x = (x_1, x_2, \ldots)$ for which $\sum_{n=1}^{\infty} |x_n|^p < \infty$. The the ℓ_p-**norm** of the sequence x is the number $\|x\|_p = \left(\sum_{n=1}^{\infty} |x_n|^p\right)^{1/p}$.

As p becomes larger, the larger values of x_n tend to dominate in the calculation of the ℓ_p-norm and indeed, $\lim_{p \to \infty} \|x\|_p = \sup\{|x_n|\}$. This brings us to ℓ_∞. This space is defined to be the set of all real sequences $x = (x_1, x_2, \ldots)$ satisfying $\sup\{|x_n|\} < \infty$. This supremum is called the ℓ_∞-**norm** of x and is denoted $\|x\|_\infty$. This norm is also called the **supremum norm** or sometimes the **uniform norm**, because a sequence of sequences converges uniformly to a limiting sequence in ℓ_∞ if and only if it converges in this norm.

All of these spaces are vector spaces under the usual (pointwise) addition and scalar multiplication. Furthermore, these spaces are nested. If $p \leqslant q$, then $\ell_p \subset \ell_q$.

There are a couple of other sequence spaces worth noting. The space of all convergent sequences is denoted c. The space of all sequences converging to zero is denoted c_0. Finally the collection of all sequences with only finitely many nonzero terms is denoted φ. All of these collections are vector spaces too, and for $1 \leqslant p < \infty$ we have the following vector subspace inclusions:

$$\varphi \subset \ell_p \subset c_0 \subset c \subset \ell_\infty \subset \mathbb{R}^{\mathbb{N}}.$$

Chapter 16 discusses the properties of these spaces at length.

The space ℓ_∞ plays a major role in the neoclassical theory of growth. Under commonly made assumptions in the one sector growth model, capital/labor ratios are uniformly bounded over time. If there is an exhaustible resource in fixed supply, then ℓ_1 may be an appropriate setting for time series.

Spaces of functions

One way to think of $\mathbb{R}^n$ is as the set of all real functions on $\{1, \ldots, n\}$. If we replace $\{1, \ldots, n\}$ by an arbitrary set X, the set of all real functions on X, denoted $\mathbb{R}^X$, is a natural generalization of $\mathbb{R}^n$. In fact, sequence spaces are a special case of function spaces, where X is the set of natural numbers $\{1, 2, 3, \ldots\}$. When X has a topological structure (see Chapter 2), it may be acceptable to restrict attention to $C(X)$, the continuous real functions on X.

Function spaces arise in models of uncertainty. In this case X represents the set of *states of the world*. Functions on X are then state-contingent variables. In statistical modeling it is common practice to denote the set of states by Ω and to endow it with additional structure, namely a σ-algebra Σ and a probability measure μ. In this case it is natural to consider the L_p-spaces. For $1 \leqslant p < \infty$, $L_p(\mu)$ is defined to be the collection of all (μ-equivalence classes of) μ-measurable functions f for which $\int_\Omega |f|^p \, d\mu < \infty$. (These terms are all explained in Chapter 11. It is okay to think of these integrals as $\int_0^1 |f(x)|^p \, dx$ for now.) The number $\|f\|_p = \left(\int_\Omega |f|^p \, d\mu\right)^{1/p}$ is the L_p**-norm** of f. The L_∞-norm is defined by

$$\|f\|_\infty = \operatorname{ess\,sup} f = \sup \{t : \mu(\{x : |f(x)| \geqslant t\}) > 0\}.$$

This norm is also known as the **essential supremum** of f. The space L_∞ is the space of all μ-measurable functions with finite essential supremum. Chapter 13 covers the L_p-spaces.

Spaces of measures

Given a vector x in $\mathbb{R}^n$ and a subset A of indices $\{1, \ldots, n\}$ define the set function $x(A) = \sum_{i \in A} x_i$. If $A \cap B = \varnothing$, then $x(A \cup B) = x(A) + x(B)$. In this way we can think of $\mathbb{R}^n$ as the collection of additive functions on the subsets of $\{1, \ldots, n\}$. The natural generalization of $\mathbb{R}^n$ from this point of view is to consider the spaces of measures or charges on an algebra of sets. (These terms are all defined in Chapter 11.) Spaces of measures on topological spaces can inherit some of the properties from the underlying space. For instance, the space of Borel probability measures on a compact metrizable space is naturally a compact metrizable space. Results of this sort are discussed in Chapters 12 and 15.

The compactness properties of spaces of measures makes them good candidates for commodity spaces for models of commodity differentiation. They are also central to models of stochastic dynamics, which are discussed in Chapter 19.

Spaces of sets

Since set theory can be used as the foundation of almost all mathematics, spaces of sets subsume everything else. In Chapter 3 we discuss natural ways of topologizing spaces of subsets of metrizable spaces. These results are also used in Chapter 17 to discuss continuity and measurability of correspondences. The topology of **closed convergence** of sets has proven to be useful as a way of topologizing preferences and demand correspondences. Topological spaces of sets have also been used in the theory of incentive contracts.

Prerequisites

The main prerequisite is what is often called "mathematical sophistication." This is hard to define, but it includes the ability to manipulate abstract concepts, and an understanding of the notion of "proof."

We assume that you know the basic facts about the standard model of the real numbers. These include the fact that between any two distinct real numbers there is a rational number and also an irrational number. (You can see that we already assume you know what these are. It was only a few centuries ago that this knowledge was highly protected.) We take for granted that the real numbers are complete. We assume you know what it means for sequences and series of real numbers to converge. We trust you are familiar with naïve set theory and its notation. We assume that you are familiar with arguments using induction. We hope that you are familiar with the basic results about metric spaces. Aliprantis and Burkinshaw [13, Chapter 1], Dieudonné [97, Chapter 3], and Rudin [292, Chapter 2] are excellent expositions of the theory of metric spaces. It would be nice, but not necessary, if you had heard of the Lebesgue integral; we define it in Chapter 11. We assume that you are familiar with the concept of a vector space. A good brief reference for vector spaces is Apostol [17]. A more detailed reference is Halmos [147].

Chapter 1

Odds and ends

One purpose of this chapter is to standardize some terminology and notation. In particular, Definition 1.1 defines what we mean by the term "function space," and Section 1.4 introduces a number of kinds of binary relations. We also use this chapter to present some useful odds and ends that should be a part of everyone's mathematical tool kit, but which don't conveniently fit anywhere else. We introduce correspondences and the notion of the evaluation duality. Our presentation is informal and we do not prove many of our claims. We also feel free to get ahead of ourselves and refer to definitions and examples that appear much later on.

We do prove a few theorems including Szpilrajn's Extension Theorem 1.9 for partial preorders, the existence of a Hamel basis (Theorem 1.8), and the Knaster–Tarski Fixed Point Theorem 1.10. These are presented as applications of Zorn's Lemma 1.7. Example 1.4 uses a standard cardinality argument to show that the lexicographic order cannot be represented by a numerical function.

We also try to present the flavor of the subtleties of modern set theory without actually proving the results. We do however prove Cantor's Diagonal Theorem 1.5 and describe Russell's Paradox. We mention some of the more esoteric aspects of the Axiom of Choice in Section 1.11 in order to convince you that you really do want to put up with it, and all it entails, such as non-measurable sets (Corollary 10.42). We also introduce the ordinals in Section 1.13.

1.1 Numbers

Leopold Kronecker is alleged to have remarked that, "God made the integers, all the rest is the work of man."[1] The **natural numbers** are $1, 2, 3, \ldots$, etc., and the set of natural numbers is denoted $\mathbb{N}$. (Some authors consider zero to be a natural number as well, and there are times we may do likewise.) We do not attempt to develop a construction of the real numbers, or even the natural numbers here. A very readable development may be found in E. Landau [221] or C. D. Aliprantis and O. Burkinshaw [13, Chapter 1].

---

[1] According to E. T. Bell [36, p. 477].

We use the symbol $\mathbb{R}$ to denote the set of real numbers, and may refer to the set of real numbers as the **real line**. We use the standard symbols $\mathbb{Z}$ for the integers, and $\mathbb{Q}$ for the rational numbers. We take for granted many of the elementary properties of the real numbers. For instance: Between any two distinct real numbers there are both a rational number and an irrational number. Any nonempty bounded set of real numbers has both an infimum and a supremum. Any nonempty set of nonnegative integers has a least element.

We have occasion to use the **extended real number** system $\mathbb{R}^*$. This is the set of real numbers together with the entities ∞ (infinity) and $-\infty$ (negative infinity). These have the property that $-\infty < r < \infty$ for any real number $r \in \mathbb{R}$. They also satisfy the following arithmetic conventions:

$$r + \infty = \infty \quad \text{and} \quad r - \infty = -\infty;$$
$$\infty \cdot r = \infty \text{ if } r > 0 \quad \text{and} \quad \infty \cdot r = -\infty \text{ if } r < 0;$$
$$\infty \cdot 0 = 0;$$

for any real r. The combination $\infty - \infty$ of symbols has no meaning. The symbols ∞ and $-\infty$ are not really meant to be used for arithmetic, they are only used to avoid awkward expressions involving infima and suprema.[2]

1.2   Sets

Informally, a set is a collection of objects. In most versions of set theory, these objects are themselves sets. Even numbers are viewed as sets. We employ the following commonly used set theoretic notation. We expect that this is familiar material, and only mention it to make sure we are all using the same notation. For variety's sake, we may use the term **family** or **collection** in place of the term **set**. The expression $x \in A$ means that x **belongs** to the set A, and $x \notin A$ means that it does not. We may also say that x is a **member** of A, a **point** in A, or an **element** of A, or that A **contains** x if $x \in A$. Two sets are equal if they have the same members. The symbol $\varnothing$ denotes the **empty set**, the set with no members. The expression $X \setminus A$ denotes the **complement** of A in X, that is, the set $\{x \in X : x \notin A\}$. When the reference set X is understood, we may simply write A^c.

The symbols $A \subset B$ or $B \supset A$ mean that the set A is a **subset** of the set B or B is a **superset** of A, that is, $x \in A$ implies $x \in B$. We also say in this case that B

---

[2] Do not confuse the extended reals with "nonstandard" models of the real numbers. Nonstandard models of the real numbers contain *infinitesimals* (positive numbers that are smaller than every standard positive real number) and *infinitely large* numbers (numbers that are larger than every standard real number), yet nevertheless obey all the rules of real arithmetic (in an appropriately formulated language). See, for instance, R. F. Hoskins [169], A. E. Hurd and P. A. Loeb [173], or K. D. Stroyan and W. A. J. Luxemburg [326] for a good introduction to nonstandard analysis.

includes A.[3] In particular, $A \subset B$ allows for the possibility that $A = B$. If we wish to exclude the possibility that $A = B$ we say that A is a **proper subset** of B, or that B **properly includes** A, or write $A \subsetneq B$.

The **union** of A and B, $\{x : x \in A \text{ or } x \in B\}$, is denoted $A \cup B$. Their **intersection**, $\{x : x \in A \text{ and } x \in B\}$, is denoted $A \cap B$. We say that A and B are **disjoint** if $A \cap B = \varnothing$ and that A **meets** B if $A \cap B \neq \varnothing$. If $\mathcal{A}$ is a set of sets, then $\bigcup \mathcal{A}$ or $\bigcup \{A : A \in \mathcal{A}\}$ or $\bigcup_{A \in \mathcal{A}} A$ denotes the union of all the sets in $\mathcal{A}$, that is, $\{x : x \in A \text{ for some } A \in \mathcal{A}\}$. In particular, $\bigcup \varnothing = \varnothing$. If $\mathcal{A}$ is nonempty, then $\bigcap \mathcal{A}$ denotes the intersection of all sets in $\mathcal{A}$. (There is a serious difficulty assigning meaning to $\bigcap \varnothing$, so we leave it undefined. The problem is that the conditional $(A \in \varnothing \implies x \in A)$ is vacuously true for any x, but there is no set that contains every x. See Section 1.11 below.) We let $A \triangle B$ denote the **symmetric difference** of A and B, defined by $A \triangle B = (A \setminus B) \cup (B \setminus A)$.

The **power set** of a set X is the collection of all subsets of X, denoted 2^X. Nonempty subsets of 2^X are usually called families of sets, and they are often written as an **indexed family**, that is, in the form $\{A_i\}_{i \in I}$, where I is called the **index set**. Given a nonempty subset $\mathcal{C}$ of 2^X, we can write it as an indexed family with $I = \mathcal{C}$ and $A_i = i$ for each $i \in I$. For any nonempty family $\{A_i\}_{i \in I}$ of subsets of a set X, we have the following useful identities known as **de Morgan's laws**:

$$\left(\bigcup_{i \in I} A_i \right)^c = \bigcap_{i \in I} A_i^c \quad \text{and} \quad \left(\bigcap_{i \in I} A_i \right)^c = \bigcup_{i \in I} A_i^c.$$

The **Cartesian product** $\prod_{i \in I} A_i$ of a family $\{A_i\}_{i \in I}$ of sets is the collection of all I-tuples $\{x_i\}_{i \in I}$, where of course, each x_i satisfies $x_i \in A_i$. Each set A_i is a **factor** in the product.[4] We may also write

$$A_1 \times A_2 \times \cdots \times A_n$$

for the Cartesian product $\prod_{i=1}^{n} A_i$. In the product $A \times A$ of a set with itself, the set $\{(x, x) : x \in A\}$ is called the **diagonal**. Some useful identities are

$$A \times \left(\bigcap_{i \in I} B_i \right) = \bigcap_{i \in I} (A \times B_i) \quad \text{and} \quad A \times \left(\bigcup_{i \in I} B_i \right) = \bigcup_{i \in I} (A \times B_i).$$

Also, for subsets A and B of a vector space, we define the **algebraic sum** $A + B$ to be $\{a + b : a \in A, b \in B\}$. The **scalar multiple** αA is defined to be $\{\alpha a : a \in A\}$ for any scalar α. Note that a careful reading of the definition implies $A + \varnothing = \varnothing$ for any set A.

---

[3] Ideally, one would never say "A contains B" (meaning $B \in A$) when one intends "A includes B" (meaning $B \subset A$), but it happens, and usually no harm is done.

[4] Note that $A \times (B \times C)$, $(A \times B) \times C$, and $A \times B \times C$ are three distinct sets, cf. K. J. Devlin [91], but there is an obvious identification, so we shan't be picky about distinguishing them.

1.3   Relations, correspondences, and functions

Given two sets X and Y, we can form the Cartesian product $X \times Y$, which is the collection of ordered pairs of elements from X and Y. (We assume you know what ordered pairs are and do not give a formal definition.) A **relation** between members of X and members of Y can be thought of as a subset of $X \times Y$.[5] A relation between members of X is called a **binary relation** on X. For a binary relation R on a set X, that is, $R \subset X \times X$, it is customary to write $x \, R \, y$ rather than $(x, y) \in R$.

A near synonym for relation is **correspondence**, but the connotation is much different. We think of a correspondence φ from X to Y as associating to each x in X a subset $\varphi(x)$ of Y, and we write $\varphi \colon X \twoheadrightarrow Y$. The **graph** of φ, denoted $\mathrm{Gr}\,\varphi$ is $\{(x, y) \in X \times Y : y \in \varphi(x)\}$. The space X is the **domain** of the correspondence and Y is the **codomain**. Given a subset $A \subset X$, the **image** $\varphi(A)$ of A under φ is defined by $\varphi(A) = \bigcup \{\varphi(x) : x \in A\}$. The **range** of φ is the image of X itself. We may occasionally call Y the **range space** of φ. When the range space and the domain are the same, we say that a point x is a **fixed point** of the correspondence φ if $x \in \varphi(x)$. We have a lot more to say about correspondences in Chapters 17 and 18.

A special kind of relation is a **function**. A relation R between X and Y is a function if $(x, y) \in R$ and $(x, z) \in R$ imply $y = z$. A function is sometimes called a **mapping** or **map**. We think of a function f from X into Y as "mapping" each point x in X to a point $f(x)$ in Y, and we write $f \colon X \to Y$. We may also write $x \mapsto f(x)$ to refer to the function f. The **graph** of f, denoted $\mathrm{Gr}\,f$ is $\{(x, y) \in X \times Y : y = f(x)\}$. As with correspondences, the space X is the **domain** of the function and Y is the **codomain**. Given a subset $A \subset X$, the **image** of A under f is $f(A) = \{f(x) : x \in A\}$. The **range** of f is the image of X itself. When the range space and the domain are the same, we say that a point x is a **fixed point** of the function f if $x = f(x)$.

The graph of a function f is also the graph of a singleton-valued correspondence φ defined by $\varphi(x) = \{f(x)\}$, and vice versa. Clearly f and φ represent the same relation, but their values are not exactly the same objects.

A **partial function** from X to Y is a function from a subset of X to Y. If $f \colon X \to Y$ and $A \subset X$, then $f|_A$ is the **restriction** of f to A. That is, $f|_A$ has domain A, and for each $x \in A$, $f|_A(x) = f(x)$. We also say that f is an **extension** of $f|_A$.

A function $x \colon \mathbb{N} \to X$, from the natural numbers to the set X, is called a **sequence** in the set X. The traditional way to denote the value $x(n)$ is x_n, and it is called the nth **term of the sequence**. Using an abused (standard) notation, we shall denote the sequence x by $\{x_n\}$, and we shall consider it both as a function and

---

[5] Some authors, e.g., N. Bourbaki [62] and K. J. Devlin [91] pointedly make a distinction between a relation, which is a linguistic notion, and the set of ordered pairs that stand in that relation to each other, which is a set theoretic construct. In practice, there does not seem to be a compelling reason to be so picky.

as its range—a subset of X. A **subsequence** of a sequence $\{x_n\}$ is a sequence $\{y_n\}$ for which there exists a strictly increasing sequence $\{k_n\}$ of natural numbers (that is, $1 \leqslant k_1 < k_2 < k_3 < \cdots$) such that $y_n = x_{k_n}$ holds for each n.

The **indicator function** (or **characteristic function**) χ_A of a subset A of X is defined by $\chi_A(x) = 1$ if $x \in A$ and $\chi_A(x) = 0$ if $x \notin A$. The set of all functions from X to Y is denoted Y^X. Recall that the power set of X is denoted 2^X. This is also the notation for the set of all functions from X into $2 = \{0, 1\}$. The rationale for this is that every subset A of X can be identified with its characteristic function χ_A, which assumes only the values 0 and 1.

If $f \colon X \to Y$ and $g \colon Y \to Z$, the **composition** of g with f, denoted $g \circ f$, is the function from X to Z defined by the formula $(g \circ f)(x) = g(f(x))$. We may also draw the accompanying sort of diagram to indicate that $h = g \circ f$. We sometimes say that this **diagram commutes** as another way of saying $h = g \circ f$.

More generally, for any two relations $R \subset X \times Y$ and $S \subset Y \times Z$, the composition relation $S \circ R$ is defined by

$$S \circ R = \{(x, z) \in X \times Z : \exists y \in Y \text{ with } (x, y) \in R \text{ and } (y, z) \in S\}.$$

A function $f \colon X \to Y$ is **one-to-one**, or an **injection**, if for every y in the range space, there is at most one x in the domain satisfying $y = f(x)$. The function f maps X **onto** Y, or is a **surjection**, if for every y in Y, there is some x in X with $f(x) = y$. A **bijection** is a one-to-one onto function. A bijection may sometimes be referred to as a **one-to-one correspondence**. The **inverse image**, or simply **inverse**, of a subset A of Y under f, denoted $f^{-1}(A)$, is the set of x with $f(x) \in A$. If f is one-to-one, the inverse image of a singleton is either a singleton or empty, and there is a function $g \colon f(X) \to X$, called the **inverse** of f, that satisfies $x = g(y)$ if and only if $f(x) = y$. The inverse function is usually denoted f^{-1}. Note that we may write $f^{-1}(y)$ to denote the inverse image of the singleton $\{y\}$ even if the function f is not one-to-one.

You should verify that the inverse image preserves the set theoretic operations. That is,

$$f^{-1}\left(\bigcap_{i \in I} A_i\right) = \bigcap_{i \in I} f^{-1}(A_i), \qquad f^{-1}\left(\bigcup_{i \in I} A_i\right) = \bigcup_{i \in I} f^{-1}(A_i),$$
$$f^{-1}(A \setminus B) = f^{-1}(A) \setminus f^{-1}(B).$$

1.4 A bestiary of relations

There are many conditions placed on binary relations in various contexts, and we summarize a number of them here. Some we have already mentioned above. We gather them here largely to standardize our terminology. Not all authors use the same terminology that we do. Each of these definitions should be interpreted as if prefaced by the appropriate universal quantifiers "for every x, y, z," etc. The symbol $\neg$ indicates negation, and a compound expression such as $x R y$ and $y R z$ may be abbreviated $x R y R z$.

A binary relation R on a set X is:

- **reflexive** if $x R x$.

- **irreflexive** if $\neg(x R x)$.

- **symmetric** if $x R y$ implies $y R x$. Note that this does not imply reflexivity.

- **asymmetric** if $x R y$ implies $\neg(y R x)$. An asymmetric relation is irreflexive.

- **antisymmetric** if $x R y$ and $y R x$ imply $x = y$. An antisymmetric relation may or may not be reflexive.

- **transitive** if $x R y$ and $y R z$ imply $x R z$.

- **complete**, or **connected**, if either $x R y$ or $y R x$ or both. Note that a complete relation is reflexive.

- **total**, or **weakly connected**, if $x \neq y$ implies either $x R y$ or $y R x$ or both. Note that a total relation may or may not be reflexive. Some authors call a total relation complete.

- a **partial order** if it is reflexive, transitive, and antisymmetric. Some authors (notably J. L. Kelley [198]) do not require a partial order to be reflexive.

- a **linear order** if it is total, transitive, and antisymmetric; a total partial order, if you will. It obeys the following **trichotomy law**: For every pair x, y exactly one of $x R y$, $y R x$, or $x = y$ holds.

- an **equivalence relation** if it is reflexive, symmetric, and transitive.

- a **preorder**, or **quasiorder**, if it is reflexive and transitive. An antisymmetric preorder is a partial order.

- the **symmetric part** of the relation S if $x R y \iff (x S y \ \& \ y S x)$.

- the **asymmetric part** of the relation S if $x R y \iff (x S y \ \& \ \neg y S x)$.

- the **transitive closure** of the relation S when $x R y$ whenever either $x S y$ or there is a finite set $\{x_1, \ldots, x_n\}$ such that $x S x_1 S x_2 \cdots x_n S y$. The transitive closure of S is the intersection of all the transitive relations (as sets of ordered pairs) that include S. (Note that the relation $X \times X$ is transitive and includes S, so we are not taking the intersection of the empty set.)

1.5 Equivalence relations

Equivalence relations are among the most important. As defined above, an **equivalence relation** on a set X is a reflexive, symmetric, and transitive relation, often denoted $\sim$. Here are several familiar equivalence relations.

- Equality is an equivalence relation.

- For functions on a measure space, almost everywhere equality is an equivalence relation.

- In a semimetric space (X, d), the relation defined by $x \sim y$ if $d(x, y) = 0$ is an equivalence relation.

- Given any function f with domain X, we can define an equivalence relation $\sim$ on X by $x \sim y$ whenever $f(x) = f(y)$.

Given an equivalence relation $\sim$ on a set X we define the **equivalence class** $[x]$ of x by $[x] = \{y : y \sim x\}$. If $x \sim y$, then $[x] = [y]$; and if $x \nsim y$, then $[x] \cap [y] = \emptyset$. The $\sim$-equivalence classes thus partition X into disjoint sets. The collection of $\sim$-equivalence classes of X is called the **quotient of X modulo $\sim$**, often written as $X/\sim$. The function $x \mapsto [x]$ is called the **quotient mapping**. In many contexts, we **identify** the members of an equivalence class. What we mean by this is that we write X instead of $X/\sim$, and we write x instead of $[x]$. Hopefully, you (and we) will not become confused and make any mistakes when we do this. As an example, if we identify elements of a semimetric space as described above, the quotient space becomes a true metric space in the obvious way. In fact, all the L_p-spaces are quotient spaces defined in this manner.

A **partition** $\{D_i\}_{i \in I}$ of a set X is a collection of nonempty subsets of X satisfying $D_i \cap D_j = \emptyset$ for $i \neq j$ and $\bigcup_{i \in I} D_i = X$. Every partition defines an equivalence relation on X by letting $x \sim y$ if $x, y \in D_i$ for some i. In this case, the equivalence classes are precisely the sets D_i.

1.6 Orders and such

A **partial order** (or **partial ordering**, or simply **order**) is a reflexive, transitive, and antisymmetric binary relation. It is traditional to use a symbol like $\geqslant$ to denote a partial order. The expressions $x \geqslant y$ and $y \leqslant x$ are synonyms. A set X equipped with a partial order is a **partially ordered set**, sometimes called a **poset**. Two elements x and y in a partially ordered set are **comparable** if either $x \geqslant y$ or $y \geqslant x$ (or both, in which case $x = y$). A **total order** or **linear order** $\geqslant$ is a partial order where every two elements are comparable. That is, a total order is a partial order that is total. A **chain** in a partially ordered set is a subset that is totally ordered—any two elements of a chain are comparable. In a partially ordered set the notation $x > y$ means $x \geqslant y$ and $x \neq y$. The **order interval** $[x, y]$ is the set $\{z \in X : x \leqslant z \leqslant y\}$. Note that if $y \not\geqslant x$, then $[x, y] = \varnothing$.

Let $(X, \geqslant)$ be a partially ordered set. An **upper bound** for a set $A \subset X$ is an element $x \in X$ satisfying $x \geqslant y$ for all $y \in A$. An element x is a **maximal element** of X if there is no y in X for which $y > x$. Similarly, a **lower bound** for A is an $x \in X$ satisfying $y \geqslant x$ for all $y \in A$. **Minimal elements** are defined analogously. A **greatest element** of A is an $x \in A$ satisfying $x \geqslant y$ for all $y \in A$. **Least elements** are defined in the obvious fashion. Clearly a nonempty subset of X has at most one greatest element and a greatest element if it exists is maximal. If the partial order is complete, then a maximal element is also the greatest. The **supremum** of a set is its least upper bound and the **infimum** is its greatest lower bound. The supremum and infimum of a set need not exist. We write $x \vee y$ for the supremum, and $x \wedge y$ for the infimum, of the two point set $\{x, y\}$. For linear orders, $x \vee y = \max\{x, y\}$ and $x \wedge y = \min\{x, y\}$. A **lattice** is a partially ordered set in which every pair of elements has a supremum and an infimum. It is easy to show (by induction) that every finite set in a lattice has a supremum and an infimum. A **sublattice** of a lattice is a subset that is closed under pairwise infima and suprema. A **complete lattice** is a lattice in which every nonempty subset A has a supremum $\bigvee A$ and an infimum $\bigwedge A$. In particular, a complete lattice itself has an infimum, denoted 0, and a supremum denoted 1. The monograph by D. M. Topkis [331] provides a survey of some of the uses of lattices in economics.

A function $f : X \to Y$ between two partially ordered sets is **monotone** if $x \geqslant y$ in X implies $f(x) \geqslant f(y)$ in Y. Some authors use the term **isotone** instead. The function f is **strictly monotone** if $x > y$ in X implies $f(x) > f(y)$ in Y. Monotone functions are also called **increasing** or **nondecreasing function.**[6] We may also say that f is **decreasing** or **nonincreasing** if $x \geqslant y$ in X implies $f(y) \geqslant f(x)$ in Y. Strictly decreasing functions are defined in the obvious way.

1.7   Real functions

A function whose range space is the real numbers is called a **real function** or a **real-valued function**. A function whose range space is the extended real numbers is called an **extended real function**. If an extended real function satisfies $f(x) = 0$ for all x in a set A, we say that f **vanishes** on A. Or if $x \notin B$ implies $f(x) = 0$, we say that f **vanishes outside** B. For traditional reasons we also use the term **functional** to indicate a real linear or sublinear function on a vector space. (These terms are defined in Chapter 5.)

The **epigraph** of an (extended) real function f on a set X, denoted epi f, is the set in $X \times \mathbb{R}$ defined by epi $f = \{(x, \alpha) \in X \times \mathbb{R} : \alpha \geqslant f(x)\}$. That is, epi f is the set of points lying on or above the graph of f. Notice that if $f(x) = \infty$, then the

---

[6] We use this terminology despite the fact, as D. M. Topkis [331] points out, the negation of "f is increasing" is not "f is nonincreasing." Do you see why?

pair (x, ∞) does not belong to the epigraph of f. Consequently the epigraph of the constant function $f = \infty$ is the empty set. The **hypograph** or **subgraph** of f is the set $\{(x, \alpha) \in X \times \mathbb{R} : \alpha \leqslant f(x)\}$ of points lying on or below the graph of f.

There are various operations on functions with common domain and range that may be performed **pointwise**. For instance, if $f, g \colon X \to \mathbb{R}$, then the function $f + g$ from X to $\mathbb{R}$, defined by $(f + g)(x) = f(x) + g(x)$, is the pointwise sum of f and g.

Real-valued functions can also be ordered pointwise. We say that the function f **dominates** g or $f \geqslant g$ **pointwise** if $f(x) \geqslant g(x)$ for every $x \in X$.[7] Unless otherwise stated, for any two real functions $f, g \colon X \to \mathbb{R}$, the symbols $f \vee g$ and $f \wedge g$ denote the pointwise maximum and minimum of the functions f and g,

$$(f \vee g)(x) = \max\{f(x), g(x)\} \quad \text{and} \quad (f \wedge g)(x) = \min\{f(x), g(x)\}.$$

The **pointwise supremum** of a family $\{f_i : i \in I\}$ of real functions on a set X is defined by $(\sup_i f_i)(x) = \sup\{f_i(x) : i \in I\}$ for each $x \in X$. Similarly the **pointwise infimum** of $\{f_i : i \in I\}$ is given by $(\inf_i f_i)(x) = \inf\{f_i(x) : i \in I\}$ for each $x \in X$.

Likewise, we say that a sequence $\{f_n\}$ of real functions **converges pointwise** to f if $f_n(x) \to f(x)$ for every $x \in X$. (More generally we can define pointwise convergence when the range is any topological space.) Pointwise lim sup and lim inf are defined in a like fashion.

In some applied areas, the term "function space" is applied to any vector space of functions on a common domain, especially if it is infinite dimensional, but in this volume we reserve the term for particular kinds of vector spaces—those that are also closed under pointwise suprema and infima.

1.1 Definition *A set E of real functions on a nonempty set X is a **function space** if it is a vector subspace of $\mathbb{R}^X$ under pointwise addition and scalar multiplication, and it is closed under finite pointwise suprema and infima. That is, if $f, g \in \mathcal{F}$ and $\alpha \in \mathbb{R}$, then the functions $f + g$, αf, $|f|$, $f \vee g$, and $f \wedge g$ also belong to $\mathcal{F}$.*

1.8 Duality of evaluation

There is a peculiar symmetry between a family $\mathcal{F}$ of real functions on a set X and the set X itself. Namely, each point of X can be identified with a real function on $\mathcal{F}$. That is, if $\mathcal{F} \subset \mathbb{R}^X$, then X can be identified with a subset of $\mathbb{R}^{\mathcal{F}}$. It works like this. For each x in X, define the real function e_x on $\mathcal{F}$ by $e_x(f) = f(x)$. This real function is called the **evaluation** functional at x. The function $x \mapsto e_x$ maps X into $\mathbb{R}^{\mathcal{F}}$. To emphasize the symmetry of the roles played by X and $\mathcal{F}$, we sometimes write $\langle x, f \rangle$ for $f(x)$. The mapping $\langle \cdot, \cdot \rangle \colon X \times \mathcal{F} \to \mathbb{R}$ is called the **evaluation duality**,

---

[7] In economics, domination may mean $f(x) \geqslant g(x)$ for every x and $f(x) > g(x)$ for at least one x.

or simply the **evaluation**. This notion of duality and the resultant symmetry of points and functions is extremely important in understanding infinite dimensional vector spaces; see Section 5.14.

1.9 Infinit**X**ies

It is astonishing to the uninitiated that mathematicians, at least since the time of G. Cantor, are able to distinguish different "sizes" of infinity. By reading on, you will be able to as well. The notion of size is called **cardinality**, or occasionally **power**. We say that a set A has the **same cardinality** as B if there is a one-to-one correspondence (that is, bijection) between A and B.[8] We also say that A has **cardinality at least as large** as B if there is a one-to-one correspondence between B and a *subset* of A. The next theorem known as either the Schröder–Bernstein Theorem (as in [149]) or the Cantor–Bernstein Theorem (see [77]), simplifies proving that two sets have the same cardinality. You only have to prove that each has cardinality as large as the other. For a proof of this result, see for instance, P. R. Halmos [149, Section 22, pp. 86–89].

1.2 Theorem (Cantor–Schröder–Bernstein)   *Given sets A and B, if A has cardinality at least as large as B and B has cardinality at least as large as A, then A and B have the same cardinality.*

   In other words, if there is a bijection between A and a subset of B, and a bijection between B and a subset of A, then there is a bijection between A and B.

The definition of size as cardinality is quite satisfactory for finite sets, but is a bit unsettling for infinite sets. For instance, the integers are in one-to-one correspondence with the even integers via the correspondence $n \leftrightarrow 2n$. But only "half" the integers are even. Nonetheless, cardinality has proven to be the most useful notion of size for sets. In fact, a useful definition of an infinite set is one that can be put into one-to-one correspondence with a proper subset of itself.

Sets of the same cardinality as a set $\{1, \dots, n\}$ of natural numbers for some $n \in \mathbb{N}$ are **finite**. A set of the same cardinality as the set $\mathbb{N}$ of natural numbers itself are called **countably infinite**. Sets that are either finite or countably infinite are called **countable**. (Other sets are **uncountable**.) We freely use the following properties of countable sets.

- Subsets of countable sets are countable.

- Countable unions of countable sets are countable.

- Finite Cartesian products of countable sets are countable.

---

[8] Those who talk about the power of a set (not to be confused with its power set) will say that two sets having the same cardinality are **equipotent**.

In particular, the set of rational numbers is countable. (Why?) The following fact is an immediate consequence of those above.

- The set of all finite subsets of a countable set is again countable.

We use the countability of the rationals to jump ahead and prove the following well-known and important result.

1.3 Theorem (Discontinuities of increasing functions)   *Let I be an interval in* $\mathbb{R}$ *and let* $f: I \to \mathbb{R}$ *be nondecreasing, that is,* $x > y$ *implies* $f(x) \geqslant f(y)$. *Then* f *has at most countably many points of discontinuity.*

Proof: For each x, since f is nondecreasing,

$$\sup\{f(y) : y < x\} = f(x_-) \leqslant f(x) \leqslant f(x_+) = \inf\{f(y) : y > x\}.$$

Clearly f is continuous at x if and only if $f(x_-) = f(x) = f(x_+)$. So if x is a point of discontinuity, then there is a rational number q_x satisfying $f(x_-) < q_x < f(x_+)$. Furthermore if x and y are points of discontinuity and $x < y$, then $q_x < q_y$. (Why?) Thus f has at most countably many points of discontinuity.   ∎

Not every infinite set is countable; some are larger. G. Cantor showed that the set of real numbers is not countable using a technique now referred to as the **Cantor diagonal process**. It works like this. Suppose the unit interval $[0, 1]$ were countable. Then we could list the decimal expansion of the reals in $[0, 1]$ in order. We now construct a real number that does not appear on the list by romping down the diagonal and making sure our number is different from each number on the list. One way to do this is to choose a real number b whose decimal expansion $0.b_1b_2b_3 \ldots$ satisfies $b_n = 7$ unless $a_{n,n} = 7$ in which case we choose $b_n = 3$. In this way, b differs from every number on the list. This shows that it is impossible to enumerate the unit interval with the integers. It also shows that $\mathbb{N}^{\mathbb{N}}$, the set of all sequences of natural numbers, is uncountable.

$\mathbb{N}$	$\mathbb{R}$
1	$0.a_{11}a_{12}a_{13}\ldots$
2	$0.a_{21}a_{22}a_{23}\ldots$
3	$0.a_{31}a_{32}a_{33}\ldots$
4	$0.a_{41}a_{42}a_{43}\ldots$
$\vdots$	$\vdots$ $\ddots$

A corollary of the uncountability of the reals is that there are well behaved linear orderings that have no real-valued representation.

1.4 Example (An order with no utility)   Define the linear order $\geqslant$ on $\mathbb{R}^2$ by $(x_1, x_2) \geqslant (y_1, y_2)$ if and only if either $x_1 > y_1$ or $x_1 = y_1$ and $x_2 \geqslant y_2$. (This order is called the **lexicographic order** on the plane.) A **utility** for this order is a function $u: \mathbb{R}^2 \to \mathbb{R}$ satisfying $x \geqslant y$ if and only if $u(x) \geqslant u(y)$. Now suppose by way of contradiction that this order has a utility. Then for each real number x, we have $u(x, 1) > u(x, 0)$. Consequently there must be some rational number r_x satisfying $u(x, 1) > r_x > u(x, 0)$. Furthermore, if $x > y$, then $r_x > r_y$. Thus

$x \leftrightarrow r_x$ is a one-to-one correspondence between the real numbers and a set of rational numbers, implying that the reals are countable. This contradiction proves the claim.   ∎

The cardinality of the set of real numbers $\mathbb{R}$ is called the cardinality of the **continuum**, written card $\mathbb{R} = \mathfrak{c}$. Here are some familiar sets with cardinality $\mathfrak{c}$.

•   The intervals $[0, 1]$ and $(0, 1)$ (and as a matter of fact any nontrivial subinterval of $\mathbb{R}$).

•   The Euclidean spaces $\mathbb{R}^n$.

•   The set of irrational numbers in any nontrivial subinterval of $\mathbb{R}$.

•   The collection of all subsets of a countably infinite set.

•   The set $\mathbb{N}^{\mathbb{N}}$ of all sequences of natural numbers.

For more about the cardinality of sets see, for instance, T. Jech [185].

1.10   The Diagonal Theorem and Russell's Paradox

The diagonal process used by Cantor to show that the real numbers are not countable can be viewed as a special case of the following more general argument.

1.5 Cantor's Diagonal Theorem   *Let X be a set and let $\varphi \colon X \twoheadrightarrow X$ be a correspondence. Then the set $A = \{x \in X : x \notin \varphi(x)\}$ of non-fixed points of φ is not a value of φ. That is, there is no x satisfying $\varphi(x) = A$.* [9]

Proof: Assume by way of contradiction that there is some $x_0 \in X$ satisfying $\varphi(x_0) = A$. If x_0 is not a fixed point of φ, that is, $x_0 \notin \varphi(x_0)$, then by definition of A, we have $x_0 \in A = \varphi(x_0)$, a contradiction. On the other hand, if x_0 is a fixed point of φ, that is, $x_0 \in \varphi(x_0)$, then by definition of A, we have $x_0 \notin A = \varphi(x_0)$, also a contradiction. Hence A is not the value of φ at any point.   ∎

Russell's Paradox is a clever argument devised by Bertrand Russell as an attack on the validity of the proof of the Diagonal Theorem. It goes like this. Let S be the set of all sets, and let $\varphi \colon S \twoheadrightarrow S$ be defined by $\varphi(A) = \{B \in S : B \in A\}$ for every $A \in S$. Since $\varphi(A)$ is just the set of members of A, we have $\varphi(A) = A$. That is, φ is the identity on S, so the set of its values is just S again. By Cantor's Diagonal Theorem, the set $C = \{A \in S : A \notin \varphi(A)\}$ is not a value of φ, so it cannot be a set, which is a contradiction.

---

[9] Descriptive set theorists state the theorem as "A is not in the range of φ," but they think of φ as a function from X to its power set 2^X. For them the range is a subset of 2^X, namely $\{\varphi(x) : x \in X\}$, but by our definition, the range is a subset of X, namely $\bigcup \{\varphi(x) : x \in X\}$.

The paradox was resolved not by repudiating the Diagonal Theorem, but by the realization that S, the collection of all sets, cannot itself be a set. What this means is that we have to be very much more careful about deciding what is a set and what is not a set.

1.11 The axiom of choice and axiomatic set theory

In Section 1.2, we were sloppy, even for us, but we were hoping you would not notice. For instance, we took it for granted that the union of a set of sets was a set, and that I-tuples (whatever they are) existed. Russell's Paradox tells us we should worry if these really are sets. Well maybe not we, but someone should worry. If you are worried, we recommend P. R. Halmos [149], or A. Shen and N. K. Vereshchagin [303] for "naïve set theory." For an excellent exposition of "axiomatic set theory," we recommend K. J. Devlin [92] or T. Jech [185].

Axiomatic set theory is viewed by many happy and successful people as a subject of no practical relevance. Indeed you may never have been exposed to the most popular axioms of set theory, the **Zermelo–Frankel (ZF) set theory**. For your edification we mention that ZF set theory proper has eight axioms. For instance, the Axiom of Infinity asserts the existence of an infinite set. There is also a ninth axiom, the **Axiom of Choice**, and ZF set theory together with this axiom is often referred to as ZFC set theory. We shall not list the others here, but suffice it to say that the first eight axioms are designed so that the collection of objects that we call sets is closed under certain set theoretic operations, such as unions and power sets. They were also designed to ward off Russell's Paradox.

The ninth axiom of ZFC set theory, the Axiom of Choice, is a seemingly innocuous set theoretic axiom with much hidden power.

1.6 Axiom of Choice   *If $\{A_i : i \in I\}$ is a nonempty set of nonempty sets, then there is a function $f : I \to \bigcup_{i \in I} A_i$ satisfying $f(i) \in A_i$ for each $i \in I$. In other words, the Cartesian product of a nonempty set of nonempty sets is itself a nonempty set.*

The function f, whose existence the axiom asserts, chooses a member of A_i for each i. Hence the term "Axiom of Choice." This axiom is both consistent with and independent of ZF set theory proper. That is, if the Axiom of Choice is dropped as an axiom of set theory, it cannot be proven by using the remaining eight axioms that the Cartesian product of nonempty sets is a nonempty set. Furthermore, adding the Axiom of Choice does not make the axioms of ZF set theory inconsistent. (A collection of axioms is inconsistent if it is possible to deduce both a statement P and its negation $\neg P$ from the axioms.)

There has been some debate over the desirability of assuming the Axiom of Choice. (G. Moore [251] presents an excellent history of the Axiom of Choice and the controversy surrounding it.) Since there may be no way to describe the choice function, why should we assume it exists? Further, the Axiom of Choice has some unpleasant consequences. The Axiom of Choice makes it possible, for instance, to prove the existence of non-Lebesgue measurable sets of real numbers (Corollary 10.42). R. Solovay [316] has shown that by dropping the Axiom of Choice, it is possible to construct models of set theory in which all subsets of the real line are Lebesgue measurable. Since measurability is a major headache in integration and probability theory, it would seem that dropping the Axiom of Choice would be desirable. Along the same lines is the **Banach–Tarski Paradox** due to S. Banach and A. Tarski [32]. They prove, using the Axiom of Choice, that the unit ball U in $\mathbb{R}^3$ can be partitioned into two disjoint sets X and Y with the property that X can be partitioned into five disjoint sets, which can be reassembled (after translation and rotation) to make a copy of U, and the same is true of Y. That is, the ball can be cut up into pieces and reassembled to make two balls of the same size! (These pieces are obviously not Lebesgue measurable. Worse yet, this paradox shows that it is impossible to define a finitely additive volume in any reasonable manner on $\mathbb{R}^3$.) For a proof of this remarkable result, see, e.g., T. Jech [184, Theorem 1.2, pp. 3–6].

On the other hand, dropping the Axiom of Choice also has some unpleasant side effects. For example, without some version of the Axiom of Choice, our previous assertion that the countable union of countable sets is countable ceases to be true. Its validity can be restored by assuming the Countable Axiom of Choice, a weaker assumption that says only that a countable product of sets is a set. Without the Countable Axiom of Choice, there exist infinite sets that have no countably infinite subset. (See, for instance, T. Jech [184, Section 2.4, pp. 20–23].)

From our point of view, the biggest problem with dropping the Axiom of Choice is that some of the most useful tools of analysis would be thrown out with it. J. L. Kelley [197] has shown that the Tychonoff Product Theorem 2.61 would be lost. Most proofs of the Hahn–Banach Extension Theorem 5.53 make use of the Axiom of Choice, but it is not necessary. The Hahn–Banach theorem, which is central to linear analysis, can be proven using the Prime Ideal Theorem of Boolean Algebra, see W. A. J. Luxemburg [232]. The Prime Ideal Theorem is equivalent to the Ultrafilter Theorem 2.19, which we prove using Zorn's Lemma 1.7 (itself equivalent to the Axiom of Choice). J. D. Halpern [152] has shown that the Ultrafilter Theorem does not imply the Axiom of Choice. Nevertheless, M. Foreman and F. Wehrung [126] have shown that if the goal is to eliminate non-measurable sets, then we have to discard the Hahn–Banach Extension Theorem. That is, any superset of the ZF axioms strong enough to prove the Hahn–Banach theorem is strong enough to prove the existence of non-measurable sets. We can learn to live with non-measurable sets, but not without the Hahn–Banach theorem. So we might as well assume the Axiom of Choice. For more on the Axiom of Choice, we recommend the monograph by P. Howard and J. E. Rubin [170]. In addition, P. R. Halmos [149] and J. L. Kelley [198, Chapter 0] have extended discussions of the Axiom of Choice.

1.12   Zorn's Lemma

A number of propositions are equivalent to the Axiom of Choice. One of these is Zorn's Lemma, due to M. Zorn [350]. That is, Zorn's Lemma is a theorem if the Axiom of Choice is assumed, but if Zorn's Lemma is taken as an axiom, then the Axiom of Choice becomes a theorem.

1.7 Zorn's Lemma   *If every chain in a partially ordered set X has an upper bound, then X has a maximal element.*

We indicate the power of Zorn's Lemma by employing it to prove a number of useful results from mathematics and economics. In addition to the results that we present in this section, we also use Zorn's Lemma to prove the Ultrafilter Theorem 2.19, the Tychonoff Product Theorem 2.61, the Hahn–Banach Extension Theorem 5.53, and the Krein–Milman Theorem 7.68.

The first use of Zorn's Lemma is the well-known fact that vector spaces possess Hamel bases. Recall that a **Hamel basis** or simply a **basis** of a vector space V is a linearly independent set B (every finite subset of B is linearly independent) such that for each nonzero $x \in V$ there are $b_1, \dots, b_k \in B$ and nonzero scalars $\alpha_1, \dots, \alpha_k$ (all uniquely determined) such that $x = \sum_{i=1}^{k} \alpha_i b_i$.

1.8 Theorem   *Every nontrivial vector space has a Hamel basis.*

Proof: Let V be a nontrivial vector space, that is, $V \neq \{0\}$. Let X denote the collection of all linearly independent subsets of V. Since $\{x\} \in X$ for each $x \neq 0$, we see that $X \neq \varnothing$. Note that X is partially ordered by set inclusion. In addition, note that an element of X is maximal if and only if it is a basis. (Why?) Now if $\mathcal{C}$ is a chain in X, then $A = \bigcup_{C \in \mathcal{C}} C$ is a linearly independent subset of V, so A belongs to X and is an upper bound for $\mathcal{C}$. By Zorn's Lemma 1.7, X has a maximal element. Thus V has a basis.   ∎

As another example of the use of Zorn's Lemma, we present the following result, essentially due to E. Szpilrajn [327]. It is used to prove the key results in the theory of revealed preference, see M. K. Richter [283, Lemma 2, p. 640]. The proof of the result is not hard, but we present it in agonizing detail because the argument is so typical of how to use Zorn's Lemma.

It is always possible to extend any binary relation R on a set X to the total relation S defined by $x \, S \, y$ for all x, y. But this is not very interesting since it destroys any asymmetry present in R. Let us say that the binary relation S on a set X is a **compatible extension** of the relation R if S extends R and preserves the asymmetry of R. That is, $x \, R \, y$ implies $x \, S \, y$, and together $x \, R \, y$ and $\neg(y \, R \, x)$ imply $\neg(y \, S \, x)$.

1.9 Theorem   (Total extension of preorders)   *Any preorder has a compatible extension to a total preorder.*

Proof: Let R be a preorder (reflexive transitive binary relation) on the set X. Let $\mathcal{E}$ be the set of preorders that compatibly extend R, and let $\mathcal{E}$ be partially ordered by inclusion (as subsets of $X \times X$). Note that $\mathcal{E}$ contains R and so is nonempty.

Let $\mathcal{C}$ be a nonempty chain in $\mathcal{E}$. We claim the relation $U = \bigcup \{S : S \in \mathcal{C}\}$ is an upper bound for $\mathcal{C}$ in $\mathcal{E}$. Clearly U is reflexive and extends R. To see that U is transitive, suppose $x \, U \, y$ and $y \, U \, z$. Then $x \, S_1 \, y$ and $y \, S_2 \, z$ for some $S_1, S_2 \in \mathcal{C}$. Since $\mathcal{C}$ is a chain, $S_1 \subset S_2$ or $S_2 \subset S_1$, say $S_1 \subset S_2$. Then $x \, S_2 \, y \, S_2 \, z$, so $x \, S_2 \, z$ by transitivity of S_2. Thus $x \, U \, z$. Moreover U is a compatible extension of R. For suppose that $x \, R \, y$ and $\neg(y \, R \, x)$. Then $\neg(y \, S \, x)$ for any S in $\mathcal{E}$, so $\neg(y \, U \, x)$. Thus U is a reflexive and transitive compatible extension of R, and U is also an upper bound for $\mathcal{C}$ in $\mathcal{E}$. Since $\mathcal{C}$ is an arbitrary chain in $\mathcal{E}$, Zorn's Lemma 1.7 asserts that $\mathcal{E}$ has a maximal element.

We now show that any preorder in $\mathcal{E}$ that is not total cannot be maximal in $\mathcal{E}$. So fix a compatible extension S in $\mathcal{E}$, and suppose that S is not total. Then there is a pair $\{x, y\}$ of distinct elements such that neither $x \, S \, y$ nor $y \, S \, x$. Define the relation $T = S \cup \{(x, y)\}$, and let W be the transitive closure of T. Clearly W is a preorder and extends R. We now verify that W is a compatible extension of S.

Suppose by way of contradiction that $u \, S \, v$ and $\neg(v \, S \, u)$, but $v \, W \, u$ for some u, v belonging to X. By the definition of transitive closure, $v \, W \, u$ means

$$v = u_0 \, T \, u_1 \, T \cdots T \, u_n \, T \, u_{n+1} = u.$$

for some $u_1, \ldots, u_n$. Since T differs from S only by (x, y), either (i) we can replace T by S everywhere above or (ii) one of the (u_i, u_{i+1}) pairs must be (x, y). Case (i) implies $v \, S \, u$ by transitivity, a contradiction. In case (ii), by omitting terms if necessary, we may assume that $(x, y) = (u_i, u_{i+1})$ only once. Then starting with $y = u_{i+1}$ we have $y = u_{i+1} \, T \cdots T \, u_n = u \, T \, v = u_0 \, T \, u_1 \, T \cdots T \, u_i = x$. Now we may replace T by S everywhere, and conclude by transitivity that $y \, S \, x$, another contradiction. Therefore W is a compatible extension of R, and since it properly includes S, we see that S cannot be maximal in $\mathcal{E}$.

Thus any maximal compatible extension of R is a total preorder. ∎

Next is the fixed point theorem of B. Knaster [211] and A. Tarski. Let $(X, \geqslant_X)$ and $(Y, \geqslant_Y)$ be partially ordered sets. Recall that a function $f : X \to Y$ is **monotone** if $x \geqslant_X z$ implies $f(x) \geqslant_Y f(z)$. Recall that for f mapping X into itself, a **fixed point** of f is a point x satisfying $f(x) = x$.

1.10 Knaster–Tarski Fixed Point Theorem   *Let $(X, \geqslant)$ be a partially ordered set with the property that every chain in X has a supremum. Let $f : X \to X$ be monotone, and assume that there exists some a in X with $a \leqslant f(a)$. Then the set of fixed points of f is nonempty and has a maximal fixed point.*

Proof: Consider the partially ordered subset

$$P = \{x \in X : x \leqslant f(x)\}.$$

The set P contains a so it is nonempty. Now suppose C is a chain in P, and b is its supremum in X. Since $c \leqslant b$ for every $c \in C$, we see that $f(c) \leqslant f(b)$. Since $c \leqslant f(c)$ for $c \in C$, it follows that $f(b)$ is an upper bound for C. Since b is the least such upper bound, we have $b \leqslant f(b)$. Therefore, $b \in P$. Thus the supremum of any chain in P belongs to P. Then by Zorn's Lemma 1.7, P has a maximal element, call it x_0.

Now $x_0 \leqslant f(x_0)$, since x_0 is in P. Since f is monotone, $f(x_0) \leqslant f(f(x_0))$. But this means that $f(x_0)$ belongs to P. Since x_0 is a maximal element of P, we see that $x_0 = f(x_0)$.

Furthermore, if x is a fixed point of f, then $x \in P$. This shows that x_0 is a maximal fixed point of f.   ∎

We point out that the hypotheses can be weakened so that only the subset $P \cap \{x \in X : x \geqslant a\}$ is required to have the property that chains have suprema. The proof is the same. The hypothesis that there exists at least one a with $a \leqslant f(a)$ is necessary. (Why?)

There is a related fixed point theorem, also due to A. Tarski [329]. It strengthens the hypotheses to require $(X, \geqslant)$ to be a complete lattice, and draws the stronger conclusion that the set of fixed points is also a complete lattice. Recall that the infimum of a complete lattice is denoted 0, and the supremum is denoted 1. Also, if A is a subset of X, by $(A, \geqslant)$ we mean the partially ordered set A where $\geqslant$ is just the restriction of the order on X to A.

1.11 Tarski Fixed Point Theorem *If $(X, \geqslant)$ is a complete lattice, and $f : X \to X$ is monotone, then the set F of fixed points of f is nonempty and $(F, \geqslant)$, is itself a complete lattice.*

Proof: As in the proof of Theorem 1.10, let $P = \{x \in X : x \leqslant f(x)\}$, put $\overline{x} = \bigvee P$ (note that $0 \in P$) and conclude that $f(\overline{x}) = \overline{x}$. Since $F \subset P$, we have that $\overline{x} = \bigvee F$. A similar argument shows that $\underline{x} = \bigwedge \{x \in X : x \geqslant f(x)\}$ satisfies $\underline{x} = \bigwedge F \in F$.

To prove that $(F, \geqslant)$ is a complete lattice, fix a nonempty subset A of F, and let $\overline{a}$ be the supremum of A (in X). Now the order interval $I = [\overline{a}, 1] = \{x \in X : \overline{a} \leqslant x\}$ is also a complete lattice in its own right. We show next that f maps I into itself.

To see this, observe that if $x \in A$, then $x \leqslant \overline{a}$, so $f(x) \leqslant f(\overline{a})$. But $x = f(x)$, so we have $x \leqslant f(\overline{a})$. Thus $f(\overline{a})$ is also an upper bound for A, so $\overline{a} \leqslant f(\overline{a})$. Hence if z belongs to I, that is, if $\overline{a} \leqslant z$, we have $f(\overline{a}) \leqslant f(z)$, and $\overline{a} \leqslant f(\overline{a}) \leqslant f(z)$, which implies that $f(z)$ also belongs to I. Therefore f maps I into itself.

Let $\hat{f}$ denote the restriction of f to I, and let $\hat{F}$ denote the (nonempty) set of fixed points of $\hat{f}$. By the first part of the proof, $\underline{z} = \bigwedge \hat{F}$ is a fixed point of $\hat{f}$ and so of f. Since $\underline{z}$ belongs to I, it is an upper bound for A that lies in F. Indeed it is the least upper bound of A that lies in F: for if b is an upper bound for A, then $b \in I$, so if b is also a fixed point of f, then $b \in \hat{F}$, so $\underline{z} \leqslant b$. Therefore, $\underline{z}$ is the supremum of A in $(F, \geqslant)$. A similar argument shows that A has an infimum in F as well. In other words, $(F, \geqslant)$ is a complete lattice.   ∎

Some care must be taken in the interpretation of this result. The theorem does *not* assert that the set F of fixed points is a sublattice of X. It may well be that the supremum of a set in the lattice $(F, \geqslant)$ is not the same as its supremum in the lattice $(X, \geqslant)$. For example, let $X = \{0, 1, a, b, b'\}$ and define the partial order $\geqslant$ by $1 \geqslant a \geqslant b \geqslant 0$ and $1 \geqslant a \geqslant b' \geqslant 0$ (and all the other comparisons implied by transitivity and reflexivity). Note that b and b' are not comparable. Define the monotone function $f \colon X \to X$ by $f(x) = x$ for $x \neq a$ and $f(a) = 1$. The set of F of fixed points of f is $\{0, b, b', 1\}$, which is a complete lattice. Let $B = \{b, b'\}$ and note that $\bigvee B = 1$ when B viewed as a subset of F, but $\bigvee B = a$, when B viewed as a subset of X.

In a converse direction, any incomplete lattice has a fixed point-free monotone function into itself. For a proof, see A. C. Davies [81]. Tarksi's Theorem has been extended to cover increasing correspondences by R. E. Smithson [314] and X. Vives [336]. See F. Echenique [113] for more constructive proofs of these and related results.

1.13   Ordinals

We now apply Zorn's Lemma to the proof of the Well Ordering Principle, which is yet another equivalent of the Axiom of Choice.

1.12 Definition   *A set X is **well ordered** by the linear order $\leq$ if every nonempty subset of X has a first element. An element x of A is **first** in A if $x \leq y$ for all $y \in A$.*

*An **initial segment** of $(X, \leq)$ is any set of the form $I(x) = \{y \in X : y \leq x\}$.*

*An **ideal** in a well ordered set X is a nonempty subset A of X such that for each $a \in A$ the initial segment $I(a)$ is included in A.*

1.13 Well Ordering Principle   *Every nonempty set can be well ordered.*

Proof: Let X be a nonempty set, and let

$$\mathcal{X} = \{(A, \leq_A) : A \subset X \text{ and } \leq_A \text{ well orders } A\}.$$

Note that $\mathcal{X}$ is nonempty, since every finite set is well ordered by any linear order. Define the partial order $\geqslant$ on $\mathcal{X}$ by $(A, \leq_A) \geqslant (B, \leq_B)$ if B is an ideal in A and $\leq_A$ extends $\leq_B$. If $\mathcal{C}$ is a chain in $\mathcal{X}$, set $C = \bigcup\{A : (A, \leq_A) \in \mathcal{C}\}$, and define $\leq_C$ on C by $x \leq_C y$ if $x \leq_A y$ for some $(A, \leq_A) \in \mathcal{C}$. Then $\leq_C$ is a well defined order on C, and $(C, \leq_C)$ belongs to $\mathcal{X}$ (that is, $\leq_C$ well orders C) and is an upper bound for $\mathcal{C}$. (Why?) Therefore, by Zorn's Lemma 1.7, the partially ordered set $\mathcal{X}$ has a maximal element $(A, \leq)$. We claim that $A = X$, so that X is well ordered by $\leq$. For if there is some $x \notin A$, extend $\leq$ to $A \cup \{x\}$ by $y \leq x$ for all $y \in A$. This extended relation well orders $A \cup \{x\}$ and A is an ideal in $A \cup \{x\}$ (why?), contradicting the maximality of $(A, \leq)$.   ∎

We now prove the existence of a remarkable and useful well ordered set.

1.14 Theorem   *There is an ordered set $(\Omega, \leq)$ satisfying the following properties.*

1. *Ω is uncountable and well ordered by $\leq$.*

2. *Ω has a greatest element ω_1.*

3. *If $x < \omega_1$, then the initial segment $I(x)$ is countable.*

4. *If $x < \omega_1$, then $\{y \in \Omega : x \leq y \leq \omega_1\}$ is uncountable.*

5. *Every nonempty subset of Ω has a least upper bound.*

6. *A nonempty subset of $\Omega \backslash \{\omega_1\}$ has a least upper bound in $\Omega \backslash \{\omega_1\}$ if and only if it is countable. In particular, the least upper bound of every uncountable subset of Ω is ω_1.*

Proof: Let $(X, \leq)$ be an uncountable well ordered set, and consider the set A of elements x of X such that the initial segment $I(x) = \{y \in X : y \leq x\}$ is uncountable. Without loss of generality we may assume A is nonempty, for if A is empty, append a point y to X, and extend the ordering $\leq$ by $x \leq y$ for all $x \in X$. This order well orders $X \cup \{y\}$. Under the extension, A is now nonempty. The set A has a first element, traditionally denoted ω_1. Set $\Omega = I(\omega_1)$, the initial segment generated by ω_1. Clearly Ω is an uncountable well ordered set with greatest element ω_1.

The proofs of the other properties except (6) are straightforward, and we leave them as exercises. So suppose $C = \{x_1, x_2, \ldots\}$ is a countable subset of $\Omega \backslash \{\omega_1\}$. Then $\bigcup_{n=1}^{\infty} I(x_n)$ is countable, so there is some $x < \omega_1$ not belonging to this union. Such an x is clearly an upper bound for C so its least upper bound b (which exists by (5)), satisfies $b \leq x < \omega_1$. For the converse, observe that if $b < \omega_1$ is a least upper bound for a set C, then C is included in the countable set $I(b)$.   ∎

The elements of Ω are called **ordinals**, and ω_1 is called the **first uncountable ordinal**. The set $\Omega_0 = \Omega \backslash \{\omega_1\}$ is the set of **countable ordinals**. Also note that we can think of the natural numbers $\mathbb{N} = \{1, 2, \ldots\}$ as a subset of Ω: Identify 1 with the first element of Ω, and recursively identify n with the first element of $\Omega \backslash \{1, 2, \ldots, n-1\}$. In interval notation we may write $\Omega = [1, \omega_1]$ and $\Omega_0 = [1, \omega_1)$.

The first element of $\Omega \backslash \mathbb{N}$ is denoted ω_0. It is the **first infinite ordinal**. [10] Clearly, $n < \omega_0$ for each $n \in \mathbb{N}$. The names are justified by the fact that if we take any other well ordered uncountable set with a greatest element and find the first uncountable initial segment $\Omega' = [1', \omega']$, then there is a strictly monotone function f from Ω onto Ω'. To establish the existence of such a function f argue as follows. Let

$$\mathfrak{X} = \{(x, g) \mid x \in \Omega \text{ and } g \colon I(x) \to \Omega' \text{ is strictly monotone and has range } I(g(x))\}.$$

---

[10] Be aware that some authors use Ω to denote the first uncountable ordinal and ω to denote the first infinite ordinal.

If $\mathbb{N} = \{1, 2, \ldots\}$ and $\mathbb{N}' = \{1', 2', \ldots\}$ are the natural numbers of Ω and Ω' respectively, and $g\colon \mathbb{N} \to \Omega'$ is defined by $g(n) = n'$, then $(n, g) \in \mathcal{X}$ for each $n \in \mathbb{N}$. This shows that $\mathcal{X}$ is nonempty. Next, define a partial order $\succcurlyeq$ on $\mathcal{X}$ by $(x, g) \succcurlyeq (y, h)$ if $x \ge y$ and $g = h$ on $I(y)$. Now let $\{(x_\alpha, g_\alpha)\}_{\alpha \in A}$ be a chain in $\mathcal{X}$. Put $x = \sup_{\alpha \in A} x_\alpha$ in Ω and define $g\colon I(x) \to \Omega'$ by $g(y) = g_\alpha(y)$ if $y < x_\alpha$ for some α and $g(x) = \sup_{\alpha \in A} g(x_\alpha)$. Notice that g is well defined, strictly monotone, and satisfies $g(I(x)) = I(g(x))$ and $(x, g) \succcurlyeq (x_\alpha, g_\alpha)$ for each $\alpha \in A$. This shows that every chain in $\mathcal{X}$ has an upper bound. By Zorn's lemma, $\mathcal{X}$ has a maximal element, say (x, f). We now leave it as an exercise to you to verify that $x = \omega_1$ and that $f(\omega_1) = \omega_1'$. You should also notice that f is uniquely determined and, in fact, $f(x)$ is the first element of the set $\Omega' \setminus \{f(y) : y < x\}$.

In the next chapter we make use of the following result.

1.15 Interlacing Lemma   *Suppose $\{x_n\}$ and $\{y_n\}$ are interlaced sequences in Ω_0. That is, $x_n \le y_n \le x_{n+1}$ for all n. Then both sequences have the same least upper bound in Ω_0.*

Proof: By Theorem 1.14 (6), each sequence has a least upper bound in Ω_0. Call the least upper bounds x and y respectively. Since $y_n \ge x_n$ for all n, we have $y \ge x$. Since $x_{n+1} \ge y_n$ for all n, we have $x \ge y$. Thus $x = y$. ∎

As an aside, here is how the Well Ordering Principle implies the Axiom of Choice. Let $\{A_i : i \in I\}$ be a nonempty family of nonempty sets. Well order $\bigcup_{i \in I} A_i$ and let $f(i)$ be the first element of A_i. Then f is a choice function.

Chapter 2

Topology

We begin with a chapter on what is now known as general topology. Topology is the abstract study of convergence and approximation. We presume that you are familiar with the notion of convergence of a sequence of real numbers, and you may even be familiar with convergence in more general normed or metric spaces. Recall that a sequence $\{x_n\}$ of real numbers converges to a real number x if $\{|x_n-x|\}$ converges to zero. That is, for every $\varepsilon > 0$, there is some n_0 such that $|x_n - x| < \varepsilon$ for all $n \geqslant n_0$. In metric spaces, the general notion of the distance between two points (given by the *metric*) plays the role of the absolute difference between real numbers, and the theory of convergence and approximation in metric spaces is not all that different from the theory of convergence and approximation for real numbers. For instance, a sequence $\{x_n\}$ of points in a metric space converges to a point x if the distance $d(x_n, x)$ between x_n and x converges to zero as a sequence of real numbers. That is, if for every $\varepsilon > 0$, there is an n_0 such that $d(x_n, x) < \varepsilon$ for all $n \geqslant n_0$. However, metric spaces are inadequate to describe approximation and convergence in more general settings. A very real example of this is given by the notion of pointwise convergence of real functions on the unit interval. It turns out there is no way to define a metric on the space of all real functions on the interval $[0, 1]$ so that a sequence $\{f_n\}$ of functions converges pointwise to a function f if and only if the distance between f_n and f converges to zero. Nevertheless, the notion of pointwise convergence is extremely useful, so it is imperative that a general theory of convergence should include it.

There are many equivalent ways we could develop a general theory of convergence.[1] In some ways, the most natural place to start is with the notion of a *neighborhood* as a primitive concept. A neighborhood of a point x is a collection of points that includes all those "sufficiently close" to x. (In metric spaces, "sufficiently close" means within some positive distance ε.) We could define the collection of all neighborhoods and impose axioms on the family of neighborhoods. Instead of this, we start with the concept of an open set. An *open* set is a set that is a neighborhood of all its points. It is easier to impose axioms on

---

[1] The early development of topology used many different approaches to capture the notion of approximation: closure operations, proximity spaces, L-spaces, uniform spaces, etc. Some of these notions were discarded, while others were retained because of their utility.

the family of open sets than it is to impose them directly on neighborhoods. The family of all open sets is called a *topology*, and a set with a topology is called a *topological space*.

Unfortunately for you, a theory of convergence for topological spaces that is adequate to deal with pointwise convergence has a few quirks. Most prominent is the inadequacy of using sequences to describe continuity of functions. A function is continuous if it carries points sufficiently close in the domain to points sufficiently close in the range. For metric spaces, continuity of f is equivalent to the condition that the sequence $\{f(x_n)\}$ converges to $f(x)$ whenever the sequence $\{x_n\}$ converges to x. This no longer characterizes continuity in the more general framework of topological spaces. Instead, we are forced to introduce either *nets* or *filters*. A net is like a sequence, except that instead of being indexed by the natural numbers, the index set can be much larger. Two particularly important techniques for indexing nets include indexing the net by the family of neighborhoods of a point, and indexing the net by the class of all finite subsets of a set.

There are offsetting advantages to working with general topological spaces. For instance, we can define topologies to make our favorite functions continuous. These are called *weak* topologies. The topology of pointwise convergence is actually a weak topology, and weak topologies are fundamental to understanding the equilibria of economies with an infinite dimensional commodity space.

Another important topological notion is compactness. Compact sets can be approximated arbitrarily well by finite subsets. (In Euclidean spaces, the compact sets are the closed and bounded sets.) Two of the most important theorems in this chapter are the Weierstrass Theorem 2.35, which states that continuous functions achieve their maxima on compact sets, and the Tychonoff Product Theorem 2.61, which asserts that the product of compact sets is compact in the product topology (the topology of pointwise convergence). This latter result is the basis of the Alaoglu Theorem 5.105, which describes a general class of compact sets in infinite dimensional spaces.

Liberating the notions of neighborhood and convergence from their metric space setting often leads to deeper insights into the structure of approximation methods. The idea of weak convergence and the keystone Tychonoff Product Theorem are perhaps the most important contributions of general topology to analysis—although at least one of us has heard the complaint that "topology is killing analysis." We collect a few fundamental topological definitions and results here. In the interest of brevity, we have included only material that we use later on, and have neglected other important and potentially useful results. We present no discussion of algebraic or differential topology, and have omitted discussion of quotient topologies, projective and inductive limits, metrizability theorems, extension theorems, and a variety of other topics. For more detailed treatments of general topology, there are a number of excellent standard references, including Dugundji [106], Kelley [198], Kuratowski [218], Munkres [256], and Willard [342]. Willard's historical notes are especially thorough.

2.1   Topological spaces

Having convinced you of the need for a more general approach, we start, as promised, with the definition of a topology. It captures most of the important properties of the family of open sets in a metric space, with one exception, the Hausdorff property, which we define presently.

2.1 Definition   *A **topology** τ on a set X is a collection of subsets of X satisfying:*

1. $\varnothing, X \in \tau$.

2. *τ is closed under finite intersections.*

3. *τ is closed under arbitrary unions.*

*A nonempty set X equipped with a topology τ is called a **topological space**, and is denoted (X, τ), (or simply X when no confusion should arise). We call a member of τ an **open set** in X. The complement of an open set is a **closed set**. A set that is both closed and open is called a **clopen set**.*

A set may be both open and closed, or it may be neither. In particular, both $\varnothing$ and X are both open and closed. The family of closed sets has the following properties, which are dual to the properties of the open sets. Prove them using de Morgan's laws.

- Both $\varnothing$ and X are closed.

- A finite union of closed sets is closed.

- An arbitrary intersection of closed sets is closed.

2.2 Example   (Topologies)   The following examples illustrate the variety of topological spaces.

1. The **trivial topology** or **indiscrete topology** on a set X consists of only X and $\varnothing$. These are also the only closed sets.

2. The **discrete topology** on a set X consists of all subsets of X. Thus every set is both open and closed.

3. A **semimetric** d on a space X is a real-valued function on $X \times X$ that is nonnegative, symmetric, satisfies $d(x, x) = 0$ for every x, and in addition satisfies the **triangle inequality**, $d(x, z) \leqslant d(x, y) + d(y, z)$. A **metric** is a semimetric that has the property that $d(x, y) = 0$ implies $x = y$. A pair (X, d), where d is a metric on X, is called a **metric space**.

   Given a semimetric d, define $B_\varepsilon(x) = \{y : d(x, y) < \varepsilon\}$, the **open ε-ball** around x. A set U is open in the **semimetric topology** generated by d if for each point x in U there is an $\varepsilon > 0$ satisfying $B_\varepsilon(x) \subset U$. The triangle inequality guarantees that each open ball is an open set. A topological space X is **metrizable** if there exists a metric d on X that generates the topology of X.

The **discrete metric**, defined by $d(x, y) = 1$ if $x \neq y$ and $d(x, y) = 0$ if $x = y$, generates the discrete topology. The zero semimetric, defined by $d(x, y) = 0$ for all x, y, generates the trivial topology.

4. The metric $d(x, y) = |x - y|$ defines a topology on the real line $\mathbb{R}$. Unless we state otherwise, $\mathbb{R}$ is assumed to have this topology.

   Every open interval (a, b) is an open set in this topology. Further, every open set is a countable union of disjoint open intervals (where the end points ∞ and $-\infty$ are allowed). To see this, note that every point in an open set must be contained in a maximal open interval, every open interval contains a rational number, and the rational numbers are countable.

5. The **Euclidean metric** on $\mathbb{R}^n$, $d(x, y) = \left[\sum_{i=1}^n (x_i - y_i)^2\right]^{1/2}$, defines its usual topology, also called the **Euclidean topology**. The Euclidean topology is also generated by the alternative metrics $d'(x, y) = \sum_{i=1}^n |x_i - y_i|$ and $d''(x, y) = \max_i |x_i - y_i|$.

6. The extended real line $\mathbb{R}^* = [-\infty, \infty] = \mathbb{R} \cup \{-\infty, \infty\}$ has a natural topology too. It consists of all subsets U such that for each $x \in U$:

   a. If $x \in \mathbb{R}$, then there exists some $\varepsilon > 0$ with $(x - \varepsilon, x + \varepsilon) \subset U$;

   b. If $x = \infty$, then there exists some $y \in \mathbb{R}$ with $(y, \infty] \subset U$; and

   c. If $x = -\infty$, then there exists some $y \in \mathbb{R}$ such that $[-\infty, y) \subset U$.

7. A different, and admittedly contrived, topology on $\mathbb{R}$ consists of all sets A such that for each x in A, there is a set of the form $U \setminus C \subset A$, where U is open in the usual topology, C is countable, and $x \in U \setminus C$.

8. Let $\mathbb{N} = \{1, 2, \ldots\}$. The collection of sets consisting of the empty set and all sets containing 1 is a topology on $\mathbb{N}$. The closed sets are $\mathbb{N}$ and all sets not containing 1.

9. Again let $\mathbb{N} = \{1, 2, \ldots\}$ and set $U_n = \{n, n + 1, \ldots\}$. Then the empty set and all the U_ns comprise a topology on $\mathbb{N}$. The closed sets are just the initial segments $\{1, 2, \ldots, n\}$ and $\mathbb{N}$ itself. ∎

We have just seen that a nontrivial set X can have many different topologies. The family of all topologies on X is partially ordered by set inclusion. If $\tau' \subset \tau$, that is, if every τ'-open set is also τ-open, then we say that τ' is **weaker** or **coarser** than τ, and that τ is **stronger** or **finer** than τ'.

The intersection of a family of topologies on a set is again a topology. (Why?) If $\mathcal{A}$ is an arbitrary nonempty family of subsets of a set X, then there exists a smallest (with respect to set inclusion) topology that includes $\mathcal{A}$. It is the intersection of all topologies that include $\mathcal{A}$. (Note that the discrete topology always includes $\mathcal{A}$.) This topology is called the **topology generated by** $\mathcal{A}$ and consists precisely of $\varnothing$, X and all sets of the form $\bigcup_\alpha V_\alpha$, where each V_α is a finite intersection of sets from $\mathcal{A}$.

A **base** for a topology τ is a subfamily $\mathcal{B}$ of τ such that each $U \in \tau$ is a union of members of $\mathcal{B}$. Equivalently, $\mathcal{B}$ is a base for τ if for every $x \in X$ and every open set U containing x, there is a basic open set $V \in \mathcal{B}$ satisfying $x \in V \subset U$. Conversely, if $\mathcal{B}$ is a family of sets that is closed under finite intersections and $\bigcup \mathcal{B} = X$, then the family τ of all unions of members of $\mathcal{B}$ is a topology for which $\mathcal{B}$ is a base. A subfamily $\mathcal{S}$ of a topology τ is a **subbase** for τ if the collection of all finite intersections of members of $\mathcal{S}$ is a base for τ. Note that if $\varnothing$ and X belong to a collection $\mathcal{S}$ of subsets, then $\mathcal{S}$ is a subbase for the topology it generates. A topological space is called **second countable** if it has a countable base. (Note that a topology has a countable base if and only if it has a countable subbase.)

If Y is a subset of a topological space (X, τ), then an easy argument shows that the collection τ_Y of subsets of Y, defined by

$$\tau_Y = \{V \cap Y : V \in \tau\},$$

is a topology on Y. This topology is called the **relative topology** or the **topology induced by** τ on Y. When $Y \subset X$ is equipped with its relative topology, we call Y a **(topological) subspace** of X. A set in τ_Y is called **(relatively) open** in Y. For example, since $X \in \tau$ and $Y \cap X = Y$, then Y is relatively open in itself. Note that the relatively closed subsets of Y are of the form

$$Y \setminus (Y \cap V) = Y \setminus V = Y \cap (X \setminus V),$$

where $V \in \tau$. That is, the relatively closed subsets of Y are the restrictions of the closed subsets of X to Y. Also note that for a semimetric topology, the relative topology is derived from the same semimetric restricted to the subset at hand. Unless otherwise stated, a subset Y of X carries its relative topology.

Part of the definition of a topology requires that a finite intersection of open sets is also an open set. However, a countable intersection of open sets need not be an open set. For instance, $\{0\} = \bigcap_{n=1}^\infty (-\frac{1}{n}, \frac{1}{n})$ is a countable intersection of open sets in $\mathbb{R}$ that is not open. Similarly, although finite unions of closed sets are closed sets, an arbitrary countable union of closed sets need not be closed; for instance, $(0, 1] = \bigcup_{n=1}^\infty [\frac{1}{n}, 1]$ is a countable union of closed sets in $\mathbb{R}$ that is neither open nor closed. The sets that are countable intersections of open sets or countable unions of closed sets are important enough that they have been given two special, albeit curious, names.

2.3 Definition   *A subset of a topological space is:*

- *a $\mathcal{G}_\delta$-set, or simply a $\mathcal{G}_\delta$, if it is a countable intersection of open sets.*

- *an $\mathcal{F}_\sigma$-set, or simply an $\mathcal{F}_\sigma$, if it is a countable union of closed sets.* [2]

The example $(0,1] = \bigcup_{n=1}^{\infty} [\frac{1}{n}, 1] = \bigcap_{n=1}^{\infty} (0, 1 + \frac{1}{n})$ shows that a set can be simultaneously a $\mathcal{G}_\delta$- and an $\mathcal{F}_\sigma$-set.

2.2   Neighborhoods and closures

Let (X, τ) be a topological space, and let A be any subset of X. The topology τ defines two sets intimately related to A. The **interior** of A, denoted A°, is the largest (with respect to inclusion) open set included in A. (It is the union of all open subsets of A.) The interior of a nonempty set may be empty. The **closure** of A, denoted $\overline{A}$, is the smallest closed set including A; it is the intersection of all closed sets including A. It is not hard to verify that $A \subset B$ implies $A^\circ \subset B^\circ$ and $\overline{A} \subset \overline{B}$. Also, it is obvious that a set A is open if and only if $A = A^\circ$, and a set B is closed if and only if $B = \overline{B}$. Consequently, for any set A, $\overline{(\overline{A})} = \overline{A}$ and $(A^\circ)^\circ = A^\circ$.

2.4 Lemma   *For any subset A of a topological space, $A^\circ = (\overline{A^c})^c$.*

Proof: Clearly,

$$A^\circ \subset A \implies A^c \subset (A^\circ)^c \implies \overline{A^c} \subset \overline{(A^\circ)^c} = (A^\circ)^c \implies A^\circ \subset (\overline{A^c})^c.$$

Also, $A^c \subset \overline{A^c}$ implies $(\overline{A^c})^c \subset A$. Since $(\overline{A^c})^c$ is an open set and A° is the largest open set included in A, we see that $A^\circ = (\overline{A^c})^c$.   ∎

The following property of the closure of the union of two sets easy to prove.

2.5 Lemma   *If A and B are subsets of a topological space, then $\overline{A \cup B} = \overline{A} \cup \overline{B}$.*

A **neighborhood** of a point x is any set V containing x in its interior. In this case we say that x is an **interior point** of V. According to our definition, a neighborhood need not be an open set, but some authors define neighborhoods to be open.

2.6 Lemma   *A set is open if and only if it is a neighborhood of each of its points.*

---

[2] This terminology seems to be derived from the common practice of using G to denote open sets and F for closed sets. The use of F probably comes from the French *fermé*, and G follows F. The letter σ probably comes from the word sum, which was often the way unions were described. According to H. L. Royden [290, p. 53], the letter δ is for the German *durchschnitt*.

The collection of all neighborhoods of a point x, called the **neighborhood base**, or **neighborhood system**, at x, is denoted $\mathcal{N}_x$. It is easy to verify that $\mathcal{N}_x$ satisfies the following properties.

1. $X \in \mathcal{N}_x$.

2. For each $V \in \mathcal{N}_x$, we have $x \in V$ (so $\varnothing \notin \mathcal{N}_x$).

3. If $V, U \in \mathcal{N}_x$, then $V \cap U \in \mathcal{N}_x$.

4. If $V \in \mathcal{N}_x$ and $V \subset W$, then $W \in \mathcal{N}_x$.

2.7 Definition   *A topology on X is called **Hausdorff** (or **separated**) if any two distinct points can be separated by disjoint neighborhoods of the points. That is, for each pair $x, y \in X$ with $x \neq y$ there exist neighborhoods $U \in \mathcal{N}_x$ and $V \in \mathcal{N}_y$ such that $U \cap V = \varnothing$.*

It is easy to see that singletons are closed sets in a Hausdorff space. (Why?) Topologies defined by metrics are Hausdorff. The trivial topology and the topologies in Examples 2.2.8 and 2.2.9 are not Hausdorff.

A **neighborhood base** at x is a collection $\mathcal{B}$ of neighborhoods of x with the property that if U is any neighborhood of x, then there is a neighborhood $V \in \mathcal{B}$ with $V \subset U$. A topological space is called **first countable** if every point has a countable neighborhood base.[3] Every semimetric space is first countable: the balls of radius $\frac{1}{n}$ around x form a countable neighborhood base at x. Clearly every second countable space is also first countable, but the converse is not true. (Consider an uncountable set with the discrete metric.)

A point x is a **point of closure** or **closure point** of the set A if every neighborhood of x meets A. Note that $\overline{A}$ coincides with the set of all closure points of A. A point x is an **accumulation point** (or a **limit point**, or a **cluster point**) of A if for each neighborhood V of x we have $(V \setminus \{x\}) \cap A \neq \varnothing$.

To see the difference between closure points and limit points, consider the subset $A = [0, 1) \cup \{2\}$ of $\mathbb{R}$. Then 2 is a closure point of A in $\mathbb{R}$, but not a limit point. The point 1 is both a closure point and a limit point of A.

We say that x is a **boundary point** of A if each neighborhood V of x satisfies both $V \cap A \neq \varnothing$ and $V \cap A^c \neq \varnothing$. Clearly, accumulation and boundary points of A belong to its closure $\overline{A}$. Let A' denote the set of all accumulation points of A (called the **derived set** of A) and ∂A denote the **boundary** of A, the set of all boundary points of A. We have the following identities:

$$\overline{A} = A^{\circ} \cup \partial A \quad \text{and} \quad \partial A = \partial A^c = \overline{A} \cap \overline{A^c}.$$

From the above identities, we see that a set A is closed if and only if $A' \subset A$ (and also if and only if $\partial A \subset A$). In other words, we have the following result.

---

[3] Now you know why the term "second countable" exists.

2.8 Lemma   *A set is closed if and only if it contains all its limit points.*

To illustrate this morass of definitions, again let $A = [0, 1) \cup \{2\}$ be viewed as a subset of $\mathbb{R}$. Then the boundary of A is $\{0, 1, 2\}$ and its derived set is $[0, 1]$. The closure of A is $[0, 1] \cup \{2\}$ and its interior is $(0, 1)$. Also note that the boundary of the set of rationals in $\mathbb{R}$ is the entire real line.

A subset A of a topological space X is **perfect** (in X) if it is closed and every point in A is an accumulation point of A. In particular, every neighborhood of a point x in A contains a point of A different from x. The space X is perfect if all of its points are accumulation points. A point $x \in A$ is an **isolated point** of A if there is a neighborhood V of x with $(V \setminus \{x\}) \cap A = \emptyset$. That is, if $\{x\}$ is a relatively open subset of A. A set is perfect if and only if it is closed and has no isolated points. Note that if A has no isolated points, then its closure, $\overline{A}$, is perfect in X. (Why?) Also, note that the empty set is perfect.

2.3   Dense subsets

A subset D of a topological space X is **dense** (in X) if $\overline{D} = X$. In other words, a set D is dense if and only if every nonempty open subset of X contains a point in D. In particular, if D is dense in X and x belongs to X, then every neighborhood of x contains a point in D. This means that any point in X can be approximated arbitrarily well by points in D. A set N is **nowhere dense** if its closure has empty interior. A topological space is **separable** if it includes a countable dense subset.

2.9 Lemma   *Every second countable space is separable.*

Proof: Let $\{B_1, B_2, \ldots\}$ be a countable base for the topology, and pick $x_i \in B_i$ for each i. Then $\{x_1, x_2, \ldots\}$ is dense. (Why?)   ∎

The converse is true for metric spaces (Lemma 3.4), but not in general.

2.10 Example   (A separable space with no countable base)   We give two examples of separable spaces that do not have countable bases. The first example is highly artificial, but easy to understand. The second example is both natural and important, but it requires some material that we do not cover till later.

1. Let X be an uncountable set and fix $x_0 \in X$. Take the topology consisting of the empty set and all sets containing x_0, cf. Example 2.2 (8). The set $\{x_0\}$ is dense in X, so X is separable. Furthermore, each set of the form $\{x_0, x\}$, $x \in X$, is open, so there is no countable base.

2. In order to understand this example you need some knowledge of weak topologies (Section 2.13) and the representation of linear functionals on

sequence spaces (see Chapter 16). The example is the space ℓ_1 of all absolutely summable real sequences equipped with the weak topology $\sigma(\ell_1, \ell_\infty)$.
The countable set of all eventually zero sequences with rational components
is a dense subset of ℓ_1 (why?), so $(\ell_1, \sigma(\ell_1, \ell_\infty))$ is a separable Hausdorff
space. However, $\sigma(\ell_1, \ell_\infty)$ is not first countable; see Theorem 6.26.   ∎

2.4 Nets

A **sequence** in X is a function from the natural numbers $\mathbb{N} = \{1, 2, \ldots\}$ into X.
We usually think of a sequence as a subset of X indexed by $\mathbb{N}$. A net is a direct
generalization of the notion of a sequence. Instead of the natural numbers, the
index set can be more general. The key issue is that the index set have a sense
of direction. A **direction** $\geq$ on a (not necessarily infinite) set D is a reflexive
transitive binary relation with the property that each pair has an upper bound.
That is, for each pair $\alpha, \beta \in D$ there exists some $\gamma \in D$ satisfying $\gamma \geq \alpha$ and $\gamma \geq \beta$.
Note that a direction need not be a partial order since we do not require it to be
antisymmetric. In practice, though, most directions are partial orders. Also note
that for a direction, every finite set has an upper bound. A **directed set** is any set
D equipped with a direction $\geq$. Here are a few examples.

1. The set of all natural numbers $\mathbb{N} = \{1, 2, \ldots\}$ with the direction $\geq$ defined
   by $m \geq n$ whenever $m \geqslant n$.

2. The set $(0, \infty)$ under the direction $\geq$ defined by $x \geq y$ whenever $x \geqslant y$.

3. The set $(0, 1)$ under the direction $\geq$ defined by $x \geq y$ whenever $x \leqslant y$.

4. The neighborhood system $\mathcal{N}_x$ of a point x in a topological space under the
   direction $\geq$ defined by $V \geq W$ whenever $V \subset W$. (The fact that the neighborhood system of a point is a directed set is the reason nets are so useful.)

5. The collection Φ of all finite subsets of a set X under the direction $\geq$ defined
   by $A \geq B$ whenever $A \supset B$.

If D is a directed set, then it is customary to denote the direction of D by $\geqslant$
instead of $\geq$. The context in which the symbol $\geqslant$ is employed indicates whether
or not it represents the direction of a set. If A and B are directed sets, then their
Cartesian product $A \times B$ is also a directed set under the **product direction** defined
by $(a, b) \geqslant (c, d)$ whenever $a \geqslant c$ and $b \geqslant d$. As a matter of fact, if $\{D_i : i \in I\}$ is an
arbitrary family of directed sets, then their Cartesian product $D = \prod_{i \in I} D_i$ is also
a directed set under the product direction defined by $(a_i)_{i \in I} \geqslant (b_i)_{i \in I}$ whenever
$a_i \geqslant b_i$ for each $i \in I$. Unless otherwise indicated, the Cartesian product of a
family of directed sets is directed by the product direction.

2.11 Definition   *A **net** in a set X is a function x: D → X, where D is a directed set. The directed set D is called the **index set** of the net and the members of D are indexes.*

In particular, sequences are nets. It is customary to denote the function $x(\cdot)$ simply by $\{x_\alpha\}$ and the directed set is understood. However, in case the index set D must be emphasized, the net is denoted $\{x_\alpha\}_{\alpha \in D}$. Moreover, we abuse notation slightly and write $\{x_\alpha\} \subset X$ for a net $\{x_\alpha\}$ in X. Observe that any directed set D is a net in itself under the identity function.

A net $\{x_\alpha\}$ in a topological space (X, τ) **converges** to some point x if it is eventually in every neighborhood of x. That is, if for each neighborhood V of x there exists some index α_0 (depending on V) such that $x_\alpha \in V$ for all $\alpha \geqslant \alpha_0$. We say that x is the **limit** of the net, and write $x_\alpha \to x$ or $x_\alpha \xrightarrow{\tau} x$. Note that in a metric space $x_\alpha \to x$ if and only if $d(x_\alpha, x) \to 0$. In Hausdorff spaces limits are unique.

2.12 Theorem   *A topological space is Hausdorff if and only if every net converges to at most one point.*

Proof: It is clear that in a Hausdorff space every net has at most one limit. (Why?) For the converse, assume that in a topological space X every net has at most one limit, and suppose by way of contradiction that X is not Hausdorff. Then there exist $x, y \in X$ with $x \neq y$ and such that for each $U \in \mathcal{N}_x$ and each $V \in \mathcal{N}_y$ we have $U \cap V \neq \emptyset$. For each $(U, V) \in \mathcal{N}_x \times \mathcal{N}_y$ let $x_{U,V} \in U \cap V$ and note that the net $\{x_{U,V}\}_{(U,V) \in \mathcal{N}_x \times \mathcal{N}_y}$ converges to both x and y, a contradiction.   ∎

While in metric spaces sequences suffice to describe closure points of sets (and several other properties as well), nets must be used to describe similar properties in general topological spaces.

2.13 Example (Sequences are not enough)   Recall the unusual topology on $\mathbb{R}$ described in Example 2.2.7. Sets of the form $U \setminus C$, where U is open in the usual topology and C is countable, constitute a base for this topology. In this topology, the only sequences converging to a point x are sequences that are eventually constant!

Note that the closure of $(0, 1)$ in this topology is still $[0, 1]$, but that no sequence in $(0, 1)$ converges to either 0 or 1. (If $\{x_1, x_2, \ldots\}$ is a sequence in $(0, 1)$, then $(0, 2) \setminus \{x_1, x_2, \ldots\}$ is a neighborhood of 1 containing no point of the sequence.)   ∎

This example is admittedly a contrived example. For more natural examples where nets are necessary, see Example 2.64, and Theorems 6.38 and 16.36.

2.14 Theorem   *A point belongs to the closure of a set if and only if it is the limit of a net in the set.*

Proof: Let x be a closure point of A. If $V \in \mathcal{N}_x$, then $V \cap A \neq \varnothing$, so there exists some $x_V \in V \cap A$. Then, $\{x_V\}_{V \in \mathcal{N}_x}$ is a net (where $\mathcal{N}_x$ is directed by $V \geqslant W$ whenever $V \subset W$) and $x_V \to x$.

For the converse, note that if a net $\{x_\alpha\}$ in A satisfies $x_\alpha \to x$, then x is clearly a closure point of A. ∎

The notion of subnet generalizes the notion of a subsequence.

2.15 Definition   *A net $\{y_\lambda\}_{\lambda \in \Lambda}$ is a **subnet** of a net $\{x_\alpha\}_{\alpha \in A}$ if there is a function $\varphi: \Lambda \to A$ satisfying*

1. $y_\lambda = x_{\varphi_\lambda}$ *for each $\lambda \in \Lambda$, where φ_λ stands for $\varphi(\lambda)$; and*

2. *for each $\alpha_0 \in A$ there exists some $\lambda_0 \in \Lambda$ such that $\lambda \geqslant \lambda_0$ implies $\varphi_\lambda \geqslant \alpha_0$.*

The following examples illustrate the definition of subnet.

- Every subsequence of a sequence is a subnet.

- Define the sequence $\{x_n\}$ of natural numbers by $x_n = n^2 + 1$. Then the net $\{y_{m,n}\}_{(m,n) \in \mathbb{N} \times \mathbb{N}}$ of natural numbers defined by $y_{m,n} = m^2 + 2mn + n^2 + 1$, is a subnet of the sequence $\{x_n\}$. To see this consider the function $\varphi: \mathbb{N} \times \mathbb{N} \to \mathbb{N}$ defined by $\varphi(m, n) = m + n$. But note that the net $\{y_{m,n}\}$ is not a subsequence of $\{x_n\}$.

- Consider the nets $\{y_\lambda\}_{\lambda \in (0,1)}$ and $\{x_\alpha\}_{\alpha \in (1,\infty)}$ defined by:

  ○ $y_\lambda = 1/\lambda$, where $(0, 1)$ is directed by $\lambda \geq \mu \iff \lambda \leqslant \mu$; and

  ○ $x_\alpha = \alpha$, where $(1, \infty)$ is directed by $\alpha \geq \beta \iff \alpha \geqslant \beta$.

Then, $\{y_\lambda\}$ is a subnet of $\{x_\alpha\}$ and conversely. To see this, consider the invertible function $\varphi: (0, 1) \to (1, \infty)$ defined by $\varphi(\lambda) = 1/\lambda$.

Subnets are associated with limit points of nets. An element x in a topological space is a **limit point** of a net $\{x_\alpha\}$ if for each neighborhood V of x and each index α there exists some $\beta \geqslant \alpha$ such that $x_\beta \in V$. The (possibly empty) set of all limit points of $\{x_\alpha\}$ is denoted $\mathrm{Lim}\,\{x_\alpha\}$.

2.16 Theorem   *In a topological space, a point is a limit point of a net if and only if it is the limit of some subnet.*

Proof: Let x be a limit point of a net $\{x_\alpha\}_{\alpha \in A}$ in some topological space. For each $(\alpha, V) \in A \times \mathcal{N}_x$ (where $A \times \mathcal{N}_x$ is directed by the product direction), pick some $\varphi_{\alpha,V} \in A$ with $\varphi_{\alpha,V} \geqslant \alpha$ and $x_{\varphi_{\alpha,V}} \in V$. Now define the net $\{y_{\alpha,V}\}$ by $y_{\alpha,V} = x_{\varphi_{\alpha,V}}$, and note that $\{y_{\alpha,V}\}_{(\alpha,V) \in A \times \mathcal{N}_x}$ is a subnet of $\{x_\alpha\}$ that converges to x.

For the converse, assume that in a topological space a subnet $\{y_\lambda\}_{\lambda \in \Lambda}$ of a net $\{x_\alpha\}_{\alpha \in A}$ converges to some point x. Fix $\alpha_0 \in A$ and a neighborhood V of x and let

$\varphi \colon \Lambda \to A$ be the mapping appearing in the definition of the subnet. Also, pick some $\lambda_0 \in \Lambda$ satisfying $y_\lambda \in V$ for each $\lambda \geqslant \lambda_0$. Next, choose some $\lambda_1 \in \Lambda$ such that $\varphi_\lambda \geqslant \alpha_0$ for each $\lambda \geqslant \lambda_1$. If $\lambda_2 \in \Lambda$ satisfies $\lambda_2 \geqslant \lambda_1$ and $\lambda_2 \geqslant \lambda_0$, then the index $\beta = \varphi_{\lambda_2}$ satisfies $\beta \geqslant \alpha_0$ and $x_\beta = x_{\varphi_{\lambda_2}} = y_{\lambda_2} \in V$, so that x is a limit point of the net $\{x_\alpha\}$.   ∎

2.17 Lemma   *In a topological space, a net converges to a point if and only if every subnet converges to that same point.*

Proof: Let $\{x_\alpha\}$ be a net in the topological space X converging to x. Clearly, for every subnet $\{y_\lambda\}$ of $\{x_\alpha\}$ we have $y_\lambda \to x$. For the converse, assume that every subnet of $\{x_\alpha\}$ converges to x, and assume by way of contradiction that $\{x_\alpha\}$ does not converge to x. Then, there exists a neighborhood V of x such that for any index $\alpha \in A$ there exists some $\varphi_\alpha \geqslant \alpha$ with $x_{\varphi_\alpha} \notin V$. Now if $y_\alpha = x_{\varphi_\alpha}$, then $\{y_\alpha\}_{\alpha \in A}$ is a subnet of $\{x_\alpha\}$ that fails to converge to x. This is a contradiction, so $x_\alpha \to x$, as desired. Note that limits do not need to be unique for this result.   ∎

As with sequences, every bounded net $\{x_\alpha\}$ of real numbers has a largest and a smallest limit point. The largest limit point of $\{x_\alpha\}$ is called the **limit superior**, written $\limsup_\alpha x_\alpha$, and the smallest is called the **limit inferior**, written $\liminf_\alpha x_\alpha$. It is not difficult to show that

$$\liminf_\alpha x_\alpha = \sup_\alpha \inf_{\beta \geqslant \alpha} x_\beta \leqslant \limsup_\alpha x_\alpha = \inf_\alpha \sup_{\beta \geqslant \alpha} x_\beta.$$

Also, note that $x_\alpha \to x$ in $\mathbb{R}$ if and only if

$$x = \liminf_\alpha x_\alpha = \limsup_\alpha x_\alpha.$$

2.5   Filters

The canonical example of a filter (and the reason filters are important in topology) is the neighborhood system $\mathcal{N}_x$ of a point x in a topological space. We introduce filters not to maximize the number of new concepts, but because they are genuinely useful in their own right, see for instance, Theorem 2.86.

2.18 Definition   *A **filter** on a set X is a family $\mathcal{F}$ of subsets of X satisfying:*

1. *$\varnothing \notin \mathcal{F}$ and $X \in \mathcal{F}$;*

2. *If $A, B \in \mathcal{F}$, then $A \cap B \in \mathcal{F}$; and*

3. *If $A \subset B$ and $A \in \mathcal{F}$, then $B \in \mathcal{F}$.*

*A **free filter** is a filter $\mathcal{F}$ with empty intersection, that is, $\bigcap_{A \in \mathcal{F}} A = \varnothing$. Filters that are not free are called **fixed**.*

Here are two more examples of filters.

- Let X be an arbitrary set, and let S be a nonempty subset of X. Then the collection of sets

$$\mathcal{F} = \{A \subset X : S \subset A\}$$

is a filter. Note that this filter is fixed.

- Let X be an infinite set and consider the collection $\mathcal{F}$ of cofinite sets. (A set is **cofinite** if it is the complement of a finite set.) That is,

$$\mathcal{F} = \{A \subset X : A^c \text{ is a finite set}\}.$$

Observe that $\mathcal{F}$ is a free filter.

A filter $\mathcal{G}$ is a **subfilter** of another filter $\mathcal{F}$ if $\mathcal{F} \subset \mathcal{G}$. In this case we also say that $\mathcal{G}$ is **finer** than $\mathcal{F}$. Note that despite the term *sub*filter, this partial order on filters is the opposite of inclusion. A filter $\mathcal{U}$ is an **ultrafilter** if $\mathcal{U}$ has no proper subfilter. That is, $\mathcal{U}$ is an ultrafilter if $\mathcal{U} \subset \mathcal{G}$ for a filter $\mathcal{G}$ implies $\mathcal{U} = \mathcal{G}$.

2.19 Ultrafilter Theorem   *Every filter is included in at least one ultrafilter. Consequently, every infinite set has a free ultrafilter.*

Proof: Let $\mathcal{F}$ be a filter on a set X, and let $\mathcal{C}$ be the nonempty collection of all subfilters of $\mathcal{F}$. That is,

$$\mathcal{C} = \{\mathcal{G} : \mathcal{G} \text{ is a filter and } \mathcal{F} \subset \mathcal{G}\}.$$

The collection $\mathcal{C}$ is partially ordered by inclusion. Given a chain $\mathcal{B}$ in $\mathcal{C}$, the family $\{A : A \in \mathcal{G} \text{ for some } \mathcal{G} \in \mathcal{B}\}$ is a filter that is an upper bound for $\mathcal{B}$ in $\mathcal{C}$. Thus the hypotheses of Zorn's Lemma 1.7 are satisfied, so $\mathcal{C}$ has a maximal element. Note that every maximal element of $\mathcal{C}$ is an ultrafilter including $\mathcal{F}$.

For the last part, note that if X is an infinite set, then

$$\mathcal{F} = \{A \subset X : A^c \text{ is finite}\}$$

is a free filter. Any ultrafilter that includes $\mathcal{F}$ is a free ultrafilter.   ∎

Several useful properties of ultrafilters are included in the next three lemmas.

2.20 Lemma   *Every fixed ultrafilter on a set X is of the form*

$$\mathcal{U}_x = \{A \subset X : x \in A\}$$

for a unique $x \in X$.

Proof: Let $\mathcal{U}$ be a fixed ultrafilter on X and let $x \in \bigcap_{A \in \mathcal{U}} A$. Then the family $\mathcal{U}_x = \{A \subset X : x \in A\}$ is a filter on X satisfying $\mathcal{U} \subset \mathcal{U}_x$. Hence $\mathcal{U} = \mathcal{U}_x$.   ∎

A nonempty collection $\mathcal{B}$ of subsets of a set X is a **filter base** if

1. $\varnothing \notin \mathcal{B}$; and

2. if $A, B \in \mathcal{B}$, then there exists some $C \in \mathcal{B}$ with $C \subset A \cap B$. (That is, $\mathcal{B}$ is directed by $\subset$.)

Every filter is, of course, a filter base. On the other hand, if $\mathcal{B}$ is a filter base for a set X, then the collection of sets

$$\mathcal{F}_{\mathcal{B}} = \{A \subset X : B \subset A \text{ for some } B \in \mathcal{B}\}$$

is a filter, called the **filter generated by** $\mathcal{B}$. For instance, the open neighborhoods at a point x of a topological space form a filter base $\mathcal{B}$ satisfying $\mathcal{F}_{\mathcal{B}} = \mathcal{N}_x$ (the filter of all neighborhoods at x).

2.21 Lemma   *An ultrafilter $\mathcal{U}$ on a set X satisfies the following:*

1. *If $A_1 \cup \cdots \cup A_n \in \mathcal{U}$, then $A_i \in \mathcal{U}$ for some i.*

2. *If $A \cap B \neq \varnothing$ for all $B \in \mathcal{U}$, then $A \in \mathcal{U}$.*

Proof: (1) Let $\mathcal{U}$ be an ultrafilter on X and let $A \cup B \in \mathcal{U}$. If $A \notin \mathcal{U}$, then the collection of sets $\mathcal{F} = \{C \subset X : A \cup C \in \mathcal{U}\}$ is a filter satisfying $B \in \mathcal{F}$ and $\mathcal{U} \subset \mathcal{F}$. Hence, $\mathcal{F} = \mathcal{U}$, so $B \in \mathcal{U}$. The general case follows by induction.

(2) Assume that $A \cap B \neq \varnothing$ for all $B \in \mathcal{U}$. If $\mathcal{B} = \{A \cap B : B \in \mathcal{U}\}$, then $\mathcal{B}$ is a filter base and the filter $\mathcal{F}$ it generates satisfies $\mathcal{U} \subset \mathcal{F}$ and $A \in \mathcal{F}$. Since $\mathcal{U}$ is an ultrafilter, we see that $\mathcal{F} = \mathcal{U}$, so $A \in \mathcal{U}$.   ∎

2.22 Lemma   *If $\mathcal{U}$ is a free ultrafilter on a set X, then $\mathcal{U}$ contains no finite subsets of X. In particular, only infinite sets admit free ultrafilters.*

Proof: We first note that a free filter $\mathcal{U}$ contains no singletons. For if $\{x\} \in \mathcal{U}$, then $\{x\} \cap A \neq \varnothing$ for each $A \in \mathcal{U}$, so $x \in A$ for each $A \in \mathcal{U}$. Hence $\bigcap_{A \in \mathcal{U}} A \neq \varnothing$, a contradiction.

Now for an ultrafilter $\mathcal{U}$, if the finite set $\{x_1, \ldots, x_n\} = \bigcup_{i=1}^{n} \{x_i\}$ belongs to $\mathcal{U}$, then by Lemma 2.21 (1) we have $\{x_i\} \in \mathcal{U}$ for some i, contrary to the preceding observation. Hence, no finite subset of X can be a member of $\mathcal{U}$.   ∎

We now come to the definition of convergence for filters. A filter $\mathcal{F}$ in a topological space **converges** to a point x, written $\mathcal{F} \to x$, if $\mathcal{F}$ includes the neighborhood filter $\mathcal{N}_x$ at x, that is, $\mathcal{N}_x \subset \mathcal{F}$. Similarly, a filter base $\mathcal{B}$ converges to some point x, denoted $\mathcal{B} \to x$, if the filter generated by $\mathcal{B}$ converges to x. Clearly, $\mathcal{N}_x \to x$ for each x.

An element x in a topological space is a **limit point** of a filter $\mathcal{F}$ whenever $x \in \overline{A}$ for each $A \in \mathcal{F}$. The set of all limit points of $\mathcal{F}$ is denoted $\text{Lim } \mathcal{F}$. Clearly, $\text{Lim } \mathcal{F} = \bigcap_{A \in \mathcal{F}} \overline{A}$. As with nets, the limit points of a filter are precisely the limits of its subfilters.

2.23 Theorem   *In a topological space, a point is a limit point of a filter if and only if there exists a subfilter converging to it.*

Proof: Let x be a limit point of a filter $\mathcal{F}$ in a topological space. That is, let $x \in \bigcap_{A \in \mathcal{F}} \overline{A}$. Then, the collection of sets

$$\mathcal{B} = \{V \cap A : V \in \mathcal{N}_x \text{ and } A \in \mathcal{F}\}$$

is a filter base. Moreover, if $\mathcal{G}$ is the filter it generates, then both $\mathcal{F} \subset \mathcal{G}$ and $\mathcal{N}_x \subset \mathcal{G}$. That is, $\mathcal{G}$ is a subfilter of $\mathcal{F}$ converging to x.

For the converse, assume that $\mathcal{G}$ is a subfilter of $\mathcal{F}$ (that is, $\mathcal{F} \subset \mathcal{G}$) satisfying $\mathcal{G} \to x$ (that is, $\mathcal{N}_x \subset \mathcal{G}$). Then each $V \in \mathcal{N}_x$ and each $A \in \mathcal{F}$ both belong to $\mathcal{G}$. Consequently, $V \cap A \neq \varnothing$. Therefore, $x \in \bigcap_{A \in \mathcal{F}} \overline{A}$.   ∎

We state without proof the following characterization of convergence.

2.24 Lemma   *In a topological space, a filter converges to a point if and only if every subfilter converges to that same point.*

2.6   Nets and Filters

There is an intimate connection between nets and filters. Let $\{x_\alpha\}_{\alpha \in D}$ be a net in a topological space X. For each α define the **section** or **tail** $F_\alpha = \{x_\beta : \beta \geq \alpha\}$ and consider the family of sets $\mathcal{B} = \{F_\alpha : \alpha \in D\}$. It is a routine matter to verify that $\mathcal{B}$ is a filter base. The filter $\mathcal{F}$ generated by $\mathcal{B}$ is called the **section filter** of $\{x_\alpha\}$ or the **filter generated by the net** $\{x_\alpha\}$.

The net $\{x_\alpha\}_{\alpha \in D}$ and its section filter $\mathcal{F}$ have the same limit points. That is, $\mathrm{Lim}\, \{x_\alpha\} = \mathrm{Lim}\, \mathcal{F}$. Indeed, if $x \in \mathrm{Lim}\, \{x_\alpha\}$, then x is (by Theorem 2.16) the limit of some subnet $\{y_\lambda\}$ of $\{x_\alpha\}$. A simple argument shows that the filter $\mathcal{G}$ generated by $\{y_\lambda\}$ is a subfilter of $\mathcal{F}$ and $\mathcal{G} \to x$. Conversely, if $x \in \mathrm{Lim}\, \mathcal{F}$, then for each index α and each $V \in \mathcal{N}_x$ we have $V \cap F_\alpha \neq \varnothing$. Thus if we choose some $y_{\alpha,V} \in V \cap F_\alpha$, then $\{y_{\alpha,V}\}_{(\alpha,V) \in D \times \mathcal{N}_x}$ defines a subnet of $\{x_\alpha\}$ satisfying $y_{\alpha,V} \to x$, so $x \in \mathrm{Lim}\, \{x_\alpha\}$.

Next, consider an arbitrary filter $\mathcal{F}$ in a topological space X and then define the set $D = \{(a, A) : A \in \mathcal{F} \text{ and } a \in A\}$. The set D has a natural direction $\geq$ defined by $(a, A) \geq (b, B)$ whenever $A \subset B$, so the formula $x_{a,A} = a$ defines a net in X, called the **net generated by the filter** $\mathcal{F}$. Observe that the section $F_{a,A} = A$, so the filter generated by the net $\{x_{a,A}\}$ is precisely $\mathcal{F}$. In particular, we have $\mathrm{Lim}\, \{x_{a,A}\} = \mathrm{Lim}\, \mathcal{F}$.

This argument establishes the following important equivalence result for nets and filters.

2.25 Theorem   (Equivalence of nets and filters)   *In a topological space, a net and the filter it generates have the same limit points. Similarly, a filter and the net it generates have the same limit points.*

2.7 Continuous functions

One of the most important duties of topologies is defining the class of continuous functions.

2.26 Definition *A function $f: X \to Y$ between topological spaces is **continuous** if $f^{-1}(U)$ is open in X for each open set U in Y.*

*We say that f is **continuous at the point** x if $f^{-1}(V)$ is a neighborhood of x whenever V is an open neighborhood of $f(x)$.*

In a metric space, continuity at a point x reduces to the familiar ε-δ definition: For each $\varepsilon > 0$, the ε-ball at $f(x)$ is a neighborhood of $f(x)$. The inverse image of the ball is a neighborhood of x, so for some $\delta > 0$, the δ-ball at x is in the inverse image. That is, if y is within δ of x, then $f(y)$ is within ε of $f(x)$. The next two theorems give several other characterizations of continuity.

2.27 Theorem *For a function $f: X \to Y$ between topological spaces the following statements are equivalent.*

1. *f is continuous.*

2. *f is continuous at every point.*

3. *If C is a closed subset of Y, then $f^{-1}(C)$ is a closed subset of X.*

4. *If B is an arbitrary subset of Y, then $f^{-1}(B^\circ) \subset \left[f^{-1}(B) \right]^\circ$.*

5. *If A is an arbitrary subset of X, then $f(\overline{A}) \subset \overline{f(A)}$.*

6. *$f^{-1}(V)$ is open in X for each V in some subbase for the topology on Y.*

Proof: (1) $\implies$ (2)   This is obvious.

(2) $\implies$ (3)   Let C be a closed subset of Y and let $x \in [f^{-1}(C)]^c = f^{-1}(C^c)$. So $f(x) \in C^c$. Since C^c is an open set, the continuity of f at x guarantees the existence of some neighborhood V of x such that $y \in V$ implies $f(y) \in C^c$. The latter implies $V \subset f^{-1}(C^c)$, so $f^{-1}(C^c)$ is a neighborhood of all of its points. Thus $f^{-1}(C^c)$ is open, which implies that $f^{-1}(C) = [f^{-1}(C^c)]^c$ is closed.

(3) $\implies$ (4)   Let B be a subset of Y. Since B° is open, the set $(B^\circ)^c$ is closed, so by hypothesis $\left[f^{-1}(B^\circ) \right]^c = f^{-1}((B^\circ)^c)$ is also closed. This means that $f^{-1}(B^\circ)$ is open, and since $f^{-1}(B^\circ) \subset f^{-1}(B)$ is true, we see that $f^{-1}(B^\circ) \subset \left[f^{-1}(B) \right]^\circ$.

(4) $\implies$ (5)   Let A be an arbitrary subset of X and let $y \in f(\overline{A})$. Then, there exists some $x \in \overline{A}$ with $y = f(x)$. If V is an open neighborhood of y, then $f^{-1}(V) = f^{-1}(V^\circ) \subset [f^{-1}(V)]^\circ$, so $f^{-1}(V) = \left[f^{-1}(V) \right]^\circ$, proving that $f^{-1}(V)$ is an open neighborhood of x. Since $x \in \overline{A}$, we see that $f^{-1}(V) \cap A \neq \varnothing$, so $V \cap f(A) \neq \varnothing$. Therefore $y \in \overline{f(A)}$.

(5) $\implies$ (6)   Let V be an open subset of Y. Put $A = \left[f^{-1}(V)\right]^c = f^{-1}(V^c)$ and note that from

$$f(\overline{A}) \subset \overline{f(A)} = \overline{f(f^{-1}(V^c))} \subset \overline{V^c} = V^c,$$

we see that $\overline{A} \subset f^{-1}(V^c) = A$. Since $A \subset \overline{A}$ is trivially true, we infer that $A = \overline{A}$, so that A is a closed set. Hence, $f^{-1}(V) = A^c$ is open.

(6) $\implies$ (1)   This is straightforward.   ∎

Given a filter base $\mathcal{B}$ in a set X and a function $f : X \to Y$, notice that the collection of sets $f(\mathcal{B}) = \{f(B) : B \in \mathcal{B}\}$ is a filter base in Y.

Continuity is often more easily expressed in terms of convergence of nets and filters.

2.28 Theorem   *For a function $f : X \to Y$ between two topological spaces and point x in X the following statements are equivalent.*

1. *The function f is continuous at x.*

2. *If a net $x_\alpha \to x$ in X, then $f(x_\alpha) \to f(x)$ in Y.*

3. *If a filter $\mathcal{F} \to x$ in X, then $f(\mathcal{F}) \to f(x)$ in Y.*

Proof: (1) $\implies$ (3)   Let $\mathcal{F} \to x$. That is, let $\mathcal{N}_x \subset \mathcal{F}$. The continuity of f at x guarantees that $f^{-1}(V) \in \mathcal{N}_x$ for each $V \in \mathcal{N}_{f(x)}$. Hence, $f^{-1}(V) \in \mathcal{F}$ for each $V \in \mathcal{N}_{f(x)}$. But then from $f(f^{-1}(V)) \subset V$, we see that $\mathcal{N}_{f(x)}$ is included in the filter generated by $f(\mathcal{F})$. Thus $f(\mathcal{F}) \to f(x)$.

(3) $\implies$ (2)   Assume that a net $\{x_\alpha\}_{\alpha \in A}$ satisfies $x_\alpha \to x$. If for each α we define $F_\alpha = \{x_\beta : \beta \geqslant \alpha\}$, then the filter base $\mathcal{B} = \{F_\alpha : \alpha \in A\}$ converges to x, so by hypothesis, $f(\mathcal{B}) \to f(x)$. This implies that if V is an arbitrary neighborhood of $f(x)$, then there exists some index α_0 satisfying $f(F_{\alpha_0}) \subset V$. Hence, $f(x_\alpha) \in V$ for all $\alpha \geqslant \alpha_0$, so $f(x_\alpha) \to f(x)$.

(2) $\implies$ (1)   Assume (2) and assume by way of contradiction that f is not continuous at x. Then there is an open neighborhood V of $f(x)$ such that $f^{-1}(V)$ is not a neighborhood of x—that is, $x \notin \left[f^{-1}(V)\right]^\circ$. By Lemma 2.4 we have $x \in \overline{\left[f^{-1}(V)\right]^c}$, so (by Theorem 2.14) there exists a net $\{x_\alpha\}$ in $\left[f^{-1}(V)\right]^c = f^{-1}(V^c)$ such that $x_\alpha \to x$. So by hypothesis, $f(x_\alpha) \to f(x)$. Since $\{f(x_\alpha)\} \subset V^c$, which is closed, $f(x) \in V^c$, a contradiction.   ∎

The preceding two theorems have the following useful corollary for real functions, which we present without proof.

2.29 Corollary   *If $f, g : X \to \mathbb{R}$ are continuous real functions on a topological space, then the following real functions are also continuous: $\alpha f + \beta g$, fg, $\min\{f, g\}$, $\max\{f, g\}$, $|f|$, where α, β are real numbers. If $g(x) \neq 0$ for all x, then $\frac{f}{g}$ is also continuous.*

Another simple consequence of the definition of continuity is the following lemma.

2.30 Lemma   *The composition of continuous functions between topological spaces is continuous.*

Two topological spaces X and Y are called **homeomorphic** if there is a one-to-one continuous function f from X onto Y such that f^{-1} is continuous too. The function f is called a **homeomorphism**. The homeomorphism defines a one-to-one correspondence between the points of the spaces and the open sets of the two spaces. From the topological point of view two homeomorphic spaces are identical—only the names of the points have been changed. Any topological property, that is, any property defined in terms of the topology, possessed by one space is also possessed by the other. There is a well-known line that claims that a topologist is someone who cannot tell the difference between a coffee cup and a donut (since they are homeomorphic). As another example, the open unit interval $(0, 1)$ and the whole real line $\mathbb{R}$ are homeomorphic. (Can you find a homeomorphism?) It is a nontrivial exercise to verify that Euclidean spaces of different dimensions are *not* homeomorphic.

A mapping $f\colon X \to Y$ between two topological spaces is an **embedding** if $f\colon X \to f(X)$ is a homeomorphism. In this case we can think of X as a topological subspace of Y by identifying X with its image $f(X)$.

2.8   Compactness

We have already seen that the definition of a topology is sufficiently weak to allow some pathetic topologies, for example, the trivial topology. In order to prove any interesting results we need additional hypotheses on the topology.

An **open cover** of a set K is a collection of open sets whose union includes K. A subset K of a topological space is **compact** if every open cover of K includes a finite subcover. That is, K is compact if every family $\{V_i : i \in I\}$ of open sets satisfying $K \subset \bigcup_{i \in I} V_i$ has a finite subfamily $V_{i_1}, \ldots, V_{i_n}$ such that $K \subset \bigcup_{j=1}^{n} V_{i_j}$. A topological space is called a **compact space** if it is a compact set. A subset of a topological space is called **relatively compact** if its closure is compact.[4]

For the trivial topology every set is compact; for the discrete topology only finite sets are compact. It is easily seen that every subset in Example 2.2.9 is compact. The well-known Heine–Borel Theorem 3.30 below is often mistaken for the definition of compactness. It states that a subset of $\mathbb{R}^n$ is compact if and only if it is closed and bounded. This result is false in more general metric spaces. For instance, consider an infinite set with the discrete metric.

---

[4] Note that relative compactness unfortunately has nothing to do with the relative topology on a set. Indeed, a set is compact in its relative topology if and only if it is compact. Nevertheless, the terminology is standard.

A family of sets has the **finite intersection property** if every finite subfamily has a nonempty intersection. Every filter has the finite intersection property, and an ultrafilter is a maximal family with the finite intersection property. Compactness can also be characterized in terms of the finite intersection property.

2.31 Theorem   *For a topological space X, the following are equivalent.*

1. *X is compact.*

2. *Every family of closed subsets of X with the finite intersection property has a nonempty intersection.*

3. *Every net in X has a limit point (or, equivalently, every net has a convergent subnet).*

4. *Every filter in X has a limit point, (or, equivalently, every filter has a convergent subfilter).*

5. *Every ultrafilter in X is convergent.*

Proof: (1) $\iff$ (2)   Assume that X is compact, and let $\mathcal{E}$ be a family of closed subsets of X. If $\bigcap_{E \in \mathcal{E}} E = \varnothing$, then $X = \bigcup_{E \in \mathcal{E}} E^c$, therefore $\{E^c : E \in \mathcal{E}\}$ is an open cover of X. Thus there exist $E_1, \ldots, E_n \in \mathcal{E}$ satisfying $X = \bigcup_{i=1}^n E_i^c$. This implies $\bigcap_{i=1}^n E_i = \varnothing$, so $\mathcal{E}$ does not have the finite intersection property. Thus, if $\mathcal{E}$ possesses the finite intersection property, then $\bigcap_{E \in \mathcal{E}} E \neq \varnothing$.

For the converse, assume that (2) is true and that $\mathcal{V}$ is an open cover of X. Then $\bigcap_{V \in \mathcal{V}} V^c = \varnothing$, so the finite intersection property must be violated. That is, there exist $V_1, \ldots, V_n \in \mathcal{V}$ satisfying $\bigcap_{j=1}^n V_j^c = \varnothing$, or $X = \bigcup_{j=1}^n V_j$, which proves that X is compact.

(3) $\iff$ (4)   This equivalence is immediate from Theorem 2.25.

(4) $\iff$ (5)   This equivalence follows from Theorems 2.23 and 2.19.

(4) $\iff$ (2)   Assume first that $\mathcal{G}$ is a family of closed subsets of X with the finite intersection property. Then $\mathcal{G}$ is a filter base, so by hypothesis the filter $\mathcal{F}$ it generates has a limit point. Now note that $\bigcap_{G \in \mathcal{G}} G = \bigcap_{A \in \mathcal{F}} \overline{A} = \text{Lim } \mathcal{F} \neq \varnothing$.

For the converse, assume that (2) is true and that $\mathcal{F}$ is a filter on X. Then the family of closed sets $\mathcal{G} = \{\overline{A} : A \in \mathcal{F}\}$ satisfies the finite intersection property, so $\text{Lim } \mathcal{F} = \bigcap_{A \in \mathcal{F}} \overline{A} \neq \varnothing$.   ∎

A subset A of a topological space is **sequentially compact** if every sequence in A has a subsequence converging to an element of A. A topological space X is **sequentially compact** if X itself is a sequentially compact set.

In many ways compactness can be viewed as a topological generalization of finiteness. There is an informal principle that compact sets behave like points in many instances. We list a few elementary properties of compact sets.

- Finite sets are compact.

- Finite unions of compact sets are compact.

- Closed subsets of compact sets are compact.

- If $K \subset Y \subset X$, then K is a compact subset of X if and only if K is a compact subset of Y (in the relative topology).

We note the following result, which we use frequently without any special mention. It is an instance of how compact sets act like points.

2.32 Lemma   *If K is a compact subset of a Hausdorff space, and $x \notin K$, then there are disjoint open sets U and V with $K \subset U$ and $x \in V$. In particular, compact subsets of Hausdorff spaces are closed.*

Proof: Since X is Hausdorff, for each y in K, there are disjoint open neighborhoods U_y of y and V_y of x. The U_ys cover K, so there is a finite subfamily $U_{y_1}, \ldots, U_{y_n}$ covering K. Now note that the disjoint open sets $U = \bigcup_{i=1}^n U_{y_i}$ and $V = \bigcap_{i=1}^n V_{y_i}$ have the desired properties.   ∎

Compact subsets of non-Hausdorff spaces need not be closed.

2.33 Example (A compact set that is not closed)   Let X be a set with at least two elements, endowed with the indiscrete topology. Any singleton is compact, but X is the only nonempty closed set.   ∎

2.34 Theorem   *Every continuous function between topological spaces carries compact sets to compact sets.*

Proof: Let $f: X \to Y$ be a continuous function between two topological spaces, and let K be a compact subset of X. Also, let $\{V_i : i \in I\}$ be an open cover of $f(K)$. Then $\{f^{-1}(V_i) : i \in I\}$ is an open cover of K. By the compactness of K there exist indexes $i_1, \ldots, i_n$ satisfying $K \subset \bigcup_{j=1}^n f^{-1}(V_{i_j})$. Hence,

$$f(K) \subset f\left(\bigcup_{j=1}^n f^{-1}(V_{i_j})\right) = \bigcup_{j=1}^n f(f^{-1}(V_{i_j})) \subset \bigcup_{j=1}^n V_{i_j},$$

which shows that $f(K)$ is a compact subset of Y.   ∎

Since a subset of the real line is compact if and only if it is closed and bounded, the preceding lemma yields the following fundamental result.

2.35 Corollary (Weierstrass)   *A continuous real-valued function defined on a compact space achieves its maximum and minimum values.*

A function $f: X \to Y$ between topological spaces is **open** if it carries open sets to open sets ($f(U)$ is open whenever U is), and **closed** if it carries closed sets to closed sets ($f(F)$ is closed whenever F is). If f has an inverse, then f^{-1} is continuous if and only if f is open (and also if and only if f is closed).

The following is a simple but very useful result.

2.36 Theorem   *A one-to-one continuous function from a compact space onto a Hausdorff space is a homeomorphism.*

Proof: Assume that $f: X \to Y$ satisfies the hypotheses. If C is a closed subset of X, then C is a compact set, so by Theorem 2.34 the set $f(C)$ is also compact. Since Y is Hausdorff, it follows that $f(C)$ is also a closed subset of Y. That is, f is a closed function. Now note that $(f^{-1})^{-1}(C) = f(C)$, and by Theorem 2.27, the function $f^{-1}: Y \to X$ is also continuous.   ∎

We close with an example of a compact Hausdorff space whose unusual properties are exploited in Examples 12.9 and 14.13.

2.37 Example (Space of ordinals)   The set $\Omega = [1, \omega_1]$ of ordinals is a Hausdorff topological space with its **order topology**. A subbase for this topology consists of all sets of the form $\{y \in \Omega : y < x\}$ or $\{y \in \Omega : y > x\}$ for some $x \in \Omega$. Recall that any increasing sequence in Ω has a least upper bound. The least upper bound is also the limit of the sequence in the order topology.

The topological space Ω is compact. To see this, let $\mathcal{V}$ be an open cover of Ω. Since ω_1 is contained in some open set, then for some ordinal $x_0 < \omega_1$ the interval $(x_0, \omega_1] = \{y \in \Omega : x_0 < y \le \omega_1\}$ is included in some member of the cover. Let x_1 be the first such ordinal, and let $V_1 \in \mathcal{V}$ satisfy $(x_1, \omega_1] \subset V_1$. By the same reasoning, unless $x_1 = 1$ there is a first ordinal $x_2 < x_1$ with $(x_2, x_1]$ included in some $V_2 \in \mathcal{V}$. Proceeding inductively, as long as $x_{n-1} \ne 1$, we can find $x_n < x_{n-1}$, the first ordinal with $(x_n, x_{n-1}] \subset V_n \in \mathcal{V}$. We claim that $x_n = 1$ for some n, so this process stops. Otherwise the set $\{x_1 > x_2 > \cdots\}$ has no first element. Thus $V_1, \ldots, V_n$ cover Ω with the possible exception of the point 1, which belongs to some member of $\mathcal{V}$.

Note that Ω is not separable: Let C be any countable subset of Ω, and let b be the least upper bound of $C \setminus \{\omega_1\}$. Then any x with $b < x < \omega_1$ cannot lie in the closure of C, so C is not dense. A consequence of this is that Ω is not metrizable, since by Lemma 3.26 below, a compact metrizable space must be separable.   ∎

2.9   Nets vs. sequences

So far, we have seen several similarities between nets and sequences, and you may be tempted to think that for most practical purposes nets and sequences behave alike. This is a mistake. We warn you that there are subtle differences between nets and sequences that you need to be careful of. The most important of them is highlighted by the following theorem and example.

2.38 Theorem   *In a topological space, if a sequence $\{x_n\}$ converges to a point x, then the set $\{x, x_1, x_2, \ldots\}$ of all terms of the sequence together with the limit point x is compact.*

Proof: Let $\{U_i\}_{i \in I}$ be an open cover of $S = \{x, x_1, x_2, \ldots\}$. Pick some index i_0 with $x \in U_{i_0}$ and note that there exists some m such that $x_n \in U_{i_0}$ for all $n > m$. Now for each $1 \leqslant k \leqslant m$ pick an index i_k with $x_k \in U_{i_k}$ and note that $S \subset \bigcup_{k=0}^{m} U_{i_k}$, which shows that S is compact.   ∎

Nets need not exhibit this property.

2.39 Example   (A convergent net without compact tails)   Let D be the set of rational numbers in the interval $(0, 1)$, directed by the usual ordering $\geqslant$ on the real numbers. It defines a net $\{x_\alpha\}_{\alpha \in D}$ in the compact metric space $[0, 1]$ by letting $x_\alpha = \alpha$. Clearly, $x_\alpha \to 1$ in $[0, 1]$. If $\alpha_0 \in D$, then note that

$$\{x_\alpha : \alpha \geqslant \alpha_0\} \cup \{1\} = \{r \in [\alpha_0, 1] : r \text{ is a rational number}\},$$

which fails to be compact (or even closed) for any $\alpha_0 \in D$.

It is also interesting to note that for any $\alpha_0 \in D$, every real number $z \in [\alpha_0, 1)$ is an accumulation point of the set $\{x_\alpha : \alpha \geqslant \alpha_0\}$. However, note that there is no subnet of $\{x_\alpha\}$ that converges to z. (Every subnet of $\{x_\alpha\}$ converges to 1.)   ∎

Whenever possible, it is desirable to replace nets with sequences, and theorems to this effect are very useful. One case that allows us to replace nets with sequences is the case of a first countable topology (each point has a countable neighborhood base). This class of spaces includes all metric spaces.

2.40 Theorem   *Let X be a first countable topological space.*

1. *If A is a subset of X, then x belongs to the closure of A if and only if there is a sequence in A converging to x.*

2. *A function $f : X \to Y$, where Y is another topological space, is continuous if and only if $x_n \to x$ in X implies $f(x_n) \to f(x)$ in Y.*

Proof: (1)   Let $x \in \overline{A}$. Let $\{V_1, V_2, \ldots\}$ be a countable base for the neighborhood system $\mathcal{N}_x$ at x. Since $x \in \overline{A}$, we have $\left(\bigcap_{k=1}^{n} V_k\right) \cap A \neq \varnothing$ for each n. Pick $x_n \in \left(\bigcap_{k=1}^{n} V_k\right) \cap A$ and note that $x_n \to x$.

(2)   If $f : X \to Y$ is continuous, then $x_n \to x$ implies $f(x_n) \to f(x)$. For the converse, assume that $x_n \to x$ in X implies $f(x_n) \to f(x)$ in Y and let $A \subset X$. By Theorem 2.27 (5), it suffices to show that $f(\overline{A}) \subset \overline{f(A)}$. So let $x \in \overline{A}$. By part (1), there exists a sequence $\{x_n\} \subset A$ satisfying $x_n \to x$. By hypothesis, $f(x_n) \to f(x)$, so $f(x) \in \overline{f(A)}$.   ∎

2.10   Semicontinuous functions

A function $f \colon X \to [-\infty, \infty]$ on a topological space X is:

- **lower semicontinuous** if for each $c \in \mathbb{R}$ the set $\{x \in X : f(x) \leqslant c\}$ is closed (or equivalently, the set $\{x \in X : f(x) > c\}$ is open).

- **upper semicontinuous** if for each $c \in \mathbb{R}$ the set $\{x \in X : f(x) \geqslant c\}$ is closed (or equivalently, the set $\{x \in X : f(x) < c\}$ is open).

Clearly, a function f is lower semicontinuous if and only $-f$ is upper semicontinuous, and vice versa. Also, a real function is continuous if and only if it is both upper and lower semicontinuous.

2.41 Lemma   *The pointwise supremum of a family of lower semicontinuous functions is lower semicontinuous. Similarly, the pointwise infimum of a family of upper semicontinuous functions is upper semicontinuous.*

Proof: We prove the lower semicontinuous case only. To this end, let $\{f_\alpha\}$ be a family of lower semicontinuous functions defined on a topological space X, and let $f(x) = \sup_\alpha f_\alpha(x)$ for each $x \in X$. From the identity

$$\{x \in X : f(x) \leqslant c\} = \bigcap_\alpha \{x \in X : f_\alpha(x) \leqslant c\},$$

we see that $\{x \in X : f(x) \leqslant c\}$ is closed for each $c \in \mathbb{R}$.   ∎

The next characterization of semicontinuity is sometimes used as a definition. Later, in Corollary 2.60, we present another characterization of semicontinuity.

2.42 Lemma   *Let $f \colon X \to [-\infty, \infty]$ be a function on a topological space. Then:*

f is lower semicontinuous if and only if $x_\alpha \to x \implies \liminf\limits_\alpha f(x_\alpha) \geqslant f(x)$.

f is upper semicontinuous if and only if $x_\alpha \to x \implies \limsup\limits_\alpha f(x_\alpha) \leqslant f(x)$.

When X is first countable, nets can be replaced by sequences.

Proof: We establish the lower semicontinuous case. So assume first that f is lower semicontinuous, and let $x_\alpha \to x$ in X. If $f(x) = -\infty$, then the desired inequality is trivially true. So suppose $f(x) > -\infty$. Fix $c < f(x)$ and note that (by the lower semicontinuity of f) the set $V = \{y \in X : f(y) > c\}$ is open. Since $x \in V$, there is some α_0 such that $x_\beta \in V$ for all $\beta \geqslant \alpha_0$, that is, $f(x_\beta) > c$ for all $\beta \geqslant \alpha_0$. Hence,

$$\liminf_\alpha f(x_\alpha) = \sup_\alpha \inf_{\beta \geqslant \alpha} f(x_\beta) \geqslant \inf_{\beta \geqslant \alpha_0} f(x_\beta) \geqslant c$$

for all $c < f(x)$. This implies that $\liminf_\alpha f(x_\alpha) \geqslant f(x)$.

Now assume that $x_\alpha \to x$ in X implies $\liminf_\alpha f(x_\alpha) \geq f(x)$, and let $c \in \mathbb{R}$. Consider the set $F = \{x \in X : f(x) \leq c\}$, and let $\{y_\alpha\}$ be a net in F satisfying $y_\alpha \to y$ in X. Then, from the inequality $f(y_\alpha) \leq c$ for each α, we obtain $f(y) \leq \liminf_\alpha f(y_\alpha) \leq c$, so $y \in F$. That is, F is closed, and hence f is lower semicontinuous. ∎

The following result generalizes Weierstrass' Theorem (Corollary 2.35) on the extreme values of continuous functions.

2.43 Theorem   *A real-valued lower semicontinuous function on a compact space attains a minimum value, and the nonempty set of minimizers is compact. Similarly, an upper semicontinuous function on a compact set attains a maximum value, and the nonempty set of maximizers is compact.*

Proof: Let X be a compact space and let $f : X \to \mathbb{R}$ be lower semicontinuous. Put $A = f(X)$, and for each c in A, put $F_c = \{x \in X : f(x) \leq c\}$. Since f is lower semicontinuous, the nonempty set F_c is closed. Furthermore, the family $\{F_c : c \in A\}$ has the finite intersection property. (Why?) Since X is compact, $\bigcap_{c \in A} F_c$ is compact and nonempty. But this is just the set of minimizers of f. ∎

We can generalize this result to maximal elements of binary relations. Let $\geq$ be a total preorder, that is, a reflexive total transitive binary relation, on a topological space X. Say that $\geq$ is **continuous** if $\geq$ is a closed subset of $X \times X$. Let us say that $\geq$ is **upper semicontinuous** if $\{x \in X : x \geq y\}$ is closed for each y. In particular, if $\geq$ is continuous, then it is upper semicontinuous. The following theorem can be strengthened, but it is useful enough.

2.44 Theorem (Maxima of binary relations)   *An upper semicontinuous total preorder on a compact space has a greatest element.*

Proof: Let X be compact, and for each y, let $F(y) = \{x \in X : x \geq y\}$. Then $\{F(y) : y \in X\}$ is a family of nonempty closed sets with the finite intersection property. (Why?) Therefore $F = \bigcap_{y \in X} F(y)$ is nonempty. Clearly, $x \in F$ implies $x \geq y$ for every $y \in X$. ∎

2.11 Separation properties

There are several "separation" properties in addition to the Hausdorff property that an arbitrary topological space may or may not satisfy. Let us say that two nonempty sets are **separated by open sets**, if they are included in disjoint open sets, and that they are **separated by continuous functions** if there is a real continuous function taking on values only in $[0, 1]$ that assumes the value zero on one set and the value one on the other. Clearly separation by continuous functions implies separation by open sets.

2.45 Definition   *A topological space X is:*

• **regular** *if every nonempty closed set and every singleton disjoint from it can be separated by open sets.*

• **completely regular** *if every nonempty closed set and every singleton disjoint from it can be separated by continuous functions.*

• **normal** *if every pair of disjoint nonempty closed sets can be separated by open sets.*

The next two results are the main reason that normal spaces are important. Their proofs are similar and involve a cumbersome recursive construction of families of closed sets.

2.46 Urysohn's Lemma   *For a topological space X, the following statements are equivalent.*

1. *The space X is normal.*

2. *Every pair of nonempty disjoint closed subsets of X can be separated by a continuous function.*

3. *If C is a closed subset of X and $f: C \to [0, 1]$ is continuous, then there is a continuous extension $\hat{f}: X \to [0, 1]$ of f satisfying*

$$\sup_{x \in X} \hat{f}(x) = \sup_{x \in C} f(x).$$

For a proof, see, e.g., [13, Theorem 10.5, p. 81]. In particular, Urysohn's Lemma implies that every normal Hausdorff space is completely regular.

2.47 Tietze Extension Theorem   *Let C be a closed subset of a normal topological space X, and let $f: C \to \mathbb{R}$ be continuous. Then there exists a continuous extension of f to X.*

For a proof, see, e.g., [13, Theorem 10.6, p. 84]. Unfortunately, we cannot guarantee that if A and B are disjoint closed subsets of a normal space that there is a continuous function f satisfying $A = f^{-1}(1)$ and $B = f^{-1}(0)$. A topological space that has this property is called **perfectly normal**.[5] Clearly perfectly normal spaces are normal. We shall see (Corollary 3.21) that every metric space is perfectly normal.

---

[5] Our definition is the one used by K. Kuratowski [218]. S. Willard [342] requires in addition that the space be T_1 (see the end of this section for the T_1 property). J. L. Kelley [198] and N. Bourbaki [61] define a space to be perfectly normal if it is normal and every closed set is a $\mathcal{G}_\delta$. For Hausdorff spaces the definitions agree, cf. [14, Problem 10.9, p. 96] or [342, Exercise 15C, p. 105].

2.48 Theorem　　*Every compact Hausdorff space is normal, and therefore completely regular.*

Proof: Let X be a compact Hausdorff space and let E and F be disjoint nonempty closed subsets of X. Then both E and F are compact. Choose a point $x \in E$. By Lemma 2.32 for each $y \in F$, there exist disjoint open sets V_y and U_y with $y \in V_y$ and $E \subset U_y$. Since $\{V_y : y \in F\}$ is an open cover of F, which is compact, there exist $y_1, \ldots, y_k \in F$ such that $F \subset \bigcup_{i=1}^{k} V_{y_i}$. Now note that the open sets $V = \bigcup_{i=1}^{k} V_{y_i}$ and $U = \bigcap_{i=1}^{k} U_{y_i}$ satisfy $E \subset U$, $F \subset V$, and $U \cap V = \varnothing$.　■

We can modify the proof of Theorem 2.48 in order to prove a slightly stronger result. Before we can state the result we need the following definition. A topological space is a **Lindelöf space** if every open cover has a countable subcover. Clearly every second countable space is a Lindelöf space.

2.49 Theorem　　*Every regular Lindelöf space is normal.*

Proof: Let A and B be nonempty disjoint closed subsets of a Lindelöf space X. The regularity of X implies that for each $x \in A$ there exists an open neighborhood V_x of x such that $\overline{V_x} \cap B = \varnothing$. Similarly, for each $y \in B$ there exists an open neighborhood W_y of y such that $\overline{W_y} \cap A = \varnothing$. Clearly the collection of open sets $\{V_x : x \in A\} \cup \{W_y : y \in B\} \cup \{X \setminus A \cup B\}$ covers X. Since X is a Lindelöf space, there exist a countable subcollection $\{V_n\}$ of $\{V_x\}_{x \in A}$ and a countable subcollection $\{W_n\}$ of $\{W_y\}_{y \in B}$ such that $A \subset \bigcup_{n=1}^{\infty} V_n$ and $B \subset \bigcup_{n=1}^{\infty} W_n$.

Now for each n let $V_n^* = V_n \setminus \bigcup_{i=1}^{n} \overline{W_i}$ and $W_n^* = W_n \setminus \bigcup_{i=1}^{n} \overline{V_i}$. Then the sets V_n^* and W_n^* are open, $V_n^* \cap W_m^* = \varnothing$ for all n and m, $A \subset \bigcup_{n=1}^{\infty} V_n^* = V$, and $B \subset \bigcup_{n=1}^{\infty} W_n^* = W$. To finish the proof note that $V \cap W = \varnothing$.　■

In addition to the properties already mentioned, there is another classification of topological spaces that you may run across, but which we eschew. A topological space is called a T_0-**space** if for each pair of distinct points, there is a neighborhood of one of them that does not contain the other. A T_1-**space** is one in which for each pair of distinct points, each has a neighborhood that does not contain the other. This is equivalent to each singleton being closed. A T_2-**space** is another name for a Hausdorff space. A T_3-**space** is a regular T_1-space. A T_4-**space** is a normal T_1-space. Finally, a $T_{3\frac{1}{2}}$-**space** or a **Tychonoff space** is a completely regular T_1-space.[6]

Here are some of the relations among the properties: Every Hausdorff space is T_1, and every T_1-space is T_0. A regular or normal space need not be Hausdorff: consider any two point set with the trivial topology. Every normal T_1-space is Hausdorff. A Tychonoff space is Hausdorff. For other separation axioms see A. Wilansky [340].

---

[6] If we had our way, the Hausdorff property would be part of the definition of a topology, and life would be much simpler.

2.12 Comparing topologies

The following two lemmas are trivial applications of the definitions, but they are included for easy reference. We feel free to refer to these results without comment. The proofs are left as an exercise.

2.50 Lemma   *For two topologies τ' and τ on a set X the following statements are equivalent.*

1. *τ' is weaker than τ, that is, $\tau' \subset \tau$.*

2. *The identity mapping $x \mapsto x$, from (X, τ) to (X, τ'), is continuous.*

3. *Every τ'-closed set is also τ-closed.*

4. *Every τ-convergent net is also τ'-convergent to the same point.*

5. *The τ-closure of any subset is included in its τ'-closure.*

2.51 Lemma   *If τ' is weaker than τ, then each of the following holds.*

1. *Every τ-compact set is also τ'-compact.*

2. *Every τ' continuous function on X is also τ continuous.*

3. *Every τ-dense set is also τ'-dense.*

When we have a choice of what topology to put on a set, there is the following rough tradeoff. The finer the topology, the more open sets there are, so that more functions are continuous. On the other hand, there are also more insidious open covers of a set, so there tend to be fewer compact sets. There are a number of useful theorems involving continuous functions and compact sets. One is the Weierstrass Theorem 2.35, which asserts that a real continuous function on a compact set attains its maximum and minimum. The Brouwer–Schauder–Tychonoff Fixed Point Theorem 17.56 says that a continuous function from a compact convex subset of a locally convex linear space into itself has a fixed point. Another example is a Separating Hyperplane Theorem 5.79 that guarantees the existence of a continuous linear functional strongly separating a compact convex set from a disjoint closed convex set in a locally convex linear space.

2.13 Weak topologies

There are two classes of topologies that by and large include everything of interest. The first and most familiar is the class of topologies that are generated by a metric. The second class is the class of weak topologies.

Let X be a nonempty set, let $\{(Y_i, \tau_i)\}_{i \in I}$ be a family of topological spaces and for each $i \in I$ let $f_i \colon X \to Y_i$ be a function. The **weak topology** or **initial topology** on X generated by the family of functions $\{f_i\}_{i \in I}$ is the weakest topology on X that makes all the functions f_i continuous. It is the topology generated by the family of sets

$$\{f_i^{-1}(V) : i \in I \text{ and } V \in \tau_i\}.$$

Another subbase for this topology consists of

$$\{f_i^{-1}(V) : i \in I \text{ and } V \in S_i\},$$

where S_i is a subbase for τ_i. Let w denote this weak topology. A base for the weak topology can be constructed out of the finite intersections of sets of this form. That is, the collection of sets of the form $\bigcap_{k=1}^{n} f_{i_k}^{-1}(V_{i_k})$, where each V_{i_k} belongs to τ_{i_k} and $\{i_1, \ldots, i_n\}$ is an arbitrary finite subset of I, is a base for the weak topology. The next lemma is an important tool for working with weak topologies.

2.52 Lemma   *A net satisfies $x_\alpha \xrightarrow{w} x$ for the weak topology w if and only if $f_i(x_\alpha) \xrightarrow{\tau_i} f_i(x)$ for each $i \in I$.*

Proof: Since each f_i is w-continuous, if $x_\alpha \xrightarrow{w} x$, then $f_i(x_\alpha) \xrightarrow{\tau_i} f_i(x)$ for all $i \in I$. Conversely, let $V = \bigcap_{k=1}^{n} f_{i_k}^{-1}(V_{i_k})$ be a basic neighborhood of x, where each $V_{i_k} \in \tau_{i_k}$. For each k, if $f_{i_k}(x_\alpha) \xrightarrow{\tau_{i_k}} f_{i_k}(x)$, then there is α_{i_k} such that $\alpha \geq \alpha_{i_k}$ implies $x_\alpha \in f_{i_k}^{-1}(V_{i_k})$. Pick $\alpha_0 \geq \alpha_{i_k}$ for all k. Then $\alpha \geq \alpha_0$ implies $x_\alpha \in V$. That is, $x_\alpha \xrightarrow{w} x$.   ∎

An important special case is the weak topology generated by a family of real functions. For a family $\mathcal{F}$ of real functions on X, the weak topology generated by $\mathcal{F}$ is denoted $\sigma(X, \mathcal{F})$. It is easy to see that a subbase for $\sigma(X, \mathcal{F})$ can be found by taking all sets of the form

$$U(f, x, \varepsilon) = \{y \in X : |f(y) - f(x)| < \varepsilon\},$$

where $f \in \mathcal{F}$, $x \in X$, and $\varepsilon > 0$.

We say that a family $\mathcal{F}$ of real functions on X is **total**, or **separates points** in X, if $f(x) = f(y)$ for all f in $\mathcal{F}$ implies $x = y$. Another way to say the same thing is that $\mathcal{F}$ separates points in X if for every $x \neq y$ there is a function f in $\mathcal{F}$ satisfying $f(x) \neq f(y)$. The weak topology $\sigma(X, \mathcal{F})$ is Hausdorff if and only if $\mathcal{F}$ is total.

Here is a subtle point about weak topologies. Let $\mathcal{F}$ be a family of real-valued functions on a set X. Every subset $A \subset X$ has a relative topology induced by the $\sigma(X, \mathcal{F})$ weak topology on X. It also has its own weak topology, the $\sigma(A, \mathcal{F}|_A)$ topology, where $\mathcal{F}|_A$ is the family of restrictions of the functions in $\mathcal{F}$ to A. Are these topologies the same? Conveniently the answer is yes.

2.53 Lemma   (Relative weak topology)   *Let $\mathcal{F}$ be a family of real-valued functions on a set X, and let A be a subset of X. The $\sigma(A, \mathcal{F}|_A)$ weak topology on A is the relative topology on A induced by the $\sigma(X, \mathcal{F})$ weak topology on X.*

Proof: Use Lemma 2.52 to show that the convergent nets in each topology are the same. This implies that the identity is a homeomorphism.   ∎

We employ the following standard notation throughout this monograph:

• $\mathbb{R}^X$ denotes the vector space of real-valued functions on a nonempty set X.

• $C(X)$ denotes the vector space of continuous real-valued functions on the topological space (X, τ). We may occasionally use the abbreviation C for $C(X)$ when X is clear from the context. We also use the common shorthand $C[0, 1]$ for $C([0, 1])$, the space of continuous real functions on the unit interval $[0, 1]$.

• $C_b(X)$ is the space of bounded continuous real functions on (X, τ). It is a vector subspace of $C(X)$.[7]

•   The **support** of a real function $f \colon X \to \mathbb{R}$ on a topological space is the closure of the set $\{x \in X : f(x) \neq 0\}$, denoted supp f. That is,

$$\operatorname{supp} f = \overline{\{x \in X : f(x) \neq 0\}}.$$

$C_c(X)$ denotes the vector space of all continuous real-valued functions on X with compact support.

The vector space $\mathbb{R}^X$ coincides, of course, with the vector space $C(X)$ when X is equipped with the discrete topology.

We now make a simple observation about weak topologies.

2.54 Lemma   *The weak topology on the topological space X generated by $C(X)$ is the same as the weak topology generated by $C_b(X)$.*

Proof: Consider a subbasic open set $U(f, x, \varepsilon) = \{y \in X : |f(y) - f(x)| < \varepsilon\}$, where $f \in C(X)$. Define the function $g \colon X \to \mathbb{R}$ by

$$g(z) = \min\{f(x) + \varepsilon, \max\{f(x) - \varepsilon, f(z)\}\}.$$

Then $g \in C_b(X)$ and $U(g, x, \varepsilon) = U(f, x, \varepsilon)$. Thus $\sigma(X, C_b)$ is as strong as $\sigma(X, C)$. The converse is immediate. Therefore $\sigma(X, C_b) = \sigma(X, C)$.   ∎

We can use weak topologies to characterize completely regular spaces.

2.55 Theorem   *A topological space (X, τ) is completely regular if and only if $\tau = \sigma(X, C(X)) = \sigma(X, C_b(X))$.*

---

[7] The notation C^* is used in some specialties for denoting C_b.

Proof: For any topological space (X, τ), we have $\sigma(X, C) \subset \tau$.

Assume first that (X, τ) is completely regular. Let x belong to the τ-open set U. Pick $f \in C(X)$ satisfying $f(x) = 0$ and $f(U^c) = \{1\}$. Then $\{y \in X : f(y) < 1\}$ is a $\sigma(X, C)$-open neighborhood of x included in U. Thus U is also $\sigma(X, C)$-open, so $\sigma(X, C) = \tau$.

Suppose now that $\tau = \sigma(X, C)$. Let F be closed and $x \notin F$. Since F^c is $\sigma(X, C)$-open, there is a neighborhood $U \subset F^c$ of x of the form

$$U = \bigcap_{i=1}^{m} \{y \in X : |f_i(y) - f_i(x)| < 1\},$$

where each $f_i \in C(X)$. For each $1 \leqslant i \leqslant m$ let $g_i(z) = \min\{1, |f_i(z) - f_i(x)|\}$ and $g(z) = \max_i g_i(z)$. Then g continuously maps X into $[0, 1]$, and satisfies $g(x) = 0$ and $g(F) = \{1\}$. Thus X is completely regular. ∎

2.56 Corollary *The completely regular spaces are precisely those whose topology is the weak topology generated by a family of real functions.*

Proof: If (X, τ) is completely regular, then $\tau = \sigma(X, C(X))$.

Conversely, suppose $\tau = \sigma(X, \mathcal{F})$ for a family $\mathcal{F}$ of real functions. Then $\mathcal{F} \subset C(X)$, so $\tau = \sigma(X, \mathcal{F}) \subset \sigma(X, C(X))$. But on the other hand, τ always includes $\sigma(X, C(X))$. Thus $\tau = \sigma(X, C(X))$, so by Theorem 2.55, (X, τ) is completely regular. ∎

The next easy corollary of Theorem 2.55 and Lemma 2.52 characterizes convergence in completely regular spaces.

2.57 Corollary *If X is completely regular, then a net $x_\alpha \to x$ in X if and only if $f(x_\alpha) \to f(x)$ for all $f \in C_b(X)$.*

For additional results on completely regular spaces see Chapter 3 of the excellent book by L. Gillman and M. Jerison [138].

2.14 The product topology

Let $\{(X_i, \tau_i)\}_{i \in I}$ be a family of topological spaces and let $X = \prod_{i \in I} X_i$ denote its Cartesian product. A typical element x of the product may also be denoted $(x_i)_{i \in I}$ or simply (x_i). For each $j \in I$, the **projection** $P_j \colon X \to X_j$ is defined by $P_j(x) = x_j$. The **product topology** τ, denoted $\prod_{i \in I} \tau_i$, is the weak topology on X generated by the family of projections $\{P_i : i \in I\}$. That is, τ is the weakest topology on X that makes each projection P_i continuous. A subbase for the product topology consists of all sets of the form $P_j^{-1}(V_j) = \prod_{i \in I} V_i$ where $V_i = X_i$ for all $i \neq j$ and V_j is open in X_j. A base for the product topology consists of all sets of the form

$$V = \prod_{i \in I} V_i,$$

where $V_i \in \tau_i$ and $V_i = X_i$ for all but finitely many i.

From this, we see that a net $\{(x_i^\alpha)_{i \in I}\}$ in X satisfies $(x_i^\alpha)_{i \in I} \xrightarrow{\tau} (x_i)_{i \in I}$ in X if and only if $x_i^\alpha \xrightarrow[\alpha]{\tau_i} x_i$ in X_i for each $i \in I$. Unless otherwise stated, the Cartesian product of a family of topological spaces is endowed with its product topology. A function $f : \prod_{i \in I} X_i \to Y$ is called **jointly continuous** if it is continuous with respect to the product topology.

In particular, note that if $(X_1, \tau_1), \ldots, (X_n, \tau_n)$ are topological spaces, then a base for the product topology on $X = X_1 \times \cdots \times X_n$ consists of all sets of the form $V = V_1 \times \cdots \times V_n$, where $V_i \in \tau_i$ for each i. Also, if $y_\alpha \to y$ in Y and $z_\beta \to z$ in Z, then the product net $(y_\alpha, z_\beta)_{(\alpha, \beta)} \to (y, z)$ in $Y \times Z$. We also point out that the Euclidean metric on $\mathbb{R}^n$ induces the product topology on $\mathbb{R}^n$, viewed as the product of n copies of $\mathbb{R}$ with its usual topology.

Recall that the **graph** $\mathrm{Gr}\, f$ of the function $f : X \to Y$ is the set

$$\mathrm{Gr}\, f = \{(x, y) \in X \times Y : y = f(x)\}.$$

Sometimes the closedness of $\mathrm{Gr}\, f$ in the product space $X \times Y$ characterizes the continuity of the function f. An important case is presented next.

2.58 Closed Graph Theorem   *A function from a topological space into a compact Hausdorff space is continuous if and only if its graph is closed.*

Proof: If $f : X \to Y$ is continuous and Y is Hausdorff, then $\mathrm{Gr}\, f$ is a closed subset of $X \times Y$: Suppose $(x_\alpha, y_\alpha) \to (x, y)$, where $y_\alpha = f(x_\alpha)$. Since f is continuous, $y_\alpha = f(x_\alpha) \to f(x)$. Since $y_\alpha \to y$ and Y is Hausdorff, we conclude $y = f(x)$. In other words, the graph of f is closed.

For the converse, assume that $\mathrm{Gr}\, f$ is a closed subset of $X \times Y$ and let $x_\alpha \to x$ in X. Suppose by way of contradiction that $f(x_\alpha) \nrightarrow f(x)$. Then there exists a neighborhood V of $f(x)$ and a subnet of $\{f(x_\alpha)\}$ (which by relabeling we also denote by $\{f(x_\alpha)\}$) satisfying $f(x_\alpha) \notin V$ for all α. The compactness of Y guarantees that $\{f(x_\alpha)\}$ has a convergent subnet, which we again denote by $\{f(x_\alpha)\}$, so we may assume $f(x_\alpha) \to y$ for some y in Y.

Thus $(x_\alpha, f(x_\alpha)) \to (x, y)$ in $X \times Y$, so from the closedness of $\mathrm{Gr}\, f$, we see that $y = f(x)$. However, this contradicts $f(x_\alpha) \notin V$ for each α. Thus $f(x_\alpha) \to f(x)$, which shows that f is continuous. ∎

The preceding result may fail if we do not assume that Y is compact.

2.59 Example (Closed graph may not imply continuity)   Define $f : \mathbb{R} \to \mathbb{R}$ by $f(x) = 1/x$ if $x \neq 0$ and $f(0) = 0$, and note that its graph is closed while f is not continuous. Of course, the range is not compact.

An even more dramatic example is this one of a function with closed graph that is discontinuous everywhere. Let $X = [0, 1]$ equipped with the Euclidean topology and let $Y = [0, 1]$ equipped with the discrete topology. Both X and Y are

Hausdorff spaces (in fact, they are complete metric spaces) with X compact and Y non-compact. Letting $I: X \to Y$ be the identity mapping, it is easy to see that I has closed graph and is discontinuous at every point. ∎

The following related result is an immediate consequence of Lemma 2.42 and the definition of the product topology.

2.60 Corollary   *An extended real-valued function f on a topological space X is lower semicontinuous if and only if its epigraph $\{(x, c) \in X \times \mathbb{R} : c \geqslant f(x)\}$ is closed in $X \times \mathbb{R}$. An extended real-valued function is upper semicontinuous if and only if its hypograph is closed.*

We now come to one of the most important compactness results in mathematics. It is known as the Tychonoff Product Theorem and asserts that an arbitrary Cartesian product of compact spaces is compact.

2.61 Tychonoff Product Theorem   *The product of a family of topological spaces is compact in the product topology if and only if each factor is compact.*

Proof: Let $\{X_i : i \in I\}$ be a family of topological spaces. If $X = \prod_{i \in I} X_i$ is compact, then $X_i = P_i(X)$ is also compact for each i; see Theorem 2.34.

For the converse, assume that each X_i is a compact space and let $\mathcal{U}$ be an ultrafilter on X. By Theorem 2.31, we have to show that $\mathcal{U}$ converges in X. To this end, start by observing that for each i the collection $\mathcal{U}_i = \{P_i(U) : U \in \mathcal{U}\}$ is a filter base of X_i. So by Theorem 2.31, we see that $\bigcap_{U \in \mathcal{U}} \overline{P_i(U)} \neq \varnothing$. For each i fix some $x_i \in \bigcap_{U \in \mathcal{U}} \overline{P_i(U)}$ and let $x = (x_i)_{i \in I} \in X$. We claim that $\mathcal{N}_x \subset \mathcal{U}$. To see this, note that if V_i is an arbitrary neighborhood of x_i in X_i, then $V_i \cap P_i(U) \neq \varnothing$ for each $U \in \mathcal{U}$. Rewriting the latter, we see that $P_i^{-1}(V_i) \cap U \neq \varnothing$ for all $U \in \mathcal{U}$, which (in view of Lemma 2.21) implies that $P_i^{-1}(V_i) \in \mathcal{U}$. From the definition of the product topology, it follows that each neighborhood of x belongs to $\mathcal{U}$, that is, $\mathcal{N}_x \subset \mathcal{U}$. In other words, $\mathcal{U} \to x$ in X, as desired. ∎

The following handy result is a consequence of the Tychonoff Product Theorem. It is used in the proof of Theorem 17.28 on products of correspondences.

2.62 Theorem   *Let $\{X_i\}_{i \in I}$ be a family of topological spaces, and for each i let K_i be a compact subset of X_i. If G is an open subset of $\prod_{i \in I} X_i$ including $\prod_{i \in I} K_i$, then there exists a basic open set $\prod_{i \in I} V_i$ (where V_i is open in X_i, and $V_i = X_i$ for all but a finite number of indexes i) such that $\prod_{i \in I} K_i \subset \prod_{i \in I} V_i \subset G$.*

Proof: Assume first that the family consists of two topological spaces, say X_1 and X_2. Since $K_1 \times K_2$ is a compact subset of $X_1 \times X_2$ and G is a union of basic open sets, there exists a finite collection of basic open sets $\{U_1 \times V_1, \ldots, U_n \times V_n\}$ such that $K_1 \times K_2 \subset \bigcup_{j=1}^{n} U_j \times V_j \subset G$. Now for each $x \in K_1$, let $U_x = \bigcap_{x \in U_j} U_j$

and note that U_x is an open neighborhood of x. Similarly, for every $y \in K_2$ set $V_y = \bigcap_{y \in V_j} V_j$. Observe that for each (x, y), the neighborhood $U_x \times V_y$ is included in one of the original $U_i \times V_i$. (Why?) From the compactness of K_1 and K_2, there exist elements $x_1, \ldots, x_m \in K_1$ and $y_1, \ldots, y_\ell \in K_2$ with $K_1 \subset \bigcup_{j=1}^{m} U_{x_j}$ and $K_2 \subset \bigcup_{r=1}^{\ell} V_{y_r}$. Next, note that the open sets $U = \bigcup_{j=1}^{m} U_{x_j}$ and $V = \bigcup_{r=1}^{\ell} V_{y_r}$ satisfy

$$K_1 \times K_2 \subset U \times V \subset \bigcup_{j=1}^{n} U_j \times V_j \subset G.$$

So the conclusion is true for a family of two topological spaces. By induction, the claim is true for any finite family of topological spaces. (Why?) For the general case, pick a finite collection $\{\prod_{i \in I} V_i^j\}_{j=1,\ldots,k}$ of basic open sets such that $K = \prod_{i \in I} K_i \subset \bigcup_{j=1}^{k} (\prod_{i \in I} V_i^j) \subset G$. (This is possible since K is compact by the Tychonoff Product Theorem 2.61.) This implies that the general case can be reduced to that of a finite family of topological spaces. We leave the remaining details as an exercise.   ∎

2.15   Pointwise and uniform convergence

For a nonempty set X, the product topology on $\mathbb{R}^X$ is also called the **topology of pointwise convergence** on X because a net $\{f_\alpha\}$ in $\mathbb{R}^X$ satisfies $f_\alpha \to f$ in $\mathbb{R}^X$ if and only if $f_\alpha(x) \to f(x)$ in $\mathbb{R}$ for each $x \in X$.

Remarkably, if $\mathcal{F}$ is a set of real-valued functions on X, we can also regard X as a set of real-valued functions on $\mathcal{F}$. Each $x \in X$ can be regarded as an **evaluation functional** $e_x : \mathcal{F} \to \mathbb{R}$, where $e_x(f) = f(x)$. As such, there is also a weak topology on $\mathcal{F}$, $\sigma(\mathcal{F}, X)$. This topology is identical to the relative topology on $\mathcal{F}$ as a subset of $\mathbb{R}^X$ endowed with the product topology. We also note the following important result.

2.63 Lemma   *If $\mathcal{F}$ is a total family of real functions on a set X, the function $x \mapsto e_x$, mapping $(X, \sigma(X, \mathcal{F}))$ into $\mathbb{R}^{\mathcal{F}}$ with its product topology, is an embedding.*

Proof: Since $\mathcal{F}$ is a total, the mapping $x \mapsto e_x$ is one-to-one. The rest is just a restatement of Lemma 2.52, using the observation that the product topology on $\mathbb{R}^{\mathcal{F}}$ is the topology of pointwise convergence on $\mathcal{F}$.   ∎

From the Tychonoff Product Theorem 2.61, it follows that a subset $\mathcal{F}$ of $\mathbb{R}^X$ is compact in the product topology if and only if it is closed and pointwise bounded. Since a subset of $\mathcal{F}$ is compact in $\mathcal{F}$ if and only if it is compact in $\mathbb{R}^X$, we see that a subset of $\mathcal{F}$ is weakly compact (compact in the product topology) if and only if it is pointwise bounded and contains the pointwise limits of its nets.

We are now in a position to give a natural example of the inadequacy of sequences. They cannot describe the product topology on an uncountable product.

2.64 Example   Let $[0, 1]^{[0,1]}$ be endowed with its product topology, the topology of pointwise convergence. Let F denote the family of indicator functions of finite subsets of $[0, 1]$. Recall that the indicator function χ_A of a set A is defined by

$$\chi_A(x) = \begin{cases} 1 & \text{if } x \in A, \\ 0 & \text{if } x \notin A. \end{cases}$$

Then $\mathbf{1}$, the function that is identically one, is not the pointwise limit of any sequence in F: Let χ_{A_n} be a sequence in F. Then $A = \bigcup_{n=1}^{\infty} A_n$ is countable, so there is some point x not belonging to A. Since $\chi_{A_n}(x) = 0$ for all n, the sequence does not converge pointwise to $\mathbf{1}$.

However there is a net in F that converges pointwise to $\mathbf{1}$: Take the family $\mathcal{F}$ of all finite subsets of $[0, 1]$ directed upward by inclusion—that is, $A \geqslant B$ if $A \supset B$. Then the net $\{\chi_A : A \in \mathcal{F}\}$ converges pointwise to $\mathbf{1}$. (Do you see why?)   ∎

A net $\{f_\alpha\}$ in $\mathbb{R}^X$ converges **uniformly** to a function $f \in \mathbb{R}^X$ whenever for each $\varepsilon > 0$ there exists some index α_0 (depending upon ε alone) such that

$$\left| f_\alpha(x) - f(x) \right| < \varepsilon$$

for each $\alpha \geqslant \alpha_0$ and each $x \in X$. Clearly, uniform convergence implies pointwise convergence, but the converse is not true.

2.65 Theorem   *The uniform limit of a net of continuous real functions is continuous.*

Proof: Let $\{f_\alpha\}$ be a net of continuous real functions on a topological space X that converges uniformly to a function $f \in \mathbb{R}^X$. Suppose $x_\lambda \to x$ in X. We now show that $f(x_\lambda) \to f(x)$.

Let $\varepsilon > 0$ be given, and pick some α_0 satisfying $|f_\alpha(y) - f(y)| < \varepsilon$ for all $\alpha \geqslant \alpha_0$ and all $y \in X$. Since f_{α_0} is a continuous function, there exists some λ_0 such that $|f_{\alpha_0}(x_\lambda) - f_{\alpha_0}(x)| < \varepsilon$ for all $\lambda \geqslant \lambda_0$. Hence, for $\lambda \geqslant \lambda_0$ we have

$$
\begin{aligned}
\left| f(x_\lambda) - f(x) \right| \\
\leqslant \left| f(x_\lambda) - f_{\alpha_0}(x_\lambda) \right| + \left| f_{\alpha_0}(x_\lambda) - f_{\alpha_0}(x) \right| + \left| f_{\alpha_0}(x) - f(x) \right| \\
< \varepsilon + \varepsilon + \varepsilon = 3\varepsilon.
\end{aligned}
$$

Thus, $f(x_\lambda) \to f(x)$, so f is a continuous function.   ∎

Here is a simple sufficient condition for a net to converge uniformly.

2.66 Dini's Theorem   *If a net of continuous real functions on a compact space converges monotonically to a continuous function pointwise, then the net converges uniformly.*

Proof: Let $\{f_\alpha\}$ be a net of continuous functions on the compact space X satisfying $f_\alpha(x) \downarrow f(x)$ for each $x \in X$, where f is continuous. Replacing f_α by $f_\alpha - f$ we may assume that f is identically zero.

Let $\varepsilon > 0$. For each $x \in X$ pick an index α_x such that $0 \leqslant f_{\alpha_x}(x) < \varepsilon$. By the continuity of f_{α_x} there is an open neighborhood V_x of x such that $0 \leqslant f_{\alpha_x}(y) < \varepsilon$ for all $y \in V_x$. Since $\alpha \geqslant \alpha_x$ implies $f_\alpha \leqslant f_{\alpha_x}$, we see that $0 \leqslant f_\alpha(y) < \varepsilon$ for each $\alpha \geqslant \alpha_x$ and all $y \in V_x$.

From $X = \bigcup_{x \in X} V_x$ and the compactness of X, we see that there exist $x_1, \ldots, x_k$ in X with $X = \bigcup_{i=1}^{k} V_{x_i}$. Now choose some index α_0 satisfying $\alpha_0 \geqslant \alpha_{x_i}$ for all $i = 1, \ldots, k$ and note that $\alpha \geqslant \alpha_0$ implies $0 \leqslant f_\alpha(y) < \varepsilon$ for all $y \in X$. That is, the net $\{f_\alpha\}$ converges uniformly to zero. ∎

2.16   Locally compact spaces

A topological space is **locally compact** if every point has a compact neighborhood.[8] The existence of a single compact neighborhood at each point is enough to guarantee many more.

2.67 Theorem (Compact neighborhood base) *In a locally compact Hausdorff space, every neighborhood of a point includes a compact neighborhood of the point. Consequently, in a locally compact Hausdorff space, each point has a neighborhood base of compact neighborhoods.*

Proof: Let G be an open neighborhood of x and let W be a compact neighborhood of x. If $W \subset G$, we are done, so assume $A = W \cap G^c \neq \varnothing$. For each $y \in A$ choose an open neighborhood U_y of y and an open neighborhood W_y of x satisfying $W_y \subset W$ and $U_y \cap W_y = \varnothing$. Since A (= $W \cap G^c$) is compact, there exist $y_1, \ldots, y_k \in A$ such that $A \subset \bigcup_{i=1}^{k} U_{y_i}$. Put $V = \bigcap_{i=1}^{k} W_{y_i}$ and $U = \bigcup_{i=1}^{k} U_{y_i}$. Now V is an open neighborhood of x, and we claim that $\overline{V}$ is compact and included in G.

To see this, note first that $\overline{V} \subset W$ implies that $\overline{V}$ is compact. Now, since U and V are both open and $V \cap U = \varnothing$, it follows that $\overline{V} \cap U = \varnothing$. Consequently, from

$$\overline{V} \cap G^c = \overline{V} \cap (W \cap G^c) = \overline{V} \cap A \subset \overline{V} \cap U = \varnothing,$$

we see that $\overline{V} \cap G^c = \varnothing$. Hence $\overline{V} \subset G$ is a compact neighborhood of x. ∎

Every compact space is locally compact. In fact, the following corollary is easily seen to be true.

2.68 Corollary *The intersection of an open subset with a closed subset of a locally compact Hausdorff space is locally compact.*

In particular, every open subset and every closed subset of a locally compact Hausdorff space is locally compact.

---

[8] Some authors require that a locally compact space be Hausdorff.

The next result is another useful corollary.

2.69 Corollary   *If K is a compact subset of a locally compact Hausdorff space, and G is an open set including K, then there is an open set V with compact closure satisfying $K \subset V \subset \overline{V} \subset G$.*

Proof: By Theorem 2.67, each point x in K has an open neighborhood V_x with compact closure satisfying $x \in V_x \subset \overline{V}_x \subset G$. Since K is compact there is a finite subcollection $\{V_{x_1}, \ldots, V_{x_n}\}$ of these sets covering K. Then $V = \bigcup_{i=1}^{n} V_{x_i}$ is the desired open set. (Why?) ∎

A **compactification** of a Hausdorff space X is a compact Hausdorff space Y where X is homeomorphic to a dense subset of Y, so we may treat X as an actual dense subset of Y. Note that if X is already compact, then it is closed in any Hausdorff space including it, so any compactification of a compact Hausdorff space is the space itself. The locally compact Hausdorff spaces are open sets in all of their compactifications. The details follow.

2.70 Theorem   *Let $\hat{X}$ be a compactification of a Hausdorff space X. Then X is locally compact if and only if X is an open subset of $\hat{X}$.*

In particular, if X is a locally compact Hausdorff space, then X is an open subset of any of its compactifications.

Proof: Let $(\hat{X}, \hat{\tau})$ be a compactification of a Hausdorff space (X, τ). If X is an open subset of $\hat{X}$, then it follows from Corollary 2.68 that X is locally compact. For the converse, assume that (X, τ) is locally compact and fix $x \in X$. Choose a compact τ-neighborhood U of x and then pick an open τ-neighborhood V of x such that $V \subset U$. Now select $W \in \hat{\tau}$ such that $V = W \cap X$ and note that

$$W = W \cap \hat{X} = W \cap \overline{X} \subset \overline{W \cap X} = \overline{V} \subset \overline{U} = U \subset X.$$

This shows that x in a $\hat{\tau}$-interior point of X, so $X \in \hat{\tau}$. ∎

2.71 Corollary   *Only locally compact Hausdorff spaces can possibly be compactified with a finite number of points.*

The simplest compactification of a noncompact locally compact Hausdorff space is its one-point compactification. It is obtained by appending a point ∞, called the **point at infinity**, that does not belong to the space X, and we write X_∞ for $X \cup \{\infty\}$. We leave the proof of the next theorem as an exercise.

2.72 Theorem   (One-point compactification)   *Let (X, τ) be a noncompact locally compact Hausdorff space and let $X_\infty = X \cup \{\infty\}$, where $\infty \notin X$. Then the collection*

$$\tau_\infty = \tau \cup \{X_\infty \setminus K : K \subset X \text{ is compact}\}$$

is a topology on X_∞. Moreover, (X_∞, τ_∞) is a compact Hausdorff space and X is an open dense subset of X_∞, that is, X_∞ is a compactification of X.

The space (X_∞, τ_∞) is called the **Alexandroff one-point compactification** of X. As an example, the one-point compactification $\mathbb{R}_\infty$ of the real numbers $\mathbb{R}$ is homeomorphic to a circle.

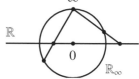

One such homeomorphism is described by mapping the "north pole" $(0, 1)$ on the unit circle in $\mathbb{R}^2$ to ∞ and every other point (x, y) on the circle is mapped to the point on the x-axis where the ray through (x, y) from ∞ crosses the axis. See Figure 2.1. Mapmakers have long known that the one-point compactification

Figure 2.1. $\mathbb{R}_\infty$ is a circle.

of $\mathbb{R}^2$ is the sphere. (Look up stereographic projection in a good dictionary.)

It is immediate from Theorem 2.72 that a subset F of X is closed in X_∞ if and only if F is compact. We also have the following observation.

2.73 Lemma   *For a subset A of X, the set $A \cup \{\infty\}$ is closed in X_∞ if and only if A is closed in X.*

Proof: To see this, just note that $X_\infty \setminus (A \cup \{\infty\}) = X \setminus A$.   ∎

The one-point compactification allows us to prove the following.

2.74 Corollary   *In a locally compact Hausdorff space, nonempty compact sets can be separated from disjoint nonempty closed sets by continuous functions. In particular, every locally compact Hausdorff space is completely regular.*

Proof: Let A be a nonempty compact subset and B a nonempty closed subset of a locally compact Hausdorff space X satisfying $A \cap B = \varnothing$. Then A is a compact (and hence closed) subset of the one-point compactification X_∞ of X. Let $C = B \cup \{\infty\}$. Then C is a closed subset of X_∞ (why?) and $A \cap C = \varnothing$.

Since X_∞ is a compact Hausdorff space, it is normal by Theorem 2.48. Now by Theorem 2.46 there exists a continuous function $f : X_\infty \to [0, 1]$ satisfying $f(x) = 1$ for all $x \in A$ and $f(y) = 0$ for all $y \in C$. Clearly, the restriction of f to X has the desired properties.   ∎

2.75 Example (Topology of the extended reals)   The extended real numbers $\mathbb{R}^* = [-\infty, \infty]$ are naturally topologized as a two-point compactification of the space $\mathbb{R}$ of real numbers. A neighborhood base of ∞ is given by the collection of intervals of the form $(c, \infty]$ for $c \in \mathbb{R}$, and the intervals $[-\infty, c)$ constitute a neighborhood base for $-\infty$. Note that a sequence $\{x_n\}$ in $\mathbb{R}^*$ converges to ∞ if for every $n \in \mathbb{N}$, there exits an n_0 such that for all $n \geqslant n_0$ we have $x_n > m$. You should verify that this is indeed a compact space, that it is first countable, and that $\mathbb{R}$ is a dense subspace of $\mathbb{R}^*$. In fact by Theorem 3.40 it is metrizable. You should further check that an extended real-valued function that is both upper and lower semicontinuous is continuous with respect to this topology.   ∎

A topological space is σ-**compact**, if it is the union of a countable family of compact sets.[9] For instance, every Euclidean space is σ-compact.

2.76 Lemma   *A second countable locally compact Hausdorff space has a countable base of open sets with compact closures. Consequently, it is σ-compact.*

Proof: Let X satisfy the hypotheses of the theorem and fix a countable base $\mathcal{B}$ for X. Consider the countable collection $\mathcal{B}_1 = \{G \in \mathcal{B} : \overline{G}$ is compact$\}$. Now let $x \in U$ with U open. By Theorem 2.67 there exists an open neighborhood V of x with compact closure satisfying $\overline{V} \subset U$. Since $\mathcal{B}$ is a base, there exists some $G \in \mathcal{B}$ such that $x \in G$ and $G \subset V$. But then $\overline{G} \subset \overline{V}$ shows that $\overline{G}$ is compact. That is, $G \in \mathcal{B}_1$. Therefore, $\mathcal{B}_1$ is a countable base with the desired properties.   ∎

A topological space X is **hemicompact** if it can be written as the union of a sequence $\{K_n\}$ of compact sets such that every compact set K of X is included in some K_n. This is actually a stronger condition than σ-compactness.

2.77 Corollary   *If X is a locally compact σ-compact Hausdorff space, then there exists a sequence $\{K_1, K_2, \ldots\}$ of compact sets with $K_n \subset K_{n+1}^\circ$ for each n, and $X = \bigcup_{n=1}^\infty K_n = \bigcup_{n=1}^\infty K_n^\circ$. In particular, X is hemicompact.*

Proof: Let $X = \bigcup_{n=1}^\infty C_n$, where each C_n is compact. By Corollary 2.69 there is a compact set K_1 with $C_1 \subset K_1^\circ \subset X$. Recursively define K_n so that $K_{n-1} \cup C_n$, which is compact, lies in the interior of K_n. Then $X = \bigcup_{n=1}^\infty C_n = \bigcup_{n=1}^\infty K_n = \bigcup_{n=1}^\infty K_n^\circ$. Furthermore, given any compact $K \subset X$, the open cover $\{K_n^\circ\}$ must have a finite subcover. Since the K_n°s are nested, one of them actually includes K. So X is hemicompact.   ∎

2.17   The Stone–Čech compactification

While the one-point compactification is easy to describe, it is not satisfactory in one important respect. The space of continuous functions on the one-point compactification can be very different from the space of bounded continuous functions on the underlying topological space. It is true that every continuous real function on X_∞ defines a bounded continuous real function on X. However, not every bounded continuous function on X extends to a continuous function on X_∞. For example, the sine function cannot be extended from $\mathbb{R}$ to $\mathbb{R}_\infty$. The next example presents an extreme case.

2.78 Example $(C(X_\infty)$ **vs.** $C_b(X))$ Let X be an uncountable set endowed with the discrete topology. Then every real function is continuous on X. Nearly the

---

[9] Some authors, notably Dugundji [106] also require local compactness as part of the definition of σ-compactness. Others do not. Be careful.

opposite is true of X_∞. If a real function is continuous on X_∞, the value at all but countably many points is the same as the value at the point ∞.

To see this, recall that open neighborhoods of ∞ are complements of compact subsets of X. Since X has the discrete topology, only finite sets are compact. Now let $f\colon X_\infty \to \mathbb{R}$ be continuous and set $c = f(\infty)$. Then $f^{-1}((c - \frac{1}{n}, c + \frac{1}{n}))$ is a neighborhood of ∞ for each $n > 0$. That is, only finitely many points of X have values of f outside $(c - \frac{1}{n}, c + \frac{1}{n})$. Letting $n \to \infty$, we conclude that at most countably many points of X have f values different from c. ∎

Completely regular Hausdorff (Tychonoff) spaces possess a compactification that avoids this defect. It is known as the Stone–Čech compactification. Its description is a wee bit complicated. Let X be a completely regular Hausdorff space and define the mapping $\varepsilon\colon X \to \mathbb{R}^{C_b(X)}$ by

$$\varepsilon(x) = e_x,$$

which associates to each x the evaluation functional at x. As usual, we topologize $\mathbb{R}^{C_b(X)}$ with the product topology. (That is, the topology of pointwise convergence on C_b). It is easy to see that ε is one-to-one, and from Lemma 2.63 we see that ε is actually an embedding. Thus X, identified with $\varepsilon(X)$, can be viewed as a topological subspace of $\mathbb{R}^{C_b(X)}$.

For each $f \in C_b(X)$, choose a real number $M_f > 0$ satisfying $|f(x)| \leqslant M_f$ for each $x \in X$. It is then clear that

$$\varepsilon(X) \subset \prod_{f \in C_b(X)} [-M_f, M_f] = Q.$$

By the Tychonoff Product Theorem 2.61, the set Q is a compact subset of $\mathbb{R}^{C_b(X)}$. Therefore, the closure $\overline{\varepsilon(X)}$ of $\varepsilon(X)$ is likewise a compact subset of $\mathbb{R}^{C_b(X)}$. In other words, $\overline{\varepsilon(X)}$ is a compactification of X. This compactification is called the **Stone–Čech compactification** of X and is denoted βX.

2.79 Theorem (**Extension property**)   *Let X be a completely regular Hausdorff space. If Y is a compact Hausdorff space and $g\colon X \to Y$ is a continuous mapping, then g extends uniquely to a continuous mapping from the Stone–Čech compactification βX to Y.*

Proof: Since Y is a compact Hausdorff space, it is a completely regular Hausdorff space (Theorem 2.48). Let $\varepsilon_X\colon X \to \mathbb{R}^{C_b(X)}$ and $\varepsilon_Y\colon Y \to \mathbb{R}^{C_b(Y)}$ be the embeddings of X and Y, respectively, via evaluation functionals, as described above. Then $\beta X = \overline{\varepsilon_X(X)}$ and $\beta Y = \overline{\varepsilon_Y(Y)}$. Since Y is compact, notice that $\varepsilon_Y(Y)$ is a compact subset of $\mathbb{R}^{C_b(Y)}$, so $\beta Y = \varepsilon_Y(Y)$.

Now note that if $h \in C_b(Y)$, then $h \circ g \in C_b(X)$. So define the mapping $\Gamma\colon \mathbb{R}^{C_b(X)} \to \mathbb{R}^{C_b(Y)}$ by

$$\Gamma\mu(h) = \mu(h \circ g)$$

for each $h \in C_b(Y)$, where we use the notation $\Gamma\mu$ rather than $\Gamma(\mu)$ to denote the value of Γ at $\mu \in \mathbb{R}^{C_b(X)}$. We claim that Γ is a continuous function. To see this, let $\{\mu_\alpha\}$ be a net in $\mathbb{R}^{C_b(X)}$ and suppose $\mu_\alpha \to \mu$ pointwise on $C_b(X)$. This means that $\mu_\alpha(f) \to \mu(f)$ in $\mathbb{R}$ for each f in $C_b(X)$. In particular, $\mu_\alpha(h \circ g) \to \mu(h \circ g)$ for each $h \in C_b(Y)$. Thus

$$\Gamma\mu_\alpha(h) = \mu_\alpha(h \circ g) \to \mu(h \circ g) = \Gamma\mu(h),$$

or $\Gamma\mu_\alpha \to \Gamma\mu$ pointwise on $C_b(Y)$. Thus Γ is continuous.

Now notice that for $x \in X$,

$$\Gamma e_x(h) = e_x(h \circ g) = h(g(x)) = e_{g(x)}(h)$$

for every $h \in C_b(Y)$, so identifying x with $\varepsilon_X(x)$ and $g(x)$ with $\varepsilon_Y(g(x))$, we have

$$\Gamma(x) = g(x).$$

That is, Γ extends g. Using Theorem 2.27 (5), we see that

$$\Gamma(\beta X) = \Gamma(\overline{\varepsilon_X(X)}) \subset \overline{\Gamma(\varepsilon_X(X))} \subset \overline{\varepsilon_Y(Y)} = \varepsilon_Y(Y).$$

Thus, Γ is the unique continuous extension of g to all of βX.   ∎

There are a number of important corollaries.

2.80 Corollary (Uniqueness)   *Let K be a compactification of a completely regular Hausdorff space X and suppose that whenever Y is a compact Hausdorff space and $g: X \to Y$ is continuous, then g has a unique continuous extension from K to Y. Then K is homeomorphic to βX.*

Proof: Take $Y = \beta X$ in Theorem 2.79.   ∎

It is a good mental workout to imagine an element of $\beta X = \overline{\varepsilon(X)}$ that does not belong to $\varepsilon(X)$. For a real function μ on $C_b(X)$ to belong to $\overline{\varepsilon(X)}$, there must be a net $\{x_\alpha\}$ in X with $e_{x_\alpha} \to \mu$ pointwise on C_b. That is, for each $f \in C_b(X)$, we have $f(x_\alpha) \to \mu(f)$. If $\{x_\alpha\}$ converges, say to x, since ε is an embedding, we conclude $\mu = e_x$, which belongs to $\varepsilon(X)$. Thus if μ belongs to $\overline{\varepsilon(X)} \setminus \varepsilon(X)$ it cannot be the case that the net $\{x_\alpha\}$ converges. On the other hand, $\{x_\alpha\}$ must have a limit point in any compactification of X. Let x_0 be a limit point of $\{x_\alpha\}$ in βX. Then μ acts like an evaluation at x_0.

Thus we can think of the Stone–Čech compactification βX as adding limit points to all the nets in X in such a way that every f in $C_b(X)$ extends continuously to βX.[10] Indeed it is characterized by this extension property.

2.81 Corollary   *Let K be a compactification of a completely regular Hausdorff space X and suppose that every bounded continuous real function on X has a (unique) continuous extension from X to K. Then K is homeomorphic to βX.*

---

[10] Professional topologists express this with the phrase "X is C^*-embedded in βX."

Proof: Given any $f \in C_b(X)$, let $\hat{f}$ denote its continuous extension to K. Since the restriction of a continuous function on K is a bounded continuous function on X, the mapping $f \mapsto \hat{f}$ from $C_b(X)$ to $C(K)$ is one-to-one and onto.

Define the mapping φ from K into $\mathbb{R}^{C_b(X)}$ by $\varphi_x(f) = \hat{f}(x)$. Observe that φ is continuous. Furthermore φ is one-to-one. To see this, suppose $\varphi_x = \varphi_y$, that is, $\hat{f}(x) = \hat{f}(y)$ for every $f \in C_b(X)$. Then $f(x) = f(y)$ for every $f \in C(K)$. But $C(K)$ separates points of K (why?), so $x = y$. Consequently, φ is a homeomorphism from K to $\varphi(K)$ (Theorem 2.36).

Treating X as a dense subset of K, observe that if x belongs to X, then φ_x is simply the evaluation at x, so by definition, $\overline{\varphi(X)}$ is the Stone–Čech compactification of X. Since X is dense, $\varphi(X) \subset \varphi(K) \subset \overline{\varphi(X)}$. But $\varphi(K)$ is compact and therefore closed. Thus $\varphi(K) = \overline{\varphi(X)}$, and we are done. ∎

We take this opportunity to describe the Stone–Čech compactification of the space $\Omega_0 = \Omega \setminus \{\omega_1\}$ of countable ordinals. Recall that it is an open subset of the compact Hausdorff space Ω of ordinals, and thus locally compact. We start with the following peculiar property of continuous functions on Ω_0.

2.82 Lemma (Continuous functions on Ω_0) *Any continuous real function on $\Omega_0 = \Omega \setminus \{\omega_1\}$ is constant on some tail of Ω_0. That is, if f is a continuous real function Ω_0, there is an ordinal $x \in \Omega_0$ such that $y \geq x$ implies $f(y) = f(x)$.*

Proof: We start by making the following observation. If $f : \Omega_0 \to \mathbb{R}$ is continuous, and $a > b \in \mathbb{R}$, then at least one of $[f \geq a]$ or $[f \leq b]$ is countable. To see this, suppose that both are uncountable. Pick $x_1 \in \Omega_0$ so that $f(x_1) \geq a$. Since the initial segment $I(x_1)$ is countable, there is some $y_1 > x_1$ with $f(y_1) \leq b$. Proceeding in this fashion we can construct two interlaced sequences satisfying $x_n < y_n < x_{n+1}$, $f(x_n) \geq a$, and $f(y_n) \leq b$ for all n. By the Interlacing Lemma 1.15, these sequences have a common least upper bound z, which must then be the limit of each sequence. Since f is continuous, we must have $f(z) = \lim f(x_n) \geq a$ and $f(z) = \lim f(y_n) \leq b$, a contradiction. Therefore at least one set is countable.

Since Ω_0 is uncountable, there is some (possibly negative) integer k, such that the set $[k \leq f \leq k + 1]$ is uncountable. Since $[f \geq k]$ and $[f \leq k + 1]$ are uncountable, by the observation above we see that for each positive n, the sets $[f \leq k - \frac{1}{n}]$ and $[f \geq k + 1 + \frac{1}{n}]$ are countable. So except for countably many x, we have $k \leq f(x) \leq k + 1$. Let $I_1 = [k, k + 1]$. Now divide I_1 in half. Then either $[k \leq f \leq k + \frac{1}{2}]$ or $[k + \frac{1}{2} \leq f \leq k + 1]$ is uncountable. (Both sets may be uncountable, for instance, if f is constant with value $k + \frac{1}{2}$.) Without loss of generality, assume $[k \leq f \leq k + \frac{1}{2}]$ is uncountable, and set $I_2 = [k, k + \frac{1}{2}]$. Observe that $\{x \in \Omega_0 : f(x) \notin I_2\}$ is countable. Proceeding in this way we can find a nested sequence $\{I_n\}$ of closed real intervals, with the length of I_n being $\frac{1}{2^n}$, and having the property that $\{x \in \Omega_0 : f(x) \notin I_n\}$ is countable. Let a denote the unique point in $\bigcap_{n=1}^{\infty} I_n$. Then $\{x \in \Omega_0 : f(x) \neq a\}$ is countable. By Theorem 1.14 (6), this set has a least upper bound b. Now pick any $x > b$. Then $y \geq x$ implies $f(y) = a$. ∎

We now come to the compactifications of Ω_0.

2.83 Theorem   (Compactification of Ω_0)   *The compact Hausdorff space Ω can be identified with both the Stone–Čech compactification and the one-point compactification of Ω_0.*

Proof: The identification with the one-point compactification is straightforward. Now note that by Lemma 2.82, every continuous real function on Ω_0 has a unique continuous extension to Ω. Thus by Corollary 2.81, we can identify Ω with the Stone–Čech Compactification of Ω_0.   ∎

There are some interesting observations that follow from this. Since Ω is compact, this means that every continuous real function on Ω_0 is bounded, even though Ω_0 is not compact. (The open cover $\{[1, x) : x \in \Omega_0\}$ has no finite sub-cover.) Since every initial segment of Ω_0 is countable, we also see that every continuous real function on Ω takes on only countably many values.

We observed above that $f \mapsto \hat{f}$ from $C_b(X)$ into $C(\beta X)$ is one-to-one and onto. In addition, for $f, g \in C_b(X)$ it is easy to see that:

1. $(f + g)\widehat{} = \hat{f} + \hat{g}$ and $(\alpha f)\widehat{} = \alpha \hat{f}$ for all $\alpha \in \mathbb{R}$;

2. $(\max\{f, g\})\widehat{} = \max\{\hat{f}, \hat{g}\}$ and $(\min\{f, g\})\widehat{} = \min\{\hat{f}, \hat{g}\}$; and

3. $\|f\|_\infty = \sup\{|f(x)| : x \in X\} = \sup\{|f(x)| : x \in \beta X\} = \|\hat{f}\|_\infty$.

In Banach lattice terminology (see Definition 9.16), these properties are summarized as follows.

2.84 Corollary   *If X is a completely regular Hausdorff space, then the mapping $f \mapsto \hat{f}$ is a lattice isometry from $C_b(X)$ onto $C(\beta X)$. That is, under this identification, $C_b(X) = C(\beta X)$.*

Getting ahead of ourselves a bit, we note that $C_b(X)$ is an AM-space with unit, so by Theorem 9.32 it is lattice isometric to $C(K)$ for some compact Hausdorff space K. According to Corollary 2.84 the space K is just the Stone–Čech compactification βX.

Unlike the one-point compactification, which is often very easy to describe, the Stone–Čech compactification can be very difficult to get a handle on. For instance, the Stone–Čech compactification of $(0, 1]$ is not homeomorphic to $[0, 1]$. The real function $\sin(\frac{1}{x})$ is bounded and continuous on $(0, 1]$, but cannot be extended to a continuous function on $[0, 1]$. However, for discrete spaces, such as the natural numbers $\mathbb{N}$, there is an interesting interpretation of the Stone–Čech compactification described in the next section.

2.18   Stone–Čech compactification of a discrete set

In this section we characterize the Stone–Čech compactification of a discrete space. Any discrete space X is metrizable by the discrete metric, and hence completely regular and Hausdorff. Thus it has a Stone–Čech compactification βX. Since every set is open in a discrete space, every such space X is **extremally disconnected**, that is, it has the property that the closure of every open set is itself open. It turns out that βX inherits this property.

2.85 Theorem   *For an infinite discrete space X:*

1. *If A is a subset of X, then $\overline{A}$ is an open subset of βX, where the bar denotes the closure in βX.*

2. *If $A, B \subset X$ satisfy $A \cap B = \varnothing$, then $\overline{A} \cap \overline{B} = \varnothing$.*

3. *The space βX is extremally disconnected.*

Proof: (1 & 2)   Let $A \subset X$. Put $C = X \setminus A$ and note that $A \cap C = \varnothing$. Define $f : X \rightarrow [0,1]$ by $f(x) = 1$ if $x \in A$ and $f(x) = 0$ if $x \in C$. Clearly, f is continuous, so it extends uniquely to a continuous function $\hat{f} : \beta X \rightarrow [0,1]$. From $A \cup C = X$, we get $\overline{A} \cup \overline{C} = \beta X$. (Do you see why?) It follows that $\overline{A} = \hat{f}^{-1}(\{1\})$ and $\overline{C} = \hat{f}^{-1}(\{0\})$. Therefore, $\overline{A} \cap \overline{C} = \varnothing$, and $\overline{A}$ is open. Now if $B \subset X$ satisfies $A \cap B = \varnothing$, then $B \subset C$, so $\overline{A} \cap \overline{B} = \varnothing$.

(3)   Let V be an open subset of βX. By (1), the set $\overline{V \cap X}$ is an open subset of βX. Note that if $x \in \overline{V}$ and W is an open neighborhood of x, then $W \cap V \neq \varnothing$, so $W \cap V \cap X \neq \varnothing$, or $x \in \overline{V \cap X}$. Therefore, $\overline{V} = \overline{V \cap X}$, so that $\overline{V}$ is open.   ∎

Let $\mathfrak{U}$ denote the set of all ultrafilters on X. That is,

$$\mathfrak{U} = \{\mathcal{U} : \mathcal{U} \text{ is an ultrafilter on } X\}.$$

As we already know, ultrafilters on X are either fixed or free. Every $x \in X$ gives rise to a unique fixed ultrafilter $\mathcal{U}_x$ on X via the formula

$$\mathcal{U}_x = \{A \subset X : x \in A\},$$

and every fixed ultrafilter on X is of the form $\mathcal{U}_x$.

Now let $\mathcal{U}$ be a free ultrafilter on X. Then $\mathcal{U}$ is a filter base in βX. Thus the filter $\mathcal{F}$ it generates has a limit point in βX (Theorem 2.31). That is, we have $\bigcap_{F \in \mathcal{F}} \overline{F} = \bigcap_{A \in \mathcal{U}} \overline{A} \neq \varnothing$. We claim that this intersection is a singleton. To see this, assume that there exist $x, y \in \bigcap_{A \in \mathcal{U}} \overline{A}$ with $x \neq y$. Then the collections

$$\mathcal{B}_x = \{V \cap A : V \in \mathcal{N}_x, A \in \mathcal{U}\} \quad \text{and} \quad \mathcal{B}_y = \{W \cap B : W \in \mathcal{N}_y, B \in \mathcal{U}\},$$

are both filter bases on X. Since the filters they generate include the ultrafilter $\mathcal{U}$, it follows that $\mathcal{B}_x \cup \mathcal{B}_y \subset \mathcal{U}$. Since βX is a Hausdorff space, there exist $V \in \mathcal{N}_x$

and $W \in \mathcal{N}_y$ such that $V \cap W = \emptyset$. This implies $\emptyset \in \mathcal{U}$, a contradiction. Hence, $\bigcap_{A \in \mathcal{U}} \overline{A}$ is a singleton.

Conversely, if $x \in \beta X \setminus X$, then the collection

$$\mathcal{B} = \{V \cap X : V \in \mathcal{N}_x\} \qquad\qquad (\star)$$

of subsets of X is a filter base on X. By Zorn's Lemma there exists an ultrafilter $\mathcal{U}$ on X including $\mathcal{B}$. Then $\mathcal{U}$ is a free ultrafilter (on X) satisfying $\bigcap_{A \in \mathcal{U}} \overline{A} = \{x\}$. (Why?) In other words, every point of $\beta X \setminus X$ is the limit point of a free ultrafilter on X.

It turns out that every point of $\beta X \setminus X$ is the limit point of exactly one free ultrafilter on X. To see this, let $\mathcal{U}_1$ and $\mathcal{U}_2$ be two free ultrafilters on X such that $x \in \bigcap_{A \in \mathcal{U}_1} \overline{A} = \bigcap_{B \in \mathcal{U}_2} \overline{B}$. If $A \in \mathcal{U}_1$, then $A \in \mathcal{U}_2$. Otherwise, $A \notin \mathcal{U}_2$ implies $X \setminus A \in \mathcal{U}_2$, so (by Theorem 2.85) $x \in \overline{A} \cap \overline{X \setminus A} = \emptyset$, a contradiction. So $\mathcal{U}_1 \subset \mathcal{U}_2$. Similarly, $\mathcal{U}_2 \subset \mathcal{U}_1$, and hence $\mathcal{U}_1 = \mathcal{U}_2$.

For each point $x \in \beta X \setminus X$, we denote by $\mathcal{U}_x$ the unique free ultrafilter on the set X—whose filter base is given by $(\star)$—having x as its unique limit point. Thus, we have established a one-to-one mapping $x \mapsto \mathcal{U}_x$ from βX onto the set $\mathfrak{U}$ of all ultrafilters on X, where the points of X correspond to the fixed ultrafilters and the points of $\beta X \setminus X$ to the free ultrafilters.

We can describe the topology on βX in terms of $\mathfrak{U}$: For each subset A of X, let

$$\mathfrak{U}_A = \{\mathcal{U} \in \mathfrak{U} : A \notin \mathcal{U}\}.$$

The collection $\mathcal{A} = \{\mathfrak{U}_A : A \subset X\}$ enjoys the following properties.

a. $\mathfrak{U}_\emptyset = \mathfrak{U}$ and $\mathfrak{U}_X = \emptyset$.

b. $\mathfrak{U}_A \cap \mathfrak{U}_B = \mathfrak{U}_{A \cup B}$ and $\mathfrak{U}_A \cup \mathfrak{U}_B = \mathfrak{U}_{A \cap B}$.

From properties (a) and (b), we see that $\mathcal{A}$ is a base for a topology τ. This topology is called the **hull-kernel topology**.[11] The topological space $(\mathfrak{U}, \tau)$ is referred to as the **ultrafilter space** of X.

The ultrafilter space is a Hausdorff space. To see this, let $\mathcal{U}_1 \neq \mathcal{U}_2$. Then there exists some $A \in \mathcal{U}_1$ with $A \notin \mathcal{U}_2$ (or vice versa), so $B = X \setminus A \notin \mathcal{U}_1$. Hence $\mathcal{U}_2 \in \mathfrak{U}_A$ and $\mathcal{U}_1 \in \mathfrak{U}_B$, while $\mathfrak{U}_A \cap \mathfrak{U}_B = \mathfrak{U}_{A \cup B} = \mathfrak{U}_X = \emptyset$.

And now we have the main result of this section: The ultrafilter space with the hull-kernel topology is homeomorphic to the Stone–Čech compactification of X.

2.86 Theorem   *For a discrete space X, the mapping $x \mapsto \mathcal{U}_x$ is a homeomorphism from βX onto $\mathfrak{U}$. So βX can be identified with the ultrafilter space $\mathfrak{U}$ of X.*

---

[11] See, e.g., W. A. J. Luxemburg and A. C. Zaanen [235, Chapter 1] for an explanation of the name.

Proof: We first demonstrate continuity. Let $\mathfrak{U}_A$ for some $A \subset X$ be a basic neighborhood of $\mathfrak{U}_x$ in $\mathfrak{U}$. We need to find a neighborhood N of x in βX such that $y \in N$ implies that $\mathfrak{U}_y \in \mathfrak{U}_A$. Since $\mathfrak{U}_x \in \mathfrak{U}_A$, we have $A \notin \mathfrak{U}_x$. Thus $B = X \setminus A \in \mathfrak{U}_x$ (why?), and consequently $x \in \overline{B}$. Now $\overline{B}$ is open in βX by Theorem 2.85. Also $A \cap B = \varnothing$, so $\overline{A} \cap \overline{B} = \varnothing$, again by Theorem 2.85. Thus $y \in \overline{B}$ implies $y \notin \overline{A}$, so $A \notin \mathfrak{U}_y$. (Why?) That is, $\mathfrak{U}_y \in \mathfrak{U}_A$. Thus $\overline{B}$ is our neighborhood.

By Theorem 2.36 the mapping $x \mapsto \mathfrak{U}_x$ is a homeomorphism. ∎

The Stone–Čech compactification of a general completely regular Hausdorff space X can be described in terms of so-called $\mathcal{Z}$-ultrafilters. A **$\mathcal{Z}$-set** is the zero set of a bounded continuous function. That is, a set of the form $\{x \in X : f(x) = 0\}$, where $f \in C_b(X)$. It is not hard to see that the intersection of two $\mathcal{Z}$-sets is another $\mathcal{Z}$-set. In a discrete space, every set is a $\mathcal{Z}$-set. A **$\mathcal{Z}$-filter** is a collection of $\mathcal{Z}$-sets that satisfy the definition of a filter, where only $\mathcal{Z}$-sets are allowed. That is, a collection $\mathcal{F}$ of $\mathcal{Z}$-sets is a $\mathcal{Z}$-filter if:

1. $\varnothing \notin \mathcal{F}$ and $X \in \mathcal{F}$;

2. If $A, B \in \mathcal{F}$, then $A \cap B \in \mathcal{F}$; and

3. If $A \subset B$, B is a $\mathcal{Z}$-set, and $A \in \mathcal{F}$, then $B \in \mathcal{F}$.

A **$\mathcal{Z}$-ultrafilter** is a maximal $\mathcal{Z}$-filter. The $\mathcal{Z}$-ultrafilter space, topologized with the hull-kernel topology, can be identified with the Stone–Čech compactification. See L. Gillman and M. Jerison [138, Chapter 6] for details. Further results may be found in the survey by R. C. Walker [338].

2.19 Paracompact spaces and partitions of unity

If $\mathcal{V} = \{V_i\}_{i \in I}$ and $\mathcal{W} = \{W_\alpha\}_{\alpha \in A}$ are covers of a set, then we say that $\mathcal{W}$ is a **refinement** of $\mathcal{V}$ if for each $\alpha \in A$ there is some $i \in I$ with $W_\alpha \subset V_i$. A collection of subsets $\{V_j\}_{j \in J}$ of a topological space is **locally finite** if each point has a neighborhood that meets at most finitely many V_j.

2.87 Definition   *A Hausdorff space is **paracompact** if every open cover of the space has an open locally finite refinement cover.*

An immediate consequence of the preceding definition is the following.

2.88 Lemma   *Every compact Hausdorff space is paracompact.*

The concept of a "partition of unity" is closely related to paracompactness. Partitions of unity define "moving" convex combinations, and are the basic tools for proving selection theorems and fixed point theorems; see, e.g., Theorems 17.63 and 17.54.

2.89 Definition   *A **partition of unity** on a set X is a family $\{f_i\}_{i \in I}$ of functions from X into $[0, 1]$ such that at each $x \in X$, only finitely many functions in the family are nonzero and*

$$\sum_{i \in I} f_i(x) = 1,$$

where by convention the sum of an arbitrary collection of zeros is zero.

*A partition of unity is **subordinated** to a cover $\mathcal{U}$ of X if each function vanishes outside some member of $\mathcal{U}$. For a topological space, a partition of unity is called **continuous** if each function is continuous, and is **locally finite** if every point has a neighborhood on which all but finitely many of the functions vanish.* [12]

We remark that if $\{f_i\}_{i \in I}$ is a locally finite partition of unity subordinated to the cover $\mathcal{U}$, then there is a locally finite partition of unity subordinated to $\mathcal{U}$ and indexed by $\mathcal{U}$: For each i pick $U_i \in \mathcal{U}$ such that f_i vanishes on U_i^c. For each $U \in \mathcal{U}$, define f_U by $f_U = \sum_{\{i : U_i = U\}} f_i$, where we set $f_U = 0$ if $\{i : U_i = U\} = \emptyset$. Note that f_U is continuous if each f_i is. We leave it as an exercise to verify that this indeed defines the desired partition of unity.

Here is the relationship between paracompactness and partitions of unity.

2.90 Theorem   *A Hausdorff space X is paracompact if and only if every open cover of X has a continuous locally finite partition of unity subordinated to it.*

Proof: One direction is easy. If $\{f_U\}_{U \in \mathcal{U}}$ is a continuous locally finite partition of unity subordinated to the open cover $\mathcal{U}$, then the collection $\{V_U\}_{U \in \mathcal{U}}$, where $V_U = \{x \in X : f_U(x) > 0\}$, is a locally finite refinement of $\mathcal{U}$.

The proof of the converse proceeds along the lines of the proof of Urysohn's Lemma 2.46. That is, it is very technical and not especially enlightening. See J. Dugundji [106, Theorem 4.2, p. 170] for details.   ∎

A consequence of the preceding result is the following.

2.91 Theorem   *Every paracompact space is normal.*

Proof: Let A and B be disjoint closed sets and consider the open cover $\{A^c, B^c\}$. By Theorem 2.90 there is a finite continuous partition of unity $\{f_{A^c}, f_{B^c}\}$ subordinated to it. Clearly $f_{A^c} = 1$ on B and $f_{A^c} = 0$ on A.   ∎

However, a normal Hausdorff space need not be paracompact; see for example, S. Willard [342, Example 20.11, p. 147].

The next result guarantees the existence of locally finite partitions of unity subordinate to a given open cover.

---

[12] When X is an open subset of some Euclidean space $\mathbb{R}^n$, then there are also C^∞-partitions of unity. For details, see e.g., J. Horváth [168, pp. 166–169].

2.92 Lemma *Let $\mathcal{U}$ be an open cover of a compact Hausdorff space X. Then there is a locally finite family $\{f_U\}_{U\in\mathcal{U}}$ of real functions such that:*

1. *$f_U : X \to [0,1]$ is continuous for each U.*

2. *f_U vanishes on U^c.*

3. *$\sum_{U\in\mathcal{U}} f_U(x) = 1$ for all $x \in X$.*

That is, $\{f_U\}_{U\in\mathcal{U}}$ is a continuous locally finite partition of unity subordinated to $\mathcal{U}$.

Proof: For each x pick a neighborhood $U_x \in \mathcal{U}$ of x. By Theorem 2.48, the space X is normal, so by Urysohn's Lemma 2.46, for each x there is a continuous real function $g_x : X \to [0,1]$ satisfying $g_x = 0$ on U_x^c and $g_x(x) = 1$. The set $V_x = \{z \in X : g_x(z) > 0\}$ is an open neighborhood of x, so $\{V_x : x \in X\}$ is an open cover of X. Thus there is a finite subcover $\{V_{x_1}, \ldots, V_{x_n}\}$. Observe that $g_{x_j}(z) > 0$ for each $z \in V_{x_j}$ and vanishes outside U_{x_j}. Define g by $g(z) = \sum_{j=1}^{n} g_{x_j}(z)$ and note that $g(z) > 0$ for every $z \in X$. Replacing g_{x_j} by g_{x_j}/g, we can assume that $\sum_{j=1}^{n} g_{x_j}(z) = 1$ for each $z \in X$.

Finally, put $f_U = \sum_{\{i:U_{x_i}=U\}} g_{x_i}$ (if $\{i : U_{x_i} = U\} = \varnothing$, we let $f_U = 0$), and note that the family $\{f_U\}_{U\in\mathcal{U}}$ of real functions satisfies the desired properties. ∎

Theorem 3.22 below shows that metric spaces are paracompact.

Chapter 3

Metrizable spaces

In Chapter 2 we introduced topological spaces to handle problems of convergence that metric spaces could not. Nevertheless, every sane person would rather work with a metric space if they could. The reason is that the metric, a real-valued function, allows us to analyze these spaces using what we know about the real numbers. That is why they are so important in real analysis. We present here some of the more arcane results of the theory of metric spaces. Most of this material can be found in some form in K. Kuratowski's [218] tome. Many of these results are the work of Polish mathematicians in the 1920s and 1930s. For this reason, a complete separable metric space is called a *Polish space*.

Here is a guide to the major points of interest in the territory covered in this chapter. The distinguishing features of the theory of metric spaces, which are absent from the general theory of topology, are the notions of *uniform continuity* and *completeness*. These are not topological notions, in that there may be two *equivalent* metrics inducing the same topology, but they may have different uniformly continuous functions, and one may be complete while the other isn't. Nevertheless, if a topological space is *completely metrizable*, there are some topological consequences. One of these is the Baire Category Theorem 3.47, which asserts that in a completely metrizable space, the countable intersection of open dense sets is dense. Complete metric spaces are also the home of the Contraction Mapping Theorem 3.48, which is one of the fundamental theorems in the theory of dynamic programming (see the book by N. Stokey, R. E. Lucas, and E. C. Prescott [322].)

Lemma 3.23 embeds an arbitrary metric space in the Banach space of its bounded continuous real-valued functions. This result is useful in characterizing complete metric spaces. By the way, all the Euclidean spaces are complete.

In a metric space, it is easy to show that second countability and separability are equivalent (Lemma 3.4). The Urysohn Metrization Theorem 3.40 asserts that every second countable regular Hausdorff is metrizable, and that this property is equivalent to being embedded in the *Hilbert cube*. This leads to a number of properties of separable metrizable spaces. Another useful property is that in metric spaces, a set is compact if and only if it is sequentially compact (Theorem 3.28).

We also introduce the compact metric space called the *Cantor set*. It can be viewed as a subset of the unit interval, but every compact metric space is the image of the Cantor set under a continuous function. In the same vein, we study the *Baire space* of sequences of natural numbers. It is a Polish space, and every Polish space is a continuous image of it. It is also the basis for the study of *analytic sets*, which we describe in Section 12.5.

We also discuss topologies for spaces of subsets of a metric space. The most straightforward way to topologize the collection of nonempty closed subsets of a metric space is through the Hausdorff metric. Unfortunately, this technique is not topological. That is, the topology on the space of closed subsets may be different for different compatible metrics on the underlying space (Example 3.86). However, restricted to the compact subsets, the topology is independent of the compatible metric (Theorem 3.91). Since every locally compact separable metrizable space has a metrizable compactification (Corollary 3.45), for this class of spaces there is a nice topological characterization of the *topology of closed convergence* on the space of closed subsets (Corollary 3.95). Once we have a general method for topologizing subsets, our horizons are greatly expanded. For example, since binary relations are just subsets of Cartesian products, they can be topologized in a useful way; see A. Mas-Colell [240]. As another example, F. H. Page [268] uses a space of sets in order to prove the existence of an optimal incentive contract.

Finally, we conclude with a discussion of the space $C(X, Y)$ of continuous functions from a compact space into a metrizable space under the topology of uniform convergence. It turns out that this topology depends only on the topology of Y and not on any particular metric (Lemma 3.98). The space $C(X, Y)$ is complete if Y is complete, and separable if Y is separable; see Lemmas 3.97 and 3.99.

3.1   Metric spaces

Recall the following definition from Chapter 2.

3.1 Definition   *A **metric** (or **distance**) on a set X is a function $d\colon X \times X \to \mathbb{R}$ satisfying the following four properties:*

1. *Positivity: $d(x, y) \geqslant 0$ and $d(x, x) = 0$ for all $x, y \in X$.*

2. *Discrimination: $d(x, y) = 0$ implies $x = y$.*

3. *Symmetry: $d(x, y) = d(y, x)$ for all $x, y \in X$.*

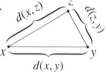

4. *The Triangle Inequality: $d(x, y) \leqslant d(x, z) + d(z, y)$ for all $x, y, z \in X$.*

*A **semimetric** on X is a function $d\colon X \times X \to \mathbb{R}$ satisfying (1), (3), and (4). Obviously, every metric is a semimetric. If d is a metric on a set X, then the pair (X, d) is called a **metric space**, and similarly if d is a semimetric, then (X, d) is a **semimetric space**.*

If d is a semimetric, then the binary relation defined by $x \sim y$ if $d(x, y) = 0$ is an equivalence relation, and d defines a metric $\hat{d}$ on the set of equivalence classes by $\hat{d}([x], [y]) = d(x, y)$. For this reason we deal mostly with metric spaces. *Be aware that when we define a concept for metric spaces, there is nearly always a corresponding notion for semimetric spaces, even if we do not explicitly mention it.* The next definition is a good example.

For a nonempty subset A of a metric space (X, d) its **diameter** is defined by

$$\operatorname{diam} A = \sup\{d(x, y) : x, y \in A\}.$$

A set A is **bounded** if $\operatorname{diam} A < \infty$, while A is **unbounded** if $\operatorname{diam} A = \infty$. If $\operatorname{diam} X < \infty$, then X is **bounded** and d is called a **bounded metric**. Similar terminology applies to semimetrics.

In a semimetric space (X, d) the **open ball** centered at a point $x \in X$ with radius $r > 0$ is the subset $B_r(x)$ of X defined by

$$B_r(x) = \{y \in X : d(x, y) < r\}.$$

The **closed ball** centered at a point $x \in X$ with radius $r > 0$ is the subset $C_r(x)$ of X defined by

$$C_r(x) = \{y \in X : d(x, y) \leqslant r\}.$$

3.2 Definition *Let (X, d) be a semimetric space. A subset A of X is **d-open** (or simply **open**) if for each $a \in A$ there exists some $r > 0$ (depending on a) such that $B_r(a) \subset A$.*

You should verify that the collection of subsets

$$\tau_d = \{A \subset X : A \text{ is } d\text{-open}\}$$

is a topology on X, called the **topology generated** or **induced** by d. When d is a metric, we call τ_d the **metric topology** on (X, d). A topological space (X, τ) is **metrizable** if the topology τ is generated by some metric. A metric generating a topology is called **compatible** or **consistent** with the topology. Two metrics generating the same topology are **equivalent**.

We have already seen a number of examples of metrizable spaces and compatible metrics in Example 2.2. There are always several metrics on any given set that generate the same topology. Let (X, d) be a metric space. Then $2d$ is also a metric generating the same topology. More interesting is the metric $\hat{d}(x, y) = \min\{d(x, y), 1\}$. It too generates the same open sets as d, but X is bounded under $\hat{d}$. In fact, notice that the $\hat{d}$-diameter of X is less than or equal to 1. A potential drawback of $\hat{d}$ is that the families of balls of radius r around x are different for d and $\hat{d}$. (For instance, $\{x \in \mathbb{R} : |x| < 2\}$ is a ball of radius 2 around 0 in the usual metric on $\mathbb{R}$, but in the truncated metric it is not a ball of any finite radius.)

Lemma 3.6 below describes a bounded metric that avoids this criticism. The point of this lemma is that for most anything topological that we want to do with a metric space, it is no restriction to assume that its metric takes on values only in the unit interval $[0, 1]$.

The following lemma summarizes some of the basic properties of metric and semimetric topologies. The proofs are straightforward applications of the definitions. You should be able to do them without looking at the hints.

3.3 Lemma (Semimetric topology)   *Let (X, d) be a semimetric space. Then:*

1. *The topology τ_d is Hausdorff if and only if d is a metric.*

2. *A sequence $\{x_n\}$ in X satisfies $x_n \xrightarrow{\tau_d} x$ if and only if $d(x_n, x) \to 0$.*

3. *Every open ball is an open set.*

4. *The topology τ_d is first countable.*

5. *A point x belongs to the closure $\overline{A}$ of a set A if and only if there exists some sequence $\{x_n\}$ in A with $x_n \to x$.*

6. *A closed ball is a closed set.*

7. *The closure of the open ball $B_r(x)$ is included in the closed ball $C_r(x)$. But the inclusion may also be proper.*

8. *If (X, d_1) and (Y, d_2) are semimetric spaces, the product topology on $X \times Y$ is generated by the semimetric*

    $$\rho((x, y), (u, v)) = d_1(x, u) + d_2(y, v).$$

    It is also generated by $\max\{d_1(x, u), d_2(y, v)\}$ and $(d_1(x, u)^2 + d_2(y, v)^2)^{1/2}$.

9. *For any four points u, v, x, y, the semimetric obeys*

    $$|d(x, y) - d(u, v)| \leqslant d(x, u) + d(y, v).$$

10. *The real function $d \colon X \times X \to \mathbb{R}$ is jointly continuous.*

Hints: The proofs of (1) and (2) are straightforward, and (5) follows from (4).

   (3)   Let y belong to the open ball $B_r(x)$. Put $\varepsilon = r - d(x, y) > 0$. If $z \in B_\varepsilon(y)$, then the triangle inequality implies $d(x, z) \leqslant d(x, y) + d(y, z) < d(x, y) + \varepsilon = r$. So $B_\varepsilon(y) \subset B_r(x)$, which means that $B_r(x)$ is a τ_d-open set.

   (4)   The countable family of open neighborhoods $\{B_{1/n}(x) : n \in \mathbb{N}\}$ is a base for the neighborhood system at x.

   (6)   Suppose $y \notin C_r(x)$. Then $\varepsilon = d(x, y) - r > 0$, so by the triangle inequality, $B_\varepsilon(y)$ is an open neighborhood of y disjoint from $C_r(x)$. This shows that the complement of $C_r(x)$ is open.

(7) Now $B_r(x) \subset C_r(x)$, so $\overline{B_r(x)} \subset \overline{C_r(x)} = C_r(x)$. For an example of proper inclusion consider the open ball of radius one under the discrete metric.

(8) Think about $\mathbb{R}^2$.

(9) The triangle inequality implies $d(x, y) \leqslant d(x, u) + d(u, v) + d(v, y)$ so

$$d(x, y) - d(u, v) \leqslant d(x, u) + d(y, v).$$

By symmetry, we obtain the result.

(10) Suppose $(x_n, y_n) \to (x, y)$ in the product topology. Then $x_n \to x$ and $y_n \to y$ in X. That is, $d(x_n, x) \to 0$ and $d(y_n, y) \to 0$. But then from (9) we get

$$|d(x_n, y_n) - d(x, y)| \leqslant d(x_n, x) + d(y_n, y) \to 0,$$

so that $d(x_n, y_n) \to d(x, y)$. ■

Although for general topological spaces the property of second countability is stronger than separability, for metrizable spaces the two properties coincide. The next result will be used again and again, often without explicit reference.

3.4 Lemma *A metrizable space is separable if and only if it is second countable.*

Proof: Let (X, τ) be a metrizable topological space and let d be a metric generating τ. First assume X is separable, and let A be a countable dense subset. Then the collection $\{B_{1/n}(x) : x \in A, \ n \in \mathbb{N}\}$ of d-open balls is a countable base for the topology τ. The converse is proven in Lemma 2.9. ■

For a general topological space, second countability is clearly inherited by its subspaces, whereas separability may not be. For metrizable spaces, separability is inherited.

3.5 Corollary *Every subset of a separable metrizable space is separable.*

3.2 Completeness

A **Cauchy sequence** in a metric space (X, d) is a sequence $\{x_n\}$ such that for each $\varepsilon > 0$ there exists some n_0 (depending upon ε) satisfying $d(x_n, x_m) < \varepsilon$ for all $n, m \geqslant n_0$, or equivalently, if $\lim_{n,m \to \infty} d(x_n, x_m) = 0$, or also equivalently, if $\lim_{n \to \infty} \mathrm{diam}\{x_n, x_{n+1}, \ldots\} = 0$. A metric space (X, d) is **complete** if every Cauchy sequence in X converges in X, in which case we also say that d is a **complete metric** on X.

Note that whether a sequence is Cauchy or a space is complete depends on the metric, not just the topology. It is possible for two metrics to induce the same topology, even though one is complete and the other is not. See Example 3.32.

A topological space X is **completely metrizable** if there is a consistent metric d for which (X, d) is complete. A separable topological space that is completely metrizable is called a **Polish space**. Such a topology is called a **Polish topology**.

Here are some important examples of complete metric spaces.

- The space $\mathbb{R}^n$ with the Euclidean metric $d(x, y) = \left[\sum_{i=1}^{n}(x_i - y_i)^2\right]^{1/2}$ is a complete metric space.

- The discrete metric is always complete.

- Let Y be a nonempty subset of a complete metric space (X, d). Then $(Y, d|_Y)$ is a complete metric space if and only if Y is a closed subset of X.

- If X is a nonempty set, then the vector space $B(X)$ of all bounded real functions on X is a complete metric space under the **uniform metric** defined by

$$d(f, g) = \sup_{x \in X} |f(x) - g(x)|.$$

It is clear that a sequence $\{f_n\}$ in $B(X)$ is d-convergent to $f \in B(X)$ if and only if it converges uniformly to f. First let us verify that d is indeed a metric on $B(X)$. Clearly, d satisfies the positivity, discrimination, and symmetry properties of a metric.

To see that d satisfies the triangle inequality, note that if $f, g, h \in B(X)$, then for each $x \in X$ we have

$$|f(x) - g(x)| \leqslant |f(x) - h(x)| + |h(x) - g(x)| \leqslant d(f, h) + d(h, g).$$

Therefore, $d(f, g) = \sup_{x \in X} |f(x) - g(x)| \leqslant d(f, h) + d(h, g)$.

Now we establish that $(B(X), d)$ is complete. To this end, let $\{f_n\}$ be a d-Cauchy sequence in $B(X)$. This means that for each $\varepsilon > 0$ there exists some k such that

$$|f_n(x) - f_m(x)| \leqslant d(f_n, f_m) < \varepsilon \qquad (\star)$$

for all $x \in X$ and all $n, m \geqslant k$. In particular, $\{f_n(x)\}$ is a Cauchy sequence of real numbers for each $x \in X$. Let $\lim f_n(x) = f(x) \in \mathbb{R}$ for each $x \in X$. To finish the proof we need to show that f is bounded and so belongs to $B(X)$, and that $d(f_n, f) \to 0$. Pick some $M > 0$ such that $|f_k(x)| \leqslant M$ for each $x \in X$, and then use $(\star)$ to see that

$$|f(x)| \leqslant \lim_{m \to \infty} |f_m(x) - f_k(x)| + |f_k(x)| \leqslant \varepsilon + M$$

for each $x \in X$, so f belongs to $B(X)$. Now another glance at $(\star)$ yields

$$|f_n(x) - f(x)| = \lim_{m \to \infty} |f_n(x) - f_m(x)| \leqslant \varepsilon$$

for all $n \geqslant k$. Hence $d(f_n, f) = \sup_{x \in X} |f_n(x) - f(x)| \leqslant \varepsilon$ for all $n \geqslant k$. This shows that $(B(X), d)$ is a complete metric space.

• If X is a topological space, then the vector space $C_b(X)$ of all bounded continuous real functions on X is a complete metric space under the uniform metric. (Recall that Theorem 2.65 implies that the uniform limit of a sequence of continuous functions is continuous.)

• More generally, let X be any nonempty set and define $d: \mathbb{R}^X \times \mathbb{R}^X \to \mathbb{R}$ by

$$d(f, g) = \sup_{x \in X} \min\{1, |f(x) - g(x)|\}.$$

Then $(\mathbb{R}^X, d)$ is a complete metric space, and a net $\{f_\alpha\}$ in $\mathbb{R}^X$ converges uniformly to $f \in \mathbb{R}^X$ if and only if $d(f_\alpha, f) \to 0$.

3.6 Lemma   *Let (X, d) be an arbitrary metric space. Then the metric ρ defined by $\rho(x, y) = \frac{d(x,y)}{1+d(x,y)}$ is a bounded equivalent metric taking values in $[0, 1)$.*

Moreover, d and ρ have the same Cauchy sequences, and (X, d) is complete if and only if (X, ρ) is complete.

Proof: The proof is left as an exercise. Here is a generous hint: $d(x, y) \leqslant \varepsilon$ if and only if $\rho(x, y) \leqslant \varepsilon/(1 + \varepsilon)$.   ∎

The next result is a profoundly useful fact about complete metric spaces. Let us say that a sequence $\{A_n\}$ of nonempty sets has **vanishing diameter** if

$$\lim_{n \to \infty} \operatorname{diam} A_n = 0.$$

3.7 Cantor's Intersection Theorem   *In a complete metric space, if a decreasing sequence of nonempty closed subsets has vanishing diameter, then the intersection of the sequence is a singleton.*

Proof: Let $\{F_n\}$ be a decreasing sequence (that is, $F_{n+1} \subset F_n$ holds for each n) of nonempty closed subsets of the complete metric space (X, d), and assume that $\lim_{n \to \infty} \operatorname{diam} F_n = 0$. The intersection $F = \bigcap_{n=1}^{\infty} F_n$ cannot have more that one point, for if $a, b \in F$, then $d(a, b) \leqslant \operatorname{diam} F_n$ for each n, so $d(a, b) = 0$, which implies $a = b$.

To see that F is a nonempty set, for each n pick some $x_n \in F_n$. Since $d(x_n, x_m) \leqslant \operatorname{diam} F_n$ for $m \geqslant n$, the sequence $\{x_n\}$ is Cauchy. Since X is complete there is some $x \in X$ with $x_n \to x$. But x_m belongs to F_n for all $m \geqslant n$, and each F_n is closed, so $\lim_{m \to \infty} x_m = x$ belongs to F_n for each n.   ∎

Continuous images may preserve the vanishing diameter property.

3.8 Lemma   *Let $\{A_n\}$ be a sequence of subsets in a metric space (X, d) such that $\bigcap_{n=1}^{\infty} A_n$ is nonempty. If $f: (X, d) \to (Y, \rho)$ is a continuous function and $\{A_n\}$ has vanishing d-diameter, then $\{f(A_n)\}$ has vanishing ρ-diameter.*

Proof: Since $\{A_n\}$ has vanishing diameter and $\bigcap_{n=1}^{\infty} A_n$ is nonempty, the intersection $\bigcap_{n=1}^{\infty} A_n$ must be some singleton $\{x\}$. Let $\varepsilon > 0$ be given. Since f is continuous, there is some $\delta > 0$ such that $d(z, x) < \delta$ implies $\rho(f(z), f(x)) < \varepsilon$. Also there is some n_0 such that for all $n \geqslant n_0$, if $z \in A_n$, then $d(z, x) < \delta$. Thus for $n \geqslant n_0$, the image $f(A_n)$ is included in the ball of ρ-radius ε around $f(x)$, so ρ- diam $f(A_n) \leqslant 2\varepsilon$. This shows that $\{f(A_n)\}$ has vanishing ρ-diameter—and also that $\bigcap_{n=1}^{\infty} f(A_n) = \{f(x)\}$.   ∎

Note that the hypothesis that $\bigcap_{n=1}^{\infty} A_n$ is nonempty is necessary. For instance, consider $X = (0, 1]$ and $Y = \mathbb{R}$ with their usual metrics, let $A_n = (0, \frac{1}{n}]$, and let $f(x) = \sin \frac{1}{x}$. Then for each n, the image $f(A_n) = [-1, 1]$, which does not have vanishing diameter.

3.3   Uniformly continuous functions

Some aspects of metric spaces are not topological, but depend on the particular compatible metric. These properties include its uniformly continuous functions and Cauchy sequences. A function $f \colon (X, d) \to (Y, \rho)$ between two metric spaces is **uniformly continuous** if for each $\varepsilon > 0$ there exists some $\delta > 0$ (depending only on ε) such that $d(x, y) < \delta$ implies $\rho(f(x), f(y)) < \varepsilon$. Any uniformly continuous function is obviously continuous. An important property of uniformly continuous functions is that they map Cauchy sequences into Cauchy sequences. (The proof of this is a simple exercise.)

A function $f \colon (X, d) \to (Y, \rho)$ between metric spaces is **Lipschitz continuous** if there is some real number c such that for every x and y in X,

$$\rho(f(x), f(y)) \leqslant c d(x, y).$$

The number c is called a **Lipschitz constant** for f. Clearly every Lipschitz continuous function is uniformly continuous.

The set $X \times X$ has a natural metric ρ given by $\rho((x, y), (u, v)) = d(x, u) + d(y, v)$. The metric d can be viewed as a function from the metric space $(X \times X, \rho)$ to $\mathbb{R}$. Viewed this way, d is Lipschitz continuous with Lipschitz constant 1 (and hence it is also a uniformly continuous function). This fact, which follows immediately from Property (9) of Lemma 3.3, may be used throughout this book without any specific reference.

An **isometry** between metric spaces (X, d) and (Y, ρ) is a one-to-one function φ mapping X into Y satisfying

$$d(x, y) = \rho(\varphi(x), \varphi(y))$$

for all $x, y \in X$. If in addition φ is surjective, then (X, d) and (Y, ρ) are **isometric**. If two metric spaces are isometric, then any property expressible in terms of metrics holds in one if and only if it holds in the other. Notice that isometries are uniformly continuous, indeed Lipschitz continuous.

Given a metric space (X, d), denote by $U_d(X)$ or more simply, U_d, the collection of all bounded d-uniformly continuous real-valued functions on X. The set U_d is a function space (recall Definition 1.1) that includes the constant functions.

In general, two different equivalent metrics determine different classes of uniformly continuous functions. For example, $x \mapsto \frac{1}{x}$ is not uniformly continuous on $(0, 1)$ under the usual metric, but it is uniformly continuous under the equivalent metric d defined by $d(x, y) = \left| \frac{1}{x} - \frac{1}{y} \right|$.

The example just given is a particular instance of the following lemma on creating new metric spaces out of old ones. The proof of the lemma is a straightforward application of the definitions and is left as an exercise.

3.9 Lemma *Let $\varphi \colon (X, d) \to Y$ be one-to-one and onto. Then φ induces a metric ρ on Y by $\rho(x, y) = d(\varphi^{-1}(x), \varphi^{-1}(y))$. Furthermore, $\varphi \colon (X, d) \to (Y, \rho)$ is an isometry. The metric ρ is also known as $d \circ \varphi^{-1}$.*

On the other hand, if $\varphi \colon Y \to (X, d)$, then φ induces a semimetric ρ on Y by $\rho(x, y) = d(\varphi(x), \varphi(y))$. If φ is one-to-one, then it is an isometry onto its range.

The bounded uniformly continuous functions form a complete subspace of the space of bounded continuous functions.

3.10 Lemma *If X is metrizable and ρ is a compatible metric on X, then the vector space $U_\rho(X)$ of all bounded ρ-uniformly continuous real functions on X is a closed subspace of $C_b(X)$. Thus $U_\rho(X)$ equipped with the uniform metric is a complete metric space in its own right.* [1]

The next theorem asserts that every uniformly continuous partial function can be uniquely extended to a uniformly continuous function on the closure of its domain simply by taking limits. The range space is assumed to be complete.

3.11 Lemma (Uniformly continuous extensions) *Let A be a nonempty subset of (X, d), and let $\varphi \colon (A, d) \to (Y, \rho)$ be uniformly continuous. Assume that (Y, ρ) is complete. Then φ has a unique uniformly continuous extension $\hat{\varphi}$ to the closure $\overline{A}$ of A. Moreover, the extension $\hat{\varphi} \colon \overline{A} \to Y$ is given by*

$$\hat{\varphi}(x) = \lim_{n \to \infty} \varphi(x_n)$$

for any $\{x_n\} \subset A$ satisfying $x_n \to x$.

In particular, if $Y = \mathbb{R}$, then $\|\varphi\|_\infty = \|\hat{\varphi}\|_\infty$.

---

[1] In the terminology of Section 9.5, $U_\rho(X)$ is a closed Riesz subspace of $C_b(X)$, and is also an AM-space with unit the constant function one.

Proof: Let $x \in \overline{A}$ and pick a sequence $\{x_n\}$ in A converging to x. Since $\{x_n\}$ converges, it is d-Cauchy. Since φ is uniformly continuous, $\{\varphi(x_n)\}$ is ρ-Cauchy. Since Y is ρ-complete, there is some $y \in Y$ such that $\varphi(x_n) \to y$.

This y is independent of the particular sequence $\{x_n\}$. To see this, let $\{z_n\}$ be another sequence in A converging to x. Interlace the terms of $\{z_n\}$ and $\{x_n\}$ to form the sequence $\{z_1, x_1, z_2, x_2, \ldots\}$ converging to x. Then $\{\varphi(z_1), \varphi(x_1), \varphi(z_2), \varphi(x_2), \ldots\}$ is again ρ-Cauchy and since $\{\varphi(x_n)\}$ is a subsequence, the limit is again y. The latter implies that $\varphi(z_n) \to y$. Thus, setting $\hat{\varphi}(x) = y$ is well defined.

To see that $\hat{\varphi}$ is uniformly continuous on $\overline{A}$, let $\varepsilon > 0$ be given and pick $\delta > 0$ so that if $x, y \in A$ and $d(x, y) < \delta$, then $\rho(\varphi(x), \varphi(y)) < \varepsilon$. Now suppose $x, y \in \overline{A}$ and $d(x, y) < \delta$. Pick sequences $\{x_n\}$ and $\{y_n\}$ in A converging to x and y respectively. From $|d(x_n, y_n) - d(x, y)| \leqslant d(x_n, x) + d(y_n, y)$, we see that $d(x_n, y_n) \to d(x, y)$, so eventually $d(x_n, y_n) < \delta$. Thus $\rho(\varphi(x_n), \varphi(y_n)) < \varepsilon$ eventually, so

$$\rho(\hat{\varphi}(x), \hat{\varphi}(y)) = \lim_{n \to \infty} \rho(\varphi(x_n), \varphi(y_n)) \leqslant \varepsilon.$$

The uniqueness of the extension is obvious.   ∎

It is interesting to note that with an appropriate change of the metric of the domain of a continuous function between metric spaces the function becomes Lipschitz continuous.

3.12 Lemma   *If $f: (X, d) \to (Y, \rho)$ is a continuous function between metric spaces, then there exists an equivalent metric d_1 on X such that $f: (X, d_1) \to (Y, \rho)$ is Lipschitz (and hence uniformly) continuous.*

More generally, if $\mathcal{F}$ is a countable family of continuous functions from (X, d) to (Y, ρ), then there exists an equivalent metric d_2 on X and an equivalent metric ρ_1 on Y such that for each $f \in \mathcal{F}$ the function $f: (X, d_2) \to (Y, \rho_1)$ is Lipschitz (and hence uniformly) continuous.

Proof: The metric d_1 is defined by $d_1(x, y) = d(x, y) + \rho(f(x), f(y))$. The reader should verify that d_1 is indeed a metric on X such that $d_1(x_n, x) \to 0$ holds in X if and only if $d(x_n, x) \to 0$. This shows that the metric d_1 is equivalent to d. Now notice that the inequality $\rho(f(x), f(y)) \leqslant d_1(x, y)$ guarantees that the function $f: (X, d_1) \to (Y, \rho)$ is Lipschitz continuous.

The general case can be established in a similar manner. To see this, consider a countable set $\mathcal{F} = \{f_1, f_2, \ldots\}$ of continuous functions from (X, d) to (Y, ρ). Next, introduce the equivalent metric ρ_1 on Y by $\rho_1(u, v) = \frac{\rho(u,v)}{1+\rho(u,v)}$. Subsequently, define the function $d_2: X \times X \to \mathbb{R}$ by

$$d_2(x, y) = d(x, y) + \sum_{n=1}^{\infty} \frac{1}{2^n} \rho_1(f_n(x), f_n(y)),$$

and note that d_2 is a metric on X that is equivalent to d. In addition, for each n we have the inequality $\rho_1(f_n(x), f_n(y)) \leqslant 2^n d_2(x, y)$. This shows that each function $f_n: (X, d_2) \to (Y, \rho_1)$ is Lipschitz continuous.   ∎

3.4 Semicontinuous functions on metric spaces

On metric spaces, upper and lower semicontinuous functions are pointwise limits of monotone sequences of Lipschitz continuous functions.

3.13 Theorem *Let $f: (X, d) \to \mathbb{R}$ be bounded below. Then f is lower semicontinuous if and only if it is the pointwise limit of an increasing sequence of Lipschitz continuous functions.*

Similarly, if $g: (X, d) \to \mathbb{R}$ is bounded above, then g is upper semicontinuous if and only if it is the pointwise limit of a decreasing sequence of Lipschitz continuous functions.

Proof: We give a constructive proof of the first part. The second part follows from the first applied to $-f$. Let $f: X \to \mathbb{R}$ be lower semicontinuous and bounded from below. For each n, define $f_n: X \to \mathbb{R}$ by

$$f_n(x) = \inf\{f(y) + nd(x, y) : y \in X\}.$$

Clearly, $f_n(x) \leqslant f_{n+1}(x) \leqslant f(x)$ for each x. Moreover, observe that

$$|f_n(x) - f_n(z)| \leqslant nd(x, z),$$

which shows that each f_n is Lipschitz continuous.

Let $f_n(x) \uparrow h(x) \leqslant f(x)$ for each x. Now fix x and let $\varepsilon > 0$. For each n pick some $y_n \in X$ with

$$f(y_n) \leqslant f(y_n) + nd(x, y_n) \leqslant f_n(x) + \varepsilon. \qquad (\star)$$

If $f(u) \geqslant M > -\infty$ for all $u \in X$, then it follows from $(\star)$ that

$$0 \leqslant d(x, y_n) \leqslant \frac{f_n(x) + \varepsilon - f(y_n)}{n} \leqslant \frac{f(x) + \varepsilon - M}{n}$$

for each n, and this shows that $y_n \to x$. Using the lower semicontinuity of f and the inequality $f(y_n) \leqslant f_n(x) + \varepsilon$, we see that

$$f(x) \leqslant \liminf_{n \to \infty} f(y_n) \leqslant \lim_{n \to \infty} [f_n(x) + \varepsilon] = h(x) + \varepsilon$$

for each $\varepsilon > 0$. So $f(x) \leqslant h(x)$, and hence $f(x) = h(x) = \lim_{n \to \infty} f_n(x)$.

The converse follows immediately from Lemma 2.41. ∎

3.14 Corollary *Let (X, d) be a metric space, and let F be a closed subset of X. Then there is a sequence $\{f_n\}$ of Lipschitz continuous functions taking values in $[0, 1]$ satisfying $f_n(x) \downarrow \chi_F(x)$ for all $x \in X$.*

Proof: The indicator function of a closed set F is upper semicontinuous. So there exists a sequence $\{f_n\}$ of Lipschitz continuous functions from X to $\mathbb{R}$ satisfying $f_n(x) \downarrow \chi_F(x)$ for each $x \in X$. If we let $g_n = f_n \wedge \mathbf{1}$, then the sequence $\{g_n\}$ satisfies the desired properties.   ∎

3.15 Corollary   *Let (X, d) be a metric space and $f : X \to \mathbb{R}$ a bounded continuous function. Then there exist sequences of bounded Lipschitz continuous functions $\{g_n\}$ and $\{h_n\}$ with $g_n(x) \uparrow f(x)$ and $h_n(x) \downarrow f(x)$ for all $x \in X$.*

Proof: A continuous function is both upper and lower semicontinuous, so invoke Theorem 3.13.   ∎

3.5   Distance functions

For a nonempty set A in a semimetric space (X, d), the **distance function** $d(\cdot, A)$ on X is defined by
$$d(x, A) = \inf\{d(x, y) : y \in A\}.$$

The **ε-neighborhood** $N_\varepsilon(A)$ of a nonempty subset A of X is defined by

$$N_\varepsilon(A) = \{x \in X : d(x, A) < \varepsilon\}.$$

Note that $N_\varepsilon(A)$ depends on the metric d, but our notation does not indicate this. We shall try not to confuse you. Also observe that

$$\overline{A} = \{x \in X : d(x, A) = 0\} = \bigcap_{\varepsilon > 0} N_\varepsilon(A).$$

3.16 Theorem   *Distance functions are Lipschitz continuous.*

Proof: If $x, y \in X$ and $z \in A$, then $d(x, A) \leqslant d(x, z) \leqslant d(x, y) + d(y, z)$. Therefore $d(x, A) - d(x, y) \leqslant d(y, z)$ for every $z \in A$. This implies $d(x, A) - d(x, y) \leqslant d(y, A)$, or $d(x, A) - d(y, A) \leqslant d(x, y)$. By symmetry, we have $d(y, A) - d(x, A) \leqslant d(x, y)$, so

$$|d(x, A) - d(y, A)| \leqslant d(x, y). \tag{$\star$}$$

That is, $d(\cdot, A) : X \to \mathbb{R}$ is Lipschitz continuous with Lipschitz constant 1.   ∎

3.17 Corollary   *For $\varepsilon > 0$, the ε-neighborhood $N_\varepsilon(A)$ of a nonempty subset A (of a semimetric space) is an open set.*

3.18 Corollary   *For $\varepsilon > 0$ and a nonempty set A, we have*

$$N_\varepsilon(\overline{A}) = N_\varepsilon(A).$$

Proof: Clearly $N_\varepsilon(A) \subset N_\varepsilon(\overline{A})$. For the reverse inclusion, let $y \in N_\varepsilon(\overline{A})$. Then there is some $x \in \overline{A}$ (so $d(x, A) = 0$) satisfying $d(x, y) < \varepsilon$. By equation ($\star$) in the proof of Theorem 3.16, we have $d(y, A) < \varepsilon$, or in other words $y \in N_\varepsilon(A)$. ∎

3.19 Corollary *In a metrizable space, every closed set is a $\mathcal{G}_\delta$, and every open set is an $\mathcal{F}_\sigma$.*

Proof: Let F be a closed subset of (X, d), and put $G_n = \{x \in X : d(x, F) < 1/n\}$. Since the distance function is continuous, G_n is open, and clearly $F = \bigcap_{n=1}^\infty G_n$. Thus F is a $\mathcal{G}_\delta$. Since the complement of an open set is closed, de Morgan's laws imply that every open set is an $\mathcal{F}_\sigma$. ∎

We can now show that a metric space is perfectly normal.

3.20 Lemma *If (X, d) is a metric space and A and B are disjoint nonempty closed sets, then the continuous function $f : X \to [0, 1]$, defined by*

$$f(x) = \frac{d(x, A)}{d(x, A) + d(x, B)},$$

satisfies $f^{-1}(0) = A$ and $f^{-1}(1) = B$.

Moreover, if $\inf\{d(x, y) : x \in A$ and $y \in B\} > 0$, then the function f is Lipschitz continuous, and hence d-uniformly continuous.

Proof: The first assertion is obvious. For the second, assume that there exists some $\delta > 0$ such that $d(x, y) \geqslant \delta$ for all $x \in A$ and all $y \in B$. Then, for any $z \in X$, $a \in A$, and $b \in B$, $\delta \leqslant d(a, b) \leqslant d(a, z) + d(z, b)$, so $d(z, A) + d(z, B) \geqslant \delta > 0$ for each $z \in X$. Now use the inequalities

$$
\begin{aligned}
|f(x) - f(y)| &= \left| \frac{d(x, A)}{d(x, A) + d(x, B)} - \frac{d(y, A)}{d(y, A) + d(y, B)} \right| \\
&= \frac{|[d(y, A) + d(y, B)]d(x, A) - [d(x, A) + d(x, B)]d(y, A)|}{[d(x, A) + d(x, B)][d(y, A) + d(y, B)]} \\
&= \frac{|[d(x, A) - d(y, A)]d(x, B) + [d(y, B) - d(x, B)]d(x, A)|}{[d(x, A) + d(x, B)][d(y, A) + d(y, B)]} \\
&\leqslant \frac{[d(x, B) + d(x, A)]d(x, y)}{[d(x, A) + d(x, B)][d(y, A) + d(y, B)]} \leqslant \frac{d(x, y)}{\delta},
\end{aligned}
$$

to see that f is indeed Lipschitz continuous. ∎

3.21 Corollary *Every metrizable space is perfectly normal.*

Using distance functions we can establish the following useful result.

3.22 Theorem *Every metrizable space is paracompact.*

Proof: Let X be a metrizable space and let d be a compatible metric. Also, let $X = \bigcup_{i \in I} V_i$ be an open cover of X. Without loss of generality, we can assume that I is an infinite set. We must show that the open cover $\{V_i\}_{i \in I}$ has an open locally finite refinement cover.

For $\varepsilon > 0$ and any nonempty subset A of X, recall that the ε-neighborhood $N_\varepsilon(A) = \{x \in X : d(x, A) < \varepsilon\}$ of A is open, and define

$$E_\varepsilon(A) = \{x \in X : B_\varepsilon(x) \subset A\} = \{x \in X : d(x, A^c) \geqslant \varepsilon\}.$$

Note that $E_\varepsilon(A)$ is closed (but possibly empty). Moreover, we have the following easily verified properties:

$$E_\varepsilon(A) \subset A \subset N_\varepsilon(A) \quad \text{and} \quad N_\varepsilon(E_\varepsilon(A)) \subset A. \tag{1}$$

$$\text{If } x \in E_\varepsilon(A) \text{ and } y \in X \setminus A, \text{ then } d(x, y) \geqslant \varepsilon. \tag{2}$$

$$\text{If } x \in X \text{ satisfies } B_\varepsilon(x) \cap E_\varepsilon(A) \neq \varnothing, \text{ then } x \in E_\varepsilon(A). \tag{3}$$

For simplicity, for each n and any nonempty subset A of X write $N_n(A) = N_{1/2^n}(A)$ and $E_n(A) = E_{1/2^n}(A)$.

Next, let $\geq$ be a well-order of the index set I; such a well-order always exists by Theorem 1.13. Using "transfinite induction,"[2] for each $n \in \mathbb{N}$ and each $i \in I$ we define the set

$$S_i^n = E_n\left(V_i \setminus \bigcup_{j<i} S_j^n\right). \tag{$\star$}$$

We claim that

$$X = \bigcup_{n=1}^{\infty} \bigcup_{i \in I} S_i^n. \tag{4}$$

To see this, let $x \in X$ and put $i_0 = \min\{i \in I : x \in V_i\}$ and then choose some n such that $B_{1/2^n}(x) \subset V_{i_0}$ and note that $x \in S_{i_0}^n$. Indeed, if $x \notin S_{i_0}^n$, then from the definition of $S_{i_0}^n$, it follows that $B_{1/2^n}(x) \cap [\bigcup_{j<i} S_j^n] \neq \varnothing$. This implies $B_{1/2^n}(x) \cap S_j^n \neq \varnothing$ for some $j < i_0$. But then, from (3) and (1), we get $x \in V_j$, which is impossible. Hence, (4) holds.

Next we define the sets $C_i^n = \overline{N_{n+3}(S_i^n)}$ and $U_i^n = N_{n+2}(S_i^n)$; of course, if $S_i^n = \varnothing$, then $C_i^n = U_i^n = \varnothing$. Clearly, C_i^n is closed, U_i^n is open and $C_i^n \subset U_i^n$. Now if $j > i$, then note that

$$S_j^n \subset V_j \setminus \bigcup_{\ell<j} S_\ell^n \subset X \setminus S_i^n.$$

So if $x \in S_j^n$ and $y \in S_i^n$, then $y \notin V_j \setminus \bigcup_{\ell<j} S_\ell^n$, and (2) yields:

$$\text{If } i \neq j, \, x \in S_j^n \text{ and } y \in S_i^n, \text{ then } d(x, y) \geqslant 1/2^n. \tag{5}$$

---

[2] The term **transfinite induction** refers to the following procedure: If i_0 is the first element of I we let $S_{i_0}^n = E_n(V_{i_0})$. Likewise, if i_1 is the first element of $I \setminus \{i_0\}$, then we let $S_{i_1}^n = E_n(V_{i_1} \setminus S_{i_0}^n)$. Now if we consider the set $J = \{i \in I : S_j^n \text{ is defined by } (\star) \text{ for all } j < i \text{ and all } n\}$, then we claim that $J = I$. If $I \setminus J \neq \varnothing$, then let j be the first element of $I \setminus J$ and note that according to $(\star)$ the set S_j^n is defined for all n, a contradiction.

Now let $j > i$, $x \in U_j^n$ and $y \in U_i^n$. Pick $u \in S_j^n$ and $v \in S_i^n$ so that $d(x, u) < 1/2^{n+2}$ and $d(y, v) < 1/2^{n+2}$ and note that from (5) we get

$$\frac{1}{2^n} \leqslant d(u, v) \leqslant d(u, x) + d(x, y) + d(y, v) < d(x, y) + \frac{1}{2^{n+1}}.$$

This implies:

$$\text{If } i \neq j, \ x \in U_j^n \text{ and } y \in U_i^n, \text{ then } d(x, y) > 1/2^{n+1}. \tag{6}$$

Next, for each fixed n consider the family of closed sets $\{C_i^n\}_{i \in I}$. We claim that for each $x \in X$ the open ball $B = B_{\frac{1}{2^{n+2}}}(x)$ intersects at most one of the sets $\{C_i^n\}_{i \in I}$. To see this, assume that for $i \neq j$ we have $y \in B \cap C_i^n$ and $z \in B \cap C_j^n$. Now a glance at (6) yields

$$\frac{1}{2^{n+1}} < d(y, z) \leqslant d(y, x) + d(x, z) < \frac{1}{2^{n+2}} + \frac{1}{2^{n+2}} = \frac{1}{2^{n+1}},$$

a contradiction. This implies (how?) that for each n the set $C_n = \bigcup_{i \in I} C_i^n$ is closed.

Finally, for each n and $i \in I$ define the sets:

$$W_i^1 = U_i^1 \quad \text{and} \quad W_i^n = U_i^n \setminus \bigcup_{k=1}^{n-1} C_k \quad \text{if } n > 1.$$

Clearly, each W_i^n is an open set. We claim that the family of open sets $\{W_i^n\}_{(n,i) \in \mathbb{N} \times I}$ is an open locally finite refinement cover of $\{V_i\}_{i \in I}$. We establish this claim by steps.

Step I: $\{W_i^n\}_{(n,i) \in \mathbb{N} \times I}$ *is a refinement of* $\{V_i\}_{i \in I}$.

To see this, note that $W_i^n \subset U_i^n = N_{n+2}(S_i^n) \subset N_n(S_i^n) \subset N_n(E_n(V_i)) \subset V_i$.

Step II: $\{W_i^n\}_{(n,i) \in \mathbb{N} \times I}$ *covers X, that is,* $X = \bigcup_{n=1}^{\infty} \bigcup_{i \in I} W_i^n$.

Fix $x \in X$. From $S_i^n \subset C_i^n$ and (4), we see that the family $\{C_i^n\}_{(n,i) \in \mathbb{N} \times I}$ covers X. Put $k = \min\{n \in \mathbb{N} : x \in C_i^n \text{ for some } i\}$. Assume that $x \notin W_i^1$. If $x \in C_i^k \subset U_i^k$, then $k > 1$ and $x \notin C_n$ for each $n < k$. Hence $x \in W_i^k$.

Step III: $\{W_i^n\}_{(n,i) \in \mathbb{N} \times I}$ *is locally finite*.

Fix $x \in X$. According to (4) there exists some n and $i_0 \in I$ such that $x \in S_{i_0}^n$. Now note that

$$B_{1/2^{n+3}}(x) \subset N_{n+3}(S_{i_0}^n) \subset \overline{N_{n+3}(S_{i_0}^n)} = C_{i_0}^n \subset C_n.$$

This implies $B_{1/2^{n+3}}(x) \cap W_i^k = \varnothing$ for all $k > n$ and all $i \in I$.

Next, fix $1 \leqslant k \leqslant n$ and assume that $B_{1/2^{n+3}}(x) \cap U_i^k \neq \varnothing$ for some $i \in I$. Then $B_{1/2^{n+3}}(x) \cap U_j^k = \varnothing$ for all $j \neq i$. To see this, assume that for $i \neq j$ there exist $y \in B_{1/2^{n+3}}(x) \cap U_i^k$ and $z \in B_{1/2^{n+3}}(x) \cap U_j^k$. But then from (6) we get $1/2^{n+1} \leqslant d(y, z) \leqslant d(y, x) + d(x, z) < 1/2^{n+3}$, which is impossible. This shows that $B_{1/2^{n+3}}(x)$ intersects at most n of the $\{U_i^k : 1 \leqslant k \leqslant n \text{ and } i \in I\}$. It follows that $B_{1/2^{n+3}}(x)$ intersects at most n of the sets W_i^k. ∎

3.6 Embeddings and completions

An **isometric embedding** of the metric space (X, d) in the metric space (Y, ρ) is simply an isometry $f : X \to Y$.

3.23 Embedding Lemma   *Every metric space can be isometrically embedded in its space of bounded uniformly continuous real functions.*

Proof: Let (X, d) be a metric space. Fix an arbitrary point $a \in X$ as a reference, and for each x define the function θ_x by

$$\theta_x(y) = d(x, y) - d(a, y).$$

For the uniform continuity of θ_x note that

$$|\theta_x(y) - \theta_x(z)| \leqslant |d(x, y) - d(x, z)| + |d(a, y) - d(a, z)| \leqslant 2d(y, z).$$

To see that θ_x is bounded, use the inequality $d(x, y) \leqslant d(x, a) + d(a, y)$ and the definition of the function θ_x to see that $\theta_x(y) \leqslant d(x, a)$. Likewise the inequality $d(a, y) \leqslant d(a, x) + d(x, y)$ implies $-\theta_x(y) = d(a, y) - d(x, y) \leqslant d(x, a)$. Furthermore, these inequalities hold exactly for $y = a$ and $y = x$ respectively. Consequently we have $\|\theta_x\|_\infty = \sup_y |\theta_x(y)| = d(x, a)$.

Next, observe that

$$\begin{aligned} |\theta_x(y) - \theta_z(y)| &= |d(x, y) - d(a, y) - [d(z, y) - d(a, y)]| \\ &= |d(x, y) - d(z, y)| \leqslant d(x, z) \end{aligned}$$

for all $y \in X$. Also $|\theta_x(z) - \theta_z(z)| = d(x, z)$. Thus,

$$\|\theta_x - \theta_z\|_\infty = \sup_{y \in X} |\theta_x(y) - \theta_z(y)| = d(x, z)$$

for all $x, z \in X$. That is, θ is an isometry.   ∎

Note that for the special case when d is a bounded metric on X, the mapping $x \mapsto d(x, \cdot)$ is an isometry from X into $C_b(X)$.

A complete metric space (Y, ρ) is the **completion** of the metric space (X, d) if there exists an isometry $\varphi : (X, d) \to (Y, \rho)$ satisfying $\overline{\varphi(X)} = Y$. It is customary to identify X with $\varphi(X)$ and consider X to be a dense subset of Y. The next result justifies calling Y *the* completion of X rather than *a* completion of X.

3.24 Theorem   *Every metric space has a completion. It is unique up to isometry, that is, any two completions are isometric.*

Proof: Since $C_b(X)$ is a complete metric space in the metric induced by its norm, Lemma 3.23 shows that a completion exists, namely $\overline{\theta(X)}$.

To prove the uniqueness of the completion up to isometry, let both (Y_1, ρ_1) and (Y_2, ρ_2) be completions of (X, d) with isometries $\varphi_i \colon (X, d) \to (Y_i, \rho_i)$. Then the function $\varphi = \varphi_1 \circ \varphi_2^{-1} \colon (\varphi_2(X), \rho_2) \to (\varphi_1(X), \rho_1)$ is an isometry and hence is uniformly continuous. By Lemma 3.11, φ has a uniformly continuous extension $\hat{\varphi}$ to the closure Y_2 of $\varphi_2(X)$. Routine arguments show that $\hat{\varphi} \colon (Y_2, \rho_2) \to (Y_1, \rho_1)$ is a surjective isometry. That is, (Y_2, ρ_2) and (Y_1, ρ_1) are isometric. ∎

3.25 Theorem   *The completion of a separable metric space is separable.*

Proof: Let Y be the completion of a metric space X and let $\varphi \colon X \to Y$ be an isometry such that $\overline{\varphi(X)} = Y$. If A is a countable dense subset of X, then (in view of Theorem 2.27 (5)) the countable subset $\varphi(A)$ of Y satisfies $\varphi(X) = \varphi(\overline{A}) \subset \overline{\varphi(A)}$, so $Y = \overline{\varphi(X)} = \overline{\varphi(A)}$. ∎

3.7   Compactness and completeness

A subset A of a metric space X is **totally bounded** if for each $\varepsilon > 0$ there exists a finite subset $\{x_1, \ldots, x_n\} \subset X$ that is **ε-dense** in A, meaning that the collection of ε-balls $B_\varepsilon(x_i)$ covers A. Note that if a set is totally bounded, then so are its closure and any subset. Any metric for which the space X is totally bounded is also called a **totally bounded metric**.

Every compact metric space is obviously totally bounded. It is easy to see that a totally bounded metric space is separable.

3.26 Lemma   *Every totally bounded metric space is separable.*

Proof: If (X, d) is totally bounded, then for each n pick a finite subset F_n of X such that $X = \bigcup_{x \in F_n} B_{1/n}(x)$, and then note that the set $F = \bigcup_{n=1}^{\infty} F_n$ is countable and dense. ∎

This implies that every compact metric space is separable, but that is not necessarily true of nonmetrizable compact topological spaces. (Can you think of a nonseparable compact topological space?) For the next result, recall that a topological space is sequentially compact if every sequence has a convergent subsequence.

3.27 Lemma   *Let (X, d) be a sequentially compact metric space, and let $\{V_i\}_{i \in I}$ be an open cover of X. Then there exists some $\delta > 0$, called the **Lebesgue number** of the cover, such that for each $x \in X$ we have $B_\delta(x) \subset V_i$ for at least one i.*

Proof: Assume by way of contradiction that no such δ exists. Then for each n there exists some $x_n \in X$ satisfying $B_{1/n}(x_n) \cap V_i^c \neq \varnothing$ for each $i \in I$. If x is the limit point of some subsequence of $\{x_n\}$, then it is easy to see (how?) that $x \in \bigcap_{i \in I} V_i^c = \left(\bigcup_{i \in I} V_i \right)^c = \varnothing$, a contradiction.   ∎

The next two results sharpen the relationship between compactness and total boundedness.

3.28 Theorem   (Compactness of metric spaces)   *For a metric space the following are equivalent:*

1. *The space is compact.*

2. *The space is complete and totally bounded.*

3. *The space is sequentially compact. That is, every sequence has a convergent subsequence.*

Proof: Let (X, d) be a metric space.

(1) $\implies$ (2)   Since $X = \bigcup_{x \in X} B_\varepsilon(x)$, there exist $x_1, \ldots, x_k$ in X such that $X = \bigcup_{i=1}^{k} B_\varepsilon(x_i)$. That is, X is totally bounded. To see that X is also complete, let $\{x_n\}$ be a Cauchy sequence in X, and let $\varepsilon > 0$ be given. Pick n_0 so that $d(x_n, x_m) < \varepsilon$ whenever $n, m \geq n_0$. By Theorem 2.31, the sequence $\{x_n\}$ has a limit point, say x. We claim that $x_n \to x$. Indeed, if we choose $k \geq n_0$ such that $d(x_k, x) < \varepsilon$, then for each $n \geq n_0$, we have

$$d(x_n, x) \leq d(x_n, x_k) + d(x_k, x) < \varepsilon + \varepsilon = 2\varepsilon,$$

proving $x_n \to x$. That is, X is also complete.

(2) $\implies$ (3)   Fix a sequence $\{x_n\}$ in X. Since X is totally bounded, there must be infinitely many terms of the sequence in a closed ball of radius $1/2$. (Why?) This ball is totally bounded too, so it must also include a closed set of diameter less than $\frac{1}{4}$ that contains infinitely many terms of the sequence. By induction, construct a decreasing sequence of closed sets with vanishing diameter, each of which contains infinitely many terms of the sequence. Use this and the Cantor Intersection Theorem 3.7 to construct a convergent subsequence.

(3) $\implies$ (1)   By Lemma 3.27, there is some $\delta > 0$ such that for each $x \in X$ we have $B_\delta(x) \subset V_i$ for at least one i. We claim that there exist $x_1, \ldots, x_k \in X$ such that $X = \bigcup_{i=1}^{k} B_\delta(x_i)$. To see this, assume by way of contradiction that this is not the case. Fix $y_1 \in X$. Since the claim is false, there exists some $y_2 \in X$ such that $d(y_1, y_2) \geq \delta$. Similarly, since $X \neq B_\delta(y_1) \cup B_\delta(y_2)$, there exists some $y_3 \in X$ such that $d(y_1, y_3) \geq \delta$ and $d(y_2, y_3) \geq \delta$. So by an inductive argument, there exists a sequence $\{y_n\}$ in X satisfying $d(y_n, y_m) \geq \delta$ for $n \neq m$. However, any such sequence $\{y_n\}$ cannot have any convergent subsequence, contrary to our hypothesis. Hence there exist $x_1, \ldots, x_k \in X$ such that $X = \bigcup_{i=1}^{k} B_\delta(x_i)$.

Finally, for each $1 \leqslant j \leqslant k$ choose an index i_j such that $B_\delta(x_j) \subset V_{i_j}$. Then $X = \bigcup_{j=1}^k V_{i_j}$, proving that X is compact. ∎

3.29 Corollary *A metric space is totally bounded if and only if its completion is compact.*

Proof: Clearly compact metric spaces are totally bounded and so are their subsets. Conversely, if (X, d) is totally bounded, then so is its completion. (Why?) But totally bounded complete metric spaces are compact. ∎

It is easy to see that any bounded subset of $\mathbb{R}^n$ is totally bounded, which yields the following classical result as an easy corollary.

3.30 Heine–Borel Theorem *Subsets of $\mathbb{R}^n$ are compact if and only if they are closed and bounded.*

Another easy consequence of Theorem 3.28 is the following useful result.

3.31 Corollary *Every continuous function from a compact metric space to a metric space is uniformly continuous.*

Proof: Let $f: (X, d) \to (Y, \rho)$ be a continuous function between metric spaces with (X, d) compact, and let $\varepsilon > 0$ be given. For each x, let V_x be the inverse image under f of $B_{\varepsilon/2}(f(x))$. Then $u, v \in V_x$ implies $\rho(f(u), f(v)) < \varepsilon$. By Theorem 3.28, the space (X, d) is also sequentially compact so by Lemma 3.27, there exists $\delta > 0$ such that for each $v \in X$, we have $B_\delta(v) \subset V_x$ for some x. Thus $u \in B_\delta(v)$ implies $\rho(f(u), f(v)) < \varepsilon$. That is, f is uniformly continuous ∎

While a metric space is compact if and only if it is complete and totally bounded, neither total boundedness nor completeness is a topological property. It is perfectly possible that a metrizable space can be totally bounded in one compatible metric and complete in a different compatible metric, yet not be compact.

3.32 Example (Completeness vs. total boundedness) Consider the set $\mathbb{N}$ of natural numbers with its usual (discrete) topology. This is clearly not a compact space, and a sequence is convergent if and only if it is eventually constant.

The discrete topology is induced by the discrete metric d, where, as you may recall, $d(n, m) = 1$ if $n \neq m$ and $d(n, n) = 0$. Clearly $(\mathbb{N}, d)$ is not totally bounded. But $(\mathbb{N}, d)$ is complete, since only eventually constant sequences are d-Cauchy.

On the other hand, the discrete topology on $\mathbb{N}$ is also induced by the bounded metric $\rho(n, m) = \left| \frac{1}{n} - \frac{1}{m} \right|$. (To see this, for each n let $r_n = 1/n(n+1)$, and notice that $B_{r_n}(n) = \{n\}$.) But $(\mathbb{N}, \rho)$ is not complete, as the sequence $\{1, 2, 3, \ldots\}$ is ρ-Cauchy, but it is not eventually constant, and so has no limit.

However $(\mathbb{N}, \rho)$ is totally bounded: Let $\varepsilon > 0$ be given, and pick some natural number k such that $1/k < \varepsilon/2$ and note that $B_\varepsilon(k) \supset \{k, k+1, k+2, \ldots\}$. Therefore, $\mathbb{N} = \bigcup_{n=1}^k B_\varepsilon(n)$, proving that $(\mathbb{N}, \rho)$ is totally bounded. ∎

The next three results deal with subsets of metric spaces that are completely metrizable given their induced topologies.

3.33 Lemma   *If the relative topology of a subset of a metric space is completely metrizable, then the subset is a $\mathcal{G}_\delta$.*

Proof: Let X be a subset of a metric space (Y, d) such that X admits a metric ρ that is consistent with the relative topology on X and for which (X, ρ) is complete.

Heuristically, X is $\bigcap_{n=1}^{\infty} \{y \in Y : d(y, X) < 1/n\} \cap \{y \in Y : \rho(y, X) < 1/n\}$. But this makes no sense, since $\rho(y, x)$ is not defined for $y \in Y \setminus X$. So what we need is a way to include points in Y that would be both d-close and ρ-close to X if ρ were defined on Y. Recall that any open set U in X is the intersection of X with an open subset V of Y. The idea is to consider open sets V where $V \cap X$ is ρ-small. To this end, for each n let

$$Y_n = \{y \in Y : \text{there is an open set } V \text{ in } Y \text{ with } y \in V \text{ and } \rho\text{-diam}\,(X \cap V) < 1/n\},$$

and put

$$G_n = \{y \in Y : d(y, X) < 1/n\} \cap Y_n.$$

First, we claim that each G_n is an open subset of Y. Indeed, if $y \in G_n$, then pick the open subset V of Y with $y \in V$ and $\rho\text{-diam}\,(X \cap V) < \frac{1}{n}$ and note that the open neighborhood $W = V \cap \{z \in Y : d(z, X) < \frac{1}{n}\}$ of y in Y satisfies $W \subset G_n$. To complete the proof, we shall show that $X = \bigcap_{n=1}^{\infty} G_n$.

First let x belong to X and fix n. Then $U = \{y \in X : \rho(y, x) < 1/3n\}$ is an open subset of X. So there exists an open subset V of Y with $U = X \cap V$. It follows that $x \in V$ and $\rho\text{-diam}\,(X \cap V) < 1/n$, so $x \in G_n$. Since n is arbitrary, $X \subset \bigcap_{n=1}^{\infty} G_n$.

For the reverse inclusion, let $y \in \bigcap_{n=1}^{\infty} G_n$. Then $d(y, X) = 0$, so $y \in \overline{X}$. In particular, there exists a sequence $\{x_n\}$ in X such that $x_n \to y$. For each n pick an open subset V_n of Y with $y \in V_n$ and $\rho\text{-diam}\,(X \cap V_n) < 1/n$. Since $X \cap V_n$ is an open subset of X, it follows that for each n there exists some k_n such that $x_m \in V_n$ for all $m \geqslant k_n$. From $\rho\text{-diam}\,(X \cap V_n) < 1/n$, we see that $\{x_n\}$ is a ρ-Cauchy sequence, and since (X, ρ) is complete, $\{x_n\}$ is ρ-convergent to some $z \in X$. It follows that $y = z \in X$, so $X = \bigcap_{n=1}^{\infty} G_n$, as desired.   ∎

For complete metric spaces the converse of Lemma 3.33 is also true.

3.34 Alexandroff's Lemma   *Every $\mathcal{G}_\delta$ in a complete metric space is completely metrizable.*

Proof: Let (Y, d) be a complete metric space, and assume that $X \neq Y$ is a $\mathcal{G}_\delta$. (The case $X = Y$ is trivial.) Then there exists a sequence $\{G_n\}$ of open sets satisfying $G_n \neq Y$ for each n and $X = \bigcap_{n=1}^{\infty} G_n$. (We want $G_n \neq Y$ so that $G_n^c = Y \setminus G_n$ is nonempty, so $0 < d(x, G_n^c) < \infty$ for all $x \in X$.) Next, define the metric ρ on X by

$$\rho(x, y) = d(x, y) + \sum_{n=1}^{\infty} \min \left\{ \frac{1}{2^n}, \left| \frac{1}{d(x, G_n^c)} - \frac{1}{d(y, G_n^c)} \right| \right\}.$$

Since each mapping $x \mapsto d(x, G_n^c)$ is continuous, a direct calculation shows that ρ is a metric equivalent to d on X. To finish, we show that (X, ρ) is complete.

To this end, let $\{x_n\}$ be a ρ-Cauchy sequence in X. It should be clear that $\{x_n\}$ is also a d-Cauchy sequence in Y, and since (Y, d) is complete, there is some $y \in Y$ such that $d(x_n, y) \to 0$. In particular, $d(x_n, G_k^c) \xrightarrow[n \to \infty]{} d(y, G_k^c)$ for each k. Also, from $\lim_{n,m \to 0} \rho(x_n, x_m) = 0$, we see that

$$\left| \frac{1}{d(x_n, G_k^c)} - \frac{1}{d(x_m, G_k^c)} \right| \xrightarrow[n,m \to \infty]{} 0,$$

so $\lim_{n \to \infty} 1/d(x_n, G_k^c)$ exists in $\mathbb{R}$ for each k. Since $\lim_{n \to \infty} d(x_n, G_k^c) = d(y, G_k^c)$, it follows that $d(y, G_k^c) > 0$, so $y \in G_k$ for each k. Therefore, y belongs to $\bigcap_{k=1}^{\infty} G_k = X$, and hence (since ρ is equivalent to d on X) we see that $\rho(x_n, y) \to 0$, as desired. ∎

The next corollary is immediate.

3.35 Corollary *Every open subset of a complete metric space is completely metrizable.*

3.8 Countable products of metric spaces

In this section, we consider a countable collection $\{X_1, X_2, \ldots\}$ of nonempty topological spaces. The Cartesian product of the sequence $\{X_n\}$ is denoted X, so $X = \prod_{n=1}^{\infty} X_n$.

3.36 Theorem *The product topology on X is metrizable if and only if each topological space X_n is metrizable.*

Proof: Assume first that each X_n is metrizable, and let d_n be a consistent metric on X_n. Define a metric d on the product space X by

$$d((x_n), (y_n)) = \sum_{n=1}^{\infty} \frac{1}{2^n} \cdot \frac{d_n(x_n, y_n)}{1 + d_n(x_n, y_n)}.$$

It is a routine matter to verify that d is indeed a metric on X, and that a net $\{x_\alpha\}$ in X satisfies $d(x_\alpha, x) \to 0$, where $x_\alpha = (x_n^\alpha)$ and $x = (x_n)$, if and only if $d_n(x_n^\alpha, x_n) \xrightarrow[\alpha]{} 0$ for each n. This shows that the product topology and the topology generated by d coincide.

For the converse, fix X_k and let d be a compatible metric on X. Also, for each n fix some $u_n \in X_n$. Now for $x \in X_k$ define $\hat{x} = (x_1, x_2, \ldots) \in X$ by $x_k = x$ and $x_n = u_n$ for $n \neq k$. Next, define a metric d_k on X_k via the formula

$$d_k(x, y) = d(\hat{x}, \hat{y}).$$

Note that d_k is indeed a metric on X_k. Since d-convergence in X is equivalent to pointwise convergence, it is a routine matter to verify that the metric d_k generates the topology of X_k.   ∎

The next result follows from similar arguments to those employed in the proof of Theorem 3.36.

3.37 Theorem   *The product of a countable collection of topological spaces is completely metrizable if and only if each factor is completely metrizable.*

Countable products of separable metrizable spaces are also separable.

3.38 Theorem   *The product of a countable collection of metrizable topological spaces is separable if and only if each factor is separable.*

Proof: Let $\{(X_n, d_n)\}$ be a sequence of separable metric spaces. As we saw in the proof of Theorem 3.36, the product topology on X is generated by the metric

$$d((x_n), (y_n)) = \sum_{n=1}^{\infty} \frac{1}{2^n} \cdot \frac{d_n(x_n, y_n)}{1 + d_n(x_n, y_n)}.$$

Now for each n let D_n be a countable dense subset of X_n. Also, for each n fix some $u_n \in D_n$. Now note that the set

$$D = \{(x_n) \in X : x_n \in D_n \text{ for each } n \text{ and } x_n = u_n \text{ eventually}\},$$

is a countable dense subset of X.

The converse follows by noting that the continuous image of a separable topological space is separable. (Use Theorem 2.27 (5).)   ∎

3.39 Corollary   *The product of a sequence of Polish spaces is a Polish space.*

3.9   The Hilbert cube and metrization

The **Hilbert cube** $\mathcal{H}$ is the set of all real sequences with values in $[0, 1]$. That is, $\mathcal{H} = [0, 1]^{\mathbb{N}}$. It is compact in the product topology by the Tychonoff Product Theorem 2.61, and it is easy to see that the metric

$$d_{\mathcal{H}}((x_n), (y_n)) = \sum_{n=1}^{\infty} \frac{1}{2^n} |x_n - y_n|,$$

induces the product topology on $\mathcal{H}$. The Hilbert cube "includes" every separable metrizable space. Indeed, we have the following theorem characterizing separable metrizable spaces.

3.40 Urysohn Metrization Theorem   *For a Hausdorff space X, the following are equivalent.*

1. *X can be embedded in the Hilbert cube.*

2. *X is a separable metrizable space.*

3. *X is regular and second countable.*

Proof: (1) $\implies$ (2) By Corollary 3.5, any subset of a separable metrizable space is separable.

(2) $\implies$ (3) Lemma 3.20 shows that a metrizable space is completely regular, and Lemma 3.4 shows that a separable metrizable space is second countable.

(3) $\implies$ (1) By Theorem 2.49, X is normal. Let $\mathcal{B}$ be a countable base of nonempty subsets of X, and let $\mathcal{C} = \{(U, V) : \overline{U} \subset V \text{ and } U, V \in \mathcal{B}\}$. The normality of X implies that $\mathcal{C}$ is nonempty. Since $\mathcal{C}$ is countable, let $\{(U_1, V_1), (U_2, V_2), \ldots\}$ be an explicit enumeration of $\mathcal{C}$. Now for each n pick a continuous real function f_n with values in $[0, 1]$ satisfying $f_n(\overline{U}) = \{1\}$ and $f_n(V^c) = \{0\}$. Note that since X is Hausdorff, the family $\{f_n\}$ separates points.

Define $\varphi \colon X \to \mathcal{H}$ by

$$\varphi(x) = (f_1(x), f_2(x), \ldots).$$

(If $\mathcal{C}$ is actually finite, fill out the sequence with zero functions.) Since $\{f_n\}$ separates points, φ is one-to-one. Since each f_n is continuous, so is φ. To show that φ is an embedding, we need to show that φ^{-1} is continuous. So suppose $\varphi(x_\alpha) \to \varphi(x)$, and let W be a neighborhood of x. Then $x \in U_n \subset \overline{U}_n \subset V_n \subset W$ for some n (why?), so $f_n(x) = 1$. Since $\varphi(x_\alpha) \to \varphi(x)$, we have $f_n(x_\alpha) \to f_n(x)$ for each n, so for large enough α we have $f_n(x_\alpha) > 0$. But this implies $x_\alpha \in V_n \subset W$ for large enough α. Thus $x_\alpha \to x$, so φ^{-1} is continuous.   ∎

3.41 Corollary   *Every separable metrizable topological space admits a compatible metric that is totally bounded. Consequently, every separable metrizable space has a metrizable compactification—the completion of this totally bounded metric space.*

Proof: Let X be a separable metrizable space. By the Urysohn Metrization Theorem 3.40, there is an embedding $\varphi \colon X \to \mathcal{H}$. Define a metric ρ on X by

$$\rho(x, y) = d_{\mathcal{H}}(\varphi(x), \varphi(y)).$$

The Hilbert cube $(\mathcal{H}, d_{\mathcal{H}})$ is a compact metric space, and hence is totally bounded. The metric ρ inherits this property.   ∎

We mention here that this compactification is not in general the same as the Stone–Čech compactification, which is usually not metrizable. To see this, you can verify that the compactification described in the proof of Corollary 3.41 of $(0, 1]$ is $[0, 1]$. But recall that the Stone–Čech compactification of $(0, 1]$ is nearly indescribable. However, it is true that every completely metrizable space is a $\mathcal{G}_\delta$ in its Stone–Čech compactification. See, e.g., [342, Theorem 24.13, p.180].

3.42 Corollary   *Every Polish space is a $\mathcal{G}_\delta$ in some metrizable compactification.*

Proof: This follows from Lemma 3.33 and Corollary 3.41.   ∎

3.43 Corollary   *The continuous image of a compact metric space in a Hausdorff space is metrizable.*

Proof: Let $f: X \to Y$ be continuous, where X is a compact metric space and Y is Hausdorff. Replacing Y by $f(X)$, we can assume without loss of generality that $Y = f(X)$. Thus Y is compact as the continuous image of the compact set X (Theorem 2.34). Hence by Theorem 2.48, Y is normal and so regular. By the Urysohn Metrization Theorem 3.40, we need only show that Y is second countable.

For any open set G in X, its complement is closed and thus compact, so $f(G^c)$ is compact and thus closed. Therefore each set of the form $Y \setminus f(G^c)$ is open in Y if G is open in X. Now let $\mathcal{B}$ be a countable base for X, and let $\mathcal{F}$ be the collection of finite unions of members of $\mathcal{B}$. We claim that $\{Y \setminus f(G^c) : G \in \mathcal{F}\}$ is a countable base for Y. It is clearly countable since $\mathcal{B}$ is. To see that it forms a base for Y, suppose that W is open in Y and that $y \in W$. Since Y is Hausdorff, the nonempty set $f^{-1}(y)$ is closed in X, and so compact (why?). Thus $f^{-1}(y)$ is covered by some finite subfamily of $\mathcal{B}$, so there is some G belonging to $\mathcal{F}$ with $f^{-1}(y) \subset G \subset f^{-1}(W)$. (Why?) Since $f^{-1}(y) \subset G$, we must have $y \notin f(G^c)$. But then $y \in Y \setminus f(G^c) \subset f(G) \subset W$, and the proof is finished.   ∎

3.10   Locally compact metrizable spaces

We are now in a position to discuss metrizability of the one-point compactification of a metrizable space.

3.44 Theorem (Metrizability of X_∞)   *The one-point compactification X_∞ of a noncompact locally compact Hausdorff space X is metrizable if and only if X is second countable.*

Proof: If X_∞ is metrizable, then since it is compact, it is separable, and so second countable. This implies that X itself is second countable.

For the converse, if X is a locally compact second countable Hausdorff space, then Lemma 2.76 and Corollary 2.77 imply that we can write $X = \bigcup_{n=1}^\infty K_n$, where

$K_n \subset K_{n+1}^\circ$, and each K_n is compact. Furthermore X is hemicompact, that is, every compact subset K is included in some K_n.

Thus the collection $\{X_\infty \setminus K_n : n = 1, 2, \ldots\}$ is a countable base at ∞. This in turn implies that X_∞ is second countable. Since X_∞ is also regular (being compact and Hausdorff), it follows from the Urysohn Metrization Theorem 3.40 that X_∞ is indeed a metrizable space. ∎

Since a separable metrizable space is second countable, we have the following.

3.45 Corollary   *The one-point compactification of a noncompact locally compact separable metrizable space is metrizable.*

3.11 The Baire Category Theorem

The notion of **Baire category** captures a topological notion of "sparseness" for subsets of a topological space X. Recall that a subset A of X is **nowhere dense** if it is not dense in any open subset of X, that is, $(\overline{A})^\circ = \varnothing$. A subset A of X is of **first (Baire) category**, or **meager**, if it is a countable union of nowhere dense sets. A subset of X is of **second (Baire) category** if it is not of first category. A **Baire space** (not to be confused with *the* Baire space $\mathbb{N}^\mathbb{N}$, described in Section 3.14) is a topological space in which nonempty open sets are not meager. The next result characterizes Baire spaces.

3.46 Theorem   *For a topological space X the following are equivalent.*

1. *X is a Baire space.*

2. *Every countable intersection of open dense sets is also dense.*

3. *If $X = \bigcup_{n=1}^\infty F_n$ and each F_n is closed, then the open set $\bigcup_{n=1}^\infty (F_n)^\circ$ is dense.*

Proof: (1) $\implies$ (2)   First note that if G is an open dense set, then its complement G^c is nowhere dense. To see this note that G^c is itself closed, so it suffices to show that G^c has empty interior. Now by Lemma 2.4, $(G^c)^\circ = (\overline{G})^c$, which is empty since G is dense.

Assume X is a Baire space and let $\{G_n\}$ be a sequence of open dense subsets of X. Set $A = \bigcap_{n=1}^\infty G_n$ and suppose $A \cap U = \varnothing$ for some nonempty open set U. Then $X = (A \cap U)^c = A^c \cup U^c$, so

$$U = X \cap U = A^c \cap U = \left[\bigcap_{n=1}^\infty G_n\right]^c \cap U = \bigcup_{n=1}^\infty (G_n^c \cap U).$$

This shows that U is a meager set, which is impossible. So A is dense in X.

(2) $\implies$ (3) Let $\{F_n\}$ be a sequence of closed sets with $X = \bigcup_{n=1}^{\infty} F_n$ and consider the open set $G = \bigcup_{n=1}^{\infty} F_n^{\circ}$. For each n, let $E_n = F_n \setminus F_n^{\circ}$, and note that E_n is a nowhere dense closed set. In particular, the set $E = \bigcup_{n=1}^{\infty} E_n$ is meager.

Since E_n is closed and nowhere dense, each E_n^c is an open dense set. By hypothesis, $E^c = \bigcap_{n=1}^{\infty} E_n^c$ is also dense. Now notice that

$$G^c = X \setminus G = \bigcup_{n=1}^{\infty} F_n \setminus \bigcup_{n=1}^{\infty} F_n^{\circ} \subset \bigcup_{n=1}^{\infty} (F_n \setminus F_n^{\circ}) = E,$$

so $E^c \subset G$. Since E^c is dense, G is dense, as desired.

(3) $\implies$ (1) Let G be a nonempty open set. If G is meager, then G can be written as a countable union $G = \bigcup_{n=1}^{\infty} A_n$, where $(\overline{A_n})^{\circ} = \varnothing$ for each n. Then

$$X = G^c \cup \overline{A}_1 \cup \overline{A}_2 \cup \overline{A}_3 \cup \cdots$$

is a countable union of closed sets, so by hypothesis the open set

$$(G^c)^{\circ} \cup (\overline{A}_1)^{\circ} \cup (\overline{A}_2)^{\circ} \cup (\overline{A}_3)^{\circ} \cup \cdots = (G^c)^{\circ}$$

is dense in X. From $(G^c)^{\circ} \subset G^c$, we see that G^c is also dense in X. In particular, $G \cap G^c \neq \varnothing$, which is impossible. Hence G is not meager, so X is a Baire space. ∎

The class of Baire spaces includes all completely metrizable spaces.

3.47 Baire Category Theorem   *A completely metrizable space is a Baire space.*

Proof: Let d be a complete compatible metric on the space X. Now let $\{G_n\}$ be a sequence of open dense subsets of X and put $A = \bigcap_{n=1}^{\infty} G_n$. By Theorem 3.46, it suffices to show that A is a dense subset of X, or that $B_r(x) \cap A \neq \varnothing$ for each $x \in X$ and $r > 0$. So fix $x \in X$ and $r > 0$.

Since G_1 is open and dense in X, there exist $y_1 \in X$ and $0 < r_1 < 1$ such that $C_{r_1}(y_1) \subset B_r(x) \cap G_1$, where you may recall that $B_r(x)$ denotes the open ball of radius r around x and $C_r(x)$ is the corresponding closed ball. Similarly, since G_2 is open and dense in X, we have $B_{r_1}(y_1) \cap G_2 \neq \varnothing$, so there exist $y_2 \in X$ and $0 < r_2 < 1/2$ such that $C_{r_2}(y_2) \subset B_{r_1}(y_1) \cap G_2$. Proceeding inductively, we see that there exists a sequence $\{y_n\}$ in X and a sequence $\{r_n\}$ of positive real numbers satisfying

$$C_{r_{n+1}}(y_{n+1}) \subset B_{r_n}(y_n) \cap G_{n+1} \subset C_{r_n}(y_n) \quad \text{and} \quad 0 < r_n < \tfrac{1}{n}$$

for each n. Now the Cantor Intersection Theorem 3.7 guarantees that $\bigcap_{n=1}^{\infty} C_{r_n}(y_n)$ is a singleton. From $\bigcap_{n=1}^{\infty} C_{r_n}(y_n) \subset B_r(x) \cap A$, we see that $B_r(x) \cap A \neq \varnothing$. ∎

A well-known application of the Baire Category Theorem is a proof of the existence of continuous functions on $[0, 1]$ that are nowhere differentiable, see, for example, [14, Problem 9.28, p. 89]. We shall use it in the proof of the Uniform Boundedness Principle 6.14.

3.12   Contraction mappings

A Lipschitz continuous function $f: X \to X$ on the metric space (X, d) is called a
contraction if it has a Lipschitz constant strictly less than 1. That is, there exists
a constant $0 \leqslant c < 1$ (called a **modulus of contraction**) such that

$$d(f(x), f(y)) \leqslant cd(x, y)$$

for all $x, y \in X$.

Recall that a fixed point of a function $f: X \to X$ is an x satisfying $f(x) = x$.
The next theorem is an important existence theorem. It asserts the existence of a
fixed point for a contraction mapping on a complete metric space, and is known
as the **Contraction Mapping Theorem** or as the **Banach Fixed Point Theorem**.
This theorem plays a fundamental role in the theory of dynamic programming, see
E. V. Denardo [90].

3.48 Contraction Mapping Theorem   *Let (X, d) be a complete metric space
and let $f: X \to X$ be a contraction. Then f has a unique fixed point x. Moreover,
for any choice x_0 in X, the sequence defined recursively by*

$$x_{n+1} = f(x_n), \ n = 0, 1, 2, \ldots,$$

converges to the fixed point x and

$$d(x_n, x) \leqslant c^n d(x_0, x)$$

for each n.

Proof: Let $0 \leqslant c < 1$ be a modulus of contraction for f. Suppose $f(x) = x$ and
$f(y) = y$. Then
$$d(x, y) = d(f(x), f(y)) \leqslant cd(x, y),$$
so $d(x, y) = 0$. That is, $x = y$. Thus f can have at most one fixed point.

To see that f has a fixed point, pick any point $x_0 \in X$, and then define a
sequence $\{x_n\}$ inductively by the formula

$$x_{n+1} = f(x_n), \ n = 0, 1, \ldots.$$

For $n \geqslant 1$, we have

$$d(x_{n+1}, x_n) = d(f(x_n), f(x_{n-1})) \leqslant cd(x_n, x_{n-1}),$$

and by induction, we see that $d(x_{n+1}, x_n) \leqslant c^n d(x_1, x_0)$. Hence, for $n > m$ the
triangle inequality yields

$$d(x_n, x_m) \leqslant \sum_{k=m+1}^{n} d(x_k, x_{k-1}) \leqslant \sum_{k=m+1}^{n} c^{k-1} d(x_1, x_0)$$

$$\leqslant \sum_{k=m+1}^{\infty} c^{k-1} d(x_1, x_0) = \frac{c^m}{1-c} \cdot d(x_1, x_0),$$

which implies that $\{x_n\}$ is a d-Cauchy sequence. Since by completeness $x_n \to x$ for some x, the continuity of f implies

$$x = \lim_{n\to\infty} x_{n+1} = \lim_{n\to\infty} f(x_n) = f(x),$$

so x is the unique fixed point of f. (The last inequality follows from the relation

$$d(x_{n+1}, x) = d(f^{n+1}(x_0), f^{n+1}(x)) \leqslant cd(f^n(x_0), f^n(x)) = cd(x_n, x)$$

and an easy inductive argument.) ∎

3.49 Corollary   *Let $f : (X, d) \to (X, d)$ be a contraction on a complete metric space. If C is an f-invariant nonempty closed subset of X, that is, $f(C) \subset C$, then the unique fixed point of f belongs to $f(C)$.*

Proof: Clearly, $f : (C, d) \to (C, d)$ is a contraction. Since C is closed, (C, d) is a complete metric space. So by the Contraction Mapping Theorem, there exists some $c \in C$ such that $f(c) = c$. Since c is the only fixed point of f, we infer that $c = f(c) \in f(C)$. ∎

3.50 Corollary   *Let $f : (X, d) \to (X, d)$ be a function on a complete metric space. If for some k, the kth iterate $f^k : X \to X$ is a contraction, then f has a unique fixed point.*

Proof: Assume that for some k and some $0 \leqslant c < 1$, we have

$$d(f^k(x), f^k(y)) \leqslant cd(x, y)$$

for all $x, y \in X$. By the Contraction Mapping Theorem, there exists a unique fixed point x of f^k. From

$$d(f(x), x) = d(f(f^k(x)), f^k(x)) = d(f^k(f(x)), f^k(x)) \leqslant cd(f(x), x),$$

we obtain $0 \leqslant (1 - c)d(f(x), x)) \leqslant 0$. Hence, $d(f(x), x) = 0$, so $f(x) = x$. That is, x is also a fixed point of f.

Now if $f(y) = y$, then clearly $f^k(y) = y$, so $y = x$. Hence, x is the only fixed point of f. ∎

There is a variation of the contraction mapping theorem that does not require the completeness of the domain. By Lemma 3.11 a contraction $f : (X, d) \to (\hat{X}, d)$ has a unique continuous extension $\hat{f}$ to the completion $(\hat{X}, d)$ of (X, d). It follows that $\hat{f} : \hat{X} \to \hat{X}$ is also a contraction with the same modulus of contraction. This proves the next result.

3.51 Theorem (Generalized Contraction Mapping Theorem) *If (X, d) is a metric space with completion $\hat{X}$ and $f: X \to X$ is a contraction mapping, then there exists a unique fixed point $\hat{x}$ of $\hat{f}: \hat{X} \to \hat{X}$, where $\hat{f}$ is the unique continuous extension of f to $\hat{X}$. Moreover, if c is a modulus of contraction for f and $x_0 \in X$ is any point, then the sequence $\{x_n\}$ defined recursively by*

$$x_{n+1} = f(x_n), \quad n = 0, 1, 2, \ldots,$$

converges to the fixed point $\hat{x}$ and $d(x_n, \hat{x}) \leqslant c^n d(x_0, \hat{x})$ for each n.

A simple example illustrates the result. Let $\mathfrak{I}$ be the set of irrational numbers, equipped with the usual metric $d(x, y) = |x - y|$. The completion of (X, d) is $\mathbb{R}$. Now consider the contraction mapping $f: \mathfrak{I} \to \mathfrak{I}$ defined by $f(x) = x/2$. The unique fixed point of f is 0, which does not lie in $\mathfrak{I}$, but in its completion $\mathbb{R}$.

For compact metric spaces, we need only functions that are "almost" contractions in order to prove a fixed point theorem.

3.52 Theorem *If a function $f: X \to X$ on a compact metric space (X, d) satisfies $d(f(x), f(y)) < d(x, y)$ for all $x \neq y$, then f has a unique fixed point.*

Proof: It should be clear that f has at most one fixed point. To see that f has a fixed point define the function $\varphi: X \to \mathbb{R}$ by $\varphi(x) = d(x, f(x))$. Clearly, φ is continuous and so (since X is compact) there exists some $x_0 \in X$ such that $\varphi(x_0) = \min_{x \in X} \varphi(x)$. Now note that if $y = f(x_0)$ satisfies $y \neq x_0$, then we have $\varphi(y) = d(y, f(y)) < d(x_0, f(x_0)) = \varphi(x_0)$, which is impossible. Hence $f(x_0) = x_0$ so that x_0 is a fixed point for f. ∎

This result depends crucially on compactness. For instance, consider the function $f: (0, 1) \to (0, 1)$ defined by $f(x) = \frac{1}{2}x$.

As an application of contraction mappings, we present a fundamental result in the theory of dynamic programming due to D. Blackwell [48].

3.53 Blackwell's Theorem *Let X be a nonempty set and let $B(X)$ denote the complete metric space of all bounded real functions equipped with the uniform metric, that is, $d(f, g) = \sup_{x \in X} |f(x) - g(x)|$. Let L be a closed linear subspace of $B(X)$ that includes the constant functions. Assume that $T: L \to L$ is a (not necessarily linear) mapping such that:*

1. *T is monotone in the sense that $f \leqslant g$ implies $T(f) \leqslant T(g)$, and*

2. *there exists some constant $0 \leqslant \beta < 1$ such that for each constant function c we have $T(f + c) \leqslant T(f) + \beta c$.*

Then T has a unique fixed point.

Proof: We shall prove that T is a contraction with modulus of contraction β. Then L, as a closed subset of the complete metric space $B(X)$, is complete, and the conclusion follows from the Contraction Mapping Theorem 3.48.

So let $f, g \in L$ and consider the constant function $c(x) = d(f, g)$ for each $x \in X$. By the definition of d we have $f \leqslant g + c$ and $g \leqslant f + c$. Now (1) implies $T(f) \leqslant T(g + c)$ and (2) implies $T(g + c) \leqslant T(g) + \beta c$, which together imply $T(f) - T(g) \leqslant \beta c$. Similarly, $T(g) - T(f) \leqslant \beta c$. Thus $|T(f)(x) - T(g)(x)| \leqslant \beta c$ for each $x \in X$, so

$$d(T(f), T(g)) = \sup_{x \in X} |T(f)(x) - T(g)(x)| \leqslant \beta c = \beta d(f, g).$$

Therefore T is a contraction with modulus of contraction β, as desired.   ∎

3.13   The Cantor set

The Cantor set, named for G. Cantor, has long been a favorite of mathematicians because it is a rich source of counterexamples. There are several ways of describing it. We begin with the simplest.

3.54 Definition   *The **Cantor set** is the countable product $\Delta = \{0, 1\}^{\mathbb{N}}$, where the two-point set $\{0, 1\}$ has the discrete topology.*

Two remarks are in order. First, we can replace the set $\{0, 1\}$ by any two point set; the choice of the two point set often simplifies proofs. Second, the formula

$$d(a, b) = \sum_{n=1}^{\infty} \frac{|a_n - b_n|}{3^n},$$

where $a = (a_1, a_2, \ldots)$ and $b = (b_1, b_2, \ldots)$, defines a metric that generates the product topology on Δ. Also, the Tychonoff Product Theorem 2.61 implies that the Cantor set is compact. It is thus a compact metric space. Indeed, we shall see below that it is in some sense the most fundamental compact metric space.

The Cantor set can also be identified with a closed subset of $[0, 1]$. It can be constructed by repeatedly re-moving open "middle-third" in-tervals. Start with $C_0 = [0, 1]$ and subdivide it into three equal subintervals $[0, \frac{1}{3}], (\frac{1}{3}, \frac{2}{3}), [\frac{2}{3}, 1]$, and remove the open middle in-terval (here $(\frac{1}{3}, \frac{2}{3})$) and let $C_1 = [0, \frac{1}{3}] \cup [\frac{2}{3}, 1]$. Now we proceed inductively. If C_n consists of 2^n closed subintervals, subdivide each into three subintervals of equal length and delete from each one of them the open middle subinterval. The union of the remaining 2^{n+1} closed subintervals is C_{n+1}. By this process, the Cantor set is then the compact set

$$C = \bigcap_{n=1}^{\infty} C_n.$$

Or in yet other words, it can be thought of as the set of real numbers in $[0, 1]$ that have a ternary expansion that does not use the digit 1, that is,

$$C = \left\{ \sum_{n=1}^{\infty} \frac{a_n}{3^n} : a_n = 0 \text{ or } a_n = 2 \right\}.$$

3.55 Lemma   *The Cantor set Δ is homeomorphic to C.*

Proof: Define $\varphi \colon \Delta \to C$ by $\varphi(a_1, a_2, \ldots) = \sum_{n=1}^{\infty} 2a_n/3^n$. Then φ is continuous, one-to-one, and surjective, so by Theorem 2.36, φ is a homeomorphism.   ∎

Viewed as a subset of the unit interval, it is easy to see that C includes no intervals. The sum of the lengths of the omitted intervals is 1, so the Cantor set has total "length" zero. Moreover, every point that belongs to C is the limit of other points in C. The Cantor diagonal process can be used to show that the Cantor set is also uncountable. Summing up we have the following.

3.56 Lemma   *The Cantor set C is an uncountable, perfect, and nowhere dense set of Lebesgue measure zero.*

Notably, the Cantor set is homeomorphic to a countable power of itself.

3.57 Lemma   *The Cantor set Δ is homeomorphic to $\Delta^{\mathbb{N}}$.*

Proof: Write $\mathbb{N} = \bigcup_{k=1}^{\infty} \mathbb{N}_k$, where each $\mathbb{N}_k$ is a countably infinite subset of $\mathbb{N}$, and $\mathbb{N}_k \cap \mathbb{N}_m = \varnothing$ whenever $k \neq m$.[3] Write $\mathbb{N}_k = \{n_1^k, n_2^k, \ldots\}$, where $n_1^k < n_2^k < \cdots$. Also, for $a = (a_1, a_2, \ldots) \in \Delta^{\mathbb{N}}$, let $a_k = (a_1^k, a_2^k, \ldots)$ and put $b_{n_i^k} = a_i^k$. Now define the function $\psi \colon \Delta^{\mathbb{N}} \to \Delta$ by

$$\psi(a_1, a_2, \ldots) = (b_1, b_2, \ldots).$$

It follows that ψ is one-to-one, surjective, and continuous. By Theorem 2.36, ψ is also a homeomorphism.   ∎

More amazing is the list of spaces that are continuous images of the Cantor set. The next series of results shows that *every* compact metric space is the image of the Cantor set under some continuous function!

---

[3] One way of constructing (by induction) such a partition is as follows. Start with $\mathbb{N}_1 = \{1, 3, 5, \ldots\}$ and assume that $\mathbb{N}_k$ has been selected so that $\mathbb{N} \setminus \mathbb{N}_k = \{n_1, n_2, n_3, \ldots\}$ is countably infinite, where $n_1 < n_2 < \cdots$. To complete the inductive argument put $\mathbb{N}_{k+1} = \{n_1, n_3, n_5, \ldots\}$.

3.58 Lemma   *Both the closed interval* $[0, 1]$ *and the Hilbert cube* $\mathcal{H}$ *are continuous images of the Cantor set.*

Proof: Let $\Delta = \{0, 1\}^{\mathbb{N}}$ and define $\theta: \Delta \to [0, 1]$ by

$$\theta(a) = \sum_{n=1}^{\infty} \frac{\alpha_n}{2^n},$$

where $a = (\alpha_1, \alpha_2, \ldots)$. Clearly, θ is continuous and since every number in $[0, 1]$ has a dyadic expansion θ is also surjective, but not one-to-one (since the dyadic expansion need not be unique).

Next, define $\varphi: \Delta^{\mathbb{N}} \to \mathcal{H}$ by $\varphi(a_1, a_2, \ldots) = (\theta(a_1), \theta(a_2), \ldots)$. An easy verification shows that φ is continuous and surjective. Now invoke Lemma 3.57 to see that $\mathcal{H}$ is a continuous image of Δ.   ∎

A nonempty set A in a topological space X is a **retract** of X if there is a continuous function $f: X \to A$ that leaves each point of A fixed.[4] That is, $f(x) = x$ for all $x \in A$. The map f is called a **retraction** of X onto A. Note that if A is a retract of X and $A \subset B \subset X$, then A is also a retract of B under the retraction $f|_B$.

3.59 Lemma   *Any nonempty closed subset of* Δ *is a retract of* Δ.

Proof: Let K be a nonempty closed, and hence compact, subset of Δ. For each point $x = (x_1, x_2, \ldots)$ in the Cantor set $\Delta = \{0, 2\}^{\mathbb{N}}$ there exists a unique element $f(x) = y = (y_1, y_2, \ldots) \in K$ minimizing $d(x, \cdot)$ over K. That is,

$$d(x, f(x)) = \sum_{n=1}^{\infty} \frac{|x_n - y_n|}{3^n} = d(x, K) = \inf\{d(x, z) : z \in K\}.$$

(For the uniqueness of the point y, we use the fact that $\sum_{n=1}^{\infty} \frac{a_n}{3^n} = \sum_{n=1}^{\infty} \frac{b_n}{3^n}$ with $a_n, b_n \in \{0, 2\}$ implies $a_n = b_n$ for each n.) Clearly, $f(x) = x$ for each $x \in K$, and we claim that f is also continuous.

Suppose $x_n \to x$, but $f(x_n) \not\to f(x)$. Since K is compact, by passing to a subsequence if necessary (how?), we can assume that $f(x_n) \to y$ for some $y \in K$. By Theorem 3.16,

$$d(x, f(x)) = d(x, K) = \lim_{n \to \infty} d(x_n, K) = \lim_{n \to \infty} d(x_n, f(x_n)) = d(x, y).$$

Since $f(x)$ is the unique minimizer of $d(x, \cdot)$ in K, we have $y = f(x)$, a contradiction. Therefore f is continuous.   ∎

3.60 Theorem   *Every compact metrizable space is a continuous image of the Cantor set.*

---

[4] Another way of expressing this is by saying that A is the range of continuous **projection** f on X, that is, $f \circ f = f$ and $f(X) = A$.

Proof: Let X be a compact metrizable space. By the Urysohn Metrization Theorem 3.40, X is homeomorphic to a closed subset Y of the Hilbert cube $\mathcal{H}$. Let $\varphi: Y \to X$ be such a homeomorphism. By Lemma 3.58 there exists a continuous mapping ψ from Δ onto $\mathcal{H}$. So $\psi^{-1}(Y)$ is a closed subset of Δ. By Lemma 3.59 there is a continuous retraction $f: \Delta \to \psi^{-1}(Y)$ satisfying $f(z) = z$ for each $z \in \psi^{-1}(Y)$. Schematically,

$$\Delta \xrightarrow{\ f\ } \psi^{-1}(Y) \xrightarrow{\ \psi\ } Y \xrightarrow{\ \varphi\ } X,$$

so $\varphi \circ \psi \circ f$ is a continuous function from Δ onto X. ∎

3.14 The Baire space $\mathbb{N}^{\mathbb{N}}$

Another fundamental metric space is the **Baire space** $\mathcal{N} = \mathbb{N}^{\mathbb{N}}$ of functions from $\mathbb{N}$ into $\mathbb{N}$ (or sequences of natural numbers), endowed with its product topology. Since the discrete metric on $\mathbb{N}$ is complete, $\mathbb{N}$ is a Polish space. Corollary 3.39 shows that $\mathcal{N}$ is Polish too. We denote typical elements of $\mathcal{N}$ by $\boldsymbol{m}, \boldsymbol{n}$, etc.

Recall that a base for the product topology on $\mathcal{N}$ is given by products of open subsets on $\mathbb{N}$, all but finitely many of which are $\mathbb{N}$. Since $\mathbb{N}$ is discrete, a moment's reflection should convince you that the collection of sets of the form

$$U_{n_1,\ldots,n_m} = \{n_1\} \times \{n_2\} \times \cdots \times \{n_m\} \times \mathbb{N} \times \mathbb{N} \times \cdots , \qquad (\star)$$

where $n_1, \ldots, n_m$ and m are natural numbers, is a base for the topology on $\mathcal{N}$. Note that this base is countable.

At this point, it is convenient to introduce a new bit of notation. Recall that a **finite sequence** in A is any ordered n-tuple of elements of A. The collection of all finite sequences in a set A is traditionally denoted $A^{<\mathbb{N}}$. That is,

$$A^{<\mathbb{N}} = \bigcup_{n=1}^{\infty} A^n.$$

We are particularly interested in $\mathbb{N}^{<\mathbb{N}}$, the finite sequences in $\mathbb{N}$. Keep in mind the important fact that $\mathbb{N}^{<\mathbb{N}}$ is a countable set, whereas $\mathcal{N} = \mathbb{N}^{\mathbb{N}}$ is uncountable. We have just argued that the collection $\{U_s : s \in \mathbb{N}^{<\mathbb{N}}\}$ of open sets defined by $(\star)$ is a base for the product topology on $\mathcal{N}$.

The next result asserts that countable products of $\mathcal{N}$ all look alike.

3.61 Theorem (Products of $\mathcal{N}$) *The Polish spaces $\mathcal{N}^2, \mathcal{N}^3, \ldots,$ and $\mathcal{N}^{\mathbb{N}}$ are all homeomorphic to $\mathcal{N}$.*

Proof: We prove only that $\mathcal{N}$ and $\mathcal{N}^{\mathbb{N}}$ are homeomorphic. You can easily see how to modify this proof to show that $\mathcal{N}$ and $\mathcal{N}^n$ are homeomorphic.

Choose an infinite countable partition $\{N_1, N_2, \ldots\}$ of $\mathbb{N}$ such that each N_k is countably infinite. That is, $\mathbb{N} = \bigcup_{i=1}^{\infty} N_i$, $N_i \cap N_j = \varnothing$ for $i \neq j$, and each N_i is infinite—see the footnote in the proof of Lemma 3.57. Let $N_i = \{k_1^i, k_2^i, \ldots\}$ for each i. Now consider the function $f \colon \mathcal{N} \to \mathcal{N}^{\mathbb{N}}$ defined by $f(\boldsymbol{n}) = (\boldsymbol{n}_1, \boldsymbol{n}_2, \ldots)$, where $\boldsymbol{n}_i(j) = \boldsymbol{n}(k_j^i)$. Clearly, f is surjective and one-to-one. We leave the task of verifying that f is a homeomorphism as an exercise.   ∎

There is a very useful natural metric on $\mathcal{N}$ described in the next lemma. In fact, the metric is an **ultrametric**. A metric d is an ultrametric, if it satisfies the **ultrametric inequality**,

$$d(x, y) \leqslant \max\{d(x, z), d(z, y)\},$$

for all x, y, and z.

3.62 Lemma   *The function* $\mathsf{t} \colon \mathcal{N} \times \mathcal{N} \to [0, 1]$*, defined by*

$$\mathsf{t}(\boldsymbol{n}, \boldsymbol{m}) = \begin{cases} 0 & \text{if } \boldsymbol{n} = \boldsymbol{m} \\ \frac{1}{k} & \text{if } \boldsymbol{n} \neq \boldsymbol{m}, \text{ and } k = \min\{j : \boldsymbol{n}(j) \neq \boldsymbol{m}(j)\}, \end{cases}$$

is a complete ultrametric for the product topology on $\mathcal{N}$.

Proof: The function t clearly satisfies symmetry, positivity, and discrimination. The triangle inequality follows easily from the ultrametric inequality, so we shall prove that. Let $\boldsymbol{n}$, $\boldsymbol{m}$, and $\boldsymbol{p}$ belong to $\mathcal{N}$ and let $k_1 = \min\{k : \boldsymbol{n}(k) \neq \boldsymbol{p}(k)\}$, $k_2 = \min\{k : \boldsymbol{m}(k) \neq \boldsymbol{p}(k)\}$, and assume without loss of generality that $k_1 \leqslant k_2$. Then for $j \leqslant k_1$,

$$\boldsymbol{n}(j) = \boldsymbol{p}(j) = \boldsymbol{m}(j),$$

so $\min\{j : \boldsymbol{n}(j) \neq \boldsymbol{m}(j)\} \geqslant k_1$. Hence

$$\mathsf{t}(\boldsymbol{n}, \boldsymbol{m}) \leqslant 1/k_1 = \mathsf{t}(\boldsymbol{n}, \boldsymbol{p}) = \max\{\mathsf{t}(\boldsymbol{n}, \boldsymbol{p}), \mathsf{t}(\boldsymbol{p}, \boldsymbol{m})\}.$$

We now show that t induces the product topology. To see this, observe that the open ball $B_{1/m}(\boldsymbol{n})$ of radius $1/m$ around $\boldsymbol{n}$ is the set of all sequences $\boldsymbol{m}$ that agree with $\boldsymbol{n}$ for the first m terms. This set also happens to be a basic open set of the form given by $(\star)$, namely

$$B_{1/m}(\boldsymbol{n}) = U_{\boldsymbol{n}(1), \ldots, \boldsymbol{n}(m)} = \{\boldsymbol{n}(1)\} \times \cdots \times \{\boldsymbol{n}(m)\} \times \mathbb{N} \times \mathbb{N} \times \cdots.$$

Finally we show that t is complete. Suppose $\{\boldsymbol{m}_n\}$ is a t-Cauchy sequence. That is, for each natural number k, there is some n_k such that for all $n \geqslant n_k$, $\mathsf{t}(\boldsymbol{m}_n, \boldsymbol{m}_{n_k}) < 1/k$. In other words, $\boldsymbol{m}_n(j) = \boldsymbol{m}_{n_k}(j)$ for $j \leqslant k$ and $n \geqslant n_k$. Define $\boldsymbol{m}$ by $\boldsymbol{m}(k) = \boldsymbol{m}_{n_k}(k)$ and observe that $\mathsf{t}(\boldsymbol{m}_n, \boldsymbol{m}) < 1/k$ for all $n \geqslant n_k$. Thus $\{\boldsymbol{m}_n\}$ t-converges to $\boldsymbol{m}$, which shows that t is complete.   ∎

We shall refer to t as the **tree metric**. (See A. S. Kechris [196, § 2] for a discussion of trees that justifies this terminology.)

There are two natural orderings on the Baire space $\mathcal{N}$. The usual pointwise ordering of functions is one. This is the partial order defined by $n \leqslant m$ if for all k, $n(k) \leqslant m(k)$. The Baire space is also naturally linearly ordered by the **lexicographic order**. This complete order is defined by $n \leqslant_{\text{lex}} m$ if $n = m$ or else if there is some k such that $n(j) = m(j)$ for $j < k$ and $n(k) < m(k)$. The lexicographic ordering is an extension of the pointwise ordering. That is, $n \leqslant m$ implies $n \leqslant_{lex} m$.

3.63 Lemma   *Every closed subset of the Baire space $\mathcal{N}$ has a least element in the lexicographic order.*

Proof: Let F be a nonempty closed subset of the Baire space $\mathcal{N}$. Let $F_0 = F$ and define a sequence $\{F_n\}$ of nonempty closed subsets of $\mathcal{N}$ inductively by

$$F_n = \{m \in F_{n-1} : m(n) = \min\{r(n) : r \in F_{n-1}\}\},$$

$n = 1, 2, \ldots$. For each n, F_n is closed, $F_n \subset F_{n-1}$, and t- diam $F_n \leqslant 1/n$. So by the Cantor Intersection Theorem 3.7, the intersection of the sequence is a singleton, say $\bigcap_{n=1}^{\infty} F_n = \{p\}$. Clearly p is the least element of F with respect to $\leqslant_{lex}$.   ∎

As with the Cantor set, a nonempty closed subset F of $\mathcal{N}$ is a retract of $\mathcal{N}$, and hence a retract of every set that includes F.

3.64 Lemma   *Every nonempty closed subset of $\mathcal{N}$ is a retract of $\mathcal{N}$.*

Proof: The proof is constructive. Let F be a nonempty closed subset of $\mathcal{N}$ and fix $n \in \mathcal{N}$. Put $F_0^n = F$ and construct a sequence of nonempty closed subsets of F inductively as follows.

Suppose the nonempty closed sets $F_k^n \subset F_{k-1}^n \subset \cdots \subset F_0^n$ have been constructed. If there exists an $m \in F_k^n$ with $m(k+1) = n(k+1)$, then let

$$F_{k+1}^n = \{m \in F_k^n : m(k+1) = n(k+1)\},$$

otherwise let

$$F_{k+1}^n = \{m \in F_k^n : m(k+1) = \min\{r(k+1) : r \in F_k^n\}\}.$$

Each F_k^n is a nonempty closed subset of the closed set F, and an easy inductive argument shows that t- diam $F_k^n \leqslant 1/k$ for each $k \geqslant 1$. It follows from the Cantor Intersection Theorem 3.7 that $\bigcap_{k=1}^{\infty} F_k^n$ is a singleton, say $\bigcap_{k=1}^{\infty} F_k^n = \{f(n)\}$.

This process defines a function $f : \mathcal{N} \to F$ via $n \mapsto f(n)$. To finish the proof, we shall show that f is a retraction of $\mathcal{N}$ onto F.

First, it is clear from the construction that $f(\boldsymbol{n}) = \boldsymbol{n}$ for each $\boldsymbol{n}$ belonging to F. Now suppose $t(\boldsymbol{n}, \boldsymbol{m}) = 1/k$. Then $\boldsymbol{n}(j) = \boldsymbol{m}(j)$ for $1 \leqslant j \leqslant k$, so

$$F_j^n = F_j^m \text{ for all } j = 1, \ldots, k.$$

From this we obtain $f(\boldsymbol{n})(j) = f(\boldsymbol{m})(j)$ for all $1 \leqslant j \leqslant k$, so

$$t(f(\boldsymbol{n}), f(\boldsymbol{m})) \leqslant 1/k = t(\boldsymbol{n}, \boldsymbol{m}),$$

proving that f is Lipschitz continuous with Lipschitz constant 1. ∎

3.65 Corollary　*Let F be a nonempty closed subset of $\mathcal{N}$, and let f be a continuous function from F into a topological space Y. Then f extends to a continuous function $\hat{f}$ on $\mathcal{N}$ with $\hat{f}(\mathcal{N}) = f(F)$.*

Proof: Set $\hat{f} = f \circ g$, where g is a retraction of $\mathcal{N}$ onto F. ∎

Just as Theorem 3.60 showed that every compact metrizable set is a continuous image of the Cantor set, every Polish space is a continuous image of $\mathcal{N}$. The idea of the proof is to use each sequence of natural numbers to assign a complete "address" or "postal code" to each point. Each postal code consists of an infinite sequence of natural numbers. We start by dividing the space into countably many regions indexed by natural numbers. Each region is dividend into countably many subregions, and so on, *ad infinitum*. The first integer in the postal code determines the region in which the point resides, the second designates the subregion, etc. The sub-regions are chosen to have vanishing diameter, so that the sequence pins down the point exactly. The tree metric tells us how many stages of subdivision two addresses have in common, so there is a continuous map from postal codes to points. A Polish space is second countable, which makes it "small" enough for $\mathcal{N}$ to specify complete addresses for all of its points.

3.66 Theorem　*Every Polish space is a one-to-one continuous image of a closed subset of $\mathcal{N}$.*

Proof: Let X be a Polish space, and fix a complete compatible metric d. We start with the following claim: If A is an $\mathcal{F}_\sigma$-set in X and $\varepsilon > 0$, then A is the union of a sequence of nonempty closed sets, each having diameter no more than ε. To see this, let $A = \bigcup_{n=1}^\infty F_n$, where each F_n is closed. Let $\{x_1, x_2, \ldots\}$ be a countable dense subset of X and let C_m be the closed ball of radius $\varepsilon/2$ around x_m. Let $D_{n,m} = F_n \cap C_m$. Each $D_{n,m}$ is closed and has diameter no more than ε, and $A = \bigcup_{n=1}^\infty \bigcup_{m=1}^\infty D_{n,m}$.

We next claim: If A is an $\mathcal{F}_\sigma$-set in X and $\varepsilon > 0$, then A is the union of a sequence $\{B_n\}$ of pairwise disjoint $\mathcal{F}_\sigma$-sets, each having diameter no more than ε, and such that $\overline{B_n} \subset A$ for each n. In light of the previous claim, we can write $A = \bigcup_{n=1}^\infty C_n$, where each C_n is closed with diameter no more than ε. Now let

$B_1 = C_1$ and $B_{n+1} = C_{n+1} \setminus \bigcup_{m=1}^{n} C_m$ for $n \geqslant 1$. Since $B_n \subset C_n$ we have diam $B_n \leqslant \varepsilon$ and $\overline{B_n} \subset C_n \subset A$. Now observe that the set $D = \bigcup_{m=1}^{n} C_m$ is closed so its complement, being open, is an $\mathcal{F}_\sigma$. Therefore each B_n is also an $\mathcal{F}_\sigma$. Furthermore the sets B_n are pairwise disjoint (although possibly empty) and $\bigcup_{n=1}^{\infty} B_n = A$.

Thus, we can write $X = \bigcup_{n=1}^{\infty} A_n$, where $\{A_n\}$ is a pairwise disjoint sequence of $\mathcal{F}_\sigma$-sets, each having diameter no more than $1/2$. But for each n_1, A_{n_1} can also be written as the union of a sequence $\{A_{n_1,n}\}$ of pairwise disjoint $\mathcal{F}_\sigma$-sets, each of diameter no more than $1/4$, and such that $\overline{A_{n_1,n}} \subset A_{n_1}$ for each n. Repeat this process inductively, so that for every finite sequence $(n_1, \ldots, n_m)$ of natural numbers we have selected an $\mathcal{F}_\sigma$-subset $A_{n_1,\ldots,n_m}$ of X with diam $A_{n_1,\ldots,n_m} \leqslant 1/2^m$, such that for each $j \neq k$, $A_{n_1,\ldots,n_m,j}$ and $A_{n_1,\ldots,n_m,k}$ are disjoint, $\overline{A_{n_1,\ldots,n_m,n}} \subset A_{n_1,\ldots,n_m}$ for each n, and

$$A_{n_1,\ldots,n_m} = \bigcup_{k=1}^{\infty} A_{n_1,\ldots,n_m,k}. \qquad (\star\star)$$

Now let $\boldsymbol{n}$ belong to $\mathcal{N}$ and notice that the sequence $\{\overline{A_{n(1),\ldots,n(k)}} : k = 1, 2, \ldots\}$ of closed sets satisfies

$$\overline{A_{n(1)}} \supset \overline{A_{n(1),n(2)}} \supset \cdots \supset \overline{A_{n(1),\ldots,n(k)}} \supset \cdots$$

and has vanishing diameter. Since X is d-complete, it follows from the Cantor Intersection Theorem 3.7 that $\bigcap_{k=1}^{\infty} \overline{A_{n(1),\ldots,n(k)}}$ is a singleton, provided $A_{n(1),\ldots,n(k)}$ is nonempty for each k.

Let $F = \{\boldsymbol{n} \in \mathcal{N} : A_{n(1),\ldots,n(k)} \neq \varnothing$ for all $k\}$. The set F is closed, for if $\boldsymbol{n}$ belongs to F^c, there is some k for which $A_{n(1),\ldots,n(k)} = \varnothing$. Now if $\mathrm{t}(\boldsymbol{n}, \boldsymbol{m}) \leqslant 1/k$, then $\boldsymbol{m}(i) = \boldsymbol{n}(i)$ for each $i = 1, \ldots, k$, so $A_{m(1),\ldots,m(k)} = A_{n(1),\ldots,n(k)} = \varnothing$, which shows that $B_{1/k}(\boldsymbol{n}) \subset F^c$. That is, F^c is open.

Further, we can define a mapping $f \colon F \to X$, via

$$\boldsymbol{n} \mapsto f(\boldsymbol{n}), \quad \text{where } \{f(\boldsymbol{n})\} = \bigcap_{k=1}^{\infty} \overline{A_{n(1),\ldots,n(k)}} = \bigcap_{k=1}^{\infty} A_{n(1),\ldots,n(k)}.$$

Since each point of X belongs to some A_{n_1}, and so to some A_{n_1,n_2}, etc., it is easy to see that F is nonempty and the mapping $f \colon F \to X$ is surjective. Since $A_{n_1,n_2,\ldots,n_m,j}$ and $A_{n_1,n_2,\ldots,n_m,k}$ are disjoint for $j \neq k$, it follows that f is one-to-one.

The continuity of f is easy to verify. To see this, fix $\boldsymbol{n}$ in $\mathcal{N}$. If $\boldsymbol{m}$ in F satisfies $\mathrm{t}(\boldsymbol{m}, \boldsymbol{n}) < 1/k$, then $\boldsymbol{m}(j) = \boldsymbol{n}(j)$ for all $j \leqslant k$, so both $f(\boldsymbol{n})$ and $f(\boldsymbol{m})$ belong to $A_{n(1),\ldots,n(k)} = A_{m(1),\ldots,m(k)}$, which has diameter $\leqslant 1/2^k$. Therefore f is continuous at $\boldsymbol{n}$, and the proof is finished. $\blacksquare$

3.67 Corollary *Every Polish space is a continuous image of $\mathcal{N}$.*

Proof: This follows from Theorem 3.66 and Corollary 3.65. $\blacksquare$

A family of subsets of some set X indexed by $\mathbb{N}^{<\mathbb{N}}$, such as the one used in the proof of Theorem 3.66, is called a **Suslin scheme**. That is, a Suslin scheme is a function $A\colon \mathbb{N}^{<\mathbb{N}} \to 2^X$. Given a Suslin scheme $\{A_s : s \in \mathbb{N}^{<\mathbb{N}}\}$, the **Suslin $\mathcal{A}$ operation** defines a set, called the **nucleus** of the scheme, by

$$\mathcal{A}(A) = \bigcup_{n \in \mathcal{N}} \bigcap_{k=1}^{\infty} A_{n(1),\dots,n(k)}.$$

The nucleus of the Suslin scheme in the proof of Theorem 3.66 is the Polish space X. When the correspondence $n \mapsto \bigcap_{k=1}^{\infty} A_{n(1),\dots,n(k)}$ is actually a function, it is called the **associated map** of the Suslin scheme.

A Suslin scheme is **regular** if for each $n_1, \dots, n_m$ and k we have

$$A_{n_1,\dots,n_m} \supset A_{n_1,\dots,n_m,k}.$$

A **Lusin scheme** is a regular Suslin scheme that also satisfies

$$A_{n_1,\dots,n_m,k} \cap A_{n_1,\dots,n_m,j} = \varnothing$$

whenever $k \neq j$. Finally, a Suslin scheme $\{A_s : s \in \mathbb{N}^{<\mathbb{N}}\}$ in a metric space has **vanishing diameter** if for each n in $\mathcal{N}$, $\lim_{k \to \infty} \operatorname{diam} A_{n(1),\dots,n(k)} = 0$.

In terms of our address analogy, a Suslin scheme is a general way of assigning postal codes, where the set A_s is a region of the space determined by the initial finite sequence s of the postal code. A regular Suslin schemes says that the regions are arranged hierarchically: a region specified by an initial sequence of length $m+1$ is a subset of the region specified by its initial sequence of length m. A Lusin scheme requires that each stage, each region is partitioned into disjoint regions by its addressing scheme. A good postal system would probably use a Lusin scheme.

As an aside, we mention the surprising fact that $\mathcal{N}$ is also homeomorphic to the space $\mathcal{I}$ of irrational numbers in $(0, 1)$ (with their usual metric topology). There are two ways (at least) to prove this. The "classical" proof uses some results from the theory of infinite continued fractions, which is not widely taught these days. A proof along these lines may be found in D. P. Bertsekas and S. E. Shreve [39, Proposition 7.5, pp. 109–112]. For an elementary introduction to continued fractions, see C. D. Olds [263]. We shall use Suslin schemes along the lines proposed by A. W. Miller [250, Theorem 1.1, p. 5].

3.68 Theorem (The space of irrationals) *The Baire space $\mathcal{N}$ is homeomorphic to the space $\mathcal{I}$ of irrationals in $(0, 1)$.*

Proof: In this proof, we construct a Lusin scheme whose associated map is a homeomorphism. Start by enumerating the rationals in $(0, 1)$ as $q_1, q_2, \dots$. The Lusin scheme is constructed inductively as follows.

In step 1, take a doubly infinite sequence $\cdots < r_{-1} < r_0 < r_1 < \cdots$ of rationals belonging to $(0, 1)$ indexed by the (negative as well as positive) integers $\mathbb{Z}$ such that (i) $r_{n+1} - r_n < 1/2$ for each n, (ii) $\inf_n r_n = 0$ and $\sup_n r_n = 1$, and (iii) q_1 is one of the r_n. Enumerate the intervals (r_{n-1}, r_n), $n \in \mathbb{Z}$ as $U_1, U_2, U_3, \ldots$. Note that $\bigcup_{n=1}^\infty U_n = (0, 1) \setminus \{r_n : n \in \mathbb{Z}\}$.

For the induction step, assume that for some $m > 1$ we have constructed for each $(m-1)$-tuple $(n_1, \ldots, n_{m-1})$ of natural numbers an open subinterval $U_{n_1, \ldots, n_{m-1}}$ of $(0, 1)$ with rational endpoints and of length at most $1/2^{m-1}$. Assume also that the rational number q_{m-1} does not belong to any one of the open intervals $U_{n_1, \ldots, n_{m-1}}$.

Now for each such open subinterval $U_{n_1, \ldots, n_{m-1}} = (q, p)$ take a doubly infinite sequence $\cdots < r_{-1} < r_0 < r_1 < \cdots$ of rational numbers belonging to (q, p) such that (i) $r_{n+1} - r_n < 1/2^m$ for each n, (ii) $\inf_n r_n = q$ and $\sup_n r_n = p$, and (iii) in case q_m belongs to (q, p) make sure that q_m is an endpoint of some interval (r_{n-1}, r_n). Enumerate the intervals (r_{n-1}, r_n) as $U_{n_1, \ldots, n_{m-1}, 1}, U_{n_1, \ldots, n_{m-1}, 2}, \ldots$, and note that

$$\bigcup_{n=1}^\infty U_{n_1, \ldots, n_{m-1}, n} = U_{n_1, \ldots, n_{m-1}} \setminus \{r_n : n \in \mathbb{Z}\}.$$

By construction, this Lusin scheme is regular, has vanishing diameter, and for any $\boldsymbol{n}$ in $\mathbb{N}$, and any k in $\mathbb{N}$, $\overline{U_{n(1), \ldots, n(k), n(k+1)}} \subset U_{n(1), \ldots, n(k)}$ and $q_k \notin U_{n(1), \ldots, n(k)}$. Therefore

$$\bigcap_{k=1}^\infty U_{n(1), \ldots, n(k)} = \bigcap_{k=1}^\infty \overline{U_{n(1), \ldots, n(k)}},$$

and this intersection is a singleton irrational. Call this irrational $f(\boldsymbol{n})$.

It is clear by the construction employed that the mapping $f \colon \mathbb{N} \to \mathbb{J}$ defined this way is one-to-one and onto. We leave it as exercise to show that f is a homeomorphism. Hint: The mapping f is continuous just as in the proof of Theorem 3.66, and the sets $\mathbb{J} \cap U_{n_1, n_2, \ldots, n_m}$ form a base for the topology on $\mathbb{J}$. ∎

Since we can write $\mathbb{J} = \bigcap_{q \in \mathbb{Q}} \{q\}^c \cap (0, 1)$, we see that $\mathbb{J}$ is a $\mathcal{G}_\delta$ in $\mathbb{R}$. Therefore Alexandroff's Lemma 3.34 implies that $\mathbb{J}$ is completely metrizable, even though the usual metric is not complete. The result above shows how to metrize $\mathbb{J}$ with the complete tree metric.

We close this section with the following observation.

3.69 Theorem *The Baire space $\mathbb{N}$ is not σ-compact.*

Proof: It suffices to show that for any sequence $\{K_n\}$ of compact subsets of $\mathbb{N}$, there is some point $\boldsymbol{m}$ that does not belong to $\bigcup_{n=1}^\infty K_n$. To see this, we use a variant of Cantor's diagonal process. Let π_k be the projection of $\mathbb{N}$ onto its kth factor, that is, $\pi_k(\boldsymbol{n}) = n(k)$. This mapping is continuous, and so achieves its maximum on each compact set K_n. Pick an element $\boldsymbol{m}$ of $\mathbb{N}$ so that for each n we have $m(n) > \max_{K_n} \pi_n$. Then $\boldsymbol{m}$ does not belong to K_n for any n. ∎

3.15 Uniformities

Metric spaces are special cases of what are called **uniform spaces**. We present here a brief discussion of uniform spaces. We do not prove any results, as they can be found in any standard topology text. For instance, see S. Willard [342, Chapter 9, pp. 238–277].

Before we define uniformities, consider a metric space (X, d). Let

$$U(\varepsilon) = \{(x, y) \in X \times X : d(x, y) < \varepsilon\}.$$

Then each $U(\varepsilon)$ is a binary relation on X and is an open subset of $X \times X$. The family $\{U(\varepsilon) : \varepsilon > 0\}$ has the following properties:

1. $\bigcap_{\varepsilon > 0} U(\varepsilon) = \{(x, x) : x \in X\}$, the diagonal of $X \times X$.

2. $U(\varepsilon_1) \cap U(\varepsilon_2) = U(\min\{\varepsilon_1, \varepsilon_2\})$.

3. $U(\varepsilon) = U^{-1}(\varepsilon)$, where $U^{-1}(\varepsilon)$ is the usual relational inverse. That is, $U^{-1}(\varepsilon) = \{(x, y) : (y, x) \in U(\varepsilon)\}$.

4. $U(\varepsilon/2) \circ U(\varepsilon/2) \subset U(\varepsilon)$, where $\circ$ denotes the usual composition of relations (see Section 1.3).

5. $B_\varepsilon(x) = \{y \in X : (x, y) \in U(\varepsilon)\}$.

Note that the collection $\{U(\varepsilon) : \varepsilon > 0\}$ is a filter base. That is, it does not contain the empty set, and it has the finite intersection property.

It is possible to phrase the definition of uniform continuity using these sets. That is, a function $f : X \to Y$ between metric spaces is uniformly continuous if for every $\varepsilon > 0$, there is a $\delta > 0$ such that $(x, y) \in U_X(\delta)$ implies $(f(x), f(y)) \in U_Y(\varepsilon)$. Also a sequence $\{x_n\}$ is Cauchy if for every $\varepsilon > 0$, there is some n such that $k, m > n$ implies $(x_k, x_m) \in U(\varepsilon)$. Uniform spaces were introduced to generalize these notions.

Let us therefore define a **diagonal uniformity**, or simply a **uniformity**, on a nonempty set X to be a nonempty collection $\mathcal{U}$ of subsets of $X \times X$ such that:

1. $U \in \mathcal{U}$ implies $\{(x, x) \in X \times X : x \in X\} \subset U$.

2. $U_1, U_2 \in \mathcal{U}$ implies $U_1 \cap U_2 \in \mathcal{U}$.

3. $U \in \mathcal{U}$ implies that $V \circ V \subset U$ for some $V \in \mathcal{U}$.

4. $U \in \mathcal{U}$ implies that $V^{-1} = \{(x, y) \in X \times X : (y, x) \in V\} \subset U$ for some $V \in \mathcal{U}$.

5. $U \in \mathcal{U}$ and $U \subset V$ imply $V \in \mathcal{U}$.

Members of $\mathcal{U}$ are called **surroundings** or **entourages**. Note that a uniformity is a filter. A **base** for a uniformity $\mathcal{U}$ is a filter base for $\mathcal{U}$ in $X \times X$. For a metric space (X, d), the collection of $U(\varepsilon)$s mentioned above is a base for the **metric uniformity** on X. A **uniform space** is simply a space equipped with a uniformity.

A uniformity $\mathcal{U}$ on a nonempty set X defines a topology as follows. Given a set $U \in \mathcal{U}$, put $U[x] = \{y \in X : (x, y) \in U\}$. Then the collection $\{U[x] : U \in \mathcal{U}\}$ is a neighborhood base at x. The topology corresponding to this neighborhood base is called the topology generated by $\mathcal{U}$. A set G is open in this topology if and only if for every $x \in G$ there is a set $U \in \mathcal{U}$ with $U[x] \subset G$. The topology is Hausdorff if and only if $\bigcap \mathcal{U} = \{(x, x) : x \in X\}$, in which case we say that $\mathcal{U}$ is **separating**. A function $f : (X, \mathcal{U}_X) \to (Y, \mathcal{U}_Y)$ between uniform spaces is **uniformly continuous** if for every $U \in \mathcal{U}_Y$ there is a $V \in \mathcal{U}_X$ such that $(x, z) \in V$ implies $(f(x), f(z)) \in U$. Every uniformly continuous function is continuous with respect to the topologies generated by the uniformities.

Cauchy nets are defined as we indicated earlier, so it is possible to discuss completeness for uniform spaces. Not all uniform spaces are generated by a metric. For instance, the trivial uniformity $\{X \times X\}$ generates the trivial topology on X, which is not metrizable unless X has only one point. The following results are worth noting.

- A uniformity is generated by a semimetric if and only it has a countable base; see S. Willard [342, Theorem 38.3, p.257]. Consequently, a uniformity is generated by a metric if and only it has a countable base and is separating; see [342, Corollary 38.4, p.258].

- A topology is generated by a uniformity if and only if it is completely regular; see [342, Theorem 38.2, p. 256]. A completely regular topology τ on the space X is generated by the uniformity with base given by the finite intersections of sets of the form $\{(x, y) \in X \times X : |f(x) - f(y)| < \varepsilon\}$ where f is a bounded τ-continuous real function and $\varepsilon > 0$.

- The metrics d and ρ on a set X generate the same uniformity if there exist positive constants c and C satisfying $cd(x, y) \leqslant \rho(x, y) \leqslant Cd(x, y)$ for all $x, y \in X$.

3.16   The Hausdorff distance

We now take a look at ways to topologize the collection of nonempty closed subsets of a metrizable space. There are three popular ways to do this, the *Vietoris topology*, the *Fell topology* or *topology of closed convergence*, and the *Hausdorff metric*. In the next few sections we describe these topologies and the relations among them. We also briefly discuss the *Wijsman topology*. For a more in-depth study we recommend G. A. Beer [35]. We start with the Hausdorff distance.

3.70 Definition   *Let (X, d) be a semimetric space. For each pair of nonempty subsets A and B of X, define*

$$h_d(A, B) = \max\left\{\sup_{a \in A} d(a, B), \sup_{b \in B} d(b, A)\right\}.$$

*The extended real number $h_d(A, B)$ is the **Hausdorff distance** between A and B relative to the semimetric d. The function h_d is the **Hausdorff semimetric** induced by d. By convention, $h_d(\varnothing, \varnothing) = 0$ and $h_d(A, \varnothing) = \infty$ for $A \neq \varnothing$.*

While h_d depends on d, we may omit the subscript when d is clear from the context.

We can also define the Hausdorff distance in terms of neighborhoods of sets. Recall our definition of the ε-neighborhood of a nonempty subset A of the semimetric space (X, d) as the set

$$N_\varepsilon(A) = \{x \in X : d(x, A) < \varepsilon\}.$$

Recall that $\bigcap_{\varepsilon > 0} N_\varepsilon(A) = \overline{A}$ and note that

$$N_\varepsilon(\textstyle\bigcup_{i \in I} A_i) = \bigcup_{i \in I} N_\varepsilon(A_i).$$

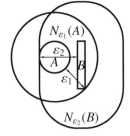

$\varepsilon_1 = \sup_{b \in B} d(b, A),$
$\varepsilon_2 = \sup_{a \in A} d(a, B)$

Figure 3.1.

3.71 Lemma   *If A and B are nonempty subsets of a semimetric space (X, d), then*

$$h(A, B) = \inf\{\varepsilon > 0 : A \subset N_\varepsilon(B) \text{ and } B \subset N_\varepsilon(A)\}.$$

Proof: If $\{\varepsilon > 0 : A \subset N_\varepsilon(B) \text{ and } B \subset N_\varepsilon(A)\} = \varnothing$, then for each $\varepsilon > 0$, either there is some $a \in A$ with $d(a, B) \geqslant \varepsilon$ or there is some $b \in B$ with $d(b, A) \geqslant \varepsilon$. This implies $h(A, B) \geqslant \varepsilon$ for each $\varepsilon > 0$, so $h(A, B) = \infty$. (Recall that $\inf \varnothing = \infty$.)

Now suppose $\delta = \inf\{\varepsilon > 0 : A \subset N_\varepsilon(B) \text{ and } B \subset N_\varepsilon(A)\} < \infty$. If ε satisfies $A \subset N_\varepsilon(B)$ and $B \subset N_\varepsilon(A)$, then $d(a, B) < \varepsilon$ for all $a \in A$ and $d(b, A) < \varepsilon$ for each $b \in B$, so $h(A, B) \leqslant \varepsilon$. Thus $h(A, B) \leqslant \delta$. On the other hand, if $\varepsilon > h(A, B)$, then obviously $A \subset N_\varepsilon(B)$ and $B \subset N_\varepsilon(A)$, so indeed $h(A, B) = \delta$. (See Figure 3.1.) ∎

The function h has all the properties of a semimetric except for the fact that it can take on the value ∞.

3.72 Lemma   *If (X, d) is a semimetric space, then h is an "extended" semimetric on 2^X. That is, $h: 2^X \times 2^X \to \mathbb{R}^*$ is an extended real-valued function such that for all A, B, C in 2^X, the following properties are satisfied.*

1. $h(A, B) \geqslant 0$ and $h(A, A) = 0$.

2. $h(A, B) = h(B, A)$.

3. $h(A, B) \leqslant h(A, C) + h(C, B)$.

4. $h(A, B) = 0$ *if and only if* $\overline{A} = \overline{B}$.

Proof: Except for the triangle inequality, these claims follow immediately from the definition.

For the triangle inequality, if any of A, B, or C is empty, the result is trivial, so assume each is nonempty. Note that for $a \in A$, $b \in B$, and $c \in C$, we have $d(a, B) \leqslant d(a, b) \leqslant d(a, c) + d(c, b)$, so

$$d(a, B) \leqslant d(a, c) + d(c, B) \leqslant d(a, c) + h(C, B).$$

Taking the infimum on the right with respect to $c \in C$, we get

$$d(a, B) \leqslant d(a, C) + h(C, B) \leqslant h(A, C) + h(C, B).$$

So $\sup_{a \in A} d(a, B) \leqslant h(A, C) + h(C, B)$. Therefore, by symmetry,

$$h(A, B) = \max\Big\{\sup_{a \in A} d(a, B), \sup_{b \in B} d(b, A)\Big\} \leqslant h(A, C) + h(C, B).$$

This completes the proof.   ∎

The following properties of the Hausdorff distance are easy to verify.

3.73 Lemma   *Let A and B be nonempty subsets of the semimetric space* (X, d).

1. *If both A and B are nonempty and d-bounded, then* $h(A, B) < \infty$. *(However, it is possible that* $h(A, B) < \infty$ *even if both A and B are unbounded, e.g., let A and B be parallel lines in* $\mathbb{R}^2$.)

2. *If A is d-bounded and* $h(A, B) < \infty$, *then B is d-bounded.*

3. *If A is d-unbounded and* $h(A, B) < \infty$, *then B is d-unbounded.*

4. *If A is d-bounded and B is d-unbounded, then* $h(A, B) = \infty$.

We can also characterize the Hausdorff distance in terms of distance functions.

3.74 Lemma   *Let* (X, d) *be a semimetric space. Then for any nonempty subsets A and B of X,*

$$h(A, B) = \sup_{x \in X} |d(x, A) - d(x, B)|.$$

Proof: Let A and B be two nonempty subsets of X. Then for each $a \in A$ and each $b \in B$, we have $d(x, A) - d(x, b) \leqslant d(x, a) - d(x, b) \leqslant d(a, b)$. Hence,

$$d(x, A) - d(x, b) \leqslant \inf_{a \in A} d(a, b) = d(b, A) \leqslant h(A, B)$$

for each $b \in B$. It follows that $d(x, A) - d(x, B) \leqslant h(A, B)$. By the symmetry of the situation, $|d(x, A) - d(x, B)| \leqslant h(A, B)$, and consequently

$$\sup_{x \in X} |d(x, A) - d(x, B)| \leqslant h(A, B).$$

If $b \in B$, then $d(b, A) = |d(b, A) - d(b, B)| \leqslant \sup_{x \in X} |d(x, A) - d(x, B)|$, so

$$\sup_{b \in B} d(b, A) \leqslant \sup_{x \in X} |d(x, A) - d(x, B)|.$$

Likewise $\sup_{a \in A} d(a, B) \leqslant \sup_{x \in X} |d(x, A) - d(x, B)|$, so the reverse inequality

$$h(A, B) \leqslant \sup_{x \in X} |d(x, A) - d(x, B)|$$

is also true.   ∎

You might ask at this point whether there are points $a \in A$ and $b \in B$ satisfying $h(A, B) = d(a, b)$. If A and B are not closed, we should not expect this to happen but the following example shows that even for closed sets this may not be the case.

3.75 Example   (Hausdorff distance not attained [259])   In ℓ_2, the Banach space of square summable sequences, the set $B = \{e_1, e_2, \ldots\}$ of unit coordinate vectors is closed. Let $x = (-1, -\frac{1}{2}, \ldots, -\frac{1}{n}, \ldots)$, and put $A = B \cup \{x\}$. Clearly, A is also closed. Then $\sup_{b \in B} d(b, A) = 0$ (as $B \subset A$), so $h(A, B) = d(x, B)$. Now

$$d(x, e_n) = \left\| x - e_n \right\|_2 = \left[(1 + \tfrac{1}{n})^2 + \sum_{i \neq n} \tfrac{1}{i^2} \right]^{\frac{1}{2}} = \left(1 + \tfrac{2}{n} + \sum_{i=1}^{\infty} \tfrac{1}{i^2} \right)^{\frac{1}{2}} = \left(1 + \tfrac{\pi^2}{6} + \tfrac{2}{n} \right)^{\frac{1}{2}}.$$

So $h(A, B) = \inf_n d(x, e_n) = (1 + \tfrac{\pi^2}{6})^{\frac{1}{2}}$, while $d(x, e_n) > (1 + \tfrac{\pi^2}{6})^{\frac{1}{2}} = h(A, B)$ for each n.   ∎

Note that the above example involved a set that was closed but not compact. For compact sets, we have the following.

3.76 Lemma   *Let (X, d) be a semimetric space, and let A and B be nonempty subsets of X.*

1. *For every $\varepsilon > 0$ and every element $a \in A$, there exists some $b \in B$ satisfying $d(a, b) < h(A, B) + \varepsilon$.*

2. *If B is compact, then for each $a \in A$ there exists some $b \in B$ satisfying $d(a, b) \leqslant h(A, B)$.*

3. *If A and B are both compact, then there exist $a \in A$ and $b \in B$ such that $d(a, b) = h(A, B)$.*

Proof: (1) This is immediate from the definition of the Hausdorff distance.

(2) Since the real function $x \mapsto d(a, x)$ is continuous, it achieves its minimum value $d(a, B)$ over the compact set B at some point $b \in B$. But then, we have $d(a, b) = d(a, B) \leqslant h(A, B)$.

(3) Assume $h(A, B) = \sup_{a \in A} d(a, B)$. Since $x \mapsto d(x, B)$ is continuous and A is compact, there exists some $a \in A$ with $d(a, B) = h(A, B)$. Since the function $x \mapsto d(a, x)$ is also continuous and B is compact, there exists some $b \in B$ satisfying $d(a, b) = \min_{x \in B} d(a, x) = d(a, B) = h(A, B)$. ∎

3.17 The Hausdorff metric topology

When d is a metric, and A and B are closed, then $h(A, B) = 0$ if and only if $A = B$. It is thus natural to use the "extended" metric h to define a Hausdorff topology at least on the collection of closed sets. We start by introducing some notation. Given a metric space (X, d),

- $\mathcal{F}$ denotes the collection of *nonempty* closed subsets of X,

- $\mathcal{F}_d$ denotes the collection of *nonempty* d-bounded closed subsets of X, and

- $\mathcal{K}$ denotes the collection of *nonempty* compact subsets of X.

If d is a bounded metric, then $\mathcal{F}_d$ coincides with $\mathcal{F}$. Of course no reference to d is needed in the definition of $\mathcal{K}$, since compactness, unlike boundedness, is a topological property. Should the need arise, we may write $\mathcal{F}(X)$, etc., to indicate the underlying space.

For $F \in \mathcal{F}$ and $\varepsilon > 0$, define

$$B_\varepsilon(F) = \{C \in \mathcal{F} : h(C, F) < \varepsilon\},$$

which by analogy to a genuine metric, we call the **open ε-ball centered at** F. Note well the difference between $N_\varepsilon(F) = \{x \in X : d(x, F) < \varepsilon\}$, which is a subset of X, and $B_\varepsilon(F) = \{C \in \mathcal{F} : h(C, F) < \varepsilon\}$, a subset of $\mathcal{F}$. Clearly $C \in B_\varepsilon(F)$ implies $C \subset N_\varepsilon(F)$, but not vice versa.

The next result is straightforward.

3.77 Lemma *The collection of balls $B_\varepsilon(F)$, where $F \in \mathcal{F}$ and $0 < \varepsilon < \infty$, forms a base for a first countable Hausdorff topology on $\mathcal{F}$.*

This topology is called the **Hausdorff metric topology** (even when h assumes the value ∞) and is denoted τ_h.

Lemma 3.73 implies that both $\mathcal{F}_d$ and $\mathcal{F} \setminus \mathcal{F}_d$ are τ_h-open, and hence both are clopen. It is possible to add the empty set as an isolated point of $\mathcal{F} \cup \{\varnothing\}$. The set X can be naturally viewed as a subset of $\mathcal{F}$.

3.78 Lemma   *Let (X, d) be a metric space. Then the mapping $x \mapsto \{x\}$ embeds X isometrically as a closed subset of $(\mathcal{F}_d, \mathsf{h})$, and hence as a closed subset of $\mathcal{F}$.*

Proof: Note that $\mathsf{h}(\{x\}, \{y\}) = d(x, y)$ for all $x, y \in X$, so $x \mapsto \{x\}$ is an isometry. To see that X is closed in $\mathcal{F}_d$, assume that $\mathsf{h}(\{x_n\}, A) \to 0$. If $x \in A$ (recall that A is nonempty), then from $d(x_n, x) \leqslant \mathsf{h}(\{x_n\}, A)$ we get $\mathsf{h}(\{x_n\}, \{x\}) = d(x_n, x) \to 0$. Thus $A = \{x\}$, so X is closed in $\mathcal{F}_d$.   ∎

We now discuss two criteria for convergence in $(\mathcal{F}, \tau_\mathsf{h})$. Lemma 3.74 immediately implies the following.

3.79 Corollary   *Let (X, d) be a metric space. Then $F_n \xrightarrow{\tau_\mathsf{h}} F$ in $\mathcal{F}$ if and only if the sequence $\{d(\cdot, F_n)\}$ of real functions converges uniformly to $d(\cdot, F)$ on X.*

The following notion of convergence of sets is defined solely in terms of the topology on X and does not depend on any particular metric.

3.80 Definition   *Let $\{E_n\}$ be a sequence of subsets of a topological space X. Then:*

1. *A point x in X belongs to the **topological lim sup**, denoted $\mathrm{Ls}\, E_n$, if for every neighborhood V of x there are infinitely many n with $V \cap E_n \neq \varnothing$.*

2. *A point x in X belongs to the **topological lim inf**, denoted $\mathrm{Li}\, E_n$, if for every neighborhood V of x, we have $V \cap E_n \neq \varnothing$ for all but finitely many n.*

3. *If $\mathrm{Li}\, E_n = \mathrm{Ls}\, E_n = E$, then the set E is called the **closed limit** of the sequence $\{E_n\}$.*[5]

Note that the definition of the closed limit is actually topological. It depends only on the topology and not on any particular metric. Clearly, $\mathrm{Li}\, E_n \subset \mathrm{Ls}\, E_n$. We leave it as an exercise to prove the following lemma. (Hint: A set is closed if and only if its complement is open, and a point is in the closure of a set if and only if every neighborhood of the point meets the set.)

3.81 Lemma   *Let $\{E_n\}$ be a sequence of subsets of a topological space X. Then both $\mathrm{Li}\, E_n$ and $\mathrm{Ls}\, E_n$ are closed sets, and moreover*

$$\mathrm{Ls}\, E_n = \bigcap_{m=1}^{\infty} \overline{\bigcup_{k=m}^{\infty} E_k}.$$

---

[5] F. Hausdorff [155, §28.2, p. 168] uses the terms "closed upper limit" and "closed lower limit." The terminology here is adapted from W. Hildenbrand [158]. The topological lim sup and lim inf of a sequence are different from the set theoretic lim sup and lim inf, defined by

$$\limsup E_n = \bigcap_{m=1}^{\infty} \bigcup_{k=m}^{\infty} E_k \quad \text{and} \quad \liminf E_n = \bigcup_{m=1}^{\infty} \bigcap_{k=m}^{\infty} E_k.$$

The next result, which appears in F. Hausdorff [155, p. 171], shows that a limit with respect to the Hausdorff metric is also the closed limit.

3.82 Theorem   (Closed convergence in $\mathcal{F}$)   *If (X, d) is a metric space and $F_n \xrightarrow{\tau_h} F$ in $\mathcal{F}$, then $F = \mathrm{Li}\, F_n = \mathrm{Ls}\, F_n$.*

Proof: Let $F_n \to F$ in the Hausdorff metric topology τ_h. Since $\mathrm{Li}\, F_n \subset \mathrm{Ls}\, F_n$, it suffices to show $\mathrm{Ls}\, F_n \subset F \subset \mathrm{Li}\, F_n$.

Let x belong to F, and pick $\varepsilon > 0$. Then $h(F_n, F) < \varepsilon$ for large enough n. In that case, there is some $x_n \in F_n$ with $d(x_n, x) < \varepsilon$. That is, $B_\varepsilon(x) \cap F_n \neq \varnothing$ for all large enough n. Therefore, $F \subset \mathrm{Li}\, F_n$.

Now let $x \in \mathrm{Ls}\, F_n$ and let $\varepsilon > 0$ be given. Then $B_\varepsilon(x) \cap F_n \neq \varnothing$ for infinitely many n. In particular, for infinitely many $x_n \in F_n$ we have $d(x, x_n) < \varepsilon$. Since $d(x_n, F) \leqslant h(F_n, F)$ and $h(F_n, F) \to 0$, we have $d(x_m, F) < \varepsilon$ and $d(x, x_m) < \varepsilon$ for some m. Pick $y \in F$ with $d(x_m, y) < \varepsilon$. From $d(x, y) \leqslant d(x, x_m) + d(x_m, y) < 2\varepsilon$, it follows that $B_{2\varepsilon}(x) \cap F \neq \varnothing$ for each $\varepsilon > 0$, which shows that $x \in \overline{F} = F$. Therefore, $\mathrm{Ls}\, F_n \subset F$.   ∎

The converse of Theorem 3.82 is false unless X is compact. In general, the closed limit of a sequence of closed sets need not be a Hausdorff metric limit. But if X is compact, see Theorem 3.93.

3.83 Example   (Closed limit vs. Hausdorff metric limit)   Consider $\mathbb{N}$ with the discrete metric d. Let $F_n = \{1, 2, \ldots, n\}$. Then $\mathrm{Ls}\, F_n = \mathrm{Li}\, F_n = \mathbb{N}$, but $h(F_n, \mathbb{N}) = 1$ for all n. Thus, the closed limit of a sequence need not be a limit in the Hausdorff metric.   ∎

We can use Lemma 3.74 to isometrically embed the metric space $(\mathcal{F}_d, \tau_h)$ of d-bounded nonempty closed sets into the space $C_b(X)$ of bounded continuous real function equipped with the sup metric. Now unless d is a bounded metric, the distance function $d(\cdot, A)$ need not be bounded, but we can find a bounded function that has the right properties. Fix x_0 in the metric space (X, d). For each nonempty subset A of X define the $f_A \colon X \to \mathbb{R}$ by

$$f_A(x) = d(x, A) - d(x, x_0).$$

If A is d-bounded, then f_A is bounded: Since $d(x, A) \leqslant d(x, x_0) + d(x_0, A)$ and $d(x, x_0) \leqslant d(x, A) + \mathrm{diam}\, A + d(x_0, A)$, for any x we have

$$|f_A(x)| \leqslant d(x_0, A) + \mathrm{diam}\, A.$$

Also note that f_A is Lipschitz continuous. In fact,

$$|f_A(x) - f_A(y)| \leqslant |d(x, A) - d(y, A)| + |d(y, x_0) - d(x, x_0)| \leqslant 2d(x, y),$$

see the proof of Theorem 3.16.

3.84 Theorem (Kuratowski)   *Let (X, d) be a metric space. Then the mapping $A \mapsto f_A$ isometrically embeds $(\mathcal{F}_d, h)$ into $U_d(X) \subset C_b(X)$.*

Proof: This follows from $f_A(x) - f_B(x) = d(x, A) - d(x, B)$ and Lemma 3.74. ∎

Note that the completion of $\mathcal{F}_d$ is simply the closure of $\mathcal{F}_d$ in $C_b(X)$. The topological space $(\mathcal{F}, h)$ inherits several important metric properties from (X, d).

3.85 Theorem (Completeness and compactness)   *Let (X, d) be a metric space. Then:*

1. *$(\mathcal{F}, \tau_h)$ is separable $\iff$ $(\mathcal{F}, \tau_h)$ is totally bounded $\iff$ (X, d) is totally bounded.*

2. *$(\mathcal{F}_d, \tau_h)$ is complete if and only if (X, d) is complete.*

3. *$(\mathcal{F}, \tau_h)$ is Polish $\iff$ $(\mathcal{F}, \tau_h)$ is compact $\iff$ (X, d) is compact.*

Proof: (1)   If $\mathcal{F}$ is h-totally bounded, then (X, d) is totally bounded, since (X, d) can be isometrically embedded in $(\mathcal{F}_d, h)$ (Lemma 3.78).

Assume that the metric space (X, d) is totally bounded. Let $\varepsilon > 0$ and let $\{x_1, \ldots, x_n\}$ be an $\varepsilon/2$-dense subset of X, and let C_i denote the closed ball centered at x_i with radius $\varepsilon/2$. For any $C \in \mathcal{F} = \mathcal{F}_d$, the set $F = \bigcup\{C_i : C_i \cap C \neq \varnothing\}$ is closed and satisfies $h(C, F) \leqslant \varepsilon$. This shows that the finite set comprising all finite unions from $\{C_1, \ldots, C_n\}$ is ε-dense in $\mathcal{F}$, so $(\mathcal{F}, \tau_h)$ is totally bounded, and therefore separable.

To show that the separability of $(\mathcal{F}, \tau_h)$ implies the total boundedness of (X, d), proceed by contraposition. If (X, d) is not totally bounded, then for some $\varepsilon > 0$ there is an infinite subset A of X satisfying $d(x, y) > 3\varepsilon$ for all distinct x, y in A. If E and F are distinct nonempty finite subsets of A, then $h(E, F) \geqslant 3\varepsilon$. In particular, the uncountable family $\{B_\varepsilon(F)\}$ of open balls, where F runs over the nonempty finite subsets of A, is pairwise disjoint. This implies that $(\mathcal{F}, \tau_h)$ cannot be separable.

(2)   If $(\mathcal{F}_d, \tau_h)$ is complete, then (X, d) is complete since by Lemma 3.78, X can be isometrically identified with a closed subset of $\mathcal{F}_d$.

Next assume that (X, d) is complete, and let $\{C_n\}$ be an h-Cauchy sequence in $\mathcal{F}_d$. We must show that $C_n \xrightarrow{h} C$ for some $C \in \mathcal{F}_d$. By passing to a subsequence, we can assume without loss of generality that $h(C_n, C_{n+1}) < 1/2^{n+1}$ for each n. Then $h(C_k, C_n) < 1/2^k$ for all $n > k$. From Theorem 3.82, the limit C, if it exists in $\mathcal{F}_d$, it must coincide with $\operatorname{Ls} C_n$. So put

$$C = \bigcap_{m=1}^{\infty} \overline{\bigcup_{r \geqslant m} C_r}.$$

Clearly, C (as an intersection of closed sets) is a closed set. First, let us check that C is nonempty. In fact, we shall establish that for each $b \in C_k$ there exists some $c \in C$ with $d(b, c) \leqslant 1/2^{k-1}$ (so $\sup_{b \in C_k} d(b, C) \leqslant 1/2^{k-1}$). To this end, fix k and $b \in C_k$. From

$$h(C_n, C_{n+1}) = \max\left\{ \sup_{a \in C_{n+1}} d(a, C_n), \sup_{x \in C_n} d(x, C_{n+1}) \right\} < 1/2^{n+1}$$

and an easy induction argument, we see that there exists a sequence $\{c_n\}$ in X such that $c_1 = c_2 = \cdots = c_k = b \in C_k$, $c_n \in C_n$ for $n > k$ and $d(c_n, c_{n+1}) < 1/2^{n+1}$ for all n. It easily follows that $\{c_n\}$ is a d-Cauchy sequence in X, so (by the d-completeness of X) there exists some $c \in X$ such that $d(c_n, c) \to 0$. Now note that $c \in C$ (so $C \neq \varnothing$) and that for $n > k$, we have

$$d(b, c_n) = d(c_k, c_n) \leqslant \sum_{i=k}^{n-1} d(c_i, c_{i+1}) \leqslant \sum_{i=k}^{n-1} 1/2^{i+1} \leqslant 1/2^k < 1/2^{k-1}.$$

Hence, $d(b, C) \leqslant d(b, c) = \lim_{n \to \infty} d(b, c_n) \leqslant 1/2^{k-1}$ for each $b \in C_k$.

Now let $x \in C$ and k be fixed. Then, there exists some $n > k$ and some $a \in C_n$ with $d(x, a) < 1/2^k$. From $h(C_k, C_n) < 1/2^k$, we see that $d(a, C_k) < 1/2^k$, so there exists some $b \in C_k$ with $d(a, b) < 1/2^k$. Therefore,

$$d(x, C_k) \leqslant d(x, b) \leqslant d(x, a) + d(a, b) < 1/2^{k-1},$$

so $\sup_{x \in C} d(x, C_k) \leqslant 1/2^{k-1}$. In other words, we have shown that

$$h(C_k, C) = \max\left\{ \sup_{b \in C_k} d(b, C), \sup_{x \in C} d(x, C_k) \right\} \leqslant 1/2^{k-1}$$

for $k = 1, 2, \ldots$. This shows that $C \in \mathcal{F}_d$ and $C_n \to C$ in $\mathcal{F}_d$.

(3)   The equivalences follow immediately from the preceding parts by taking into account that a metric space is compact if and only if it is complete and totally bounded (Theorem 3.28).   ∎

The fact that $(\mathcal{F}, \tau_h)$ can fail to be Polish, even when (X, d) is Polish and d is a bounded complete metric is mildly disturbing. There is however another topology on $\mathcal{F}$ that is Polish. The **Wijsman topology** τ_W is the weak topology on $\mathcal{F}$ generated by the family of functions $F \mapsto d(x, F)$ as x ranges over X. That is, $F_n \xrightarrow{\tau_W} F$ if and only if $d(\cdot, F_n) \to d(\cdot, F)$ pointwise. For the Hausdorff metric topology, Corollary 3.79 asserts that $F_n \xrightarrow{\tau_h} F$ if and only if $d(\cdot, F_n) \to d(\cdot, F)$ uniformly. Thus τ_W is weaker than τ_h. The Wijsman topology agrees with the Hausdorff metric topology when X is compact. G. A. Beer [34] proves that the Wijsman topology on $\mathcal{F}$ is Polish whenever X is Polish. The method of proof is to embed $\mathcal{F}$ in $C_b(X)$ via $F \mapsto d(\cdot, F)$, except that in this construction $C_b(X)$ is endowed with the topology of pointwise convergence, not the topology of uniform convergence. See [34] for the somewhat intricate details.

The Hausdorff metric has another disturbing defect. Unless X is compact, the Hausdorff metric topology on $\mathcal{F}$ depends on the actual metric d, not just on the topology of X. That is, it may be that d and ρ are equivalent bounded metrics on X, so that the bounded closed sets are the same for both metrics, but h_d and h_ρ may not be equivalent metrics on $\mathcal{F}$. However, Theorem 3.91 below shows that the Hausdorff metric topology on $\mathcal{K}$ is topological, so this defect is not an issue if X is compact. Here is an example.

3.86 Example (Hausdorff metric is not topological)   Consider the bounded metrics d and ρ on $\mathbb{N}$ defined by

$$d(n,m) = \begin{cases} 0 & \text{if } n = m, \\ 1 & \text{if } n \neq m, \end{cases} \qquad \text{and} \qquad \rho(n,m) = \left| \tfrac{1}{n} - \tfrac{1}{m} \right|.$$

Both metrics generate the discrete topology on $\mathbb{N}$. Thus, $\mathcal{F}$ is just the collection of nonempty subsets of $\mathbb{N}$.

For each n, define $F_n = \{1, 2, \ldots, n\}$. It is easy to see that $\mathsf{h}_d(F_n, \mathbb{N}) = 1$ for all n. On the other hand, for $k \notin F_n$, we have $\rho(k, F_n) = \tfrac{1}{n} - \tfrac{1}{k}$. Consequently,

$$\mathsf{h}_\rho(F_n, \mathbb{N}) = \sup_k \rho(k, F_n) = \lim_{k \to \infty} \left(\tfrac{1}{n} - \tfrac{1}{k} \right) = \frac{1}{n}.$$

Thus, $F_n \xrightarrow{\mathsf{h}_\rho} \mathbb{N}$. So the Hausdorff metrics h_d and h_ρ are not equivalent.   ∎

This example made use of two metrics that generate different uniformities. If two equivalent bounded metrics generate the same uniformity, then the induced Hausdorff metrics are also equivalent. That is, the Hausdorff metric topology depends only on the uniformity induced by the metric.

3.87 Theorem   *Suppose X is metrizable with bounded compatible metrics d and ρ that generate the same uniformity $\mathcal{U}$. Then the corresponding Hausdorff metrics h_d and h_ρ are equivalent on $\mathcal{F}$.*

Proof: Let F belong to $\mathcal{F}$. It suffices to show that for every $\varepsilon > 0$, there is $\delta > 0$ so that the h_d-ball of radius 2ε at F includes the h_ρ-ball of radius δ at F. Let $U_d(\varepsilon) = \{(x, y) \in X \times X : d(x, y) < \varepsilon\}$ be an entourage in $\mathcal{U}$. Since ρ generates $\mathcal{U}$, there is some $\delta > 0$ with $U_\rho(2\delta) = \{(x, y) \in X \times X : \rho(x, y) < 2\delta\} \subset U_d(\varepsilon)$. Suppose now that $\mathsf{h}_\rho(F, C) < \delta$. Then by Lemma 3.71, $F \subset N_{2\delta}^\rho(C)$ and $C \subset N_{2\delta}^\rho(F)$. Now note that $N_\varepsilon^d(F) = \{y \in X : (x, y) \in U_d(\varepsilon) \text{ for some } x \in F\}$. Thus we see that $F \subset N_\varepsilon^d(C)$ and $C \subset N_\varepsilon^d(F)$, so $\mathsf{h}_d(C, F) \leqslant \varepsilon$. Thus $B_\delta^\rho(F) \subset B_{2\varepsilon}^d(F)$, as desired.   ∎

We now give conditions under which the collection $\mathcal{K}$ of nonempty compact sets is a closed subset of $\mathcal{F}$.

3.88 Theorem   *For a metric space (X, d):*

1. *The collection $\mathcal{F}_{tb}$ of all nonempty totally d-bounded closed sets is closed in $\mathcal{F}$.*

2. *If in addition X is d-complete, then the collection $\mathcal{K}$ of nonempty compact sets is closed in $\mathcal{F}$.*

Proof: (1)   Suppose F belongs to the closure of $\mathcal{F}_{tb}$ in $\mathcal{F}$. Let $\varepsilon > 0$. Pick some $C \in \mathcal{F}_{tb}$ with $h(C, F) < \varepsilon/2$. Since C is d-totally bounded, there is a finite subset $\{x_1, \ldots, x_m\}$ of X satisfying $C \subset \bigcup_{i=1}^{m} B_{\varepsilon/2}(x_i)$.

Now let x belong to F. From $d(x, C) \leqslant h(C, F) < \varepsilon/2$, it follows that there is some $c \in C$ satisfying $d(x, c) < \varepsilon/2$. Next select some i satisfying $d(x_i, c) < \varepsilon/2$, and note that $d(x, x_i) < \varepsilon$. Therefore, $x \in \bigcup_{i=1}^{m} B_\varepsilon(x_i)$, so $F \subset \bigcup_{i=1}^{m} B_\varepsilon(x_i)$. This shows that $F \in \mathcal{F}_{tb}$. Thus $\mathcal{F}_{tb}$ is h-closed in $\mathcal{F}$.

(2)   Since X is d-complete, so is every closed subset. Since every compact set is totally bounded, part (1) and Theorem 3.28 imply that the limit of any sequence of compact sets is also compact.   ∎

3.18   Topologies for spaces of subsets

The next result describes a topology on the power set of a topological space.

3.89 Definition   *For any nonempty subset A of a set X, define*

$$A^u = \{B \in 2^X \setminus \{\varnothing\} : B \subset A\} \quad and \quad A^\ell = \{B \in 2^X : A \cap B \neq \varnothing\}.$$

We also define $\varnothing^u = \varnothing^\ell = \varnothing$.

Clearly, $A^u = 2^A \setminus \{\varnothing\} \subset A^\ell$ for each nonempty subset A. Moreover, notice that $A^u \cap B^u = (A \cap B)^u$ and $(A \cap B)^\ell \subset A^\ell \cap B^\ell$ for all subsets A and B.

Let X be a topological space. The collection of sets of the form

$$G_0^u \cap G_1^\ell \cap \cdots \cap G_n^\ell$$

where $G_0, \ldots, G_n$ are open subsets of X, is closed under finite intersections. Since $X^u = X^\ell = 2^X \setminus \{\varnothing\}$, it thus forms a base for a topology on the power set of X. This topology is known variously as the **exponential topology**, e.g. [218, § 17], or the **Vietoris topology**, e.g., [158, 209]. We are most interested in the relativization of this topology to the space $\mathcal{F}$ of nonempty closed subsets or the space $\mathcal{K}$ of nonempty compact subsets of a metrizable space. In this case, the term Vietoris topology seems more common, so we shall denote the topology τ_V. For more general results see K. Kuratowski [218, § 17–18, 42–44].

3.90 Corollary **(Finite sets are dense)** *If D is a dense subset of a Hausdorff topological space X, then the set $\mathcal{D}$ of all finite subsets of D is dense in the Vietoris topology on 2^X. Consequently, if X is separable, then so are $(2^X, \tau_V)$ and $(\mathcal{F}, \tau_V)$.*

Proof: To see that $\mathcal{D}$ is dense, let $\mathcal{U} = G_0^u \cap G_1^\ell \cap \cdots \cap G_n^\ell$ be a nonempty basic open set in τ_V. It is easy to see that $G_0 \cap G_i \neq \varnothing$ for $i = 1, \ldots, n$. Since D is dense, for each $i = 1, \ldots, n$ there is some $x_i \in D$ belonging to $G_0 \cap G_i$. But then the finite (and closed) subset $\{x_1, \ldots, x_n\}$ of X belongs to $\mathcal{U}$. Therefore $\mathcal{D}$ is dense. To prove separability, note that if D is countable, then $\mathcal{D}$ is also countable. ∎

When X is metrizable, the Vietoris topology τ_V and the Hausdorff metric topology τ_h coincide when relativized to $\mathcal{K}$.

3.91 Theorem *Let X be a metrizable space, and let d be any compatible metric. Then the Vietoris topology and the Hausdorff metric topology coincide on $\mathcal{K}$, the space of nonempty compact subsets of X. Consequently, all compatible metrics on X generate the same Hausdorff metric topology on $\mathcal{K}$.*

Proof: We start by showing that for each open subset G of X, the sets G^u and G^ℓ are both open in the Hausdorff metric topology on $\mathcal{K}$. (Of course, this is relativized to $\mathcal{K}$, so $G^u = \{K \in \mathcal{K} : K \subset G\}$, etc.) Since $\varnothing^u = \varnothing^\ell = \varnothing$ and $X^u = X^\ell = \mathcal{K}$, we can suppose that G is a nonempty proper open subset of X. First, we establish that G^u and G^ℓ are τ_h-open subsets of $\mathcal{K}$.

Suppose first that $C \in G^u$. That is, the set C is compact and $C \subset G$. Put $\varepsilon = \min_{x \in C} d(x, G^c) > 0$. If $K \in B_\varepsilon(C)$, then $K \subset G$. That is, $B_\varepsilon(C) \subset G^u$. This shows that G^u is an open subset of $\mathcal{K}$.

Now suppose $C \in G^\ell$. That is, C is compact and $C \cap G \neq \varnothing$. Fix some $x \in C \cap G$. Then there exists some $\varepsilon > 0$ such that $B_\varepsilon(x) \subset G$. We claim that $B_\varepsilon(C) \subset G^\ell$. To see this, let $K \in B_\varepsilon(C)$. That is, K is compact and $h(C, K) < \varepsilon$. From $d(x, K) \leqslant h(C, K)$, it follows that there exists some $y \in K$ with $d(x, y) < \varepsilon$, so $y \in G$. That is, $G \cap K \neq \varnothing$, or in other words $K \in G^\ell$. Hence $B_\varepsilon(C) \cap \mathcal{K} \subset G^\ell$, which implies that G^ℓ is τ_h-open.

Next we show that any open ball in the Hausdorff metric topology is Vietoris-open. So let C be a nonempty compact subset of X, and let $\varepsilon > 0$. We need to show that there is some τ_V-open set $\mathcal{U}$ satisfying $C \in \mathcal{U} \subset B_\varepsilon(C)$.

To establish this, let $G_0 = N_{\varepsilon/2}(C) = \{x \in X : d(x, C) < \varepsilon/2\}$. Since C is compact, there is a finite subset $\{x_1, \ldots, x_n\}$ of C with $C \subset \bigcup_{i=1}^n B_{\varepsilon/2}(x_i)$. Put $G_i = B_{\varepsilon/2}(x_i)$ and then let $\mathcal{U} = G_0^u \cap G_1^\ell \cap \cdots \cap G_n^\ell$. Clearly, $C \in \mathcal{U}$. Now suppose that $K \in \mathcal{U}$. That is, K is a compact subset of X satisfying $K \subset G_0$ and $K \cap G_i \neq \varnothing$ for each $i = 1, \ldots, n$. From $K \subset G_0 = N_{\varepsilon/2}(C)$, we see that $\sup_{x \in K} d(x, C) < \varepsilon$. On the other hand, since each $x \in C$ belongs to some $G_i = B_{\varepsilon/2}(x_i)$, which contains points from K, we see that $\sup_{x \in C} d(x, K) < \varepsilon$. Therefore, $h(C, K) < \varepsilon$. Thus, $C \in \mathcal{U} \subset B_\varepsilon(C)$, and the proof is finished. ∎

There is a weakening of the Vietoris topology due to by J. M. G. Fell [122], called the **Fell topology**. It has a base given by sets of the form

$$(K^c)^u \cap G_1^\ell \cap \cdots \cap G_n^\ell,$$

where K is compact and $G_1, \ldots, G_n$ are open subsets of X.

3.92 Lemma *Let X be a locally compact Hausdorff topological space. Then the Fell topology on $\mathcal{F}$ is a Hausdorff topology.*

Proof: Let $F_1, F_2 \in \mathcal{F}$ satisfy $F_1 \neq F_2$. We can assume that there exists a point $x_0 \in F_1 \setminus F_2$. Pick an open neighborhood G of x_0 whose closure $K = \overline{G}$ is compact such that $K \cap F_2 = \varnothing$ (see Theorem 2.67). Set $\mathcal{U} = \{F \in \mathcal{F} : F \subset K^c\}$ and $\mathcal{V} = \{F \in \mathcal{F} : F \cap G \neq \varnothing\}$. Then $\mathcal{U}$ and $\mathcal{V}$ are open in the Fell topology, $F_2 \in \mathcal{U}$, $F_1 \in \mathcal{V}$, and $\mathcal{U} \cap \mathcal{V} = \varnothing$. ∎

For a locally compact Polish space the Fell topology is also called the **topology of closed convergence**, denoted τ_C. The reason is that (as we shall see in Corollary 3.95 below) in this case, closed limits are also limits in $(\mathcal{K}, \tau_C)$.

When the underlying space X is a compact metric space, the Hausdorff metric topology on $\mathcal{K} = \mathcal{F}$ coincides with the Fell topology and also with the Vietoris topology. In this case, the converse of Theorem 3.82 is true for the space $\mathcal{K}$. This is a consequence of the characterization of the Hausdorff metric topology in Theorem 3.91.

3.93 Theorem *If X is a compact metric space, then τ_C coincides with the Hausdorff metric topology, and*

$$K_n \xrightarrow{\tau_C} K \text{ in } \mathcal{F} (= \mathcal{K}) \quad \text{if and only if} \quad K = \operatorname{Li} K_n = \operatorname{Ls} K_n.$$

Proof: Let X be a compact metric space. Then $\mathcal{F} = \mathcal{K}$ and the Vietoris and Fell topologies coincide on $\mathcal{F}$ (since the complement of any open set is compact). So by Theorem 3.91 they agree with the Hausdorff metric topology on $\mathcal{F}$ for any compatible metric. It thus follows from Theorem 3.82 that if $K_n \xrightarrow{\tau_C} K$, then we have $K = \operatorname{Li} K_n = \operatorname{Ls} K_n$.

Now suppose K is a nonempty compact subset satisfying $K = \operatorname{Li} K_n = \operatorname{Ls} K_n$, where $\{K_n\} \subset \mathcal{K}$. To show that $K_n \to K$ in the topology of closed convergence, it suffices to prove that for every neighborhood of K of the form G^u and every neighborhood of the form G^ℓ, where G is open in X, eventually K_n lies in G^u and in G^ℓ.

So consider first the case that $K \in G^\ell$, where G is open. That is, $K \cap G \neq \varnothing$. Fix some $x \in K \cap G$. Then $x \in K = \operatorname{Li} K_n$ implies $G \cap K_n \neq \varnothing$ for all n sufficiently large. That is, $K_n \in G^\ell$ for all n sufficiently large.

Next consider the case that $K \in G^u$. That is, $K \subset G$, where $G \neq X$ is a nonempty open set. Since K is compact, the continuous function $x \mapsto d(x, G^c)$

attains its minimum over K, say $\min_{x\in K} d(x, G^c) = \varepsilon > 0$. Now we claim that $\bigcup_{n=m}^{\infty} K_n \subset G$ for some m. For if this is not the case, then for each m there exists some $x_m \in F_m = \overline{\bigcup_{n=m}^{\infty} K_n}$ with $x_m \notin G$, so $d(x_m, K) \geqslant \varepsilon$. If $x \in X$ is an accumulation point of the sequence $\{x_m\}$, then $d(x, K) \geqslant \varepsilon$ too. But since the F_ms are closed and nested, $x \in F_m$ for each m. That is, $x \in \bigcap_{n=1}^{\infty} \overline{\bigcup_{n=m}^{\infty} K_n} = \operatorname{Ls} K_n = K$, a contradiction. Thus for some m, if $n \geqslant m$, then $K_n \subset F_m \subset G$, so $K_n \in G^u$. This completes the proof. ∎

Example 3.83 shows that compactness of X is essential in the above theorem. Nevertheless, we can extend this analysis to the closed sets of a locally compact separable metrizable space X. By Corollary 3.45, the one-point compactification X_∞ of X is metrizable. Therefore, by Theorem 3.91, there is a topological characterization of the space $\mathcal{F}_\infty = \mathcal{K}_\infty$ of nonempty compact subsets of the one-point compactification X_∞. We use this to define a topology on $\mathcal{F}$ that depends only on the topology of X.

3.94 Lemma *Let X be a noncompact locally compact separable metrizable space. Let $\mathcal{F}$ denote the set of all nonempty closed subsets of X, and let $\mathcal{F}_\infty$ be the space of all nonempty closed subsets of X_∞ equipped with its Hausdorff metric topology. Then the mapping $\theta\colon (\mathcal{F}, \tau_C) \to \mathcal{F}_\infty = \mathcal{K}_\infty$, defined by*

$$\theta(F) = F \cup \{\infty\},$$

is an embedding of $(\mathcal{F}, \tau_C)$ as a closed subspace of $\mathcal{F}_\infty$.

Proof: In this proof, let the symbols A^u and A^ℓ be relativized to $\mathcal{K}_\infty$. That is, $A^u = \{K \in \mathcal{K}_\infty : K \subset A\}$, etc. Now note that $\theta(\mathcal{F}) = \{K \in \mathcal{K}_\infty : \infty \in K\}$. Consequently,

$$\mathcal{F}_\infty \setminus \theta(\mathcal{F}) = \{K \in \mathcal{K}_\infty : K \subset X\} = X^u.$$

But $X = X_\infty \setminus \{\infty\}$ is open in X_∞, so X^u is open in $\mathcal{K}_\infty$ by Theorem 3.91, which means that $\theta(\mathcal{F})$ is closed (and hence compact) in $\mathcal{K}_\infty$.

Clearly θ is one-to-one. We claim that it is an embedding. By Theorem 2.36, it is enough to show that θ is an open mapping.

It suffices to show that θ carries every basic set for τ_C to an open set in $\mathcal{K}_\infty$. But this follows from Theorem 3.91 by observing that for each basic τ_C-open set

$$\mathcal{U} = \{F \in \mathcal{F} : F \subset K^c \text{ and } F \cap G_i \neq \varnothing, \ i = 1, \ldots, n\},$$

we have $\theta(\mathcal{U}) = (X_\infty \setminus K)^u \cap G_1^\ell \cap \cdots \cap G_n^\ell$. ∎

And now here is the basic theorem concerning the topology of closed convergence for locally compact separable metrizable spaces.

3.95 Corollary (**Closed convergence in $\mathcal{F}$**)   *If X is a locally compact Polish space, then $(\mathcal{F}, \tau_C)$ is compact and metrizable.*

Moreover, $F_n \xrightarrow{\tau_C} F$ if and only if $F = \mathrm{Li}\, F_n = \mathrm{Ls}\, F_n$.

Proof: If X is compact, then this is Theorem 3.91. So assume that X is not compact. By Theorem 3.85 (3), the space $\mathcal{K}_\infty$ is compact and metrizable, and so is the closed subspace $\theta(\mathcal{F})$, which is (by Lemma 3.94) homeomorphic to $\mathcal{F}$.

Now assume that a sequence $\{F_n\}$ in $\mathcal{F}$ satisfies $F = \mathrm{Li}\, F_n = \mathrm{Ls}\, F_n$ for some $F \in \mathcal{F}$. We shall show that $F_n \xrightarrow{\tau_C} F$ in $\mathcal{F}$. Let K be a compact subset of X such that $F \subset K^c$. We claim that $F_n \subset K^c$ for all n sufficiently large. For if this were not the case, then $F_n \cap K \neq \varnothing$ for infinitely many n. Since K is compact, it follows that there exists some $x \in K \cap \mathrm{Ls}\, F_n = K \cap F \subset K \cap K^c = \varnothing$, a contradiction. On the other hand, if $x \in F \cap G = (\mathrm{Li}\, F_n) \cap G$ for some open set G, then G is an open neighborhood of x, so $G \cap F_n \neq \varnothing$ for all n sufficiently large. The above show that if $\mathcal{U}$ is a basic neighborhood for τ_C, then $F_n \in \mathcal{U}$ for all n sufficiently large. That is, $F_n \xrightarrow{\tau_C} F$.

For the converse, assume that $F_n \xrightarrow{\tau_C} F$ in $\mathcal{F}$. Then $\theta(F_n) \to \theta(F)$ in K_∞, so by Theorem 3.82 we have $\mathrm{Li}\, \theta(F_n) = \mathrm{Ls}\, \theta(F_n) = \theta(F)$. Now the desired conclusion follows from the identities $\mathrm{Li}\, F_n = X \cap \mathrm{Li}\, \theta(F_n)$ and $\mathrm{Ls}\, F_n = X \cap \mathrm{Ls}\, \theta(F_n)$. ∎

As an aside, Corollary 3.95 easily shows that in Example 3.83,

$$\{1, 2, \ldots, n\} \xrightarrow[n \to \infty]{\tau_C} \mathbb{N}.$$

3.19   The space $C(X, Y)$

In this section we discuss the topology of uniform convergence of functions on a compact topological space. So fix a compact space X and a metrizable space Y. Let $C(X, Y)$ denote the set of all continuous functions from X to Y. That is,

$$C(X, Y) = \{f \in Y^X : f \text{ is continuous}\}.$$

If ρ is a compatible metric on Y, then the formula

$$d_\rho(f, g) = \sup_{x \in X} \rho(f(x), g(x))$$

defines a metric on $C(X, Y)$. The verification of the metric properties are straightforward. Since X is compact, we have $d_\rho(f, g) < \infty$ for each $f, g \in C(X, Y)$. Thus, we have the following result.

3.96 Lemma (**Metrizability of $C(X, Y)$**)   *If X is a compact space, Y is a metrizable space, and ρ is a compatible metric on Y, then $(C(X, Y), d_\rho)$ is a metric space.*

This metric characterizes the **topology of d-uniform convergence** on X of functions in $C(X, Y)$. Since d-uniform convergence of a sequence of functions implies pointwise convergence, the topology of uniform convergence is stronger than the topology of pointwise convergence (Lemma 2.50).

The next result characterizes the completeness of $(C(X, Y), d_\rho)$.

3.97 Lemma (Completeness of $C(X, Y)$)   *Let X be a compact space, let Y be a metrizable space, and let ρ be a compatible metric on Y. Then the metric space $(C(X, Y), d_\rho)$ is d_ρ-complete if and only if Y is ρ-complete.*

Proof: For simplicity, write d for d_ρ. Assume first that $(C(X, Y), d)$ is d-complete, and let $\{y_n\}$ be a ρ-Cauchy sequence in Y. For each n consider the constant function $f_n(x) = y_n$ for each $x \in X$. Then $\{f_n\}$ is a d-Cauchy sequence, so there exists a function $f \in C(X, Y)$ such that $d(f_n, f) \to 0$. Now for each $x_0 \in X$, we have $\rho(y_n, f(x_0)) \leqslant d(f_n, f) \to 0$. That is, $y_n \to f(x_0)$, so Y is ρ-complete.

Conversely, suppose that Y is ρ-complete, and let $\{f_n\}$ be a d-Cauchy sequence in $C(X, Y)$. Then, for each $\varepsilon > 0$ there exists some n_0 such that $\rho(f_n(x), f_m(x)) < \varepsilon$ for each $x \in X$ and all $n, m \geqslant n_0$. In other words, $\{f_n(x)\}$ is a ρ-Cauchy sequence in Y. If $\rho(f_n(x), f(x)) \to 0$ for each $x \in X$, then (as in the proof of Theorem 2.65) we see that $f \in C(X, Y)$ and $d(f_n, f) \to 0$. ∎

The next result shows that the topology on $C(X, Y)$ induced by d_ρ depends only on the topology of Y, not on the particular metric ρ. As a result, we can view $C(X, Y)$ as a topological space without specifying a metric for Y, and we can refer simply to the **topology of uniform convergence** on $C(X, Y)$.

3.98 Lemma (Equivalent metrics on $C(X, Y)$)   *Let X be a compact space and let Y be a metrizable space. If ρ_1 and ρ_2 are compatible metrics on Y, then d_{ρ_1} and d_{ρ_2} are equivalent metrics on $C(X, Y)$. That is, d_{ρ_1} and d_{ρ_2} generate the same topology on $C(X, Y)$.*

Proof: Let ρ_1 and ρ_2 be two compatible metrics on Y. Also, let a sequence $\{f_n\}$ in $C(X, Y)$ satisfy $d_{\rho_1}(f_n, f) \to 0$ for some $f \in C(X, Y)$. To complete the proof, it suffices to show that $d_{\rho_2}(f_n, f) \to 0$.

To this end, assume by way of contradiction that $d_{\rho_2}(f_n, f) \not\to 0$. So by passing to a subsequence if necessary, we can suppose that there exists some $\varepsilon > 0$ such that $d_{\rho_2}(f_n, f) > \varepsilon$ for each n. Next, pick a sequence $\{x_n\}$ in X satisfying $\rho_2(f_n(x_n), f(x_n)) > \varepsilon$ for each n. The compactness of X guarantees the existence of a subnet $\{x_{n_\alpha}\}$ of the sequence $\{x_n\}$ such that $x_{n_\alpha} \to x$ holds in X. Since $f \in C(X, Y)$, we see that $f(x_{n_\alpha}) \to f(x)$. This implies $\rho_1(f(x_{n_\alpha}), f(x)) \to 0$ and $\rho_2(f(x_{n_\alpha}), f(x)) \to 0$. Moreover, from

$$\rho_1(f_{n_\alpha}(x_{n_\alpha}), f(x)) \leqslant \rho_1(f_{n_\alpha}(x_{n_\alpha}), f(x_{n_\alpha})) + \rho_1(f(x_{n_\alpha}), f(x))$$
$$\leqslant d_{\rho_1}(f_{n_\alpha}, f) + \rho_1(f(x_{n_\alpha}), f(x)) \to 0$$

and the equivalence of ρ_1 and ρ_2, we see that $\rho_2(f_{n_\alpha}(x_{n_\alpha}), f(x)) \to 0$. But then

$$0 < \varepsilon < \rho_2(f_{n_\alpha}(x_{n_\alpha}), f(x_{n_\alpha})) \leqslant \rho_2(f_{n_\alpha}(x_{n_\alpha}), f(x)) + \rho_2(f(x_{n_\alpha}), f(x)) \to 0,$$

which is impossible, and the proof is finished.   ■

From now on in this section $C(X, Y)$ is endowed with the topology of uniform convergence. It is worth noting that if Y is a normed space, then under the usual pointwise algebraic operations, $C(X, Y)$ is a vector space that becomes a normed space under the norm $\|f\| = \sup_{x \in X} \|f(x)\|$. If Y is a Banach space, then Lemma 3.97 shows that $C(X, Y)$ is a Banach space too.

3.99 Lemma (Separability of $C(X, Y)$)   *If X is compact and metrizable, and Y is separable and metrizable, then the metrizable space $C(X, Y)$ is separable.*

Proof: Fix compatible metrics ρ_1 for X and ρ for Y, respectively, and let $d = d_\rho$ denote the metric generating the topology on $C(X, Y)$. Since a metrizable space is separable if and only if it is second countable, it suffices to show that $C(X, Y)$ has a countable base.

For each compact subset K of X and each open subset V of Y let

$$U_{K,V} = \{f \in C(X, Y) : f(K) \subset V\}.$$

We claim that each $U_{K,V}$ is an open subset of $C(X, Y)$.

To see this, let h belong to $U_{K,V}$. Then $h(K) \subset V$, so for each point $x \in K$ there is some $\varepsilon_x > 0$ such that the ball $B_{2\varepsilon_x}(h(x))$ is included in V. Since $h(K)$ is compact and $h(K) \subset \bigcup_{x \in K} B_{\varepsilon_x}(h(x))$, there is a finite subset $\{x_1, \ldots, x_n\}$ of K such that $h(K) \subset \bigcup_{i=1}^{n} B_{\varepsilon_{x_i}}(h(x_i))$. Let $\varepsilon = \min\{\varepsilon_{x_1}, \ldots, \varepsilon_{x_n}\}$. Now assume that $g \in C(X, Y)$ satisfies $d(h, g) < \varepsilon$. Then, given $x \in K$ pick some i satisfying $\rho(h(x), h(x_i)) < \varepsilon_{x_i}$ and note that the inequalities

$$\rho(g(x), h(x_i)) \leqslant \rho(g(x), h(x)) + \rho(h(x), h(x_i)) < 2\varepsilon_{x_i}$$

imply $g(x) \in B_{2\varepsilon_{x_i}}(h(x_i)) \subset V$, so $g(K) \subset V$. Therefore $B_\varepsilon(h) \subset U_{K,V}$, proving that $U_{K,V}$ is an open subset of $C(X, Y)$.

Next, fix a countable dense subset $\{z_1, z_2, \ldots\}$ of X and let $\{C_1, C_2, \ldots\}$ be an enumeration of the countable collection of closed (hence compact) ρ_1-balls with centers at the points z_i and rational radii. Now pick a countable base $\{V_1, V_2, \ldots\}$ for the topology on Y. To finish the proof, we establish that the countable collection of all finite intersections of the open sets U_{C_i, V_j} $(i, j = 1, 2, \ldots)$ is a base for the topology on $C(X, Y)$.

To this end, let W be an open subset of $C(X, Y)$ and let $f \in W$. Pick $\delta > 0$ so that $B_{2\delta}(f) = \{g \in C(X, Y) : d(f, g) < 2\delta\} \subset W$. Next, write $Y = \bigcup_{n=1}^{\infty} W_n$, where each $W_n \in \{V_1, V_2, \ldots\}$ and has ρ-diameter less than δ. Subsequently, we can write each $f^{-1}(W_n)$ as a union of open ρ_1-balls having centers at appropriate

z_i and rational radii such that the corresponding closed balls with the same centers and radii also lie in $f^{-1}(W_n)$. From $X = \bigcup_{n=1}^{\infty} f^{-1}(W_n)$ and the compactness of X, we infer that there exists a finite collection $C_{m_1}, \ldots, C_{m_k}$ of these closed balls satisfying $X = \bigcup_{i=1}^{k} C_{m_i}$. For each i choose some ℓ_i such that $C_{m_i} \subset f^{-1}(V_{\ell_i})$.

Now let $g \in \bigcap_{i=1}^{k} U_{C_{m_i}, V_{\ell_i}}$. For $x \in X$, choose some i with $x \in C_{m_i}$, and note that $f(x), g(x) \in V_{\ell_i}$. Since V_{ℓ_i} has ρ-diameter less than δ, we have $\rho(f(x), g(x)) < \delta$. Hence $d(f, g) \leqslant \delta < 2\delta$, which implies $g \in B_{2\delta}(f) \subset W$. As a consequence, $f \in \bigcap_{i=1}^{k} U_{C_{m_i}, V_{\ell_i}} \subset W$, and the proof is finished. ∎

The metrizable space $C(X, Y)$ need not be compact even if both X and Y are compact metric spaces.

3.100 Example (C(X, Y) is not compact)   Let $X = Y = [0, 1]$ and consider the sequence $\{f_n\}$ in $C(X, Y)$ defined by $f_n(x) = x^n$. Then $\{f_n\}$ converges pointwise to the discontinuous function f defined by $f(1) = 1$ and $f(x) = 0$ for $0 \leqslant x < 1$. This implies that $\{f_n\}$ does not have any uniformly convergent subsequence in $C(X, Y)$, so the Polish space $C([0, 1], [0, 1])$ is not compact. ∎

Chapter 4

Measurability

A major motivation for studying measurable structures is that they are at the foundations of probability and statistics. Suppose we wish to assign probabilities to various *events*. Given events A and B it is natural to consider the events "A and B," "A or B," and the event "not A." If we model events as sets of *states of the world*, then the family of events should be closed under intersections, unions, and complements. It should also include the set of all states of the world. Such a family of sets is called an *algebra* of sets. If we also wish to discuss the "law of averages," which has to do with the average behavior over an infinite sequence of trials, then it is useful to add closure under countable intersections to our list of desiderata. An algebra that is closed under countable intersections is a σ-*algebra*. A set equipped with a σ-algebra of subsets is a *measurable space* and elements of this σ-algebra are called *measurable sets*. In Chapter 10, we discuss the measurability of sets with respect to a *measure*. In that chapter, we show that a measure μ induces a σ-algebra of μ-measurable sets. The reason we do not start with a measure here is that in statistical decision theory events have their own interpretation independent of any measure, and since probability is a purely subjective notion, there is no "correct" measure that deserves special stature in defining measurability.

The first part of this chapter deals with the properties of algebras, σ-algebras, and the related classes of *semirings*, *monotone classes*, and *Dynkin systems*. This means that the ratio of definitions to results is uncomfortably high in this chapter, but these concepts are necessary. The major result in this area is Dynkin's Lemma 4.11. Semirings are important because the class of *measurable rectangles* in a product of measurable spaces is a semiring (Lemma 4.42). The σ-algebra generated by the collection of measurable rectangles is called the *product σ-algebra*.

When the underlying space has a topological structure, we may wish all the open and closed sets to be measurable. The smallest σ-algebra of sets that contains all open sets is called the *Borel σ-algebra* of the topological space. Corollaries 4.15, 4.16, and 4.17 give other characterizations of the Borel algebra. Unless otherwise specified, we view every topological space as measurable space where the σ-algebra of measurable sets is the Borel σ-algebra. The product σ-algebra of two Borel σ-algebras is the Borel σ-algebra of the product topology provided both spaces are second countable (Theorem 4.44).

A function between measurable spaces is a *measurable function* if for every measurable set in its range, the inverse image is a measurable set in the domain. (In probability theory, real-valued measurable functions are known as *random variables*.) Section 4.5 deals with properties of measurable functions: A measurable function from a measurable space into a second countable Hausdorff space (with its Borel σ-algebra) has a graph that is measurable in the product σ-algebra (Theorem 4.45). When the range space is the set of real numbers (with the Borel σ-algebra), the class of measurable functions is a vector lattice of functions closed under pointwise limits of sequences (Theorem 4.27). (It is not generally closed under pointwise limits of nets.) If the range space is metrizable, then the class of measurable functions is closed under pointwise limits (Lemma 4.29). Also, when the range is separable and metrizable, a function is measurable if and only if it is the pointwise limit of a sequence of *simple* measurable functions. This result cannot be generalized too far. Example 4.31 presents a pointwise convergent sequence of Borel measurable functions from a compact metric space (the unit interval) into a compact (nonmetrizable) Hausdorff space whose limit is not Borel measurable. For separable metrizable spaces, the class of bounded Borel measurable real functions is obtained by taking monotone limits of bounded continuous real functions (Theorem 4.33).

A *Carathéodory function* is a function from the product of a measurable space S and a topological space X into a topological space Y that is measurable in one variable and continuous in the other. If the topological spaces are metrizable, then under certain conditions a Carathéodory function is *jointly measurable*, that is, measurable with respect to the product σ-algebra on $S \times X$ (Theorem 4.51). Under stronger conditions (Theorem 4.55) Carathéodory functions characterize the measurable functions from S to $C(X, Y)$ (continuous functions from X to Y).

For Polish spaces, there are some remarkable results concerning Borel sets that are related to the Baire space $\mathcal{N} = \mathbb{N}^{\mathbb{N}}$. Given a Polish space and a Borel subset, there is a stronger Polish topology (generating the same Borel σ-algebra) for which the given Borel set is actually closed (Lemma 4.56). Similarly given a Borel measurable function from a Polish space into a second countable space there is a stronger Polish topology (generating the same Borel σ-algebra) for which the given function is actually continuous. This means that for many proofs we may assume that a Borel set is actually closed or that a Borel measurable function is actually continuous. We use this technique to show every Borel subset of a Polish space is the *one-to-one* continuous image of a closed subset of $\mathcal{N}$ (Theorem 4.60).

It is easy to see that every function f into a measurable space defines a smallest σ-algebra $\sigma(f)$ on its domain for which it is measurable. Theorem 4.41 asserts that a real-valued function is $\sigma(f)$-measurable if and only if it can be written as a function of f. It is also easy to see that every continuous function between topological spaces is Borel measurable (Corollary 4.26). But what is the smallest σ-algebra for which every continuous function is measurable? In general, this σ-algebra is smaller than the Borel σ-algebra, and is called the *Baire σ-algebra*.

Example 4.66 gives a dramatic example of the difference. The Baire σ-algebra can be missing some very important sets. But for locally compact Polish spaces (such as the Euclidean space $\mathbb{R}^n$), the two σ-algebras coincide (Lemma 4.65). The Baire σ-algebra figures prominently in the classical representation of certain positive functionals as integrals, see, e.g., Theorem 14.16.

4.1 Algebras of sets

We start by describing algebras and σ-algebras, which are the nicest families of sets that we deal with in connection with measure theory and probability. If we think of random events as being described by sentences, it makes sense to consider connecting these sentences with "and," "or," and "not" to make new events. These correspond to the set operations of intersection, union, and complementation. Algebras are families that are closed under these operations.

4.1 Definition   *A nonempty family $\mathcal{A}$ of subsets of a set X is an **algebra** of sets if it is closed under finite unions and complementation. That is,*

$$A, B \in \mathcal{A} \quad \implies \quad [A \cup B \in \mathcal{A} \quad and \quad A^c = X \setminus A \in \mathcal{A}].$$

A σ-algebra is an algebra that is also closed under countable unions. That is, $\{A_n\} \subset \mathcal{A}$ implies $\bigcup_{n=1}^{\infty} A_n \in \mathcal{A}.$ [1]

In probability theory, an algebra of sets is often called a **field**, and a σ-algebra is then a σ**-field**. Some French authors use the term *tribu* for a σ-field and it is sometimes translated as "tribe."

Clearly, every algebra $\mathcal{A}$ contains $\varnothing$ and X. Indeed, since $\mathcal{A}$ is nonempty, there exists some $A \in \mathcal{A}$, so $A^c \in \mathcal{A}$. Hence, $X = A \cup A^c \in \mathcal{A}$ and $\varnothing = X^c \in \mathcal{A}$. Thus the simplest example of an algebra, indeed of a σ-algebra, is $\{\varnothing, X\}$, which is the smallest (with respect to inclusion) algebra of subsets of X. The largest possible algebra (or σ-algebra) of subsets of X is 2^X, the collection of all subsets of X.

Every algebra is closed under finite intersections and every σ-algebra is closed under countable intersections. As a matter of fact, when a nonempty family $\mathcal{A}$ of subsets of a set X is closed under complementation, then $\mathcal{A}$ is an algebra (resp. a σ-algebra) if and only if it is closed under finite intersections (resp. countable intersections). These claims easily follow from de Morgan's laws.

Every nonempty collection $\mathcal{C}$ of subsets of a set X is included in the σ-algebra 2^X. It is also clear that the intersection of any nonempty family of σ-algebras is a σ-algebra. Therefore, the intersection of all σ-algebras that include $\mathcal{C}$ is the smallest σ-algebra including $\mathcal{C}$. This σ-algebra is called, as you might expect, the

---

[1] The σ in this definition is a mnemonic for (infinite) sequences.

σ-algebra generated by $\mathcal{C}$.[2] The σ-algebra generated by $\mathcal{C}$ is denoted $\sigma(\mathcal{C})$. In other words,

$$\sigma(\mathcal{C}) = \bigcap \{\mathcal{A} \subset 2^X : \mathcal{C} \subset \mathcal{A} \text{ and } \mathcal{A} \text{ is a } \sigma\text{-algebra}\}.$$

Notice that if $\mathcal{A} = \sigma(\mathcal{C})$ and $\mathcal{F} = \{A^c : A \in \mathcal{C}\}$, then $\sigma(\mathcal{F}) = \mathcal{A}$ too.

The σ-algebra generated by a family is characterized as follows.

4.2 Theorem   *If $\mathcal{C}$ is a nonempty collection of subsets of a set X, then $\sigma(\mathcal{C})$ is the smallest family $\mathcal{A}$ of subsets of X that includes $\mathcal{C}$ and satisfies:*

  i. *if $A \in \mathcal{C}$, then $A^c \in \mathcal{A}$,*

  ii. *$\mathcal{A}$ is closed under countable intersections, and*

  iii. *$\mathcal{A}$ is closed under countable disjoint unions.*

Before we present its proof, let us consider why this theorem is nontrivial. Note first that property (i) does not say that $\mathcal{A}$ is closed under complementation. It says that $\mathcal{A}$ includes the complements of sets in $\mathcal{C}$. Also, (iii) does not imply that $\mathcal{A}$ is closed under countable unions. Here is a simple example.

4.3 Example   (Disjoint unions vs. unions)   Consider the countable family $\mathcal{C} = \{\{1, n\} \subset \mathbb{N} : n > 1\}$ of subsets of the natural numbers $\mathbb{N}$. Since no pair of elements of $\mathcal{C}$ is disjoint, it is vacuously closed under countable disjoint unions. On the other hand, it is not closed under countable unions, since $\mathbb{N} = \bigcup \mathcal{C}$ is itself a countable union, and $\mathbb{N} \notin \mathcal{C}$.   ∎

Thus it is conceivable that a set could satisfy (i)–(iii), yet not be a σ-algebra. Indeed, let $X = \{0, 1\}$, $\mathcal{C} = \{\{0\}\}$, and $\mathcal{A} = \{\{1\}\}$. Then $\mathcal{A}$ satisfies (i)–(iii), but it does not include $\mathcal{C}$, and is not an algebra. In particular, $\mathcal{A}$ is not closed under complementation.

Proof of Theorem 4.2: Let $\mathcal{A}$ be the smallest family of sets that includes $\mathcal{C}$ and satisfies (i)–(iii). (Note that such a smallest family exists, as the family of all subsets of X has these properties, and the intersection of an arbitrary set of families with these properties also has these properties.) Since $\sigma(\mathcal{C})$ also satisfies (i)–(iii), we have $\mathcal{A} \subset \sigma(\mathcal{C})$. Let $\mathcal{F} = \{A \in \mathcal{A} : A^c \in \mathcal{A}\}$. Then $\mathcal{F}$ is closed under complementation, and by (i) we have $\mathcal{C} \subset \mathcal{F} \subset \mathcal{A}$.

It suffices to show that $\mathcal{F}$ is a σ-algebra. For then $\sigma(\mathcal{C}) \subset \mathcal{F}$, and therefore $\mathcal{A} = \mathcal{F} = \sigma(\mathcal{C})$ (since $\mathcal{F} \subset \mathcal{A} \subset \sigma(\mathcal{C})$). We do this in steps.

---

[2] In fact, let P denote any set of properties for a family of subsets of X. (The set P of properties might define the class of σ-algebras, or it might define monotone classes, or it might define a kind of class for which we have not coined a name.) Let $\mathcal{C}$ be a family of subsets of X. When we refer to the family of subsets satisfying P generated by $\mathcal{C}$, we mean the unique family $\mathcal{F}$ satisfying P and also (i) $\mathcal{C} \subset \mathcal{F}$, and (ii) if $\mathcal{E}$ satisfies P and $\mathcal{C} \subset \mathcal{E}$, then $\mathcal{F} \subset \mathcal{E}$.

If such smallest family exists, it is often $\bigcap \{\mathcal{E} \subset 2^X : \mathcal{C} \subset \mathcal{E} \text{ and } \mathcal{E} \text{ satisfies } P\}$. But there are certain classes for which such a smallest member may fail to exist. See the discussion of semirings below.

Step I: If $A, B \in \mathcal{F}$, then $A \setminus B \in \mathcal{F}$.

Let $A, B \in \mathcal{F}$. Then $A^c, B^c \in \mathcal{F}$. Since the family $\mathcal{A}$ is closed under countable intersections, we see that $A \setminus B = A \cap B^c \in \mathcal{A}$. But $\mathcal{A}$ is also closed under countable disjoint unions, so from the identity $(A \setminus B)^c = A^c \cup (A \cap B)$ we have $(A \setminus B)^c \in \mathcal{A}$. Therefore, $A \setminus B \in \mathcal{F}$.

Step II: The family $\mathcal{F}$ is closed under finite unions, and so an algebra.

Let $A, B \in \mathcal{F}$. This means that A, B, A^c, and B^c all belong to $\mathcal{A}$. Clearly, $(A \cup B)^c = A^c \cap B^c \in \mathcal{A}$. From and Step I and the disjoint union

$$(A \setminus B) \cup (A \cap B) \cup (B \setminus A) = A \cup B$$

we see that $A \cup B \in \mathcal{A}$. Therefore, $A \cup B \in \mathcal{F}$.

Step III: The algebra $\mathcal{F}$ is a σ-algebra.

Let $\{A_n\}$ be a sequence in $\mathcal{F}$. Define a sequence of pairwise disjoint sets recursively by $B_1 = A_1$, and $B_n = A_n \setminus (A_1 \cup \cdots \cup A_{n-1})$ for $n > 1$. Since $\mathcal{F}$ is an algebra of sets, each B_n belongs to $\mathcal{F}$, and by (iii) the countable disjoint union $\bigcup_{n=1}^{\infty} B_n = \bigcup_{n=1}^{\infty} A_n$ belongs to $\mathcal{A}$. Thus, we have shown that if $\{A_n\} \subset \mathcal{F}$, then $\bigcup_{n=1}^{\infty} A_n \in \mathcal{A}$.

Now let $\{A_n\} \subset \mathcal{F}$. Then by the preceding argument, $\bigcup_{n=1}^{\infty} A_n \in \mathcal{A}$. Moreover, since $\{A_n^c\} \subset \mathcal{A}$ and $\mathcal{A}$ is closed under countable intersections, we have $\left(\bigcup_{n=1}^{\infty} A_n\right)^c = \bigcap_{n=1}^{\infty} A_n^c \in \mathcal{A}$. Thus $\bigcup_{n=1}^{\infty} A_n \in \mathcal{F}$, so $\mathcal{F}$ is a σ-algebra. ∎

4.2 Rings and semirings of sets

While the class of σ-algebras, or at least algebras, captures the properties that we want the family of "events" to have, it is sometimes easier, especially when describing a measure, to start with a family of sets that has less structure and look at the σ-algebra it generates. That is the object of the Carathéodory Extension Procedure 10.23 in Chapter 10. In fact many mathematicians work with a measure theory where the underlying family of events is a ring.

4.4 Definition   *A nonempty collection $\mathcal{R}$ of subsets of a set X is a **ring** if it is closed under pairwise unions and relative complementation. That is,*

$$A, B \in \mathcal{R} \quad \Longrightarrow \quad [A \cup B \in \mathcal{R} \quad and \quad A \setminus B \in \mathcal{R}].$$

*A **σ-ring** is a ring that is also closed under countable unions. That is, $\{A_n\} \subset \mathcal{R}$ implies $\bigcup_{n=1}^{\infty} A_n \in \mathcal{R}$.*

Since a ring $\mathcal{R}$, being nonempty by definition, contains some set A, it follows that $\varnothing = A \setminus A \in \mathcal{R}$, so the empty set belongs to every ring. Thus the simplest example of a ring, in fact a σ-ring, is just $\{\varnothing\}$. From the identities

$$A \cap B = A \setminus (A \setminus B) \quad \text{and} \quad A \triangle B = (A \setminus B) \cup (B \setminus A),$$

we see that every ring is closed under pairwise intersections and symmetric differences. On the other hand, from the identities

$$A \cup B = (A \bigtriangleup B) \bigtriangleup (A \cap B) \quad \text{and} \quad A \setminus B = A \bigtriangleup (A \cap B),$$

it follows that a nonempty family $\mathcal{R}$ of subsets of a set X that is closed under symmetric differences and finite intersections is a ring. In other words, a nonempty family $\mathcal{R}$ of subsets of a set X is a ring if and only if it is closed under symmetric differences and finite intersections.[3]

Every algebra is a ring, but a ring need not contain X, and so may fail to be an algebra. A ring that does contain X is an algebra. A σ-ring is always closed under countable intersections. To see this, let $\{A_n\}$ be a sequence in a σ-ring $\mathcal{R}$ and let $A = \bigcap_{n=1}^{\infty} A_n$. Then $A_1 \setminus A = \bigcup_{n=1}^{\infty}(A_1 \setminus A_n) \in \mathcal{R}$, so $A = A_1 \setminus (A_1 \setminus A) \in \mathcal{R}$.

If $\mathcal{R}$ is a ring, then the collection $\{A \subset X : A \in \mathcal{R} \text{ or } A^c \in \mathcal{R}\}$ is an algebra. We leave it to you to verify that any nonempty family $\mathcal{C}$ of subsets of X generates a smallest ring and a smallest σ-ring that includes it.

We now turn to collections of sets that are slightly less well behaved than rings, but which arise naturally in the study of Cartesian products and the theory of integration.

4.5 Definition   *A* **semiring** *$\mathcal{S}$ is a nonempty family of subsets of a set X satisfying the properties.*

1. $\varnothing \in \mathcal{S}$.

2. *If $A, B \in \mathcal{S}$, then $A \cap B \in \mathcal{S}$.*

3. *If $A, B \in \mathcal{S}$, then there exist pairwise disjoint sets $C_1, \ldots, C_n \in \mathcal{S}$ such that* $A \setminus B = \bigcup_{i=1}^{n} C_i$.

Any family of pairwise disjoint subsets of a set together with the empty set (in particular any partition of a set together with the empty set) is automatically a semiring. Another important example of a semiring is the collection $\mathcal{S}$ of all half-open rectangles in $\mathbb{R}^n$ defined by

$$\mathcal{S} = \{[a_1, b_1) \times \cdots \times [a_n, b_n) : a_i, b_i \in \mathbb{R} \text{ for each } i = 1, \ldots, n\},$$

---

[3] Rings of sets are commutative rings in the algebraic sense, where symmetric difference is addition and intersection is multiplication. That is, $(\mathcal{R}, \bigtriangleup)$ is an Abelian group under addition: (i) $\bigtriangleup$ is associative, $A \bigtriangleup (B \bigtriangleup C) = (A \bigtriangleup B) \bigtriangleup C = \{x \in X : x \text{ belongs to exactly one of } A, B, C\}$. (ii) $\varnothing$ is a zero, $A \bigtriangleup \varnothing = A$. (iii) Every A has an inverse, since $A \bigtriangleup A = \varnothing$. (iv) $\bigtriangleup$ is commutative, $A \bigtriangleup B = B \bigtriangleup A$. Being a ring further requires: (v) $\cap$ is associative, $A \cap (B \cap C) = (A \cap B) \cap C$. (vi) The distributive law, $A \cap (B \bigtriangleup C) = (A \cap B) \bigtriangleup (A \cap C)$. For a commutative ring we need: (vii) $\cap$ is commutative, $A \cap B = B \cap A$. (This definition of commutative ring is that of I. N. Herstein [157, pp. 83–84], but other definitions of ring are in use, see e.g., S. MacLane and G. Birkhoff [238, p. 85].)

When X belongs to $\mathcal{R}$, then X is a unit, $A \cap X = X \cap A = A$. Unfortunately, even in this case $\mathcal{R}$ is not an algebraist's field, since $A \cap A^c = \varnothing$. (Unless $\mathcal{R} = \{\varnothing, X\}$.)

where $[a_i, b_i) = \varnothing$ if $b_i \leqslant a_i$. The collection $\mathcal{S}$ is a semiring but not a ring. This semiring plays an important role in the theory of Lebesgue measure on $\mathbb{R}^n$.

One of the useful properties of semirings is this: If $\mathcal{S}_X$ and $\mathcal{S}_Y$ are semirings of subsets of X and Y, respectively, then the family of rectangles

$$\{A \times B : A \in \mathcal{S}_X \text{ and } B \in \mathcal{S}_Y\}$$

is a semiring of subsets of $X \times Y$ called the **product semiring** (Lemma 4.42 below). The product semiring is denoted $\mathcal{S}_X \times \mathcal{S}_Y$. Do not confuse this with the Cartesian product $\{(A, B) : A \in \mathcal{S}_X \text{ and } B \in \mathcal{S}_Y\}$. Even if $\mathcal{A}_1$ and $\mathcal{A}_2$ are σ-algebras, their product $\mathcal{A}_1 \times \mathcal{A}_2$ need not be an algebra, although it is always a semiring.

Unlike the other kinds of classes of families of sets we have described, the intersection of a collection of semirings need not be a semiring. For example, let $X = \{0, 1, 2\}$, $\mathcal{S}_1 = \{\varnothing, X, \{0\}, \{1\}, \{2\}\}$, and $\mathcal{S}_2 = \{\varnothing, X, \{0\}, \{1, 2\}\}$. Then $\mathcal{S}_1$ and $\mathcal{S}_2$ are semirings (in fact, $\mathcal{S}_2$ is an algebra), but their intersection

$$\mathcal{C} = \mathcal{S}_1 \cap \mathcal{S}_2 = \{\varnothing, X, \{0\}\}$$

is not a semiring as $X \setminus \{0\} = \{1, 2\}$ is not a union of sets in $\mathcal{C}$. Thus we cannot say that there is a smallest semiring including $\mathcal{C}$. Each of $\mathcal{S}_1$ and $\mathcal{S}_2$ is a minimal, but not smallest, semiring including $\mathcal{C}$.

If $\mathcal{S}$ is a semiring of sets, then the family $\mathcal{R}$ of all finite unions of members of $\mathcal{S}$ is the ring generated by $\mathcal{S}$. Consequently, a semiring closed under finite unions is a ring. The following schematic diagram summarizes the relationships among the various families of sets.

4.6 Example   To keep these notions straight, and to show that none of the converse implications hold, consider an uncountable set. Then:

1. The family of singleton subsets together with the empty set is a semiring but not a ring.

2. The family of all finite subsets is a ring but neither an algebra nor a σ-ring. (Remember, the empty set is finite.)

3. The family of all subsets that are either finite or have finite complement is an algebra but neither a σ-algebra nor a σ-ring.

4. The family of countable subsets is a σ-ring but not an algebra.

5. The family of all subsets that are either countable or have countable complement is a σ-algebra. It is the σ-algebra generated by the singletons.   ∎

We close the section by presenting two technical properties of semirings that are of use in later chapters.

4.7 Lemma   *For a semiring $\mathcal{S}$ we have the following.*

1. *If $A_1, \ldots, A_n, A \in \mathcal{S}$, then the set $A \setminus \bigcup_{i=1}^{n} A_i$ can be written as a union of a pairwise disjoint finite subset of $\mathcal{S}$.*

2. *If $\{A_n\}$ is a sequence in $\mathcal{S}$, then there exists a pairwise disjoint sequence $\{C_k\}$ in $\mathcal{S}$ satisfying $\bigcup_{n=1}^{\infty} A_n = \bigcup_{k=1}^{\infty} C_k$ and such that for each k there exists some n with $C_k \subset A_n$.*

Proof: (1)   The proof is by induction. The case $n = 1$ follows from the definition of semiring. So assume the claim true for n, and let $A_1, \ldots, A_n, A_{n+1}, A$ belong to $\mathcal{S}$. By the induction hypothesis, there are pairwise disjoint sets $C_1, \ldots, C_k$ in $\mathcal{S}$ such that $A \setminus \bigcup_{i=1}^{n} A_i = \bigcup_{j=1}^{k} C_j$. Clearly,

$$A \setminus \bigcup_{i=1}^{n+1} A_i = \left(A \setminus \bigcup_{i=1}^{n} A_i \right) \setminus A_{n+1} = \left(\bigcup_{j=1}^{k} C_j \right) \setminus A_{n+1} = \bigcup_{j=1}^{k} (C_j \setminus A_{n+1}).$$

Now for each j, pick a pairwise disjoint set $\{D_1^j, \ldots, D_{k_j}^j\}$ included in $\mathcal{S}$ satisfying $C_j \setminus A_{n+1} = \bigcup_{r=1}^{k_j} D_r^j$. Then $\{D_r^j : j = 1, \ldots, k,\ r = 1, \ldots, k_j\}$ is a finite pairwise disjoint subset of $\mathcal{S}$, and $A \setminus \bigcup_{i=1}^{n+1} A_i = \bigcup_{j=1}^{k} \bigcup_{r=1}^{k_j} D_r^j$.

(2)   Let $\{A_n\}$ be a sequence in $\mathcal{S}$ and put $A = \bigcup_{n=1}^{\infty} A_n$. Let $B_1 = A_1$ and $B_{n+1} = A_{n+1} \setminus \bigcup_{i=1}^{n} A_i$ for each $n \geq 1$. Then $B_i \cap B_j = \varnothing$ for $i \neq j$ and $A = \bigcup_{n=1}^{\infty} B_n$. By part (1) each B_n can be written as a union of a finite pairwise disjoint family of members of $\mathcal{S}$. Now notice that the union of all these pairwise disjoint families of $\mathcal{S}$ gives rise to a pairwise disjoint sequence $\{C_k\}$ of $\mathcal{S}$ that satisfies the desired properties.   ∎

4.8 Lemma   *Let $\mathcal{S}$ be a semiring and let $A_1, \ldots, A_n$ belong to $\mathcal{S}$. Then there exists a finite family $\{C_1, \ldots, C_k\}$ of pairwise disjoint members of $\mathcal{S}$ such that:*

1. *Each C_i is a subset of some A_j; and*

2. *Each A_j is a union of a subfamily of the family $\{C_1, \ldots, C_k\}$.*

Proof: The proof is by induction. For $n = 1$, the claim is trivial. So assume our claim to be true for any n members of $\mathcal{S}$ and let $A_1, \ldots, A_n, A_{n+1} \in \mathcal{S}$. For the sets $A_1, \ldots, A_n$ there exist—by the induction hypothesis—pairwise disjoint sets $C_1, \ldots, C_k \in \mathcal{S}$ satisfying (1) and (2). Now consider the finite family of pairwise disjoint subsets of $\mathcal{S}$

$$\{C_1 \cap A_{n+1}, C_1 \setminus A_{n+1}, \ldots, C_k \cap A_{n+1}, C_k \setminus A_{n+1}, A_{n+1} \setminus \bigcup_{i=1}^{k} C_i\}.$$

By the definition of the semiring we can write each $C_i \setminus A_{n+1}$ $(i = 1, \ldots, k)$ as a union of a pairwise disjoint finite family of members of $\mathcal{S}$. Likewise, by Lemma 4.7, the set $A_{n+1} \setminus \bigcup_{i=1}^{k} C_i$ can be written as a union of a pairwise disjoint finite family of members of $\mathcal{S}$. The sets in these unions together with the $C_i \cap A_{n+1}$ $(i = 1, \ldots, k)$ make a pairwise disjoint finite family of members of $\mathcal{S}$ that satisfies properties (1) and (2) for the family $A_1, \ldots, A_n, A_{n+1}$.   ∎

4.3 Dynkin's lemma

A σ-algebra is usually most conveniently described in terms of a generating family. In this section we study families of sets possessing certain monotonicity properties that are of interest mostly for technical reasons relating to the σ-algebras they generate. As usual, the notation $A_n \uparrow A$ means $A_n \subset A_{n+1}$ for each n and $A = \bigcup_{n=1}^{\infty} A_n$, and $A_n \downarrow A$ means $A_{n+1} \subset A_n$ for each n and $A = \bigcap_{n=1}^{\infty} A_n$. The most useful families are the Dynkin systems.

4.9 Definition   *A **Dynkin system** or a **λ-system**[4] is a nonempty family $\mathcal{A}$ of subsets of a set X with the following properties:*

1. *$X \in \mathcal{A}$.*

2. *If $A, B \in \mathcal{A}$ and $A \subset B$, then $B \setminus A \in \mathcal{A}$.*

3. *If a sequence $\{A_1, A_2, \ldots\} \subset \mathcal{A}$ satisfies $A_n \uparrow A$, then $A \in \mathcal{A}$.*

A **π-system** is a nonempty family of subsets of a set that is closed under finite intersections. The property of being a π-system and a Dynkin system characterizes σ-algebras.

4.10 Lemma   *A nonempty family of subsets of a set X is a σ-algebra if and only if it is both a π-system and a Dynkin system.*

Proof: Clearly, a σ-algebra is both a Dynkin system and a π-system. For the converse, let $\mathcal{A}$ be a Dynkin system that is also closed under finite intersections (that is, a π-system). Note that $\mathcal{A}$ is closed under complementation ($A^c = X \setminus A$), so $\mathcal{A}$ is in fact an algebra. To see that it is a σ-algebra, suppose $A = \bigcup_{n=1}^{\infty} A_n$ with $\{A_n\} \subset \mathcal{A}$. Letting $B_n = \bigcup_{k=1}^{n} A_k \in \mathcal{A}$, and noting that $B_n \uparrow A$, we get $A \in \mathcal{A}$.   ∎

---

[4] D. P. Bertsekas and S. E. Shreve [39, p. 133] use the term Dynkin system, while P. Billingsley [43] and E. B. Dynkin [111] himself use the term λ-system. B. Fristedt and L. Gray [129, pp. 724–725] use the term **Sierpiński class** as they attribute Dynkin's Lemma 4.11 below to W. Sierpiński [306], though they credit Dynkin with popularizing it. R. M. Blumenthal and R. K. Getoor [52] use the term d-system.

Notice that a Dynkin system that is not a π-system need not be an algebra. For example, consider $X = \{1, 2, 3, 4\}$. Then

$$\mathcal{A} = \{\varnothing, \{1, 2\}, \{3, 4\}, \{1, 3\}, \{2, 4\}, X\}$$

is a Dynkin system that is neither an algebra nor a π-system.

The following result is a key result in establishing measurability properties and is known as Dynkin's π-λ Lemma or simply as Dynkin's Lemma.

4.11 Dynkin's Lemma   *If $\mathcal{A}$ is a Dynkin system and a nonempty family $\mathcal{F} \subset \mathcal{A}$ is closed under finite intersections, then $\sigma(\mathcal{F}) \subset \mathcal{A}$. That is, if $\mathcal{F}$ is a π-system, then $\sigma(\mathcal{F})$ is the smallest Dynkin system that includes $\mathcal{F}$.*

Proof: Let $\mathcal{A}$ be a Dynkin system and let a nonempty family $\mathcal{F} \subset \mathcal{A}$ be closed under finite intersections. Denote by $\mathcal{D}$ the smallest Dynkin system that includes $\mathcal{F}$ (that is, the intersection of the collection of all Dynkin systems that include $\mathcal{F}$). It suffices to show that $\sigma(\mathcal{F}) \subset \mathcal{D} \subset \mathcal{A}$.

To this end, let

$$\mathcal{A}_1 = \{A \in \mathcal{D} : A \cap F \in \mathcal{D} \text{ for all } F \in \mathcal{F}\}.$$

It easy to see that $\mathcal{A}_1$ is a Dynkin system including $\mathcal{F}$, so $\mathcal{A}_1 = \mathcal{D}$. Now let

$$\mathcal{A}_2 = \{A \in \mathcal{D} : A \cap B \in \mathcal{D} \text{ for all } B \in \mathcal{D}\}.$$

Again, $\mathcal{A}_2$ is a Dynkin system including $\mathcal{F}$, so $\mathcal{A}_2 = \mathcal{D}$, which means that $\mathcal{D}$ is closed under finite intersections. By Lemma 4.10, $\mathcal{D}$ is a σ-algebra, and since it includes $\mathcal{F}$, we have $\sigma(\mathcal{F}) \subset \mathcal{D}$ (in fact, $\sigma(\mathcal{F}) = \mathcal{D}$), as desired.   ∎

Monotone classes also are closely related to σ-algebras and Dynkin systems.

4.12 Definition   *A **monotone class** is a nonempty family $\mathcal{M}$ of subsets of a set X such that if a sequence $\{A_n\}$ in $\mathcal{M}$ satisfies $A_n \uparrow A$ or $A_n \downarrow A$, then $A \in \mathcal{M}$.*

The following diagram summarizes some of these relationships.

$$\sigma\text{-algebra} \implies \text{Dynkin system} \implies \text{monotone class}$$

The last implication requires a bit of thought. Let $\{A_n\}$ be a sequence in a Dynkin system $\mathcal{A}$ satisfying $A_n \downarrow A$. If $B_n = A_n^c \in \mathcal{A}$, then $B_n \uparrow A^c$ so $A^c \in \mathcal{A}$, which implies $A = (A^c)^c \in \mathcal{A}$.

A monotone class need not be a Dynkin system. For instance, if $X = \{0, 1\}$, then the family $\{X, \{1\}\}$ is a monotone class but not a Dynkin system. Clearly the intersection of a collection of monotone classes is again a monotone class. Thus, every nonempty family $\mathcal{C}$ of sets is included in a smallest monotone class, namely the intersection of all monotone classes including it—this is the monotone class generated by $\mathcal{C}$.

4.13 Monotone Class Lemma *If $\mathcal{A}$ is an algebra, then $\sigma(\mathcal{A})$ is the smallest monotone class including $\mathcal{A}$, that is, $\sigma(\mathcal{A})$ is the monotone class generated by $\mathcal{A}$. In particular, an algebra $\mathcal{A}$ is a monotone class if and only if it is a σ-algebra.*

Proof: Let $\mathcal{M}$ be the smallest monotone class including $\mathcal{A}$. It is easy to see that $\mathcal{A} \subset \mathcal{M} \subset \sigma(\mathcal{A})$. Let $\mathcal{C} = \{B \in \mathcal{M} : B \setminus A \in \mathcal{M}$ for each $A \in \mathcal{A}\}$. Then $\mathcal{C}$ is a monotone class (why?) including the algebra $\mathcal{A}$, and hence $\mathcal{M} = \mathcal{C}$. That is, $B \setminus A \in \mathcal{M}$ for each $B \in \mathcal{M}$ and all $A \in \mathcal{A}$. Now let

$$\mathcal{D} = \{B \in \mathcal{M} : M \setminus B \in \mathcal{M} \text{ for each } M \in \mathcal{M}\}.$$

Again, $\mathcal{D}$ is a monotone class that (by the above) satisfies $\mathcal{A} \subset \mathcal{D}$. Thus, $\mathcal{D} = \mathcal{M}$. This shows that $\mathcal{M}$ is a Dynkin system. Since the algebra $\mathcal{A}$ is closed under finite intersections, by Dynkin's Lemma 4.11, $\sigma(\mathcal{A}) \subset \mathcal{M}$, so $\mathcal{M} = \sigma(\mathcal{A})$. ∎

4.4   The Borel σ-algebra

The most important example of a σ-algebra is the σ-algebra of subsets of a topological space generated by its open sets.

4.14 Definition   *The **Borel σ-algebra** of a topological space (X, τ) is $\sigma(\tau)$, the σ-algebra generated by the family τ of open sets.[5] Members of the Borel σ-algebra are **Borel sets**. The Borel σ-algebra is denoted $\mathcal{B}_X$, or simply $\mathcal{B}$.*

The σ-algebra $\mathcal{B}$ is also generated by the closed sets of X. As a consequence of Dynkin's Lemma 4.11 we have another characterization of the Borel sets.

4.15 Corollary   *The Borel σ-algebra is the smallest Dynkin system containing the open sets. It is also the smallest Dynkin system containing the closed sets.*

The next result gives another characterization of the Borel σ-algebra. It follows immediately from Theorem 4.2.

4.16 Corollary   *The Borel σ-algebra of a topological space is the smallest family of sets containing all the open sets and all the closed sets that is closed under countable intersections and countable disjoint unions.*

---

[5] Be warned that there are several slightly different definitions of the Borel sets in use. For instance, K. Kuratowski [219] defines the Borel sets to be the members of the smallest family of sets including the closed sets that is closed under countable unions and countable intersections. For metric spaces this definition is equivalent to ours by Corollary 4.18. (Interestingly, in [218] he uses the same definition we do.) P. R. Halmos [148] defines the Borel sets of a locally compact Hausdorff space to be the members of the smallest σ-ring containing every compact set. This differs significantly from our definition—on an uncountable discrete space, only countable sets are Borel sets under this definition. For σ-compact spaces the two definitions agree.

Here is a slightly different characterization of the Borel sets of a metric space.

4.17 Corollary   *The Borel σ-algebra of a metrizable space is the smallest family of sets that includes the open sets and is closed under countable intersections and countable disjoint unions.*

Proof: By Corollary 3.19, every closed set is a $\mathcal{G}_\delta$, so every family of sets including the open sets that is closed under countable intersections must include the closed sets. Now apply Corollary 4.16.   ■

To get a similar result for a family containing the closed sets, we assume closure under all countable unions, not only disjoint ones.

4.18 Corollary   *The Borel σ-algebra of a metrizable space is the smallest family of sets that includes the closed sets and is closed under countable intersections and countable unions.*

Proof: In a metrizable space every open set is an $\mathcal{F}_\sigma$. So every family of sets including the closed sets that is closed under countable unions must include the open sets, and the conclusion follows from Corollary 4.16.   ■

Since every closed set is a Borel set, the closure of *any* set is a Borel set. Likewise, the interior of any set is a Borel set, and the boundary of any set is a Borel set. In a Hausdorff space every point is closed, so every countable set is a Borel set. Also in a Hausdorff space every compact set is closed, so every compact set is a Borel set.

Unless otherwise stated, the real line $\mathbb{R}$ is tacitly understood to be equipped with the σ-algebra of its Borel sets $\mathcal{B}_{\mathbb{R}}$. For the real line almost any class of intervals generates the Borel σ-algebra. We leave the proof to you.

4.19 Lemma   *Consider the following families of intervals in $\mathbb{R}$:*

$$\mathcal{C}_1 = \{(a,b) : a < b\}, \quad \mathcal{C}_2 = \{[a,b] : a < b\}, \quad \mathcal{C}_3 = \{[a,b) : a < b\},$$
$$\mathcal{C}_4 = \{(a,b] : a < b\}, \quad \mathcal{C}_5 = \{(a,\infty) : a \in \mathbb{R}\}, \quad \mathcal{C}_6 = \{(-\infty,b) : b \in \mathbb{R}\},$$
$$\mathcal{C}_7 = \{[a,\infty) : a \in \mathbb{R}\}, \quad \mathcal{C}_8 = \{(-\infty,b] : b \in \mathbb{R}\}.$$

Then $\sigma(\mathcal{C}_1) = \sigma(\mathcal{C}_2) = \cdots = \sigma(\mathcal{C}_8) = \mathcal{B}_{\mathbb{R}}$.

4.20 Lemma   *If X is a topological space and Y is a subset of X, then the Borel sets of Y (where Y has the relative topology) are the restrictions of the Borel sets of X to Y. That is,*

$$\mathcal{B}_Y = \{B \cap Y : B \in \mathcal{B}_X\}.$$

Proof: Let $\mathcal{C} = \{B \cap Y : B \in \mathcal{B}_X\}$. Clearly, $\mathcal{C}$ is a σ-algebra containing the open subsets of Y, so $\mathcal{B}_Y \subset \mathcal{C}$.

Now let $\mathcal{A} = \{B \in \mathcal{B}_X : B \cap Y \in \mathcal{B}_Y\}$. Then $\mathcal{A}$ is a σ-algebra containing the open subsets of X, so $\mathcal{A} = \mathcal{B}_X$. That is, $B \in \mathcal{B}_X$ implies $B \cap Y \in \mathcal{B}_Y$, so $\mathcal{C} = \{B \cap Y : B \in \mathcal{B}_X\} \subset \mathcal{B}_Y$. Thus $\mathcal{B}_Y = \mathcal{C}$, as claimed.   ■

Certain classes of Borel sets have been given special names. Recall that a countable intersection of open sets is called a $\mathcal{G}_\delta$-set. A countable union of closed sets is called an $\mathcal{F}_\sigma$-set. Similarly, an $\mathcal{F}_{\sigma\delta}$ is a countable intersection of $\mathcal{F}_\sigma$-sets, and so on *ad infinitum*. All these kinds of sets are Borel sets, of course. You may be tempted to believe that any Borel set may be obtained by applying the operations of countable union, countable intersection, and complementation to a family of open sets some finite or maybe countable number of times. This is not the case for uncountable metric spaces. We won't go into details here, but if you are interested, consult K. Kuratowski [218, Section 30, pp. 344-373].

One trick for proving something is a Borel set is to write a description of the set involving universal (for all) and existential (there exists) quantifiers. This can be converted into a sequence of set theoretic operations involving unions and intersections. (In fact, the well-known Polish notation [using $\vee$ to mean "for all" and $\wedge$ to mean "there exists"] is designed to emphasize this.) For an example of this technique, see the proof of Theorem 4.28 below. We also use this trick in the proof of Lemma 7.63 below to show that the set of extreme points of a metrizable compact convex set is a $\mathcal{G}_\delta$-set.

The theory of Borel sets is most satisfying for Polish spaces. The reason is that in metric spaces convergence and closure can be described using (countable) sequences, since each point has a countable neighborhood base. Completeness allows convergence to be phrased in terms of the Cauchy property. Adding separability introduces another source of countable operations—the countable base for the topology.

4.5 Measurable functions

Let $\mathcal{A}_X$ and $\mathcal{A}_Y$ be nonempty families of subsets of X and Y, respectively. A function $f: X \to Y$ is $(\mathcal{A}_X, \mathcal{A}_Y)$-**measurable** if $f^{-1}(A)$ belongs to $\mathcal{A}_X$ for each A in $\mathcal{A}_Y$. It is a good time now to remind you of the bracket notation for inverse images, namely $[f \in A]$ means $f^{-1}(A)$. We may say that f is **measurable** when $\mathcal{A}_X$ and $\mathcal{A}_Y$ are understood. Usually, $\mathcal{A}_X$ and $\mathcal{A}_Y$ will be σ-algebras, but the definition makes sense with arbitrary families. However, we do reserve the term "measurable space" for a set equipped with a σ-algebra.

4.21 Definition *A measurable space is a pair (X, Σ), where X is a set and Σ is a σ-algebra of subsets of X.*

When either X or Y is a topological space, it is by default a measurable space equipped with its Borel σ-algebra. In particular, in the special case of a real function $f: (X, \mathcal{A}) \to \mathbb{R}$, we say that f is $\mathcal{A}$-measurable if it is $(\mathcal{A}, \mathcal{B}_\mathbb{R})$-measurable. When both X and Y are topological spaces, we say that f is **Borel measurable** if f is $(\mathcal{B}_X, \mathcal{B}_Y)$-measurable. We may also in this case simply say that f is a **Borel function**.

Compositions of measurable functions are measurable. The proof is trivial.

4.22 Lemma   *Let $(X, A_X) \xrightarrow{f} (Y, A_Y) \xrightarrow{g} (Z, A_Z)$ be measurable. Then the composition $g \circ f \colon (X, A_X) \to (Z, A_Z)$ is also measurable.*

Now note that taking inverse images preserves σ-algebras. To formulate this proposition, for a function $f \colon X \to Y$ and a family $\mathcal{F}$ of subsets of Y, define

$$f^{-1}(\mathcal{F}) = \{f^{-1}(A) : A \in \mathcal{F}\}.$$

Note that if $\mathcal{F}$ is a σ-algebra, then $f^{-1}(\mathcal{F})$ is also a σ-algebra.

4.23 Lemma   *If $f \colon X \to Y$ is a function between two sets and $\mathcal{F}$ is a nonempty family of subsets of Y, then*

$$\sigma(f^{-1}(\mathcal{F})) = f^{-1}(\sigma(\mathcal{F})).$$

Proof: Observe first that $f^{-1}(\sigma(\mathcal{F}))$ is a σ-algebra of subsets of X including $f^{-1}(\mathcal{F})$, so $\sigma(f^{-1}(\mathcal{F})) \subset f^{-1}(\sigma(\mathcal{F}))$. For the reverse inclusion, let

$$\mathcal{A} = \left\{A \in \sigma(\mathcal{F}) : f^{-1}(A) \in \sigma(f^{-1}(\mathcal{F}))\right\}.$$

Note that $\mathcal{A}$ is a σ-algebra of subsets of X that includes $\mathcal{F}$, so $\sigma(\mathcal{F}) = \mathcal{A}$. Consequently, $f^{-1}(\sigma(\mathcal{F})) = f^{-1}(\mathcal{A}) \subset \sigma(f^{-1}(\mathcal{F}))$, which gives the desired identity.   ∎

The following consequence of Lemma 4.23 shows that we do not have to check each inverse image to verify measurability.

4.24 Corollary   *Let $f \colon (X, \Sigma_X) \to (Y, \Sigma_Y)$ be a function between measurable spaces, and let $\mathcal{C}$ generate Σ_Y, that is, $\sigma(\mathcal{C}) = \Sigma_Y$. Then f is measurable if and only if $f^{-1}(C) \in \Sigma_X$ for each $C \in \mathcal{C}$.*

We use the next results frequently without any reference. Their proofs follow from Lemma 4.19 and Corollary 4.24.

4.25 Corollary   *For a function $f \colon (X, \Sigma) \to \mathbb{R}$ on a measurable space, let $\mathcal{C}$ be any one of the families of intervals described in Lemma 4.19. Then f is measurable if and only if $f^{-1}(I)$ belongs to Σ for each $I \in \mathcal{C}$.*

Since the open sets generate the Borel σ-algebra we have the following.

4.26 Corollary   *Every continuous function between topological spaces is Borel measurable.*

4.6 The space of measurable functions

The collection of all measurable real-valued functions is closed under most interesting pointwise operations. In particular, it is a function space in the sense of Definition 1.1.

4.27 Theorem *Let (X, Σ) be a measurable space. For any pair $f, g \colon X \to \mathbb{R}$ of Σ-measurable functions and any $\alpha \in \mathbb{R}$, the following functions are measurable:*

$$\alpha f, \quad f + g, \quad fg, \quad f/g, \quad f \vee g, \quad f \wedge g, \quad f^+, \quad f^-, \quad \text{and} \quad |f|.$$

Also, if $\{f_n\}$ is a sequence of Σ-measurable functions, then

$$\lim f_n, \quad \limsup f_n, \quad \text{and} \quad \liminf f_n$$

are Σ-measurable, provided they are defined and real-valued (finite).

   To summarize, the collection of Σ-measurable real-valued functions is a function space and an algebra that is closed under pointwise sequential limits.

Proof: We shall give hints for some of the proofs and leave the rest as an exercise, or see, e.g., [13, Section 16]. We start by noting some identities:

$$f^+ = \frac{|f| + f}{2}, \quad f^- = \frac{|f| - f}{2},$$

$$f \vee g = \frac{f + g + |f + g|}{2}, \quad f \wedge g = \frac{f + g - |f + g|}{2},$$

$$fg = \frac{(f + g)^2 - f^2 - g^2}{2}.$$

Thus we only need to show measurability of αf, $f + g$, $|f|$, and f^2 to get the rest.

   Measurability of the sum may require the most cleverness, so here is a hint. By Lemma 4.19 and Corollary 4.24, it is enough to show that for each $\alpha \in \mathbb{R}$, the set $[f + g > \alpha] = \{x \in X : f(x) + g(x) > \alpha\}$ belongs to Σ. Observe that $f(x) + g(x) > \alpha$ if and only if there is a rational number q satisfying $f(x) > q > \alpha - g(x)$. Thus $[f + g > \alpha] = \bigcup_{q \in \mathbb{Q}} [f > q] \cap [g > \alpha - q]$. You can handle it from here.

   To see that limits are measurable, let $f \colon X \to \mathbb{R}$ be a real function, and suppose there is a sequence $\{f_n\}$ of Σ-measurable functions satisfying $f_n(x) \to f(x)$ for each $x \in X$. Then note that

$$[f > \alpha] = \bigcup_{k=1}^{\infty} \bigcap_{n=k}^{\infty} [f_n > \alpha + \tfrac{1}{k}]$$

for each $\alpha \in \mathbb{R}$, so f is Σ-measurable. ∎

   Even if a sequence of measurable real functions does not converge, we can at least say something about the set of points where it does converge.

4.28 Hahn's Theorem   *Let (X, Σ) be a measurable space and let $\{f_n\}$ be a sequence of Σ-measurable real-valued functions. Then the set of points at which the sequence $\{f_n\}$ converges (in $\mathbb{R}$) is a Σ-measurable set.*

Proof: Recall that the sequence $\{f_1(x), f_2(x), \ldots\}$ of real numbers converges if and only if it is a Cauchy sequence. That is, for every n, there is an m, such that for every k we have $|f_{m+k}(x) - f_m(x)| < \frac{1}{n}$. Thus the set C of points at which the sequence $\{f_1, f_2, \ldots\}$ of functions converges pointwise can be written as

$$C = \left\{ x \in X : \forall n \, \exists m \, \forall k \, |f_{m+k}(x) - f_m(x)| < \tfrac{1}{n} \right\}$$
$$= \bigcap_{n=1}^{\infty} \bigcup_{m=1}^{\infty} \bigcap_{k=1}^{\infty} [|f_{m+k} - f_m| < \tfrac{1}{n}].$$

Since each f_n is Σ-measurable, Theorem 4.27 implies that $[|f_{m+k} - f_m| < \frac{1}{n}]$ belongs to Σ. Thus, C is derived from countable unions and intersections of sets in Σ, and hence $C \in \Sigma$.   ∎

We can generalize some of these results from real-valued functions to functions taking values in a metric space. In particular, when the range is a metric space, the pointwise limit of a sequence of measurable functions is measurable.

4.29 Lemma   *The pointwise limit of a sequence of measurable functions from a measurable space into a metrizable space is measurable.*

Proof: Let $\{f_n\}$ be a sequence of measurable functions from (X, Σ) into the metrizable space Y. Suppose f is the pointwise limit of $\{f_n\}$. Let F be a nonempty closed subset of Y. By Corollary 4.24, it suffices to show that $f^{-1}(F) \in \Sigma$.

Choose a compatible metric d for Y and set $G_n = \{y \in Y : d(y, F) < \frac{1}{n}\}$ for each n. Then each G_n is open and $\bigcap_{n=1}^{\infty} G_n = F$. We claim that

$$f^{-1}(F) = \bigcap_{k=1}^{\infty} \bigcup_{m=1}^{\infty} \bigcap_{n=m}^{\infty} f_n^{-1}(G_k).$$

Given the claim, since each G_k is open and f_n is measurable, $f_n^{-1}(G_k)$ is a measurable set, so $f^{-1}(F)$ is also measurable.

To prove the claim, suppose first that $x \in f^{-1}(F)$, that is, $f(x)$ belongs to F. Since $f_n(x) \to f(x)$, and for each k the set G_k is a neighborhood of $f(x)$, there is some m such that for all $n \geqslant m$ we have $f_n(x) \in G_k$. That is, x belongs to $\bigcap_{k=1}^{\infty} \bigcup_{m=1}^{\infty} \bigcap_{n=m}^{\infty} f_n^{-1}(G_k)$.

For the reverse inclusion, assume x belongs to $\bigcap_{k=1}^{\infty} \bigcup_{m=1}^{\infty} \bigcap_{n=m}^{\infty} f_n^{-1}(G_k)$. That is, for every k, the point $f_n(x)$ eventually lies in G_k. Thus $f(x) = \lim_n f_n(x) \in \overline{G_k}$, so $f(x)$ belongs to $\bigcap_{k=1}^{\infty} \overline{G_k}$. But $\overline{G_{k+1}} \subset G_k$, so we may replace $\overline{G_k}$ by G_k in the intersection, that is, $f(x) \in \bigcap_{k=1}^{\infty} G_k = F$. This establishes the claim.   ∎

If the range is a separable metric space, then the next lemma, reduces the question of measurability of a function into a metric space to the measurability of real-valued functions.

4.30 Lemma   *Let $f: (S, \Sigma) \to (X, d)$ be a function from a measurable space into a separable metric space and for each $x \in X$ define the function $\theta_x: S \to \mathbb{R}$ by $\theta_x(s) = d(x, f(s))$. Then f is measurable if and only if the real function θ_x is measurable for each $x \in X$.*

Proof: Since X is separable, every open subset of X is the union of a countable family of open d-balls. So f is measurable if and only if $f^{-1}(B_\varepsilon(x))$ belongs to Σ for each $x \in X$ and each $\varepsilon > 0$. But

$$f^{-1}(B_\varepsilon(x)) = \{s \in S : d(x, f(s)) < \varepsilon\} = [\theta_x < \varepsilon].$$

So f is measurable if and only if θ_x is measurable for each $x \in X$.   ∎

These results are more subtle than it might appear. For instance, the conclusion no longer follows if we drop the metrizability assumption on the range, even if the range is compact. The next example may be found, e.g., in R. M. Dudley [103].

4.31 Example (Limit not measurable) Let $S = I = [0, 1]$. Then I^I, the space of functions (measurable or not) from I into I, endowed with its product topology is a compact space that is not metrizable. For each n define the function $\varphi_n: S \to I^I$ by $\varphi_n(s)(x) = (1 - n|s - x|)^+$. Note that $s \mapsto \varphi_n(s)$ is continuous from S into I^I, and therefore Borel measurable. Furthermore $\varphi_n \xrightarrow[n \to \infty]{\text{pointwise}} \varphi$, where $\varphi(s) = \chi_{\{s\}} \in I^I$ is the indicator function of the singleton $\{s\}$.

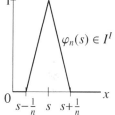

For each $s \in S$ there is an open subset U_s of I^I such that $\varphi^{-1}(U_s) = \{s\}$, for example, let $U_s = \{f \in I^I : f(s) > 0\}$. Now let A be a non-Borel subset of S and put $V = \bigcup_{s \in A} U_s$. (We show in Corollary 10.42 below that a non-Borel subset of I exists.) Then V is open in I^I, but $\varphi^{-1}(V) = A$, so φ is not measurable.   ∎

We now turn attention to the relation between the space of bounded Borel measurable functions and bounded continuous functions.

4.32 Definition   *The collection of all bounded Borel measurable real functions defined on the topological space X is denoted $B_b(X)$.*

It is easy to see that with the usual (everywhere) pointwise algebraic and lattice operations $B_b(X)$ is a function space. The next result shows how the space $C_b(X)$ of bounded continuous real functions lies in $B_b(X)$.

4.33 Theorem   *Let X be a metrizable space, and let $\mathcal{F}$ be a vector subspace of $B_b(X)$ including $C_b(X)$. Then $\mathcal{F} = B_b(X)$ if and only if $\mathcal{F}$ is closed under monotone sequential pointwise limits in $B_b(X)$. (That is, if and only if $\{f_n\} \subset \mathcal{F}$ and $f_n \uparrow f \in B_b(X)$ imply $f \in \mathcal{F}$.)*

Proof: Let X be a metrizable space, and let $\mathcal{F}$ be a vector subspace of $B_b(X)$ including $C_b(X)$ and containing its monotone sequential pointwise limits in $B_b(X)$. Consider the family $\mathcal{A}$ of all Borel sets whose indicators lie in $\mathcal{F}$. That is, we let $\mathcal{A} = \{A \in \mathcal{B}_X : \chi_A \in \mathcal{F}\}$. We shall show that $\mathcal{A}$ is a Dynkin system containing the closed sets. Consequently, by Lemma 4.11, $\mathcal{A}$ contains all the Borel sets, so $\mathcal{A} = \mathcal{B}_X$. It follows that $\mathcal{F}$ contains all the simple Borel measurable functions. By Theorem 4.36 below, every $0 \leqslant f \in B_b(X)$ is a pointwise limit of an increasing sequence of simple functions, so it follows that $\mathcal{F} = B_b(X)$.

We now show that $\mathcal{A}$ is a Dynkin system containing the closed sets. Corollary 3.14 states that the indicator function of every closed set is a decreasing pointwise limit of a sequence of bounded continuous functions, so $\mathcal{A}$ contains all closed sets. In particular, $X \in \mathcal{A}$. If $A \subset B$ and $A, B \subset \mathcal{F}$, then $\chi_{B \setminus A} = \chi_B - \chi_A \in \mathcal{F}$, so that $\mathcal{A}$ is closed under proper set differences. Also $\mathcal{A}$ is closed under increasing countable unions, since $A_n \uparrow A$ if and only if $\chi_{A_n}(x) \uparrow \chi_A(x)$ for each x. This shows that $\mathcal{A}$ is a Dynkin system containing the closed sets, and the proof is complete.   ∎

4.7   Simple functions

In this section we present some useful technical results on approximation of measurable functions by simple functions. A simple function is a function that assumes only finitely many distinct values. We also require as part of the definition a kind of measurability condition. Let $\mathcal{A}$ be an algebra of subsets of a set X and let Y be an arbitrary set. A function $\varphi \colon (X, \mathcal{A}) \to Y$ is a **simple function**, or more specifically $\mathcal{A}$**-simple**, if φ takes a finite number of values, say $y_1, \dots, y_n$, and $A_i = \varphi^{-1}(\{y_i\}) \in \mathcal{A}$ for each $i = 1, \dots, n$. We may on occasion be redundant and use the term measurable simple function. Note that no measurable structure is assumed for the space Y, but if Y is a measurable space where singletons are measurable, a simple function is measurable in the usual sense.

If Y is a vector space (or $\mathbb{R}$), then we may write φ in its **standard representation** $\varphi = \sum_{i=1}^n y_i \chi_{A_i}$, where the y_i are nonzero and distinct. If φ assumes the value zero, this just means that $\bigcup_{i=1}^n A_i$ is not all of X. By convention, the standard representation of the constant function zero is $\chi_\varnothing$.

The collection of real-valued simple functions is a function space in the sense of Definition 1.1, namely a vector space of functions closed under pointwise lattice operations.

4.34 Lemma   *If $\mathcal{A}$ is an algebra of subsets of a set X, then the collection of simple real-valued functions is a function space.*

For functions taking on values in a vector space we have the following result.

4.35 Lemma   *Assume that A is an algebra of subsets of a set X and that Y is a vector space. Then the collection of A-simple functions from X to Y under the pointwise operations is a vector space.*

One reason that simple functions are important is that they are pointwise dense in the vector space of all measurable functions.

4.36 Theorem   *Let A be an algebra of subsets of X, and let $A_{\mathbb{R}}$ denote the algebra generated by the half-open intervals $[a, b)$. If $f : X \to [0, \infty)$ is an $(A, A_{\mathbb{R}})$-measurable function, then there exists a sequence $\{\varphi_n\}$ of nonnegative simple functions such that $\varphi_n(x) \uparrow f(x)$ for all $x \in X$.*

Proof: Suppose f is $(A, A_{\mathbb{R}})$-measurable. Break up the range of f as follows. Given n, partition the interval $[0, n)$ into $n2^n$ half-open intervals of length $1/2^n$, and for each $1 \leqslant i \leqslant n2^n$ let

$$A_i^n = \left[\tfrac{i-1}{2^n} \leqslant f < \tfrac{i}{2^n} \right] = \{ x \in X : \tfrac{i-1}{2^n} \leqslant f(x) < \tfrac{i}{2^n} \} \in A.$$

Define the A-measurable simple function $\varphi_n = \sum_{i=1}^{n2^n} \tfrac{i-1}{2^n} \chi_{A_i^n}$. For x with $f(x) < n$, we have $\varphi_n(x) = \tfrac{i-1}{2^n} \leqslant f(x) < \tfrac{i}{2^n}$ for some i, and $\varphi_n(x) = 0$ otherwise. This construction guarantees that $\varphi_n(x) \uparrow f(x)$ for each $x \in X$. ∎

The real-valued measurable functions on a measurable space are precisely the pointwise limits of sequences of simple functions.

4.37 Corollary   *If (X, Σ) is a measurable space, then a real-valued function $f : X \to \mathbb{R}$ is Σ-measurable if and only if there exists a sequence $\{\varphi_n\}$ of simple functions satisfying $f_n(x) \to f(x)$ for each $x \in X$.*

Proof: Note that $f = f^+ - f^-$, so use Lemma 4.34 and Theorem 4.36. ∎

We can extend theses results from the case of a real-valued function to a function taking values in a separable metric space.

4.38 Theorem   *For a function $f : (X, \Sigma) \to (Y, d)$, from a measurable space into a separable metric space, we have the following.*

1. *The function f is measurable if and only if it is the pointwise d-limit of a sequence of $(\Sigma, \mathcal{B}_Y)$-measurable simple functions.*

2. *If in addition, (Y, d) is totally bounded, then f is a measurable function if and only if it is the d-uniform limit of a sequence of $(\Sigma, \mathcal{B}_Y)$-measurable simple functions.*

Proof: We establish (2) first. Start by noticing that if f is the pointwise limit of a sequence of simple functions, then by Corollary 4.29, f is $(\Sigma, \mathcal{B}_Y)$-measurable.

Now assume that f is $(\Sigma, \mathcal{B}_Y)$-measurable and let $\varepsilon > 0$. Since Y is a totally bounded metric space, there exist $y_1, \ldots, y_k \in Y$ such that $Y = \bigcup_{i=1}^k B_\varepsilon(y_i)$. Put $A_1 = B_\varepsilon(y_1)$ and $A_{n+1} = B_\varepsilon(y_{n+1}) \setminus \bigcup_{i=1}^n B_\varepsilon(y_i)$ for $n = 1, \ldots, k-1$. Then each A_i is a Borel subset of Y, $A_i \cap A_j = \varnothing$ for $i \neq j$ and $Y = \bigcup_{i=1}^k A_i$. Clearly, $X = \bigcup_{i=1}^k f^{-1}(A_i)$ and $f^{-1}(A_i) \cap f^{-1}(A_j) = \varnothing$ if $i \neq j$. Now define $\varphi \colon X \to Y$ by letting $\varphi(x) = y_i$ for each $x \in f^{-1}(A_i)$. Then φ is a simple function of X and satisfies $d(f(x), \varphi(x)) < \varepsilon$ for each $x \in X$. From this, it easily follows that f is the d-uniform limit of a sequence of $(\Sigma, \mathcal{B}_Y)$-measurable simple functions.

Next assume that f is $(\Sigma, \mathcal{B}_Y)$-measurable. Since (Y, d) is separable, there exists (by Corollary 3.41) a totally bounded metric ρ on Y that is equivalent to d. But then, by the preceding conclusion, f is the ρ-uniform limit of a sequence of $(\Sigma, \mathcal{B}_Y)$-measurable simple functions, say $\{\varphi_n\}$. Clearly, this sequence $\{\varphi_n\}$ of $(\Sigma, \mathcal{B}_Y)$-measurable simple functions d-converges pointwise to f. This completes the proof of the theorem. ∎

If the range space is not separable, then a Borel function (even a continuous function) need not be the pointwise limit of a sequence of simple functions. Below is an example that depends on the following lemma on simple functions, which is interesting in its own right. This lemma does not require that the simple functions be measurable in any sense, only that they have a finite range.

4.39 Lemma   *The pointwise limit of a sequence of simple (finite range) functions from a set to a metrizable space has a separable range.*

Proof: Let X be a set, Y a metrizable space and let $\{\varphi_n\}$ be a sequence of simple functions from X to Y. Assume that $\varphi_n(x) \to f(x)$ in Y for each $x \in X$. Put $A = \bigcup_{n=1}^\infty \varphi_n(X)$, and note that A is countable. Since A is dense in its closure $\overline{A}$, this implies that $\overline{A}$ is separable. From $\varphi_n(x) \to f(x)$ for each $x \in X$, we see that $f(X) \subset \overline{A}$. Since every subset of a separable metrizable space is separable (Corollary 3.5), the range $f(X)$ of f is separable. ∎

4.40 Example   Let $X = \ell_2([0, 1])$, the (real) Hilbert space with respect to the set $[0, 1]$. This means that $\ell_2([0, 1])$ consists of all functions $f \colon [0, 1] \to \mathbb{R}$ satisfying $f(\lambda) \neq 0$ for at most countably many $\lambda \in [0, 1]$ and $\sum_{\lambda \in [0,1]} |f(\lambda)|^2 < \infty$.[6] The inner product of two functions $f, g \in \ell_2([0, 1))$ is defined by

$$(f, g) = \sum_{\lambda \in [0,1]} f(\lambda)g(\lambda).$$

---

[6] If $\{\lambda_1, \lambda_2, \ldots\} \subset [0, 1]$ is a countable set for which $f(\lambda) = 0$ for all $\lambda \notin \{\lambda_1, \lambda_2, \ldots\}$, then as usual we let $\sum_{\lambda \in [0,1]} |f(\lambda)|^2 = \sum_{n=1}^\infty |f(\lambda_n)|^2$.

The distance d on X is defined by means of the norm as follows:

$$d(f,g) = \|f - g\| = \Big(\sum_{\lambda \in [0,1]} |f(\lambda) - g(\lambda)|^2 \Big)^{\frac{1}{2}}.$$

The space (X, d) is a complete non-separable metric space. To see that X is non-separable, for each $\lambda \in [0, 1]$ let e_λ denote the function defined by $e_\lambda(x) = 0$ if $x \neq \lambda$ and $e_\lambda(\lambda) = 1$. Then $e_\lambda \in X$ and $d(e_\lambda, e_\mu) = \sqrt{2}$ for $\lambda \neq \mu$. So the uncountable family of open balls $\{B(e_\lambda, 1/2)\}_{\lambda \in [0,1]}$ is pairwise disjoint. Thus no countable subset of X can be dense.

Now define the function $F \colon X \to X$ by $F(f)(\lambda) = f(\lambda^2)$ for each $f \in X$ and all $\lambda \in [0, 1]$. It is easy to see that F is one-to-one and surjective. Moreover, a moment's thought reveals that $d(F(f), F(g)) = d(f, g)$ for all $f, g \in X$, that is, F is a surjective isometry. In particular, F is continuous—and hence Borel measurable.

However, F cannot be the pointwise limit of any sequence of simple functions; since otherwise, according to Lemma 4.39, its range X should be separable, which is not the case. ∎

4.8 The σ-algebra induced by a function

If $f \colon X \to (Y, \Sigma)$ is a function and Σ is a σ-algebra of subsets of Y, then it is easy to see that $\sigma(f) = \{f^{-1}(A) : A \in \Sigma\}$ is a σ-algebra of subsets of X, known as the **σ-algebra induced by** f. It turns out that a real function that is $\sigma(f)$-measurable can actually be written as a function of f, a fact that is of extreme importance in the theory of conditional expectations in probability.

4.41 Theorem *Let (Y, Σ) be a measurable space, $f \colon X \to (Y, \Sigma)$, and $g \colon X \to \mathbb{R}$. Then the function g is $\sigma(f)$-measurable if and only if there exists a Σ-measurable function $h \colon Y \to \mathbb{R}$ such that $g = h \circ f$.*

Proof: The theorem is illustrated by this commuting diagram. Clearly g is Σ-measurable if such an h exists. For the converse assume that g is $\sigma(f)$-measurable. The existence of such a Σ-measurable function h is established in steps.

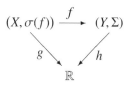

Step I: Assume that g is a $\sigma(f)$-measurable simple function.

Let $g = \sum_{i=1}^{n} a_i \chi_{A_i}$ be the standard representation of g, where the A_i are pairwise disjoint subsets belonging to $\sigma(f)$. For each i choose $B_i \in \Sigma$ such that $A_i = f^{-1}(B_i)$. The Σ-measurable simple function $h = \sum_{i=1}^{n} a_i \chi_{B_i}$ is easily seen to satisfy $h \circ f = g$.

Step II: The general case.

Since g is $\sigma(f)$-measurable, by Corollary 4.37 there is a sequence $\{\varphi_n\}$ of $\sigma(f)$-measurable simple functions satisfying $\varphi_n(x) \rightarrow g(x)$ for each $x \in X$. Now, by Step I, for each n there exists a Σ-measurable function $h_n \colon Y \rightarrow \mathbb{R}$ such that $h_n \circ f = \varphi_n$. Next, let

$$L = \{y \in Y : \lim_{n \to \infty} h_n(y) \text{ exists in } \mathbb{R}\}.$$

From $h_n(f(x)) = \varphi_n(x) \rightarrow g(x)$, we have $f(X) \subset L$. Put $h(y) = \lim_{n \to \infty} h_n(y)$ for $y \in L$ and $h(y) = 0$ for $y \notin L$. Since $f(X) \subset L$, we see that $h \circ f = g$. Now Hahn's Theorem 4.28 implies that L belongs to Σ, so $h_n \chi_L$ is Σ-measurable, and $h_n \chi_L \rightarrow h$, so h is Σ-measurable. ∎

4.9   Product structures

For each $i = 1, \ldots, n$, let $\mathcal{S}_i$ be a semiring of subsets of the set X_i. A subset of the product $X_1 \times X_2 \times \cdots \times X_n$ is a **measurable rectangle** if it is of the form $A_1 \times A_2 \times \cdots \times A_n$ where each A_i belongs to $\mathcal{S}_i$.

4.42 Lemma   *The family of measurable rectangles is a semiring.*

Proof: To see this, verify the identities

$$(A \times B) \cap (C \times D) = (A \cap C) \times (B \cap D); \quad \text{and}$$

$$A \times B \setminus C \times D = [(A \setminus C) \times B] \cup [(A \cap C) \times (B \setminus D)],$$

and use induction on n. ∎

This semiring is called the **product semiring** of $\mathcal{S}_1, \mathcal{S}_2, \ldots, \mathcal{S}_n$, and is denoted $\mathcal{S}_1 \times \mathcal{S}_2 \times \cdots \times \mathcal{S}_n$.[7] The product of σ-algebras is defined in a slightly different fashion.

4.43 Definition   *Let Σ_i be a σ-algebra of subsets of X_i ($i = 1, \ldots, n$). The* ***product σ-algebra*** *$\Sigma_1 \otimes \Sigma_2 \otimes \cdots \otimes \Sigma_n$ is the σ-algebra generated by the semiring $\Sigma_1 \times \Sigma_2 \times \cdots \times \Sigma_n$ of measurable rectangles. That is,*

$$\Sigma_1 \otimes \Sigma_2 \otimes \cdots \otimes \Sigma_n = \sigma(\Sigma_1 \times \Sigma_2 \times \cdots \times \Sigma_n).$$

One of the useful properties of the Borel σ-algebra is that, in an important class of cases, the Borel σ-algebra of a product of topological spaces is the product of their Borel σ-algebras. Not surprisingly, second countability is important.

---

[7] This is at odds with the standard Cartesian product notation. However, this is the notation used by most authors and we retain it. You should not have any problem understanding its meaning from the context of the discussion.

4.44 Theorem   *For any two topological spaces X and Y:*

1. $\mathcal{B}_X \otimes \mathcal{B}_Y \subset \mathcal{B}_{X \times Y}$.

2. *If X and Y are second countable, then* $\mathcal{B}_X \otimes \mathcal{B}_Y = \mathcal{B}_{X \times Y}$.

Proof: (1)   For each subset A of X, let

$$\Sigma(A) = \{B \subset Y : A \times B \in \mathcal{B}_{X \times Y}\}.$$

Then $\Sigma(A)$ satisfies the following properties.

   a. $\varnothing \in \Sigma(A)$. To see this, note that $A \times \varnothing = \varnothing \in \mathcal{B}_{X \times Y}$.

   b. If $B, C \in \Sigma(A)$, then $B \setminus C \in \Sigma(A)$. Indeed, if $B, C \in \Sigma(A)$, then observe that $A \times (B \setminus C) = (A \times B) \setminus (A \times C) \in \mathcal{B}_{X \times Y}$.

   c. $\Sigma(A)$ is closed under countable unions. To see this, note that if $\{B_n\}$ is a sequence in $\Sigma(A)$, then $A \times (\bigcup_{n=1}^{\infty} B_n) = \bigcup_{n=1}^{\infty} A \times B_n \in \mathcal{B}_{X \times Y}$.

The above three properties show that $\Sigma(A)$ is a σ-ring. It is a σ-algebra if $Y \in \Sigma(A)$.

Next note that for any open subset G of X, $U \in \Sigma(G)$ for every open subset U of Y. Since Y itself is open, if $G \subset X$ is open, then $\Sigma(G)$ is a σ-algebra of subsets of Y that includes τ_Y. Thus $\mathcal{B}_Y \subset \Sigma(G)$ whenever G is open.

Now let

$$\mathcal{A} = \{A \subset X : \mathcal{B}_Y \subset \Sigma(A)\}.$$

As we just remarked, $\mathcal{A}$ includes τ_X.

Also note that $\mathcal{A}$ is closed under complementation. To see this, let A belong to $\mathcal{A}$ and let B belong to $\mathcal{B}_Y$. Then $A \times B \in \mathcal{B}_{X \times Y}$. But X is open, so $X \times B \in \mathcal{B}_{X \times Y}$ too. Therefore so $A^c \times B = (A \times B)^c \cap (X \times B)$ belongs to $\mathcal{B}_{X \times Y}$. That is, $B \in \Sigma(A^c)$. Since B is an arbitrary Borel set, we have $\mathcal{B}_Y \subset \Sigma(A^c)$. In other words $A^c \in \mathcal{A}$. (We shall see in Lemma 4.45 below that $A \times B \in \mathcal{B}_{X \times Y}$ implies $B \in \mathcal{B}_Y$, so $\Sigma(A) = \mathcal{B}_Y$ for any $A \in \mathcal{A}$.)

Finally, if $\{A_n\} \subset \mathcal{A}$, and $B \in \mathcal{B}_Y$, we have $A_n \times B \in \mathcal{B}_{X \times Y}$ for each n. Using the fact that $\bigcup_{n=1}^{\infty} (A_n \times B) = (\bigcup_{n=1}^{\infty} A_n) \times B$, we see that $B \in \Sigma(\bigcup_{n=1}^{\infty} A_n)$. That is, $\mathcal{A}$ is closed under countable unions. Therefore $\mathcal{A}$ is a σ-algebra including τ_X, so $\mathcal{B}_X \subset \mathcal{A}$.

Thus, we have shown that for any Borel subsets A of X and B of Y the rectangle $A \times B$ belongs to $\mathcal{B}_{X \times Y}$. Therefore $\mathcal{B}_X \otimes \mathcal{B}_Y \subset \mathcal{B}_{X \times Y}$.

(2)   If both X and Y are second countable, then every open subset of $X \times Y$ is a countable union of subsets of the form $U \times V$, where U is open in X and V is open in Y. Consequently, $\mathcal{B}_X \otimes \mathcal{B}_Y \supset \mathcal{B}_{X \times Y}$, so we indeed have equality.   ∎

The next result gives a sufficient condition for the graph of a function to be a measurable set in the product σ-algebra.

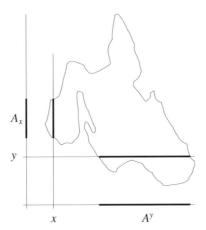

Figure 4.1. Sections of a set.

4.45 Theorem   *Let (X, Σ) be a measurable space, and let Y be a second count-able Hausdorff space. If $f \colon X \to Y$ is $(\Sigma, \mathcal{B}_Y)$-measurable, then the graph of f is $\Sigma \otimes \mathcal{B}_Y$-measurable. That is, $\mathrm{Gr}\, f \in \Sigma \otimes \mathcal{B}_Y$.*

Proof: Let $U_1, U_2, \ldots$ be a countable base for Y. Then $f(x) \neq y$ if and only if there is some U_n for which $f(x) \in U_n$ and $y \notin U_n$. Thus

$$(\mathrm{Gr}\, f)^c = \bigcup_{n=1}^{\infty} f^{-1}(U_n) \times (U_n)^c,$$

which, since f is measurable, is a countable union of measurable rectangles, and so belongs to $\Sigma \otimes \mathcal{B}_Y$. Therefore $\mathrm{Gr}\, f$ belongs to $\Sigma \otimes \mathcal{B}_Y$.   ∎

When X and Y are Polish spaces equipped with their Borel σ-algebras, then the converse is true, but the proof must wait until Theorem 12.28. We now turn our attention to sections of subsets of product spaces.

If A is a subset of a Cartesian product $X \times Y$, then for each $x \in X$ and $y \in Y$ the x- and y-sections of A are defined by

$$A_x = \{y \in Y : (x, y) \in A\} \quad \text{and} \quad A^y = \{x \in X : (x, y) \in A\}.$$

Clearly, for each $x \in X$ and each $y \in Y$, we have

$$\varnothing_x = \varnothing^y = \varnothing, \quad (X \times Y)_x = Y, \quad \text{and } (X \times Y)^y = X.$$

It is easy to see that the sections of a collection $\{A_i\}_{i \in I}$ of subsets of $X \times Y$ satisfy the following properties:

$$\Big(\bigcup_{i \in I} A_i\Big)_x = \bigcup_{i \in I} (A_i)_x \quad \text{and} \quad \Big(\bigcap_{i \in I} A_i\Big)_x = \bigcap_{i \in I} (A_i)_x, \text{ and}$$

$$\Big(\bigcup_{i \in I} A_i\Big)^y = \bigcup_{i \in I} (A_i)^y \quad \text{and} \quad \Big(\bigcap_{i \in I} A_i\Big)^y = \bigcap_{i \in I} (A_i)^y.$$

From these identities, it follows that if $\mathcal{A}$ is a σ-algebra of subsets of X, then the family $\{A \subset X \times Y : A^y \in \mathcal{A} \text{ for all } y \in Y\}$ is a σ-algebra of subsets of $X \times Y$.

Now let (X, Σ_X) and (Y, Σ_Y) be measurable spaces. A subset A of $X \times Y$ has **measurable sections** if $A_x \in \Sigma_Y$ for each $x \in X$ and $A^y \in \Sigma_X$ for each $y \in Y$.

4.46 Lemma   *Every set in the product σ-algebra $\Sigma_X \otimes \Sigma_Y$ has measurable sections.*

Proof: Consider the family $\mathcal{A}$ of subsets of $X \times Y$ given by

$$\mathcal{A} = \{A \subset X \times Y : A_x \in \Sigma_Y \text{ and } A^y \in \Sigma_X \text{ for all } x \in X \text{ and } y \in Y\}.$$

Then, as mentioned above, $\mathcal{A}$ is a σ-algebra. Moreover, from

$$(A \times B)_x = \begin{cases} B & \text{if } x \in A \\ \varnothing & \text{if } x \notin A \end{cases} \quad \text{and} \quad (A \times B)^y = \begin{cases} A & \text{if } y \in B \\ \varnothing & \text{if } y \notin B, \end{cases}$$

we see that every measurable rectangle belongs to $\mathcal{A}$, that is, $\Sigma_X \times \Sigma_Y \subset \mathcal{A}$. Since $\mathcal{A}$ is a σ-algebra, we get $\Sigma_X \otimes \Sigma_Y = \sigma(\Sigma_X \times \Sigma_Y) \subset \mathcal{A}$.   ∎

The converse is not true in general. In fact, W. Sierpiński [305] shows that there exists a non-Borel (in fact, a non-Lebesgue measurable) subset A of $\mathbb{R}^2$, called a **Sierpiński set**, whose intersection with each straight line of the plane consists of at most two points. (See also B. R. Gelbaum and J. M. H. Olmsted [134, p. 130] and M. Frantz [127].) Clearly, any Sierpiński set has measurable sections but is not a Borel subset of $\mathbb{R}^2$.

Just as sets have sections, so do functions. If $f: X \times Y \to Z$ is a function, then for each $x \in X$ the symbol f_x denotes the function $f_x: Y \to Z$ defined by $f_x(y) = f(x, y)$ for each $y \in Y$. Similarly, the function $f^y: X \to Z$ is defined by $f^y(x) = f(x, y)$ for each $x \in X$.

4.47 Definition   *Let (X, Σ_X), (Y, Σ_Y), and (Z, Σ_Z) be measurable spaces. We say a function $f: X \times Y \to Z$ is:*

1. *jointly measurable if it is $(\Sigma_X \otimes \Sigma_Y, \Sigma_Z)$-measurable.*

2. *measurable in x if $f^y: (X, \Sigma_X) \to (Z, \Sigma_Z)$ is measurable for each $y \in Y$.*

3. *measurable in y if $f_x: (Y, \Sigma_Y) \to (Z, \Sigma_Z)$ is measurable for each $x \in X$.*

4. *separately measurable if it is both measurable in x and measurable in y.*

Jointly measurable functions are separately measurable.

4.48 Theorem   *Let (X, Σ_X), (Y, Σ_Y), and (Z, Σ_Z) be measurable spaces. Then every jointly measurable function $f: X \times Y \to Z$ is separately measurable.*

Proof: If $y \in Y$ is fixed, then note that for each $A \in \Sigma_Z$, we have

$$(f^y)^{-1}(A) = \{x \in X : f_y(x) = f(x, y) \in A\} = [f^{-1}(A)]^y \in \Sigma_X,$$

where the last membership holds by virtue of Lemma 4.46. This shows that $f^y: X \to Z$ is measurable for each $y \in Y$.   ∎

Separate measurability does not imply joint measurability. For instance, the indicator function for a Sierpiński set is separately measurable, but fails to be jointly measurable. However, functions into a product σ-algebra are measurable if and only if each component is.

4.49 Lemma   *Let (X, Σ), (X_1, Σ_1) and (X_2, Σ_2) be measurable spaces, and let $f_1: X \to X_1$ and $f_2: X \to X_2$. Define $f: X \to X_1 \times X_2$ by*

$$f(x) = (f_1(x), f_2(x)).$$

Then $f: (X, \Sigma) \to (X_1 \times X_2, \Sigma_1 \otimes \Sigma_2)$ is measurable if and only if the two functions $f_1: (X, \Sigma) \to (X_1, \Sigma_1)$ and $f_2: (X, \Sigma) \to (X_2, \Sigma_2)$ are both measurable.

Proof: Start by observing that if $A \subset X_1$ and $B \subset X_2$ are arbitrary, then

$$f^{-1}(A \times B) = f_1^{-1}(A) \cap f_2^{-1}(B). \qquad (\star)$$

Now assume that both $f_1: (X, \Sigma) \to (X_1, \Sigma_1)$ and $f_2: (X, \Sigma) \to (X_2, \Sigma_2)$ are measurable and let A belong to Σ_1 and B belong to Σ_2. Then, from $(\star)$ it easily follows that $f^{-1}(A \times B) \in \Sigma$. Since the rectangles $A \times B$ with $A \in \Sigma_1$ and $B \in \Sigma_2$ generate the product σ-algebra, it follows from Corollary 4.24 that the function $f: (X, \Sigma) \to (X_1 \times X_2, \Sigma_1 \otimes \Sigma_2)$ is measurable.

For the converse, assume $f: (X, \Sigma) \to (X_1 \times X_2, \Sigma_1 \otimes \Sigma_2)$ is measurable and let A belong to Σ_1. Then $A \times X_2$ belongs to $\Sigma_1 \otimes \Sigma_2$, so by $(\star)$,

$$f_1^{-1}(A) = f_1^{-1}(A) \cap X = f_1^{-1}(A) \cap f_2^{-1}(X_2) = f^{-1}(A \times X_2) \in \Sigma.$$

This shows that the function $f_1: (X, \Sigma) \to (X_1, \Sigma_1)$ is measurable. Similarly, the function $f_2: (X, \Sigma) \to (X_2, \Sigma_2)$ is also measurable.   ∎

4.10 Carathéodory functions

In this section we shall discuss a special class of useful functions that are continuous in one variable and measurable in another. They are known as Carathéodory functions.

4.50 Definition   *Let (S, Σ) be a measurable space, and let X and Y be topological spaces. A function $f \colon S \times X \to Y$ is a **Carathéodory function** if:*

1. *for each $x \in X$, the function $f^x = f(\cdot, x) \colon S \to Y$ is $(\Sigma, \mathcal{B}_Y)$-measurable; and*

2. *for each $s \in S$, the function $f_s = f(s, \cdot) \colon X \to Y$ is continuous.*

Carathéodory functions have the virtue of being jointly measurable in many important cases.

4.51 Lemma (Carathéodory functions are jointly measurable)   *Let (S, Σ) be a measurable space, X a separable metrizable space, and Y a metrizable space. Then every Carathéodory function $f \colon S \times X \to Y$ is jointly measurable.*

Proof: Let d and ρ be compatible metrics on X and Y respectively. Let $\{x_1, x_2, \ldots\}$ be a countable dense subset of X and observe that since $f(s, \cdot)$ is continuous, $f(s, x)$ belongs to the closed set F if and only if for each n there is some x_m with $d(x, x_m) < \frac{1}{n}$ and $\rho(f(s, x_m), F) < \frac{1}{n}$. This easily implies

$$f^{-1}(F) = \bigcap_{n=1}^{\infty} \bigcup_{m=1}^{\infty} \{s \in S : f(s, x_m) \in N_{\frac{1}{n}}(F)\} \times B_{\frac{1}{n}}(x_m),$$

where $N_{\frac{1}{n}}(F) = \{y \in Y : \rho(y, F) < \frac{1}{n}\}$. Since f is measurable in s, and $N_{\frac{1}{n}}(F)$ is open (and hence Borel), $\{s \in S : f(s, x_m) \in N_{\frac{1}{n}}(F)\}$ is measurable. Thus $f^{-1}(F)$ is measurable.   ∎

The theorem above relies on the separability of the space X. In fact, R. O. Davies and J. Dravecký [82] show that the theorem may fail when X is not separable. The next result is technical, but it is used later. Note the role that separability of the spaces Y_i plays.

4.52 Lemma   *Let (S, Σ) be a measurable space, X, Y_1, and Y_2 be separable metrizable spaces, and Z be a topological space. If $f_i \colon S \times X \to Y_i$, $i = 1, 2$, are Carathéodory functions, and $g \colon Y_1 \times Y_2 \to Z$ is Borel measurable, then the composition $h \colon S \times X \to Z$ defined by $h(s, x) = g(f_1(s, x), f_2(s, x))$ is jointly measurable.*

Proof: By Lemma 4.51 each f_i is $(\Sigma \otimes \mathcal{B}_X, \mathcal{B}_{Y_i})$-measurable from $S \times X$ into Y_i. Therefore, by Lemma 4.49, the function $(s, x) \mapsto (f_1(s, x), f_2(s, x))$ (from $S \times X$ to $Y_1 \times Y_2$) is $(\Sigma \otimes \mathcal{B}_X, \mathcal{B}_{Y_1} \otimes \mathcal{B}_{Y_2})$-measurable. Now g is $(\mathcal{B}_{Y_1 \times Y_2}, \mathcal{B}_Z)$-measurable, and since each Y_i is separable, $\mathcal{B}_{Y_1} \otimes \mathcal{B}_{Y_2} = \mathcal{B}_{Y_1 \times Y_2}$ by Theorem 4.44, so the composition $h = g \circ (f_1, f_2)$ is measurable, as desired. ∎

There is a one-to-one correspondence between functions from S into Y^X and functions from $S \times X$ into Y. In certain cases we can identify Carathéodory functions with measurable functions from S into $C(X, Y)$, where $C(X, Y)$ is endowed with its topology of uniform convergence. But first we must describe the Borel σ-algebra of $C(X, Y)$.

Recall that for each $x \in X$ the **evaluation functional** at x is the function $e_x \colon C(X, Y) \to Y$ defined by $e_x(f) = f(x)$ for each $f \in C(X, Y)$. Clearly each evaluation functional is continuous on $C(X, Y)$, and therefore Borel measurable.

4.53 Lemma *Assume that X is compact and metrizable and that Y is separable and metrizable. Then the family*

$$\mathcal{C} = \{e_x^{-1}(F) : x \in X \text{ and } F \text{ is closed in } Y\}$$
$$= \left\{\{f \in C(X, Y) : f(x) \in F\} : x \in X \text{ and } F \text{ is closed in } Y\right\}$$

generates the Borel σ-algebra on $C(X, Y)$, that is, $\mathcal{B}_{C(X,Y)} = \sigma(\mathcal{C})$.

In other words, the Borel σ-algebra $\mathcal{B}_{C(X,Y)}$ is the smallest σ-algebra on $C(X, Y)$ for which all the evaluations are Borel measurable functions.

Proof: As mentioned before the theorem, every set in $\mathcal{C}$ is a Borel subset of $C(X, Y)$. It remains to show that $\mathcal{B}_{C(X,Y)} \subset \sigma(\mathcal{C})$.

For the reverse inclusion, it suffices to show that every open set in $C(X, Y)$ is contained in $\sigma(\mathcal{C})$. Now (by Lemma 3.99) $C(X, Y)$ is separable, so let D be a countable dense subset of $C(X, Y)$. Let ρ be a compatible metric on Y. Then the topology on $C(X, Y)$ is generated by the metric

$$d(f, g) = \sup_{x \in X} \rho(f(x), g(x)).$$

(Lemma 3.98 asserts that the topology on $C(X, Y)$ is independent of the compatible metric ρ.) Thus every open subset of $C(X, Y)$ is a countable union of closed sets of the form $\{F_{\frac{1}{n}}(f) : f \in D,\ n \in \mathbb{N}\}$, where $F_\varepsilon(f) = \{g \in C(X, Y) : d(g, f) \leqslant \varepsilon\}$. It suffices to show that every such $F_\varepsilon(f)$ belongs to $\sigma(\mathcal{C})$.

So fix $f \in C(X, Y)$, $x \in X$, and $\varepsilon > 0$. Let A be the closed ball in Y centered at $f(x)$ with radius $\varepsilon > 0$, that is, $A = \{y \in Y : \rho(f(x), y) \leqslant \varepsilon\}$. Then

$$\{g \in C(X, Y) : \rho(f(x), g(x)) \leqslant \varepsilon\} = \{g \in C(X, Y) : e_x(g) = g(x) \in A\}$$
$$= e_x^{-1}(A) \in \mathcal{C}.$$

Now let $\{x_1, x_2, \ldots\}$ be a countable dense subset of X. Then

$$F_\varepsilon(f) = \bigcap_{n=1}^{\infty} \{g \in C(X, Y) : \rho(f(x_n), g(x_n)) \leqslant \varepsilon\} \in \sigma(\mathcal{C}).$$

Therefore $\mathcal{B}_{C(X,Y)} = \sigma(\mathcal{C})$, as claimed. ∎

4.54 Corollary   *Assume that X is compact and metrizable and that Y is separable and metrizable. Let $\mathcal{F}$ be any family of sets generating the Borel σ-algebra of Y. Then the family*

$$\mathcal{C} = \{e_x^{-1}(F) : x \in X \text{ and } F \in \mathcal{F}\}$$

generates the Borel σ-algebra on $C(X, Y)$.

As usual, to simplify notation, for $f : S \to Y^X$ write f_s for $f(s)$. Given $f : S \times X \to Y$ define $\hat{f} : S \to Y^X$ by $\hat{f}_s(x) = f(s, x)$. Similarly, for $g : S \to Y^X$ define $\overline{g} : S \times X \to Y$ by $\overline{g}(s, x) = g_s(x)$. Observe that $f = \overline{\hat{f}}$ and $g = \hat{\overline{g}}$. That is, $f \mapsto \hat{f}$ and $g \mapsto \overline{g}$ are inverses. Under these mappings, Carathéodory functions are the same as Borel measurable functions into $C(X, Y) \subset Y^X$, when Y is a separable metric space, X is compact and metrizable, and $C(X, Y)$ is endowed with its topology of uniform convergence.

4.55 Theorem (Measurable functions into $C(X, Y)$)   *Let (S, Σ) be a measurable space, X a compact metrizable space, (Y, d) a separable metric space, and let $C(X, Y)$ be endowed with the topology of d-uniform convergence.*

1. *If $f : S \times X \to Y$ is a Carathéodory function, then $\hat{f}$ maps S into $C(X, Y)$ and is Borel measurable.*

2. *If $g : S \to C(X, Y)$ is Borel measurable, then $\overline{g}$ is a Carathéodory function.*

Proof: (1)   Let $f : S \times X \to Y$ be a Carathéodory function. By Lemma 4.53 it suffices to show that $\hat{f}^{-1}(B) \in \Sigma$ for each set B of the form

$$B = e_x^{-1}(F) = \{h \in C(X, Y) : d(h(x), F) = 0\},$$

where F is an arbitrary closed subset of Y. To this end, define $\theta : S \times X \to \mathbb{R}$ by $\theta(s, x) = d(f(s, x), F)$. By Lemma 4.52, θ is jointly measurable, so θ^x defined by $\theta^x(s) = \theta(s, x)$ is measurable. Then

$$\hat{f}^{-1}(B) = \{s \in S : d(f(s, x), F) = 0\} = (\theta^x)^{-1}(0),$$

which belongs to Σ, so $\hat{f}$ is Borel measurable.

(2)   Let $g : S \to C(X, Y)$ be Borel measurable, and define $\overline{g} : S \times X \to Y$ by $\overline{g}(s, x) = g_s(x)$. Clearly $\overline{g}$ is continuous in x for each s. To see that $\overline{g}(\cdot, x)$

is Borel measurable, let U be an open subset of Y. Now the pointwise open set $G = \{h \in C(X, Y) : h(x) \in U\}$ is an open subset of $C(X, Y)$. But

$$\{s \in S : \overline{g}(s, x) \in U\} = \{s \in S : g_s \in G\} = g^{-1}(G) \in \Sigma,$$

since g is Borel measurable. Thus $\overline{g}$ is a Carathéodory function. ∎

We can view the evaluation as a function $e\colon C(X, Y) \times X \to Y$ defined via $e(f, x) = f(x)$. It is easy to verify that under the hypotheses of Theorem 4.55 that e is continuous in f and x separately and thus jointly measurable (as a Carathéodory function). R. J. Aumann [25] provides additional results on e viewed this way. In particular, he shows that more generally e may fail to be jointly measurable.

4.11 Borel functions and continuity

In this section we present some relationships between Borel measurability and continuity of functions on Polish spaces. It turns out that a Polish domain can be re-topologized to make a given Borel function continuous, and yet retain the same Borel σ-algebra. Later on, we prove Lusin's Theorem 12.8, which asserts that even with the given topology, a Borel measurable function is "almost" continuous in a measure theoretic sense. We start with a couple of simple lemmas on Polish topologies taken from A. S. Kechris [196, Lemmas 13.2, 13.3, p. 82].

4.56 Lemma   *Let (X, τ) be a Polish topological space, and let F be a closed subset of X. Then there is a Polish topology $\tau_F \supset \tau$ on X such that F is τ_F-clopen and $\sigma(\tau_F) = \sigma(\tau)$.*

Proof: Let F be a closed subset of X, and let τ_F be the topology generated by $\tau \cup \{F\}$. This is the smallest topology including τ for which F is open and hence clopen. The τ_F-open sets are precisely those of the form $V \cup (W \cap F)$ where $V, W \in \tau$. Consequently a net $\{x_\alpha\}$ in X satisfies $x_\alpha \xrightarrow{\tau_F} x$ in X if and only if (i) $x_\alpha \xrightarrow{\tau} x$ and (ii) if $x \in F$, then $x_\alpha \in F$ for all α large enough. Further, since $F^c \in \tau$, it is easy to see that $\sigma(\tau) = \sigma(\tau_F)$.

It remains to show that τ_F is completely metrizable. Consider the Polish topological space $X \times \{0, 1\}$; it is convenient to view this space as a *disjoint union* of X with itself. If $G = F \times \{0\} \cup F^c \times \{1\}$, then $G = \bigcap_{n=1}^{\infty}[N_{\frac{1}{n}}(F) \times \{0\} \cup F^c \times \{1\}]$ is a $\mathcal{G}_\delta$ in $X \times \{0, 1\}$. Therefore by Alexandroff's Lemma 3.34, it is completely metrizable. But there is an obvious homeomorphism between G and (X, τ_F), namely $(x, n) \mapsto x$. This completes the proof. ∎

4.57 Lemma   *Let (X, τ) be Polish, and let $\{\tau_n\}$ be a sequence of Polish topologies on X with $\tau_n \supset \tau$ for each n. Then the topology τ_∞ generated by $\bigcup_{n=1}^{\infty} \tau_n$ is Polish. Further, if $\sigma(\tau) = \sigma(\tau_n)$ for each n, then $\sigma(\tau) = \sigma(\tau_\infty)$.*

Proof: Clearly, the product topological space $Y = \prod_{n=1}^{\infty}(X, \tau_n)$ (with the product topology) is a Polish space. The mapping $f: X \to Y$, defined by $f(x) = (x, x, \ldots)$, is one-to-one. Moreover, $f(X)$ is closed, and so Polish. (Indeed, if a net $\{x_\alpha\}$ in X satisfies $f(x_\alpha) = (x_\alpha, x_\alpha, \ldots) \to (y_1, y_2, \ldots)$ in Y, then $x_\alpha \xrightarrow{\tau_n} y_n$ for each n, and, in particular, $x_\alpha \xrightarrow{\tau} y_n$ for each n. This implies $y_1 = y_2 = \cdots$, so $(y_1, y_2, \ldots) \in f(X)$, proving that $f(X)$ is closed in Y.) Since a net $\{x_\alpha\}$ in X satisfies $x_\alpha \xrightarrow{\tau_\infty} x$ if and only if $x_\alpha \xrightarrow{\tau_n} x$ for each n, it follows that f is a homeomorphism onto its range. Therefore (X, τ_∞) is Polish.

Now assume $\sigma(\tau) = \sigma(\tau_n)$ for each n, and let $\{U_k^n : k \in \mathbb{N}\}$ be a countable base for τ_n. Clearly, $\sigma(\{U_k^n : k \in \mathbb{N}\}) = \sigma(\tau_n) = \sigma(\tau)$ for each n. This implies $\sigma(\tau_\infty) = \sigma(\{U_k^n : n, k \in \mathbb{N}\}) = \sigma(\tau)$. ∎

These lemmas enable us to prove the following result.

4.58 Lemma *Let $\mathcal{C}$ be a countable family of Borel subsets of a Polish space (X, τ). Then there is a Polish topology $\tau^* \supset \tau$ on X with the same Borel σ-algebra, that is, $\sigma(\tau^*) = \sigma(\tau)$, and for which each set in $\mathcal{C}$ is τ^*-clopen.*

Proof: This is another of those results where it is easier to characterize the family of sets satisfying a property than it is to prove that any given set has the property.

Let $\mathcal{A}$ be the family of all subsets A of X such that there is a Polish topology $\tau_A \supset \tau$ satisfying $\sigma(\tau) = \sigma(\tau_A)$ and for which A is τ_A-clopen. By Lemma 4.56, $\mathcal{A}$ includes the closed sets. Now observe that if A is τ_A-clopen, then so is A^c, which shows that $\mathcal{A}$ is closed under complementation. Lemma 4.57 guarantees that $\mathcal{A}$ is closed under countable unions. Since $\varnothing, X \in \mathcal{A}$, it follows that $\mathcal{A}$ is in fact a σ-algebra. Therefore $\mathcal{A}$ includes the Borel sets of X.

Now assume that $\mathcal{C} = \{B_1, B_2, \ldots\}$ is a countable family of Borel sets. By the preceding case, for each n, there exists a Polish topology τ_n on X such that $\tau_n \supset \tau$, $\sigma(\tau_n) = \sigma(\tau)$, and B_n is τ_n-clopen. But then, by Lemma 4.57 again, the topology τ^* generated by $\bigcup_{n=1}^{\infty} \tau_n$ is a Polish topology on X satisfying $\tau^* \supset \tau$ and $\sigma(\tau^*) = \sigma(\tau)$, and clearly every member of $\mathcal{C}$ is τ^*-clopen. ∎

The following remarkable theorem can be found in A. S. Kechris [196, Theorem 13.11, p. 84], who attributes the basic ideas to K. Kuratowski.

4.59 Theorem *Let $\mathcal{F}$ be a countable family of Borel functions from a Polish space (X, τ) to a second countable topological space Y. Then there exists a Polish topology $\tau^* \supset \tau$ on X such that $\sigma(\tau^*) = \sigma(\tau)$, and $f: (X, \tau^*) \to Y$ is continuous for each $f \in \mathcal{F}$.*

Proof: Let $\{V_1, V_2, \ldots\}$ be a countable base for Y. Then the family of Borel sets $\mathcal{C} = \{f^{-1}(V_n) : f \in \mathcal{F} \text{ and } n \in \mathbb{N}\}$ is countable. So by Lemma 4.58, there exists a Polish topology τ^* on X such that $\tau^* \supset \tau$ and each member of $\mathcal{C}$ is τ^*-clopen. Clearly each $f: (X, \tau^*) \to Y$ is continuous. ∎

4.60 Theorem *Every Borel subset of a Polish space is the one-to-one continuous image of a closed subset of the Baire space* $\mathcal{N}$.

Proof: Let B be a Borel subset of the Polish space (X, τ). By Lemma 4.58 there is a Polish topology τ_B finer than τ for which B is closed. In particular (B, τ_B) is a Polish space, so by Theorem 3.66, B is the one-to-one continuous image of a closed subset of $\mathcal{N}$. But such a function is also continuous onto (B, τ). ∎

4.61 Corollary *Every nonempty Borel subset of a Polish space is a continuous image of* $\mathcal{N}$.

Proof: This follows from the preceding theorem by observing that every closed subset of $\mathcal{N}$ is a retract of $\mathcal{N}$ (Lemma 3.64). ∎

Two additional interesting and important results must be postponed until we have some more machinery. One is that a function is Borel measurable if and only if its graph is Borel set (Theorem 12.28). The other (Theorem 12.29) asserts that the one-to-one image of a Borel set under a Borel function is a Borel set.

4.12   The Baire σ-algebra

Corollary 4.26 implies that every real-valued continuous function between topological spaces is Borel measurable. In particular, for every topological space X, every function in $C_c(X)$, the space of continuous real functions with compact support, is measurable with respect to the Borel σ-algebra. The **Baire σ-algebra** of X, denoted $\mathcal{B}aire(X)$, is defined to be the smallest σ-algebra on X for which all members of $C_c(X)$ are measurable. That is, the Baire σ-algebra is the σ-algebra of subsets of X generated by the family of sets

$$\{f^{-1}(V) : f \in C_c(X) \text{ and } V \text{ is open in } X\}.$$

Members of this σ-algebra are called **Baire sets**. The Baire σ-algebra is most interesting when X is a locally compact Hausdorff space.

Some authors use another definition of the Baire sets, so let $\mathcal{B}aire^*$ denote the smallest σ-algebra for which all members of $C(X)$, the continuous real functions, are measurable.[8] This is also the σ-algebra generated by $C_b(X)$, the bounded continuous real functions. Clearly, $\mathcal{B}aire \subset \mathcal{B}aire^*$.

We start our investigation of the Baire sets by noting the following properties of locally compact Hausdorff spaces.

---

[8] Be warned that as with the Borel sets, there are different definitions of the Baire sets in common use. Dudley [104] defines the Baire sets to be what we call the Baire* sets. Halmos [148] defines the Baire sets to be the members of the σ-ring generated by the nonempty compact $\mathcal{G}_\delta$s. See Royden [290, Section 13.1, pp. 331–334], whose terminology we adopt, for an extended discussion of the various definitions. For locally compact separable metrizable spaces (such as $\mathbb{R}^n$) all these definitions agree.

4.62 Lemma   *For a compact subset K of a locally compact Hausdorff space X we have the following.*

1. *If F is a closed set satisfying $K \cap F = \varnothing$, then there is a function $f \in C_c(X)$ with $0 \leqslant f(x) \leqslant 1$ for all x, $f(x) = 0$ for $x \in F$, and $f(x) = 1$ for $x \in K$.*

2. *If G is an open set satisfying $K \subset G$, then there exist an open Baire set U and a compact $\mathcal{G}_\delta$-set D such that $K \subset U \subset D \subset G$.*

Proof: (1)   By Corollary 2.69 there is an open set W having compact closure and satisfying $K \subset W \subset \overline{W} \subset F^c$. By Corollary 2.74 there is a continuous real function f with $0 \leqslant f \leqslant 1$ satisfying $f(x) = 1$ for each $x \in K$, and $f(x) = 0$ for $x \in W^c$. The support of f lies in the compact set $\overline{W}$, so f belongs to $C_c(X)$ and has the desired properties.

(2)   By part (1) there is some $f \in C_c(X)$ satisfying $f(x) = 1$ for each $x \in K$ and $f(x) = 0$ for $x \in G^c$. Then $D = [f \geqslant \frac{1}{2}]$ is a compact $\mathcal{G}_\delta$ (why?) and $U = [f > \frac{1}{2}]$ is an open Baire set, and $K \subset U \subset D \subset G$.   ∎

4.63 Corollary   *In a locally compact Hausdorff space, the open Baire sets constitute a base for the topology.*

Proof: Let U be an open neighborhood of the point x. By Lemma 4.62 here is some $f \in C_c$ with $f(x) = 1$ and $f(y) = 0$ for $y \in U^c$. Then $V = [f > \frac{1}{2}]$ is an open Baire set satisfying $x \in V \subset U$.   ∎

We now present another characterization of the Baire sets.

4.64 Lemma   *The Baire σ-algebra of a locally compact Hausdorff space is the σ-algebra generated by the family of compact $\mathcal{G}_\delta$-sets.*

Proof: Let X be a locally compact Hausdorff space. First we show that every Baire set belongs to the σ-algebra $\mathcal{A}$ generated by the compact $\mathcal{G}_\delta$-sets. By definition, $\mathcal{B}aire$ is the σ-algebra generated by the family of sets of the form $[f \geqslant \alpha]$ for $f \in C_c(X)$. So let f belong to $C_c(X)$, and note that $-f$ also belongs to $C_c(X)$. Since $[f \geqslant \alpha] = \bigcap_{n=1}^{\infty}[f > \alpha - \frac{1}{n}]$, we see that $[f \geqslant \alpha]$ is always a $\mathcal{G}_\delta$. Further, for $\alpha > 0$, $[f \geqslant \alpha]$ is a closed subset of the support of f, so it is compact. Thus $[f \geqslant \alpha] \in \mathcal{A}$ whenever $\alpha > 0$. Now observe that for $\alpha < 0$ we have $0 < -\alpha + \frac{\alpha}{2n} < -\alpha$ and that

$$[f \geqslant \alpha] = [f < \alpha]^c = [-f > -\alpha]^c = \left(\bigcap_{n=1}^{\infty}[-f \geqslant -\alpha + \tfrac{\alpha}{2n}] \right)^c \in \mathcal{A}.$$

Also observe that $[f \geqslant 0] = \bigcap_{n=1}^{\infty}[f \geqslant -\frac{1}{n}] \in \mathcal{A}$. This shows that $\mathcal{A}$ is a σ-algebra containing every set of the form $[f \geqslant \alpha]$ where $f \in C_c(X)$. Therefore, $\mathcal{B}aire \subset \mathcal{A}$.

Next we show that $\mathcal{A} \subset \mathcal{B}aire$. So let $K = \bigcap_{n=1}^{\infty} G_n$ be a compact $\mathcal{G}_\delta$, with each G_n an open set. By Lemma 4.62 (2), for each n there is an open Baire set V_n such that $K \subset V_n \subset G_n$. This implies $K = \bigcap_{n=1}^{\infty} V_n$, so K is a Baire set. Thus $\mathcal{A} \subset \mathcal{B}aire$, and the proof is complete.   ∎

We mention without proof that every compact Baire set is actually a compact $\mathcal{G}_\delta$. See P. R. Halmos [148, Theorem D, p. 221] for a proof, but be aware that his definition is different from ours in a way that does not matter for this proposition. The next lemma relates the Baire sets and the Borel sets.

4.65 Lemma (Baire and Borel sets)   *For a topological space X:*

1. *Every Baire set is a Borel set. Indeed, $\mathcal{B}aire \subset \mathcal{B}aire^* \subset \mathcal{B}_X$.*

2. *If X is locally compact, separable, and metrizable, then*

$$\mathcal{B}aire = \mathcal{B}aire^* = \mathcal{B}_X.$$

3. *If X is metrizable, then $\mathcal{B}aire^* = \mathcal{B}_X$.*

Proof: (1)   As mentioned above, this follows from Corollary 4.26.

(2)   Let X be a separable locally compact metrizable space with compatible metric d. It suffices to show that every closed set is a Baire set. Now Lemma 2.76 implies that X is a countable union of compact sets. Therefore each closed set is likewise a countable union of compact sets. So it suffices to show that each compact set is a Baire set. But every compact set in a metric space is a $\mathcal{G}_\delta$ by Corollary 3.19, so by Lemma 4.64 a Baire set too.

(3)   The distance functions $d(\cdot, F)$, where F is closed, are continuous, and thus $F = \{x \in X : d(x, F) = 0\} \in \mathcal{B}aire^*$. This implies that $\mathcal{B}aire^*$ includes all the closed sets, and therefore all the Borel sets.   ∎

The Baire and Borel σ-algebras may be different in general. Here is a slightly complicated but very interesting example, showing that the Baire sets may not include all the interesting Borel sets.

4.66 Example (Baire vs. Borel sets)   Let X_∞ be the one-point compactification of an uncountable discrete space X. Then X_∞ is a compact Hausdorff space. Furthermore, every subset of X is open in X_∞, and $\{\infty\}$ is closed, so every subset of X_∞ is a Borel set. That is, $\mathcal{B}_{X_\infty} = 2^{X_\infty}$.

The Baire sets of X_∞ are more difficult to describe. The only compact subsets of X_∞ that are subsets of X are finite. These sets are also open, so they are $\mathcal{G}_\delta$-sets. Now note that any set that contains ∞ is closed, since its complement, as a subset of X, is open. Since X_∞ is compact, any set containing ∞ must be compact too.

Now recall that any open set that contains ∞ must be the complement of a compact (that is, finite) subset of X. Thus any $\mathcal{G}_\delta$ that contains ∞ must be the complement of a countable subset of X.

Therefore the compact $\mathcal{G}_\delta$-sets in X_∞ are the finite subsets of X and the complements (in X_∞) of countable subsets of X. It follows by Lemma 4.64 that the Baire σ-algebra of X_∞ comprises the countable subsets of X and their complements. In particular, neither X nor $\{\infty\}$ is a Baire set, and any uncountable Baire set contains ∞.   ∎

Note that in a second countable topological space any base for the topology also generates the Borel σ-algebra. The reason is that a base must contain a countable base, so every open set must belong to the σ-algebra generated by the base. For locally compact Hausdorff spaces, the σ-algebra generated by any base (countable or not) includes the Baire σ-algebra.

4.67 Lemma   *Let $\mathcal{V}$ be a subbase for the topology on a locally compact Hausdorff space. Then $\mathcal{B}aire \subset \sigma(\mathcal{V}) \subset \mathcal{B}$.*

Proof: Since the family of finite intersections from $\mathcal{V}$ is a base, $\sigma(\mathcal{V})$ includes a base for the topology. It suffices to prove that $\sigma(\mathcal{V})$ contains every compact $\mathcal{G}_\delta$. So suppose $K = \bigcap_{n=1}^{\infty} G_n$ is a compact $\mathcal{G}_\delta$, where each G_n is open. Since K is compact, for each n, there is an open set V_n such that $K \subset V_n \subset G_n$ and V_n is a finite union of basic open sets in $\mathcal{V}$. Therefore $V_n \in \mathcal{V}$, and $K = \bigcap_{n=1}^{\infty} V_n$, so $K \in \sigma(\mathcal{V})$.   ∎

For an important class of spaces that includes all the Euclidean spaces, the Baire σ-algebra of a product of two spaces is the product of their Baire σ-algebras.

4.68 Theorem   *If X and Y are second countable locally compact Hausdorff spaces, then*

$$\mathcal{B}aire(X \times Y) = \mathcal{B}aire(X) \otimes \mathcal{B}aire(Y).$$

Proof: For locally compact Hausdorff spaces, the Baire σ-algebra is generated by the compact $\mathcal{G}_\delta$-sets. Also note that if X and Y are locally compact, then $X \times Y$ is locally compact. (Why?) Now define

$$\Sigma(A) = \{B \subset Y : A \times B \in \mathcal{B}aire(X \times Y)\}.$$

As in the proof of Theorem 4.44 (1), $\Sigma(A)$ is a σ-ring for any A. Now suppose K is a compact $\mathcal{G}_\delta$ in X. If C is a compact $\mathcal{G}_\delta$ in Y, then $K \times C$ is a compact $\mathcal{G}_\delta$ in $X \times Y$. (Why?) Thus $\Sigma(K)$ contains every compact $\mathcal{G}_\delta$. Furthermore, if Y is second countable, it follows from Corollary 2.77 and Lemma 4.65 that Y belongs to $\Sigma(K)$. Thus $\Sigma(K)$ is a σ-algebra that includes $\mathcal{B}aire(Y)$.

Now put $\mathcal{A} = \{A \subset X : \mathcal{B}aire(Y) \subset \Sigma(A)\}$. This set is closed under complementation and countable intersections, and we have just shown that it contains every compact $\mathcal{G}_\delta$. Consequently, $\mathcal{B}aire(X) \otimes \mathcal{B}aire(Y) \subset \mathcal{B}aire(X \times Y)$.

For the reverse inclusion, observe that Lemma 4.62 implies that sets of the form $U \times V$, where U is an open Baire set in X and V is an open Baire set of Y constitute a base for $X \times Y$. Since each such $U \times V$ belongs to $\mathcal{B}aire(X) \times \mathcal{B}aire(Y)$, by Lemma 4.67, we see that $\mathcal{B}aire(X) \otimes \mathcal{B}aire(Y)$ includes $\mathcal{B}aire(X \times Y)$, so we have equality.   ∎

If we take the Baire sets to be the members of the σ-ring generated by the compact $\mathcal{G}_\delta$-sets, then the hypothesis of second countability may be dropped from the theorem above.

Chapter 5

Topological vector spaces

One way to think of functional analysis is as the branch of mathematics that studies the extent to which the properties possessed by finite dimensional spaces generalize to infinite dimensional spaces. In the finite dimensional case there is only one natural linear topology. In that topology every linear functional is continuous, convex functions are continuous (at least on the interior of their domains), the convex hull of a compact set is compact, and nonempty disjoint closed convex sets can always be separated by hyperplanes. On an infinite dimensional vector space, there is generally more than one interesting topology, and the topological dual, the set of continuous linear functionals, depends on the topology. In infinite dimensional spaces convex functions are not always continuous, the convex hull of a compact set need not be compact, and nonempty disjoint closed convex sets cannot generally be separated by a hyperplane. However, with the right topology and perhaps some additional assumptions, each of these results has an appropriate infinite dimensional version.

Continuous linear functionals are important in economics because they can often be interpreted as prices. Separating hyperplane theorems are existence theorems asserting the existence of a continuous linear functional separating disjoint convex sets. These theorems are the basic tools for proving the existence of efficiency prices, state-contingent prices, and Lagrange multipliers in Kuhn–Tucker type theorems. They are also the cornerstone of the theory of linear inequalities, which has applications in the areas of mechanism design and decision theory. Since there is more than one topology of interest on an infinite dimensional space, the choice of topology is a key modeling decision that can have economic as well as technical consequences.

The proper context for separating hyperplane theorems is that of linear topologies, especially locally convex topologies. The classic works of N. Dunford and J. T. Schwartz [110, Chapter V], and J. L. Kelley and I. Namioka, *et al.* [199], as well as the more modern treatments by R. B. Holmes [166], H. Jarchow [181], J. Horváth [168], A. P. Robertson and W. J. Robertson [287], H. H. Schaefer [293], A. E. Taylor and D. C. Lay [330], and A. Wilansky [341] are good references on the general theory of linear topologies. R. R. Phelps [278] gives an excellent treatment of convex functions on infinite dimensional spaces. For applications to problems of optimization, we recommend J.-P. Aubin and I. Ekeland [23], I. Ekeland and R. Temam [115], I. Ekeland and T. Turnbull [116], and R. R. Phelps [278].

Here is the road map for this chapter. We start by defining a *topological vector space* (tvs) as a vector space with a topology that makes the vector operations continuous. Such a topology is *translation invariant* and can therefore be characterized by the neighborhood base at zero. While the topology may not be metrizable, there is a base of neighborhoods that behaves in some ways like the family of balls of positive radius (Theorem 5.6). In particular, if V is a neighborhood of zero, it includes another neighborhood W such that $W + W \subset V$. So if we think of V as an ε-ball, then W is like the $\varepsilon/2$-ball.

There is a topological characterization of finite dimensional topological vector spaces. (Finite dimensionality is an algebraic, not topological property.) A Hausdorff tvs is finite dimensional if and only if it is locally compact (Theorem 5.26). There is a unique Hausdorff linear topology on any finite dimensional space, namely the Euclidean topology (Theorem 5.21). Any finite dimensional subspace of a Hausdorff tvs is closed (Corollary 5.22) and *complemented* (Theorem 5.89) in locally convex spaces.

There is also a simple characterization of metrizable topological vector spaces. A Hausdorff tvs is metrizable if and only if there is a countable neighborhood base at zero (Theorem 5.10).

Without additional structure, these spaces can be quite dull. In fact, it is possible to have an infinite dimensional metrizable tvs where zero is the only continuous linear functional (Theorem 13.31). The additional structure comes from convexity. A set is *convex* if it includes the line segments joining any two of its points. A real function f is convex if its epigraph, $\{(x, \alpha) : \alpha \geqslant f(x)\}$, is convex. All linear functionals are convex. A convex function on an open convex set is continuous if it is bounded above on a neighborhood of a point (Theorem 5.43). Thus linear functions are continuous if and only if they are bounded on a neighborhood of zero. When zero has a base of convex neighborhoods, the space is *locally convex*. These are the spaces we really want. A convex neighborhood of zero gives rise to a convex homogeneous function known as its *gauge*. The gauge function of a set tells for each point how much the set must be enlarged to include it. In a normed space, the *norm* is the gauge of the unit ball. Not all locally convex spaces are normable, but the family of gauges of symmetric convex neighborhoods of zero, called *seminorms*, are a good substitute. The best thing about locally convex spaces is that they have lots of continuous linear functionals. This is a consequence of the seemingly innocuous Hahn–Banach Extension Theorem 5.53. The most important consequence of the Hahn–Banach Theorem is that in a locally convex space, there are *hyperplanes* that strictly separate points from closed convex sets that don't contain them (Corollary 5.80). As a result, every closed convex set is the intersection of all closed *half spaces* including it.

Another of the consequences of the Hahn–Banach Theorem is that the set of continuous linear functionals on a locally convex space separates points. The collection of continuous linear functionals on X is known as the (topological) *dual space*, denoted X'. Now each $x \in X$ defines a linear functional on X' by $x(x') = x'(x)$. Thus we are led to the study of *dual pairs* $\langle X, X' \rangle$ of spaces and their associated *weak topologies*. These weak topologies are locally convex. The weak topology on X' induced by X is called the *weak** topology on X'. The most familiar example of a dual pair is probably the pairing of functions and measures—each defines a linear functional via the integral $\int f \, d\mu$, which is linear in f for a fixed μ, and linear in μ for a fixed f. (The weak topology induced on probability measures by this duality with continuous functions is the topology of convergence in distribution that is used in Central Limit Theorems.) Remarkably, in a dual pair $\langle X, X' \rangle$, any subspace of X' that separates the points of X is weak* dense in X' (Corollary 5.108).

G. Debreu [84] introduced dual pairs in economics in order to describe the duality between commodities and prices. According to this interpretation, a dual pair $\langle X, X' \rangle$ represents the commodity-price duality, where X is the commodity space, X' is the price space, and $\langle x, x' \rangle$ is the value of the bundle x at prices x'. This is the basic ingredient of the Arrow–Debreu–McKenzie model of general economic equilibrium; see [9].

If we put the weak topology on X generated by X', then X' is the set of all continuous linear functionals on X (Theorem 5.93). Given a weak neighborhood V of zero in X, we look at all the linear functionals that are bounded on this neighborhood. Since they are bounded, they are continuous and so lie in X'. We further normalize them so that they are bounded by unity on V. The resulting set is called the *polar* of V, denoted V°. The remarkable Alaoglu Theorem 5.105 asserts that V° is compact in the weak topology X generates on X'. Its proof relies on the Tychonoff Product Theorem 2.61. The useful Bipolar Theorem 5.103 states the polar of the polar of a set A is the closed convex circled hull of A.

We might ask what other topologies besides the weak topology on X give X' as the dual. The Mackey–Arens Theorem 5.112 answers this question. The answer is that for a topology on X to have X' as its dual, there must be a base at zero consisting of the duals of a family of weak* compact convex circled subsets of X'. Thus the topology generated by the polars of all the weak* compact convex circled sets in X' is the strongest topology on X for which X' is the dual. This topology is called the *Mackey topology* on X, and it has proven to be extremely useful in the study of infinite dimensional economies. It was introduced to economics by T. F. Bewley [40]. The usefulness stems from the fact that once the dual space of continuous linear functionals has been fixed, the Mackey topology allows the greatest number of continuous real (nonlinear) functions.

There are entire volumes devoted to the theory of topological vector spaces, so we cannot cover everything in one chapter. Chapter 6 describes the additional properties of spaces where the topology is derived from a norm. Chapter 7 goes into more depth on the properties of convex sets and functions. Convexity involves a strange synergy between the topological structure of the space and its algebraic structure. A number of results there are special to the finite dimensional case. Another important aspect of the theory is the interaction of the topology and the order structure of the space. Chapter 8 covers *Riesz spaces*, which are partially ordered topological vector spaces where the partial order has topological and algebraic restrictions modeled after the usual order on $\mathbb{R}^n$. Chapter 9 deals with normed partially ordered spaces.

5.1 Linear topologies

Recall that a (real) **vector space** or (real) **linear space** is a set X (whose elements are called vectors) with two operations: addition, which assigns to each pair of vectors x, y the vector $x + y$, and scalar multiplication, which assigns to vector x and each scalar (real number) α the vector αx. There is a special vector 0. These operations satisfy the following properties: $x + y = y + x$, $(x + y) + z = x + (y + z)$, $x + 0 = x$, $x + (-1)x = 0$, $1x = x$, $\alpha(\beta x) = (\alpha\beta)x$, $\alpha(x + y) = \alpha x + \alpha y$, and $(\alpha + \beta)x = \alpha x + \beta x$. (There are also complex vector spaces, where the scalars are complex numbers, but we won't have occasion to refer to them.)

A subset of a vector space is called a **vector subspace** or **(linear subspace)** if it is a vector space in its own right under the induced operations. The **(linear)** **span** of a subset is the smallest vector subspace including it. A function $f : X \to Y$ between two vector spaces is **linear** if it satisfies

$$f(\alpha x + \beta z) = \alpha f(x) + \beta f(z)$$

for every $x, z \in X$ and $\alpha, \beta \in \mathbb{R}$. Linear functions between vector spaces are usually called **linear operators**. A linear operator from a vector space to the real line is called a **linear functional**.

A topology τ on a vector space X is called a **linear topology** if the operations addition and scalar multiplication are τ-continuous. That is, if $(x, y) \mapsto x + y$ from $X \times X$ to X and $(\alpha, x) \mapsto \alpha x$ from $\mathbb{R} \times X$ to X are continuous. Then (X, τ) is called a **topological vector space** or **tvs** for short. (A topological vector space may also be called a **linear topological space**, especially in older texts.) A tvs need not be a Hausdorff space.

A mapping $\varphi : L \to M$ between two topological vector spaces is a **linear homeomorphism** if φ is one-to-one, linear, and $\varphi : L \to \varphi(L)$ is a homeomorphism. The linear homeomorphism φ is also called an **embedding** and $\varphi(L)$ is referred to as a **copy** of L in M. Two topological vector spaces are **linearly homeomorphic** if there exists a linear homeomorphism from one onto the other.

5.1 Lemma *Every vector subspace of a tvs with the induced topology is a topological vector space in its own right.*

Products of topological vector spaces are topological vector spaces.

5.2 Theorem *The product of a family of topological vector spaces is a tvs under the pointwise algebraic operations and the product topology.*

Proof: Let $\{(X_i, \tau_i)\}_{i \in I}$ be a family of topological vector spaces and let $X = \prod_{i \in I} X_i$ and $\tau = \prod_{i \in I} \tau_i$. We show only that addition on X is continuous and leave the case of scalar multiplication as an exercise.

Let $(x_i^\alpha) \xrightarrow[\alpha]{\tau} (x_i)$ and $(y_i^\lambda) \xrightarrow[\lambda]{\tau} (y_i)$ in X. Then $x_i^\alpha \xrightarrow[\alpha]{\tau_i} x_i$ and $y_i^\lambda \xrightarrow[\lambda]{\tau_i} y_i$ in X_i for each i, so also $x_i^\alpha + y_i^\lambda \xrightarrow[\alpha,\lambda]{\tau_i} x_i + y_i$ in X_i for each i. Since the product topology on X is the topology of pointwise convergence, we see that

$$(x_i^\alpha) + (y_i^\lambda) = (x_i^\alpha + y_i^\lambda) \xrightarrow[\alpha,\lambda]{\tau} (x_i + y_i) = (x_i) + (y_i),$$

and the proof is finished. ∎

Linear topologies are **translation invariant**. That is, a set V is open in a tvs X if and only if the translation $a + V$ is open for all a. Indeed, the continuity of addition implies that for each $a \in X$, the function $x \mapsto a + x$ is a linear homeomorphism. In particular, every neighborhood of a is of the form $a + V$, where V is a neighborhood of zero. In other words, the neighborhood system at zero determines the neighborhood system at every point of X by translation. Also note that the mapping $x \mapsto \alpha x$ is linear homeomorphism for any $\alpha \neq 0$. In particular, if V is a neighborhood of zero, then so is αV for all $\alpha \neq 0$.

The most familiar linear topologies are derived from norms. A **norm** on a vector space is a real function $\| \cdot \|$ satisfying

1. $\|x\| \geqslant 0$ for all vectors x, and $\|x\| = 0$ implies $x = 0$.

2. $\|\alpha x\| = |\alpha|\, \|x\|$ for all vectors x and all scalars α.

3. $\|x + y\| \leqslant \|x\| + \|y\|$ for all vectors x and y.

A neighborhood base at zero consists of all sets of the form $\{x : \|x\| < \varepsilon\}$ where ε is a positive number. The **norm topology** for a norm $\|\cdot\|$ is the metrizable topology generated by the metric $d(x, y) = \|x - y\|$.

The next lemma presents some basic facts about subsets of topological vector spaces. Most of the proofs are straightforward.

5.3 Lemma *In a topological vector space:*

1. *The algebraic sum of an open set and an arbitrary set is open.*

2. *Nonzero multiples of open sets are open.*

3. *If B is open, then for any set A we have $\overline{A} + B = A + B$.*

4. *The algebraic sum of a compact set and a closed set is closed. (However, the algebraic sum of two closed sets need not be closed.)*

5. *The algebraic sum of two compact sets is compact.*

6. *Scalar multiples of closed sets are closed.*

7. *Scalar multiples of compact sets are compact.*

8. *A linear functional is continuous if and only if it is continuous at 0.*

Proof: We shall prove only parts (3) and (4).

(3) Clearly $A + B \subset \overline{A} + B$. For the reverse inclusion, let $y \in \overline{A} + B$ and write $y = x + b$ where $x \in \overline{A}$ and $b \in B$. Then there is an open neighborhood V of zero such that $b + V \subset B$. Since $x \in \overline{A}$, there exists some $a \in A \cap (x - V)$. Then $y = x + b = a + b + (x - a) \in a + b + V \subset A + B$.

(4) Let A be compact and B be closed, and let a net $\{x_\alpha + y_\alpha\}$ in $A + B$ satisfy $x_\alpha + y_\alpha \to z$. Since A is compact, we can assume (by passing to a subnet) that $x_\alpha \to x \in A$. The continuity of the algebraic operations yields

$$y_\alpha = (x_\alpha + y_\alpha) - x_\alpha \to z - x = y.$$

Since B is closed, $y \in B$, so $z = x + y \in A + B$, proving that $A + B$ is closed.   ∎

5.4 Example (Sum of closed sets)   To see that the sum of two closed sets need not be closed, consider the closed sets $A = \{(x, y) : x > 0, y \geq \frac{1}{x}\}$ and $B = \{(x, y) : x < 0, y \geq -\frac{1}{x}\}$ in $\mathbb{R}^2$. While A and B are closed, neither is compact, and $A + B = \{(x, y) : y > 0\}$ is not closed.   ∎

5.2 Absorbing and circled sets

We now describe some special algebraic properties of subsets of vector spaces. The **line segment** joining vectors x and y is the set $\{\lambda x + (1 - \lambda)y : 0 \leq \lambda \leq 1\}$.

5.5 Definition   *A subset A of a vector space is:*

• *__convex__ if it includes the line segment joining any pair of its points.*

• *__absorbing__ (or __radial__) if for any x some multiple of A includes the line segment joining x and zero. That is, if there exists some $\alpha_0 > 0$ satisfying $\alpha x \in A$ for every $0 \leq \alpha \leq \alpha_0$.*

Equivalently, A is absorbing if for each vector x there exists some $\alpha_0 > 0$ such that $\alpha x \in A$ whenever $-\alpha_0 \leq \alpha \leq \alpha_0$.

• *__circled__ (or __balanced__) if for each $x \in A$ the line segment joining x and $-x$ lies in A. That is, if for any $x \in A$ and any $|\alpha| \leq 1$ we have $\alpha x \in A$.*

• *__symmetric__ if $x \in A$ implies $-x \in A$.*

• *__star-shaped about zero__ if it includes the line segment joining each of its points with zero. That is, if for any $x \in A$ and any $0 \leq \alpha \leq 1$ we have $\alpha x \in A$.*

Circled and absorbing, but not convex. Star-shaped, but neither symmetric nor convex. Circled, but neither absorbing nor convex.

Figure 5.1. Shapes of sets in $\mathbb{R}^2$.

Note that an absorbing set must contain zero, and any set including an absorbing set is itself absorbing. For any absorbing set A, the set $A \cap (-A)$ is nonempty, absorbing, and symmetric. Every circled set is symmetric. Every circled set is star-shaped about zero, as is every convex set containing zero. See Figure 5.1 for some examples.

Let X be a topological vector space. For each fixed scalar $\alpha \neq 0$ the mapping $x \mapsto \alpha x$ is a linear homeomorphism, so αV is a neighborhood of zero whenever V is and $\alpha \neq 0$. Now if V is a neighborhood of zero, then the continuity of the function $(\alpha, x) \mapsto \alpha x$ at $(0, 0)$ guarantees the existence of a neighborhood W at zero and some $\alpha_0 > 0$ such that $x \in W$ and $|\alpha| \leqslant \alpha_0$ imply $\alpha x \in V$. Thus, if $U = \bigcup_{|\alpha| \leqslant \alpha_0} \alpha W$, then U is a neighborhood of zero, $U \subset V$, and U is circled. Moreover, from the continuity of the addition map $(x, y) \mapsto x + y$ at $(0, 0)$, we see that there is a neighborhood W of zero such that $x, y \in W$ implies $x + y \in V$, that is, $W + W \subset V$. Also note that since $W + W \subset V$, it follows that $\overline{W} \subset V$. (For if $x \in \overline{W}$, then $x - W$ is a neighborhood of x, so $(x - W) \cap W \neq \varnothing$ implies $x \in W + W \subset V$.)

Since the closure of an absorbing circled set remains absorbing and circled (why?), we have just shown that zero has a neighborhood base consisting of closed, absorbing, and circled sets. We cannot conclude that zero has a neighborhood base consisting of convex sets. If the tvs does have a neighborhood base at zero of convex sets, it is called a **locally convex space**.

The following theorem establishes the converse of the results above and characterizes the structure of linear topologies.

5.6 Structure Theorem *If (X, τ) is a tvs, then there is a neighborhood base $\mathcal{B}$ at zero such that:*

1. *Each $V \in \mathcal{B}$ is absorbing.*

2. *Each $V \in \mathcal{B}$ is circled.*

3. *For each $V \in \mathcal{B}$ there exists some $W \in \mathcal{B}$ with $W + W \subset V$.*

4. *Each $V \in \mathcal{B}$ is closed.*

Conversely, if a filter base $\mathcal{B}$ on a vector space X satisfies properties (1), (2), and (3) above, then there exists a unique linear topology τ on X having $\mathcal{B}$ as a neighborhood base at zero.

Proof: If τ is a linear topology, then by the discussion preceding the theorem, the collection of all τ-closed circled neighborhoods of zero satisfies the desired properties. For the converse, assume that a filter base $\mathcal{B}$ of a vector space X satisfies properties (1), (2), and (3).

We have already mentioned that a linear topology is translation invariant and so uniquely determined by the neighborhoods of zero. So define τ to be the collection of all subsets A of X satisfying

$$A = \{x \in A : \exists\, V \in \mathcal{B} \text{ such that } x + V \subset A\}. \qquad (\star)$$

Then clearly $\varnothing, X \in \tau$ and the collection τ is closed under arbitrary unions. If $A_1, \ldots, A_k \in \tau$ and $x \in A_1 \cap \cdots \cap A_k$, then for each $i = 1, \ldots, k$ there exists some $V_i \in \mathcal{B}$ such that $x + V_i \subset A_i$. Since $\mathcal{B}$ is a filter base, there exists some $V \in \mathcal{B}$ with $V \subset V_1 \cap \cdots \cap V_k$. Now note that $x + V \subset A_1 \cap \cdots \cap A_k$ and this proves that $A_1 \cap \cdots \cap A_k \in \tau$. Therefore, we have established that τ is a topology on X.

The next thing we need to observe is that if for each subset A of X we let

$$A^\sharp = \{x \in A : \exists\, V \in \mathcal{B} \text{ such that } x + V \subset A\},$$

then $A^\sharp$ coincides with the τ-interior of A, that is, $A^\circ = A^\sharp$. If $x \in A^\circ$, then by $(\star)$ and the fact that A° is τ-open, there exists some $V \in \mathcal{B}$ such that $x + V \subset A^\circ \subset A$, so $x \in A^\sharp$. Therefore, $A^\circ \subset A^\sharp$. To see that equality holds, it suffices to show that $A^\sharp$ is τ-open. To this end, $y \in A^\sharp$. Pick some $V \in \mathcal{B}$ such that $y + V \subset A$. By (3) there exists some $W \in \mathcal{B}$ such that $W + W \subset V$. Now if $w \in W$, then we have $y + w + W \subset y + W + W \subset y + V \subset A$, so that $y + w \in A^\sharp$ for each $w \in W$, that is, $y + W \subset A^\sharp$. This proves that $A^\sharp$ is τ-open, so $A^\circ = A^\sharp$.

Now it easily follows that for each $x \in X$ the collection $\{x + V : V \in \mathcal{B}\}$ is a τ-neighborhood base at x. Next we shall show that the addition map $(x, y) \mapsto x + y$ is a continuous function. To see this, fix $x_0, y_0 \in X$ and a set $V \in \mathcal{B}$. Choose some $U \in \mathcal{B}$ with $U + U \subset V$ and note that $x \in x_0 + U$ and $y \in y_0 + U$ imply $x + y \in x_0 + y_0 + V$. Consequently, the addition map is continuous at (x_0, y_0) and therefore is a continuous function.

Finally, let us prove the continuity of scalar multiplication. Fix $\lambda_0 \in \mathbb{R}$ and $x_0 \in X$ and let $V \in \mathcal{B}$. Pick some $W \in \mathcal{B}$ such that $W + W \subset V$. Since W is an absorbing set there exists some $\varepsilon > 0$ such that for each $-\varepsilon < \delta < \varepsilon$ we have $\delta x_0 \in W$. Next, select a natural number $n \in \mathbb{N}$ with $|\lambda_0| + \varepsilon < n$ and note that if $\lambda \in \mathbb{R}$ satisfies $|\lambda - \lambda_0| < \varepsilon$, then $\left|\frac{\lambda}{n}\right| \leqslant \frac{|\lambda_0| + \varepsilon}{n} < 1$. Now since W is (by (2)) a circled set, for each $\lambda \in \mathbb{R}$ with $|\lambda - \lambda_0| < \varepsilon$ and all $x \in x_0 + \frac{1}{n}W$ we have

$$\lambda x = \lambda_0 x_0 + (\lambda - \lambda_0)x_0 + \lambda(x - x_0) \in \lambda_0 x_0 + W + \tfrac{\lambda}{n}W \subset \lambda_0 x_0 + W + W \subset \lambda_0 x_0 + V.$$

This shows that multiplication is a continuous function at (λ_0, x_0). ∎

In a topological vector space the interior of a circled set need not be a circled set; see, for instance, the third set in Figure 5.1. However, the interior of a circled neighborhood V of zero is automatically an open circled set. To see this, note first that 0 is an interior point of V. Now let $x \in V^\circ$ and fix some nonzero $\lambda \in \mathbb{R}$ with $|\lambda| \leqslant 1$. Pick some neighborhood W of zero with $x + W \subset V$ and note that the neighborhood λW of zero satisfies $\lambda x + \lambda W = \lambda(x + W) \subset \lambda V \subset V$. Therefore $\lambda x \in V^\circ$ for each $|\lambda| \leqslant 1$, so V° is a circled set. This conclusion yields the following.

5.7 Lemma   *In a topological vector space the collection of all open and circled neighborhoods of zero is a base for the neighborhood system at zero.*

If τ is a linear topology on a vector space and $\mathcal{N}$ denotes the τ-neighborhood system at zero, then the set $K_\tau = \bigcap_{V \in \mathcal{N}} V$ is called the **kernel of the topology** τ. From Theorem 5.6 it is not difficult to see that K_τ is a closed vector subspace. The vector subspace K_τ is the trivial subspace $\{0\}$ if and only if τ is a Hausdorff topology. The proof of the next result is straightforward and is left for the reader.

5.8 Lemma   *A linear topology τ on a vector space is Hausdorff if and only if its kernel K_τ is trivial (and also if and only if $\{0\}$ is a τ-closed set).*

Property (3) of the Theorem 5.6 allows to use "$\varepsilon/2$ arguments" even when we don't have a metric. As an application of this result, we offer another instance of the informal principle that compact sets behave like points.

5.9 Theorem   *Let K be a compact subset of a topological vector space X, and suppose $K \subset U$, where U is open. Then there exist an open neighborhood W of zero and a finite subset Φ of K such that*

$$K \subset \Phi + W \subset K + W \subset U.$$

Proof: Since $K \subset U$, for each $x \in K$, there is open neighborhood W_x of zero such that $x + W_x + W_x \subset U$. Since K is compact, there is a finite set $\{x_1, \ldots, x_n\}$ with $K \subset \bigcup_{i=1}^n (x_i + W_{x_i})$. Let $W = \bigcap_{i=1}^n W_{x_i}$ and note that W is an open neighborhood of zero. Since the open sets $x_i + W_{x_i}$, $i = 1, \ldots, n$, cover K, given $y \in K$, there is an x_i satisfying $y \in x_i + W_{x_i}$. For this x_i we have $y + W \subset x_i + W_{x_i} + W_{x_i} \subset U$, and from this we see that $K + W \subset U$.

Now from $K \subset K + W = \bigcup_{y \in K}(y + W) \subset U$ and the compactness of K, it follows that there exists a finite subset $\Phi = \{y_1, \ldots, y_m\}$ of K such that $K \subset \bigcup_{j=1}^m (y_j + W)$. Now note that $K \subset \Phi + W \subset K + W \subset U$. ∎

5.3   Metrizable topological vector spaces

A metric d on a vector space is said to be **translation invariant** if it satisfies
$d(x + a, y + a) = d(x, y)$ for all x, y, and a. Every metric induced by a norm
is translation invariant, but the converse is not true (see Example 5.78 below).
For Hausdorff topological vector spaces, the existence of a compatible translation
invariant metric is equivalent to first countability.

5.10 Theorem   *A Hausdorff topological vector space is metrizable if and only
if zero has a countable neighborhood base. In this case, the topology is generated
by a translation invariant metric.*

Proof: Let (X, τ) be a Hausdorff tvs. If τ is metrizable, then τ has clearly a countable neighborhood base at zero. For the converse, assume that τ has a countable
neighborhood base at zero. Choose a countable base $\{V_n\}$ of circled neighborhoods of zero such that $V_{n+1} + V_{n+1} + V_{n+1} \subset V_n$ holds for each n. Now define the
function $\rho \colon X \to [0, \infty)$ by

$$\rho(x) = \begin{cases} 1 & \text{if } x \notin V_1, \\ 2^{-k} & \text{if } x \in V_k \setminus V_{k+1}, \\ 0 & \text{if } x = 0. \end{cases}$$

Then it is easy to check that for each $x \in X$ we have the following:

1. $\rho(x) \geqslant 0$ and $\rho(x) = 0$ if and only if $x = 0$.

2. $x \in V_k$ for some k if and only if $\rho(x) \leqslant 2^{-k}$.

3. $\rho(x) = \rho(-x)$ and $\rho(\lambda x) \leqslant \rho(x)$ for all $|\lambda| \leqslant 1$.

4. $\lim_{\lambda \to 0} \rho(\lambda x) = 0$.

We also note the following property.

- $x_n \xrightarrow{\tau} 0$ *if and only if* $\rho(x_n) \to 0$.

    Now by means of the function ρ we define the function $\pi \colon X \to [0, \infty)$ via the
formula:
$$\pi(x) = \inf \left\{ \sum_{i=1}^{n} \rho(x_i) : x_1, \ldots, x_n \in X \text{ and } \sum_{i=1}^{n} x_i = x \right\}.$$
The function π satisfies the following properties.

a. $\pi(x) \geqslant 0$ *for each* $x \in X$.

b. $\pi(x + y) \leqslant \pi(x) + \pi(y)$ *for all* $x, y \in X$.

c. $\frac{1}{2}\rho(x) \leqslant \pi(x) \leqslant \rho(x)$ *for each* $x \in X$ *(so* $\pi(x) = 0$ *if and only if* $x = 0$*)*.

Property (a) follows immediately from the definition of π. Property (b) is straightforward. The proof of (c) will be based upon the following property:

$$\text{If } \sum_{i=1}^{n} \rho(x_i) < \tfrac{1}{2^m}, \text{ then } \sum_{i=1}^{n} x_i \in V_m. \tag{$\star$}$$

To verify $(\star)$, we use induction on n. For $n = 1$ we have $\rho(x_1) < 2^{-m}$, and consequently $x_1 \in V_{m+1} \subset V_m$ is trivially true. For the induction step, assume that if $\{x_i : i \in I\}$ is any collection of at most n vectors satisfying $\sum_{i \in I} \rho(x_i) < 2^{-m}$ for some $m \in \mathbb{N}$, then $\sum_{i \in I} x_i \in V_m$.

Suppose that $\sum_{i=1}^{n+1} \rho(x_i) < \tfrac{1}{2^m}$ for some $m \in \mathbb{N}$. Clearly, we have $\rho(x_i) \leqslant \tfrac{1}{2^{m+1}}$, so $x_i \in V_{m+1}$ for each $1 \leqslant i \leqslant n+1$. We now distinguish the following two cases.

Case 1: $\sum_{i=1}^{n+1} \rho(x_i) < \tfrac{1}{2^{m+1}}$

Clearly $\sum_{i=1}^{n} \rho(x_i) < \tfrac{1}{2^{m+1}}$, so by the induction hypothesis $\sum_{i=1}^{n} x_i \in V_{m+1}$. Thus

$$\sum_{i=1}^{n+1} x_i = \sum_{i=1}^{n} x_i + x_{n+1} \in V_{m+1} + V_{m+1} \subset V_m.$$

Case 2: $\sum_{i=1}^{n+1} \rho(x_i) \geqslant \tfrac{1}{2^{m+1}}$

Let $1 \leqslant k \leqslant n + 1$ be the largest k such that $\sum_{i=k}^{n+1} \rho(x_i) \geqslant \tfrac{1}{2^{m+1}}$. If $k = n + 1$, then $\rho(x_{n+1}) = \tfrac{1}{2^{m+1}}$, so from $\sum_{i=1}^{n+1} \rho(x_i) < \tfrac{1}{2^m}$ we have $\sum_{i=1}^{n} \rho(x_i) < \tfrac{1}{2^{m+1}}$. But then, as in Case 1, we get $\sum_{i=1}^{n+1} x_i \in V_m$.

Thus, we can assume that $k < n + 1$. Assume first that $k > 1$. From the inequalities $\sum_{i=1}^{n+1} \rho(x_i) < \tfrac{1}{2^m}$ and $\sum_{i=k}^{n+1} \rho(x_i) \geqslant \tfrac{1}{2^{m+1}}$, we obtain $\sum_{i=1}^{k-1} \rho(x_i) < \tfrac{1}{2^{m+1}}$. So our induction hypothesis yields $\sum_{i=1}^{k-1} x_i \in V_{m+1}$. Also, by the choice of k we have $\sum_{i=k+1}^{n+1} \rho(x_i) < \tfrac{1}{2^{m+1}}$, and thus by our induction hypothesis also we have $\sum_{i=k+1}^{n+1} x_i \in V_{m+1}$. Therefore, in this case we obtain

$$\sum_{i=1}^{n+1} x_i = \sum_{i=1}^{k-1} x_i + x_k + \sum_{i=k+1}^{n+1} x_i \in V_{m+1} + V_{m+1} + V_{m+1} \subset V_m.$$

If $k = 1$, then we have $\sum_{i=2}^{n+1} \rho(x_i) < \tfrac{1}{2^{m+1}}$, so $\sum_{i=2}^{n+1} x_i \in V_{m+1}$. This implies $\sum_{i=1}^{n+1} x_i = x_1 + \sum_{i=2}^{n+1} x_i \in V_{m+1} + V_{m+1} \subset V_m$. This completes the induction and the proof of $(\star)$.

Next, we verify (c). To this end, let $x \in X$ satisfy $\rho(x) = 2^{-m}$ for some $m \geqslant 0$. Also, assume by way of contradiction that the vectors $x_1, \ldots, x_k$ satisfy $\sum_{i=1}^{k} x_i = x$ and $\sum_{i=1}^{k} \rho(x_i) < \tfrac{1}{2}\rho(x) = 2^{-m-1}$. But then, from $(\star)$ we get $x = \sum_{i=1}^{k} x_i \in V_{m+1}$, so $\rho(x) \leqslant 2^{-m-1} < 2^{-m} = \rho(x)$, which is impossible. This contradiction, establishes the validity of (c).

Finally, for each $x, y \in X$ define $d(x, y) = \pi(x - y)$ and note that d is a translation invariant metric that generates τ. ∎

Even if a tvs is not metrizable, it is nonetheless uniformizable by a translation invariant uniformity. For a proof of this result, stated below, see, for example, H. H. Schaefer [293, §1.4, pp. 16–17].

5.11 Theorem   *A topological vector space is uniformizable by a unique translation invariant uniformity. A base for the uniformity is the collection of sets of the form $\{(x, y) : x - y \in V\}$ where V ranges over a neighborhood base $\mathcal{B}$ at zero.*

A **Cauchy net** in a topological vector space is a net $\{x_\alpha\}$ such that for each neighborhood V of zero there is some α_0 such that $x_\alpha - x_\beta \in V$ for all $\alpha, \beta \geq \alpha_0$. Every convergent net is Cauchy. (Why?) Similarly, a filter $\mathcal{F}$ on a topological vector space is called a **Cauchy filter** if for each neighborhood V of zero there exists some $A \in \mathcal{F}$ such that $A - A \subset V$. Convergent filters are clearly Cauchy. From the discussion in Section 2.6, it is easy to see that a filter is Cauchy if and only if the net it generates is a Cauchy net (and that a net is Cauchy if and only if the filter it generates is Cauchy).

A topological vector space (X, τ) is **topologically complete**, or simply **complete** (and τ is called a **complete topology**), if every Cauchy net is convergent, or equivalently, if every Cauchy filter is convergent.

The proof of the next lemma is straightforward and is omitted.

5.12 Lemma   *Let $\{(X_i, \tau_i)\}_{i \in I}$ be a family of topological vector spaces, and let $X = \prod_{i \in I} X_i$ endowed with the product topology $\tau = \prod_{i \in I} \tau_i$. Then (X, τ) is τ-complete if and only if each factor (X_i, τ_i) is τ_i-complete.*

If a linear topology τ on a vector space X is generated by a translation invariant metric d, then (X, d) is a complete metric space if and only if (X, τ) is topologically complete as defined above, that is, (X, d) is a complete metric space if and only if every τ-Cauchy sequence in X is τ-convergent. Not every consistent metric of a metrizable topological vector space is translation invariant. For instance, consider the three metrics d_1, d_2, and d_3 on $\mathbb{R}$ defined by:

$$d_1(x, y) = |x - y|,$$
$$d_2(x, y) = |x - y| + \left|\tfrac{1}{1+|x|} - \tfrac{1}{1+|y|}\right|, \text{ and}$$
$$d_3(x, y) = \left|e^{-x} - e^{-y}\right|.$$

Then d_1, d_2, and d_3 are equivalent metrics, d_1 is complete and translation invariant, d_2 is complete but not translation invariant, and d_3 is neither complete nor translation invariant.

5.13 Definition   *A **completely metrizable topological vector space** is a topologically complete metrizable topological vector space. In other words, a completely metrizable tvs is a topologically complete tvs having a countable neighborhood base at zero.*

Note that (according to Theorem 5.10) every completely metrizable topological vector space admits a compatible translation invariant complete metric. Clearly, the class of completely metrizable topological vector spaces includes the class of Banach spaces.

A complete Hausdorff topological vector space Y is called a **topological completion** or simply a **completion** of another Hausdorff topological vector space X if there is a linear homeomorphism $T: X \to Y$ such that $T(X)$ is dense in Y; identifying X with $T(X)$, we can think of X as a subspace of Y. This leads to the next result, which appears in many places; see, for instance, J. Horváth [168, Theorem 1, p. 131].

5.14 Theorem   *Every Hausdorff topological vector space has a unique (up to linear homeomorphism) topological completion.*

The concept of uniform continuity makes sense for functions defined on subsets of topological vector spaces. A function $f: A \to Y$, where A is a subset of a tvs X and Y is another tvs, is **uniformly continuous** if for each neighborhood V of zero in Y there exists a neighborhood W of zero in X such that $x, y \in A$ and $x - y \in W$ imply $f(x) - f(y) \in V$. You should notice that if X is a tvs, then both addition $(x, y) \mapsto x + y$, from $X \times X$ to X, and scalar multiplication $(\alpha, x) \mapsto \alpha x$, from $\mathbb{R} \times X$ to X, are uniformly continuous.

The analogue of Lemma 3.11 can now be stated as follows—the proof is left as an exercise.

5.15 Theorem   *Let A be a subset of a tvs, let Y be a complete Hausdorff topological vector space, and let $f: A \to Y$ be uniformly continuous. Then f has a unique uniformly continuous extension to the closure $\overline{A}$ of A.*

5.4   The Open Mapping and Closed Graph Theorems

In this section we prove two basic theorems of functional analysis, the Open Mapping Theorem and the Closed Graph Theorem. We do this in the setting of completely metrizable topological vector spaces. For more on these theorems and extensions to general topological vector spaces we recommend T. Husain [174]. We start by recalling the definition of an operator.

5.16 Definition   *A function $T: X \to Y$ between two vector spaces is a **linear operator** (or simply an **operator**) if*

$$T(\alpha x + \beta y) = \alpha T(x) + \beta T(y)$$

*for all $x, y \in X$ and all scalars $\alpha, \beta \in \mathbb{R}$. When Y is the real line $\mathbb{R}$, we call T a **linear functional**.*

It is common to denote the vector $T(x)$ by Tx, and we do it quite often. If $T: X \to Y$ is not a linear operator, then T is referred to as a **nonlinear operator**. The following lemma characterizes continuity of linear operators.

5.17 Lemma (Continuity at zero) *An operator $T: X \to Y$ between topological vector spaces is continuous if and only if it is continuous at zero (in which case it is uniformly continuous).*

Proof: Everything follows from the identity $T(x) - T(y) = T(x - y)$.   ∎

Recall that a function between topological spaces is called an **open mapping** if it carries open sets to open sets.

5.18 The Open Mapping Theorem   *A surjective continuous operator between completely metrizable topological vector spaces is an open mapping.*

Proof: Let $T: (X_1, \tau_1) \to (X_2, \tau_2)$ be a surjective continuous operator between completely metrizable topological vector spaces and let U be a circled τ_1-closed neighborhood of zero. It suffices to show that the set $T(U)$ is a τ_2-neighborhood of zero. We first establish the following claim.

●   *For any τ_1-neighborhood W of zero in X_1 there exists a τ_2-neighborhood V of zero in X_2 satisfying $V \subset \overline{T(W)}$.*

To see this, let W and W_0 be circled τ_1-neighborhoods of zero that satisfy $W_0 + W_0 \subset W$. From $X_1 = \bigcup_{n=1}^{\infty} nW_0$ and the fact that T is surjective, it follows that $X_2 = T(X_1) = \bigcup_{n=1}^{\infty} nT(W_0)$. Therefore, by the Baire Category Theorem 3.47, for some n the set $\overline{nT(W_0)} = n\overline{T(W_0)}$ must have an interior point. This implies that there exists some $y \in \overline{T(W_0)}$ and some circled τ_2-neighborhood V of zero with $y + V \subset \overline{T(W_0)}$. Since $\overline{T(W_0)}$ is symmetric, we see that $v - y \in \overline{T(W_0)}$ for each $v \in V$. Thus, if $v \in V$, then it follows from $v = (v-y)+y \in \overline{T(W_0)}+\overline{T(W_0)} \subset \overline{T(W)}$ that $v \in \overline{T(W)}$, so $V \subset \overline{T(W)}$.

Now pick a countable base $\{W_n\}$ at zero for τ_1 consisting of τ_1-closed circled sets satisfying $W_{n+1} + W_{n+1} \subset W_n$ for all $n = 1, 2, \ldots$ and $W_1 + W_1 \subset U$. The claim established above and an easy inductive argument guarantee the existence of a countable base $\{V_n\}$ at zero for τ_2 consisting of circled and τ_2-closed sets satisfying $V_{n+1} + V_{n+1} \subset V_n$ and $V_n \subset \overline{T(W_n)}$ for all $n = 1, 2, \ldots$. We finish the proof by showing that $V_1 \subset T(U)$.

To this end, let $y \in V_1$. From $V_1 \subset \overline{T(W_1)}$ and the fact that $y + V_2$ is a τ_2-neighborhood of y, it follows that there exists some $w_1 \in W_1$ with $y - T(w_1) \in V_2$, so $y - T(w_1) \in \overline{T(W_2)}$. Now by an inductive argument, we can construct a sequence $\{w_n\}$ in X_1 such that for each $n = 1, 2, \ldots$ we have $w_n \in W_n$ and

$$y - \sum_{i=1}^{n} T(w_i) = y - T\Big(\sum_{i=1}^{n} w_i\Big) \in V_{n+1}. \qquad (\star)$$

Next, let $x_n = \sum_{i=1}^{n} w_i$ and note that from

$$x_{n+p} - x_n = \sum_{i=n+1}^{n+p} w_i \in W_{n+1} + W_{n+2} + \cdots + W_{n+p} \subset W_n,$$

we see that $\{x_n\}$ is a τ_1-Cauchy sequence. Since (X_1, τ_1) is τ_1-complete, there is some $x \in X_1$ such that $x_n \xrightarrow{\tau_1} x$. Rewriting $(\star)$ as $y - T(x_n) \in V_{n+1}$ for each n, we see that $y - T(x_n) \xrightarrow{\tau_2} 0$ in X_2. On the other hand, the continuity of T yields $T(x_n) \xrightarrow{\tau_2} T(x)$, and from this we get $y = T(x)$.

Finally, from $x_n = \sum_{i=1}^{n} w_i \in W_1 + W_2 + \cdots + W_n \subset W_1 + W_1 \subset U$ and the τ_1-closedness of U, we easily infer that $x \in U$, so $y = T(x) \in T(U)$. In other words $V_1 \subset T(U)$, and the proof is finished. ∎

5.19 Corollary *A surjective continuous one-to-one operator between completely metrizable topological vector spaces is a homeomorphism.*

Recall that the **graph** of a function $f: A \to B$ is simply the subset of the Cartesian product $A \times B$ defined by

$$\text{Gr} f = \{(a, f(a)) : a \in A\}.$$

Notice that if $T: X \to Y$ is an operator between vector spaces, then the graph $\text{Gr} T$ of T is a vector subspace of $X \times Y$.

5.20 The Closed Graph Theorem *An operator between completely metrizable topological vector spaces is continuous if and only if it has closed graph.*

Proof: Assume that $T: (X_1, \tau_1) \to (X_2, \tau_2)$ is an operator between completely metrizable topological vector spaces such that its graph $\text{Gr} T = \{(x, T(x)) : x \in X_1\}$ is a closed subspace of $X_1 \times X_2$. It follows that $\text{Gr} T$ (with the induced product topology from $X_1 \times X_2$) is also a completely metrizable topological vector space. Since the mapping $S : \text{Gr} T \to X_1$ defined by $S(x, T(x)) = x$ is a surjective continuous one-to-one operator, it follows from Corollary 5.19 that S is a homeomorphism. In particular, the operator $x \mapsto (x, T(x)) = S^{-1}(x)$, from X_1 to $\text{Gr} T$, is continuous. Since the projection $P_2: X_1 \times X_2 \to X_2$, defined by $P_2(x_1, x_2) = x_2$, is continuous it follows that the operator $T = P_2 S^{-1}$ is likewise continuous. ∎

5.5 Finite dimensional topological vector spaces

This section presents some distinguishing properties of finite dimensional vector spaces. Recall that the **Euclidean norm** $\|\cdot\|_2$ on $\mathbb{R}^n$ is defined by $\|x\|_2 = (\sum_{i=1}^{n} x_i^2)^{\frac{1}{2}}$. It generates the **Euclidean topology**. Remarkably, this is the only Hausdorff linear topology on $\mathbb{R}^n$. In particular, any two norms on a finite dimensional vector space are equivalent: Two norms $\| \cdot \|$ and $\| \cdot \|$ on a vector space X are **equivalent** if they generate the same topology. In view of Theorem 6.17, this occurs if and only if there exist two positive constants K and M satisfying $K\|x\| \leqslant \|x\| \leqslant M\|x\|$ for each $x \in X$.

5.21 Theorem   *Every finite dimensional vector space admits a unique Hausdorff linear topology, namely the complete Euclidean topology.*

Proof: Let $X = \mathbb{R}^n$, let τ_1 be a Hausdorff linear topology on X, and let τ denote the linear topology generated by the Euclidean norm $\| \cdot \|_2$. Clearly, (X, τ) is topologically complete.

We know that a net $\{x_\alpha = (x_1^\alpha, \ldots, x_n^\alpha)\}$ in $\mathbb{R}^n$, satisfies $x_\alpha \xrightarrow{\|\cdot\|_2} 0$ if and only if $x_i^\alpha \xrightarrow[\alpha]{} 0$ in $\mathbb{R}$ for each i. Thus, if $x_\alpha \xrightarrow{\|\cdot\|_2} 0$, then since addition and scalar multiplication are τ_1-continuous,

$$x_\alpha = \sum_{i=1}^n x_i^\alpha e_i \xrightarrow[\alpha]{\tau_1} \sum_{i=1}^n 0 e_i = 0,$$

where as usual, e_i denotes the ith coordinate unit vector of $\mathbb{R}^n$. Thus, the identity $I \colon (X, \tau) \to (X, \tau_1)$ is continuous and so $\tau_1 \subset \tau$.

Now let $B = \{x \in X : \|x\|_2 < 1\}$. Since $S = \{x \in X : \|x\|_2 = 1\}$ is τ-compact, it follows from $\tau_1 \subset \tau$ that S is also τ_1-compact. Therefore (since τ_1 is Hausdorff) S is τ_1-closed. Since $0 \notin S$, we see that there exists a circled τ_1-neighborhood V of zero such that $V \cap S = \varnothing$. Since V is circled, we have $V \subset B$: For if there exists some $x \in V$ such that $x \notin B$ (that is, $\|x\|_2 \geqslant 1$), then $\frac{x}{\|x\|_2} \in V \cap S$, a contradiction.

Thus, B is a τ_1-neighborhood of zero. Since scalar multiples of B form a τ-neighborhood base at zero, we see that $\tau \subset \tau_1$. Therefore $\tau_1 = \tau$.   ∎

When we deal with finite dimensional vector spaces, we shall assume tacitly (and without any specific mention) that they are equipped with their Euclidean topologies and all topological notions will be understood in terms of Euclidean topologies.

The remaining results in this section are consequences of Theorem 5.21.

5.22 Corollary   *A finite dimensional vector subspace of a Hausdorff topological vector space is closed.*

Proof: Let Y be a finite dimensional subspace of a Hausdorff topological vector space (X, τ), and let $\{y_\alpha\}$ be a net in Y satisfying $y_\alpha \xrightarrow{\tau} x$ in X. Therefore it is a Cauchy net in X, and hence also in Y. By Theorem 5.21, τ induces the Euclidean topology on Y. Since Y (with its Euclidean metric) is a complete metric space, it follows that $y_\alpha \xrightarrow{\tau} y$ in Y. Since τ is Hausdorff, we see that $x = y \in Y$, so Y is a closed subspace of X.   ∎

5.23 Corollary   *Every Hamel basis of an infinite dimensional completely metrizable topological vector space is uncountable.*

Proof: Let $\{e_1, e_2, \ldots\}$ be a countable Hamel basis of an infinite dimensional completely metrizable tvs X. For each n let X_n be the finite dimensional vector subspace generated by $\{e_1, \ldots, e_n\}$. By Theorem 5.21 each X_n is closed. Now note that $X = \bigcup_{n=1}^{\infty} X_n$ and then use the Baire Category Theorem 3.47 to conclude that some X_n has a nonempty interior. This implies $X = X_n$ for some n, which is impossible. ∎

5.24 Corollary   *Let $v_1, v_2, \ldots, v_m$ be linearly independent vectors in a Hausdorff topological vector space (X, τ). For each n let $x_n = \sum_{i=1}^{m} \lambda_i^n v_i$. If $x_n \xrightarrow{\tau} x$ in X, then there exist $\lambda_1, \ldots, \lambda_m$ such that $x = \sum_{i=1}^{m} \lambda_i v_i$ (that is, x is in the linear span of $\{v_1, \ldots, v_m\}$) and $\lambda_i^n \xrightarrow[n\to\infty]{} \lambda_i$ for each i.*

Proof: Let Y be the linear span of $\{v_1, \ldots, v_m\}$. By Corollary 5.22, Y is a closed vector subspace of X, so $x \in Y$. That is, there exist scalars $\lambda_1, \ldots, \lambda_m$ such that $x = \sum_{i=1}^{m} \lambda_i v_i$.

Now for each $y = \sum_{i=1}^{m} \alpha_i v_i \in Y$, let $\|y\| = \sum_{i=1}^{m} |\alpha_i|$. Then $\| \cdot \|$ is a norm on Y, and thus (by Theorem 5.21) the topology induced by τ on Y coincides with the topology generated by the norm $\| \cdot \|$ on Y. Now note that

$$\|x_n - x\| = \left\| \sum_{i=1}^{m} \lambda_i^n v_i - \sum_{i=1}^{m} \lambda_i v_i \right\| = \sum_{i=1}^{m} |\lambda_i^n - \lambda_i| \xrightarrow[n\to\infty]{} 0$$

if and only if $\lambda_i^n \xrightarrow[n\to\infty]{} \lambda_i$ for each i. ∎

A **ray** in a vector space X is the set of nonnegative multiples of some vector, that is, a set of the form $\{\alpha v : \alpha \geqslant 0\}$, where $v \in X$. It is **trivial** if it contains only zero. We may also refer to a translate of such a set as a ray or a **half line**. A **cone** is a set of rays, or in other words a set that contains every nonnegative multiple of each of its members. That is, C is a cone if $x \in C$ implies $\alpha x \in C$ for every $\alpha \geqslant 0$.[1] In particular, we consider linear subspaces to be cones. A cone is **pointed** if it includes no lines. (A **line** is a translate of a one-dimensional subspace, that is, a set of the form $\{x + \alpha v : \alpha \in \mathbb{R}\}$, where $x, v \in X$ and $v \neq 0$.)

Let S be a nonempty subset of a vector space. The **cone generated by** S is the smallest cone that includes S and is thus $\{\alpha x : \alpha \geqslant 0 \text{ and } x \in S\}$. The **convex cone generated by** S is the smallest convex cone generated by S. You should verify that it consists of all nonnegative linear combinations from S.

5.25 Corollary   *In a Hausdorff topological vector space, the convex cone generated by a finite set is closed.*

---

[1] Some authors, notably R. T. Rockafellar [288] and G. Choquet [76], define a cone to be a set closed under multiplication by strictly positive scalars. The point zero may or may not belong to such a cone. Other authorities, e.g., W. Fenchel [123] and D. Gale [133] use our definition.

Proof: Let $S = \{x_1, x_2, \ldots, x_k\}$ be a nonempty finite subset of a Hausdorff topological vector space X. Then the convex cone K generated by S is given by

$$K = \left\{ \sum_{i=1}^{k} \lambda_i x_i : \lambda_i \geq 0 \text{ for each } i \right\}.$$

Now fix a nonzero $x = \sum_{i=1}^{k} \lambda_i x_i \in K$. We claim that there is a linearly independent subset T of S and nonnegative scalars $\{\beta_t : t \in T\}$ such that $x = \sum_{t \in T} \beta_t t$.

To see this, start by noticing that we can assume that $\lambda_i > 0$ for each i; otherwise drop the terms with $\lambda_i = 0$. Now if the set S is linearly independent, then there is nothing to prove. So assume that S is linearly dependent. This means that there exist scalars $\alpha_1, \ldots, \alpha_k$, not all zero, such that $\sum_{i=1}^{k} \alpha_i x_i = 0$. We can assume that $\alpha_i > 0$ for some i; otherwise multiply them by -1. Now let $\mu = \max\{\frac{\alpha_i}{\lambda_i} : i = 1, \ldots, k\}$, and notice that $\mu > 0$. In particular, we have $\lambda_i \geq \frac{1}{\mu}\alpha_i$ for each i and $\lambda_i = \frac{1}{\mu}\alpha_i$ for some i. This implies that

$$x = \sum_{i=1}^{k} \lambda_i x_i = \sum_{i=1}^{k} \lambda_i x_i - \frac{1}{\mu} \sum_{i=1}^{k} \alpha_i x_i = \sum_{i=1}^{k} (\lambda_i - \frac{1}{\mu}\alpha_i) x_i$$

is a linear combination of the x_i with nonnegative coefficients, and one of them is zero. In other words, we have shown that if the set S is not a linearly independent set, then we can write x as a linear combination with positive coefficients of at most $k-1$ vectors of S. Our claim can now be completed by repeating this process.

Now assume that a sequence $\{y_n\}$ in K satisfies $y_n \to y$ in X. Since the collection of all linearly independent subsets of S is a finite set, by the above discussion, there exist a linearly independent subset of S, say $\{z_1, \ldots, z_m\}$, and a subsequence of $\{y_n\}$, which we shall denote by $\{y_n\}$ again, such that

$$y_n = \sum_{i=1}^{m} \mu_i^n z_i$$

with all coefficients μ_i^n nonnegative. It follows from Corollary 5.24 that y belongs to K, so K is closed. $\blacksquare$

There are no infinite dimensional locally compact Hausdorff topological vector spaces. This is essentially due to F. Riesz.

5.26 Theorem (F. Riesz) *A Hausdorff topological vector space is locally compact if and only if is finite dimensional.*

Proof: Let (X, τ) be a Hausdorff topological vector space. If X is finite dimensional, then τ coincides with the Euclidean topology and since the closed balls are compact sets, it follows that (X, τ) is locally compact.

For the converse assume that (X, τ) is locally compact and let V be a τ-compact neighborhood of zero. From $V \subset \bigcup_{x \in V}(x + \frac{1}{2}V)$, we see that there exists a finite subset $\{x_1, \ldots, x_k\}$ of V such that

$$V \subset \bigcup_{i=1}^{k}(x_i + \tfrac{1}{2}V) = \{x_1, \ldots, x_k\} + \tfrac{1}{2}V. \qquad (\star)$$

Let Y be the linear span of $x_1, \ldots, x_k$. From $(\star)$, we get $V \subset Y + \frac{1}{2}V$. This implies $\frac{1}{2}V \subset \frac{1}{2}(Y + \frac{1}{2}V) = Y + \frac{1}{2^2}V$, so $V \subset Y + (Y + \frac{1}{2^2}V) = Y + \frac{1}{2^2}V$. By induction we see that

$$V \subset Y + \frac{1}{2^n}V \qquad (\star\star)$$

for each n. Next, fix $x \in V$. From $(\star\star)$, it follows that for each n there exist $y_n \in Y$ and $v_n \in V$ such that $x = y_n + \frac{1}{2^n}v_n$. Since V is τ-compact, there exists a subnet $\{v_{n_\alpha}\}$ of the sequence $\{v_n\}$ such that $v_{n_\alpha} \xrightarrow{\tau} v$ in X (and clearly $\frac{1}{2^{n_\alpha}} \to 0$ in $\mathbb{R}$). So

$$y_{n_\alpha} = x - \tfrac{1}{2^{n_\alpha}}v_{n_\alpha} \xrightarrow{\tau} x - 0v = x.$$

Since (by Corollary 5.22) Y is a closed subspace, $x \in Y$. That is, $V \subset Y$. Since V is also an absorbing set, it follows that $X = Y$, so that X is finite dimensional. $\blacksquare$

5.6 Convex sets

Recall that a subset of a vector space is **convex** if it includes the line segment joining any two of its points. Or in other words, a set C is convex if whenever $x, y \in C$, the line segment $\{\alpha x + (1 - \alpha)y : \alpha \in [0, 1]\}$ is included in C. By induction, a set C is convex if and only if for every finite subset $\{x_1, \ldots, x_n\}$ of C and nonnegative scalars $\{\alpha_1, \ldots, \alpha_n\}$ with $\sum_{i=1}^{n} \alpha_i = 1$, the linear combination $\sum_{i=1}^{n} \alpha_i x_i$ lies in C. Such a linear combination is called a **convex combination**, and the coefficients may be called **weights**.

The next lemma presents some elementary properties of convex sets.

5.27 Lemma *In any vector space:*

1. *The sum of two convex sets is convex.*

2. *Scalar multiples of convex sets are convex.*

3. *A set C is convex if and only if $\alpha C + \beta C = (\alpha + \beta)C$ for all nonnegative scalars α and β.*

4. *The intersection of an arbitrary family of convex sets is convex.*

5. *A convex set containing zero is circled if and only if it is symmetric.*

6. *In a topological vector space, both the interior and the closure of a convex set are convex.*

Proof: We prove only the first part of the last claim and leave the proofs of everything else as an exercise.

Let C be a convex subset of a tvs and let $0 \leqslant \alpha \leqslant 1$. Since C° is an open set, the set $\alpha C^\circ + (1 - \alpha)C^\circ$ is likewise open. (Why?) The convexity of C implies $\alpha C^\circ + (1 - \alpha)C^\circ \subset C$. Since C° is the largest open set included in C, we see that $\alpha C^\circ + (1 - \alpha)C^\circ \subset C^\circ$. This shows that C° is convex.   ∎

In topological vector spaces we can say a little bit more about the interior and closure of a convex set.

5.28 Lemma   *If C is a convex subset of a tvs, then:*

$$0 < \alpha \leqslant 1 \quad \Longrightarrow \quad \alpha C^\circ + (1 - \alpha)\overline{C} \subset C^\circ. \qquad (\star)$$

In particular, if $C^\circ \neq \varnothing$, then:

1. *The interior of C is dense in $\overline{C}$, that is, $\overline{C^\circ} = \overline{C}$.*

2. *The interior of $\overline{C}$ coincides with the interior of C, that is, $\overline{C}^\circ = C^\circ$.*

Proof: The case $\alpha = 1$ in $(\star)$ is immediate. So let C be convex, $x \in C^\circ$, $y \in \overline{C}$, and let $0 < \alpha < 1$. Choose an open neighborhood U of zero such that $x + U \subset C$. Since $y - \frac{\alpha}{1-\alpha}U$ is a neighborhood of y, there is some $z \in C \cap (y - \frac{\alpha}{1-\alpha}U)$, so that $(1 - \alpha)(y - z)$ belongs to αU. Since C is convex, the (nonempty) open set $V = \alpha(x + U) + (1 - \alpha)z = \alpha x + \alpha U + (1 - \alpha)z$ lies entirely in C. Moreover, from

$$\alpha x + (1 - \alpha)y = \alpha x + (1 - \alpha)(y - z) + (1 - \alpha)z \in \alpha x + \alpha U + (1 - \alpha)z = V \subset C,$$

we see that $\alpha x + (1 - \alpha)y \in C^\circ$. This proves $(\star)$, and letting $\alpha \to 0$ proves (1).

For (2), fix $x_0 \in C^\circ$ and $x \in \overline{C}^\circ$. Pick a neighborhood W of zero satisfying $x + W \subset \overline{C}$. Since W is absorbing, there is some $0 < \varepsilon < 1$ such that $\varepsilon(x - x_0) \in W$, so $x + \varepsilon(x - x_0) \in \overline{C}$. By $(\star)$, we have $x - \varepsilon(x - x_0) = \varepsilon x_0 + (1 - \varepsilon)x \in C^\circ$. But then, using $(\star)$ once more, we obtain $x = \frac{1}{2}[x - \varepsilon(x - x_0)] + \frac{1}{2}[x + \varepsilon(x - x_0)] \in C^\circ$. Therefore, $\overline{C}^\circ \subset C^\circ \subset \overline{C}^\circ$ so that $\overline{C}^\circ = C^\circ$.   ∎

Note that a convex set with an empty interior may have a closure with a nonempty interior. For instance, any dense (proper) vector subspace has this property.

The **convex hull** of a set A, denoted $\mathrm{co}\, A$, consists precisely of all convex combinations from A. That is,

$$\mathrm{co}\, A = \Big\{x : \exists\, x_i \in A,\ \alpha_i \geqslant 0\, (1 \leqslant i \leqslant n),\ \sum_{i=1}^{n} \alpha_i = 1,\ \text{and}\ x = \sum_{i=1}^{n} \alpha_i x_i \Big\}.$$

The convex hull $\mathrm{co}\, A$ is the smallest convex set including A and by Lemma 5.27 (4) is the intersection of all convex sets that include A. In a topological vector space, the **closed convex hull** of a set A, denoted $\overline{\mathrm{co}}\,A$, is the smallest closed convex set including A. By Lemma 5.27 (6) it is the closure of $\mathrm{co}\,A$, that is, $\overline{\mathrm{co}}\,A = \overline{\mathrm{co}\,A}$.

The next lemma presents further results on the relationship between topological and convexity properties. The **convex circled hull** of a subset A of a vector space is the smallest convex and circled set that includes A. It is the intersection of all convex and circled sets that include A. The **closed convex circled hull** of A is the smallest closed convex circled set including A. It is the closure of the convex circled hull of A.

5.29 Lemma   *For nonempty convex sets $A_1, \ldots, A_n$ in a tvs we have:*

1. *The convex hull of the union $\bigcup_{i=1}^{n} A_i$ satisfies*

$$\mathrm{co}\Big(\bigcup_{i=1}^{n} A_i\Big) = \Big\{\sum_{i=1}^{n} \lambda_i x_i : \lambda_i \geq 0,\ x_i \in A_i,\ and\ \sum_{i=1}^{n} \lambda_i = 1\Big\}.$$

   In particular, if each A_i is also compact, then $\mathrm{co}(\bigcup_{i=1}^{n} A_i)$ is compact.

2. *If, in addition, each A_i is circled, then the convex circled hull of the union $\bigcup_{i=1}^{n} A_i$ is the set*

$$\Big\{\sum_{i=1}^{n} \lambda_i x_i : \lambda_i \in \mathbb{R},\ x_i \in A_i,\ and\ \sum_{i=1}^{n} |\lambda_i| \leq 1\Big\}.$$

   Furthermore, if each A_i is also compact, then the convex circled hull of $\bigcup_{i=1}^{n} A_i$ is compact.

Proof: Let X be a vector space and let $A_1, \ldots, A_n$ be convex subsets of X. You can easily verify that the indicated sets coincide with the convex and convex circled hull of the union $\bigcup_{i=1}^{n} A_i$, respectively.

Now let X be equipped with a linear topology. Consider the compact sets

$$C = \Big\{\lambda \in \mathbb{R}_+^n : \sum_{i=1}^{n} \lambda_i = 1\Big\} \quad and \quad K = \Big\{\lambda \in \mathbb{R}^n : \sum_{i=1}^{n} |\lambda_i| \leq 1\Big\}.$$

Define the continuous function $f : \mathbb{R}^n \times A_1 \times \cdots \times A_n \to X$ by

$$f(\lambda, x_1, \ldots, x_n) = \sum_{i=1}^{n} \lambda_i x_i.$$

The compactness assertions follow from the fact that the continuous image of a compact set is compact, and the observations that

$$\mathrm{co}\Big(\bigcup_{i=1}^{n} A_i\Big) = f(C \times A_1 \times \cdots \times A_n)$$

and that $f(K \times A_1 \times \cdots \times A_n)$ is the convex circled hull of $\bigcup_{i=1}^{n} A_i$.   ∎

The convexity of the sets A_i is crucial for the results in Lemma 5.29; see Example 5.34 below. Here are some straightforward corollaries. The convex hull of a finite set is called a **polytope**.

5.30 Corollary   *Every polytope in a tvs is compact.*

5.31 Corollary   *The convex circled hull of a compact convex subset of a tvs is compact.*

Proof: Note that if C is a compact convex set, then its convex circled hull coincides with the convex circled hull of $C \cup (-C)$.   ∎

In finite dimensional vector spaces, the convex hull of a set is characterized by the celebrated Carathéodory convexity theorem.

5.32 Carathéodory's Convexity Theorem   *In an n-dimensional vector space, every vector in the convex hull of a nonempty set can be written as a convex combination using no more than n+1 vectors from the set.*

Proof: Let A be a nonempty subset of some n-dimensional vector space, and let $x \in \operatorname{co} A$. Consider the nonempty set of natural numbers

$$S = \{\ell \in \mathbb{N} : x \text{ is a convex combination of some } \ell \text{ vectors from } A\},$$

and let $k = \min S$. We must show that $k \leqslant n+1$.

Assume by way of contradiction that $k > n+1$. Pick $x_1, \ldots, x_k \in A$ and positive constants $\alpha_1, \ldots, \alpha_k$ with $\sum_{i=1}^k \alpha_i = 1$ and $x = \sum_{i=1}^k \alpha_i x_i$. Since $k - 1 > n$, the $k - 1$ vectors $x_2 - x_1, x_3 - x_1, \ldots, x_k - x_1$ of the n-dimensional vector space X must be linearly dependent. Consequently, there exist scalars $\lambda_2, \lambda_3, \ldots, \lambda_k$, not all zero, such that $\sum_{i=2}^k \lambda_i(x_i - x_1) = 0$. Letting $c_1 = -\sum_{i=2}^k \lambda_i$ and $c_i = \lambda_i$ ($i = 2, 3, \ldots, k$), we see that not all the c_i are zero and satisfy

$$\sum_{i=1}^k c_i x_i = 0 \quad \text{and} \quad \sum_{i=1}^k c_i = 0.$$

Without loss of generality we can assume that $c_j > 0$ for some j. Next, put $c = \min\{\alpha_i/c_i : c_i > 0\}$, and pick some m with $\alpha_m/c_m = c > 0$. Note that

1. $\alpha_i - c c_i \geqslant 0$ for each i and $\alpha_m - c c_m = 0$; and

2. $\sum_{i=1}^k (\alpha_i - c c_i) = 1$ and $x = \sum_{i=1}^k (\alpha_i - c c_i) x_i$.

The above shows that x can be written as a convex combination of fewer than k vectors of A, contrary to the definition of k.   ∎

Since continuous images of compact sets are compact, Carathéodory's theorem immediately implies the following. (Cf. proof of Lemma 5.29.)

5.33 Corollary   *The convex hull and the convex circled hull of a compact subset of a finite dimensional vector space are compact sets.*

The convex hull of a compact subset of an infinite dimensional topological vector space need not be a compact set.

5.34 Example (Noncompact convex hull)   Consider ℓ_2, the space of all square summable sequences. For each n let $u_n = (\underbrace{0,\ldots,0}_{n-1},\frac{1}{n},0,0,\ldots)$. Observe that $\|u_n\|_2 = \frac{1}{n}$, so $u_n \to 0$. Consequently,

$$A = \{u_1, u_2, u_3, \ldots\} \cup \{0\}$$

is a norm compact subset of ℓ_2. Since $0 \in A$, it is easy to see that

$$\mathrm{co}\, A = \Big\{ \sum_{i=1}^{k} \alpha_i u_i : \alpha_i \geqslant 0 \text{ for each } i \text{ and } \sum_{i=1}^{k} \alpha_i \leqslant 1 \Big\}.$$

In particular, each vector of $\mathrm{co}\, A$ has only finitely many nonzero components. We claim that $\mathrm{co}\, A$ is not norm compact. To see this, set

$$x_n = \big(\tfrac{1}{2}, \tfrac{1}{2}\cdot\tfrac{1}{2^2}, \tfrac{1}{3}\cdot\tfrac{1}{2^3}, \ldots, \tfrac{1}{n}\cdot\tfrac{1}{2^n}, 0, 0, \ldots\big) = \sum_{i=1}^{n} \tfrac{1}{2^i} u_i,$$

so $x_n \in \mathrm{co}\, A$. Now $x_n \xrightarrow{\|\cdot\|_2} x = \big(\tfrac{1}{2}, \tfrac{1}{2}\tfrac{1}{2^2}, \tfrac{1}{3}\cdot\tfrac{1}{2^3}, \ldots, \tfrac{1}{n}\cdot\tfrac{1}{2^n}, \tfrac{1}{n+1}\cdot\tfrac{1}{2^{n+1}}, \ldots\big)$ in ℓ_2. But $x \notin \mathrm{co}\, A$, so $\mathrm{co}\, A$ is not even closed, let alone compact.

In this example, the convex hull of a compact set failed to be closed. The question remains as to whether the closure of the convex hull is compact. In general, the answer is no. To see this, let X be the space of sequences that are eventually zero, equipped with the ℓ_2-norm. Let A be as above, and note that $\overline{\mathrm{co}}\, A$ (where the closure is taken in X, not ℓ_2) is not compact either. To see this, observe that the sequence $\{x_n\}$ defined above has no convergent subsequence (in X).   ∎

However there are three important cases when the closed convex hull of a compact set is compact. The first is when the compact set is a finite union of compact convex sets. This is just Lemma 5.29. The second is when the space is completely metrizable and locally convex. This includes the case of all Banach spaces with their norm topologies. Failure of completeness is where the last part of Example 5.34 goes awry. The third case is a compact set in the weak topology on a Banach space; this is the Krein–Šmulian Theorem 6.35 ahead. Here is the proof for the completely metrizable locally convex case.

5.35 Theorem (Closed convex hull of a compact set)   *In a completely metrizable locally convex space, the closed convex hull of a compact set is compact.*

Proof: Let K be compact subset of a completely metrizable locally convex space X. By Theorem 5.10 the topology is generated by some compatible complete metric d. By Theorem 3.28, it suffices to prove that $\overline{\text{co}}\, K$ is d-totally bounded. So let $\varepsilon > 0$ be given. By local convexity there is a convex neighborhood V of zero satisfying $V + V \subset B_\varepsilon$, the d-open ball of radius ε at zero. Since K is compact, there is a finite set Φ with $K \subset \Phi + V$. Clearly, $\text{co}\, K \subset \text{co}\, \Phi + V$. (Why?) By Corollary 5.30, co Φ is compact, so there is a finite set F satisfying $\text{co}\, \Phi \subset F + V$. Therefore

$$\text{co}\, K \subset \text{co}\, \Phi + V \subset F + V + V \subset F + B_\varepsilon.$$

Thus co K, and hence $\overline{\text{co}}\, K$, is d-totally bounded.   ∎

Note that the proof above does not require the entire space to be completely metrizable. The same argument works provided $\overline{\text{co}}\, K$ lies in a subset of a locally convex space that is completely metrizable.

Finally, we shall present a case where the convex hull of the union of two closed convex sets is closed. But first, we need a definition.

5.36 Definition   *A subset A of a topological vector space (X, τ) is **(topologically) bounded**, or more specifically **τ-bounded**, if for each neighborhood V of zero there exists some $\lambda > 0$ such that $A \subset \lambda V$.*

Observe that for a normed space, the topologically bounded sets coincide with the norm bounded sets. Also, notice that if $\{x_\alpha\}$ is a topologically bounded net in a tvs and $\lambda_\alpha \to 0$ in $\mathbb{R}$, then $\lambda_\alpha x_\alpha \to 0$.

5.37 Lemma   *If A and B are two nonempty convex subsets of a Hausdorff topological vector space such that A is compact and B is closed and bounded, then $\text{co}(A \cup B)$ is closed.*

Proof: Let $z_\alpha = (1 - \lambda_\alpha)x_\alpha + \lambda_\alpha y_\alpha \to z$, where $0 \leqslant \lambda_\alpha \leqslant 1$, $x_\alpha \in A$, and $y_\alpha \in B$ for each α. By passing to a subnet, we can assume that $x_\alpha \to x \in A$ and $\lambda_\alpha \to \lambda \in [0, 1]$. If $\lambda > 0$, then $y_\alpha \to \frac{z-(1-\lambda)x}{\lambda} = y \in B$, and consequently $z = (1 - \lambda)x + \lambda y \in \text{co}(A \cup B)$.

Now consider the case $\lambda = 0$. The boundedness of B implies $\lambda_\alpha y_\alpha \to 0$, so $z_\alpha = (1 - \lambda_\alpha)x_\alpha + \lambda_\alpha y_\alpha \to x$. Since the space is Hausdorff, $z = x \in \text{co}(A \cup B)$.   ∎

5.7   Convex and concave functions

The interaction of the algebraic and topological structure of a topological vector space is manifested in the properties of the important class of convex functions. The definition is purely algebraic.

5.38 Definition   *A function $f: C \to \mathbb{R}$ on a convex set C in a vector space is:*

- *convex if $f(\alpha x + (1 - \alpha)y) \leqslant \alpha f(x) + (1 - \alpha)f(y)$ for all $x, y \in C$ and all $0 \leqslant \alpha \leqslant 1$.*

- *strictly convex if $f(\alpha x + (1 - \alpha)y) < \alpha f(x) + (1 - \alpha)f(y)$ for all $x, y \in C$ with $x \neq y$ and all $0 < \alpha < 1$.*

- *concave if $-f$ is a convex function.*

- *strictly concave if $-f$ is strictly convex.*

Note that a real function f on a convex set is convex if and only if

$$f\left(\sum_{i=1}^{n} \alpha_i x_i\right) \leqslant \sum_{i=1}^{n} \alpha_i f(x_i)$$

for every convex combination $\sum_{i=1}^{n} \alpha_i x_i$.
You may verify the following lemma.

5.39 Lemma   *A function $f: C \to \mathbb{R}$ on a convex subset of a vector space is convex if and only if its **epigraph**, $\{(x, \alpha) \in C \times \mathbb{R} : \alpha \geqslant f(x)\}$, is convex. Similarly, f is concave if and only if its **hypograph**, $\{(x, \alpha) \in C \times \mathbb{R} : \alpha \leqslant f(x)\}$, is convex.*

Some important properties of convex functions are immediate consequences of the definition. There is of course a corresponding lemma for concave functions. We omit it.

5.40 Lemma   *The collection of convex functions on a fixed convex set has the following properties.*

1. *Sums and nonnegative scalar multiples of convex functions are convex.*

2. *The (finite) pointwise limit of a net of convex functions is convex.*

3. *The (finite) pointwise supremum of a family of convex functions is convex.*

The next simple inequality is useful enough that it warrants its own lemma. It requires no topology.

5.41 Lemma   *Let $f: C \to \mathbb{R}$ be a convex function, where C is a convex subset of a vector space. Let x belong to C and suppose z satisfies $x + z \in C$ and $x - z \in C$. Let $\delta \in [0, 1]$. Then*

$$\left|f(x + \delta z) - f(x)\right| \leqslant \delta \max\{f(x + z) - f(x), f(x - z) - f(x)\}$$

Proof: Now $x + \delta z = (1 - \delta)x + \delta(x + z)$, so $f(x + \delta z) \leqslant (1 - \delta)f(x) + \delta f(x + z)$. Rearranging terms yields

$$f(x + \delta z) - f(x) \leqslant \delta\left[f(x + z) - f(x)\right], \tag{1}$$

and replacing z by $-z$ gives

$$f(x - \delta z) - f(x) \leqslant \delta\left[f(x - z) - f(x)\right]. \tag{2}$$

Also, since $x = \frac{1}{2}(x + \delta z) + \frac{1}{2}(x - \delta z)$, we have $f(x) \leqslant \frac{1}{2}f(x + \delta z) + \frac{1}{2}f(x - \delta z)$. Multiplying by two and rearranging terms we obtain

$$f(x) - f(x + \delta z) \leqslant f(x - \delta z) - f(x). \tag{3}$$

Combining (2) and (3) yields

$$f(x) - f(x + \delta z) \leqslant f(x - \delta z) - f(x) \leqslant \delta\left[f(x - z) - f(x)\right].$$

This in conjunction with (1) yields the conclusion of the lemma. ∎

5.42 Theorem (Local continuity of convex functions) *Let $f : C \to \mathbb{R}$ be a convex function, where C is a convex subset of a topological vector space. If f is bounded above on a neighborhood of an interior point of C, then f is continuous at that point.*

Proof: Assume that for some $x \in C$ there exist a circled neighborhood V of zero and some $M > 0$ satisfying $x + V \subset C$ and $f(y) < f(x) + M$ for each $y \in x + V$. Fix $\varepsilon > 0$ and choose some $0 < \delta \leqslant 1$ so that $\delta M < \varepsilon$. But then if $y \in x + \delta V$, then from Lemma 5.41 it follows that for each $y \in x + \delta V$ we have $|f(y) - f(x)| < \varepsilon$. This shows that f is continuous at x. ∎

Amazingly, continuity at a single point implies global continuity for convex functions on open sets.

5.43 Theorem (Global continuity of convex functions) *For a convex function $f : C \to \mathbb{R}$ on an open convex subset of a topological vector space, the following statements are equivalent.*

1. *f is continuous on C.*

2. *f is upper semicontinuous on C.*

3. *f is bounded above on a neighborhood of each point in C.*

4. *f is bounded above on a neighborhood of some point in C.*

5. *f is continuous at some point in C.*

Proof: (1) $\implies$ (2)   Obvious.

(2) $\implies$ (3)   Assume that f is upper semicontinuous and $x \in C$. Then the set $\{y \in C : f(y) < f(x) + 1\}$ is an open neighborhood of x on which f is bounded.

(3) $\implies$ (4)   This is trivial.

(4) $\implies$ (5)   This is Theorem 5.42.

(5) $\implies$ (1)   Suppose f is continuous at the point x, and let y be any other point in C. Since scalar multiplication is continuous, $\{\beta \in \mathbb{R} : x + \beta(y - x) \in C\}$ includes an open neighborhood of 1. This implies that there exist $z \in C$ and $0 < \lambda < 1$ such that $y = \lambda x + (1 - \lambda)z$.

Also, since f is continuous at x, there is a circled neighborhood V of zero such that $x + V \subset C$
and f is bounded above on $x + V$, say by μ. We
claim that f is bounded above on $y + \lambda V$. To see
this, let $v \in V$. Then $y + \lambda v = \lambda(x+v)+(1-\lambda)z \in C$.
The convexity of f thus implies

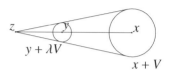

$$f(y + \lambda v) \leqslant \lambda f(x + v) + (1 - \lambda)f(z) \leqslant \lambda\mu + (1 - \lambda)f(z).$$

That is, f is bounded above by $\lambda\mu + (1 - \lambda)f(z)$ on $y + \lambda V$. So by Theorem 5.42, f is continuous at y.   ∎

If the topology of a tvs is generated by a norm, continuity of a convex function at an interior point implies local Lipschitz continuity. The proof of the next result is adapted from A. W. Roberts and D. E. Varberg [285].

5.44 Theorem   *Let $f : C \to \mathbb{R}$ be convex, where C is a convex subset of a normed tvs. If f is continuous at the interior point x of C, then f is Lipschitz continuous on a neighborhood of x. That is, there exists $\delta > 0$ and $\mu > 0$, such that $B_\delta(x) \subset C$ and for $y, z \in B_\delta(x)$, we have*

$$|f(y) - f(z)| \leqslant \mu \|y - z\|.$$

Proof: Since f is continuous at x, there exists $\delta > 0$ such that $B_{2\delta}(x) \subset C$ and $w, z \in B_{2\delta}(x)$ implies $|f(w) - f(z)| < 1$. Given distinct y and z in $B_\delta(x)$, let $\alpha = \|y - z\|$ and let $w = y + \frac{\delta}{\alpha}(y - z)$, so $\|w - y\| = \frac{\delta}{\alpha}\|y - z\| = \delta$. Then w belongs to $B_{2\delta}(x)$ and we may write y as the convex combination $y = \frac{\alpha}{\alpha+\delta}w + \frac{\delta}{\alpha+\delta}z$. Therefore

$$f(y) \leqslant \frac{\alpha}{\alpha + \delta}f(w) + \frac{\delta}{\alpha + \delta}f(z).$$

Subtracting $f(z)$ from each side gives

$$f(y) - f(z) \leqslant \frac{\alpha}{\alpha + \delta}[f(w) - f(z)] < \frac{\alpha}{\alpha + \delta}.$$

Switching the roles of y and z allows us to conclude

$$|f(y) - f(z)| < \frac{\alpha}{\alpha + \delta} < \frac{\alpha}{\delta} = \frac{1}{\delta}\|y - z\|,$$

so $\mu = 1/\delta$ is the desired Lipschitz constant. ∎

We also point out that strictly convex functions on infinite dimensional spaces are quite special. In order for a continuous function to be strictly convex on a compact convex set, the relative topology of the set must be metrizable. This result relies on facts about metrizability of uniform spaces that we do not wish to explore, but if you are interested, see G. Choquet [76, p. II-139].

5.8   Sublinear functions and gauges

A real function f defined on a vector space is **subadditive** if

$$f(x + y) \leqslant f(x) + f(y)$$

for all x and y. Recall that a nonempty subset C of a vector space is a cone if $x \in C$ implies $\alpha x \in C$ for every $\alpha \geqslant 0$. A real function f defined on a cone C is **positively homogeneous** if

$$f(\alpha x) = \alpha f(x)$$

for every $\alpha \geqslant 0$. Clearly, if f is positively homogeneous, then $f(0) = 0$ and f is completely determined by its values on any absorbing set. In other words, two positively homogeneous functions are equal if and only if they agree on an absorbing set.

5.45 Definition   *A real function on a vector space is **sublinear** if it is both positively homogeneous and subadditive, or equivalently, if it is both positively homogeneous and convex.*

To see the equivalence in the definition above, observe that for a subadditive positively homogeneous function f we have

$$f(\lambda x + (1 - \lambda)y) \leqslant f(\lambda x) + f((1 - \lambda x)) = \lambda f(x) + (1 - \lambda)f(x),$$

so f is convex. Conversely, to see that a positively homogeneous convex function is subadditive, note that

$$f(x) + f(y) = \tfrac{1}{2}f(2x) + \tfrac{1}{2}f(2y) \leqslant f\left(\tfrac{1}{2}2x + \tfrac{1}{2}2y\right) = f(x + y).$$

Clearly every linear functional is sublinear, and so too is every norm. An important subclass of sublinear functions consists of functions called seminorms, which satisfy most of the properties norms, and which turn out to be crucial to the study of locally convex spaces.

5.46 Definition   *A **seminorm** is a subadditive function $p \colon X \to \mathbb{R}$ on a vector space satisfying*

$$p(\alpha x) = |\alpha| p(x)$$

for all $\alpha \in \mathbb{R}$ and all $x \in X$. [2]
   *A seminorm p that satisfies $p(x) = 0$ if and only if $x = 0$ is called a **norm**.*

Note that every seminorm is indeed sublinear, and every sublinear function satisfying $p(-x) = p(x)$ for all x is a seminorm. In particular, if f is a linear functional, then $p(x) = |f(x)|$ defines a seminorm. A seminorm p defines a semimetric d via $d(x, y) = p(x - y)$. If p is a norm, then the semimetric is actually a metric.
   We now state some simple properties of sublinear functions. The proofs are left as exercises.

5.47 Lemma (Sublinearity)   *If $p \colon X \to \mathbb{R}$ is sublinear, then:*

1. $p(0) = 0$.

2. *For all x we have $-p(x) \leqslant p(-x)$. Consequently p is linear if and only if $p(-x) = -p(x)$ for all $x \in X$.*

3. *The function q defined by $q(x) = \max\{p(x), p(-x)\}$ is a seminorm.*

4. *If p is a seminorm, then $p(x) \geqslant 0$ for all x.*

5. *If p is a seminorm, then the set $\{x : p(x) = 0\}$ is a linear subspace.*

We now come to the important class of Minkowski functionals, or gauges.

5.48 Definition *The **gauge**, [3] or the **Minkowski functional**, p_A, of a subset A of a vector space is defined by*

$$p_A(x) = \inf\{\alpha > 0 : x \in \alpha A\},$$

where, by convention, $\inf \varnothing = \infty$. In other words, $p_A(x)$ is the smallest factor by which the set A must be enlarged to contain the point x.

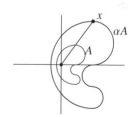

Figure 5.2. The gauge of A.

The next lemma collects a few elementary properties of gauges. The proof is left as an exercise.

---

[2] Be assured at once that, as we shall see in the following result, every seminorm $p \colon X \to \mathbb{R}$ satisfies $p(x) \geqslant 0$ for each $x \in X$.

[3] Dunford and Schwartz [110, p. 411] use the term support functional instead of gauge. We however have another, more standard, use in mind for the term support functional.

5.49 Lemma　　*For nonempty subsets B and C of a vector space X:*

1. $p_{-C}(x) = p_C(-x)$ *for all* $x \in X$.

2. *If C is symmetric, then* $p_C(x) = p_C(-x)$ *for all* $x \in X$.

3. $B \subset C$ *implies* $p_C \leqslant p_B$.

4. *If C includes a subspace M, then* $p_C(x) = 0$ *for all* $x \in M$.

5. *If C is star-shaped about zero, then*

$$\{x \in X : p_C(x) < 1\} \subset C \subset \{x \in X : p_C(x) \leqslant 1\}.$$

6. *If X is a tvs and C is closed and star-shaped about zero, then*

$$C = \{x \in X : p_C(x) \leqslant 1\}.$$

7. *If B and C are star-shaped about zero, then* $p_{B \cap C} = p_B \vee p_C$, *where as usual* $[p_B \vee p_C](x) = \max\{p_B(x), p_C(x)\}$.

Absorbing sets are of interest in part because any positively homogeneous function is completely determined by its values on any absorbing set.

5.50 Lemma　　*For a nonnegative function* $p: X \to \mathbb{R}$ *on a vector space we have the following.*

1. *p is positively homogeneous if and only if it is the gauge of an absorbing set—in which case for every subset A of X satisfying*

$$\{x \in X : p(x) < 1\} \subset A \subset \{x \in X : p(x) \leqslant 1\}$$

   we have $p_A = p$.

2. *p is sublinear if and only if it is the gauge of a convex absorbing set C, in which case we may take* $C = \{x \in X : p(x) \leqslant 1\}$.

3. *p is a seminorm if and only if it is the gauge of a circled convex absorbing set C, in which case we may take* $C = \{x \in X : p(x) \leqslant 1\}$.

4. *When X is a tvs, p is a continuous seminorm if and only if it is the gauge of a unique closed, circled and convex neighborhood V of zero, namely* $V = \{x \in X : p(x) \leqslant 1\}$.

5. *When X is finite dimensional, p is a norm if and only if it is the gauge of a unique circled, convex and compact neighborhood V of zero, namely* $V = \{x \in X : p(x) \leqslant 1\}$.

Proof: (1)   If $p = p_A$ for some absorbing subset A of X, then it is easy to see that p is positively homogeneous. For the converse, assume that p is positively homogeneous, and let A be any subset of X satisfying

$$\{x \in X : p(x) < 1\} \subset A \subset \{x \in X : p(x) \leqslant 1\}.$$

Clearly, A is an absorbing set, so $p_A : X \to \mathbb{R}$ is a nonnegative real-valued positively homogeneous function.

Now fix $x \in X$. If some $\alpha > 0$ satisfies $x \in \alpha A$, then pick some $u \in A$ such that $x = \alpha u$ and note that $p(x) = p(\alpha u) = \alpha p(u) \leqslant \alpha$. From this, we easily infer that $p(x) \leqslant p_A(x)$. On the other hand, the positive homogeneity of p implies that for each $\beta > p(x)$ we have $\frac{x}{\beta} \in A$ or $x \in \beta A$, so $p_A(x) \leqslant \beta$ for all $\beta > p(x)$. Hence $p_A(x) \leqslant p(x)$ is also true. Therefore $p_A(x) = p(x)$ for all $x \in X$.

(2)   Let $p = p_C$, the gauge of the absorbing convex set C. Clearly p_C is nonnegative and positively homogeneous. For the subadditivity of p_C, let $\alpha, \beta > 0$ satisfy $x \in \alpha C$ and $y \in \beta C$. Then $x + y \in \alpha C + \beta C = (\alpha + \beta)C$, so $p_C(x+y) \leqslant \alpha + \beta$. Taking infima yields $p_C(x + y) \leqslant p_C(x) + p_C(y)$, so p_C is subadditive. For the converse, assume that p is a sublinear function. Let $C = \{x \in X : p(x) \leqslant 1\}$ and note that C is convex and absorbing. Now a glance at part (1) shows that $p = p_C$.

(3)   Repeat the arguments of the preceding part.

(4)   If p is a continuous seminorm, then the set $V = \{x \in X : p(x) \leqslant 1\}$ is a closed, circled and convex neighborhood of zero such that $p = p_V$. Conversely, if V is a closed, circled and convex neighborhood of zero and $p = p_V$, then p_V is (by part (3)) a seminorm. But then $p_V \leqslant 1$ on V and Theorem 5.43 guarantee that p is continuous.

For the uniqueness of the set V, assume that W is any other closed, circled and convex neighborhood of zero satisfying $p = p_V = p_W$. If $x \in W$, then $p(x) = p_W(x) \leqslant 1$, so $x \in V$. Therefore, $W \subset V$. For the reverse inclusion, let $x \in V$. This implies $p_W(x) = p_V(x) \leqslant 1$. If $p_W(x) < 1$, then pick $0 \leqslant \alpha < 1$ and $w \in W$ such $x = \alpha w$. Since W is circled, $x \in W$. On the other hand, if $p_W(x) = 1$, then pick a sequence $\{\alpha_n\}$ of real numbers and a sequence $\{w_n\} \subset W$ satisfying $\alpha_n \downarrow 1$ and $x = \alpha_n w_n$ for each n. But then $w_n = \frac{x}{\alpha_n} \to x$ and the closedness of W yield $x \in W$. Thus, $V \subset W$ is also true, so $W = V$.

(5)   If p is a norm, then p generates the Euclidean topology on X, so the set $V = \{x \in X : p(x) \leqslant 1\}$ is circled, convex and compact neighborhood of zero and satisfies $p = p_V$. Its uniqueness should be obvious. On the other hand, if $p = p_V$, where $V = \{x \in X : p(x) \leqslant 1\}$ is a circled, convex and compact neighborhood of zero, then it is not difficult to see that the seminorm p is indeed a norm.   ∎

The continuity of a sublinear functional is determined by its behavior near zero. Recall that a real function $f : D \to \mathbb{R}$ on a subset of a tvs is uniformly continuous on D if for every $\varepsilon > 0$, there is a neighborhood V of zero such that $|f(x) - f(y)| < \varepsilon$ whenever $x, y \in D$ satisfy $x - y \in V$.

5.51 Lemma   *A sublinear function on a tvs is (uniformly) continuous if and only if it is bounded on some neighborhood of zero.*[4]

Proof: Let $h: X \to \mathbb{R}$ be a sublinear function on a tvs. Note that h is bounded on $h^{-1}((-1, 1))$, which is a neighborhood of zero if h is continuous.

For the converse, continuity follows from Theorem 5.43, but uniform continuity is easy to prove directly. Assume that $|h(x)| < M$ for each x in some circled neighborhood V of zero. Note that for any x and y we have

$$h(x) = h(x - y + y) \leqslant h(x - y) + h(y),$$

so $h(x) - h(y) \leqslant h(x - y)$. In a similar fashion, $h(y) - h(x) \leqslant h(y - x)$. Thus, $|h(x) - h(y)| \leqslant \max\{h(x - y), h(y - x)\}$. So if $x - y \in \frac{\varepsilon}{M} V$, then $|h(x) - h(y)| < \varepsilon$, which shows that h is uniformly continuous. ∎

The next result elaborates on Lemma 5.50.

5.52 Theorem (Semicontinuity of gauges)   *A nonnegative sublinear function on a topological vector space is:*

1. *Lower semicontinuous if and only if it is the gauge of an absorbing closed convex set.*

2. *Continuous if and only if it is the gauge of a convex neighborhood of zero.*

Proof: Let $p: X \to \mathbb{R}$ be a nonnegative sublinear function on a tvs.

(1) Suppose first that the function p is lower semicontinuous on X. Then $C = \{x \in X : p(x) \leqslant 1\}$ is absorbing, closed and convex. By Lemma 5.50, $p = p_C$, the gauge of C.

Let C be an arbitrary absorbing, closed and convex subset of X. Then for $0 < \alpha < \infty$ the lower contour set $\{x \in X : p_C(x) \leqslant \alpha\} = \alpha C$ (why?), which is closed. The set $\{x \in X : p_C(x) \leqslant 0\} = \bigcap_{\alpha > 0} \alpha C$, which is closed, being the intersection of closed sets. Finally, $\{x \in X : p_C(x) \leqslant \alpha\}$ for $\alpha < 0$ is empty. Thus, p_C is lower semicontinuous.

(2) If p is continuous, then the set $C = \{x \in X : p(x) \leqslant 1\}$ includes the set $\{x \in X : p(x) < 1\}$, which is open. Thus C is a (closed) convex neighborhood of zero, and $p = p_C$. On the other hand, if C is a neighborhood of zero and $p = p_C$, then $p_C \leqslant 1$ on C, so by Lemma 5.51 it is continuous. ∎

---

[4] By Theorem 7.24, every sublinear function on a finite dimensional vector space is continuous, since it is convex.

5.9   The Hahn–Banach Extension Theorem

Let X^* denote the vector space of all linear functionals on the linear space X. The space X^* is called the **algebraic dual** of X to distinguish it from the **topological dual** X', the vector space of all continuous linear functionals on a tvs X.[5]

The algebraic dual X^* is in general very large. To get a feeling for its size, fix a Hamel basis $\mathcal{H}$ for X. Every $x \in X$ has a unique representation $x = \sum_{h \in \mathcal{H}} \lambda_h h$, where only a finite number of the λ_h are nonzero; see Theorem 1.8. If $f^* \in X^*$, then $f^*(x) = \sum_{h \in \mathcal{H}} \lambda_h f^*(h)$, so the action of f^* on X is completely determined by its action on $\mathcal{H}$. This implies that every $f \in \mathbb{R}^{\mathcal{H}}$ gives rise to a (unique) linear functional f^* on X via the formula $f^*(x) = \sum_{h \in \mathcal{H}} \lambda_h f(h)$. The mapping $f \mapsto f^*$ is a linear isomorphism from $\mathbb{R}^{\mathcal{H}}$ onto X^*, so X^* can be identified with $\mathbb{R}^{\mathcal{H}}$.[6] In general, when we use the term dual space, we mean the topological dual.

One of the most important and far-reaching results in analysis is the following seemingly mild theorem. It is usually stated for the case where p is sublinear, but this more general statement is as easy to prove. Recall that a real-valued function f **dominates** a real-valued function g on A if $f(x) \geqslant g(x)$ for all $x \in A$.

5.53 Hahn–Banach Extension Theorem   *Let X be a vector space and let $p\colon X \to \mathbb{R}$ be any convex function. Let M be a vector subspace of X and let $f\colon M \to \mathbb{R}$ be a linear functional dominated by p on M. Then there is a (not generally unique) linear extension $\hat{f}$ of f to X that is dominated by p on X.*

---

[5] Be warned! Some authors use X' for the algebraic dual and X^* for the topological dual.

[6] This depends on the fact that any two Hamel bases $\mathcal{H}$ and $\mathcal{H}'$ of X have the same cardinality. From elementary linear algebra, we know that this is true if $\mathcal{H}$ is finite. We briefly sketch the proof of this claim when $\mathcal{H}$ and $\mathcal{H}'$ are infinite. The proof is based upon the fact that $\mathcal{H} \times \mathbb{N}$ has the same cardinality as $\mathcal{H}$. To see this, let $\mathcal{X}$ be the set of all pairs (S, f), where S is a nonempty subset of $\mathcal{H}$ and the function $f\colon S \times \mathbb{N} \to S$ is one-to-one and surjective. Since $\mathcal{X}$ contains the countable subsets of $\mathcal{H}$, the set $\mathcal{X}$ is nonempty. On $\mathcal{X}$ we define a partial order $\geqslant$ by letting $(S, f) \geqslant (T, g)$ whenever $S \supset T$ and $f = g$ on T. It is not difficult to see that $\geqslant$ is indeed a partial order on $\mathcal{X}$ and that every chain in $\mathcal{X}$ has an upper bound. By Zorn's Lemma 1.7, $\mathcal{X}$ has a maximal element, say (R, φ). We claim that $\mathcal{H} \setminus R$ is a finite set. Otherwise, if $\mathcal{H} \setminus R$ is an infinite set, then $\mathcal{H} \setminus R$ must include a countable subset A. Let $R' = R \cup A$ and fix any one-to-one and surjective function $g\colon A \times \mathbb{N} \to A$. Now define $\psi\colon R' \times \mathbb{N} \to R'$ by $\psi(r, n) = \varphi(r, n)$ if $(r, n) \in R \times \mathbb{N}$ and $\psi(a, n) = g(a, n)$ if $(a, n) \in A \times \mathbb{N}$. But then we have $(R', \psi) \in \mathcal{X}$ and $(R', \psi) > (R, \varphi)$, contrary to the maximality property of (R, φ). Therefore, $\mathcal{H} \setminus R$ is a finite set. Next, pick a countable set Y of R and fix a one-to-one and surjective function $h\colon [(\mathcal{H} \setminus R) \cup Y] \times \mathbb{N} \to [\varphi(Y \times \mathbb{N}) \cup (\mathcal{H} \setminus R)]$ and then define the function $\theta\colon \mathcal{H} \times \mathbb{N} \to \mathcal{H}$ by $\theta(x, n) = \varphi(x, n)$ if $(x, n) \in (R \setminus Y) \times \mathbb{N}$ and $\theta(x, n) = h(x, n)$ if $(x, n) \in [(\mathcal{H} \setminus R) \cup Y] \times \mathbb{N}$. Clearly, $\theta\colon \mathcal{H} \times \mathbb{N} \to \mathcal{H}$ is one-to-one and surjective.

For each $x \in \mathcal{H}$ there exists a unique nonempty finite subset $\mathcal{H}'(x) = \{y_1^x, \ldots, y_{k_x}^x\}$ of $\mathcal{H}'$ and nonzero scalars $\lambda_1^x, \ldots, \lambda_{k_x}^x$ such that $x = \sum_{i=1}^{k_x} \lambda_i^x y_i^x$. Since $\mathcal{H}$ and $\mathcal{H}'$ are Hamel bases, it follows that $\mathcal{H}' = \bigcup_{x \in \mathcal{H}} \mathcal{H}'(x)$. Now define the function $\alpha\colon \mathcal{H} \times \mathbb{N} \to \mathcal{H}'$ by $\alpha(x, n) = y_1^x$ if $n > k_x$ and $\alpha(x, n) = y_n^x$ if $1 \leqslant n \leqslant k_x$. Clearly, α is surjective and from this we infer that there exists a one-to-one function $\beta\colon \mathcal{H}' \to \mathcal{H} \times \mathbb{N}$. But then the scheme $\mathcal{H}' \overset{\beta}{\longrightarrow} \mathcal{H} \times \mathbb{N} \overset{\theta}{\longrightarrow} \mathcal{H}$ shows that $\mathcal{H}$ has cardinality at least as large as $\mathcal{H}'$. By symmetry, $\mathcal{H}'$ has cardinality at least as large as $\mathcal{H}$ and a glance at the classical Cantor–Schröder–Bernstein Theorem 1.2 shows that $\mathcal{H}$ and $\mathcal{H}'$ have the same cardinality.

Proof: The proof is an excellent example of what is known as transfinite induction. It has two parts. One part says that an extension of f whose domain is not all of X can be extended to a larger subspace and still satisfy $\hat{f} \leqslant p$. The second part says that this is enough to conclude that we can extend f all the way to X and still satisfy $\hat{f} \leqslant p$.

Let $f \leqslant p$ on the subspace M. If $M = X$, then we are done. So suppose there exists $v \in X \setminus M$. Let N be the linear span of $M \cup \{v\}$. For each $x \in N$ there is a unique decomposition of x of the form $x = z + \lambda v$ where $z \in M$. (To see the uniqueness, suppose $x = z_1 + \lambda_1 v = z_2 + \lambda_2 v$. Then $z_1 - z_2 = (\lambda_2 - \lambda_1)v$. Since $z_1 - z_2 \in M$ and $v \notin M$, it must be the case that $\lambda_2 - \lambda_1 = 0$. But then $\lambda_1 = \lambda_2$ and $z_1 = z_2$.)

Any linear extension $\hat{f}$ of f to N must satisfy $\hat{f}(z + \lambda v) = f(z) + \lambda \hat{f}(v)$. Thus what we need to show is that we can choose $c = \hat{f}(v) \in \mathbb{R}$ so that $\hat{f} \leqslant p$ on N. That is, we must demonstrate the existence of a real number c satisfying

$$f(z) + \lambda c \leqslant p(z + \lambda v) \tag{1}$$

for all $z \in M$ and all $\lambda \in \mathbb{R}$. It is a routine matter to verify that (1) is true if and only if there exists some real number c satisfying

$$\tfrac{1}{\lambda}[f(x) - p(x - \lambda v)] \leqslant c \leqslant \tfrac{1}{\mu}[p(y + \mu v) - f(y)] \tag{2}$$

for all $x, y \in M$ and all $\lambda, \mu > 0$. Now notice that (2) is true for some $c \in \mathbb{R}$ if and only if

$$\sup_{x \in M, \lambda > 0} \tfrac{1}{\lambda}[f(x) - p(x - \lambda v)] \leqslant \inf_{y \in M, \mu > 0} \tfrac{1}{\mu}[p(y + \mu v) - f(y)], \tag{3}$$

which is equivalent to

$$\tfrac{1}{\lambda}[f(x) - p(x - \lambda v)] \leqslant \tfrac{1}{\mu}[p(y + \mu v) - f(y)] \tag{4}$$

for all $x, y \in M$ and $\lambda, \mu > 0$. Rearranging terms, we see that (4) is equivalent to

$$f(\mu x + \lambda y) \leqslant \mu p(x - \lambda v) + \lambda p(y + \mu v) \tag{5}$$

for all $x, y \in M$ and all $\lambda, \mu > 0$. Thus, an extension of f to all of N exists if and only if (5) is valid. For the validity of (5) note that if $x, y \in M$ and $\lambda, \mu > 0$, then

$$
\begin{aligned}
f(\mu x + \lambda y) &= (\lambda + \mu) f\left(\frac{\mu}{\lambda + \mu} x + \frac{\lambda}{\lambda + \mu} y\right) \\
&\leqslant (\lambda + \mu) p\left(\frac{\mu}{\lambda + \mu} x + \frac{\lambda}{\lambda + \mu} y\right) \\
&= (\lambda + \mu) p\left(\frac{\mu}{\lambda + \mu}[x - \lambda v] + \frac{\lambda}{\lambda + \mu}[y + \mu v]\right) \\
&\leqslant (\lambda + \mu)\left[\frac{\mu}{\lambda + \mu} p(x - \lambda v) + \frac{\lambda}{\lambda + \mu} p(y + \mu v)\right] \\
&= \mu p(x - \lambda v) + \lambda p(y + \mu v).
\end{aligned}
$$

This shows that as long as there is some $v \notin M$, there is an extension of f to a larger subspace containing v that satisfies $\hat{f} \leq p$.

To conclude the proof, consider the set of all pairs (g, N) of partial extensions of f such that: N is a linear subspace of X with $M \subset N$, $g \colon N \to \mathbb{R}$ is a linear functional, $g|_M = f$, and $g(x) \leq p(x)$ for all $x \in N$. On this set, we introduce the partial order $(h, L) \geq (g, N)$ whenever $L \supset N$ and $h|_N = g$; note that this relation is indeed a partial order.

It is easy to verify that if $\{(g_\alpha, N_\alpha)\}$ is a chain, then the function g defined on the linear subspace $N = \bigcup_\alpha N_\alpha$ by $g(x) = g_\alpha(x)$ for $x \in N_\alpha$ is well defined and linear, $g(x) \leq p(x)$ for all $x \in N$, and $(g, N) \geq (g_\alpha, N_\alpha)$ for each α. By Zorn's Lemma 1.7, there is a maximal extension $\hat{f}$ satisfying $\hat{f} \leq p$. By the first part of the argument, $\hat{f}$ must be defined on all of X. ∎

The next result tells us when a sublinear functional is actually linear.

5.54 Theorem   *A sublinear function $p \colon X \to \mathbb{R}$ on a vector space is linear if and only if it dominates exactly one linear functional on X.*

Proof: First let $p \colon X \to \mathbb{R}$ be a sublinear functional on a vector space. If p is linear and $f(x) \leq p(x)$ for all $x \in X$ and some linear functional $f \colon X \to \mathbb{R}$, then $-f(x) = f(-x) \leq p(-x) = -p(x)$, so $p(x) \leq f(x)$ for all $x \in X$, that is, $f = p$.

Now assume that p dominates exactly one linear functional on X. Note that p is linear if and only if $p(-x) = -p(x)$ for each $x \in X$. So if we assume by way of contradiction that p is not linear, then there exists some $x_0 \neq 0$ such that $-p(-x_0) < p(x_0)$. Let $M = \{\lambda x_0 : \lambda \in \mathbb{R}\}$, the vector subspace generated by x_0, and define the linear functionals $f, g \colon M \to \mathbb{R}$ by $f(\lambda x_0) = \lambda p(x_0)$ and $g(\lambda x_0) = -\lambda p(-x_0)$. From $f(x_0) = p(x_0)$ and $g(x_0) = -p(-x_0)$, we see that $f \neq g$. Next, notice that $f(z) \leq p(z)$ and $g(z) \leq p(z)$ for each $z \in M$, that is, p dominates both f and g on the subspace M. Now by the Hahn–Banach Theorem 5.53, the two distinct linear functionals f and g have linear extensions to all of X that are dominated by p, a contradiction. ∎

5.10   Separating hyperplane theorems

There is a geometric interpretation of the Hahn–Banach Theorem that is more useful. Assume that X is a vector space. Taking a page from the statisticians' notational handbook, let $[f = \alpha]$ denote the level set $\{x : f(x) = \alpha\}$, and $[f > \alpha]$ denote $\{x : f(x) > \alpha\}$, etc. A **hyperplane** is a set of the form $[f = \alpha]$, where f is a nonzero linear functional on X and α is a real number. (Note well that it is a crucial part of the definition that f be nonzero.) A hyperplane defines two **strict half spaces**, $[f > \alpha]$ and $[f < \alpha]$, and two **weak half spaces**, $[f \geq \alpha]$ and $[f \leq \alpha]$. A set in a vector spaces is a **polyhedron** if it is the intersection of finitely many weak half spaces.

Figure 5.3. Strong separation.

Figure 5.4. These sets cannot be
separated by a hyperplane.

The hyperplane $[f = \alpha]$ **separates** two sets A and B if either $A \subset [f \leqslant \alpha]$ and
$B \subset [f \geqslant \alpha]$ or if $B \subset [f \leqslant \alpha]$ and $A \subset [f \geqslant \alpha]$. We say that the hyperplane
$[f = \alpha]$ **properly separates** A and B if it separates them and $A \cup B$ is not included
in H. A hyperplane $[f = \alpha]$ **strictly separates** A and B if it separates them and in
addition, $A \subset [f > \alpha]$ and $B \subset [f < \alpha]$ or vice-versa. We say that $[f = \alpha]$ **strongly
separates** A and B if there is some $\varepsilon > 0$ with $A \subset [f \leqslant \alpha]$ and $B \subset [f \geqslant \alpha + \varepsilon]$
or vice-versa. We may also say that the linear functional f itself separates the sets
when some hyperplane $[f = \alpha]$ separates them, etc. (Note that this terminology is
inconsistent with the terminology of Chapter 2 regarding separation by continuous
functions. Nevertheless, it should not lead to any confusion.)

It is obvious—but we shall spell it out anyhow, because it is such a useful
trick—that if $[f = \alpha]$ separates two sets, then so does $[-f = -\alpha]$, but the sets are
in the opposite half spaces. This means we can take our choice of putting A in
$[f \geqslant \alpha]$ or in $[f \leqslant \alpha]$.

5.55 Lemma   *A hyperplane $H = [f = \alpha]$ in a topological vector space is either
closed or dense, but not both; it is closed if and only if f is continuous, and dense
if and only if f is discontinuous.*

Proof: If e satisfies $f(e) = \alpha$ and $H_0 = [f = 0]$, then $H = e + H_0$. This shows that
we can assume that $\alpha = 0$. If f is continuous, then clearly H_0 is closed. Also, if
H_0 is dense, then f cannot be continuous (otherwise f is the zero functional).

Now assume that H_0 is closed and let $x_\lambda \to 0$. Also, fix some u with $f(u) = 1$.
If $f(x_\lambda) \nrightarrow 0$, then (by passing to a subnet if necessary) we can assume that
$|f(x_\lambda)| \geqslant \varepsilon$ for each λ and some $\varepsilon > 0$. Put $y_\lambda = u - \frac{f(u)}{f(x_\lambda)} x_\lambda$ and note that $y_\lambda \in H_0$
for each λ and $y_\lambda \to u$. So $u \in H_0$, which is impossible. Thus $f(x_\lambda) \to 0$, so f is
continuous.

Next, suppose that f is discontinuous. Then there exist a net $\{x_\lambda\}$ and some
$\varepsilon > 0$ satisfying $x_\lambda \to 0$ and $|f(x_\lambda)| \geqslant \varepsilon$ for each λ. If x is arbitrary, then put
$z_\lambda = x - \frac{f(x)}{f(x_\lambda)} x_\lambda \in H_0$ and note that $z_\lambda \to x$. So H_0 (and hence H) is dense, and
the proof is finished.   ∎

Ordinary separation is a weak notion because it does not rule out that both sets might actually lie in the hyperplane. The following example illustrates some of the possibilities.

5.56 Example (Kinds of separation)   Consider the plane $\mathbb{R}^2$ and set $f(x, y) = y$. Put $A_1 = \{(x, y) : y > 0 \text{ or } (y = 0 \text{ and } x > 0)\}$ and $B_1 = -A_1$. Also define $A_2 = \{(x, y) : x > 0 \text{ and } y \geq \frac{1}{x}\}$ and $B_2 = \{(x, y) : x > 0 \text{ and } y \leq -\frac{1}{x}\}$. Then the hyperplane $[f = 0]$ separates A_1 and B_1 and strictly separates A_2 and B_2. But the sets A_1 and B_1 cannot be strictly separated, while the sets A_2 and B_2 cannot be strongly separated. ∎

The following simple facts are worth pointing out, and we may use these facts without warning.

5.57 Lemma   *If a linear functional f separates the sets A and B, then f is bounded above or below on each set. Consequently, if say A is a linear subspace, then f is identically zero on A.*

Likewise, if B is a cone, then f can take on values of only one sign on B and the opposite sign on A.

Proof: Suppose $f(x) \neq 0$ for some x in the subspace A. For any real number λ define $x_\lambda = \frac{\lambda}{f(x)}x$. Then x_λ also belongs to A and $f(x_\lambda) = \lambda$, which contradicts the fact f is bounded on A.

For the case where B is a cone, observe that either $\lambda f(b) = f(\lambda b) \leq f(a)$ holds for all $b \in B$, $a \in A$ and $\lambda \geq 0$ or $\lambda f(b) \geq f(a)$ for all $b \in B$, $a \in A$ and $\lambda \geq 0$. This implies either $f(b) \leq 0 \leq f(a)$ for all $b \in B$ and $a \in A$ or $f(b) \geq 0 \geq f(a)$ for all $b \in B$ and $a \in A$. ∎

We may say that a linear functional **annihilates** a subspace when it is bounded, and hence zero, on the subspace.

Another cheap trick stems from the following observation. In a vector space, for nonempty sets A and B we have:

$$A \cap B = \varnothing \quad \Longleftrightarrow \quad 0 \notin A - B.$$

We use this fact repeatedly.

The first important separation theorem is a plain vanilla separating hyperplane theorem—it holds in arbitrary linear spaces and requires no topological assumptions. Instead, a purely algebraic property is assumed.

5.58 Definition   *A point x in a vector space is an **internal point** of a set B if there is an absorbing set A such that $x + A \subset B$, or equivalently if the set $B - x$ is absorbing.*

In other words, a point x is an internal point of a set B if and only if for each vector u there exists some $\alpha_0 > 0$ depending on u such that $x + \alpha u \in B$ whenever $|\alpha| \leqslant \alpha_0$.

5.59 Example (Internal point vs. interior point) It should be clear that interior points are internal points. We shall show later (see Lemma 5.60) that a vector in a convex subset of a finite dimensional vector space is an internal point if and only if it is an interior point. However, in infinite dimensional topological vector spaces an internal point of a convex set need not be an interior point. For an example, let $X = C[0, 1]$, the vector space of all continuous real-valued functions defined on $[0, 1]$. On X we consider the two norms $\|f\|_\infty = \max_{x \in [0,1]} |f(x)|$ and $\|f\| = \int_0^1 |f(x)| \, dx$, and let τ_∞ and τ be the Hausdorff linear topologies generated by $\|\cdot\|_\infty$ and $\|\cdot\|$, respectively. If $C = \{f \in C[0, 1] : \|f\|_\infty < 1\}$, then C is a convex set and has 0 as a τ_∞-interior point. In particular, 0 is an internal point of C. Now notice that 0 is not a τ-interior point of C. ∎

As mentioned in the preceding example, in finite dimensional vector spaces the internal points of a convex set are precisely the interior points of the set.

5.60 Lemma *Let C be a nonempty convex subset of a finite dimensional vector space X. Then a vector of C is an internal point of C if and only if it is an interior point of C (for the Euclidean topology on X).*

Proof: Let x_0 be an internal point of C. Replacing C by $C - x_0$, we can assume that $x_0 = 0$. It is easy to see that there exists a basis $\{e_1, \ldots, e_k\}$ of X such that $\pm e_i \in C$ for all $i = 1, \ldots, k$. Now note that the norm $\left\| \sum_{i=1}^k \alpha_i e_i \right\| = \sum_{i=1}^k |\alpha_i|$ must be equivalent to the Euclidean norm; see Theorem 5.21. If $x = \sum_{i=1}^k \alpha_i e_i \in B_1(0)$, then x can be written as a convex combination of the collection of vectors $\{0, \pm e_1, \ldots, \pm e_k\}$ of C, so since C is convex we have $x \in C$. Thus $B_1(0) \subset C$ so that 0 is an interior point of C. (For more details see also the proof of Theorem 7.24.) ∎

We are now ready for the fundamental separating hyperplane theorem.

5.61 Basic Separating Hyperplane Theorem *Two nonempty disjoint convex subsets of a vector space can be properly separated by a nonzero linear functional, provided one of them has an internal point.*

Proof: Let A and B be disjoint nonempty convex sets in a vector space X, and suppose A has an internal point. Then the nonempty convex set $A - B$ has an internal point. Let z be an internal point of $A - B$. Clearly, $z \neq 0$ and the set $C = A - B - z$ is nonempty, convex, absorbing, and satisfies $-z \notin C$. (Why?) By part (2) of Lemma 5.50, the gauge p_C of C is a sublinear function.

We claim that $p_C(-z) \geqslant 1$. Indeed, if $p_C(-z) < 1$, then there exist $0 \leqslant \alpha < 1$ and $c \in C$ such that $-z = \alpha c$. Since $0 \in C$, it follows that $-z = \alpha c + (1 - \alpha)0 \in C$, a contradiction. Hence $p_C(-z) \geqslant 1$.

Let $M = \{\alpha(-z) : \alpha \in \mathbb{R}\}$, the one-dimensional subspace generated by $-z$, and define $f : M \to \mathbb{R}$ by $f(\alpha(-z)) = \alpha$. Clearly, f is linear and moreover $f \leqslant p_C$ on M, since for each $\alpha \geqslant 0$ we have $p_C(\alpha(-z)) = \alpha p_C(-z) \geqslant \alpha = f(\alpha(-z))$, and $\alpha < 0$ yields $f(\alpha(-z)) < 0 \leqslant p_C(\alpha(-z))$. By the Hahn–Banach Extension Theorem 5.53, f extends to $\hat{f}$ defined on all of X satisfying $\hat{f}(x) \leqslant p_C(x)$ for all $x \in X$. Note that $\hat{f}(z) = -1$, so $\hat{f}$ is nonzero.

To see that $\hat{f}$ separates A and B let $a \in A$ and $b \in B$. Then we have

$$\hat{f}(a) = \hat{f}(a - b - z) + \hat{f}(z) + \hat{f}(b) \leqslant p_C(a - b - z) + \hat{f}(z) + \hat{f}(b)$$
$$= p_C(a - b - z) - 1 + \hat{f}(b) \leqslant 1 - 1 + \hat{f}(b) = \hat{f}(b).$$

This shows that the nonzero linear functional $\hat{f}$ separates the convex sets A and B.

To see that the separation is proper, let $z = a - b$, where $a \in A$ and $b \in B$. Since $\hat{f}(z) = -1$, we have $\hat{f}(a) \neq \hat{f}(b)$, so A and B cannot lie in the same hyperplane. ∎

5.62 Corollary   *Let A and B be two nonempty disjoint convex subsets of a vector space X. If there exists a vector subspace Y including A and B such that either A or B has an internal point in Y, then A and B can be properly separated by a nonzero linear functional on X.*

Proof: By Theorem 5.61 there is a nonzero linear functional f on Y that properly separates A and B. Now note that any linear extension of f to X is a nonzero linear functional on X that properly separates A and B. ∎

5.11   Separation by continuous functionals

Theorem 5.61 makes no mentions of any topology. In this section we impose topological hypotheses and draw topological conclusions. The next lemma gives a topological condition that guarantees the existence of internal points, which is a prerequisite for applying the Basic Separating Hyperplane Theorem 5.61. It is a consequence of the basic Structure Theorem 5.6 and although we have mentioned it before, we state it again in order to emphasize its importance.

5.63 Lemma   *In a topological vector space, every neighborhood of zero is an absorbing set. Consequently, interior points are internal.*

Note that the converse of this is not true. In a topological vector space there can be absorbing sets with empty interior. For example, the unit ball in an infinite dimensional normed space is a very nice convex absorbing set, but it has empty interior in the weak topology, see Corollary 6.27.

The next lemma gives a handy criterion for continuity of a linear functional on a topological vector space. It generalizes the result for Banach spaces that linear functionals are bounded if and only if they are continuous.

5.64 Lemma   *If a linear functional on a tvs is bounded either above or below on a neighborhood of zero, then it is continuous.*

Proof: If f is linear, then both f and $-f$ are convex, so the conclusion follows from Theorem 5.43. Or more directly, if $f \leqslant M$ on a symmetric neighborhood V of zero, then $x - y \in \frac{\varepsilon}{M} V$ implies $|f(x) - f(y)| = |f(x - y)| \leqslant \frac{\varepsilon}{M} M = \varepsilon$.   ∎

The proof of the next result is left as an exercise.

5.65 Lemma   *A nonzero continuous linear functional on a topological vector space properly separates two nonempty sets if and only if it properly separates their closures.*

Some more separation properties of linear functionals are contained in the next lemma.

5.66 Lemma   *If A is a nonempty subset of a tvs X and a nonzero linear functional f on X satisfies $f(x) \geqslant \alpha$ for all $x \in A$, then $f(x) > \alpha$ for all $x \in A^\circ$ (and so if $A^\circ \neq \varnothing$, then f is continuous).*

In particular, in a tvs, if a nonzero linear functional separates two nonempty sets, one of which has an interior point, then it is continuous and properly separates the two sets.

Proof: Assume that $x_0 + V \subset A$, where V is a circled neighborhood of zero. If $f(x_0) = \alpha$, then for each $v \in V$ we have $\alpha \pm f(v) = f(x_0 \pm v) \geqslant \alpha$. Consequently, $\pm f(v) \geqslant 0$ or $f(v) = 0$ for all $v \in V$. Since V is absorbing, the latter yields $f(y) = 0$ for all $y \in X$, that is, $f = 0$, which is impossible. Hence $f(x) > \alpha$ holds for all $x \in A^\circ$. Now from $f(v) \geqslant \alpha - f(x_0)$ for all $v \in V$, it follows from Lemma 5.64 that f is continuous.

For the last part, let A and B be two nonempty subsets of a tvs X with $A^\circ \neq \varnothing$ and assume that there exist a linear functional f on X and some $\alpha \in \mathbb{R}$ satisfying $f(a) \geqslant \alpha \geqslant f(b)$ for all $a \in A$ and all $b \in B$. By the first part, f is continuous and $f(a) > \alpha$ for all $a \in A^\circ$. The latter shows that f properly separates A and B (so f also property separates $\overline{A}$ and $\overline{B}$).   ∎

We now come to a basic topological separating hyperplane theorem.

5.67 Interior Separating Hyperplane Theorem   *In any tvs, if the interiors of a convex set A is nonempty and is disjoint from another nonempty convex set B, then $\overline{A}$ and $\overline{B}$ can be properly separated by a nonzero continuous linear functional.*

Moreover, the pairs of convex sets $(A, \overline{B})$, $(\overline{A}, B)$, and (A, B) likewise can be properly separated by the same nonzero continuous linear functional.

Proof: Assume that A and B are two nonempty convex subsets of a tvs X such that $A° \neq \varnothing$ and $A° \cap B = \varnothing$. By Lemma 5.28 we know that $\overline{A°} = \overline{A}$. Now, according to Theorem 5.61, there exists a nonzero linear functional f on X that properly separates $A°$ and B. But then (by Lemma 5.66) f is continuous and properly separates $\overline{A°} = \overline{A}$ and $\overline{B}$. ∎

5.68 Corollary *In any tvs, if the interior of two convex sets are nonempty and disjoint, then their closures (and so the convex sets themselves) can be properly separated by a nonzero continuous linear functional.*

The hypothesis that one of the sets must have a nonempty interior cannot be dispensed with. The following example, due to J. W. Tukey [332], presents two disjoint nonempty closed convex subsets of a Hilbert space that cannot be separated by a continuous linear functional.

5.69 Example (Inseparable disjoint closed convex sets) In ℓ_2, the Hilbert space of all square summable sequences, let

$$A = \{x = (x_1, x_2, \ldots) \in \ell_2 : x_1 \geqslant n|x_n - n^{-\frac{2}{3}}| \text{ for } n = 2, 3, \ldots\}.$$

The sequence v with $v_n = n^{-\frac{2}{3}}$ lies in ℓ_2 and belongs to A, so A is nonempty. Clearly A is convex. It is also easy to see that A is norm closed. Let

$$B = \{x = (x_1, 0, 0, \ldots) \in \ell_2 : x_1 \in \mathbb{R}\}.$$

The set B is clearly nonempty, convex, and norm closed. Indeed, it is a straight line, a one-dimensional subspace.

Observe that A and B are disjoint. To see this note that if x belongs to B, then $n|x_n - n^{-\frac{2}{3}}| = n^{\frac{1}{3}} \xrightarrow[n]{} \infty$, so x cannot lie in A.

We now claim that A and B cannot be separated by any nonzero continuous linear functional on ℓ_2. In fact, we prove the stronger result that $A - B$ is dense in ℓ_2. To see this, fix any $z = (z_1, z_2, \ldots)$ in ℓ_2 and let $\varepsilon > 0$. Choose k so that $\sum_{n=k+1}^{\infty} n^{-\frac{4}{3}} < \varepsilon^2/4$ and $\sum_{n=k+1}^{\infty} z_n^2 < \varepsilon^2/4$.

Now consider the vector $x = (x_1, x_2, \ldots) \in A$ defined by

$$x_n = \begin{cases} \max\limits_{1 \leqslant i \leqslant k} i|z_i - i^{-\frac{2}{3}}| & \text{if } n = 1, \\ z_n & \text{if } 2 \leqslant n \leqslant k, \\ n^{-\frac{2}{3}} & \text{if } n > k. \end{cases}$$

Let $y = (x_1 - z_1, 0, 0, \ldots) \in B$ and note that the vector $x - y \in A - B$ satisfies

$$\left\| z - (x - y) \right\| = \left[\sum_{n=k+1}^{\infty} (z_n - n^{-\frac{2}{3}})^2 \right]^{\frac{1}{2}} \leqslant \left[\sum_{n=k+1}^{\infty} z_n^2 \right]^{\frac{1}{2}} + \left[\sum_{n=k+1}^{\infty} n^{-\frac{4}{3}} \right]^{\frac{1}{2}} < \varepsilon.$$

That is, $A - B$ is dense, so A cannot be separated from B by a continuous linear functional. (Why?) ∎

As an application of the Interior Separating Hyperplane Theorem 5.67, we shall present a useful result on concave functions due to K. Fan, I. Glicksberg, and A. J. Hoffman [120]. It takes the form of an **alternative**, that is, an assertion that exactly one of two mutually incompatible statements is true. We shall see more alternatives in the sequel.

5.70 Theorem (The Concave Alternative)   *Let* $f_1, \ldots, f_m \colon C \to \mathbb{R}$ *be concave functions defined on a nonempty convex subset of some vector space. Then exactly one of the following two alternatives is true.*

1. *There exists some* $x \in C$ *such that* $f_i(x) > 0$ *for each* $i = 1, \ldots, m$.

2. *There exist nonnegative scalars* $\lambda_1, \ldots, \lambda_m$, *not all zero, such that*

$$\sum_{i=1}^{m} \lambda_i f_i(x) \leqslant 0$$

for each $x \in C$.

Proof: It is easy to see that both statements cannot be true. Now consider the subset of $\mathbb{R}^m$:

$$A = \{ y \in \mathbb{R}^m : \exists\, x \in C \text{ such that } y_i \leqslant f_i(x) \text{ for each } i \}.$$

Clearly A is nonempty. To see that A is convex, let $y, z \in A$, and pick $x_1, x_2 \in C$ satisfying $y_i \leqslant f_i(x_1)$ and $z_i \leqslant f_i(x_2)$ for each i. Now if $0 \leqslant \alpha \leqslant 1$, then the concavity of the functions f_i implies

$$\alpha y_i + (1 - \alpha) z_i \leqslant \alpha f_i(x_1) + (1 - \alpha) f_i(x_2) \leqslant f_i(\alpha x_1 + (1 - \alpha) x_2)$$

for each i. Since $\alpha x_1 + (1 - \alpha) x_2 \in C$, the inequalities show that $\alpha y + (1 - \alpha) z \in A$. That is, A is a convex subset of $\mathbb{R}^m$. Now notice that if (1) is not true, then the convex set A is disjoint from the interior of the convex set $\mathbb{R}^m_+$. So, according to Theorem 5.67 there exists a nonzero vector $\lambda = (\lambda_1, \ldots, \lambda_m)$ such that

$$\lambda \cdot y = \sum_{i=1}^{m} \lambda_i y_i \geqslant \sum_{i=1}^{m} \lambda_i f_i(x)$$

for all $y \in \mathbb{R}^m_+$ and all $x \in C$. Clearly, $\sum_{i=1}^{m} \lambda_i f_i(x) \leqslant 0$ for all $x \in C$ and $\lambda \cdot y \geqslant 0$ for all $y \in \mathbb{R}^m_+$. The latter yields $\lambda_i \geqslant 0$ for each i and the proof is complete.   ∎

5.12   Locally convex spaces and seminorms

To obtain a separating hyperplane theorem with a stronger conclusion than proper separation, we need stronger hypotheses. One such hypothesis is that the linear space be a locally convex space.

5.71 Definition   *Recall that a topological vector space is **locally convex**, or is a* **locally convex space**, *if every neighborhood of zero includes a convex neighborhood of zero.*[7]

A **Fréchet space** *is a completely metrizable locally convex space.*

Since in a topological vector space the closure of a convex set is convex, the Structure Theorem 5.6 implies that in a locally convex space the closed convex circled neighborhoods of zero form a neighborhood base at zero. Next notice that the convex hull of a circled set is also circled. From this and the fact that the interior of a convex (resp. circled) neighborhood of zero is a convex (resp. circled) neighborhood of zero, it follows that in a locally convex space the collection of all open convex circled neighborhoods of zero is also a neighborhood base at zero.

In other words, we have the following result.

5.72 Lemma   *In a locally convex space:*

1. *The collection of all the closed, convex and circled neighborhoods of zero is a neighborhood base at zero.*

2. *The collection of all open, convex and circled neighborhoods of zero is a neighborhood base at zero.*

It turns out that the locally convex topologies are precisely the topologies derived from families of seminorms. Let X be a vector space. For a seminorm $p \colon X \to \mathbb{R}$ and $\varepsilon > 0$, let us write

$$V_p(\varepsilon) = \{x \in X : p(x) \leqslant \varepsilon\},$$

the closed ε-ball of p centered at zero. Now let $\{p_i\}_{i \in I}$ be a family of seminorms on X. Then the collection $\mathcal{B}$ of all sets of the form

$$V_{p_1}(\varepsilon) \cap \cdots \cap V_{p_n}(\varepsilon), \quad \varepsilon > 0,$$

is a filter base of convex sets that satisfies conditions (1), (2), and (3) of the Structure Theorem 5.6. Consequently, $\mathcal{B}$ induces a unique locally convex topology on X having $\mathcal{B}$ as a neighborhood base at zero. This topology is called the locally convex topology **generated by the family of seminorms** $\{p_i\}_{i \in I}$. A family $\mathcal{F}$ of seminorms is **saturated** if $p, q \in \mathcal{F}$ implies $p \vee q \in \mathcal{F}$. If a family of seminorms is saturated, then it follows from Lemmas 5.50 and 5.49 (7) that a neighborhood base at zero is given by the collection of all $V_p(\varepsilon)$, no intersections required.

In the converse direction, let τ be a locally convex topology on a vector space X, and let $\mathcal{B}$ denote the neighborhood base at zero consisting of all circled convex closed neighborhoods of zero. Then, for each $V \in \mathcal{B}$ the gauge p_V is a seminorm on X. An easy argument shows that the family of seminorms $\{p_V\}_{V \in \mathcal{B}}$ is a saturated family generating τ. Thus, we have the following important characterization of locally convex topologies.

---

[7] Many authors define a locally convex space to be Hausdorff as well.

5.73 Theorem (Seminorms and local convexity) *A linear topology on a vector space is locally convex if and only if it is generated by a family of seminorms.*

In particular, a locally convex topology is generated by the family of gauges of the convex circled closed neighborhoods of zero.

Here is a simple example of a locally convex space.

5.74 Lemma *For any nonempty set X, the product topology on $\mathbb{R}^X$ is a complete locally convex Hausdorff topology.*

Proof: Note that the product topology is generated by the family of seminorms $\{p_x\}_{x \in X}$, where $p_x(f) = |f(x)|$.   ∎

If X is countable, then $\mathbb{R}^X$ is a completely metrizable locally convex space, that is, $\mathbb{R}^X$ is a Fréchet space. The metrizable locally convex spaces are characterized by the following result whose proof follows from Theorem 5.10 and 5.73.

5.75 Lemma *A Hausdorff locally convex space (X, τ) is metrizable if and only if τ is generated by a sequence $\{q_n\}$ of seminorms—in which case the topology τ is generated by the translation invariant metric d given by*

$$d(x, y) = \sum_{n=1}^{\infty} \frac{1}{2^n} \cdot \frac{q_n(x-y)}{1 + q_n(x-y)}.$$

Recall that a subset A of a topological vector space (X, τ) is **(topologically) bounded**, or more specifically **τ-bounded**, if for each neighborhood V of zero there exists some $\lambda > 0$ such that $A \subset \lambda V$.

The proof of the following simple lemma is left as an exercise.

5.76 Lemma *If a family of seminorms $\{p_i\}_{i \in I}$ on a vector space X generates the locally convex topology τ, then:*

1. *τ is Hausdorff if and only if $p_i(x) = 0$ for all $i \in I$ implies $x = 0$.*

2. *A net $\{x_\alpha\}$ satisfies $x_\alpha \xrightarrow{\tau} x$ if and only if $p_i(x_\alpha - x) \to 0$ for each i.*

3. *A subset A of X is τ-bounded if and only if $p_i(A)$ is a bounded subset of real numbers for each i.*

A locally convex space is **normable** if its topology is generated by a single norm.

5.77 Theorem (Normability) *A locally convex Hausdorff space is normable if and only if it has a bounded neighborhood of zero.*

Proof: If V is a convex, circled, closed, and bounded neighborhood of zero, then note that p_V is a norm that generates the topology.   ∎

Here is a familiar example of a completely metrizable locally convex space that is not normable.

5.78 Example ($\mathbb{R}^{\mathbb{N}}$ **is not normable**)   According to Lemma 5.74 the product topology τ on $\mathbb{R}^{\mathbb{N}}$ is a Hausdorff locally convex topology that is generated by the countable collection $\{p_1, p_2, \ldots\}$ of seminorms, where $p_n(x) = |x_n|$ for each $x = (x_1, x_2, \ldots) \in \mathbb{R}^{\mathbb{N}}$. But then, by Lemma 5.75, the topology τ is also completely metrizable—and, indeed, is generated by the complete translation invariant metric $d(x, y) = \sum_{n=1}^{\infty} 2^{-n} \frac{|x_n - y_n|}{1 + |x_n - y_n|}$. In other words, it is a Fréchet space. However, the product topology τ is not normable: Let

$$V = \{x = (x_1, x_2, \ldots) \in \mathbb{R}^{\mathbb{N}} : |x_{n_i}| < \varepsilon \text{ for all } i = 1, \ldots, k\}$$

be a basic τ-neighborhood of zero and choose n such that $n \neq n_i$ for all $i = 1, \ldots, k$. Then it is easy to see that $\sup p_n(V) = \infty$. This shows that no τ-neighborhood of zero can be τ-bounded and therefore, by Theorem 5.77, τ is not normable.   ∎

Not every tvs is locally convex. Theorems 13.31 and 13.43 show some of the surprises lurking in infinite dimensional spaces. Sometimes, zero is the only continuous linear functional!

5.13   Separation in locally convex spaces

In locally convex spaces, we have the following strong separating hyperplane theorem. (For a sharper version of this result holding for Banach spaces see Corollary 7.47.)

5.79 Strong Separating Hyperplane Theorem   *For disjoint nonempty convex subsets of a (not necessarily Hausdorff) locally convex space, if one is compact and the other closed, then there is a nonzero continuous linear functional strongly separating them.*

Proof: Let A and B satisfy the hypotheses. By Lemma 5.3, $A - B$ is a nonempty closed convex set, and it does not contain zero. Thus its complement is an open neighborhood of zero, and since the space is locally convex, there is a circled convex open neighborhood V of zero disjoint from $A - B$. Since V is open, the Interior Separating Hyperplane Theorem 5.67 guarantees that there is a nonzero continuous linear functional f separating V and $A - B$. That is, $f(v) \leqslant f(a) - f(b)$ for all $v \in V$, $a \in A$, and $b \in B$. Since f is nonzero and V is absorbing, f cannot vanish on V. Therefore there exists some $v_0 \in V$ with $f(v_0) > 0$. Now if $\varepsilon = f(v_0)$ and $\alpha = \sup_{b \in B} f(b)$, then note that $f(a) \geqslant \alpha + \varepsilon > \alpha \geqslant f(b)$ for all a in A and b in B. That is, f strongly separates A and B.   ∎

We state some easy consequences.

5.80 Corollary (Separating points from closed convex sets)   *In a locally convex space, if K is a nonempty closed convex set and z ∉ K, then there exists a nonzero continuous linear functional strongly separating K and z.*

5.81 Corollary (Non-dense vector subspaces)   *A vector subspace of a locally convex space fails to be dense if and only if there exists a nonzero continuous linear functional that vanishes on it.*

5.82 Corollary (The dual separates points)   *The topological dual of a locally convex space separates points if and only if the topology is Hausdorff.*

Proof: Let (X, τ) be a locally convex space. If the topological dual X' separates points and $x \neq y$ pick some $f \in X'$ satisfying $f(x) < f(y)$ and note that if $f(x) < c < f(y)$, then the open half spaces $[f < c]$ and $[f > c]$ are disjoint open neighborhoods of x and y.

Conversely, if τ is a Hausdorff topology, then singletons are closed and compact, so the separation of points follows immediately from Corollary 5.80. ∎

This last result stands in marked contrast to Theorem 13.31, where it is shown that zero is the only continuous linear functional on $L_p(\mu)$ for $0 < p < 1$. Of course, these spaces are not locally convex.

Closed convex sets can be characterized in terms of closed half spaces. Consequently they are determined by the dual space. (For a sharper version of the second part of the next theorem that is valid for Banach spaces see Corollary 7.48.)

5.83 Corollary (Closed convex sets)   *In a locally convex space, if a convex set is not dense, then its closure is the intersection of all (topologically) closed half spaces that include it.*

In particular, in a locally convex space X, every proper closed convex subset of X is the intersection of all closed half spaces that include it.

Proof: Let A be a non-dense convex subset of a locally convex space. Recall that a closed half space is a set of the form $[f \leq \alpha] = \{x : f(x) \leq \alpha\}$, where f is a nonzero continuous linear functional. If $a \notin \overline{A}$, then according to Corollary 5.80 there exist a nonzero continuous linear functional g and some scalar α satisfying $\overline{A} \subset [g \leq \alpha]$ and $g(a) > \alpha$. This implies that $\overline{A}$ is the intersection of all closed half spaces including A. ∎

Note that if a convex set is dense in the space X, then its closure, X, is not included in any half space, so we cannot omit the qualification "not dense" in the theorem above. The last corollary takes the form of an alternative.

5.84 Corollary (The Convex Cone Alternative) *If C is a convex cone in a locally convex space (X, τ), then for each $x \in X$ one of the following two mutually exclusive alternatives holds.*

1. *The point x belongs to the τ-closure of C, that is, $x \in \overline{C}$.*

2. *There exists a τ-continuous linear functional f on X satisfying*

$$f(x) > 0 \qquad and \qquad f(c) \leqslant 0 \text{ for all } c \in C.$$

Proof: It is easy to check that statements (1) and (2) are mutually exclusive. Assume $x \notin \overline{C}$. Then, by the Strong Separating Hyperplane Theorem 5.79, there exist a nonzero τ-continuous linear functional f on X and some constant α satisfying $f(x) > \alpha$ and $f(c) \leqslant \alpha$ for all $c \in \overline{C}$. Since C is a cone, it follows that $\alpha \geqslant 0$ and $f(c) \leqslant 0$ for all $c \in \overline{C}$. Consequently, $f(x) > 0$ and $f(c) \leqslant 0$ for all $c \in C$. In other words, we have shown that if $x \notin \overline{C}$, then (2) is true and the proof is finished. ∎

A special case of this result is known as Farkas' Lemma. It and its relatives are instrumental to the study of linear programming and decision theory.

5.85 Corollary (Farkas' Lemma [121]) *If A is a real $m \times n$ matrix and b is a vector in $\mathbb{R}^m$, then one of the following mutually exclusive alternatives holds.*

1. *There exists a vector $\lambda \in \mathbb{R}_+^n$ such that $b = A\lambda$.*

2. *There exists a nonzero vector $a \in \mathbb{R}^m$ satisfying*

$$a \cdot b > 0 \qquad and \qquad A^t a \leqslant 0.$$

Here, as usual, λ is an n-dimensional column vector, and A^t denotes the $n \times m$ transpose matrix of A.

Proof: By Corollary 5.25 the convex cone C in $\mathbb{R}^m$ generated by the n columns of A is closed. Statement (1) is equivalent to $b \in C$. Corollary 5.84 says that either (1) holds or else there is a linear functional (represented by a nonzero vector a) such that $a \cdot b > 0$ and $a \cdot c \leqslant 0$ for all $c \in C$. But $a \cdot c \leqslant 0$ for all $c \in C$ if and only if $A^t a \leqslant 0$. But this is just (2). ∎

Recall that a seminorm p on a vector space X dominates a linear functional f if $f(x) \leqslant p(x)$ for each $x \in X$. This is equivalent to $|f(x)| \leqslant p(x)$ for each $x \in X$.

5.86 Lemma (Continuous linear functionals) *A linear functional on a tvs is continuous if and only if it is dominated by a continuous seminorm.*

Proof: Let (X, τ) be a tvs and let f be a linear functional on X. If $|f(x)| \leqslant p(x)$ for all $x \in X$ and some τ-continuous seminorm p, then it easily follows that $\lim_{x \to 0} f(x) = 0$, which shows that f is τ-continuous.

For the converse, simply note that if f is a τ-continuous linear functional, then $x \mapsto |f(x)|$ is a τ-continuous seminorm dominating f.   ∎

5.87 Theorem (Dual of a subspace)   *If (X, τ) is a locally convex space and Y is a vector subspace of X, then every τ-continuous linear functional on Y (endowed with the relative topology) extends to a (not necessarily unique) τ-continuous linear functional on X.*

In particular, the continuous linear functionals on Y are precisely the restrictions to Y of the continuous linear functionals on X.

Proof: Let $f : Y \to \mathbb{R}$ be a continuous linear functional. Pick some convex and circled τ-neighborhood V of zero satisfying $|f(y)| \leqslant 1$ for each y in $V \cap Y$. From part (3) of Lemma 5.50 we see that p_V is a continuous seminorm and it is easy to check that $|f(y)| \leqslant p_V(y)$ for all $y \in Y$. By the Hahn–Banach Theorem 5.53 there exists an extension $\hat{f}$ of f to all of X satisfying $|\hat{f}(x)| \leqslant p_V(x)$ for all $x \in X$. By Lemma 5.86, $\hat{f}$ is τ-continuous, and we are done.   ∎

As an application of the preceding result, we shall show that every finite dimensional vector subspace of a locally convex Hausdorff space is complemented.

5.88 Definition   *A vector space X is the **direct sum** of two subspaces Y and Z, written $X = Y \oplus Z$, if every $x \in X$ has a unique decomposition of the form $x = y + z$, where $y \in Y$ and $z \in Z$.*

*A closed vector subspace Y of a topological vector space X is **complemented** in X if there exists another closed vector subspace Z such that $X = Y \oplus Z$.*

5.89 Theorem   *In a locally convex Hausdorff space every finite dimensional vector subspace is complemented.*

Proof: Let (X, τ) be a locally convex Hausdorff space and let Y be a finite dimensional vector subspace of X. Pick a basis $\{y_1, \ldots, y_k\}$ for Y and consider the linear functionals $f_i \colon Y \to \mathbb{R}$ $(i = 1, \ldots, k)$ defined by $f_i(\sum_{j=1}^{k} \lambda_j y_j) = \lambda_i$.

Clearly, each $f_i \colon (Y, \tau) \to \mathbb{R}$ is continuous. By Theorem 5.87, each f_i has a τ-continuous extension to all of X, which we again denote f_i. Now consider the continuous projection $P \colon X \to X$ defined by

$$P(x) = \sum_{i=1}^{k} f_i(x) y_i.$$

That is, P projects x onto the space spanned by $\{y_1, \ldots, y_k\}$. Now define the closed vector subspace $Z = \{x - P(x) : x \in X\}$ of X, and note that Z satisfies $Y \oplus Z = X$. ∎

5.14   Dual pairs

Dual pairs are an extremely useful way of obtaining locally convex spaces.

5.90 Definition   *A **dual pair** (or a **dual system**) is a pair $\langle X, X' \rangle$ of vector spaces together with a bilinear functional $(x, x') \mapsto \langle x, x' \rangle$, from $X \times X'$ to $\mathbb{R}$, that separates the points of X and X'. That is:*

1. *The mapping $x' \mapsto \langle x, x' \rangle$ is linear for each $x \in X$.*

2. *The mapping $x \mapsto \langle x, x' \rangle$ is linear for each $x' \in X'$.*

3. *If $\langle x, x' \rangle = 0$ for each $x' \in X'$, then $x = 0$.*

4. *If $\langle x, x' \rangle = 0$ for each $x \in X$, then $x' = 0$.*

Each space of a dual pair $\langle X, X' \rangle$ can be interpreted as a set of linear functionals on the other. For instance, each $x \in X$ defines the linear functional $x' \mapsto \langle x, x' \rangle$. Conditions (1) and (2) are the ones required for the definition of a **bilinear functional**. The bilinear functional $(x, x') \mapsto \langle x, x' \rangle$ is also called the **duality** (or the **bilinearity**) of the dual pair. Recall that a family $\mathcal{F}$ of linear functionals on X is **total** if it separates the points of X: $f(x) = f(y)$ for all $f \in \mathcal{F}$ implies $x = y$. Conditions (3) and (4) in the definition of a dual pair require that each space separates the points of the other. One way to obtain a dual pair is to start with a vector space X, and choose an arbitrary total subspace X' of the algebraic dual X^*. Then it is readily seen that $\langle X, X' \rangle$ is a dual pair under the evaluation duality $(x, x') \mapsto x'(x)$.

Here are some familiar examples of dual pairs.

- $\langle \mathbb{R}^n, \mathbb{R}^n \rangle$ under the duality $\langle x, y \rangle = \sum_{i=1}^{n} x_i y_i$.

- $\langle L_p(\mu), L_q(\mu) \rangle$, $1 \leqslant p, q \leqslant \infty$, $\frac{1}{p} + \frac{1}{q} = 1$ and $\langle f, g \rangle = \int f g \, d\mu$.

- $\langle C[0, 1], ca[0, 1] \rangle$ under the duality $\langle f, \mu \rangle = \int_0^1 f(x) \, d\mu(x)$.

- $\langle \ell_\infty, \ell_1 \rangle$ under the duality $\langle x, y \rangle = \sum_{i=1}^{\infty} x_i y_i$.

- $\langle X, X^* \rangle$, where X is an arbitrary vector space, X^* is its algebraic dual, and $\langle x, x^* \rangle = x^*(x)$.

Since we can consider X to be a vector subspace of $\mathbb{R}^{X'}$, X inherits the product topology of $\mathbb{R}^{X'}$. This topology is referred to as the **weak topology** on X and is denoted $\sigma(X, X')$, or simply w. Since the product topology on $\mathbb{R}^{X'}$ is a locally convex Hausdorff topology, the weak topology $\sigma(X, X')$ is likewise Hausdorff and locally convex. Observe that $x_\alpha \xrightarrow{w} x$ in X if and only if $\langle x_\alpha, x' \rangle \to \langle x, x' \rangle$ in $\mathbb{R}$ for each $x' \in X'$. For this reason the weak topology is also known as the **topology**

of pointwise convergence on X'. A family of seminorms that generates the weak topology $\sigma(X, X')$ is $\{p_{x'} : x' \in X'\}$, where

$$p_{x'}(x) = |\langle x, x' \rangle|, \quad x \in X.$$

The locally convex Hausdorff topology $\sigma(X', X)$ is defined in a similar manner. It is generated by the family of seminorms $\{p_x : x \in X\}$, where $p_x(x') = |\langle x, x' \rangle|$ for each $x' \in X'$. The topology $\sigma(X', X)$ is known as the **weak* topology** on X' and is denoted simply by w^*. Observe that $x'_\alpha \xrightarrow{w^*} x'$ in X' if and only if $\langle x, x'_\alpha \rangle \to \langle x, x' \rangle$ in $\mathbb{R}$ for each $x \in X$.

We next establish that the topological dual of $(X, \sigma(X, X'))$ really is X'. The value of this result is that if we start with a vector space X, we can take any total vector subspace $\mathcal{F}$ of X^* and find a topology on X, namely $\sigma(X, \mathcal{F})$, that makes $\mathcal{F}$ the topological dual of X. That is, we get to pick the dual! To do this, we need a lemma. The **kernel** of a linear functional f on a vector space X is the vector subspace defined by

$$\ker f = \{x \in X : f(x) = 0\} = f^{-1}(\{0\}).$$

5.91 Fundamental Theorem of Duality   *Let $f, f_1, \ldots, f_n$ be linear functionals on a vector space X. Then f lies in the span of $f_1, \ldots, f_n$ (that is, $f = \sum_{i=1}^n \lambda_i f_i$ for some scalars $\lambda_1, \ldots, \lambda_n$) if and only if $\bigcap_{i=1}^n \ker f_i \subset \ker f$.*

Proof: If $f = \sum_{i=1}^n \lambda_i f_i$, then clearly $\bigcap_{i=1}^n \ker f_i \subset \ker f$. For the converse, assume that $\bigcap_{i=1}^n \ker f_i \subset \ker f$. Then $T: X \to \mathbb{R}^n$ via $T(x) = (f_1(x), \ldots, f_n(x))$ is a linear operator. Since $\bigcap_{i=1}^n \ker f_i \subset \ker f$, if $(f_1(x), \ldots, f_n(x)) = (f_1(y), \ldots, f_n(y))$, then $f_i(x - y) = 0$ for each i and so $f(x) = f(y)$. Thus the linear functional $\varphi: T(X) \to \mathbb{R}$ defined by $\varphi(f_1(x), \ldots, f_n(x)) = f(x)$ is well defined. Now note that φ extends to all of $\mathbb{R}^n$, so there exist scalars $\lambda_1, \ldots, \lambda_n$ such that $\varphi(\alpha_1, \ldots, \alpha_n) = \sum_{i=1}^n \lambda_i \alpha_i$. Thus $f(x) = \sum_{i=1}^n \lambda_i f_i(x)$ for each $x \in X$, as desired.   ∎

As with many result on separating hyperplanes, it is possible to recast the conclusion of this theorem as an alternative. We leave it to you to figure out why the next result is equivalent to Theorem 5.91.

5.92 Corollary   *If $f, f_1, \ldots, f_n$ are linear functionals on a vector space X, then either there exist scalars $\lambda_1, \ldots, \lambda_n$ such that $f(x) = \sum_{i=1}^n \lambda_i f_i(x)$ for all $x \in X$, or else there exists an x such that $f_1(x) = \cdots = f_n(x) = 0$ and $f(x) > 0$.*

The Fundamental Theorem of Duality deserves its name because of its role in the next result, which asserts that spaces in a dual pair are each other's duals.

5.93 Theorem   (Dual pairs are weakly dual)   *If $\langle X, X' \rangle$ is a dual pair, then the topological dual of the tvs $(X, \sigma(X, X'))$ is X'. That is, if $f: X \to \mathbb{R}$ is a $\sigma(X, X')$-continuous linear functional, then there exists a unique $x' \in X'$ such that $f(x) = \langle x, x' \rangle$ for each $x \in X$. Similarly, we have $(X', \sigma(X', X))' = X$.*

Proof: Let $f: X \to \mathbb{R}$ be a $\sigma(X, X')$-continuous linear functional. The continuity of f at zero implies the existence of a basic $\sigma(X, X')$-neighborhood of zero that f maps into $[-1, 1]$. That is, there exist $x'_1, \ldots, x'_n \in X'$ and some $\varepsilon > 0$ such that $|\langle x, x'_i \rangle| \leqslant \varepsilon$ for $i = 1, \ldots, n$ implies $|f(x)| \leqslant 1$. So if $x \in \bigcap_{i=1}^{n} \ker x'_i$, then $\langle \alpha x, x'_i \rangle = 0$ for each i and α. Hence $\alpha |f(x)| \leqslant 1$ for each α, so $f(x) = 0$. Consequently, $\bigcap_{i=1}^{n} \ker x'_i \subset \ker f$. By the Fundamental Theorem of Duality 5.91, there are $\lambda_1, \ldots, \lambda_n$ such that $f = \sum_{i=1}^{n} \lambda_i x'_i \in X'$. Uniqueness follows from the fact that X is total. ∎

Theorem 5.93 states that every dual pair $\langle X, X' \rangle$ is obtained from a locally convex Hausdorff space (X, τ) and its topological dual X'. An obvious consequence of Theorem 5.93 is stated next.

5.94 Corollary *Let X'_1 and X'_2 be total subspaces of X^*. Then $\sigma(X, X'_1)$ is weaker than $\sigma(X, X'_2)$ if and only if $X'_1 \subset X'_2$.*

We also have the following consequence of Theorem 5.93.

5.95 Corollary *For a collection of linear functionals $f, f_1, \ldots, f_n$ on a vector X, the following statements are equivalent.*

1. *There exist nonnegative scalars $\lambda_1, \ldots, \lambda_n$ such that $f = \sum_{j=1}^{n} \lambda_j f_j$.*

2. *If $x \in X$ satisfies $f_j(x) \leqslant 0$ for each $j = 1, \ldots, n$, then $f(x) \leqslant 0$.*

Proof: Clearly, (1) $\Longrightarrow$ (2). For the converse, assume (2). We must show that f belongs to the convex cone C generated by $f_1, \ldots, f_n$ in X^*. Consider the dual pair $\langle X, X^* \rangle$ and note that (by Corollary 5.25) the cone C is $\sigma(X^*, X)$-closed. So, if $f \notin C$, then by Corollary 5.84 there exists a $\sigma(X^*, X)$-continuous linear functional on X^* (which according to Theorem 5.93 must be some point $x \in X$) satisfying $f(x) > 0$ and $f_j(x) \leqslant 0$ for each $j = 1, \ldots, n$, contrary to the validity of (2). This contradiction establishes that $f \in C$, as desired. ∎

5.15 Topologies consistent with a given dual

Since the set of $\sigma(X, X')$-continuous linear functionals is precisely X', and for infinite dimensional spaces there are many subspaces X' of the algebraic dual X^* that separate points in X, we have some latitude in the choice of a dual. This presents another topological tradeoff. By enlarging X' we get a stronger (finer) $\sigma(X, X')$-topology on X, so more sets are closed. With more continuous linear functionals, we can separate more sets in X. The tradeoff is that, for a stronger topology, there are fewer compact sets.

Enlarging the dual is not the only way to obtain a stronger topology. There are topologies on X stronger than $\sigma(X, X')$ that still give X' as the dual.

5.96 Definition   *A locally convex topology τ on X is **consistent** (or **compatible**) with the dual pair $\langle X, X' \rangle$ if $(X, \tau)' = X'$. Consistent topologies on X' are defined similarly.*

The following lemma is immediate consequence of Corollary 5.82.

5.97 Lemma   *Every topology consistent with a dual pair is Hausdorff.*

By Theorem 5.93, for every locally convex Hausdorff space (X, τ) both the weak topology $\sigma(X, X')$ and τ are consistent with the dual pair $\langle X, X' \rangle$. It takes several sections to characterize consistent topologies, so we first mention some simple results. The first is an immediate consequence of Corollary 5.83. It states that for a given dual pair, we may speak of a *closed convex set* without specifying the compatible topology to which we are referring.

5.98 Theorem   (Closed convex sets)   *All locally convex topologies consistent with a given dual pair have the same collection of closed convex sets.*

5.99 Corollary   *The set of lower semicontinuous convex functions on a space is the same in all topologies consistent with the dual pair.*

Proof: By Corollary 2.60, a real function on X is lower semicontinuous function if and only if its epigraph is closed, and by Lemma 5.39 is convex if and only if its epigraph is convex. By Theorem 5.98, if these sets are closed in one consistent topology, then they are closed in all consistent topologies.   ∎

Recall that a subset A of a topological vector space (X, τ) is topologically bounded if for each neighborhood V of zero there exists some $\lambda > 0$ such that $A \subset \lambda V$. We prove later (Theorem 6.20) that all consistent topologies share the same bounded sets. It thus suffices to characterize the weakly bounded sets.

5.100 Lemma   (Weakly bounded sets)   *A subset A of X is weakly bounded if and only if it is pointwise bounded. That is, if and only if for every $x' \in X'$ the set $\{\langle x, x' \rangle : x \in A\}$ is bounded in $\mathbb{R}$. Likewise, $B \subset X'$ is weak* bounded if and only if for every $x \in X$ the set $\{\langle x, x' \rangle : x' \in B\}$ is bounded.*

Proof: We prove only the first part. Recall that $\sigma(X, X')$ is generated by the family of seminorms $\{p_{x'} : x' \in X'\}$, where $p_{x'}(x) = |\langle x, x' \rangle|$. Thus by Lemma 5.76, A is bounded if and only if $p_{x'}(A) = \{|\langle x, x' \rangle| : x \in A\}$ is bounded in $\mathbb{R}$.   ∎

Viewing $A \subset X$ as a family of linear functionals on the space X', we see that $\sigma(X, X')$-boundedness is just pointwise boundedness on X'. Likewise a set $B \subset X'$ is $\sigma(X', X)$-bounded if and only if it is pointwise bounded as a set of linear functionals on X.

5.16 Polars

The construct of the polar of a set is fundamental in describing the collection of all consistent topologies for a dual pair. It captures some of the features of the unit ball in the dual of a normed space.

5.101 Definition *For a dual pair $\langle X, X' \rangle$, the **absolute polar**, or simply the **polar**,[8] A° of a nonempty subset A of X, is the subset of X' defined by*

$$A^\circ = \{x' \in X' : |\langle x, x' \rangle| \leqslant 1 \text{ for all } x \in A\}.$$

Similarly, if B is a nonempty subset of X', then its polar is the subset of X defined by

$$B^\circ = \{x \in X : |\langle x, x' \rangle| \leqslant 1 \text{ for all } x' \in B\}.$$

*The **bipolar** of a subset A of X or X' is the set $(A^\circ)^\circ$ written simply as $A^{\circ\circ}$. The bipolar of a subset of X' is defined in a similar manner.*

*The **one-sided polar** of $A \subset X$, denoted $A^\odot$, is defined by*

$$A^\odot = \{x' \in X' : \langle x, x' \rangle \leqslant 1 \text{ for all } x \in A\}.$$

Likewise, for a nonempty subset B of X' we let

$$B^\odot = \{x \in X : \langle x, x' \rangle \leqslant 1 \text{ for all } x' \in B\}.$$

The canonical example of an absolute polar is the polar of the unit ball in a normed space—it is the unit ball in the dual; see Section 6.5. Other examples are shown in Figure 5.5. You should also observe that the basic neighborhoods of zero in the $\sigma(X, X')$ topology are the absolute polars of finite subsets of X'; see Section 7.14.

The next result summarizes some elementary properties of polar sets.

---

[8] **WARNING:** This definition of the polar of a set is *not* universally used. German authors, e.g., G. Köthe [214, p. 245], who introduced the concept, or H. H. Schaefer [293], define the polar of a subset A of X to be $\{x' \in X' : \langle x, x' \rangle \leqslant 1 \text{ for all } x \in A\}$, and use the term **absolute polar** for what we here call the polar. Many French authors, for instance, N. Bourbaki [63] or G. Choquet [76], define the polar of the set A to be the set $\{x' \in X' : \langle x, x' \rangle \geqslant -1 \text{ for all } x \in A\}$. (Although C. Castaing and M. Valadier [75] use the German definition.) The French polar is the negative of the German polar. The absolute polar of A is the German polar of the circled hull of A. Over time, English speaking authors have replaced the term *absolute polar* by the simpler term *polar*. For circled sets, the two concepts agree. (Why?) Consequently, the definition of $\mathfrak{S}$-topologies below is independent of the definition that is used.

However, when it comes to cones, there is a dramatic difference among the definitions. The absolute polar of a cone C is the vector subspace of all linear functionals that annihilate the cone, that is, $\{x' \in X' : \langle x, x' \rangle = 0 \text{ for all } x \in C\}$, while the Köthe one-sided polar (if we may use the expression) is the (generally much larger) set $\{x' \in X' : \langle x, x' \rangle \leqslant 0 \text{ for all } x \in C\}$. Things are complicated by the fact that English speaking authors might use the term *polar cone* in either the Köthe or Bourbaki sense.

$$A \qquad\qquad A^{\circ} \qquad\qquad A^{\circ\circ}$$

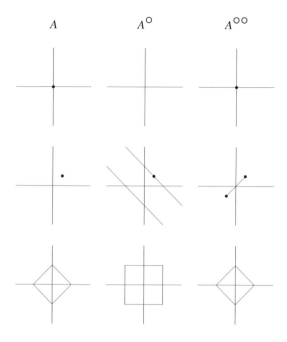

Figure 5.5. Some examples of absolute polars in $\mathbb{R}^2$.

5.102 Lemma **(Properties of polars)**   *Let $\langle X, X' \rangle$ be a dual pair, let A, B be nonempty subsets of X, and let $\{A_i\}$ be a family of nonempty subsets of X. Then:*

1. *If $A \subset B$, then $A^{\circ} \supset B^{\circ}$ and $A^{\odot} \supset B^{\odot}$.*

2. *If $\varepsilon \neq 0$, then $(\varepsilon A)^{\circ} = \frac{1}{\varepsilon} A^{\circ}$ and $(\varepsilon A)^{\odot} = \frac{1}{\varepsilon} A^{\odot}$.*

3. *$\bigcap A_i^{\circ} = (\bigcup A_i)^{\circ}$ and $\bigcap A_i^{\odot} = (\bigcup A_i)^{\odot}$.*

4. *The absolute polar A° is nonempty, convex, circled, $\sigma(X', X)$-closed, and contains zero.*

5. *The one-sided polar $A^{\odot}$ is nonempty, convex, $\sigma(X', X)$-closed, and contains zero.*

6. *If A is absorbing, then both A° and $A^{\odot}$ are $\sigma(X', X)$-bounded.*

7. *The set A is $\sigma(X, X')$-bounded if and only if A° is absorbing.*

The corresponding dual statements are true for subsets of X'.

Proof: We prove the claims only for the absolute polars.

    (1)   Obvious.

    (2)   Just apply the definition.

    (3)   Just think about it.

    (4)   Clearly, A° contains zero, is convex, and circled. To see that A° is also w^*-closed, let the net $\{x'_\alpha\}$ in A° satisfy $x'_\alpha \xrightarrow{\sigma(X',X)} x'$ in X' and let $x \in A$. From $|\langle x, x'_\alpha\rangle| \leq 1$ for each α, we get $|\langle x, x'\rangle| = \lim_\alpha |\langle x, x'_\alpha\rangle| \leq 1$, so $x' \in A^\circ$.

    (5)   Repeat the proof of part (4) above.

    (6)   Assume that A is an absorbing set and fix $x \in X$. Choose a scalar $\alpha > 0$ such that $\pm\alpha x \in A$. If $x' \in A^\circ$, then $\alpha|\langle x, x'\rangle| = |\langle \alpha x, x'\rangle| \leq 1$, so $|\langle x, x'\rangle| \leq \frac{1}{\alpha}$ for all $x' \in A^\circ$. So A° is $\sigma(X', X)$-bounded; see Lemma 5.100. Similarly, if $x' \in A^\circ$, then $\pm\alpha\langle x, x'\rangle = \langle \pm\alpha x, x'\rangle \leq 1$, and consequently $|\alpha\langle x, x'\rangle| \leq 1$ or $|\langle x, x'\rangle| \leq \frac{1}{\alpha}$ for all $x' \in A^\circ$. Hence A° is also $\sigma(X', X)$-bounded.

    (7)   Assume first that A is $\sigma(X, X')$-bounded and let $x' \in X'$. According to Lemma 5.100 there exists some $\lambda > 0$ such that $|\langle x, x'\rangle| \leq \lambda$ for each $x \in A$. If $\alpha_0 = \frac{1}{\lambda}$, then $|\langle x, \alpha x'\rangle| \leq 1$ for each $x \in A$ and all $0 \leq \alpha \leq \alpha_0$. Hence $\alpha x' \in A^\circ$ for all $0 \leq \alpha \leq \alpha_0$, so that A° is an absorbing set.

For the converse, suppose that A° is absorbing. By part (6) the set $A^{\circ\circ}$ is $\sigma(X, X')$-bounded, so A (as a subset of $A^{\circ\circ}$) is likewise $\sigma(X, X')$-bounded. ∎

The following fundamental result is known as the **Bipolar Theorem**.

5.103 Bipolar Theorem   *Let $\langle X, X'\rangle$ be a dual pair, and let A be a nonempty subset of X.*

1. *The bipolar $A^{\circ\circ}$ is the convex circled $\sigma(X, X')$-closed hull of A. Hence if A is convex, circled, and $\sigma(X, X')$-closed, then $A = A^{\circ\circ}$.*

2. *The one-sided bipolar $A^{\odot\odot}$ is the convex $\sigma(X, X')$-closed hull of $A \cup \{0\}$. Hence if A is convex, $\sigma(X, X')$-closed, and contains zero, then $A = A^{\odot\odot}$.*

Corresponding results hold for subsets of X'.

Proof: By Lemma 5.102(4) the set $A^{\circ\circ}$ is convex, circled, and $\sigma(X, X')$-closed. Clearly, $A \subset A^{\circ\circ}$. Let B be the convex circled $\sigma(X, X')$-closed hull of A,

$$B = \bigcap\{C : C \text{ is convex, circled, and } \sigma(X, X')\text{-closed with } A \subset C\}.$$

Clearly $B \subset A^{\circ\circ}$. For the reverse inclusion, suppose there exists some $a \in A^{\circ\circ}$ with $a \notin B$. By the Separation Corollary 5.80 and Theorem 5.93 there exist some $x' \in X'$ and some $\alpha > 0$ satisfying $|\langle x, x'\rangle| \leq \alpha$ for each $x \in B$ and $\langle a, x'\rangle > \alpha$. Replacing x' by $\frac{x'}{\alpha}$, we can assume that $\alpha = 1$. This implies $x' \in A^\circ$. However, $\langle a, x'\rangle > 1$ yields $a \notin A^{\circ\circ}$, a contradiction. Therefore $B = A^{\circ\circ}$.

By Lemma 5.102(4), the set $A^{\odot\odot}$ is convex, and $\sigma(X, X')$-closed. Furthermore, it is clear that $A \cup \{0\} \subset A^{\odot\odot}$. Let C denote the $\sigma(X, X')$-closed convex hull of $A \cup \{0\}$. Then $C \subset A^{\infty}$. Suppose $x \in A^{\infty} \setminus C$. Then by the Separation Corollary 5.80 and Theorem 5.93 there exist some $x' \in X'$ and some $\alpha \neq 0$ satisfying $\langle y, x' \rangle \leqslant \alpha$ for each $y \in C$ and $\langle x, x' \rangle > \alpha$. Since $0 \in C$, we must have $\alpha > 0$. Replacing x' by $\frac{x'}{\alpha}$, we can assume that $\alpha = 1$. This implies $x' \in A^{\circ}$. However, $\langle x, x' \rangle > 1$ yields $x \notin A^{\infty}$, a contradiction. Therefore $C = A^{\infty}$. ∎

5.104 Corollary *For any family $\{A_i\}$ of nonempty convex, circled, and $\sigma(X, X')$-closed subsets of X, the polar $(\bigcap A_i)^{\circ}$ is the convex circled $\sigma(X', X)$-closed hull of the set $\bigcup A_i^{\circ}$.*

Proof: From the Bipolar Theorem 5.103 each $A_i^{\circ\circ} = A_i$, so Lemma 5.102(3) implies the identity

$$\left(\bigcup A_i^{\circ}\right)^{\circ\circ} = \left[\left(\bigcup A_i^{\circ}\right)^{\circ}\right]^{\circ} = \left(\bigcap A_i^{\circ\circ}\right)^{\circ} = \left(\bigcap A_i\right)^{\circ}.$$

Applying the Bipolar Theorem 5.103 once more, note that $\left(\bigcup A_i^{\circ}\right)^{\circ\circ}$ is the convex circled $\sigma(X', X)$-closed hull of $\bigcup A_i^{\circ}$, and we are done. ∎

Now we come to the **Alaoglu Compactness Theorem**, due to L. Alaoglu [5], which is one of the most useful theorems in analysis. It describes one of the primary sources of compact sets in an infinite dimensional setting.

5.105 Alaoglu's Compactness Theorem *Let V be a neighborhood of zero for some locally convex topology on X consistent with the dual pair $\langle X, X' \rangle$. Then its polar V° is a weak* compact subset of X'.*

Similarly, if W is a neighborhood of zero for some consistent locally convex topology on X', then its polar W° is a weakly compact subset of X.

Proof: It suffices to prove the first part, since the proof of the second just interchanges the roles of X and X'. So let V be a neighborhood of zero for some consistent locally convex topology τ on X. Recall that $\sigma(X', X)$ is the topology of pointwise convergence on X. That is, it is the topology on X' induced by the product topology on $\mathbb{R}^X$ (where each $x' \in X'$ is identified with a linear function on X). By the Tychonoff Product Theorem 2.61, a subset of $\mathbb{R}^X$ is compact if and only if it is pointwise closed and pointwise bounded.

To establish that V° is pointwise bounded, pick $x \in X$. Since V is a neighborhood of zero, there is some $\lambda_x > 0$ such that $x \in \lambda_x V$. But then $|\langle x, x' \rangle| \leqslant \lambda_x$ for each $x' \in V^{\circ}$. (Why?) Thus V° is pointwise bounded.

To see that V° is closed in $\mathbb{R}^X$, let $\{x'_\alpha\}$ be a net in V° satisfying $x'_\alpha \to f$ in $\mathbb{R}^X$. That is, $\langle x, x'_\alpha \rangle \to f(x)$ for each $x \in X$. It is easy to see that f is linear, and that $|f(x)| \leqslant 1$ for each $x \in V$. By Lemma 5.64, f is τ-continuous, so $f \in X'$, and in particular $f \in V^{\circ}$. Therefore, V° is closed in $\mathbb{R}^X$ too. ∎

We close this section on polars with a discussion of a closely related notion.

5.106 Definition   *Let $\langle X, X' \rangle$ be a dual pair, and let A be a nonempty subset of X. The* **annihilator** $A^\perp$ *of A is the set of linear functionals in X' that vanish on A.*

That is, $A^\perp = \{x' \in X' : \langle x, x' \rangle = 0$ for all $x \in A\}$. Clearly the annihilator of A is a weak*-closed linear subspace of X'. The annihilator of a subset of X' is similarly defined.

If A is a vector subspace of X (or X'), then the annihilator of A coincides with its absolute polar (why?). That is, if A is a vector subspace, then $A^\perp = A^\circ$. If A is not a vector subspace, then it is easy to see that $A^\perp$ coincides with the absolute polar of the vector subspace spanned by A. The following result is an immediate consequence of the Bipolar Theorem 5.103.

5.107 Theorem   *Let $\langle X, X' \rangle$ be a dual pair and let M be a linear subspace of X. Then $M^\perp = M^\odot = M^\circ$. If M is weakly closed, then $M^{\perp\perp} = M$.*
An analogous result holds for linear subspaces of X'.

The next result is another important consequence of the Bipolar Theorem. It gives a simple test to show that a subspace is dense.

5.108 Corollary (Weak* dense subspaces)   *Let $\langle X, X' \rangle$ be a dual pair and let Y' be a linear subspace of X'. Then the following are equivalent.*

1. *Y' is total. That is, Y' separates points of X.*

2. *$(Y')^\perp = \{0\}$.*

3. *Y' is weak* dense in X'.*

The corresponding symmetric result is true for subspaces of X.

Proof: (1) $\implies$ (2)   Obvious from the definitions.
   (2) $\implies$ (3)   From Theorem 5.107 we see that $(Y')^{\perp\perp}$ is the w^*-closure of Y' in X. But $(Y')^{\perp\perp} = \{0\}^\perp = X'$, so Y' is w^*-dense.
   (3) $\implies$ (1)   Suppose $y'(x) = 0$ for all $y' \in Y'$. Let x' belong to X' and let $\{y'_\alpha\}$ be a net in Y' with $y'_\alpha \xrightarrow{w^*} x'$. Then $x'(x) = \lim y'_\alpha(x) = 0$. Since x' is arbitrary, and X' separates points of X, we see that $x = 0$. This proves that Y' is total.   ∎

5.109 Corollary (Separation by a dense subspace)   *Let $\langle X, X' \rangle$ be a dual pair and suppose C and K are nonempty disjoint weakly compact convex subsets of X. Let Y' be a weak*-dense subspace of X'. Then there exists $y' \in Y'$ that strongly separates C and K.*

Proof: By Corollary 5.94, the topology $\sigma(X, Y')$ is weaker than $\sigma(X, X')$, so by Lemma 2.51 both C and K are $\sigma(X, Y')$-compact. By Corollary 5.108, Y' is total, so $\langle X, Y' \rangle$ is a dual pair. Consequently the topology $\sigma(X, Y')$ is Hausdorff, so C

and K are also $\sigma(X, Y')$-closed. Theorem 5.93 asserts that Y' is the dual of X under its $\sigma(X, Y')$ topology, so the desired conclusion follows from the Strong Separating Hyperplane Theorem 5.79.   ∎

The above result does not hold if C is closed but not compact. For instance, suppose Y' is weak* dense in X', pick x' in $X' \setminus Y'$ and set $C = \ker x' = [x' = 0]$. Let K be a singleton $\{x\}$ with $x'(x) = 1$. If y' strongly separates x from C, we must have $y'(z) = 0$ for all $z \in C$. But then the Fundamental Theorem of Duality 5.91 implies $y' = \alpha x'$ for some $\alpha \neq 0$, so $y' \notin Y'$.

The next simple result is important for understanding weak topologies. Let L be a linear subspace of a vector space X. We say that L has **codimension** m if it is the complement of an m-dimensional subspace. That is, if we can write $X = M \oplus L$, where M has dimension m. The annihilator of an m-dimensional subspace is a subspace of codimension m.

5.110 Theorem   *Let $\langle X, X' \rangle$ be a dual pair and let M be an m-dimensional linear subspace of X. Then $M^{\perp}$ has codimension m. That is, X' is the direct sum of $M^{\perp}$ and an m-dimensional subspace.*

The corresponding result holds for finite dimensional subspaces of X'.

Proof: Let $\{x_1, \ldots, x_m\}$ be a basis for M. For each k, define the continuous linear functional f_k on M by $f_k(\sum_{j=1}^{m} \lambda_j x_j) = \lambda_k$, and consider a continuous linear extension x'_k to X, as in the proof of Theorem 5.89. Then $x'_k(x_j) = 1$ if $j = k$, and $x'_k(x_j) = 0$ if $k \neq j$. This implies that $\{x'_1, \ldots, x'_m\}$ is linearly independent. (Why?) Let L be the m-dimensional span of $\{x'_1, \ldots, x'_m\}$. We claim that $X' = M^{\perp} \oplus L$.

Clearly, $x' \in M^{\perp} \cap L$ implies $x' = 0$. To see that $X' = M^{\perp} \oplus L$, let $x' \in X'$. Put $y' = \sum_{j=1}^{m} x'(x_j) x'_j \in L$ and $z' = x' - y'$. Then an easy argument shows that $z' \in M^{\perp}$, so $x' = z' + y' \in M^{\perp} \oplus L$.   ∎

5.17 $\mathfrak{S}$-topologies

We now take the polar route to characterizing consistent locally convex topologies for a dual pair $\langle X, X' \rangle$. We start with an arbitrary nonempty $\sigma(X', X)$-bounded subset A of X'. By Lemma 5.100, the formula

$$q_A(x) = \sup\{|\langle x, x' \rangle| : x' \in A\}$$

defines a seminorm on X. Furthermore $\{x \in X : q_A(x) \leqslant 1\} = A^{\circ}$, and we have the identity

$$q_A(x) = \sup\{|\langle x, x' \rangle| : x' \in A\} = \inf\{\alpha > 0 : x \in \alpha A^{\circ}\} = p_{A^{\circ}}(x).$$

To see that $q_A = p_{A^{\circ}}$ fix x in X. If x belongs to αA°, then write $x = \alpha y$ with $\alpha > 0$ and $y \in A^{\circ}$. Note that $|\langle x, x' \rangle| = \alpha |\langle y, x' \rangle| \leqslant \alpha$ for all $x' \in A$. Hence $q_A(x) \leqslant \alpha$, from which we see that $q_A(x) \leqslant p_{A^\circ}(x)$. To prove the reverse inequality, note that x/β belongs to A° for every $\beta > q_A(x)$. Thus $p_{A^\circ}(\frac{x}{\beta}) = p_{A^\circ}(x)/\beta \leqslant 1$, so $p_{A^\circ}(x) \leqslant \beta$ for all $\beta > q_A(x)$. Hence $p_{A^\circ}(x) \leqslant q_A(x)$, and we are done.

In other words, q_A is a seminorm that coincides with the gauge A°. By the Bipolar Theorem 5.103, the set $A^{\circ\circ}$ is the convex circled $\sigma(X', X)$-closed hull of A. Since $A^\circ = (A^{\circ\circ})^\circ$, we see that $q_A = q_{A^{\circ\circ}}$.

Now let $\mathfrak{S}$ be a family of nonempty $\sigma(X', X)$-bounded subsets of X'.[9] The corresponding **$\mathfrak{S}$-topology** on X is the locally convex topology generated by the family of seminorms $\{q_A : A \in \mathfrak{S}\}$. Equivalently, it is the topology generated by the neighborhood subbase $\{\varepsilon A^\circ : A \in \mathfrak{S} \text{ and } \varepsilon > 0\}$ at zero. Thus we may expand $\mathfrak{S}$ to $\widehat{\mathfrak{S}} = \{\varepsilon A : A \in \mathfrak{S} \text{ and } \varepsilon > 0\}$ and still generate the same topology on X. In other words, the neighborhood base at zero for the $\mathfrak{S}$-topology consists of all sets of the form

$$A_1^{\,\circ} \cap \cdots \cap A_n^{\,\circ},$$

where $A_1, \ldots, A_n \in \widehat{\mathfrak{S}}$. Also note that since $q_A = q_{A^{\circ\circ}}$, we may restrict attention to families of convex circled sets. The $\mathfrak{S}$-topology is Hausdorff if and only if the span of the set $\bigcup_{A \in \mathfrak{S}} A$ is $\sigma(X', X)$-dense in X'. (Why?)

Since $x_\alpha \xrightarrow{\mathfrak{S}} x$ in X if and only if $q_A(x_\alpha - x) \to 0$ for every $A \in \mathfrak{S}$, and $q_A(x_\alpha - x) \to 0$ for every $A \in \mathfrak{S}$ if and only if $\{x_\alpha\}$ converges uniformly to x on each member of $\mathfrak{S}$, the $\mathfrak{S}$-topology is also called the **topology of uniform convergence** on members of $\mathfrak{S}$.

Remarkably, every consistent locally convex topology on X (or on X') is an $\mathfrak{S}$-topology. This important result is known as the Mackey–Arens Theorem. It finally answers the question of what topologies are consistent for a given dual pair. The next lemma breaks out a major part of the proof. It doesn't tell you anything new—it is just a peculiar, but useful, way of rewriting Lemma 5.64, which says that a linear functional is continuous if and only if it is bounded on some neighborhood of zero. (Recall that X^*, the algebraic dual of X, is the vector space of all real linear functions on X, continuous or not.)

5.111 Lemma   *If τ is a linear topology of a vector space X and $\mathcal{B}$ is a neighborhood base at zero, then the topological dual of (X, τ) is $\bigcup_{V \in \mathcal{B}} V^\bullet$, where $V^\bullet$ is the polar of V taken with respect to the dual pair $\langle X, X^* \rangle$.*

Proof: Let x' be τ-continuous. Clearly, $W = (x')^{-1}([-1, 1])$ is a τ-neighborhood of zero and if $V \in \mathcal{B}$ satisfies $V \subset W$, then $x' \in V^\bullet$.

Conversely, if $x' \in V^\bullet$, it is bounded on V and so τ-continuous.   ∎

The next result is due to G. W. Mackey [237] and R. Arens [19]. It characterizes all the linear topologies consistent with a dual pair.

---

[9] We use the symbol $\mathfrak{S}$ because it is well established. For those of you who don't know how to pronounce it, $\mathfrak{S}$ is an upper case "S" in the old German fraktur alphabet.

5.112 Mackey–Arens Theorem *A locally convex topology τ on X is consistent with the dual pair $\langle X, X' \rangle$ if and only if τ is the $\mathfrak{S}$-topology for a family $\mathfrak{S}$ of convex, circled, and $\sigma(X', X)$-compact subsets of X' with $\bigcup_{A \in \mathfrak{S}} A = X'$.*

Proof: First we show that a consistent topology is an $\mathfrak{S}$-topology. Let τ be a consistent topology and let $\mathcal{B}$ be the neighborhood base of all the convex, circled, τ-closed τ-neighborhoods of zero. Let $\mathfrak{S} = \{V^{\circ} : V \in \mathcal{B}\}$. By Alaoglu's Theorem 5.105, each V° is $\sigma(X', X)$-compact. Further, each is convex and circled, and $\bigcup_{V \in \mathcal{B}} V^{\circ} = X'$. The Bipolar Theorem 5.103 implies $V^{\circ\circ} = V$ for each $V \in \mathcal{B}$, so we have $\{A^{\circ} : A \in \mathfrak{S}\} = \{V^{\circ\circ} : V \in \mathcal{B}\} = \mathcal{B}$. Therefore τ is the $\mathfrak{S}$-topology.

The converse is only a bit trickier. We must deal with both the $\langle X, X' \rangle$ and $\langle X, X^* \rangle$ dual pairs. Keep in mind that the $\sigma(X, X^*)$-topology on X is stronger than the $\sigma(X, X')$-topology. Furthermore, the $\sigma(X', X)$-topology on X' is the relativization to $X' \subset X^*$ of the $\sigma(X^*, X)$-topology on X^* (Lemma 2.53). For this proof, let A° denote the polar of A with respect to $\langle X, X' \rangle$, and let $A^{\bullet}$ denote the polar with respect to $\langle X, X^* \rangle$. Observe that for a set $A \subset X' \subset X^*$, we have $A^{\circ} = A^{\bullet}$. (Why?)

Now suppose that τ is an $\mathfrak{S}$-topology for a family $\mathfrak{S}$ of convex, circled, and $\sigma(X', X)$-compact subsets of X' with $\bigcup_{A \in \mathfrak{S}} A = X'$. Without loss of generality, we can assume that $\varepsilon A \in \mathfrak{S}$ for each $\varepsilon > 0$ and all $A \in \mathfrak{S}$. Then the family $\mathcal{B}$ of all finite intersections of the form

$$V = (A_1)^{\circ} \cap \cdots \cap (A_n)^{\circ} = (A_1)^{\bullet} \cap \cdots \cap (A_n)^{\bullet}, \qquad (\star)$$

where $A_1, \ldots, A_n \in \mathfrak{S}$, is a neighborhood base at zero for τ.

Let $X^{\#} \subset X^*$ denote the topological dual of (X, τ). By Lemma 5.111, we know that $X^{\#} = \bigcup_{V \in \mathcal{B}} V^{\bullet}$. If $x' \in X'$, then $x' \in A$ for some $A \in \mathfrak{S}$, so $|\langle x, x' \rangle| \leqslant 1$ for all $x \in A^{\circ}$. Thus x' is bounded on A°, a τ-neighborhood of zero, so $x' \in X^{\#}$. Therefore, $X' \subset X^{\#}$.

To show that $X^{\#} \subset X'$, let V be a basic τ-neighborhood as in $(\star)$. It suffices to show that $V^{\bullet} \subset X'$. By Lemma 5.102(3), $V^{\bullet} = (\bigcup_{i=1}^{n} A_i)^{\bullet\bullet}$. By the Bipolar Theorem 5.103, the set $(\bigcup_{i=1}^{n} A_i)^{\bullet\bullet}$ is the convex circled $\sigma(X^*, X)$-closed hull of $\bigcup_{i=1}^{n} A_i$. Now use Lemma 5.29 (2) to see that the convex circled hull C of $\bigcup_{i=1}^{n} A_i$ in X^* is precisely the set

$$C = \Big\{ \sum_{i=1}^{n} \lambda_i x_i' : \lambda_i \in \mathbb{R}, \ x_i' \in A_i \ (i = 1, \ldots, n), \text{ and } \sum_{i=1}^{n} |\lambda_i| \leqslant 1 \Big\},$$

which is a subset of X'. Since by assumption each A_i is $\sigma(X', X)$-compact, each is also a $\sigma(X^*, X)$-compact subset of X^*. So again by Lemma 5.29 (2), the set C is $\sigma(X^*, X)$-compact, and so $\sigma(X^*, X)$-closed. Consequently $V^{\bullet} = C \subset X'$, and the proof is finished.   ∎

5.18   The Mackey topology

Observe that the weak topology $\sigma(X, X')$ is the $\mathfrak{S}$-topology for the collection $\mathfrak{S} = \{\{x'\} : x' \in X'\}$. The weak topology $\sigma(X, X')$ is the smallest locally convex topology on X consistent with $\langle X, X' \rangle$. The largest consistent locally convex topology on X is by Theorem 5.112 the $\mathfrak{S}$-topology for the family $\mathfrak{S}$ consisting of all convex, circled, and $\sigma(X', X)$-compact subsets of X'. This important topology is called the **Mackey topology** and denoted $\tau(X, X')$. The Mackey topology $\tau(X', X)$ is defined analogously.

The Mackey–Arens Theorem 5.112 can be restated in terms of the weak and Mackey topologies as follows.

5.113 Theorem   *A locally convex topology τ on X is consistent with the dual pair $\langle X, X' \rangle$ if and only if $\sigma(X, X') \subset \tau \subset \tau(X, X')$.*

Similarly, locally convex topology τ' on X' is consistent with the dual pair $\langle X, X' \rangle$ if and only if $\sigma(X', X) \subset \tau' \subset \tau(X', X)$.

Even though the Mackey topology is defined in terms of circled subsets of X', we have the following lemma.

5.114 Lemma   (Mackey neighborhoods)   *If a nonempty subset K of X' is $\sigma(X', X)$-compact, then the one-sided polar $K^{\odot}$ is a convex $\tau(X, X')$ (Mackey) neighborhood of zero in X. Conversely, the one-sided polar $V^{\odot}$ of an arbitrary $\tau(X, X')$-neighborhood V of zero is nonempty, convex, and $\sigma(X', X)$-compact.*

Proof: Suppose first that K is a nonempty w^*-compact subset of X'. Let C be the convex circled hull of K. By Corollary 5.31, C is weak* compact. Thus $C^{\odot} = C^{\mathbf{O}}$ is a Mackey neighborhood of zero. But since $K \subset C$, we have $K^{\odot} \supset C^{\odot}$, so $K^{\odot}$ is a Mackey neighborhood too.

Conversely, if V is a Mackey neighborhood of zero, then there is a basic neighborhood $W \subset V$ of the form $W = A^{\mathbf{O}}$, where A is a nonempty convex circled $\sigma(X', X)$-compact subset of X'. Note that $W^{\mathbf{O}} = W^{\odot} \supset V^{\odot}$ since W is circled. Now $W^{\mathbf{O}}$ is $\sigma(X', X)$-compact by Alaoglu's Theorem 5.105, and $V^{\odot}$ is convex and $\sigma(X', X)$-closed by Lemma 5.102(4). Therefore $V^{\odot}$ is $\sigma(X', X)$-compact.   ∎

5.19   The strong topology

There is another important topology on X. It is the $\mathfrak{S}$-topology generated by the family $\mathfrak{S}$ of all $\sigma(X', X)$-bounded subsets of X'. It is known as the **strong topology** and is denoted $\beta(X, X')$. In general, the strong topology $\beta(X, X')$ is *not* consistent with the dual pair $\langle X, X' \rangle$. The dual strong topology $\beta(X', X)$ is defined analogously.

If (X, τ) is a locally convex Hausdorff space, then the **double dual** of (X, τ) is the topological dual of $(X', \beta(X', X))$ and is denoted X''. It is customary to consider X'' equipped with the strong topology $\beta(X'', X')$.

Recall that every $x \in X$ defines a linear functional $\hat{x}$ on X', the evaluation at x, via $\hat{x}(x') = x'(x)$. If $B = \{x\}$, then B is a bounded subset of X, and on the $\beta(X', X)$-neighborhood B° of zero we have

$$|\hat{x}(x')| = |x'(x)| \leqslant 1 \quad \text{for all} \quad x' \in B^{\circ}.$$

By Lemma 5.64, $\hat{x}$ is $\beta(X', X)$-continuous, that is, $\hat{x} \in X''$. Since X' separates the points of X (Corollary 5.82), we see that $x \mapsto \hat{x}$ is a linear isomorphism, so X identified with its image can be viewed as a vector subspace of its double dual X''. A locally convex Hausdorff space is called **semi-reflexive** if $X'' = X$.

Chapter 6

Normed spaces

This chapter studies some of the special properties of normed spaces. All finite dimensional spaces have a natural norm, the Euclidean norm. On a finite dimensional vector space, the Hausdorff linear topology the norm generates is unique (Theorem 5.21). The Euclidean norm makes $\mathbb{R}^n$ into a complete metric space. A normed space that is complete in the metric induced by its norm is called a *Banach space*. Here is an overview of some of the more salient results in this chapter.

The norm topology on a vector space X defines a topological dual X', giving rise to a natural dual pair $\langle X, X' \rangle$. Thus we may refer to the weak topology on a normed space without specifying a dual pair. In such cases, it is understood that X is paired with its norm dual. Since a finite dimensional space has only one Hausdorff linear topology, the norm topology and the weak topology must be the same. This is not true in infinite dimensional normed spaces. On an infinite dimensional normed space, the weak topology is strictly weaker than the norm topology (Theorem 6.26). The reason for this is that every basic weak neighborhood includes a nontrivial linear subspace—the intersection of the kernels of a finite collection of continuous linear functionals. This linear subspace is of course unbounded in norm, so no norm bounded set can be weakly open (Corollary 6.27). This fact leads to some surprising conclusions. For instance, in an infinite dimensional normed space, zero is always in the weak closure of the unit sphere $\{x : \|x\| = 1\}$ (Corollary 6.29). In fact, in infinite dimensional normed spaces, there always exist nets converging weakly to zero, but wandering off to infinity in norm (Lemma 6.28). Also, the weak topology on an infinite dimensional normed space is never metrizable (Theorem 6.26). Despite this, it is possible for the weak topology to be metrizable when restricted to bounded subsets, such as the unit ball (Theorems 6.30 and 6.31). It also turns out that on a normed space, there is no stronger topology with the same dual. That is, the norm topology is the Mackey topology for the natural dual pair (Theorem 6.23).[1]

---

[1] The natural duality of a normed space with its norm dual is not always the most useful pairing. Two important examples are the normed spaces $B_b(X)$ of bounded Borel measurable functions on a metrizable space, and the space $L_\infty(\mu)$ of μ-essentially bounded functions. (Both include ℓ_∞ as a special case.) The dual of B_b is the space of bounded charges, but the pairing $\langle B_b, ca \rangle$ of B_b with countably additive measures is more common. See Section 14.1 for a discussion of this pair. Similarly, the dual of L_∞ is larger than L_1, but the pairing $\langle L_\infty, L_1 \rangle$ is more useful. This can be confusing at times.

Linear operators are linear functions from one vector space into another. An important special case is when the range is the real line, which is a Banach space under the absolute value norm. Norms on the domain and the range allow us to define the boundedness of an operator. An operator is *bounded* if it maps norm bounded sets into norm bounded sets. Boundedness is equivalent to norm continuity of an operator, which is equivalent to uniform continuity (Lemmas 5.17 and 6.4). The Open Mapping Theorem 5.18 shows that if a bounded operator between Banach spaces is surjective, then it carries open sets to open sets. The *operator norm* of a bounded operator $T : X \to Y$ is defined by $\|T\| = \sup\{\|T(x)\| : \|x\| \leqslant 1\}$. This makes the vector space $L(X, Y)$ of all continuous linear operators from X into Y a normed space. It is a Banach space if Y is (Theorem 6.6). In particular, the topological dual of a normed space is also a Banach space. The Uniform Boundedness Principle 6.14 says that a family of bounded linear operators from a Banach space to a normed space is bounded in the norm on $L(X, Y)$ if and only if it is a pointwise bounded family. This is used to prove that for general dual pairs, all consistent topologies have the same bounded sets (Theorem 6.20).

There are many ways to recognize the continuity of a linear operator between normed spaces. One of these is via the Closed Graph Theorem 5.20, which states that a linear operator between Banach spaces is continuous if and only if its graph is closed. Another useful fact is that a linear operator is continuous in the norm topology if and only it is continuous in the weak topology (Theorem 6.17). Any pointwise limit of a sequence of continuous linear operators on a Banach space is a continuous operator (Corollary 6.19). Every operator T from X to Y, defines an *(algebraic) adjoint operator* T^* from Y^* to X^* by means of the formula $T^*y^* = y^* \circ T$, where X^* and Y^* are the algebraic duals of X and Y respectively. A useful result is that an operator T is continuous if and only if its adjoint carries Y' into X' (Theorem 6.43). Finally, we point out that the evaluation duality $\langle x, x' \rangle$, while jointly norm continuous, is not jointly weak-weak* continuous for infinite dimensional spaces (Theorems 6.37 and 6.38).

The topological dual of a normed space is a Banach space under the operator norm. Alaoglu's Compactness Theorem 6.21 asserts that the unit ball in the dual of a normed space is weak* compact. Since the dual X' of a normed space X is a Banach space, its dual X'' is a Banach space too, called the *second dual* of X. In general, there is a natural isometric embedding of X as a $\sigma(X'', X')$-dense subspace of X'' (Theorem 6.24), and in some cases the two coincide. In this case we say that X is *reflexive*. A Banach space is reflexive if and only if its closed unit ball is weakly compact (Theorem 6.25).

There are some useful results about weak compactness in normed spaces. Recall that for any metric space, a set is compact if and only if it is sequentially com-

---

For instance, the Mackey topology $\tau(\ell_\infty, \ell_1)$ for the dual pair $\langle \ell_\infty, \ell_1 \rangle$ is *not* the norm topology on ℓ_∞: it is weaker. In this chapter at least, we do not deal with other pairings. But when it comes to applying these theorems, make sure you know your dual.

pact (Theorem 3.28). The celebrated Eberlein–Šmulian Theorem 6.34 asserts that in a normed space, a set is weakly compact if and only if it is weakly sequentially compact. Theorem 5.35 implies that the closed convex hull of a norm compact subset of a Banach space is norm compact. The Krein–Šmulian Theorem 6.35 says that the closed convex hull of a weakly compact subset of a Banach space is weakly compact. James' Theorem 6.36 says that a weakly closed bounded subset of a Banach space is weakly compact if and only if every continuous linear functional achieves its maximum on the set.

A linear operator from X to Y induces in a natural way another linear operator from the dual Y^* to the dual X^*. This is called the *adjoint* operator. (In finite dimensional spaces the matrix representation of the adjoint is the transpose of the matrix representation of the original operator.) We make heavy use of adjoints in Chapter 19.

We conclude with an introduction to *Hilbert spaces*, which are Banach spaces where the norm is derived from an *inner product*. An inner product maps pairs of vectors into the real numbers. The inner product of x and y is denoted (x, y). Every Euclidean space is a Hilbert space, and the inner product is the familiar vector dot product. One of the most important properties of a Hilbert space is that it is self-dual. That is, every continuous linear functional corresponds to the inner product with a vector y, that is, it is of the form $x \mapsto (y, x)$ (Corollary 6.55). The other important concept that an inner product allows is that of *orthogonality*. Two vectors are orthogonal if their inner product is zero. Convex sets in Hilbert spaces also have the *nearest point property* (Theorem 6.53).

6.1 Normed and Banach spaces

The class of Banach spaces is a special class of both complete metric spaces and locally convex spaces. A **normed space** is a vector space[2] X equipped with a norm $\|\cdot\|$. Recall that a norm is a function $\|\cdot\| \colon X \to \mathbb{R}$ that satisfies the properties:

1. $\|x\| \geqslant 0$ for all $x \in X$, and $\|x\| = 0$ if and only if $x = 0$.

2. $\|\alpha x\| = |\alpha| \|x\|$ for all $\alpha \in \mathbb{R}$ and all $x \in X$.

3. $\|x + y\| \leqslant \|x\| + \|y\|$ for all $x, y \in X$.

Property (3) is known as the **triangle inequality**. The norm induces a metric d via the formula $d(x, y) = \|x - y\|$. Properties (2) and (3) guarantee that a ball of radius r around zero is convex, so the topology generated by this metric is a locally convex Hausdorff topology. It is known as the **norm topology** on X. The triangle inequality easily implies

$$\big| \|x\| - \|y\| \big| \leqslant \|x - y\|$$

---

[2] Remember, in this book, we only consider real vector spaces.

for all x, y. This readily shows that the norm (as a real function $x \mapsto \|x\|$ on X) is a uniformly continuous function.

A subset of a normed space is **norm bounded** if it is bounded in the metric induced by the norm. Equivalently, a set A is norm bounded if there is some real constant M such that $\|x\| \leqslant M$ for all $x \in A$.

The **closed unit ball** U of a normed space X is the set of vectors of norm no greater than one. That is,

$$U = \{x \in X : \|x\| \leqslant 1\}.$$

Clearly U is norm bounded, convex, circled, and norm (hence weakly) closed. (Why?) The **open unit ball** is $\{x \in X : \|x\| < 1\}$.

6.1 Definition   *A **Banach space** is a normed space that is also a complete metric space under the metric induced by its norm.*

Banach spaces are the most important class of locally convex spaces, and are often studied without reference to the general theory. Here is a list of some familiar Banach spaces.

•   The Euclidean space $\mathbb{R}^n$ with its Euclidean norm. A special case is the real line $\mathbb{R}$ with the absolute value norm.

•   The $L_p(\mu)$-space ($1 \leqslant p < \infty$) with the L_p-norm defined by

$$\|f\|_p = \left(\int |f|^p \, d\mu \right)^{\frac{1}{p}}.$$

•   The $L_\infty(\mu)$-space with the norm $\|f\|_\infty = \operatorname{ess\,sup} |f|$.

•   The vector space c_0 of all real sequences converging to zero, with the sup norm $\|x\|_\infty = \sup\{|x_n| : n = 1, 2, \ldots\}$.

•   The vector space $ba(\mathcal{A})$ of bounded charges on an algebra $\mathcal{A}$ of subsets of a set Ω, with the total variation norm $\|\mu\| = |\mu|(\Omega)$. (See Theorem 10.53.)

•   The vector space $C_b(\Omega)$ of all bounded continuous real functions on a topological space Ω, with the sup norm $\|f\|_\infty = \sup\{|f(\omega)| : \omega \in \Omega\}$.

•   The vector space $C^k[a, b]$ of all k-continuously differentiable real functions on an interval $[a, b]$ with the norm

$$\|f\| = \|f\|_\infty + \|f'\|_\infty + \cdots + \|f^{(k)}\|_\infty.$$

6.2   Linear operators on normed spaces

In this section, we discuss some basic properties of continuous operators acting between normed spaces. The proof of the next lemma is left as an exercise.

6.2 Lemma   *If* $T: X \to Y$ *is an operator between normed spaces, then*

$$\sup_{\|x\|\leq 1} \|Tx\| = \min\{M \geq 0 : \|Tx\| \leq M\|x\| \text{ for all } x \in X\},$$

where we adhere to the convention $\min \varnothing = \infty$. *If the normed space* X *is nontrivial* (*that is,* $X \neq \{0\}$), *then we also have*

$$\sup_{\|x\|\leq 1} \|Tx\| = \sup_{\|x\|=1} \|Tx\|.$$

We are now in a position to define the norm of an operator.

6.3 Definition   *The **norm of an operator** $T: X \to Y$ between normed spaces is the nonnegative extended real number $\|T\|$ defined by*

$$\|T\| = \sup_{\|x\|\leq 1} \|Tx\| = \min\{M \geq 0 : \|Tx\| \leq M\|x\| \text{ for all } x \in X\}.$$

If $\|T\| = \infty$, *we say that T is an **unbounded operator**, while in case $\|T\| < \infty$, we say that T is a **bounded operator**.*

Consequently, an operator $T: X \to Y$ between normed spaces is bounded if and only if there exists some positive real number $M > 0$ satisfying the inequality $\|T(x)\| \leq M\|x\|$ for all $x \in X$. Another way of stating the boundedness of an operator is this: An operator $T: X \to Y$ is bounded if and only if it carries the closed (or open) unit ball of X to a norm bounded subset of Y. The following simple result follows immediately from Lemma 6.2 and the definition of the operator norm. It is used often without any special mention. Its proof is straightforward and is omitted.

6.4 Lemma (Boundedness and continuity)   *For a bounded operator $T: X \to Y$ between normed spaces the following hold true.*

1. *For each $x \in X$ we have $\|Tx\| \leq \|T\| \cdot \|x\|$.*

2. *The operator T is continuous if and only if it is bounded.*

Now let X and Y be two normed spaces. If T and S are linear operators from X into Y, then you can easily verify the following properties of the operator norm.

- $\|T\| \geq 0$ and $\|T\| = 0$ if and only if $T = 0$.

- $\|\alpha T\| = |\alpha| \cdot \|T\|$ for each $\alpha \in \mathbb{R}$.

- $\|S + T\| \leq \|S\| + \|T\|$.

Consequently, we have the following fact.

6.5 Lemma   *The vector space $L(X, Y)$ of all bounded operators from X to Y is a normed vector space.*

We write $L(X)$ for $L(X, X)$. Clearly, $T_n \to T$ in $L(X, Y)$ implies $T_n x \to T x$ in Y for each $x \in X$. The normed space $L(X, Y)$ is a Banach space exactly when Y is a Banach space. The details follow.

6.6 Theorem   *For normed spaces X and Y we have:*

1. *If Y is a Banach space, then $L(X, Y)$ is also a Banach space.*

2. *If X is nontrivial and the normed space $L(X, Y)$ is a Banach space, then Y is likewise a Banach space.*

Proof: (1)   Assume first that Y is a Banach space and let $\{T_n\}$ be a Cauchy sequence in $L(X, Y)$. Then, for each $x \in X$ we have

$$\left\| T_n x - T_m x \right\| \leqslant \| T_n - T_m \| \cdot \|x\|. \qquad (\star)$$

Now let $\varepsilon > 0$. Pick some n_0 such that $\| T_n - T_m \| < \varepsilon$ for all $n, m \geqslant n_0$. From $(\star)$, we see that $\| T_n x - T_m x \| \leqslant \varepsilon \|x\|$ for all $n, m \geqslant n_0$ and each x. So $\{T_n x\}$ is a Cauchy sequence in Y for each $x \in X$. Therefore, if $T x = \lim_{n \to \infty} T_n x$, then T defines a linear operator from X to Y and $\| T_n x - T x \| \leqslant \varepsilon \|x\|$ for each x and all $n \geqslant n_0$. This implies $T \in L(X, Y)$ and that $T_n \to T$ in $L(X, Y)$. (Why?)

(2)   Assume that $L(X, Y)$ is a Banach space, and let $\{y_n\}$ be a Cauchy sequence in Y. Since $X \neq \{0\}$, there exists a continuous nonzero linear functional f on X. Now for each n consider the operator T_n in $L(X, Y)$ defined by $T_n(x) = f(x) y_n$. It is easy to see that $\{T_n\}$ is a Cauchy sequence in $L(X, Y)$. So if $T_n \to T$ in $L(X, Y)$ and $x_0 \in X$ satisfies $f(x_0) = 1$, then $y_n = T_n(x_0) \to T(x_0)$ in Y. This shows that Y is a Banach space. ∎

6.3   The norm dual of a normed space

It is time now to discuss some important properties of the first and second duals of a normed space.

6.7 Definition   *The **norm dual** X' of a normed space $(X, \| \cdot \|)$ is Banach space $L(X, \mathbb{R})$. The operator norm on X' is also called the **dual norm**, also denoted $\| \cdot \|$. That is,*

$$\|x'\| = \sup_{\|x\| \leqslant 1} |x'(x)| = \sup_{\|x\| = 1} |x'(x)|.$$

The dual space is indeed a Banach space by Theorem 6.6.

6.8 Theorem   *The norm dual of a normed space is a Banach space.*

The next result is a nifty corollary of the Hahn–Banach Extension Theorem.

6.9 Lemma   (Norm preserving extension)   *A continuous linear functional defined on a subspace of a normed space can be extended to a continuous linear functional on the entire space while preserving its original norm.*

Proof: Let Y be a subspace of a normed space X and let $f: Y \to \mathbb{R}$ be a continuous linear functional. Let

$$M = \sup\{|f(y)| : y \in Y \text{ and } \|y\| \leq 1\} < \infty$$

and note that $|f(y)| \leq M \cdot \|y\|$ for each $y \in Y$. Clearly, the norm $p(x) = M \cdot \|x\|$ is a sublinear mapping on X. Any extension $\hat{f}$ of f to all of X satisfying $\hat{f}(x) \leq p(x)$ for each $x \in X$ has the desired properties.   ∎

The norm dual of X' is called the **second dual** (or the **double dual**) of X and is denoted X''. The normed space X can be embedded isometrically in X'' in a natural way. Each $x \in X$ gives rise to a norm-continuous linear functional $\hat{x}$ on X' via the formula

$$\hat{x}(x') = x'(x) \text{ for each } x' \in X'.$$

6.10 Lemma   *For each $x \in X$, we have $\|\hat{x}\| = \|x\| = \max_{\|x'\| \leq 1} |x'(x)|$, where $\|\hat{x}\|$ is the operator norm of $\hat{x}$ as a linear functional on the normed space X'.*

Proof: By definition, $\|\hat{x}\| = \sup_{\|x'\| \leq 1} |\hat{x}(x')|$. But

$$|\hat{x}(x')| = |x'(x)| \leq \|x'\| \cdot \|x\|,$$

so $\|\hat{x}\| = \sup_{\|x'\| \leq 1} |\hat{x}(x')| \leq \|x\|$.
   Now let $V = \{\alpha x : \alpha \in \mathbb{R}\}$ and let $f: V \to \mathbb{R}$ by $f(\alpha x) = \alpha \|x\|$. If $p(y) = \|y\|$, then $f(\alpha x) \leq p(\alpha x)$ and from the Hahn–Banach Extension Theorem 5.53, we can extend f to all of X in such a way that $f(y) \leq p(y) = \|y\|$ for each $y \in X$. It follows that $f \in X'$, $\|f\| \leq 1$, and $f(x) = \|x\|$. Therefore,

$$\|\hat{x}\| = \sup_{\|x'\| \leq 1} |x'(x)| \geq f(x) = \|x\|.$$

Thus $\|\hat{x}\| = \sup_{\|x'\| \leq 1} |x'(x)| = \max_{\|x'\| \leq 1} |x'(x)| = \|x\|$.   ∎

6.11 Corollary   *The mapping $x \mapsto \hat{x}$ from X into X'' is a linear isometry (a linear operator and an isometry), so X can be identified with a subspace $\hat{X}$ of X''.*

The closure $\overline{\hat{X}}$ of $\hat{X}$ in X'' (which is a closed vector subspace of X'') is the norm completion of X. That is, $\overline{\hat{X}}$ is the completion of X when X is equipped with the metric induced by the norm. Therefore, we have proven the following.

6.12 Theorem   *The norm completion of a normed space is a Banach space.*

When the linear isometry $x \mapsto \hat{x}$ from a Banach space X into its double dual X'' is surjective, the Banach space is called reflexive. That is, we have the following definition.

6.13 Definition   *A Banach space is called **reflexive** if $X = \hat{X} = X''$.*

6.4   The uniform boundedness principle

Let X and Y be two normed spaces. A family of operators $\mathcal{A}$ of $L(X, Y)$ is **pointwise bounded** if for each $x \in X$ there exists some $M_x > 0$ such that $\|T(x)\| \leqslant M_x$ for each $T \in \mathcal{A}$. The following important theorem is known as the **Uniform Boundedness Principle**.

6.14 Uniform Boundedness Principle   *Let X be a Banach space, let Y be a normed space, and let $\mathcal{A}$ be a nonempty subset of $L(X, Y)$. Then $\mathcal{A}$ is norm bounded if and only if it is pointwise bounded.*

Proof: If there exists some $M > 0$ satisfying $\|T\| \leqslant M$ for each $T \in \mathcal{A}$, then $\|Tx\| \leqslant \|T\| \cdot \|x\| \leqslant M\|x\|$ for each $x \in X$ and all $T \in \mathcal{A}$.

For the converse, assume that $\mathcal{A}$ is pointwise bounded. For each n define

$$C_n = \{x \in X : \|Tx\| \leqslant n \text{ for all } T \in \mathcal{A}\},$$

Each C_n is norm closed, and since $\mathcal{A}$ is pointwise bounded, $X = \bigcup_{n=1}^{\infty} C_n$. Taking into account that X is complete, it follows from Theorem 3.46 and the Baire Category Theorem 3.47 that some C_k has a nonempty interior. So there exist $a \in C_k$ and $r > 0$ such that $\|y - a\| \leqslant r$ implies $y \in C_k$. Now let $T \in \mathcal{A}$ and let $x \in X$ satisfy $\|x\| \leqslant 1$. From $\|(a + rx) - a\| \leqslant r$, it follows that $a + rx \in C_k$, so

$$r\|Tx\| = \|T(rx)\| = \|T(a + rx) - T(a)\| \leqslant \|T(a + rx)\| + \|T(a)\| \leqslant 2k.$$

Therefore, $\|Tx\| \leqslant \frac{2k}{r} = M$ for all $T \in \mathcal{A}$ and all $x \in X$ with $\|x\| \leqslant 1$. It follows that $\|T\| = \sup_{\|x\| \leqslant 1} \|Tx\| \leqslant M$ for each $T \in \mathcal{A}$, and the proof is finished.   ∎

Since $X' = L(X, \mathbb{R})$, we have the following important special case of the Uniform Boundedness Principle for a collection of continuous linear functionals.

6.15 Corollary   *A nonempty set in the dual of a Banach space is norm bounded if and only if it is pointwise bounded.*

A subset A of a normed space X, viewed as a subset of X'', is **pointwise bounded** if for each $x' \in X'$ there exists a constant $M_{x'} > 0$ (depending upon x') such that $|x'(a)| \leqslant M_{x'}$ for each $a \in A$.

6.16 Corollary   *A nonempty subset of a normed vector space is norm bounded if and only if it is pointwise bounded.*

Proof: If A is a subset of a normed space X, embed X naturally in its double dual X'' and apply Corollary 6.15 to A as a subset of the double dual X''. ∎

For linear operators norm continuity and weak continuity are equivalent.

6.17 Theorem (Norm and weak continuity) *A linear operator between two normed spaces is norm continuous if and only if it is weakly continuous. That is, $T : X \to Y$ is norm continuous if and only if T is continuous when X has its $\sigma(X, X')$-topology and Y has its $\sigma(Y, Y')$-topology.*

Proof: First let T be norm continuous. Note that if $y' \in Y'$, then $y' \circ T \in X'$. So if $x_\alpha \xrightarrow{w} 0$ and $y' \in Y'$, then $y'(Tx_\alpha) = (y' \circ T)(x_\alpha) \to 0$. That is, $Tx_\alpha \xrightarrow{w} 0$ in Y.

Now let T be weakly continuous and assume by way of contradiction that T is unbounded. Then there exists a sequence $\{x_n\}$ in X satisfying $\|x_n\| \leqslant 1$ and $\|Tx_n\| \geqslant n^2$ for each n. Clearly, $\|\frac{x_n}{n}\| \to 0$, so $\frac{x_n}{n} \xrightarrow{w} 0$. Hence, $T(\frac{x_n}{n}) \xrightarrow{w} 0$ in Y and, in particular, the sequence $\{T(\frac{x_n}{n})\}$ is pointwise bounded. By Corollary 6.16, $\{T(\frac{x_n}{n})\}$ is also norm bounded, contrary to $\|T(\frac{x_n}{n})\| \geqslant n$ for each n. Therefore, T must be a bounded (and hence continuous) operator. ∎

Another useful consequence of the Uniform Boundedness Principle is that the pointwise limit of a family of continuous operators is continuous.

6.18 Corollary *Assume that X is a Banach space and Y is a normed space. If a sequence $\{T_n\} \subset L(X, Y)$ satisfies $T_n x \xrightarrow{w} Tx$ in Y for each $x \in X$, then T is a continuous operator.*

Proof: Clearly, the mapping $T : X \to Y$ defined by $Tx = w\text{-}\lim_{n \to \infty} T_n x$ is a linear operator. Next, let $\mathcal{A} = \{T_1, T_2, \ldots\}$. Since the sequence $\{T_n x\}$ is weakly convergent for each x, we see that $\{T_n x\}$ is a norm bounded sequence for each x (see Corollary 6.16). So by the Uniform Boundedness Principle 6.14, there exists some $M > 0$ such that $\|T_n\| \leqslant M$ for each n. Now note that if $\|x\| \leqslant 1$ and $y' \in Y'$, then

$$|\langle T_n x, y' \rangle| \leqslant \|y'\| \cdot \|T_n\| \cdot \|x\| \leqslant M\|y'\|$$

for each n. This implies $|\langle Tx, y' \rangle| \leqslant M\|y'\|$ for each $\|x\| \leqslant 1$ and all $y' \in Y'$. Therefore $\|T(x)\| = \sup_{\|y'\| \leqslant 1} |\langle Tx, y' \rangle| \leqslant M$ for all $x \in X$ with $\|x\| \leqslant 1$, and thus $\|T\| = \sup_{\|x\| \leqslant 1} \|T(x)\| \leqslant M$. This shows that $T \in L(X, Y)$. ∎

6.19 Corollary *If a sequence of continuous linear functionals on a Banach space converges pointwise, then the pointwise limit is a continuous linear functional.*

The Uniform Boundedness Principle can also be employed to establish that all consistent topologies on a dual pair have the same bounded sets. This result is due to G. Mackey [237]. The proof here uses a clever trick to make a subspace of the dual into a Banach space, so that Corollary 6.16 can be applied.

6.20 Theorem (Mackey) *Given a dual pair $\langle X, X' \rangle$, all consistent topologies on X have the same bounded sets.*

Proof: Clearly, every $\tau(X, X')$-bounded subset of X is bounded with respect every consistent topology on X. We must establish that every weakly bounded subset of X is Mackey-bounded. To this end, let A be a $\sigma(X, X')$-bounded subset of X, and let C be a nonempty, convex, circled and weak∗ compact subset of X'. We must show that there exists some $\lambda > 0$ such that $\lambda A \subset C^\circ$.

Consider the subset $E = \bigcup_{n=1}^{\infty} nC$ of X'. Since C is convex and circled, E is a vector subspace of X'. Let $\| \cdot \|$ denote the gauge of C restricted to E. That is,

$$\|x'\| = \inf\{\alpha > 0 : x' \in \alpha C\}, \quad x' \in E.$$

Clearly, $\| \cdot \|$ is a seminorm on E, and we claim that $\| \cdot \|$ is in fact a norm. To see this, assume that $\|x'\| = 0$. This implies that for each n there exists some $0 \leqslant \varepsilon_n < \frac{1}{n}$ and $y'_n \in C$ such that $x' = \varepsilon_n y'_n$. Since C is w^*-compact, there exists a subnet $\{y'_{n_\alpha}\}$ of the sequence $\{y'_n\}$ satisfying $y'_{n_\alpha} \xrightarrow{w^*} y'$ in X'. From $\varepsilon_{n_\alpha} \to 0$, we see that $x' = w^*\text{-}\lim_\alpha \varepsilon_{n_\alpha} y'_{n_\alpha} = 0y' = 0$.

Next, we claim that the closed unit ball under $\| \cdot \|$ is precisely C. Clearly, $\|x'\| \leqslant 1$ for each $x' \in C$. On the other hand, if $\|x'\| \leqslant 1$, then $x' \in (1 + \frac{1}{n})C$ for each n, so for each n we can write $x' = (1 + \frac{1}{n})z'_n$ with $z'_n \in C$. If $z' \in C$ is a weak∗ limit of $\{z'_n\}$, then $x' = z' \in C$. Thus, $C = \{x' \in E : \|x'\| \leqslant 1\}$.

Our next assertion is that $(E, \| \cdot \|)$ is a Banach space. To see this, let $\{x'_n\} \subset E$ be a $\| \cdot \|$-Cauchy sequence. This means that for each $\varepsilon > 0$ there exists some n_0 such that $x'_n - x'_m \in \varepsilon C$ for all $n, m \geqslant n_0$. By passing to a subsequence, we can assume that $x'_{n+1} - x'_n \in \frac{1}{2^{n+1}}C$ for each n. Using once more that C is convex and circled, we see that

$$x'_n = x'_1 + \sum_{i=1}^{n-1}(x'_{i+1} - x'_i) \in x'_1 + \Big(\sum_{i=1}^{n-1} \tfrac{1}{2^{i+1}} \Big)C \subset x'_1 + C$$

for each n. Since $x'_1 + C$ is w^*-compact, the sequence $\{x'_n\}$ has a w^*-accumulation point $x' \in x'_1 + C \subset E$. Also, from

$$x'_{n+k} - x'_n = \sum_{i=n}^{n+k-1}(x'_{i+1} - x'_i) \in \Big(\sum_{i=n}^{n+k-1} \tfrac{1}{2^{i+1}} \Big)C \subset \tfrac{1}{2^n}C,$$

we see that $x' - x'_n \in \frac{1}{2^n}C$ for each n. Thus $\|x'_n - x'\| \leqslant \frac{1}{2^n}$ for each n, which implies that $(E, \| \cdot \|)$ is a Banach space.

Next, note that since C is w^*-compact, every $x \in X$ (as a linear functional on X') is bounded on C. In particular, A can be viewed as a collection of continuous linear functionals on E. By our hypothesis, A is a pointwise bounded collection of continuous linear functionals on E. So by Corollary 6.16, there exists some $\lambda > 0$ such that $\|x\| = \sup_{x' \in C} |\langle x, x' \rangle| \leqslant \frac{1}{\lambda}$ for each $x \in A$. Thus, $|\langle \lambda x, x' \rangle| \leqslant 1$ for each $x \in A$ and each $x' \in C$. In other words, $\lambda A \subset C^\circ$, as desired. ∎

6.5 Weak topologies on normed spaces

In this section, we discuss some important properties of the weak and weak∗ topologies on normed spaces. From now on in this chapter, whenever we refer to a normed space X, we implicitly consider the dual pair $\langle X, X' \rangle$, where X' is the norm dual of X. For instance, when we refer to the weak topology on a normed space X, we mean the $\sigma(X, X')$-topology.

Recall that the closed unit ball of a normed space X is denoted

$$U = \{x \in X : \|x\| \leqslant 1\}.$$

Similarly, the closed unit balls U' and U'' of X' and X'' are defined by

$$U' = \{x' \in X' : \|x'\| \leqslant 1\} \quad \text{and} \quad U'' = \{x'' \in X'' : \|x''\| \leqslant 1\}.$$

Note that U' is norm bounded, convex, circled, and weak∗ closed. (Why?)

It is easy to see from $|x'(x)| \leqslant \|x'\| \cdot \|x\|$ that

$$U^{\circ} = U' \quad \text{and} \quad (U')^{\circ} = U^{\circ\circ} = U.$$

Since, by the definition of X', the norm topology on X is consistent with the dual pair $\langle X, X' \rangle$, we have the following very important special case of Alaoglu's Compactness Theorem 5.105.

6.21 Alaoglu's Theorem *The closed unit ball of the norm dual of a normed space is weak∗ compact. Consequently, a subset of the norm dual of a normed space is weak∗ compact if and only if it is weak∗ closed and norm bounded.*

Be warned that though the closed unit ball in X' is weak∗ compact, the closed unit sphere, $\{x' : \|x'\| = 1\}$, need *not* be weak∗ compact. This is because the norm on X' is not weak∗ continuous, so the unit sphere is not even weak∗ closed, except in the finite dimensional case (see Corollary 6.29 below). However, the dual norm is always weak∗ lower semicontinuous.

6.22 Lemma (Semicontinuity of norms) *If X is a normed space, then the norm function $x \mapsto \|x\|$ is weakly lower semicontinuous on X, and the dual norm function $x' \mapsto \|x'\|$ is weak∗ lower semicontinuous on X'.*

Proof: It is easy to prove these statements directly, but we offer the following clever proofs, which merit study.

First, we consider the norm on X. Since $x \mapsto \|x\|$ is norm continuous, it is also lower semicontinuous. Since the norm is a convex function, Corollary 5.99 implies it is lower semicontinuous in every topology consistent with the dual pair $\langle X, X' \rangle$. In particular, it is weakly lower semicontinuous.

Now for the dual norm. The argument above cannot be used, since X is not generally the norm dual of X'. But by definition, each x is a weak∗ continuous linear functional on X', and hence lower semicontinuous. Since the supremum of a family of lower semicontinuous functions is lower semicontinuous (Lemma 2.41), $x' \mapsto \|x'\| = \sup_{\|x\| \leqslant 1} \langle x, x' \rangle$ is weak* lower semicontinuous. ∎

A consequence of Alaoglu's Theorem 6.21 is that for a normed space X the Mackey topology $\tau(X, X')$ coincides with the norm topology on X.

6.23 Corollary (Norm topology is Mackey) *For a normed space X, the Mackey topology, the strong topology, and the norm topology are the same.*

Proof: Let X be a normed space with norm dual X'. Since the Mackey topology is the strongest locally convex topology consistent with $\langle X, X' \rangle$, it must be at least as strong as the norm topology. On the other hand, the unit ball U' in X' is convex, circled, and by Alaoglu's Theorem 6.21, $\sigma(X', X)$-compact. From the definition of the Mackey topology, the polar of U' is a Mackey neighborhood of zero. But $(U')^\circ$ is the closed unit ball U of X. Therefore, the norm topology is as strong the Mackey topology.

It also follows from Lemma 6.10 that a set in X' is $\sigma(X', X)$-bounded if and only if it is norm bounded. Thus norm convergence implies convergence in the strong topology, so the two are equal. ∎

Theorem 6.21 also sheds some light on the embedding of X into X''.

6.24 Theorem (Embedding X in X'') *For a normed space X:*

1. *The topology $\sigma(X'', X')$ induces $\sigma(X, X')$ on X.*

2. *The closed unit ball U of X is $\sigma(X'', X')$-dense in the closed unit ball U'' of X''.*

3. *The vector space X is $\sigma(X'', X')$-dense in X''.*

Proof: (1) This is just Lemma 2.53.

(2) By Alaoglu's Theorem 6.21, U'' is $\sigma(X'', X')$-compact. So if $\overline{U}$ is the $\sigma(X'', X')$-closure of U in X'', then $\overline{U} \subset U''$.

For the reverse inclusion, assume by way of contradiction that there exists some $x'' \in U'' \setminus \overline{U}$. Since $\overline{U}$ is convex and $\sigma(X'', X')$-compact, Corollary 5.80 and Theorem 5.93 imply that there exists some $x' \in X'$ strictly separating x'' and $\overline{U}$. That is, there is some $c > 0$ such that $x''(x') > c$ and $x'(x) \leqslant c$ for all $x \in U$. In particular, we have $\|x'\| = \sup_{x \in U} x'(x) \leqslant c$. But then, we have $c < x''(x') \leqslant \|x''\| \cdot \|x'\| \leqslant 1 \cdot c = c$, which is impossible. Hence $\overline{U} = U''$.

(3) This follows immediately from part (2). ∎

Since the norm topology on X' is not in general consistent with the dual pair $\langle X, X' \rangle$, it follows that the closed unit ball $U = (U')^\circ$ need not be weakly compact. However, as we show next, U is weakly compact if and only if X is reflexive.

6.25 Theorem (Reflexive Banach spaces) *For a Banach space X, the following statements are equivalent.*

1. *The Banach space X is reflexive.*

2. *The closed unit ball of X is weakly compact.*

3. *The dual Banach space X' is reflexive.*

Proof: (1) $\iff$ (2) Assume first that X is reflexive. Then $U = U''$ and by Alaoglu's Theorem 6.21 the closed unit ball is $\sigma(X'', X')$-compact. So by Theorem 6.24 (1), the closed unit ball U is weakly compact. Conversely, if U is weakly compact, then it follows from Theorem 6.24 (2) that $U = U''$. Hence, $X = X''$.

(3) $\iff$ (1) Clearly, (1) implies (3). Next, assume that X' is reflexive. We know that X is a norm-closed subspace of X'', so X is also $\sigma(X'', X''')$-closed. Since $X''' = X'$, we see that X is $\sigma(X'', X')$-closed. However, by Theorem 6.24 (3), we know that X is also $\sigma(X'', X')$-dense in X''. Therefore, $X = X''$. ∎

6.6 Metrizability of weak topologies

Finite dimensionality can be characterized in terms of weak topologies.

6.26 Theorem (Finite dimensional spaces) *For a normed space X the following are equivalent.*

1. *The vector space X is finite dimensional.*

2. *The weak and norm topologies on X coincide.*

3. *The weak topology on X is metrizable.*

4. *The weak topology is first countable.*

Proof: A finite dimensional space has only one Hausdorff linear topology (Theorem 5.21), so (1) $\implies$ (2). The implications (2) $\implies$ (3) $\implies$ (4) are obvious. It remains to be shown that (4) $\implies$ (1).

So suppose that the weak topology $\sigma(X, X')$ is first countable. Choose a sequence $\{x'_n\}$ in X' such that the sequence of weak neighborhoods $\{V_1, V_2, \ldots\}$, where

$$V_n = \{x \in X : |x'_i(x)| \leq 1 \text{ for } i = 1, \ldots, n\},$$

is a countable base at zero for $\sigma(X, X')$. (Why?)

Now assume by way of contradiction that X is not finite dimensional. We claim that $\bigcap_{i=1}^n \ker x'_i \neq \{0\}$ for each n. For suppose $\bigcap_{i=1}^n \ker x'_i = \{0\}$ for some n. Then $\{0\} = \bigcap_{i=1}^n \ker x'_i \subset \ker x'$ for each $x' \in X'$. By the Fundamental Theorem of Duality 5.91, the functionals $x'_1 \ldots, x'_n$ span X', which implies that X'

is finite dimensional. Consequently, X'' is finite dimensional. (Why?) Since X can be considered to be a vector subspace of X'', X itself is finite dimensional, a contradiction.

Thus, for each n there exists some nonzero $x_n \in \bigcap_{i=1}^{n} \ker x_i'$, which we can normalize so $\|x_n\| = n$. Clearly, $x_n \in V_n$ for each n so $x_n \xrightarrow{w} 0$. In particular, $\{x_n\}$ is pointwise bounded. (Why?) By Corollary 6.15, $\{x_n\}$ is a norm bounded sequence, contrary to $\|x_n\| = n$ for each n. Therefore, X must be finite dimensional.   ∎

For a finite dimensional vector space, we need the hypothesis that the space is Hausdorff to guarantee uniqueness of the topology; see Theorem 5.21. After all, any single nonzero element of the dual generates a weak topology that is not Hausdorff (unless the space is one-dimensional). These topologies are distinct if the generating members of the dual are independent.

6.27 Corollary   *The weak interior of every closed or open ball in an infinite dimensional normed space is empty.*

Proof: Let X be an infinite dimensional normed space, and assume by way of contradiction that there exists a weak neighborhood W of zero and some $u \in U$ such that $u + W \subset U$, where U is the closed unit ball of X. If $w \in W$, then $\|\frac{1}{2}w\| = \frac{1}{2}\|(u+w)-u\| \leqslant 1$, so $\frac{1}{2}W \subset U$. This means that U is a weak neighborhood of zero, so (by Theorem 6.26) X is finite dimensional, a contradiction. Hence the closed unit ball U of X has an empty weak interior.   ∎

Another immediate consequence of Theorem 6.26 is that in an infinite dimensional normed space, the weak topology is strictly weaker than the norm topology. So in this case, there must exist a net $\{x_\alpha\}$ with $x_\alpha \xrightarrow{w} 0$ and $\|x_\alpha\| \nrightarrow 0$. The next lemma exhibits such a net.

6.28 Lemma   *Every infinite dimensional normed space admits a net $\{x_\alpha\}$ satisfying $x_\alpha \xrightarrow{w} 0$ and $\sup\{\|x_\beta\| : \beta \geqslant \alpha\} = \infty$ for each α.*

Proof: Let X be an infinite dimensional normed space and let A denote the collection of all nonempty finite subsets of the norm dual X'. The set A is directed by the set inclusion, $\alpha \geqslant \beta$ whenever $\alpha \supset \beta$. As in the proof of Theorem 6.26, for each $\alpha = \{x_1', \ldots, x_n'\}$ there exists some $x_\alpha \in \bigcap_{i=1}^{n} \ker x_i'$ such that $\|x_\alpha\| = |\alpha|$ (the cardinality of α). Now note that the net $\{x_\alpha\}_{\alpha \in A}$ satisfies the desired properties.   ∎

Note that this line of argument does not guarantee that we can find a *sequence* (rather than a net) converging weakly to zero, but not converging in norm. Indeed, ℓ_1 has the property that if a sequence converges weakly to zero, then it converges to zero in norm; see [12, Theorem 13.1, p. 200]. (This property is called the **Schur property**.) In the same vein we have the following remarkable property.

6.29 Corollary   *In any infinite dimensional normed space, the closed unit sphere is weakly dense in the closed unit ball.*

Proof: Fix $u \in U$ with $\|u\| < 1$ and then alter the proof of Lemma 6.28 by scaling the x_α so that $\|x_\alpha + u\| = 1$ and $x_\alpha \xrightarrow{w} 0$. This implies $x_\alpha + u \xrightarrow{w} u$.   ∎

The next two results deal with separability and metrizability properties of the weak and weak∗ topologies. When we say that a set A is τ-metrizable for some topology τ, we mean that the topological space $(A, \tau|_A)$, where $\tau|_A$ is the relativiza-tion of τ to A, is metrizable. It is quite possible for a subset of a normed space to be weakly metrizable even if the whole space is not. The simplest example is a finite set, which is metrizable by the discrete metric. We now present some more interesting cases.

6.30 Theorem   *A normed space is separable if and only if the closed unit ball of its dual space is w^*-metrizable.*

Proof: Let X be a normed space with unit ball U. First assume that X is separable, so there exists a countable dense set $\{x_1, x_2, \ldots\}$ in U. Let

$$d(x', y') = \sum_{n=1}^{\infty} \frac{1}{2^n} \cdot \left| x'(x_n) - y'(x_n) \right|.$$

Since each x_n lies in U, it follows that $d(x', y') \leqslant \|x'\| + \|y'\|$. Now observe that d is a metric on X'. We claim that d generates w^* on U'. Indeed d induces w^* on any norm bounded subset of X'.

To see this, consider the identity mapping $I : (U', w^*) \to (U', d)$. Since U' is w^*-compact (Alaoglu's Theorem 6.21), it suffices to show that I is continuous (Theorem 2.36). To this end, let $\{x'_\alpha\}$ be a net in U' satisfying $x'_\alpha \xrightarrow{w^*} x'$ and let $\varepsilon > 0$. Fix some k such that $\sum_{n=k+1}^{\infty} \frac{1}{2^n} < \varepsilon$. Since each $x'_\alpha \in U'$ and $x_n \in U$, we have $|x'_\alpha(x_n) - x'(x_n)| \leqslant 2$, so

$$d(x'_\alpha, x') \leqslant \sum_{n=1}^{k} \left| x'_\alpha(x_n) - x'(x_n) \right| + 2\varepsilon.$$

Since $x'_\alpha(x_n) \xrightarrow{\alpha} x'(x_n)$, we see that $\limsup_\alpha d(x'_\alpha, x') \leqslant 2\varepsilon$ for all $\varepsilon > 0$. Thus $\lim_\alpha d(x'_\alpha, x') = 0$, as desired. Since every bounded subset of X' lies in a closed ball, which is w^*-compact by Alaoglu's Theorem 6.21, the preceding argument shows that d metrizes w^* on every bounded subset of X'.

For the converse, assume that (U', w^*) is a compact metrizable space. Choose a sequence $\{x_n\}$ in X such that the w^*-neighborhoods of zero

$$V_n = \{x' \in U' : |x'(x_i)| \leqslant 1 \text{ for all } 1 \leqslant i \leqslant n\}, \ n = 1, 2, \ldots,$$

satisfy $\bigcap_{n=1}^{\infty} V_n = \{0\}$. (Why is this possible?) Let Y denote the closure of the linear subspace generated by $\{x_1, x_2, \ldots\}$. We claim that $Y = X$. If $Y \neq X$, then by Corollary 5.80 there exists some nonzero $x' \in U'$ that vanishes on Y. This implies $x' \in V_n$ for each n, so $x' = 0$, which is a contradiction. Hence $Y = X$. Now note that the set of all finite linear combinations of $\{x_1, x_2, \ldots\}$ with rational coefficients is a countable dense subset of X.   ∎

In a similar fashion, we can establish the following result.

6.31 Theorem   *The dual X' of a Banach space X is separable if and only if the unit ball of X is weakly metrizable.*

Proof: See [12, Theorem 10.8, p. 153].   ∎

The next result describes one more interesting metrizability property of the weak topology.

6.32 Theorem   *If the dual X' of a normed space X includes a countable total set, then every weakly compact subset of X is weakly metrizable.*

Proof: Let $\{x_1', x_2', \ldots\}$ be a countable total subset of X'. We can assume that $\|x_n'\| \leqslant \frac{1}{2^n}$ for each n. Notice that the formula

$$d(x, y) = \sum_{n=1}^{\infty} \left| x_n'(x - y) \right|$$

defines a metric on X.

Now let W be a weakly compact subset of X. We claim that the metric d induces the topology $\sigma(X, X')$ on W. To see this, consider the identity mapping $I: (W, w) \to (W, d)$. In view of Theorem 2.36, it suffices to show that I is continuous. To this end, let $x_\alpha \xrightarrow{w} x$ in W and let $\varepsilon > 0$. Since W is norm bounded (why?), there exists some k such that $\sum_{n=k+1}^{\infty} |x_n'(x_\alpha - x)| < \varepsilon$. This implies $d(x_\alpha, x) \leqslant \sum_{n=1}^{k} |x_n'(x_\alpha - x)| + \varepsilon$ for each α, from which it follows that $\limsup_\alpha d(x_\alpha, x) \leqslant \varepsilon$ for each $\varepsilon > 0$. Thus, $\lim_\alpha d(x_\alpha, x) = 0$, so the identity is (w, d)-continuous, and the proof is finished.   ∎

We close the section by stating four important theorems dealing with weak compactness in normed spaces. Recall that a subset of a topological space is **relatively compact** if its closure is compact.

6.33 Grothendieck's Theorem [143]   *A subset A of a Banach space X is relatively weakly compact if and only if for each $\varepsilon > 0$ there exists a weakly compact set W such that $A \subset W + \varepsilon U$, where U denotes the closed unit ball of X.*

Proof: See [12, Theorem 10.17, p. 159].   ∎

6.34 Eberlein–Šmulian Theorem [112, 315]   *In the weak topology on a normed space, compactness and sequential compactness coincide. That is, a subset A of a normed space X is relatively weakly compact (respectively, weakly compact) if and only if every sequence in A has a weakly convergent subsequence in X (respectively, in A).*

Proof: See [12, Theorem 10.13, p. 156].   ∎

6.35 Krein–Šmulian Theorem [216]   *In a Banach space, both the convex circled hull and the convex hull of a relatively weakly compact set are relatively weakly compact sets.*

Proof: See [12, Theorem 10.15, p. 158].   ∎

The next theorem is extremely deep.

6.36 James' Theorem [178]   *A nonempty weakly closed bounded subset of a Banach space is weakly compact if and only if every continuous linear functional attains a maximum on the set.*

Proof: See [166, Section 19, pp. 157–161].   ∎

Corollary 7.81 asserts that if $F \subset X'$ is finite, then every continuous linear functional in $\operatorname{co} F$ attains its maximum on F°. This result does not generalize from finite sets to the closed unit ball of X'. To see this, observe that since the closed unit ball U of X is the polar of the closed unit ball in the dual, if every functional in U' attains its maximum, then James' Theorem 6.36 implies that the closed unit ball U is weakly compact, so by Theorem 6.25 the space must be reflexive. We show later on that ℓ_1, for instance, is not reflexive.

6.7   Continuity of the evaluation

From the point of view of economic theory, one of the main differences between finite and infinite dimensional vector spaces is the continuity of the evaluation map. Let $\langle X, X' \rangle$ be a dual pair, and consider the evaluation map $(x, x') \mapsto \langle x, x' \rangle$. If X is finite dimensional, then the evaluation map is (jointly) continuous. Since finite dimensional spaces have only one Hausdorff linear topology, the choice of topology is not an issue. For normed spaces, the evaluation is jointly continuous for the norm topologies. As we are about to see, giving one of the spaces its weak topology destroys the global joint continuity of the evaluation, but it survives on compact sets.

6.37 Theorem   *Let X be a normed space with norm dual X'. Then the duality $(x, x') \mapsto \langle x, x' \rangle$, from $X \times X'$ to $\mathbb{R}$, is jointly norm continuous.*

Proof: It suffices to prove continuity at zero. By Lemma 6.4, if $\|x_n\| \to 0$ and $\|x_n'\| \to 0$, then $|\langle x_n, x_n' \rangle| \leqslant \|x_n\| \cdot \|x_n'\| \to 0$.   ∎

With the weak topology on an infinite dimensional space things are different.

6.38 Theorem   *Let X be an infinite dimensional normed space with norm dual X'. Then the evaluation $(x, x') \mapsto \langle x, x' \rangle$ from $X \times X'$ to $\mathbb{R}$ is not jointly continuous if either space is given its weak topology for the dual pair and the other space its norm topology.*

Proof: We first consider the case where X is given its $\sigma(X, X')$-topology and X' its norm topology. As in the proof of Lemma 6.28, we can find a net $\{x_\alpha\}$ indexed by the finite subsets of X' with $x_\alpha \xrightarrow{\sigma(X,X')} 0$ and $\|x_\alpha\| = |\alpha|$ (the cardinality of α).

Next, for each α, there exists a continuous linear functional $f_\alpha \in X'$ with $\|f_\alpha\| \leqslant 1$ satisfying $f_\alpha(x_\alpha) = \|x_\alpha\| = |\alpha|$; cf. Lemma 6.10. Now let $x_\alpha' = \frac{f_\alpha}{|\alpha|}$, and note that $\|x_\alpha'\| \to 0$. By construction, the equality $\langle x_\alpha, x_\alpha' \rangle = 1$ holds for each α. But $(x_\alpha, x_\alpha') \xrightarrow{\sigma(X,X') \times \|\cdot\|} (0, 0)$, so the evaluation is not jointly continuous.

Next we consider the case where X' is endowed with its $\sigma(X', X)$-topology and X its norm topology. In this case, just as before, we construct a net $\{x_\alpha'\}$ indexed by finite subsets of X such that $x_\alpha' \xrightarrow{w^*} 0$ and $\|x_\alpha'\| = |\alpha|$. Now use the fact that $\|x'\| = \sup\{\langle x, x' \rangle : \|x\| \leqslant 1\}$ to find y_α satisfying $\langle y_\alpha, x_\alpha' \rangle \geqslant \frac{1}{2}|\alpha|$ and $\|y_\alpha\| \leqslant 1$. Put $x_\alpha = \frac{2}{|\alpha|} y_\alpha$. Then $\|x_\alpha\| \to 0$ and $\langle x_\alpha, x_\alpha' \rangle \geqslant 1$ for all α. So the evaluation is not jointly continuous whenever X is given its norm topology and X' is given the weak∗ topology.   ∎

Note that we may replace the norm topology in the preceding theorem with any weaker topology and the evaluation still fails to be jointly continuous. However, the evaluation is jointly continuous on certain restricted subsets.

6.39 Theorem   *Let $\langle X, X' \rangle$ be a dual pair and τ be a consistent topology on X. Let V be a τ-neighborhood of zero. Then the evaluation $\langle \cdot, \cdot \rangle$ restricted to $X \times V^\circ$ is jointly continuous in the $\tau \times \sigma(X', X)$-topology.*

Proof: Fix $\varepsilon > 0$ and let $x_\alpha \xrightarrow{\tau} x$ and $x_\alpha' \xrightarrow{\sigma(X',X)} x'$, where $\{x_\alpha'\} \subset V^\circ$. Then

$$|\langle x_\alpha, x_\alpha' \rangle - \langle x, x' \rangle| \leqslant |\langle x_\alpha, x_\alpha' \rangle - \langle x, x_\alpha' \rangle| + |\langle x, x_\alpha' \rangle - \langle x, x' \rangle|.$$

Since $x_\alpha \xrightarrow{\tau} x$, eventually $x_\alpha - x \in \frac{\varepsilon}{2}V$, so $|\langle x_\alpha, x_\alpha' \rangle - \langle x, x_\alpha' \rangle| \leqslant \frac{\varepsilon}{2}$, since each $x_\alpha' \in V^\circ$. Since $x_\alpha' \xrightarrow{\sigma(X',X)} x'$, eventually $|\langle x, x_\alpha' \rangle - \langle x, x' \rangle| < \frac{\varepsilon}{2}$. Therefore, eventually $|\langle x_\alpha, x_\alpha' \rangle - \langle x, x' \rangle| < \varepsilon$.   ∎

6.40 Corollary   *Let X be a Banach space and B a norm bounded subset of X'. Then the evaluation $\langle \cdot, \cdot \rangle$ restricted to $X \times B$ is jointly continuous, where X has its norm topology and B has its w^*-topology.*

There is a dual version of Corollary 6.40, and we leave its proof as an exercise.

6.41 Theorem   *Let B be a norm bounded subset of the Banach space X. Then the evaluation $(x, x') \mapsto \langle x, x' \rangle$ restricted to $B \times X'$ is jointly continuous when B is endowed with the weak topology and X' with its norm topology.*

6.8   Adjoint operators

The study of operators plays an important role in functional analysis and its applications. Here we discuss briefly a few concepts and results associated with (linear) operators. These results are employed extensively in Chapter 19.

Let $T: X \to Y$ be an operator between two vector spaces and let X^* and Y^* denote the algebraic duals of X and Y respectively. Every $y^* \in Y^*$ gives rise to a real function T^*y^* on X defined pointwise via the formula $T^*y^* = y^* \circ T$. Clearly T^*y^* is linear and so belongs to X^*. It is also easy to verify that the mapping $y^* \mapsto T^*y^*$ from Y^* to X^* is linear, that is, $T^*(\alpha y^* + \beta z^*) = \alpha T(y^*) + \beta T(z^*)$ for all $y^*, z^* \in Y^*$ and all $\alpha, \beta \in \mathbb{R}$. Thus, the operator $T: X \to Y$ defines a companion operator $T^*: Y^* \to X^*$ via the formula $T^*y^*(x) = y^*(Tx)$ for all $y^* \in Y^*$ and all $x \in X$.

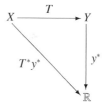

6.42 Definition   *The operator T^* is called the **algebraic adjoint** of T and is defined by $T^*y^* = y^* \circ T$, or equivalently via the duality identity*

$$\langle x, T^*y^* \rangle = \langle Tx, y^* \rangle,$$

where $x \in X$ and $y^* \in Y^*$.

The next result offers a very simple criterion for deciding whether a linear operator is weakly continuous. You only have to check that its adjoint carries continuous functionals into continuous functionals.

6.43 Theorem (**Weak continuity and adjoints**)   *Let $\langle X, X' \rangle$ and $\langle Y, Y' \rangle$ be dual pairs (of not necessarily normed spaces) and let $T: X \to Y$ be a linear operator, where X and Y are endowed with their weak topologies. Then T is (weakly) continuous if and only if the algebraic adjoint T^* satisfies $T^*(Y') \subset X'$.*

Proof: If $T^*(Y') \subset X'$ and $x_\alpha \xrightarrow{\sigma(X,X')} 0$, then for each $y' \in Y'$ we have

$$\langle Tx_\alpha, y' \rangle = \langle x_\alpha, T^*y' \rangle \to 0,$$

which shows that $Tx_\alpha \xrightarrow{\sigma(Y,Y')} 0$. That is, T is weakly continuous.

For the converse, assume that T is weakly continuous. Let $x_\alpha \xrightarrow{\sigma(X,X')} 0$ and $y' \in Y'$. Then $\langle x_\alpha, T^*y' \rangle = \langle Tx_\alpha, y' \rangle \to 0$, so T^*y' is $\sigma(X, X')$-continuous on X. By Theorem 5.93, T^*y' belongs to X'. Thus, $T^*(Y') \subset X'$.   ∎

6.44 Definition   *Let $\langle X, X' \rangle$ and $\langle Y, Y' \rangle$ be dual pairs and let $T : X \to Y$ be a weakly continuous operator. Then the adjoint $T^* : Y^* \to X^*$ restricted to Y' is called the* **topological adjoint** (*or simply the* **adjoint** *of T*) *and is denoted T'.*

Now consider a continuous operator $T : X \to Y$ between two normed vector spaces. Then, by Theorems 6.17 and 6.43, we see that $T^*(Y') \subset X'$ (where X' and Y' are the norm duals of X and Y, respectively), so $T' = T^*|_{Y'}$ maps Y' into X'. In this case, T' is simply called the (**norm**) **adjoint** of T. It is easy to see that T and T' have the same norm. Indeed,

$$\|T'\| = \sup_{\|y'\| \leqslant 1} \|T'y'\| = \sup_{\|y'\| \leqslant 1} \left(\sup_{\|x\| \leqslant 1} |\langle x, T'y' \rangle| \right)$$

$$= \sup_{\|x\| \leqslant 1} \left(\sup_{\|y'\| \leqslant 1} |\langle Tx, y' \rangle| \right) = \sup_{\|x\| \leqslant 1} \|Tx\| = \|T\|.$$

In other words, for normed spaces the mapping $T \mapsto T'$ (where T' is the norm adjoint of T) is a linear isometry from $L(X, Y)$ into $L(Y', X')$.

The adjoint of the operator $T' : Y' \to X'$ is called the **second adjoint** of T and is denoted T''. Therefore, the second adjoint $T'' : X'' \to Y''$ satisfies the duality identity

$$\langle y', T''x'' \rangle = \langle T'y', x'' \rangle, \quad y' \in Y', \ x'' \in X''.$$

In particular, if $T : X \to Y$ is a continuous operator between normed spaces, and if we consider X and Y to be embedded in the natural way in X'' and Y'' respectively, then $T''x = Tx$ for each $x \in X$. In other words, the second adjoint operator $T'' : X'' \to Y''$ is the (unique) norm-continuous linear extension of T.

6.9   Projections and the fixed space of an operator

Given a linear operator T from a vector space X into itself, an **eigenvector** of T is a nonzero vector $x \in X$ for which there exists a scalar λ satisfying $Tx = \lambda x$. The scalar λ is an **eigenvalue** of T associated with x. By definition, the vector zero is never an eigenvector. Even in the finite dimensional case, real eigenvalues and eigenvectors need not exist. The **eigenspace** of T associated with the eigenvalue λ is the span of the eigenvectors associated with λ. If X is finite dimensional, it is especially useful to count the dimension of the eigenspace associated with an eigenvalue, which is the called the multiplicity of the eigenvalue. Fixed points of T are eigenvectors associated with the eigenvalue 1, and the eigenspace associated with 1 is called the **fixed space** $\mathcal{F}_T$ of T. That is,

$$\mathcal{F}_T = \{x \in X : T(x) = x\}.$$

If X is a normed space, then the fixed space of a continuous operator is a closed subspace of X.

6.45 Lemma   *If $T: X \to X$ is a continuous linear operator, then the fixed space of the adjoint operator T' is given by*

$$\mathcal{F}_{T'} = \{x' \in X' : x' \text{ vanishes on the range of } I - T\},$$

where I is the identity operator on X.

Proof: The conclusion follows easily from the observation that

$$\langle x - Tx, x' \rangle = \langle x, x' - T'x' \rangle$$

for all $x \in X$ and all $x' \in X'$.   ∎

When we compose any mapping with itself, it is traditional to use a superscript to indicate the fact. Thus $T^2 = T \circ T$, and $T^3 = T \circ T \circ T$, etc. A mapping is **idempotent** if it satisfies $T^2 = T$. Idempotent operators are also called projections.

6.46 Definition   *A linear operator $P: X \to X$ on a vector space is called a* **projection** *if $P^2 = P$. Equivalently, an operator P is a projection if and only if P coincides with the identity operator on the range of P.*

Clearly the fixed space of a projection is its range. Continuous projections are associated with complemented closed subspaces. Recall that a closed subspace Y of a Banach space is **complemented** if there exists another closed subspace Z such that $X = Y \oplus Z$; the closed subspace Z is called a **complement** of Y.

6.47 Theorem   *A closed subspace Y of a Banach space X is complemented if and only if it is the range of a continuous projection on X.*

Proof: If $P: X \to X$ is a continuous projection having range Y, then consider the closed subspace $Z = (I - P)(X)$ and note that $X = Y \oplus Z$.

For the converse, assume that $X = Y \oplus Z$, where Z is a closed subspace of X. For each $x \in X$, there exist $x_1 \in Y$ and $x_2 \in Z$ (both uniquely determined) such that $x = x_1 + x_2$. Now define the operator $P: X \to X$ by $Px = x_1$. Clearly, $P^2 = P$ and P has range Y.

To finish the proof, we must show that P is continuous. For this, it suffices to establish that P has closed graph (Theorem 5.20). So assume $x_n \to x$ and $Px_n \to y$. Since Y is closed, we get $y \in Y$. On the other hand, $x_n - Px_n \in Z$ for each n, so $x_n - Px_n \to x - y$. Now the closedness of Z implies $z = x - y \in Z$. That is, we have written $x = y + z$ with $y \in Y$ and $z \in Z$. By the definition of P, we get $y = Px$, and the proof is finished.   ∎

6.10 Hilbert spaces

Hilbert spaces form a special class of Banach spaces where the norm is derived from an *inner product*. All finite dimensional Euclidean spaces are Hilbert spaces.

6.48 Definition   *A (real) inner product space is a real vector space X equipped with a function $(\cdot, \cdot): X \times X \to \mathbb{R}$ such that:*

1. $(x, x) \geqslant 0$ *for all $x \in X$ and $(x, x) = 0$ if and only if $x = 0$.*

2. $(x, y) = (y, x)$ *for all $x, y \in X$.*

3. $(\alpha x + \beta y, z) = \alpha(x, z) + \beta(y, z)$ *for all $x, y, z \in X$ and all $\alpha, \beta \in \mathbb{R}$.*

We shall see that the function $x \mapsto \|x\| = \sqrt{(x, x)}$, from X to $[0, \infty)$, is a norm, the **norm induced by the inner product**. To prove this we need to establish an inequality known as the Cauchy–Schwarz inequality.

6.49 Lemma (Cauchy–Schwarz Inequality)   *If X is an inner product space, then for all $x, y \in X$ we have*

$$|(x, y)| \leqslant \|x\| \cdot \|y\|.$$

Equality holds if and only if the vectors x and y are linearly dependent.

Proof: Let $x, y \in X$ be nonzero and define the quadratic function $Q: \mathbb{R} \to \mathbb{R}$ by

$$Q(\lambda) = \|x + \lambda y\|^2 = \|y\|^2 \lambda^2 + 2(x, y)\lambda + \|x\|^2.$$

Clearly, $Q(\lambda) \geqslant 0$ for each $\lambda \in \mathbb{R}$. So the discriminant of the quadratic is nonpositive, that is, $4|(x, y)|^2 - 4\|x\|^2 \cdot \|y\|^2 \leqslant 0$ or $|(x, y)| \leqslant \|x\| \cdot \|y\|$. Equality holds if and only if the quadratic has a zero, that is, if and only if there exists some $\lambda \in \mathbb{R}$ such that $x + \lambda y = 0$, which is, of course, equivalent to having the vectors x and y linearly dependent. ∎

With the help of the Cauchy–Schwarz inequality we can show that the function $x \mapsto \|x\|$ is indeed a norm.

6.50 Lemma   *If X is an arbitrary inner product space, then the real function $x \mapsto \|x\| = \sqrt{(x, x)}$ is a norm on X.*

Proof: The only thing that needs verification is the triangle inequality. To this end, let $x, y \in X$ and then use the Cauchy–Schwarz inequality to get

$$\|x + y\|^2 = \|x\|^2 + 2(x, y) + \|y\|^2 \leqslant \|x\|^2 + 2\|x\| \cdot \|y\| + \|y\|^2 = (\|x\| + \|y\|)^2.$$

This implies $\|x + y\| \leqslant \|x\| + \|y\|$. ∎

Inner product spaces also satisfy an identity known as the parallelogram law.

6.51 Lemma (Parallelogram Law)   *If X is an inner product space, then for each $x, y \in X$ we have*

$$\|x + y\|^2 + \|x - y\|^2 = 2\|x\|^2 + 2\|y\|^2.$$

Proof: Note that $\|x+y\|^2 = \|x\|^2 + 2(x, y) + \|y\|^2$ and $\|x-y\|^2 = \|x\|^2 - 2(x, y) + \|y\|^2$. Adding these two identities yields $\|x + y\|^2 + \|x - y\|^2 = 2\|x\|^2 + 2\|y\|^2$. ∎

The parallelogram law, which is a simple consequence of the Pythagorean Theorem, asserts that the sum of the squares of the lengths of the diagonals of a parallelogram is equal to the sum of the squares of the lengths of the sides. Consider the parallelogram with vertices $0, x, y, x + y$. Its diagonals are the segments $[0, x + y]$ and $[x, y]$, and their lengths are $\|x + y\|$ and $\|x - y\|$. It has two sides of length $\|x\|$ and two of length $\|y\|$.

In fact, a norm on a vector space is induced by an inner product if and only if it satisfies the parallelogram law; see for instance [14, Problem 32.10, p. 303].

The definition of a Hilbert space is next.

6.52 Definition   *An inner product space is called a **Hilbert space** if it is a Banach space under the norm induced by its inner product. (That is, the induced metric is complete.)*

The two classical examples of Hilbert spaces are the spaces $\mathbb{R}^n$ equipped with the Euclidean inner product $(x, y) = \sum_{i=1}^{n} x_i y_i$ and the Banach space ℓ_2 of all square summable sequences of real numbers with the inner product $(x, y) = \sum_{i=1}^{\infty} x_i y_i$.

We now come to a basic result regarding closed convex subsets of Hilbert spaces.

6.53 Theorem (Nearest Point Theorem)   *If C is a nonempty closed convex subset of a Hilbert space H, then for each $x \in H$ there exists a unique vector $\pi_C(x) \in C$ satisfying $\|x - \pi_C(x)\| \leqslant \|x - y\|$ for all $y \in C$.*

Proof: We can assume that $x = 0$. Put $d = \inf_{u \in C} \|u\|$ and then select a sequence $\{u_n\} \subset C$ such that $\|u_n\| \to d$. From the parallelogram law

$$\|u_n - u_m\|^2 = 2\|u_n\|^2 + 2\|u_m\|^2 - 4\left\|\tfrac{u_n + u_m}{2}\right\|^2$$
$$\leqslant 2\|u_n\|^2 + 2\|u_m\|^2 - 4d^2 \xrightarrow[n,m \to \infty]{} 0,$$

we see that $\{u_n\}$ is a Cauchy sequence. If $u_n \to u$ in H, then $u \in C$ and $\|u\| = d$. This establishes the existence of a point in C nearest zero.

For the uniqueness of the nearest point, assume that some $v \in C$ satisfies $\|v\| = d$. Then using the parallelogram law once more, we get

$$0 \leqslant \|u - v\|^2 = 2\|u\|^2 + 2\|v\|^2 - 4\left\|\tfrac{u+v}{2}\right\|^2 \leqslant 2d^2 + 2d^2 - 4d^2 = 0,$$

so $\|u - v\| = 0$ or $u = v$.   ∎

Figure 6.1.

The point $\pi_C(x)$ of C nearest x is called the **metric projection** of x on C, and the mapping $\pi_C : H \to C$ is referred to as the **(metric) projection of H onto C**. The geometrical illustration of the nearest point is shown in Figure 6.1. The properties of the nearest point and the projection are included in the next result. When C is a linear subspace, then the metric projection is a projection in the sense of Definition 6.46.

6.54 Lemma   *For a nonempty closed convex subset C of a Hilbert space H the metric projection mapping $\pi_C : H \to C$ satisfies the following properties.*

a. *If $x \in C$, then $\pi_C(x) = x$.*

b. *If $x \notin C$, then $\pi_C(x) \in \partial C$.*

c. *For each $x \in H$ and each $y \in C$ we have $(x - \pi_C(x), y - \pi_C(x)) \leqslant 0$. In other words, the hyperplane through $\pi_C(x)$ defined by $x - \pi_C(x)$,*

$$\{z \in H : (x - \pi_C(x), z) = (x - \pi_C(x), \pi_C(x))\},$$

strongly separates x from C.

d. *For all $x, y \in H$ we have $\|\pi_C(x) - \pi_C(y)\| \leqslant \|x - y\|$. In particular, π_C is uniformly continuous and C is a retract of H under π_C.*

e. *If C is a closed vector subspace of H, then for each $x \in H$ the vector $x - \pi_C(x)$ is orthogonal to C, that is, $(x - \pi_C(x), c) = 0$ for all $c \in C$. In this case $\pi_C(x)$ is also called the **orthogonal projection** of x on C.*

Proof: (a)   If $x \in C$, then clearly $\pi_C(x) = x$.

(b)   Let $x \notin C$ and note that the point $x_\lambda = \lambda x + (1 - \lambda)\pi_C(x) \in H$ satisfies $\|x - x_\lambda\| = (1 - \lambda)\|x - \pi_C(x)\| < \|x - \pi_C(x)\|$ for each $0 < \lambda < 1$. So if $\pi_C(x)$ were an interior point of C, then for small λ the vector x_λ belongs to C and is closer to x than $\pi_C(x)$, a contradiction. Hence, $\pi_C(x) \in \partial C$.

(c)   Let $x \in H$, $y \in C$ and $0 < \alpha < 1$. From the definition of $\pi_C(x)$, we get

$$\begin{aligned}
\left\|x - \pi_C(x)\right\|^2 &\leqslant \left\|x - [\alpha y + (1 - \alpha)\pi_C(x)]\right\|^2 \\
&= \left\|[x - \pi_C(x)] - \alpha[y - \pi_C(x)]\right\|^2 \\
&= \left\|x - \pi_C(x)\right\|^2 - 2\alpha(x - \pi_C(x), y - \pi_C(x)) + \alpha^2\left\|y - \pi_C(x)\right\|^2.
\end{aligned}$$

This implies $0 \leqslant -2\alpha(x - \pi_c(x), y - \pi_c(x)) + \alpha^2 \|y - \pi_c(x)\|^2$ or

$$-2(x - \pi_c(x), y - \pi_c(x)) + \alpha \|y - \pi_c(x)\|^2 \geqslant 0.$$

Letting $\alpha \downarrow 0$ yields $-2(x - \pi_c(x), y - \pi_c(x)) \geqslant 0$ or $(x - \pi_c(x), y - \pi_c(x)) \leqslant 0$.

(d) Let $x, y \in H$. Replacing y in (c) with $\pi_c(y) \in C$, we get

$$(x - \pi_c(x), \pi_c(y) - \pi_c(x)) \leqslant 0. \qquad (\star)$$

Exchanging x and y in $(\star)$ yields $(y - \pi_c(y), \pi_c(x) - \pi_c(y)) \leqslant 0$ or

$$(\pi_c(y) - y, \pi_c(y) - \pi_c(x)) \leqslant 0. \qquad (\star\star)$$

Adding $(\star)$ and $(\star\star)$, we obtain

$$\begin{aligned}
0 &\geqslant (x - \pi_c(x), \pi_c(y) - \pi_c(x)) + (\pi_c(y) - y, \pi_c(y) - \pi_c(x)) \\
&= (x - y, \pi_c(y) - \pi_c(x)) + \|\pi_c(y) - \pi_c(x)\|^2.
\end{aligned}$$

From this and the Cauchy–Schwarz inequality, we get

$$\begin{aligned}
\|\pi_c(y) - \pi_c(x)\|^2 &\leqslant (y - x, \pi_c(y) - \pi_c(x)) \\
&\leqslant \|y - x\| \cdot \|\pi_c(y) - \pi_c(x)\|.
\end{aligned}$$

This implies $\|\pi_c(x) - \pi_c(y)\| \leqslant \|x - y\|$ for all $x, y \in H$.

(e) Fix $c \in C$. Since C is a vector subspace, we have $\pi_c(x) + \lambda(\pm c) \in C$ for all $\lambda \in \mathbb{R}$. It follows that

$$\begin{aligned}
\|x - \pi_c(x)\|^2 &\leqslant \|x - [\pi_c(x) + \lambda(\pm c)]\|^2 \\
&= \|x - \pi_c(x)\|^2 \mp 2\lambda(x - \pi_c(x), c) + \lambda^2 \|c\|^2
\end{aligned}$$

or $\pm 2\lambda(x - \pi_c(x), c) \leqslant \lambda^2 \|c\|^2$ for all $\lambda \in \mathbb{R}$. So $\pm(x - \pi_c(x), c) \leqslant \frac{\lambda}{2} \|c\|^2$ for all $\lambda > 0$ and by letting $\lambda \downarrow 0$ we get $\pm(x - \pi_c(x), c) \leqslant 0$ or $(x - \pi_c(x), c) = 0$ for all $c \in C$. ∎

As a first application of the preceding result we shall obtain a characterization of the norm dual of a Hilbert space.

6.55 Corollary (F. Riesz) *If H is a Hilbert space and $f \in H'$, then there exists a unique vector $y_f \in H$ such that $f(x) = (x, y_f)$ holds for all $x \in H$. Moreover, the mapping $f \mapsto y_f$, from H' to H, is a surjective linear isometry (so subject to this linear isometry we can write $H' = H$).*

Proof: Let $f \in H'$. If $f(x) = (x, y) = (x, z)$ for all $x \in H$, then $(x, y - z) = 0$ for all $x \in H$. Letting $x = y - z$ we get $\|y - z\|^2 = 0$ or $y = z$. This establishes the uniqueness of the representing vector y_f.

For the existence of the vector y_f we consider two cases. If $f = 0$, then clearly $y_f = 0$ is the desired vector. So we can assume $f \neq 0$. In this case, if $C = \ker f$, then C is a proper closed subspace of H, so by part (e) of Lemma 6.54 there exists a unit vector $z \in H$ satisfying $(u, z) = 0$ for all $u \in C$. Now notice that $z \notin C$ and that for each $x \in H$ we have $x - \frac{f(x)}{f(z)} z \in C$. This implies $(x - \frac{f(x)}{f(z)} z, z) = 0$ or $f(x) = (x, f(z)z)$ for all $x \in H$, so $y_f = f(z)z$. The rest of the proof can be completed by using the Cauchy–Schwarz inequality and is left for the reader. $\blacksquare$

For any nonempty subset A of a Hilbert space H its **orthogonal complement** $A^\perp$ is the set

$$A^\perp = \{x \in H : (x, a) = 0 \text{ for all } a \in A\}.$$

Note that the orthogonal complement of a set A is simply the annihilator of the set A as introduced in Definition 5.106. Clearly, $A^\perp$ is a closed vector subspace of H. The orthogonal complement of $A^\perp$ is denoted $A^{\perp\perp}$, that is, $A^{\perp\perp} = (A^\perp)^\perp$. When A itself is a closed linear subspace, then the orthogonal complement is a complementary subspace in the sense of Definition 5.88. That is, the Hilbert space is the direct sum of any closed linear subspace and its orthogonal complement. Moreover, for a linear subspace, its orthogonal complement is also its polar. These two basic properties of orthogonal complements are summarized in the next result.

6.56 Lemma   *If M is a closed vector subspace of a Hilbert space H, then*

1. $M^{\perp\perp} = M$, *and*

2. *The orthogonal complement $M^\perp$ of M is indeed a complement of M, that is,* $M \oplus M^\perp = H$.

Proof: (1)   Clearly, $M \subset M^{\perp\perp}$. If $M \neq M^{\perp\perp}$, then (by part (e) of Lemma 6.54) there exists a non zero vector $u \in M^{\perp\perp}$ such that $u \in M^\perp$. It follows that $(u, u) = 0$ or $u = 0$, which is a contradiction. Hence $M = M^{\perp\perp}$.

(2)   A straightforward verification shows that $M \oplus M^\perp$ is a closed vector subspace. If $M \oplus M^\perp \neq H$, then there exists some nonzero vector $v \in H$ such that v is orthogonal to $M \oplus M^\perp$. It follows that $v \in M^\perp \cap M^{\perp\perp} = \{0\}$, which is impossible. Hence $M \oplus M^\perp = H$. $\blacksquare$

6.57 Corollary   *Every proper linear subspace of a finite dimensional vector space is a finite intersection of hyperplanes—and therefore a polyhedron.*

Proof: Let M be a proper linear subspace of a finite dimensional vector space X. If $\{a_1, \ldots, a_k\}$ is a Hamel basis of $M^\perp$, then

$$M = M^{\perp\perp} = \{x \in X : (x, a_i) = 0 \text{ for all } i = 1, \ldots, k\}.$$

This shows that M is a finite intersection of hyperplanes. $\blacksquare$

Chapter 7

Convexity

This chapter provides an introduction to *convex analysis*, the properties of convex sets and functions. We start by taking the convexity of the epigraph to be the definition of a convex function, and allow convex functions to be extended-real valued. Any real-valued convex function on a convex set can be extended to the entire vector space by setting it to ∞ where it was previously undefined. The set of points where a convex function does not assume the value ∞ is its *effective domain*. If the effective domain is not empty and the convex function does not assume the value $-\infty$, then it is a *proper* convex function.

By Theorem 5.98 the collection of closed convex sets is the same for all topologies consistent with a given dual pair. Consequently, if a convex function is lower semicontinuous in one consistent topology, then it is lower semicontinuous in every consistent topology. If a convex function is continuous on its effective domain, and the domain is closed, then its extension is lower semicontinuous everywhere. Thus lower semicontinuous proper convex functions are especially interesting. A lower semicontinuous proper convex function is the pointwise supremum of the continuous *affine functions* it dominates (Theorem 7.6).

One of the main themes of this chapter is the maximization of linear functions over subsets of a locally convex space. This is also a recurring theme in economics, where linear functionals are interpreted as prices, and profit maximization and cost minimization are key concepts. The *support functional* of a set assigns to each continuous linear functional its supremum over the set. This supremum may be ∞, which is a prime motive for allowing convex functions to be extended valued. The support functional of any set and its closed convex hull are identical. Since a closed convex subset of a locally convex space is the intersection of the closed half spaces that include it, it is characterized by its support functionals, which encapsulates this information. Thus there is a one-to-one correspondence between closed convex sets and support functionals. Convex sets are partially ordered by inclusion and support functions are ordered pointwise, and the correspondence between them preserves the order structure. See Theorems 7.52 and 7.51 and the following discussion. Even the Hausdorff metric on the space of closed convex subsets of a normed space can be defined in terms of support functionals (Lemma 7.58).

Points at which a nonzero linear functional attains a maximum over a set are *support points* of the set. The associated hyperplane on which the support point lies is called a *supporting hyperplane.* The support point is *proper* if the set does not lie wholly in the supporting hyperplane. Support points must be boundary points, but not every boundary point need be a support point, even for closed convex sets. Indeed, Example 7.9, which is due to V. Klee, provides an example of a nonempty closed convex set in an infinite dimensional Fréchet space that has no support points whatsoever. (In other words, no nonzero continuous linear functional attains a maximum on this set.) However, there are important cases for which support points are plentiful. If a closed convex set has a nonempty interior, then every boundary point is a proper support point (Lemma 7.7). In a finite dimensional space, every point on the *relative boundary* is a proper support point. We also present the Bishop–Phelps Theorem 7.43, which asserts that in a Banach space the set of support points of a closed convex set is a dense subset of the boundary.

We already remarked that a lower semicontinuous convex function is the pointwise supremum of the affine functions it dominates. If it agrees with one of these affine functions at some point, then the graph of the affine function is a supporting hyperplane to the epigraph (Lemma 7.11). The linear functional defining the affine functional is called a *subgradient.* It is easy to see that a convex function attains a minimum at a point only if the zero functional is a subgradient (Lemma 7.10). The collection of subgradients of a convex function at a point in the effective domain is a (possibly empty) weak∗ compact convex set, called the *subdifferential.* One reason for this terminology is that the *one-sided directional derivative* of a convex function defines a positively homogeneous convex functional, and the set of linear functionals it dominates is the subdifferential (Theorem 7.16). The subgradients of the support functional of a convex set at a particular linear functional in the dual space are the maximizers of the linear functional (Theorem 7.57). The Brøndsted–Rockafellar Theorem 7.50, using an argument similar to the Bishop–Phelps Theorem, shows that in a Banach space, a convex function has a subgradient on a dense subset of its effective domain.

Section 7.5 refines the conditions for the existence of a supporting hyperplane in terms of the existence of cones with particular properties. C. D. Aliprantis, R. Tourky, and N. C. Yannelis [16] provide a survey of their use in economics, where they are called *properness* conditions. A supporting hyperplane is a particular kind of separating hyperplane, so these results also refine our separating hyperplane theorems (Lemma 7.20). In finite dimensional spaces, there are further refinements of the separating hyperplane theorems. In a finite dimensional space, any two nonempty disjoint convex sets can be properly separated (Theorem 7.30). Indeed in a finite dimensional space, two nonempty convex sets can be properly separated if and only if their *relative interiors* are disjoint (Theorem 7.35).

Section 7.6 gives additional properties of proper convex functions on finite dimensional spaces. They are continuous on the relative interiors of their effective domains (Theorem 7.24). We mention without proof that a convex function is (Fréchet) differentiable almost everywhere in its effective domain (Theorem 7.26) and possesses a kind of second differential almost everywhere (Theorem 7.28).

Example 5.34 provides an example of a compact subset of a tvs whose closed convex hull is not compact. Theorem 5.35 asserts that the closed convex hull of a compact set is compact for the special case of completely metrizable locally convex spaces. This includes all the finite dimensional spaces. The Krein–Milman Theorem 7.68 asserts that, in a locally convex space, compact convex sets are the closure of the convex hull of their *extreme points*. An extreme point of a convex set is one that can be deleted and still leave a convex set.

Thus there are two useful ways to describe a compact convex set: as the closed convex hull of its extreme points, and as the intersection of all closed half spaces that include it. One might be tempted to say that we do not really know such a set unless we know both descriptions. K. C. Border [58] provides an example in economics of the use of both descriptions of a compact convex set to characterize the set of implementable *auctions*.

The Bauer Maximum Principle 7.69 is closely related to the Krein–Milman Theorem, and returns to theme of maximizing linear functionals. It asserts that a continuous convex function on a compact convex set achieves its maximum at an extreme point of the set. In fact, it is enough for the function to be *explicitly quasiconvex* (Corollary 7.75). If the maximizer of a linear functional is unique, then it is an *exposed point*. In finite dimensional spaces, the set of exposed points is dense in the set of extreme points (Theorem 7.89).

The convex hull of a finite set is called a *polytope*. Polytopes are always compact. The intersection of finitely many closed half spaces is a *polyhedron*. Every basic weak neighborhood of zero is a polyhedron by definition. But it can be written as the sum of a polytope plus a linear subspace (Theorem 7.80). While we do not discuss it here, in finite dimensional spaces, every compact polyhedron is a polytope. For more on this, we recommend the book by G. M. Ziegler [349].

It is often possible to efficiently characterize a compact convex set in terms of its extreme points. For instance, S. Brumelle and R. Vickson [71] have applied the Krein–Milman Theorem to characterize stochastic dominance relations; see also K. C. Border [57]. M. Berliant [38] has applied it to the problem of equilibrium pricing of land.

While there are dozens of excellent books devoted to the various aspects of convex analysis, we have space to mention only a few favorites. The classics on the finite dimensional case are the mimeographed notes on W. Fenchel's [123] lectures and the comprehensive tome by R. T. Rockafellar [288]. As the former is hard to find, and many of our colleagues find the latter difficult going, we highly recommend the treatment by J.-B. Hiriart-Urruty and C. Lemaréchal [163, 164, 165]. The appendix to D. W. Katzner's [195] monograph is brief, but remarkably informative as well. The infinite dimensional case is treated by J. R. Giles [136], R. R. Phelps [278], and A. W. Roberts and D. E. Varberg [284]. D. Gale [133] and

H. Nikaidô [262] address different problems in economics. G. M. Ziegler [349] is devoted to polytopes and polyhedral convexity. The works of M. Florenzano and C. Le Van [125], I. Ekeland and R. Temam [115], I. Ekeland and T. Turnbull [116], and J. Stoer and C. Witzgall [321] are devoted to convex analysis and optimization.

7.1   Extended-valued convex functions

Surprisingly, the definition of convex (or concave) function that we have been using to date is not the most useful for the analysis of convex sets and functions. So far we have allowed convex functions to be defined on convex subsets of a vector space. It is often more useful to require that a convex (or concave) function be defined everywhere. We can do this by considering concave and convex to be extended-real valued. If we take a function f that is convex in the sense of Definition 5.38, defined on the convex set C, we may extend it to the entire vector space X by defining it to be ∞ outside C. By Lemma 5.39 a function is convex if and only if its epigraph is a convex set. We now take this to be the definition for extended-valued functions defined on the entire vector space.

7.1 Definition (Extended convex functions)   *An extended-real valued function* $f: X \to \mathbb{R}^* = [-\infty, \infty]$ *on a vector space* X *is* **convex** *if its epigraph*

$$\text{epi } f = \{(x, \alpha) \in X \times \mathbb{R} : \alpha \geqslant f(x)\}$$

is a convex subset of the vector space $X \times \mathbb{R}$.

*The **effective domain** of a convex function* f *is the set* $\{x \in X : f(x) < \infty\}$ *and is denoted* dom f. *A convex function is **proper** if its effective domain is nonempty and additionally, it never assumes the value* $-\infty$.

Similarly, an extended-real valued function $f: X \to \mathbb{R}^*$ *on a vector space* X *is* **concave** *if its hypograph*

$$\text{hypo } f = \{(x, \alpha) \in X \times \mathbb{R} : \alpha \leqslant f(x)\}$$

is convex.

*The **effective domain** of a concave function* f *is the set* $\{x \in X : f(x) > -\infty\}$. *A concave function is **proper** if its effective domain is nonempty and it never assumes the value* ∞.

Positive homogeneity of an extended-real function can be defined in the usual fashion, provided we remember the convention that $0 \cdot \infty = 0$.

While we concentrate on convex functions in the remainder of the chapter, you should keep in mind that for every theorem about convex functions there is a corresponding result for concave functions, where the epigraph is replaced by the hypograph, and various inequalities are reversed.

We now turn your attention to a few simple facts.

•   Linear functions are both concave and convex.

•   Any real-valued convex function defined on a nonempty convex subset C of X may be regarded as a proper convex function on all of X by putting $f(x) = \infty$ for $x \notin C$. If f is continuous on C and C is closed in X, then f is lower semicontinuous as an extended real-valued function on X.

•   Note well that the epigraph of an extended-real valued function is a subset of $X \times \mathbb{R}$, not a subset of $X \times \mathbb{R}^*$. As a result, $x \in \mathrm{dom}\, f$ if and only if $(x, f(x)) \in \mathrm{epi}\, f$. In other words, the effective domain of an extended-real valued function f is the projection on X of its epigraph. We shall use this fact without any special mention.

•   The effective domain of a convex function is a convex set.

•   The constant function $f = -\infty$ is convex (its epigraph is $X \times \mathbb{R}$), but not proper, and the constant function $g = \infty$ is also convex (its epigraph is the empty set, which is convex), but not proper. These functions are also concave.

•   The function $g \colon \mathbb{R} \to \mathbb{R}^*$ defined by $g(x) = 0$ for $x = \pm 1$, $g(x) = \infty$ for $|x| > 1$, and $g(x) = -\infty$ for $|x| < 1$ is an example of a nontrivial improper convex function given by Rockafellar [288].

•   If a convex function is proper, then its epigraph is a nonempty proper subset of $X \times \mathbb{R}$.

•   Let f be an extended real-valued function on X. *If is f finite at x and continuous at x, then in fact x belongs to the interior of the effective domain of f.*

•   A convex function need not be finite at all points of continuity. The proper convex function f defined by $f(x) = 1/x$ for $x > 0$, and $f(x) = \infty$ for $x \leqslant 0$ is continuous everywhere, even at zero.

•   Given a convex set C in X, the function $\delta_C \colon X \to \mathbb{R}^*$ defined by $\delta_C(x) = 0$ for $x \in C$ and $\delta_C(x) = \infty$ for $x \notin C$ is a convex function, called the **(convex) indicator** function of C. If C is nonempty, then δ_C is proper.

7.2   Lower semicontinuous convex functions

From Corollary 2.60, an extended-valued proper convex function f on a topological vector space is lower semicontinuous if and only if its epigraph is a closed (and convex) subset of $X \times \mathbb{R}$. (Similarly a function is an upper semicontinuous proper concave function if and only if its hypograph is closed and convex.) In a locally convex space, by Corollary 5.83, every closed convex proper subset of the space is the intersection of all the closed half spaces that include it. Recall that if

f is a proper convex function then epi f is a proper subset of $X \times \mathbb{R}$. Thus the epigraph of a proper lower semicontinuous convex function is the intersection of all the closed half spaces that include it. We now relate certain closed half spaces in $X \times \mathbb{R}$ to the epigraphs of certain functions. We make use of the following simple fact, the proof of which we leave as an exercise.

- The topological dual of $X \times \mathbb{R}$ is $X' \times \mathbb{R}$ under the duality

$$\langle (x, \alpha), (x', \lambda) \rangle = x'(x) + \lambda \alpha.$$

The functions with closed half spaces for epigraphs are the affine functions.

7.2 Definition   *A function $f : X \to \mathbb{R}$ on a vector space is **affine** if it is of the form $f(x) = x^*(x) + c$ for some linear function $x^* \in X^*$ and some real c.*

Clearly every linear functional is affine, and every affine function is both convex and concave.

Let us refer to a typical element in $X \times \mathbb{R}$ as a point (x, α), where $x \in X$ and $\alpha \in \mathbb{R}$. We may call x the "vector component" and α the "real component." A closed hyperplane in $X \times \mathbb{R}$ is defined in terms of a continuous linear functional $(x', \lambda) \in X' \times \mathbb{R}$. If the real component $\lambda = 0$, we say the hyperplane is **vertical**. If the hyperplane is not vertical, by homogeneity we can normalize λ to be -1.

7.3 Lemma   *Any non-vertical hyperplane in $X \times \mathbb{R}$ is the graph of some affine function on X. The graph of any affine function on X is some non-vertical hyperplane in $X \times \mathbb{R}$.*

Proof: The non-vertical hyperplane $\{(x, \alpha) \in X \times \mathbb{R} : \langle (x^*, \lambda), (x, \alpha) \rangle = c\}$ where $\lambda \neq 0$ is the graph of the affine function $g(x) = (-1/\lambda)x^*(x) + c/\lambda$. On the other hand, the graph of the affine function $x \mapsto x^*(x) + c$ is the non-vertical hyperplane $\{(x, \alpha) \in X \times \mathbb{R} : \langle (x^*, -1), (x, \alpha) \rangle = -c\}$. ∎

It follows from Lemmas 2.41 and 5.40 that the pointwise supremum of a family of lower semicontinuous affine functions on a topological vector space is lower semicontinuous and convex. Similarly, the pointwise infimum of a family of upper semicontinuous affine functions is upper semicontinuous and concave. This suggests the following definition.

7.4 Definition   *Let C be a nonempty closed convex subset of the topological vector space X, and let $f : C \to \mathbb{R}$. Define the extended real functions $\hat{f}$ and $\check{f}$ on C by*

$$\hat{f}(x) = \inf\{g(x) : g \geqslant f \text{ and } g \text{ is affine and continuous}\}$$

and

$$\check{f}(x) = \sup\{g(x) : g \leqslant f \text{ and } g \text{ is affine and continuous}\},$$

*where the conventions $\sup \varnothing = -\infty$ and $\inf \varnothing = \infty$ apply. The function $\hat{f}$ is called the **concave envelope** of f, and $\check{f}$ is called the **convex envelope** of f.*

Clearly $\check{f} \leqslant f \leqslant \hat{f}$. As we remarked above, the convex envelope of a function is convex and lower semicontinuous. In locally convex Hausdorff spaces, lower semicontinuous proper convex functions agree with their convex envelope. Before we prove the theorem, we first prove a useful lemma.

7.5 Lemma *Let X be a locally convex Hausdorff space, and let $f \colon X \to \mathbb{R}^*$ be a lower semicontinuous proper convex function. If x belongs to the effective domain of f and $\alpha < f(x)$, then there exists a continuous affine function g satisfying $g(x) = \alpha$ and $g \ll f$, where $g \ll f$ means $g(y) < f(y)$ for all $y \in X$.*

Proof: Not that the epigraph of f is a nonempty closed convex proper subset of $X \times \mathbb{R}$, and by construction (x, α) does not belong to epi f. Thus by the Separating Hyperplane Theorem 5.80 there is a nonzero linear functional (x', λ) in the dual space $X' \times \mathbb{R}$ and $\varepsilon > 0$ satisfying

$$x'(x) + \lambda\alpha + \varepsilon < x'(y) + \lambda\beta \quad \text{for every } (y, \beta) \in \text{epi } f. \qquad (\star)$$

Since this inequality holds for β arbitrarily large, we must have $\lambda \geqslant 0$.

Since x belongs to the effective domain of f, we have $f(x) < \infty$. Then evaluating $(\star)$ at $(y, \beta) = (x, f(x))$ we rule out $\lambda = 0$. Now dividing by $\lambda > 0$ in $(\star)$, we see that the function $g \colon X \to \mathbb{R}$ defined by

$$g(z) = \frac{\langle x - z, x' \rangle}{\lambda} + \alpha$$

satisfies $g(x) = \alpha$, and for all $y \in \text{dom } f$ we have $g(y) + \varepsilon < f(y)$. For $y \notin \text{dom } f$, we have $g(y) < \infty = f(y)$. ∎

7.6 Theorem *Let X be a locally convex Hausdorff space, and let $f \colon X \to \mathbb{R}^*$ be a lower semicontinuous proper convex function. Then for each x we have*

$$f(x) = \sup\{g(x) : g \ll f \text{ and } g \text{ is affine and continuous}\}.$$

Consequently, $f = \check{f}$.

Proof: Fix x and let $\alpha \in \mathbb{R}$ satisfy $\alpha < f(x)$. (Since f is proper, we cannot have $f(x) = -\infty$, so such a real α exists.) It suffices to show that there is a continuous affine function g with $g \ll f$ and $g(x) \geqslant \alpha$. (Why?)

There are two cases to consider. The first is that x belongs to the effective domain of f. This is covered by Lemma 7.5 directly.

In case x is not in the effective domain, we may still proceed as in the proof of Lemma 7.5 to show that here exists a nonzero linear functional (x', λ) in the dual space $X' \times \mathbb{R}$ and $\varepsilon > 0$ satisfying $(\star)$ with $\lambda \geqslant 0$. However, we may not conclude that $\lambda > 0$. So suppose that $\lambda = 0$. Then $(\star)$ becomes

$$x'(x) + \varepsilon < x'(y) \quad \text{for every } y \in \text{dom } f.$$

Define the affine function h by $h(z) = \langle x - z, x' \rangle + \varepsilon/2$ and observe that $h(x) > 0$ and for $y \in \mathrm{dom}\, f$ we have $h(y) < 0$.

Next pick some $\bar{y} \in \mathrm{dom}\, f$, and use Lemma 7.5 to find an affine function $\bar{g}$ satisfying $\bar{g} \ll f$ (and $\bar{g}(\bar{y}) = f(\bar{y}) - 1$, which is irrelevant for our purpose). Now consider affine functions of the form

$$g(z) = \gamma h(z) + \bar{g}(z), \quad \text{where } \gamma > 0.$$

For $y \in \mathrm{dom}\, f$ we have $h(y) < 0$ and $\bar{g}(y) < f(y)$, so $g(y) < f(y)$. For y not in the effective domain of f, we have $g(y) < \infty = f(y)$. But $h(x) > 0$, so for γ large enough, $g(x) > \alpha$, as desired. ∎

A remark is in order. We know that the epigraph of a lower semicontinuous proper convex function is a proper closed convex subset of $X \times \mathbb{R}$. Therefore it is the intersection of all the closed half-spaces that include it. The theorem refines this to the intersection of all the closed half spaces corresponding to non-vertical hyperplanes.

7.3 Support points

One of the recurring themes of this chapter is the characterization of the maxima and minima of linear functionals over nonempty convex sets.

Let A be a nonempty subset of a topological vector space X, and let f be a nonzero continuous linear functional on X. If f attains either its maximum or its minimum over A at the point $x \in A$, we say that f **supports** A at x and that x is a **support point** of A.[1] Letting $\alpha = f(x)$ we may also say that the hyperplane $[f = \alpha]$ supports A at x. If A is not wholly included in the hyperplane we say it is **properly supported** at x. Finally, we may also say that the associated closed half space 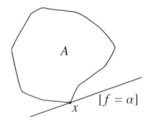 that includes A (that is, $[f \leqslant \alpha]$ for a maximum at x or $[f \geqslant \alpha]$ for a minimum at x) supports A at x. Since f is minimized whenever $-f$ is maximized, and vice versa, we are free to choose our support points to be either maximizers or minimizers, whichever is more convenient. Mathematicians, e.g., [278], tend to define support points as maximizers, while economists, e.g., [244], may define them as minimizers.

It is clear that only boundary points of A can be support points, however not every boundary point need be a support point. Theorem 7.36 below shows that in the finite dimension case, every point in a convex set that does not belong to

---

[1] N. Dunford and J. T. Schwartz [110, Definition V.9.4 p. 447] refer to such a linear functional f as a **tangent functional**.

its relative interior is a support point. (We postpone this result in part because we need to explain the relative interior.) But in the infinite dimensional case, not every boundary point of A need be a support point even if A is closed and convex, but we do have the following important case. We shall later prove the Bishop–Phelps Theorem 7.43, which asserts that in a Banach space the set of support points of a closed convex set is a dense subset of its boundary.

7.7 Lemma   *Let C be a nonempty convex subset of a tvs X with nonempty interior. If x is a boundary point of C that belongs to C, then C is properly supported at x.*

Proof: Since C° (the interior of C) is a nonempty open convex set and x does not belong to C°, there exists a nonzero continuous linear functional f satisfying $f(x) \leqslant f(y)$ for all $y \in C^\circ$, see Theorem 5.67. But the interior of C is dense in C (Lemma 5.28), so in fact $f(x) \leqslant f(y)$ for all $y \in C$. That is, f supports C at the point x. By Theorem 5.66 the support is proper.   ∎

The next example shows how the above conclusion may fail for a convex set with empty interior.

7.8 Example (A boundary point that is not a support point) Consider the set ℓ_1^+ of nonnegative sequences in ℓ_1, the Banach space of all summable sequences under the $\| \cdot \|_1$-norm. It is clearly a closed convex cone. However, its interior is empty. To see this, note that for each $\varepsilon > 0$ and every $x = (x_1, x_2, \ldots) \in \ell_1^+$, there is some n_0 such that $x_{n_0} < \varepsilon$. Define $y = (y_1, y_2, \ldots)$ by $y_i = x_i$ for $i \neq n_0$, and $y_{n_0} = -\varepsilon$. Then y does not belong to ℓ_1^+, but $\|x - y\| < 2\varepsilon$. Since ε is arbitrary we see that ℓ_1^+ has empty interior (quite unlike the finite dimensional case).

Thus every point in ℓ_1^+ is a boundary point. But no strictly positive sequence in ℓ_1^+ is a support point. To see this we make use of the fact that the dual space of ℓ_1 is ℓ_∞, the space of bounded sequences (see Theorem 16.20 below). Let $x = (x_1, x_2, \ldots)$ be an element of ℓ_1 such that $x_i > 0$ holds for each i, and suppose some nonzero $y = (y_1, y_2, \ldots) \in \ell_\infty$ satisfies

$$\sum_{i=1}^\infty y_i x_i \leqslant \sum_{i=1}^\infty y_i z_i$$

for all $z = (z_1, z_2, \ldots) \in \ell_1^+$. Letting $z_k = x_k + 1$ and $z_i = x_i$ for $i \neq k$ yields $y_k \geqslant 0$ for all k. Since y is nonzero, we must have $y_k > 0$ for some k, so $\sum_{i=1}^\infty y_i x_i > 0$. But then $z = 0$ implies $0 = \sum_{i=1}^\infty y_i z_i \geqslant \sum_{i=1}^\infty y_i x_i > 0$, which is impossible. Thus x cannot be a support point.

On the other hand, if some $x \in \ell_1^+$ has $x_k = 0$ for some k, then the nonzero continuous linear functional $e_k \in \ell_\infty$ satisfies

$$0 = e_k(x) \leqslant e_k(z)$$

for all $z \in \ell_1^+$. This means that e_k supports the set ℓ_1^+ at x. Moreover, note that the collection of such x is norm dense in ℓ_1^+. ∎

V. Klee [208] presents an even more remarkable example of a proper closed convex set in $\mathbb{R}^{\mathbb{N}}$, that has no support points whatsoever. Recall that $\mathbb{R}^{\mathbb{N}}$ (the space of all real sequences) with its product topology is a Fréchet space, but not a Banach space (see Example 5.78). The following example presents Klee's construction.

7.9 Example (A proper closed convex set with no support points)   Let V_m denote the closed linear subspace of $\mathbb{R}^{\mathbb{N}}$ for which all the terms after the first m are zero, and let π_m be the natural projection of $\mathbb{R}^{\mathbb{N}}$ onto V_m, that is, π_m maps $x = (x_1, x_2, \ldots)$ to $(x_1, \ldots, x_m, 0, 0, \ldots)$. Let d_m be the Euclidean metric on V_m, that is, $d_m(x, y) = [\sum_{i=1}^{m} (x_i - y_i)^2]^{1/2}$, and let U_m be the open unit ball in V_m, that is, $U_m = \{x \in V_m : d_m(0, x) < 1\}$. Finally, let $f(\alpha) = \frac{\alpha}{1+\alpha}$ for $\alpha \geqslant 0$.

We now construct a sequence of convex sets inductively. Let $C_1 = V_1^+$, that is, the set of sequences x such that $x_1 \geqslant 0$ and $x_i = 0$ for $i > 1$. Now define C_{m+1} inductively by

$$C_{m+1} = \{x \in V_{m+1} : x_{m+1} \geqslant 0 \text{ and } \pi_m(x) \in C_m + 2^{-m} f(x_{m+1}) U_m\}.$$

Note that for $\alpha > 0$, the set $C_m + \alpha U_m$ is $\{x \in V_m : d_m(\pi_m(x), C_m) < \alpha\} = N_\alpha(C_m)$. In particular, we have:

- If $x \in C_{m+1}$, then there is $\tilde{x} \in C_m$ with $d_m(\tilde{x}, \pi_m(x)) \leqslant 2^{-m} f(x_{m+1})$. The inequality is strict if $x_{m+1} > 0$.

To get a feel for this construction, Figure 7.1 depicts C_1, C_2, and $\pi_2(C_3)$ in V_2 identified with the plane. The set V_1 is identified with the x_1-axis, and the set $C_1 + 2^{-1} U_1$ is identified with the open interval $(-1/2, \infty)$ of the x_1-axis, which is open in V_1. Notice how the boundary of C_2 asymptotes to the vertical line at $x_1 = -1/2$. This guarantees that the projection of $\overline{C_2}$ onto V_1 is open in V_1.) The set C_3 extends out of the page, its cross section starting at C_2 for $x_3 = 0$, and asymptotically approaching the wall coming out of the page along the boundary of $\pi_2(C_3)$ as $x_3 \to \infty$.

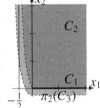

Figure 7.1.

There are three properties of this family of sets that are immediate. Namely, for each $m = 1, 2, \ldots$ we have:

$$V_m^+ \subset C_m \tag{1}$$

$$C_m \subset C_{m+1} \tag{2}$$

$$\pi_m(C_{m+1}) = C_m + 2^{-m} U_m \tag{3}$$

The proof of (1) proceeds by induction on m. For $m = 1$ we have $V_1^+ = C_1$. For the induction step, assume that $V_m^+ \subset C_m$. If $x \in V_{m+1}^+$, then $x_{m+1} \geqslant 0$ and $\pi_m(x) \in V_m^+ \subset C_m$. Obviously $C_m \subset C_m + 2^{-m} f(x_{m+1}) U_m$. Thus $x \in C_{m+1}$, proving $V_{m+1}^+ \subset C_{m+1}$.

For (2), let $x \in C_m$. In particular, $x \in V_m \subset V_{m+1}$ and $x_{m+1} = 0$, so $f(x_{m+1}) = 0$. Thus $\pi_m(x) = x \in C_m = C_m + \{0\} = C_m + 2^{-m} f(x_{m+1})U_m$. That is, $x \in C_{m+1}$.

We now verify (3). If $x \in C_{m+1}$, then $\pi_m(x) \in C_m + 2^{-m} f(x_{m+1})U_m \subset C_m + 2^{-m} U_m$, so $\pi_m(C_{m+1}) \subset C_m + 2^{-m} U_m$. For the reverse inclusion, let $y \in C_m + 2^{-m} U_m$ and note that $\pi_m(y) = y$. Pick $u \in C_m$ and $v \in U_m$ such that $y = u + 2^{-m}v$. Since $v \in U_m$, there exists some β with $d_m(0, v) < \beta < 1$. Pick $\alpha > 0$ such that $f(\alpha) = \beta$, let $w = \frac{v}{\beta} \in U_m$ and note that $v = f(\alpha)w$. Setting $x = (y_1, \ldots, y_m, \alpha, 0, 0, \ldots) \in V_{m+1}$, we have $x_{m+1} = \alpha > 0$ and

$$\pi_m(x) = \pi_m(y) = y = u + 2^{-m} f(\alpha)w \in C_m + 2^{-m} f(x_{m+1})U_m.$$

In other words, $x \in C_{m+1}$, so $\pi_m(x) = y \in \pi_m(C_{m+1})$.

The sequence of sets $\{C_m\}$ also satisfies the following list of properties.

Property (I): *Each C_m is a convex subset of $\mathbb{R}^{\mathbb{N}}$.*

The proof is by induction on m. Clearly C_1 is convex. For the induction step, we assume that C_m is convex and must show that C_{m+1} is convex. To this end, let $x, y \in C_{m+1}$, fix $0 < \lambda < 1$, and put $z = \lambda x + (1 - \lambda)y$. Clearly, $z \in V_{m+1}$ and $z_{m+1} \geqslant 0$. Since $x, y \in C_m$, by the induction hypothesis we have $z \in C_m$ and so $z \in C_{m+1}$. Next, pick $\tilde{x}, \tilde{y} \in C_m$ so that $d_m(\tilde{x}, \pi_m(x)) \leqslant 2^{-m} f(x_{m+1})$ and $d_m(\tilde{y}, \pi_m(y)) \leqslant 2^{-m} f(y_{m+1})$. We can assume that either $x_{m+1} > 0$ or $y_{m+1} > 0$, so at least one of these inequalities is strict. By our induction hypothesis the vector $u = \lambda\tilde{x} + (1 - \lambda)\tilde{y}$ belongs to C_m. Now the linearity of π_m and the sublinearity of d_m imply

$$
\begin{aligned}
d_m(u, \pi_m(z)) &= d_m(\lambda\tilde{x} + (1 - \lambda)\tilde{y}, \lambda\pi_m(x) + (1 - \lambda)\pi_m(y)) \\
&\leqslant \lambda d_m(\tilde{x}, \pi_m(x)) + (1 - \lambda)d_m(\tilde{y}, \pi_m(y)) \\
&< 2^{-m}[\lambda f(x_{m+1}) + (1 - \lambda)f(y_{m+1})] \\
&\leqslant 2^{-m} f(\lambda x_{m+1} + (1 - \lambda)y_{m+1}),
\end{aligned}
$$

where the last inequality follows from the concavity of the function f. This implies $d_m(u, \pi_m(z)) < 2^{-m} f(z_{m+1})$, and from this we get $z \in C_{m+1}$, and the convexity of C_{m+1} has been established.

Property (II): *For all k and m we have*

$$\pi_m(C_{m+k}) = C_m + \sum_{n=m}^{m+k-1} 2^{-n} U_m = C_m + 2^{-(m-1)}(1 - 2^{-k})U_m.$$

The proof works by induction on k. The case $k = 1$ is just the definition of C_{m+1}. Assume the result is true for all m and some k. Then using (3) and the induction hypothesis, we see that

$$\pi_m(C_{m+k+1}) = \pi_m(\pi_{m+k}(C_{m+k+1})) = \pi_m(C_{m+k} + 2^{-(m+k)}U_{m+k})$$

$$= \pi_m(C_{m+k}) + \pi_m(2^{-(m+k)}U_{m+k}) = C_m + \sum_{n=m}^{m+k-1} 2^{-n}U_m + 2^{-(m+k)}U_m$$

$$= C_m + \sum_{n=m}^{m+k} 2^{-n}U_m.$$

Now consider the following subset of $\mathbb{R}^{\mathbb{N}}$:

$$C = \bigcup_{m=1}^{\infty} C_m.$$

We shall prove that the closure $\overline{C}$ of C in $\mathbb{R}^{\mathbb{N}}$ is a nonempty proper closed convex set with no support points. To do this we need a few more results.

Property (III): *The set $\overline{C}$ includes $\mathbb{R}_+^{\mathbb{N}}$.*

Indeed, for any $x \in \mathbb{R}_+^{\mathbb{N}}$, we have $\pi_m(x) \in V_m^+ \subset C_m \subset C$, and therefore from $\pi_m(x) \to x$ in $\mathbb{R}^{\mathbb{N}}$ we get $x \in \overline{C}$.

Property (IV): *If $x \in C$, then for each m we have*

$$\pi_m(x) \in C_m + [1 + f(x_{m+1} + 2^{-m})]2^{-m}U_m.$$

Clearly $\pi_m(x) = (\pi_m \circ \pi_{m+1})(x)$. Since $x \in C$, it belongs to some C_j. If $j \leqslant m$, then $x \in C_m$, and if $x \in C_{m+1}$, then the definition proves the conclusion. The only challenging case is $j > m + 1$. Then by Property (II),

$$\pi_{m+1}(x) \in \pi_{m+1}(C_j) = C_{m+1} + \sum_{n=m+1}^{j-1} 2^{-n}U_{m+1} \subset C_{m+1} + 2^{-m}U_{m+1}.$$

Thus there exists some $\tilde{x} \in C_{m+1}$ with $d_{m+1}(\tilde{x}, \pi_{m+1}(x)) < 2^{-m}$. In particular, we have $\tilde{x}_{m+1} < x_{m+1} + 2^{-m}$. Since $\tilde{x} \in C_{m+1}$ we have

$$\pi_m(\tilde{x}) \in C_m + 2^{-m}f(\tilde{x}_{m+1})U_m \subset C_m + 2^{-m}f(x_{m+1} + 2^{-m})U_m.$$

Now note the equalities $\pi_m(x) = \pi_m(\tilde{x}) + \pi_m(x - \tilde{x})$, $\pi_m(x - \tilde{x}) = \pi_m(\pi_{m+1}(x - \tilde{x}))$, $\pi_{m+1}(x - \tilde{x}) = \pi_{m+1}(x) - \tilde{x} \in 2^{-m}U_{m+1}$, and $\pi_m(2^{-m}U_{m+1}) = 2^{-m}U_m$. Putting it all together yields

$$\pi_m(x) \in C_m + 2^{-m}f(x_{m+1} + 2^{-m})U_m + 2^{-m}U_m$$

$$= C_m + [1 + f(x_{m+1} + 2^{-m})]2^{-m}U_m.$$

Property (V): *The nonempty set C is convex and for each m we have*

$$\pi_m(C) = \pi_m(\overline{C}) = C_m + 2^{-(m-1)}U_m.$$

In particular, $\pi_m(\overline{C})$ is an open subset of V_m.

The convexity of C follows from (2) and the convexity of each C_m. For the remainder, we first show that $\pi_m(C) = C_m + 2^{-(m-1)}U_m$. Property (II) implies that $\pi_m(C_i) \subset C_m + 2^{-(m-1)}U_m$ for all $i \geqslant m$. Since $C_i \subset C_m$ for each $1 \leqslant i < m$, we easily see that

$$\pi_m(C) = \bigcup_{i=1}^{\infty} \pi_m(C_i) \subset C_m + 2^{-(m-1)}U_m.$$

For the reverse inclusion, let $x = c + 2^{-(m-1)}v$ with $c \in C_m$ and $v \in U_m$. Now notice that from $d_m(0, v) < 1$ and $\lim_{k \to \infty}(1 - 2^{-k}) = 1$, it follows that there exists some k such that the vector $w = \frac{v}{1-2^{-k}}$ belongs to U_m. But then from Property (II) we have

$$x = c + 2^{-(m-1)}v = c + 2^{-(m-1)}(1 - 2^{-k})w$$

$$\in C_m + 2^{-(m-1)}(1 - 2^{-k})U_m = \pi_m(C_{m+k}) \subset \pi_m(C).$$

Thus, $x \in \pi_m(C)$ and so $C_m + 2^{-(m-1)}U_m \subset \pi_m(C)$ is also true.

Now we claim that the closure also satisfies

$$\pi_m(\overline{C}) = C_m + 2^{-(m-1)}U_m.$$

Clearly $\pi_m(\overline{C})$ includes $\pi_m(C) = C_m + 2^{-(m-1)}U_m$, so it suffices to show the opposite inclusion. To this end, let $x \in \overline{C}$ and pick a sequence $\{x_n\} \subset C$ with $x_n \to x$ in $\mathbb{R}^\mathbb{N}$, where $x_n = (x_{n,1}, x_{n,2} \ldots)$ and $x = (x_1, x_2, \ldots)$. Then there is some n_0 such that for all $n \geqslant n_0$ we have $x_{n,m+1} < x_{m+1} + 1$. Then by Property (IV), since $x_n \in C$, we must have

$$\pi_m(x_n) \in C_m + \alpha 2^{-m}U_m,$$

where $\alpha = 1 + f(x_{m+1} + 1 + 2^{-m})$ satisfies $0 < \alpha < 2$. Thus

$$\pi_m(x) = \lim_{n \to \infty} \pi_m(x_n) \in \overline{C_m + \alpha 2^{-m}U_m}.$$

We have reduced the problem to showing $\overline{C_m + \alpha 2^{-m}U_m} \subset C_m + 2^{-(m-1)}U_m$. To prove this, first note that the closed unit ball $\overline{U}_m$ of V_m is a compact subset of $\mathbb{R}^\mathbb{N}$. Consequently, according to Lemma 5.3 (4), the set $\overline{C}_m + \alpha 2^{-m}\overline{U}_m$ is closed. Also note that $\alpha 2^{-m}\overline{U}_m \subset 2^{-(m-1)}U_m$ since $\alpha < 2$. Finally note that Lemma 5.3 (3) assures us that $\overline{C}_m + 2^{-(m-1)}U_m = C_m + 2^{-(m-1)}U_m$. So we have

$$\overline{C_m + \alpha 2^{-m}U_m} \subset \overline{\overline{C}_m + \alpha 2^{-m}\overline{U}_m} = \overline{C}_m + \alpha 2^{-m}\overline{U}_m$$

$$\subset \overline{C}_m + 2^{-(m-1)}U_m = C_m + 2^{-(m-1)}U_m,$$

and we are done with the proof of Property (V).

Property (VI): *The nonempty closed convex set $\overline{C}$ is a proper subset of $\mathbb{R}^{\mathbb{N}}$.*

From Property (IV) we know that $\pi_1(\overline{C}) = C_1 + U_1$. Now observe that the point $x = (-1, 0, 0, \ldots)$ does not belong to $C_1 + U_1$, but $\pi_1(x) = x$ so $x \notin \overline{C}$.

Property (VII): *The proper closed convex subset $\overline{C}$ of $\mathbb{R}^{\mathbb{N}}$ has no support points.*

We prove in Theorem 16.3 that the dual space of $\mathbb{R}^{\mathbb{N}}$ is $\varphi = \bigcup_{n=1}^{\infty} V_n$, the space of sequences with only finitely many nonzero terms. That is, given a sequence p in φ the mapping $x \mapsto \sum_{n=1}^{\infty} p_n x_n$ is a continuous linear function on $\mathbb{R}^{\mathbb{N}}$ and all continuous linear functionals are of this form. Now any p in φ belongs to some V_m, so maximizing p over $\overline{C}$ is the same as maximizing p over $\pi_m(\overline{C})$, which is an open set in V_m. Therefore a nonzero p has no maximum (or minimum). In other words $\overline{C}$ has no support points.   ∎

7.4   Subgradients

We now turn in a roundabout way to characterizing certain support points of the epigraph of a convex function.

Given a dual pair $\langle X, X' \rangle$ and a convex function f on X, we say that $x' \in X'$ is a **subgradient** of f at x if it satisfies the following **subgradient inequality**

$$f(y) \geqslant f(x) + x'(y - x)$$

for every $y \in X$. The set of subgradients at x is the **subdifferential** of f, denoted $\partial f(x)$. It may be that $\partial f(x)$ is empty, but if it is nonempty, we say that f is **subdifferentiable** at x. (For a concave function f, if $x' \in X'$ satisfies the reverse subgradient inequality, that is, if $f(y) \leqslant f(x) + x'(y - x)$ for all y, then we say that x' is a **supergradient** of f at x, and refer to the collection of them as the **superdifferential**, also denoted $\partial f(x)$.[2])

An immediate consequence of the definition is the following result, which we shall see later can be interpreted as a kind of "first order condition" for a minimum.

7.10 Lemma   *A convex function f is minimized at x if and only if $0 \in \partial f(x)$.*

The proof follows immediately by setting $x' = 0$ in the subgradient inequality. A byproduct of this result is a sufficient condition (x minimizes f) for $\partial f(x)$ to be nonempty. Note that if f is proper (that is, if its effective domain is nonempty and f never takes on the value $-\infty$), by considering $y \in \operatorname{dom} f$, we see that the subgradient inequality can only be satisfied if $f(x) < \infty$, that is, if $x \in \operatorname{dom} f$. We shall not be very interested in subgradients of improper functions.

---

[2] Some authors, notably Rockafellar [288, p. 308], use "subgradient" to mean both subgradient and supergradient, and subdifferential to mean both subdifferential and superdifferential. Rockafellar does suggest that our terminology is more appropriate than the terminology he actually uses.

Another way to phrase the subgradient inequality is that x' is a subgradient of f at $x \in \operatorname{dom} f$ if f dominates the affine function $g(y) = x'(y) - x'(x) + f(x)$, which agrees with f at x. Now the graph of an affine function is a nonvertical hyperplane in $X \times \mathbb{R}$, so the subgradient inequality implies that this hyperplane supports the epigraph of f at $(x, f(x))$. In fact, we have the following lemma.

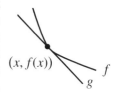

7.11 Lemma *The functional x' is a subgradient of the proper convex function f at $x \in \operatorname{dom} f$ if and only if $(x, f(x))$ maximizes the linear functional $(x', -1)$ over* epi f.

Proof: We leave the proof as an exercise, but here is a generous hint: Rewrite the subgradient inequality $f(y) \geq f(x) + x'(y - x)$ as

$$\langle (x, f(x)), (x', -1) \rangle \geq \langle (y, f(y)), (x', -1) \rangle,$$

for all $y \in \operatorname{dom} f$. ∎

As mentioned earlier, the subdifferential may be empty, that is, there may be no subgradient at x. In fact, A. Brøndsted and R. T. Rockafellar [64] give an example of a lower semicontinuous proper convex function defined on a Fréchet space that is nowhere subdifferentiable. Their example is based on the set in Example 7.9. On a positive note, we do have the following simple sufficient condition.

7.12 Theorem *Let $\langle X, X' \rangle$ be a dual pair, and let f be a proper convex function on X. If x is an interior point of $\operatorname{dom} f$ and if f is $\sigma(X, X')$-continuous at x, then f has a subgradient at x.*

Proof: By Theorem 5.43, there is a neighborhood V of x on which f is bounded above by some real number $c < \infty$ and continuous. Then $V \times (c, \infty)$ is an open subset of $X \times \mathbb{R}$ included in the epigraph of f, so epi f has nonempty interior. But $(x, f(x))$ is not an interior point of epi f as $(x, f(x) - \varepsilon)$ does not belong to the epigraph for any $\varepsilon > 0$. Therefore by Lemma 7.7, the point $(x, f(x))$ is a proper support point of epi f.

We just need to show that the supporting hyperplane is nonvertical. So suppose by way of contradiction that $(x, f(x))$ maximizes the nonzero linear functional $(x', 0)$ over epi f, or equivalently x maximizes x' over $\operatorname{dom} f$. Since x is in the interior of $\operatorname{dom} f$, we must have that x' is identically zero, a contradiction. ∎

The subdifferential $\partial f(x)$, being defined in terms of weak linear inequalities, is a weak*-closed convex set (possibly empty). In fact, it is weak* compact.

7.13 Theorem *Let $\langle X, X' \rangle$ be a dual pair, and let f be a proper convex function on X. If $x \in \operatorname{dom} f$, then $\partial f(x)$ is a weak* compact convex subset of X'.*

Proof: The convexity of $\partial f(x)$ is easy. For compactness, as in the proof of Alaoglu's Theorem 5.105, we rely on the Tychonoff Product Theorem 2.61. It thus suffices to show that $\partial f(x)$ is weak∗ closed and pointwise bounded. Writing y as $x + v$ (where $v = y - x$) the subgradient inequality implies that $\partial f(x)$ is the intersection

$$\bigcap_{v \in X} \{x' \in X' : x'(v) \leqslant f(x + v) - f(x)\}$$

of weak∗-closed sets, so it is weak∗ closed. Now we need to find bounds on $x'(v)$ for each $v \in X$. By the subgradient inequality, if $x' \in \partial f(x)$, we have $x'(v) \leqslant f(x+v) - f(x)$. For a lower bound, observe that the subgradient inequality implies $x'(-v) \leqslant f(x - v) - f(x)$. But $x'(-v) = -x'(v)$ so $x'(v) \geqslant f(x) - f(x - v)$. Thus for any $v \in X$, we have

$$x' \in \partial f(x) \implies f(x) - f(x - v) \leqslant x'(v) \leqslant f(x + v) - f(x).$$

That is, $\partial f(x)$ is pointwise bounded, and the proof is complete. ∎

We now relate the subdifferential to the directional derivatives of f. We prove next that for a convex function, the difference quotient $\frac{f(x+\lambda v)-f(x)}{\lambda}$ is nonincreasing as $\lambda \downarrow 0$, so it has a limit in $\mathbb{R}^*$, although it may be $-\infty$.

7.14 Lemma   *Let f be a proper convex function, let x belong to* dom f, *let v belong to X, and let $0 < \mu < \lambda$. Then the difference quotients satisfy*

$$\frac{f(x + \mu v) - f(x)}{\mu} \leqslant \frac{f(x + \lambda v) - f(x)}{\lambda}.$$

In particular, $\lim_{\lambda \downarrow 0} \frac{f(x+\lambda v)-f(x)}{\lambda}$ *exists in* $\mathbb{R}^*$.

Proof: The point $x + \mu v$ is the convex combination $\frac{\mu}{\lambda}(x + \lambda v) + \frac{\lambda - \mu}{\lambda}x$, so by convexity

$$f(x + \mu v) \leqslant \frac{\mu}{\lambda}f(x + \lambda v) + \frac{\lambda - \mu}{\lambda}f(x).$$

Dividing by $\mu > 0$ and rearranging yields the desired inequality. ∎

Define the **one-sided directional derivative** $d^+ f(x): X \to \mathbb{R}^*$ [3] at x by

$$d^+ f(x)(v) = \lim_{\lambda \downarrow 0} \frac{f(x + \lambda v) - f(x)}{\lambda}.$$

Remarkably, if f is subdifferentiable at x, then this limit is finite. To see this, rewrite the subgradient inequality

$$f(y) \geqslant f(x) + x'(y - x)$$

---

[3] This is the notation used by Phelps [278]. Fenchel [123] and Rockafellar [288] write $f'(x; v)$. Neither one is very pretty.

as

$$x'(v) \leqslant \frac{f(x + \lambda v) - f(x)}{\lambda}, \quad \text{where } y = x + \lambda v.$$

In this case, the difference quotient is bounded below by $x'(v)$ for any $x' \in \partial f(x)$, so the limit is finite.

We now show that $d^+ f(x)$ is a positively homogeneous convex function.

7.15 Theorem   *Let f be a proper convex function on the tvs X. The directional derivative mapping $v \mapsto d^+ f(x)(v)$ from X into $\mathbb{R}^*$ satisfies the following properties.*

  a. *The function $d^+ f(x)$ is a positively homogeneous convex function (that is, sublinear) and its effective domain is a convex cone.*

  b. *If f is continuous at $x \in \operatorname{dom} f$, then $v \mapsto d^+ f(x)(v)$ is continuous and finite-valued.*

Proof: It is easy to see that the function $v \mapsto d^+ f(x)(v)$ is homogeneous, as $\frac{f(x+\lambda \alpha v)-f(x)}{\lambda} = \alpha \frac{f(x+\lambda \alpha v)-f(x)}{\alpha \lambda}$, and so $d^+ f(x)(\alpha v) = \alpha d^+ f(x)(v)$. This also shows that the effective domain is a cone. For convexity, observe that

$$\frac{f(x + \lambda(\alpha v + (1 - \alpha)w) - f(x)}{\lambda} = \frac{f(\alpha(x + \lambda v) + (1 - \alpha)(x + \lambda w)) - f(x)}{\lambda}$$

$$\leqslant \frac{\alpha f(x + \lambda v) + (1 - \alpha)f(x + \lambda w)) - f(x)}{\lambda}$$

$$= \alpha \frac{f(x + \lambda v) - f(x)}{\lambda} + (1 - \alpha)\frac{f(x + \lambda w) - f(x)}{\lambda},$$

and letting $\lambda \downarrow 0$ yields $d^+ f(x)(\alpha v + (1 - \alpha)w) \leqslant \alpha d^+ f(x)(v) + (1 - \alpha)d^+ f(x)(w)$.
By Lemma 5.41, we have

$$|f(x + \lambda v) - f(x)| \leqslant \lambda \max\{f(x + v) - f(x), f(x - v) - f(x)\}$$

for $0 < \lambda \leqslant 1$. So let $\varepsilon > 0$ be given. If f is continuous at x, there exists an absorbing circled neighborhood V of 0 such that $v \in V$ implies $|(f(x+v)-f(x)| < \varepsilon$. We thus have

$$|d^+ f(x)(v)| \leqslant \frac{|f(x + \lambda v) - f(x)|}{\lambda} \leqslant \max\{f(x + v) - f(x), f(x - v) - f(x)\} < \varepsilon.$$

(Why?) That is, $d^+ f(x)$ is bounded above on V. So by Lemma 5.51, it is continuous. ∎

When the sublinear function $v \mapsto d^+ f(x)(v)$ is actually linear (and finite-valued), it is called the **Gâteaux derivative** of f at x.

7.16 Theorem   *For a proper convex function f and a point $x \in$ dom f, the following are equivalent.*

1. *x' is a subgradient of f at x.*

2. *$(x', -1)$ is maximized over epi f at $(x, f(x))$.*

3. *$x' \leqslant d^+ f(x)$.*

Proof: The equivalence of (1) and (2) is just Lemma 7.11. To see the equivalence of (2) and (3), first note that any point y can be written as $x + \lambda v$ with $v = y - x$ and $\lambda = 1$. So (2) can be rewritten as

$$\langle (x', -1), (x + \lambda v, f(x + \lambda v)) \rangle \leqslant \langle (x', -1), (x, f(x)) \rangle \qquad (\star\star)$$

for all $x + \lambda v \in$ dom f. Now note that $(\star\star)$ is equivalent to the inequalities

$$x'(x + \lambda v) - f(x + \lambda v) \leqslant x'(x) - f(x)$$
$$x'(\lambda v) \leqslant f(x + \lambda v) - f(x)$$
$$x'(v) \leqslant \frac{f(x + \lambda v) - f(x)}{\lambda}.$$

In light of Lemma 7.14, his shows that (2) $\iff$ (3). $\blacksquare$

In light of Theorem 5.54, which states that a sublinear functional is linear if and only if it dominates exactly one linear functional (itself), we have the following corollary.

7.17 Corollary   *For a proper convex function f, the subdifferential $\partial f(x)$ is a singleton if and only if $d^+ f(x)$ is the Gâteaux derivative of f at x.*

7.5   Supporting hyperplanes and cones

This section refines the characterization of support points. Recall that a cone is a nonempty subset of a vector space that is closed under multiplication by nonnegative scalars. However, we define an **open cone**, to be a nonempty open subset of a topological vector space closed under multiplication by strictly positive scalars. An open cone contains the point zero only if it is the whole space. It is often convenient to translate a cone or an open cone around the vector space, so let us say that a nonempty subset of a vector space is a **cone with vertex x** if it is of the form $x + C$ where C is a bona fide cone (with vertex zero). Every linear subspace is a cone and each of its points is a vertex of the subspace in this sense. Other definitions regarding cones have obvious generalizations to cones with vertex x.

Every closed half space is a cone with a possibly nonzero vertex, and they are the largest closed cones except for the entire space. The next fundamental result on supporting hyperplanes of cones is due to V. Klee [205, 206].

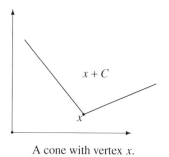

A cone with vertex x.

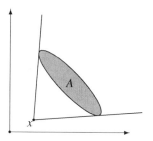

The cone with vertex x generated by A.

7.18 Lemma (Klee) *In a locally convex space a convex cone is supported at its vertex if and only if the cone is not dense.*

Proof: Let C be a convex cone with vertex x in a locally convex space. If C is supported at x, it lies in some closed half space and is thus not dense.

For the converse, assume that C is not dense. Without loss of generality we may assume that C is a cone with vertex 0. Now there exists some $x_0 \notin \overline{C}$. Since $\overline{C}$ is a closed convex set, the Separation Corollary 5.80 guarantees the existence of some nonzero continuous linear functional f satisfying $f(x_0) < f(x)$ for all $x \in C$. By the remarks following Example 5.56, it follows that C lies in the closed half space $[f \geqslant 0]$, and so is supported at zero. ∎

You may have some difficulty thinking of a dense convex cone other than the entire vector space. Indeed in finite dimensional Hausdorff spaces, the whole space is the only dense cone. In infinite dimensional spaces, however, there can be dense proper subspaces, which are thus also dense cones. For instance, the set of polynomials is dense in the space of continuous functions on the unit interval. The Stone–Weierstrass Theorem 9.13 gives the existence of many dense subspaces.

The next theorem gives several characterizations of support points. Some of them have been used in economics, where they are called **properness conditions**. C. D. Aliprantis, R. Tourky, and N. C. Yannelis [16] discuss the role of the conditions in general economic equilibrium theory.

7.19 Theorem *Let C be a convex subset of a locally convex space and let x be a boundary point of C. If $x \in C$, then the following statements are equivalent.*

1. *The set C is supported at x.*

2. *There is a non-dense convex cone with vertex x that includes C, or equivalently, the convex cone with vertex x generated by C is not dense.*

3. *There is an open convex cone K with vertex x such that $K \cap C = \varnothing$.*

4. *There exist a nonzero vector v and a neighborhood V of zero such that $x - \alpha v + z \in C$ with $\alpha > 0$ implies $z \notin \alpha V$.*

Proof: (1) $\implies$ (2) Any closed half space that supports C at x is a closed convex cone with vertex x that is not dense and includes C.

(2) $\implies$ (3) Let $\hat{K}$ be a non-dense convex cone with vertex x that includes C. By Lemma 7.18, x is a point of support of $\hat{K}$. Now if f is a nonzero continuous linear functional attaining its maximum over $\hat{K}$ at x, then the open half space $K = [f > f(x)]$ is an open convex cone with vertex x that satisfies $K \cap C = \emptyset$.

(3) $\implies$ (4) Let K be an open convex cone with vertex 0 that satisfies $(x + K) \cap C = \emptyset$. Fix a vector $w \in K$ and choose a neighborhood V of zero such that $w + V \subset K$. Put $v = -w \neq 0$. We claim that v and V satisfy the desired properties. To see this, assume that $x - \alpha v + z \in C$ with $\alpha > 0$. If $z = \alpha u$ for some $u \in V$, then

$$x - \alpha v + z = x + \alpha(w + u) \in x + K,$$

so $x - \alpha v + z \in (x + K) \cap C$, which is a contradiction. Hence $z \notin \alpha V$.

(4) $\implies$ (1) We can assume that the neighborhood V of zero is open and convex. The given condition guarantees that the open convex cone K generated by $-v + V$ with vertex zero, that is,

$$K = \{\alpha(-v + w) : \alpha > 0 \text{ and } w \in V\},$$

satisfies $(x + K) \cap C = \emptyset$. Then, by the Interior Separating Hyperplane Theorem 5.67, there exists a nonzero continuous linear functional f separating $x + K$ and C. That is, f satisfies $f(x + k) \leqslant f(y)$ for all $k \in K$ and all $y \in C$. Since K is a cone, we have $f(x + \alpha k) \leqslant f(y)$ for all $\alpha > 0$ and each $k \in K$, and by letting $\alpha \downarrow 0$, we get $f(x) \leqslant f(y)$ for all $y \in C$. ∎

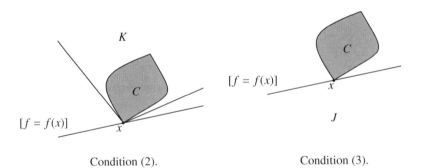

Condition (2).   Condition (3).

Figure 7.2. Theorem 7.19

The geometry of Theorem 7.19 is shown in Figure 7.2. We can rephrase a separating hyperplane theorem in terms of cones. Recall that the cone generated by S is the smallest cone that includes S and is thus $\{\alpha x : \alpha \geqslant 0 \text{ and } x \in S\}$.

7.20 Lemma (Separation of sets and cones) *Two nonempty subsets A and B of a locally convex space can be separated if and only if the convex cone generated by the set A − B is not dense.*

Proof: Suppose that a nonzero continuous linear functional f separates two nonempty sets A and B, that is, assume that $f(b) \leqslant f(a)$ for all $a \in A$ and all $b \in B$. Then $A - B \subset \{x : f(x) \geqslant 0\} = H$, so the closed convex cone generated by $A - B$ lies in the non-dense closed cone H.

Next, assume that the convex cone C generated by $A - B$ is not dense. Then by Lemma 7.18, there exists a nonzero continuous linear functional f satisfying $f(x) \geqslant 0$ for all $x \in C$. This implies $f(a) \geqslant f(b)$ for all $a \in A$ and all $b \in B$. ∎

7.6 Convex functions on finite dimensional spaces

In this section we gather several important properties of convex functions on finite dimensional spaces. For a more detailed account see the definitive volume by R. T. Rockafellar [288]. The first result is a refinement of Lemma 7.14 for the one-dimensional case.

7.21 Theorem *For a function $f : I \to \mathbb{R}$ on an interval in $\mathbb{R}$ the following statements are equivalent.*

1. *The function f is convex.*

2. *If $x_1, x_2, x_3 \in I$ satisfy $x_1 < x_2 < x_3$, then*

$$f(x_2) \leqslant \frac{x_3 - x_2}{x_3 - x_1} f(x_1) + \frac{x_2 - x_1}{x_3 - x_1} f(x_3).$$

3. *If $x_1, x_2, x_3 \in I$ satisfy $x_1 < x_2 < x_3$, then*

$$\frac{f(x_2) - f(x_1)}{x_2 - x_1} \leqslant \frac{f(x_3) - f(x_1)}{x_3 - x_1} \leqslant \frac{f(x_3) - f(x_2)}{x_3 - x_2}.$$

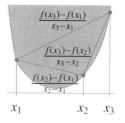

Proof: (1) $\implies$ (2) This follows from the fact that x_2 is the convex combination $\frac{x_3 - x_2}{x_3 - x_1} x_1 + \frac{x_2 - x_1}{x_3 - x_1} x_3$.

(2) $\implies$ (3) Observe that

$$f(x_2) - f(x_1) \leqslant \left(\frac{x_3 - x_2}{x_3 - x_1} - 1 \right) f(x_1) + \frac{x_2 - x_1}{x_3 - x_1} f(x_3)$$

$$= \frac{x_2 - x_1}{x_3 - x_1} [f(x_3) - f(x_1)],$$

and the first inequality follows. The second inequality uses a similar argument.

$(3) \implies (1)$ Assume $x_1 < x_3$ and $0 < \alpha < 1$ and put $x_2 = \alpha x_1 + (1 - \alpha)x_3$. Then $\alpha = \frac{x_3 - x_2}{x_3 - x_1}$ and $\beta = 1 - \alpha = \frac{x_2 - x_1}{x_3 - x_1}$. Clearly, $x_1 < x_2 < x_3$, so the first inequality yields

$$f(x_2) - f(x_1) \leqslant \frac{x_2 - x_1}{x_3 - x_1}[f(x_3) - f(x_1)]$$

$$= \beta f(x_3) - \beta f(x_1),$$

or $f(x_2) \leqslant \alpha f(x_1) + \beta f(x_3)$. ∎

The next result is an immediate consequence of the preceding.

7.22 Theorem *For a convex function $f \colon I \to \mathbb{R}$ defined on a real interval:*

1. *The function f is continuous at every interior point of I.*

2. *The left and right derivatives exist and are finite at each interior point of I. Moreover, if f_ℓ and f_r denote the left and right derivatives of f, respectively, and $x < y$ are interior points of I, then*

$$f_\ell(x) \leqslant f_r(x) \leqslant f_\ell(y) \leqslant f_r(y).$$

   In particular, the left and right derivatives are both nondecreasing on the interior of I.

3. *The function f is differentiable everywhere on the interior of I except for at most countably many points.*

Proof: We shall prove only (3): Given (2), the only way f can fail to be differentiable at x is if $f_\ell(x) < f_r(x)$ in which case there exists some rational number q_x satisfying $f_\ell(x) < q_x < f_r(x)$. It follows from (2) that if $x < y$ are both points of nondifferentiability, then $q_x < q_y$. The conclusion follows from the countability of the rational numbers. ∎

Another consequence of Theorem 7.21 is that a convex function on finite a dimensional space is subdifferentiable at every interior point of its domain.

7.23 Theorem *Every convex function $f \colon I \to \mathbb{R}$ defined on an open subinterval of $\mathbb{R}$ is subdifferentiable at every point of I.*

Proof: Let $a \in I$ and let m_r and m_ℓ denote the right and left derivatives of f at a, respectively. From Theorem 7.22 (1), we get $-\infty < m_\ell \leqslant m_r < \infty$. Now pick any number m such that $m_\ell \leqslant m \leqslant m_r$ and use part (3) of Theorem 7.21 to conclude that $f(x) \geqslant f(a) + m(x - a)$ for each $x \in I$. In other words $\partial f(x) = [m_\ell, m_r]$. ∎

Part (3) of Theorem 7.22 can be generalized as follows.

7.24 Theorem *In a finite dimensional vector space, every convex function is continuous on the relative interior of its domain.*

Proof: Let $f: C \to \mathbb{R}$ be a convex function defined on a convex subset C of a finite dimensional vector space X, and let F be the flat generated by C. Then $F = x_0 + M$ where M is a k-dimensional subspace of X and $x_0 \in \mathrm{ri}(C)$. Pick a basis $\{e_1, \ldots, e_k\}$ of M such that $x_0 \pm e_i \in C$ holds for each $i = 1, \ldots, k$. Now notice that the function $\|\cdot\|: M \to \mathbb{R}$ defined by $\left\|\sum_{i=1}^k \alpha_i e_i\right\| = \sum_{i=1}^k |\alpha_i|$ is a norm on M. So by Theorem 5.21 the norm $\|\cdot\|$ must generate the Euclidean topology of M. In particular, the set $V = \{x \in M : \|x\| < 1\}$ is a neighborhood of zero. Now notice that if $x \in x_0 + V$, then we can write $x = x_0 + \sum_{i=1}^k \alpha_i e_i$ with $\sum_{i=1}^k |\alpha_i| < 1$, so from

$$x = x_0 + \sum_{i=1}^k \alpha_i e_i = \sum_{\{i:\alpha_i>0\}} \alpha_i(x_0 + e_i) + \sum_{\{i:\alpha_i<0\}} (-\alpha_i)(x_0 - e_i) + \left(1 - \sum_{i=1}^k |\alpha_i|\right)x_0$$

and the convexity of C, we see that $x \in C$. Therefore, $x_0 + V \subset C$. Next, if we let $\mu = \max\{f(x_0), f(x_0 \pm e_1), f(x_0 \pm e_2), \ldots, f(x_0 \pm e_k)\}$, then using the convexity of f for each $x \in x_0 + V$ we have

$$f(x) \leqslant \sum_{\{i:\alpha_i>0\}} \alpha_i f(x_0 + e_i) + \sum_{\{i:\alpha_i<0\}} (-\alpha_i)f(x_0 - e_i) + \left(1 - \sum_{i=1}^k |\alpha_i|\right)f(x_0) \leqslant \mu.$$

This shows that f is bounded above on the neighborhood $x_0 + V$ of x_0. By Theorem 5.42 the function f is continuous at x_0. ∎

A convex function on a convex subset of an infinite dimensional topological vector space need not be continuous on the interior of its domain. For instance, any discontinuous linear functional on an infinite dimensional topological vector space provides such an example.

We now state but do not prove some useful theorems on the differentiability of convex functions on $\mathbb{R}^n$. But first we must recall the definition of differentiability for functions between normed spaces (such as $\mathbb{R}^n$ and $\mathbb{R}$). Let X and Y be normed spaces. Recall that $L(X, Y)$ is the normed space of continuous linear operators from X into Y.

A function $f: U \to Y$, where U is an open subset of X, is **differentiable** at some vector $x \in X$ if there exists an operator $T \in L(X, Y)$ such that for all vectors $v \in X$ for v sufficiently close to zero we have[4]

$$f(x + v) = f(x) + T(v) + o(\|v\|).$$

---

[4] The notation $o(\|v\|)$ stands for any remainder $r(v)$ satisfying $\lim_{v \to 0} \frac{r(v)}{\|v\|} = 0$. In particular, $\frac{f(x+\lambda v)-f(x)}{\lambda}$ converges uniformly as $\lambda \to 0$ to $T(v)$ on every ball at zero. Consequently, a Fréchet derivative is also a Gâteaux derivative, but not necessarily vice-versa.

It is easy to see that such a bounded operator T is uniquely determined and is called the **Fréchet derivative** or the **differential** of f at x, and is traditionally denoted $Df(x)$. The value of this operator at $v \in X$ is denoted $Df(x)(v)$. (For the case $X = \mathbb{R}^n$ and $Y = \mathbb{R}$, we can identify $L(\mathbb{R}^n, \mathbb{R})$ with $\mathbb{R}^n$ and $Df(x)$ with the gradient of f at x.)

If f is differentiable at every point in U, the mapping $x \mapsto Df(x)$ from U into $L(X, Y)$ is itself a function from an open subset of a normed space to a normed space. If this mapping is differentiable at x, its differential, which belongs to $L(X, L(X, Y))$, is called the **second differential** of f at x, denoted $D^2 f(x)$. (Thus for each $v \in X$, $D^2 f(x)(v)$ is an operator in $L(X, Y)$. Its value at $w \in X$ is $D^2 f(x)(v)(w)$. For the case $X = \mathbb{R}^n$ and $Y = \mathbb{R}$, we can identify $L(\mathbb{R}^n, L(\mathbb{R}^n, \mathbb{R}))$ with $\mathbb{R}^{n^2}$ and $D^2 f(x)$ with the **Hessian matrix** $\left[\frac{\partial^2 f(x)}{\partial x_i \partial x_j}\right]$ of f at x via $D^2 f(x)(v)(w) = \sum_{i=1}^n \sum_{j=1}^n \frac{\partial^2 f(x)}{\partial x_i \partial x_j} v_i w_j$.)

We already know from Theorem 7.12 that the subdifferential is nonempty at interior points and that when it is a singleton, it consists of the Gâteaux derivative (Theorem 7.16). The next theorem asserts that for convex functions on a finite dimensional space, a singleton subdifferential consists of the Fréchet derivative. A complete proof may be found in R. T. Rockafellar [288, Theorem 25.1, p. 242].

7.25 Theorem   *Let C be an open convex subset of $\mathbb{R}^n$ and let $f \colon C \to \mathbb{R}$ be a convex function. Then f is differentiable at x if and only if the subdifferential $\partial f(x)$ is a singleton, in which case the lone subgradient is in fact the differential of f at x.*

The next result shows that the derivative a convex function defined on an open subset of a finite dimensional space exists almost everywhere.[5]

7.26 Theorem   *If C is an open convex subset of $\mathbb{R}^n$ and $f \colon C \to \mathbb{R}$ is a convex function, then:*

   a. *the set A of points where f is differentiable is a dense $\mathcal{G}_\delta$ subset of C,*

   b. *its complement has Lebesgue measure zero, and*

   c. *the function $x \mapsto Df(x)$ from A to $L(\mathbb{R}^n, \mathbb{R})$ is continuous.*

A proof of this can be found in R. T. Rockafellar [288, Theorem 25.5 and Corollary 25.5.1, p. 246] or W. Fenchel [123, Theorems 33 and 34, pp. 86–87]. The proof that a convex function is differentiable almost everywhere has two parts. The first part uses Lemma 7.22 (3) and Fubini's Theorem 11.27 to show that a convex function has partial derivatives almost everywhere, and the second part shows that if all the partial derivative of a convex function (on a finite dimensional space) exist at a point, then it is in fact differentiable at that point.

---

[5] We discuss Lebesgue measure and sets of measure zero in Chapter 10. The phrase "almost everywhere" means "except possibly on a set of Lebesgue measure zero."

7.27 Theorem   *Let $f: C \to \mathbb{R}$ be a twice differentiable real function on an open convex subset of $\mathbb{R}^n$. Then f is convex if and only its Hessian matrix is positive semidefinite[6] everywhere in C.*

For a proof of the above result, see C. D. Aliprantis [7, Problem 1.1.2, p. 3] or J.-B. Hiriart-Urruty and C. Lemaréchal [163, Theorem 4.3.1, p. 190]. We can also say something about the almost everywhere existence of the second differential. Let us say that a correspondence $\varphi: \mathbb{R}^n \twoheadrightarrow \mathbb{R}^m$ is **differentiable** at x if there is a linear mapping $T: \mathbb{R}^n \to \mathbb{R}^m$ (that is, an $m \times n$ matrix) satisfying

$$y_v = y + T(v) + o(\|v\|), \quad \text{for all } y \in \varphi(x), \ y_v \in \varphi(x + v).$$

If φ is differentiable at x, then it is singleton-valued at x, and the linear mapping T is unique and is called the **derivative** of φ at x.

The following theorem is due to A. D. Alexandroff [6]. This statement is based on R. Howard [171, Theorems 6.1 and 7.1] and J.-B. Hiriart-Urruty and C. Lemaréchal [163, Theorem 4.3.4, p. 192]. See also the enlightening discussion in Sections I.5.1–2 (pp. 30–33) and IV.4.3 (pp. 190–193) of [163].

7.28 Theorem (Alexandroff's Theorem)   *If $f: \mathbb{R}^n \to \mathbb{R}$ is a proper convex function, and $\operatorname{dom} f$ has nonempty interior, then there exists a subset A of the interior of $\operatorname{dom} f$ such that:*

1. *The set $(\operatorname{dom} f)^\circ \setminus A$ has Lebesgue measure zero.*

2. *Both f and ∂f are differentiable at every point of A.*

3. *The derivative T of ∂f at each point of A is symmetric, positive definite, and satisfies the "second order Taylor expansion formula"*

$$f(x + v) = f(x) + Df(x)(v) + \tfrac{1}{2}T(x)(v) \cdot v + o(\|v\|^2).$$

7.7   Separation and support in finite dimensional spaces

In this section we prove results on support points that are special to the finite dimensional case.

7.29 Lemma   *In a finite dimensional vector space, if zero does not belong to a nonempty convex set C, then the convex cone generated by C is not dense.*

Proof: Let C be a nonempty convex subset of a finite dimensional vector space such that $0 \notin C$. Then the convex cone generated by C is

$$K = \{\lambda x : \lambda \geqslant 0 \text{ and } x \in C\}.$$

Let $\{e_1, e_2, \ldots, e_k\}$ be a maximal collection of linearly independent vectors that

---

[6] Recall that an $n \times n$ symmetric matrix M is **positive semidefinite** if $x^t M x \geqslant 0$ holds for each $x \in \mathbb{R}^n$. Equivalently, M is positive semidefinite if its eigenvalues are real and nonnegative.

lie in C. Then, it is easy to see that for each $x \in C$ there exist scalars $\alpha_1, \ldots, \alpha_k$ such that $x = \sum_{i=1}^{k} \alpha_i e_i$. To finish the proof, we shall show that if $v = \sum_{i=1}^{k} e_i$, then the vector $-v$ does not belong to the closure of K.

To see this, assume by way of contradiction that $-v \in \overline{K}$. Then there is a sequence $\{x_n\}$ in C and a sequence of positive scalars $\{\lambda_n\}$ such that $\lambda_n x_n \to -v$. Now for each n write $x_n = \sum_{i=1}^{k} \alpha_i^n e_i$, and note that $\lambda_n x_n = \sum_{i=1}^{k} \lambda_n \alpha_i^n e_i \to \sum_{i=1}^{k} (-1) e_i$. Since $e_1, \ldots, e_k$ are linearly independent, it follows that $\lim_{n \to \infty} \lambda_n \alpha_i^n = -1$ for each $1 \leqslant i \leqslant k$; see Corollary 5.24. This implies that for some n we must have $\alpha_i^n < 0$ for each $1 \leqslant i \leqslant k$. But then, for that n, the convexity of C implies

$$0 = \frac{1}{1 - \sum_{i=1}^{k} \alpha_i^n} x_n + \sum_{i=1}^{k} \frac{-\alpha_i^n}{1 - \sum_{i=1}^{k} \alpha_i^n} e_i \in C,$$

which is a contradiction. Hence, $-v \notin \overline{K}$, so K is not dense. ∎

In finite dimensional vector spaces disjoint convex sets can always be properly separated. (Compare this with Example 5.69.)

7.30 The Finite Dimensional Separating Hyperplane Theorem *In a finite dimensional vector space any two nonempty disjoint convex sets can be properly separated by a nonzero linear functional.*

Proof: Let A and B be two nonempty disjoint convex subsets of a finite dimensional vector space X. We assume that X is equipped with its Euclidean topology so that X is a Hausdorff topological vector space. The convex set $A - B$ does not contain 0 and then use Lemma 7.29 to conclude that the convex cone generated by $A - B$ is not dense. Now a glance at Lemma 7.20 guarantees that A and B can be separated.

The proof that A and B can be properly separated is by induction on the dimension of X. For $n = 1$, we can assume that $X = \mathbb{R}$ in which case A and B are intervals or singletons. Also, we can suppose that there exists some $\alpha \in \mathbb{R}$ such that $a \leqslant \alpha \leqslant b$ for all $a \in A$ and all $b \in B$. Now it is easy to see that the linear functional $f \colon \mathbb{R} \to \mathbb{R}$ defined by $f(x) = x$ properly separates A and B. For the induction step assume that the conclusion is true for all vector spaces of dimension n and let A and B be two nonempty disjoint convex subsets of some $(n{+}1)$-dimensional vector space X. By the first part, there exists some nonzero linear functional $f \colon X \to \mathbb{R}$ and some $\alpha \in \mathbb{R}$ such that $A \subset [f \leqslant \alpha]$ and $B \subset [f \geqslant \alpha]$. If $A \cup B$ is not a subset of $[f = \alpha]$, then f properly separates A and B. So we consider the case $A \cup B \subset [f = \alpha] = u + H$, where $H = \{x \in X : f(x) = 0\}$ and $f(u) = \alpha$. Clearly, H is a vector subspace of X of dimension n and the nonempty disjoint convex sets $A - u$ and $B - u$ are both subsets of H. But then, according to our induction hypothesis, there exists a nonzero linear functional $g \colon H \to \mathbb{R}$ that properly separates $A - u$ and $B - u$. Now note that any linear extension h of g to all of X is a linear functional that properly separates $A - u$ and $B - u$. It follows that h properly separates A and B. This completes the induction step and the proof of the theorem. ∎

For the sake of completeness, we mention without proof an additional result, due to V. Klee [207], which deals with strict separation in finite dimensional spaces. Recall that a hyperplane $[f = \alpha]$ strictly separates two sets if one lies in the half space $[f < \alpha]$ and the other lies in the half space $[f > \alpha]$.

7.31 Theorem (Strict separation in finite dimension spaces) *For two disjoint nonempty closed subsets of a finite dimensional vector space, if neither includes a ray in its boundary, then they can be strictly separated by a hyperplane.*

There is a nice characterization of separation of convex sets in finite dimensional vector spaces that will be presented in the sequel. To do this, we need some preliminary discussion regarding affine subspaces. So let X be a vector space. A translation $x + Y$ of a vector subspace Y of X is called an **affine subspace** or a **flat**. The **dimension** of an affine subspace $x + Y$ is simply the dimension of the vector subspace Y.

7.32 Lemma *Every nonempty subset A of a vector space is included in a smallest (with respect to inclusion) affine subspace—called the **affine subspace** (or the **flat**) **generated** by A and denoted F_A. The **dimension** of A is the dimension of the flat F_A.*

We present the proof and several properties of the flat F_A below. So fix a nonempty subset A of a vector space X. For each $a \in A$ let M_a denote the linear span of $A - a$.

- *If $a, b \in A$, then $M_a = M_b$, $a + M_a = b + M_b$, and $A \subset a + M_a$.*

Clearly $A \subset a + M_a$. Let $a, b, c \in A$. Then we may write $c - a = (c - b) - (a - b)$, where both $c - b$ and $a - b$ belong to $A - b$, so $c - a$ belongs to M_b. Since c is arbitrary, we have $A - a \subset M_b$, so $M_a \subset M_b$. By symmetry, $M_b \subset M_a$ is also true so that $M_a = M_b$. We may now drop the subscripts from M_a, etc. Since a and c are arbitrary members of A, we have also shown that $a - c \in M$ whenever $a, c \in A$.

To see that $a + M = b + M$, consider a typical element $a + u$ of $a + M$, where of course $u \in M$. Now write $a + u = b - (a - b) + u$. Since both $a - b$ and u belong to M, so does their sum, which shows that $a + u \in b + M$. Hence $a + M \subset b + M$. Similarly, $b + M \subset a + M$, so $a + M = b + M$.

A consequence of this is that if $x + Y = u + V$, where Y and V are vector subspaces, then $Y = V$, so every affine subspace is the translation of a unique vector subspace. (Let $A = x + Y = u + V$.)

Thus, the span of $A - a$ is independent of the vector $a \in A$, so we shall refer to it simply as M. It turns out that $a + M$ is F_A, the smallest flat that includes A.

- *If a vector subspace Y of X satisfies $A \subset x + Y$ and $a \in A$, then $M \subset Y$ and $a + M \subset x + Y$. In particular, the flat generated by A is precisely $a + M$, that is, $F_A = a + M$ for each $a \in A$.*

Since $A \subset x + Y$, there is some $y_1 \in Y$ with $a = x + y_1$. If $z \in M$, then $z = u - a$ for some $u \in a + M \subset x + Y$. Thus there is some $y_2 \in Y$ such that $u = x + y_2$, so $z = (x + y_2) - (x - y_1) = y_1 + y_2 \in Y$, showing that $M \subset Y$. Also, for any $z \in M \subset Y$, we may write $a + z = x + y_1 + z \in x + Y$, so $a + M \subset x + Y$.

Thus $a + M$ is included in any flat that includes A, so $a + M = F_A$.

- *If A is not a singleton and $a \in A$, then any maximal collection of linearly independent vectors of $A - a$ (which must exist by Zorn's lemma) is a Hamel basis for M_a.*

Let $\mathcal{H}$ be a maximal collection of linearly independent vectors of $A - a$. Then every vector in $A - a$ must be a linear combination of a finite collection of vectors of $\mathcal{H}$. This easily implies that $\mathcal{H}$ is a Hamel basis of M_a.

Now assume that X is a topological vector space. Then F_A is automatically topologized with the induced topology from X. The **relative interior** ri(A) of A is the interior of the set A considered as a subset of F_A. We let Intr($A - a$) denote the interior of the set $A - a$ viewed as a subset of the topological vector subspace M_a (where the topology of M_a is the one induced from X). Since the mapping $x \mapsto x + a$, from M_a to F_A, is an onto homeomorphism, it is clear that

$$\mathrm{ri}(A) = a + \mathrm{Intr}(A - a).$$

7.33 Lemma   *If X is a finite dimensional vector space and A is a nonempty convex subset of X, then its relative interior ri(A) is a nonempty convex set that is dense in A.*

Proof: If $A = \{a\}$, then $M_a = \{0\}$ and ri(A) $= \{a\}$. So we can assume that A is not a singleton. Fix $a \in A$ and then pick a basis $\{e_1, \dots, e_k\}$ of M_a such that $\{e_1, \dots, e_k\} \subset A - a$. Since $0 \in A - a$, we see that each vector of the form $\sum_{i=1}^{k} \lambda_i e_i$ with $\lambda_i \geq 0$ for each i and $\sum_{i=1}^{k} \lambda_i \leq 1$ lies in $A - a$.

Now notice that the function $\| \sum_{i=1}^{k} \alpha_i e_i \| = \sum_{i=1}^{k} |\alpha_i|$ is a norm on M_a. So the topology it generates coincides with the topology induced on M_a by the topology of X. We claim that the vector $e = \frac{1}{2k} \sum_{i=1}^{k} e_i \in A - a$ is an interior point of $A - a$. Indeed, if we pick $0 < \varepsilon < \frac{1}{2k}$, then it is not difficult to verify that the open ball centered at e with radius ε lies entirely in $A - a$. Thus Intr($A - a$) is a non empty convex subset of M_a that is dense in $A - a$; see Lemma 5.27 (6) and Lemma 5.28. To complete the proof now note that ri(A) $= a + \mathrm{Intr}(A - a)$. ∎

Notice that the flat generated by a nonempty convex subset A of a finite dimensional vector space X is equal to X if and only if A has an interior point in X (in which case the relative interior of A and the interior of A in X coincide).

Using Theorem 7.24 and the preceding result we also have the following.

7.34 Lemma   *If A is a nonempty convex subset of a finite dimensional vector space, then every convex function defined on A is continuous on* ri(A).

And now we are ready to characterize the separation of convex sets in finite dimensional vector spaces.

7.35 Separation of Convex Sets in Finite Dimensional Spaces   *In a finite dimensional vector space two nonempty convex sets can be properly separated if and only if their relative interiors are disjoint.*

Proof: Let A and B be nonempty convex subsets of a finite dimensional vector space X. Assume that there exists a nonzero linear functional $f: X \to \mathbb{R}$ that properly separates A and B, that is, for some $\alpha \in \mathbb{R}$ we have $A \subset [f \leqslant \alpha]$ and $B \subset [f \geqslant \alpha]$, and $A \cup B$ does not lie in $[f = \alpha]$.

Consider the case where either A or B is a singleton, say $A = \{a\}$. Clearly, ri(A) = $\{a\}$ = A. Now assume by way of contradiction that ri(A) $\cap$ ri(B) $\neq \varnothing$ or that $a \in$ ri(B). This implies $f(a) = \alpha$; otherwise $f(a) < \alpha$ yields ri(A) $\cap$ ri(B) = $\varnothing$. Now notice that f does not vanish on M_a (the linear span of $B-a$). Indeed, if $f = 0$ on M_a, then $A \cup B \subset [f = \alpha]$, which is impossible. From $f(b - a) \geqslant \alpha - f(a)$ for all $b \in B$ and Lemma 5.66, it follows that $f(x) > \alpha - f(a)$ for all $x \in$ Intr($B - a$). This implies $f(b) > \alpha$ for all $b \in a +$ Intr($B - a$) = ri(B). In particular, we get $f(a) > \alpha = f(a)$, which is absurd. Hence ri(A) $\cap$ ri(B) = $\varnothing$ must be true if either A or B is a singleton.

Next, suppose that neither A nor B is a singleton. Fix some $a \in A$ and some $b \in B$. If $f = 0$ on $M_a \cup M_b$ (where M_a is the linear span of $A - a$ and M_b the linear span of $B - b$), then we have $f(x) = f(a) \leqslant \alpha \leqslant f(b) = f(y)$ for all $x \in A$ and all $y \in B$. Since $A \cup B$ is not included in $[f = \alpha]$, it follows that either $f(a) < \alpha$ or $f(b) > \alpha$. Now notice that in either case we have $A \cap B = \varnothing$, so ri(A) $\cap$ ri(B) = $\varnothing$.

The preceding discussion shows that we can assume that either f does not vanish on M_a or on M_b. Without loss of generality we can suppose that f does not vanish on M_a. From $A \subset [f \leqslant \alpha]$ we get $A - a \subset [f \leqslant \alpha - f(a)]$, so Lemma 5.66 implies Intr($A - a$) $\subset [f < \alpha - f(a)]$, where Intr($A - a$) is the interior of $A - a$ in M_a. Hence, ri(A) = $a +$ Intr($A - a$) $\subset [f < \alpha]$. But then it easily follows that ri(A) $\cap$ ri(B) = $\varnothing$ is also true in this case.

For the converse, assume that ri(A) $\cap$ ri(B) = $\varnothing$. Then ri(A) and ri(B) are nonempty disjoint convex subsets of X, so by Theorem 7.30 they can be properly separated by a nonzero linear functional, say f. Since ri(A) and ri(B) are dense in A and B, respectively, it follows that f also properly separates A and B. ∎

We can now characterize support points in finite dimensional spaces.

7.36 Finite Dimensional Supporting Hyperplane Theorem   *Let C be a nonempty convex subset of a finite dimensional Hausdorff space and let x belong to C. Then there is a linear functional properly supporting C at x if and only if x ∉ ri(C).*

Proof: ($\Longrightarrow$) Assume that the linear functional f properly supports C at x. Then it properly supports $C - x$ at zero. Letting M denote the linear span of $C - x$, we have that $f|_M$ supports $C - x$ at zero. Lemma 5.66 then implies that $0 \notin (C - x)^\circ$ (relative to M) so that $x \notin \mathrm{ri}(C)$.

($\Longleftarrow$) Assume that $x \in C \setminus \mathrm{ri}(C)$. Then $\mathrm{ri}(\{x\}) = \{x\}$ and $\mathrm{ri}(\mathrm{ri}(C)) = \mathrm{ri}(C)$ are disjoint nonempty convex sets, so by Theorem 7.35, there is a linear functional properly separating them. This functional also properly supports C at x. ∎

7.8   Supporting convex subsets of Hilbert spaces

This section presents a few properties of the support points of a closed convex subset of a Hilbert space(Section 6.10). As before, we are only interested in real Hilbert spaces. Recall that for a point x and a nonempty closed convex set C in a Hilbert space the metric projection $\pi_C(x)$ is the unique point in C that is nearest x (Theorem 6.53).

7.37 Lemma   *For a nonempty proper closed convex subset C of a Hilbert space we have the following.*

1. *A point $c_0 \in \partial C$ is a support point of C if and only if there exists some $x \notin C$ such that $\pi_C(x) = c_0$ (in which case the vector $c_0 - x$ supports the convex set C at c_0).*

2. *The set of support points of C is dense in the boundary ∂C of C.*

Proof: (1)   Fix $c_0 \in \partial C$. Assume first that there exists some $x \notin C$ such that $\pi_C(x) = c_0$. According Lemma 6.54(c), we know that $(x - \pi_C(x), y - \pi_C(x)) \leqslant 0$, that is, $(x - c_0, y - c_0) \leqslant 0$ for each $y \in C$. Since $x \neq c_0$, the vector $p = c_0 - x \neq 0$, and the last inequality can be rewritten as $(p, c_0) \leqslant (p, y)$ for all $y \in C$. This shows that p supports C at c_0.

For the converse, assume that c_0 is a support point of C. So there exists a nonzero vector $p \in H$ such that $(p, c_0) \leqslant (p, c)$ holds for each $c \in C$. We claim that the vector $x = c_0 - p$ does not belong to C. Indeed, if $x \in C$, then from $(p, c_0) \leqslant (p, x) = (p, c_0 - p) = (p, c_0) - \|p\|^2$, we get $0 < \|p\|^2 \leqslant 0$, a contradiction. Hence, $x \notin C$. Now note that for each $c \in C$ we have $(p, c_0 - c) \leqslant 0$, so

$$
\begin{aligned}
\|x - c\|^2 &= \|(c_0 - p) - c\|^2 = \|(c_0 - c) - p\|^2 \\
&= \|c_0 - c\|^2 - 2(p, c_0 - c) + \|p\|^2 \\
&\geqslant \|p\|^2 = \|x - c_0\|^2
\end{aligned}
$$

holds for each $c \in C$. This shows that $\pi_C(x) = c_0$.

(2)   Let $c_0 \in \partial C$ and pick a sequence $\{x_n\}$ of points such that $x_n \notin C$ for each n and $x_n \to c_0$. By part (1), each $\pi_C(x_n)$ is a support point of C. By Lemma 6.54 (d), the function π_C is continuous, so $\pi_C(x_n) \to \pi_C(c_0) = c_0$. ∎

We can adapt Example 7.8 to show that not every boundary point of a closed convex subset of an infinite dimensional Hilbert space is a support point.

7.38 Example   Let $C = \ell_2^+$, the positive cone of the Hilbert space ℓ_2. It is easy to see that C is a nonempty closed convex subset of ℓ_2. Moreover, $C^\circ = \varnothing$, so every point of C is a boundary point. We claim that a point $c = (c_1, c_2, \ldots) \in C$ is a support point of C if and only if $c_i = 0$ for some i.

To see this, let $c = (c_1, c_2, \ldots) \in C$. Assume first that $c_i = 0$ for some i. It is easy to see that the basic unit vector e_i supports C at c. For the converse, assume that c is a support point of C. So there exists a nonzero vector $p \in \ell_2$ such that $(p, x) \geq (p, c)$ for all $x \in C$. Since $x \in C$ implies $x + c \in C$, we conclude that $(p, x) \geq 0$ for each $x \in \ell_2^+$. From this, it easily follows that the vector $p = (p_1, p_2, \ldots)$ satisfies $p_i \geq 0$ for each i.

Since $p \neq 0$, we infer that $p_i > 0$ for some i. Therefore, from $0 \in C$, we get $0 \leq p_i c_i \leq (p, c) \leq (p, 0) = 0$, so $c_i = 0$. Hence, the support points of C are the nonnegative vectors having at least one coordinate equal to zero. In particular, the latter shows that not every boundary point of C is a support point.

Another way of seeing this is as follows. Let $c = (c_1, c_2, \ldots) \in \ell_2^+$ be a vector satisfying $c_i > 0$ for each i. We claim that $\pi_C(x) \neq c$ for all $x \notin C$. Indeed, if $x \notin C$, then $x_k < 0$ must be the case for some k. If we consider the vector $y = (y_1, y_2, \ldots)$ defined by $y_i = c_i$ if $x_i \geq 0$ and $y_i = 0$ if $x_i < 0$, then note that the vector y belongs to C and satisfies $\|x - y\| < \|x - c\|$. This shows that $\pi_C(x) \neq c$ for all $x \notin C$. ∎

7.9 The Bishop–Phelps Theorem

We have seen that not every boundary point of a nonempty closed convex set need be a support point. Lemma 7.37 shows that for Hilbert spaces, the set of support points is at least dense in the boundary. E. Bishop and R. R. Phelps [46] extend this result to Banach spaces. In this section we present a proof of this remarkable result. The results in this section are taken from [46]. (See also R. R. Phelps [278, Theorem 3.18, p. 48] and G. J. O. Jameson [179, Theorem 3.8.14, p. 127].)

To understand the Bishop–Phelps Theorem we need a discussion of certain cones. Let X be a Banach space. For $f \in X'$ with $\|f\| = 1$ and $0 < \delta < 1$, define

$$K(f, \delta) = \{x \in X : f(x) \geq \delta\|x\|\}.$$

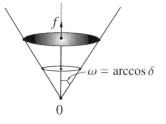

Figure 7.3. The cone $K(f, \delta)$

It is straightforward to verify that $K(f, \delta)$ is a closed convex pointed cone having a nonempty interior.[7] In a Euclidean space, it can be described as the cone with major axis

---

[7] The interior of the convex pointed cone $K(f, \delta)$ is the convex set $\{x \in X : f(x) > \delta\|x\|\}$. To see

the half line determined by the unit vector f and having angle $\omega = \arccos \delta$, see Figure 7.3. We point out now that the definition of this cone does not have an analogue outside normed spaces, which is one reason the Bishop–Phelps Theorem cannot be generalized to Fréchet spaces.

As we shall see, the cones $K(f, \delta)$ are related to the support points of convex sets. We need several properties that will be stated in terms of lemmas below.

7.39 Lemma   *Let C be a nonempty convex subset of a Banach space and assume that $c \in \partial C$. Then C is supported at c if and only if there exists some cone of the form $K(f, \delta)$ satisfying $C \cap [c + K(f, \delta)] = \{c\}$.*

Proof: Assume that some continuous linear functional g of norm one supports C at c, that is, $g(x) \leqslant g(c)$ for each $x \in C$. Then $C \cap [c + K(g, \delta)] = \{c\}$ for each $0 < \delta < 1$.

For the converse, assume that $C \cap [c + K(f, \delta)] = \{c\}$ for some continuous linear functional of norm one and some $0 < \delta < 1$. If $K = c + K(f, \delta)$, then $C \cap K^\circ = \varnothing$, and by Theorem 7.19 (3) the set C is supported at c. ∎

7.40 Lemma   *Let f and g be norm one linear functionals on a Banach space X. If for some $0 < \varepsilon < 1$ we have $\left\| g|_{\ker f} \right\| \leqslant \varepsilon$, that is, if the norm of g restricted to the kernel of f is no more than ε, then either $\|f + g\| \leqslant 2\varepsilon$ or $\|f - g\| \leqslant 2\varepsilon$.*

Proof: Let $\left\| g|_{\ker f} \right\| \leqslant \varepsilon$. By Lemma 6.9, there exists a continuous linear extension h of $g|_{\ker f}$ to all of X with $\|h\| \leqslant \varepsilon$. Since $\ker f \subset \ker(g - h)$, there exists some scalar α such that $g - h = \alpha f$; see Theorem 5.91. Now note that

$$|\alpha| = \|\alpha f\| = \|g - h\| \leqslant \|g\| + \|h\| \leqslant 1 + \varepsilon,$$

and that $0 < 1 - \varepsilon \leqslant \|g\| - \|h\| \leqslant \|g - h\| = |\alpha|$. So $\alpha \neq 0$ and $|1 - |\alpha|| \leqslant \varepsilon$.

Now if $\alpha > 0$, then $|1 - \alpha| = |1 - |\alpha||$, and thus

$$\|g - f\| = \|h + (\alpha - 1)f\| \leqslant \|h\| + |1 - \alpha| \leqslant 2\varepsilon.$$

On the other hand, if $\alpha < 0$, then $|1 + \alpha| = |1 - |\alpha||$, so

$$\|g + f\| = \|h + (1 + \alpha)f\| \leqslant \|h\| + |1 + \alpha| \leqslant 2\varepsilon,$$

and the proof is finished. ∎

7.41 Lemma   *Let f and g be bounded linear functionals of norm one on a Banach space, and let $0 < \varepsilon < \frac{1}{2}$ be given. If g is a positive linear functional with respect to the cone $K(f, \frac{\varepsilon}{2+\varepsilon})$, that is, if $g(x) \geqslant 0$ for all $x \in K(f, \frac{\varepsilon}{2+\varepsilon})$, then $\|f - g\| \leqslant 2\varepsilon$.*

---

this, notice that if some $x_0 \in X$ satisfies $f(x_0) = \delta\|x_0\|$, then the vector x_0 cannot be an interior point of $K(f, \delta)$. Indeed, if any such x_0 is an interior point of $K(f, \delta)$, then there exists some $\eta > 0$ such that $f(x_0) + f(y) = f(x_0 + y) \geqslant \delta\|x_0 + y\| \geqslant \delta\|x_0\| - \delta\|y\|$ for all $\|y\| \leqslant \eta$. This implies $|f(y)| \leqslant \delta\|y\|$ for all $\|y\| \leqslant \eta$ and consequently for all $y \in X$. So $\|f\| \leqslant \delta$, which is impossible.

Proof: Assume that f, g and $0 < \varepsilon < \frac{1}{2}$ satisfy the stated properties. Start by choosing some unit vector $x_0 \in X$ such that $f(x_0) > \frac{1+\varepsilon}{2+\varepsilon}$.

Next, note that for each $y \in \ker f$ satisfying $\|y\| \leq \frac{1}{\varepsilon}$ we have

$$\|x_0 \pm y\| \leq \|x_0\| + \|y\| \leq 1 + \frac{1}{\varepsilon} = \frac{1+\varepsilon}{\varepsilon} \leq \frac{2+\varepsilon}{\varepsilon} f(x_0) = \frac{2+\varepsilon}{\varepsilon} f(x_0 \pm y),$$

so $f(x_0 \pm y) \geq \frac{\varepsilon}{2+\varepsilon}\|x_0 \pm y\|$. Therefore, $x_0 \pm y \in K(f, \frac{\varepsilon}{2+\varepsilon})$. Now the positivity of g on $K(f, \frac{\varepsilon}{2+\varepsilon})$ implies $g(x_0 \pm y) \geq 0$, so $|g(y)| \leq g(x_0) \leq 1$ for all $y \in \ker f$ with $\|y\| \leq \frac{1}{\varepsilon}$. The latter easily yields $\left\|g|_{\ker f}\right\| \leq \varepsilon$. Now a glance at Lemma 7.40 guarantees that either $\|f - g\| \leq 2\varepsilon$ or $\|f + g\| \leq 2\varepsilon$ is true.

To see that $\|f - g\| \leq 2\varepsilon$ is true, note that $2\varepsilon < 1$ implies that there exists some unit vector $x \in X$ such that $f(x) > 2\varepsilon = 2\varepsilon\|x\| \geq \frac{\varepsilon}{2+\varepsilon}\|x\|$. Thus $x \in K(f, \frac{\varepsilon}{2+\varepsilon})$, so $g(x) \geq 0$. Consequently,

$$\|f + g\| \geq f(x) + g(x) \geq f(x) > 2\varepsilon.$$

This proves that $\|f - g\| \leq 2\varepsilon$ must be the case.   ∎

7.42 Lemma   *Let f be a norm one linear functional on a Banach space X, and let $0 < \delta < 1$ be given. If D is a nonempty closed bounded subset of X, then for each $d \in D$ there exists some $m \in D$ satisfying $D \cap [m + K(f, \delta)] = \{m\}$ and $m - d \in K(f, \delta)$.*

Proof: Define a partial order on D by $x \geq y$ if $x - y \in K(f, \delta)$. Now fix $u \in D$ and consider the nonempty set $D_d = \{x \in D : x \geq d\}$. Notice that a vector $m \in D$ satisfies $D \cap [m + K(f, \delta)] = \{m\}$ and $m - d \in K(f, \delta)$ if and only if m is a maximal element in D_d with respect to $\geq$. So to complete the proof we must show that the partially ordered set $(D_d, \geq)$ has a maximal element. By Zorn's Lemma, it suffices to prove that every chain in D_d has an upper bound in D_d.

To this end, let $\mathcal{C}$ be a chain in D_u. If some $u \in \mathcal{C}$ satisfies $u \geq v$ for all $v \in \mathcal{C}$, then we are done. So assume that for each $u \in \mathcal{C}$ there exists some $v \in \mathcal{C}$ with $v > u$. If we let $A = \mathcal{C}$ and $x_\alpha = \alpha$ for each $\alpha \in \mathcal{C}$, we can identify $\mathcal{C}$ with the increasing net $\{x_\alpha\}$. Since D_u is norm bounded, it follows that $\{f(x_\alpha)\}$ is an increasing bounded net of real numbers, and hence a Cauchy net. Since for any α and β we have either $x_\alpha \geq x_\beta$ or $x_\beta \geq x_\alpha$, it follows that $\delta\|x_\alpha - x_\beta\| \leq |f(x_\alpha) - f(x_\beta)|$ for all α and β. This implies that $\{x_\alpha\}$ is a Cauchy net in X. Since X is a Banach space this net converges in X, say to some $m \in X$. Clearly, $m \in D$ and since the cone $K(f, \delta)$ is closed, we get $m \geq x_\alpha$ for each α (see also the footnote in the proof of Theorem 8.43). That is, m is an upper bound of the chain $\mathcal{C}$, and the proof is finished.   ∎

We are now ready to state and prove the Bishop–Phelps Theorem.

7.43 Theorem (Bishop–Phelps) *For a closed convex subset C of a Banach space X we have the following.*

1. *The set of support points of C is dense in the boundary of C.*

2. *If in addition C is norm bounded, then the set of bounded linear functionals on X that support C is dense in X'.*

Proof: (1) Fix some $x_0 \in \partial C$ and let $\varepsilon > 0$. Choose some $y_0 \notin C$ such that $\|x_0 - y_0\| < \frac{\varepsilon}{2}$. By the Separation Corollary 5.80 there is a nonzero continuous linear functional f satisfying $f(y_0) > f(z)$ for all $z \in C$. Without loss of generality we may normalize f so that its norm is one, that is, $\|f\| = \sup\{f(z) : \|z\| \leqslant 1\} = 1$.

Now let $K = K(f, \frac{1}{2})$ and let $D = C \cap (x_0 + K)$, which is a nonempty closed convex set. If $x \in D$, then $x - x_0 \in K$, so

$$\tfrac{1}{2}\|x - x_0\| \leqslant f(x - x_0) = f(x) - f(x_0) < f(y_0) - f(x_0) \leqslant \|y_0 - x_0\| < \tfrac{\varepsilon}{2}.$$

Hence, $\|x - x_0\| < \varepsilon$ and thus $D \subset B_\varepsilon(x_0)$. In particular, D is a norm bounded set. According to Lemma 7.42, there exists some $m \in D$ such that $D \cap (m + K) = \{m\}$ and $m - x_0 \in K$.

Clearly, $m \in C \cap (m + K)$. Now fix $x \in C \cap (m + K)$. Then there exists some $k \in K$ such that $x = m + k = x_0 + (m - x_0) + k \in x_0 + K$, and so $x \in C \cap (x_0 + K) = D$. This implies $x \in D \cap (m + K) = \{m\}$, that is, $x = m$. Consequently $C \cap (m + K) = \{m\}$ and so (by Lemma 7.39) m is a support point of C satisfying $\|x_0 - m\| < \varepsilon$.

(2) Now assume that C is a nonempty, closed, norm bounded, and convex subset of a Banach space X. Fix $f \in X'$ with $\|f\| = 1$ and let $\varepsilon > 0$. Pick some $0 < \delta < \frac{1}{2}$ satisfying $2\delta < \varepsilon$. It suffices to show that there exists a norm one linear functional $g \in X'$ that supports C satisfying $\|f - g\| \leqslant 2\delta$.

For simplicity, let $K = K(-f, \frac{\delta}{2+\delta})$. By Lemma 7.42 there exists some $m \in C$ such that $C \cap (m + K^\circ) = \varnothing$. So by the Separation Theorem 5.67, there exists some bounded linear functional $h \in X'$ of norm one satisfying $h(c) \leqslant h(m + k)$ for all $c \in C$ and all $k \in K$. This implies that h attains its maximum at $m \in C$ and that $h(x) \geqslant 0$ for each $x \in K$. Now a glance at Lemma 7.41 guarantees that $\|f - h\| \leqslant 2\delta$. So h is a norm one bounded linear functional that supports C at m and satisfies $\|f - h\| \leqslant 2\delta < \varepsilon$. ∎

We remind you that Example 7.9 shows that the Bishop–Phelps Theorem cannot be extended to Fréchet spaces. The proof relies heavily on properties of the cones $K(f, \delta)$, which are cones by virtue of the homogeneity of the norm. There is no corresponding construction without a norm. We also point out that the situation in complex Banach spaces is also different. V. I. Lomonosov [229] has exhibited a bounded closed convex subset of a complex Banach space with no support points whatever.

Let us illustrate the Bishop–Phelps Theorem with some examples. For the first part of the Bishop–Phelps Theorem we consider a closed convex set from a Banach lattice.

7.44 Example   The closed convex set we have in mind is the positive cone of a Banach lattice. So let E be a Banach lattice without order units—for instance let $E = \ell_1$ or $E = L_1[0, 1]$. In this case, we know (Corollary 9.41) that E^+ has an empty interior, so $\partial E^+ = E^+$. Also, recall that a vector $x \in E^+$ is *strictly positive* if $f(x) > 0$ for each $0 < f \in E'_+$.

Now let $x_0 \in E^+$ be a point of support of E^+. This means that there exists a nonzero continuous linear functional $f \in E'$ satisfying $f(x_0) \leqslant f(x)$ for all $x \in E^+$. Since E^+ is a cone, it easily follows that $f(x) \geqslant 0$ for each $x \in E^+$, that is, $f \in E'_+$. From $0 \in E^+$ we get $f(x_0) = 0$. In other words, we have shown that a vector $x_0 \in E^+$ is a support point of E^+ if and only if there exists a nonzero positive linear functional f satisfying $f(x_0) = 0$. In particular, no strictly positive vector is a support point of E^+. The Bishop–Phelps Theorem in this case asserts that if e is a strictly positive vector, then for each $\varepsilon > 0$ there exists a support point x_0 of E^+ such that $\|e - x_0\| < \varepsilon$.[8] (It is also interesting to recall that the set of strictly positive vectors is either empty or dense in E^+.)   ∎

7.45 Example   Before presenting an example for the second part of the Bishop–Phelps Theorem, let us make a comment. If D is any nonempty weakly compact subset of a Banach space X, then every continuous linear functional on X attains a maximum value on D, that is, it supports D. Since every nonempty bounded convex closed subset of a Banach space is also weakly closed (Theorem 5.98), the second part of the Bishop–Phelps Theorem is really a new result when C is a nonempty bounded weakly closed convex subset of a Banach space that is not weakly compact.

We now invoke James' Theorem 6.36, which states that: *A nonempty bounded weakly closed subset of a Banach space is weakly compact if and only if every continuous linear functional attains a maximum on the set.* Also, recall that Theorem 6.25 asserts that: *A Banach space is reflexive if and only if its closed unit ball is weakly compact.*

Thus, the closed unit ball U of a non-reflexive Banach space X is an example of a nonempty bounded weakly closed convex subset of a Banach space that cannot be supported by each bounded linear functional. However, by the second part of the Bishop–Phelps Theorem the bounded linear functionals that support U are dense in X'.   ∎

7.46 Theorem (Bishop–Phelps) *Assume that A and B are two nonempty subsets of a Banach space X satisfying the following properties.*

    a. *A is closed and convex.*

---

[8] To prove directly that the set S of all support points of E^+ is dense in E^+ argue as follows. Let $x \in E^+$ and assume by way of contradiction that x does not belong to the norm closure of S. Then there exists some $\delta > 0$ such that $B_\delta(x) \cap S = \varnothing$. Now let $y \in B_\delta(x)$. From $|x - y^+| = |x^+ - y^+| \leqslant |x - y|$, it follows that $y^+ \in B_\delta(x)$, so $y^+ \notin S$, which means that $y^+ \gg 0$. In particular, y^+ is a weak unit. The latter, in view of $y^+ \wedge y^- = 0$, implies $y^- = 0$ or $y = y^+ - y^- = y^+ \in E^+$. Therefore, $B_\delta(x) \subset E^+$, contrary to the fact that E^+ has no interior points.

b. *B is bounded.*

c. *There exists some $f \in X'$ with $\|f\| = 1$ such that $\sup f(A) < \inf f(B)$.*

Then for each $\varepsilon > 0$ we can find some $g \in X'$ with $\|g\| = 1$ and some $a \in A$ so that $\|f - g\| \leqslant 2\varepsilon$ and $g(a) = \sup g(A) < \inf g(B)$.

Proof: It is enough to consider $0 < \varepsilon < \frac{1}{2}$. Let $\alpha = \sup f(A)$, $\beta = \inf f(B)$, and then fix γ such that $\alpha < \gamma < \beta$. Now consider the nonempty bounded set $V = B + (\beta - \gamma)U$ and note that $\inf f(V) = [\inf f(B)] - (\beta - \gamma) = \gamma$. Now let $\delta = \frac{2+\varepsilon}{\varepsilon}$ and then choose $u \in A$ such that

$$\alpha - f(u) < \frac{\gamma - \alpha}{2\delta}.$$

Now fix some $\theta > \max\{\frac{\gamma-\alpha}{2}, \sup_{y \in V} \|y - u\|\}$, put $k = \frac{2\delta\theta}{\gamma-\alpha}$ and note that $1 < \delta < k$. By Lemma 7.42 there exists some $a_0 \in A$ such that $A \cap [a_0 + K(f, \frac{1}{k})] = \{a_0\}$ and $a_0 - u \in K(f, \frac{1}{k})$. We claim that

$$V \subset a_0 + K(f, \tfrac{1}{k}). \tag{$\star$}$$

To see this, note that for each $y \in V$ we have

$$\|y - a_0\| \leqslant \|y - u\| + \|a_0 - u\| < \theta + \|a_0 - u\| \leqslant \theta + kf(a_0 - u)$$
$$\leqslant \theta + k[\alpha - f(u)] < \theta + \frac{k(\gamma - \alpha)}{2\delta} = 2\theta$$
$$< 2\delta\theta = k(\gamma - \alpha) \leqslant kf(y - a_0).$$

Next, pick a nonzero linear functional $g \in X'$ with $\|g\| = 1$ such that

$$g(a_0) = \sup g(A) \leqslant \inf g(a_0 + K(f, \tfrac{1}{k})) \leqslant \inf g(V)$$
$$= [\inf g(B)] - (\beta - \gamma) < \inf g(B).$$

Moreover, from $\inf g(a_0 + K(f, \frac{1}{k})) \geqslant g(a_0)$, it follows that the linear functional g is $K(f, \frac{1}{k})$-positive. Since $\frac{1}{k} < \frac{1}{\delta} = \frac{\varepsilon}{2+\varepsilon}$ implies $K(f, \frac{\varepsilon}{2+\varepsilon}) \subset K(f, \frac{1}{k})$, we see that g is also $K(f, \frac{\varepsilon}{2+\varepsilon})$-positive. But then a glance at Lemma 7.41 guarantees that $\|f - g\| \leqslant 2\varepsilon$, and the proof is finished. ∎

This theorem has a number of interesting applications. The first one is a sharper Banach space version of the Strong Separating Hyperplane Theorem 5.79.

7.47 Corollary *Assume that A and B are two nonempty disjoint convex subsets of a Banach space X such that A is closed and B is weakly (and in particular norm) compact. Then there exist a non-zero linear functional $g \in X'$ and vectors $a_0 \in A$ and $b_0 \in B$ such that*

$$\sup g(A) = g(a_0) < g(b_0) = \inf g(B).$$

Proof: By Theorem 5.79 there is a nonzero linear functional $f \in X'$ satisfying $\sup f(A) < \inf f(B)$. Without loss of generality we may assume $\|f\| = 1$. The hypotheses of Theorem 7.46 are satisfied, so there exist $g \in X'$ with $\|g\| = 1$ and some $a_0 \in A$ satisfying $g(a_0) = \sup g(A) < \inf g(B)$. Since B is weak* compact there is a point $b_0 \in B$ satisfying $g(b_0) = \inf g(B)$. ∎

The next result follows immediately from the preceding corollary and is a much stronger version of Corollary 5.83 that is valid for Banach spaces.

7.48 Corollary   *Every proper nonempty convex closed subset of a Banach space is the intersection of all closed half spaces that support it.*

7.49 Corollary   *Every proper nonempty closed subset of a separable Banach space is the intersection of a countable collection of closed half spaces that support it.*

Proof: Let C be a proper nonempty convex closed subset of a separable Banach space X and let $\{x_1, x_2, \ldots\}$ be a countable subset of $X \setminus C$ that is norm dense in $X \setminus C$. For each n let $d_n = d(x_n, C) > 0$, the distance of x_n from C, and note that $C \cap (x_n + \frac{d_n}{2} U) = \varnothing$. Now, according to Theorem 7.46, for each n there exist some nonzero $g_n \in X'$ and some $y_n \in C$ such that

$$g(y_n) = \sup g_n(C) < \inf g_n(x_n + \tfrac{d_n}{2} U). \qquad (\star)$$

Next, take any $x \in X \setminus C$ and put $d = d(x, C) > 0$. Choose some x_k such that $\|x - x_k\| < \frac{d}{3}$. This implies that for each $c \in C$ we have

$$\|c - x_k\| \geqslant \|c - x\| - \|x - x_k\| \geqslant d - \frac{d}{3} = \frac{2}{3} d.$$

Thus $d_k = \inf_{c \in C} \|c - x_k\| \geqslant \frac{2}{3} d$, which implies that $\|x - x_k\| < \frac{1}{2} d_k$. Consequently $x \in x_k + \frac{d_k}{2} U$ and from $(\star)$ we get $g_k(x) > \sup g_k(C)$ or

$$-g_k(x) < -g_k(y_k) = \inf[-g_k(C)],$$

and this leads to the desired conclusion. ∎

Another result that is closely related to the Bishop–Phelps Theorem is due to A. Brøndsted and R. T. Rockafellar [64].

7.50 Theorem (Brøndsted–Rockafellar)   *Let $f : X \to \mathbb{R}^*$ be a lower semicontinuous proper convex function on a Banach space. Then the set of points at which f is subdifferentiable is dense in* dom f.

The proof is subtler than you might think—after all, we have already remarked that [64] contains an example of a nowhere subdifferentiable lower semicontinuous proper convex function on a Fréchet space. The proof uses constructions similar to those in the proof of the Bishop–Phelps Theorem. See R. R. Phelps [278] for a complete proof and a discussion of the relationships between the two theorems.

7.10   Support functionals

Recall that for a given pair $\langle X, X' \rangle$ all consistent topologies on X (or X') have the same closed convex sets and the same lower semicontinuous sublinear functions. Moreover, every proper closed convex subset C of X' is the intersection of all the closed half spaces that include it. A convenient way to summarize information on the half spaces including C is via its support functional. For a nonempty subset C of X', define the **support functional**[9] $h_C \colon X \to \mathbb{R}^*$ of C by

$$h_C(x) = \sup\{\langle x, x' \rangle : x' \in C\}.$$

Note that this supremum may be ∞ if C is not compact. Given an extended real-valued sublinear function $h \colon X \to (-\infty, \infty]$, define

$$C_h = \{x' \in X' : \langle x, x' \rangle \leqslant h(x) \text{ for all } x \in X\}.$$

That is, C_h is the set of linear functionals that are dominated by h. The support functional h of a nonempty set is a proper convex function since $h(0) = 0$. Under the usual convention that $\sup \varnothing = -\infty$, if we apply the definition of the support functional to the empty set, we obtain the constant function $h_\varnothing = -\infty$, which is an improper convex function that fails to be positively homogeneous, since $h_\varnothing(0) = -\infty \neq 0$.

7.51 Theorem   *Let $\langle X, X' \rangle$ be a dual pair, and let C be a nonempty, closed, convex subset of X'. Then the support functional $h_C \colon X \to (-\infty, \infty]$ is a proper sublinear and lower semicontinuous functional.*

Conversely, if $h \colon X \to (-\infty, \infty]$ is a proper lower semicontinuous sublinear function, then C_h is a nonempty closed convex subset of X'.

Furthermore, we have the duality $C = C_{h_C}$ and $h = h_{C_h}$.

Proof: Let C be a nonempty, convex and closed subset of X'. For each $x' \in C$ and all $x, y \in X$, we have

$$|\langle x + y, x' \rangle| \leqslant |\langle x, x' \rangle| + |\langle y, x' \rangle| \leqslant h_C(x) + h_C(y).$$

Hence $h_C(x + y) \leqslant h_C(x) + h_C(y)$, so h_C is subadditive. Clearly, $h_C(\alpha x) = \alpha h_C(x)$ for all $\alpha \geqslant 0$. Properness follows from the nonemptiness of C.

Since h_C is the supremum of the family C of continuous linear functionals, it is lower semicontinuous by Lemma 2.41.

Next, we establish that $C = C_{h_C} = \{x' \in X' : \langle x, x' \rangle \leqslant h_C(x) \text{ for all } x \in X\}$. Note first that $C \subset C_{h_C}$. Now note that by Corollary 5.80 on separating points

---

[9] For reasons we do not wish to go into here (involving duality of functions), most authors in the field of convex analysis employ the notation $\delta^*(x \mid C)$ rather than $h_C(x)$ for the support functional of the set C.

from closed convex sets, we have that if $y' \notin C$ there exists some $x \in X$ such that $\langle x, y' \rangle > \sup\{\langle x, x' \rangle : x' \in C\} = h_C(x)$, so $y' \notin C_{h_C}$. Thus $C_{h_C} = C$.

For the second part of the theorem, assume that h is a lower semicontinuous sublinear function. Then $C_h = \bigcap_{x \in X}\{x' \in X' : \langle x, x' \rangle \leqslant h(x)\}$ is obviously a weak∗-closed convex subset of X'. It is nonempty by the Hahn–Banach Extension Theorem 5.53.

Finally, to complete the proof, we need to show that $h = h_{C_h}$, or in other words we need to show that

$$h(x) = \sup\{\langle x, x' \rangle : x' \in X' \text{ and } \langle y, x' \rangle \leqslant h(y) \text{ for all } y \in X\}.$$

That is, we need to show that f is the pointwise supremum of the linear functions that it dominates. By Theorem 7.6 we know that h is the pointwise supremum of all the affine functions that it dominates. But h is proper and homogeneous, so $h(0) = 0$, which implies that h dominates the affine function $x \mapsto x'(x) + \alpha$, then $\alpha = 0$, in which case h dominates the linear function x'. This shows that $h = h_{C_h}$. ∎

If, in addition, the set C is weak∗ compact, we can say more, namely that its support functional is finite and Mackey continuous. Recall that the polar C° of a set C, is the convex set $\{x' \in X' : |\langle x, x' \rangle| \leqslant 1 \text{ for all } x \in C\}$.

7.52 Theorem   *Let $\langle X, X' \rangle$ be a dual pair, and let K be a nonempty weak∗-compact convex subset of X'. Then the support functional h_K is a proper $\tau(X, X')$-continuous sublinear function on X.*

Conversely, if $h \colon X \to \mathbb{R}$ is a $\tau(X, X')$-continuous sublinear function, then K_h is a nonempty weak∗ compact convex subset of X'.

Furthermore, we have the duality $K = K_{h_K}$ and $h = h_{K_h}$.

Proof: Assume first that K is a nonempty weak∗-compact convex subset of X'. By Theorem 7.51 all that remains to be proven is that h_K is Mackey continuous. To see this, let C be the convex circled hull of K. By Corollary 5.31, we know that C is a weak∗-compact, convex, and circled subset of X'. So from the definition of the Mackey topology, its polar C° is a $\tau(X, X')$-neighborhood of zero. Now for $x \in C^\circ$ and $x' \in K$, we have $|\langle x, x' \rangle| \leqslant 1$, so $|h_K(x)| \leqslant 1$ for each $x \in C^\circ$. By Lemma 5.51, h_K is Mackey continuous.

Now assume that h is a $\tau(X, X')$-continuous sublinear function on X. By Theorem 7.51 all that remains to be proven is that K_h is weak∗ compact. Since K_h is weak∗ closed, it suffices to show that it is included in a weak∗-compact set. Now by the Mackey continuity of h at zero, there exists a nonempty, convex, circled, and w^*-compact subset C of X' such that $|h(x)| \leqslant 1$ for each $x \in C^\circ$. But then for each $x \in C^\circ$ and $x' \in K_h$, we have $\pm\langle x, x' \rangle \leqslant \max\{h(x), h(-x)\} \leqslant 1$, so $|\langle x, x' \rangle| \leqslant 1$ for each $x \in C^\circ$. It follows that $x' \in C^{\circ\circ} = C$. Thus $K_h \subset C$, so K_h is a weak∗-compact subset of X'. ∎

The following corollary appears in K. Back [29].

7.53 Corollary   *Let $\langle X, X' \rangle$ be a dual pair and let K be a nonempty weak*-compact convex set in X'. Assume $0 \notin K$. Then $\{x \in X : \langle x, x' \rangle < 0 \text{ for all } x' \in K\}$ is a nonempty Mackey-open convex cone.*

Proof: Observe that

$$\{x \in X : \langle x, x' \rangle < 0 \text{ for all } x' \in K\} = \{x \in X : h_K(x) < 0\}.$$

Since $0 \notin K$, the Separating Hyperplane Theorem 5.80 shows that this set is nonempty. (Why?) Theorem 7.52 implies that it is Mackey open, and clearly it is a convex cone.   ∎

We take this opportunity to point out the following simple results.

7.54 Lemma   *For a dual pair $\langle X, X' \rangle$ we have the following.*

1. *The support functional of a singleton $\{x'\}$ in X' is x' itself.*

2. *The support functional of the sum of two nonempty subsets F and C of X' satisfies $h_{F+C} = h_F + h_C$.*

3. *Let $\{K_n\}$ be a decreasing sequence of nonempty weak* compact subsets of the space X'. If $K = \bigcap_{n=1}^{\infty} K_n$, then $K \neq \varnothing$ and the sequence $\{h_{K_n}\}$ of support functionals satisfies $h_{K_n}(x) \downarrow h_K(x)$ for each $x \in X$.*

Proof: We prove only the third claim. So let $\{K_n\}$ be a sequence of nonempty weak* compact subsets of X' satisfying $K_{n+1} \subset K_n$ for each n. Then $K = \bigcap_{n=1}^{\infty} K_n$ is nonempty. Clearly $h_K(x) \leqslant h_{K_n}(x)$ for all n, so $h_K(x) \leqslant \inf_n h_{K_n}(x)$ for $x \in X$.

For the reverse inequality, fix $x \in X$. Then by the weak* compactness of K_n, for each n there exists some $x'_n \in K_n$ satisfying $x'_n(x) = h_{K_n}(x)$. Since $\{x'_n\} \subset K_1$, it follows that the sequence $\{x'_n\}$ has a weak* limit point x' in X'. It follows (how?) that $x' \in K$ and $h_K(x) \geqslant x'(x) = \inf_n h_{K_n}(x)$. Therefore $h_K(x) = \inf_n h_{K_n}(x)$ for each $x \in X$.   ∎

We already know that if the support functional of a convex weak* compact set C dominates a continuous linear functional x', then x' belongs to C. The same is true of the linear part of an affine function.

7.55 Lemma   *Let $\langle X, X' \rangle$ be a dual pair, and let C be a weak*-closed convex subset of X' with support functional h_C. If $g = x' + c$ is a continuous affine function satisfying $g \leqslant h_C$, then $x' \in C$ and $c \leqslant 0$.*

Proof: The cases $C = \emptyset$ is trivial: No affine g satisfies $g \leqslant h_\emptyset = -\infty$. So suppose that C is nonempty. Then $h_C(0) = 0$. Let g be a continuous affine function satisfying $g \leqslant h_C$. Write $g(x) = x'(x) + c$, where $x' \in X'$ and $c \in \mathbb{R}$. Now fix x in X. By hypothesis $g(\lambda x) = x'(\lambda x) + c \leqslant h_C(\lambda x)$ for every λ. Therefore $-c \geqslant x'(\lambda x) - h_C(\lambda x) = \lambda[x'(x) - h_C(x)]$ for all $\lambda > 0$. This implies $x'(x) \leqslant h_C(x)$. Since x is arbitrary, $x' \leqslant h_C$. Theorem 7.51 now implies that $x' \in C$. Since $c = x'(0) + c \leqslant h_C(0) = 0$, we have $c \leqslant 0$. ∎

We can now describe the support functional of the intersection of two closed convex sets.

7.56 Theorem   *Let $\langle X, X' \rangle$ be a dual pair, and let A and B be weak*-closed convex subsets of X' with $A \cap B \neq \emptyset$. Then the support functional of $A \cap B$ is the convex envelope of* $\min\{h_A, h_B\}$.

Proof: Let $h_C : X \to \mathbb{R}^*$ denote the support functional of $C = A \cap B$. By Theorem 7.51, h_C is an extended real-valued lower semicontinuous sublinear functional on X, and clearly $h_C \leqslant \min\{h_A, h_B\}$. Therefore by Theorem 7.6 it suffices to prove that if g is a continuous affine function satisfying $g \leqslant \min\{h_A, h_B\}$, then $g \leqslant h_C$. So suppose g is such a function and write $g(x) = x'(x) + c$, where $x' \in X'$ and $c \in \mathbb{R}$. By Lemma 7.55 we conclude that $c \leqslant 0$ and $x' \in A \cap B = C$, so by Theorem 7.51, $x' \leqslant h_C$. Therefore $g(x) = x'(x) + c \leqslant x'(x) \leqslant h_C(x)$, and we are finished. ∎

Note that if one of A or B is weak* compact, then the preceding theorem applies even if A and B are disjoint: The support function $h_\emptyset(x)$ of the empty set at x is the supremum of the empty set, which is $-\infty$ by convention. The convex envelope of $\min\{h_A, h_B\}$ is the supremum of the continuous affine functions that it dominates. Suppose that $g(x) = x'(x) + c$ is a continuous affine function satisfying $g \leqslant h_A$ and $g \leqslant h_B$. Since $h_A(0) = h_B(0) = 0$, we must have $c \leqslant 0$. Since A and B are disjoint and one is compact, they can be strongly separated by some x in X. That is, $y'(x) \geqslant \alpha$ for $y' \in A$ and $y'(x) < \alpha - \varepsilon$ for $y' \in B$ and some $\varepsilon > 0$. Therefore $h_A(-x) \leqslant -\alpha$ and $h_B(x) \leqslant \alpha - \varepsilon$. Then for any $\lambda > 0$, we have

$$g(-\lambda x) = x'(-\lambda x) + c \leqslant h_A(-\lambda x) \leqslant -\lambda\alpha$$

and $g(\lambda x) = x'(\lambda x) + c \leqslant h_B(\lambda x) < \lambda(\alpha - \varepsilon)$. By rearranging terms, we get $\lambda(\alpha - \varepsilon) - c \geqslant x'(\lambda x) \geqslant \lambda\alpha + c$. Thus we conclude $c \leqslant -\frac{\lambda\varepsilon}{2}$ for every $\lambda > 0$, which is impossible. In other words there can be no continuous affine function g satisfying $g \leqslant \min\{h_A, h_B\}$. Taking the supremum over the empty set implies that the convex envelope of $\min\{h_A, h_B\}$ is the constant $-\infty$, which is just the support functional of the empty set.

We now point out that the family of weak* compact convex subsets of X' partially ordered by inclusion is a lattice. (That is, every pair of sets has both an infimum and a supremum.) The infimum of A and B, $A \wedge B$, is just $A \cap B$, and the supremum $A \vee B$ is $co(A \cup B)$. (Recall that Lemma 5.29 guarantees that $co(A \cup B)$ is compact.) Likewise, the collection of continuous sublinear functions on X under the pointwise ordering is a lattice with $f \vee g = \max\{f, g\}$, and $f \wedge g$ is the convex envelope of $\min\{f, g\}$. (Here we include the constant $-\infty$ as an honorary member of the family.) Now consider the surjective one-to-one mapping $A \mapsto h_A$ between these two lattices. It follows from Lemma 7.54 and Theorem 7.56 that this mapping preserves the algebraic and lattice operations in the following sense:

- $h_{A \vee B} = h_A \vee h_B$ and $h_{A \wedge B} = h_A \wedge h_B$.

- $A \subset B$ implies $h_A \leqslant h_B$.

- $h_{A+B} = h_A + h_B$ and $h_{\lambda A} = \lambda h_A$ for $\lambda > 0$.

We close with a characterization of the subdifferentiability of a support function. To simplify notation, we refer to the support function h_C as simply h.

7.57 Theorem *Let $\langle X, X' \rangle$ be a dual pair, and let C be a nonempty weak*-closed convex subset of X' with support functional $h: X \to \mathbb{R}^*$, that is,*

$$h(x) = \sup\{\langle x, x' \rangle : x' \in C\}.$$

Then $x' \in X'$ is a subgradient of h at x if and only if $\langle x, x' \rangle = h(x)$, that is, x' maximizes x over C.

Proof: Assume first that x' is a subgradient of h at x, that is, it satisfies the subgradient inequality

$$h(y) \geqslant h(x) + \langle y - x, x' \rangle \qquad (\star)$$

for all $y \in X$. If $x' \notin C$, by Corollary 5.80, there is some $y \in X$ with $\langle y, x' \rangle > h(y)$. Then for $\lambda > 0$ large enough, $\langle \lambda y, x' \rangle - h(\lambda y) > \langle x, x' \rangle - h(x)$, which violates $(\star)$. By contraposition, we may conclude that $x' \in C$. Now evaluate $(\star)$ at $y = 0$ to get $\langle x, x' \rangle \geqslant h(x)$. But since $x' \in C$, we must have $\langle x, x' \rangle \leqslant h(x)$, so in fact $\langle x, x' \rangle = h(x)$.

For the converse, assume that $x' \in C$ and $\langle x, x' \rangle = h(x)$. Since $h(y) \geqslant \langle y, x' \rangle$ for any y by definition, combining these two facts yields $(\star)$. ∎

7.11 Support functionals and the Hausdorff metric

We offer for your consideration a characterization of the Hausdorff metric on the space of closed bounded convex subsets of a normed space. Following C. Castaing and M. Valadier [75, Theorem II-18, p. 49], we start with a seminorm.

Let X be a locally convex space, and fix a continuous seminorm p on X. Let U denote the closed unit ball $\{x \in X : p(x) \leqslant 1\}$, and let d denote the semimetric induced by p. Let $\mathcal{C}$ denote the collection of all closed and p-bounded nonempty convex subsets of X. Let ρ_d denote the Hausdorff semimetric on $\mathcal{C}$ induced by d. That is, $\rho(A, B) = \max\{\sup_{x \in A} d(x, B), \sup_{x \in B} d(x, A)\}$. Recall that the support functional $h_C \colon X' \to \mathbb{R}^*$ of a nonempty subset C of X is given by $h_C(x') = \sup\{x'(x) : x \in C\}$.

7.58 Lemma   (Hausdorff semimetric and support functionals)   *Let X be a locally convex space, and let p be a continuous seminorm on X with induced semimetric d. Let U denote the closed unit ball $\{x \in X : p(x) \leqslant 1\}$. Then for any two nonempty closed and p-bounded convex subsets A and B of X we have*

$$\rho_d(A, B) = \sup\{|h_A(x') - h_B(x')| : x' \in U^\circ\}.$$

Proof: Observe that since A and B are closed and convex, $A \subset B$ if and only if $h_A \leqslant h_B$. (See the remarks at the end of Section 7.10.) Recall Lemma 3.71, which implies that

$$\rho_d(A, B) = \inf\{\varepsilon > 0 : B \subset A + \varepsilon U \text{ and } A \subset B + \varepsilon U\}. \qquad (\star)$$

Also, recall that $h_{B+\varepsilon U} = h_B + \varepsilon h_U$ (Lemma 7.54 (2)). Therefore, recalling the homogeneity of support functionals and rearranging terms, $A \subset B + \varepsilon U$ if and only if $h_A(x') - h_B(x') \leqslant \varepsilon h_U(x')$ for all $x' \in U^\circ$. Thus $\rho_d(A, B) \leqslant \varepsilon$ if and only if $|h_A(x') - h_B(x')| \leqslant \varepsilon$ for all $x' \in U^\circ$. This equivalence coupled with $(\star)$ proves the desired formula.   ∎

7.59 Corollary   (Hausdorff metric on convex sets)   *For nonempty norm-closed and bounded convex subsets A and B of a normed space we have*

$$\rho(A, B) = \sup_{x' \in U'} |h_A(x') - h_B(x')|,$$

where ρ is the Hausdorff metric induced by the norm and U' is the closed unit ball of X'.

Proof: This follows from Lemma 7.58 by recalling that $U^\circ = U'$.   ∎

In certain instances, the space of convex nonempty w^*-closed sets is itself a closed subspace of the space of nonempty w^*-closed sets.

7.60 Theorem   *Let X be a separable normed space and let $\mathcal{F}$ denote the compact metrizable space of all nonempty w^*-closed subsets of the compact metrizable space (U', w^*). Then the collection of nonempty convex w^*-closed subsets of U' is a closed subset of $\mathcal{F}$.*

Proof: Start by recalling that if $\{x_1, x_2, \ldots\}$ is a dense subset of the closed unit ball U of X, then the formula $d(x', y') = \sum_{m=1}^{\infty} \frac{1}{2^m} |x'(x_m) - y'(x_m)|$ defines a metric on U' that generates the w^*-topology on U'; see the proof of Theorem 6.30.

Now let $\{C_n\}$ be a sequence of convex nonempty w^*-closed subsets of U' satisfying $C_n \to F$ in $(\mathcal{F}, \rho_d)$ and let $\varepsilon > 0$. Then for all sufficiently large n we have $F \subset N_\varepsilon(C_n)$ and $C_n \subset N_\varepsilon(F)$; see Lemma 3.71. Now $N_\varepsilon(C_n)$ is convex (why?), so $\overline{\text{co}}\,F \subset N_{2\varepsilon}(C_n)$, and since $C \subset N_{2\varepsilon}(F)$ we certainly have $C \subset N_{2\varepsilon}(\overline{\text{co}}\,F)$. But this shows that $C_n \to \overline{\text{co}}\,F$, so $F = \overline{\text{co}}\,F$. Thus the collection of all nonempty, convex and w^*-closed subsets of U' is a closed (and hence compact) subset of $(\mathcal{F}, \rho_d)$. ∎

7.12 Extreme points of convex sets

Many different sets may have the same closed convex hull. In this section we partially characterize the minimal such set—the set of extreme points. In a sense, the extreme points of a convex set characterize all the members.

7.61 Definition   An *extreme subset* of a (not necessarily convex) subset C of a *vector space, is a nonempty subset F of C with the property that if x belongs to F it cannot be written as a convex combination of points of C outside F. That is, if* $x \in F$ and $x = \alpha y + (1 - \alpha)z$, where $0 < \alpha < 1$ and $y, z \in C$, then $y, z \in F$. A point *x is an **extreme point** of C if the singleton $\{x\}$ is an extreme set. The set of extreme points of C is denoted $\mathcal{E}(C)$.*

That is, a vector x is an extreme point of C if it cannot be written as a strict convex combination of distinct points in C. A **face** of a convex set C is a convex extreme subset of C. Here are some examples.

- The extreme points of a closed disk are all the points on its circumference.

- The set of extreme points of a convex set is an extreme set—if it is nonempty.

- In $\mathbb{R}^n$, the extreme points of a convex polyhedron are its vertexes. All its faces and edges are extreme sets.

- The rays of a pointed closed convex cone that are extreme sets are called **extreme rays**. For instance, the nonnegative axes are the extreme rays of the usual positive cone in $\mathbb{R}^n$.

The following useful property is easy to verify.

7.62 Lemma   *A point a in a convex set C is an extreme point if and only if $C \setminus \{a\}$ is a convex set.*

In general, the set of extreme points of a convex set K may be empty, and if nonempty, need not be closed. For instance, the set C of all strictly positive functions on the unit interval is a convex subset of $\mathbb{R}^{[0,1]}$ without extreme points. To see this, let f be strictly positive. Then, $g = \frac{1}{2}f$ is also strictly positive and distinct from f, but $f = \frac{1}{2}g + \frac{1}{2}(f + g)$, proving that f cannot be an extreme point of C. As an example of a compact convex set for which the set of extreme points is not closed, consider the subset of $\mathbb{R}^3$

$$A = \{(x, y, 0) \in \mathbb{R}^3 : x^2 + y^2 \leqslant 1\} \cup \{(0, -1, 1), (0, -1, -1)\}.$$

The convex hull of A is compact, but the set of extreme points of A is

$$\{(x, y, 0) \in \mathbb{R}^3 : x^2 + y^2 = 1\} \cup \{(0, -1, 1), (0, -1, -1)\} \setminus \{(0, -1, 0)\},$$

which is not closed. See Figure 7.4.

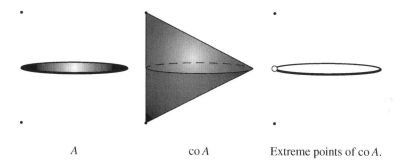

A co A Extreme points of co A.

Figure 7.4. The set of extreme points of co A is not closed.

You should verify the following properties of extreme subsets.

1. An extreme subset of an extreme subset of a set C is an extreme subset of C.

2. A nonempty intersection of a collection of extreme subsets of a set C is an extreme subset of the set C.

While the set of extreme points of a set K is not necessarily closed, if K is compact and the topology of K is metrizable, then it is easy to see that it is a $\mathcal{G}_\delta$, a countable intersection of open sets. Although most weak topologies of interest are not metrizable, Theorems 6.30 and 6.31 show that restricted to norm bounded subsets of duals (resp. preduals) of separable Banach spaces, the weak∗ (resp. weak) topology is metrizable. Thus the next lemma does have some important applications. Unfortunately, in general, the set of extreme points of a convex set need not even be a Borel set; see E. Bishop and K. DeLeeuw [45], and J. E. Jayne and C. A. Rogers [182].

7.63 Lemma   *If K is a metrizable compact convex subset of a topological vector space, then the set of extreme points of K is a $\mathcal{G}_\delta$ in K.*

Proof: Define $f\colon K \times K \to K$ by $f(x,y) = \frac{x+y}{2}$. Then a point is not extreme if and only if it is the image under f of a pair (x,y) with $x \neq y$. Now let d be a metric for K, and note that $x \neq y$ if and only if there is some n for which $d(x,y) \geqslant \frac{1}{n}$. Letting D_n denote the compact set $\{(x,y) \in K \times K : d(x,y) \geqslant \frac{1}{n}\}$, we see that the set of nonextreme points of K is $\bigcup_{n=1}^{\infty} f(D_n)$. Thus

$$\mathcal{E}(K) = K \setminus \bigcup_{n=1}^{\infty} f(D_n) = \bigcap_{n=1}^{\infty} K \setminus f(D_n).$$

Since continuous images of compact sets are compact, and compact subsets of metric spaces are closed, each $K \setminus f(D_n)$ is open in K. ∎

The extreme points of a convex set are of interest primarily because of the Krein–Milman Theorem and its generalizations. The Krein–Milman Theorem asserts that a compact convex subset K of a locally convex Hausdorff space is the closed convex hull of its extreme points. That is, the convex hull of the set of extreme points is dense in K. This means that if every extreme point of K has some property P, and if P is preserved by taking limits and convex combinations, then every point in K also enjoys property P. For instance to show that a compact convex set K lies in the polar of a set A, it is enough to show that every extreme point lies in the polar of A.

7.64 Lemma   *The set of maximizers of a convex function is either an extreme set or is empty. Likewise, the set of minimizers of a concave function is either an extreme set or is empty.*

Proof: Let $f\colon C \to \mathbb{R}$ be convex. Suppose f achieves a maximum on C. Put $M = \max\{f(x) : x \in C\}$ and let $F = \{x \in C : f(x) = M\}$. Suppose that $x = \alpha y + (1-\alpha)z \in F$, $0 < \alpha < 1$, and $y, z \in C$. If $y \notin F$, then $f(y) < M$, so

$$\begin{aligned} M = f(x) &= f(\alpha y + (1-\alpha)z) \leqslant \alpha f(y) + (1-\alpha)f(z) \\ &< \alpha M + (1-\alpha)M = M, \end{aligned}$$

a contradiction. Hence $y, z \in F$, so F is an extreme subset of C. ∎

The following lemma is the basic result concerning the existence of extreme points.

7.65 Lemma   *In a locally convex Hausdorff space, every compact extreme subset of a set C contains an extreme point of C.*

Proof: Let C be a subset of some locally convex Hausdorff space and let F be a compact extreme subset of C. Consider the collection of sets

$$\mathcal{F} = \{G \subset F : G \text{ is a compact extreme subset of } C\}.$$

Since $F \in \mathcal{F}$, we have $\mathcal{F} \neq \varnothing$, and $\mathcal{F}$ is partially ordered by set inclusion. The compactness of F (as expressed in terms of the finite intersection property) guarantees that every chain in $\mathcal{F}$ has a nonempty intersection. Clearly, the intersection of extreme subsets of C is an extreme subset of C if it is nonempty. Thus, Zorn's Lemma applies, and yields a minimal compact extreme subset of C included in F, call it G. We claim that G is a singleton. To see this, assume by way of contradiction that there exist $a, b \in G$ with $a \neq b$. By the Separation Corollary 5.82 there is a continuous linear functional f on X such that $f(a) > f(b)$. Let M be the maximum value of f on G. Arguing as in the proof of Lemma 7.64, we see that the compact set $G_0 = \{c \in G : f(c) = M\}$ is an extreme subset of G (and hence of C) and $b \notin G_0$, contrary to the minimality of G. Hence G must be a singleton. Its unique element is an extreme point of C lying in F.   ∎

Since every nonempty compact subset C is itself an extreme subset of C, we have the following immediate consequence of Lemma 7.65.

7.66 Corollary   *Every nonempty compact subset of a locally convex Hausdorff space has an extreme point.*

7.67 Theorem   *Every nonempty compact subset of a locally convex Hausdorff space is included in the closed convex hull of its extreme points.*

Proof: Let C be a nonempty compact subset of a locally convex Hausdorff space X, and let B denote the closed convex hull of its extreme points. We claim that $C \subset B$. Suppose by way of contradiction that there is some $a \in C$ with $a \notin B$. By Corollary 7.66 the set B is nonempty. So by the Separation Corollary 5.80 there exists a continuous linear functional f on X with $f(a) > f(b)$ for all $b \in B$. Let A be the set of maximizers of f over C. Clearly, A is a nonempty compact extreme subset of C, and $A \subset C \setminus B$. By Lemma 7.65, A contains an extreme point of C. But then, $A \cap B \neq \varnothing$, a contradiction. Hence $C \subset B$, as claimed.   ∎

The celebrated Krein–Milman Theorem [215] is now a consequence of the preceding result.

7.68 The Krein–Milman Theorem   *In a locally convex Hausdorff space X each nonempty convex compact subset is the closed convex hull of its extreme points.*
    If X is finite dimensional, then every nonempty convex compact subset is the convex hull of its extreme points.

Proof: Only the second part needs proof. The proof will be done by induction on the dimension n of X. For $n = 1$ a nonempty convex compact subset of $\mathbb{R}$ is either a point or a closed interval, in which case the conclusion is obvious. For the induction step, assume that the result is true for all nonempty convex compact subsets of finite dimensional vector spaces of dimension less than or equal to n.

This implies that the result is also true for all nonempty convex compact subsets of affine subspaces of dimension less than or equal to n. Now assume that C is a nonempty convex compact subset of an $(n+1)$-dimensional vector space X and let $\mathcal{E}$ be the collection of all extreme points of C. By the "Krein–Milman" part, we have $\overline{co}\,\mathcal{E} = C$.

If the affine subspace generated by C is of dimension less that $n + 1$, then the conclusion follows from our induction hypothesis. So we can assume that the affine subspace generated by C is X itself. This means that the interior of C is nonempty. In particular, co $\mathcal{E}$ must have a nonempty interior. Otherwise, if co $\mathcal{E}$ has an empty interior, then $\overline{co}\,\mathcal{E}$ has dimension less than $n + 1$, contrary to $\overline{co}\,\mathcal{E} = C$, as desired.

Now let x belong to C. If $x \in C^\circ$, then from Lemma 5.28 it follows that $x \in C^\circ = (\overline{co}\,\mathcal{E})^\circ = (co\,\mathcal{E})^\circ \subset co\,\mathcal{E}$. On the other hand, if $x \in \partial C$, then (by Lemma 7.7) there exists a nonzero $f \in X^*$ satisfying $f(x) \leqslant f(a)$ for all $a \in C$. If we let $F = \{a \in C : f(a) = f(x)\} = C \cap [f = f(x)]$, then F is a compact face of C that lies in the n-dimensional flat $[f = f(x)]$. By the induction hypothesis x is a convex combination of extreme points of F. Now notice that every extreme point of F is an extreme point of C, and from this we get $x \in co\,\mathcal{E}$. Thus, $C \subset co\,\mathcal{E}$, so $C = co\,\mathcal{E}$. ∎

Pay careful attention to the statement of the Krein–Milman Theorem. It does *not* state that the closed convex hull of a compact set is compact. Indeed, that is not necessarily true, see Example 5.34. Rather it says that if a convex set is compact, then it is the closed convex hull of its extreme points. Furthermore, the hypothesis of local convexity cannot be dispensed with. J. W. Roberts [286] gives an example of a compact convex subset of the completely metrizable tvs $L_{\frac{1}{2}}[0, 1]$ that has no extreme points.

We know that continuous functions always achieve their maxima and minima over nonempty compact sets. In a topological vector space we can say more. A continuous convex function on a nonempty compact convex set will always have at least one maximizer that is an extreme point of the set. This result is known as the **Bauer Maximum Principle**. Note that this result does not claim that all maximizers are extreme points.

7.69 Bauer Maximum Principle   *If C is a compact convex subset of a locally convex Hausdorff space, then every upper semicontinuous convex function on C has a maximizer that is an extreme point.*

Proof: Let f be an upper semicontinuous convex function on the nonempty, compact, and convex set. By Theorem 2.43 the set F of maximizers of f is nonempty and compact. By Lemma 7.64 it is an extreme set. But then Lemma 7.65 implies that F contains an extreme point of C. ∎

The following corollary gives two immediate consequences of the Bauer Maximum Principle.

7.70 Corollary   *If C is a nonempty compact convex subset of a locally convex Hausdorff space, then:*

1. *Every lower semicontinuous concave function on C has a minimizer that is an extreme point of C.*

2. *Every continuous linear functional has a maximizer and a minimizer that are extreme points of C.*

7.13   Quasiconvexity

There are generalizations of convexity for functions that are commonly applied in economic theory and operations research.

7.71 Definition   *A real function $f: C \to \mathbb{R}$ on a convex subset C of a vector space is:*

- *quasiconvex if $f(\alpha x + (1 - \alpha)y) \leqslant \max\{f(x), f(y)\}$ for all $x, y \in C$ and all $0 \leqslant \alpha \leqslant 1$.*

- *strictly quasiconvex if $f(\alpha x + (1 - \alpha)y) < \max\{f(x), f(y)\}$ for all $x, y \in C$ with $x \neq y$ and all $0 < \alpha < 1$.*

- *quasiconcave if $-f$ is a quasiconvex function. Explicitly, f is quasiconcave if $f(\alpha x + (1 - \alpha)y) \geqslant \min\{f(x), f(y)\}$ for all $x, y \in C$ and all $0 \leqslant \alpha \leqslant 1$.*

- *strictly quasiconcave if $-f$ is strictly quasiconvex.*

Then next lemma is a simple consequence of the definitions.

7.72 Lemma   *Every convex function is quasiconvex (and every concave function is quasiconcave).*

Characterizations of quasiconvexity are given in the next lemma.

7.73 Lemma   *For a real function $f: C \to \mathbb{R}$ on a convex set, the following statements are equivalent:*

1. *The function f is quasiconvex.*

2. *For each $\alpha \in \mathbb{R}$, the strict lower contour set $\{x \in C : f(x) < \alpha\}$ is a (possibly empty) convex set.*

3. *For each $\alpha \in \mathbb{R}$, the lower contour set $\{x \in C : f(x) \geqslant \alpha\}$ is a (possibly empty) convex set.*

We omit the proof, and note that there is of course an analogous result for quasiconcave functions and upper contour sets. On a topological vector space, convex functions have a fair amount of built-in continuity. We note that Theorem 5.98 on closed convex sets implies the following generalization of Corollary 5.99.

7.74 Corollary   *All locally convex topologies consistent with a given dual pair have the same lower semicontinuous quasiconvex functions.*

Proof: If f is quasiconvex, then $\{x : f(x) \leqslant \alpha\}$ is convex for each α. By Theorem 5.98, if these sets are closed in one consistent topology, then they are closed in all consistent topologies.   ∎

Note that an even stronger version of the Bauer Maximum Principle is true. Let us call a real function g **explicitly quasiconvex**, if it is quasiconvex and in addition, $g(x) < g(y)$ implies $g(\lambda x + (1 - \lambda)y) < g(y)$ for $0 < \lambda < 1$. (The latter condition does not imply quasiconvexity, as the function $g(x) = 0$ for $x \neq 0$ and $g(0) = 1$ demonstrates.)

7.75 Corollary   *Let C be a nonempty compact convex subset of a locally convex Hausdorff space. Every upper semicontinuous explicitly quasiconvex function has a maximizer on C that is an extreme point of C.*

Proof: Let $f \colon C \to \mathbb{R}$ be an upper semicontinuous explicitly quasiconvex function. By Theorem 2.43 the set F of maximizers of f is nonempty and compact. Put $M = \max\{f(x) : x \in C\}$, so $F = \{x \in C : f(x) = M\}$.

We wish to show that F is an extreme subset of C, that is, if x belongs to F, and $x = \alpha y + (1 - \alpha)z$, where $0 < \alpha < 1$ and $y, z \in C$, then both y and z belong to F. If say $y \notin F$, then $f(y) < M = f(x)$, so by quasiconvexity we have $M = f(x) \leqslant \max\{f(y), f(z)\}$, which implies $f(x) = f(z) = M > f(y)$. On the other hand, since f is explicitly quasiconvex, and $f(y) < f(z)$, we must also have $f(x) < f(z)$, a contradiction. Therefore F is an extreme set.

By Lemma 7.65, F contains an extreme point of C.   ∎

Explicit quasiconvexity is defined analogously, and a similar result holds.

7.14   Polytopes and weak neighborhoods

In this section we discuss the relation between weak topologies and finite systems of linear inequalities. Given a dual pair $\langle X, X' \rangle$, each linear functional $x' \in X'$ and each real number α give rise to a **linear inequality** of the form $x'(x) \leqslant \alpha$. The **solution set** of this inequality is the collection of $x \in X$ that satisfy the inequality. That is, $\{x \in X : x'(x) \leqslant \alpha\}$. This set is a $\sigma(X, X')$-closed half space in X. Similarly, each $x \in X$ and α define a linear inequality on X'. Its solution set is a weak∗-closed half space in X'. Due to the symmetry of the role of X and X' in a dual pair, everything we say about inequalities on X has a corresponding statement about linear inequalities on X'. We do not explicitly mention these results, you can figure them out yourself.

A finite system of linear inequalities is defined by a finite set $\{x'_1, \ldots, x'_m\} \subset X'$ and a corresponding set $\{\alpha_1, \ldots, \alpha_m\}$ of reals. The solution set of the system is $\{x \in X : x'_i(x) \leq \alpha_i, i = 1, \ldots, m\}$. The solution set of a finite system of linear inequalities is the intersection of finitely many weakly closed half spaces.

Recall that a **polyhedron** in X is a finite intersection of weakly closed half spaces. That is, a polyhedron is the solution set of a finite system of linear inequalities on X.

Clearly the polar (one-sided or absolute) of a finite subset of X' is a polyhedron. Thus there is a base of weak neighborhoods of zero consisting of polyhedra. In a finite dimensional space, it is possible for a polyhedron to be compact. The Fundamental Theorem of Duality 5.91 implies that this cannot happen in an infinite dimensional space (see the proof of Theorem 6.26). Nevertheless we show (Theorem 7.80) that polars of finite sets do have some salient properties.

Recall that a **polytope** in a vector space is the convex hull of a finite set. The next lemma sets forth the basic properties of polytopes.

7.76 Lemma   *In a topological vector space, the convex hull of a finite set F is compact, and its set of extreme points is nonempty and included in F. That is, $\mathcal{E}(\operatorname{co} F) \neq \varnothing$ and $\mathcal{E}(\operatorname{co} F) \subset F$.*

Proof: Let $F = \{x_1, \ldots, x_n\}$ be a finite subset of a topological vector space. By Corollary 5.30, the convex hull of F is compact.

Now let $x = \sum_{i=1}^n \lambda_i x_i$, where $0 \leq \lambda_i \leq 1$ for each i and $\sum_{i=1}^n \lambda_i = 1$, belong to $\operatorname{co} F$. Assume that $x \neq x_i$ for each i. This implies that $0 < \lambda_j < 1$ for some j. In particular, the point $y = \sum_{i \neq j} \frac{\lambda_i}{1-\lambda_j} x_i$ belongs to $\operatorname{co} F$. Therefore we have

$$x = \lambda_j x_j + (1 - \lambda_j) \sum_{i \neq j} \frac{\lambda_i}{1 - \lambda_j} x_i = \lambda_j x_j + (1 - \lambda_j)y,$$

which shows that x cannot be an extreme point of $\operatorname{co} F$. In other words, the extreme points of $\operatorname{co} F$ are among the points of F.

To see that $\operatorname{co} F$ has extreme points, notice first that $\operatorname{co} F \subset M$, where M is the finite dimensional vector subspace generated by F. If M is equipped with its Euclidean topology (which is locally convex), then $\operatorname{co} F$ is a compact subset of M, so by the Krein–Milman Theorem 7.68 it is also the convex hull (in M) of its extreme points. Thus $\mathcal{E}(\operatorname{co} F) \neq \varnothing$.   ∎

Scalar products and sums of polytopes are also polytopes.

7.77 Lemma   *The algebraic sum of two polytopes is a polytope.*

Proof: If $A = \mathrm{co}\{x_1,\ldots,x_n\}$ and $B = \mathrm{co}\{y_1,\ldots,y_m\}$, then you can verify that $A + B = \mathrm{co}\{x_i + y_j : i = 1,\ldots,n,\ j = 1,\ldots,m\}$. Generous hint: If $x = \sum_{i=1}^{n} \lambda_i x_i$ and $y = \sum_{j=1}^{m} \alpha_j y_j$, then $x+y = \sum_{i=1}^{n} \sum_{j=1}^{m} \lambda_i \alpha_j (x_i+y_j)$ is a convex combination. ∎

In the finite dimensional case, it is well-known that the solution set of a finite system of linear inequalities has finitely many extreme points. (If it is half space it has no extreme points.) We prove this in a general framework via an elegant argument taken from H. Nikaidô [262, p. 40].

7.78 Lemma *Let X be a (not necessarily locally convex) topological vector space, and let $x_1',\ldots,x_m'$ belong to X' and $\alpha_1,\ldots,\alpha_m$ belong to $\mathbb{R}$. Then the solution set*

$$S = \{x \in X : x_i'(x) \leqslant \alpha_i \text{ for each } i = 1,\ldots,m\}$$

is a closed convex set and has at most 2^m extreme points.

Proof: The solution set S is clearly closed and convex. With regard to extreme points, start by defining a mapping A from S to the set of all subsets of $\{1,\ldots,m\}$, via

$$A(x) = \{i \in \{1,\ldots,m\} : x_i'(x) < \alpha_i\}.$$

That is, $A(x)$ is the set of "slack" inequalities at x. We shall show that the mapping $x \mapsto A(x)$ is one-to-one on $\mathcal{E}(S)$. Since there are 2^m distinct subsets of $\{1,\ldots,m\}$, this establishes the claim. [10] To this end, suppose $x, y \in \mathcal{E}(S)$ satisfy $A(x) = A(y)$. We must show that $x = y$.

Suppose first that $A(x) = A(y) = \varnothing$. Then $x_i'(x) = x_i'(y) = \alpha_i$ for all i, so $x_i'(x - y) = 0$ for all i. Therefore, $x_i'(y + 2(x - y)) = \alpha_i$ for all i, so $y + 2(x - y) \in S$. Now from $x = \frac{1}{2}y + \frac{1}{2}[y + 2(x - y)]$ and the fact that x is an extreme point, we see that $x = y$.

Now suppose that $A(x) = A(y) = B \neq \varnothing$. In this case, we let

$$\varepsilon = \min\left\{\frac{\alpha_i - x_i'(x)}{\alpha_i - x_i'(y)} : i \in B\right\} > 0.$$

Then $\varepsilon[\alpha_i - x_i'(y)] \leqslant \alpha_i - x_i'(x)$ for each $i = 1,\ldots,m$. (If i does not belong to B, then $\alpha_i - x_i'(x) = \alpha_i - x_i'(y) = 0$.) Suppose first that $\varepsilon \geqslant 1$. This implies $\alpha_i - x_i'(y) \leqslant \alpha_i - x_i'(x)$, so $x_i'(x - y) \leqslant 0$ for all i. Therefore $y + 2(x - y)$ satisfies $x_i'(y+2(x-y)) \leqslant \alpha_i$ for all i, so $y+2(x-y) \in S$. In particular, $x = \frac{1}{2}y+\frac{1}{2}[y+2(x-y)]$, which shows that $x = y$. Now suppose $0 < \varepsilon < 1$. Then $x_i'(x - \varepsilon y) \leqslant (1 - \varepsilon)\alpha_i$, or $x_i'(\frac{1}{1-\varepsilon}(x - \varepsilon y)) \leqslant \alpha_i$ for each i. Therefore $z = \frac{1}{1-\varepsilon}(x - \varepsilon y) \in S$. But then $x = \varepsilon y + (1 - \varepsilon)z$, so again $x = y$. ∎

---

[10] With more work, we can show that there at most $2^m - 1$ extreme points, because except for the trivial case $X = \{0\}$, it can never happen that $A(x) = \{1,\ldots,m\}$ for an extreme point x.

And now we come to a basic result regarding linear inequalities. It states that if the set of solutions to a finite system of linear inequalities is compact, then it is a polytope. That is, every compact polyhedron is a polytope.

7.79 Theorem (Solutions of Linear Inequalities)   *Let $\langle X, X' \rangle$ be a dual pair, and let $x'_1, \ldots, x'_m$ belong to X' and $\alpha_1, \ldots, \alpha_m$ belong to $\mathbb{R}$. If the solution set*

$$S = \{x \in X : x'_i(x) \leqslant \alpha_i \text{ for each } i = 1, \ldots, m\}$$

is $\sigma(X, X')$-compact and nonempty, then it is a polytope, and X is finite dimensional.

Moreover, a nonempty convex compact subset of a finite dimensional vector space is a polyhedron if and only if it is a polytope.

Proof: If the solution set S is $\sigma(X, X')$-compact and nonempty, then the Krein–Milman Theorem 7.68 implies that S is the $\sigma(X, X')$-closed convex hull of its set of extreme points. But, by Lemma 7.78, the solution set S has a finite number of extreme points, so it is a polytope (see Corollary 5.30).

To see that X is finite dimensional, let $M = \bigcap_{i=1}^{m} \ker x'_i$, which is a linear subspace of X. Note that $S + M = S$, which is $\sigma(X, X')$-compact, so $M \subset S - S$ must be $\sigma(X, X')$-compact. The only way that M can be $\sigma(X, X')$-compact, is if $M = \{0\}$. (Why?) But then, for any $x' \in X'$, we have $M \subset \ker x'$ so by the Fundamental Theorem of Duality 5.91, the functionals $x'_1, \ldots, x'_m$ span X', which implies that X' is finite dimensional. Consequently, X'' is finite dimensional. (Why?) Since X can be considered a vector subspace of X'', X is itself finite dimensional.

For the last part, we need only to show that every polytope is a polyhedron. So let $A = \text{co}\{a_1, \ldots, a_k\}$ be a polytope in a finite dimensional vector space X. We can assume that zero is an interior point of A. (Why?) By part (5) of Lemma 5.102 the one sided polar $A^\circ = \{x' \in X' : x'(a_i) \leqslant 1 \text{ for } i = 1, \ldots, k\}$ is $\sigma(X', X)$-bounded and $\sigma(X', X)$-closed. Since X' is finite dimensional, A° is $\sigma(X', X)$-compact. So by the previous part A° is a polytope and from this and the Bipolar Theorem 5.103, we see that $A = A^{\circ\circ}$ is a polyhedron. ∎

Actually, more is known. In a finite dimensional space, every polyhedron is the sum of a linear subspace, a polyhedral cone, and a polytope. (Any of these pieces may contain only zero.) For a comprehensive treatment of polyhedra in finite dimensional spaces, see for example, D. Gale [133, Chapter 2], J. Stoer and C. Witzgall [321, Chapter 2], or G. M. Ziegler [349]. See also the excellent book by M. Florenzano and C. Le Van [125].

We can now examine some of the finer points of the structure of basic weak neighborhoods of zero. Recall that a base of weak neighborhoods is given by the polars of finite subsets of X'. These polars are infinite "polyhedral prisms."

7.80 Theorem (Basic Weak Neighborhoods)   *Let $\langle X, X' \rangle$ be a dual pair and let F be a finite subset of X', let M be the finite dimensional subspace spanned by F, and let $V = F^{\circ}$ be its (absolute) polar. Then*

$$V = C \oplus M^{\perp},$$

where C is a polytope containing zero. That is, every x in V has a unique decomposition of the form $x = x_C + x_M$, where $x_C \in C$ and $x_M \in M^{\perp}$.

Proof: First consider the trivial case $F = \{0\}$. Then $M^{\perp} = X$ and $V = X = C \oplus M^{\perp}$, where $C = \{0\}$, a polytope. So we can assume that F contains a nonzero vector and M has dimension at least one. By Theorem 5.110 we can write $X = L \oplus M^{\perp}$, where L is finite dimensional and has the same dimension as M. Set $C = L \cap V$. Clearly, C is convex and $0 \in C$. From $X = L \oplus M^{\perp}$, it easily follows that $V = C \oplus M^{\perp}$.

   We claim that C is a polytope. First note that C is the set of solutions to the following finite system of linear inequalities:

$$C = \{x \in L : \pm x'(x) \leqslant 1 \text{ for each } x' \in F\}.$$

Clearly, C is a closed subset of L. Since C lies in the finite dimensional subspace L, it suffices to prove that C is bounded in L, where we now assume that L is equipped with its Euclidean norm $\| \cdot \|$. Suppose by way of contradiction that C is not bounded. Then for each n there is some $y_n \in C$ satisfying $\|y_n\| \geqslant n$. Let $x_n = \frac{y_n}{\|y_n\|} \in L$, so $\|x_n\| = 1$ for each n. Since the unit sphere of L is compact, we can assume by passing to a subsequence that there exists some $x \in L$ with $\|x\| = 1$ and $x_n \to x$. Then for $x' \in F$, we have

$$|\langle x_n, x' \rangle| = \tfrac{1}{n} \cdot \tfrac{n}{\|y_n\|} \cdot |\langle y_n, x' \rangle| \leqslant \tfrac{1}{n} \cdot 1 \cdot 1 = \tfrac{1}{n},$$

so $\langle x, x' \rangle = \lim_{n \to \infty} \langle x_n, x' \rangle = 0$ for each $x' \in F$. Therefore $\langle x, x' \rangle = 0$ for all x' in $M = \text{span } F$. That is, $x \in M^{\perp}$. So $x \in M^{\perp} \cap L = \{0\}$, contrary to $\|x\| = 1$. This contradiction completes the proof of the theorem.   ∎

7.81 Corollary   *Let $\langle X, X' \rangle$ be a dual pair and let F be a finite subset of X'. Then every $x' \in \text{co } F$ attains a maximum and a minimum on $V = F^{\circ}$.*

Proof: By Theorem 7.80, we can write $V = C \oplus M^{\perp}$, where M is the linear span of F, and C is a polytope. Then for any x' in $\text{co } F$ (or any $x' \in M$ for that matter) and any $x = x_C + x_M \in C \oplus M^{\perp}$, we have $x'(x) = x'(x_C)$. Since C is compact (why?), x' attains a maximum (and a minimum) on C and hence on V.   ∎

   The next result on one-sided polars is used to prove Theorem 17.41.

7.82 Lemma   *Let $\langle X, X' \rangle$ be a dual pair. Let K be a polytope in X and assume $0 \in K$. Let V be a basic closed $\sigma(X, X')$-neighborhood of zero, that is, V is the absolute polar of a finite subset of X'. Then the one-sided polar $(K + V)^{\odot}$ is a polytope included in $V^{\odot}$.*

Proof: Start by noting that we can write $V = F^{\circ}$, where $F = \{x_1', \ldots, x_n'\}$ is a symmetric finite subset of X'. (Why?) The Bipolar Theorem 5.103 thus implies

$$V^{\circ} = V^{\odot} = \text{co } F.$$

Since $0 \in K$, we see that $V \subset K + V$, so $(K + V)^{\odot} \subset V^{\odot} = V^{\circ} = \text{co } F$, which is w^*-compact. Thus, the one-sided polar $(K + V)^{\odot}$ is w^*-compact and convex. By Theorem 7.79 it suffices to show that $(K + V)^{\odot}$ is the solution set of a finite system of linear inequalities defined by points of X.

To this end, let M be the linear span of F. By Theorem 7.80, we can write $V = C \oplus M^{\perp}$, where C is a polytope. We claim that

$$(K + V)^{\odot} = \{x' \in M : \langle x, x' \rangle \leqslant 1 \text{ for all } x \in K + C\}. \qquad (\star)$$

To see this, let $S = \{x' \in M : \langle x, x' \rangle \leqslant 1 \text{ for all } x \in K + C\}$. Assume first that $x' \in (K + V)^{\odot} \subset M$. If $x \in K + C$, then $x \in K + C + M^{\perp} = K + V$, so $\langle x, x' \rangle \leqslant 1$ for each $x \in K + C$. This shows that $(K + V)^{\odot} \subset S$. For the reverse inclusion, suppose $x' \in S$. That is, $x' \in M$ and $\langle x, x' \rangle \leqslant 1$ for each $x \in K + C$. This implies $\langle x, x' \rangle \leqslant 1$ for each $x \in K + C + M^{\perp} = K + V$, which means that $x' \in (K + V)^{\odot}$. Thus, $S \subset (K + V)^{\odot}$, so $(K + V)^{\odot} = S$.

By Lemma 7.77, the sum $K + C$ is a polytope. In fact, if $C = \text{co}\{z_1, \ldots, z_k\}$ and $K = \text{co}\{x_1, \ldots, x_m\}$, then $K + C = \text{co}\{x_i + z_j : i = 1, \ldots, m, \ j = 1, \ldots, k\}$. By the Bauer Maximum Principle 7.69 any $x' \in (K + V)^{\odot}$ achieves its maximum at an extreme point of $K + C$, which by Lemma 7.76 must be one of the points $x_i + z_j$. Therefore, from $(\star)$, we see that $(K + V)^{\odot}$ is the solution set in the finite dimensional space M to the finite system of linear inequalities: $\langle x_i + z_j, x' \rangle \leqslant 1$, $i = 1, \ldots, m; \ j = 1, \ldots, k$. That is,

$$(K + V)^{\odot} = \{x' \in M : \langle x_i + z_j, x' \rangle \text{ for all } i = 1, \ldots, m, \ j = 1, \ldots, k\},$$

and the proof is finished. ∎

7.15 Exposed points of convex sets

In this section, we shall discuss some special kinds of extreme points of convex sets—exposed and strongly exposed points. We begin with the definition.

7.83 Definition *Let A be a nonempty convex set in a tvs (X, τ). A point $e \in A$ is:*

* *an **exposed point** of A if it is the unique maximizer (or minimizer) over A of a nonzero continuous linear functional. That is, if there exists a nonzero continuous linear functional $x' \in X'$ such that $x'(e) > x'(a)$ for all $a \in A \setminus \{e\}$. We say that the linear functional x' **exposes** the point e.*

- a **strongly exposed point** of A *if there exists a nonzero continuous linear functional* $x' \in X'$ *that supports A at e and such that any net* $\{a_\lambda\}$ *in A satisfying* $x'(a_\lambda) \to x'(e)$ *converges to e, that is,* $a_\lambda \xrightarrow{\tau} e$. *We say that* x' **strongly exposes** e.

Some remarks are in order.

- It is clear that every exposed point is an extreme point. However, an extreme point need not be an exposed point; see Figure 7.5, which shows the union of a rectangle and a half disk. The indicated point, where the disk meets the rectangle, is extreme, but not exposed, since any linear function that attains its minimum at that point also attains its minimum along the entire bottom of the rectangle.

Figure 7.5.

- Strongly exposed points of convex subsets of Hausdorff topological vector spaces are automatically exposed points. Indeed, if x' strongly exposes a point e of a convex subset A of a Hausdorff tvs and $x'(e) = x'(a)$ holds for some $a \in A$ with $a \neq e$, then the constant sequence $a_n = a$ satisfies $x'(a_n) \to x'(e)$ while $\{a_n\}$ fails to converge to e.

One way to understand strongly exposed points is to consider the case of a completely metrizable tvs such as a Banach space. In this case, x' strongly exposes the point e in A, with say $x'(e) = \alpha > x'(a)$ for all points $a \in A$, if and only if $\operatorname{diam} A \cap [x' \geqslant \alpha - 1/n] \to 0$ as $n \to \infty$.

- An exposed point need not be a strongly exposed point. For example, consider $C[0, 1]$, the Banach lattice of all continuous real functions on $[0, 1]$. Let $A = [0, \mathbf{1}]$, the convex set of all $x \in C[0, 1]$ satisfying $0 \leqslant x \leqslant \mathbf{1}$. Since $0 < \int_0^1 x(t)\, dt$ holds for all $0 < x \in C[0, 1]$, it follows that Lebesgue measure exposes 0, so 0 is an exposed point of A. We claim that 0 is not a strongly exposed point of A. To see this, let μ be any nonzero measure on $[0, 1]$. If $x_n \in C[0, 1]$ is the function whose graph consists of the line segments joining the points $(0, 0)$, $(\frac{1}{2n}, 1)$, $(\frac{1}{n}, 0)$, and $(1, 0)$, then $\|x_n\|_\infty = 1$ for each n and $x_n(t) \to 0$ for each $t \in [0, 1]$. So by the Dominated Convergence Theorem 11.21, we have $\int_{[0,1]} x_n(t)\, d\mu(t) \to \int_{[0,1]} 0\, d\mu(t) = 0$, and from this we see that μ cannot strongly expose 0.

Unfortunately we cannot draw a simple picture to illustrate the difference between exposed and strongly exposed points since in finite dimensional vector spaces they are the same.

7.84 Lemma *Let C be a nonempty closed convex subset of a finite dimensional vector space X. Then a point $e \in C$ is exposed by a nonzero linear functional if and only if it strongly exposes the point e. In particular, the sets of exposed and strongly exposed points of C coincide.*

Proof: Fix a norm $\| \cdot \|$ on X and let f be a nonzero linear functional on X that exposes a point $e \in C$. Also, let $\{x_n\}$ be sequence in C satisfying $f(x_n) \to f(e)$. We need to show that $x_n \to e$.

We first claim that $\{x_n\}$ is a norm bounded sequence. To see this, suppose by way of contradiction that $\{x_n\}$ is not norm bounded. By passing to a subsequence if needed, we can assume that $\|x_n\| > n$ for each n. Let $y_n = (1 - \frac{1}{\|x_n\|})e + \frac{1}{\|x_n\|}x_n \in C$. Passing to one more subsequence if necessary, we can assume that $\frac{x_n}{\|x_n\|} \to x$. Clearly, $\|x\| = 1$, so in particular, $x \neq 0$. It follows that $y_n \to e + x$ and the closedness of C guarantees that $e + x \in C$. Since f exposes e, it follows from $e \neq e + x$ that $f(e) > f(e + x)$ or $f(x) < 0$. However, from $f(x_n) \to f(e)$, we get $f(x) = \lim f(\frac{x_n}{\|x_n\|}) = \lim \frac{f(x_n)}{\|x_n\|} = 0$, which is a contradiction. Consequently, $\{x_n\}$ is a norm bounded sequence.

Now let $\{y_n\}$ be a subsequence of $\{x_n\}$. Since $\{y_n\}$ is bounded and X is finite dimensional, there is a convergent subsequence $\{z_n\}$ of $\{y_n\}$ with limit point z belonging to C. Then $f(z_n) \to f(z)$, but by hypothesis $f(z_n) \to f(e)$. By the definition of exposure, this implies $z = e$. Thus, every subsequence of $\{x_n\}$ has a subsequence that converges to e, and hence $x_n \to e$, as desired. ∎

Every vector on the unit sphere of a Hilbert space is a strongly exposed point of the closed unit ball. The details follow.

7.85 Lemma *Let H be a Hilbert space and let $U = \{x \in H : \|x\| \leqslant 1\}$ be its closed unit ball. Then every boundary point of U is a strongly exposed point and if $x \in \partial U$, that is, if $\|x\| = 1$, then the vector x is the only unit vector that strongly exposes the point x.*

Proof: From the Cauchy–Schwarz inequality we have $(x, y) \leqslant \|x\| \cdot \|y\| \leqslant 1$ for each $y \in U$, so $(x, x) = 1 \geqslant (x, y)$. This shows that x supports U at x. Suppose some unit vector z satisfies $\|z\| = 1$ and $(z, x) \geqslant (z, y)$ for all $y \in U$. Then evaluating this at $y = z$ yields $1 \geqslant (z, x) \geqslant (z, z) = 1$ and thus $(x, z) = 1$. Consequently $|(x, z)| = \|x\| \cdot \|z\|$ and hence $z = \lambda x$. From $(x, z) = 1$, we get $\lambda = 1$, that is, $z = x$. This establishes the uniqueness of the supporting unit vector.

To see that x strongly exposes x assume that a sequence $\{x_n\}$ in U satisfies $(x, x_n) \to (x, x) = 1$. From $|(x, x_n)| \leqslant \|x\| \cdot \|x_n\| = \|x_n\| \leqslant 1$, we get $\|x_n\| \to 1$, so $\|x_n - x\|^2 = \|x_n\|^2 - 2(x, x_n) + \|x\|^2 \to 1 - 2 + 1 = 0$. ∎

7.86 Corollary *Let H be a Hilbert space and let $C(a, r)$ be the closed ball centered at $a \in H$ with radius r, that is, $C(a, r) = \{x \in H : \|x - a\| \leqslant r\}$. Then every boundary point of $C(a, r)$ is a strongly exposed point and if $c \in \partial C(a, r)$, that is, if $\|c - a\| = r$, then (up to a positive multiple) the vector $c - a$ is the only vector that strongly exposes the point x.*

7.87 Corollary   *Let C be a nonempty convex subset of a Hilbert space H. If for some point $a \in H$, the point $c \in C$ is the farthest point in C from a, that is, $\|x - a\| \leqslant \|c - a\|$ for all $x \in C$, then c is a strongly exposed point of C.*

Proof: Let $B = \{y \in H : \|y - a\| \leqslant \|c - a\|\}$, the closed ball of radius $\|c - a\|$ centered at a. Then (by Corollary 7.86) the vector $c - a$ strongly exposes c in the convex set B. Since $C \subset B$, it follows that $c - a$ also strongly exposes c in C.   ∎

The final results of the section deal with a density property of the strongly exposed points in finite dimensional vector spaces.

7.88 Lemma   *Let C be a nonempty convex subset of a tvs X and let G be a nonempty open convex subset of X. Letting $\mathrm{Exp}(S)$ denote the collection of exposed points of a convex set S, we have $\mathrm{Exp}(G \cap C) = G \cap \mathrm{Exp}(C)$.*

Proof: Start by observing that the inclusion $G \cap \mathrm{Exp}(C) \subset \mathrm{Exp}(G \cap C)$ is obvious. For the reverse inclusion, let $e \in \mathrm{Exp}(G \cap C)$. Pick some $f \in X'$ that exposes e over $G \cap C$. We claim that f also exposes e over C. If this is not the case, then there exists some $y \in C$ satisfying $y \neq e$ and $f(y) \geqslant f(e)$. In particular, we have $f(\alpha y + (1 - \alpha)e) \geqslant f(e)$ for all $0 < \alpha < 1$. Since $e \in G$ and $\lim_{\alpha \downarrow 0}[\alpha y + (1 - \alpha)e] = e$, there exists some $0 < \alpha_0 < 1$ such that the vector $z = \alpha_0 y + (1 - \alpha_0)e \in C$ satisfies $z \in G$ and $z \neq e$. But then we have $z \in G \cap C$, $z \neq e$ and $f(z) \geqslant f(e)$, contradicting the fact that f exposes e over $G \cap C$. Hence $e \in G \cap \mathrm{Exp}(C)$, so $\mathrm{Exp}(G \cap C) \subset G \cap \mathrm{Exp}(C)$.   ∎

We also have the following density result due to S. Straszewicz [325].

7.89 Theorem (Straszewicz)   *In a finite dimensional vector space, the set of exposed points (and hence the set of strongly exposed points) of a nonempty closed convex subset is dense in the set of its extreme points.*

Proof: We assume first that C is a nonempty compact convex subset of some $\mathbb{R}^n$; we consider $\mathbb{R}^n$ equipped with its Euclidean norm so that $\mathbb{R}^n$ is a Hilbert space. For each $u \in \mathbb{R}^n$ let $F_u = \{a \in A : \|x - u\| \leqslant \|a - u\|$ for all $x \in A\}$, that is, F_u consists of the vectors in C that are farthest from u. Since C is nonempty and compact, each F_u is nonempty (and closed), and according to Corollary 7.87 it consists of strongly exposed vectors of C. Put $F = \bigcup_{u \in \mathbb{R}^n} F_u$, and we claim that $\overline{\mathrm{co}}\, F = C$.

To see this, suppose by way of contradiction that there exists some u belonging to $C \setminus \overline{\mathrm{co}}\, F$. Let v be the metric projection of u onto $\overline{\mathrm{co}}\, F$ (that is, let v be the vector in $\overline{\mathrm{co}}\, F$ nearest u) and let $w = \frac{1}{2}(u + v)$. Next, consider the sequence of open balls $\{U_n\}$ with centers at the vectors $u_n = w + n(v - w)$ and radii $\|u_n - w\| = n\|v - w\|$. Since for each $x \neq 0$ we have $\bigcup_{n=1}^{\infty} B_{n\|x\|}(nx) = \{y \in \mathbb{R}^n : x \cdot y > 0\}$ (why?), it follows that

$$\bigcup_{n=1}^{\infty} U_n = w + \{y \in \mathbb{R}^n : (v - w) \cdot y > 0\} = \{y \in \mathbb{R}^n : (v - w) \cdot (y - w) > 0\}.$$

Now if $x \in \overline{\text{co}}\, F$, then (by part (c) of Lemma 6.54) we have $(u - v) \cdot (x - v) \leqslant 0$. Given that $u - v = 2(w - v)$, it follows that we have $(v - w) \cdot (v - x) \leqslant 0$ for all $x \in \overline{\text{co}}\, F$. So if $x \in \overline{\text{co}}\, F$, then

$$
\begin{aligned}
(v - w) \cdot (x - w) &= (v - w) \cdot [(v - w) + (x - v)] \\
&= \|v - w\|^2 - (v - w) \cdot (v - x) \\
&> -(v - w) \cdot (v - x) \geqslant 0,
\end{aligned}
$$

and so $x \in \bigcup_{n=1}^{\infty} U_n$ for each $x \in \overline{\text{co}}\, F$. Thus, $\overline{\text{co}}\, F \subset \bigcup_{n=1}^{\infty} U_n$, and by the compactness of $\overline{\text{co}}\, F$ we infer that $\overline{\text{co}}\, F \subset U_k$ for some k. Now notice that the vector $u \in C$ satisfies $\|u - u_k\| > \|u_k - w\|$. This implies $F_{u_k} = F_{u_k} \cap F = \varnothing$, which is impossible. This contradiction establishes that $\overline{\text{co}}\, F = C$.

Now assume that e is an extreme point of C. From the preceding conclusion we have $e \in \overline{\text{co}}\, F$. So there exists a sequence $\{e_n\} \subset \text{co}\, F$ such that $e_n \to e$. By Carathédory's Theorem 5.32 we can write $e_n = \sum_{i=0}^{n} \lambda_n^i e_n^i$, where $\{e_n^i\} \subset F$ and $\{\lambda_n^i\} \subset [0, 1]$ satisfies $\sum_{i=0}^{n} \lambda_n^i = 1$. By the compactness of C and $[0, 1]$, we can assume (by passing to a subsequence) that $\lambda_n^i \xrightarrow[n \to \infty]{} \lambda_i \geqslant 0$ and $e_n^i \xrightarrow[n \to \infty]{} e_i \in C$. It follows that $e = \sum_{i=0}^{n} \lambda_i e_i$ with $\sum_{i=0}^{n} \lambda_i = 1$. Since e is an extreme point of C, we conclude that $e = e_i$ for some i and so e is the limit of a sequence of strongly exposed points.

Finally, we consider the general case, that is, assume that C is a nonempty closed convex subset of $\mathbb{R}^n$. Fix an extreme point e of C and let $\varepsilon > 0$. Put $C_\varepsilon(e) = \{x \in \mathbb{R}^n : \|x - e\| \leqslant \varepsilon\}$ and $G = B_\varepsilon(e)$. Clearly, e is an extreme point of the nonempty compact convex set $C_\varepsilon(e) \cap C$. By the preceding case, there exists an exposed point x_0 of the set $C_\varepsilon(e) \cap C$ such that $x_0 \in G \cap [C_\varepsilon(e) \cap C] = G \cap C$. It follows that x_0 is an exposed point of $G \cap C$, and from $\text{Exp}(G \cap C) = G \cap \text{Exp}(C)$, we see that x_0 is also an exposed point of C satisfying $\|e - x_0\| < \varepsilon$. ∎

7.90 Corollary *Every extreme point of a polytope in a locally convex Hausdorff space is a strongly exposed point.*

Proof: It follows from Theorem 7.89 and the fact that every polytope lies in a finite dimensional vector space and has a finite number of extreme points (see Lemma 7.76). ∎

Chapter 8

Riesz spaces

A *Riesz space* is a real vector space equipped with a partial order satisfying the following properties. Inequalities are preserved by adding the same vector to each side, or by multiplying both sides by the same positive scalar. Each pair $\{x, y\}$ of vectors has a *supremum* or least upper bound, denoted $x \vee y$. Thus Riesz spaces mimic some of the order properties possessed by the real numbers. However, the real numbers possess other properties not shared by all Riesz spaces, such as order completeness and the Archimedean property. To further complicate matters, the norm of a real number coincides with its absolute value. In more general normed Riesz spaces the norm and absolute value are different.

Riesz spaces capture the natural notion of *positivity* for functions on ordered vector spaces. For the special class of *Banach lattices*, every continuous linear functional is the difference of two positive linear functionals. As a result, many results proven for positive functionals extend to continuous functionals.

The abstraction of the order properties frees them from the details of any particular space and makes it easier to prove general theorems about Riesz spaces in a straightforward fashion. Without this general theory, even special cases are difficult. For example, the well-known Hahn–Jordan and Lebesgue Decomposition Theorems are difficult theorems of measure theory yet are special cases of general results from the theory of Riesz spaces. Conveniently, most spaces used in economic analysis are Riesz spaces, see for instance, [9, 10, 137, 190, 243].

The importance of ordered vector spaces in economic analysis stems from the fact that often there is a natural ordering on commodity vectors for which "more is better." That is, preferences are monotonic in the order on the commodity space. In this case, a reasonable requirement is that equilibrium prices be positive. Furthermore, in Riesz spaces, the order interval defined by the social endowment corresponds roughly to the Edgeworth box. For symmetric Riesz pairs, order intervals are weakly compact, so that the order structure provides a source of compact sets.

This chapter is a brief introduction to the basic theory of Riesz spaces. For a more thorough treatment we recommend C. D. Aliprantis and O. Burkinshaw [12, 15], W. A. J. Luxemburg and A. C. Zaanen [235], P. Meyer-Nieberg [247], H. H. Schaefer [294], and A. C. Zaanen [347].

8.1 Orders, lattices, and cones

Recall that a **partially ordered set** $(X, \geqslant)$ is a set X equipped with a partial order $\geqslant$. That is, $\geqslant$ is a transitive, reflexive, antisymmetric relation. The notation $y \leqslant x$ is, of course, equivalent to $x \geqslant y$. Also, $x > y$ means $x \geqslant y$ and $x \neq y$.[1] The expression "x **dominates** y" means $x \geqslant y$, and we say "x **strictly dominates** y" whenever $x > y$.

Recall that a partially ordered set $(X, \geqslant)$ is a **lattice** if each pair of elements $x, y \in X$ has a supremum (or least upper bound) and an infimum (or greatest lower bound). An element z is the **supremum** of a pair of elements $x, y \in X$ if

i. z is an upper bound of the set $\{x, y\}$, that is, $x \leqslant z$ and $y \leqslant z$; and

ii. z is the least such bound, that is, $x \leqslant u$ and $y \leqslant u$ imply $z \leqslant u$.

The **infimum** of two elements is defined similarly. We denote the supremum and infimum of two elements $x, y \in X$ by $x \vee y$, and $x \wedge y$ respectively. That is,

$$x \vee y = \sup\{x, y\} \quad \text{and} \quad x \wedge y = \inf\{x, y\}.$$

The functions $(x, y) \mapsto x \vee y$ and $(x, y) \mapsto x \wedge y$ are the **lattice operations** on X. In a lattice, every finite nonempty set has a supremum and an infimum. If $\{x_1, \ldots, x_n\}$ is a finite subset of a lattice, then we write

$$\sup\{x_1, \ldots, x_n\} = \bigvee_{i=1}^{n} x_i \quad \text{and} \quad \inf\{x_1, \ldots, x_n\} = \bigwedge_{i=1}^{n} x_i.$$

Recall that a subset C of a vector space E is a **pointed convex cone** if:

a. C is a cone: $\alpha C \subset C$ for all $\alpha \geqslant 0$ (equivalently, $\alpha \geqslant 0$ and $x \in C$ imply $\alpha x \in C$);

b. C is convex: which given (a) amounts to $C + C \subset C$ (equivalently, $x, y \in C$ implies $x + y \in C$); and

c. C is pointed: $C \cap (-C) = \{0\}$.

A pointed convex cone C induces a partial order $\geqslant$ on E defined by $x \geqslant y$ whenever $x - y \in C$. The partial order induced by a pointed convex cone C is compatible with the algebraic structure of E in the sense that it satisfies the following two properties:

1. $x \geqslant y$ implies $x + z \geqslant y + z$ for each $z \in E$; and

---

[1] Note that this notation is at odds with the notation often used by economists for the usual order on $\mathbb{R}^n$, where $x > y$ means $x_i > y_i$ for all i, $x \geq y$ means $x_i \geqslant y_i$ for all i, and $x \geqslant y$ means $x \geq y$ and $x \neq y$.

2. $x \geqslant y$ implies $\alpha x \geqslant \alpha y$ for each $\alpha \geqslant 0$.

In the converse direction, if $\geqslant$ is a partial order on a real vector space E that satisfies properties (1) and (2), then the subset $C = \{x \in E : x \geqslant 0\}$ of E is a pointed convex cone, which induces the order $\geqslant$ on E. (We recommend you verify this as an exercise.)

An **ordered vector space** E is a real vector space with an order relation $\geqslant$ that is compatible with the algebraic structure of E in the sense that it satisfies properties (1) and (2). In an ordered vector space E, the set $\{x \in E : x \geqslant 0\}$ is a pointed convex cone, called the **positive cone** of E, denoted E^+ (or E_+). Any vector in E^+ is called **positive** The cone E^+ is also called the **nonnegative cone** of E.

8.2   Riesz spaces

An ordered vector space that is also a lattice is called a **Riesz space** or a **vector lattice**. The geometric interpretation of the lattice structure on a Riesz space is shown in Figure 8.1

Figure 8.1. The geometry of sup and inf.

For a vector x in a Riesz space, the **positive part** x^+, the **negative part** x^-, and the **absolute value** $|x|$ are defined by

$$x^+ = x \vee 0, \quad x^- = (-x) \vee 0, \quad \text{and} \quad |x| = x \vee (-x).$$

We list here two handy identities that are used all the time without any special mention:

$$x = x^+ - x^- \quad \text{and} \quad |x| = x^+ + x^-.$$

Also note that $|x| = 0$ if and only if $x = 0$.

8.1 Example   (Riesz spaces)   Many familiar spaces are Riesz spaces, as the following examples show.

1. The Euclidean space $\mathbb{R}^n$ is a Riesz space under the usual ordering where $x = (x_1, \ldots, x_n) \geqslant y = (y_1, \ldots, y_n)$ whenever $x_i \geqslant y_i$ for each $i = 1, \ldots, n$. The infimum and supremum of two vectors x and y are given by

$$x \vee y = (\max\{x_1, y_1\}, \ldots, \max\{x_n, y_n\})$$

and
$$x \wedge y = (\min\{x_1, y_1\}, \ldots, \min\{x_n, y_n\}).$$

2. Both the vector space $C(X)$ of all continuous real functions and the vector space $C_b(X)$ of all bounded continuous real functions on the topological space X are Riesz spaces when the ordering is defined pointwise. That is, $f \geqslant g$ whenever $f(x) \geqslant g(x)$ for each $x \in X$. The lattice operations are:

$$(f \vee g)(x) = \max\{f(x), g(x)\} \quad \text{and} \quad (f \wedge g)(x) = \min\{f(x), g(x)\}.$$

3. The vector space $L_p(\mu)$ ($0 \leqslant p \leqslant \infty$) is a Riesz space under the almost everywhere pointwise ordering. That is, $f \geqslant g$ in $L_p(\mu)$ if $f(x) \geqslant g(x)$ for μ-almost every x. The lattice operations are given by

$$(f \vee g)(x) = \max\{f(x), g(x)\} \quad \text{and} \quad (f \wedge g)(x) = \min\{f(x), g(x)\}.$$

4. Let $ba(\mathcal{A})$ denote the vector space of all signed charges of bounded variation on a given algebra $\mathcal{A}$ of subsets of a set X. Under the ordering defined by $\mu \geqslant \nu$ whenever $\mu(A) \geqslant \nu(A)$ for each $A \in \mathcal{A}$, $ba(\mathcal{A})$ is a Riesz space. Its lattice operations are given by

$$(\mu \vee \nu)(A) = \sup\{\mu(B) + \nu(A \setminus B) : B \in \mathcal{A} \text{ and } B \subset A\}$$

and

$$(\mu \wedge \nu)(A) = \inf\{\mu(B) + \nu(A \setminus B) : B \in \mathcal{A} \text{ and } B \subset A\}.$$

For details see Theorem 10.53.

5. The vector spaces ℓ_p ($0 < p \leqslant \infty$) and c_0 are Riesz spaces under the usual pointwise ordering. For details see Chapter 16

6. A slightly less familiar example of a Riesz space, but one that has applications to the theory of financial options, is the space of piecewise linear functions on an interval of the real line, with the usual pointwise ordering.

7. Lest you think that every ordered linear space you can imagine is a Riesz space, we offer for your consideration the case of the vector space of all differentiable functions on the real line, under the usual pointwise ordering. Clearly, the pointwise supremum of two differentiable functions need not be differentiable, but this fact alone does not mean that there is not a smallest differentiable function dominating any given pair of differentiable functions. Nonetheless, in general, there is no supremum to an arbitrary pair of differentiable functions. To convince yourself of this, consider the functions $f(x) = x$ and $g(x) = -x$.

8. Every function space in the sense of Definition 1.1 is a Riesz space under the pointwise ordering. ∎

8.3 Order bounded sets

A subset A of a Riesz space is **order bounded, from above** if there is a vector u (called an **upper bound** of A) that dominates each element of A, that is, satisfying $a \leqslant u$ for each $a \in A$. Sets order bounded from below are defined similarly. Notice that a subset A of a Riesz space is order bounded from above (resp. below) if and only if $-A$ is order bounded from below (resp. above). A subset A of a Riesz space is **order bounded** if A is both order bounded from above and from below. A **box** or an **order interval**, is any set of the form

$$[x, y] = \{z : x \leqslant z \leqslant y\}.$$

If x and y are incomparable, then $[x, y] = \varnothing$. Observe that a set is order bounded if and only if it fits in a box.

A nonempty subset of a Riesz space has a **supremum** (or a **least upper bound**) if there is an upper bound u of A such that $a \leqslant v$ for all $a \in A$ implies $u \leqslant v$. Clearly, the supremum, if it exists, is unique, and is denoted $\sup A$. The **infimum** (or **greatest lower bound**) of a nonempty subset A is defined similarly, and is denoted $\inf A$. (Recall that any nonempty bounded subset of real numbers has both an infimum and a supremum—this is the Completeness Axiom.) If we index $A = \{a_i : i \in I\}$, then we may employ the standard lattice notation

$$\sup A = \bigvee_{i \in I} a_i \qquad \text{and} \qquad \inf A = \bigwedge_{i \in I} a_i.$$

Keep in mind that a subset of a Riesz space can have at most one supremum and at most one infimum. Note also that if a set A has a supremum, then the set $-A = \{-a : a \in A\}$ has an infimum, and

$$\inf(-A) = -\sup A.$$

A net $\{x_\alpha\}$ in a Riesz space is **decreasing**, written $x_\alpha \downarrow$, if $\alpha \geqslant \beta$ implies $x_\alpha \leqslant x_\beta$. The symbol $x_\alpha \uparrow$ indicates an **increasing net**, while $x_\alpha \uparrow \leqslant x$ (resp. $x_\alpha \downarrow \geqslant x$) denotes an increasing (resp. decreasing) net that is order bounded from above (resp. below) by x. The notation $x_\alpha \downarrow x$ means that $x_\alpha \downarrow$ and $\inf\{x_\alpha\} = x$. The meaning of $x_\alpha \uparrow x$ is similar. Some basic properties of increasing nets are listed below. You can verify these properties as exercises; there are corresponding statements for decreasing nets.

- If $x_\alpha \uparrow x$ and $y_\beta \uparrow y$, then $x_\alpha + y_\beta \uparrow x + y$;

- If $x_\alpha \uparrow x$, then $\lambda x_\alpha \uparrow \lambda x$ for $\lambda > 0$, and $\lambda x_\alpha \downarrow \lambda x$ for $\lambda < 0$;

- If $x_\alpha \uparrow x$ and $y_\beta \uparrow y$, then $x_\alpha \vee y_\beta \uparrow x \vee y$ and $x_\alpha \wedge y_\beta \uparrow x \wedge y$.

A subset A of a Riesz space is **directed upward** (resp. **downward**), written $A \uparrow$ (resp. $A \downarrow$), if for each pair $a, b \in A$ there exists some $c \in A$ satisfying $a \vee b \leqslant c$ (resp. $a \wedge b \geqslant c$). That is, A is directed upward if and only if $(A, \geqslant)$ is a directed set. The symbol $A \uparrow a$ means $A \uparrow$ and $\sup A = a$ (and similarly, $A \downarrow a$ means $A \downarrow$ and $\inf A = a$). You can easily see that upward directed sets and increasing nets are for all practical purposes equivalent. However, in certain situations it is more convenient to employ upward directed sets than increasing nets.

8.4   Order and lattice properties

There are two important additional properties that the real numbers exhibit, but that a Riesz space E may or may not possess.

8.2 Definition   *A Riesz space E is **Archimedean** if whenever $0 \leqslant nx \leqslant y$ for all $n = 1, 2, \ldots$ and some $y \in E^+$, then $x = 0$. Equivalently, E is Archimedean if $\frac{1}{n}x \downarrow 0$ for each $x \in E^+$.*

*A Riesz space E is **order complete**, or **Dedekind complete** if every nonempty subset that is order bounded from above has a supremum. (Equivalently, if every nonempty subset that is bounded from below has an infimum).*

Note that the Archimedean property described here is different from the property often used in connection with the real numbers. The alternative "Archimedean property" is that for any nonzero x and any y, there exists an n satisfying $|y| \leqslant n|x|$. In the case of the real numbers, these two properties are equivalent, but they are not equivalent in general, as the next example shows.

8.3 Example   (The Archimedean property)   Let $C(0, 1)$ denote the vector space of all continuous functions on the open interval $(0, 1)$. It is an Archimedean Riesz space under the usual pointwise ordering. To see this, suppose $0 \leqslant f$. Then $\frac{1}{n}f(x) \downarrow 0$ in $\mathbb{R}$ for each x, so $\frac{1}{n}f \downarrow 0$ in $C(0, 1)$.

Now consider $f(x) = \frac{1}{x}$ and $g(x) = 1$ for all x. Observe that there is no n for which $f \leqslant ng$, so the alternative Archimedean property is not satisfied.   ∎

A moment's thought reveals that for any set A, the order $\geqslant$ on the set S of suprema of finite subsets of A is a direction: for each pair $x, y \in S$, we have $x \leqslant x \vee y$, $y \leqslant x \vee y$, and $x \vee y \in S$. Furthermore, S has the same upper bounds as A. This observation implies that a Riesz space is order complete if and only if $0 \leqslant x_\alpha \uparrow \leqslant x$ implies that $\sup\{x_\alpha\}$ exists (and also if and only if $x_\alpha \downarrow \geqslant x \geqslant 0$ implies that $\inf\{x_\alpha\}$ exists).

8.4 Lemma   *Every order complete Riesz space is Archimedean.*

Proof: Suppose $0 \leqslant nx \leqslant y$ for each $n = 1, 2, \ldots$ and some x, y in an order complete Riesz space E. Then $0 \leqslant x \leqslant \frac{1}{n} y$ for each n, so by the order completeness of E, $\frac{1}{n} y \downarrow z \geqslant x$ for some z. It follows that $\frac{2}{n} y = 2 \frac{1}{n} y \downarrow 2z$ and also $\frac{2}{n} y \downarrow z$. Hence, $2z = z$, so $z = 0$. From $0 \leqslant x \leqslant z = 0$, we see that $x = 0$. ∎

The converse is false—an Archimedean Riesz space need not be order complete. As the next example shows, $C[0, 1]$ is Archimedean but is not order complete.

8.5 Example (*$C[0, 1]$ is not order complete*)   Consider the sequence of piecewise linear functions in $C[0, 1]$ defined by

$$f_n(x) = \begin{cases} 1 & \text{if } 0 \leqslant x \leqslant \frac{1}{2} - \frac{1}{n}, \\ -n(x - \frac{1}{2}) & \text{if } \frac{1}{2} - \frac{1}{n} < x < \frac{1}{2}, \\ 0 & \text{if } \frac{1}{2} \leqslant x \leqslant 1. \end{cases}$$

Then $0 \leqslant f_n \uparrow \leqslant 1$ in $C[0, 1]$, where 1 is the constant function one, but $\{f_n\}$ does not have a supremum in $C[0, 1]$ (why?); see Figure 8.2.

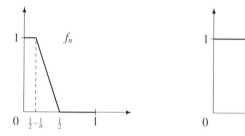

Figure 8.2. Example 8.5.

Incidentally, notice that $f_n(x) \uparrow f(x)$ for each $x \in [0, 1]$ implies that $f_n \uparrow f$ in the lattice sense. On the other hand, $f_n \uparrow f$ in the lattice sense does not imply that $f_n(x) \uparrow f(x)$ for each $x \in [0, 1]$. For example, define g_n by

$$g_n(x) = \begin{cases} 1 & \text{if } 0 \leqslant x \leqslant 1 - \frac{1}{n}, \\ n(1 - x) & \text{if } 1 - \frac{1}{n} < x \leqslant 1. \end{cases}$$

Notice that $g_n \uparrow 1$ in the lattice sense, while $g_n(1) = 0$ for all n. See Figure 8.2. ∎

Two Riesz spaces E and F are **lattice isomorphic,** (or **Riesz isomorphic** or simply **isomorphic**) if there exists a one-to-one, onto, lattice preserving linear operator $T: E \to F$. That is, besides being linear, one-to-one, and surjective, T also satisfies the identities

$$T(x \vee y) = T(x) \vee T(y) \qquad \text{and} \qquad T(x \wedge y) = T(x) \wedge T(y)$$

for all $x, y \in E$. From the point of view of Riesz space theory, two isomorphic Riesz spaces cannot be distinguished.

Remarkably, every Archimedean Riesz space is lattice isomorphic to an appropriate function space; for a proof, see [235, Chapter 7] Since the lattice operations in a function space are defined pointwise, this result implies the following remarkable fact.

8.6 Theorem   *Every lattice identity that is true for real numbers is also true in every Archimedean Riesz space.*

For instance, you can easily verify the following lattice identities for real numbers.

1. $x \wedge y = -[(-x) \vee (-y)]$ and $x \vee y = -[(-x) \wedge (-y)]$.

2. $x + y \vee z = (x + y) \vee (x + z)$ and $x + y \wedge z = (x + y) \wedge (x + z)$.

3. For $\alpha \geqslant 0$, $(\alpha x) \vee (\alpha y) = \alpha(x \vee y)$ and $(\alpha x) \wedge (\alpha y) = \alpha(x \wedge y)$.

4. $x + y = x \vee y + x \wedge y$.

5. $x = x^+ - x^-$, $|x| = x^+ + x^-$ and $x^+ \wedge x^- = 0$.

6. $|\alpha x| = |\alpha||x|$.

7. $|x - y| = x \vee y - x \wedge y$.

8. $x \vee y = \frac{1}{2}(x + y + |x - y|)$ and $x \wedge y = \frac{1}{2}(x + y - |x - y|)$.

9. $|x| \vee |y| = \frac{1}{2}\big||x + y| + |x - y|\big|$ and $|x| \wedge |y| = \frac{1}{2}\big||x + y| - |x - y|\big|$.

10. $|x + y| \vee |x - y| = |x| + |y|$.

By the above claim, all these lattice identities are true in any Archimedean Riesz space—and in fact, in any Riesz space. Similarly, for lattice inequalities we have:

8.7 Corollary   *If a lattice inequality is true for real numbers, then it is true in any Riesz space.*

For instance, for arbitrary vectors x, y, and z in a Riesz space, we have:

a. $|x + y| \leqslant |x| + |y|$ and $\big||x| - |y|\big| \leqslant |x - y|$.

b. $|x \vee y - z \vee y| \leqslant |x - z|$ and $|x \wedge y - z \wedge y| \leqslant |x - z|$.

c. $|x^+ - y^+| \leqslant |x - y|$ and $|x^- - y^-| \leqslant |x - y|$.

d. If $x, y, z \geqslant 0$, then $x \wedge (y + z) \leqslant x \wedge y + x \wedge z$.

For more about lattice identities and inequalities in Riesz spaces see [12, 235].

To see how one can prove directly the above identities and inequalities we shall prove the first part of (1) and (2) and (4). For the first part of (1) let $a = -[(-x) \vee (-y)]$. Adding $x + a$ to both sides of the inequality $-x \leqslant -a$ we get $a \leqslant x$ and similarly $a \leqslant y$. Thus $a \leqslant x \wedge y$. Now assume that $b \leqslant x$ and $b \leqslant y$. It follows that $-x \leqslant -b$ and $-y \leqslant -b$. Therefore $(-x) \vee (-y) \leqslant -b$ or $b \leqslant a$. Thus $x \wedge y = a$.

For the first part of (2) let $a = x + (y \vee z)$ and $b = (x + y) \vee (x + z)$. From $a - x = y \vee z$ we get $y \leqslant a - x$ and $z \leqslant a - x$ or $x + y \leqslant a$ and $x + z \leqslant a$. Therefore, $b \leqslant a$. Now note that $x + y \leqslant b$ and $x + z \leqslant b$ imply $y \leqslant b - x$ and $z \leqslant b - x$. Consequently, $y \vee z \leqslant b - x$ or $a = x + (y \vee z) \leqslant b$. Thus $a = b$.

For (4) start by observing that $y - y \wedge x \geqslant 0$ implies $x \leqslant x + y - x \wedge y$ and similarly $y \leqslant x + y - x \wedge y$. Hence $x \vee y \leqslant x + y - x \wedge y$ or $x \vee y + x \wedge y \leqslant x + y$. Next, use that $y - x \vee y \leqslant 0$ to get $x \geqslant x + y - x \vee y$ and likewise $y \geqslant x + y - x \vee y$. Therefore, $x \wedge y \geqslant x + y - x \vee y$ or $x \vee y + x \wedge y \geqslant x + y$. It follows that $x \vee y + x \wedge y = x + y$. Incidentally, letting $y = 0$ in (4) and using (1), we get

$$x^+ - x^- = x \wedge 0 - [(-x) \vee 0] = x \vee 0 + x \wedge 0 = x + 0 = x.$$

Just as a vector subspace of a vector space is a subset that is closed under linear combinations, a vector subspace F of a Riesz space E is a **Riesz subspace** if for each $x, y \in F$ the vector $x \vee y$ (taken in E) belongs to F (and, of course, the vector $x \wedge y = -(-x) \vee (-y)$ also belongs to F). In other words, a vector subspace is a Riesz subspace if and only if it is closed under the lattice operations on E.

A **Dedekind completion** of a Riesz space E is an order complete Riesz space $\hat{E}$ having a Riesz subspace F that is lattice isomorphic to E (hence F can be identified with E) satisfying

$$\hat{x} = \sup\{x \in F : x \leqslant \hat{x}\} = \inf\{y \in F : \hat{x} \leqslant y\}$$

for each $\hat{x} \in \hat{E}$. Only Archimedean Riesz spaces can have Dedekind completions, and the converse is also true.

8.8 Theorem   *Every Archimedean Riesz space has a unique (up to lattice isomorphism) Dedekind completion.*

Proof: See [235, Section 32]. ∎

8.5   The Riesz decomposition property

Riesz spaces satisfy an important property known as the **Riesz Decomposition Property**.

8.9 Riesz Decomposition Property   *In a Riesz space, if the vector y satisfies $|y| \leqslant \left|\sum_{i=1}^n x_i\right|$, then there exist vectors $y_1, \ldots, y_n$ such that $y = \sum_{i=1}^n y_i$ and $|y_i| \leqslant |x_i|$ for each i. If y is positive, then the y_is can be chosen to be positive too.*

Proof: We prove the result for $n = 2$, and leave the completion of the proof by induction as an exercise. So assume $|y| \leqslant |x_1 + x_2|$.

Let $y_1 = [(-|x_1|) \vee y] \wedge |x_1|$. Clearly, $-|x_1| \leqslant y_1 \leqslant |x_1|$ or $|y_1| \leqslant |x_1|$. Also, note that if $y \geqslant 0$, then $0 \leqslant y_1 \leqslant y$. Next, put $y_2 = y - y_1$ and note that $y = y_1 + y_2$ and that $0 \leqslant y$ implies $y_2 \geqslant 0$. To finish the proof, we must show that $|y_2| \leqslant |x_2|$. To this end, start by observing that $|y| \leqslant |x_1 + x_2| \leqslant |x_1| + |x_2|$ implies $-|x_1| - |x_2| \leqslant y \leqslant |x_1| + |x_2|$ or $-|x_2| \leqslant |x_1| + y$ and $y - |x_1| \leqslant |x_2|$. So $-|x_2| \leqslant (|x_1| + y) \wedge 0$ and $(y - |x_1|) \vee 0 \leqslant |x_2|$. Now from

$$\begin{aligned} y_2 &= y - [(-|x_1|) \vee y] \wedge |x_1| \\ &= y + [|x_1| \wedge (-y)] \vee (-|x_1|) \\ &= [(|x_1| + y) \wedge 0] \vee (y - |x_1|), \end{aligned}$$

we see that $-|x_2| \leqslant y_2 \leqslant |x_2|$ or $|y_2| \leqslant |x_2|$. ∎

A. Mas-Colell has suggested the following economic interpretation of the Riesz Decomposition Property. Interpret each $x_i \geqslant 0$ as the vector of holdings of person i. Then $x = \sum_{i=1}^{n} x_i$ represents the total wealth of the economy. Think of the vector y $(0 \leqslant y \leqslant x)$ as a tax. The Riesz Decomposition Property says that in a Riesz space, if the tax is feasible in the aggregate, then there is a feasible way to distribute the tax among the individuals.

8.6   Disjointness

The notion of disjointness in Riesz spaces is much different from set-theoretic disjointness. It is also different from orthogonality in an inner product space.

8.10 Definition   *Two vectors x and y in a Riesz space are mutually* **disjoint**, *or* **orthogonal** *written $x \perp y$, if $|x| \wedge |y| = 0$.*

Note that if x and y are disjoint vectors, then so are αx and βy for all scalars α and β, as $0 \leqslant |\alpha x| \wedge |\beta y| \leqslant [(|\alpha| + |\beta|)|x|] \wedge [(|\alpha| + |\beta|)|y|] = 0$. As usual, we say that a set A of vectors is pairwise disjoint if each pair of distinct vectors in A is disjoint.

We saw earlier that $x = x^+ - x^-$ is a decomposition of a vector x in a Riesz space as a difference of two positive disjoint vectors. This decomposition is unique:

8.11 Theorem   *In a Riesz space, $x = y - z$ and $y \wedge z = 0$ imply $y = x^+$ and $z = x^-$.*

Proof: Note that $x^+ = x \vee 0 = (y - z) \vee 0 = y \vee z - z = (y + z - y \wedge z) - z = y$. Similarly, $x^- = z$. ∎

The next theorem characterizes disjoint vectors.

8.12 Theorem   *For vectors x and y in a Riesz space the following statements are equivalent.*

1. $x \perp y$, *that is,* $|x| \wedge |y| = 0$.

2. $|x + y| = |x - y|$.

3. $|x + y| = |x| \vee |y|$.

Consequently, if $\{x_1, \ldots, x_n\}$ is a finite pairwise disjoint set of vectors, then $\left|\sum_{i=1}^n x_i\right| = \sum_{i=1}^n |x_i| = \bigvee_{i=1}^n |x_i|$.

Proof: We present a proof using the lattice identities listed in Section 8.4.
(1) $\Longrightarrow$ (2)   It follows from $|x| \wedge |y| = \frac{1}{2}\big||x + y| - |x - y|\big|$.
(2) $\Longrightarrow$ (3)   Note that $|x| \vee |y| = \frac{1}{2}(|x + y| + |x - y|) = |x + y|$.
(3) $\Longrightarrow$ (1)   From $|x+y| = |x| \vee |y| = \frac{1}{2}(|x+y|+|x-y|)$, we get $|x+y| = |x-y|$.
So $|x| \wedge |y| = \frac{1}{2}\big||x + y| - |x - y|\big| = 0$.
For the last part, note first that $x_1 \perp x_2$ implies

$$|x_1 + x_2| = |x_1| \vee |x_2| = |x_1| + |x_2| - |x_1| \wedge |x_2| = |x_1| + |x_2|.$$

To complete the proof, observe that $x_n \perp x_1 + \cdots + x_{n-1}$, and induct on n.   ∎

8.7   Riesz subspaces and ideals

Recall that a vector subspace F of a Riesz space E is a **Riesz subspace** if it is closed under the lattice operations on E. From the identity $x \vee y = \frac{1}{2}(x+y+|x-y|)$, we see that a vector subspace F is a Riesz subspace if and only if $x \in F$ implies $|x| \in F$. Riesz subspaces of Archimedean Riesz spaces are likewise Archimedean.

For example, the collection of piecewise linear functions on $[0, 1]$ is a Riesz subspace of $C[0, 1]$, the Riesz space of continuous real functions on $[0, 1]$. In turn, $C[0, 1]$ is a Riesz subspace of $B[0, 1]$, the function space of bounded functions on $[0, 1]$. In its own right, this space is a Riesz subspace of $\mathbb{R}^{[0,1]}$, the function space of all real-valued functions on $[0, 1]$.

A subset S of a Riesz space is called **solid** if $|y| \leqslant |x|$ and $x \in S$ imply $y \in S$. A solid vector subspace of a Riesz space is called an **ideal** Since every solid set contains the absolute values of its elements, we see that every ideal is a Riesz subspace. However, a Riesz subspace need not be an ideal. For instance, $C[0, 1]$ is a Riesz subspace of $\mathbb{R}^{[0,1]}$, but it is not an ideal. On the other hand, the ℓ_p-spaces are ideals in $\mathbb{R}^{\mathbb{N}}$.

8.13 Theorem   *A Riesz subspace F is an ideal if and only if $0 \leqslant x \leqslant y$ and $y \in F$ imply $x \in F$.*

Proof: We prove the "if" part only. So assume that F is a Riesz subspace such that $0 \leqslant x \leqslant y$ and $y \in F$ imply $x \in F$. Now let $|x| \leqslant |y|$ with $y \in F$. Since F is a Riesz subspace, we have $|y| \in F$. From $0 \leqslant x^+ \leqslant |y|$ and $0 \leqslant x^- \leqslant |y|$, we get $x^+, x^- \in F$. Thus $x = x^+ - x^- \in F$, so F is an ideal. ∎

8.14 Lemma   *An ideal in an order complete Riesz space is order complete.*

Proof: Let A be an ideal in an order complete Riesz space E, and suppose the net $\{x_\alpha\}$ satisfies $0 \leqslant x_\alpha \uparrow \leqslant x$ in A. Since E is order complete, there exists some $y \in E$ with $x_\alpha \uparrow y$. Clearly, $0 \leqslant y \leqslant x$. Since $x \in A$ and A is an ideal, $y \in A$. It follows that $x_\alpha \uparrow y$ in A, so A is order complete. ∎

Every subset S of a Riesz space E is included in a smallest ideal. The existence of such a smallest ideal follows from the fact that E itself is an ideal including S, and the fact that the intersection of a family of ideals is an ideal (why?). The **ideal generated** by S is the intersection of all ideals that include S. A moment's thought shows that the ideal generated by S consists of all vectors $x \in E$ for which there exist a finite number of vectors $x_1, \ldots, x_n \in S$ and positive scalars $\lambda_1, \ldots, \lambda_n$ such that $|x| \leqslant \sum_{i=1}^n \lambda_i |x_i|$. A **principal ideal** is an ideal generated by a singleton. The principal ideal generated by $\{x\}$ in a Riesz space E is denoted E_x. Clearly,

$$E_x = \{ y \in E : \exists \lambda > 0 \text{ with } |y| \leqslant \lambda |x| \}.$$

An element $e > 0$ in a Riesz space E is an **order unit**, or simply a **unit**, if for each $x \in E$ there exists a $\lambda > 0$ such that $|x| \leqslant \lambda e$. Equivalently e is a unit if its principal ideal E_e is all of E. Units and principal ideals reappear in later sections, particularly Section 9.4.

8.8   Order convergence and order continuity

A net $\{x_\alpha\}$ in a Riesz space E **converges in order** (or is **order convergent**) to some $x \in E$, written $x_\alpha \xrightarrow{o} x$, if there is a net $\{y_\alpha\}$ (with the same directed set) satisfying $y_\alpha \downarrow 0$ and $|x_\alpha - x| \leqslant y_\alpha$ for each α. A function $f : E \to F$ between two Riesz spaces is **order continuous** if $x_\alpha \xrightarrow{o} x$ in E implies $f(x_\alpha) \xrightarrow{o} f(x)$ in F.

A net can have at most one order limit. Indeed, if $x_\alpha \xrightarrow{o} x$ and $x_\alpha \xrightarrow{o} y$, then pick two nets $\{y_\alpha\}$ and $\{z_\alpha\}$ with $|x_\alpha - x| \leqslant y_\alpha \downarrow 0$ and $|x_\alpha - y| \leqslant z_\alpha \downarrow 0$ for each α, and note that $0 \leqslant |x - y| \leqslant |x_\alpha - x| + |x_\alpha - y| \leqslant y_\alpha + z_\alpha \downarrow 0$ implies $|x - y| = 0$, or $x = y$. Here are some simple properties of order convergent nets.

8.15 Lemma (Order convergence)   *If $x_\alpha \xrightarrow{o} x$ and $y_\beta \xrightarrow{o} y$, then:*

1. $x_\alpha + y_\beta \xrightarrow{o} x + y$.

2. $x_\alpha^+ \xrightarrow{o} x^+$, $x_\alpha^- \xrightarrow{o} x^-$, *and* $|x_\alpha| \xrightarrow{o} |x|$.

3. $\lambda x_\alpha \xrightarrow{o} \lambda x$ for each $\lambda \in \mathbb{R}$.

4. $x_\alpha \vee y_\beta \xrightarrow{o} x \vee y$ and $x_\alpha \wedge y_\beta \xrightarrow{o} x \wedge y$.

5. If $x_\alpha \leqslant y_\alpha$ for all $\alpha \geqslant \alpha_0$, then $x \leqslant y$.

The **limit superior** and **limit inferior** of an order bounded net $\{x_\alpha\}$ in an order complete Riesz space are defined by the formulas

$$\limsup_\alpha x_\alpha = \bigwedge_\alpha \bigvee_{\beta \geqslant \alpha} x_\beta \qquad \text{and} \qquad \liminf_\alpha x_\alpha = \bigvee_\alpha \bigwedge_{\beta \geqslant \alpha} x_\beta.$$

Note that $\liminf_\alpha x_\alpha \leqslant \limsup_\alpha x_\alpha$. (Why?) The limit superior and limit inferior characterize order convergence in order complete Riesz spaces.

8.16 Theorem (Order convergence) *An order bounded net $\{x_\alpha\}$ in an order complete Riesz space satisfies $x_\alpha \xrightarrow{o} x$ if and only if*

$$x = \liminf_\alpha x_\alpha = \limsup_\alpha x_\alpha.$$

Proof: Assume $x_\alpha \xrightarrow{o} x$. Then there is another net $\{y_\alpha\}$ such that $|x_\alpha - x| \leqslant y_\alpha \downarrow 0$. Now note that for each $\beta \geqslant \alpha$, we have $x_\beta = (x_\beta - x) + x \leqslant y_\beta + x \leqslant y_\alpha + x$, so $\bigvee_{\beta \geqslant \alpha} x_\beta \leqslant y_\alpha + x$. Hence,

$$\limsup_\alpha x_\alpha = \bigwedge_\alpha \bigvee_{\beta \geqslant \alpha} x_\beta \leqslant \bigwedge_\alpha (y_\alpha + x) = x.$$

Similarly, $\liminf_\alpha x_\alpha \geqslant x$, so $x = \limsup_\alpha x_\alpha = \liminf_\alpha x_\alpha$.

For the converse, note that if $x = \limsup_\alpha x_\alpha = \liminf_\alpha x_\alpha$, then by letting $y_\alpha = \bigvee_{\beta \geqslant \alpha} x_\beta - \bigwedge_{\gamma \geqslant \alpha} x_\gamma$, we get $y_\alpha \downarrow 0$ and $|x_\alpha - x| \leqslant y_\alpha$ for each α. This shows that $x_\alpha \xrightarrow{o} x$. ∎

The next result is obvious but bears pointing out. It says that in a wide class of spaces where pointwise convergence makes sense, order convergence and pointwise convergence coincide. We leave the proof as an exercise.

8.17 Lemma   *An order bounded sequence $\{f_n\}$ in some $L_p(\mu)$ space satisfies $f_n \xrightarrow{o} f$ if and only if $f_n(x) \to f(x)$ in $\mathbb{R}$ for μ-almost all x.*

Similarly, an order bounded net $\{f_\alpha\}$ in $\mathbb{R}^X$ satisfies $f_\alpha \xrightarrow{o} f$ if and only if $f_\alpha(x) \to f(x)$ in $\mathbb{R}$ for all $x \in X$.

However, norm convergence and order convergence do not generally coincide.

8.18 Example (Order convergence vs. norm convergence)   The sequence $\{u_n\}$ in ℓ_∞ defined by $u_n = (\underbrace{1, \ldots, 1}_{n} 0, 0, \ldots)$ converges pointwise and in order to

$\mathbf{1} = (1, 1, \ldots)$, but not in norm. ∎

8.9 Bands

A subset S of a Riesz space is **order closed** if $\{x_\alpha\} \subset S$ and $x_\alpha \xrightarrow{o} x$ imply $x \in S$. A solid set A is order closed if and only if $0 \leqslant x_\alpha \uparrow x$ and $\{x_\alpha\} \subset A$ imply $x \in A$. To see this, assume the condition is satisfied and let a net $\{x_\alpha\}$ in A satisfy $x_\alpha \xrightarrow{o} x$. Pick a net $\{y_\alpha\}$ with $y_\alpha \downarrow 0$ and $|x_\alpha - x| \leqslant y_\alpha$ for each α. Then $(|x| - y_\alpha)^+ \leqslant |x_\alpha|$ for each α, so $(|x| - y_\alpha)^+ \in A$ for each α. Now the relation $(|x| - y_\alpha)^+ \uparrow |x|$ coupled with our condition yields $|x| \in A$, so $x \in A$.

An order closed ideal is called a **band**. By the above, an ideal A is a band if and only if $\{x_\alpha\} \subset A$ and $0 \leqslant x_\alpha \uparrow x$ imply $x \in A$. Here are two illustrative examples of bands.

• If V is an open subset of a completely regular topological space X, then the vector space

$$B = \{f \in C(X) : f(x) = 0 \text{ for all } x \in V\}$$

is a band in the Riesz space $C(X)$. (Why?)

• If E is a measurable set in a measure space (X, Σ, μ) and $0 \leqslant p \leqslant \infty$, then the vector space

$$C = \{f \in L_p(\mu) : f(x) = 0 \text{ for } \mu\text{-almost all } x \in E\}$$

is a band in the Riesz space $L_p(\mu)$.

If S is a nonempty subset of a Riesz space E, then its **disjoint complement** S^d, defined by

$$S^d = \{x \in E : |x| \wedge |y| = 0 \text{ for all } y \in S\},$$

is necessarily a band. This follows immediately from the order continuity of the lattice operations. We write S^{dd} for $(S^d)^d$.

The **band generated by a subset** D of a Riesz space E is the intersection of all bands that include D. Here are two important bands generated by special sets.

• The band generated by a singleton $\{x\}$ is called a **principal band**, denoted B_x. Note that

$$B_x = \{y \in E : |y| \wedge n|x| \uparrow_n |y|\}.$$

• The band generated by an ideal A is given by

$$\{x \in E : \exists \text{ a net } \{x_\alpha\} \subset A \text{ with } 0 \leqslant x_\alpha \uparrow |x|\}.$$

8.19 Theorem (Double disjoint complement of a band) *In an Archimedean Riesz space every band B satisfies $B = B^{dd}$. Also, the band generated by any set S is precisely S^{dd}.*

Proof: Let B be a band in an Archimedean Riesz space E. Then $B \subset B^{dd}$. To see that $B^{dd} \subset B$, fix $0 < x \in B^{dd}$ and let $D = \{y \in B : 0 \leqslant y \leqslant x\}$. Obviously $D \uparrow$. We claim that $D \uparrow x$.

To see this, assume by way of contradiction that there exists some z in E^+ satisfying $y \leqslant z < x$ for all $y \in D$. From $0 < x - z \in B^{dd}$, we infer that $x - z \notin B^d$ (keep in mind that $B^d \cap B^{dd} = \{0\}$). So there exists some $0 < v \in B$ such that $u = v \wedge (x - z) > 0$. Then $u \in B$ and $0 < u \leqslant x$, so $u \in D$. Consequently, $2u = u + u \leqslant z + (x - z) = x$, and thus $2u \in D$. By induction, we see that $0 < nu \leqslant x$ for each n, contrary to the Archimedean property of E. Thus $D \uparrow x$. Since B is a band, $x \in B$. Therefore, $B = B^{dd}$. ∎

A vector $e > 0$ in a Riesz space E is called a **weak unit** if the principal band $B_e = E$. This differs from an order unit, which has the property that its principal ideal is E. For instance, the constant function **1** is an order unit in $C[0, 1]$, but only a weak unit in $L_1[0, 1]$. If E is Archimedean, a vector $e > 0$ is a weak unit if and only if $x \perp e$ implies $x = 0$. (Why?)

Recall that a vector space L is the direct sum of two vector subspaces M and N, written $L = M \oplus N$, if every $x \in L$ has a unique decomposition $x = y + z$ with $y \in M$ and $z \in N$. This decomposition defines two linear mappings $x \mapsto y$ and $x \mapsto z$, the projections onto M and N. A band B in a Riesz space E is a **projection band** if $E = B \oplus B^d$. F. Riesz has shown that in an order complete Riesz space, every band is a projection band.

8.20 Theorem (F. Riesz) *Every band B in an order complete Riesz space E is a projection band. That is, $E = B \oplus B^d$.*

Proof: Try it as an exercise, or see [15, Theorem 1.46, p. 21]. ∎

An important example of a band is the set of countably additive measures on a σ-algebra of sets. This is a band in the Riesz space of charges (Theorem 10.56). Its disjoint complement is the set of all purely finitely additive charges. The band generated by a signed measure μ is the collection of signed measures that are absolutely continuous with respect to μ (Theorem 10.61). The **Lebesgue Decomposition Theorem** is nothing but the fact that this band is a projection band.

8.10 Positive functionals

A linear functional $f: E \to \mathbb{R}$ on a Riesz space E is:

- **positive** if $x \geqslant 0$ implies $f(x) \geqslant 0$.

- **strictly positive** if $x > 0$ implies $f(x) > 0$.

- **order bounded** if f carries order bounded subsets of E to bounded subsets of $\mathbb{R}$ (or, equivalently, if $f([x, y])$ is a bounded subset of $\mathbb{R}$ for each box $[x, y]$).

Amazingly, there are Riesz spaces that have no strictly positive linear functional!

8.21 Example (No strictly positive functionals) On $\mathbb{R}^N$, the Riesz space of all real sequences, there are no strictly positive linear functionals. This is because any positive linear functional on $\mathbb{R}^N$ is representable by a sequence with only finitely many nonzero terms (Theorem 16.3). ∎

A Riesz space E has the **countable sup property** if every subset of E that has a supremum in E includes a countable subset having the same supremum in E.

8.22 Theorem *If a Riesz space E admits a strictly positive linear functional, then E is Archimedean and has the countable sup property.*

Proof: Let $f: E \to \mathbb{R}$ be a strictly positive linear functional. If $0 \leqslant y \leqslant \frac{1}{n}x$ for all n and some $x, y \in E^+$, then $0 \leqslant f(y) \leqslant \frac{1}{n}f(x)$ for all n, so $f(y) = 0$. The strict positivity of f implies $y = 0$. Hence, E is Archimedean.

Next, let $\sup A = a$. Replacing A by the set of all finite suprema of the elements of A, we can assume that $A \uparrow a$. Now pick a sequence $\{a_n\} \subset A$ with $a_n \uparrow$ and $f(a_n) \uparrow s = \sup\{f(x) : x \in A\} < \infty$. Clearly, if $x \in A$, then $f(x \vee a_n) \uparrow s$. We claim that $a_n \uparrow a$. To see this, let $a_n \leqslant b$ for each n. Then, for each $x \in A$, we have

$$0 \leqslant f((x-b)^+) \leqslant f((x-a_n)^+) = f(x \vee a_n) - f(a_n) \to s - s = 0,$$

so $f((x-b)^+) = 0$. The strict positivity of f implies $(x-b)^+ = 0$ or $x \leqslant b$ for each $x \in A$. Hence $a \leqslant b$, proving that $\sup\{a_n\} = a$. ∎

The Riemann integral is a strictly positive linear functional on $C[0,1]$, and so is the Lebesgue integral on $L_p[0,1]$ ($1 \leqslant p \leqslant \infty$). So $C[0,1]$ and $L_p[0,1]$ ($1 \leqslant p \leqslant \infty$) are Riesz spaces with the countable sup property. The Riesz space $\mathbb{R}^{[0,1]}$ does not have the countable sup property. (Why?)

Every linear functional is additive, that is, $f(x+y) = f(x) + f(y)$. In a Riesz space, a sort of converse result is also true.

8.23 Lemma (Kantorovich) *If E is a Riesz space and $f: E^+ \to \mathbb{R}^+$ is additive, then f extends uniquely to a positive linear functional $\hat{f}$ on E. Moreover, the unique positive linear extension is given by the formula[2]*

$$\hat{f}(x) = f(x^+) - f(x^-).$$

Proof: Clearly, any linear extension of f must satisfy the formula defining $\hat{f}$. To complete the proof we must show that $\hat{f}$ is linear. To see that $\hat{f}$ is additive, let $x, y \in E$. Then

$$(x+y)^+ - (x+y)^- = x + y = x^+ - x^- + y^+ - y^-,$$

---

[2] The proof below shows that in actuality we have the following stronger result: *Let E and F be two Riesz spaces with F Archimedean. If a function $T: E^+ \to F^+$ is additive, then T has a unique positive linear extension $\hat{T}: E \to F$ given by the formula $\hat{T}(x) = T(x^+) - T(x^-)$.*

so $(x + y)^+ + x^- + y^- = x^+ + y^+ + (x + y)^-$. Using the fact that f is additive on E^+, we obtain $f((x + y)^+) + f(x^-) + f(y^-) = f(x^+) + f(y^+) + f((x + y)^-)$, or

$$\hat{f}(x + y) = f((x + y)^+) - f((x + y)^-)$$
$$= [f(x^+) - f(x^-)] + [f(y^+) - f(y^-)] = \hat{f}(x) + \hat{f}(y).$$

Also,

$$\hat{f}(-x) = f((-x)^+) - f((-x)^-) = f(x^-) - f(x^+) = -\hat{f}(x).$$

Moreover, since $f(kx) = kf(x)$ for each natural number k and $x \in E^+$, for each rational number $r = \frac{m}{n}$ with $m, n \in \mathbb{N}$ and each $x \in E^+$ we have

$$rf(x) = \tfrac{m}{n} f(x) = \tfrac{m}{n} f(\tfrac{nx}{n}) = \tfrac{m}{n} n f(\tfrac{x}{n}) = m f(\tfrac{x}{n}) = f(\tfrac{m}{n} x) = f(rx).$$

Next notice that $0 \leqslant x \leqslant y$ implies $f(x) \leqslant f(x) + f(y - x) = f(y)$.

The above observations show that in order to establish the homogeneity of $\hat{f}$ it suffices to show that $f(\lambda x) = \lambda f(x)$ for each $\lambda > 0$ and each $x \in E^+$. So let $\lambda > 0$ and $x \in E^+$. Pick two sequences $\{r_n\}$ and $\{t_n\}$ of rational numbers such that $0 \leqslant r_n \uparrow \lambda$ and $t_n \downarrow \lambda$. From $0 \leqslant r_n x \leqslant \lambda x \leqslant t_n x$, we see that

$$r_n f(x) = f(r_n x) \leqslant f(\lambda x) \leqslant f(t_n x) = t_n f(x),$$

and by letting $n \to \infty$, we obtain $f(\lambda x) = \lambda f(x)$.   ∎

Clearly, every positive linear functional is monotone ($f(x) \geqslant f(y)$ whenever $x \geqslant y$), and so order bounded. It is also straightforward that the set of all order bounded linear functionals on a Riesz space E (under the usual algebraic operations) is a vector space. This vector space is denoted $E^\sim$ and is called the **order dual** of E. The order dual $E^\sim$ becomes an ordered vector space under the ordering $f \geqslant g$ if $f(x) \geqslant g(x)$ for each $x \in E^+$.

F. Riesz has shown that the order dual of any Riesz space is, in fact, an order complete Riesz space.

8.24 Theorem (F. Riesz)   *The order dual $E^\sim$ of any Riesz space E is an order complete Riesz space. Its lattice operations are given by*

$$(f \vee g)(x) = \sup\{f(y) + g(z) : y, z \in E^+ \text{ and } y + z = x\}$$

and

$$(f \wedge g)(x) = \inf\{f(y) + g(z) : y, z \in E^+ \text{ and } y + z = x\}$$

for all $f, g \in E^\sim$ and all $x \in E^+$. In particular, for $f \in E^\sim$ and $x \in E^+$, we have:

1. *$f^+(x) = \sup\{f(y) : 0 \leqslant y \leqslant x\};$*

2. $f^-(x) = \sup\{-f(y) : 0 \leqslant y \leqslant x\} = -\inf\{f(y) : 0 \leqslant y \leqslant x\};$

3. $|f|(x) = \sup\{f(y) : |y| \leqslant x\} = \sup\{|f(y)| : |y| \leqslant x\};$ *and*

4. $|f(x)| \leqslant |f|(|x|).$

Moreover, $f_\alpha \uparrow f$ *holds in* $E^\sim$ *if and only if* $f_\alpha(x) \uparrow f(x)$ *for each* $x \in E^+$.

Proof: We prove the supremum formula and leave everything else as an exercise.
So let $f, g \in E^\sim$. Define $h: E^+ \to \mathbb{R}$ by

$$h(x) = \sup\{f(y) + g(x - y) : 0 \leqslant y \leqslant x\}.$$

We claim that h is additive. To see this, let $u, v \in E^+$. Then for arbitrary $0 \leqslant u_1 \leqslant u$
and $0 \leqslant v_1 \leqslant v$, we have

$$[f(u_1) + g(u - u_1)] + [f(v_1) + g(v - v_1)]$$
$$= f(u_1 + v_1) + g(u + v - (u_1 + v_1)) \leqslant h(u + v),$$

from which we deduce that $h(u) + h(v) \leqslant h(u + v)$. Now if $0 \leqslant y \leqslant u + v$, then by
the Riesz Decomposition Property 8.9 there exist $y_1, y_2 \in E^+$ such that $y = y_1 + y_2$,
$0 \leqslant y_1 \leqslant u$, and $0 \leqslant y_2 \leqslant v$. Consequently,

$$f(y) + g((u + v) - y)) = [f(y_1) + g(u - y_1)] + [f(y_2) + g(v - y_2)] \leqslant h(u) + h(v).$$

This implies $h(u + v) \leqslant h(u) + h(v)$. Therefore, $h(u + v) = h(u) + h(v)$ for all
$u, v \in E^+$.

Now, by Lemma 8.23, h has a unique positive linear extension $\hat{h}$ to all of E.
Clearly, $f(x) \leqslant \hat{h}(x)$ and $g(x) \leqslant \hat{g}(x)$ for all $x \in E^+$. Moreover, if $\theta \in E^\sim$ satisfies
$f \leqslant \theta$ and $g \leqslant \theta$, then $0 \leqslant y \leqslant x$ implies

$$f(y) + g(x - y) \leqslant \theta(y) + \theta(x - y) = \theta(x),$$

so $\hat{h}(x) = h(x) \leqslant \theta(x)$ for each $x \in E^+$. Therefore, $\hat{h} = f \vee g$ in $E^\sim$.   ∎

Since f^+ and f^- are positive, we have the following.

8.25 Corollary   *Every order bounded linear functional is the difference of two
positive linear functionals.*

8.26 Definition   *A linear functional* $f: E \to \mathbb{R}$ *is*

- *order continuous (or a normal integral) if* $f(x_\alpha) \to 0$ *in* $\mathbb{R}$ *whenever the net*
$x_\alpha \xrightarrow{o} 0$ *in* E.

- *σ-order continuous (or an integral) if* $f(x_n) \to 0$ *in* $\mathbb{R}$ *whenever the sequence* $x_n \xrightarrow{o} 0$ *in* E.

Clearly every order continuous linear functional is σ-order continuous but the converse is false. The Lebesgue integral on the Riesz space $B_b[0, 1]$ of all bounded measurable functions on $[0, 1]$ is σ-order continuous, but not order continuous. (See the discussion on page 415.)

8.27 Lemma   *On an Archimedean Riesz space, every σ-order continuous linear functional is order bounded.*

Proof: Let f be a σ-order continuous linear functional on the Archimedean Riesz space E, and suppose by way of contradiction that f is not order bounded. Then there is a sequence $\{x_n\}$ in a box $[-x, x]$ satisfying $|f(x_n)| > n^2$. Since $|\frac{1}{n}x_n| \leqslant \frac{1}{n}x$ and E is Archimedean, $\frac{1}{n}x_n \xrightarrow{o} 0$, and therefore $f(\frac{1}{n}x_n) \to 0$. But by hypothesis, $|f(\frac{1}{n}x_n)| > n$, a contradiction.   ∎

The set of all order continuous linear functionals is a vector subspace of $E^\sim$, denoted $E_n^\sim$. It is called the **order continuous dual** of E. Similarly, the vector space of all σ-order continuous linear functionals is called the **σ-order continuous dual** of E, denoted $E_c^\sim$. T. Ogasawara has shown that both the order continuous and the σ-order continuous duals are bands of the order dual $E^\sim$.

8.28 Theorem (Ogasawara)   *Both the order and σ-order continuous duals of a Riesz space are bands in its order dual.*

Proof: See [12, Theorem 4.4, p. 44].   ∎

Since (by Theorem 8.24) $E^\sim$ is an order complete Riesz space, it follows from Theorems 8.28 and 8.20 that $E_n^\sim$ is a projection band in $E^\sim$, so $E^\sim = E_n^\sim \oplus (E_n^\sim)^d$. The band $(E_n^\sim)^d$ is denoted $E_s^\sim$ and its members are called **singular functionals**. So $E^\sim = E_n^\sim \oplus E_s^\sim$, which means that every order bounded linear functional $f \in E^\sim$ can be written uniquely in the form $f = g + h$, where $g \in E_n^\sim$ (called the **order continuous component** of f) and $h \in E_s^\sim$ (called the **singular component** of f).

8.29 Example (Riesz spaces and their duals)   Here are some familiar Riesz spaces and their duals.

• If $E = C[0, 1]$, then $E^\sim = ca[0, 1]$, the set of all (countably additive) Borel signed measures on $[0,1]$. It can be shown that $E_c^\sim = E_n^\sim = \{0\}$, and $E_s^\sim = E^\sim$. We emphasize: There is no nonzero σ-order continuous linear functional on the Riesz space $C[0, 1]$!, For a proof, see [234, Example 24.5(ii) p. 674].

• If $E = \ell_\infty$, then $E_c^\sim = E_n^\sim = \ell_1$, which can be identified with the vector space of all signed measures of bounded variation on the positive integers. Its complement $E_s^\sim$ is the vector space consisting of all purely finitely additive bounded signed charges. For details, see Section 16.7.

• If $E = L_p(\mu)$ for some $1 < p < \infty$, then $E^\sim = E_c^\sim = E_n^\sim = L_q(\mu)$ (where $\frac{1}{p} + \frac{1}{q} = 1$) and $E_s^\sim = \{0\}$.

• If $E = \mathbb{R}^\mathbb{N}$, the Riesz space of all real sequences, then $E^\sim = E_c^\sim = E_n^\sim$ is the Riesz space of all sequences that have only finitely many nonzero components, and $E_s^\sim = \{0\}$; see Theorem 16.3.
■

8.11   Extending positive functionals

The Hahn–Banach Extension Theorem 5.53 has a natural generalization to Riesz space-valued functions. As in the real case, a function $p \colon X \to E$ from a vector space to a partially ordered vector space is **sublinear**, if p is **subadditive** that is, $p(x + y) \leqslant p(x) + p(y)$ for all $x, y \in X$, and **positively homogeneous**, that is, $p(\alpha x) = \alpha p(x)$ for all $x \in X$ and all scalars $\alpha \geqslant 0$.

We can now state a more general form of the Hahn–Banach Extension Theorem. Its proof is a Riesz space analogue of the proof of Theorem 5.53; see [12, Theorem 2.1, p. 21].

8.30 Vector Hahn–Banach Extension Theorem   *Let X be vector space and let $p \colon X \to E$ be a convex (or in particular, a sublinear) function from X to an order complete Riesz space. If M be a vector subspace of X and $T \colon M \to E$ is a linear operator satisfying $T(x) \leqslant p(x)$ for each $x \in M$, then there exists a linear extension $\hat{T}$ of T to all of X satisfying $\hat{T}(x) \leqslant p(x)$ for all $x \in X$.*

Recall that a function $f \colon X \to Y$ between partially ordered sets is **monotone** if $x \leqslant y$ implies $f(x) \leqslant f(y)$.

8.31 Theorem   *Let F be a Riesz subspace of a Riesz space E and let $f \colon F \to \mathbb{R}$ be a positive linear functional. Then f extends to a positive linear functional on all of E if and only if there is a monotone sublinear function $p \colon E \to \mathbb{R}$ satisfying $f(x) \leqslant p(x)$ for all $x \in F$.*

Proof: One direction is simple. If g is a positive extension of f to E, just let $p(x) = g(x^+)$.

For the converse, suppose there is a monotone sublinear function $p \colon E \to \mathbb{R}$ with $f(x) \leqslant p(x)$ for $x \in F$. By the Hahn–Banach Theorem 5.53 there is a linear extension g of f to E satisfying $g(x) \leqslant p(x)$ for all $x \in E$. Observe that if $x \geqslant 0$, then $-g(x) = g(-x) \leqslant p(-x) \leqslant p(0) = 0$, which implies $g(x) \geqslant 0$. So g is a positive extension of f.
■

Let M be a vector subspace of a partially ordered vector space E. We say that M **majorizes** E if for each $x \in E$, there is some $y \in M$ with $x \leqslant y$.

8.32 Theorem (**Kantorovich [193]**)   *If M is a vector subspace of a Riesz space E that majorizes E, then every positive linear functional on M extends to a positive linear functional on E.*

Proof: Let M be a majorizing subspace of a Riesz space E, and let $f : M \to \mathbb{R}$ be a positive linear functional. Define the mapping $p : E \to \mathbb{R}$ by

$$p(x) = \inf\{f(y) : y \in M \text{ and } x \leqslant y\}.$$

Notice that the positivity of f and the majorization by M guarantee that p is indeed real-valued. Now an easy verification shows that p is a sublinear mapping satisfying $f(x) = p(x)$ for all $x \in M$.

By the Hahn–Banach Theorem 5.53 there exists a linear extension g of f to all of E satisfying $g(x) \leqslant p(x)$ for all $x \in E$. In particular, for $x \geqslant 0$ we have $-x \leqslant 0 \in M$, so $-g(x) = g(-x) \leqslant p(-x) \leqslant f(0) = 0$, or $g(x) \geqslant 0$. Thus, g is a positive extension of f to all of E. ∎

Since any subspace containing a unit is a majorizing subspace, the following result is a special case of Theorem 8.32 (cf. L. Nachbin [257, Theorem 7, p. 119]).

8.33 Corollary   *If a vector subspace M of a Riesz space E contains an order unit, then every positive linear functional on M extends to a (not necessarily unique) positive linear functional on E.*

For an application of the preceding result, notice that the Riesz space of continuous functions $C_b(X)$ majorizes $B(X)$, the Riesz space of all bounded real functions on X. By Theorem 8.32 every positive linear functional on $C_b(X)$ extends to a positive linear functional on $B(X)$.

The **double order dual** of a Riesz space E is the order dual of $E^\sim$, denoted $E^{\sim\sim}$. Every vector x in a Riesz space E gives rise to an order bounded linear functional $\hat{x}$ on $E^\sim$ via the formula

$$\hat{x}(f) = f(x), \quad f \in E^\sim.$$

In fact, an easy argument shows that $\hat{x}$ is order continuous on $E^\sim$. Thus, $x \mapsto \hat{x}$, is a linear mapping from E into $E^{\sim\sim}$. It turns out to be lattice preserving, as we shall see. That is, it also satisfies

$$\widehat{x \vee y} = \hat{x} \vee \hat{y} \quad \text{and} \quad \widehat{x \wedge y} = \hat{x} \wedge \hat{y}$$

for all $x, y \in E$. In case $E^\sim$ separates the points of E, the mapping $x \mapsto \hat{x}$ is also one-to-one, so E (identified with its image $\hat{E}$) can be considered a Riesz subspace of its double order dual $E^{\sim\sim}$. This is a special case of the next theorem (for $F = E^\sim$).

8.34 Theorem   *Let E be a Riesz space, and let F be an ideal in the order dual $E^\sim$ that separates the points of E. Then the mapping $x \mapsto \hat{x}$ from E to $F^\sim$, where*

$$\hat{x}(f) = f(x), \quad f \in F^\sim,$$

is a lattice isomorphism onto its range. Hence E identified with its image $\hat{E}$ can be viewed as a Riesz subspace of $F^\sim$.

Proof: Clearly, $x \mapsto \hat{x}$ is a linear isomorphism onto its range. To see that it is also lattice preserving, it suffices to show that $\widehat{x^+} = \hat{x}^+$ for each $x \in E$. To this end, let $x \in E$ be fixed and let $f \in F^+$ be arbitrary. Then

$$\hat{x}^+(f) = \sup\{\hat{x}(g) : g \in F \text{ and } 0 \leqslant g \leqslant f\} \leqslant f(x^+) = \widehat{x^+}(f).$$

Now let $Y = \{\lambda x : \lambda \in \mathbb{R}\}$, and define the seminorm $p : E \to \mathbb{R}$ by $p(z) = f(z^+)$. Clearly, Y is a Riesz subspace of E and if we define $h : Y \to \mathbb{R}$ by $h(\lambda x) = \lambda f(x^+)$, then $h(y) \leqslant p(y)$ for each $y \in Y$. By Theorem 8.30, h has a linear extension to all of E (which we denote by h again) such that $h(z) \leqslant p(z)$ for each $z \in E$. It follows that $0 \leqslant h \leqslant f$, so $h \in F$. Moreover,

$$\widehat{x^+}(f) = f(x^+) = h(x) = \hat{x}(h) \leqslant \hat{x}^+(f),$$

and hence $\widehat{x^+}(f) = \hat{x}^+(f)$ for all $f \in F$. Therefore, $\widehat{x^+} = \hat{x}^+$.   ∎

8.35 Corollary   *Let E be a Riesz space, and let F be an ideal in the order dual $E^\sim$ that separates the points of E. Then a vector $x \in E$ satisfies $x \geqslant 0$ if and only if $f(x) \geqslant 0$ for each $0 \leqslant f \in F$.*

8.12   Positive operators

In this section, we discuss some basic properties of positive operators that are used in later chapters. For detailed accounts of the theory of positive operators you can consult the books by Aliprantis and Burkinshaw [12], Schaefer [294], and Zaanen [347].

8.36 Definition   *A **positive operator** $T : E \to F$ between ordered vector spaces is a linear operator that maps positive vectors to positive vectors. That is, T is positive if $x \geqslant 0$ in E implies $T(x) \geqslant 0$ in F.*

The definition of order continuity is analogous to the one for real functions.

8.37 Definition   *A positive operator $T : E \to F$ between Riesz spaces is:*

- *σ-**order continuous** if $x_n \downarrow 0$ in E implies $T(x_n) \downarrow 0$ in F.*

- ***order continuous** if $x_\alpha \downarrow 0$ in E implies $T(x_\alpha) \downarrow 0$ in F.*

Obviously, a positive order continuous operator is automatically σ-order continuous. The converse is false—can you produce an example?

If $T: E \to F$ is a positive operator between two Riesz spaces, then its **order adjoint** $T^\sim: F^\sim \to E^\sim$ between the order duals is defined as $T^\sim f = f \circ T$ and so satisfies the familiar duality identity

$$\langle x, T^\sim f \rangle = \langle Tx, f \rangle = f(Tx),$$

where $x \in E$ and $f \in F^\sim$. The same formula can be used to define the order adjoint for a general order bounded operator. An operator between Riesz spaces is **order bounded** if it carries order bounded sets in the domain to order bounded sets in the range. A positive operator is always order bounded.

8.38 Theorem *The order adjoint of a positive operator is an order continuous positive operator.*

Proof: Let $T: E \to F$ be a positive operator between Riesz spaces. Clearly, $T^\sim: F^\sim \to E^\sim$ is a positive operator. Now suppose $f_\alpha \downarrow 0$ in $F^\sim$. That is, $f_\alpha(u) \downarrow 0$ in $\mathbb{R}$ for each $u \in F^+$. So for each $x \in E^+$ we have

$$\langle x, T^\sim f_\alpha \rangle = \langle Tx, f_\alpha \rangle = f_\alpha(Tx) \downarrow 0.$$

Thus $T^\sim f_\alpha \downarrow 0$ in $E^\sim$, so $T^\sim: F^\sim \to E^\sim$ is order continuous. ∎

The next result characterizes order continuity and σ-order continuity of positive operators in terms of their behavior on the order continuous and σ-order continuous duals.

8.39 Theorem *For a positive operator $T: E \to F$ between Riesz spaces:*

1. *If T is σ-order continuous, then $T^\sim(F_c^\sim) \subset E_c^\sim$. Conversely, if $F_c^\sim$ separates the points of F and $T^\sim(F_c^\sim) \subset E_c^\sim$, then T is σ-order continuous.*

2. *If T is order continuous, then $T^\sim(F_n^\sim) \subset E_n^\sim$. Conversely, if $F_n^\sim$ separates the points of F and $T^\sim(F_n^\sim) \subset E_n^\sim$, then T is order continuous.*

Proof: We prove only (1). Suppose that T is σ-order continuous. Let $f \in F_c^\sim$ and assume $x_n \downarrow 0$. Then $Tx_n \downarrow 0$, so

$$\langle x_n, T^\sim f \rangle = \langle Tx_n, f \rangle = f(Tx_n) \downarrow 0,$$

which means that $T^\sim f$ is σ-order continuous.

For the converse, assume $T^\sim(F_c^\sim) \subset E_c^\sim$ and let $x_n \downarrow 0$ in E. Also, choose y to satisfy $0 \leqslant y \leqslant Tx_n$ for each n. Then, for each $0 \leqslant f \in F_c^\sim$ we have

$$0 \leqslant f(y) \leqslant f(Tx_n) = \langle Tx_n, f \rangle = \langle x_n, T^\sim f \rangle \downarrow 0,$$

so $f(y) = 0$ for each $f \in F_c^\sim$. Since $F_c^\sim$ separates the points of F, we get $y = 0$. Thus, $Tx_n \downarrow 0$, and this shows that T is σ-order continuous. ∎

8.13   Topological Riesz spaces

Recall that a subset A of a Riesz space is **solid** if $|y| \leq |x|$ and $x \in A$ imply $y \in A$.
The **solid hull** of a subset B of a Riesz space E, denoted sol (B), is the smallest
solid set that includes B. Note that

$$\text{sol}(B) = \{y \in E : \exists\, x \in B \text{ with } |y| \leq |x|\}.$$

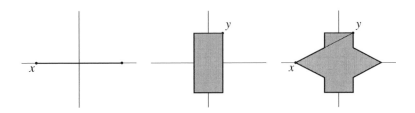

Figure 8.3. The solid hulls of $\{x\}$, $\{y\}$, and the segment $\overline{xy}$.

Clearly, every solid set is circled. The solid hull of a convex set need not be a
convex set; see Figure 8.3. But:

8.40 Lemma   *The convex hull of a solid set is solid.*

Proof: Let A be a solid set, and suppose $|y| \leq \left|\sum_{i=1}^{n} \lambda_i x_i\right|$, where $\lambda_i > 0$ and $x_i \in A$
for each i and $\sum_{i=1}^{n} \lambda_i = 1$. By the Riesz Decomposition Property (Theorem 8.9)
there exist $y_1, \ldots, y_n$ with $|y_i| \leq |\lambda_i x_i| = \lambda_i |x_i|$ for each i and $y = \sum_{i=1}^{n} y_i$. If $z_i = \frac{1}{\lambda_i} y_i$,
then $|z_i| \leq |x_i|$, so $z_i \in A$ for each i. Therefore, $y = \sum_{i=1}^{n} y_i = \sum_{i=1}^{n} \lambda_i z_i \in \text{co}\, A$, so
co A is solid.   ∎

A linear topology τ on a Riesz space E is **locally solid**, (and (E, τ) is called
a **locally solid Riesz space**) if τ has a base at zero consisting of solid neighborhoods. The local solidness of a linear topology is intrinsically related to the
uniform continuity of the lattice operations of the Riesz space. Recall that the
mappings $(x, y) \mapsto x \vee y$, $(x, y) \mapsto x \wedge y$, $x \mapsto x^+$, $x \mapsto x^-$, and $x \mapsto |x|$ are called
the **lattice functions** or the **lattice operations**, on E. Also recall that a function
$f : E \to F$ between two topological vector spaces is **uniformly continuous** if for
each neighborhood W of zero in F there is a neighborhood V of zero in E such
that $x - y \in V$ implies $f(x) - f(y) \in W$. Notice that the uniform continuity of any
one of the lattice functions guarantees the uniform continuity of the others. This
property is tied up with local solidness.

8.41 Theorem   *A linear topology τ on a Riesz space is locally solid if and only
if the lattice operations are uniformly continuous with respect to τ.*

Proof: You should verify that the uniform continuity of any one of the lattice operations implies the uniform continuity of the other lattice operations. Let τ be a linear topology on a Riesz space. If τ is locally solid, then the inequality $|x^+ - y^+| \leqslant |x - y|$ implies that the lattice operation $x \mapsto x^+$ is uniformly continuous.

For the converse, assume that the lattice operation $x \mapsto x^+$ is uniformly continuous, and let U be a τ-neighborhood of zero. We must demonstrate the existence of a solid τ-neighborhood U_1 of zero such that $U_1 \subset U$.

Start by choosing a circled neighborhood V of zero with $V + V \subset U$. Next, using uniform continuity, pick a neighborhood V_1 of zero such that $x - y \in V_1$ implies $x^+ - y^+ \in V$, and then choose another neighborhood V_2 of zero such that $V_2 + V_2 \subset V_1$. Again using the uniform continuity of $x \mapsto x^+$, select a circled neighborhood W of zero such that $x - y \in W$ implies $x^+ - y^+ \in V_2$. To finish the proof, we show that $\text{sol}(W) \subset U$.

To this end, assume $|v| \leqslant |w|$ where $w \in W$. Since $w - 0 = w \in W$, the choice of V_2 implies that $w^+ \in V_2$, and similarly $w^- \in V_2$. Consequently, we have $|w| = w^+ + w^- \in V_2 + V_2 \subset V_1$. Now from $v^+ - (v^+ - |w|) = |w| \in V_1$, we see that $v^+ = (v^+)^+ - (v^+ - |w|)^+ \in V$, and similarly $v^- \in V$. Then $v = v^+ - v^- \in V + V \subset U$, which shows that $\text{sol}(W) \subset U$, as desired.   ∎

8.42 Lemma   *In a locally solid Riesz space, the closure of a solid set is solid.*

Proof: Let A be a solid subset of a locally solid Riesz space (E, τ), and suppose $|x| \leqslant |y|$ with $y \in \overline{A}$. Pick a net $\{y_\alpha\}$ in A such that $y_\alpha \xrightarrow{\tau} y$. Put $z_\alpha = [(-|y_\alpha|) \vee x] \wedge |y_\alpha|$ and note that $|z_\alpha| \leqslant |y_\alpha|$ for each α. Since A is solid, $\{z_\alpha\} \subset A$. Now the continuity of the lattice operations implies

$$z_\alpha = [(-|y_\alpha|) \vee x] \wedge |y_\alpha| \xrightarrow{\tau} [(-|y|) \vee x] \wedge |y| = x,$$

so $x \in \overline{A}$. Therefore, $\overline{A}$ is solid.   ∎

Some elementary (but important) relationships between the order structure and the topological structure on a locally solid Hausdorff Riesz space are listed in the next theorem.

8.43 Theorem   *In a locally solid Hausdorff Riesz space (E, τ):*

1. *The positive cone E^+ is τ-closed.*

2. *If $x_\alpha \uparrow$ and $x_\alpha \xrightarrow{\tau} x$, then $x_\alpha \uparrow x$. That is, $x = \sup\{x_\alpha\}$.*

3. *The Riesz space E is Archimedean.*

4. *Every band in E is τ-closed.*

Proof: (1)   From the lattice identity $x = x^+ - x^-$, we see that

$$E^+ = \{x \in E : x^- = 0\}.$$

To see that E^+ is τ-closed, note that the lattice operation $x \mapsto x^-$ is a (uniformly) continuous function.

(2)   Let $x_\alpha \uparrow$ and $x_\alpha \xrightarrow{\tau} x$. Since $x_\alpha - x_\beta \in E^+$ for each $\alpha \geq \beta$, we see that for each β the net $\{x_\alpha - x_\beta : \alpha \geq \beta\}$ in E^+ satisfies $x_\alpha - x_\beta \xrightarrow[\alpha \geq \beta]{\tau} x - x_\beta$. Since E^+ is τ-closed, $x - x_\beta \in E^+$ for each β. This shows that x is an upper bound of the net $\{x_\alpha\}$. To see that x is the least upper bound of $\{x_\alpha\}$, assume that $x_\alpha \leq y$ for each α. Then, $y - x_\alpha \in E^+$ for each α and $y - x_\alpha \xrightarrow{\tau} y - x$ imply $y - x \in E^+$, or $y \geq x$. [3]

(3)   If $x \in E^+$, then $\frac{1}{n}x \downarrow$ and $\frac{1}{n}x \xrightarrow{\tau} 0$. By (2), we see that $\frac{1}{n}x \downarrow 0$.

(4)   Let D be an arbitrary nonempty subset of E. Then its disjoint complement $D^d = \{x \in E : |x| \wedge |y| = 0 \text{ for all } y \in D\}$ is τ-closed. Indeed, if $\{x_\alpha\}$ is a net in D^d satisfying $x_\alpha \xrightarrow{\tau} x$ and $y \in D$, then (by continuity of the lattice operations), $|x| \wedge |y| = \lim_\alpha |x_\alpha| \wedge |y| = 0$. This shows that $x \in D^d$, so D^d is τ-closed.

To see that a band A is τ-closed use the fact that (since E is Archimedean) $A = A^{dd} = (A^d)^d$; see Theorem 8.19.   ∎

8.44 Definition   *A **locally convex-solid topology** is a locally convex topology τ on a Riesz space E that is also locally solid, and (E, τ) is called a **locally convex-solid Riesz space***

From Lemmas 8.40 and 8.42, we see that a topology τ on a Riesz space is locally convex-solid if and only if τ has a base at zero consisting of neighborhoods that are simultaneously closed, solid, and convex.

8.45 Definition   *A seminorm p on a Riesz space is a **lattice seminorm** (or a **Riesz seminorm**) if $|x| \leq |y|$ implies $p(x) \leq p(y)$ or, equivalently, if*

1. *p is **absolute**, $p(x) = p(|x|)$ for all x; and*

2. *p is **monotone** on the positive cone, $0 \leq x \leq y$ implies $p(x) \leq p(y)$.*

The gauge of an absorbing, convex, and solid subset A of a Riesz space is always a lattice seminorm. Indeed, if $|x| \leq |y|$, then the seminorm p_A satisfies

$$p_A(x) = \inf\{\alpha > 0 : x \in \alpha A\} = \inf\{\alpha > 0 : |x| \leq \alpha A\}$$
$$\leq \inf\{\alpha > 0 : |y| \in \alpha A\} = p_A(y).$$

8.46 Theorem   *A linear topology on a Riesz space is locally convex-solid if and only if it is generated by a family of lattice seminorms.*

---

[3] This proof actually shows the following more general result. *If E is a partially ordered vector space whose cone is closed for a linear topology τ (not necessarily Hausdorff), then $x_\alpha \uparrow$ and $x_\alpha \xrightarrow{\tau} x$ imply $x_\alpha \uparrow x$.*

Proof: Let τ be a locally convex-solid topology on a Riesz space and let $\mathcal{B}$ be a base at zero consisting of all the τ-closed, convex, and solid neighborhoods. Then $\{p_V : V \in \mathcal{B}\}$ is a family of lattice seminorms generating the topology τ. ∎

8.47 Example (Locally convex-solid Riesz spaces)   Here are some familiar locally convex-solid Riesz spaces.

1. For a compact Hausdorff space K, the Riesz space $C(K)$ with the topology generated by the sup norm is a locally convex-solid Riesz space. Notice that the sup norm

$$\|f\|_\infty = \sup\{|f(x)| : x \in K\}$$

   is indeed a lattice norm.

2. The Riesz space $\mathbb{R}^X$ of all real functions defined on a nonempty set X equipped with the product topology is a locally convex-solid Riesz space. The product topology is generated by the family $\{p_x : x \in X\}$ of lattice seminorms, where $p_x(f) = |f(x)|$.

3. The Riesz space $L_0(\mu)$ of equivalence classes of μ-measurable real functions on a finite measure space (X, Σ, μ) with the metric topology of convergence in measure is a locally solid Riesz space that *fails* to be locally convex if μ is nonatomic; see Theorem 13.41 (3). The **topology of convergence in measure** is generated by the metric

$$d(f, g) = \int_X \frac{|f-g|}{1+|f-g|} \, d\mu.$$

4. The Riesz space $ba(\mathcal{A})$ of all signed charges of bounded variation on an algebra $\mathcal{A}$ of subsets of some set X becomes a locally convex-solid Riesz space when equipped with the topology generated by the total variation lattice norm $\|\mu\| = |\mu|(X)$. For details see Theorem 10.53 ∎

Not all locally convex topologies on a Riesz space are locally solid. Except in the finite dimensional case, the weak topology on a Banach lattice is not locally convex-solid; see [15, Theorem 2.38, p. 65].

As usual, the topological dual of a topological vector space X is denoted X', and its members are designated with primes. For instance, x', y', etc., denote elements of X'. The topological dual of a locally solid Riesz space E is an ideal in the order dual $E^\sim$.

8.48 Theorem   *If (E, τ) is a locally solid Riesz space, then its topological dual E' is an ideal in the order dual $E^\sim$. In particular, E' is order complete.*

Proof: Assume $|x'| \leqslant |y'|$ with $y' \in E'$ and let $x_\alpha \xrightarrow{\tau} 0$. Fix $\varepsilon > 0$ and for each α pick some y_α with $|y_\alpha| \leqslant |x_\alpha|$ and $|y'|(|x_\alpha|) \leqslant |y'|(y_\alpha)| + \varepsilon$. The local solidness of τ implies $y_\alpha \xrightarrow{\tau} 0$, and from the inequalities

$$|x'(x_\alpha)| \leqslant |x'|(|x_\alpha|) \leqslant |y'|(|x_\alpha|) \leqslant |y'|(y_\alpha)| + \varepsilon,$$

we see that $\limsup_\alpha |x'(x_\alpha)| \leqslant \varepsilon$ for each $\varepsilon > 0$. Therefore, $x'(x_\alpha) \to 0$, so $x' \in E'$. This shows that E' is an ideal in $E^\sim$. ∎

Every nonempty subset A of the order dual $E^\sim$ of a Riesz space E gives rise to a natural locally convex-solid topology on E via the family $\{p_{x'} : x' \in A\}$ of lattice seminorms, where

$$p_{x'}(x) = |x'|(|x|) = \langle |x|, |x'| \rangle.$$

This locally convex-solid topology on E is called the **absolute weak topology** generated by A, denoted $|\sigma|(E, A)$.

Similarly, if A is a nonempty subset of E, and E' is a Riesz subspace of $E^\sim$, then the family of lattice seminorms $\{p_x : x \in A\}$, where

$$p_x(x') = |x'|(|x|) = \langle |x|, |x'| \rangle,$$

defines a locally convex-solid topology on E'. This topology is called the **absolute weak* topology** on E' generated by A, denoted $|\sigma|(E', A)$.

8.49 Theorem (Kaplan) *If E is a Riesz space, and A is a subset of the order dual $E^\sim$, then the topological dual of the locally convex-solid Riesz space $(E, |\sigma|(E, A))$ coincides with the ideal generated by A in $E^\sim$.*

Proof: Let $I(A)$ be the ideal generated by A in $E^\sim$ and let E' denote the topological dual of $(E, |\sigma|(E, A))$. Since (by Theorem 8.48) E' is an ideal in $E^\sim$ and $A \subset E'$ (why?), we see that $I(A) \subset E'$. Now if $x' \in E'$, then there exist $x_1', \ldots, x_n' \in E'$ and positive scalars $\lambda_1, \ldots, \lambda_n$ satisfying $|x'(x)| \leqslant \sum_{i=1}^n \lambda_i \langle |x|, |x_i'| \rangle$ for each $x \in E$. This implies that $|x'| \leqslant \sum_{i=1}^n \lambda_i |x_i'|$ (why?), or $x' \in I(A)$. Therefore $E' = I(A)$ as claimed. ∎

If (E, τ) is a locally convex-solid Hausdorff Riesz space, then by Theorem 8.48 E' is an ideal in $E^\sim$, so by Theorem 8.49, the absolute weak topology $|\sigma|(E, E')$ is a locally convex-solid topology on E consistent with $\langle E, E' \rangle$. (Why?) In particular, we have

$$\sigma(E, E') \subset |\sigma|(E, E') \subset \tau(E, E'),$$

where, as you may recall, the Mackey topology $\tau(E, E')$ is the strongest consistent topology. As a matter of fact, the absolute weak topology $|\sigma|(E, E')$ is the weakest locally convex-solid topology on E that is consistent with the duality $\langle E, E' \rangle$. Also, note that $x_\alpha \xrightarrow{|\sigma|(E, E')} 0$ in E if and only if $x'(|x_\alpha|) \to 0$ for each $0 \leqslant x' \in E'$.

8.14 The band generated by E'

In this section, (E, τ) denotes a (not necessarily Hausdorff) locally convex-solid Riesz space. By Theorem 8.48, we know that E' (the topological dual of (E, τ)) is an ideal in the order dual $E^\sim$. The next result, due to W. A. J. Luxemburg [233, Theorem 5.3, p. 127], characterizes the band generated by E' in topological terms.

8.50 Theorem (**Luxemburg**) *The band generated by E' in $E^\sim$ is precisely the set of all order bounded linear functionals that are τ-continuous on the order intervals of E.*

Proof: We start by considering the set $\mathcal{B}$ of order bounded linear functionals

$$\mathcal{B} = \{f \in E^\sim : f \text{ is } \tau\text{-continuous on the order intervals of } E\}.$$

The proof consists of showing three claims.

- The set $\mathcal{B}$ is a band in $E^\sim$, and $E' \subset \mathcal{B}$.

Clearly $\mathcal{B}$ is a vector subspace of $E^\sim$ satisfying $E' \subset \mathcal{B}$. To see that $\mathcal{B}$ is an ideal, suppose $|f| \leqslant |g|$ with $g \in \mathcal{B}$. Also, suppose an order bounded net $\{x_\alpha\}$ satisfies $x_\alpha \xrightarrow{\tau} 0$, and let $\varepsilon > 0$. For each α pick some y_α with $|y_\alpha| \leqslant |x_\alpha|$ so that

$$|f(x_\alpha)| \leqslant |f|(|x_\alpha|) \leqslant |g(y_\alpha)| + \varepsilon.$$

(This follows from $|g|(|x_\alpha|) = \sup\{|g(y)| : |y| \leqslant |x_\alpha|\}$.) Since $|y_\alpha| \leqslant |x_\alpha|$ and $x_\alpha \xrightarrow{\tau} 0$, local solidness of τ implies $y_\alpha \xrightarrow{\tau} 0$, so $g(y_\alpha) \to 0$. Hence $\limsup_\alpha |f(x_\alpha)| \leqslant \varepsilon$ for all $\varepsilon > 0$. This implies $f(x_\alpha) \to 0$. That is, $f \in \mathcal{B}$. It is easy to see that $\mathcal{B}$ is order closed, so $\mathcal{B}$ is a band.

- Let $0 \leqslant f \in (E')^{\mathrm{d}}$. If $x > 0$ satisfies $f(x) > 0$, then for each $0 < \varepsilon < f(x)$ the convex set $S_\varepsilon = \{y \in [0, x] : f(y) \leqslant \varepsilon\}$ is τ-dense in $[0, x]$.

If S_ε is not τ-dense in $[0, x]$, then x does not belong to the τ-closure $\overline{S}_\varepsilon$ of S_ε. (Why?) By Separating Hyperplane Theorem 5.80 there exists some $g \in E'$ such that $g(x) > 1$ and $g(y) \leqslant 1$ for each $y \in S_\varepsilon$. Replacing g by g^+, we can assume that $g \geqslant 0$. From $g \wedge f = 0$, we infer that there exists a sequence $\{x_n\} \subset [0, x]$ such that $f(x_n) + g(x - x_n) \to 0$. It follows that $x_n \in S_\varepsilon$ for all $n \geqslant n_0$, so $g(x) = g(x_n) + g(x - x_n) \leqslant 1 + g(x - x_n) \to 1$, contradicting $g(x) > 1$. Consequently, S_ε is τ-dense in $[0, x]$.

- If $f \in \mathcal{B} \cap (E')^{\mathrm{d}}$, then $f = 0$ (and hence $\mathcal{B} = (E')^{\mathrm{dd}}$, the band generated by E' in $E^\sim$).

Let $0 \leqslant f \in \mathcal{B} \cap (E')^{\mathrm{d}}$ and assume by way of contradiction that there exists some $x > 0$ with $f(x) = 1$. Let $\varepsilon = \frac{1}{2}$ in the previous claim, and then select a net $\{x_\alpha\} \subset S_\varepsilon$ such that $x_\alpha \xrightarrow{\tau} x$. From $f \in \mathcal{B}$, we see that $f(x_\alpha) \to f(x) = 1$. However, this contradicts $f(x_\alpha) \leqslant \frac{1}{2}$ for each α, and the proof is complete. ∎

We close the section by illustrating Theorem 8.50 with an example.

8.51 Example (Topological continuity on boxes)   Let $E = \varphi$, the Riesz space of all real sequences that are eventually zero. Let τ be the locally convex-solid topology generated on E by the lattice norm

$$\|(x_1, x_2, \ldots)\|_\infty = \sup\{|x_i| : i = 1, 2, \ldots\}.$$

It is easy to see (Theorem 16.3) that $E^\sim = \mathbb{R}^N$ (the Riesz space of all sequences). The topological dual E' coincides with ℓ_1. The band generated by ℓ_1 coincides with $\mathbb{R}^N$. A moment's thought reveals that every sequence in $\mathbb{R}^N$ defines a linear functional on E that is indeed τ-continuous on the order intervals of E. ∎

8.15   Riesz pairs

Riesz pairs play an important role in economics as models for commodity spaces and their associated price space.

8.52 Definition   *A **Riesz pair** $\langle E, E' \rangle$ is a dual pair of Riesz spaces, where E' is an ideal in the order dual $E^\sim$.*

8.53 Lemma   *In a Riesz pair, the polar of a solid set is solid.*

Proof: Let $\langle E, E' \rangle$ be a Riesz pair and let A be a nonempty solid subset of E. To see that A° is solid, assume that $x', y' \in E'$ satisfy $|x'| \leqslant |y'|$ and $y' \in A^\circ$. Then, for each $x \in A$ we have

$$\begin{aligned}
|\langle x, x' \rangle| \leqslant \langle |x|, |x'| \rangle &\leqslant \langle |x|, |y'| \rangle \\
&= \sup\{|\langle y, y' \rangle| : y \in E \text{ and } |y| \leqslant |x|\} \\
&= \sup\{|\langle y, y' \rangle| : y \in A \text{ and } |y| \leqslant |x|\} \leqslant 1,
\end{aligned}$$

which shows that $x' \in A^\circ$.

Now let B be a nonempty solid subset of E' and consider the Riesz pair $\langle E', (E')^\sim \rangle$. By the preceding case, the polar $B^\bullet$ of B in $(E')^\sim$, that is, the set

$$B^\bullet = \{x'' \in (E')^\sim : |\langle x', x'' \rangle| \leqslant 1 \text{ for all } x' \in B\}$$

is a solid subset of $(E')^\sim$. If we consider E as a Riesz subspace of $(E')^\sim$ (embedded under its natural embedding as in Theorem 8.34), then we see that

$$B^\circ = \{x \in E : |\langle x, x' \rangle| \leqslant 1 \text{ for all } x' \in B\} = B^\bullet \cap E.$$

This easily implies that B° is a solid subset of E. ∎

Riesz pairs possess a number of special properties. If $\langle E, E' \rangle$ is a Riesz pair, then the strong topology, $\beta(E', E)$ on E' is locally convex-solid. (Why?) Moreover, it is clear from Lemma 8.53 that if $\mathfrak{S}$ is the collection of all convex, solid, and $\sigma(E', E)$-compact subsets of E' (note that $\{[-x', x'] : 0 \leqslant x' \in E'\} \subset \mathfrak{S}$), then the $\mathfrak{S}$-topology is a locally convex-solid topology on E. It is known as the **absolute Mackey topology** denoted $|\tau|(E, E')$. The absolute Mackey topology $|\tau|(E, E')$ is the largest locally convex-solid topology on E that is consistent with $\langle E, E' \rangle$. Thus

$$\sigma(E, E') \subset |\sigma|(E, E') \subset |\tau|(E, E') \subset \tau(E, E').$$

A locally convex-solid topology τ on E is consistent with the Riesz pair $\langle E, E' \rangle$ if and only if

$$|\sigma|(E, E') \subset \tau \subset |\tau|(E, E').$$

In a Riesz pair $\langle E, E' \rangle$, a positive vector $x \in E^+$ is **strictly positive**, written $x \gg 0$, if $\langle x, x' \rangle > 0$ for each $0 < x' \in E'$. Equivalently, $x \in E^+$ is strictly positive if x acts as a strictly positive linear functional on E' when considered as a member of $(E')^\sim$. A strictly positive vector is also called a **quasi-interior point**.

8.54 Theorem   *In a Riesz pair $\langle E, E' \rangle$, a vector $x \in E^+$ is strictly positive if and only if the principal ideal E_x is weakly dense in E.*

Proof: Assume first that x is strictly positive. If E_x is not weakly dense in E, then choose z not belonging to the weak closure of E_x. By Separating Hyperplane Theorem 5.80 we can separate the weak closure of E_x from z by a nonzero linear functional $x' \in E'$. Since E_x is a linear subspace, $\langle y, x' \rangle = 0$ for all $y \in E_x$. Note that $x' \neq 0$ implies $|x'| > 0$, so by the strict positivity of x, we have $\langle x, |x'| \rangle > 0$. On the other hand, from Theorem 8.24, it follows that

$$\langle x, |x'| \rangle = \sup\{\langle y, x' \rangle : |y| \leqslant x\} = 0.$$

This is a contradiction. Hence, E_x must be weakly dense in E.

For the converse, assume that E_x is weakly dense in E, and choose $0 < x' \in E'$. If $\langle x, x' \rangle = 0$, then $x' = 0$ on E_x, and consequently (by the weak denseness of E_x) $x' = 0$ on E, contrary to $x' > 0$. Hence, $\langle x, x' \rangle > 0$, so x is a strictly positive vector. ∎

The next result describes extensions of positive functionals on ideals.

8.55 Theorem   *Let $\langle E, E' \rangle$ be a Riesz pair, let τ be a consistent locally convex topology on E, and let J be an ideal in E. If $f : J \to \mathbb{R}$ is a positive τ-continuous linear functional, then f has a positive τ-continuous linear extension to all of E. Moreover, the formula*

$$f_J(x) = \sup\{f(y) : y \in J \text{ and } 0 \leqslant y \leqslant x\}, \ x \in E^+,$$

defines a positive τ-continuous linear extension of f to all of E such that:

1. $f_J(x) = 0$ for all $x \in J^d$; and

2. f_J is the **minimal extension** of f in the sense that if $0 \leqslant x' \in E'$ is another extension of f, then $f_J \leqslant x'$.

Proof: By Theorem 5.87, f has a τ-continuous linear extension to all of E, say g. Then we claim that g^+ is a τ-continuous positive linear extension of f to all of E. Indeed, since J is an ideal, $0 \leqslant y \leqslant x \in J$ implies $y \in J$. So for $0 \leqslant x \in J$ we have

$$g^+(x) = \sup\{g(y) : y \in E \text{ and } 0 \leqslant y \leqslant x\}$$
$$= \sup\{f(y) : y \in J \text{ and } 0 \leqslant y \leqslant x\} = f(x).$$

Next, consider the formula

$$f_J(x) = \sup\{f(y) : y \in J \text{ and } 0 \leqslant y \leqslant x\}, \ x \in E^+.$$

First we claim that f_J is additive on E^+. To see this, let $x, y \in E^+$. If $u, v \in J$ satisfy $0 \leqslant u \leqslant x$ and $0 \leqslant v \leqslant y$, then $u + v \in J$ and $0 \leqslant u + v \leqslant x + y$. So $f(u) + f(v) = f(u + v) \leqslant f_J(x + y)$, which implies $f_J(x) + f_J(y) \leqslant f_J(x + y)$.

For the reverse inequality, let $w \in J$ satisfy $0 \leqslant w \leqslant x + y$. Then, by the Riesz Decomposition Property, there exist $w_1, w_2 \in E$ such that $0 \leqslant w_1 \leqslant x$, $0 \leqslant w_2 \leqslant y$, and $w = w_1 + w_2$. Since J is an ideal, w_1, w_2 belong to J. So $f(w) = f(w_1) + f(w_2) \leqslant f_J(x) + f_J(y)$, which implies $f_J(x + y) \leqslant f_J(x) + f_J(y)$. Thus, $f_J(x + y) = f_J(x) + f_J(y)$. By Lemma 8.23, f_J defines a positive linear functional on E which is a positive linear extension of f to all of E.

Next note that if $0 \leqslant x \in J^d$ and $y \in J$ satisfy $0 \leqslant y \leqslant x$, then $y \in J \cap J^d = \{0\}$. So $\{y \in J : 0 \leqslant y \leqslant x\} = \{0\}$, and hence $f_J(x) = 0$ for each $x \in J^d$.

Now let $0 \leqslant h \in E'$ be any positive linear extension of f. If $x \in E^+$ and $y \in J$ satisfy $0 \leqslant y \leqslant x$, then $f(y) = h(y) \leqslant h(x)$, so

$$f_J(x) = \sup\{f(y) : y \in J \text{ and } 0 \leqslant y \leqslant x\} \leqslant h(x).$$

Finally, by the first part f has a positive extension $0 \leqslant g \in E'$, so it follows that $0 \leqslant f_J \leqslant g$. Since J is an ideal in E', we have $f_J \in E'$, and the proof is finished. ∎

8.16 Symmetric Riesz pairs

Recall that a Riesz pair is a dual pair $\langle E, E' \rangle$ of Riesz spaces where E' is an ideal in $E^\sim$. A **symmetric Riesz pair** is a Riesz pair where E is an ideal in $(E')^\sim$ (or, equivalently, if E is an ideal in $(E')_n^\sim$), where E is embedded in $(E')^\sim$ via the lattice isomorphism $x \mapsto \hat{x}$ defined by $\hat{x}(x') = \langle x, x' \rangle$ for each $x' \in E'$. Equivalently, $\langle E, E' \rangle$ is a symmetric Riesz pair if and only if $\langle E', E \rangle$ is a Riesz pair. Here is a list of some important symmetric Riesz pairs.

- $\langle \mathbb{R}^n, \mathbb{R}^n \rangle$.

- $\langle \ell_\infty, \ell_1 \rangle$, and in general $\langle L_\infty(\mu), L_1(\mu) \rangle$, when μ is σ-finite.

- $\langle \ell_1, \ell_\infty \rangle$, and in general $\langle L_1(\mu), L_\infty(\mu) \rangle$, when μ is σ-finite.

- $\langle L_p(\mu), L_q(\mu) \rangle$; $1 < p, q < \infty$, $\frac{1}{p} + \frac{1}{q} = 1$.

- $\langle \mathbb{R}^X, \varphi \rangle$, where φ denotes the Riesz space of all real functions on X that vanish outside finite subsets of X.

- $\langle c_0, \ell_1 \rangle$.

The Riesz pairs of the form $\langle C(K), ca(K) \rangle$ are not generally symmetric.

Symmetric Riesz pairs are intimately related to the weak compactness of order intervals, as the following discussion explains. Remember that if $\langle E, E' \rangle$ is a Riesz pair, then $\sigma((E')^\sim, E')$ is the restriction of the pointwise topology on $\mathbb{R}^{E'}$ to $(E')^\sim$ and that $\sigma((E')^\sim, E')$ induces the weak topology $\sigma(E, E')$ on E.

8.56 Lemma *In a Riesz pair $\langle E, E' \rangle$ every order interval in E' is $\sigma(E', E)$-compact.*

Proof: Let $0 \leqslant x' \in E'$. Clearly, the order interval $[0, x']$ as a subset of $\mathbb{R}^E$ is pointwise bounded. Moreover, we claim that $[0, x']$ is pointwise closed. To see this, assume that a net $\{x'_\alpha\}$ in $[0, x']$ satisfies $x'_\alpha(x) \to f(x)$ for each $x \in E$ and some $f \in \mathbb{R}^E$. Then f is a linear functional, and from $0 \leqslant x'_\alpha(x) \leqslant x'(x)$ for each $x \in E^+$, we see that f is a positive linear functional satisfying $0 \leqslant f \leqslant x'$. Since E' is an ideal in $E^\sim$, we see that $f \in [0, x']$. In other words, the order interval $[0, x']$ is pointwise bounded and closed. By Tychonoff's Product Theorem 2.61, $[0, x]$ is $\sigma(E', E)$-compact. ∎

If $\langle E, E' \rangle$ is a Riesz pair and $x \in E^+$, then let $[[0, x]]$ denote the order interval determined by x when considered as an element of $(E')^\sim$. That is,

$$[[0, x]] = \{x'' \in (E')^\sim : 0 \leqslant x'' \leqslant x\}.$$

As usual, $[0, x] = \{y \in E : 0 \leqslant y \leqslant x\}$.

8.57 Lemma *If $\langle E, E' \rangle$ is a Riesz pair, then for each $x \in E^+$ the order interval $[0, x]$ is $\sigma((E')^\sim, E')$-dense in $[[0, x]]$. In particular, for $x \in E^+$, the order interval $[0, x]$ is weakly compact if and only if $[0, x] = [[0, x]]$.*

Proof: Clearly $\langle E', (E')^\sim \rangle$ is a Riesz pair, so Lemma 8.56 implies that $[[0, x]]$ is $\sigma((E')^\sim, E')$-compact. Let $\overline{[0, x]}$ denote the $\sigma((E')^\sim, E')$-closure of $[0, x]$ in $(E')^\sim$. Clearly, $\overline{[0, x]} \subset [[0, x]]$. If $\overline{[0, x]} \neq [[0, x]]$, then there is some x'' in $[[0, x]]$ with

$x'' \notin \overline{[0, x]}$. By Separating Hyperplane Theorem 5.80, there exists some $x' \in E'$ such that $x''(x') > 1$ and $x'(y) \leqslant 1$ for each $y \in [0, x]$. Thus

$$(x')^+(x) = \sup\{x'(y) : y \in E \text{ and } 0 \leqslant y \leqslant x\} \leqslant 1.$$

This implies that $x''(x') \leqslant x''((x')^+) \leqslant x((x')^+) = (x')^+(x) \leqslant 1$, which contradicts $x''(x') > 1$. Hence, $\overline{[0, x]} = [[0, x]]$.

The last part of the theorem follows from the fact that $\sigma((E')^\sim, E')$ induces the topology $\sigma(E, E')$ on E.   ∎

8.58 Definition   *A linear topology τ on a Riesz space E is called **order continuous** (resp. **σ-order continuous**) if $x_\alpha \xrightarrow{o} 0$ (resp. $x_n \xrightarrow{o} 0$) implies $x_\alpha \xrightarrow{\tau} 0$ (resp. $x_n \xrightarrow{\tau} 0$).*

If τ is locally solid, then τ is order continuous if and only if $x_\alpha \downarrow 0$ in E implies $x_\alpha \xrightarrow{\tau} 0$. Also, notice that if τ is an order continuous locally solid topology on a Riesz space E, then the topological dual E' of (E, τ) is in fact an ideal in the order continuous dual $E_n^\sim$.

We also have the following density theorem due to H. Nakano.

8.59 Theorem (Nakano)   *If $E' \subset (E')_n^\sim$, then for every $0 < x''$ in $(E')_n^\sim$ there exists some $y \in E$ such that $0 < y \leqslant x''$.*

Proof: See [12, Theorem 5.5, p. 59].   ∎

The following important theorem characterizes symmetric Riesz pairs.

8.60 Theorem   *For a Riesz pair $\langle E, E' \rangle$ the following statements are equivalent.*

1. *$\langle E, E' \rangle$ is a symmetric Riesz pair.*

2. *The absolute weak* topology $|\sigma|(E', E)$ is consistent with $\langle E, E' \rangle$.*

3. *The order intervals of E are $\sigma(E, E')$-compact.*

4. *E is order complete, and every consistent locally convex-solid topology on E is order continuous.*

5. *E is order complete, and the weak topology $\sigma(E, E')$ is order continuous.*

6. *E is order complete, and $E' \subset E_n^\sim$.*

Proof: (1) $\implies$ (2)   By Theorem 8.49, the topological dual of $(E', |\sigma|(E', E))$ coincides with the ideal generated by E in $(E')^\sim$. Since $\langle E, E' \rangle$ is a symmetric Riesz pair, this ideal is just E, so $|\sigma|(E', E)$ is consistent with the dual pair $\langle E, E' \rangle$.

(2) $\implies$ (3)   Theorem 8.49 informs us that E is again an ideal in $(E')^\sim$. In particular, we have $[0, x] = [[0, x]]$ for each $x \in E^+$. By Lemma 8.56, every order interval of E is weakly compact.

(3) $\implies$ (4) By Lemma 8.57, we know that $[0, x] = [[0, x]]$ for each $x \in E^+$, and this shows that E is an ideal in $(E')^\sim$. In addition, by Lemma 8.14, E is an order complete Riesz space.

Next let τ be a consistent locally convex-solid topology on E and assume $x_\alpha \downarrow 0$ in E. We can suppose that $0 \leqslant x_\alpha \leqslant x$ for all α and some $x \in E^+$. Also, let V be a solid τ-neighborhood of zero.

Since $[0, x]$ is weakly compact, by passing to a subnet we may assume that $x_\alpha \xrightarrow{w} y$ in E. Since E^+ is $\sigma(E, E^+)$-closed, it follows from the footnote to the proof of Theorem 8.43 that $y = 0$. Therefore, $x_\alpha \xrightarrow{w} 0$. In particular, zero belongs to the weakly (and hence to the τ-) closed convex hull of $\{x_\alpha\}$. So there exist indexes $\alpha_1, \ldots, \alpha_n$ and positive constants $\lambda_1, \ldots, \lambda_n$ such that $\sum_{i=1}^n \lambda_i = 1$ and $\sum_{i=1}^n \lambda_i x_{\alpha_i} \in V$. Next fix some α_0 such that $\alpha_0 \geqslant \alpha_i$ for each $i = 1, \ldots, n$. Now if $\alpha \geqslant \alpha_0$, then $0 \leqslant x_\alpha = \sum_{i=1}^n \lambda_i x_\alpha \leqslant \sum_{i=1}^n \lambda_i x_{\alpha_i} \in V$. Since V is solid, $x_\alpha \in V$ for each $\alpha \geqslant \alpha_0$. That is, $x_\alpha \xrightarrow{\tau} 0$.

(4) $\implies$ (5) Let $x_\alpha \downarrow 0$ in E. Note that the absolute weak topology $|\sigma|(E, E')$ is a consistent locally convex-solid topology on E. Consequently, $f(x_\alpha) \downarrow 0$ for each $0 \leqslant f \in E'$. This easily implies that $x_\alpha \xrightarrow{w} 0$.

(5) $\implies$ (6) If $x_\alpha \downarrow 0$ and $0 \leqslant f \in E'$, then the order continuity of $\sigma(E, E')$ implies $f(x_\alpha) \downarrow 0$, which shows that $E' \subset (E')^\sim_n$.

(6) $\implies$ (1) Assume that $0 \leqslant x'' \leqslant x$ in $(E')^\sim$ with $x \in E$. Consider the set $U = \{u \in E : 0 \leqslant u \leqslant x''\}$. Let $z = \sup U$ in E and $z^* = \sup U$ in $(E')^\sim$ (The suprema exist since E and $(E')^\sim$ are both order complete.) Moreover, $z, z^* \in (E')^\sim_n$ and $z^* \leqslant z$ in $(E')^\sim$. (Why?) We claim that $z = z^*$. To see this, assume by way of contradiction that $z^* < z$. Then by Nakano's Theorem 8.59, there exists some $v \in E$, with $0 < v \leqslant z - z^*$, so $u \in U$ implies $0 < u \leqslant z^* \leqslant z - v < z$, contrary to $z = \sup U$ in E. Therefore $z = z^* \leqslant x''$.

We claim that $x'' = z \in E$. To see this, assume by way of contradiction that $z < x''$. Again, by Nakano's Theorem there exists some $u \in E$ such that $0 < u \leqslant x'' - z$, so $z < z + u \leqslant x''$, contrary to $z = \sup U$.

These arguments show that $[0, x] = [[0, x]]$ for each $x \in E^+$, which means that E is an ideal in $(E')^\sim$. That is, $\langle E, E' \rangle$ is a symmetric Riesz pair, and the proof of the theorem is finished. ∎

8.61 Corollary *If $\langle E, E' \rangle$ is a symmetric Riesz pair, then $\langle E', E \rangle$ is also a symmetric Riesz pair.*

Proof: Assume that $\langle E, E' \rangle$ is a symmetric Riesz pair. By Theorem 8.60(3), the order intervals of E are weakly compact. Lemma 8.57 implies that E is an ideal in $(E')^\sim$, so $\langle E', E \rangle$ is a Riesz pair. Since, by Lemma 8.56, the intervals of E' are $\sigma(E', E)$-compact, it follows from Theorem 8.60(3) that $\langle E', E \rangle$ is in fact a symmetric Riesz pair. ∎

8.62 Corollary *If E is an order complete Riesz space and the order continuous dual $E^\sim_n$ separates points, then $\langle E, E^\sim_n \rangle$ is a symmetric Riesz pair.*

Chapter 9

Banach lattices

Recall that a lattice norm is norm is a norm that is monotone in the absolute value of a vector (Definition 8.45). Normed Riesz spaces are simply Riesz spaces equipped with lattice norms. By Theorem 8.46, such spaces are locally convex-solid. If the norm is also complete, the space is a *Banach lattice*. Of course, the metric induced by a lattice norm need not be complete, but if it is complete there are surprising consequences. For instance, positive operators between Banach lattices must be continuous. Not every Riesz space can be fitted with a complete lattice norm, but if it can, the norm is unique to positive multiple. A *Fréchet lattice* is a completely metrizable locally solid Riesz space.

In this chapter we start with some examples of Fréchet and Banach lattices and develop some of their basic properties. We continue with a discussion of lattice isometries between Banach lattices and order continuous norms. Of key interest for its wide range of applications is the fact that a Banach lattice and its norm dual form a symmetric Riesz pair if and only if the Banach lattice has order continuous norm. A Banach lattice has *order continuous norm* if every decreasing net that order converges to zero also converges to zero in norm. The other important fact about Fréchet lattices and Banach lattices is that every *positive linear functional* is automatically continuous (Theorem 9.6). Also, for Fréchet lattices the topological and order duals coincide (Theorem 9.11).

We also present, but do not prove, two versions of the Stone–Weierstrass Theorem (Theorems 9.12 and 9.13). These theorems describe dense subspaces of the space of continuous functions a compact space. The lattice version gives conditions for a Riesz subspace to be dense.

There are two important special classes of Banach lattices: the AL-spaces and the AM-spaces. AL-spaces are abstract versions of the $L_1(\mu)$-spaces, while AM-spaces are the abstract versions of the $C(K)$-spaces (K compact Hausdorff). Remarkably, the AL- and AM-spaces are mutually dual. A Banach lattice is an AL-space (resp. an AM-space) if and only if its norm dual is an AM-space (resp. an AL-space). Principal ideals in Banach lattices are the prime examples of AM-spaces. One interesting fact, especially for economists, is that the positive cone of a Banach lattice has nonempty norm interior if and only if it is an AM-space with unit. In AM-spaces, the Stone–Weierstrass Theorem 9.13 provides a plethora of dense subspaces.

In finite dimensional Euclidean spaces, the positive cone of the space is big in the sense that it has a nonempty interior. In infinite dimensional spaces, the interior of the positive cone of a Banach lattice is often empty. Section 9.6 shows that if the positive cone is nonempty, then the space can be represented as dense subset of a $C(X)$ space.

Next we discuss the properties of positive projections and contractions in Riesz subspaces, and close the chapter with a discussion of the space of functions of bounded variation. This is a space with at least two natural order structures.

9.1 Fréchet and Banach lattices

Recall that a lattice norm $\| \cdot \|$ has the property that $|x| \leqslant |y|$ in E implies $\|x\| \leqslant \|y\|$. A Riesz space equipped with a lattice norm is called a **normed Riesz space**. A complete normed Riesz space is called a **Banach lattice**

9.1 Example   (Normed Riesz spaces)   Here are some familiar examples of normed Riesz spaces and Banach lattices.

•   The Euclidean spaces $\mathbb{R}^n$ with their Euclidean norms are all Banach lattices.

•   If K is a compact space, then the Riesz space $C(K)$ of all continuous real functions on K under the sup norm

$$\|f\|_\infty = \sup\{|f(x)| : x \in K\}$$

is a Banach lattice.

•   If X is a topological space, then $C_b(X)$, the Riesz space of all bounded real continuous functions on X, under the lattice norm

$$\|f\|_\infty = \sup\{|f(x)| : x \in X\}$$

is a Banach lattice.

•   The Riesz space $C[0, 1]$ under the L_1 lattice norm

$$\|f\| = \int_0^1 |f(x)|\, dx$$

is a normed Riesz space, but *not* a Banach lattice.

•   If X is an arbitrary nonempty set, then the Riesz space $B(X)$ of all bounded real functions on X under the lattice norm

$$\|f\|_\infty = \sup\{|f(x)| : x \in X\}$$

is a Banach lattice.

•   The Riesz spaces $L_p(\mu)$, $1 \leqslant p < \infty$, (and hence the ℓ_p-spaces) are all Banach lattices when equipped with their L_p-norms

$$\|f\|_p = \Big(\int |f|^p \, d\mu \Big)^{\frac{1}{p}}.$$

Similarly, the $L_\infty(\mu)$-spaces are all Banach lattices with their essential sup norms; see Theorem 13.5.

•   If $\mathcal{A}$ is an algebra of subsets of X, then the Riesz space $ba(\mathcal{A})$ of all signed charges of bounded variation is a Banach lattice under the total variation norm

$$\|\mu\| = |\mu|(X).$$

See Theorem 10.53 for details.

•   The Riesz space c_0 of all real sequences converging to zero (null sequences) under the sup norm

$$\|(x_1, x_2, \ldots)\|_\infty = \sup\{|x_n| : n = 1, 2, \ldots\}$$

is a Banach lattice.

∎

The Fréchet lattices are defined as follows.

9.2 Definition   A **Fréchet lattice** is a completely metrizable locally solid Riesz space.

The next result characterizes completeness in metrizable locally solid Riesz spaces.

9.3 Theorem   A metrizable locally solid Riesz space is topologically complete (that is, a Fréchet lattice) if and only if every increasing positive Cauchy sequence is convergent.

In particular, a normed Riesz space is a Banach lattice if and only if every increasing positive Cauchy sequence is norm convergent.

Proof: Assume that E is a metrizable locally solid Riesz space in which every increasing positive Cauchy sequence is topologically convergent. Let $\{x_n\}$ be a Cauchy sequence. We must show that $\{x_n\}$ has a convergent subsequence.

To this end, start by fixing a countable base $\{V_n\}$ at zero consisting of solid sets satisfying $V_{n+1} + V_{n+1} \subset V_n$ for each n. Also, (by passing to a subsequence) we can assume $x_{n+1} - x_n \in V_{n+1}$ for each n, so by solidness $(x_{n+1} - x_n)^+ \in V_{n+1}$ for each n. Next, define the two increasing positive sequences $\{y_n\}$ and $\{z_n\}$ by

$$y_n = \sum_{i=1}^{n} (x_{i+1} - x_i)^+ \quad \text{and} \quad z_n = \sum_{i=1}^{n} (x_{i+1} - x_i)^-,$$

and note that $x_n = x_1 + y_{n-1} - z_{n-1}$ for each $n \geqslant 2$. From

$$y_{n+p} - y_n = \sum_{i=n+1}^{n+p} (x_{i+1} - x_i)^+ \in V_{n+1} + V_{n+2} + \cdots + V_{n+p+1} \subset V_n,$$

we see that $\{y_n\}$ is a Cauchy sequence. Similarly, $\{z_n\}$ is a Cauchy sequence. If $y_n \to y$ and $z_n \to z$, then $x_n \to x_1 + y - z$. ∎

9.4 Lemma   *Both the norm dual and the norm completion of a normed Riesz space are Banach lattices.*

Proof: Let E be a normed Riesz space. We shall show that its norm dual E' is a Banach lattice—we already know from Theorem 8.48 that E' is an ideal in $E^\sim$. It remains to be shown that the norm of E' is a lattice norm. To this end, let $|x'| \leqslant |y'|$ in E'. From

$$|x'(x)| \leqslant |x'|(|x|) \leqslant |y'|(|x|) = \sup\{|y'(y)| : |y| \leqslant |x|\},$$

we see that

$$\|x'\| = \sup_{\|x\| \leqslant 1} |x'(x)| \leqslant \sup_{\|x\| \leqslant 1} \sup_{|y| \leqslant |x|} |y'(y)| \leqslant \sup_{\|y\| \leqslant 1} |y'(y)| = \|y'\|.$$

For the other assertion, note that the norm completion of E coincides with the closure of E in the Banach lattice E''. ∎

In particular, every Banach lattice is a Fréchet lattice, but the converse is not true. For instance, for $0 < p < 1$ the Riesz space $L_p[0, 1]$ is a Fréchet lattice under the distance $d(f, g) = \int_0^1 |f(x) - g(x)| \, dx$, but it does not admit any lattice norm; see Theorem 13.31.

The proof of the next result is left as an exercise.

9.5 Lemma   *The topological completion of a metrizable locally solid Riesz space is a Fréchet lattice.*

And now we come to a remarkable result. Positive operators on a Fréchet lattice are continuous.

9.6 Theorem (Continuity of positive operators)   *Every positive operator from a Fréchet lattice into a locally solid Riesz space is continuous. In particular, every positive real linear functional on a Fréchet lattice is continuous.*

Proof: Let (E, τ) be a Fréchet lattice, let F be a locally solid Riesz space, and let $T: E \to F$ be a positive operator. Assume by way of contradiction that T is not continuous. Then there exist a sequence $\{x_n\}$ in E and a neighborhood W of zero in F such that $x_n \xrightarrow{\tau} 0$ and $Tx_n \notin W$ for each n. Pick a countable base $\{V_n\}$ of solid τ-neighborhoods of zero satisfying $V_{n+1} + V_{n+1} \subset V_n$ for each n. By passing to a subsequence of $\{x_n\}$, we can suppose that $x_n \in \frac{1}{n}V_n$ (or $nx_n \in V_n$) for each n. Next, for each n let $y_n = \sum_{i=1}^{n} i|x_i|$, and note that

$$y_{n+p} - y_n = \sum_{i=n+1}^{n+p} i|x_i| \in V_{n+1} + V_{n+2} + \cdots + V_{n+p} \subset V_n.$$

Therefore $\{y_n\}$ is a τ-Cauchy sequence, so $y_n \xrightarrow{\tau} y$ for some y in E. By Theorem 8.43 (2), we have $y_n \uparrow y$. Hence, $0 \leqslant y_n \leqslant y$ for each n. Now the positivity of T implies

$$|Tx_n| \leqslant T|x_n| = \tfrac{1}{n}T(n|x_n|) \leqslant \tfrac{1}{n}Ty_n \leqslant \tfrac{1}{n}Ty,$$

which shows that $Tx_n \to 0$ in F, contrary to $Tx_n \notin W$ for each n. Consequently, T must be a continuous operator. ∎

The hypothesis of topological completeness in the preceding theorem cannot be dropped. As the next example shows, a positive operator on a normed Riesz space need not be continuous.

9.7 Example (Discontinuous positive operator) Let φ denote the order complete Riesz space of all real sequences that are eventually zero. The Riesz space φ is a normed Riesz space under the sup norm $\|\cdot\|_\infty$, where as usual

$$\|x\|_\infty = \sup\{|x_n| : n = 1, 2, \ldots\}.$$

Now consider the linear functional $f : E \to \mathbb{R}$ defined by

$$f(x_1, x_2, \ldots) = \sum_{n=1}^{\infty} x_n.$$

Clearly, f is a positive linear functional, but it fails to be norm continuous. To see this, let $u_n = (1, \ldots, 1, 0, 0, \ldots) \in E$, where the 1s occupy the first n coordinates. Then $\|u_n\|_\infty = 1$ and $f(u_n) = n$ for each n. Consequently,

$$\|f\| = \sup_{\|x\|_\infty \leqslant 1} |f(x)| \geqslant \sup_n f(u_n) = \infty,$$

so $\|f\| = \infty$. Thus, f is not continuous. ∎

Theorem 9.6 has a number of important consequences.

9.8 Corollary *If (E, τ) is a Fréchet lattice, then τ is the finest locally solid topology on E.*

Proof: If τ_1 is an arbitrary locally solid topology on E, then the identity operator $I : (E, \tau) \to (E, \tau_1)$ is a positive operator. Hence, by Theorem 9.6, I must be continuous, so $\tau_1 \subset \tau$. ∎

An immediate consequence of the preceding corollary is the following uniqueness property.

9.9 Corollary *A Riesz space admits at most one metrizable locally solid topology that makes it a Fréchet lattice.*

Specializing this result to Banach lattices yields the following.

9.10 Corollary *Any two lattice norms that make a Riesz space into a Banach lattice are equivalent.*

For Fréchet lattices the topological and order duals coincide.

9.11 Theorem (Order dual of a Fréchet lattice) *The topological dual and the order dual of a Fréchet lattice E (in particular, of a Banach lattice E) coincide. That is, $E' = E^\sim$.*

Proof: By Theorem 8.48, we know that E' is an ideal in the order dual $E^\sim$. On the other hand, by Theorem 9.6, every positive linear functional on E is continuous. Since each linear functional in $E^\sim$ is the difference of two positive linear functionals, we see that $E' = E^\sim$. ∎

9.2   The Stone–Weierstrass Theorem

There are two results known as the Stone–Weierstrass Approximation Theorems that present conditions under which a vector subspace of $C(X)$ is uniformly dense. One is a lattice-theoretic statement, the other is algebraic. We state the lattice version first. For a proof see [13, Theorem 11.3, p. 88].

9.12 Stone–Weierstrass Theorem (Lattice Version) *Let X be a compact space. A Riesz subspace of $C(X)$ that separates the points of X and contains the constant function **1** is uniformly dense in $C(X)$.*

For an illustration of Theorem 9.12, let $X = [0, 1]$ and consider the Riesz subspace of $C[0, 1]$ consisting of all piecewise linear continuous functions on the interval $[0, 1]$.

An **algebra of functions** (not to be confused with an algebra of sets) is a linear space of real functions that is closed under (pointwise) multiplication. Recall that if (X, d) is a metric space, then U_d, the vector space of bounded real-valued and d-uniformly continuous functions on X, is always a uniformly closed algebra of functions.

The algebraic version of the Stone–Weierstrass Theorem follows.

9.13 Stone–Weierstrass Theorem (Algebraic Version) *An algebra $\mathcal{A}$ of real-valued continuous functions on a compact space X that separates the points of X and contains the constant function **1** is uniformly dense in $C(X)$.*

Proof: To prove the result, one must establish that the uniform closure $\overline{\mathcal{A}}$ of $\mathcal{A}$ is a Riesz subspace of $C(X)$ and then apply Theorem 9.12. For details, see [13, Theorem 11.5, p. 89]. ∎

The Stone–Weierstrass Theorem can be used to characterize the metrizability of a compact Hausdorff topological space in terms of the separability of its space of continuous functions.

9.14 Theorem *A compact Hausdorff space X is metrizable if and only if $C(X)$ is a separable Banach lattice.*

Proof: Let X be a compact Hausdorff space. Assume first that X is metrizable, and let d be a consistent metric on X. Fix a countable dense subset $\{x_1, x_2, \ldots\}$ of X and for each n define $f_n(x) = d(x, x_n)$. Let $\mathcal{F} = \{\mathbf{1}, f_1, f_2, \ldots\}$, where $\mathbf{1}$ is the constant function one. The set $\mathcal{F}$ separates the points of X. (Why?) Now let $\mathcal{F}_1$ denote the (countable) set of all finite products of the functions in $\mathcal{F}$. Next, let $\mathcal{A}$ be the set of all continuous functions that are (finite) linear combinations of the elements of $\mathcal{F}_1$. Then $\mathcal{A}$ is an algebra that separates the points of X and contains the constant function one. By the Stone–Weierstrass Theorem 9.13, the algebra of functions $\mathcal{A}$ is uniformly dense in $C(X)$. Since the finite linear combinations from $\mathcal{F}_1$ with rational coefficients is a countable uniformly dense subset of $C(X)$, we see that $C(X)$ is a separable Banach lattice.

For the converse, note that if $C(X)$ is separable, then (by Theorem 6.30) the closed unit ball U' of the dual of $C(X)$ is w^*-compact and metrizable. Since X can be identified with a closed subset of U', (the embedding $x \mapsto \delta_x$ is a homeomorphism by Corollary 2.57; see also Theorem 15.8), we see that X is a metrizable topological space. ∎

9.3 Lattice homomorphisms and isometries

We now discuss lattice properties of operators. As usual, if $T: X \to Y$ is a linear operator between vector spaces, then for brevity we write Tx rather than $T(x)$.

9.15 Theorem *For a linear operator $T: E \to F$ between Riesz spaces, the following statements are equivalent.*

1. $T(x \vee y) = T(x) \vee T(y)$ *for all* $x, y \in E$.

2. $T(x \wedge y) = T(x) \wedge T(y)$ *for all* $x, y \in E$.

3. $T(x^+) = (Tx)^+$ *for all* $x \in E$.

4. $T(x^-) = (Tx)^-$ *for all* $x \in E$.

5. $T(|x|) = |Tx|$ *for all* $x \in E$.

6. *If* $x \wedge y = 0$ *in* E, *then* $Tx \wedge Ty = 0$ *in* F.

Proof: The proof is a direct application of the lattice identities in Riesz spaces. To indicate how to prove this result, we establish the equivalence of (1) and (5). So assume first that (1) is true. Then

$$T|x| = T(x \vee (-x)) = T(x) \vee T(-x) = T(x) \vee (-T(x)) = |Tx|.$$

Now assume that (5) is true. From $x \vee y = \frac{1}{2}(x + y + |x - y|)$, we see that

$$T(x \vee y) = \tfrac{1}{2}(Tx + Ty + T|x - y|) = \tfrac{1}{2}(Tx + Ty + |Tx - Ty|) = Tx \vee Ty.$$

For more details, see [12, Theorem 7.2, p. 88]. ∎

9.16 Definition   *A linear operator $T: E \to F$ between Riesz spaces is a **lattice homomorphism** (or a **Riesz homomorphism**) if T satisfies any one of the equivalent statements of Theorem 9.15*
     *A lattice homomorphism that is also one-to-one is a **lattice isomorphism** (or a **Riesz isomorphism**).*

Every lattice homomorphism $T: E \to F$ is a positive operator; indeed, if $x \geqslant 0$, then $Tx = T(x^+) = (Tx)^+ \geqslant 0$. Also notice that if $T: E \to F$ is a lattice homomorphism, then the range $T(E)$ is a Riesz subspace of F. In case $T: E \to F$ is a lattice isomorphism, then $T(E)$ and E can be considered to be identical Riesz spaces. Two Riesz spaces E and F are **lattice isomorphic** if there is a lattice isomorphism from E onto F.

9.17 Theorem   *Let $T: E \to F$ be a linear operator between Riesz spaces that is one-to-one and onto. Then T is a lattice isomorphism if and only if both T and T^{-1} are positive operators. That is, T is a lattice isomorphism provided $x \geqslant 0$ in E if and only if $Tx \geqslant 0$ in F.*

Proof: If T is a lattice isomorphism, then clearly both T and T^{-1} are positive operators. For the converse, assume that T and T^{-1} are both positive operators and let $x, y \in E$.
     From $x \leqslant x \vee y$ and $y \leqslant x \vee y$, we get $Tx \leqslant T(x \vee y)$ and $Ty \leqslant T(x \vee y)$ or $Tx \vee Ty \leqslant T(x \vee y)$. The same arguments applied to Tx, Ty, and the operator T^{-1} in place of x, y and T show that $x \vee y \leqslant T^{-1}(Tx \vee Ty)$. Applying T, we get $T(x \vee y) \leqslant Tx \vee Ty$. Hence, $T(x \vee y) = T(x) \vee T(y)$, so T is a lattice isomorphism. ∎

A linear operator $T: X \to Y$ between normed spaces is a **linear homeomorphism** if $T: X \to T(X)$ is a homeomorphism. (Or equivalently, if there exist positive constants K and M such that

$$K\|x\| \leqslant \|T(x)\| \leqslant M\|x\|$$

for each $x \in X$). Two normed spaces X and Y are **linearly homeomorphic** if there is a linear homeomorphism from X onto Y. A linear operator $T: X \to Y$ between normed spaces that satisfies $\|T(x)\| = \|x\|$ for all $x \in X$ is a **linear isometry** Two normed spaces X and Y are **linearly isometric** if there exists a linear isometry from X onto Y. Clearly, every linear isometry is a linear homeomorphism.

A lattice isomorphism $T \colon E \to F$ between normed Riesz spaces is:

- A **lattice homeomorphism** if T is also a linear homeomorphism.

- A **lattice isometry** if T is also a linear isometry.

9.18 Definition   *Two normed Riesz spaces E and F are **lattice isometric** if there is a lattice isometry from E onto F.*

From the point of view of Riesz spaces, two lattice isometric normed Riesz spaces are identical.

9.19 Lemma   *A lattice isomorphism $T \colon E \to F$ between normed Riesz spaces is a lattice isometry if and only if $\|Tx\| = \|x\|$ for each $x \in E^+$.*

Proof: If $\|Tx\| = \|x\|$ for each $x \in E^+$, then for each $z \in E$ we have

$$\|Tz\| = \big\|\,|Tz|\,\big\| = \big\|\,|T|z|\,\big\| = \big\|\,|z|\,\big\| = \|z\|,$$

which proves that T is a lattice isometry.   ∎

9.4   Order continuous norms

We now discuss an important connection between the topological and order structures of a Banach lattice. This connection is usually known as the "order continuity of the norm."

9.20 Definition   *A lattice norm $\|\cdot\|$ on a Riesz space is:*

- *order continuous if $x_\alpha \downarrow 0$ implies $\|x_\alpha\| \downarrow 0$.*

- *σ-order continuous if $x_n \downarrow 0$ implies $\|x_n\| \downarrow 0$.*

Obviously, order continuity implies σ-order continuity. The converse is false, even for Banach lattices.

9.21 Example   **(Order continuity of the norm)**   Let X be an uncountable discrete space, and let X_∞ be the one-point compactification of X. We claim the sup norm on $C(X_\infty)$ is σ-order continuous, but not order continuous.

Recall from Example 2.78 that if a function is continuous on X_∞, the value at all but countably many points is the same as the value at ∞. Next note that for any point x in X, the indicator function $\chi_{\{x\}}$ is a continuous function on X_∞. This implies that a net $f_\alpha \downarrow 0$ in $C(X_\infty)$ if and only if $f_\alpha(x) \downarrow 0$ for each x in X. For if $f_\alpha(x) \downarrow \varepsilon > 0$, then $\varepsilon\chi_{\{x\}}$ is a lower bound of $\{f_\alpha\}$.

Now suppose $f_n \downarrow 0$ in $C(X_\infty)$. Then $f_n(x) \downarrow 0$ for each x in X. Further, by the above discussion, the set $\bigcup_{n=1}^\infty \{x \in X : f_n(x) \neq f_n(\infty)\}$ is countable. Since X is uncountable, there is some x_0 in X satisfying $f_n(x_0) = f_n(\infty)$ for all n. Since $f_n(x_0) \downarrow 0$, we have $f_n(\infty) \downarrow 0$ too. Thus $f_n(x) \downarrow 0$ for each x in X_∞. It now follows from Dini's Theorem 2.66 that $f_n \downarrow 0$ uniformly on X_∞, that is, $\|f_n\|_\infty \downarrow 0$. In other words, $C(X_\infty)$ has σ-order continuous norm.

To see that $\|\cdot\|_\infty$ is not order continuous, consider the directed family of all finite subsets of X, directed upward by inclusion. For each finite subset F of X, set $f_F = \mathbf{1} - \chi_F$ (where $\mathbf{1}$ is the constant function one). Then $\{f_F\}$ is a net in $C(X_\infty)$ satisfying $f_F \downarrow 0$ and $\|f_F\|_\infty = 1$ for each F.   ∎

The norm of a Banach lattice is, of course, order continuous if and only if the locally solid topology it generates is order continuous. The order continuity of the norm has several useful characterizations. They are listed in the next theorem, which is the Banach lattice version of Theorem 8.60.

9.22 Theorem   *For a Banach lattice E the following statements are equivalent.*

1. *$\langle E, E' \rangle$ is a symmetric Riesz pair, where E' is the norm dual of E.*

2. *E has order continuous norm.*

3. *E has σ-order continuous norm and is order complete.*

4. *$E' = E_n^\sim$.*

5. *E is an ideal in its double norm dual E''.*

6. *The boxes of E are $\sigma(E, E')$-compact.*

7. *Every order bounded disjoint sequence in E converges in norm to zero.*

Proof: See [12, Theorems 12.9 and 12. 13].   ∎

Two immediate consequences are worth pointing out.

9.23 Corollary   *A reflexive Banach lattice has order continuous norm.*

9.24 Corollary   *A Banach lattice with order continuous norm is order complete.*

The Banach lattices c_0 (with supremum norm), the $L_p(\mu)$ spaces ($1 \leqslant p < \infty$), and $ba(\mathcal{A})$ all have order continuous norms; see Theorem 13.7. In general, the Banach lattices $C(K)$ (for K compact Hausdorff), and $L_\infty(\mu)$ do not have order continuous norms. Banach lattices with order continuous norms admit plenty of "locally" strictly positive linear functionals.

9.25 Theorem *If E is a Banach lattice with order continuous norm and E_x is a principal ideal, then there exists a positive linear functional on E that is strictly positive on E_x.*

Proof: See [12, Theorem 12.14, p. 183]. ∎

9.5 AM- and AL-spaces

In this section we consider normed spaces satisfying additional algebraic and lattice theoretic properties.

9.26 Definition *A lattice norm on a Riesz space is:*

- *an **M-norm** if $x, y \geqslant 0$ implies $\|x \vee y\| = \max\{\|x\|, \|y\|\}$.*

- *an **L-norm** if $x, y \geqslant 0$ implies $\|x + y\| = \|x\| + \|y\|$.*

A normed Riesz space equipped with an M-norm (resp. an L-norm) is called an **M-space** (resp. an **L-space**.) A norm complete M-space is an **AM-space** Similarly, a norm complete L-space is an **AL-space**.[1] You can easily verify that the norm completion of an M-space (resp. an L-space) is an AM-space (resp. an AL-space). AM-spaces and AL-spaces are dual to each other.

9.27 Theorem *A Banach lattice is an AL-space (resp. an AM-space) if and only if its dual is an AM-space (resp. an AL-space).*

Proof: See [12, Theorem 12.22, p. 188]. ∎

The $C(K)$-spaces and $L_\infty(\mu)$-spaces are AM-spaces, while the $L_1(\mu)$-spaces are AL-spaces. Also, the Banach lattice $ba(\mathcal{A})$ is an AL-space; see Theorem 10.53. Remarkably, every principal ideal in an arbitrary Banach lattice has the structure of an AM-space with unit.

9.28 Theorem *If E is either a Banach lattice or an order complete Riesz space, then for each $x \in E$ the principal ideal E_x, equipped with the norm*

$$\|y\|_\infty = \inf\{\lambda > 0 : |y| \leqslant \lambda|x|\} = \min\{\lambda \geqslant 0 : |y| \leqslant \lambda x\},$$

is an AM-space, with unit $|x|$.[2]

---

[1] The term AL-space is an abbreviation for "abstract Lebesgue space." The term M-space is a mnemonic for "maximum," but its use may come from the fact that M follows L in the Latin alphabet.

[2] Actually, this conclusion is true for the class of all Archimedean uniformly complete Riesz spaces; see [235, Theorem 45.4, p. 308]. In general, on every principal ideal E_x of an Archimedean Riesz space the $\|\cdot\|_\infty$-norm is an M-norm.

Proof: Let $0 < x \in E$. We leave it as an exercise to verify that the formula

$$\|y\|_\infty = \inf\{\lambda > 0 : |y| \leq \lambda x\} = \min\{\lambda \geq 0 : |y| \leq \lambda x\}$$

defines a lattice norm on the principal ideal E_x. Next we show that $\|\cdot\|_\infty$ is an M-norm. To this end, let $0 \leq y, z \in E_x$ and put $m = \max\{\|y\|_\infty, \|z\|_\infty\}$. Clearly, $m \leq \|y \vee z\|_\infty$. From $y \leq \|y\|_\infty x$ and $z \leq \|z\|_\infty x$, we see that $y \vee z \leq mx$, so $\|y \vee z\|_\infty \leq m$. Therefore, $\|y \vee z\|_\infty = \max\{\|y\|_\infty, \|z\|_\infty\}$. It is clear that x is a unit for E_x.

Next we show that $(E_x, \|\cdot\|_\infty)$ is a Banach lattice. Let $\{y_n\}$ be a positive increasing $\|\cdot\|_\infty$-Cauchy sequence in E_x. By Theorem 9.3, it suffices to show that $\{y_n\}$ is $\|\cdot\|_\infty$-convergent in E_x.

To this end, fix $\varepsilon > 0$ and then choose n_0 such that $\|y_n - y_m\|_\infty < \varepsilon$ for all $n, m \geq n_0$. Then for $n, m \geq n_0$ we have

$$|y_n - y_m| \leq \|y_n - y_m\|_\infty x < \varepsilon x. \qquad (\star)$$

From $(\star)$, we see that $y_n \leq y_{n_0} + \varepsilon x$ for all $n \geq n_0$, so $\{y_n\}$ is order bounded in E_x. Thus, if E is order complete, then there exists some $y \in E_x$ with $y_n \uparrow y$. On the other hand, if E is a Banach lattice, then it follows from $(\star)$ that $\{y_n\}$ is a norm Cauchy sequence in E. So if $\|y_n - y\| \to 0$, then (from Theorem 8.43 (2)) $y_n \uparrow y$ in E (so $y \in E_x$). Thus, in either case, there exists some $y \in E_x$ with $y_n \uparrow y$. Since $0 \leq y_n - y_m \uparrow_{n \geq m} y - y_m$, it follows from $(\star)$ that $0 \leq y - y_m \leq \varepsilon x$ for all $m \geq n_0$, or $\|y_m - y\|_\infty \leq \varepsilon$ for all $m \geq n_0$, as desired. ∎

Recall that a vector $e > 0$ in a Riesz space E is an **order unit**, or simply a **unit**, if for each $x \in E$ there exists some $\lambda > 0$ such that $|x| \leq \lambda e$. Equivalently e is a unit if its principal ideal E_e is all of E. (This differs from a weak order unit, which has the property that its principal band is E.) Now assume that E is a Banach lattice. Since by Theorem 9.28 the principal ideal E_e, under the lattice norm

$$\|x\|_\infty = \inf\{\lambda > 0 : |x| \leq \lambda e\}$$

is a Banach lattice, it follows from Theorem 9.6 that the two norms $\|\cdot\|_\infty$ and $\|\cdot\|$ are equivalent. In addition, the $\|\cdot\|_\infty$-closed unit ball of E coincides with the box $[-e, e]$. From now on when we use the phrase **an AM-space with unit** we mean a Banach lattice with a unit e whose norm is the $\|\cdot\|_\infty$-norm defined above.

9.29 Lemma   *If E is an AM-space with unit e, then for every $x' \in E'$ we have*

$$\|x'\| = \left\|\,|x'|\,\right\| = |x'|(e).$$

Proof: We know that the closed unit ball of E coincides with the box $[-e, e]$. So if $x' \in E'$, then $\|x'\| = \left\|\,|x'|\,\right\| = \sup_{x \in [-e,e]} |x'(x)| = |x'|(e)$. ∎

The norm dual of an AL-space is an AM-space with unit.

9.30 Theorem   *The norm dual of an AL-space is an AM-space with unit e', where e' is the linear functional defined by the norm. That is,*

$$e'(x) = \|x^+\| - \|x^-\|.$$

Proof: Let E be an AL-space and define $e' : E \to \mathbb{R}$ by $e'(x) = \|x^+\| - \|x^-\|$. By Lemma 8.23, e' is a positive (and hence continuous) linear functional on E. Moreover, for each $x \in E^+$ and each $x' \in E'$, we have

$$|x'|(x) \leqslant \|x'\| \cdot \|x\| = \|x'\| e'(x).$$

That is, $|x'| \leqslant \|x'\| e'$. Thus, e' is an order unit of E'. Now note that the closed unit ball of E' coincides with the box $[-e', e']$.   ∎

The next theorem shows that units are preserved in double duals.

9.31 Theorem   *If E is an AM-space with unit e, then E'' is also an AM-space with the same unit e.*

Proof: Let E be an AM-space with unit a e satisfying $\|e\| = 1$ and let U'' denote the closed unit ball of E''. Put $[\![-e, e]\!] = \{x'' \in E'' : -e \leqslant x'' \leqslant e\}$ and note that $[\![-e, e]\!] \subset U''$. Now assume $x'' \in U''$. Then $|x''| \in U''$, and for each $0 \leqslant x' \in E'$ we have

$$|x''|(x') \leqslant \|x''\| \cdot \|x'\| \leqslant \|x'\| = x'(e) = e(x').$$

Therefore, $|x''| \leqslant e$ or $x'' \in [\![-e, e]\!]$. This shows that $U'' \subset [\![-e, e]\!]$ is also true, and so $U'' = [\![-e, e]\!]$. Consequently, E'' is likewise an AM-space with unit e.   ∎

The final remarkable results of this section state that an AM-space with unit is lattice isometric to some $C(K)$-space, and that an AL-space is lattice isometric to an $L_1(\mu)$-space.

9.32 Theorem (Kakutani–Bohnenblust–M. Krein–S. Krein)   *A Banach lattice is an AM-space with unit if and only if it is lattice isometric to $C(K)$ for some compact Hausdorff space K. The space K is unique up to homeomorphism.*

Proof: We only sketch the proof. Let E be a Banach lattice with unit e. Also let $U'_+ = \{x' \in U' : x' \geqslant 0\}$, the positive part of the closed unit ball U' of E'. Then E is lattice isometric to $C(K)$, where

$$K = \{x' \in U'_+ : x' \text{ is an extreme point of } U'_+ \text{ with } \|x'\| = x'(e) = 1\}$$
$$= \{x' \in U'_+ : x' \text{ is a lattice homomorphism with } \|x'\| = x'(e) = 1\}$$

is equipped with the weak* topology. (The hard part is showing the equality of these two sets.) It is clear from the second characterization of K that it is a weak*-closed subset of U, so K is a nonempty weak*-compact set. Now notice that each

$x \in E$ defines (via evaluation) a unique continuous real function on K. So with this identification E is a Riesz subspace of $C(K)$. Moreover, since U' is the closed convex hull of K, it follows (why?) that E is in fact a $\|\cdot\|_\infty$-closed Riesz subspace of $C(K)$. Since E separates the points of K and contains the constant function one (here the unit e acts as the constant function one on K), the Stone–Weierstrass Theorem 9.12 below implies that E is norm dense in $C(K)$. Therefore E coincides with $C(K)$. For details see [12, Theorem 12.28, p. 194].   ∎

Thus by this result and Theorem 9.28, every principal ideal in a Banach lattice is lattice isometric to a $C(K)$-space. The next representation result is more delicate, and we omit its proof.

9.33 Theorem (Kakutani) *A Banach lattice is an AL-space if and only if it is lattice isometric to an $L_1(\mu)$-space.*

Proof: See [12, Theorem 12.26, p. 192].   ∎

A special case of Theorem 9.33, is that the Banach lattice $ba(\mathcal{A})$ of all signed charges on an algebra of sets is lattice isometric to some $L_1(\mu)$-space.

Every AL-space has order continuous norm. Indeed, if E is an AL-space and $\{x_n\}$ is a positive order bounded disjoint sequence (that is, $x_n \wedge x_m = 0$ for $n \neq m$ and $x_n \leq x$ for all n and some $x \in E^+$), then from $\sum_{n=1}^k x_n = \bigvee_{n=1}^k x_n \leq x$, we see that $\sum_{n=1}^\infty \|x_n\| = \lim_{k\to\infty} \sum_{n=1}^k \|x_n\| = \lim_{k\to\infty} \left\|\sum_{n=1}^k x_n\right\| \leq \|x\| < \infty$, so $\lim \|x_n\| = 0$. By Theorem 9.22 (1) and (6), we infer that E has order continuous norm. Thus, by Theorem 9.22 again, every AL-space is an ideal in its double dual. In fact, a stronger conclusion is true and is presented next.

9.34 Theorem *If E is an AL-space, then E is a band in E''. In fact, $E = (E')_n^\sim$. Consequently, $E'' = E \oplus E^d$.*

Proof: Let E be an AL-space. We first show that E is a band in E''. Note that since E and E'' are AL-spaces, both E and E'' have order continuous norms. In particular, by Theorem 9.22, E is an ideal in E''. Now assume $0 \leq x_\alpha \uparrow x''$ in E'' with $\{x_\alpha\} \subset E$. The order continuity of the norm on E'' implies that $\{x_\alpha\}$ is a norm Cauchy net in E'' (and hence in E). If $\|x_\alpha - x\| \to 0$ in E, then $x_\alpha \uparrow x$ (Theorem 8.43 (2)), so (since E is an ideal in E'') $x_\alpha \uparrow x$ in E''. Hence, $x'' = x \in E$, and therefore E is a band in E''.

To see that $E = (E')_n^\sim$, consider the symmetric Riesz pair $\langle E', (E')_n^\sim \rangle$. By Theorem 8.60 (6), the absolute weak topology $|\sigma|((E')_n^\sim, E')$ is a consistent topology. So if E is not $|\sigma|((E')_n^\sim, E')$-dense in $(E')_n^\sim$, then there exists (by Corollary 5.81) some nonzero $x' \in E'$ that vanishes on E, a contradiction. Thus, E is $|\sigma|((E')_n^\sim, E')$-dense in $(E')_n^\sim$.[3] By Theorem 8.43 (4)), E is $|\sigma|((E')_n^\sim, E')$-closed, so $E = (E')_n^\sim$.   ∎

---

[3] This conclusion is a general result. That is, the same proof shows that if $\langle E, E' \rangle$ is a Riesz pair, then E is always $|\sigma|((E')_n^\sim, E')$-dense in $(E')_n^\sim$.

In particular, notice that if E is an AL-space, then every $x'' \in E''$ can be written uniquely in the form $x'' = x + y$, where $x \in E$ and $y \in E^{\mathrm{d}}$. The decomposition $x'' = x + y$ is known as the **Yosida–Hewitt decomposition** of x''. [4]

A Banach lattice that is a band in its double dual is known as a **KB-space** (an abbreviation for Kantorovich–Banach space). This class of KB-spaces enjoys certain remarkable properties. For instance:

9.35 Theorem   *In a KB-space the solid hull of a relatively weakly compact set is relatively weakly compact.*

For a proof of this and additional results on KB-spaces, see [12, Section 14]. It follows that if E is a KB-space, then $\langle E', E \rangle$ is a symmetric Riesz pair. Since every $\sigma(E, E')$-compact subset of E has a relatively $\sigma(E, E')$-compact solid hull, the Mackey topology $\tau(E', E)$ is locally convex-solid, that is, $|\tau|(E', E) = \tau(E', E)$. The following result is a special case of this conclusion having important applications, e.g., [40, 243].

9.36 Theorem   *If μ is a σ-finite measure, then for the symmetric Riesz pair $\langle L_\infty(\mu), L_1(\mu) \rangle$ the Mackey topology $\tau(L_\infty, L_1)$ is locally convex-solid.*

Consequently, in this case the Mackey and absolute Mackey topologies coincide. That is, $\tau(L_\infty, L_1) = |\tau|(L_\infty, L_1)$.

We say that a Banach space X has the **Dunford–Pettis Property** whenever $x_n \xrightarrow{\sigma(X,X')} x$ in X and $x'_n \xrightarrow{\sigma(X',X'')} x'$ in X' imply $\langle x_n, x'_n \rangle \to \langle x, x' \rangle$. In other words, a Banach space X has the Dunford–Pettis Property if and only if the evaluation mapping $(x, x') \mapsto \langle x, x' \rangle$ is sequentially $(\sigma(X, X'), \sigma(X', X''))$-continuous.

9.37 Theorem (Grothendieck)   *Every AL-space and AM-space possesses the Dunford–Pettis Property.*

Proof: See [12, Theorem 19.6, p. 336].   ∎

9.38 Theorem   *Every reflexive Banach space with the Dunford–Pettis property is finite dimensional—so every reflexive AL- or AM-space is finite dimensional.*

Proof: Let X be a reflexive Banach space with the Dunford–Pettis property. Then the closed unit ball U of X is weakly compact (Theorem 6.25). We shall prove that U is norm compact. This allows us to use Theorem 5.26 to conclude that X is finite dimensional.

Let $\{x_n\}$ be a sequence in U. Since U is weakly compact, the Eberlein–Šmulian Theorem 6.34 asserts that $\{x_n\}$ has a weakly convergent subsequence. Thus we can assume that $x_n \xrightarrow{w} x$. Also, replacing $\{x_n\}$ by $\{x_n - x\}$, we can assume that

---

[4] K. Yosida and E. Hewitt [346] decomposed charges into a countably additive part and a purely finitely additive part. See Definition 10.57

$x_n \xrightarrow{w} 0$. To complete the proof, we show that $\|x_n\| \to 0$. Indeed, if $\|x_n\| \nrightarrow 0$, then there exist some $\varepsilon > 0$ and a subsequence of $\{x_n\}$ (which we denote $\{x_n\}$ again) satisfying $\|x_n\| > \varepsilon$ for each n. So for each n there exists some $x'_n \in X'$ with $\|x'_n\| = 1$ and $|x'_n(x_n)| > \varepsilon$. Since X' is also reflexive, by passing to a subsequence, we can assume $x'_n \xrightarrow{w} x'$ in X'. But then the Dunford–Pettis property implies $x'_n(x_n) \to 0$, contrary to $|x'_n(x_n)| > \varepsilon$ for each n. Therefore, $\|x_n\| \to 0$. ∎

9.39 Corollary   *An AL-space is lattice homeomorphic to an AM-space if and only if it is finite dimensional.*

Proof: If an AL-space E is lattice homeomorphic to an AM-space, then the dual AM-space E' with unit is lattice homeomorphic to an AL-space. (Why?) In particular, its closed unit ball (which is an order interval) is weakly compact (recall that AL-spaces have order continuous norms), so E' is reflexive. By Theorem 9.38, E' (and hence E) is finite dimensional. ∎

For more about the Dunford–Pettis Property see [12, Section 19].

9.6   The interior of the positive cone

A variety of applications of separating hyperplane theorems in economics assume the existence of interior points in the positive cone of an ordered vector space. In this section we establish that if the positive cone of a topological Riesz space has a nonempty interior, then the Riesz space is essentially a Riesz subspace of some $C(K)$-space.

We start with a characterization of the interior points of the positive cone.

9.40 Theorem   *If τ is a linear topology τ on an ordered vector space E, then a vector $e \in E^+$ is a τ-interior point of E^+ if and only if the box $[-e, e]$ is a τ-neighborhood of zero. In particular, interior points of E^+ are order units of E.*

Proof: Assume that a symmetric τ-neighborhood V of zero satisfies $e + V \subset E^+$. We claim that $V \subset [-e, e]$. To see this, suppose $v \in V$. Then $e + v \in E^+$, that is, $e + v \geqslant 0$, so $v \geqslant -e$. On the other hand, since V is a symmetric neighborhood, we have $-v \in V$, so $e - v \geqslant 0$. Hence, $v \leqslant e$ too and the inclusion $V \subset [-e, e]$ is established.

For the converse, suppose that If $V = [-e, e]$ is a τ-neighborhood of zero. Then the identity $e + V = e + [-e, e] = [0, 2e] \subset E^+$ shows that e is a τ-interior point of E^+.

The last part follows immediately from the fact that neighborhoods of zero for linear topologies are absorbing sets. ∎

9.41 Corollary *If an ordered vector space E does not have an order unit, then its positive cone E^+ has empty interior in any linear topology.*

By Corollary 9.39, only finite dimensional AL-spaces have order units, so the positive cone of an infinite dimensional AL-space has empty interior. The nonemptiness of the interior of the positive cone imposes a severe restriction on the lattice structure of the space.

9.42 Theorem *If the positive cone of an Archimedean Riesz space E has a nonempty interior in some Hausdorff linear topology, then E is lattice isomorphic to a Riesz subspace of $C(K)$ for some compact Hausdorff topological space K. Moreover, we can choose K so that the Riesz subspace is uniformly dense in $C(K)$.*

Proof: Let τ be a linear topology on an Archimedean Riesz space E and let e be a τ-interior point of E^+. By Theorem 9.40, the box $V = [-e, e]$ is a τ-neighborhood of zero and e is an order unit. Next, we present two different ways to view E as a Riesz subspace of some $C(K)$-space.

First, consider the Dedekind completion E^* of E. Then e is also an order unit for E^* and since E^* is now order complete, under the lattice norm

$$\|x\|_\infty = \inf\{\lambda > 0 : |x| \leqslant \lambda e\},$$

E^* is an AM-space having e as unit. By Theorem 9.32, E^* is lattice isomorphic to a $C(K)$ for some compact Hausdorff space K, where the space K is unique up to a homeomorphism. An easy argument now shows that the Riesz space E can be identified with a Riesz subspace of $C(K)$, where the vector e corresponds to the constant function **1** on K. Also, you should note that in this case K is **extremally disconnected.**, That is, the closure of every open set is always an open set; see [235, Theorem 43.11 p. 288].

Another way of proving the last part of the theorem is to observe that

$$\|x\|_\infty = \inf\{\lambda > 0 : |x| \leqslant \lambda e\},$$

is a lattice norm on E satisfying $\|x \vee y\|_\infty = \max\{\|x\|_\infty, \|y\|_\infty\}$ for each $x, y \in E^+$. That is, $(E, \|\cdot\|_\infty)$ is an M-space. The norm completion $\hat{E}$ of the normed Riesz space $(E, \|\cdot\|_\infty)$ is an AM-space having e as unit. Hence, by Theorem 9.32, $\hat{E}$ is lattice isomorphic to some $C(K)$-space, and consequently E is lattice isomorphic to a uniformly dense Riesz subspace of $C(K)$. ∎

9.43 Corollary *If a Riesz space E is not lattice isomorphic to a Riesz subspace of any $C(K)$-space, then the positive cone E^+ has an empty interior with respect to any linear topology on E.*

9.7 Positive projections

This section takes up the properties of positive operators that are also projections (Definition 6.46) or contractions.

9.44 Definition   *A continuous operator $T\colon X \to X$ on a normed space is a* **contraction operator** *if $\|T\| \leqslant 1$.* [5]

9.45 Definition   *A Banach lattice has* **strictly monotone norm** *if $0 \leqslant x < y$ implies $\|x\| < \|y\|$.*

The L_p-spaces for $1 \leqslant p < \infty$ have strictly monotone norm while L_∞ does not have strictly monotone norm.

9.46 Theorem   *The fixed space of every positive contraction operator on a Banach lattice with strictly monotone norm is a closed Riesz subspace.*

Proof: Let $T\colon E \to E$ be a positive contraction operator on a Banach lattice with strictly monotone norm. Suppose $x \in \mathcal{F}_T$, that is, $Tx = x$. Then we have $|x| = |Tx| \leqslant T|x|$. If $|x| < T|x|$, then the strict monotonicity of the norm implies

$$\|x\| = \big\||x|\big\| < \big\|T|x|\big\| \leqslant \|T\| \cdot \big\||x|\big\| \leqslant \|x\|,$$

a contradiction. Hence, $|x| = T|x|$, so $|x| \in \mathcal{F}_T$. Thus $\mathcal{F}_T$ is a Riesz subspace. ∎

The following theorem of H. H. Schaefer [294, Proposition 11.5, p. 214] exhibits a remarkable property of positive projections.

9.47 Theorem   (Schaefer)   *Let $P\colon E \to E$ be a positive projection on a Riesz space, that is, $P \geqslant 0$ and $P^2 = P$. Then the range $F = P(E)$ of P satisfies the following properties.*

1. *The vector space F with the induced ordering from E is a Riesz space, which is not necessarily a Riesz subspace of E. Its lattice operations $\vee_F$ and $\wedge_F$ are given by*

$$x \vee_F y = P(x \vee y), \ \ x \wedge_F y = P(x \wedge y), \ \text{and} \ |x|_F = P(|x|).$$

2. *If P is strictly positive, then F is a Riesz subspace of E.*

3. *If E is a Banach lattice, then the norm*

$$\||x\|| = \big\||x|_F\big\| = \big\|P|x|\big\|, \ \ x \in F,$$

is a lattice norm on F. Moreover, $\|| \cdot \||$ is equivalent to $\| \cdot \|$ and $(F, \|| \cdot \||)$ is a Banach lattice.

---

[5] Notice that this terminology is inconsistent with the terminology of Chapter 3. The alternative is to call these operators **nonexpansive**, but the terminology is traditional.

Proof: (1) Clearly, $E^+ \cap F$ is a cone. We must show that this cone induces a lattice ordering on F. We prove only the supremum formula. The other lattice operations can be proven in a similar manner.

To this end, let $x, y \in F$. Then $x \leqslant x \vee y$ and $y \leqslant x \vee y$ imply $x = Px \leqslant P(x \vee y)$ and $y \leqslant P(x \vee y)$. That is, $P(x \vee y)$ is an upper bound in F for the set $\{x, y\}$. To see that this is the least upper bound, assume $x \leqslant z$ and $y \leqslant z$ for some $z \in F$. Then, $x \vee y \leqslant z$, and the positivity of P implies $P(x \vee y) \leqslant Pz = z$.

(2) Let $x \in F$. Then $|x| = |Px| \leqslant P|x|$, so $P|x| - |x| \geqslant 0$. Consequently, $P(P|x| - |x|) = 0$. Since P is strictly positive, we see that $|x| = P|x| = |x|_F$. In other words, F is a Riesz subspace of E.

(3) We first show that F is a (norm-) closed subspace of E. To this end, assume that a sequence $\{x_n\} \subset F$ satisfies $x_n \to x$ in E. The positivity of P guarantees that the operator P is norm continuous (Theorem 9.6), therefore $x_n = P(x_n) \to P(x)$. Hence, $P(x) = x$, and consequently $x \in F$.

Clearly, $||| \cdot |||$ is a lattice norm. Moreover, for each $x \in F$, we have

$$\|x\| = \|Px\| = \big\||Px|\big\| \leqslant \big\|P|x|\big\| = |||x||| \leqslant \|P\| \cdot \|x\|,$$

which means that the two norms $||| \cdot |||$ and $\| \cdot \|$ are equivalent on F. ∎

The preceding result presents some examples of lattice subspaces. A **lattice subspace**, is a vector subspace of an ordered vector space that is a Riesz space under the induced ordering. Theorem 9.47 simply asserts that the range of a positive projection on a Riesz space is a lattice subspace. For more about lattice subspaces see Y. A. Abramovich C. D. Aliprantis, and I. A. Polyrakis [2], I. A. Polyrakis [281, 282], and their references.

9.8 The curious AL-space BV_0

The Banach lattice of functions of bounded variation is important in financial economics because it is the smallest vector space of functions containing all the increasing functions. Increasing functions are the natural candidates for utility functions for wealth and play a crucial role in the definition of stochastic dominance [57]. Furthermore, since the zero point of a utility function is irrelevant to the preference it induces, there is no generality lost in considering only functions vanishing at a given point.

Throughout this section $[a, b]$ denotes a fixed (finite) closed interval of $\mathbb{R}$. For arbitrary real numbers $x < y$, we let $\mathbb{P}[x, y]$ denote the set of all partitions of $[x, y]$. A **partition** of $[x, y]$ is a finite set of points $\{x_0, x_1, \ldots, x_n\}$ such that $x = x_0 < x_1 < \cdots < x_n = y$. For each function $f \in \mathbb{R}^{[a,b]}$, we associate three extended real-valued functions P_f, N_f, and T_f defined by

$$P_f(x) = \sup\Big\{\sum_{i=1}^{n} [f(x_i) - f(x_{i-1})]^+ : \{x_0, x_1, \ldots, x_n\} \in \mathbb{P}[a, x]\Big\},$$

$$N_f(x) = \sup\Big\{\sum_{i=1}^{n} [f(x_i) - f(x_{i-1})]^- : \{x_0, x_1, \ldots, x_n\} \in \mathbb{P}[a, x]\Big\},$$

$$T_f(x) = \sup\Big\{\sum_{i=1}^{n} |f(x_i) - f(x_{i-1})| : \{x_0, x_1, \ldots, x_n\} \in \mathbb{P}[a, x]\Big\}.$$

These functions are called the **positive variation**, the **negative variation**, and the **total variation** of f on $[a, x]$. Notice that P_f, N_f, and T_f are increasing[6] and $T_f = N_f + P_f$.

A function $f \in \mathbb{R}^{[a,b]}$ is of **bounded variation** if T_f is real-valued (that is, $T_f(b) < \infty$). It is not difficult to verify that every function of bounded variation is bounded. The collection of all functions of bounded variation on $[a, b]$ is denoted $BV[a, b]$ or simply BV. Routine arguments guarantee that under the usual point-wise ordering, $f \geqslant g$ if $f(x) \geqslant g(x)$ for all $x \in [a, b]$, BV is a Riesz space that is also closed under pointwise multiplication. As a matter of fact, BV is a function space. The properties of BV are summarized in the next theorem, whose proof is omitted.

9.48 Theorem   *The vector space BV of all functions of bounded variation on $[a, b]$ is a Riesz space under the pointwise algebraic and lattice operations. Moreover, BV with the sup norm is an M-space.*

We note that $BV[a, b]$ is not an AM-space. In particular, it is not complete under the sup norm. For instance, let $g(x) = x^2 \cos(\frac{1}{x^2})$ for $x > 0$, and consider the functions in $\mathbb{R}^{[0,1]}$ defined by

$$f(x) = \begin{cases} 0 & \text{if } x = 0, \\ g(x) & \text{if } 0 < x \leqslant 1, \end{cases} \quad \text{and} \quad f_n(x) = \begin{cases} 0 & \text{if } 0 \leqslant x < \frac{1}{n}, \\ g(x) & \text{if } \frac{1}{n} < x \leqslant 1. \end{cases}$$

Then you can verify that each f_n is of bounded variation ($f_n \in BV[0, 1]$) and $|f_n(x) - f(x)| \leqslant \frac{1}{n^2}$ for each n and all $x \in [0, 1]$. So $\{f_n\}$ converges uniformly to f, but f fails to be of bounded variation on $[0, 1]$. (Why?) The norm completion of $BV[a, b]$ is its norm closure in the AM-space of all bounded real-valued functions on $[a, b]$.

An important Riesz subspace of BV that we isolate and study is denoted $BV_0[a, b]$ or simply BV_0. This is the Riesz subspace of BV consisting of all real-valued functions on $[a, b]$ that vanish at a. That is,

$$BV_0[a, b] = \{f \in BV[a, b] : f(a) = 0\}.$$

---

[6] We use the term "increasing" synonymously with nondecreasing.

We may identify BV_0 with a quotient space of BV, where two functions are identified if they differ by a constant function. If $f \in BV_0$, then $f = P_f - N_f$, from which it follows that every function from BV_0 is the difference of two increasing functions. In addition, observe that if $f \in BV_0$ is an increasing function, then $f = P_f$. By the above, BV_0 is an M-space under the pointwise algebraic and lattice operations and the sup norm. And now we come to the interesting part of this section:

- BV_0 can be renormed and reordered so that it becomes an AL-space!

Introduce the ordering $\geq$ defined by $f \geq g$ if $f - g$ is an increasing function. It is easy to verify that $(BV_0, \geq)$ is indeed a partially ordered vector space where the positive cone is the cone of all increasing real-valued functions on $[a, b]$ that vanish at zero. We now show that the order $\geq$ makes BV_0 a Riesz space.

9.49 Lemma   *The ordered vector space $(BV_0, \geq)$ is a Riesz space whose lattice operations for each $f \in BV_0$ are given by*

$$f^+ = P_f, \quad f^- = N_f, \quad and \quad |f| = T_f.$$

Proof: It suffices to show that $f^+ = P_f$ under this ordering. That is, we must show that P_f is the least upper bound of 0 and f in $(BV_0, \geq)$.

To this end, observe that $P_f \geq 0$ is trivially true. Also, from $P_f - f = N_f$, we see that $P_f - f$ is an increasing function. Hence, $P_f \geq 0$ and $P_f \geq f$.

To see that P_f is the least upper bound of 0 and f, let $g \in BV_0$ satisfy $g \geq f$ and $g \geq 0$. That is, both functions g and $g - f$ are increasing. We must show that $g \geq P_f$ or that $g - P_f$ is an increasing function. To see this, first let $a \leqslant u < v \leqslant b$. Since $g - f$ is increasing, $(g - f)(v) \geqslant (g - f)(u)$, or $g(v) - g(u) \geqslant f(v) - f(u)$. Since g is increasing, $g(v) - g(u) \geqslant [f(v) - f(u)]^+$. Thus, if $a \leqslant x < y \leqslant b$, then for any partition $\{x_0, x_1, \ldots, x_n\}$ of $[x, y]$, we have

$$g(y) - g(x) = \sum_{i=1}^{n} [g(x_i) - g(x_{i-1})] \geqslant \sum_{i=1}^{n} [f(x_i) - f(x_{i-1})]^+.$$

Taking suprema over all partitions yields $g(y) - g(x) \geqslant P_f(y) - P_f(x)$. This implies that $g - P_f$ is an increasing function, and the proof is finished.   ∎

The supremum and infimum of two functions $f, g \in BV_0$ can be computed from the formulas $f \vee g = (f - g)^+ + g$ and $f \wedge g = -[(-f) \vee (-g)]$. To obtain these formulas, let us introduce some notation. Let $x \in [a, b]$ and let $P = \{x_0, x_1, \ldots, x_n\}$ be a partition of $[a, x]$. Then for each f in $\mathbb{R}^{[a,b]}$, we write $\Delta_i f = f(x_i) - f(x_{i-1})$. Also, for each pair $f, g \in \mathbb{R}^{[a,b]}$ we let

$$S_P^{f,g}(x) = \sum_{i=1}^{n} [(\Delta_i f) \vee (\Delta_i g)] \quad and \quad R_P^{f,g}(x) = \sum_{i=1}^{n} [(\Delta_i f) \wedge (\Delta_i g)].$$

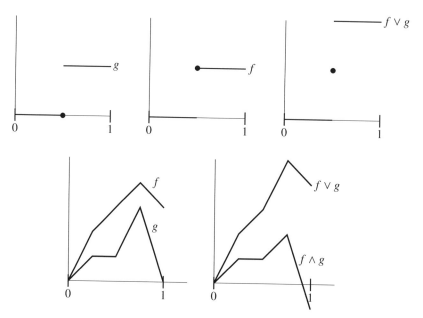

Figure 9.1. $BV_0[0, 1]$ as an AL-space.

Some examples are shown in Figure 9.1.

Now we can express the familiar lattice formulas $f \vee g = (f - g)^+ + g$ and $f \wedge g = -[(-f) \vee (-g)]$ as follows. (The proof is left to you.)

9.50 Lemma   *If $f, g \in BV_0[a, b]$, then their sup and inf in $BV_0[a, b]$ satisfy*

$$(f \vee g)(x) = \sup_{P \in \mathbb{P}[a,x]} S_P^{f,g} \qquad and \qquad (f \wedge g)(x) = \inf_{P \in P[a,x]} R_P^{f,g}$$

for each $x \in [a, b]$.

On the Riesz space $(BV_0, \geq)$, the total variation introduces an L-norm $\| \cdot \|$ defined for each $f \in BV_0$ by

$$\|f\| = T_f(b).$$

Notice that if $f, g \geq 0$, then we indeed have

$$\|f + g\| = T_{f+g}(b) = T_f(b) + T_g(b) = \|f\| + \|g\|.$$

(We leave the verification that this is an L-norm to you.) The L-norm $\| \cdot \|$ is known as the **total variation norm**. (We point out that $\| \cdot \|$ is a seminorm on all of BV as the norm of any constant function is zero.)

9.51 Theorem   *The vector space BV_0 under the pointwise algebraic operations, the ordering $\geq$, and the total variation norm is an AL-space.*

Proof: The only thing that remains to be shown is that BV_0 is $\|\cdot\|$-complete. Since $\|\cdot\|$ is a lattice norm, it suffices to show that every increasing positive Cauchy sequence converges; see Theorem 9.3.

So assume that a Cauchy sequence $\{f_n\}$ satisfies $f_{n+1} \geq f_n \geq 0$ for each n. Then all the f_n and all the $f_{n+p} - f_n$ are increasing functions, so

$$0 \leqslant f_{n+p}(x) - f_n(x) \leqslant f_{n+p}(b) - f_n(b) = \|f_{n+p} - f_n\|,$$

which implies that $\{f_n\}$ is a uniformly Cauchy sequence. If f is its uniform limit, then f and all the $f - f_n$ are increasing functions and $\|f_n - f\| = f(b) - f_n(b) \to 0$. The proof of the theorem is now complete. ∎

We let BV_0^ℓ (resp. BV_0^r) denote the vector subspace of all left (resp. right) continuous functions of BV_0.

9.52 Theorem   *Both BV_0^ℓ and BV_0^r are bands in BV_0 (and hence they are both AL-spaces in their own right).*

Proof: We establish only that BV_0^ℓ is a band—the other case is similar. Assume first that $0 \leq f \leq g$ and $g \in BV_0^\ell$. Fix $x \in [a,b]$. Since f is increasing, we have $f(x-) \leqslant f(x)$, where $f(x-)$ means $\lim_{y \uparrow x} f(y)$. Similarly, the increasingness of $g - f$ implies $(g - f)(x-) \leqslant (g - f)(x)$. This fact, in view of the left continuity of g, implies $g(x) - f(x-) \leqslant g(x) - f(x)$. Hence, $f(x) \leqslant f(x-)$, so $f(x) = f(x-)$, which means that f is left continuous at each x, so $f \in BV_0^\ell$. Since $|f| = T_f$ is a left continuous function whenever f is (why?), it follows from Theorem 8.13 that BV_0^ℓ is an ideal in BV_0.

To see that BV_0^ℓ is a band, let $0 \leq f_\alpha \uparrow f$ in BV_0 with $\{f_\alpha\} \subset BV_0^\ell$. The order continuity of the norm in BV_0 (every AL-space has order continuous norm, see the discussion following Theorem 9.33) implies that $\{f_\alpha\}$ is norm convergent to f. In particular, as in the proof of Theorem 9.51, we see that the net $\{f_\alpha\}$ converges uniformly to f and from this, it easily follows (how?) that f is left continuous. ∎

The indicator function $f = \chi_{[\frac{1}{2},1]} \in BV_0[0,1]$ is not left continuous, but it is orthogonal (in the vector lattice sense of Definition 8.10) to $BV_0^\ell[0,1]$. To see this, let $0 \leq g \in BV_0^\ell[0,1]$. Now for $0 < x < \frac{1}{2}$ consider the partition $P = \{0, x, \frac{1}{2}, 1\}$ and note that

$$0 \leq (f \wedge g)(1) \leqslant S_P^{f,g} = \min\{1, g(\tfrac{1}{2}) - g(x)\} \xrightarrow[x \uparrow \frac{1}{2}]{} 0,$$

or $(f \wedge g)(1) = 0$. Since $f \wedge g$ is increasing (keep in mind that the infimum is taken in $BV_0[0,1]$), we see that $f \wedge g = 0$ in $BV_0[0,1]$. Thus, f belongs to $(BV_0^\ell[0,1])^{\mathrm{d}}$.

Finally, note that BV_0^ℓ and BV_0^r are projection bands since BV_0 is order complete. So we have the direct sum decompositions

$$BV_0 = BV_0^\ell \oplus [BV_0^\ell]^{\mathrm{d}} = BV_0^r \oplus [BV_0^r]^{\mathrm{d}}.$$

For instance, if $f \in BV_0[0, 1]$ is defined by

$$f(x) = \begin{cases} 0 & \text{if } 0 \leqslant x < \frac{1}{2}, \\ 1 & \text{if } x = \frac{1}{2}, \\ 2 & \text{if } \frac{1}{2} < x \leqslant 1, \end{cases}$$

then $f = \chi_{(\frac{1}{2},1]} + \chi_{[\frac{1}{2},1]} \in BV_0^\ell \oplus [BV_0^\ell]^d$. (Can you find the decomposition of f as $f = f_1 + f_2 \in BV_0^r \oplus [BV_0^r]^d$?) For more about BV_0^ℓ and BV_0^r see Theorem 10.62.

Chapter 10

Charges and measures

In Chapter 4 we introduced the concept of σ-algebra of sets to capture the properties of events in probability theory. We used the traditional terminology of referring to sets belonging to such a σ-algebra as *measurable sets*. While we have good pedagogical reasons for introducing the material in this order, it is not obvious what a σ-algebra has to do with measurement of anything. In this chapter we hope to remedy this omission. Historically, mathematicians were interested in generalizing the notions of length, area, and volume. The most useful generalization of these concepts is the notion of a *measure*. In its abstract form a measure is a set function with additivity properties that reflect the properties of length, area, and volume. A *set function* is an extended real function defined on a collection of subsets of an underlying *measurable space*. (We also impose the restriction that a measure assumes at most one of the values ∞ and $-\infty$.) In this chapter we consider set functions that have some of the properties ascribed to area. The main property is *additivity*. The area of two regions that do not overlap is the sum of their areas. A *charge* is any nonnegative set function that is additive in this sense. A *measure* is a charge that is countably additive. That is, the area of a sequence of disjoint regions is the infinite series of their areas. A *probability measure* is a measure that assigns measure one to the entire space. Charges and measures are intimately entwined with *integration*, which we take up in Chapter 11. But here we study them in their own right.

The reason we are interested in charges and measures is that in probability theory and economics, the underlying measurable space has a natural interpretation in terms of states of the world, or in some economic models, as the space of attributes of consumers and/or commodities. See, for instance, M. Berliant [38], W. Hildenbrand [158], L. E. Jones [187, 188, 189] or A. Mas-Colell [241] for a representative sample of this literature. When the underlying measurable space has an interpretation, the set functions also have natural interpretations, such as probability, population, or resource endowments. Thus measures are natural ways to describe the parameters of our models.

On the other hand, due to the Riesz Representation Theorems (see Chapter 14), measure theory can be approached as a branch of the theory of positive operators on Banach lattices, and indeed this approach is often adopted by mathematicians interested more in the theory than its interpretation. The Radon–Nikodym Theorem 13.18 and the Kakutani Representation Theorem 9.33 show that spaces of measures play a fundamental role in the theory of Banach lattices.

There are too many treatises on measure theory and integration to mention any significant fraction of them. Halmos [148] is a classic. Aliprantis and Burkinshaw [13], Royden [290], and Rudin [291] provide very readable introductions to the Lebesgue measure and its applications. Billingsley [43], Doob [99], and Dudley [104] elaborate on the role of measure theory in the theory of probability. Neveu [261] contains a number of results that do not seem to appear elsewhere. Luxemburg [233] has a very nice brief treatment of (finitely additive) charges, while Bhaskara Rao and Bhaskara Rao [41] present a detailed analysis of them.

Here is a guide to the main points of this chapter. As we said above, much of the interest in measures stems from interest in integration. The modern Lebesgue–Daniell approach to integration differs from the ancient Archimedes–Riemann approach in the following way. The Riemann integral is calculated by dividing the domain of a function into manageable regions (intervals), approximating the value of the function on each region, summing, and passing to the limit as the size of the regions goes to zero. The Lebesgue approach starts by partitioning the range of the function into small pieces, finding regions in the domain on which the function is approximately constant (these regions may be quite complicated), measuring the size of these regions, summing and passing to the limit as the size of pieces in the range goes to zero. In order to pursue this approach, we need to be able to measure complicated pieces of the domain. Furthermore, when we look for places where the value of a function is nearly constant, we are looking at the inverse image of a small interval. Thus we want the collection of sets that we can measure to include the inverse image of every real interval.

At this point you may ask, *why can't we measure all subsets of the domain?* The answer to this question is quite subtle and takes us into the realm of axiomatic set theory, and the **Problem of Measure**. The Problem of Measure is this: Given any set, is there a probability measure defined on its power set such that every singleton has measure zero? [1] Clearly the answer to this question can only depend on the cardinality of the set. The cardinality of a set is said to be a **measurable cardinal** if the answer to this question is yes. If X is countable, then countable additivity implies that no such probability measure exists. But what if X is uncountable? This is where the set theory comes in. The **Continuum Hypothesis** asserts that $\mathfrak{c}$, the cardinality of the continuum (that is, the cardinality of $[0, 1]$), is the smallest uncountable cardinal. So the Continuum Hypothesis asserts that any uncountable set must have cardinality at least $\mathfrak{c}$. The Continuum Hypothesis, like the Axiom of Choice, is one of those agnostic axioms of ZF set theory—you may assume it without creating contradictions, yet you cannot prove it, even using the

---

[1] The Ultrafilter Theorem 2.19 implies that for any infinite set there is a probability charge that assigns mass zero to each point. Every free ultrafilter defines a charge assuming only the values zero and one; see Lemma 16.35

Axiom of Choice. (See for instance, the classic by P. J. Cohen [77]) It is beyond our scope here, but under the Continuum Hypothesis, S. Banach and K. Kuratowski [31] have shown that there is no probability measure defined on the power set of [0, 1] with the property that it assigns measure zero to every singleton. That is, $\mathfrak{c}$ is not a measurable cardinal.[2]

Since it is often natural to assign measure zero to each point on the real line, we have a choice. Either we scale back our ambitions and not try to measure every subset, or we can give up countable additivity and work with charges.[3] Each approach has its limitations, which trickle out over the next few chapters. It is this limitation though that brings us back to σ-algebras as the natural classes of measurable sets.

We start with a measure defined on a semiring of sets. The reason we choose a semiring is that the collection of half-open intervals on a line is a semiring, and length (Lebesgue measure) is one of the main applications. Once we have a measure defined on a semiring of sets, we can define an *outer measure* on the class of all sets, called the *Carathéodory extension of the measure*. (An outer measure differs from a measure in that it is only *countably subadditive*.) This construction also generates a σ-algebra of sets that we might reasonably call measurable (Definition 10.17 and Theorem 10.20), as the outer measure is actually a measure when restricted to this σ-algebra. This is a potential source of confusion. We may start out with a σ-algebra of sets that we can a priori measure. Any measure μ defined on this σ-algebra generates a new (generally larger) σ-algebra of μ-measurable sets. We try to be careful about distinguishing these σ-algebras.

The Carathéodory extension procedure is used in Section 10.6 to define the *Lebesgue measure* of subsets of the line. We start out with the semiring of half-open intervals, on which length is the measure, and extend it to the σ-algebra of *Lebesgue measurable sets*.

We also consider the vector spaces of all *signed* charges and measures of bounded variation on a fixed algebra. (The vector operations are defined setwise.) These spaces turn out to be AL-spaces under the *total variation* norm (Theorems 10.53 and 10.56).

---

[2] An interesting question is whether there are any measurable cardinals at all. The question is still open, but it is known that if there are any measurable cardinals, they must be so large that you will never encounter them; see T. Jech [185, Chapter 5]. Surprisingly there are some straightforward questions regarding probability measures on metric spaces whose answers depend on the existence of measurable cardinals. For instance, the question of whether the support of a Borel probability measure is separable is such a question; see P. Billingsley [43]. For a short proof of this result, see R. M. Dudley [104, Appendix D, pp. 415–416]. We do however prove, using only the Axiom of Choice, the simpler result that it is impossible to find a translation invariant measure defined on all subsets of $\mathbb{R}$ that assigns each interval its length (Theorem 10.41). There are, however, translation invariant charges on the power set of $\mathbb{R}$ that assign each interval its length. See S. Banach [30] or L. V. Kantorovich and G. P. Akilov [194, Theorem 9, p. 154].

[3] Even restricting attention to charges does not enable us to measure all the subsets of $\mathbb{R}^3$ in a way that is invariant under both translation and rotation. This observation is due to F. Hausdorff [154]. The famous Banach–Tarski Paradox [32] (see page 14) is a refinement of his work.

10.1   Set functions

We mentioned in Chapter 4 that from the point of view of probability, the most interesting families of sets are σ-algebras, but that semirings arise naturally in certain applications.

The following properties of set functions on semirings intuitively ought to be satisfied by any notion of length, area, or volume. As usual, we denote the extended real numbers (the reals together with the elements ∞ and $-\infty$) by $\mathbb{R}^*$.

10.1 Definition   *A set function $\mu\colon \mathcal{S} \to \mathbb{R}^*$ on a semiring is:*

- **monotone** *if $A \subset B$ and $A, B \in \mathcal{S}$ imply $\mu(A) \leqslant \mu(B)$.*

- **additive** *if for each finite family $\{A_1, \ldots, A_n\}$ of pairwise disjoint sets in $\mathcal{S}$ with $\bigcup_{i=1}^n A_i \in \mathcal{S}$, we have $\mu(\bigcup_{i=1}^n A_i) = \sum_{i=1}^n \mu(A_i)$.*

- **σ-additive** *if for each countable family $\{A_n\}$ of pairwise disjoint sets in $\mathcal{S}$ with $\bigcup_{n=1}^\infty A_n \in \mathcal{S}$ we have $\mu(\bigcup_{n=1}^\infty A_n) = \sum_{n=1}^\infty \mu(A_n)$.*

- **subadditive** *if $\{A_1, \ldots, A_n\} \subset \mathcal{S}$ and $\bigcup_{i=1}^n A_i \in \mathcal{S}$ imply*

$$\mu\left(\bigcup_{i=1}^n A_i\right) \leqslant \sum_{i=1}^n \mu(A_i).$$

- **σ-subadditive** *if $\{A_n\} \subset \mathcal{S}$ and $\bigcup_{n=1}^\infty A_n \in \mathcal{S}$ imply*

$$\mu\left(\bigcup_{n=1}^\infty A_n\right) \leqslant \sum_{n=1}^\infty \mu(A_n).$$

A σ-additive set function is also called **countably additive**. We may also call a set function **finitely additive** if it is additive, but not necessarily σ-additive. Similar terminology also applies to subadditive set functions.

10.2 Definition   *A set function $\mu\colon \mathcal{S} \to [-\infty, \infty]$ on a semiring is:*

- *A **signed charge** if μ is additive, assumes at most one of the values $-\infty$ and ∞, and $\mu(\varnothing) = 0$. A signed charge that assumes only nonnegative values is called a **charge***

- *A **signed measure** if μ is σ-additive, assumes at most one of the values $-\infty$ and ∞, and $\mu(\varnothing) = 0$. If a signed measure assumes only nonnegative values, then it is called a **measure**. [4]*

---

[4] It may seem more natural to call any signed charge a measure and then specialize to say countably additive measures or positive measures. In fact, many authors refer to charges as finitely additive measures. The terminology we use has the virtue of brevity. Not every author follows these conventions, so beware.

An important example of a measure is **counting measure**. Under counting measure, if a set is finite with n elements, its measure is n. If a set is infinite, its counting measure is ∞. Counting measure is important because (as we shall see) summation of a series is the same as integration with respect to counting measure. Thus theorems about integration apply directly to infinite series.

10.3 Lemma   *Every charge (and hence every measure) is monotone and subadditive. Moreover, every measure is σ-subadditive.*

Proof: Let $\mu\colon \mathcal{S} \to [0, \infty]$ be a charge on the semiring $\mathcal{S}$. For the monotonicity of μ, assume $A, B \in \mathcal{S}$ satisfy $A \subset B$. By the definition of a semiring, there exist pairwise disjoint sets $C_1, \ldots, C_k \in \mathcal{S}$ such that $B \setminus A = \bigcup_{i=1}^{k} C_i$. Clearly, the family $\{A, C_1, \ldots, C_k\}$ of $\mathcal{S}$ is pairwise disjoint and satisfies $B = A \cup C_1 \cup \cdots \cup C_k$. So

$$\mu(A) \leqslant \mu(A) + \mu(C_1) + \cdots + \mu(C_k) = \mu(A \cup C_1 \cup \cdots \cup C_k) = \mu(B).$$

For the last two claims we need the following simple property: *If $A \in \mathcal{S}$ and $A_1, \ldots, A_k \in \mathcal{S}$ satisfy $\bigcup_{i=1}^{k} A_i \subset A$ and $A_i \cap A_j = \emptyset$ for each $i \neq j$, then $\sum_{i=1}^{k} \mu(A_i) \leqslant \mu(A)$.*
To see this, use Lemma 4.7 to write $A \setminus \bigcup_{i=1}^{k} A_i = \bigcup_{j=1}^{n} D_j$, where each D_j belongs to $\mathcal{S}$ and $D_i \cap D_j = \emptyset$ for $i \neq j$, and then note that the disjoint union $A = A_1 \cup \cdots \cup A_k \cup D_1 \cup \cdots \cup D_n$ implies

$$\sum_{i=1}^{k} \mu(A_i) \leqslant \sum_{i=1}^{k} \mu(A_i) + \sum_{j=1}^{n} \mu(D_j) = \mu(A).$$

For the subadditivity of μ, let $A_1, \ldots, A_n \in \mathcal{S}$ satisfy $A = \bigcup_{i=1}^{n} A_i \in \mathcal{S}$. Put $B_1 = A_1$ and $B_k = A_k \setminus \bigcup_{i=1}^{k-1} A_i$ for $k \geqslant 2$. Then $B_i \cap B_j = \emptyset$ for $i \neq j$ and $A = \bigcup_{k=1}^{n} B_k$. Moreover, by Lemma 4.7, we can write $B_k = \bigcup_{j=1}^{m_k} C_j^k$, where the C_j^k belong to $\mathcal{S}$ and are pairwise disjoint. From $B_k \subset A_k$ and the property stated above, we have $\sum_{j=1}^{m_k} \mu(C_j^k) \leqslant \mu(A_k)$. Now taking into consideration the disjoint union $A = \bigcup_{k=1}^{n} \bigcup_{j=1}^{m_k} C_j^k$, we see that

$$\mu(A) = \sum_{k=1}^{n} \sum_{j=1}^{m_k} \mu(C_j^k) \leqslant \sum_{k=1}^{n} \mu(A_k).$$

If μ is a measure and $A = \bigcup_{n=1}^{\infty} A_n$, where $\{A, A_1, \ldots\} \subset \mathcal{S}$, then define the sets B_n as above and repeat the preceding arguments to obtain $\mu(A) \leqslant \sum_{n=1}^{\infty} \mu(A_n)$.   ∎

We also point out that every charge is "subtractive" in the following sense.

10.4 Lemma   *If $\mu\colon \mathcal{S} \to [0, \infty]$ is a charge and $A, B \in \mathcal{S}$ satisfy $A \subset B$, $B \setminus A \in \mathcal{S}$ and $\mu(B) < \infty$, then*

$$\mu(B \setminus A) = \mu(B) - \mu(A).$$

Proof: The claim follows from the disjoint union $B = A \cup (B \setminus A)$.   ∎

We may occasionally use the pleonasm "(countably additive) measure" in place of "measure" as a reminder of the fact that what we are about to say may not be true of charges. Clearly, every measure is a charge, but the converse is not true, as the following example makes clear.

10.5 Example (**Not all charges are measures**)   Consider the Banach lattice ℓ_∞ of all bounded real sequences on the set of positive integers $\mathbb{N} = \{1, 2, \ldots\}$. Let c denote the majorizing Riesz subspace of ℓ_∞ consisting of all convergent sequences, and let Lim: $c \to \mathbb{R}$ be the positive linear functional defined by

$$\text{Lim}(x) = \lim_{n \to \infty} x_n.$$

By Theorem 8.32 there exists a positive linear extension of Lim to ℓ_∞, which we again denote Lim.

Now, let $\mathcal{A}$ denote the σ-algebra of all subsets of $\mathbb{N}$, the power set of $\mathbb{N}$. If we define $\mu \colon \mathcal{A} \to [0, 1]$ by

$$\mu(A) = \text{Lim}(\chi_A),$$

then μ is a charge that fails to be σ-additive. (Why?)   ∎

10.6 Definition   *A charge $\mu \colon \mathcal{A} \to [0, \infty]$ on an algebra is **finite** (or **totally finite**) if $\mu(X) < \infty$.*

The next example shows that a signed charge may take on only finite values, yet nevertheless its range may be unbounded.

10.7 Example (**An unbounded finite-valued signed charge**)   Let $\mathcal{A}$ be the algebra consisting of all finite subsets of $\mathbb{N}$ and their complements. Define the signed charge μ on $\mathcal{A}$ by setting $\mu(A) = n$ if A is finite with n elements and $\mu(A) = -\mu(A^c)$ if A is infinite. Note that μ is finite-valued and its range is the set of all integers.   ∎

Measures satisfy the following important continuity properties.

10.8 Theorem   *For a measure $\mu \colon \mathcal{S} \to [0, \infty]$ defined on a semiring and a sequence $\{A_n\}$ in $\mathcal{S}$, we have the following.*

1. *If $A_n \uparrow A$ and $A \in \mathcal{S}$, then $\mu(A_n) \uparrow \mu(A)$.*

2. *If $A_n \downarrow A$, $A \in \mathcal{S}$, and $\mu(A_k) < \infty$ for some k, then $\mu(A_n) \downarrow \mu(A)$.*

Proof: (1)   Let $\{A_n\}$ be a sequence in $\mathcal{S}$ satisfying $A_n \subset A_{n+1}$ for each n and assume $A = \bigcup_{n=1}^\infty A_n$ belongs to $\mathcal{S}$. If $\mu(A_n) = \infty$ for some n, then $\mu(A_n) \uparrow \mu(A)$ is trivial. So assume $\mu(A_n) < \infty$ for each n. Letting $A_0 = \varnothing$, we may write each set $A_k \setminus A_{k-1}$ as a finite pairwise disjoint union of sets in $\mathcal{S}$, say $C_1^k, \ldots, C_{m_k}^k$.

This guarantees that $A_k = A_{k-1} \cup C_1^k \cup \cdots \cup C_{m_k}^k$, so the additivity of μ yields $\sum_{i=1}^{m_k} \mu(C_i^k) = \mu(A_k) - \mu(A_{k-1})$. Now using the σ-additivity of μ we obtain

$$\mu(A) = \sum_{k=1}^{\infty} \sum_{i=1}^{m_k} \mu(C_i^k) = \sum_{k=1}^{\infty} [\mu(A_k) - \mu(A_{k-1})]$$

$$= \lim_{n \to \infty} \sum_{k=1}^{n-1} [\mu(A_k) - \mu(A_{k-1})] = \lim_{n \to \infty} \mu(A_n).$$

(2) Without loss of generality, we can consider the case $\mu(A_1) < \infty$. Then $A_1 \setminus A = \bigcup_{n=1}^{\infty}(A_n \setminus A_{n+1})$. Once again we may write each set $A_n \setminus A_{n+1}$ as a finite pairwise disjoint union of sets $C_1^n, \ldots, C_{m_n}^n$ in $\mathcal{S}$. So by the σ-additivity of μ we get

$$\mu(A_1) - \mu(A) = \sum_{k=1}^{\infty} \sum_{i=1}^{m_k} \mu(C_i^k) = \lim_{n \to \infty} \sum_{k=1}^{n-1} [\mu(A_k) - \mu(A_{k+1})]$$

$$= \mu(A_1) - \lim_{n \to \infty} \mu(A_n),$$

which shows that $\mu(A_n) \downarrow \mu(A)$. ∎

We also note the following useful simple result whose easy proof is left as an exercise.

10.9 Lemma *A finite charge μ on an algebra is countably additive if and only if it satisfies $\mu(A_n) \downarrow 0$ whenever $A_n \downarrow \varnothing$.*

The next result gives a necessary and sufficient condition for two finite measures to be equal. We do not have to verify that their values are the same on every set in the σ-algebra, it is enough to check values on a generating family closed under finite intersections.

10.10 Theorem *Assume that μ and ν are finite measures on a σ-algebra Σ of subsets of X such that $\mu(X) = \nu(X)$. Let $\mathcal{C}$ generate Σ and assume that $\mathcal{C}$ is closed under finite intersections. If $\mu(A) = \nu(A)$ for all $A \in \mathcal{C}$, then $\mu = \nu$.*

Proof: Let $\mathcal{D} = \{A \in \Sigma : \mu(A) = \nu(A)\}$. It is easy to see that $\mathcal{D}$ is a Dynkin system satisfying $\mathcal{C} \subset \mathcal{D}$. By Dynkin's Lemma 4.11, we get $\Sigma = \sigma(\mathcal{C}) \subset \mathcal{D} \subset \Sigma$. Hence $\Sigma = \mathcal{D}$, so $\mu = \nu$. ∎

10.11 Corollary *Two finite measures on the Borel σ-algebra of a topological space coincide if they agree on the open sets or on the closed sets.*

10.12 Example The assumption of finiteness in Theorem 10.10 cannot be dropped. The family of subsets $C_n = \{n, n+1, \ldots\}$ of $\mathbb{N}$ generates the power set σ-algebra of $\mathbb{N}$, and is closed under finite intersections. Let μ be the counting measure and $\nu = 2\mu$. Then $\mu(C_n) = \nu(C_n) = \infty$ for each n, but μ and ν are distinct. ∎

We now give sufficient conditions for a finite charge to be countably additive, and so a measure. The conditions are based on a property related to the topological property of compactness. Recall that a family of sets has the **finite intersection property** if the intersection of every finite subfamily is nonempty. Let us call a family $\mathcal{C}$ of subsets of X a **compact class** if every sequence $\{C_n\}$ in $\mathcal{C}$ with the finite intersection property has a nonempty intersection. For instance, the family of compact sets in a Hausdorff topological space is a compact class (Theorem 2.31). An equivalent restatement is that $\mathcal{C}$ is a compact class if $\bigcap_{n=1}^{\infty} C_n = \varnothing$ (where $\{C_n\} \subset \mathcal{C}$) implies there is some m for which $C_1 \cap \cdots \cap C_m = \varnothing$. For more results on compact classes, see the monograph by J. Pfanzagl and P. Pierlo [276]

The next result is taken from J. Neveu [261, Proposition I.6.2, p. 27]. It states that if a charge on an algebra is "tight" relative to a compact class, then it is countably additive.

10.13 Theorem   *Let μ be a finite charge on an algebra $\mathcal{A}$ of subsets of X. Let $\mathcal{C}$ be a compact subclass of $\mathcal{A}$, and suppose that for every $A \in \mathcal{A}$ we have*

$$\mu(A) = \sup\{\mu(C) : C \in \mathcal{C} \text{ and } C \subset A\}.$$

Then μ is countably additive on $\mathcal{A}$.

Proof: Let $A_n \downarrow \varnothing$, where each A_n belongs to $\mathcal{A}$. By Lemma 10.9 it suffices to show that $\mu(A_n) \downarrow 0$. To this end, let $\varepsilon > 0$.

For each n choose $C_n \in \mathcal{C}$ satisfying $C_n \subset A_n$ and $\mu(A_n) \leqslant \mu(C_n) + \frac{\varepsilon}{2^n}$. Observe that

$$\left(\bigcap_{i=1}^{n} A_i\right) \setminus \left(\bigcap_{i=1}^{n} C_i\right) \subset \bigcup_{i=1}^{n}\left(A_i \setminus C_i\right). \tag{$\star$}$$

Now let $\mathcal{F}$ be the collection of all finite intersections of sets from $\mathcal{C}$. That is, $\mathcal{F}$ consists of all sets of the form $E_1 \cap \cdots \cap E_n$ for some n, where each E_i belongs to $\mathcal{C}$. Obviously $\mathcal{F}$ is also a compact class. Let $K_n = \bigcap_{i=1}^{n} C_i$, which belongs to $\mathcal{F} \cap \mathcal{A}$. Observe that $K_n = \bigcap_{i=1}^{n} C_i \subset \bigcap_{i=1}^{n} A_i = A_n$, so $K_n \downarrow \varnothing$. Since $\mathcal{F}$ is a compact class, there is some m for which $K_m = \varnothing$.

Since the A_ns are nested, for $n \geqslant m$ equation $(\star)$ reduces to

$$A_n \subset \bigcup_{i=1}^{n}(A_i \setminus C_i).$$

Consequently for $n \geqslant m$ we have

$$\mu(A_n) \leqslant \sum_{i=1}^{n} \mu(A_i \setminus C_i) \leqslant \varepsilon.$$

This proves that $\lim_{n\to\infty} \mu(A_n) = 0$.   ∎

10.2   Limits of sequences of measures

In this section we list two important results that deal with setwise limits of sequences of finite measures defined on a common σ-algebra.

10.14 Theorem   (Vitali–Hahn–Saks)   *Let $\{\mu_n\}$ be a sequence of finite measures defined on a common σ-algebra Σ. If for each $A \in \Sigma$ the sequence $\{\mu_n(A)\}$ converges in $\mathbb{R}$, then the formula*

$$\mu(A) = \lim_{n\to\infty} \mu_n(A)$$

defines a finite measure on Σ.

Proof: See N. Dunford and J. T. Schwartz [110, pp. 158–159] or C. D. Aliprantis and O. Burkinshaw [14, Problem 37.5, p. 356].   ∎

The next theorem is harder to prove than it seems.

10.15 Theorem   (Dieudonné)   *Let $\{\mu_n\}$ be a sequence of finite measures defined on the Borel sets of a Polish space. If $\{\mu_n(G)\}$ converges in $\mathbb{R}$ for every open set G, then $\{\mu_n(B)\}$ converges for every Borel set B (so by Theorem 10.14 above $\lim_n \mu_n$ defines a finite Borel measure).*

Proof: See J. K. Brooks [65] or J. K. Brooks and R. V. Chacon [66].   ∎

10.3   Outer measures and measurable sets

In this section, we discuss the basic properties of what are known as outer measures. Outer measures were introduced by C. Carathéodory [72].

10.16 Definition   *An **outer measure** μ on a set X is a nonnegative extended real set function defined on the power set of X that is monotone, σ-subadditive, and satisfies $\mu(\varnothing) = 0$.*

In other words, a nonnegative extended real set function μ defined on the power set of a set X is an outer measure whenever

1. $\mu(\varnothing) = 0$,

2. $A \subset B$ implies $\mu(A) \leqslant \mu(B)$, and

3. $\mu(\bigcup_{n=1}^{\infty} A_n) \leqslant \sum_{n=1}^{\infty} \mu(A_n)$ for each sequence $\{A_n\}$ of subsets of X.

While an outer measure μ is defined on the power set of X, there is an especially useful class of sets determined by μ, called μ-measurable sets, on which μ is actually a measure. This is the subject of Theorem 10.20 below.

10.17 Definition   *Let μ be an outer measure on a set X. Then a subset A of X is called μ-measurable (or more simply **measurable**) if*

$$\mu(S) = \mu(S \cap A) + \mu(S \cap A^c)$$

for each subset S of X. The collection of all μ-measurable subsets is denoted Σ_μ. That is,

$$\Sigma_\mu = \{A \subset X : \mu(S) = \mu(S \cap A) + \mu(S \cap A^c) \text{ for each subset } S \text{ of } X\}.$$

The next result is an easy consequence of the subadditivity property of outer measures.

10.18 Lemma   *A subset A of X is μ-measurable if and only if*

$$\mu(S) \geqslant \mu(S \cap A) + \mu(S \cap A^c)$$

for each subset S.

A μ-**null set** (or simply a null set) is a set A with $\mu(A) = 0$. The monotonicity of μ implies that any subset of a null set is also a null set. The next result is a straightforward consequence of Lemma 10.18

10.19 Lemma   *Every μ-null set is μ-measurable.*

The next theorem elucidates a fundamental relationship between outer measures and measures. It asserts that the collection Σ_μ of all measurable sets is a σ-algebra and that μ restricted to Σ_μ is a measure.

10.20 Theorem (Carathéodory)   *If μ is an outer measure on a set X, then the family Σ_μ of μ-measurable sets is a σ-algebra and μ restricted to Σ_μ is a measure.*

Proof: Clearly, $\varnothing, X \in \Sigma_\mu$, and Σ_μ is closed under complementation. First, we show that Σ_μ is an algebra.

Since Σ_μ is closed under complementation, it suffices to show that Σ_μ is closed under finite unions. To this end, let $A, B \in \Sigma_\mu$. Fix a subset S of X and let $C = A \cup B$. Using the fact that $C = A \cup (A^c \cap B)$ and $C^c = A^c \cap B^c$, we see that

$$\begin{aligned}
\mu(S) &\leqslant \mu(S \cap C) + \mu(S \cap C^c) \\
&\leqslant [\mu(S \cap A) + \mu(S \cap (A^c \cap B))] + \mu((S \cap A^c) \cap B^c) \\
&= \mu(S \cap A) + [\mu((S \cap A^c) \cap B) + \mu((S \cap A^c) \cap B^c)] \\
&= \mu(S \cap A) + \mu(S \cap A^c) = \mu(S),
\end{aligned}$$

which implies $\mu(S) = \mu(S \cap C) + \mu(S \cap C^c)$ for each subset S of X. Thus, $C = A \cup B \in \Sigma_\mu$, so Σ_μ is an algebra.

Now we claim that $\mu \colon \Sigma_\mu \to [0, \infty]$ is additive. As a matter of fact, we shall prove that if $A_1, \ldots, A_k \in \Sigma_\mu$ are pairwise disjoint (that is, $A_i \cap A_j = \emptyset$ for $i \neq j$), then

$$\mu\left(S \cap \left[\bigcup_{n=1}^{k} A_n\right]\right) = \sum_{n=1}^{k} \mu(S \cap A_n) \qquad (\star)$$

for each subset S of X. Indeed, if $A, B \in \Sigma_\mu$ satisfy $A \cap B = \emptyset$ and $S \subset X$, then the measurability of A yields

$$\mu(S \cap (A \cup B)) = \mu(S \cap (A \cup B) \cap A) + \mu(S \cap (A \cup B) \cap A^c)$$
$$= \mu(S \cap A) + \mu(S \cap B).$$

The general case can be established easily by induction. Letting $S = X$ in $(\star)$, we see that μ is additive.

Next, to see that Σ_μ is a σ-algebra, it suffices to establish that Σ_μ is closed under pairwise disjoint countable unions. To this end, let $\{A_n\}$ be a pairwise disjoint sequence in Σ_μ. Put $A = \bigcup_{n=1}^{\infty} A_n$ and $B_k = \bigcup_{n=1}^{k} A_n$ for each k. Now if S is an arbitrary subset of X, then by $(\star)$ and the monotonicity of μ we obtain

$$\mu(S) = \mu(S \cap B_k) + \mu(S \cap B_k^c) \geqslant \mu(S \cap B_k) + \mu(S \cap A^c)$$
$$= \sum_{n=1}^{k} \mu(S \cap A_n) + \mu(S \cap A^c)$$

for each k. This combined with the σ-subadditivity of μ yields

$$\mu(S) \geqslant \sum_{n=1}^{\infty} \mu(S \cap A_n) + \mu(S \cap A^c) \geqslant \mu(S \cap A) + \mu(S \cap A^c),$$

from which it follows that $\mu(S) = \mu(S \cap A) + \mu(S \cap A^c)$. Thus $A \in \Sigma_\mu$, so Σ_μ is a σ-algebra. Moreover, for each k we have

$$\sum_{n=1}^{k} \mu(A_n) = \mu\left(\bigcup_{n=1}^{k} A_n\right) \leqslant \mu\left(\bigcup_{n=1}^{\infty} A_n\right) \leqslant \sum_{n=1}^{\infty} \mu(A_n),$$

so $\mu(\bigcup_{n=1}^{\infty} A_n) = \sum_{n=1}^{\infty} \mu(A_n)$. That is, μ is σ-additive on Σ_μ. ∎

10.4   The Carathéodory extension of a measure

Sometimes we start with a measure defined on a small semiring of sets and wish to extend it to a larger class of sets. For instance in Section 10.6 below, Lebesgue measure is constructed by defining it on the half open intervals and extending it to the class of Lebesgue measurable sets. Another example is the construction of product measures in Section 10.7 below. A general method for extending measures was developed by C. Carathéodory and is known as the **Carathéodory Extension Procedure**. We start with the following definition.

10.21 Definition   *Consider a measure $\mu \colon \mathcal{S} \to [0, \infty]$ defined on a semiring of subsets of the set X. The measure μ generates a nonnegative extended real-valued set function μ^* defined on the power set of X via the formula*

$$\mu^*(A) = \inf\Big\{\sum_{n=1}^{\infty} \mu(A_n) : \{A_n\} \subset \mathcal{S} \text{ and } A \subset \bigcup_{n=1}^{\infty} A_n\Big\},$$

where the usual convention $\inf \varnothing = \infty$ *applies. This new set function μ^* generated by μ as above is called the **Carathéodory extension** of μ.*

We shall soon show that the Carathéodory extension is an outer measure, but before we can fully state the main result we need another definition.

10.22 Definition   *A measure μ on a semiring $\mathcal{S}$ of subsets of X is **σ-finite** if there exists a sequence $\{A_n\}$ in $\mathcal{S}$ (which can be taken to be pairwise disjoint) such that $X = \bigcup_{n=1}^{\infty} A_n$ and $\mu(A_n) < \infty$ for each n.*
    *A measure μ on a semiring is **finite** if $\mu^*(X) < \infty$.*[5]

It is important to notice that not every semiring admits a σ-finite measure. For instance, if X is uncountable and $\mathcal{S}$ is the semiring of singleton sets together with the empty set, then no measure on $\mathcal{S}$ can be σ-finite, since no countable collection of sets in $\mathcal{S}$ has union equal to X. Thus the assumption that a measure is σ-finite is a joint assumption on the measure and the semiring.

Notice also that a measure on a semiring is σ-finite (resp. finite) if and only if there exists a sequence $\{A_n\}$ of sets satisfying $X = \bigcup_{n=1}^{\infty} A_n$ and $\mu^*(A_n) < \infty$ for each n (resp. $\sum_{n=1}^{\infty} \mu^*(A_n) < \infty$).

We now state the main result of this section. Parts of it were proven already in Section 10.3 on outer measures. We prove the rest in a series of lemmas.

10.23 Carathéodory Extension Procedure Theorem   *Let $\mathcal{S}$ be a semiring of subsets of X and let $\mu \colon \mathcal{S} \to [0, \infty]$ be a measure on $\mathcal{S}$. Define the Carathéodory extension μ^* of μ via the formula*

$$\mu^*(A) = \inf\Big\{\sum_{n=1}^{\infty} \mu(A_n) : \{A_n\} \subset \mathcal{S} \text{ and } A \subset \bigcup_{n=1}^{\infty} A_n\Big\}.$$

Say that a set A is μ-measurable if

$$\mu^*(S) = \mu^*(S \cap A) + \mu^*(S \cap A^c)$$

for each subset S of X. Then:

---

[5] Some authors use the term **totally finite** to indicate that X belongs to the semiring (so it is a semialgebra) and that $\mu(X) < \infty$ rather than $\mu^*(X) < \infty$.

1. *The set function μ^* is an outer measure on X.*

2. *The extension μ^* truly is an extension of μ. That is, $\mu^*(A) = \mu(A)$ for every A belonging to the semiring $\mathcal{S}$.*

3. *The collection Σ_μ of μ-measurable subsets of X is a σ-algebra, and μ^* is a measure when restricted to Σ_μ.*

4. *Every set belonging to the semiring $\mathcal{S}$ is μ-measurable. In other words, $\mathcal{S} \subset \sigma(\mathcal{S}) \subset \Sigma_\mu$.*

5. *Intermediate extensions are compatible in the following sense: If Σ is a semiring with $\mathcal{S} \subset \Sigma \subset \Sigma_\mu$, and ν is the restriction of μ^* to Σ, then $\nu^* = \mu^*$. In particular, $(\mu^*)^* = \mu^*$.*

6. *If A is μ-measurable, then there exists some $B \in \sigma(\mathcal{S})$ with $A \subset B$ and $\mu^*(B) = \mu^*(A)$.*

7. *If μ is σ-finite and $A \in \Sigma_\mu$, then there exists some null set C such that $A \cap C = \varnothing$ and $A \cup C \in \sigma(\mathcal{S})$.*

8. *If μ is σ-finite and Σ is a semiring with $\mathcal{S} \subset \Sigma \subset \Sigma_\mu$, then μ^* is the unique extension of μ to a measure on Σ.*

We now present the pieces of this fundamental result.

10.24 Lemma *The Carathéodory extension of a measure is an outer measure.*

Proof: Let $\mu \colon \mathcal{S} \to [0, \infty]$ be a measure. Clearly, $\mu^*(\varnothing) = 0$ and $A \subset B$ implies $\mu^*(A) \leqslant \mu^*(B)$. We must establish the σ-subadditivity of μ^*.

To this end, let $\{A_n\}$ be a sequence of subsets of X. If $\sum_{n=1}^{\infty} \mu^*(A_n) = \infty$, then there is nothing to prove. So assume $\sum_{n=1}^{\infty} \mu^*(A_n) < \infty$ and let $\varepsilon > 0$. For each n pick a sequence $\{B_k^n : k = 1, 2, \ldots\} \subset \mathcal{S}$ satisfying $A_n \subset \bigcup_{k=1}^{\infty} B_k^n$ and $\sum_{k=1}^{\infty} \mu(B_k^n) < \mu^*(A_n) + \frac{\varepsilon}{2^n}$. Now note that $\bigcup_{n=1}^{\infty} A_n$ is a subset of $\bigcup_{n=1}^{\infty} \bigcup_{k=1}^{\infty} B_k^n$. Therefore,

$$\mu^*\Big(\bigcup_{n=1}^{\infty} A_n\Big) \leqslant \sum_{n=1}^{\infty} \sum_{k=1}^{\infty} \mu(B_k^n) \leqslant \sum_{n=1}^{\infty} [\mu^*(A_n) + \tfrac{\varepsilon}{2^n}] = \sum_{n=1}^{\infty} \mu^*(A_n) + \varepsilon$$

for each $\varepsilon > 0$. This implies $\mu^*(\bigcup_{n=1}^{\infty} A_n) \leqslant \sum_{n=1}^{\infty} \mu^*(A_n)$. ∎

The Carathéodory extension μ^* of μ is also known as the **outer measure generated by μ** and, as the next result shows, it is indeed an extension of the measure.

10.25 Lemma *The outer measure μ^* generated by μ is an extension of μ. That is, $\mu^*(A) = \mu(A)$ for each $A \in \mathcal{S}$.*

Proof: Let $A \in \mathcal{S}$. From $A = A \cup \varnothing \cup \varnothing \cdots$, we see that

$$\mu^*(A) \leqslant \mu(A) + 0 + 0 + \cdots = \mu(A).$$

For the reverse inequality, assume $A \subset \bigcup_{n=1}^{\infty} A_n$ with $A_n \in \mathcal{S}$ for each n, so that $A = \bigcup_{n=1}^{\infty} A_n \cap A$. By Lemma 10.3, we have $\mu(A) \leqslant \sum_{n=1}^{\infty} \mu(A_n \cap A) \leqslant \sum_{n=1}^{\infty} \mu(A_n)$. This easily implies $\mu(A) \leqslant \mu^*(A)$, so $\mu^*(A) = \mu(A)$. ∎

We now formalize the notion of measurability with respect to a measure.

10.26 Definition   *A set is **μ-measurable** if it is measurable with respect to the outer measure μ^* in the sense of Definition 10.17 That is, A is μ-measurable if*

$$\mu^*(S) = \mu^*(S \cap A) + \mu^*(S \cap A^c)$$

*for every subset S of X. By Theorem 10.20 the collection of μ measurable sets is a σ-algebra, which we denote Σ_μ (rather than Σ_{μ^*}). A real function $f: X \to \mathbb{R}$ is **μ-measurable** if $f: (X, \Sigma_\mu) \to (\mathbb{R}, \mathcal{B}_\mathbb{R})$ is measurable.*

In practice we often drop the μ and refer to sets and functions as measurable. The next lemma simplifies the verification of measurability of a set A. Read it carefully and compare it to the definition above so that you are sure that you understand the difference between the two statements.

10.27 Lemma (μ-measurability)   *Let $\mathcal{S}$ a semiring of subsets of a set X, and let $\mu: \mathcal{S} \to [0, \infty]$ be a measure on $\mathcal{S}$. Then a subset A of X is μ-measurable if and only if $\mu(S) = \mu^*(S \cap A) + \mu^*(S \cap A^c)$ for each $S \in \mathcal{S}$.*

Proof: If A is μ-measurable, then by definition, $\mu^*(S) = \mu^*(S \cap A) + \mu^*(S \cap A^c)$ for every subset S of X, and since μ agrees with μ^* on the semiring $\mathcal{S}$ it follows that $\mu(S) = \mu^*(S \cap A) + \mu^*(S \cap A^c)$ for each $S \in \mathcal{S}$.

For the converse, let $A \subset X$ satisfy $\mu(S) = \mu^*(S \cap A) + \mu^*(S \cap A^c)$ for each $S \in \mathcal{S}$. Fix a subset B of X. We need to show that $\mu^*(B) = \mu^*(B \cap A) + \mu^*(B \cap A^c)$. By Lemma 10.18, it suffices to show that $\mu^*(B) \geqslant \mu^*(B \cap A) + \mu^*(B \cap A^c)$.

If $\mu^*(B) = \infty$, the inequality is obvious. So assume $\mu^*(B) < \infty$, and let $\varepsilon > 0$. Pick a sequence $\{S_n\}$ in $\mathcal{S}$ satisfying $B \subset \bigcup_{n=1}^{\infty} S_n$ and $\sum_{n=1}^{\infty} \mu(S_n) < \mu^*(B) + \varepsilon$. But then the monotonicity and σ-subadditivity of μ^* imply

$$\mu^*(B \cap A) + \mu^*(B \cap A^c) \leqslant \sum_{n=1}^{\infty} [\mu^*(S_n \cap A) + \mu^*(S_n \cap A^c)]$$
$$= \sum_{n=1}^{\infty} \mu(S_n) \; < \; \mu^*(B) + \varepsilon$$

for all $\varepsilon > 0$, and the desired inequality follows. ∎

We are now ready to show that every set belonging to the semiring $\mathcal{S}$ is μ-measurable.

10.28 Corollary *Every set in $\mathcal{S}$ is μ-measurable. That is, $\mathcal{S} \subset \Sigma_\mu$.*

Proof: Let $A \in \mathcal{S}$. If $S \in \mathcal{S}$, then we can write $S \cap A^c = S \setminus A = \bigcup_{i=1}^n C_i$, where $C_i \in \mathcal{S}$ for each i and $C_i \cap C_j = \varnothing$ for $i \neq j$. By the σ-subadditivity of μ^*, we have $\mu^*(S \cap A^c) \leqslant \sum_{i=1}^n \mu(C_i)$. Now note that the disjoint union $S = (S \cap A) \cup C_1 \cup \cdots \cup C_n$ implies

$$\mu(S) = \mu(S \cap A) + \sum_{i=1}^n \mu(C_i) \geqslant \mu(S \cap A) + \mu^*(S \cap A^c),$$

for each $S \in \mathcal{S}$, and the conclusion follows from Lemma 10.27. $\blacksquare$

In other words, every measure μ extends to a measure on the σ-algebra Σ_μ of its measurable sets. In particular, note that every measure extends to a measure on the σ-algebra $\sigma(\mathcal{S})$ generated by $\mathcal{S}$. What happens if we repeat the Carathéodory extension procedure on μ^*? The answer is that we get μ^* again. That is, $(\mu^*)^* = \mu^*$. The details are included in the next lemma.

10.29 Lemma *Let $\mu \colon \mathcal{S} \to [0, \infty]$ be a measure on a semiring and let Σ be another semiring such that $\mathcal{S} \subset \Sigma \subset \Sigma_\mu$. If ν denotes the restriction of μ^* to Σ, then $\nu^* = \mu^*$. In particular, we have $(\mu^*)^* = \mu^*$.*

Proof: Let Σ be a semiring satisfying $\mathcal{S} \subset \Sigma \subset \Sigma_\mu$ and let ν denote the restriction of μ^* to Σ. Fix a subset A of X. Since $\mathcal{S} \subset \Sigma$, it is immediate that $\nu^*(A) \leqslant \mu^*(A)$. If $\nu^*(A) = \infty$, then $\nu^*(A) = \mu^*(A)$ is obvious. So assume $\nu^*(A) < \infty$.

Pick a sequence $\{A_n\}$ in Σ satisfying $A \subset \bigcup_{n=1}^\infty A_n$. Then the monotonicity and σ-subadditivity of μ^* imply

$$\mu^*(A) \leqslant \mu^*\left(\bigcup_{n=1}^\infty A_n\right) \leqslant \sum_{n=1}^\infty \mu^*(A_n) = \sum_{n=1}^\infty \nu(A_n).$$

This implies $\mu^*(A) \leqslant \nu^*(A)$. Therefore $\nu^*(A) = \mu^*(A)$. $\blacksquare$

10.30 Lemma *If A is μ-measurable, then there exists some $B \in \sigma(\mathcal{S})$ satisfying $A \subset B$ and $\mu^*(B) = \mu^*(A)$.*

Proof: If $\mu^*(A) = \infty$, then let $B = X$. So assume $\mu^*(A) < \infty$. It follows (how?) from the definition of μ^* that

$$\mu^*(A) = \inf\{\mu^*(B) : B \in \sigma(\mathcal{S}) \text{ and } A \subset B\}.$$

So for each k there is $B_k \in \sigma(\mathcal{S})$ with $A \subset B_k$, and $\mu^*(B_k) < \mu^*(A) + \frac{1}{k}$. Now if $B = \bigcap_{k=1}^\infty B_k$, then $B \in \sigma(\mathcal{S})$, $A \subset B$, and $\mu^*(B) = \mu^*(A)$. $\blacksquare$

For the remaining results we need to assume that μ is σ-finite. In this case, the measurable sets Σ_μ coincide up to null sets with the sets of the σ-algebra $\sigma(\mathcal{S})$ generated by $\mathcal{S}$.

10.31 Lemma   *If μ is σ-finite on $\mathcal{S}$ and $A \in \Sigma_\mu$, then there exists some null set C such that $A \cap C = \varnothing$ and $A \cup C \in \sigma(\mathcal{S})$.*

Proof: Let A belong to Σ_μ. Since μ is σ-finite, we can write $X = \bigcup_{n=1}^\infty X_n$, where $X_n \in \mathcal{S}$ and $\mu(X_n) < \infty$ for each n. By Lemma 10.30, for each n there exists some $B_n \in \sigma(\mathcal{S})$ with $A \cap X_n \subset B_n$ and $\mu^*(B_n) = \mu^*(A \cap X_n) < \infty$. So if we let $C_n = B_n \setminus A \cap X_n$, then $\mu^*(C_n) = 0$. Now put $B = \bigcup_{n=1}^\infty B_n$, which belongs to $\sigma(\mathcal{S})$, and note that $A \subset B$. Further $B \setminus A \subset \bigcup_{n=1}^\infty C_n$, so letting $C = B \setminus A$ we see $\mu^*(C) = 0$. Moreover, $A \cup C = A \cup (B \setminus A) = B$ belongs to $\sigma(\mathcal{S})$. ∎

Is the extension of a measure μ to a measure on Σ_μ unique? The answer is "no" in general. Here is a simple example.

10.32 Example (A measure with uncountably many extensions)   Consider $X = \{0, 1\}$, $\mathcal{S} = \{\varnothing, \{0\}\}$, and define $\mu \colon \mathcal{S} \to [0, \infty]$ by $\mu(\varnothing) = 0$ and $\mu(\{0\}) = 1$. Note that $\sigma(\mathcal{S}) = 2^X = \{\varnothing, \{0\}\{1\}, X\}$. Since 1 does not belong to any member of $\mathcal{S}$, we have $\mu^*(\{1\}) = \inf \varnothing = \infty$. In particular, observe that μ is not σ-finite. Now notice that for any $0 \leqslant \alpha \leqslant \infty$, the set function $\nu \colon 2^X \to [0, \infty]$, defined by $\nu(\varnothing) = 0$, $\nu(\{0\}) = 1$, $\nu(\{1\}) = \alpha$, and $\nu(X) = 1 + \alpha$, is a measure that agrees with μ on $\mathcal{S}$. This shows that μ has uncountably many extensions. ∎

However, if μ is σ-finite, then the Carathéodory extension μ^* is the unique extension of μ to a measure on Σ_μ.

10.33 Lemma   *Let $\mu \colon \mathcal{S} \to [0, \infty]$ be a σ-finite measure on a semiring, and let Σ be a semiring satisfying $\mathcal{S} \subset \Sigma \subset \Sigma_\mu$. Then μ^* is the unique extension of μ to a measure on Σ.*

Proof: Let $\mu \colon \mathcal{S} \to [0, \infty]$ be a σ-finite measure and let Σ be a semiring satisfying $\mathcal{S} \subset \Sigma \subset \Sigma_\mu$. Also, let $\nu \colon \Sigma \to [0, \infty]$ be an extension of μ to a measure on Σ. Let ν^* denote the Carathéodory extension of ν. If A is an arbitrary subset of X and a sequence $\{A_n\} \subset \mathcal{S}$ satisfies $A \subset \bigcup_{n=1}^\infty A_n$, then

$$\nu^*(A) \leqslant \sum_{n=1}^\infty \nu^*(A_n) = \sum_{n=1}^\infty \nu(A_n) = \sum_{n=1}^\infty \mu(A_n),$$

so $\nu^*(A) \leqslant \mu^*(A)$ for each subset A of X.

So in order to establish that $\nu = \mu^*$ on Σ, it suffices (in view of the σ-finiteness of μ) to show that $\mu^*(A) \leqslant \nu(A)$ for each $A \in \Sigma$ with $\mu^*(A) < \infty$. (Why?) So let $A \in \Sigma$ with $\mu^*(A) < \infty$ and fix $\varepsilon > 0$. Pick a sequence $\{A_n\}$ in $\mathcal{S}$ satisfying

$A \subset \bigcup_{n=1}^{\infty} A_n$ and $\sum_{n=1}^{\infty} \mu(A_n) < \mu^*(A) + \varepsilon$. Put $B = \bigcup_{n=1}^{\infty} A_n$. By Lemma 4.7, there exists a pairwise disjoint sequence $\{C_n\}$ in $\mathcal{S}$ such that $B = \bigcup_{n=1}^{\infty} C_n \in \sigma(\Sigma)$. Since μ^* and ν^* are both measures on $\sigma(\Sigma)$ that agree with μ on $\mathcal{S}$, we see that

$$\mu^*(B) = \sum_{n=1}^{\infty} \mu^*(C_n) = \sum_{n=1}^{\infty} \mu(C_n) = \sum_{n=1}^{\infty} \nu(C_n) = \nu^*(B).$$

Moreover, by the discussion at the beginning of the proof,

$$\nu^*(B \setminus A) \leqslant \mu^*(B \setminus A) = \mu^*(B) - \mu^*(A) \leqslant \sum_{n=1}^{\infty} \mu(A_n) - \mu^*(A) < \varepsilon.$$

So $\mu^*(A) \leqslant \mu^*(B) = \nu^*(B) = \nu(A) + \nu^*(B \setminus A) < \nu(A) + \varepsilon$ for each $\varepsilon > 0$, which shows that $\mu^*(A) \leqslant \nu(A)$. Thus, $\nu(A) = \mu^*(A)$ for each $A \in \Sigma$. ∎

10.5 Measure spaces

According to the Carathéodory Extension Theorem 10.23, we can always extend any measure on a semiring to the σ-algebra it generates. Accordingly, the following definition seems appropriate.

10.34 Definition *A **measure space** is a triplet (X, Σ, μ), where Σ is a σ-algebra of subsets of X and $\mu \colon \Sigma \to [0, \infty]$ is a measure. If $\mu(X) = 1$, then μ is a **probability measure** and we may call (X, Σ, μ) a **probability space**.*

*A measure space (X, Σ, μ) is **complete** if Σ is equal to Σ_μ, the collection of all μ-measurable sets. In this case we say that μ is a **complete measure**.*

It follows from Lemma 10.29 that the Carathéodory extension of any measure μ when restricted to Σ_μ is a complete measure. This restriction is called the **completion** of μ.

The phrase μ-**almost everywhere** (abbreviated μ-a.e. or simply a.e.) means "everywhere except possibly for a set A with $\mu^*(A) = 0$," where μ^* is the outer measure generated by μ. For instance, we say that two functions $f, g \colon X \to \mathbb{R}$ are μ-almost everywhere equal, written $f = g$ a.e., if $\mu^*(\{x : f(x) \neq g(x)\}) = 0$. Or we may say $f_n \to f$ μ-almost everywhere if $\mu^*(\{x : f_n(x) \nrightarrow f(x)\}) = 0$. The notation $f_n \uparrow$ a.e. means $f_n \leqslant f_{n+1}$ a.e. for each n. (The French use the abbreviation p.p., which stands for *presque partout*. Statisticians and probabilists write a.s., for "almost surely," when μ is a probability measure.)

Let (X, Σ, μ) be a measure space and let $f \colon X \to \mathbb{R}$ be a function. For brevity, we say that f is Σ-measurable instead of $(\Sigma, \mathcal{B}_{\mathbb{R}})$-measurable and Σ_μ-measurable instead of $(\Sigma_\mu, \mathcal{B}_{\mathbb{R}})$-measurable. Since $\Sigma \subset \Sigma_\mu$, every Σ-measurable function is Σ_μ-measurable. In the converse direction, we have the following result.

10.35 Theorem   *Let (X, Σ, μ) be a σ-finite measure space and consider a Σ_μ-measurable function $f: X \to \mathbb{R}$. Then there exists a Σ-measurable function $g: X \to \mathbb{R}$ such that $f = g$ μ-a.e.*

Proof: We can assume $f(x) \geqslant 0$ for each $x \in X$ (otherwise, we apply the arguments below to f^+ and f^- separately). If $f = \chi_A$ for some $A \in \Sigma_\mu$, then by Lemma 10.30 there exists a μ-null set C such that $B = A \cup C \in \Sigma$. So if $g = \chi_B$, then g is Σ-measurable and $f = g$ μ-a.e. It follows that if φ is a Σ_μ-simple function, then there is a Σ-simple function ψ satisfying $\psi = \varphi$ μ-a.e.

Now, by Theorem 4.36, there exists a sequence $\{\varphi_n\}$ of Σ_μ-simple functions satisfying $0 \leqslant \varphi_n(x) \uparrow f(x)$ for each $x \in X$. For each n fix a Σ-simple function ψ_n such that $\psi_n = \varphi_n$ μ-a.e. So as above, for each n there exists a μ-null set $A_n \in \Sigma$ with $\psi_n(x) = \varphi_n(x)$ for all $x \notin A_n$. Put $A = \bigcup_{n=1}^{\infty} A_n \in \Sigma$, and note that A is a μ-null set. Moreover, we have $\psi_n \chi_{A^c}(x) \uparrow f\chi_{A^c}(x)$ for each x. If $g = f\chi_{A^c}$, then (by Theorem 4.27) g is Σ-measurable and $g = f$ μ-a.e. Indeed, the above argument shows that there is a μ-null set N belonging to Σ (not just Σ_μ) such that $g(x) = f(x)$ for all $x \notin N$.   ∎

10.36 Theorem   *Let (X, Σ, μ) be a measure space and let $f: X \to \mathbb{R}$ be a Σ-measurable function. Then either f is constant μ-a.e. or else there exists a nonzero constant c satisfying*

$$\mu([f < c]) > 0 \quad and \quad \mu([f > c]) > 0,$$

where $[f < c] = \{x \in X : f(x) < c\}$ and $[f > c] = \{x \in X : f(x) > c\}$.

Proof: Suppose $f: X \to \mathbb{R}$ is Σ-measurable and not constant μ-a.e. Assume first that $f(x) \geqslant 0$ for each $x \in X$. Let

$$c_0 = \sup\{c \in \mathbb{R} : \mu([f \leqslant c]) = 0\}.$$

Clearly, $0 \leqslant c_0 < \infty$ and $\mu([f < c_0]) = 0$. Since f is not constant μ-a.e., there exists some $c > c_0$ such that $\mu([f > c]) > 0$. (Why?) Now if k satisfies $c_0 < k < c$, then by the definition of c_0 we have $\mu([f < c]) \geqslant \mu([f \leqslant k]) > 0$, and the desired conclusion is established in this case.

In the general case, either f^+ or f^- is not equal to a constant μ-a.e. We consider only the case where f^+ is not equal to a constant μ-a.e. (The other case can be treated in a similar fashion.) By the preceding case, there exists some $c > 0$ satisfying $\mu([f^+ > c]) > 0$ and $\mu([f^+ < c]) > 0$. To finish the proof notice that $[f^+ > c] = [f > c]$ and $[f^+ < c] = [f < c]$.   ∎

10.37 Lemma   *Let (X, Σ) be a measurable space, and let $f: X \to [0, 1]$ be Σ-measurable. If μ is a measure on Σ, then either there is a set A in Σ with $f = \chi_A$ μ-a.e., or else there is a constant $0 < c < \frac{1}{2}$ with $\mu([c < f < 1 - c]) > 0$.*

Proof: For each n let $A_n = [\frac{1}{2n} < f < 1 - \frac{1}{2n}]$. If $\mu(A_n) = 0$ for each n, then from $A_n \uparrow [0 < f < 1]$, we see that $\mu([0 < f < 1]) = 0$. This shows that $f = \chi_A$ μ-a.e. for some $A \in \Sigma$.   ∎

We close the section by stating an interesting result known as **Egoroff's Theorem**, asserting that the pointwise convergence of a sequence of measurable functions on a finite measure space is "almost" uniform.

10.38 Egoroff's Theorem *If a sequence $\{f_n\}$ of measurable functions on a finite measure space (X, Σ, μ) satisfies $f_n \to f$ a.e., then for each $\varepsilon > 0$ there exists some $A \in \Sigma$ such that:*

1. *$\mu(A) < \varepsilon$; and*

2. *The sequence $\{f_n\}$ converges uniformly to f on A^c.*

Proof: See the proof of [13, Theorem 16.7, p. 125].   ∎

10.6 Lebesgue measure

One of the most important measures is Lebesgue measure on the real line, and its generalizations to Euclidean spaces. It is the unique measure on the Borel sets, whose value on every interval is its length. As we mentioned earlier, the collection $\mathcal{S}$ of all half-open intervals,

$$\mathcal{S} = \{[a, b) : a \leqslant b \in \mathbb{R}\},$$

where $[a, a) = \varnothing$, is a semiring of subsets of $\mathbb{R}$.

10.39 Theorem   *The set function $\lambda \colon \mathcal{S} \to [0, \infty)$ defined by*

$$\lambda([a, b)) = b - a$$

is a σ-finite measure on $\mathcal{S}$.

Proof: Let $[a, b) = \bigcup_{n=1}^{\infty} [a_n, b_n)$, where the sequence $\{[a_n, b_n)\}$ consists of nonempty pairwise disjoint half-open intervals. For each $a < x \leqslant b$, let

$$s_x = \sum_i (b_i - a_i).$$

where the sum (possibly an infinite series) extends over all i for which $b_i \leqslant x$; we let $s_x = 0$ if there is no such i. It is easy to see that $s_x \leqslant x - a$ (why?). Obviously $a < x < y \leqslant b$ imply $s_x \leqslant s_y$. Next, consider the nonempty set

$$A = \{x \in (a, b] : s_x = x - a\}.$$

Put $t = \sup A$ and note that $a < t \leqslant b$. Now if $x \in A$, then

$$x - a = s_x \leqslant s_t \leqslant t - a,$$

and from this it easily follows that $s_t = t - a$. That is, $t \in A$. We claim that $t = b$.

To see this, assume by way of contradiction that $a < t < b$. Then $a_m \leqslant t < b_m$ must hold for exactly one m. Since the sequence $\{[a_n, b_n)\}$ is pairwise disjoint, $b_i \leqslant t$ if and only if $b_i \leqslant a_m$. This implies that $s_t = s_{a_m}$. But then from the relation

$$t - a = s_t = s_{a_m} \leqslant a_m - a \leqslant t - a = s_t,$$

we see that $a_m \in A$, which in turn implies $b_m \in A$, contrary to $t < b_m$. Hence, $t = b$. That is,

$$\lambda([a,b)) = b - a = \sum_{n=1}^{\infty} (b_n - a_n) = \sum_{n=1}^{\infty} \lambda([a_n, b_n)),$$

and the proof is finished.   ∎

Therefore, by Lemmas 10.25 and 10.33, λ has a unique extension to Σ_λ, the σ-algebra of λ-measurable sets. We again denote this extension by λ. It is called **Lebesgue measure** on the real line. The members of Σ_λ are called **Lebesgue measurable sets**. We note that Lebesgue measure is **translation invariant** That is, $\lambda(A) = \lambda(x + A)$ for each number x and each Lebesgue measurable set A, where $x + A = \{x + y : y \in A\}$. As a matter of fact, the outer measure λ^* satisfies $\lambda^*(A) = \lambda^*(x + A)$ for each real number x and set A of real numbers.

And now we come to a natural question. *Is there a translation invariant measure defined on the power set of $\mathbb{R}$ that assigns each interval its length?* The answer is no. To see this, we need a lemma.

10.40 Lemma (Vitali) *There exists a subset A of $[0, 1]$ with the property that if $\{r_1, r_2, \ldots\}$ is any enumeration of the rationals in the interval $[-1, 1]$, then the sets $A_n = r_n + A$ ($n = 1, 2, \ldots$) satisfy*

$$A_n \cap A_m = \varnothing \text{ for } n \neq m \quad \text{and} \quad [0, 1] \subset \bigcup_{n=1}^{\infty} A_n \subset [-1, 2].$$

Proof: Define the equivalence relation $\sim$ on $[0, 1]$ by $x \sim y$ if $x - y$ is rational. Using the Axiom of Choice 1.6, let A be a set containing exactly one element from each equivalence class. Now let $\{r_1, r_2, \ldots\}$ be an enumeration of the rational numbers of the interval $[-1, 1]$ and let $A_n = r_n + A$ for each n. It is a routine matter to verify that the sequence $\{A_n\}$ satisfies the desired properties.   ∎

10.41 Theorem *There is no translation invariant measure defined on the power set of $\mathbb{R}$ that assigns each interval its length. (In fact, there is no translation invariant measure defined on the power set of $\mathbb{R}$ that assigns each nonempty bounded interval a finite positive measure.)*

Proof: Assume by way of contradiction that there exists a translation invariant measure μ defined on the power set of $\mathbb{R}$ that assigns each nonempty bounded interval a finite positive measure. Consider the set A satisfying the properties stated in Lemma 10.40. Fix an enumeration $\{r_1, r_2, \ldots\}$ of the rationals in the interval $[-1, 1]$, and define the sets $A_n = r_n + A$ for each n. Since μ is translation invariant, we have $\mu(A_n) = \mu(A)$ for each n. Moreover, note that

$$0 < \mu([0, 1]) \leqslant \mu\left(\bigcup_{n=1}^{\infty} A_n\right) = \sum_{n=1}^{\infty} \mu(A_n) = \lim_{n \to \infty} n\mu(A) \leqslant \mu([-1, 2]) < \infty.$$

However, it is easy to see that there is no number $\mu(A)$ satisfying the above property, and our conclusion follows. ∎

10.42 Corollary *There is a subset of $\mathbb{R}$ that is not Lebesgue measurable.*

Proof: If Σ_λ coincides with the power set of $\mathbb{R}$, then λ is a translation invariant measure defined on the power set of $\mathbb{R}$ that assigns each interval its length, contradicting Theorem 10.41. As a matter of fact, the set A defined in Lemma 10.40 cannot be Lebesgue measurable. (Why?) ∎

Note that since $(a, b) = \bigcup_{n=1}^{\infty} [a + \frac{b-a}{n}, b)$, the σ-algebra $\sigma(\mathcal{S})$ generated by $\mathcal{S}$ contains every open interval. It therefore contains every open set. Therefore $\sigma(\mathcal{S})$ includes $\mathcal{B}_{\mathbb{R}}$, the Borel σ-algebra of $\mathbb{R}$. Conversely, $\mathcal{B}_{\mathbb{R}} \supset \mathcal{S}$. (Why?) Thus $\sigma(\mathcal{S}) = \mathcal{B}_{\mathbb{R}}$. It follows from Theorems 10.20 and 10.25 that every Borel set is Lebesgue measurable.

We summarize the preceding discussion in the following result.

10.43 Theorem *We have $\sigma(\mathcal{S}) = \mathcal{B}_{\mathbb{R}} \subset \Sigma_\lambda$.*

Not every Lebesgue measurable set is a Borel set.

10.44 Theorem *The Cantor set, which has Lebesgue measure zero, has a subset that is not a Borel set.*

Proof: See, for example the proof of [13, Theorem 18.11, p. 143]. ∎

We mention here that *n*-**dimensional Lebesgue measure** is defined analogously using the semiring of half-open rectangles, and assigning each rectangle its n-dimensional "volume."

10.7 Product measures

Now let $\mathcal{S}_i$ be a semiring of subsets of the set X_i ($i = 1, \ldots, n$) and assume that $\mu_i \colon \mathcal{S}_i \to [0, \infty]$ is a measure on $\mathcal{S}_i$. Then on the product semiring a set function

$\mu \colon \mathcal{S}_1 \times \cdots \times \mathcal{S}_n \to [0, \infty]$ can be defined via the formula

$$\mu(A_1 \times A_2 \times \cdots \times A_n) = \prod_{i=1}^{n} \mu_i(A_i),$$

where, as usual, we adhere to the convention $0 \cdot \infty = 0$. It turns out that μ is a measure, called the **product measure** and denoted $\mu_1 \times \mu_2 \times \cdots \times \mu_n$. That is,

$$\mu_1 \times \mu_2 \cdots \times \mu_n(A_1 \times A_2 \times \cdots \times A_n) = \prod_{i=1}^{n} \mu_i(A_i).$$

10.45 Theorem   *If μ_i is a measure on the semiring $\mathcal{S}_i$ $(i = 1, \ldots, n)$, then the set function $\mu_1 \times \mu_2 \times \cdots \times \mu_n$ is a measure on the product semiring $\mathcal{S}_1 \times \mathcal{S}_2 \times \cdots \times \mathcal{S}_n$.*

Proof: See the proof of [13, Theorem 26.1, p. 205].   ∎

We note the following facts about measurable sets of a product measure.

10.46 Theorem   *Let $\mu_i \colon \mathcal{S}_i \to [0, \infty]$ be a measure on a semiring of subsets of a set X_i and let A_i be a measurable subset of X_i with $\mu^*(A_i) < \infty$ $(i = 1, \ldots, n)$. Then*

$$(\mu_1 \times \cdots \times \mu_n)^*(A_1 \times \cdots \times A_n) = \mu_1^* \times \cdots \times \mu_n^*(A_1 \times \cdots \times A_n)$$

$$= \prod_{i=1}^{n} \mu_i^*(A_i).$$

Proof: See the proof of [13, Theorem 26.2, p. 206].   ∎

10.47 Theorem   *Let $\mu_i \colon \mathcal{S}_i \to [0, \infty]$ be a measure on a semiring of subsets of a set X_i and let A_i be a measurable subset of X_i $(i = 1, \ldots, n)$. Then $A_1 \times \cdots \times A_n$ is $\mu_1 \times \cdots \times \mu_n$-measurable. That is, we have*

$$\Sigma_{\mu_1} \times \Sigma_{\mu_2} \times \cdots \times \Sigma_{\mu_n} \subset \Sigma_{\mu_1 \times \mu_2 \times \cdots \times \mu_n}.$$

Proof: See the proof of [13, Theorem 26.3, p. 206].   ∎

If each μ_i is σ-finite, then $\mu_1 \times \cdots \times \mu_n$ is also σ-finite, so $(\mu_1 \times \cdots \times \mu_n)^*$ is (by Lemma 10.33) the only extension of $\mu_1 \times \cdots \times \mu_n$ to a measure on $\Sigma_{\mu_1 \times \cdots \times \mu_n}$.

10.8   Measures on $\mathbb{R}^n$

By a "measure on $\mathbb{R}^n$" we mean a measure on the σ-algebra of the Borel sets of $\mathbb{R}^n$. In this section we study the structure of measures $\mathbb{R}^n$. For simplicity, we consider the real line first. We construct measures on $\mathbb{R}$ using the semirings

$$\mathcal{S} = \{(a, b] : a, b \in \mathbb{R}\} \quad \text{and} \quad \mathcal{S}' = \{[a, b) : a, b \in \mathbb{R}\}$$

of half-open intervals, where, as usual, $[a, b) = (a, b] = \emptyset$ if $b \leqslant a$. A real function f on $\mathbb{R}$ is **right continuous** at x if $\lim f(x_n) = f(x)$ for every sequence $x_n \downarrow x$. Similarly, f is **left continuous** at x if $\lim f(x_n) = f(x)$ for every sequence $x_n \uparrow x$.

Let $f: \mathbb{R} \to \mathbb{R}$ be nondecreasing and right continuous. Then f defines a set-valued function $\mu_f: \mathcal{S} \to [0, \infty)$ via the formula

$$\mu_f((a, b]) = f(b) - f(a)$$

for $a \leqslant b$. It turns out that μ_f is a measure on $\mathbb{R}$.

10.48 Theorem *If $f: \mathbb{R} \to \mathbb{R}$ is a nondecreasing right continuous function, then the set function μ_f is a σ-finite measure on the semiring $\mathcal{S}$. Thus the Carathéodory extension procedure can be used to extend it uniquely to a measure on the Borel σ-algebra $\mathcal{B}_{\mathbb{R}}$.*

Proof: The proof is similar to that of Theorem 10.39 and is left as an exercise. For details, see [13, Example 13.6, p. 100]. ∎

An analogous construction can be used with a nondecreasing left continuous function $f: \mathbb{R} \to \mathbb{R}$. In this case, f defines a σ-finite measure v_f on the semiring $\mathcal{S}'$ via the formula
$$v_f([a, b)) = f(b) - f(a)$$

for $a \leqslant b$. This measure again extends to a unique measure on $\mathcal{B}_{\mathbb{R}}$.

More generally, note that if $f: \mathbb{R} \to \mathbb{R}$ is a nondecreasing function then the two set functions $\mu_f: \mathcal{S} \to [0, \infty)$ and $v_f: \mathcal{S}' \to [0, \infty)$ defined by

$$\mu_f((a, b]) = f(b+) - f(a+) \quad \text{and} \quad v_f([a, b)) = f(b-) - f(a-),$$

where $f(x+) = \lim_{t \to x^+} f(t)$ and $f(x-) = \lim_{t \to x^-} f(t)$, extend to identical measures on $\mathcal{B}_{\mathbb{R}}$. (Why?)

The above discussion shows that every nondecreasing left or right continuous (or, for that matter, every nondecreasing) function defines a unique measure on $\mathbb{R}$. The converse is also true. To see this, let μ be a measure on $\mathbb{R}$ that is finite on the bounded subintervals of $\mathbb{R}$. Such a measure is called a **Borel measure**. The measure μ defines a function $f: \mathbb{R} \to \mathbb{R}$ via the formula

$$f(x) = \begin{cases} \mu((0, x]) & \text{if } x > 0, \\ -\mu((x, 0]) & \text{if } x \leqslant 0. \end{cases}$$

You can easily verify that:

  i. f is nondecreasing and right continuous; and

  ii. $\mu((a, b]) = f(b) - f(a)$.

For the right continuity of f note that $x_n \downarrow x$ implies $(0, x_n] \downarrow (0, x]$ if $x \geqslant 0$ and $(x_n, 0] \uparrow (x, 0]$ if $x < 0$.

In particular, it follows from (ii) that $\mu = \mu_f$, and consequently, we have the following important result.

10.49 Theorem   *Any Borel measure μ on $\mathbb{R}$ satisfies $\mu = \mu_f$ for a unique (up to translation by a constant) nondecreasing right continuous function f.*

Similarly, every Borel measure μ on $\mathbb{R}$ satisfies $\mu = \nu_f$, for a unique (up to translation by a constant) nondecreasing left continuous function f.

For $f(x) = x$ the measure μ_f is, of course, the classical **Lebesgue measure**.

Carrying out this identification of functions with Borel measures in $\mathbb{R}^n$ is only somewhat more difficult. Given $a, b \in \mathbb{R}^n$, let $(a, b]$ denote the half-open box $\{x \in \mathbb{R}^n : \forall i \, a_i < x_i \leqslant b_i\}$. In particular, the interval $(-\infty, b] = \{x : x \leqslant b\}$. If μ is a finite Borel measure on $\mathbb{R}^n$, then let $f(x) = \mu((-\infty, x])$. Now for $b \geqslant a$, what is the relation between $\mu((a, b])$ and the values of f? It is no longer simply $f(b) - f(a)$. Consider the case of $\mathbb{R}^2$, and write $b = (b_1, b_2)$ and $a = (a_1, a_2)$. Define $c = (b_1, a_2)$ and $d = (a_1, b_2)$. In other words, c and d are the other two corners of the box $(a, b]$. Now observe that

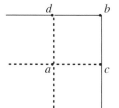

$$(a, b] = [(-\infty, b] \setminus (-\infty, d]] \setminus [(-\infty, c] \setminus (-\infty, a]].$$

Therefore

$$\mu((a, b]) = [f(b) - f(d)] - [f(c) - f(a)]. \tag{1}$$

It is easy to verify that f is continuous from above. That is, if $x_n \downarrow x$, then $f(x_n) \downarrow f(x)$.

Conversely, any $f \colon \mathbb{R}^2 \to \mathbb{R}$ that is continuous from above defines via (1) a measure on the semiring

$$\mathcal{S}_2 = \{(a_1, b_1] \times (a_2, b_2] : a_1 \leqslant b_1 \text{ and } a_2 \leqslant b_2\},$$

as long as (1) assigns nonnegative mass to each box. Thus we can apply the Carathéodory extension procedure to define a unique Borel measure satisfying (1). An identification similar to this works even if μ is not finite (as long as it is finite on bounded sets) and for dimensions greater than two. The first tricky part is figuring out a decent notation that allows us to write down the higher dimensional version of (1). In order to do this, we introduce the following **difference operator**.

Let $f \colon \mathbb{R}^n \to \mathbb{R}$ and let $h = (h_1, \ldots, h_n) \in \mathbb{R}_+^n$. Each of the 2^n corners (extreme points) of the box $(x - h, x]$ is of the form

$$x - h(\delta) = (x_1 - \delta_1 h_1, \ldots, x_n - \delta_n h_n),$$

where $h(\delta) = (\delta_1 h_1, \ldots, \delta_n h_n)$ and each δ_i is either zero or one. For each vector $\delta = (\delta_1, \ldots, \delta_n)$ of zeros and ones, let $s(\delta) = \sum_{i=1}^n \delta_i$. Then we define the difference

$$\Delta_h f(x) = \sum_\delta (-1)^{s(\delta)} f(x - h(\delta)),$$

where the sum runs over all 2^n vectors δ of zeros and ones. Then a little counting and induction should convince you that the n-dimensional equivalent of (1) is

$$\mu((a, b]) = \Delta_{b-a} f(b). \tag{2}$$

For the special case $f(x) = x_1 \cdot x_2 \cdots x_n$ you should verify that the difference $\Delta_h f(x) = h_1 \cdot h_2 \cdots h_n$, so the measure defined by (2) is ordinary Lebesgue measure on $\mathbb{R}^n$.

10.50 Theorem   *If $f : \mathbb{R}^n \to \mathbb{R}$ is continuous from above and satisfies $\Delta_h f(x) \geqslant 0$ for all $x \in \mathbb{R}^n$ and all $h \in \mathbb{R}_+^n$, then there exists a unique Borel measure μ on $\mathbb{R}^n$ satisfying (2).*

Conversely, if μ is Borel measure on $\mathbb{R}^n$, then there exists a function $f : \mathbb{R}^n \to \mathbb{R}$ (unique up to translation) that is continuous from above, satisfies $\Delta_h f(x) \geqslant 0$ for all $x \in \mathbb{R}^n$ and all $h \in \mathbb{R}_+^n$, and satisfies (2).

Proof: Given a function f, we need to verify that (2) characterizes a measure on the semiring $\mathcal{S}_n$, and applying the Carathéodory extension procedure. Given a not necessarily finite Borel measure on $\mathbb{R}^n$, we have to figure out how to define f on the various orthants of $\mathbb{R}^n$. For details see [43, Theorem 12.5, p. 149].   ∎

10.9   Atoms

In this section we consider measures with and without *atoms*, sets of positive measure that cannot be split into two smaller sets of positive measure.

10.51 Definition   *For a measure μ, a measurable set A is called an **atom**, of μ if $\mu^*(A) > 0$ and for every measurable subset B of A, either $\mu^*(B) = 0$ or $\mu^*(A \setminus B) = 0$. If μ has no atoms, then μ is **nonatomic** or **atomless**. The measure μ is **purely atomic** if there exists a countable set A such that $\mu^*(X \setminus A) = 0$ and for each $a \in A$ the singleton set $\{a\}$ is measurable with $\mu^*(\{a\}) > 0$.*

The next result states two basic properties of nonatomic measures.

10.52 Theorem   *If μ is a nonatomic measure and E is a measurable set satisfying $0 < \mu^*(E) < \infty$, then:*

1. *There exists a pairwise disjoint sequence $\{E_n\}$ of measurable subsets of E with $\mu^*(E_n) > 0$ for each n, and consequently $\mu^*(E_n) \to 0$.*

2. *For each $0 \leqslant \delta \leqslant \mu^*(E)$ there exists a measurable subset F of E with $\mu^*(F) = \delta$. Consequently, the range of μ^* is a closed interval.*

Proof: (1)   Since E is not an atom it has a measurable subset E_1 with $\mu^*(E_1) > 0$ and $\mu^*(E \setminus E_1) > 0$. Similarly, since $E \setminus E_1$ is not an atom, it has a measurable subset E_2 satisfying $\mu^*(E_2) > 0$ and $\mu^*((E \setminus E_1) \setminus E_2) = \mu^*(E \setminus (E_1 \cup E_2)) > 0$. Continuing in this way yields a pairwise disjoint sequence $\{E_n\}$ with $\mu(E_n) > 0$ for each n. Since $\sum_{n=1}^{\infty} \mu^*(E_n) = \mu^*(\bigcup_{n=1}^{\infty} E_n) \leqslant \mu^*(E) < \infty$, we see that $\mu^*(E_n) \to 0$.

(2)   We establish this by using Zorn's Lemma. Fix $0 < \delta < \mu^*(E)$. We need the following simple property: If $\mathcal{C}$ is a collection of pairwise disjoint measurable subsets of E each of which has positive measure, then $\mathcal{C}$ is a countable set. (Indeed, if $\mathcal{C}_n = \{A \in \mathcal{C} : \mu^*(A) \geqslant \frac{1}{n}\}$, then each $\mathcal{C}_n$ is finite (why?) and $\mathcal{C} = \bigcup_{n=1}^{\infty} \mathcal{C}_n$.)

Next, let $\mathcal{Z}$ be the set of all collections $\mathcal{C}$ such that $\mathcal{C}$ consists of pairwise disjoint measurable subsets of E, each one having positive measure, such that $\sum_{A \in \mathcal{C}} \mu^*(A) \leqslant \delta$. (Such a collection $\mathcal{C}$ must be countable.) By part (1) there exists (in view of $\mu^*(E_n) \to 0$) a measurable subset B of E such that $\mu^*(B) < \delta$, so $\{B\} \in \mathcal{Z}$. Thus, $\mathcal{Z}$ is nonempty and is obviously a partially ordered set under the inclusion relation $\subset$. Now if $\{\mathcal{C}_i\}_{i \in I}$ is a chain in $\mathcal{Z}$ (for each pair i, j, either $\mathcal{C}_i \subset \mathcal{C}_j$ or $\mathcal{C}_j \subset \mathcal{C}_i$), then it is easy to see (how?) that $\mathcal{C} = \bigcup_{i \in I} \mathcal{C}_i \in \mathcal{Z}$. Consequently, by Zorn's Lemma 1.7, $\mathcal{Z}$ has a maximal element, say $\mathcal{C}_0$. Put $F = \bigcup_{A \in \mathcal{C}_0} A$. Since $\mathcal{C}_0$ is countable, the set F is a measurable subset of E satisfying $\mu^*(F) \leqslant \delta$. We claim that, in fact, $\mu^*(F) = \delta$.

To see this, assume by way of contradiction that F satisfies $\mu^*(F) < \delta$. Since $\mu^*(E \setminus F) > 0$, there exists (as above) a measurable subset C of $E \setminus F$ satisfying $0 < \mu^*(C) < \delta - \mu^*(F)$. But then $\mathcal{C}_0 \cup \{C\} \in \mathcal{Z}$ (why?), contrary to the maximality property of $\mathcal{C}_0$. Hence, $\mu^*(F) = \delta$.   ∎

10.10   The AL-space of charges

Throughout this section $\mathcal{A}$ denotes an algebra (not necessarily a σ-algebra) of subsets of a set X. A **partition** of a set $A \in \mathcal{A}$ is any finite collection $\{A_1, \ldots, A_n\}$ of pairwise disjoint subsets of $\mathcal{A}$ satisfying $\bigcup_{i=1}^{n} A_i = A$. If $\mu \colon X \to [-\infty, \infty]$ is a signed charge, then the **total variation** (or simply the **variation**) of μ is defined by

$$V_\mu = \sup\Big\{ \sum_{i=1}^{n} |\mu(A_i)| : \{A_1, \ldots, A_n\} \text{ is a partition of } X \Big\}.$$

A signed charge is of **bounded variation** if $V_\mu < \infty$. Clearly, every signed charge of bounded variation is a (finite) real-valued set function.

The collection of all signed charges having bounded variation, denoted $ba(\mathcal{A})$, is called the **space of charges** on the algebra $\mathcal{A}$. (The *ba* is a mnemonic for "bounded additive.") Clearly, under the pointwise (that is to say, setwise) algebraic operations of addition and scalar multiplication,

$$(\mu + \nu)(A) = \mu(A) + \nu(A) \quad \text{and} \quad \alpha\mu(A) = (\alpha\mu)(A),$$

the space of charges $ba(\mathcal{A})$ is a vector space. In fact, as the next theorem shows, $ba(\mathcal{A})$ is an AL-space with the ordering $\geqslant$ defined setwise, $\mu \geqslant \nu$ if $\mu(A) \geqslant \nu(A)$ for all $A \in \mathcal{A}$, and norm $\|\mu\| = V_\mu$.

10.53 Theorem *If $\mathcal{A}$ is an algebra of subsets of some set X, then its space of charges $ba(\mathcal{A})$ is an AL-space. Specifically:*

1. *The lattice operations on $ba(\mathcal{A})$ are given by*

$$(\mu \vee \nu)(A) = \sup\{\mu(B) + \nu(A \setminus B) : B \in \mathcal{A} \text{ and } B \subset A\}; \text{ and}$$
$$(\mu \wedge \nu)(A) = \inf\{\mu(B) + \nu(A \setminus B) : B \in \mathcal{A} \text{ and } B \subset A\}.$$

2. *The Riesz space $ba(\mathcal{A})$ is order complete, and $\mu_\alpha \uparrow \mu$ in the lattice sense if and only if $\mu_\alpha(A) \uparrow \mu(A)$ for each $A \in \mathcal{A}$ (and $\mu_\alpha \downarrow \mu$ is, of course, equivalent to $\mu_\alpha(A) \downarrow \mu(A)$ for each $A \in \mathcal{A}$).*

3. *The total variation $\|\mu\| = V_\mu = |\mu|(X)$ is the L-norm on $ba(\mathcal{A})$.*

Proof: Note that the binary relation $\geqslant$ on $ba(\mathcal{A})$ defined by $\mu \geqslant \nu$ if $\mu(A) \geqslant \nu(A)$ for each $A \in \mathcal{A}$ is indeed an order relation under which $ba(\mathcal{A})$ is a partially ordered vector space. In addition, note that the positive cone $ba^+(\mathcal{A})$ consists precisely of all charges on $\mathcal{A}$.

First, we show that $ba(\mathcal{A})$ is a Riesz space. To see this, let $\mu, \nu \in ba(\mathcal{A})$, and for each $A \in \mathcal{A}$ let

$$\omega(A) = \sup\{\mu(B) + \nu(A \setminus B) : B \in \mathcal{A} \text{ and } B \subset A\}.$$

Clearly $\omega(A)$ is finite for each $A \in \mathcal{A}$. We claim that $\omega \in ba(\mathcal{A})$ and that $\omega = \mu \vee \nu$ in $ba(\mathcal{A})$. To see this, notice first that if $\theta \in ba(\mathcal{A})$ satisfies $\mu \leqslant \theta$, $\nu \leqslant \theta$ and $A \in \mathcal{A}$, then for each $B \in \mathcal{A}$ with $B \subset A$ we have

$$\mu(B) + \nu(A \setminus B) \leqslant \theta(B) + \theta(A \setminus B) = \theta(A),$$

so $\omega(A) \leqslant \theta(A)$ for each $A \in \mathcal{A}$. Also, $\mu \leqslant \omega$, $\nu \leqslant \omega$, and $\omega(\varnothing) = 0$ follow trivially. Thus, in order to establish that $\omega = \mu \vee \nu$, it remains to be shown that ω is finitely additive. To this end, let $A, B \in \mathcal{A}$ satisfy $A \cap B = \varnothing$. If $C, D \in \mathcal{A}$ satisfy $C \subset A$ and $D \subset B$, then

$$[\mu(C) + \nu(A \setminus C)] + [\mu(D) + \nu(B \setminus D)] = \mu(C \cup D) + \nu((A \cup B) \setminus (C \cup D))$$
$$\leqslant \omega(A \cup B),$$

so $\omega(A) + \omega(B) \leqslant \omega(A \cup B)$. For the reverse inequality, given $\varepsilon > 0$ there exists some $C \in \mathcal{A}$ with $C \subset A \cup B$ and

$$\omega(A \cup B) - \varepsilon < \mu(C) + \nu((A \cup B) \setminus C)$$
$$= [\mu(C \cap A) + \nu(A \setminus C)] + [\mu(C \cap B) + \nu(B \setminus C)]$$
$$\leqslant \omega(A) + \omega(B).$$

Since $\varepsilon > 0$ is arbitrary, $\omega(A \cup B) \leqslant \omega(A) + \omega(B)$ too. Thus $\omega(A \cup B) = \omega(A) + \omega(B)$. That is, $\omega \in ba(\mathcal{A})$.

For the order completeness of $ba(\mathcal{A})$, let $0 \leqslant \mu_\alpha \uparrow \leqslant \mu$. For each $A \in \mathcal{A}$, let $\nu(A) = \lim_\alpha \mu_\alpha(A)$. Obviously, $\nu \in ba(\mathcal{A})$ and $\mu_\alpha \uparrow \nu$ in $ba(\mathcal{A})$.

Now note that the formula $\|\mu\| = |\mu|(X)$ defines a lattice norm on $ba(\mathcal{A})$ satisfying $\|\mu\| = V_\mu$. (Why?) Clearly, for each $\mu, \nu \in ba^+(\mathcal{A})$ we have

$$\|\mu + \nu\| = (\mu + \nu)(X) = \mu(X) + \nu(X) = \|\mu\| + \|\nu\|.$$

To complete the proof, we must show that $ba(\mathcal{A})$ is norm complete. To this end, let $\{\mu_n\}$ be a Cauchy sequence. For each $A \in \mathcal{A}$, we have

$$|\mu_n(A) - \mu_m(A)| \leqslant |\mu_n - \mu_m|(A) \leqslant |\mu_n - \mu_m|(X) = \|\mu_n - \mu_m\|,$$

so $\{\mu_n(A)\}$ is a Cauchy sequences in $\mathbb{R}$ for each $A \in \mathcal{A}$. Let $\mu(A) = \lim_{n \to \infty} \mu_n(A)$ for each $A \in \mathcal{A}$. Clearly, $\mu(\varnothing) = 0$ and μ is additive on $\mathcal{A}$. Now if $A_1, \ldots, A_k$ is a partition of X, then

$$\sum_{i=1}^k |\mu(A_i)| = \lim_{n \to \infty} \sum_{i=1}^k |\mu_n(A_i)| \leqslant \limsup_{n \to \infty} \|\mu_n\| < \infty.$$

This shows that $V_\mu < \infty$, so $\mu \in ba(\mathcal{A})$. Next, note that if again $A_1, \ldots, A_k$ is a partition of X, then

$$\sum_{i=1}^k |(\mu_n - \mu)(A_i)| = \lim_{m \to \infty} \sum_{i=1}^k |(\mu_n - \mu_m)(A_i)|$$
$$\leqslant \limsup_{m \to \infty} \|\mu_n - \mu_m\|,$$

so $\|\mu_n - \mu\| \leqslant \limsup_{m \to \infty} \|\mu_n - \mu_m\|$. From this last inequality we infer that $\lim_{n \to \infty} \|\mu_n - \mu\| = 0$. Hence, $ba(\mathcal{A})$ is an AL-space. ∎

10.54 Corollary *For each $\mu \in ba(\mathcal{A})$ we have the following.*

1. *Its positive part in $ba(\mathcal{A})$ is given by:*

$$\mu^+(A) = (\mu \vee 0)(A) = \sup\{\mu(B) : B \in \mathcal{A} \text{ and } B \subset A\}.$$

2. *Its negative part in $ba(\mathcal{A})$ is given by:*

$$\mu^-(A) = [(-\mu) \vee 0](A) = -\inf\{\mu(B) : B \in \mathcal{A} \text{ and } B \subset A\}.$$

3. *Its absolute value in $ba(\mathcal{A})$ is given by*

$$|\mu|(A) = \sup\{\mu(B) - \mu(X \setminus B) : B \in \mathcal{A} \text{ and } B \subset A\}$$
$$= \sup\{|\mu(B)| + |\mu(X \setminus B)| : B \in \mathcal{A} \text{ and } B \subset A\}$$
$$= \sup\Big\{\sum_{i=1}^n |\mu(A_i)| : \{A_1, \ldots, A_n\} \text{ is a partition of } A\Big\}.$$

The following result is an easy consequence of the preceding.

10.55 Corollary   *A signed charge is of bounded variation if and only if it has bounded range.*

10.11   The AL-space of measures

The collection of all signed measures of bounded variation in $ba(\mathcal{A})$ is denoted $ca(\mathcal{A})$, where $\mathcal{A}$ is as you recall, an algebra of subsets of a set X. The notation is to remind you that these are countably additive set functions. The lattice structure of this space was thoroughly investigated by K. Yosida and E. Hewitt [346].

10.56 Theorem   *The subset $ca(\mathcal{A})$ of countably additive signed measures in $ba(\mathcal{A})$ is a projection band. That is, $ba(\mathcal{A})$ can be decomposed as*

$$ba(\mathcal{A}) = ca(\mathcal{A}) \oplus [ca(\mathcal{A})]^{\mathrm{d}}.$$

In particular, $ca(\mathcal{A})$ with the total variation norm is an AL-space.

Proof: Clearly, $ca(\mathcal{A})$ is a vector subspace of $ba(\mathcal{A})$. Next, we show that $ca(\mathcal{A})$ is a Riesz subspace. For this, it suffices (in view of Theorem 8.13) to show that $\mu \in ca(\mathcal{A})$ implies $\mu^+ \in ca(\mathcal{A})$. So let $\mu \in ca(\mathcal{A})$ and let $\{A_n\}$ be a sequence of pairwise disjoint sets in $\mathcal{A}$ such that $A = \bigcup_{n=1}^{\infty} A_n \in \mathcal{A}$. If $B \in \mathcal{A}$ satisfies $B \subset A$, then by the σ-additivity of μ, we get

$$\mu(B) = \mu\Big(\bigcup_{n=1}^{\infty} B \cap A_n\Big) = \sum_{n=1}^{\infty} \mu(B \cap A_n) \leq \sum_{n=1}^{\infty} \mu^+(A_n),$$

and consequently,

$$\mu^+(A) = \sup\{\mu(B) : B \in \mathcal{A} \text{ and } B \subset A\} \leq \sum_{n=1}^{\infty} \mu^+(A_n).$$

For the reverse inequality, let $\varepsilon > 0$. Then, from the definition of μ^+, for each n there exists some $B_n \in \mathcal{A}$ with $B_n \subset A_n$ and $\mu^+(A_n) - \frac{\varepsilon}{2^n} < \mu(B_n)$. It follows that

$$\mu^+(A) \geq \mu\Big(\bigcup_{n=1}^{k} B_n\Big) = \sum_{n=1}^{k} \mu(B_n) \geq \sum_{n=1}^{k} [\mu^+(A_n) - \tfrac{\varepsilon}{2^n}] \geq \sum_{n=1}^{k} \mu^+(A_n) - \varepsilon$$

for each k, so $\sum_{n=1}^{\infty} \mu^+(A_n) \leq \mu^+(A) + \varepsilon$ for each $\varepsilon > 0$. Putting the above together, we see that $\mu^+(A) = \sum_{n=1}^{\infty} \mu^+(A_n)$, so μ^+ is σ-additive.

To see that $ca(\mathcal{A})$ is an ideal, it is sufficient (by Theorem 8.13) to show that $0 \leq \nu \leq \mu$ and $\mu \in ca(\mathcal{A})$ imply $\nu \in ca(\mathcal{A})$. Indeed, under these hypotheses, if $\{A_n\}$ is a sequence of pairwise disjoint sets in $\mathcal{A}$ with $A = \bigcup_{n=1}^{\infty} A_n \in \mathcal{A}$, then from

$$0 \leq \nu(A) - \sum_{n=1}^{k} \nu(A_n) = \nu\Big(A \setminus \bigcup_{n=1}^{k} A_n\Big) \leq \mu\Big(A \setminus \bigcup_{n=1}^{k} A_n\Big) \downarrow_k 0,$$

it follows that $v(A) = \sum_{n=1}^{\infty} v(A_n)$. That is, $v \in ca(\mathcal{A})$.

Finally, we establish that $ca(\mathcal{A})$ is a band. So let a net $\{\mu_\alpha\}$ in $ca(\mathcal{A})$ satisfy $0 \leqslant \mu_\alpha \uparrow \mu$ and let $\{A_n\}$ be a sequence of pairwise disjoint sets in $\mathcal{A}$ satisfying $A = \bigcup_{n=1}^{\infty} A_n \in \mathcal{A}$. From

$$\sum_{n=1}^{k} \mu_\alpha(A_n) = \mu_\alpha\left(\bigcup_{n=1}^{k} A_n\right) \leqslant \mu\left(\bigcup_{n=1}^{k} A_n\right) \leqslant \mu\left(\bigcup_{n=1}^{\infty} A_n\right) = \mu(A),$$

we obtain $\sum_{n=1}^{k} \mu(A_n) = \lim_\alpha \sum_{n=1}^{k} \mu_\alpha(A_n) \leqslant \mu(A)$ for each k. Therefore we have $\sum_{n=1}^{\infty} \mu(A_n) \leqslant \mu(A)$. On the other hand, for each α we have

$$\mu_\alpha(A) = \sum_{n=1}^{\infty} \mu_\alpha(A_n) \leqslant \sum_{n=1}^{\infty} \mu(A_n),$$

so $\mu(A) = \lim_\alpha \mu_\alpha(A) \leqslant \sum_{n=1}^{\infty} \mu(A_n)$. Thus $\mu(A) = \sum_{n=1}^{\infty} \mu(A_n)$. That is, $\mu \in ca(\mathcal{A})$. Hence, $ca(\mathcal{A})$ is a band in $ba(\mathcal{A})$.

Since $ba(\mathcal{A})$ is an order complete Riesz space, it follows from Theorem 8.20 that the band $ca(\mathcal{A})$ is a projection band. ∎

10.57 Definition   *The band $[ca(\mathcal{A})]^{\mathrm{d}}$ is denoted $pa(\mathcal{A})$, and its members are called **purely finitely additive charges**. A purely additive charge is thus orthogonal (or disjoint, see Definition 8.10) to every (countably additive) measure.*

Theorem 10.56 asserts that every signed charge $\mu \in ba(\mathcal{A})$ has a unique decomposition as $\mu = \mu_c + \mu_p$, where μ_c is countably additive, called the **countably additive part** of μ and μ_p is the **purely finitely additive part** of μ. This decomposition is known as the **Yosida–Hewitt decomposition** of μ. The next lemma characterizes disjointness in $ca(\mathcal{A})$.

10.58 Lemma   *For signed measures $\mu, v \in ca(\mathcal{A})$ we have the following.*

1. *If for some $A \in \mathcal{A}$ we have $|\mu|(A) = |v|(A^c) = 0$, then $|\mu| \wedge |v| = 0$.*

2. *If $\mathcal{A}$ is a σ-algebra, and $|\mu| \wedge |v| = 0$, then there exists some $A \in \mathcal{A}$ such that $|\mu|(A) = |v|(A^c) = 0$.*

Proof: The first part follows immediately from the infimum formula. For the second part, let $\mathcal{A}$ be a σ-algebra and assume $|\mu| \wedge |v| = 0$. In particular,

$$(|\mu| \wedge |v|)(X) = \inf\{|\mu|(E) + |v|(E^c) : E \in \mathcal{A}\} = 0.$$

So for each n there exists some $E_n \in \mathcal{A}$ such that $|\mu|(E_n) + |v|(E_n^c) \leqslant 2^{-n}$. Let $A = \bigcap_{n=1}^{\infty} \bigcup_{i=n}^{\infty} E_i$. Then A belongs to the σ-algebra $\mathcal{A}$, and we have the inequality $|\mu|(A) \leqslant |\mu|(\bigcup_{i=n}^{\infty} E_i) \leqslant \sum_{i=n}^{\infty} 2^{-i} = 2^{1-n}$ for all n. Thus $|\mu|(A) = 0$.

Now $A^c = \bigcup_{n=1}^{\infty} \bigcap_{i=n}^{\infty} E_i^c$, but $\bigcap_{i=n}^{\infty} E_i^c \subset E_n^c$ and $|v|(E_n^c) \leqslant 2^{-n}$ for all n, so $|v|(\bigcap_{i=n}^{\infty} E_i^c) = 0$, which implies $|v|(A^c) = 0$. Therefore $|\mu|(A) = |v|(A^c) = 0$. ∎

10.12   Absolute continuity

We can extend the definition of absolute value to arbitrary signed charges μ via the familiar formula

$$|\mu|(A) = \sup\Big\{\sum_{i=1}^{n} |\mu(A_i)| : \{A_1, \dots, A_n\} \text{ is a partition of } A\Big\}.$$

However, in this case, notice that $|\mu|(A) = \infty$ is allowed.

With this definition in mind, the notion of absolute continuity can be formulated as follows.

10.59 Definition   *A signed charge ν is **absolutely continuous** with respect to another signed charge μ, written $\nu \ll \mu$ or $\mu \gg \nu$, if for each $\varepsilon > 0$ there exists some $\delta > 0$ such that $A \in \mathcal{A}$ and $|\mu|(A) < \delta$ imply $|\nu(A)| < \varepsilon$.*

For the countably additive case, we present the following important characterization of absolute continuity. We leave the proof as an exercise.

10.60 Lemma   *Let μ and ν be two signed measures on a σ-algebra with $|\nu|$ finite. Then $\nu \ll \mu$ if and only if $|\mu|(A) = 0$ implies $\nu(A) = 0$.*

The set of signed charges that are absolutely continuous with respect to a fixed signed charge $\mu \in ba(\mathcal{A})$ is the band generated by μ in $ba(\mathcal{A})$.

10.61 Theorem   *For each signed charge $\mu \in ba(\mathcal{A})$ the collection of all signed charges in $ba(\mathcal{A})$ that are absolutely continuous with respect to μ is the band B_μ generated by μ in $ba(\mathcal{A})$.*

*In particular, from $B_\mu \oplus B_\mu^{\mathrm{d}} = ba(\mathcal{A})$, we see that every $\nu \in ba(\mathcal{A})$ has a unique decomposition $\nu = \nu_1 + \nu_2$ (called the **Lebesgue decomposition** of ν with respect to μ), where $\nu_1 \ll \mu$ and $\nu_2 \perp \mu$.*

Proof: Assume first that $\nu \in B_\mu$ (that is, $|\nu| \wedge n|\mu| \uparrow |\nu|$) and let $\varepsilon > 0$. Then for m large enough, $(|\nu| - |\nu| \wedge m|\mu|)(X) = |\nu|(X) - |\nu| \wedge m|\mu|(X) < \varepsilon$. Put $\delta = \varepsilon/m$ and note that if $A \in \mathcal{A}$ satisfies $|\mu|(A) < \delta$, then

$$
\begin{aligned}
|\nu(A)| &\leqslant |\nu|(A) \\
&= (|\nu| - |\nu| \wedge m|\mu|)(A) + |\nu| \wedge m|\mu|(A) \\
&\leqslant (|\nu| - |\nu| \wedge m|\mu|)(X) + m|\mu|(A) \\
&< \varepsilon + \varepsilon = 2\varepsilon.
\end{aligned}
$$

That is, $\nu \ll \mu$.

For the converse, assume that $\nu \ll \mu$. From $B_\mu \oplus B_\mu^{\mathrm{d}} = ba(\mathcal{A})$, we can write $\nu = \nu_1 + \nu_2$ with $\nu_1 \in B_\mu$ and $\nu_2 \perp \mu$. From the preceding case, and $\nu_2 = \nu - \nu_1$, we infer that $v_2 \ll \mu$. We claim that $v_2 = 0$. To this end, let $B \in \mathcal{A}$ and let $\varepsilon > 0$. Since $v_2 \ll \mu$, there exists some $0 < \delta \leqslant \varepsilon$ such that $A \in \mathcal{A}$ and $|\mu|(A) < \delta$ imply $|v_2(A)| < \varepsilon$. From $|v_2| \wedge |\mu|(X) = 0$, we see that there exists some $A \in \mathcal{A}$ with $|v_2|(A) + |\mu|(A^c) < \delta$. Clearly $|\mu|(B \cap A^c) < \delta$, so $|v_2(B \cap A^c)| < \varepsilon$. It follows that

$$|v_2(B)| \leqslant |v_2(B \cap A)| + |v_2(B \cap A^c)| < |v_2|(A) + \varepsilon \leqslant 2\varepsilon$$

for each $\varepsilon > 0$, so $v_2(B) = 0$ for each $B \in \mathcal{A}$. Hence, $v_2 = 0$, which implies $v = v_1 \in B_\mu$, and the proof is finished.   ∎

Finally, let us present a connection between $BV_0[a, b]$, and $ca[a, b]$ (we write $ca[a, b]$ instead of $ca(\mathcal{B})$, where $\mathcal{B}$ is the σ-algebra of the Borel sets of $[a, b]$). Recall that $BV_0[a, b]$ is an AL-space under the total variation norm and the ordering $\geq$ defined by $f \geq g$ if $f - g$ is an increasing function (Theorem 9.51). If $0 \leq f \in BV_0$, then we can extend the function f to all of $\mathbb{R}$ by letting $f(x) = f(b)$ for $x > b$ and $f(x) = 0$ for $x < a$. By Theorem 10.49 the function f defines a measure μ_f on $\mathcal{B}_\mathbb{R}$ (which vanishes, of course, outside the interval $[a, b]$). Since every function $f \in BV_0$ is the difference of two increasing functions on $[a, b]$, it follows that every function $f \in BV_0$ defines a signed measure $\mu_f \in ca[a, b]$, where

$$\mu_f([c, d)) = f(d-) - f(c-) \quad \text{and} \quad \mu_f((c, d]) = f(d+) - f(c+).$$

Clearly, $\mu_{f+g} = \mu_f + \mu_g$ and $\mu_{\alpha f} = \alpha \mu_f$. In other words, we have defined an operator $R \colon BV_0 \to ca[a, b]$ via the formula

$$R(f) = \mu_f.$$

From Theorem 10.49, it follows that R is onto and clearly R is a positive operator. However, you should note that R is not one-to-one.

Now restricting R to BV_0^ℓ, we see that R is one-to-one, onto, and $R(f) \geqslant 0$ if and only if $f \geq 0$ in BV_0. So by Theorem 9.17, R is a lattice isomorphism. Moreover, it is not difficult to see that R is also a lattice isometry. (Why?) Therefore, we have established the following result.

10.62 Theorem   *Both AL-spaces $BV_0^\ell[a, b]$ and $BV_0^r[a, b]$ are lattice isometric to $ca[a, b]$ via $f \mapsto \mu_f$.*

Chapter 11

Integrals

In modern mathematics the process of computing areas and volumes is called *integration*. The computation of areas of curved geometrical figures originated about 2,300 years ago with the introduction by the Greek mathematician Eudoxus (ca. 365–300 B.C.E.) of the celebrated "method of exhaustion." This method also introduced the modern concept of limit. In the method of exhaustion, a convex figure is approximated by inscribed (or circumscribed) polygons—whose areas can be calculated—and then the number of vertexes of the inscribed polygons is increased until the convex region has been "exhausted." That is, the area of the convex region is computed as the limit of the areas of the inscribed polygons. Archimedes (287–212 B.C.E.) used the method of exhaustion to calculate the area of a circle and the volume of a sphere, as well as the areas and volumes of several other geometrical figures and solids. The method of exhaustion is, in fact, at the heart of all modern integration techniques.

The method of exhaustion, along with most ancient mathematics, was forgotten for almost 2,000 years until the invention of calculus by I. Newton (1642–1727) and G. Leibniz (1646–1716). The theory of integration then developed rapidly. A.-L. Cauchy (1789–1857) and G. F. B. Riemann (1826–1866) were among the first to present axiomatic abstract foundations of integration.

In the modern abstract approach to integration theory, we usually start with a *measure space* (X, Σ, μ) and the associated Riesz space L of all Σ-*step functions*. The Σ-step functions are the analogues of the inscribed (or circumscribed) polygons. If $\varphi = \sum_{i=1}^{n} a_i \chi_{A_i}$ is a Σ-step function, then the integral of φ is defined as a weighted sum of its values, the weights being the measures of the sets on which φ assumes those values. That is,

$$\int \varphi \, d\mu = \sum_{i=1}^{n} a_i \mu(A_i).$$

The integration problem now consists of finding larger classes of functions for which the integral can be defined in such a way that it preserves the fundamental properties of area and volume. This means that on the larger class (vector space) of functions the integral must remain a positive linear functional possessing a continuity property that captures the exhaustion property of Eudoxus. The measure-theoretic approach to integration was developed through the work of H. Lebesgue (1875–1941), C. Carathéodory (1873–1950), and P. J. Daniell (1889–1946). Their ideas and approach are present throughout this chapter.

An even more abstract approach to integration is as a positive operator on a Banach lattice. D. H. Fremlin [128] and K. Jacobs [177] are exemplars of this approach. In Chapter 14 we present typical results along these lines.

11.1   The integral of a step function

In this section, $\mathcal{A}$ is an algebra of subsets of a set X and $\mu\colon \mathcal{A} \to [0, \infty]$ denotes a charge. That is, μ is a nonnegative finitely additive set function defined on $\mathcal{A}$.

11.1 Definition   *A simple function $\varphi\colon X \to \mathbb{R}$ is a $\boldsymbol{\mu}$-**step function** (or simply a* **step function** *when the charge μ is well understood) if its standard representation $\varphi = \sum_{i=1}^{n} a_i \chi_{A_i}$ satisfies $\mu(A_i) < \infty$ for each i.* [1]

*A **representation** for a μ-step function φ is any expression of the form $\varphi = \sum_{j=1}^{m} b_j \chi_{B_j}$, where $B_j \in \mathcal{A}$ and $\mu(B_j) < \infty$ for each j.*

In other words, a simple function is a μ-step function if and only if the function vanishes outside of a set in $\mathcal{A}$ of finite measure. So if L denotes the collection of all μ-step functions, then a repetition of the proof of Lemma 4.34 yields the following.

11.2 Lemma   *The collection L of all μ-step functions is a Riesz space and, in fact, a function space and an algebra.*

Any satisfactory theory of integration has to treat step functions in the obvious way. That is, the integral of a step function should be a weighted sum of its values, the weights being the measures of the sets on which it assumes those values. Precisely, we have the following definition.

11.3 Definition   *Let μ be a charge on an algebra of subsets of a set X, and let $\varphi\colon X \to \mathbb{R}$ be a step function having the standard representation $\varphi = \sum_{i=1}^{n} a_i \chi_{A_i}$. The **integral** of φ (with respect to μ) is defined by*

$$\int \varphi \, d\mu = \sum_{i=1}^{n} a_i \mu(A_i).$$

---

[1] This terminology is useful, but a little bit eccentric. Many authors reserve the term "step function" for a simple function whose domain is a closed interval of the real line and has a representation in terms of indicators of intervals. It is handy though to have a term to indicate a simple function that is nonzero on a set of finite measure.

Thus the integral can be viewed as a real function on the Riesz space L of all μ-step functions. We establish next that, in fact, the integral is a positive linear functional. In order to prove this, we need to show that for any step function φ and for any representation $\varphi = \sum_{j=1}^{m} b_j \chi_{B_j}$ the value of the sum $\sum_{j=1}^{m} b_j \mu(B_j)$ coincides with the integral of φ.

11.4 Lemma   *If $\varphi = \sum_{j=1}^{m} b_j \chi_{B_j}$ is a representation of a step function φ, then*

$$\int \varphi \, d\mu = \sum_{j=1}^{m} b_j \mu(B_j).$$

Proof: Let $\varphi = \sum_{i=1}^{n} a_i \chi_{A_i}$ be the standard representation of φ. Assume first that the B_j are pairwise disjoint. Since neither the function φ nor the sum $\sum_{j=1}^{m} b_j \mu(B_j)$ changes by deleting the terms with $b_j = 0$, we can assume that $b_j \neq 0$ for each j. In such a case, we have $\bigcup_{i=1}^{n} A_i = \bigcup_{j=1}^{m} B_j$. Moreover, $a_i \mu(A_i \cap B_j) = b_j \mu(A_i \cap B_j)$ for all i and j. Indeed, if $A_i \cap B_j = \emptyset$ the equality is obvious and if $x \in A_i \cap B_j$, then $a_i = b_j = \varphi(x)$. Therefore,

$$\int \varphi \, d\mu = \sum_{i=1}^{n} a_i \mu(A_i) = \sum_{i=1}^{n} \sum_{j=1}^{m} a_i \mu(A_i \cap B_j)$$

$$= \sum_{j=1}^{m} \sum_{i=1}^{n} b_j \mu(A_i \cap B_j) = \sum_{j=1}^{m} b_j \mu(B_j).$$

Now consider the general case. By Lemma 4.8, there exist pairwise disjoint sets $C_1, \ldots, C_k \in \mathcal{A}$ such that each $B_j = \bigcup \{C_i : C_i \subset B_j\}$ and each C_i is included in some B_j. For each i and j let $\delta_i^j = 1$ if $C_i \subset B_j$ and $\delta_i^j = 0$ if $C_i \not\subset B_j$. Clearly, $\chi_{B_j} = \sum_{i=1}^{k} \delta_i^j \chi_{C_i}$ and $\mu(B_j) = \sum_{i=1}^{k} \delta_i^j \mu(C_i)$. Consequently,

$$\varphi = \sum_{j=1}^{m} b_j \chi_{B_j} = \sum_{j=1}^{m} b_j \Big[\sum_{i=1}^{k} \delta_i^j \chi_{C_i} \Big] = \sum_{i=1}^{k} \Big[\sum_{j=1}^{m} b_j \delta_i^j \Big] \chi_{C_i}.$$

So by the preceding case, we have

$$\int \varphi \, d\mu = \sum_{i=1}^{k} \Big[\sum_{j=1}^{m} b_j \delta_i^j \Big] \mu(C_i) = \sum_{j=1}^{m} b_j \Big[\sum_{i=1}^{k} \delta_i^j \mu(C_i) \Big] = \sum_{j=1}^{m} b_j \mu(B_j),$$

and the proof is finished. ∎

We are now ready to establish the linearity of the integral.

11.5 Theorem   *If μ is a charge on an algebra of sets, then the integral is a linear functional from the Riesz space L of all μ-step functions into $\mathbb{R}$. That is, for all $\varphi, \psi \in L$ and all $\alpha, \beta \in \mathbb{R}$, we have*

$$\int (\alpha\varphi + \beta\psi)\, d\mu = \alpha \int \varphi\, d\mu + \beta \int \psi\, d\mu.$$

In addition, the integral is a positive linear functional. That is, $\varphi \geq 0$ implies $\int \varphi\, d\mu \geq 0$.

Proof: Let $\varphi, \psi \in L$. Clearly, $\int (\alpha\varphi)\, d\mu = \alpha \int \varphi\, d\mu$ for each $\alpha \in \mathbb{R}$ and $\int \varphi\, d\mu \geq 0$ if $\varphi \geq 0$. For the remainder, if $\varphi = \sum_{i=1}^{n} a_i \chi_{A_i}$ and $\psi = \sum_{j=1}^{m} b_j \chi_{B_j}$, then we have the representation $\varphi + \psi = \sum_{i=1}^{n} a_i \chi_{A_i} + \sum_{j=1}^{m} b_j \chi_{B_j}$. So by Lemma 11.4, we get

$$\int (\varphi + \psi)\, d\mu = \sum_{i=1}^{n} a_i \mu(A_i) + \sum_{j=1}^{m} b_j \mu(B_j) = \int \varphi\, d\mu + \int \psi\, d\mu,$$

and the proof is finished.   ∎

The positivity of the integral can be rephrased in the following equivalent statement: If $\varphi \leq \psi$ in L, then $\int \varphi\, d\mu \leq \int \psi\, d\mu$. This property is also referred to as **monotonicity** of the integral.

For each $\mu \in ba(\mathcal{A})$ both μ^+ and μ^- are finite charges, so every simple function has integrals with respect to both μ^+ and μ^-. For every simple function φ and any $\mu \in ba(\mathcal{A})$, we define the integral $\int \varphi\, d\mu$ by

$$\int \varphi\, d\mu = \int \varphi\, d\mu^+ - \int \varphi\, d\mu^-.$$

The importance of this formula is explained a bit later.

11.2   Finitely additive integration of bounded functions

It is possible to generalize the notion of integral beyond just the step functions to the class of bounded measurable functions. For the remainder of this section fix a finite charge μ on an algebra $\mathcal{A}$ of subsets of a set X. For a bounded function $f: X \to \mathbb{R}$ define the **lower integral** of f by

$$I_*(f) = \sup\left\{ \int \varphi\, d\mu : \varphi \in L \text{ and } \varphi \leq f \right\},$$

and the **upper integral** by

$$I^*(f) = \inf\left\{ \int \psi\, d\mu : \psi \in L \text{ and } f \leq \psi \right\},$$

where L is the vector space of step functions. Clearly, $-\infty < I_*(f) \leq I^*(f) < \infty$.

We say that f is μ-**integrable** if the upper and lower integrals of f are equal. The common value is called the **integral** of f with respect to μ and is denoted $\int f\,d\mu$ [or $\int_X f\,d\mu$, or $\int_X f(x)\,d\mu(x)$, or $\int f(x)\,d\mu(x)$, or even $\int f(x)\,\mu(dx)$]. The next result characterizes integrable functions. Its easy proof is left as an exercise.

11.6 Theorem *For a bounded function $f\colon X \to \mathbb{R}$ and a finite charge μ on an algebra of subsets of X, the following statements are equivalent.*

1. *The function f is integrable.*

2. *For each $\varepsilon > 0$ there exist two step functions φ and ψ satisfying $\varphi \leqslant f \leqslant \psi$ and $\int(\psi - \varphi)d\mu < \varepsilon$.*

3. *There exist sequences $\{\varphi_n\}$ and $\{\psi_n\}$ of step functions satisfying $\varphi_n \uparrow \leqslant f$, $\psi_n \downarrow \geqslant f$, and $\int(\psi_n - \varphi_n)\,d\mu \downarrow 0$.*

The proof of the next result is also left as an exercise.

11.7 Theorem *The collection of all bounded integrable functions with respect to a finite charge is a Riesz space, and in fact, a function space. Moreover, the integral is a $\|\cdot\|_\infty$-continuous positive linear functional on the vector space of bounded integrable functions.*

Let $\mathcal{A}_{\mathbb{R}}$ denote the algebra generated in $\mathbb{R}$ by the collection of all half open intervals $\{[a,b) : a < b\}$. The proof of the next theorem involves a frequently used technique that we call "partitioning the range and taking inverse images."

11.8 Theorem *Every bounded $(\mathcal{A}, \mathcal{A}_{\mathbb{R}})$-measurable function is integrable with respect to any finite charge.*

Proof: Let $f\colon X \to \mathbb{R}$ be a bounded $(\mathcal{A}, \mathcal{A}_{\mathbb{R}})$-measurable function. Fix some $M > 0$ satisfying $-M < f(x) < M$ for each $x \in X$, and let $\varepsilon > 0$. Also, fix a partition $P = \{t_0, t_1, \ldots, t_n\}$ of $[-M, M]$ with mesh

$$|P| = \max\{t_i - t_{i-1} : i = 1, \ldots, n\} < \varepsilon.$$

Next, let $A_i = f^{-1}([t_{i-1}, t_i))$ $(i = 1, \ldots, n)$, and note that $A_i \in \mathcal{A}$ for each i, and $A_i \cap A_j = \varnothing$ whenever $i \neq j$. The two step functions

$$\varphi = \sum_{i=1}^n t_{i-1}\chi_{A_i} \quad \text{and} \quad \psi = \sum_{i=1}^n t_i\chi_{A_i}$$

satisfy $\varphi \leqslant f \leqslant \psi$. Now let μ be a finite charge on $\mathcal{A}$. Then by the finite additivity of μ, we see that

$$\int(\psi - \varphi)d\mu = \sum_{i=1}^n (t_i - t_{i-1})\mu(A_i) \leqslant \sum_{i=1}^n \varepsilon\mu(A_i) = \varepsilon\mu(X).$$

By Theorem 11.6 the function f is μ-integrable. ∎

11.3   The Lebesgue integral

Throughout this section μ denotes a measure on a semiring $\mathcal{S}$ of subsets of a set X. For such (countably additive) measures, we can extend the theory of integration to include many unbounded measurable functions as well.

Recall that a function $f \colon X \to \mathbb{R}$ is **μ-measurable** (or simply **measurable**) if f is $(\Sigma_\mu, \mathcal{B}_\mathbb{R})$-measurable. That is, f is measurable if and only if the inverse images of Borel sets under f are Lebesgue measurable sets. A **μ-step function** (or simply a **step function**) is a function $\varphi \colon X \to \mathbb{R}$ that has a representation $\varphi = \sum_{j=1}^{m} b_j \chi_{B_j}$ with $B_j \in \Sigma_\mu$ and $\mu^*(B_j) < \infty$ for each j. Again, we let $L(\mu)$, or simply L when μ is clear, denote the Riesz space of all μ-step functions.

When μ is countably additive, that is, a measure, the integral on L satisfies an important monotone continuity property. This property is the key insight of the method of exhaustion of Eudoxus.

11.9 Theorem   *Let μ be a (countably additive) measure.*

1. *If two step functions φ and ψ are equal almost everywhere, then their integrals coincide. That is, $\int \varphi \, d\mu = \int \psi \, d\mu$.*

2. *If $\{\varphi_n\}$ is a sequence of step functions with $\varphi_n \downarrow 0$ a.e., then $\int \varphi_n \, d\mu \downarrow 0$. Similarly, for step functions, $\varphi_n \uparrow \varphi$ a.e. implies $\int \varphi_n \, d\mu \uparrow \int \varphi \, d\mu$, and $\varphi_n \downarrow \varphi$ a.e. implies $\int \varphi_n \, d\mu \downarrow \int \varphi \, d\mu$.*

Proof: We establish only the second part and leave the first part as an exercise. So let $\varphi_n \downarrow 0$ a.e. Letting $A_n = \{x \in X : \varphi_{n+1}(x) > \varphi_n(x)\}$, $A_0 = \{x \in X : \varphi_n(x) \not\to 0\}$ and $A = \bigcup_{n=0}^{\infty} A_n$, we see that $\mu^*(A) = 0$. So if $\psi_n = \varphi_n \chi_{A^c}$, then $\psi_n = \varphi_n$ a.e. and $\psi_n(x) \downarrow 0$ for each x. In other words, without loss of generality we can assume from the outset that $\varphi_n(x) \downarrow 0$ for each x. Put $M = \max\{\varphi_1(x) : x \in X\}$ and $B = \{x \in X : \varphi_1(x) > 0\}$.

Now fix $\varepsilon > 0$ and for each n let $B_n = \{x \in X : \varphi_n(x) \geqslant \varepsilon\}$. From $\varphi_n(x) \downarrow 0$ for each x, we see that $B_n \downarrow \varnothing$. Consequently (by Theorem 10.8) $\mu^*(B_n) \downarrow 0$. Next note that $0 \leqslant \varphi_n \leqslant M\chi_{B_n} + \varepsilon\chi_B$ for each n. Therefore, by the monotonicity of the integral on L, we get

$$0 \leqslant \int \varphi_n \, d\mu \leqslant M \int \chi_{B_n} \, d\mu + \varepsilon \int \chi_B \, d\mu = M\mu^*(B_n) + \varepsilon\mu^*(B).$$

Thus $0 \leqslant \limsup_{n\to\infty} \int \varphi_n \, d\mu \leqslant \varepsilon\mu^*(B)$ for each $\varepsilon > 0$, which implies that $\int \varphi_n \, d\mu \downarrow 0$.   ∎

We continue with the introduction of upper functions.

11.10 Definition   *A function $f \colon X \to \mathbb{R}$ is a **μ-upper function** (or simply an **upper function**) if there exists a sequence $\{\varphi_n\}$ of step functions such that:*

1. $\varphi_n \uparrow f$ μ-a.e.; and

2. $\sup_n \int \varphi_n \, d\mu < \infty$.

Note that if $\{\psi_n\}$ is another sequence of step functions satisfying $\psi_n \uparrow f$ a.e., then for each fixed k, we have $\psi_n \wedge \varphi_k \uparrow_n f \wedge \varphi_k = \varphi_k$ a.e., so by Theorem 11.9 (2), we see that

$$\int \varphi_k \, d\mu = \lim_{n \to \infty} \int \psi_n \wedge \varphi_k \, d\mu \leqslant \lim_{n \to \infty} \int \psi_n \, d\mu$$

for each k. Hence, $\lim_{k \to \infty} \int \varphi_k \, d\mu \leqslant \lim_{n \to \infty} \int \psi_n \, d\mu$. By the symmetry of the situation, $\lim_{k \to \infty} \int \varphi_k \, d\mu = \lim_{n \to \infty} \int \psi_n \, d\mu$. In other words, the value of the limit $\lim_{n \to \infty} \int \varphi_n \, d\mu$ is independent of the sequence $\{\varphi_n\}$. This value is called the **Lebesgue integral** of f and is denoted $\int f \, d\mu$. That is,

$$\int f \, d\mu = \lim_{n \to \infty} \int \varphi_n \, d\mu.$$

Here are some of the basic properties of upper functions.

11.11 Lemma *Upper functions enjoy the following properties.*

1. *Every upper function is μ-measurable.*

2. *Every step function φ is an upper function and its Lebesgue integral coincides with $\int \varphi \, d\mu$.*

3. *If f is an upper function and if $g = f$ a.e., then g is also an upper function and $\int g \, d\mu = \int f \, d\mu$.*

4. *If f and g are upper functions, then so are $f + g$, $f \wedge g$, $f \vee g$, and αf for each $\alpha \geqslant 0$. Moreover, $\int (f + g) \, d\mu = \int f \, d\mu + \int g \, d\mu$.*

5. *If f and g are upper functions and $f \leqslant g$ a.e., then $\int f \, d\mu \leqslant \int g \, d\mu$.*

Proof: We prove (4) and (5) only. Fix sequences $\{\varphi_n\}$ and $\{\psi_n\}$ of step functions satisfying $\varphi_n \uparrow f$ and $\psi_n \uparrow g$ a.e. Now $\varphi_n + \psi_n \uparrow f + g$ a.e. and it is easy to see that $\int (\varphi_n + \psi_n) \, d\mu \uparrow \int f \, d\mu + \int g \, d\mu$, so $f + g$ is an upper function and that $\int (f + g) \, d\mu = \int f \, d\mu + \int g \, d\mu$. To see that $f \wedge g$ is an upper function, note that $\varphi_n \wedge \psi_n \uparrow f \wedge g$ a.e. and that $\lim_{n \to \infty} \int \varphi_n \wedge \psi_n \, d\mu \leqslant \lim_{n \to \infty} \int \varphi_n \, d\mu = \int f \, d\mu < \infty$. For $f \vee g$ note that $\varphi_n \vee \psi_n \uparrow f \vee g$ and that

$$\int \varphi_n \vee \psi_n \, d\mu =$$

$$\int (\varphi_n + \psi_n - \varphi_n \wedge \psi_n) \, d\mu \uparrow \int f \, d\mu + \int g \, d\mu - \int f \wedge g \, d\mu < \infty.$$

If $f \leqslant g$ a.e., then $\varphi_n \wedge \psi_n \uparrow f$ and $\varphi_n \wedge \psi_n \leqslant \psi_n$ for each n. It follows that $\int f \, d\mu = \lim_{n \to \infty} \int \varphi_n \wedge \psi_n \, d\mu \leqslant \lim_{n \to \infty} \int \psi_n \, d\mu = \int g \, d\mu$. ∎

Now we are ready to define Lebesgue integrability for general functions.

11.12 Definition   *A function $f: X \to \mathbb{R}$ is **Lebesgue integrable** if there exist two μ-upper functions $u, v: X \to \mathbb{R}$ such that $f = u - v$ a.e. The **Lebesgue integral** of f is defined by*

$$\int f \, d\mu = \int u \, d\mu - \int v \, d\mu.$$

We also use the symbols: $\int_X f \, d\mu$, $\int_X f(x) \, d\mu(x)$, $\int f(x) \, d\mu(x)$, and $\int f(x) \, \mu(dx)$. [2]

Note well that under our definition, if a function is Lebesgue integrable, then its integral is a (finite) real number—the extended numbers $\pm\infty$ are excluded as permissible values of the integral. However, in Section 11.5 below, we loosen this restriction.

It is easy to see that the value of the Lebesgue integral of a function f does not depend on the particular upper functions chosen. Indeed, if $f = u_1 - v_1 = u_2 - v_2$ a.e. for upper functions u_i and v_i, then $u_1 + v_2 = u_2 + v_1$ a.e., and is an upper function. So from

$$\int u_1 \, d\mu + \int v_2 \, d\mu = \int (u_1 + v_2) \, d\mu = \int (u_2 + v_1) \, d\mu = \int u_2 \, d\mu + \int v_1 \, d\mu,$$

it follows that $\int u_1 \, d\mu - \int v_1 \, d\mu = \int u_2 \, d\mu - \int v_2 \, d\mu$. Clearly:

• Every Lebesgue integrable function is measurable, and

• If a function f is equal almost everywhere to an integrable function g, then f is Lebesgue integrable and $\int f \, d\mu = \int g \, d\mu$.

The next theorem asserts that the Lebesgue integral is linear and monotone. Its simple direct proof is left as an exercise.

11.13 Theorem   *The Lebesgue integrable functions form a function space, and the Lebesgue integral is a positive linear functional on this function space. That is, if f and g are Lebesgue integrable functions, then the integral is:*

1. ***additive**, that is, $\int (f + g) \, d\mu = \int f \, d\mu + \int g \, d\mu$;*

2. ***homogeneous**, that is, $\int \alpha f \, d\mu = \alpha \int f \, d\mu$ for each $\alpha \in \mathbb{R}$; and*

3. ***monotone**, that is, $f \leqslant g$ a.e. implies $\int f \, d\mu \leqslant \int g \, d\mu$.*

---

[2] Unfortunately the term "Lebesgue integral" is also used to mean the (Lebesgue) integral of a function on the real line with respect to Lebesgue measure. Some authors, e.g., [148, p. 106], mean only that. It would be less ambiguous to call our Lebesgue integral an "abstract Lebesgue integral," but we stick with our terminology.

The Lebesgue integral in its general form was introduced by H. Lebesgue [225]. The present formulation of the Lebesgue integral is essentially due to P. J. Daniell [80].

The next result shows that all positive integrable functions are upper functions.

11.14 Lemma   *Any almost everywhere nonnegative integrable function is an upper function.*

Proof: Let $f: X \to \mathbb{R}$ be a nonnegative-a.e. integrable function. We can assume that $f(x) \geqslant 0$ for each $x \in X$. By Theorem 4.36, there exists a sequence $\{\varphi_n\}$ of simple functions satisfying $0 \leqslant \varphi_n(x) \uparrow f(x)$ for each x.

Next, pick two upper functions u and v such that $f = u - v$ a.e. From $f \geqslant 0$, we see that $u \geqslant v$ a.e. Since both u and v are the pointwise almost everywhere limits of sequences of step functions, there exists a sequence $\{\theta_n\}$ of step functions satisfying $\theta_n \to f$ a.e. Replacing θ_n by θ_n^+, we can assume $\theta_n \geqslant 0$ for each n. Now put $\psi_n = \varphi_n \wedge (\bigvee_{i=1}^n \theta_i)$. Then each ψ_n is a step function and $\psi_n \uparrow f$ a.e. Moreover from $\psi_n \leqslant f$, we see that $\int \psi_n \, d\mu \leqslant \int f \, d\mu < \infty$. Therefore f is an upper function. ∎

Here is a characterization of the Lebesgue integral that is used quite often as an alternate definition of the Lebesgue integral.

11.15 Theorem   *A function $f: X \to \mathbb{R}$ is Lebesgue integrable if and only if f^+ and f^- are both upper functions. Moreover, in this case we have*

$$\int f \, d\mu = \int f^+ \, d\mu - \int f^- \, d\mu. \quad [3]$$

Proof: If f^+ and f^- are both upper functions, then the function $f = f^+ - f^-$ is Lebesgue integrable, and $\int f \, d\mu = \int f^+ \, d\mu - \int f^- \, d\mu$. For the converse, assume that f is Lebesgue integrable. Then there exist two upper functions u and v such that $f = u - v$ a.e. Clearly $f^+ = u \vee v - v$ and $f^- = u \vee v - u$ a.e. Since $u \vee v$ is an upper function, we see that f^+ and f^- are both Lebesgue integrable functions. Since f^+ and f^- are also positive functions, Lemma 11.14 guarantees that both are also upper functions. ∎

The next result lists more properties of the Lebesgue integral.

11.16 Theorem   *Let $f, g, h: X \to \mathbb{R}$ be measurable. Then we have the following:*

---

[3] Many authors define the Lebesgue integral using this formula as follows. First, define the Lebesgue integral of a nonnegative measurable function f by

$$\int f \, d\mu = \sup\Bigl\{ \int \varphi \, d\mu : \varphi \text{ is a step function such that } \varphi \leqslant f \text{ a.e.} \Bigr\},$$

provided that the supremum is finite. Subsequently, say that a measurable function is Lebesgue integrable if f^+ and f^- are both Lebesgue integrable, in which case define $\int f \, d\mu = \int f^+ \, d\mu - \int f^- \, d\mu$. We leave it as an exercise to verify that this definition of the Lebesgue integral coincides with ours in the σ-finite case.

1. *If f and g are Lebesgue integrable, and $f \leqslant h \leqslant g$ a.e., then h is also Lebesgue integrable.*

2. *f is Lebesgue integrable if and only if $|f|$ is Lebesgue integrable.*

3. *If f is Lebesgue integrable and $f \geqslant 0$ a.e., then $\int f \, d\mu = 0$ if and only if $f = 0$ a.e.*

Proof: (1)   Notice that $f \leqslant h \leqslant g$ a.e. is equivalent to $0 \leqslant h - f \leqslant g - f$ a.e. This means that we can suppose $f = 0$ a.e. Pick a sequence $\{\theta_n\}$ of step functions satisfying $0 \leqslant \theta_n \uparrow g$ a.e. Also, by Theorem 4.36, there exists a sequence $\{\psi_n\}$ of simple functions satisfying $0 \leqslant \psi_n \uparrow h$ a.e. Now if $\varphi_n = \psi_n \wedge \theta_n$, then each φ_n is a step function, $\int \varphi_n \, d\mu \leqslant \int \theta_n \, d\mu \leqslant \int g \, d\mu < \infty$ for each n and $\varphi_n \uparrow h$ a.e. Thus, h is an upper function (and hence Lebesgue integrable).

(2)   If f is Lebesgue integrable, then (by Theorem 11.15) both f^+ and f^- are upper functions, so $|f| = f^+ + f^-$ is Lebesgue integrable. On the other hand, if $|f|$ is Lebesgue integrable, then from $-|f| \leqslant f \leqslant |f|$ and the first part, we see that f is also Lebesgue integrable.

(3)   Let a function $f \geqslant 0$ a.e. be Lebesgue integrable. By Lemma 11.14, f is an upper function. So there exists a sequence $\{\varphi_n\}$ of step functions such that $0 \leqslant \varphi_n \uparrow f$ a.e. If $f = 0$ a.e., then clearly $\int f \, d\mu = 0$. On the other hand, if $\int f \, d\mu = 0$, then $\int \varphi_n \, d\mu = 0$ for each n, or $\varphi_n = 0$ a.e. for each n, which implies $f = 0$ a.e. ∎

We continue with one more property of the Lebesgue integral.

11.17 Lemma   *If a sequence $\{\varphi_n\}$ of step functions satisfies $\varphi_n \uparrow \chi_A$ a.e., then $\mu^*(A) = \lim_{n\to\infty} \int \varphi_n \, d\mu$. In particular, if f is an integrable function, then every measurable set A satisfying $\chi_A \leqslant f$ a.e has finite measure.*

Proof: We can suppose that $0 \leqslant \varphi_n(x) \uparrow \chi_A(x)$ for each x. (Why?). If we let $A_n = \{x \in X : \varphi_n(x) > 0\}$, then each A_n is measurable, $\mu^*(A_n) < \infty$ for each n, and $A_n \uparrow A$. By Theorem 10.8, we know that $\mu^*(A_n) \uparrow \mu^*(A)$. Now noting that $\{\chi_{A_n}\}$ is a sequence of step functions satisfying $\chi_{A_n} \uparrow \chi_A$, it follows that

$$\mu^*(A) = \lim_{n\to\infty} \mu^*(A_n) = \lim_{n\to\infty} \int \chi_{A_n} \, d\mu = \lim_{n\to\infty} \int \varphi_n \, d\mu,$$

as claimed. ∎

We say that a function f is Lebesgue **integrable over** a measurable set A if $f\chi_A$ is Lebesgue integrable (over X). In this case we write $\int_A f \, d\mu$ for $\int f\chi_A \, d\mu$. Consequently, by the linearity of the integral, if A and B are disjoint measurable sets and f is integrable over $A \cup B$, then $\int_{A \cup B} f \, d\mu = \int_A f \, d\mu + \int_B f \, d\mu$.

11.4 Continuity properties of the Lebesgue integral

Unless otherwise stated, μ again denotes a measure on a semiring of subsets of a set X. Our purpose here is to discuss the most prominent theorems in the theory of the Lebesgue integral. In particular, we state and prove the major theorems describing when we may interchange the order of taking limits and integration. These amount to continuity properties of the integral.

Start by observing that by altering the values of a function on a set of μ-measure zero changes neither the μ-measurability of the function nor the value of its integral. This allows us to take liberties in defining a function. Specifically, we can allow a function to take on the values $+\infty$ and $-\infty$ or even be left undefined on a set of μ-measure zero. By saying that a function f "defines a μ-integrable function," we mean that the set of points where the function f assumes infinite values (or is even left undefined) has μ-measure zero. By assigning real values on this null set (for instance, we can assign the value zero at each point of this set) f becomes an integrable function.

The first theorem is a **Monotone Convergence Theorem**.

11.18 Levi's Theorem   *Assume that $\{f_n\}$ is a sequence of integrable functions such that $f_n \uparrow$ a.e. If $\lim_{n \to \infty} \int f_n \, d\mu < \infty$, then there exists a Lebesgue integrable function f such that $f_n(x) \uparrow f(x)$ for almost all x and $\int f \, d\mu = \lim_{n \to \infty} \int f_n \, d\mu$.*

Proof: Let $\{f_n\}$ be a sequence of integrable functions satisfying the properties stated in the theorem. Replacing each f_n by $f_n - f_1$, we can assume without loss of generality that $0 \leqslant f_n(x) \uparrow$ in $\mathbb{R}$ for each x. (Why?) Put $I = \lim_{n \to \infty} \int f_n \, d\mu < \infty$.

Let $f_n(x) \uparrow g(x)$ in $\mathbb{R}^*$ for each x, and let $A = \{x \in X : g(x) = \infty\}$. From $A = \bigcap_{m=1}^{\infty} \bigcup_{k=m}^{\infty} \{x \in X : f_k(x) \geqslant m\}$, we see that A is a measurable set. We now show that $\mu^*(A) = 0$.

By Lemma 11.14, each f_n is an upper function. So for each k there exists a sequence $\{\varphi_n^k\}$ of step functions satisfying $0 \leqslant \varphi_n^k(x) \uparrow_n f_k(x)$ for each x. For each n, let $\psi_n = \bigvee_{k=1}^n \varphi_n^k$. Then $\{\psi_n\}$ is a sequence of step functions satisfying $\psi_n(x) \uparrow g(x)$ for each x and $\int \psi_n \, d\mu \uparrow I$. (Why?) Now notice that for each fixed k, we have $\psi_n \wedge k\chi_A \uparrow_n k\chi_A$. A glance at Lemma 11.17 yields

$$k\mu^*(A) = \lim_{n \to \infty} \int \psi_n \wedge k\chi_A \, d\mu \leqslant \lim_{n \to \infty} \int \psi_n \, d\mu = I < \infty$$

for each k. This implies $\mu^*(A) = 0$. Now if $f = g\chi_{A^c}$, then $\psi_n \uparrow f$ a.e. and $f_n \uparrow f$ a.e., and moreover

$$\int f \, d\mu = \lim_{n \to \infty} \int \psi_n \, d\mu = \lim_{n \to \infty} \int f_n \, d\mu,$$

as desired.   ∎

You should verify that Levi's theorem can also be stated in the following equivalent "series form."

11.19 Theorem (Levi) *If $\{f_n\}$ is a sequence of nonnegative Lebesgue integrable functions such that $\sum_{n=1}^{\infty} \int f_n \, d\mu < \infty$, then the series $\sum_{n=1}^{\infty} f_n$ defines an integrable function and*

$$\int \Big[\sum_{n=1}^{\infty} f_n\Big] d\mu = \sum_{n=1}^{\infty} \int f_n \, d\mu.$$

The next result is well-known as Fatou's Lemma.

11.20 Fatou's Lemma   *If a sequence $\{f_n\}$ of Lebesgue integrable functions is bounded from below by an integrable function g (that is, $g \leqslant f_n$ a.e. for each n) and $\liminf_{n\to\infty} \int f_n \, d\mu < \infty$, then $\liminf_{n\to\infty} f_n$ defines an integrable function and*

$$\int \liminf_{n\to\infty} f_n \, d\mu \leqslant \liminf_{n\to\infty} \int f_n \, d\mu.$$

Proof: Let g be an integrable function, and $\{f_n\}$ be a sequence of integrable functions satisfying $g \leqslant f_n$ a.e. for each n and $\liminf_{n\to\infty} \int f_n \, d\mu < \infty$. Without loss of generality, we can assume that $g = 0$ and that $f_n(x) \geqslant 0$ for all n and each x. (Why?)

For each n, let $g_n(x) = \inf\{f_i(x) : i \geqslant n\}$. From $\bigwedge_{i=n}^{n+k} f_i(x) \downarrow_k g_n(x)$, we see that each g_n is a measurable function. Moreover, from $0 \leqslant g_n \leqslant f_n$, we see that each g_n is an integrable function—in fact an upper function. From

$$\liminf_{n\to\infty} f_n = \bigvee_{n=1}^{\infty} \bigwedge_{k=n}^{\infty} f_k = \bigvee_{n=1}^{\infty} g_n,$$

we see that $g_n \uparrow \liminf_{n\to\infty} f_n$. In addition, $g_n \leqslant f_n$ implies

$$\lim_{n\to\infty} \int g_n \, d\mu \leqslant \liminf_{n\to\infty} \int f_n \, d\mu < \infty.$$

Now a glance at Levi's Theorem 11.18, shows that $\liminf_{n\to\infty} f_n$ defines an integrable function that satisfies

$$\int \liminf_{n\to\infty} f_n \, d\mu = \lim_{n\to\infty} \int g_n \, d\mu \leqslant \liminf_{n\to\infty} \int f_n \, d\mu,$$

and the proof is finished. ∎

And now we come to the theorem known as the **Lebesgue Dominated Convergence Theorem**. It allows us to interchange limits and integrals and has been called "the cornerstone of the theory of integration." In the terminology of Chapter 8, it says that the Lebesgue integral is a σ-order continuous linear functional on $L_1(\mu)$. (The Banach lattice $L_1(\mu)$ is discussed in Chapter 13.)

11.21 Dominated Convergence Theorem *Assume that a sequence $\{f_n\}$ of Lebesgue integrable functions satisfies $f_n(x) \to f(x)$ for almost all x, and that $\{f_n\}$ is dominated a.e. by an integrable function g. That is, $|f_n| \leqslant g$ a.e. for each n. Then f is Lebesgue integrable and*

$$\int f \, d\mu = \lim_{n \to \infty} \int f_n \, d\mu.$$

Proof: Assume that the sequence $\{f_n\}$ of integrable functions satisfies the properties stated in the theorem. Clearly, f is measurable and $|f| \leqslant g$ a.e. By Theorem 11.16(1), f defines an integrable function.

Since $-g \leqslant f_n$ a.e. for each n, it follows from Fatou's Lemma 11.20 that $\int f \, d\mu = \int \lim_{n \to \infty} f_n \, d\mu \leqslant \liminf_{n \to \infty} \int f_n \, d\mu$. Similarly, $-g \leqslant -f_n$ a.e. for each n implies

$$-\int f \, d\mu = \int \lim_{n \to \infty} (-f_n) \, d\mu \leqslant \liminf_{n \to \infty} \left[-\int f_n \, d\mu \right] = -\limsup_{n \to \infty} \int f_n \, d\mu,$$

or $\limsup_{n \to \infty} \int f_n \, d\mu \leqslant \int f \, d\mu$. So

$$\limsup_{n \to \infty} \int f_n \, d\mu = \liminf_{n \to \infty} \int f_n \, d\mu = \int f \, d\mu,$$

which shows that indeed $\lim_{n \to \infty} f_n \, d\mu = \int f \, d\mu$. ∎

We take this opportunity to point out that these limit theorems apply only to sequences, not nets. For example, let A be a non-measurable subset of $[0, 1]$ (cf. Corollary 10.42). Let $\mathcal{F}$ be the set of finite subsets of A, directed upward by inclusion. For each $F \in \mathcal{F}$, the indicator function χ_F is (Borel) measurable and $\int \chi_A \, d\lambda = 0$, where λ is Lebesgue measure. Now the net $\{\chi_F : F \in \mathcal{F}\}$ increases upward to χ_A, which is not Lebesgue measurable. Thus none of Levi's Theorem 11.18, Fatou's Lemma 11.20, or the Dominated Convergence Theorem 11.21 can be extended to nets.

When the semiring is a σ-algebra, we can identify the Lebesgue integrable functions with Lebesgue integrable functions that are also measurable with respect to the σ-algebra.

11.22 Lemma *Let μ be a measure on a σ-algebra Σ and let $f : X \to \mathbb{R}$ be a Lebesgue integrable function. Then there is a Σ-measurable function $g : X \to \mathbb{R}$ such that $f = g$ a.e. (and hence g is also Lebesgue integrable).*

Proof: Put $A_n = \{x \in X : |f(x)| \geqslant \frac{1}{n}\}$. By Lemma 11.17, we know that $A_n \in \Sigma_\mu$ and $\mu^*(A_n) < \infty$. For each n, pick (by using Lemma 10.30) some $B_n \in \Sigma$ with $A_n \subset B_n$ and $\mu(B_n) = \mu^*(A_n)$, and put $B = \bigcup_{n=1}^{\infty} B_n \in \Sigma$. Then $f = f\chi_B$ μ-a.e., so f can be considered defined on B alone, which is a σ-finite measure space. The conclusion now follows immediately from Theorem 10.35. ∎

Finally, recall that Theorems 10.49 and 10.50 identify nondecreasing left continuous functions on $\mathbb{R}$ and regular Borel measures on $\mathbb{R}$. Given such a function f with corresponding measure μ_f, we define the **Lebesgue–Stieltjes integral** $\int g\,df$ to be $\int g\,d\mu_f$. That is, a Lebesgue–Stieltjes integral is a Lebesgue integral in disguise, and everything that applies to Lebesgue integrals applies to Lebesgue–Stieltjes integrals.

11.5   The extended Lebesgue integral

In this section μ is once more a measure on a semiring of subsets of a set X. As agreed before, a $(\Sigma_\mu, \mathcal{B}_{\mathbb{R}})$-measurable function $f\colon X \to \mathbb{R}$ is referred to simply as a **measurable function**. We know that every integrable function is measurable. On the other hand, not every measurable function is integrable; for example, the constant function one on $\mathbb{R}$ is measurable, but fails to be Lebesgue integrable.

Our purpose here is to extend the Lebesgue integral to additional functions in a reasonable manner. Consider first a nonnegative measurable function $f\colon X \to \mathbb{R}$. If f is not integrable in the sense of the preceding sections, then define the integral of f to be $+\infty$ and write $\int_X f\,d\mu = \infty$. With this convention, every nonnegative measurable function has an integral.

Now let $f\colon X \to \mathbb{R}$ be an arbitrary measurable function. If either f^+ or f^- is integrable, then define the **integral** of f to be the extended real number

$$\int_X f\,d\mu = \int_X f^+\,d\mu - \int_X f^-\,d\mu.$$

In this sense, we say that the integral of f exists—as an extended real number. We have thus assigned an integral to more functions; for instance, $\int_{\mathbb{R}} \mathbf{1}\,d\lambda = \infty$.

The next result characterizes functions that have an integral in the generalized sense.

11.23 Theorem   *The integral of a measurable function $f\colon X \to \mathbb{R}$ exists if and only if f dominates or is being dominated by an integrable function, that is, if and only if there exists an integrable function $g\colon X \to \mathbb{R}$ such that either $g \leqslant f$ a.e. or $f \leqslant g$ a.e.*

Proof: If the integral of f exists, then the inequality $-f^- \leqslant f^+ - f^- = f \leqslant f^+$ shows that either $g = -f^-$ or $g = f^+$ satisfies the desired property.

For the converse, assume that $g \leqslant f$ a.e. for some integrable function g. From the inequality $0 \leqslant f^- = (-f) \vee 0 \leqslant (-g) \vee 0 = g^-$ and Theorem 11.16 (1), it follows that f^- is integrable, so the integrable of f exists. A similar argument can be applied when $f \leqslant g$ a.e.   ∎

This conventional extension of the integral simplifies the statements of several important theorems. For instance, the Hölder and Minkowski Inequalities are true for arbitrary measurable functions (see Section 13.2) and Fatou's Lemma can be stated as follows.

- If a sequence of measurable functions $\{f_n\}$ dominates a fixed integrable function, then $\int \liminf_{n\to\infty} f_n \, d\mu \leqslant \liminf_{n\to\infty} \int f_n \, d\mu$.

Another indication of the usefulness of this convention is provided by the next result.

11.24 Jensen's Inequality   *Assume that μ is a probability measure (that is, $\mu^*(X) = 1$) and that $\varphi \colon \mathbb{R} \to \mathbb{R}$ is a convex function. Then for each Lebesgue integrable function $f \colon X \to \mathbb{R}$, we have*

$$\int_X (\varphi \circ f) \, d\mu \geqslant \varphi\Big(\int_X f \, d\mu \Big).$$

Proof: Note that since φ is continuous, the composition $\varphi \circ f$ is measurable (Lemma 4.22). Let $a = \int_X f(x) \, d\mu(x)$. By Theorem 7.23 there exists some m such that $\varphi(t) \geqslant \varphi(a) + m(t-a)$ for each $t \in \mathbb{R}$. Consequently $\varphi(f(x)) \geqslant \varphi(a) + m[f(x) - a]$ for each x. So by Theorem 11.23, the integral of $\varphi \circ f$ exists and satisfies

$$\int_X (\varphi \circ f) \, d\mu \geqslant \int_X [m(f(x) - a) + \varphi(a)] \, d\mu(x) = \varphi(a) = \varphi\Big(\int_X f \, d\mu \Big),$$

as claimed.   ∎

Given that any nonnegative measurable function has now an extended integral, we can present another characterization of the Lebesgue integral.

11.25 Theorem   *A measurable function $f \colon X \to \mathbb{R}$ on a measure space (X, Σ, μ) is Lebesgue integrable if and only if there exists a sequence $\{\varphi_n\}$ of Σ-step functions satisfying*

$$\int |f - \varphi_n| \, d\mu \to 0.$$

Moreover, in this case, $\lim_{n\to\infty} \int_A \varphi_n \, d\mu = \int_A f \, d\mu$ for each $A \in \Sigma$.

Proof: If $\int |f - \varphi_n| \, d\mu \to 0$, then $\int |f - \varphi_n| \, d\mu < \infty$ for some n, so from the identity $|f| \leqslant |f - \varphi_n| + |\varphi_n|$ and Theorem 11.16 (1), we see that f is Lebesgue integrable.

For the converse, assume that f is Lebesgue integrable, that is, both f^+ and f^- are Lebesgue integrable. By Theorem 4.36, there exist two sequences $\{\theta_n\}$ and $\{\psi_n\}$ of Σ-step functions satisfying $0 \leqslant \theta_n(x) \uparrow f^+(x)$ and $0 \leqslant \psi_n(x) \uparrow f^-(x)$ for each $x \in X$. Put $\varphi_n = \theta_n - \psi_n$ and note that each φ_n is a Σ-step function. From

$$|\varphi_n(x)| = |\theta_n(x) - \psi_n(x)| \leqslant |\theta_n(x)| + |\psi_n(x)| \leqslant f^+(x) + f^-(x) = |f(x)|,$$

we see that $|f(x) - \varphi_n(x)| \leqslant 2|f(x)|$ for each $x \in X$. Consequently, we have $|f(x) - \varphi_n(x)| \to 0$ for each $x \in X$, which (in view of the Dominated Convergence Theorem 11.21) implies $\int |f - \varphi_n| \, d\mu \to 0$. The last part is left as an exercise.   ∎

As we discuss later, the conditions of the preceding theorem are just those needed to define the Bochner integral for Banach space-valued functions; see Definition 11.42.

11.6 Iterated integrals

We present here a brief discussion of iterated integrals. For proofs and details see [13, Section 26]. Let $\mu\colon S_1 \to [0, \infty]$ and $v\colon S_2 \to [0, \infty]$ be two measures, where S_1 and S_2 are semirings of subsets of the sets X and Y, respectively. Recall that the set function $\mu \times v\colon S_1 \times S_2 \to [0, \infty]$ defined by

$$(\mu \times v)(A \times B) = \mu(A)v(B),$$

(where the convention $0 \cdot \infty = 0$ applies), turns out to be a measure on the product semiring $S_1 \times S_2$; see Theorem 10.45. Also, by Theorem 10.47 we know that $\Sigma_\mu \times \Sigma_v \subset \Sigma_{\mu \times v}$.

11.26 Definition   *For a function $f\colon X \times Y \to \mathbb{R}$, the **iterated integral** $\int\int f\,dvd\mu$ exists if*

1. *the function $f(x, \cdot)$ is v-integrable for μ-almost all x; and*

2. *the function $x \mapsto \int_Y f(x, y)\,dv(y)$ defines a μ-integrable function.*

In this case, the value of $\int\int f\,dvd\mu$ is defined by

$$\int\int f\,dvd\mu = \int_X \left[\int_Y f(x, y)\,dv(y) \right] d\mu(x).$$

The meaning of the iterated integral $\int\int f\,d\mu dv$ is analogous.

The next two theorems provide a practical way of deciding when an iterated integral exists and when we can interchange the order of integration. The first theorem, Fubini's Theorem, applies to products of arbitrary measures and requires the function be known to be integrable with respect to the product measure. The second theorem, Tonelli's Theorem, only requires that the function be measurable, but instead requires that the measures be σ-finite.

11.27 Fubini's Theorem   *Assume that $\mu\colon S_1 \to [0, \infty]$ and $v\colon S_2 \to [0, \infty]$ are measures on two semirings of subsets of the sets X and Y, respectively. If $f\colon X \times Y \to \mathbb{R}$ is a $\mu \times v$-integrable function, then both iterated integrals $\int\int f\,dvd\mu$ and $\int\int f\,d\mu dv$ exist and*

$$\int f\,d(\mu \times v) = \int\int f\,dvd\mu = \int\int f\,d\mu dv.$$

11.28 Tonelli's Theorem   *Assume that $\mu: \mathcal{S}_1 \to [0, \infty]$ and $\nu: \mathcal{S}_2 \to [0, \infty]$ are two σ-finite measures on two semirings of subsets of the sets X and Y, respectively. If $f: X \times Y \to \mathbb{R}$ is a $\mu \times \nu$-measurable function such that one of the iterated integrals $\int\int |f| \, d\nu d\mu$ and $\int\int |f| \, d\mu d\nu$ exists, then f is $\mu \times \nu$-integrable—and hence the other iterated integral also exists and*

$$\int f \, d(\mu \times \nu) = \int\int f \, d\nu d\mu = \int\int f \, d\mu d\nu.$$

We note that the last two theorems can be generalized by induction to products of any finite number of factors.

11.7   The Riemann integral

In this section, we briefly sketch the definition and basic properties of the Riemann integral of elementary calculus fame. This integral is important because when it exists, it agrees with the Lebesgue integral with respect to the familiar Lebesgue measure. Furthermore, the Fundamental Theorem of Calculus 11.33 provides a tool for computing the Riemann integral. The drawbacks of the Riemann integral are that not nearly as many functions are Riemann integrable as are Lebesgue integrable, and that it is inextricably tied to Lebesgue measure (Euclidean length, area, or volume). For simplicity, we discuss its one dimensional version and leave the details of the n-dimensional version as an exercise. Details and proofs can be found in [13, Section 23].

Throughout the section, $f: [a, b] \to \mathbb{R}$ will be a bounded function on a finite closed interval of $\mathbb{R}$. A **partition** P of $[a, b]$ is a set of points $P = \{x_0, x_1, \ldots, x_n\}$ satisfying $a = x_0 < x_1 < \cdots < x_n = b$. The **mesh** $|P|$ of P is the length of the longest of the n subintervals $[x_{i-1}, x_i]$ $(i = 1, \ldots, n)$. A partition P is **finer** than another partition Q if $Q \subset P$. In the n-dimensional case the analogue of a **closed interval** is any set of the form $I = [a_1, b_1] \times \cdots \times [a_n, b_n]$. This set coincides, of course, with the box $[\boldsymbol{a}, \boldsymbol{b}]$, where $\boldsymbol{a} = (a_1, \ldots, a_n)$ and $\boldsymbol{b} = (b_1, \ldots, b_n)$. A partition of I is any subset of the form $P_1 \times \cdots \times P_n$, where P_i is a partition for the closed interval $[a_i, b_i]$.

Given a partition $P = \{x_0, x_1, \ldots, x_n\}$, we let

$$m_i = \inf\{f(x) : x \in [x_{i-1}, x_i]\} \quad \text{and} \quad M_i = \sup\{f(x) : x \in [x_{i-1}, x_i]\}.$$

The **lower sum** of f relative to the partition P is defined by

$$\mathcal{L}_f(P) = \sum_{i=1}^{n} m_i(x_i - x_{i-1}),$$

and similarly the **upper sum** by

$$\mathcal{U}_f(P) = \sum_{i=1}^{n} M_i(x_i - x_{i-1}).$$

Clearly, $\mathcal{L}_f(P) \leqslant \mathcal{U}_f(P)$. Also, if P is a partition finer than Q, then $\mathcal{L}_f(Q) \leqslant \mathcal{L}_f(P)$ and $\mathcal{U}_f(P) \leqslant \mathcal{U}_f(Q)$. (Why?) In particular, for any two partitions P and Q, we have

$$\mathcal{L}_f(P) \leqslant \mathcal{L}_f(P \cup Q) \leqslant \mathcal{U}_f(P \cup Q) \leqslant \mathcal{U}_f(Q).$$

The **lower Riemann integral** $I_*(f)$ of f is defined by

$$I_*(f) = \sup\{\mathcal{L}_f(P) : P \text{ is a partition of } [a, b]\},$$

while the **upper Riemann integral** is

$$I^*(f) = \inf\{\mathcal{U}_f(Q) : Q \text{ is a partition of } [a, b]\}.$$

Clearly $-\infty < I_*(f) \leqslant I^*(f) < \infty$. Moreover, if $\mathbb{P}$ denotes the directed set of all partitions of $[a, b]$, then the net $\{\mathcal{L}_f(P)\}_{P \in \mathbb{P}}$ satisfies $\mathcal{L}_f(P) \uparrow_P I_*(f)$ (and similarly, $\mathcal{U}_f(P) \downarrow_P I^*(f)$).

11.29 Definition   *A bounded function $f: [a, b] \to \mathbb{R}$ is called **Riemann inte-grable** if $I_*(f) = I^*(f)$. The common value is called the **Riemann integral** of f, and is denoted $\int_a^b f(x)\, dx$.*

Here are two important characterizations of Riemann integrability.

11.30 Theorem   *For a bounded function $f: [a, b] \to \mathbb{R}$, the following statements are equivalent.*

1. *The function f is Riemann integrable.*

2. *For each $\varepsilon > 0$ there exists a partition P with $\mathcal{U}_f(P) - \mathcal{L}_f(P) < \varepsilon$.*

3. *The function f is continuous a.e. (with respect to Lebesgue measure).*

Two immediate consequences of the preceding theorem are:

•   Every continuous function on a closed interval is Riemann integrable.

•   Under the pointwise operations, the collection of all Riemann integrable functions on a closed interval is a Riesz space, and in fact, a function space and an algebra.

Given a partition $P = \{x_0, x_1, \ldots, x_n\}$ and a selection $T = \{t_1, \ldots, t_n\}$ satisfying $x_{i-1} \leqslant t_i \leqslant x_i$ for each $i = 1, \ldots, n$, define the **Riemann sum** of f corresponding to P and T by

$$R(P, T) = \sum_{i=1}^n f(t_i)(x_i - x_{i-1}).$$

Clearly, $\mathcal{L}_f(P) \leqslant R(P, T) \leqslant \mathcal{U}_f(P)$.

11.31 Theorem   *Let $f: [a,b] \to \mathbb{R}$ be a Riemann integrable function and let $\{P_n\}$ be a sequence of partitions of $[a,b]$ such that $|P_n| \to 0$. Then,*

$$\lim_{n\to\infty} \mathcal{L}_f(P_n) = \lim_{n\to\infty} \mathcal{U}_f(P_n) = \int_a^b f(x)\,dx.$$

In particular, if for each n an arbitrary selection of points T_n for P_n is chosen, then $\lim_{n\to\infty} R(P_n, T_n) = \int_a^b f(x)\,dx$.

Proof: See [13, Theorem 23.5, p. 181].   ∎

The preceding theorem allows us to compare the Riemann and Lebesgue integrals. Let $f: [a,b] \to \mathbb{R}$ be a Riemann integrable function. For each n let P_n denote the partition that subdivides $[a,b]$ into 2^n subintervals of equal length. That is, $P_n = \{x_0^n, x_1^n, \ldots, x_{2^n}^n\}$, where

$$x_i^n = a + i\tfrac{b-a}{2^n}, \ i = 0, 1, \ldots, 2^n.$$

Clearly, $P_n \subset P_{n+1}$ and $|P_n| \to 0$. Consequently, by Theorem 11.31, we have $\mathcal{L}_f(P_n) \uparrow \int_a^b f(x)\,dx$ and $\mathcal{U}_f(P_n) \downarrow \int_a^b f(x)\,dx$. Given P_n, let

$$m_i^n = \inf\{f(x) : x \in [x_{i-1}^n, x_i^n]\} \text{ and } M_i^n = \sup\{f(x) : x \in [x_{i-1}^n, x_i^n]\},$$

and then define the step functions

$$\varphi_n = \sum_{i=1}^{2^n} m_i^n \chi_{[x_{i-1}^n, x_i^n)} \quad \text{and} \quad \psi_n = \sum_{i=1}^{2^n} M_i^n \chi_{[x_{i-1}^n, x_i^n)}.$$

Clearly, $\varphi_n(x) \leqslant f(x) \leqslant \psi_n(x)$ for each $x \in [a,b)$. If λ denotes Lebesgue measure on $\mathbb{R}$, then φ_n and ψ_n are λ-step functions satisfying

$$\int \varphi_n \,d\lambda = \mathcal{L}_f(P_n) \quad \text{and} \quad \int \psi_n \,d\lambda = \mathcal{U}_f(P_n).$$

If we let $\varphi_n(x) \uparrow h(x)$ and $\psi_n(x) \downarrow g(x)$, then $h(x) \leqslant f(x) \leqslant g(x)$ for all x in $[a,b)$, and moreover both h and g are Lebesgue integrable. From the Lebesgue Dominated Convergence Theorem 11.21, we get

$$\int (g-h)\,d\lambda = \lim_{n\to\infty} \int (\psi_n - \varphi_n)\,d\lambda = \lim_{n\to\infty}[\mathcal{U}_f(P_n) - \mathcal{L}_f(P_n)] = 0,$$

and from $g-h \geqslant 0$ and Theorem 11.16, we infer that $g-h = 0$ λ-a.e., or $h = g = f$ a.e. Hence, $\varphi_n \uparrow f$ a.e., which implies that f is Lebesgue integrable (and, in fact, an upper function) and that

$$\int f \,d\lambda = \lim_{n\to\infty} \int \varphi_n \,d\lambda = \lim_{n\to\infty} \mathcal{L}_f(P_n) = \int_a^b f(x)\,dx.$$

Consequently, we have established the following important result.

11.32 Theorem   *If a function $f\colon [a, b] \to \mathbb{R}$ is Riemann integrable, then f is also Lebesgue integrable. Moreover, the two integrals coincide, that is,*

$$\int f \, d\lambda = \int_a^b f(x) \, dx.$$

The converse of the previous theorem is, of course, false. For instance, if $f\colon [0, 1] \to \mathbb{R}$ is the function $f = \chi_Q$, where Q is the set of all rational numbers of $[0, 1]$, then f is discontinuous at every point of $[0, 1]$, so (by Theorem 11.30) f is not Riemann integrable. However, since Q has Lebesgue measure zero (every countable subset of $\mathbb{R}$ has Lebesgue measure zero), we see that $f = \chi_Q = 0$ a.e., so f is Lebesgue integrable and $\int f \, d\lambda = 0$.

Finally, we present the Fundamental Theorem of Calculus, which is the basic tool for computing integrals of continuous functions. Recall that F is an **antiderivative** of f on $[a, b]$ if F is continuous on $[a, b]$, differentiable on (a, b), and $F'(x) = f(x)$ for each $x \in (a, b)$.

11.33 Fundamental Theorem of Calculus   *If $f\colon [a, b] \to \mathbb{R}$ is a continuous function and F is an antiderivative of f, then*

$$\int_a^b f(x) \, dx = F(b) - F(a).$$

For extensions of the Riemann approach to integration, see R. Henstock [156].

11.8   The Bochner integral

So far we have only examined the integral of real-valued functions. But it is clear that the definition of the integral of step functions makes sense for functions taking on values in a vector space. While integrating a function with values in a general linear space may seem like mathematical generality for its own sake, many practical problems in statistics and economics can be formulated in this fashion. We shall see in Chapter 19 that such integrals are important in the analysis of Markov processes.

Suppose Ω is a set equipped with an algebra $\mathcal{A}$ of measurable sets and a charge μ. Also, let X be a vector space. As in the real case, a function $\varphi\colon \Omega \to X$ that assumes only a finite number of values, say $x_1, \ldots, x_n$, is called an **X-simple function** if $A_i = \varphi^{-1}(\{x_i\}) \in \mathcal{A}$ for each i. As usual, the formula $\varphi = \sum_{i=1}^n x_i \chi_{A_i}$ is called the **standard representation** of φ.[4] If $\mu(A_i) < \infty$ for each nonzero x_i, then

---

[4] We write $x_i \chi_{A_i}$ instead of the correct (but awkward) notation $\chi_{A_i} x_i$. (Scalars multiply vectors from the left.)

φ is called an **X-step function**. The **integral** of an X-valued step function φ is the vector $\int \varphi\, d\mu$ in X defined via the formula

$$\int \varphi\, d\mu = \sum_{i=1}^{n} \mu(A_i) x_i.$$

As in the real case, if $\varphi = \sum_{j=1}^{m} y_j \chi_{B_j}$ is another representation of φ with $\mu(B_j) < \infty$ for each nonzero y_j, then

$$\int \varphi\, d\mu = \sum_{j=1}^{m} \mu(B_j) y_j.$$

The proof is simply a repetition of the proof of Lemma 11.4.

The technical question at hand now is how to generalize the integral of a vector-valued function beyond the case of step functions. If the vector space X is an ordered vector space, there is a hope that we could build a theory of integration based on upper functions, analogous to the development of the Lebesgue integral. Unfortunately, we know of no satisfactory theory along these lines. However, there are several useful extensions of the integral along other lines, all of which are based on the idea of reducing the question of vector integrability to integrability of real functions. All of these integrals require that μ be a measure rather than a charge. The first vector integral we discuss is the Bochner integral. The main reference for this material is J. Diestel and J. J. Uhl [96, Chapter II].

For the remainder of this chapter, unless otherwise stated, (Ω, Σ, μ) is a measure space and X a Banach space. If $f: \Omega \to X$ is a vector function, then $\|f\|$ denotes the (nonnegative) real function $\|f\|: \Omega \to \mathbb{R}$ defined by $\|f\|(\omega) = \|f(\omega)\|$ for each $\omega \in \Omega$. We call $\|f\|$ the **norm function** of f.

The vector space of all X-step functions is denoted L_X. As mentioned before, the proof of Theorem 11.4 shows that the integral is a linear operator from L_X into X. That is, we have the following result.

11.34 Theorem *The collection L_X of all X-step functions is a vector subspace of the vector space X^{Ω}. Moreover, for each $\varphi, \psi \in L_X$ and each $\alpha, \beta \in \mathbb{R}$ we have*

$$\int (\alpha\varphi + \beta\psi)\, d\mu = \alpha \int \varphi\, d\mu + \beta \int \psi\, d\mu,$$

that is, the Bochner integral is a linear operator from L_X to X.

If X is also a Banach lattice, then L_X is a Riesz space under the pointwise lattice operations and the Bochner integral is a positive operator from L_X to X.

For $\varphi \in L_X$ and $E \in \Sigma$, define $\int_E \varphi\, d\mu$, the **integral of φ over E**, by

$$\int_E \varphi\, d\mu = \int \varphi\chi_E\, d\mu.$$

The proof of the next lemma is left as an exercise.

11.35 Lemma   *If $\varphi \in L_X$ has standard representation $\varphi = \sum_{i=1}^{n} x_i \chi_{A_i}$, then the norm function $\|\varphi\|$ of φ is a real step function having standard representation $\|\varphi\| = \sum_{i=1}^{n} \|x_i\| \chi_{A_i}$. Moreover,*

$$\int \|\varphi\| d\mu = \sum_{i=1}^{n} \|x_i\| \mu(A_i) \quad and \quad \left\| \int \varphi \, d\mu \right\| \leq \int \|\varphi\| \, d\mu.$$

We are now ready to define the concept of strong measurability. The definition is simply the abstraction of Corollary 4.37.

11.36 Definition   *Let (Ω, Σ, μ) be a measure space, and let $f \colon \Omega \to X$ be a vector function. We say that f is **strongly μ-measurable** if there exists a sequence $\{\varphi_n\}$ of X-simple functions such that $\lim_{n \to \infty} \|f(\omega) - \varphi_n(\omega)\| = 0$ for μ-almost all $\omega \in \Omega$.*

Let us characterize the strongly measurable functions. Observe first that if a function $f \colon \Omega \to X$ is strongly measurable, then there exists a μ-null subset E of Ω and a separable closed subspace Y of X such that $\{f(\omega) : \omega \in \Omega \setminus E\} \subset Y$. That is, if f is strongly measurable, then for μ-almost all $\omega \in \Omega$ the value $f(\omega)$ lies in a separable closed vector subspace of X. Conversely, if f is measurable and for some μ-null set E and some separable closed subspace Y of X we have $\{f(\omega) : \omega \in \Omega \setminus E\} \subset Y$, then according to Theorem 4.38 the function $f \colon \Omega \setminus E \to Y$ is the pointwise limit of a sequence of simple functions from $\Omega \setminus E$ to Y. This easily implies that $f \colon \Omega \to X$ is strongly measurable. Thus, we have established the following result.

11.37 Lemma   *A function $f \colon \Omega \to X$ is strongly measurable if and only if it is measurable and its values $f(\omega)$ lie for μ-almost all ω in a separable closed subspace of X.*

As easy consequence of this characterization is the following result.

11.38 Lemma   *If a function $f \colon \Omega \to X$ is the μ-almost every pointwise limit of a sequence of strongly measurable functions, then f itself is strongly measurable.*

As you might expect, strongly measurable functions have measurable norm functions.

11.39 Lemma   *If $f \colon \Omega \to X$ is strongly measurable, then the real function $\|f\|$ is also measurable.*

Proof: Since $\big| \|f(\omega)\| - \|\varphi_n(\omega)\| \big| \leq \big\| f(\omega) - \varphi_n(\omega) \big\|$, we have $\|\varphi_n(\omega)\| \to \|f(\omega)\|$ for μ-almost every $\omega \in \Omega$. So by Corollary 4.37, $\|f\|$ is measurable. ∎

The collection of all strongly measurable functions from Ω to X is denoted $\mathcal{M}(\Omega, X)$. That is,

$$\mathcal{M}(\Omega, X) = \{f \in X^\Omega : f \text{ is strongly measurable}\}.$$

The verification of the following simple property is left as an exercise.

11.40 Lemma   *The collection $\mathcal{M}(\Omega, X)$ of all strongly measurable functions from Ω to X is a vector subspace of X^Ω containing all the X-step functions. That is, we have the following vector subspace inclusions:*

$$L_X \subset \mathcal{M}(\Omega, X) \subset X^\Omega.$$

Our next goal is to extend the notion of the integral from L_X to a larger subspace of $\mathcal{M}(\Omega, X)$. To do this, we need a lemma.

11.41 Lemma   *Let $f \colon \Omega \to X$ be a strongly μ-measurable function. Suppose that for two sequences $\{\varphi_n\}$ and $\{\psi_n\}$ of X-step functions the real measurable functions $\|f - \varphi_n\|$ and $\|f - \psi_n\|$ are Lebesgue integrable for each n and*

$$\lim_{n\to\infty} \int \|f - \varphi_n\|\, d\mu = \lim_{n\to\infty} \int \|f - \psi_n\|\, d\mu = 0.$$

Then for each $E \in \Sigma$ we have

$$\lim_{n\to\infty} \int_E \varphi_n\, d\mu = \lim_{n\to\infty} \int_E \psi_n\, d\mu,$$

where the last two limits are taken with respect to the norm topology on X.

Proof: Assume that the two sequences $\{\varphi_n\}$ and $\{\psi_n\}$ of X-step functions satisfy the stated property. Fix $E \in \Sigma$. From

$$\left\| \int_E \varphi_n\, d\mu - \int_E \varphi_m\, d\mu \right\| = \left\| \int_E (\varphi_n - \varphi_m)\, d\mu \right\| \leqslant \int \|\varphi_n - \varphi_m\|\, d\mu$$
$$\leqslant \int \|f - \varphi_n\|\, d\mu + \int \|f - \varphi_m\|\, d\mu,$$

we see that $\lim_{n,m\to\infty} \left\| \int_E \varphi_n\, d\mu - \int_E \varphi_m\, d\mu \right\| = 0$, which shows that the sequence $\{\int_E \varphi_n\, d\mu\}$ is a Cauchy sequence in X, so it converges in X. Similarly, the sequence $\{\int_E \psi_n\, d\mu\}$ converges in X. Now the inequality

$$\left\| \int_E \varphi_n\, d\mu - \int_E \psi_n\, d\mu \right\| \leqslant \int \|f - \varphi_n\|\, d\mu + \int \|f - \psi_n\|\, d\mu$$

easily implies $\lim_{n\to\infty} \int_E \varphi_n\, d\mu = \lim_{n\to\infty} \int_E \psi_n\, d\mu$, as claimed. ∎

The Bochner integral was introduced by S. Bochner [53] and is precisely the abstraction of Theorem 11.25.

11.42 Definition   *A strongly μ-measurable function $f: \Omega \to X$ is **Bochner integrable** if there is a sequence $\{\varphi_n\}$ of X-step functions such that the real measurable function $\|f - \varphi_n\|$ is Lebesgue integrable for each n and*

$$\lim_{n\to\infty} \int \|f - \varphi_n\| \, d\mu = 0.$$

*In this case, for each $E \in \Sigma$ the **Bochner integral** of f over E is defined by*

$$\int_E f \, d\mu = \lim_{n\to\infty} \int_E \varphi_n \, d\mu,$$

where the last limit is in the norm topology on X.

As usual, we write $\int f \, d\mu$ instead of $\int_\Omega f \, d\mu$. By Lemma 11.41, the Bochner integral is well defined, in the sense that it does not depend on the particular sequence of step functions used to approximate f. Every X-step function is Bochner integrable and if $\varphi \in L_X$ has the standard representation $\varphi = \sum_{i=1}^n x_i \chi_{A_i}$, then $\int_E \varphi \, d\mu = \sum_{i=1}^n \mu(E \cap A_i) x_i$ for each $E \in \Sigma$.

The collection of all Bochner integrable functions is a vector subspace of $\mathcal{M}(\Omega, X)$ and the Bochner integral acts as a linear operator from this space into X. The details are in the next theorem whose straightforward proof is left as an exercise.

11.43 Theorem   *If f and g are two Bochner integrable functions and $\alpha, \beta \in \mathbb{R}$, then $\alpha f + \beta g$ is also Bochner integrable and*

$$\int_E (\alpha f + \beta g) \, d\mu = \alpha \int_E f \, d\mu + \beta \int_E g \, d\mu$$

for each $E \in \Sigma$. Moreover, if X is a Banach lattice and f and g are two Bochner integrable functions satisfying $f(\omega) \leqslant g(\omega)$ for μ-almost all $\omega \in \Omega$, then

$$\int_E f \, d\mu \leqslant \int_E g \, d\mu$$

for each $E \in \Sigma$.

The definition of the Bochner integral is cumbersome to apply, but fortunately for finite measure spaces there is a manageable criterion.

11.44 Theorem   *Let (Ω, Σ, μ) be a finite measure space and let $f: \Omega \to X$ be a μ-measurable function. Then f is Bochner integrable if and only if its norm function $\|f\|$ is Lebesgue integrable, that is, $\int \|f\| \, d\mu < \infty$.*

Proof: See J. Diestel and J. J. Uhl [96, Theorem II.2.2, p. 45]. ∎

We leave the straightforward proof of the next lemma up to you.

11.45 Lemma   *Let $f: \Omega \to X$ be Bochner integrable and let Y be another Banach space. If $T: X \to Y$ is a bounded operator, then the function $Tf: \Omega \to Y$, defined by $Tf(\omega) = T(f(\omega))$, is also Bochner integrable and*

$$\int_\Omega Tf \, d\mu = T\left(\int_\Omega f \, d\mu\right).$$

There is also a Dominated Convergence Theorem for the Bochner integral.

11.46 Vector Dominated Convergence Theorem   *Let $f: \Omega \to X$ be strongly measurable and let $\{f_n\}$ be a sequence of Bochner integrable functions satisfying $\left\| f_n(\omega) - f(\omega) \right\| \to 0$ for μ-almost all $\omega \in \Omega$. If there exists a real nonnegative Lebesgue integrable function $g: \Omega \to \mathbb{R}$ such that for each n we have $\|f_n\| \leqslant g$ μ-a.e., then f is Bochner integrable and*

$$\lim_{n\to\infty} \int_E f_n \, d\mu = \int_E f \, d\mu$$

for each $E \in \Sigma$.

Proof: Clearly, $\|f\| \leqslant g$ μ-a.e., and $\|f - f_n\|$ is measurable for each n. From $\|f - f_n\| \leqslant 2g$ μ-a.e., we see that $\|f - f_n\|$ is Lebesgue integrable for each n. Moreover, from $\|f - f_n\| \to 0$ μ-a.e. and the Lebesgue Dominated Convergence Theorem 11.21, we get $\int \|f - f_n\| \, d\mu \to 0$. Next, for each n choose an X-step function φ_n with $\int \|f_n - \varphi_n\| \, d\mu < \frac{1}{n}$, and note that

$$\int \|f - \varphi_n\| \, d\mu \leqslant \int \|f - f_n\| \, d\mu + \int \|f_n - \varphi_n\| \, d\mu \to 0.$$

This implies that f is Bochner integrable and that

$$\int_E f \, d\mu = \lim_{n\to\infty} \int_E \varphi_n \, d\mu = \lim_{n\to\infty} \int_E f_n \, d\mu$$

for each $E \in \Sigma$. ∎

We close the section by presenting a useful condition for the Bochner integrability of L_1-valued functions. As usual, if $f: \Omega \to L_1(\nu)$ is a function, then we denote $f(\omega)(t)$ by $f(\omega, t)$, that is, $f(\omega)(t) = f(\omega, t)$. Note well that f may fail to be jointly measurable, but we do have the following result.

11.47 Theorem   *Let (Ω, Σ, μ) and (T, S, ν) be σ-finite measure spaces. Then for a function $f: \Omega \to L_1(\nu)$ we have the following.*

1. *If f is Bochner integrable, then there exists a $\mu \times \nu$-integrable function $F : \Omega \times T \to \mathbb{R}$ (which is uniquely determined up to a $\mu \times \nu$-null set) such that:*

   a. *for μ-almost all $\omega \in \Omega$ we have $f(\omega) = F(\omega, \cdot)$ in $L_1(\nu)$, and*

   b. *for ν-almost all $t \in T$ the real function $F(\cdot, t)$ is μ-integrable and*

   $$\left(\int_{\Omega} f \, d\mu \right)(t) = \int_{\Omega} F(\omega, t) \, d\mu(\omega).$$

   for ν-almost all $t \in T$.

2. *If $f(\cdot, \cdot)$ is $\mu \times \nu$-integrable, then f is Bochner integrable and*

   $$\left(\int_{\Omega} f \, d\mu \right)(t) = \int_{\Omega} f(\omega, t) \, d\mu(\omega).$$

   for ν-almost all $t \in T$.

Proof: See N. Dunford and J. T. Schwartz [110, Theorem 17, p. 198]. ∎

As an application, consider $f : [0, 1] \to L_1[0, 1]$ defined by $f(\omega) = \chi_{[0,\omega]}$. Then $f(\omega, t) = \chi_{[0,\omega]}(t) = \chi_A(\omega, t)$, where $A = \{(\omega, t) \in [0, 1] \times [0, 1] : t \leqslant \omega\}$. This implies that $f(\cdot, \cdot) \in L_1([0, 1] \times [0, 1])$, so

$$\left(\int_0^1 f(\omega) \, d\omega \right)(t) = \int_0^1 \chi_A(\omega, t) \, d\omega = \int_0^1 \chi_{[t,1]}(\omega) \, d\omega = 1 - t.$$

11.9   The Gelfand integral

Let (Ω, Σ, μ) be a measure space and X a Banach space with dual X'. For X'-valued functions we can introduce another notion of measurability.

11.48 Definition   *A function $f : \Omega \to X'$ is **weak* measurable** if for each $x \in X$ the function $xf : \Omega \to \mathbb{R}$, defined by*

$$xf(\omega) = \langle x, f(\omega) \rangle,$$

is measurable.

Every X'-valued strongly measurable function is weak* measurable. To see this, note first that if $\varphi = \sum_{i=1}^n x_i' \chi_{A_i}$ is an X'-simple function and $x \in X$, then

$$x\varphi(\omega) = \sum_{i=1}^n \langle x, x_i' \chi_{A_i}(\omega) \rangle = \left[\sum_{i=1}^n \langle x, x_i' \rangle \chi_{A_i} \right](\omega),$$

which shows that $x\varphi$ is indeed a real Σ-simple function. Now given a function $f\colon \Omega \to X'$, if there exists a sequence $\{\varphi_n\}$ of X'-simple functions satisfying $\|f(\omega) - \varphi_n(\omega)\| \to 0$ for almost every $\omega \in \Omega$, then

$$x\varphi_n(\omega) = \langle x, \varphi_n(\omega) \rangle \to \langle x, f(\omega) \rangle = xf(\omega)$$

for almost every $\omega \in \Omega$. That is, the sequence of real Σ-simple functions $\{x\varphi_n\}$ converges pointwise almost everywhere to xf, so xf is measurable.

For X'-valued functions, we can define a weaker notion of a vector integral, due to I. M. Gelfand [135].

11.49 Definition   *A weak* measurable function $f\colon \Omega \to X'$ is **Gelfand integrable** over a set $E \in \Sigma$ if there exists some $x'_E \in X'$ satisfying*

$$\langle x, x'_E \rangle = \int_E \langle x, f(\omega) \rangle \, d\mu(\omega)$$

*for each $x \in X$. The unique vector $x'_E \in X'$ is called the **Gelfand integral** or **weak*-integral** of f over E and is denoted $\int_E f \, d\mu$, that is, $\int_E f \, d\mu = x'_E$.*
*If the integral exists for each $E \in \Sigma$, we say that f is **Gelfand integrable**.*

Suppose that $\varphi = \sum_{i=1}^n x'_i \chi_{A_i}$ is an X'-step function. If $E \in \Sigma$ and $x \in X$, then

$$\left\langle x, \int_E \varphi \, d\mu \right\rangle = \left\langle x, \sum_{i=1}^n x'_i \mu(E \cap A_i) \right\rangle = \sum_{i=1}^n x'_i(x)\mu(E \cap A_i)$$

$$= \int_E \langle x, \varphi(\omega) \rangle \, d\mu(\omega).$$

This shows that φ is Gelfand integrable and that its Gelfand integral coincides with its Bochner integral.

Next, let $f\colon \Omega \to X'$ be Bochner integrable and let $E \in \Sigma$ be fixed. Pick a sequence $\{\varphi_n\}$ of X'-step functions satisfying $\int \|f - \varphi_n\| \, d\mu \to 0$. From

$$\langle x, f(\omega) \rangle = \langle x, f(\omega) - \varphi_n(\omega) \rangle + \langle x, \varphi_n(\omega) \rangle,$$

we see that for each $x \in X$ the real function $\omega \mapsto \langle x, f(\omega) \rangle$ is Lebesgue integrable. Moreover, for each n we have

$$\int_E \langle x, f(\omega) \rangle \, d\mu(\omega) = \int_E \langle x, f(\omega) - \varphi_n(\omega) \rangle \, d\mu(\omega) + \int_E \langle x, \varphi_n(\omega) \rangle \, d\mu(\omega). \quad (\star)$$

Now from the inequality

$$\left| \int_E \langle x, f(\omega) - \varphi_n(\omega) \rangle \, d\mu(\omega) \right| \le \|x\| \int \|f - \varphi_n\| \, d\mu \to 0,$$

the fact that $\int_E \varphi_n \, d\mu \to \int_E f \, d\mu$, and equation ($\star$), we see that

$$\int_E \langle x, f(\omega) \rangle \, d\mu(\omega) = \langle x, \int_E f \, d\mu \rangle$$

for each $x \in X$. In other words, we have established the following result.

11.50 Theorem   *If a function $f: \Omega \to X'$ is Bochner integrable, then f is Gelfand integrable and the two integrals coincide. That is, $x'_E = \int_E f \, d\mu$ for each $E \in \Sigma$.*

We can also look at the Bochner integral as a Gelfand integral. If we consider X embedded in X'' (the norm dual of X'), then every X-valued function can be also viewed as an $(X')'$-valued function. From the above discussion, if $f: \Omega \to X$ is a Bochner integrable function, then the function $f: \Omega \to X''$ is also Gelfand integrable and the two integrals coincide.

The collection of all Gelfand integrable functions is a vector space and the Gelfand integral acts on it as an X'-valued linear operator. The details are included in the next result whose straightforward proof is left as an exercise.

11.51 Theorem   *The collection of all Gelfand integrable functions is a vector subspace of $(X')^\Omega$ and the Gelfand integral is linear. That is, if f and g are Gelfand integrable over a set $E \in \Sigma$, then $\alpha f + \beta g$ $(\alpha, \beta \in \mathbb{R})$ is also Gelfand integrable over E and*

$$\int_E (\alpha f + \beta g) \, d\mu = \alpha \int_E f \, d\mu + \beta \int_E g \, d\mu.$$

The next theorem gives conditions that guarantee the Gelfand integrability of a function.

11.52 Theorem   *Let X be a Banach space and let $f: \Omega \to X'$ have the property that xf is integrable for each $x \in X$. Then f is Gelfand integrable, that is, for each E in Σ, the Gelfand integral of f over E exists.*

Proof: If the Gelfand integral of f over E exists, it can only be the linear functional $x \mapsto \int_E \langle x, f(\omega) \rangle \, d\mu(\omega)$. We need to show that this functional is continuous. To see this, we write it as the composition of two continuous linear functions. Start by noting that if $x_n \to x$ in norm, then $x_n f \to xf$ pointwise on Ω. This follows from $x_n f(\omega) = \langle x_n, f(\omega) \rangle$ and the fact that $f(\omega)$ is a continuous linear functional on X.

Next, define the linear operator $T: X \to L_1(\mu)$ by $T(x) = xf$. We claim that T has closed graph: Let $x_n \to x$, and suppose $x_n f \xrightarrow{\|\cdot\|_1} g$ in $L_1(\mu)$, that is, $\|x_n f - g\|_1 \to 0$. Then along some subsequence $\{x_{n_k}\}$, $x_{n_k} f \to g$ μ-a.e. (Theorem 13.6). But by the above remark, $x_{n_k} f \to xf$ everywhere, so $g = xf$ in $L_1(\mu)$.

That is, the graph of T is closed. By the Closed Graph Theorem 5.20, T is a continuous linear operator from X to $L_1(\mu)$. Now the mapping $g \mapsto \int_E g \, d\mu$ is a continuous linear functional on $L_1(\mu)$, so the composition is a continuous linear functional on X, that is, it is some $x'_E \in X'$. By construction, x'_E satisfies $\langle x, x'_E \rangle = \int_E \langle x, f(\omega) \rangle \, d\mu(\omega)$. ∎

The next result gives a sufficient condition for xf to be integrable.

11.53 Corollary   *If μ is a finite measure, $f : \Omega \to X'$ is weak* measurable, and the range of f is a norm bounded subset of X', then f is Gelfand integrable.*

Proof: If f is norm bounded, then for each fixed vector $x \in X$, the function $\omega \mapsto xf(\omega) = \langle x, f(\omega) \rangle$ is bounded, so it is integrable. ∎

The next theorem gives a handy and intuitive property of the Gelfand integral. It says that the Gelfand integral of a function with respect to a probability measure lies in the weak*-closed convex hull of the range of the function.

11.54 Theorem   *Let $f : \Omega \to X'$ be Gelfand integrable with respect to a probability measure μ on Σ (that is, $\mu(\Omega) = 1$) and let $A = f(\Omega)$. Then $\int_\Omega f \, d\mu$ belongs to $\overline{co} \, A$, the weak*-closed convex hull of A.*

Proof: Let $x' = \int_\Omega f \, d\mu$, $A = f(\Omega)$, and suppose $x' \notin \overline{co} \, A$. Then (by Theorem 5.79) x' is strongly separated from $\overline{co} \, A$ by some $x \in X$. That is, there exists some α satisfying $\langle x, x' \rangle > \alpha \geqslant \langle x, f(\omega) \rangle$ for all ω. This contradicts $\langle x, x' \rangle = \int_\Omega \langle x, f(\omega) \rangle \, d\mu(\omega) \leqslant \alpha \int_\Omega d\mu(\omega) = \alpha$. ∎

11.10   The Dunford and Pettis integrals

The Pettis integral and the Dunford integral are analogous to the Gelfand integral for functions taking values in X rather than in X'. Let (Ω, Σ, μ) be a measure space and X a Banach space with dual X'. A function $\varphi : \Omega \to X$ is **weakly measurable** if for every $x' \in X'$ the function $\varphi x' : \Omega \to \mathbb{R}$ defined by $\varphi x'(\omega) = \langle \varphi(\omega), x' \rangle$ is measurable. The **Pettis integral** of a weakly measurable function φ over E, if it exists, is a vector $x_E \in X$ satisfying

$$\langle x_E, x' \rangle = \int_E \langle \varphi(\omega), x' \rangle \, d\mu(\omega)$$

for each $x' \in X'$. The **Dunford integral** of φ over E is an element x''_E in the double dual X'' of X satisfying, you guessed it,

$$\langle x', x''_E \rangle = \int_E \langle \varphi(\omega), x' \rangle \, d\mu(\omega)$$

for each $x' \in X'$. A function is **Dunford integrable** (resp., **Pettis integrable**) if its Dunford (resp., Pettis) integral exists for every E in Σ.[5] It is obvious that:

Bochner integrability $\implies$ Pettis integrability $\implies$ Dunford integrability

In general no reverse implication is true. Also, note that the Gelfand integral is nothing but the Pettis integral for X'-valued functions.

A closed graph argument similar to the proof of Theorem 11.52 proves the following result on Dunford integrability. The details are omitted.

11.55 Theorem   *Let X be a Banach space and $\varphi \colon \Omega \to X$ a weakly measurable function satisfying $\varphi x' \in L_1(\mu)$ for each $x' \in X'$. Then φ is Dunford integrable.*

The case of Pettis integrals is less satisfactory. A function may be Dunford integrable but not Pettis integrable. See J. Diestel and J. J. Uhl [96, Theorem II.3.7, p. 54] for conditions guaranteeing Pettis integrability. This result is quite deep and out of place here.

The notion of Pettis and Dunford integrability can be defined in a more general setting, which is useful in its own right. It adds topological assumptions on the measure space, but the range space need not be a Banach space. Let X be a topological vector space with dual X' that separates the points of X. Let Ω be a compact Hausdorff space, and let μ be a Borel measure on Ω (that is, μ is a measure on the Borel σ-algebra of Ω). As above, we say that $f \colon \Omega \to X$ is **weakly Borel measurable** if $\omega \mapsto \langle f(\omega), x' \rangle$ is Borel measurable for each $x' \in X'$. Again, we define the **Pettis integral** over E of a weakly Borel measurable function $f \colon \Omega \to X$ as the unique vector $x_E \in X$ (if it exists) satisfying

$$\langle x_E, x' \rangle = \int_E \langle f(\omega), x' \rangle \, d\mu(\omega)$$

for each $x' \in X'$. (Uniqueness of such a vector is guaranteed since X' separates points of X.) The Dunford integral is then defined similarly as the unique $x''_E \in X''$ (if it exists), where X'' is now the topological dual of $(X', \beta(X', X))$, such that

$$\langle x', x''_E \rangle = \int_E \langle \varphi(\omega), x' \rangle \, d\mu(\omega)$$

for each $x' \in X'$.

---

[5] The Dunford integral was introduced by N. Dunford [109] and the Pettis integral by B. J. Pettis [275]

Chapter 12

Measures and topology

Chapter 10 dealt with measures and charges defined on abstract semirings or algebras of sets. In applications there is often a natural topological or metric structure on the underlying measure space. By combining topological and set theoretic notions it is possible to develop a richer and more useful theory. Some of these connections between measure theory and topology are discussed in this chapter.

One of the most useful notions involving the topological structure is *tightness*, which asserts that the measure of any measurable set can be approximated by the measure of an included compact set. Indeed if a charge (which need only be finitely additive) on the Borel sets of a Hausdorff space is tight, then it is automatically countably additive (Theorem 12.4). A somewhat stronger condition than tightness is regularity. A measure is *regular* if every compact set has finite measure, it is tight, and in addition, the measure of every set can be approximated by the measure of open sets that include it. Every finite Borel measure on a Polish space is regular (Theorem 12.7). This is not generally true for non-Polish spaces. Example 12.9 is a classic example of a non-regular Borel probability measure on a compact Hausdorff space.

There are other nice properties of Borel measures on Polish spaces. One of these is that in this case (as well as few other cases) every finite measure has a well defined *support*, or minimal closed set of full measure (Theorem 12.14). Example 12.15 shows that in general, even on a compact Hausdorff space, a Borel measure need not have a support. Lusin's Theorem 12.8 shows that when the domain is a Polish space, a Borel measurable function is continuous when restricted to a compact set whose complement has arbitrarily small positive measure. In addition, Theorem 12.22 shows the existence of *nonatomic* regular Borel measures on uncountable Polish spaces.

Section 12.5 discusses analytic subsets of Polish spaces, which are the continuous images of the Baire space $\mathcal{N}$. Every Borel set is analytic (Theorem 12.25), but not vice versa (Example 12.33). However, every analytic set is *universally measurable* (Theorem 12.41), that is, μ-measurable for any Borel probability measure μ. Analytic sets occur naturally in connection with measurable correspondences, see Theorem 18.21. They arise naturally in the study of stochastic processes (see, e.g., Dellacherie [86, 87]), dynamic programming (see, e.g., Bertsekas and Shreve [39]), and also in the theory of games with incomplete information (see, e.g., Stinchcombe and White [320]). They also appear prominently in Chapter 18 on measurable correspondences below.

Finally we prove some interesting facts about functions between Polish spaces. Theorem 12.28 asserts that a function is Borel measurable if and only if its graph is a Borel set. Theorem 12.29 says that the one-to-one image of a Borel set under a Borel measurable function is a Borel set.

The classic reference for measures on topological spaces is Halmos [148]. Some of this material is covered in standard analysis texts, such as, Aliprantis and Burkinshaw [13], Royden [290], and Rudin [291]. Billingsley [42], Neveu [261], Parthasarathy [271], and Pollard [280] concentrate on applications to probability and stochastic processes. Choquet [76] has an excellent treatment of the topological properties of spaces of Radon measures. The material on Borel functions and analytic sets derives from Kechris [196], Kuratowski [218], Lusin [231], and Parthasarathy [271]. There is also an excellent monograph by Srivastava [319].

12.1 Borel measures and regularity

In this section X is a topological space. As before, the σ-algebra of all Borel sets of X is denoted $\mathcal{B}$, or $\mathcal{B}_X$ or $\mathcal{B}orel$. Similarly, the σ-algebra of all Baire sets is denoted $\mathcal{B}aire$. The symbol $\mathcal{A}_X$ denotes the algebra generated by the open sets.

12.1 Definition   *A **(signed) Borel measure** is simply a (signed) measure defined on the Borel sets of a topological space.* [1]

*Similarly, a **(signed) Baire measure** is any (signed) measure defined on the σ-algebra $\mathcal{B}aire$ of Baire sets of a topological space.*

*A **(signed) Borel charge** is a (signed) charge that is defined either on the algebra $\mathcal{A}_X$ or σ-algebra $\mathcal{B}_X$ generated by the open sets.*

While we are more interested in charges and measures than their signed counterparts, we make the following general definitions.

12.2 Definition   *Here $\mathcal{A}$ may stand for $\mathcal{A}_X$, $\mathcal{B}_X$, or $\mathcal{B}aire(X)$. Let μ be a charge or measure on $\mathcal{A}$. The charge or measure μ is:*

- ***outer regular** if for every set A in $\mathcal{A}$,*

$$\mu(A) = \inf\{\mu(V) : V \in \mathcal{A}, \ V \ open, \ and \ A \subset V\}.$$

---

[1] Recall that in Chapter 10 we required a Borel measure to assign finite measure to every compact set. Most authors make this definition. However, for the purposes of this chapter, we do not make that requirement. We do require that what we call a regular Borel measure assign finite measure to every compact set.

- **inner regular** *if for every set A in $\mathcal{A}$,*

$$\mu(A) = \sup\{\mu(F) : F \in \mathcal{A}, \ F \ closed, \ and \ F \subset A\}.$$

- **normal** *if it is both inner and outer regular.*

- **tight** *if for every set A in $\mathcal{A}$,*

$$\mu(A) = \sup\{\mu(K) : K \in \mathcal{A}, \ K \ compact, \ and \ K \subset A\}.$$

- **regular** *if $\mu(K) < \infty$ for each compact set $K \in \mathcal{A}$ and it is both outer regular and tight.*

We say that a signed charge possesses any of these properties if both its positive and negative parts do.

A few words about these definitions are in order.[2] For any Hausdorff space, if $\mathcal{A}$ denotes either $\mathcal{A}_X$, the algebra generated by the open sets, or $\mathcal{B}_X$, the Borel σ-algebra, then the qualification that we only consider open, closed, and compact sets belonging to $\mathcal{A}$ is redundant. On the other hand, if μ is a Baire measure, then it is tight if every set can be approximated from the inside by compact Baire sets. In general, not every compact set is a Baire set, so it is not obvious that the restriction of a regular Borel measure to the Baire σ-algebra is a regular Baire measure.

In particular, if μ is regular, then for every $A \in \mathcal{A}$ with $\mu(A) < \infty$ and every $\varepsilon > 0$, there are an open set $G \in \mathcal{A}$ and a compact set $K \in \mathcal{A}$ with $K \subset A \subset G$ and $|\mu(G) - \mu(K)| < \varepsilon$.

You should notice that two outer regular Borel charges are equal if and only if they agree on the open sets. Also, tight Borel charges are equal if and only if they agree on the compact sets.

12.3 Lemma   *A finite Borel charge μ is outer regular if and only if it is inner regular (and also if and only if it is normal).*

Proof: If $\mu: \mathcal{A} \to [0, \infty)$ (where $\mathcal{A} = \mathcal{A}_X$ or $\mathcal{A} = \mathcal{B}_X$) is finite and satisfies $\mu(A) = \sup\{\mu(F) : F \ \text{closed and} \ F \subset A\}$. Then from

$$\mu(X) - \mu(A) = \mu(A^c) = \sup\{\mu(F) : F \ \text{closed and} \ F \subset A^c\}$$
$$= \mu(X) - \inf\{\mu(V) : V \ \text{open and} \ A \subset V\},$$

we see that $\mu(A) = \inf\{\mu(V) : V \ \text{open and} \ A \subset V\}$, so μ is outer regular. The converse can be proven in a similar manner.   ∎

---

[2] This terminology is not universally used. Notably, Parthasarathy [271] uses "regular" to mean what we call outer regular. Dudley [104] uses "closed regular" for what we call inner regular, and "regular" for what we call tight, unless μ is a finite measure, in which case he also uses the term "tight." Many authors use "inner regular" to mean what we call "tight." For compact Hausdorff spaces, there is no difference. Our use of the term "tight" agrees with Billingsley [42]. The term "normal measure" is our own invention, and we find it useful.

The next result is even stronger. Not only is a tight charge regular, it is countably additive.

12.4 Theorem (Tight charges are measures) *On a Hausdorff space, every tight
finite Borel charge is a regular measure.*

Proof: Let μ be a tight finite Borel charge. Since the class of compact sets is a
compact class, Theorem 10.13 implies that μ is a measure. In a Hausdorff space
every compact set is closed, so if a charge is tight, then for any Borel set A we
have

$$\mu(A) = \sup\{\mu(K) : K \text{ compact and } K \subset A\}$$
$$\leqslant \sup\{\mu(F) : F \text{ closed and } F \subset A\} \leqslant \mu(A).$$

So by Lemma 12.3, tightness implies outer regularity. Thus μ is a regular Borel
measure. ∎

On metrizable spaces, every finite measure is normal.

12.5 Theorem   *Every finite Borel measure on a metrizable space is normal.*

Proof: Let X be a metrizable space, and let μ be a finite Borel measure. Consider
the collection of sets

$$\mathcal{A} = \{A \in \mathcal{B}_X : \mu(A) = \sup\{\mu(F) : F \text{ closed and } F \subset A\}$$
$$= \inf\{\mu(G) : G \text{ open and } A \subset G\}\}.$$

We claim that $\mathcal{A}$ is a σ-algebra containing the open sets. If this claim is true, then
$\mathcal{A} = \mathcal{B}_X$, and we are done. We establish the claim in steps.

• *If $A \in \mathcal{A}$, then $A^c \in \mathcal{A}$.*

The verification of this assertion is straightforward.

• *If $\{A_n\}$ is a sequence in $\mathcal{A}$, then $\bigcup_{n=1}^\infty A_n \in \mathcal{A}$.*

Let $\{A_n\} \subset \mathcal{A}$ and let $A = \bigcup_{n=1}^\infty A_n$. Now let $\varepsilon > 0$. For each n pick an open set G_n
and a closed set F_n satisfying the inclusions $F_n \subset A_n \subset G_n$ and the inequalities
$\mu(G_n) < \mu(F_n) + \frac{\varepsilon}{2^n}$. Let $G = \bigcup_{n=1}^\infty G_n$ and $F = \bigcup_{n=1}^\infty F_n$. Then G is open, $A \subset G$,
and from $G \setminus A \subset \bigcup_{n=1}^\infty (G_n \setminus A_n)$, we see that

$$0 \leqslant \mu(G) - \mu(A) = \mu(G \setminus A) \leqslant \sum_{n=1}^\infty \mu(G_n \setminus A_n)$$
$$= \sum_{n=1}^\infty [\mu(G_n) - \mu(A_n)] < \sum_{n=1}^\infty \frac{\varepsilon}{2^n} = \varepsilon.$$

Hence $\mu(A) = \inf\{\mu(G)\} : G$ open and $A \subset G\}$.

Similarly, we get $\mu(A) < \mu(F) + \varepsilon$. Since $C_n = \bigcup_{i=1}^{n} F_i \uparrow F$, it follows from the continuity of the measure that for some n the closed set $C_n \subset A$ satisfies $\mu(A) < \mu(C_n) + \varepsilon$. Therefore,

$$\mu(A) = \sup\{\mu(F)\} : F \text{ closed and } F \subset A\}.$$

So $A \in \mathcal{A}$.

•   *The family $\mathcal{A}$ contains every open set.*

Let V be an open set. Since X is metrizable, V is an $\mathcal{F}_\sigma$-set. So there exists a sequence $\{F_n\}$ of closed sets satisfying $F_n \uparrow V$. By the continuity of the measure, we see that $\mu(F_n) \uparrow \mu(V)$, so

$$\mu(V) = \sup\{\mu(F) : F \text{ is closed and } F \subset V\}.$$

Since $\mu(V) = \inf\{\mu(G)\} : G$ open and $V \subset G\}$ is obviously true, we see that V belongs to $\mathcal{A}$.

Thus $\mathcal{A}$ is a σ-algebra that includes the open sets, so $\mathcal{A} = \mathcal{B}_X$.   ∎

The following lemma characterizes tight measures on a metrizable space.

12.6 Lemma   *Any finite Borel measure μ on a metrizable space X is tight if and only if for each $\varepsilon > 0$ there is a compact subset K of X such that $\mu(K) > \mu(X) - \varepsilon$.*

Proof: We may normalize μ to be a probability measure. Assume first that for each $0 < \varepsilon < 1$ there is a compact set K with $\mu(K) > 1 - \varepsilon$. By Theorem 12.5, μ is a normal measure. Consequently $\mu(A) = \sup\{\mu(F) : F \text{ closed and } F \subset A\}$ for each Borel set A. Therefore to show that μ is tight, it suffices to establish that

$$\mu(F) = \sup\{\mu(K) : K \text{ compact and } K \subset F\}$$

for each closed set F. (Why?) So let F be a closed set, and assume by way of contradiction that there exists some $0 < \varepsilon < 1$ such that

$$\sup\{\mu(K) : K \text{ compact and } K \subset F\} < \mu(F) - \varepsilon.$$

Now if C is a compact subset of X, then $C \cap F$ is a compact subset of F so

$$\mu(C) = \mu(C \cap F) + \mu(C \cap F^c) \leqslant \mu(F) - \varepsilon + \mu(F^c) = 1 - \varepsilon,$$

which contradicts our hypothesis. Hence, μ is tight. The converse is trivial.   ∎

12.2   Regular Borel measures

Regular Borel measures play an important role in the duality theory of spaces of
continuous functions. On Polish spaces, all finite Borel measures are regular.

12.7 Theorem   *A finite Borel measure on a Polish space is regular.*

Proof: Let X be a Polish space under the metric d, and let $\mu\colon \mathcal{B} \to [0,1]$ be a
probability measure. By Theorem 12.4 it suffices to show that μ is tight. Fix a
countable dense subset $\{x_1, x_2, \ldots\}$ of X, and for each i and n consider the closed
set $C_i^n = \{x \in X : d(x, x_i) \leqslant \frac{1}{n}\}$. Clearly, $X = \bigcup_{i=1}^{\infty} C_i^n$ for each n.
    Now fix $0 < \varepsilon < 1$ and for each n pick an integer k_n such that

$$\mu\Big(X \setminus \bigcup_{i=1}^{k_n} C_i^n\Big) < \frac{\varepsilon}{2^n}.$$

Next put $C = \bigcap_{n=1}^{\infty} \bigcup_{i=1}^{k_n} C_i^n$ and note that C is a totally bounded closed set. Since
C is also d-complete, we see that C is a compact set (Theorem 3.28). From

$$1 - \mu(C) = \mu(X \setminus C) = \mu\Big(\bigcup_{n=1}^{\infty}\Big(X \setminus \bigcup_{i=1}^{k_n} C_i^n\Big)\Big)$$
$$\leqslant \sum_{n=1}^{\infty} \mu\Big(X \setminus \bigcup_{i=1}^{k_n} C_i^n\Big) < \sum_{n=1}^{\infty} \frac{\varepsilon}{2^n} = \varepsilon,$$

we see that $\mu(C) > 1 - \varepsilon$. By Lemma 12.6 the measure μ is tight.   ∎

    As a consequence of Theorem 12.7 we obtain a remarkable theorem asserting
that Borel measurable functions between Polish spaces are "almost continuous"
in a measure-theoretic sense. For another elegant proof, and connections to the
Bochner integral, see P. A. Loeb and E. Talvila [228].

12.8 Lusin's Theorem   *Let $\mathcal{F}$ be a countable collection of Borel measurable
functions from a Polish space (X, τ) to a second countable topological space Y. If
μ is a Borel probability measure on X, then for each $\varepsilon > 0$ there exists a compact
subset K of X with $\mu(K^c) < \varepsilon$ such that the restriction of each function in $\mathcal{F}$ to K
is continuous.*

Proof: By Theorem 4.59, there exists a Polish topology $\tau^* \supset \tau$ on X such that
$\sigma(\tau^*) = \sigma(\tau) = \mathcal{B}orel$ and $f\colon (X, \tau^*) \to Y$ is continuous for each $f \in \mathcal{F}$.
    Now fix $\varepsilon > 0$. By Theorem 12.7, the measure μ is τ^*-regular. Therefore there
exists some τ^*-compact set K satisfying $\mu(X \setminus K) < \varepsilon$. Clearly, the set K is also
τ-compact and the identity mapping $I\colon (K, \tau^*) \to (K, \tau)$ is continuous. But then,
by Theorem 2.36, I is a homeomorphism, which means that τ^* and τ agree on K.
In particular, $f\colon (K, \tau) \to Y$ is continuous for each $f \in \mathcal{F}$.   ∎

In general, not every finite Borel measure is regular, even for compact Hausdorff spaces. The following example of a Borel measure that is not regular is based on P. R. Halmos [148, Exercise 10, p. 231] and W. Rudin [291, Exercise 17, pp. 58–59]. It is quite involved, but highly instructional.

12.9 Example (Nonregular Borel measure)   We present a Borel probability measure on the Borel σ-algebra of Ω, the compact Hausdorff space of ordinals (see Example 2.37), that is not regular.

Start by observing that any closed subset of Ω that does not contain ω_1 is countable. The reason is that every nonempty subset of Ω has a least upper bound. If the least upper bound is ω_1, then ω_1 is a limit point of the set. (Why?) Thus a nonempty closed set that does not contain ω_1 must have a least upper bound $b < \omega_1$, so it is a subset of the countable initial segment $I(b) = \{x \in \Omega : x \leq b\}$.

We now show that the collection of uncountable closed subsets of Ω is closed under countable intersections. That is, if $\{F_m\}$ is a countable collection of uncountable closed subsets of Ω, then the intersection $F = \bigcap_{m=1}^{\infty} F_m$ is closed and uncountable. The intersection F is clearly closed. To see that it is uncountable, start by observing that it is possible to construct an increasing sequence $\{x_n\}$ that meets each F_m infinitely often: We use an inductive argument to construct such a sequence. Start by taking a point $x_1 \neq \omega_1$ in F_1. Assume now that the points $x_1 < x_2 < \cdots < x_n < \omega_1$ have been selected so that $x_i \in F_i$ for each $1 \leq i \leq n$. Since the initial segment $I(x_n) = \{x \in \Omega : x \leq x_n\}$ is countable and F_{n+1} is uncountable, we can choose a point $x_{n+1} \in F_{n+1}$ such that $x_n < x_{n+1} < \omega_1$. Notice that every sequence so constructed has a least upper bound in Ω_0 (Theorem 1.14 (6)), which is also its limit. The limit of the sequence must belong to the intersection F, since there are subsequences included in each closed F_m. Now consider the collection L of limits of all sequences constructed as above. For any $x < \omega_1$, since $I(x)$ is countable we can construct such a sequence with $x_1 > x$, so L is not bounded by any $x < \omega_1$. Therefore $L \subset F$ is uncountable.

Let us say that a subset A of Ω is *big* if $A \cup \{\omega_1\}$ includes an uncountable closed set. Every nontrivial tail $(x, \omega_1]$ is big. A superset of a big set is big, and a countable intersection of big sets is still big. The complement of a big set is not big. (Why?) Say that a set is *small* if its complement is big. Let Σ be the family of all sets that are either big or small. The family Σ is obviously closed under complementation. It is also closed under countable intersections. To see this, note that a small set intersected with any set is small. Thus Σ is a σ-algebra.

Observe that every countable set is small: If A is countable, then $A \setminus \{\omega_1\}$ has an upper bound $b \in \Omega_0$, so $A^c \cup \{\omega_1\}$ includes the uncountable closed set $[x, \omega_1]$ for any $b < x < \omega_1$. Now every closed set is either countable or big, so Σ contains every closed set. Therefore Σ includes the Borel σ-algebra.

Define a probability measure μ on Σ by $\mu(A) = 1$ if A is big, and $\mu(A) = 0$ if A is small. We refer to this measure as the big-small measure on Ω. This set function is countably additive since countable intersections of big sets are big. In particular, this implies that no two big sets are disjoint, and that countable unions of small sets are small.

The big-small measure is not outer regular on $\mathcal{B}$: the singleton $\{\omega_1\}$ has measure zero, but every neighborhood of ω_1 has measure one since it includes a tail interval $[x, \omega_1]$ for some $x < \omega_1$, which is big.

Also note that the big-small measure μ is not tight. To see this, note first that every compact subset of Ω_0 is bounded, and so countable. (Otherwise the cover consisting of all the open sets $[1, x)$ for $x \in \Omega_0$ has no finite subcover. As an aside, this also shows that Ω_0 is not σ-compact.) Thus every compact subset of Ω_0 has μ-measure zero, while $\mu(\Omega_0) = 1$.   ∎

Measures induced from finite regular measures via continuous functions are also regular.

12.10 Theorem   *Let $f: X \to Y$ be a continuous mapping between Hausdorff spaces where X is compact. If μ is a regular Borel measure, then the measure μf^{-1} induced by μ on Y via*

$$\mu f^{-1}(A) = \mu(f^{-1}(A)),$$

is also a regular Borel measure.

Proof: Let μ be a regular Borel measure on $\mathcal{B}_X$. By Theorem 12.4, it suffices to prove that μf^{-1} is tight. Since μ is regular, for any $A \in \mathcal{B}_Y$, we have

$$\mu f^{-1}(A) = \sup\{\mu(K) : K \text{ is compact and } K \subset f^{-1}(A)\}. \qquad (\star)$$

Now let K be a compact set satisfying $K \subset f^{-1}(A)$. Then $f(K) \subset A$, so we have $K \subset f^{-1}(f(K)) \subset f^{-1}(A)$. Therefore, $\mu(K) \leqslant \mu f^{-1}(f(K)) \leqslant \mu f^{-1}(A)$. Since f is continuous, the set $C = f(K)$ is a compact subset of Y. From $(\star)$, we see that

$$\mu f^{-1}(A) = \sup\{\mu f^{-1}(C) : C \text{ compact and } C \subset A\}$$

for each $A \in \mathcal{B}_Y$. That is, μf^{-1} is tight.   ∎

12.11 Definition   *Let X be a Hausdorff space, and let $\mathcal{A}$ denote either the Borel σ-algebra $\mathcal{B}_{orel}$, the algebra $\mathcal{A}_X$ generated by the open sets, or the Baire σ-algebra $\mathcal{B}_{aire}$.*

- *$ba_n(\mathcal{A})$ is the set of normal signed Borel charges of bounded variation on $\mathcal{A}$.*

- *$ba_r(\mathcal{A})$ is the set of regular signed charges of bounded variation on $\mathcal{A}$.*

- *$ca_n(\mathcal{A})$ is the set of normal signed measures of bounded variation on $\mathcal{A}$.*

- *$ca_r(\mathcal{A}_X)$ is the set of regular signed measures of bounded variation on $\mathcal{A}$.*

Keep in mind that every charge of bounded variation is finite, so all the members of ba_r and ca_r are real-valued set functions.

12.12 Theorem *Let X be a Hausdorff space, and let $\mathcal{A}$ denote either $\mathcal{A}_X$ or $\mathcal{B}orel$. Then $ba_r(\mathcal{A})$ (resp. $ca_r(\mathcal{A})$) is a closed Riesz subspace of $ba(\mathcal{A})$ (resp. $ca(\mathcal{A})$).*

Proof: We establish the $ba_r(\mathcal{A})$ case. So let $\mu \in ba_r(\mathcal{A})$. We first show that μ^+ belongs to $ba_r(\mathcal{A})$. By Theorem 12.4, it suffices to show that μ^+ is tight. To this end, fix $A \in \mathcal{A}$ and let

$$s = \sup\{\mu^+(K) : K \text{ compact and } K \subset A\}.$$

Clearly, $s \leqslant \mu^+(A)$. Now fix $B \in \mathcal{A}$ with $B \subset A$ and let $\varepsilon > 0$. Pick a compact set $C \subset B$ such that $|\mu(C) - \mu(B)| < \varepsilon$. Then it is clear that $\mu(B) < \mu(C) + \varepsilon \leqslant s + \varepsilon$, so

$$\mu^+(A) = \sup\{\mu(B) : B \in \mathcal{A} \text{ and } B \subset A\} \leqslant s + \varepsilon$$

for all $\varepsilon > 0$. Hence $\mu^+(A) = s$, which shows that μ^+ is tight.

Next, notice that if $0 \leqslant \mu, \nu \in ba_r(\mathcal{A})$, then $\mu + \nu \in ba_r(\mathcal{A})$. (Why?) Given our discussion above, this implies that if $\mu \in ba_r(\mathcal{A})$, then $|\mu| = \mu^+ + \mu^- \in ba_r(\mathcal{A})$. This, together with the inequality

$$|(\mu + \nu)(K) - (\mu + \nu)(A)| \leqslant |\mu(K) - \mu(A)| + |\nu(K) - \nu(A)| \leqslant (|\mu| + |\nu|)(A \setminus K),$$

implies that $ba_r(\mathcal{A})$ is closed under addition. Obviously $ba_r(\mathcal{A})$ is closed under scalar multiplication. Hence, $ba_r(\mathcal{A})$ is Riesz subspace of $ba(\mathcal{A})$.

Finally, to see that $ba_r(\mathcal{A})$ is a closed subspace of $ba(\mathcal{A})$ let $\mu \in \overline{ba_r(\mathcal{A})}$. Fix $A \in \mathcal{A}$ and let $\varepsilon > 0$. Pick some $\nu \in ba_r(\mathcal{A})$ such that $\|\mu - \nu\| = |\mu - \nu|(X) < \varepsilon$ and then select a compact set $K \subset A$ with $|\nu(K) - \nu(A)| < \varepsilon$. Then

$$|\mu(K) - \mu(A)| \leqslant |\nu(K) - \mu(K)| + |\nu(K) - \nu(A)| + |\nu(A) - \mu(A)| < 3\varepsilon.$$

Thus μ is tight, so $ba_r(\mathcal{A})$ is closed. ∎

12.13 Corollary *Both $ba_r(\mathcal{A})$ and $ca_r(\mathcal{A})$ are AL-spaces in their own right.*

Similar results are true for the spaces ba_n and ca_n, and the proofs are virtually identical.

12.3 The support of a measure

Let X be a topological space. The **support** of a measure $\mu\colon \mathcal{B} \to [0, \infty]$, if it exists, is a closed set, denoted $\operatorname{supp}\mu$, satisfying:

1. $\mu((\operatorname{supp}\mu)^c) = 0$; and

2. If G is open and $G \cap \operatorname{supp}\mu \neq \varnothing$, then $\mu(G \cap \operatorname{supp}\mu) > 0$.[3]

---

[3] Many authors do not require condition (2) as part of the definition of support. The support of a measure is often defined by $\operatorname{supp}\mu = (\bigcup\{V : V \text{ open and } \mu(V) = 0\})^c$. By this definition, every measure has a (closed) support, but the support may not satisfy condition (2). See Example 12.15.

If a Borel measure has a support, then $\mu(\operatorname{supp}\mu) = \mu(X)$. A Borel measure μ cannot have more than one support. To see this, suppose that two closed sets F_1 and F_2 are supports. From $\mu(F_1^c) = 0$, we see that $F_1^c \cap F_2 = \varnothing$; otherwise, $F_1^c \cap F_2 \neq \varnothing$ implies $\mu(F_1^c) > 0$, which is a contradiction. Hence, $F_2 \subset F_1$. Similarly, $F_1 \subset F_2$, so $F_1 = F_2$.

In the same vein, a **carrier** of μ is any set $A \in \mathcal{B}$ satisfying $\mu(A^c) = 0$. In this case, we say that A **carries** μ. It is clear that a Borel measure has in general more than one carrier and its support (if it exists) is automatically a carrier.

12.14 Theorem   *Let X be a topological space, and let μ be a (not necessarily finite) Borel measure. If either X is second countable, or if μ is a tight measure, then μ has a (unique) support.*

Proof: Consider first the case where X is second countable. Let

$$G = \bigcup \{V : V \text{ open and } \mu(V) = 0\}$$

and let $S = G^c$. Then G is a countable union of open sets of measure zero, and hence has measure zero. Also, if V is open and $V \cap S \neq \varnothing$, then it follows that $\mu(V \cap S) > 0$: For if $\mu(V \cap S) = 0$, then $\mu(V) = \mu(V \cap S) + \mu(V \cap G) = 0$. Thus $V \subset G$, a contradiction. Therefore S is the support of μ.

Now consider the case where μ is tight, and define G as above. If K is a compact subset of G, then there exist open sets $V_1, \ldots, V_n$ with $\mu(V_i) = 0$ for each i and $K \subset \bigcup_{i=1}^n V_i$. It follows that $\mu(K) = 0$. Thus,

$$\mu(G) = \sup\{\mu(K) : K \text{ compact and } K \subset G\} = 0,$$

and as above, $S = G^c$ is the support of μ.   ∎

In particular, notice that every regular measure has a support. To appreciate the delicacy of the preceding result, we mention a measure that has no support.

12.15 Example (A Borel measure with no support)   Recall the big-small measure on the compact Hausdorff space Ω from Example 12.9 It is a Borel measure that has no support. To see this, observe that every ordinal $x < \omega_1$ is contained in a measure zero open set of the form $[1, y)$ where $x < y < \omega_1$. Thus the support cannot contain any point x in Ω_0. This leaves only the closed set $\{\omega_1\}$, but its complement, Ω_0, has measure one. Thus no set qualifies as the support. However, the big-small measure is carried by Ω_0, which is not closed.   ∎

We also note the following result, which we use without reference. We leave its proof as an exercise.

12.16 Lemma   *Let μ be a Borel measure on a topological space X that has a support, let $x \in \operatorname{supp}\mu$, and let $f \in C_b(X)$ satisfy $f(x) > 0$. Then for every sufficiently small neighborhood V of x, we have $\int_V f \, d\mu > 0$.*

12.17 Definition   *For any set X, δ_x, denotes the **point mass at x**, which is the probability measure on the power set of X carried by $\{x\}$. That is, $\delta_x(A) = 1$ if $x \in A$ and $\delta_x(A) = 0$ if $x \notin A$.* [4]

Note that when X is a Hausdorff space, δ_x (restricted to the Borel σ-algebra) is always a regular Borel measure having support $\{x\}$.

12.4   Nonatomic Borel measures

The main objective of this section is to show that every uncountable Polish space admits a nonatomic Borel probability measure. Recall that an atom of a measure is a measurable set of strictly positive measure that cannot be partitioned into disjoint measurable subsets of strictly positive measure. Before we can carry out the proof, we need some preliminary results.

12.18 Lemma   *Any atom of a Borel measure on a second countable Hausdorff space includes a singleton of positive measure.*

In particular, a Borel measure on a second countable Hausdorff space is nonatomic if and only if every singleton has measure zero.

Proof: Let μ be a Borel measure on a second countable Hausdorff space X. Fix a countable base $\{V_1, V_2, \ldots\}$ for the topology of X and let $A \in \mathcal{B}_X$ be an atom of μ. Let $I = \{i \in \mathbb{N} : \mu(A \cap V_i) = 0\}$ and consider the Borel set $B = A \setminus \bigcup_{i \in I} V_i$. Then B is a subset of A and $\mu(B) = \mu(A) > 0$. (Why?) In particular, $B \neq \varnothing$. We claim that B is a singleton.

To see this, suppose by way of contradiction that B contains two distinct points, say a and b. Since X is Hausdorff, there exist two disjoint basic open sets V_j and V_k such that $a \in V_j \cap A$ and $b \in V_k \cap A$. If $\mu(V_j \cap A) = 0$, then $j \in I$ contrary to $a \in B = A \setminus \bigcup_{i \in I} V_i$. Since A is an atom, we get $\mu(V_j \cap A) = \mu(A)$, and similarly $\mu(V_k \cap A) = \mu(A)$. However, since $(V_j \cap A) \cap (V_k \cap A) = \varnothing$, neither $A \cap V_j$ nor $A \setminus V_j$ has measure zero, which contradicts A being an atom. Therefore B is a singleton.   ∎

To appreciate Lemma 12.18 better, observe that the big-small measure on the ordinals we presented in Example 12.9 has the property that every set of measure one is an atom (since the only values it assumes are zero and one and an intersection of two big sets is a big set), yet every singleton set has measure zero.

12.19 Lemma   *Every separable metrizable space can be written as the disjoint union of a countable set and a perfect set. (Either of these sets may be empty.)*

---

[4] A point mass is sometimes called a **Dirac measure** or an **evaluation**.

Proof: Let (X, d) be a separable metric space, and let

$$A = \{x \in X : \text{For some } r > 0 \text{ the open ball } B_r(x) \text{ is countable}\}.$$

Since X is separable, so is A. Let $\{x_1, x_2, \ldots\}$ be a countable dense subset of A. For each n, let $\mathbb{N}_n = \{k \in \mathbb{N} : B_{1/k}(x_n) \text{ is countable}\}$. Then the open set $V = \bigcup_{n=1}^{\infty} \bigcup_{k \in \mathbb{N}_n} B_{1/k}(x_n)$ is countable, and we claim that $A \subset V$. To see this, let a belong to A. Fix some $r > 0$ such that $B_r(a)$ is countable and then select some k with $2/k < r$. Next, pick some n such that $d(a, x_n) < 1/k$ and note that $a \in B_{1/k}(x_n) \subset B_r(a)$. This implies $a \in V$.

Next, consider the closed set $P = X \setminus V$. If $x \in P$ and $r > 0$, then the open ball $B_r(x)$ has uncountably many points and, since V is countable, $B_r(x)$ contains a point in P different from x. That is, every point of P is an accumulation point of P, so P is perfect. Now note that $X = V \cup P$.   ∎

12.20 Lemma   *Every perfect set in a complete metric space includes a nonempty compact perfect set.*

Proof: Since (by definition) perfect sets are closed, let (X, d) be a perfect complete metric space. Given a set A, let $N_n(A) = \{x \in X : d(x, A) \leqslant \frac{1}{2^n}\}$. We claim that there exists a sequence $\{A_n\}$ of subsets of X such that:

1. Each A_n has 2^n elements, and

2. $A_n \subset A_{n+1}$ and $N_{n+1}(A_{n+1}) \subset N_n(A_n)$ for each n.

The proof proceeds by induction. Start by fixing two distinct points a, b in X and let $A_1 = \{a, b\}$. Now, for the inductive step, assume that A_n is a set with 2^n elements and put $\rho_n = \min\{d(u, v) : u, v \in A_n \text{ and } u \neq v\} > 0$. Since X is perfect, a ball of radius $\frac{1}{2^{n+1}}$ centered at a point in A_n contains infinitely many members of X. For each point $x \in A_n$ choose a point $y_x^n \neq x$ satisfying $d(x, y_x^n) < \frac{1}{2^{n+1}}$ and $d(x, y_x^n) < \rho_n$, and let $A_{n+1} = \bigcup_{x \in A_n} \{x, y_x^n\}$. Clearly, A_{n+1} has exactly 2^{n+1} elements and $A_n \subset A_{n+1}$. Moreover, $N_{n+1}(A_{n+1}) \subset N_n(A_n)$.

Let $K = \bigcap_{n=1}^{\infty} N_n(A_n)$. Then K is closed, and also totally bounded. Since X is complete, it follows that K is compact. Since $\bigcup_{n=1}^{\infty} A_n \subset K$, we see that K is also nonempty. Next, we claim that K is perfect.

To see this, let $x \in K$ and fix $\varepsilon > 0$. If $x \in A_k$ for some k, then $x \in A_n$ for each $n \geqslant k$, so from $d(x, y_x^n) < \frac{1}{2^{n+1}}$ it follows that for some $n \geqslant k$ the point $y_x^n \in K$ satisfies $y_x^n \neq x$ and $d(x, y_x^n) < \varepsilon$. Now assume that $x \notin A_k$ for each k. Then from $x \in N_n(A_n)$ for each n we get $d(x, A_n) \leqslant \frac{1}{2^n}$ for each n. So if we choose n so that $\frac{1}{2^n} < \varepsilon$, then any $z \in A_n \subset K$ with $d(x, z) < \frac{1}{2^n}$ satisfies $x \neq z$ and $d(x, z) < \varepsilon$. Therefore x is an accumulation point of K, so K is perfect.   ∎

You should compare the construction in the proof above to the construction of the Cantor set as a subset of the unit interval that we carried out in Section 3.13.

We can write the Cantor set as a countable intersection $\bigcap_{n=0}^{\infty} C_n$ of closed sets, where $C_0 = [0, 1]$, $C_1 = [0, \frac{1}{3}] \cup [\frac{2}{3}, 1]$, etc. (See the discussion after Definition 3.54.) Each closed set is the union of a finite collection of closed balls. The number of closed balls doubles at each stage, but their radius decreases by a factor of three. It is easy to see that the Cantor set is a compact perfect subset of $[0, 1]$.

The proof of the next result, which can be found in Y. A. Abramovich and A. W. Wickstead [3], is based on ideas of J. Feinstein. It uses the fact that the norm dual of a $C(K)$-space (K Hausdorff and compact) coincides with $ca_r(K)$; see Theorem 14.14 below. (Even though we refer to a later result, we do not introduce any circularity, and this is the appropriate section for the result.)

12.21 Theorem *If V is a nonempty open subset of a perfect locally compact Hausdorff space X, then there exists a regular Borel probability measure on X having support in V and vanishing at each singleton.*

Proof: Let V be a nonempty open subset of a perfect locally compact Hausdorff space X. The proof employs the following two properties:

a. Since X is perfect, every nonempty open set is infinite; and

b. Each point of X has a base of compact neighborhoods (Theorem 2.67).

We use induction to construct a "tree-like" sequence of compact neighborhoods, where at each stage n there are 2^n pairwise disjoint such neighborhoods.

We start the induction with $n = 0$. Fix some $x_1^0 \in V$ and then (using (b)) select a compact neighborhood $K = C_1^0$ of x_1^0 lying in V. At the nth stage of the inductive argument, assume that there are 2^n pairwise disjoint compact neighborhoods, say $C_1^n, \ldots, C_{2^n}^n$ of the points $x_1^n, \ldots, x_{2^n}^n$, respectively. We pass to the $(n+1)$-stage by obtaining two compact neighborhoods from each C_i^n as follows: For each $i = 1, \ldots, 2^n$ fix two distinct points $x_{i,1}^n$ and $x_{i,2}^n$ in the interior of C_i^n (such points always exist according to property (a)) and then choose—by using property (b)—two disjoint compact neighborhoods $C_{i,1}^n$ and $C_{i,2}^n$ lying in C_i^n of $x_{i,1}^n$ and $x_{i,2}^n$, respectively.

Now look again at the nth stage of our construction and consider the regular Borel probability measure $\mu_n = \frac{1}{2^n} \sum_{i=1}^{2^n} \delta_{x_i^n}$. Clearly $\{\mu_n\}$ is a sequence in the w^*-compact set

$$\{v \in ca_r(\mathcal{B}_K) : v \geqslant 0 \text{ and } \|v\| = v(K) = 1\}.$$

By Theorem 14.14 below, $ca_r(\mathcal{B}_K)$ is the norm dual of $C(K)$. Let $\mu \in ca_r(\mathcal{B}_K)$ be a w^*-accumulation point of $\{\mu_n\}$. To finish the proof, it suffices to show $\mu(\{x\}) = 0$ for each $x \in K$. (Clearly, $\mu(\{x\}) = 0$ for each $x \in X \setminus K$.)

So let $x \in K$. Fix n, and note that there exists at most one $i \in \{1, \ldots, 2^n\}$ such that $x \in C_i^n$. In particular, by Corollary 2.74, there exists a continuous function $f \colon X \to [0, 1]$ such that $f(x) = 1$ and $f = 0$ on C_j^n for $j \neq i$. Now note that for

$m \geqslant n$, we have $\int f \, d\mu_m \leqslant \frac{1}{2^n}$, and from this we infer that $\int f \, d\mu \leqslant \frac{1}{2^n}$. Thus, $\mu(\{x\}) \leqslant \int f \, d\mu \leqslant \frac{1}{2^n}$ for each n, so $\mu(\{x\}) = 0$.   ∎

And now we are ready to state and prove the main result of this section.

12.22 Theorem   *Every uncountable Polish space admits a nonatomic Borel probability measure.*

Proof: Let X be an uncountable Polish space. By Lemma 12.19 we can write X as the disjoint union of a countable set C and an (uncountable) perfect set P. Since P is closed, it is a Polish space in its own right. So by Lemma 12.20, P includes a nonempty compact perfect set K. By Theorem 12.21 and Lemma 12.18, K (and hence X) admits a nonatomic Borel probability measure.   ∎

For a proof of this result using the Baire Category Theorem 3.47, see K. R. Parthasarathy [271, Theorem 8.1, p. 53].

12.5   Analytic sets

Corollary 3.67 asserts that every Polish space is a continuous image of the Baire space $\mathcal{N} = \mathbb{N}^{\mathbb{N}}$. The converse is not true. As we shall presently see in Example 12.33, not every continuous image of the Baire space $\mathcal{N}$ is a separable completely metrizable space. Nonetheless, in spite of this, sets that are continuous images of $\mathcal{N}$ possess several important properties that will be discussed in this section. The results of this section make much use of Suslin schemes, so now may be a good time to review Section 3.14.

12.23 Definition   *A subset of a Polish space is **analytic**, if either it is empty or a continuous image of the Baire space $\mathcal{N}$, and **coanalytic** if its complement is analytic.*

There are two results that follow easily from the definition, and we use them without any special reference. Since the composition of continuous functions is continuous:

•   The continuous image in a Polish space of an analytic set is analytic.

From Corollary 4.61, we already know the following:

•   Every Borel subset of a Polish space is analytic.

There are several useful characterizations of analytic sets.

12.24 Theorem   *For a nonempty subset A of a Polish space X, the following statements are equivalent.*

1. *The set A is a continuous image of $\mathbb{N}$, that is, A is an analytic set.*

2. *The set A is the projection of a closed subset of $\mathbb{N} \times X$ on X.*

3. *The set A is a continuous image of a Polish space.*

4. *There is a Borel subset B of a Polish space and a Borel measurable function $f \colon B \to X$ with $f(B) = A$.*

5. *The set A is the nucleus of a regular Suslin scheme consisting of closed subsets of X and having vanishing diameter.*

Proof: (1) $\implies$ (2)   Let $f \colon \mathbb{N} \to A$ be a continuous surjection. Then Gr f is a closed subset of $\mathbb{N} \times X$ whose projection on X is A.

(2) $\implies$ (3) $\implies$ (4)   These results are straightforward, since closed subsets of Polish spaces are Polish, and continuous functions are Borel measurable.

(4) $\implies$ (1)   Assume that B is a Borel subset of a Polish space (Y, τ) and that $f \colon B \to X$ is a Borel measurable function having range A. By Lemma 4.58, there is a Polish topology $\tau_B \supset \tau$ on Y for which B is τ_B-clopen, and such that $\sigma(\tau_B) = \sigma(\tau)$. Therefore, (B, τ_B) is itself a Polish space. Now, according to Theorem 4.59, there exists a stronger Polish topology $\tau_f \supset \tau_B$ on Y such that $\sigma(\tau_f) = \sigma(\tau_B) = \sigma(\tau)$, and $f \colon (B, \tau_f) \to X$ is continuous. Next, by Theorem 4.60, there exists a continuous surjection $g \colon \mathbb{N} \to (B, \tau_f)$. This implies that the function $h = f \circ g \colon \mathbb{N} \to X$ is continuous and $h(\mathbb{N}) = A$.

(1) $\implies$ (5)   Let $f \colon \mathbb{N} \to A$ be a continuous surjection, and consider the Suslin scheme defined by

$$F_{n_1,\ldots,n_m} = \overline{f(U_{n_1,\ldots,n_m})} = \overline{f(\{n_1\} \times \cdots \times \{n_m\} \times \mathbb{N} \times \mathbb{N} \times \cdots)}.$$

This Suslin scheme is clearly regular (that is, $F_{n_1,\ldots,n_m,k} \subset F_{n_1,\ldots,n_m}$) and consists of closed sets. Moreover, Lemma 3.8 implies that it has vanishing diameter since the scheme $\{U_s : s \in \mathbb{N}^{<\mathbb{N}}\}$ has vanishing diameter (with respect to the tree metric) and f is continuous. Now to establish that $A = \mathcal{A}(F_s)$, repeat the proof of Theorem 3.66.

(5) $\implies$ (1)   Let $\{F_s : s \in \mathbb{N}^{<\mathbb{N}}\}$ be a regular Suslin scheme of closed sets with vanishing diameter whose nucleus is A. That is, $A = \bigcup_{n \in \mathbb{N}} \bigcap_{k=1}^{\infty} F_{n(1),\ldots,n(k)}$.

Since X is complete, the only way $\bigcap_{k=1}^{\infty} F_{n(1),\ldots,n(k)}$ can be empty is if for some k the set $F_{n(1),\ldots,n(k)}$ is itself empty. Let

$$\mathcal{M} = \{n \in \mathbb{N} : F_{n(1),\ldots,n(k)} \neq \emptyset \text{ for all } k\}.$$

Clearly, $A = \bigcup_{n \in \mathcal{M}} \bigcap_{k=1}^{\infty} F_{n(1),\ldots,n(k)}$. For each $n \in \mathcal{M}$ let $f(n)$ be the unique element of $\bigcap_{k=1}^{\infty} F_{n(1),\ldots,n(k)}$. As in the proof of Theorem 3.66, f is a continuous function from $\mathcal{M}$ onto A.

We now show that $\mathcal{M}$ is a closed subset of $\mathcal{N}$. Suppose $n \notin \mathcal{M}$, that is, $F_{n(1),\ldots,n(k)} = \varnothing$ for some k. Now notice that if $t(m, n) < 1/k$ (where t is the tree metric of Lemma 3.62), then $F_{m(1),\ldots,m(k)} = F_{n(1),\ldots,n(k)} = \varnothing$. Thus the complement of $\mathcal{M}$ is open.

Finally, by Lemma 3.64, $\mathcal{M}$ is a retract of $\mathcal{N}$. Hence A is a continuous image of $\mathcal{N}$, and the proof is finished. $\blacksquare$

12.25 Theorem   *The collection of analytic subsets of a Polish space is closed under countable unions and countable intersections.*

Proof: Let $\{A_1, A_2, \ldots\}$ be a countable collection of analytic subsets of the Polish space X and put $A = \bigcap_{n=1}^{\infty} A_n$. If $A = \varnothing$, then A is an analytic set, so we can assume that $A \neq \varnothing$. For each n let $f_n \colon \mathcal{N} \to A_n$ be a continuous mapping of $\mathcal{N}$ onto A_n. Now consider the metrizable space $\mathcal{N}^{\mathbb{N}}$, which is (by Theorem 3.61) homeomorphic to $\mathcal{N}$. Let D be the (nonempty) subset of $\mathcal{N}^{\mathbb{N}}$ defined by

$$D = \{(n_1, n_2, \ldots) \in \mathcal{N}^{\mathbb{N}} : f_1(n_1) = f_2(n_2) = \cdots = f_k(n_k) = \cdots\}.$$

Since each f_n is continuous, it follows that D is a closed subset of the Polish space $\mathcal{N}^{\mathbb{N}}$. Consequently, D is automatically a Polish space in its own right. Now define $f \colon D \to X$ via $f(n_1, n_2, \ldots) = f_k(n_k)$ for any k. By the construction of D, it doesn't matter which k we use, so f is well-defined and continuous. Now observe that $f(D) = \bigcap_{n=1}^{\infty} A_n = A$. By Theorem 12.24 (3), A is an analytic set.

Next we must show that the family of analytic sets is closed under countable unions. To do this, define $\mathcal{N}_k = \{(n_1, n_2, \ldots,) \in \mathcal{N} : n_1 = k\}\ (k = 1, 2, \ldots)$. Clearly, $\{\mathcal{N}_k : k \in \mathbb{N}\}$ is a partition of $\mathcal{N}$, and each $\mathcal{N}_k$ is both closed and open in $\mathcal{N}$, and each $\mathcal{N}_k$ is homeomorphic to $\mathcal{N}$. Now let $\{A_1, A_2, \ldots\}$ be a countable collection of nonempty analytic subsets of a Polish space X. For each k pick a surjective continuous function $g_k \colon \mathcal{N}_k \to A_k$. Then the function $g \colon \mathcal{N} \to X$, defined by $g|_{\mathcal{N}_k} = g_k$, is continuous and maps $\mathcal{N}$ onto the countable union $\bigcup_{n=1}^{\infty} A_n$. $\blacksquare$

The next result sheds a bit more light on the relation between Borel and analytic sets, and has some important consequences.

12.26 Lusin's Separation Theorem   *Any pair of disjoint analytic subsets of a Polish space can be separated by Borel sets. That is, if A and B are disjoint analytic subsets of a Polish space X, then there exist Borel subsets A' and B' of X satisfying $A \subset A'$, $B \subset B'$, and $A' \cap B' = \varnothing$.*

Proof: The proof will be based upon the following property (P):

(P)   *If two subsets A and B of a topological space can be written as $A = \bigcup_{n=1}^{\infty} A_n$ and $B = \bigcup_{m=1}^{\infty} B_m$, where each pair A_n and B_m can be separated by Borel sets, then the pair A and B can be also separated by Borel sets.*

Indeed, if for each n, m we choose two Borel sets $A'_{n,m}$ and $B'_{n,m}$ separating A_n and B_m, then note that the Borel sets $A' = \bigcup_{n=1}^\infty \bigcap_{m=1}^\infty A'_{n,m}$ and $B' = \bigcup_{m=1}^\infty \bigcap_{n=1}^\infty B'_{n,m}$ separate the sets A and B.

Now let A and B satisfy the hypotheses. If one is empty, say the set B, then the pair X and $\varnothing$ separates A and B. So without loss of generality assume both are nonempty. Since A and B are analytic, there are continuous surjections $f : \mathcal{N} \to A$ and $g : \mathcal{N} \to B$. Recall that for any finite sequence $s = (n_1, \ldots, n_m) \in \mathbb{N}^{<\mathbb{N}}$, the symbol U_s denotes the basic clopen subset $\{n_1\} \times \cdots \times \{n_m\} \times \mathbb{N} \times \mathbb{N} \times \cdots$ of $\mathcal{N}$, and define $A_s = f(U_s)$ and $B_s = g(U_s)$. (For simplicity, we also write $A_{n_1,\ldots,n_m}$ instead of $A_s = A_{(n_1,\ldots,n_m)}$.)

Suppose by way of contradiction that A and B cannot be separated by Borel sets. Since $A = \bigcup_{n=1}^\infty A_n$ and $B = \bigcup_{m=1}^\infty B_m$, by the above property (P), there exist n_1 and m_1 such that A_{n_1} and B_{m_1} cannot be separated by Borel sets. But $A_{n_1} = \bigcup_{n=1}^\infty A_{n_1,n}$ and $B_{m_1} = \bigcup_{m=1}^\infty B_{m_1,m}$. So by property (P) again, there must be some n_2 and m_2 such that A_{n_1,n_2} and B_{m_1,m_2} cannot be separated by Borel sets. Continuing in this way, we see that there exist $\boldsymbol{n}$ and $\boldsymbol{m}$ belonging to $\mathcal{N}$ such that for any k, the sets $A_{\boldsymbol{n}(1),\ldots,\boldsymbol{n}(k)}$ and $B_{\boldsymbol{m}(1),\ldots,\boldsymbol{m}(k)}$ cannot be separated by Borel sets. Now look at the points $x = f(\boldsymbol{n}) \in A$ and $y = g(\boldsymbol{m}) \in B$. Since A and B are disjoint, $x \neq y$. Thus there are disjoint open neighborhoods U of x and V of y in X. Since f and g are continuous, there is k large enough so that

$$A_{\boldsymbol{n}(1),\ldots,\boldsymbol{n}(k)} = f(U_{\boldsymbol{n}(1),\ldots,\boldsymbol{n}(k)}) \subset U$$

and

$$B_{\boldsymbol{m}(1),\ldots,\boldsymbol{m}(k)} = g(U_{\boldsymbol{m}(1),\ldots,\boldsymbol{m}(k)}) \subset V.$$

Thus, the open (and hence Borel) sets U and V separate $A_{\boldsymbol{n}(1),\ldots,\boldsymbol{n}(k)}$ and $B_{\boldsymbol{m}(1),\ldots,\boldsymbol{m}(k)}$, a contradiction. ∎

Recall that a subset of a Polish space is **coanalytic** if its complement is analytic. In general, disjoint coanalytic sets need not be separated by Borel sets, see, e.g., [231, pp. 260–264]. However, we have the following result.

12.27 Corollary   *A subset of a Polish space is both analytic and coanalytic if and only if it is a Borel set.*

Proof: If A is a Borel set, then so is A^c, so A is both analytic and coanalytic. Conversely, if A is both analytic and coanalytic, the Lusin Separation Theorem 12.26 says A and A^c can be separated by Borel sets, which implies they must be Borel sets themselves. ∎

This corollary leads to the following characterizations of Borel measurable functions between Polish spaces.

12.28 Theorem   *For a function $f : X \to Y$ between Polish spaces the following statements are equivalent.*

1. *f is Borel measurable.*

2. *$\mathrm{Gr}\, f$ is a Borel subset of $X \times Y$.*

3. *$\mathrm{Gr}\, f$ is an analytic subset of $X \times Y$.*

Proof: (1) $\implies$ (2) $\implies$ (3) The first implication follows from Theorem 4.45, and the second from Corollary 4.61.

(3) $\implies$ (1) Start by observing that if S is an arbitrary subset of Y, then

$$f^{-1}(S) = \pi((X \times S) \cap \mathrm{Gr}\, f),$$

where $\pi : X \times Y \to X$ is the projection of $X \times Y$ on X defined by $\pi(x, y) = x$. Clearly, π is a continuous function.

Now let B be a Borel subset of Y. Since $X \times B$ is Borel and hence analytic, and $\mathrm{Gr}\, f$ is analytic by hypothesis, their intersection is analytic in $X \times Y$ by Theorem 12.25. Therefore the projection $\pi((X \times B) \cap \mathrm{Gr}\, f) = f^{-1}(B)$ is also an analytic subset of X. Similar reasoning shows that $f^{-1}(B^c)$ is analytic in X, but $f^{-1}(B^c) = [f^{-1}(B)]^c$, so $f^{-1}(B)$ is both analytic and coanalytic. Therefore by Corollary 12.27, $f^{-1}(B)$ is a Borel subset of X, so f is Borel measurable.   ∎

The next lemma is a partial converse of Theorem 4.60.

12.29 Theorem   *Let $f : X \to Y$ be a Borel measurable function between Polish spaces. Let B be a Borel subset of X. If $f|_B$ is one-to-one, then $f(B)$ is a Borel subset of Y.*

Proof: Assume that $f : X \to Y$ is a Borel measurable function between Polish spaces and that B is a Borel subset of X such that $f|_B$ is one-to-one. By Theorem 4.59 there is a stronger Polish topology τ_f on X for which f is actually continuous. By Theorem 4.60 there is a closed subset F of $\mathcal{N}$ and a one-to-one τ_f-continuous function g from F onto B. Then $f \circ g$ is a one-to-one continuous function from F onto $f(B)$. So without loss of generality, it suffices to show that if F is a closed subset of $\mathcal{N}$ and $f : F \to Y$ is one-to-one and continuous, then $f(F)$ is a Borel subset of Y.

So consider the regular Suslin scheme $\{A_s : s \in \mathbb{N}^{<\mathbb{N}}\}$ defined by

$$A_{n_1,\dots,n_m} = f(U_{n_1,\dots,n_m} \cap F),$$

where $U_{n_1,\dots,n_m}$ is one of our basic clopen subsets of $\mathcal{N}$. Since the intersection $\bigcap_{k=1}^{\infty} U_{n(1),\dots,n(k)} = \{n\}$ is nonempty, by Lemma 3.8 the sequence $\{A_{n(1),\dots,n(m)}\}$ has vanishing diameter if n belongs to F. If n does not belong to F, then since F is closed there is some k such that $U_{n(1),\dots,n(k)} \cap F = \varnothing$, which also implies that the sequence $\{A_{n(1),...,n(m)}\}$ has vanishing diameter. Then just as in the proof of Theorem 12.24,

$$f(F) = \bigcup_{n\in F} \bigcap_{k=1}^{\infty} A_{n(1),...,n(k)}.$$

Clearly, each A_s is analytic, being the continuous image of the closed subset $U_s \cap F$ of $\mathcal{N}$. Since f is one-to-one, the sets $A_{n_1,...,n_m,j}$ and $A_{n_1,...,n_m,k}$ are disjoint whenever $j \neq k$, so $\{A_s\}$ is a Lusin scheme. By the Lusin Separation Theorem 12.26, for $j \neq k$ there exist disjoint Borel sets $B_{n_1,...,n_m,j}$ and $B_{n_1,...,n_m,k}$ with $B_{n_1,...,n_m,j} \supset A_{n_1,...,n_m,j}$ and $B_{n_1,...,n_m,k} \supset A_{n_1,...,n_m,k}$. Proceeding inductively and replacing $B_{n_1,...,n_m,j}$ by $B_{n_1,...,n_m,j} \cap B_{n_1,...,n_m}$ if necessary, we can assume that $B_{n_1,...,n_m,j} \subset B_{n_1,...,n_m}$ for each j. That is, $\{B_s\}$ is a regular Lusin scheme. Now let $\hat{B}_{n_1,...,n_m} = B_{n_1,...,n_m} \cap \overline{A}_{n_1,...,n_m}$. This guarantees that the scheme $\{\hat{B}_s\}$ has vanishing diameter, since $\{A_s\}$ does. Thus $\{\hat{B}_s\}$ is a regular Lusin scheme of Borel sets with vanishing diameter.

Put $\hat{A} = \bigcup_{n\in\mathcal{N}} \bigcap_{k=1}^{\infty} \hat{B}_{n(1),...,n(k)}$. We claim that $\hat{A}$ is a Borel set, and moreover $\hat{A} = f(F)$. To see that $\hat{A}$ is a Borel set, put

$$B = \bigcap_{m=1}^{\infty} \bigcup \{\hat{B}_{n_1,...,n_m} : (n_1,...,n_m) \in \mathbb{N}^m\}.$$

This is clearly a Borel set, being a countable intersection of countable unions of Borel sets. Furthermore, $\hat{A} = B$. This is because x belongs to B if and only for each m the point x belongs to a unique $\hat{B}_{n_1,...,n_m}$. (Uniqueness follows from disjointness.) But this is just another way of saying x belongs to $\hat{A}$.

To see that $\hat{A} = f(F)$, note that $B_{n(1),...,n(k)} \subset \overline{A}_{n(1),...,n(k)}$ by construction. So if x belongs to $\hat{A}$, then x belongs to $\mathcal{A}(\overline{A}_s)$. That is, there is some $n \in \mathcal{N}$ such that $\{x\} = \bigcap_{k=1}^{\infty} \overline{A}_{n(1),...,n(k)}$. Consequently, for this n, each $\overline{A}_{n(1),...,n(k)}$ must be nonempty, which implies that each set $A_{n(1),...,n(k)}$ is also nonempty, which in turn guarantees that for each k there exists some $n_k \in U_{n(1),...,n(k)} \cap F$. But this sequence $\{n_k\}$ is Cauchy, and so converges to some m in F. By continuity, we must have $f(m) = x$. This shows that $\hat{A} \subset f(F)$.

Finally, note that $A_{n(1),...,n(k)} \subset B_{n(1),...,n(k)}$. Therefore if $x = f(n)$ for some $n \in F$, then $\{x\} = \bigcap_{k=1}^{\infty} A_{n(1),...,n(k)}$, so $x \in \bigcap_{k=1}^{\infty} \hat{B}_{n(1),...,n(k)}$. Hence $f(F) \subset \hat{A}$. This concludes the proof. ∎

So far we have not shown that there exist analytic sets that are not Borel sets. We shall show that there is an analytic subset of $\mathcal{N}$ whose complement is not analytic, and hence by Corollary 12.27 is not Borel. In general, in any uncountable Polish space, not every analytic set is a Borel set, and the collection of analytic sets is not a σ-algebra.

We now proceed on a lengthy trip to prove this. We start with the notion of a universal set. The definition is a bit awkward, so bear with us. For each space X in some appropriate class of topological spaces, let $\mathcal{C}(X)$ be a variable symbol that stands for some distinguished class of subsets of X. We have in mind that $\mathcal{C}(X)$ may be the class of open subsets of X, or the class of analytic subsets of X, or the class of Borel sets of X, etc. We say that a subset A of $\mathcal{N} \times X$ is a **universal set for the class** $\mathcal{C}(X)$ if (i) A belongs to $\mathcal{C}(\mathcal{N} \times X)$ and (ii) the collection of sections $C_n = \{x \in X : (n, x) \in A\}$, $n \in \mathcal{N}$, coincides with $\mathcal{C}(X)$, that is, $\mathcal{C}(X) = \{C_n : n \in \mathcal{N}\}$.

In many cases it is clearer to think of a universal set as the graph of a correspondence. So let us say that a correspondence $\varphi \colon \mathcal{N} \twoheadrightarrow X$ is a **universal correspondence for** $\mathcal{C}(X)$ if (i) $\operatorname{Gr}\varphi$ belongs $\mathcal{C}(\mathcal{N} \times X)$ and (ii) the collection of images $\{\varphi(n) : n \in \mathcal{N}\}$ is equal to $\mathcal{C}(X)$. We shall also say that a topological space **admits a universal correspondence for the class** $\mathcal{C}$, if there exists a universal correspondence $\varphi \colon \mathcal{N} \twoheadrightarrow X$ for its family of subsets $\mathcal{C}(X)$. The next lemma describes a correspondence that is universal for the class of open sets of a second countable topological space.

12.30 Lemma   *Every second countable topological space admits a universal correspondence for the class of open sets.*

Proof: Let $\{U_2, U_3, \ldots\}$ be a countable base for a topological space X. For each $n \in \mathcal{N}$ let $K_n = \{k \geqslant 2 : n(k) = n(1)\}$, and define $\varphi \colon \mathcal{N} \twoheadrightarrow X$ by

$$\varphi(n) = \bigcup_{k \in K_n} U_k.$$

Note that the $\varphi(n)$ may be empty, which is good, because the empty set is an open subset of X. By construction, every value of φ is open. To show that φ is universal, we need to show that φ has open graph and that every open subset of X is a value of φ. To see the latter, let V be an open set in X and define n by $n(1) = 1$ and for $k \geqslant 2$ let

$$n(k) = \begin{cases} 1 & \text{if } U_k \subset V \\ 7 & \text{otherwise.} \end{cases}$$

Then $\varphi(n) = V$.

To see that the graph of φ is open in $\mathcal{N} \times X$, let (n, x) belong to the graph, that is, $x \in \varphi(n)$. Thus there is some $k \geqslant 2$ with $x \in U_k$ and $n(k) = n(1)$. Consider the ball $B_{\frac{1}{k}}(n)$ of radius $\frac{1}{k}$ around n (in the tree metric on $\mathcal{N}$). If m belongs to this neighborhood, then

$$m(j) = n(j) \quad \text{for } j = 1, \ldots, k.$$

In particular, $m(k) = n(k) = n(1) = m(1)$, so $U_k \subset \varphi(m)$. Therefore $B_{\frac{1}{k}}(n) \times U_k$ is a neighborhood of (n, x) included in the graph of φ, which shows that the graph is open.   ∎

By taking the complement of φ we have the following consequence.

12.31 Corollary   *Every second countable topological space admits a universal correspondence for the class of closed sets.*

We now proceed to a slightly more complicated construction.

12.32 Corollary   *Every Polish space admits a universal correspondence for the family of analytic sets.*

Proof: Let X be a Polish space. According to statement (2) of Theorem 12.24, a subset A of X is analytic if and only if there is a closed subset F of $\mathcal{N} \times X$ with $\pi_X(F) = A$. But $\mathcal{N} \times X$ is Polish, so by Corollary 12.31 there is a universal correspondence $\varphi \colon \mathcal{N} \twoheadrightarrow \mathcal{N} \times X$ for the closed subsets of $\mathcal{N} \times X$. Thus the values of $\gamma = \pi_X \circ \varphi \colon \mathcal{N} \twoheadrightarrow X$ coincide precisely with the analytic subsets of X. It remains to be shown that γ has analytic graph. Now

$$
\begin{aligned}
\operatorname{Gr}\gamma &= \{(\boldsymbol{n}, x) \in \mathcal{N} \times X : x \in \gamma(\boldsymbol{n}) = \pi_X(\varphi(\boldsymbol{n}))\} \\
&= \{(\boldsymbol{n}, x) \in \mathcal{N} \times X : (\boldsymbol{m}, x) \in \varphi(\boldsymbol{n}) \text{ for some } \boldsymbol{m} \in \mathcal{N}\} \\
&= \pi(\operatorname{Gr}\varphi),
\end{aligned}
$$

where $\pi \colon \mathcal{N} \times \mathcal{N} \times X \to \mathcal{N} \times X$ is defined by $\pi(\boldsymbol{n}, \boldsymbol{m}, x) = (\boldsymbol{n}, x)$. Clearly π is a continuous function from the Polish space $\mathcal{N} \times \mathcal{N} \times X$ onto the Polish space $\mathcal{N} \times X$, and $\operatorname{Gr}\gamma$ is the image under π of the closed set $\operatorname{Gr}\varphi$, and so analytic.   ∎

We now show that there is an analytic subset of $\mathcal{N}$ whose complement is not analytic, and hence not a Borel set.

12.33 Example (An analytic set that is not a Borel set)   By Corollary 12.32 there is a correspondence $\gamma \colon \mathcal{N} \twoheadrightarrow \mathcal{N}$ that is universal for the analytic subsets of $\mathcal{N}$. In particular, $\operatorname{Gr}\gamma$ is an analytic subset of $\mathcal{N}^2$. Let F denote the set of fixed points of γ, that is,

$$ F = \{\boldsymbol{n} \in \mathcal{N} : \boldsymbol{n} \in \gamma(\boldsymbol{n})\}. $$

Then F is an analytic subset of $\mathcal{N}$. To see this, let $D = \{(\boldsymbol{n}, \boldsymbol{n}) : \boldsymbol{n} \in \mathcal{N}\}$ be the diagonal of $\mathcal{N}^2$, which is closed, and hence analytic. By Theorem 12.25, the intersection $D \cap \operatorname{Gr}\gamma$ is analytic, so $\pi(D \cap \operatorname{Gr}\gamma)$, where π is the projection of $\mathcal{N}^2$ onto its first factor, is analytic. But $\pi(D \cap \operatorname{Gr}\gamma)$ is just F.

Now Cantor's Diagonal Theorem 1.5 asserts that there is no $\boldsymbol{n} \in \mathcal{N}$ for which $\gamma(\boldsymbol{n}) = F^c$. But since γ is universal for the analytic sets of X, the coanalytic set F^c cannot be an analytic subset of X. By Corollary 12.27 the analytic set F cannot be a Borel set either.   ∎

One good example deserves another.

12.34 Example (A closed subset of $\mathcal{N}^2$ whose projection is not Borel)   According to Example 12.33 there is an analytic subset A of $\mathcal{N}$ that is not a Borel set. Let $f \colon \mathcal{N} \to \mathcal{N}$ be continuous with $f(\mathcal{N}) = A$. Then the graph of f is a closed subset of $\mathcal{N}^2$ and its projection on the second factor is A, which is not Borel.   ∎

A. S. Kechris [196, Hint 18.17, p. 360] describes the following correspondence with closed graph and nonempty values that admits no Borel measurable selector. That is, there is no measurable function whose graph lies in the graph of the correspondence. This example is important for Chapter 18.

12.35 Example (A nonempty-valued closed correspondence with no Borel measurable selector) The idea of this example is actually quite simple. Imagine there were a "universal Borel function," that is, a function $n \mapsto f_n$ defined on $\mathcal{N}$ such that each f_n is a Borel measurable function from $\mathcal{N}$ into $\mathcal{N}$, and every Borel measurable function from $\mathcal{N}$ into $\mathcal{N}$ is of the form f_n for some n. Imagine further that there were a nonempty-valued correspondence $\varphi \colon \mathcal{N} \twoheadrightarrow \mathcal{N}$ with closed graph such that for each n, $\varphi(n)$ does not contain $f_n(n)$. Then it is clear that φ can have no Borel measurable selector. For if f were such a selector, then $f = f_k$ for some k, in which case we would have $f_k(k) \in \varphi(k)$ since f is a selector, but by construction $f_k(k) \notin \varphi(k)$, a contradiction.

Real life, alas, is not so simple, but a variation on the idea above provides the example we seek. Recall that a function $f \colon X \to Y$ between Polish spaces is Borel measurable if and only if its graph is a Borel subset of $X \times Y$ (Theorem 12.28). So by Theorem 4.60 there is a closed subset F of $\mathcal{N}$ and a one-to-one continuous function $g \colon F \to X \times Y$ such that $g(F) = \operatorname{Gr} f$. Since g is continuous, its graph is a closed subset of $\mathcal{N} \times X \times Y$. For our example, $X = Y = \mathcal{N}$, so we are naturally interested in closed subsets of $\mathcal{N}^3$, some of which define Borel measurable functions from $\mathcal{N}$ into $\mathcal{N}$.

With this point in mind, let $\psi \colon \mathcal{N} \twoheadrightarrow \mathcal{N}^3$ be a universal correspondence for the closed subsets of $\mathcal{N}^3$. By the above discussion, every Borel measurable function from $\mathcal{N}$ into $\mathcal{N}$ corresponds to some $\psi(n)$, but not every $\psi(n)$ defines a Borel measurable function.

Next fix a homeomorphism $h \colon \mathcal{N} \to \mathcal{N}^2$, and let $h_0, h_1 \colon \mathcal{N} \to \mathcal{N}$ be the functions such that $h(n) = (h_0(n), h_1(n))$. The need for this homeomorphism will be apparent only later. Now define the correspondence $\sigma \colon \mathcal{N} \twoheadrightarrow \mathcal{N}$ as follows:

$$\sigma(n) = \begin{cases} \{h_0(p)\}^c & \text{if } \exists! \, (m, p) \text{ with } (m, n, p) \in \psi(n) \\ \mathcal{N} & \text{otherwise,} \end{cases} \qquad (\star)$$

where, as usual, $\exists!$ means "there exists exactly one." This correspondence has been designed so that it has nonempty values, and whenever $\psi(n)$ defines a Borel measurable function f as described above, then $h_0(f(n))$ does not belong to $\sigma(n)$. Unfortunately, σ need not have closed graph, so we shall look for something like a subcorrespondence that does have closed graph.

We assert that $\operatorname{Gr} \sigma$ is an analytic subset of $\mathcal{N}^2$, but defer the proof of this until the end, so as not to disrupt the flow of the argument. Given that $\operatorname{Gr} \sigma$ is analytic in $\mathcal{N}^2$, it follows from Theorem 12.24 that there is a closed set C in $\mathcal{N} \times \mathcal{N}^2$ such that $\operatorname{Gr} \sigma = \pi(C)$, where $\pi \colon \mathcal{N} \times \mathcal{N}^2 \to \mathcal{N}^2$ is the projection defined by $\pi(m, n, p) = (n, p)$. In other words,

$$\sigma(n) = \{p \in \mathcal{N} : \exists \, m \in \mathcal{N} \text{ such that } (m, n, p) \in C\}. \qquad (\star\star)$$

Finally, define the correspondence $\varphi \colon \mathcal{N} \twoheadrightarrow \mathcal{N}$ by

$$\varphi(n) = \{p \in \mathcal{N} : (h_1(p), n, h_0(p)) \in C\}.$$

We claim that φ has closed graph and nonempty values, but first let us verify that φ has no Borel measurable selector.

To see this, use $(\star\star)$ to observe that:

$$p \in \varphi(n) \implies (h_1(p), n, h_0(p)) \in C \implies h_0(p) \in \sigma(n). \qquad (\star\star\star)$$

Suppose by way of contradiction that $f \colon \mathcal{N} \to \mathcal{N}$ is a Borel measurable selector for φ. By the remarks above there is a closed subset F of $\mathcal{N}$ and a one-to-one continuous function $g \colon F \to \mathcal{N}^2$ such that $g(F) = \operatorname{Gr} f$. Since g is continuous, its graph is a closed subset of $\mathcal{N}^3$. Since ψ is universal for the closed subsets of $\mathcal{N}^3$, there is some $q \in \mathcal{N}$ such that

$$\psi(q) = \operatorname{Gr} g.$$

Since g is one-to-one, there is exactly one $r \in \mathcal{N}$ for which $g(r) = (q, f(q))$. In other words, there is exactly one pair (m, p) such that the triple (m, q, p) belongs to $\operatorname{Gr} g = \psi(q)$, namely $m = r$ and $p = f(q)$. It follows from $(\star)$ that

$$\sigma(q) = \{h_0(f(q))\}^{\mathrm{c}}.$$

But f is a selector for φ, so $f(q) \in \varphi(q)$, so by $(\star\star\star)$ we have $h_0(f(q)) \in \sigma(q)$, which is a contradiction. Therefore no Borel measurable selector from φ exists.

Next, we prove that $\operatorname{Gr} \sigma$ is analytic. The proof makes use of the following fact (F):

> (F) *If X and Y are Polish spaces and F is a Borel subset of $X \times Y$, then the set $\{x \in X : \exists! \, y \in Y \text{ such that } (x, y) \in F\}$ is coanalytic in X.*

This fact is proved by A. S. Kechris [196, Theorem 18.11, pp. 123–127], and in a slightly different guise by K. Kuratowski [218, Theorem §39.VII.1, vol. I, pp. 494–495].

Now define the correspondence $\tau \colon \mathcal{N} \twoheadrightarrow \mathcal{N}$ by

$$\tau(n) = \begin{cases} \{p\} & \text{if } \exists! \, (m, p) \text{ such that } (m, n, p) \in \psi(n) \\ \varnothing & \text{otherwise.} \end{cases}$$

Then

$$\operatorname{Gr} \tau = \{(n, p) \in \mathcal{N}^2 : \exists! \, (m, p) \text{ such that } (n, m, n, p) \in \operatorname{Gr} \psi\},$$

which, since $\operatorname{Gr}\psi$ is closed, is coanalytic in light of the fact (F) above. Now observe that $\operatorname{Gr}\sigma$ is the image of the analytic set $(\operatorname{Gr}\tau)^c$ under the continuous mapping $(n, p) \to (n, h_0(p))$.

Next, we verify that φ has closed graph and nonempty values. It is easy to see that $\operatorname{Gr}\varphi$ is closed, since the function $(n, p) \mapsto (h_1(p), n, h_0(p))$ is continuous, and $\operatorname{Gr}\varphi$ is the inverse image of the closed set C.

To see that $\varphi(n)$ is nonempty for any n, recall that σ has nonempty values, so by ($\star\star$), there is some (m, p) for which $(m, n, p) \in C$. But h is a homeomorphism from $\mathcal{N}$ onto $\mathcal{N}^2$, so there is some q for which $h(q) = (h_0(q), h_1(q)) = (p, m)$. But then by definition, $p \in \varphi(n)$. (Now you know why we introduced the homeomorphism h.) This completes the construction. ∎

As an aside, we mention that in general, the σ-algebra of Borel sets does not admit a universal correspondence.

12.36 Lemma   *The Baire space $\mathcal{N}$ does not admit a universal correspondence for its class of Borel sets.*

Proof: To see this, assume by way of contradiction that there exists a universal correspondence for the class of Borel sets of $\mathcal{N}$, say $\gamma: \mathcal{N} \twoheadrightarrow \mathcal{N}$. Let D be the diagonal of $\mathcal{N} \times \mathcal{N}$. Then the set F of fixed points of γ is (if nonempty) the one-to-one image of the Borel set $D \cap \operatorname{Gr}\gamma$ under the projection mapping along the first coordinate. Thus, by Theorem 12.29, F is Borel, so F^c is also Borel. But F^c cannot be a value of γ. (Why?) Therefore, γ cannot be universal. ∎

Nonetheless, analytic sets are not too far from being Borel sets in the following sense. Let μ be a Borel probability measure on a topological space X and recall that a subset A of X is μ-measurable if it differs from a Borel set only by a set of outer μ-measure zero. The collection of μ-measurable subsets is denoted $\mathcal{B}_\mu$ and is a σ-algebra (Theorem 10.20). Sets that are μ-measurable for every Borel probability measure μ are called **universally measurable**. The collection of universally measurable sets is a σ-algebra, being the intersection of the σ-algebras $\mathcal{B}_\mu$. We show in the next section (Theorem 12.41) that every analytic subset of a Polish space is universally measurable. In general, the σ-algebra generated by the analytic sets is a proper subset of the σ-algebra of universally measurable sets. In order to prove the universal measurability of analytic sets we need to digress and develop a little bit of the theory of capacities.

12.6 The Choquet Capacity Theorem

In this section we develop just enough of the theory of capacities to show that analytic sets are universally measurable. There are several different kinds of capacity discussed in the literature, but they all share the following requirement. A

capacity on a set X is an extended real-valued set function v defined on the power set 2^X that is **monotone**. That is,

$$A \subset B \implies v(A) \leqslant v(B).$$

For our limited purpose we restrict attention to the case where X is a Hausdorff space and the capacity satisfies additional regularity properties. For lack of a better generally accepted term, we shall say that a capacity v on a Hausdorff topological space X is **nice** if it satisfies the following properties.

1. *Normalization*: $v(\varnothing) = 0$. [5]

2. *Continuity from below on* 2^X: If $A_n \uparrow A$ in 2^X, then $v(A_n) \uparrow v(A)$.

3. *Continuity from above on* $\mathcal{K}$: If a sequence $\{K_n\}$ of compact sets satisfies $K_n \downarrow K$, then $v(K_n) \downarrow v(K)$.

4. *Regularity*: If K is compact, then $v(K) < \infty$ and

$$v(K) = \inf\{v(U) : U \text{ is open and } K \subset U\}.$$

We next present the two most important examples of nice capacities.

12.37 Lemma   *If μ is a Borel probability measure on a Polish space X, then the (Carathéodory) outer measure μ^* induced by μ, defined by*

$$\mu^*(A) = \inf\{\mu(B) : B \in \mathcal{B} \text{ and } A \subset B\},$$

is a nice capacity.

Proof: Recall from the Carathéodory Extension Theorem 10.23 that μ^* so defined is an outer measure and agrees with μ on the Borel sets. It is consequently monotone and normalized. Since compact sets are Borel sets, continuity from above follows from Theorem 10.8. Since μ is (by Theorem 12.7) regular, it easily follows that μ^* satisfies regularity.

The only property that may be in doubt is continuity from below. To establish this, we use the fact that if S is an arbitrary subset of X, then there exists a Borel set E satisfying $S \subset E$ and $\mu^*(S) = \mu(E)$. Indeed, if for each n we pick a Borel set E_n with $S \subset E_n$ and $\mu(E_n) \leqslant \mu^*(S) + \frac{1}{n}$, then the Borel set $E = \bigcap_{n=1}^{\infty} E_n$ satisfies $S \subset E$ and $\mu^*(S) = \mu(E)$. (See also Lemma 10.30.)

Now let $A_n \uparrow A$. Pick a Borel set E satisfying $A \subset E$ and $\mu^*(A) = \mu(E)$. Also, for each n choose a Borel set E_n with $A_n \subset E_n$ and $\mu^*(A_n) = \mu(E_n)$. Next, consider the sequence of Borel sets $B_n = (\bigcap_{k=n}^{\infty} E_k) \cap E$, $n = 1, 2, \ldots$. Clearly, $A_n \subset B_n \subset E_n$ for each n and $B_n \uparrow B$, where $A \subset B \subset E$. By monotonicity, we

---

[5] Normalization coupled with the monotonicity of v yields $0 \leqslant v(A) \leqslant \infty$ for each $A \in 2^X$.

have $\mu^*(A_n) = \mu(B_n)$ for each n and $\mu(B) = \mu^*(A)$. Thus, by the continuity of μ from below on the Borel σ-algebra, we get

$$\mu^*(A_n) = \mu(B_n) \uparrow \mu(B) = \mu^*(A),$$

and the proof is complete.   ∎

The proof of the next lemma is left for you.

12.38 Lemma   *Let $f: X \to Y$ be a continuous function between Hausdorff spaces, and let v be a nice capacity on Y. Then the set function v_f on X defined by $v_f(A) = v(f(A))$ is a nice capacity.*

We continue with another important property of analytic sets—which is, in fact, a characterization of analytic sets. (See, e.g., [196, Theorem 25.13, p. 200].)

12.39 Lemma   *Every analytic subset A of a Polish space is the nucleus of a Suslin scheme $\{A_s : s \in \mathbb{N}^{<\mathbb{N}}\}$, that is, $A = \mathcal{A}(A_s)$, satisfying the following properties.*

1. *Each A_s is analytic.*

2. *For each finite sequence $(n_1, \ldots, n_m)$ we have*

$$A_{n_1,\ldots,n_m,k} \uparrow_k A_{n_1,\ldots,n_m}.$$

3. *For each $\boldsymbol{n}$ in $\mathcal{N}$ the set $A_{\boldsymbol{n}} = \bigcap_{k=1}^{\infty} A_{n(1),\ldots,n(k)}$ is compact.*

4. *If $A_{\boldsymbol{n}} \subset U$ for some open set U, then $A_{n(1),\ldots,n(k)} \subset U$ for some k.*

Proof: If $A = \varnothing$, then let $A_s = \varnothing$ for all s. Now assume A is a nonempty subset of a Polish space X, and let f be a continuous function from $\mathcal{N}$ onto A. For each $(n_1, \ldots, n_m)$ in $\mathbb{N}^{<\mathbb{N}}$ consider the clopen set

$$V_{n_1,\ldots,n_m} = \{\boldsymbol{n} \in \mathcal{N} : \boldsymbol{n}(k) \leqslant n_k, \ k = 1, \ldots, m\}.$$

The Tychonoff Product Theorem 2.61 implies that for $\boldsymbol{n}$ in $\mathcal{N}$ the set

$$V_{\boldsymbol{n}} = \bigcap_{k=1}^{\infty} V_{n(1),\ldots,n(k)} = \{\boldsymbol{m} \in \mathcal{N} : \boldsymbol{m}(k) \leqslant \boldsymbol{n}(k), \ k = 1, 2, \ldots\}$$

is a compact subset of $\mathcal{N}$. Next, define the Suslin scheme $\{A_s : s \in \mathbb{N}^{<\mathbb{N}}\}$ by $A_{n_1,\ldots,n_m} = f(V_{n_1,\ldots,n_m})$. We shall verify that this scheme has the desired properties.

Observe first that since each V_s is closed, each A_s is analytic. So this Suslin scheme satisfies (1). Furthermore, (2) is satisfied by construction. To establish (3), we need to verify that

$$f(V_{\boldsymbol{n}}) = \bigcap_{k=1}^{\infty} f(V_{n(1),\ldots,n(k)}) = A_{\boldsymbol{n}}. \qquad (\star)$$

To see this, suppose x belongs to $\bigcap_{k=1}^{\infty} f(V_{n(1),\ldots,n(k)})$. Then, for each k there is $n_k \in V_{n(1),\ldots,n(k)}$ such that $f(n_k) = x$. Now (using the standard diagonal process), we see that the sequence $\{n_k\}$ must have a subsequence that converges pointwise to some m in V_n, so by continuity $f(m) = x$. Thus $f(V_n) \supset \bigcap_{k=1}^{\infty} f(V_{n(1),\ldots,n(k)})$, and the other inclusion is obvious. Since each V_n is compact, it follows from $(\star)$ that each A_n is a compact set. Moreover, notice that

$$\mathcal{A}(A_s) = \bigcup_{n \in \mathcal{N}} \bigcap_{k=1}^{\infty} A_{n(1),\ldots,n(k)} = \bigcup_{n \in \mathcal{N}} f(V_n) = f\left(\bigcup_{n \in \mathcal{N}} V_n\right) = f(\mathcal{N}) = A.$$

For (4), assume A_n is included in some open set U. Suppose by way of contradiction that for each k there is some point n_k in $V_{n(1),\ldots,n(k)}$ such that $f(n_k)$ belongs to the closed set U^c. Then, using again the standard diagonal process, we see that some subsequence of $\{n_k\}$ converges pointwise to some $m \in V_n$. This implies $f(m) \in f(V_n) \cap U^c = A_n \cap U^c = \varnothing$, a contradiction. The proof is now complete. ∎

G. Choquet realized that the subadditivity property of an outer measure was irrelevant to the following result.

12.40 Choquet Capacity Theorem   *If v is a nice capacity on a Polish space, then for any analytic set A we have*

$$v(A) = \sup\{v(K) : K \text{ is compact and } K \subset A\}.$$

Proof: Let A be an analytic subset of a Polish space X and let v be a nice capacity. If $v(A) = 0$, then since $\varnothing$ is compact, we are through. So suppose $v(A) > \alpha \geqslant 0$. It suffices to find a compact set $K \subset A$ with $v(K) \geqslant \alpha$.

Let $\{A_{n_1,\ldots,n_m}\}$ be a Suslin scheme as in Lemma 12.39 associated with the analytic set A. Since $A_k \uparrow A$ and v is continuous from below, there is some n_1 such that $v(A_{n_1}) > \alpha$. Likewise there is some n_2 such that $v(A_{n_1,n_2}) > \alpha$. Continuing in this manner we construct some n in $\mathcal{N}$ such that for each k we have $v(A_{n(1),\ldots,n(k)}) > \alpha$.

Now we claim that the compact set $A_n = \bigcap_{k=1}^{\infty} A_{n(1),\ldots,n(k)}$ satisfies $v(A_n) \geqslant \alpha$. To see this, assume by way of contradiction that this is not the case, that is, suppose $v(A_n) < \alpha$. Then, by the regularity property of nice capacities there is an open set U satisfying $v(A_n) \leqslant v(U) < \alpha$. But then, by Lemma 12.39 (4), for some k we have $A_{n(1),\ldots,n(k)} \subset U$. The latter, in view of the monotonicity of v, implies

$$\alpha < v(A_{n(1),\ldots,n(k)}) \leqslant v(U) < \alpha,$$

which is impossible. This completes the proof. ∎

We mention in passing that the version of the Choquet Capacity Theorem we proved is a special case of a more general result in which the family of compact sets is replaced by an arbitrary family $\mathcal{S}$ that is closed under finite unions and countable intersections and contains $\emptyset$, and the class of analytic sets is re- placed by the family of sets obtained from $\mathcal{S}$ via the Suslin operation. See, e.g., P.-A. Meyer [246, T III.19, p. 39].

The proof that analytic sets are universally measurable is now easy.

12.41 Theorem   *Every analytic subset of a Polish space is universally measurable.*

Proof: Let μ be a Borel probability measure on a Polish space X. We know from Lemma 12.37 that the outer measure μ^* is a nice capacity on X. Let A be an analytic subset of X and fix some Borel set B with $A \subset B$ and $\mu^*(A) = \mu(B)$. By the Choquet Capacity Theorem 12.40, we have

$$\mu^*(A) = \sup\{\mu^*(K) : K \text{ is compact and } K \subset A\}.$$

Thus, for each n there is a compact subset K_n of A such that

$$\mu(B) = \mu^*(A) \leqslant \mu^*(K_n) + \tfrac{1}{n} = \mu(K_n) + \tfrac{1}{n},$$

where the last equality follows from the fact that K_n is compact and hence a Borel set. In particular, the set $E = \bigcup_{n=1}^{\infty} K_n \subset A$ is a Borel set. Now from $A \setminus E \subset B \setminus K_n$, it follows that

$$\mu^*(A \setminus E) \leqslant \mu(B \setminus K_n) = \mu(B) - \mu(K_n) < \tfrac{1}{n}$$

for each n. Thus $\mu^*(A \setminus E) = 0$, so $A \setminus E$ is a μ-measurable set. Finally, from the identity $A = E \cup (A \setminus E)$, we see that A is also μ-measurable. Since μ is arbitrary, A is universally measurable. ∎

Chapter 13

L_p-spaces

In this chapter, we introduce the classical L_p-spaces and study their basic properties. Recall that for a measure space (X, Σ, μ), two measurable real functions f and g on X are *equivalent* if they agree μ-almost everywhere. For $0 < p < \infty$, the p-norm of f is defined by

$$\|f\|_p = \left(\int |f|^p \, d\mu \right)^{\frac{1}{p}},$$

where we allow the integral to be infinite. Note that this integral depends only on the equivalence class of f. The space $L_p(\mu)$ is the collection of equivalence classes of measurable functions f for which the p-norm is finite. The space $L_\infty(\mu)$ comprises the equivalence classes of essentially bounded measurable functions, while $L_0(\mu)$ is the collection of the equivalence classes of measurable functions. With the pointwise algebraic and lattice operations all the L_p-spaces are order complete Riesz spaces. In fact, for $1 \leqslant p \leqslant \infty$, the $L_p(\mu)$-spaces are all Banach lattices (Theorem 13.5). For $0 \leqslant p < 1$ the $L_p(\mu)$-spaces are not Banach lattices, indeed they are not locally convex topological vector spaces, but they are nevertheless Fréchet lattices (Theorem 13.31). Theorem 13.11 proves the remarkable result that for a probability measure on the σ-algebra Σ, the Banach sublattices of $L_p(\Sigma)$ that contain the constant function $\mathbf{1}$ are exactly the Banach sublattices of the form $L_p(\mathcal{A})$ for some σ-subalgebra $\mathcal{A}$ of Σ.

The duals of the L_p spaces have an interesting representation—they are also L_p-spaces. Theorem 13.26, due to F. Riesz, asserts that if $1 < p, q < \infty$ satisfy $\frac{1}{p} + \frac{1}{q} = 1$, then the norm dual of $L_p(\mu)$ can be identified with $L_q(\mu)$ (and vice-versa by symmetry).

Besides norm convergence in L_p-spaces, there is another natural notion of convergence, convergence in measure. A sequence of measurable functions $\{f_n\}$ *converges in measure* to a measurable function f if for each $\varepsilon > 0$ we have

$$\lim_{n \to \infty} \mu(\{x \in X : |f_n(x) - f(x)| \geqslant \varepsilon\}) = 0.$$

Convergence in measure gives rise to the smallest Hausdorff locally solid topology on an L_p-space, and it is seldom locally convex. As a matter of fact, we establish that for finite nonatomic measure spaces, the topological dual of any $L_p(\mu)$-space with the topology of convergence in measure is trivial (Theorem 13.41).

We discuss several other topics related to L_p-spaces. For instance, we pay special attention to the Radon–Nikodym Theorem 13.18 and its applications. In particular, we use the Radon–Nikodym Theorem to prove Lyapunov's Convexity Theorem 13.33, which states that the range of finite-dimensional vector of nonatomic finite measures is a compact convex set. The chapter ends with a brief study of the extremely useful "Change of Variables" formulas.

13.1 L_p-norms

In this section (X, Σ, μ) will always be a measure space. A μ-measurable function $f \colon X \to \mathbb{R}$ is **p-integrable** (for $0 < p < \infty$) if $|f|^p$ is an integrable function. The set of p-integrable functions is denoted $L_p(\mu)$. Actually it is customary to identify functions that are equal almost everywhere. So $L_p(\mu)$ consists of equivalence classes rather than functions. We do this so the formulas below define norms and not just seminorms.

For $0 < p < \infty$, the set $L_p(\mu)$ is actually a vector space under the pointwise operations. It is clearly closed under scalar multiplication. To see that the sum of two p-integrable functions is also p-integrable, observe that for any pair a, b of real numbers, if $|a| \leqslant |b|$, then $|a + b|^p \leqslant (|a| + |b|)^p \leqslant (2|b|)^p \leqslant 2^p (|a|^p + |b|^p)$. This implies that

$$|f + g|^p \leqslant 2^p (|f|^p + |g|^p),$$

so $f + g$ is p-integrable if both f and g are.

If $f \in L_p(\mu)$, then the **L_p-norm** of f is defined by

$$\|f\|_p = \left(\int_X |f|^p \, d\mu \right)^{\frac{1}{p}}.$$

The $\|\cdot\|_\infty$-norm (or the **essential sup norm**) of a μ-measurable function $f \colon X \to \mathbb{R}$ is defined by

$$\|f\|_\infty = \inf\{M > 0 : |f(x)| \leqslant M \text{ for } \mu\text{-almost all } x\},$$

where the convention $\inf \varnothing = \infty$ applies. The collection of all equivalence classes of measurable functions f with $\|f\|_\infty < \infty$ is denoted $L_\infty(\mu)$. We let $L_0(\mu)$ denote the set of equivalence classes of measurable functions. In all cases, $L_p(\mu)$ is a vector space. The next result justifies the symbol $\|\cdot\|_\infty$ used to designate the essential sup norm.

13.1 Lemma *If (X, Σ, μ) is a finite measure space and $f \in L_\infty(\mu)$, then*

$$\lim_{p \to \infty} \|f\|_p = \|f\|_\infty.$$

Proof: Fix an arbitrary function $f \in L_\infty(\mu)$. From $|f| \leqslant \|f\|_\infty \chi_X$ a.e., we see that $\|f\|_p \leqslant \|f\|_\infty [\mu(X)]^{\frac{1}{p}}$. So

$$\limsup_{p \to \infty} \|f\|_p \leqslant \|f\|_\infty. \qquad (\star)$$

Let $\varepsilon > 0$. Then the measurable set $E = \{x \in X : |f(x)| \geqslant \|f\|_\infty - \varepsilon\}$ has positive measure. From the inequality $(\|f\|_\infty - \varepsilon)\chi_E \leqslant |f|$, it follows that $(\|f\|_\infty - \varepsilon)[\mu(E)]^{\frac{1}{p}} \leqslant \|f\|_p$. Therefore, $\|f\|_\infty - \varepsilon \leqslant \liminf_{p \to \infty} \|f\|_p$ for each $\varepsilon > 0$, which means that $\|f\|_\infty \leqslant \liminf_{p \to \infty} \|f\|_p$. This combined with $(\star)$ shows that $\lim_{p \to \infty} \|f\|_p = \|f\|_\infty$. ∎

The following useful and important observation is based on Lemma 11.22.

- *If a μ-measurable function $f: X \to \mathbb{R}$ belongs to some L_p-space, then there exists a Σ-measurable function $g: X \to \mathbb{R}$ such that $g = f$ a.e.*

The practical significance of this is that we may replace a Σ_μ-measurable function by a Σ-measurable function, and as far as integration goes, nothing changes.

13.2   Inequalities of Hölder and Minkowski

Two positive numbers $1 \leqslant p, q \leqslant \infty$ are called **conjugate exponents** if $\frac{1}{p} + \frac{1}{q} = 1$, where we adhere to the convention $\frac{1}{\infty} = 0$. Regarding conjugate exponents, we state the following important inequality known as **Hölder's Inequality**

13.2 Hölder's Inequality   *If p and q is a pair of conjugate exponents, $f \in L_p(\mu)$ and $g \in L_q(\mu)$, then $fg \in L_1(\mu)$ and*

$$\int |fg| \, d\mu \leqslant \|f\|_p \cdot \|g\|_q.$$

Proof: See [13, Theorem 31.3, p. 256]. ∎

13.3 Corollary   *If μ is finite and $0 \leqslant p < q \leqslant \infty$, then $L_q(\mu) \subset L_p(\mu)$. Moreover, if $\mu^*(X) = 1$, then for each $f \in L_q(\mu)$ we have $\|f\|_p \leqslant \|f\|_q$.*

Proof: If $q = \infty$ or $p = 0$, then the conclusion is obvious. So we may assume $0 < p < q < \infty$. Put $r = \frac{q}{p} > 1$, $s = \frac{q}{q-p} > 1$, and note that $\frac{1}{r} + \frac{1}{s} = 1$.

Now let $f \in L_q(\mu)$. Then $(|f|^p)^r = |f|^q \in L_1(\mu)$, that is, $|f|^p \in L_r(\mu)$. Since μ is a finite measure, the constant function $\mathbf{1}$ belongs to $L_s(\mu)$. So by Hölder's Inequality 13.2, we know that $|f|^p = |f|^p \cdot \mathbf{1} \in L_1(\mu)$ and

$$(\|f\|_p)^p = \int |f|^p \, d\mu = \int |f|^p \cdot \mathbf{1} \, d\mu$$
$$\leqslant \left(\int (|f|^p)^r \, d\mu \right)^{\frac{1}{r}} \cdot \left(\int \mathbf{1}^s \, d\mu \right)^{\frac{1}{s}} = (\|f\|_q)^p \cdot [\mu^*(X)]^{\frac{1}{s}}.$$

Therefore, $f \in L_p(\mu)$ and if $\mu^*(X) = 1$, then $\|f\|_p \leqslant \|f\|_q$. ∎

Minkowski's Inequality is just the triangle inequality for the L_p-norms.

13.4 Minkowski's Inequality *For $f, g \in L_p(\mu)$, where $1 \leqslant p \leqslant \infty$, we have*

$$\|f + g\|_p \leqslant \|f\|_p + \|g\|_p.$$

Proof: For $p = 1$ or $p = \infty$ the inequality is clearly true. So we can assume $1 < p < \infty$. Let $1 < q < \infty$ be such that $\frac{1}{p} + \frac{1}{q} = 1$.

We already know that if f and g belong to $L_p(\mu)$, then $f + g$ likewise belongs to $L_p(\mu)$. Next observe that since $(p-1)q = p$, it follows that $|f + g|^{p-1} \in L_q(\mu)$. Thus, by Hölder's Inequality 13.2 both functions

$$|f| \cdot |f + g|^{p-1} \quad \text{and} \quad |g| \cdot |f + g|^{p-1}$$

belong to $L_1(\mu)$, and we have the inequality

$$\int |f| \cdot |f + g|^{p-1} \, d\mu \leqslant \|f\|_p \cdot \left(\int |f + g|^{(p-1)q} \, d\mu \right)^{\frac{1}{q}}$$

$$= \|f\|_p \cdot (\|f + g\|_p)^{\frac{p}{q}}.$$

Similarly, we have

$$\int |g| \cdot |f + g|^{p-1} \, d\mu \leqslant \|g\|_p \cdot (\|f + g\|_p)^{\frac{p}{q}}.$$

So from $|f + g|^p = |f + g||f + g|^{p-1} \leqslant (|f| + |g|)|f + g|^{p-1}$, we get

$$(\|f + g\|_p)^p = \int |f + g|^p \, d\mu$$

$$\leqslant \int |f| \cdot |f + g|^{p-1} \, d\mu + \int |g| \cdot |f + g|^{p-1} \, d\mu$$

$$\leqslant \|f\|_p \cdot (\|f + g\|_p)^{\frac{p}{q}} + \|g\|_p \cdot (\|f + g\|_p)^{\frac{p}{q}}$$

$$= (\|f\|_p + \|g\|_p) \cdot (\|f + g\|_p)^{\frac{p}{q}}.$$

This easily implies

$$\|f + g\|_p = (\|f + g\|_p)^{p - \frac{p}{q}} \leqslant \|f\|_p + \|g\|_p.$$

The proof of the theorem is now complete. ∎

In each $L_p(\mu)$-space, define the partial order $f \leqslant g$ to mean $f(x) \leqslant g(x)$ for almost all x. With this ordering each $L_p(\mu)$ is a Riesz space. In fact, for $1 \leqslant p \leqslant \infty$ each $L_p(\mu)$ is a Banach lattice.

13.5 Riesz–Fischer Theorem *For $1 \leqslant p \leqslant \infty$, the Riesz space $L_p(\mu)$ equipped with the L_p-norm is a Banach lattice.*

Proof: We start by proving that each L_p-norm is a lattice norm. Actually, we prove the result for $1 \leqslant p < \infty$ and leave the case $p = \infty$ as an exercise. Let $0 \leqslant f_n \uparrow$ be a Cauchy sequence in $L_p(\mu)$. By Theorem 9.3, it suffices to show that $\{f_n\}$ converges in $L_p(\mu)$.

Since $\{f_n\}$ is Cauchy, it is easy to see that there exists some $M > 0$ such that $0 \leqslant \int (f_n)^p \, d\mu = (\|f_n\|_p)^p \uparrow \leqslant M$. By Levi's Theorem 11.18, there exists a function $0 \leqslant g \in L_1(\mu)$ such that $(f_n)^p \uparrow g$ a.e. Then $0 \leqslant f = g^{\frac{1}{p}} \in L_p(\mu)$ and from the Lebesgue Dominated Convergence Theorem 11.21, we get $\|f_n - f\|_p \to 0$. ∎

The $L_\infty(\mu)$ Banach lattices are order complete, and unless X is essentially a finite set, they do not have σ-order continuous norms. For instance, in $L_\infty[0,1]$ we have $\chi_{(0,\frac{1}{n})} \downarrow 0$, while $\|\chi_{(0,\frac{1}{n})}\|_\infty = 1$ for each n.

Although norm convergence in L_p-spaces does not imply pointwise convergence (why?), we nevertheless have the following useful result.

13.6 Theorem *If $\|f_n - f\|_p \to 0$ in some $L_p(\mu)$-space ($1 \leqslant p \leqslant \infty$), then there exist a subsequence $\{g_n\}$ of $\{f_n\}$ and a function $g \in L_p(\mu)$ satisfying $|g_n| \leqslant g$ μ-a.e. and $g_n(x) \to f(x)$ for μ-almost all x.*

Proof: For $p = \infty$ the conclusion is obvious. So assume $1 \leqslant p < \infty$. By passing to a subsequence, we can assume that $\|f_{n+1} - f_n\|_p < \frac{1}{2^n}$ for each n. By Levi's Theorem 11.18, $0 \leqslant h = \sum_{n=1}^{\infty} |f_{n+1} - f_n| \in L_p(\mu)$. (Why?) Moreover, from

$$|f_{n+k} - f_n| \leqslant \sum_{i=n}^{n+k-1} |f_{i+1} - f_i| \leqslant h, \qquad (\star)$$

we see that $\{f_n(x)\}$ is a Cauchy sequence of real numbers for μ-almost all x. Thus $f_n \to f^*$ a.e., and from $(\star)$, we get $|f_n - f^*| \leqslant h$ a.e. for all n. In particular, $|f^*| \leqslant h + |f_1|$ implies $f^* \in L_p(\mu)$. Clearly $|f_n| \leqslant h + |f^*| = g \in L_p(\mu)$ a.e. for each n. Finally, note that $|f_n - f^*| \leqslant h$ a.e. implies $\|f_n - f^*\|_p \to 0$, so $f^* = f$ μ-a.e. ∎

Every $L_p(\mu)$-space (for $1 \leqslant p < \infty$) has order continuous norm.

13.7 Theorem *For $1 \leqslant p < \infty$, the Banach lattice $L_p(\mu)$ has order continuous norm (and hence it is also order complete).*

Proof: Assume $f_\alpha \downarrow 0$ in $L_p(\mu)$, where $1 \leqslant p < \infty$. Let $\int |f_\alpha|^p \, d\mu \downarrow s \geqslant 0$. We must show that $s = 0$.

Start by picking a sequence of indexes $\{\alpha_n\}$ satisfying $\alpha_{n+1} \geqslant \alpha_n$ for each n and $\int |f_{\alpha_n}|^p \, d\mu \downarrow s$. We claim that $|f_{\alpha_n}|^p \downarrow 0$. To this end, let $|f_{\alpha_n}|^p \downarrow f \geqslant 0$ and fix some index α. For each n there exists some index β_n such that $\beta_n \geqslant \alpha$ and $\beta_n \geqslant \alpha_n$; we can assume that $\beta_{n+1} \geqslant \beta_n$ for each n. If $|f_{\beta_n}|^p \downarrow g \geqslant 0$, then $f \geqslant g$ and $\int f \, d\mu = \int g \, d\mu = s$. Hence, $g = f$, so $f = g \leqslant f_{\beta_n} \leqslant f_\alpha$ for each index α. In view of $f_\alpha \downarrow 0$, we infer that $f = 0$. Therefore, $|f_{\alpha_n}|^p \downarrow 0$, so $s = \lim_{n \to \infty} \int |f_{\alpha_n}|^p \, d\mu = 0$. ∎

13.3   Dense subspaces of L_p-spaces

We collect a few results concerning the norm denseness of certain important subspaces of L_p-spaces. The first one is immediate from the definition of L_p-spaces.

13.8 Theorem   *For each* $1 \leqslant p < \infty$ *the Riesz subspace of all step functions is norm dense in* $L_p(\mu)$. *If* μ *is finite, then the step functions are also* $\| \cdot \|_\infty$-*dense in* $L_\infty(\mu)$.

13.9 Theorem   *If* $1 \leqslant p < \infty$ *and* μ *is a regular Borel measure on a locally compact Hausdorff space* X, *then the Riesz subspace* $C_c(X)$ *of all continuous real-valued functions on* X *with compact support is norm dense in* $L_p(\mu)$.

Proof: Let $1 \leqslant p < \infty$. Since the step functions are norm dense in $L_p(\mu)$ it suffices to show that for each $A \in \mathcal{B}$ with $\mu(A) < \infty$ and each $\varepsilon > 0$ there exists some $f \in C_c(X)$ such that $\|\chi_A - f\|_p = (\int |\chi_A - f|^p \, d\mu)^{\frac{1}{p}} < \varepsilon$.

To this end, let $A \in \mathcal{B}$ satisfy $\mu(A) < \infty$, and let $\varepsilon > 0$. Since μ is a regular Borel measure, there exist a compact set K and an open set V satisfying $K \subset A \subset V$ and $\mu(V \setminus K) < \frac{\varepsilon}{2}$. By Corollary 2.69 there exists an open set W with compact closure such that $K \subset W \subset \overline{W} \subset V$, and from Corollary 2.74 there exists a function $f \in C(X)$ such that $f = 1$ on K and $f = 0$ on $X \setminus W$. Clearly, $f \in C_c(X)$. Now note that

$$\left(\int |\chi_A - f|^p \, d\mu \right)^{\frac{1}{p}} \leqslant \left(\int |\chi_A - \chi_K|^p \, d\mu \right)^{\frac{1}{p}} + \left(\int |\chi_K - f|^p \, d\mu \right)^{\frac{1}{p}}$$

$$\leqslant \left(\int \chi_{A \setminus K} \, d\mu \right)^{\frac{1}{p}} + \left(\int \chi_{V \setminus K} \, d\mu \right)^{\frac{1}{p}}$$

$$\leqslant \left(\tfrac{\varepsilon}{2} \right)^{\frac{1}{p}} + \left(\tfrac{\varepsilon}{2} \right)^{\frac{1}{p}} = 2 \left(\tfrac{\varepsilon}{2} \right)^{\frac{1}{p}} \leqslant 2 \cdot \tfrac{\varepsilon}{2} = \varepsilon,$$

and the proof is finished.   ∎

A function $f \colon \mathbb{R}^n \to \mathbb{R}$ is a C^∞-**function** if it has continuous partial derivatives of all orders. Remarkably, for $1 \leqslant p < \infty$ the C^∞-functions with compact support are norm dense in the L_p-spaces.

13.10 Theorem   *For each* $1 \leqslant p < \infty$ *the vector space of all* C^∞-*functions on* $\mathbb{R}^n$ *with compact support is norm dense in* $L_p(\mathbb{R}^n)$, *where* $\mathbb{R}^n$ *is equipped with Lebesgue measure.*

Proof: See [14, Problem 31.33, p. 292].   ∎

13.4 Sublattices of L_p-spaces

In this section we characterize the Banach sublattices of certain L_p-spaces. In particular, for the remainder of the section we assume that (X, Σ, μ) is a probability space—that is, Σ is a σ-algebra and μ is a measure on Σ satisfying $\mu(X) = 1$.

We mentioned before that every equivalence class of any $L_p(\mu)$-space contains a Σ-measurable function (see Theorem 10.35), so we can assume that all "functions" in the $L_p(\mu)$-spaces are Σ-measurable. Now let $\mathcal{A}$ be a σ-subalgebra of Σ. That is, $\mathcal{A} \subset \Sigma$ and $\mathcal{A}$ is a σ-algebra. Then, $(X, \mathcal{A}, \mu)$ is another probability space. To distinguish the L_p-spaces of the two measure spaces (X, Σ, μ) and $(X, \mathcal{A}, \mu)$, we shall write $L_p(\Sigma) = L_p(X, \Sigma, \mu)$ and $L_p(\mathcal{A}) = L_p(X, \mathcal{A}, \mu)$. We can assume that the elements of $L_p(\Sigma)$ are Σ-measurable functions and the elements of $L_p(\mathcal{A})$ are likewise $\mathcal{A}$-measurable functions. Then $L_p(\mathcal{A})$ is a closed Riesz subspace of $L_p(\Sigma)$, that is, $L_p(\mathcal{A})$ is a Banach sublattice of $L_p(\Sigma)$. This can be seen by noting that $L_p(\mathcal{A})$ is either a complete Riesz subspace of $L_p(\Sigma)$ or by employing Theorem 13.6. Remarkably, the $L_p(\mathcal{A})$ Banach sublattices are the only ones containing the constant function one.

13.11 Theorem *If (X, Σ, μ) is a probability space and $1 \leqslant p < \infty$, then the closed Riesz subspaces of $L_p(\Sigma)$ that contain the constant function $\mathbf{1}$ are exactly the Banach sublattices of the form $L_p(\mathcal{A})$ for some σ-subalgebra $\mathcal{A}$ of Σ.*

Proof: Let L be a closed Riesz subspace of some $L_p(\mu)$, where $1 \leqslant p < \infty$, such that $\mathbf{1} \in L$. Define the collection of sets $\mathcal{A} = \{A \in \Sigma : \chi_A \in L\}$. A direct verification shows that $\mathcal{A}$ is a σ-subalgebra of Σ; the order continuity of the L_p-norm is needed here to show that $\mathcal{A}$ is closed under countable unions.

Since the $\mathcal{A}$-step functions (which belong to L) are norm dense in $L_p(\mathcal{A})$, we get $L_p(\mathcal{A}) \subset L$. Now let $f \in L$. If $A = \{x \in X : f(x) > 0\}$, then $nf^+ \wedge \mathbf{1} \uparrow \chi_A$ and moreover $\{nf^+ \wedge \mathbf{1}\} \subset L$ (since L is a closed Riesz subspace). This implies $\|nf^+ \wedge \mathbf{1} - \chi_A\|_p \to 0$, from which it follows that $\chi_A \in L$, so $A \in \mathcal{A}$. Next, note that since $f - c\mathbf{1} \in L$, the preceding case implies

$$\{x \in X : f(x) > c\} = \{x \in X : (f - c\mathbf{1})(x) > 0\} \in \mathcal{A}.$$

This shows that f is $\mathcal{A}$-measurable and hence $f \in L_p(\mathcal{A})$. Thus, $L = L_p(\mathcal{A})$ and the proof is finished. ∎

A simple modification of the preceding proof yields the following L_∞-version of Theorem 13.11.

13.12 Theorem *Let (X, Σ, μ) be a probability space, and let L be a closed Riesz subspace of $L_\infty(\Sigma)$ containing the constant function $\mathbf{1}$. Then the following statements are equivalent.*

1. *$L = L_\infty(\mathcal{A})$ for some σ-subalgebra $\mathcal{A}$ of Σ.*

2. *If $\{f_n\} \subset L$, $f \in L_\infty(\Sigma)$, and $f_n(x) \uparrow f(x)$ for μ-almost all x, then $f \in L$.*

13.5 Separable L_1-spaces and measures

Let μ be a finite measure on a semiring and, as usual, let Σ_μ denote the σ-algebra of the μ-measurable sets. We say that two measurable sets A and B are μ-**equivalent** (or that $A = B$ μ-a.e.) if $\mu(A \vartriangle B) = 0$. It is easy to see that this defines an equivalence relation on Σ_μ. For simplicity, we denote the equivalence classes of Σ_μ by Σ_μ again. That is, we identify μ-equivalent sets. For instance, every μ-null is identified with the empty set.

The mapping $A \mapsto \chi_A$ is a natural embedding of Σ_μ into $L_1(\mu)$. We have $\chi_A \wedge \chi_B = \chi_{A \cap B}$, $\chi_A \vee \chi_B = \chi_{A \cup B}$, and $\chi_{A \setminus B} = \chi_A - \chi_{A \cap B}$. Identifying Σ_μ with its image in $L_1(\mu)$, we can think of Σ_μ as a subset of $L_1(\mu)$.[1] As such, under the induced metric

$$d(A, B) = \|\chi_A - \chi_B\|_1 = \int |\chi_A - \chi_B|\, d\mu = \mu(A \vartriangle B),$$

Σ_μ is a metric space and, in fact, a complete metric space.

13.13 Lemma   *The set Σ_μ is a closed subset of $L_1(\mu)$, and hence Σ_μ is a complete metric space in its own right.*

Proof: Let $\int |\chi_{A_n} - f|\, d\mu \to 0$. By Lemma 13.6, we can assume (by passing to a subsequence) that $\chi_{A_n}(x) \to f(x)$ for μ-almost all x. Thus, $f = \chi_A$ for some $A \in \Sigma_\mu$, so Σ_μ is a closed subset of $L_1(\mu)$.   ∎

The functions

$$(A, B) \mapsto A \cup B, \quad (A, B) \mapsto A \cap B, \quad \text{and} \quad (A, B) \mapsto A \setminus B,$$

from $\Sigma_\mu \times \Sigma_\mu$ to Σ_μ are all uniformly continuous. For instance, to see that the function $(A, B) \mapsto A \cup B$ is uniformly continuous, note that

$$d(A \cup B, C \cup D)$$

$$= \int |\chi_{A \cup B} - \chi_{C \cup D}|\, d\mu$$

$$= \int |\chi_A \vee \chi_B - \chi_C \vee \chi_D|\, d\mu$$

$$\leqslant \int |\chi_A \vee \chi_B - \chi_C \vee \chi_B|\, d\mu + \int |\chi_C \vee \chi_B - \chi_C \vee \chi_D|\, d\mu$$

$$\leqslant \int |\chi_A - \chi_C|\, d\mu + \int |\chi_B - \chi_D|\, d\mu$$

$$= d(A, C) + d(B, D).$$

---

[1] The set Σ_μ plays another important role as a subset of $L_1(\mu)$. An element $v \geqslant 0$ of a Riesz space is said to be a **component**, of another vector $u \geqslant 0$ if $v \wedge (u - v) = 0$. In order complete Riesz spaces, the components of an element $u > 0$ (which form a complete Boolean algebra under the induced operations) coincide with the extreme points of the order interval $[0, u]$; see [12, Theorem 3.15 p. 37]. In our case, the characteristic functions of measurable sets are exactly the components of the constant function **1**. Thus, Σ_μ coincides with the extreme points of the order interval $[0, 1]$ in $L_1(\mu)$.

13.14 Lemma *The metric space Σ_μ is separable if and only if the Banach lattice $L_1(\mu)$ is separable.*

Proof: If $L_1(\mu)$ is separable, then its subset Σ_μ is likewise separable. For the converse, assume that Σ_μ is separable. Let $\{E_1, E_2, \ldots\}$ be a dense subset of Σ_μ. Then the set of all functions that are finite linear combinations of $\{\chi_{E_1}, \chi_{E_2}, \ldots\}$ with rational coefficients is a countable dense subset of $L_1(\mu)$. (Why?) ∎

13.15 Definition *A measure μ is called **separable** if Σ_μ is a separable metric space (or, equivalently, if $L_1(\mu)$ is a separable Banach lattice).*

If μ and ν are finite measures, then the σ-algebras Σ_μ and Σ_ν are **lattice isometric** if there is a one-to-one surjective mapping $\varphi \colon \Sigma_\mu \to \Sigma_\nu$ such that for all $A, B \in \Sigma_\mu$ we have:

- $\varphi(A \cup B) = \varphi(A) \cup \varphi(B)$;

- $\varphi(A \cap B) = \varphi(A) \cap \varphi(B)$;

- $\varphi(A^c) = [\varphi(A)]^c$; and

- $\int |\chi_A - \chi_B| \, d\mu = \int |\chi_{\varphi(A)} - \chi_{\varphi(B)}| \, d\nu$.

Let Λ denote the σ-algebra of all Lebesgue measurable subsets of $[0, 1]$ and let $L_1[0, 1] = L_1([0, 1], \Lambda, \lambda)$, where λ denotes Lebesgue measure on the line. It follows from Theorem 13.9, that the continuous real functions on $[0, 1]$ are dense in $L_1[0, 1]$. Recall that, from the Stone–Weierstrass Theorem 9.13, $C([0, 1])$ has a countable uniformly dense subset. Hence Λ is separable. (Why?) Remarkably, C. Carathéodory has shown that for any nonatomic probability measure μ, Λ is the only separable Σ_μ-space.

13.16 Theorem (Carathéodory) *For a nonatomic probability measure μ the metric space Σ_μ is lattice isometric to Λ, the Lebesgue measurable sets of $[0, 1]$, if and only if μ is separable.*

Proof: See the proof of [290, Theorem 4, p. 399]. ∎

13.17 Corollary *If (X, Σ, μ) is a nonatomic probability space, then μ is separable if and only if $L_1(\mu)$ is lattice isometric to $L_1[0, 1]$. Moreover, if μ is nonatomic and separable, then $L_p(\mu)$ is lattice isometric to $L_p[0, 1]$ for each $1 \leqslant p \leqslant \infty$.*

13.6 The Radon–Nikodym Theorem

Let μ be a finite measure on a σ-algebra Σ of subsets of a set X. We shall now show that the Banach lattice $L_1(\mu)$ is lattice isometric to the band of all signed measures of bounded variation that are absolutely continuous with respect to μ.

By virtue of Theorem 10.61, we know that the vector space of all signed measures of bounded variation that are absolutely continuous with respect to μ coincides with the band B_μ generated by μ in $ba(\Sigma)$. Since $ca(\Sigma)$, the vector space of all signed measures of bounded variation, is itself a band of $ba(\Sigma)$, the band B_μ also coincides with the band generated by μ in $ca(\Sigma)$ (so B_μ consists only of signed measures).

Notice that if $f \in L_1(\mu)$, then the set function $\nu_f \colon \Sigma \to \mathbb{R}$ defined by

$$\nu_f(A) = \int_A f \, d\mu,$$

is a signed measure of bounded variation that is also absolutely continuous with respect to μ. In addition, it is not difficult to see that

$$\nu_f^+(A) = \int_A f^+ \, d\mu, \quad \nu_f^-(A) = \int_A f^- \, d\mu, \quad \text{and} \quad |\nu_f|(A) = \int_A |f| \, d\mu.$$

The celebrated Radon–Nikodym Theorem asserts that the signed measures ν_f are the only measures of bounded variation that are absolutely continuous with respect to μ.

13.18 Theorem (Radon–Nikodym) *If (X, Σ, μ) is a σ-finite measure space, and if the signed measure ν of bounded variation is absolutely continuous with respect to μ, then there exists a (μ-almost) unique μ-integrable function f satisfying*

$$\nu(A) = \int_A f \, d\mu$$

for each $A \in \Sigma$. Moreover, by Theorem 10.35, f may be taken to be Σ-measurable.

Proof: See [13, Theorem 37.8, p. 342]. Also note that if ν is a measure, then by Lemma 10.33 we have $\nu^*(A) = \int_A f \, d\mu$ for each $A \in \Sigma_\mu$.   ∎

The function f of the preceding theorem is known as the **Radon–Nikodym derivative** or the **density function** of ν with respect to μ and is denoted $\frac{d\nu}{d\mu}$. It is also customary to write $d\nu = f \, d\mu$. The Radon–Nikodym theorem implies that the mapping $f \mapsto \nu_f$, from $L_1(\mu)$ to B_μ, is one-to-one and onto. In fact, the mapping is a lattice isometry.

13.19 Theorem   *Let μ be a finite measure on a σ-algebra Σ. Then the mapping $f \mapsto \nu_f$, where*

$$\nu_f(A) = \int_A f \, d\mu,$$

is a lattice isometry from $L_1(\mu)$ onto B_μ (so under this lattice isometry the Banach lattice B_μ can be identified with the Banach lattice $L_1(\mu)$).

Proof: By the Radon–Nikodym Theorem 13.18, we know that the linear mapping $f \mapsto v_f$ is one-to-one and onto. Also, $v_f \geqslant 0$ if and only if $f \geqslant 0$, which shows that $f \mapsto v_f$ is a lattice isomorphism; see Theorem 9.17. Finally, to see that $f \mapsto v_f$ is also an isometry, note that

$$\|v_f\| = |v_f|(X) = \int_X |f| \, d\mu = \|f\|_1$$

for each $f \in L_1(\mu)$.   ∎

There is one more useful formulation of the Radon–Nikodym Theorem involving measures. It can be proven easily by employing Theorem 13.18.

13.20 Theorem (Radon–Nikodym)   *Let Σ be a σ-algebra and let μ and v be σ-finite measures on Σ. If v is absolutely continuous with respect to μ, then there exists a (μ-almost) unique Σ-measurable function $f \geqslant 0$ satisfying*

$$v^*(A) = \int_A f \, d\mu$$

for each $A \in \Sigma_\mu$, where now the values of the integral can be infinite.

We now present an example from P. R. Halmos [148, Exercise 8, p. 131] that shows that the hypothesis that μ be σ-finite cannot be dropped from the Radon–Nikodym Theorem.

13.21 Example (No Radon–Nikodym derivative)   Let X be an uncountable set and let Σ be the σ-algebra consisting of all countable subsets and their complements. Let μ be counting measure, which is not σ-finite, and define $v(A) = 0$ if A is countable and $v(A) = 1$ otherwise. Then v is absolutely continuous with respect to μ, but nevertheless has no Radon–Nikodym derivative. (Why?)   ∎

13.7   Equivalent measures

Again, in this section Σ is a σ-algebra of sets. Two measures μ and v are **equivalent**, written $\mu \equiv v$, if both $v \ll \mu$ and $\mu \ll v$. Note that two finite measures μ and v are equivalent if and only if $B_\mu = B_v$.

13.22 Lemma   *If a measure $v \in ca(\Sigma)$ is absolutely continuous with respect to a σ-finite measure μ, then $\Sigma_\mu \subset \Sigma_v$. In this case, every μ-measurable function is v-measurable.*

Proof: Let $A \in \Sigma_\mu$. By Lemma 10.31, there exists some μ-null set C satisfying $A \cap C = \varnothing$ and $A \cup C \in \Sigma \subset \Sigma_v$. From $v \ll \mu$, we get $v^*(C) = 0$, so $C \in \Sigma_v$. Consequently, $A = (A \cup C) \setminus C \in \Sigma_v$.   ∎

The next theorem justifies the formula $dv = g\,d\mu$.

13.23 Theorem   *Let μ and v be σ-finite measures on a σ-algebra Σ and let v be absolutely continuous with respect to μ. If $g = \frac{dv}{d\mu}$ is the Radon–Nikodym derivative of v with respect to μ, then for each v-integrable function f the function fg is μ-integrable and*

$$\int f\,dv = \int fg\,d\mu.$$

Proof: We can assume that $g(x) \geqslant 0$ for each x. Let

$$Y = \{x \in X : g(x) > 0\} \in \Sigma_\mu.$$

We claim that if $A \in \Sigma_v$ then $Y \cap A$ is μ-measurable. To this end, let $A \in \Sigma_v$. Since v is σ-finite, we can assume that $v^*(A) < \infty$.

Suppose first that $A \subset Y$ and $v^*(A) = 0$. By Lemma 10.30, there exists some $B \in \Sigma$ satisfying $A \subset B$ and $v(B) = 0$. Clearly, $B \cap Y \in \Sigma_\mu$. If $\mu^*(B \cap Y) > 0$, then $0 < \int_{B \cap Y} g\,d\mu = v^*(B \cap Y) \leqslant v(B) = 0$, which is impossible. Hence, $\mu^*(B \cap Y) = 0$, and from $A \subset B \cap Y$, we see that $\mu^*(A) = 0$. Thus, $A \in \Sigma_\mu$.

Next, consider the general case. Choose some $C \in \Sigma$ satisfying $A \subset C$ and $v^*(A) = v(C)$. Thus $v^*(C \setminus A) = 0$, so by the preceding case $(C \setminus A) \cap Y \in \Sigma_\mu$. Consequently, $A \cap Y = C \cap Y \setminus (C \setminus A) \cap Y \in \Sigma_\mu$.

Clearly,

$$v^*(A) = \int_A g\,d\mu = \int_{A \cap Y} g\,d\mu + \int_{A \cap Y^c} g\,d\mu = \int_{A \cap Y} g\,d\mu = v^*(A \cap Y)$$

for each $A \in \Sigma_v$. If $A \in \Sigma_v$ satisfies $v^*(A) < \infty$, then from $\chi_{A \cap Y} g = \chi_A g$ and the preceding discussion, we see that $\chi_A g$ is a μ-measurable function. From

$$\int_A \chi_A\,dv = v^*(A) = v^*(A \cap Y) = \int_{A \cap Y} g\,d\mu = \int \chi_A g\,d\mu,$$

it follows that for every v-step function φ the function φg is μ-integrable and $\int \varphi\,dv = \int \varphi g\,d\mu$.

Now let f be a v-upper function. Pick a sequence of v-step functions $\{\varphi_n\}$ satisfying $\varphi_n(x) \uparrow f(x)$ for all $x \notin A$, where $v^*(A) = 0$. Since $\varphi_n g(x) = fg(x) = 0$ for all $x \notin Y$, it follows that $\varphi_n g(x) \uparrow fg(x)$ for all $x \notin A \cap Y$. From the equalities $0 = v^*(A) = v^*(A \cap Y) = \int_{A \cap Y} g\,d\mu$, we infer that $\mu^*(A \cap Y) = 0$, so $\varphi_n g \uparrow fg$ μ-a.e. too. That is, we have

$$\varphi_n \uparrow f\ v\text{-a.e.}, \quad \varphi_n g \uparrow fg\ \mu\text{-a.e.}, \quad \text{and} \quad \int \varphi_n\,dv = \int \varphi_n g\,d\mu \text{ for each } n.$$

Taking limits, we infer that fg is a μ-integrable function and $\int f\,dv = \int fg\,d\mu$. The conclusion for an arbitrary $f \in L_1(v)$ is now immediate.   ∎

13.24 Corollary *Let μ and ν be equivalent σ-finite measures on a σ-algebra. If $f = \dfrac{d\mu}{d\nu}$ and $g = \dfrac{d\nu}{d\mu}$ are the Radon–Nikodym derivatives, then*

$$f = \frac{1}{g}\ \nu\text{-a.e.} \quad and \quad g = \frac{1}{f}\ \mu\text{-a.e.}$$

Equivalent measures also have the same space of integrable functions.

13.25 Theorem *If two σ-finite measures μ and ν on a σ-algebra are equivalent, then the Banach lattices $L_1(\mu)$ and $L_1(\nu)$ are lattice isometric.*

Proof: Let $\mu \equiv \nu$. Then, by Lemma 13.22, we get $\Sigma_\mu = \Sigma_\nu$, so μ and ν have the same measurable functions. Now note (by using Theorem 13.23) that the mapping $f \mapsto f \cdot \frac{d\mu}{d\nu}$, from $L_1(\mu)$ to $L_1(\nu)$, is an onto lattice isometry. ∎

13.8 Duals of L_p-spaces

We now characterize the duals of the L_p-spaces.

13.26 Theorem (F. Riesz) *If $1 < p, q < \infty$ are conjugate exponents ($\frac{1}{p} + \frac{1}{q} = 1$), then each $g \in L_q(\mu)$ defines a continuous linear functional F_g on $L_p(\mu)$ via the formula*

$$F_g(f) = \int fg\,d\mu.$$

Moreover, the mapping $g \mapsto F_g$ is a lattice isometry from $L_q(\mu)$ onto $L'_p(\mu)$—so the norm dual of $L_p(\mu)$ can be identified with $L_q(\mu)$.

Proof: We sketch the main idea of the proof. Let $F : L_q(\mu) \to \mathbb{R}$ be a continuous linear functional. The trick is to prove that it has such a representation. For each μ-measurable A, define $\nu(A) = F(\chi_A)$. Then it is not hard to show that ν is a signed measure on the σ-algebra of μ-measurable sets, and it is absolutely continuous with respect to μ. By the Radon–Nikodym Theorem 13.18 there is an integrable function g satisfying $F(\chi_A) = \int g\chi_A\,d\mu$ for each $A \in \Sigma_\mu$. Linearity and Levi's Theorem 11.18 show that g represents F. By evaluating F on a cleverly chosen collection, it can be shown that $g \in L_q(\mu)$. See [13, Theorems 31.16, 37.9, and 37.11] for the unsightly details. ∎

13.27 Corollary *For each $1 < p < \infty$ the Banach lattice $L_p(\mu)$ is reflexive.*

13.28 Theorem (F. Riesz) *Let (X, Σ, μ) be a σ-finite measure space. Then the mapping $F : L_\infty(\mu) \to L'_1(\mu)$, defined by*

$$F_g(f) = \int fg\,d\mu,$$

is an onto lattice isometry. That is, for σ-finite measures, the norm dual of $L_1(\mu)$ can be identified with the Banach lattice $L_\infty(\mu)$.

Proof: The idea of the proof is again to use the Radon–Nikodym Theorem to find the representing function. See [13, Theorem 37.10, p. 347] for the details. ∎

An immediate consequence of the preceding result is that the Banach lattices $L_1(\mu)$ and $L_\infty(\mu)$ are seldom reflexive.

13.29 Corollary   *If μ is σ-finite and $L_1(\mu) \neq L_\infty(\mu)$, then neither $L_1(\mu)$ nor $L_\infty(\mu)$ is reflexive.*

When $0 < p < 1$, the situation for the $L_p(\mu)$-spaces is drastically different. They are no longer Banach lattices. However, they are all Fréchet lattices.

13.30 Theorem   *For $0 < p < 1$, the $L_p(\mu)$-space is a Fréchet lattice under the metric $d(f, g) = \int |f - g|^p \, d\mu$.*

Proof: Fix $0 < p < 1$. To verify that the formula $d(f, g) = \int |f - g|^p \, d\mu$ satisfies the triangle inequality, we must employ the following elementary inequality for real numbers: If $a, b \geqslant 0$, then $(a + b)^p \leqslant a^p + b^p$. Indeed, for $a > 0$ and $b > 0$, we have

$$(a + b)^p = (a + b)(a + b)^{p-1} = a(a + b)^{p-1} + b(a + b)^{p-1}$$
$$\leqslant a \cdot a^{p-1} + b \cdot b^{p-1} = a^p + b^p.$$

Clearly, d generates a locally solid topology. Now an argument similar to that in the proof of Theorem 13.5 shows that d is a complete metric. ∎

For $0 < p < 1$ and nonatomic measures, the $L_p(\mu)$-spaces are not locally convex.

13.31 Theorem (Day [83])   *If (X, Σ, μ) is a nonatomic measure space, then for each $0 < p < 1$ we have*

$$L'_p(\mu) = L_p^\sim(\mu) = \{0\}.$$

In particular, for every nonatomic measure μ and each $0 < p < 1$ the Fréchet lattice $L_p(\mu)$ is not locally convex.

Proof: Let (X, Σ, μ) be a nonatomic measure and fix $0 < p < 1$. By Theorem 9.11, we know that $L'_p(\mu) = L_p^\sim(\mu)$. Now fix $0 \leqslant \varphi \in L_p^\sim(\mu)$. Also, we can assume that μ is a finite measure—otherwise we consider φ restricted to $L_p(E)$ for each $E \in \Sigma$ with $\mu(E) < \infty$. Since φ is continuous at zero, there exists some positive integer n such that $d(f, 0) = \int |f|^p \, d\mu \leqslant \frac{1}{n}$ implies $|\varphi(f)| \leqslant 1$. It follows that

$$|\varphi(f)| \leqslant n^{\frac{1}{p}} \|f\|_p \qquad\qquad (\star)$$

for all $f \in L_p(\mu)$.

From Corollary 13.3, we know that $L_1(\mu) \subset L_p(\mu)$. Thus, by Theorem 9.11, we see that φ restricted to $L_1(\mu)$ is continuous. Then, by Theorem 13.28, there exists some $0 \leqslant g \in L_\infty(\mu)$ such that $\varphi(f) = \int fg\,d\mu$ for each $f \in L_1(\mu)$.

Next, we claim that $g = 0$ a.e. To see this, assume by way of contradiction that $g \neq 0$ a.e. Then there exist some $\varepsilon > 0$ and some $A \in \Sigma$ with $\mu(A) \geqslant \varepsilon$ such that $g(x) \geqslant \varepsilon$ for all $x \in A$. Now if k is an arbitrary positive integer, then choose some $B \in \Sigma$ with $B \subset A$ and $0 < \mu(B) < 1/k$ (see Theorem 10.52), so using $(\star)$, we see that

$$\varepsilon\mu(B) \leqslant \int g\chi_B\,d\mu = \varphi(\chi_B) \leqslant n^{\frac{1}{p}}\|\chi_B\|_p = n^{\frac{1}{p}}[\mu(B)]^{\frac{1}{p}}.$$

Therefore, $k^{\frac{1}{p}-1} \leqslant [\mu(B)]^{1-\frac{1}{p}} \leqslant \frac{n^{\frac{1}{p}}}{\varepsilon}$ for all k, which is a contradiction. Hence, $g = 0$ a.e., so $\varphi = 0$ on $L_1(\mu)$. Since $L_1(\mu)$ is d-dense in $L_p(\mu)$ (why?), we see that $\varphi = 0$ on $L_p(\mu)$ too. ∎

13.9 Lyapunov's Convexity Theorem

The celebrated convexity theorem of A. A. Lyapunov [236] states that the range of a finite nonatomic finite dimensional vector measure is compact and convex. The Lyapunov Convexity Theorem plays a fundamental role in the formulation of the "bang-bang" principle of optimal control theory. The bang-bang principle asserts that optimal controls need only take on values that are extreme points of the admissible control set. See for instance, [144, 222, 226, 260, 264, 265, 266]. The Lyapunov convexity theorem has also been used in the study of large economies [27, 158, 335], stochastic economic equilibria [49, 142, 258], and bargaining [101]. Recently it has been applied to problems in the politics of dividing land [159, 304].

In order to prove the Lyapunov convexity theorem we make use of the following lemma, which is interesting in its own right.

13.32 Lemma   *Let $\mu_1, \ldots, \mu_n$ be finite measures (not necessarily nonatomic) on a measurable space (X, Σ) and let $\mu = \mu_1 + \cdots + \mu_n$. Then the set*

$$I = \left\{ \left(\int f\,d\mu_1, \ldots, \int f\,d\mu_n \right) : f \text{ is } \Sigma\text{-measurable and } 0 \leqslant f \leqslant 1 \ \mu\text{-a.e.} \right\}$$

is a compact convex subset of $\mathbb{R}^n$.

Proof: Notice that each μ_i is absolutely continuous with respect to the finite measure μ. Let g_i be a Σ-measurable function representing the Radon–Nikodym derivative of μ_i with respect to μ. Then $g_i = \frac{d\mu_i}{d\mu} \in L_1(\mu)$; see Theorem 13.18.

By Theorem 13.28, the dual of $L_1(\mu)$ is $L_\infty(\mu)$. Moreover the closed unit ball of $L_\infty(\mu)$ coincides with the order interval $[-1, 1]$. By Alaoglu's Theorem 6.21,

$[-1, 1]$ is weak∗-compact. It is easy to see that $P = \{f \in L_\infty(\mu) : 0 \leqslant f \leqslant 1\}$, the positive part of the unit ball of $L_\infty(\mu)$ is convex and weak∗-closed. Consequently, P is itself a weak∗-compact subset of $L_\infty(\mu)$.

Next, define the mapping $T : L_\infty(\mu) \to \mathbb{R}^n$ via the formula

$$Tf = \left(\int f \, d\mu_1, \ldots, \int f \, d\mu_n \right) = \left(\int f g_1 \, d\mu, \ldots, \int f g_n \, d\mu \right).$$

Clearly T is a linear operator. Furthermore, even though P may include functions that are not Σ-measurable, every function in P differs from a Σ-measurable function only on a set of μ-measure zero (Theorem 10.35), so $T(P) = I$. To complete the proof, it suffices to show that T is weak∗-continuous. Indeed, if $f_\alpha \xrightarrow{w^*} f$ in $L_\infty(\mu)$, that is, $\int f_\alpha g \, d\mu \to \int f g \, d\mu$ for each $g \in L_1(\mu)$, then

$$Tf_\alpha = \left(\int f_\alpha g_1 \, d\mu, \ldots, \int f_\alpha g_n \, d\mu \right) \to \left(\int f g_1 \, d\mu, \ldots, \int f g_n \, d\mu \right) = Tf,$$

and the proof is finished.   ∎

Recall that an **atom** of a measure μ on the σ-algebra Σ is a set $A \in \Sigma$ of positive measure that cannot be split into two pieces of smaller positive measure. That is, A is an atom if $\mu(A) > 0$, and for any $B \in \Sigma$ with $B \subset A$ either $\mu(B) = 0$ or $\mu(B) = \mu(A)$. For instance, any singleton is an atom of the counting measure. A measure is **nonatomic** or **atomless** if it has no atoms. For example, Lebesgue measure on $\mathbb{R}^n$ is nonatomic.

We are now ready to state and prove Lyapunov's convexity theorem. The proof presented here is due to J. Lindenstrauss [227]. It relies on the Krein–Milman Theorem 7.68. (Another, elementary, proof can be found in P. R. Halmos [145].)

13.33 Lyapunov Convexity Theorem   *If $\mu_1, \ldots, \mu_n$ are finite nonatomic measures on a measurable space (X, Σ), then the set*

$$R = \Big\{ (\mu_1(A), \ldots, \mu_n(A)) : A \in \Sigma \Big\}$$

is a compact convex subset of $\mathbb{R}^n$. Moreover, we also have that

$$R = \left\{ \left(\int f \, d\mu_1, \ldots, \int f \, d\mu_n \right) : f \text{ is } \Sigma\text{-measurable and } 0 \leqslant f \leqslant 1 \ \mu\text{-a.e.} \right\},$$

where $\mu = \mu_1 + \cdots + \mu_n$.

Proof: The proof is by induction. For $n = 1$, the conclusion follows immediately from Theorem 10.52 and Lemma 10.30. For the induction step, assume that our claim is true for any n nonatomic measures and let $\mu_1, \ldots, \mu_n, \mu_{n+1}$ be $n+1$ nonatomic measures on Σ. We establish the claim by proving that $R = I$, where

$$I = \left\{ \left(\int f \, d\mu_1, \ldots, \int f \, d\mu_{n+1} \right) : f \text{ is } \Sigma\text{-measurable and } 0 \leqslant f \leqslant 1 \ \mu\text{-a.e.} \right\},$$

which is a compact convex subset of $\mathbb{R}^n$ by Lemma 13.32. Since $R \subset I$, we must show that $I \subset R$.

To this end, let $\mu = \mu_1 + \cdots + \mu_n + \mu_{n+1}$. Also, let f be a Σ-measurable function satisfying $0 \leqslant f \leqslant 1$ μ-a.e. We must prove that there exists some $A \in \Sigma$ such that $\int f \, d\mu_i = \mu_i(A)$ for each $i = 1, \ldots, n, n+1$.

To establish this claim, we consider the nonempty, convex, and weak* compact subset $\mathcal{C}$ of $L_\infty(\mu)$ defined by

$$\mathcal{C} = \left\{ g \in [0,1] : g \text{ is } \Sigma\text{-measurable and } \int g \, d\mu_i = \int f \, d\mu_i \text{ for each } i \right\}.$$

By the Krein–Milman Theorem 7.68, $\mathcal{C}$ has an extreme point, say g. We finish the proof by proving that $g = \chi_A$ μ-a.e. for some $A \in \Sigma$.

To see this, assume by way of contradiction that $g \neq \chi_A$ μ-a.e. for each $A \in \Sigma$. Then for some $0 < \varepsilon < 1$ the set $E = \{x \in X : \varepsilon \leqslant g(x) \leqslant 1 - \varepsilon\} \in \Sigma$ satisfies $\mu(E) > 0$. (Why?) It follows that $\mu_i(E) > 0$ must hold for some i. Without loss of generality, we can assume that $\mu_{n+1}(E) > 0$. Since μ_{n+1} is nonatomic, there exists some $B \in \Sigma$ with $B \subset E$ satisfying $\mu_{n+1}(B) > 0$ and $\mu_{n+1}(E \setminus B) > 0$. Let $B_1 = B$ and $B_2 = E \setminus B$.

Apply the induction hypothesis to the measure spaces (B_1, Σ_{B_1}) and (B_2, Σ_{B_2}), where $\Sigma_{B_i} = \{A \in \Sigma : A \subset B_i\}$, to get the existence of two sets $C_1, D_1 \in \Sigma$ with $C_1 \subset B_1$ and $D_1 \subset B_2$ satisfying

$$\mu_i(C_1) = \tfrac{1}{2}\mu_i(B_1) \quad \text{and} \quad \mu_i(D_1) = \tfrac{1}{2}\mu_i(B_2)$$

for each $i = 1, \ldots, n$. In particular, for $C_2 = B_1 \setminus C_1$ and $D_2 = B_2 \setminus D_1$, we have

$$\mu_i(C_1) - \mu_i(C_2) = \mu_i(D_1) - \mu_i(D_2) = 0.$$

We can assume $|\mu_{n+1}(C_1) - \mu_{n+1}(C_2)| \leqslant |\mu_{n+1}(D_1) - \mu_{n+1}(D_2)|$. In case we have $\mu_{n+1}(D_1) - \mu_{n+1}(D_2) = 0$, we let $\alpha = \beta = \varepsilon$. Otherwise, if $\mu_{n+1}(D_1) - \mu_{n+1}(D_2) \neq 0$, then we put $\alpha = -\varepsilon$ and $\beta = \frac{\varepsilon[\mu_{n+1}(C_1) - \mu_{n+1}(C_2)]}{\mu_{n+1}(D_1) - \mu_{n+1}(D_2)}$. Either way, $0 < |\alpha| \leqslant \varepsilon$, $0 < |\beta| \leqslant \varepsilon$, and

$$\alpha[\mu_{n+1}(C_1) - \mu_{n+1}(C_2)] + \beta[\mu_{n+1}(D_1) - \mu_{n+1}(D_2)] = 0.$$

Now note that the Σ-measurable function $h = \alpha\chi_{C_1} - \alpha\chi_{C_2} + \beta\chi_{D_1} - \beta\chi_{D_2}$ is nonzero μ-a.e. and satisfies $\int h \, d\mu = 0$. Since $h = 0$ on $X \setminus E$ and $-\varepsilon \leqslant h(x) \leqslant \varepsilon$ for each $x \in E$, it follows that $0 \leqslant g \pm h \leqslant 1$. Consequently, $f \pm h \in \mathcal{C}$. But then $g = \tfrac{1}{2}(g + h) + \tfrac{1}{2}(g - h)$ contradicts the fact that g is an extreme point of $\mathcal{C}$, and the proof of the theorem is finished. ∎

The following related result is due to L. E. Dubins and E. H. Spanier [101]. One interpretation of this theorem is that it is always possible to cut a "nonatomic cake" fairly.

13.34 Theorem (Dubins–Spanier) *Let $\mu_1, \ldots, \mu_m$ be nonatomic probability measures on a measurable space (X, Σ). Given $\alpha_1, \ldots, \alpha_n \geqslant 0$ with $\sum_{j=1}^n \alpha_j = 1$, there is a partition $\{A_1, \ldots, A_n\}$ of X satisfying $\mu_i(A_j) = \alpha_j$ for all $i = 1, \ldots, m$ and $j = 1, \ldots, n$.*

Proof: Let $\mathbb{P}$ denote the set of all n-tuples $(A_1, \ldots, A_n)$, where the sets $A_1, \ldots, A_n$ belong to Σ and they form a partition of X. Given an n-tuple $P = (A_1, \ldots, A_n) \in \mathbb{P}$, define $v(P)$ to be the $m \times n$ matrix whose i, j entry is $\mu_i(A_j)$. We first show that $R = \{v(P) : P \in \mathbb{P}\}$, the range of v, is a convex set of matrices. To this end, let $(A_1, \ldots, A_n)$ and $(B_1, \ldots, B_n)$ belong to $\mathbb{P}$, and let $0 < \lambda < 1$. Define the nonatomic $2mn$-dimensional vector measure γ by $\gamma(E) = [\mu_i(E \cap A_j), \mu_i(E \cap B_j)]$. By Lyapunov's Theorem 13.33 there is a set E with $\gamma(E) = \lambda \gamma(X)$. That is, for every $i = 1, \ldots, m$ and $j = 1, \ldots, n$,

$$\mu_i(E \cap A_j) = \lambda \mu_i(A_j) \quad \text{and} \quad \mu_i(E \cap B_j) = \lambda \mu_i(B_j).$$

Define $(C_1, \ldots, C_n) \in \mathbb{P}$ by letting $C_j = (E \cap A_j) \cup (E^c \cap B_j)$. (Why is this a partition?) Then $\mu_i(C_j) = \lambda \mu_i(A_j) + (1 - \lambda)\mu_i(B_j)$. This shows that R is convex.

Now let P_j denote the partition $(A_1, \ldots, A_n)$ with $A_j = X$ and $A_k = \emptyset$ for $k \neq j$. Then $v(P_j)$ is the matrix with ones in the jth column and zeros everywhere else. Since the range R of v is convex, there is a partition $P = (A_1, \ldots, A_n)$ with $v(P) = \sum_{j=1}^n \alpha_j v(P_j)$. In other words, for every $i = 1, \ldots, m$ and $j = 1, \ldots, n$, we have $\mu_i(A_j) = \alpha_j$.   ∎

Lyapunov's theorem partially generalizes to nonatomic charges. If the algebra on which the charge is defined is nice enough, then the range of a finite dimensional nonatomic vector charge is convex, but not necessarily closed. More generally, the range can be quite perverse. For more details see T. E. Armstrong and K. Prikry [20]. Lyapunov's theorem does not generalize to infinite dimensional vector measures. That is, in general a nonatomic vector measure may have range that is neither convex nor compact. Here is a simple example due to J. J. Uhl [334].

13.35 Example (Infinite dimensional vector measure) Let $\mathcal{B}$ denote the Borel σ-algebra of the unit interval $[0, 1]$ endowed with Lebesgue measure λ. Let $\mu: \mathcal{B} \to L_1(\lambda)$ be defined by $\mu(A) = \chi_A$. Then for any sequence $\{A_1, A_2, \ldots\}$ of pairwise disjoint sets, $\sum_{i=1}^n \mu(A_i) \xrightarrow{\|\cdot\|_1} \mu(\bigcup_{i=1}^\infty ftyA_i)$, so μ is norm countably additive. It is also nonatomic. The range of μ is the collection of indicator functions of Borel sets. This set is not convex. For instance, the constant function $\frac{1}{2}\chi_{[0,1]} + \frac{1}{2}\chi_\emptyset$ does not belong to the range of μ. The range is closed, but not compact. The Borel sets $A_n = \{x \in [0, 1] : \sin(2^n \pi x) > 0\}$ satisfy $\|\chi_{A_n} - \chi_{A_m}\|_1 = \frac{1}{4}$ for $n \neq m$, so no subsequence of $\{\chi_{A_n}\}$ converges.   ∎

See J. Diestel and J. J. Uhl [96, Chapter 9] for conditions under which Lyapunov's Theorem does generalize.

13.10 Convergence in measure

As before, in this section (X, Σ, μ) will denote a measure space. *Convergence in measure* defines a metrizable linear topology on the vector space of equivalence classes of measurable functions on a finite measure space.

13.36 Definition *A sequence $\{f_n\}$ of μ-measurable functions* **converges in measure** *to a measurable function f, written $f_n \xrightarrow{\mu} f$, if for each $c > 0$,*

$$\lim_{n \to \infty} \mu^*(\{x \in X : |f_n(x) - f(x)| \geq c\}) = 0.$$

When μ is a probability measure we say that the sequence $\{f_n\}$ **converges in probability**.

You can verify easily that convergence in measure satisfies the following properties:

- If $f_n \xrightarrow{\mu} f$ and $f_n \xrightarrow{\mu} g$, then $f = g$ μ-a.e.

- If $f_n \xrightarrow{\mu} f$ and $g_n \xrightarrow{\mu} g$, then $\alpha f_n + \beta g_n \xrightarrow{\mu} \alpha f + \beta g$ for all $\alpha, \beta \in \mathbb{R}$.

- If $f_n \xrightarrow{\mu} f$, then $f_n^+ \xrightarrow{\mu} f^+$, $f_n^- \xrightarrow{\mu} f^-$, and $|f_n| \xrightarrow{\mu} |f|$.

Pointwise convergence does not imply convergence in measure. For instance, if $f_n = \chi_{(n,n+1)}$, then $f_n(x) \to 0$ for each $x \in \mathbb{R}$ while $\lambda(\{x \in \mathbb{R} : |f_n(x)| \geq 1\}) = 1$ for each n. However, on a finite measure space pointwise convergence implies convergence in measure.

13.37 Theorem *On a finite measure space, pointwise convergence implies convergence in measure.*

Proof: Assume $f_n \to f$ μ-a.e. on a finite measure space (X, Σ, μ). Put

$$A_n = \{x \in X : |f_n(x) - f(x)| \geq \varepsilon\} = \left\{x \in X : \frac{|f_n(x) - f(x)|}{1 + |f_n(x) - f(x)|} \geq \frac{\varepsilon}{1 + \varepsilon}\right\}.$$

This implies $\mu^*(A_n) \leq \frac{1+\varepsilon}{\varepsilon} \int \frac{|f_n - f|}{1 + |f_n - f|} \, d\mu \to 0$, where the last limit follows from the Dominated Convergence Theorem 11.21. Consequently, $\mu^*(A_n) \to 0$. ∎

Although convergence in measure does not imply pointwise convergence, we can always extract from any convergent in measure sequence a pointwise convergent subsequence.

13.38 Theorem *Every sequence that converges in measure has an almost everywhere pointwise convergent subsequence with the same limit.*

Proof: Assume $f_n \xrightarrow{\mu} f$. It is easy to see that there exists a strictly increasing sequence $\{k_n\}$ of natural numbers such that $\mu^*(\{x : |f_k(x) - f(x)| \geq 1/n\}) < 1/2^n$ for all $k \geq k_n$. For each n consider the measurable set $E_n = \{x : |f_{k_n}(x) - f(x)| \geq \frac{1}{n}\}$ and put $E = \bigcap_{n=1}^{\infty} \bigcup_{k=n}^{\infty} E_k$. Then

$$\mu^*(E) \leq \mu^*\left(\bigcup_{k=n}^{\infty} E_k\right) \leq \sum_{k=n}^{\infty} \mu^*(E_k) \leq 2^{1-n}$$

for each n, which implies $\mu^*(E) = 0$. On the other hand, if $x \notin E$, then there exists some n_0 such that $x \notin \bigcup_{k=n_0}^{\infty} E_k$, so $|f_{k_n}(x) - f(x)| \leq \frac{1}{n}$ for all $n \geq n_0$. Thus, $f_{k_n}(x) \to f(x)$ for all $x \notin E$, which means that $f_{k_n} \to f$ μ-almost everywhere. ∎

13.39 Theorem   *Norm convergence in any $L_p(\mu)$-space implies convergence in measure.*

Proof: Assume that $\|f_n - f\|_p \to 0$ in some $L_p(\mu)$ space with $0 < p < \infty$. If $A_n = \{x : |f_n(x) - f(x)| \geq c\}$, then $c^p \chi_{A_n} \leq |f_n - f|^p$. Therefore,

$$\mu^*(A_n) \leq \frac{1}{c^p} \int |f_n - f|^p \, d\mu \xrightarrow[n \to \infty]{} 0,$$

so $f_n \xrightarrow{\mu} f$. The case $p = \infty$ is trivial. ∎

On the vector space $L_0(\mu)$ of all equivalence classes of measurable functions on a finite measure space, convergence in measure is a metric convergence.

13.40 Lemma   *If μ is a finite measure, then convergence in measure in the vector space $L_0(\mu)$ of all equivalence classes of μ-measurable functions is equivalent to convergence with respect to the translation invariant metric*

$$d(f, g) = \int \frac{|f - g|}{1 + |f - g|} \, d\mu.$$

That is, $f_n \xrightarrow{\mu} f$ if and only if $d(f_n, f) \to 0$.

Proof: Since $0 \leq \frac{|f-g|}{1+|f-g|} \leq 1$, we have $0 \leq d(f, g) < \infty$ for all $f, g \in L_0(\mu)$. Clearly, $d(f, g) = d(g, f)$, and $d(f, g) = 0$ if and only if $f = g$ a.e. The triangle inequality follows from the following elementary property of the real numbers: If a, b, c are nonnegative real numbers and $a \leq b + c$, then $\frac{a}{1+a} \leq \frac{b}{1+b} + \frac{c}{1+c}$. (Why?)

Now let $\{f_n\}$ be a sequence of measurable functions and for each $\varepsilon > 0$ let

$$A_n = \{x \in X : |f_n(x) - f(x)| \geq \varepsilon\} = \left\{x \in X : \frac{|f_n(x) - f(x)|}{1 + |f_n(x) - f(x)|} \geq \frac{\varepsilon}{1 + \varepsilon}\right\}.$$

Assume first that $d(f_n, f) \to 0$. Fix $\varepsilon > 0$ and note that $\frac{\varepsilon}{1+\varepsilon}\chi_{A_n} \leq \frac{|f_n-f|}{1+|f_n-f|}$. This implies $\mu^*(A_n) \leq \frac{1+\varepsilon}{\varepsilon} d(f_n, f)$, so $\lim_{n\to\infty} \mu^*(A_n) = 0$, which shows that $f_n \xrightarrow{\mu} f$.

For the converse, assume $f_n \xrightarrow{\mu} f$ and let $\varepsilon > 0$. Choose some n_0 such that $\mu^*(A_n) < \varepsilon$ for all $n \geq n_0$. So

$$d(f_n, f) = \int_{A_n} \frac{|f_n - f|}{1 + |f_n - f|} \, d\mu + \int_{A_n^c} \frac{|f_n - f|}{1 + |f_n - f|} \, d\mu$$

$$\leq \mu^*(A_n) + \frac{\varepsilon}{1 + \varepsilon} \mu^*(A_n^c) \leq \varepsilon + \varepsilon \mu(X) = [1 + \mu(X)]\varepsilon,$$

for all $n \geq n_0$, which shows that $d(f_n, f) \to 0$. ∎

13.11 Convergence in measure in L_p-spaces

We start with a result that summarizes the most important properties of convergence in measure.

13.41 Theorem *For a finite measure μ we have the following.*

1. *The topology of convergence in measure defines a complete metrizable locally solid topology on $L_0(\mu)$. That is, $L_0(\mu)$ with the topology of convergence in measure is a Fréchet lattice.*

2. *The topology of convergence in measure is order continuous on $L_0(\mu)$.*

3. *If μ is also nonatomic, then $L_0'(\mu) = L_0^{\sim}(\mu) = \{0\}$, (so the topology of convergence in measure on $L_0(\mu)$ is not locally convex).*

Proof: (1) We have already proven that the topology of convergence in measure defines a linear metrizable topology on $L_0(\mu)$ that is generated by the translation invariant metric $d(f, g) = \int \frac{|f-g|}{1+|f-g|} \, d\mu$. If $|f| \leq |g|$, then $\frac{|f|}{1+|f|} \leq \frac{|g|}{1+|g|}$, so

$$d(f, 0) = \int \frac{|f|}{1 + |f|} \, d\mu \leq \int \frac{|g|}{1 + |g|} \, d\mu = d(g, 0).$$

This shows that the d-balls at zero are solid sets, so the topology of convergence in measure is locally solid.

To see that d is a complete metric, let $\{f_n\}$ be a d-Cauchy sequence. It suffices to show that $\{f_n\}$ has a convergent subsequence. By passing to a subsequence, we can assume that $d(f_n, f_m) < \frac{1}{n}$ for all $m \geq n$. For each k, m, and $\varepsilon > 0$, let

$$A_{k,m}(\varepsilon) = \{x \in X : |f_k(x) - f_m(x)| \geq \varepsilon\} = \left\{x \in X : \frac{|f_k(x) - f_m(x)|}{1 + |f_k(x) - f_m(x)|} \geq \frac{\varepsilon}{1 + \varepsilon}\right\},$$

so $\frac{\varepsilon}{1+\varepsilon} \chi_{A_{k,m}(\varepsilon)} \leq \frac{|f_k(x)-f_m(x)|}{1+|f_k(x)-f_m(x)|}$. This implies

$$\mu^*(A_{k,m}(\varepsilon)) \leq \frac{1+\varepsilon}{\varepsilon} d(f_k, f_m) < \frac{1+\varepsilon}{\varepsilon} \cdot \frac{1}{k}$$

for all $m \geqslant k$. In particular, using induction, we can choose a strictly increasing sequence $\{k_n\}$ of natural numbers such that $\mu^*(A_{k_n,m}(1/2^n)) < 1/2^n$ for all $m \geqslant k_n$. So if $g_n = f_{k_n}$, then $\{g_n\}$ is a subsequence of $\{f_n\}$ satisfying

$$\mu^*(\{x \in X : |g_{n+1}(x) - g_n(x)| \geqslant 1/2^n\}) < 1/2^n$$

for each n. Next, let $E_n = \{x : |g_{n+1}(x) - g_n(x)| \geqslant \frac{1}{2^n}\}$ and $F_n = \bigcup_{r=n}^{\infty} E_r$. Clearly, $\mu^*(E_n) < 2^{-n}$ for each n and $\mu^*(F_n) \leqslant \sum_{r=n}^{\infty} \mu^*(E_r) \leqslant 2^{-n+1}$. So if we consider the measurable set $F = \bigcap_{n=1}^{\infty} F_n$, then $\mu^*(F) = 0$.

Now if $x \notin F$, then $x \notin F_n$ for some n, or $|g_{r+1}(x) - g_r(x)| < 2^{-r}$ for all $r \geqslant n$. Therefore, for each $k \geqslant n$ and all p, we have

$$|g_{k+p}(x) - g_k(x)| \leqslant \sum_{i=k}^{\infty} |g_{i+1}(x) - g_i(x)| \leqslant 2^{-k+1},$$

which shows that $\{g_n(x)\}$ is a Cauchy sequence of real numbers for each $x \notin F$. So $g_n \to g \in L_0(\mu)$ a.e., and by Theorem 13.37, we get $d(g_n, g) \to 0$.

(2) If $f_\alpha \downarrow 0$ in $L_0(\mu)$, then $\frac{f_\alpha}{1+f_\alpha} \downarrow 0$ also holds in $L_0(\mu)$. This implies $\frac{f_\alpha}{1+f_\alpha} \downarrow 0$ in $L_1(\mu)$, so by Theorem 13.7 we get $d(f_\alpha, 0) = \int \frac{f_\alpha}{1+f_\alpha} \, d\mu \downarrow 0$. In other words, convergence in measure defines an order continuous locally solid topology.

(3) Assume here that μ is a finite and nonatomic. By Theorem 9.11, we know that $L_0'(\mu) = L_0^{\sim}(\mu)$. Let $0 \leqslant \varphi \in L_0'(\mu)$. Also, by Theorem 4.36, the Riesz space L of all μ-step functions is pointwise dense in $L_0(\mu)$, so L is d-dense in $L_0(\mu)$. (Why?) Since $L \subset L_p(\mu)$, it follows that $L_p(\mu)$ is d-dense in $L_0(\mu)$ for each $0 < p < \infty$. Now a glance at Theorem 13.31 shows that $\varphi = 0$ on each $L_p(\mu)$ with $0 < p < 1$ (recall here that since every $L_p(\mu)$-space is a Fréchet lattice every positive linear functional on an $L_p(\mu)$-space is continuous), so $\varphi = 0$ on $L_0(\mu)$. ∎

As a consequence of the preceding result, we obtain the following "non-mixing" property for certain $L_0(\mu)$-spaces.

13.42 Corollary *Let μ be σ-finite and nonatomic. If E is any Riesz space with separating order dual (in particular, if E is a Banach lattice), then there is no nontrivial positive operator from $L_0(\mu)$ to E.*

Proof: We may assume that μ is finite and nonatomic. For a positive operator $T : L_0(\mu) \to E$ and $0 \leqslant f \in E^{\sim}$, clearly $f \circ T \in L_0^{\sim}(\mu) = \{0\}$. If $E^{\sim}$ separates the points of E, then clearly we must have $T = 0$. ∎

For the statement of the next two theorems, let us write τ_p for the topology generated by the "L_p-norm" ($0 < p \leqslant \infty$) and τ_m for the topology of convergence in measure. Keep in mind that (by Theorem 9.11) τ_p is the finest locally solid topology on $L_p(\mu)$.

13.43 Theorem *If μ is finite and nonatomic, then $(L_p(\mu), \tau_m)' = \{0\}$.*

Proof: We provide a proof when $1 \leqslant p < \infty$; the case $0 < p < 1$ follows immediately from Theorem 13.31. Before starting the proof, let us put a few things together. First, since τ_m is a locally solid topology, $(L_p(\mu), \tau_m)'$ is an ideal in the order dual $L_p^\sim(\mu)$ (Theorem 8.48). Second, the order dual $L_p^\sim(\mu)$ coincides with the topological dual $(L_p(\mu), \tau_p)'$ (Theorem 9.11). Third, by Theorem 13.26, we know that $(L_p(\mu), \tau_p)' = L_q(\mu)$, where $\frac{1}{p} + \frac{1}{q} = 1$. Thus, $(L_p(\mu), \tau_m)'$ is an ideal in $L_q(\mu)$.

Now let $0 \leqslant \varphi \in (L_p(\mu), \tau_m)'$. By the above, there exists some $0 \leqslant g \in L_q(\mu)$ satisfying $\varphi(f) = \int fg \, d\mu$ for each $f \in L_p(\mu)$. Now put $A = \{x : g(x) > 0\}$, and assume by way of contradiction that $\mu^*(A) > 0$. If μ is nonatomic, then there exists a sequence $\{A_n\}$ of pairwise disjoint measurable subsets of A satisfying $\mu^*(A_n) > 0$ for each n; see Theorem 10.52. It follows that $\varphi(\chi_{A_n}) > 0$ for each n. Let $\lambda_n = \frac{1}{\varphi(\chi_{A_n})}$, and note that $f = \sum_{n=1}^{\infty} n\lambda_n \chi_{A_n} \in L_0(\mu)$. From $n\lambda_n \chi_{A_n} \leqslant f$, we see that $\lambda_n \chi_{A_n} \leqslant \frac{1}{n}f$, so $\lambda_n \chi_{A_n} \to 0$ a.e. Thus $\lambda_n \chi_{A_n} \xrightarrow{\mu} 0$ in $L_p(\mu)$, so $\varphi(\lambda_n \chi_{A_n}) \to 0$. However, this contradicts $\varphi(\lambda_n \chi_{A_n}) = 1$ for all n, and hence $g = 0$. Therefore, $(L_p(\mu), \tau_m)' = \{0\}$. ∎

The topology of convergence in measure on L_p-spaces also has an interesting minimality property.

13.44 Theorem (Aliprantis–Burkinshaw [11]) *If μ is a finite measure, then for each $0 < p < \infty$ the topology of convergence in measure restricted to $L_p(\mu)$ is the weakest locally solid Hausdorff topology. That is, if τ is any locally solid Hausdorff topology on $L_p(\mu)$, then $\tau_m \subset \tau \subset \tau_p$.*

Proof: See [11, Theorem 7, p. 169]. ∎

13.12 Change of variables

Consider two sets X and Y, and let Σ_X and Σ_Y be σ-algebras of subsets of X and Y respectively. A **measurable transformation** from X to Y is simply any measurable mapping $T : (X, \Sigma_X) \to (Y, \Sigma_Y)$. That is, $T : X \to Y$ is a measurable transformation if $T^{-1}(A) \in \Sigma_X$ for each $A \in \Sigma_Y$. If μ is a measure on Σ_X, then a measurable transformation $T : (X, \Sigma_X) \to (Y, \Sigma_Y)$ gives rise to a measure ν on Σ_Y via the formula

$$\nu(A) = \mu(T^{-1}(A)), \ A \in \Sigma_Y.$$

The measure ν is customarily denoted μT^{-1}, that is, we write $\nu = \mu T^{-1}$, and called it the **measure induced** from μ by T on Σ_Y.[2]

---

[2] When Σ is a σ-algebra of subsets of a set X and $T : X \to X$ is a measurable transformation, then a measure μ on Σ is **T-invariant** if $\mu T^{-1} = \mu$. Invariant measures play a crucial role in ergodic theory. Their existence can be demonstrated by using Banach–Mazur limits or by employing fixed point theorems; see Section 16.10

In general, if (X, Σ_X, μ) is a measure space and $T \colon X \to Y$ is an arbitrary mapping, then the collection of sets

$$\Sigma_Y = \{A \subset Y : T^{-1}(A) \in \Sigma_X\} \qquad\qquad (\star)$$

is a σ-algebra of subsets of Y and the mapping $T \colon (X, \Sigma_X) \to (Y, \Sigma_Y)$ becomes a measurable transformation. Again, μT^{-1} defines a measure on Σ_Y also called the measure induced from μ on Y by T.

The following simple (but useful) result characterizes the measurability of functions defined on Y.

13.45 Lemma *Let Σ_X be a σ-algebra of subsets of a set X, let $T \colon X \to Y$ be a function and let Σ_Y be the σ-algebra defined by $(\star)$ above. If Σ_Z is a σ-algebra of subsets of a set Z, then a function $f \colon Y \to Z$ is (Σ_Y, Σ_Z)-measurable if and only if $f \circ T$ is (Σ_X, Σ_Z)-measurable.*

Proof: Use the set identity $(f \circ T)^{-1}(A) = T^{-1}(f^{-1}(A))$.   ∎

When $T \colon X \to Y$ is a measurable transformation, we can think of the formula $y = T(x)$ as "the change of variable" from x to y via T (or as the passage from the space X to the space Y via the action T). With this interpretation in mind, we can formulate the following change of variables theorem.

13.46 Change of Variables Theorem I *Let Σ_X and Σ_Y be σ-algebras of subsets of X and Y respectively, and let $T \colon (X, \Sigma_X) \to (Y, \Sigma_Y)$ be a measurable transformation. Assume also that μ is a measure on Σ_X and let $\nu = \mu T^{-1}$ be the measure induced from μ by T on Σ_Y. For a function $f \colon Y \to \mathbb{R}$ we have:*

1. *If f is ν-integrable, then $f \circ T$ is μ-integrable and*

$$\int_Y f\, d\nu = \int_Y f\, d\mu T^{-1} = \int_X f \circ T\, d\mu.$$

2. *If ν is σ-finite, f is ν-measurable, and $f \circ T \in L_1(\mu)$, then $f \in L_1(\nu)$ and*

$$\int_Y f\, d\nu = \int_X f \circ T\, d\mu.$$

Proof: First note that for each $A \subset Y$, we have $\mu^*(T^{-1}(A)) \leqslant \nu^*(A)$. Indeed, if $A \subset \bigcup_{n=1}^{\infty} B_n$ with each $B_n \in \Sigma_Y$, then $T^{-1}(A) \subset \bigcup_{n=1}^{\infty} T^{-1}(B_n)$ and $T^{-1}(B_n) \in \Sigma_X$ for each n. Consequently,

$$\mu^*(T^{-1}(A)) \leqslant \sum_{n=1}^{\infty} \mu(T^{-1}(B_n)) = \sum_{n=1}^{\infty} \nu(B_n),$$

from which it follows that $\mu^*(T^{-1}(A)) \leqslant \nu^*(A)$. In particular, if A is a ν-null set, then $T^{-1}(A)$ is a μ-null set.

(1) Assume now that $f \in L_1(\nu)$; we can suppose that f is a ν-upper function. By Theorem 4.27 and Lemma 11.22, there is a sequence $\{\varphi_n\}$ of Σ_Y-step functions such that $\varphi_n(x) \uparrow f(x)$ for each $x \notin A$, where $\nu^*(A) = 0$. By the above, $\mu^*(T^{-1}(A)) = 0$. Now observe that $\varphi_n \circ T(x) \uparrow f \circ T(x)$ for all $x \notin T^{-1}(A)$, so $\varphi_n \circ T \uparrow f \circ T$ μ-a.e. In view of $\chi_B \circ T = \chi_{T^{-1}(B)}$ for $B \subset Y$, we see that $\{\varphi_n \circ T\}$ is a sequence of Σ_X-step functions satisfying $\int \varphi_n \circ T \, d\mu = \int \varphi_n \, d\nu$. Therefore, $f \circ T \in L_1(\nu)$ and (by taking limits) $\int f \circ T \, d\mu = \int f \, d\nu$.

(2) Next, assume that ν is σ-finite and that f is a ν-measurable function such that $f \circ T \in L_1(\mu)$. In view of $(f \circ T)^+ = f^+ \circ T$, we can suppose that $f \geqslant 0$. Since ν is σ-finite there exists a sequence $\{\varphi_n\}$ of Σ_Y-step functions such that $\varphi_n(y) \uparrow f(y)$ for each $y \notin A$, where $\nu^*(A) = 0$ (why?). By the above discussion, $\mu^*(T^{-1}(A)) = 0$. Put $\psi_n = \varphi_n \circ T$ and note that $\{\psi_n\}$ is a sequence of μ-step functions satisfying $\psi_n(x) = \varphi_n(T(x)) \uparrow f \circ T(x)$ for each $x \notin T^{-1}(A)$. Therefore $\psi_n \uparrow f \circ T$ μ-a.e. From $\int \psi_n \, d\mu = \int \varphi_n \circ T \, d\mu = \int \varphi_n \, d\nu$, we see that $f \in L_1(\nu)$ and that $\int f \, d\nu = \int f \circ T \, d\mu$. ∎

13.47 Corollary   *Let (X, Σ, μ) be a measure space, let $T : X \to Y$ be a mapping, and let ν be the measure induced from μ by T on Y. Then the mapping $f \mapsto f \circ T$, from $L_1(\nu)$ to $L_1(\mu)$, is a lattice isometry—so $L_1(\nu)$ can be considered a Banach sublattice of $L_1(\mu)$.*

Proof: Clearly, $f \mapsto f \circ T$ is a linear mapping. From $(f \circ T)^+ = f^+ \circ T$, we see that $f \mapsto f \circ T$ is a lattice homomorphism. To see that $f \mapsto f \circ T$ is also an isometry, note that if $f \in L_1(\nu)$, then Theorem 13.46 implies

$$\|f \circ T\|_1 = \int_X |f \circ T| \, d\mu = \int_X |f| \circ T \, d\mu = \int_Y |f| \, d\nu = \|f\|_1,$$

and the proof is finished. ∎

We wish to present one more change of variables theorem here. It is the classical finite dimensional change of variables formula. In order to state it in its general form, we need to recall a few definitions.

Let V be an open subset of $\mathbb{R}^n$ and let $T : V \to \mathbb{R}^n$ be a function having partial derivatives for each $x \in V$. The matrix

$$\left[\frac{\partial T_i}{\partial x_j}(x) \right] = \begin{bmatrix} \frac{\partial T_1}{\partial x_1}(x) & \cdots & \frac{\partial T_1}{\partial x_n}(x) \\ \vdots & \ddots & \vdots \\ \frac{\partial T_n}{\partial x_1}(x) & \cdots & \frac{\partial T_n}{\partial x_n}(x) \end{bmatrix}$$

is called the **Jacobian matrix** of T, and its determinant is called the **Jacobian** of T. The Jacobian determinant is denoted $J_T(x)$, that is,

$$J_T(x) = \det \left[\frac{\partial T_i}{\partial x_j}(x) \right].$$

13.48 Definition   *A function $T : V \to W$ between two open sets of some $\mathbb{R}^n$ is a*
diffeomorphism if:

1. *T is one-to-one and onto;*

2. *T is continuously differentiable;*

3. *$J_T(x) \neq 0$ for each $x \in V$; and*

4. *T is a homeomorphism (from V onto W).* [3]

We can now state the most general finite dimensional change of variables result. As usual, if X is a Lebesgue measurable subset of some $\mathbb{R}^n$, then $L_p(X)$
denotes the L_p-space for X equipped with Lebesgue measure λ.

13.49 Change of Variables Theorem II   *Let A and B be Lebesgue measurable*
subsets of some Euclidean space $\mathbb{R}^n$. Assume that there exist open sets $V \subset A$ and
$W \subset B$ such that $\lambda(A \setminus V) = \lambda(B \setminus W) = 0$. Let $T : V \to W$ be a diffeomorphism.
Then for each $f \in L_1(B)$, the function $(f \circ T) \cdot |J_T|$ (defined a.e. on A) belongs to
$L_1(A)$ and

$$\int_B f \, d\lambda = \int_A (f \circ T) \cdot |J_T| \, d\lambda.$$

Proof: See [13, Section 40].   ∎

---

[3] This property follows from the other three, but we include it in this definition in order to emphasize
its importance; see [18, Theorem 13.5, p. 371].

Chapter 14

Riesz Representation Theorems

In this chapter we discuss a well-known family of theorems, known collectively as the Riesz Representation Theorems, that assert that positive linear functionals on the classical normed Riesz space $C(X)$ of continuous real functions on X can be represented as integrals with respect to Borel measures. To make sure everything is integrable, we restrict attention either to continuous functions with compact support, $C_c(X)$, and measures that are finite on compact sets, or to finite measures and bounded continuous functions, $C_b(X)$. We also consider positive functionals on the spaces of bounded measurable real functions $B_b(X)$.

Theorem 14.9 asserts that a positive linear functional on $C_b(X)$, the space of bounded continuous real functions on X, where X is a normal Hausdorff space, has a representation as the integral with respect to a unique outer regular charge on the algebra generated by the open sets. A charge is *outer regular* if every set can be approximated (in measure) from the outside by open sets. Since $C_b(X)$ is a Banach lattice, every positive linear functional is norm continuous. Theorem 14.10 shows that the space of outer regular charges with the usual lattice operations is lattice isometric to the norm dual of $C_b(X)$.

Theorem 14.12 asserts that a positive linear functional on $C_c(X)$, the space of continuous real functions on X with compact support, where X is a locally compact Hausdorff space, has a representation as the integral with respect to a unique regular Borel measure. Indeed, positive linear functionals on $C_c(X)$ are often called *Radon measures*. A Borel measure is *regular* if it is outer regular and *tight*, meaning every Borel set can be approximated in measure from inside by compact sets. Theorem 14.14 shows that the AL-space of regular Borel measures is lattice isometric to the norm dual of $C_c(X)$. Note that every Borel measure defines a continuous linear functional on $C_c(X)$. However it is possible for two distinct Borel measures on X to define the same linear functional on $C_c(X)$ (Example 14.13). Consequently, $C_c(X)$ does not separate the points in the space of Borel measures. This means that the pairing of $C_c(X)$ with the space of Borel measures is not a dual pair. This problem is cured by restricting attention to regular Borel measures.

There are many versions of these theorems that appear in the literature, and the relations among them are not always clear. For instance, one version states that a positive linear functional on $C(X)$, where X is a compact Hausdorff space, has a representation as the integral with respect to a unique finite Baire measure. The way this result relates to Theorem 14.12 is this. We know that we only need Baire measures to be able to integrate functions in $C_c(X)$. In the locally compact case, every positive functional has a representation in terms of a Baire measure. This representation may not be unique in the space of Baire measures. The way we get a unique measure is by requiring it to have a regular extension to the Borel σ-algebra. In the smaller class of regular Borel measures the representation is unique. In the special case where X is compact, the representation is already unique in the class of Baire measures. There is still a representation as a regular Borel measure, but this may not be stated.

We also prove (Corollary 14.15) that when X is compact and metrizable, and so a special case of both locally compact and normal space, a positive linear functional on $C(X) = C_b(X) = C_c(X)$ has a representation as the integral with respect to a unique finite regular Borel measure. This is reconciled with Theorem 14.9 by showing that for compact metrizable spaces, every outer regular charge on the algebra generated by the open sets is the restriction of a unique regular Borel measure.

We also show that every continuous linear functional on the space $B_b(\Sigma)$ of bounded measurable functions on a σ-algebra has a representation as a signed charge (Lemma 14.3). These and other representation theorems are summarized in Table 14.1 on page 499.

Theorem 14.23 characterizes homomorphisms between $C(X)$ spaces, where X is compact, as *composition operators*.

14.1 The AM-space $B_b(\Sigma)$ and its dual

In this section Σ is a σ-algebra of subsets of some fixed set X.

14.1 Definition   *The collection of all bounded Σ-measurable real functions defined on X is denoted $B_b(\Sigma)$.*

Recall that if X is a topological space, then for simplicity we write $B_b(X)$ instead of $B_b(\mathcal{B}_X)$. When $B_b(\Sigma)$ is equipped with the sup norm it becomes an AM-space having unit the constant function $\mathbf{1}$. That is:

14.2 Theorem   *The Riesz space $B_b(\Sigma)$ equipped with the sup norm is a σ-order complete AM-space with unit $\mathbf{1}$.*

Next we describe the norm dual of the Banach lattice $B_b(\Sigma)$. Recall that since $B_b(\Sigma)$ is an AM-space, its norm dual $B'_b(\Sigma)$ is an AL-space and coincides with its order dual (Theorems 9.11 and 9.27).

14.3 Lemma   *If $\varphi \in B'_b(\Sigma)$, then the set function $\mu_\varphi \colon \Sigma \to \mathbb{R}$ defined by*

$$\mu_\varphi(A) = \varphi(\chi_A)$$

is a finitely additive signed measure of bounded variation. That is, $\mu_\varphi \in ba(\Sigma)$.

Proof: Clearly, $\mu_\varphi(\varnothing) = \varphi(\chi_\varnothing) = \varphi(0) = 0$. Also, if $A, B \in \Sigma$ are disjoint, then

$$\mu_\varphi(A \cup B) = \varphi(\chi_{A \cup B}) = \varphi(\chi_A + \chi_B) = \varphi(\chi_A) + \varphi(\chi_B) = \mu_\varphi(A) + \mu_\varphi(B),$$

so μ_φ is a finitely additive real-valued set function on Σ.

 To see that μ_φ is of bounded variation, note that if $\{A_1, \ldots, A_n\}$ is a partition of X, then

$$\sum_{i=1}^{n} |\mu_\varphi(A_i)| = \sum_{i=1}^{n} |\varphi(\chi_{A_i})| \leqslant \sum_{i=1}^{n} |\varphi|(\chi_{A_i})$$

$$= |\varphi|\left(\sum_{i=1}^{n} \chi_{A_i}\right) = |\varphi|(\mathbf{1}) = \|\varphi\| < \infty,$$

which implies that μ_φ is of bounded variation. ∎

 We next show that the norm (or order) dual of $B_b(\Sigma)$ coincides with $ba(\Sigma)$.

14.4 Theorem   *The mapping $\varphi \mapsto \mu_\varphi$ from $B'_b(\Sigma)$ to $ba(\Sigma)$, defined by*

$$\mu_\varphi(A) = \varphi(\chi_A),$$

is a surjective lattice isometry—so $B'_b(\Sigma) = ba(\Sigma)$.

Proof: By Lemma 14.3, we know that $\varphi \mapsto \mu_\varphi$ is indeed a mapping from $B'_b(\Sigma)$ to $ba(\Sigma)$. This mapping is clearly linear. To see that $\varphi \mapsto \mu_\varphi$ is one-to-one, assume $\mu_\varphi = 0$ for some $\varphi \in B'_b(\Sigma)$. Then, $\varphi(s) = 0$ for each Σ-simple function s. Now, as in the proof of Theorem 11.8, we can show that the vector space of all Σ-simple functions is uniformly dense in $B_b(\Sigma)$. Since φ is a $\|\cdot\|_\infty$-continuous linear functional, we infer that $\varphi = 0$, so that $\varphi \mapsto \mu_\varphi$ is one-to-one.

 Now we establish that the mapping $\varphi \mapsto \mu_\varphi$ is surjective. To see this, choose $0 \leqslant \mu \in ba(\Sigma)$, and consider the positive linear functional $\varphi \colon B_b(\Sigma) \to \mathbb{R}$ defined by $\varphi(f) = \int f \, d\mu$. (The integral always exists by virtue of Theorem 11.8.) Clearly, $\varphi_\mu = \mu$, so $\varphi \mapsto \mu_\varphi$ is surjective.

 Next, notice that $\mu_\varphi \geqslant 0$ if and only if $\varphi \geqslant 0$. So by Theorem 9.17, $\varphi \mapsto \mu_\varphi$ is a lattice isomorphism. Finally, from

$$\|\mu_\varphi\| = |\mu_\varphi|(X) = \mu_{|\varphi|}(X) = |\varphi|(\mathbf{1}) = \|\varphi\|,$$

we infer that $\varphi \mapsto \mu_\varphi$ is indeed a surjective lattice isometry. ∎

 Not surprisingly, the σ-order continuous dual of $B_b(\Sigma)$ coincides with $ca(\Sigma)$, the Riesz space of all countably additive signed measures of bounded variation.

14.5 Theorem   *The σ-order continuous dual of $B_b(\Sigma)$ coincides with $ca(\Sigma)$.*
That is, $(B_b(\Sigma))_c^{\sim} = ca(\Sigma)$. Moreover, $ca(\Sigma)$ separates the points of $B_b(\Sigma)$ and the
pair $\langle B_b(\Sigma), ca(\Sigma)\rangle$ under its natural duality

$$\langle f, \mu\rangle = \int f \, d\mu$$

is a (not necessarily symmetric) Riesz pair.

Proof: Let $0 \leqslant \varphi \in B_b'(\Sigma)$. We must show that μ_φ is a measure if and only if φ is
a σ-order continuous linear functional. To this end, assume first that φ is σ-order
continuous and let $\{A_n\}$ be a pairwise disjoint sequence in Σ. Put $A = \bigcup_{i=1}^{\infty} A_i$,
$B_n = \bigcup_{i=1}^{n} A_i$, and note that $\chi_{B_n}(x) \uparrow \chi_A(x)$ for each $x \in X$. Since φ is σ-order
continuous, it follows that

$$\sum_{i=1}^{n} \mu_\varphi(A_i) = \varphi(\chi_{B_n}) \uparrow_n \varphi(\chi_A) = \mu_\varphi(A),$$

which shows that μ_φ is σ-additive.

For the converse, assume μ_φ is a measure, and let $f_n \downarrow 0$ in $B_b(\Sigma)$. We claim
that $f_n(x) \downarrow 0$ for μ_φ-almost all x. To see this, let $f_n(x) \downarrow f(x) \geqslant 0$ for each x. Then
$f \in B_b(\Sigma)$, so $C = \{x \in X : f(x) > 0\} \in \Sigma$. If $\mu_\varphi(C) > 0$, then there exists some
$\varepsilon > 0$ such that the set

$$C_\varepsilon = \{x \in X : f(x) \geqslant \varepsilon\} \in \Sigma$$

has $\mu_\varphi(C_\varepsilon) > 0$. But then, $f_n \geqslant \varepsilon\chi_{C_\varepsilon} > 0$ for each n, contradicting $f_n \downarrow 0$ in $B_b(\Sigma)$.
So $f_n(x) \downarrow 0$ for μ_φ-almost all x. Now the Lebesgue Dominated Convergence
Theorem 11.21 implies

$$\varphi(f_n) = \int f_n \, d\mu_\varphi \downarrow 0,$$

proving that φ is σ-order continuous.

To see that $ca(\Sigma)$ separates the points of $B_b(\Sigma)$ note first that the point mass δ_x
belongs to $ca(\Sigma)$ for each $x \in X$. So if $f \neq g$, then $f(x) \neq g(x)$ for some $x \in X$,
which translates to $\int f \, d\delta_x \neq \int g \, d\delta_x$.

Finally, to see that $\langle B_b(\Sigma), ca(\Sigma)\rangle$ need not be a symmetric Riesz pair, recall
that a Riesz pair $\langle E, E'\rangle$ is symmetric if and only if the $\sigma(E, E')$ weak topology is
order continuous (Theorem 8.60). Consider the order interval $[0, 1]$ in $B_b([0, 1])$.
Let $\lambda \in ca([0, 1])$ denote Lebesgue measure. Let Φ denote the family of finite sub-
sets of $[0, 1]$. Then the net $\{\chi_\alpha\}_{\alpha \in \Phi}$ satisfies $\chi_\alpha \uparrow 1$, so $\{\chi_\alpha\}_{\alpha \in \Phi}$ is order convergent
to 1, but $\int \chi_\alpha \, d\lambda = 0$ for each α and $\int 1 \, d\lambda = 1$, so it is not weakly convergent. ∎

We finish this section with two useful properties of $B_b(X)$.

14.6 Lemma   *For a Hausdorff space X, a net $\{f_\alpha\}$ of functions in $B_b(X)$ satisfies*
$f_\alpha \downarrow 0$ (in the lattice sense) if and only if $f_\alpha(x) \downarrow 0$ in $\mathbb{R}$ for each $x \in X$.

Proof: Recall that in a Hausdorff space, singletons are closed, and hence Borel sets. If $f_\alpha \downarrow 0$ in $B_b(X)$ and $f_\alpha(x) \geqslant \varepsilon > 0$ for all α and some $x \in X$, then $f_\alpha \geqslant \varepsilon \chi_{\{x\}} > 0$ in $B_b(X)$ for each α, which is impossible. ∎

This result together with Theorem 4.33 implies that the σ-order continuous operators on $B_b(X)$ are determined by their values on $C_b(X)$.

14.7 Lemma *If S and X are metrizable, then two σ-order continuous positive operators $T_1, T_2 \colon B_b(X) \to B_b(S)$ coincide if and only if $T_1(f) = T_2(f)$ for each $f \in C_b(X)$.*

Proof: Assume that $T_1(f) = T_2(f)$ for each $f \in C_b(X)$, and let

$$\mathcal{F} = \{g \in B_b(X) : T_1(g) = T_2(g)\}.$$

Clearly $\mathcal{F}$ is a vector subspace of $B_b(X)$ satisfying $C_b(X) \subset \mathcal{F}$. Let $\{f_n\}$ be a sequence in $C_b(X)$ satisfying $f_n \uparrow f$ pointwise for some $f \in B_b(X)$. Then from $T_1(f_n) = T_2(f_n)$ and the σ-order continuity of T_1 and T_2, we have $T_1(f) = T_2(f)$. That is, $f \in \mathcal{F}$. Then $\mathcal{F} = B_b(X)$ by Theorem 4.33, so $T_1 = T_2$. ∎

The Riesz pair $\langle B_b, ca \rangle$ plays an important role in Chapter 19.

14.2 The dual of $C_b(X)$ for normal spaces

In order to describe the dual of $C_b(X)$, we start by proving that every continuous bounded function is always integrable with respect to any finite Borel charge.

14.8 Theorem *If X is a topological space and $\mu \colon \mathcal{A}_X \to [0, \infty)$ is a finite charge on the algebra $\mathcal{A}_X$ generated by the open sets of X, then $C_b(X)$ is a Riesz subspace of the Riesz space of all bounded μ-integrable real functions on X.*

Proof: If $f \colon X \to \mathbb{R}$ is continuous, then it is $(\mathcal{A}_X, \mathcal{A}_\mathbb{R})$-measurable, since

$$f^{-1}([a,b)) = f^{-1}((-\infty, b)) \cap \left[f^{-1}((-\infty, a)) \right]^c \in \mathcal{A}_X$$

The conclusion now follows from Theorem 11.8. ∎

Now we are ready to characterize the positive linear functionals on $C_b(X)$.

14.9 Theorem (Positive functionals on $C_b(X)$) *Let X be a normal Hausdorff topological space and let $\Lambda \colon C_b(X) \to \mathbb{R}$ be a positive linear functional. Then there exists a unique finite normal charge μ on the algebra $\mathcal{A}_X$ generated by the open sets satisfying $\mu(X) = \|\Lambda\| = \Lambda(\mathbf{1})$ and*

$$\Lambda(f) = \int f \, d\mu$$

for each $f \in C_b(X)$.

Proof: Let $\Lambda\colon C_b(X) \to \mathbb{R}$ be a positive linear functional, where X is Hausdorff and normal. The existence of the finite charge μ is quite involved. The uniqueness of such a measure is much simpler and we prove it first.

(*Uniqueness*) Assume that two finite regular charges μ and ν on $\mathcal{A}_X$ satisfy

$$\int f\,d\mu = \int f\,d\nu$$

for each $f \in C_b(X)$. To establish that $\mu = \nu$, it suffices to show that $\mu(C) = \nu(C)$ for each closed set C. So let C be a closed set and consider an arbitrary open set V with $C \subset V$. Since X is normal, by Urysohn's Lemma 2.46, there exists a continuous function $f\colon X \to [0,1]$ satisfying $f(x) = 1$ for each $x \in C$ and $f(x) = 0$ for all $x \in V^c$. From $\chi_C \leqslant f \leqslant \chi_V$, we see that

$$\mu(C) = \int \chi_C d\mu \leqslant \int f\,d\mu = \int f\,d\nu \leqslant \int \chi_V d\nu = \nu(V),$$

and consequently $\nu(C) = \inf\{\nu(V) : V \text{ open and } C \subset V\} \geqslant \mu(C)$. By symmetry, $\mu(C) \geqslant \nu(C)$, so $\mu(C) = \nu(C)$.

(*Existence*) We construct the charge μ in steps.

First, for each closed subset C we define

$$\mu(C) = \inf\{\Lambda(f) : f \in C_b(X) \text{ and } f \geqslant \chi_C\}.$$

Next, for each open set V, we let

$$\mu(V) = \sup\{\mu(C) : C \text{ closed and } C \subset V\}$$

(Notice that for a clopen set these formulae for μ agree.)

Finally, for an arbitrary subset A of X, we define

$$\mu^*(A) = \inf\{\mu(V) : V \text{ open and } A \subset V\}.$$

Now we prove that μ^* restricted to $\mathcal{A}_X$ has the desired properties. We accomplish this in a series of steps.

(1) $\mu(\varnothing) = 0$ *and* $\mu(X) = \|\Lambda\| = \Lambda(\mathbf{1}) < \infty$.

This follows easily from the linearity and positivity of Λ.

(2) *If C_1 and C_2 are closed, then* $\mu(C_1 \cup C_2) \leqslant \mu(C_1) + \mu(C_2)$ *with equality if* $C_1 \cap C_2 = \varnothing$.

Let C_1 and C_2 be closed. If $\chi_{C_1} \leqslant f$ and $\chi_{C_2} \leqslant g$, then $\chi_{C_1 \cup C_2} \leqslant f + g$, so $\mu(C_1 \cup C_2) \leqslant \Lambda(f + g) = \Lambda(f) + \Lambda(g)$. Taking the infimum over f and g, we get $\mu(C_1 \cup C_2) \leqslant \mu(C_1) + \mu(C_2)$.

Next, assume that $C_1 \cap C_2 = \varnothing$. By Urysohn's Lemma 2.46, there exists a continuous function $h: X \to [0, 1]$ such that $h(x) = 1$ for each $x \in C_1$ and $h(x) = 0$ for each $x \in C_2$. So if $f \in C_b(X)$ satisfies $\chi_{C_1 \cup C_2} \leqslant f$, then $\chi_{C_1} \leqslant fh$ and $\chi_{C_2} \leqslant f(1 - h)$. Therefore

$$\Lambda(f) = \Lambda(fh) + \Lambda(f(1 - h)) \geqslant \mu(C_1) + \mu(C_2).$$

So $\mu(C_1 \cup C_2) \geqslant \mu(C_1) + \mu(C_2)$, and therefore $\mu(C_1 \cup C_2) = \mu(C_1) + \mu(C_2)$.

(3) μ^* *is monotone.*

This follows immediately from the definition of μ^*.

(4) *If A is either closed or open, then* $\mu^*(A) = \mu(A)$.

If A is open, then clearly $\mu^*(A) = \mu(A)$. So assume that A is closed. If V is an open set satisfying $A \subset V$, then from the definition of $\mu(V)$, we see that $\mu(A) \leqslant \mu(V)$. Consequently,

$$\mu(A) \leqslant \inf\{\mu(V) : V \text{ open and } A \subset V\} = \mu^*(A).$$

For the reverse inequality, fix $\varepsilon > 0$ and then pick some $f \in C_b(X)$ with $\chi_A \leqslant f$ and $\Lambda(f) \leqslant \mu(A) + \varepsilon$. Next, pick $0 < \delta < 1$ and consider the sets $V = \{x \in X : f(x) > \delta\}$ and $C = \{x \in X : f(x) \geqslant \delta\} = \{x : \frac{1}{\delta}f(x) \geqslant 1\}$. Clearly, V is open, C is closed and $A \subset V \subset C$. Consequently, we have

$$\mu^*(A) \leqslant \mu(V) = \sup\{\mu(B) : B \text{ closed and } B \subset V\}$$
$$\leqslant \mu(C) \leqslant \Lambda(\tfrac{1}{\delta}f) = \tfrac{1}{\delta}\Lambda(f) \leqslant \tfrac{1}{\delta}[\mu(A) + \varepsilon].$$

Since $\varepsilon > 0$ and $0 < \delta < 1$ are arbitrary, we have $\mu^*(A) \leqslant \mu(A)$, so $\mu^*(A) = \mu(A)$.

(5) μ^* *is finitely subadditive. That is,*

$$\mu^*\left(\bigcup_{i=1}^{n} A_i\right) \leqslant \sum_{i=1}^{n} \mu^*(A_i)$$

for each finite collection $\{A_1, \ldots, A_n\}$ *of subsets of X.*

It suffices to establish the subadditivity of μ^* for $n = 2$. So let A_1 and A_2 be subsets of X and fix $\varepsilon > 0$. Pick two open sets V_1 and V_2 with $A_1 \subset V_1$, $A_2 \subset V_2$, $\mu(V_1) < \mu^*(A_1) + \varepsilon$, and $\mu(V_2) < \mu^*(A_2) + \varepsilon$. Also, fix some closed set $C \subset V_1 \cup V_2$ with $\mu(V_1 \cup V_2) < \mu(C) + \varepsilon$.

Observe that $C \setminus V_1$ and $C \setminus V_2$ are two disjoint closed sets. Since X is normal, there exist disjoint open sets W_1 and W_2 such that $C \setminus V_1 \subset W_1$ and $C \setminus V_2 \subset W_2$. Now consider the closed sets $C_1 = C \setminus W_1$ and $C_2 = C \setminus W_2$. Then $C_1 \subset V_1$, $C_2 \subset V_2$, and $C = C_1 \cup C_2$. (Why?) Therefore, using (2) and (4), we see that

$$\mu^*(A_1 \cup A_2) \leqslant \mu(V_1 \cup V_2) \leqslant \mu(C) + \varepsilon = \mu(C_1 \cup C_2) + \varepsilon$$
$$\leqslant \mu(C_1) + \mu(C_2) + \varepsilon \leqslant \mu(V_1) + \mu(V_2) + \varepsilon$$
$$\leqslant \mu^*(A_1) + \mu^*(A_2) + 3\varepsilon.$$

Since $\varepsilon > 0$ is arbitrary, we get $\mu^*(A_1 \cup A_2) \leqslant \mu^*(A_1) + \mu^*(A_2)$.

(6) *The collection $\mathcal{A}_\mu$ of all μ^*-measurable sets, that is, the collection*

$$\mathcal{A}_\mu = \{A \subset X : \mu^*(S) = \mu^*(S \cap A) + \mu^*(S \cap A^c) \text{ for all } S \subset X\},$$

is an algebra containing the open sets (so $\mathcal{A}_X \subset \mathcal{A}_\mu$).

The proof that $\mathcal{A}$ is an algebra is the same as in the proof of Theorem 10.20. So we need to show that $\mathcal{A}_\mu$ contains every open subset of X. To this end, let V be an open subset of X and let S be a subset of X. Fix $\varepsilon > 0$ and let W be an open set satisfying $S \subset W$. Now pick a closed set $C \subset V \cap W$ such that $\mu(C) > \mu(V \cap W) - \varepsilon \geqslant \mu^*(V \cap S) - \varepsilon$ and select a closed subset $K \subset W \setminus C$ with

$$\mu(K) > \mu(W \setminus C) - \varepsilon \geqslant \mu^*(W \setminus V) - \varepsilon \geqslant \mu^*(S \cap V^c) - \varepsilon.$$

Since $C \cup K \subset W$ and $C \cap K = \varnothing$, it follows from (2) that

$$\mu(W) \geqslant \mu(C \cup K) = \mu(C) + \mu(K) \geqslant \mu^*(S \cap V) + \mu^*(S \cap V^c) - 2\varepsilon,$$

for each open set W with $W \supset S$. So $\mu^*(S) \geqslant \mu^*(S \cap V) + \mu^*(S \cap V^c) - 2\varepsilon$ for each $\varepsilon > 0$. Thus, $\mu^*(S) = \mu^*(S \cap V) + \mu^*(S \cap V^c)$, so $V \in \mathcal{A}_\mu$.

(7) *μ^* is a normal charge on $\mathcal{A}_\mu$.*

Let $A_1, A_2 \in \mathcal{A}_\mu$ satisfy $A_1 \cap A_2 = \varnothing$. Put $A = A_1 \cup A_2$ and note that $A \cap A_1 = A_1$ and $A \cap A_1^c = A_2$. Now the measurability of A_1 applied to the "test set" $S = A$ yields

$$\mu^*(A_1 \cup A_2) = \mu^*(A) = \mu^*(A \cap A_1) + \mu^*(A \cap A_1^c) = \mu^*(A_1) + \mu^*(A_2).$$

This shows that μ^* is a charge. It is also clearly outer regular, so it is normal.

(8) *The restriction μ of μ^* to $\mathcal{A}_X$ satisfies $\Lambda(f) = \int f \, d\mu$ for $f \in C_b(X)$.*

Let f belong to $C_b(X)$. By translating and scaling appropriately, we can suppose that $0 \leqslant f(x) < 1$ for each x. Fix n and for each i define

$$A_i = \{x \in X : (i-1)/n \leqslant f(x) < i/n\} \quad \text{and} \quad B_i = \{x \in X : f(x) \geqslant i/n\}.$$

Clearly, each B_i is closed and each $A_i = B_{i-1} \setminus B_i$. Moreover, the $\mathcal{A}_\mu$-simple function $\varphi = \sum_{i=1}^n \frac{i}{n}\chi_{A_i}$ satisfies $\varphi \geqslant f$, so

$$\int \varphi \, d\mu = \sum_{i=1}^n \frac{i}{n}\mu(A_i) \geqslant \int f \, d\mu.$$

Next, for each i define the continuous function $\theta_i : \mathbb{R} \to [0,1]$ by $\theta(t) = 0$ if $t \leqslant \frac{i-1}{n}$, $\theta(t) = 1$ if $t \geqslant \frac{i}{n}$ and linear on the closed interval $[\frac{i-1}{n}, \frac{i}{n}]$. That is,

$$\theta_i(t) = \begin{cases} 0 & \text{if } t \leqslant \frac{i-1}{n}, \\ nt + 1 - i & \text{if } \frac{i-1}{n} < t < \frac{i}{n}, \\ 1 & \text{if } t \geqslant \frac{i}{n}. \end{cases}$$

Note that $\frac{1}{n}\sum_{i=1}^{n}\theta_i(t) = t$ for each $0 \leqslant t \leqslant 1$. Next, for each $i = 1,\ldots,n$ consider the function $f_i \in C_b(X)$ defined by $f_i(x) = \theta_i(f(x))$. By the above discussion, $\frac{1}{n}\sum_{i=1}^{n}f_i = f$. Since $f(x) \geqslant \frac{i}{n}$ for each $x \in B_i$, we see that $\chi_{B_i} \leqslant f_i$. Therefore $\Lambda(f_i) \geqslant \mu(B_i)$ for each $i = 1,\ldots,n$. Consequently,

$$
\begin{aligned}
\Lambda(f) = \frac{1}{n}\sum_{i=1}^{n}\Lambda(f_i) &\geqslant \frac{1}{n}\sum_{i=1}^{n}\mu(B_i) = \frac{1}{n}\Big[\sum_{i=1}^{n} i\mu(B_{i-1}\setminus B_i) - \mu(B_0)\Big] \\
&= \sum_{i=1}^{n}\frac{i}{n}\mu(A_i) - \frac{1}{n}\mu(B_0) = \int \varphi\,d\mu - \frac{1}{n}\mu(B_0) \\
&\geqslant \int f\,d\mu - \frac{1}{n}\mu(X)
\end{aligned}
$$

for each n. Hence, $\Lambda(f) \geqslant \int f\,d\mu$ for each $f \in C_b(X)$. Replacing f by $-f$, we get $\Lambda(f) \leqslant \int f\,d\mu$. That is, $\Lambda(f) = \int f\,d\mu$ for each $f \in C_b(X)$. ∎

We can now show that the dual of $C_b(X)$ is the AL-space $ba_n(\mathcal{A}_X)$.

14.10 Theorem (Dual of $C_b(X)$, with X normal) *Let X be a Hausdorff normal topological space and let $\mathcal{A}_X$ be the algebra generated by the open subsets of X. Then the mapping $\Lambda \colon ba_n(\mathcal{A}_X) \to C_b'(X)$, defined by*

$$
\Lambda_\mu(f) = \int f\,d\mu = \int f\,d\mu^+ - \int f\,d\mu^-,
$$

is a surjective lattice isometry. In other words, the norm dual of the AM-space $C_b(X)$ can be identified with the AL-space $ba_n(\mathcal{A}_X)$.

Proof: By Theorem 14.9, the mapping $\Lambda \colon ba_n^+(\mathcal{A}_X) \to [C_b'(X)]^+$ is additive and surjective. It follows (see the footnote to Lemma 8.23) that the mapping $\Lambda \colon ba_n(\mathcal{A}_X) \to C_b'(X)$ (as defined above) is indeed a surjective linear operator.

Next, we claim that $\Lambda_\mu \geqslant 0$ if and only $\mu \geqslant 0$. To see this, let μ in $ba_n(\mathcal{A}_X)$ satisfy $\int f\,d\mu \geqslant 0$ for each $0 \leqslant f \in C_b(X)$ and let V be an open set. Since $\mu \in ba_n(\mathcal{A}_X)$, it suffices to show that $\mu(V) \geqslant 0$. Given $\varepsilon > 0$ choose a closed set C satisfying $C \subset V$ and $\mu^-(V) - \mu^-(C) < \varepsilon$. Since X is normal and Hausdorff, there exists a continuous function $f \colon X \to [0,1]$ such that $f = 1$ on C and $f = 0$ on V^c, so $\chi_C \leqslant f \leqslant \chi_V$. Therefore,

$$
\begin{aligned}
0 \leqslant \int f\,d\mu = \int f\,d\mu^+ - \int f\,d\mu^- &\leqslant \mu^+(V) - \mu^-(C) \\
&\leqslant \mu^+(V) - \mu^-(V) + \varepsilon = \mu(V) + \varepsilon
\end{aligned}
$$

for each $\varepsilon > 0$, which shows that $\mu(V) \geqslant 0$. In particular, Λ is one-to-one.

Now Theorem 9.17 shows that Λ is a surjective lattice isomorphism, so in view of the equality

$$\|\Lambda_\mu\| = \left\||\Lambda_\mu|\right\| = \|\Lambda_{|\mu|}\| = \|\Lambda_{|\mu|}(\mathbf{1})\| = |\mu|(X) = \|\mu\|,$$

it is also an isometry. ∎

14.11 Corollary (Dual of $\ell_\infty(X)$)   *Let X be a set, and let $\ell_\infty(X)$ denote the AM-space of all bounded real functions on X. Then the norm dual of $\ell_\infty(X)$ coincides with $ba(X)$, the AL-space of all signed measures of bounded variation defined on the power set of X.*

Proof: Note that $\ell_\infty(X) = C_b(X)$ when X is equipped with the discrete topology. By Theorem 12.5 every finite charge on X is normal, so $ba_n(X) = ba(X)$. ∎

14.3   The dual of $C_c(X)$ for locally compact spaces

A **Radon measure** on a topological space X is a positive linear functional on $C_c(X)$, the Riesz space of all continuous real functions on X with compact support. The term is usually applied only to locally compact spaces. The following representation theorem, due essentially to F. Riesz, justifies the use of the term "measure," and is also known as the **Riesz–Markov Theorem**

14.12 Theorem (Positive Functionals on $C_c(X)$)   *Let X be a locally compact Hausdorff space, and let $\Lambda\colon C_c(X) \to \mathbb{R}$ be a positive linear functional. Then there exists a unique regular Borel measure μ satisfying $\|\Lambda\| = \mu(X)$ and*

$$\Lambda(f) = \int f\, d\mu$$

for all $f \in C_c(X)$.

Proof: The proof follows the same steps as the proof of Theorem 14.9 with closed sets replaced by compact sets. The key difference is that now μ^* is σ-subadditive rather than merely subadditive: Step (5) is modified by taking a sequence $\{A_n\}$ of subsets and approximating each A_n by an open set V_n that includes it so that $\mu^*\left(\bigcup_{n=1}^\infty A_n\right) \leqslant \mu^*\left(\bigcup_{n=1}^\infty V_n\right) + \varepsilon$. Next approximate $V = \bigcup_{n=1}^\infty V_n$ by an included compact set K. Since K is compact, it is actually covered by a finite subcover of $\{V_n\}$, so finite subadditivity implies countable subadditivity. (This requires some work.) It follows that the collection of μ-measurable sets is actually a σ-algebra, rather than just an algebra. The proof that every open set is measurable proceeds along the same lines. Note that the construction of μ^* guarantees that it is regular on $\mathcal{B}_X$. We leave the details as an exercise, or see [13, Theorem 38.3, p. 355]. ∎

It is important to realize that Theorem 14.12 does not say that there is a unique Borel measure representing the positive linear functional Λ and that the measure is regular. It says that there is only one regular Borel measure representing Λ. In fact, the first assertion is not true. The next example presents two distinct Borel measures representing the same functional. Since one of them cannot be regular, we use the big-small measure on the compact Hausdorff space Ω from Example 12.9. Before the example, we mention that in the special case where X is compact, there is a unique measure in the class of Baire measures. See for instance, H. L. Royden [290, Theorem 25, p. 357].

14.13 Example (Nonunique representation by Borel measures) Lemma 2.82 asserts that every continuous real function f on Ω is eventually constant. That is, there is some $x < \omega_1$ such that $y \geq x$ implies $f(y) = f(\omega_1)$. Therefore, under the (nonregular) big-small measure μ of Example 12.9 $\int_\Omega f\,d\mu = f(\omega_1)$. This defines a continuous linear functional on $C(\omega)$, namely evaluation at ω_1. This functional is also represented by another Borel measure, the point mass δ_{ω_1}, which is regular.

On $C_c(\Omega_0)$ integration with respect to μ induces the zero functional. The zero measure is of course the regular Borel measure on Ω_0 inducing the zero functional.

This example may seem to be at odds with the fact (mentioned above) that on a compact Hausdorff space the representing measure is unique in the class of Baire measures. As it turns out, on $\mathcal{B}aire(\Omega)$ the big-small measure μ and the point mass δ_{ω_1} coincide. The fact is, neither Ω_0 nor $\{\omega_1\}$ is a Baire subset of Ω. We shall not go into the proof of this here, but it follows from the fact that every compact Baire set is a compact $\mathcal{G}_\delta$-set (cf. [148, Theorem D, p. 221]). It is not hard to see that the compact set $\{\omega_1\}$ is not a $\mathcal{G}_\delta$. ∎

The sup norm on $C_c(X)$ is a lattice norm, so $C_c(X)$ equipped with the sup norm is a normed Riesz space. Of course, if X is compact, then $C_c(X)$ coincides with $C(X)$, and in this case it is a Banach lattice. We can now describe the norm dual of $C_c(X)$ for locally compact Hausdorff spaces. The proof (which is similar to that of Theorem 14.10) is left as an exercise.

14.14 Theorem (Dual of $C_c(X)$, with X locally compact) *Let X be a locally compact Hausdorff space. Then the mapping $\Lambda\colon ca_r(\mathcal{B}_X) \to C_c'(X)$, defined by*

$$\Lambda_\mu(f) = \int f\,d\mu,$$

is a surjective lattice isometry. That is, the norm dual of $C_c(X)$ can be identified with the AL-space $ca_r(\mathcal{B}_X)$ of regular signed Borel measures of bounded variation.

In particular, if X is a compact Hausdorff space, then $C_c'(X) = ca_r(\mathcal{B}_X)$.

Notice that since $C_c(X)$ (with the sup norm) need not be a Banach lattice, its norm dual need not coincide with the order dual. This means that there exist positive linear functionals that are not norm continuous. For instance, if $X = \mathbb{R}$, then Lebesgue measure λ defines the positive linear functional $f \mapsto \int f \, d\lambda$ on $C_c(\mathbb{R})$, which is not norm continuous.

Keeping in mind that on a compact metrizable space every finite Borel measure is regular (Theorem 12.7), we have the following important special case.

14.15 Corollary (Dual of $C(X)$, with X compact metrizable) *If X is a compact metrizable space, then the mapping* $\Lambda \colon ca(\mathcal{B}_X) \to C'(X)$*, defined by*

$$\Lambda_\mu(f) = \int f \, d\mu,$$

is a surjective lattice isometry. That is, if X is a compact metrizable space, then the norm dual of $C(X)$ can be identified with the AL-space $ca(X)$ of finite Borel measures on X.

In Table 14.1 on page 499 we list several common Banach lattices and their duals. For additional representation theorems see the tables in N. Dunford and J. T. Schwartz [110, IV.15, pp. 374–379] and Z. Semadeni [301, 18.5.5, p. 318].

14.4 Baire vs. Borel measures

The literature on Riesz Representation Theorems can be quite confusing. At first glance, when X is a compact Hausdorff space, Theorem 14.10 seems to contradict Theorem 14.14. According to these theorems, the dual of $C(X) = C_b(X)$ is $ba_n(\mathcal{A}_X)$ and also $ca_r(\mathcal{B}_X)$! However, this is not a contradiction because every normal signed charge of bounded variation μ on $\mathcal{A}_X$ has (in light of Theorems 14.9 and 14.12) a unique extension to a regular signed measure $\bar{\mu}$ of bounded variation on $\mathcal{B}_X$ satisfying $\int f \, d\mu = \int f \, d\bar{\mu}$ for each $f \in C(X)$. In fact, μ and $\bar{\mu}$ are the restrictions of μ^* to $\mathcal{A}_X$ and $\mathcal{B}_X$, respectively. Moreover, it is possible to show that for compact Hausdorff spaces, the mapping $\mu \mapsto \bar{\mu}$, from $ba_n(\mathcal{A}_X)$ to $ca_r(\mathcal{B}_X)$, is a surjective lattice isometry. With this identification, we have $ba_n(\mathcal{A}_X) = ca_r(\mathcal{B}_X)$. Similarly, some versions of Theorem 14.12 assert that every positive linear functional on a locally compact Hausdorff space can be represented as an integral with respect to a Baire measure rather than a regular Borel measure. Again this is not a contradiction, because the regular Borel measure restricted to the Baire σ-algebra is a Baire measure that represents the same functional. Therefore, Theorem 14.12 can be reformulated as follows.

14.16 Theorem (Dual of $C_c(X)$, with X locally compact) *If X is a locally compact Hausdorff space, then the norm dual of $C_c(X)$ can be identified with the AL-space $ca_r(\mathcal{B}_{aire})$ in the following sense:*

Normed Riesz Space		Norm Dual	Result
ℓ_p,	$1 < p < \infty$	ℓ_q	13.26
ℓ_1		ℓ_∞	13.28
c_0		ℓ_1	16.7
$L_p(\mu)$,	$1 < p < \infty$	$L_q(\mu)$	13.26
$L_1(\mu)$,	σ-finite μ	$L_\infty(\mu)$	13.28
$C_c(X)$,	X locally compact Hausdorff	$ca_r(\mathcal{B})$	14.14
$C(X)$,	X compact Hausdorff	$ca_r(\mathcal{B})$	14.14
$C(X)$,	X compact metric	$ca(\mathcal{B}) = ca_r(\mathcal{B})$	14.15
$C_b(X)$,	X normal Hausdorff	$ba_n(\mathcal{A})$	14.10
$\ell_\infty(X)$,	X discrete	$ba(X) = ba_n(X)$	13.28
$B_b(\Sigma)$,	Σ a σ-algebra	$ba(\Sigma)$	14.11

Table 14.1. A Table of Normed Riesz Spaces and their Norm Duals

Here $\mathcal{B}$ denotes the σ-algebra of Borel sets, $\mathcal{A}$ denotes the algebra generated by the open sets, and $\frac{1}{p} + \frac{1}{q} = 1$. For pairs of sequence spaces, the duality $\langle x, y \rangle = \sum_{n=1}^{\infty} x_n y_n$. For pairs of $L_p(\mu)$-spaces the duality is given by $\langle f, g \rangle = \int fg \, d\mu$. For pairs of function spaces with spaces of measures the duality is integration, $\langle f, \mu \rangle = \int f \, d\mu$.

For each continuous linear functional $\Lambda: C_c(X) \to \mathbb{R}$ there exists a unique signed measure $\mu \in ca_r(\mathcal{B}_{aire})$, which extends uniquely to a regular signed Borel measure of bounded variation on $\mathcal{B}_{orel}$, satisfying

$$\Lambda(f) = \int f \, d\mu$$

for each $f \in C_c(X)$.

Similar arguments can be applied to the restriction of regular Borel measures to the algebra $\mathcal{A}_X$ generated by the open sets for normal spaces. This reconciles Theorems 14.10 and 14.14.

14.5   Homomorphisms between $C(X)$-spaces

Here we study lattice homomorphisms between $C(X)$-spaces, where X is a compact Hausdorff space. The space $C(X)$ of continuous real functions on X equipped with its sup norms is a Banach lattice. In fact, it is an AM-space with unit $\mathbf{1}_X$, where $\mathbf{1}_X$ denotes the constant function one on X.

Throughout this section X and Y are compact Hausdorff spaces.

14.17 Lemma   *A nonzero linear functional $\theta: C(X) \to \mathbb{R}$ is a lattice homomorphism if and only if $\theta = c\delta_{x_0}$ for a unique constant $c > 0$ and a unique $x_0 \in X$.*

Proof: If $\theta = c\delta_{x_0}$ with $c \geqslant 0$, then θ is clearly a lattice homomorphism. That is, it satisfies $\theta(f \wedge g) = \min\{\theta(f), \theta(g)\}$ for each $f, g \in C(X)$.

For the converse, let θ be a lattice homomorphism. Clearly, θ is a positive linear functional. By Theorem 14.12 there exists a unique regular Borel measure μ such that $\theta(f) = \int f \, d\mu$ for each $f \in C(X)$. We claim that the support of μ consists of one point. To see this, note that if a, b belong to the support of μ and $a \neq b$, then there exist functions $f, g \in C(X)$ with $f(a) = g(b) = 1$ and $f \wedge g = 0$. Consequently $0 = \theta(f \wedge g) = \min\{\theta(f), \theta(g)\} = \min\{\int f \, d\mu, \int g \, d\mu\} > 0$, a contradiction. Hence, $\text{supp}\,\mu = \{x_0\}$, a singleton. Now note that

$$\theta(h) = \int h \, d\mu = h(x_0)\mu(\{x_0\}),$$

for each $h \in C(X)$, and the proof is finished.   ∎

Lattice homomorphisms between $C(X)$ and $C(Y)$ have a simple structure.

14.18 Theorem   *Let $T: C(X) \to C(Y)$ be a positive operator and define the function $r \in C(Y)$ by $r = T\mathbf{1}_X \geqslant 0$. Then T is a lattice homomorphism if and only* *if there exists a function $\xi\colon Y \to X$, which is uniquely determined and continuous on the set $\{y \in Y : r(y) > 0\}$, satisfying*

$$(Tf)(y) = r(y)f(\xi(y))$$

for each $f \in C(X)$.

Proof: If T is of the above form, then it is easy to check that T is indeed a lattice homomorphism. For the converse, assume that T is a lattice homomorphism. Then for each $y \in Y$, we have

$$(\delta_y \circ T)(f) = \delta_y(Tf) = (Tf)(y),$$

from which (by using that T is a lattice homomorphism) it follows that the linear functional $\delta_y \circ T\colon C(X) \to \mathbb{R}$ is a lattice homomorphism. So by Lemma 14.17, there exists a unique constant $r(y) \geqslant 0$ and some (not necessarily unique) $\xi(y) \in X$ such that

$$(Tf)(y) = (\delta_y \circ T)(y) = r(y)f(\xi(y)).$$

Clearly, $r = T\mathbf{1}_X$ and if $r(y) > 0$, then $\xi(y)$ is uniquely determined.

Now assume $r(y) > 0$ and let $\{y_\alpha\}$ be a net in Y satisfying $r(y_\alpha) > 0$ for each α and $y_\alpha \to y$. Then

$$r(y_\alpha)f(\xi(y_\alpha)) = (Tf)(y_\alpha) \to (Tf)(y) = r(y)f(\xi(y)),$$

so $f(\xi(y_\alpha)) \to f(\xi(y))$ for each $f \in C(X)$. From Corollary 2.57, we see that $\xi(y_\alpha) \to \xi(y)$, and the proof is finished. ∎

14.19 Definition *A linear operator $T\colon C(X) \to C(Y)$ is an **algebraic homomorphism** (or a **multiplicative operator**) if*

$$T(fg) = T(f)T(g)$$

for all $f, g \in C(X)$.

14.20 Lemma *Every algebraic homomorphism is a lattice homomorphism. However, the converse is false.*

Proof: Let $T\colon C(X) \to C(Y)$ be an algebraic homomorphism. Note first that T is a positive operator. Indeed, if $f \geqslant 0$, then $T(f) = T[(\sqrt{f})^2] = [T(\sqrt{f})]^2 \geqslant 0$. Now if $f \in C(X)$, then

$$|T(f)|^2 = [T(f)]^2 = T(f^2) = T(|f|^2) = [T(|f|)]^2,$$

from which it follows that $|T(f)| = T(|f|)$.

As an example of a lattice homomorphism that is not an algebraic homomorphism simply consider the lattice homomorphism $T\colon C[0, 1] \to C[0, 1]$ defined by $T(f) = 2f$. ∎

The next result characterizes algebraic homomorphisms.

14.21 Lemma   *An operator $T: C(X) \to C(Y)$ is an algebraic homomorphism if and only if there exists a unique clopen subset A of Y and a function $\xi: Y \to X$ that is continuous on A such that*

$$Tf = \chi_A \cdot (f \circ \xi)$$

for each $f \in C(X)$.

Proof: If T has the form described in the lemma, then T is clearly an algebraic homomorphism. Now assume that T is an algebraic homomorphism. By Lemma 14.20, T is a lattice homomorphism, so by Theorem 14.18, T is of the form

$$(Tf)(y) = r(y)f(\xi(y)),$$

where $T\mathbf{1}_X = r \geqslant 0$ and $\xi: Y \to X$ is continuous on $\{y \in Y : r(y) > 0\}$. Since T is an algebraic homomorphism, we have

$$r^2 = (T\mathbf{1}_X)^2 = T((\mathbf{1}_X)^2)) = T\mathbf{1}_X = r,$$

from which it follows that $r = \chi_A$ for a unique clopen subset A of Y.   ∎

14.22 Definition   *An operator $T: C(X) \to C(Y)$ is a **composition operator** if there is a continuous function $\xi: Y \to X$ satisfying*

$$Tf = f \circ \xi$$

for each $f \in C(X)$.

Clearly, every composition operator is automatically an algebraic (and hence a lattice) homomorphism. A positive operator $T: C(X) \to C(Y)$ is called a **Markov operator** if $T\mathbf{1}_X = \mathbf{1}_Y$. For Markov operators, the notions of algebraic homomorphism, lattice homomorphism and composition operator coincide. Specifically, we have the following result whose proof follows immediately from the above discussion.

14.23 Theorem   *For a Markov operator $T: C(X) \to C(Y)$ the following statements are equivalent.*

1. *T is an algebraic homomorphism.*

2. *T is a lattice homomorphism.*

3. *T is a composition operator.*

The set of Markov operators is a convex subset of the vector space of bounded operators from $C(X)$ to $C(Y)$. It turns out that its extreme points are precisely the Markov operators that are lattice homomorphisms.

14.24 Theorem   *A Markov operator $T: C(X) \to C(Y)$ is a lattice homomorphism if and only if T is an extreme point of the convex set of all Markov operators.*

Proof: Assume first that T is an extreme point of the set $\mathcal{C}$ of Markov operators. Fix a function $h \in C(X)$ such that $\mathbf{1}_X \leqslant h \leqslant 2 \cdot \mathbf{1}_X$, and then define the operators $S, R: C(X) \to C(Y)$ by

$$S(f) = \tfrac{T(hf)}{T(h)} \quad \text{and} \quad R(f) = 2T(f) - S(f).$$

Notice that $S, R \in \mathcal{C}$ and $T = \frac{1}{2}S + \frac{1}{2}R$. Since T is an extreme point of $\mathcal{C}$, it follows that $T = S$. This implies $T(hf) = T(h)T(f)$ for each $f \in C(X)$ and all $h \in C(X)$ satisfying $\mathbf{1}_X \leqslant h \leqslant 2 \cdot \mathbf{1}_X$. Now assume that $g \in C(X)$ satisfies $g \neq 0$. Then the function $h = \mathbf{1}_X + \frac{g}{2\|g\|_\infty} \in C(X)$ satisfies $\mathbf{1}_X \leqslant h \leqslant 2 \cdot \mathbf{1}_X$, so $T((\mathbf{1}_X + \frac{g}{2\|g\|_\infty})f) = T(\mathbf{1}_X + \frac{g}{2\|g\|_\infty})T(f)$ for all $f \in C(\Omega)$. This easily implies $T(gf) = T(g)T(f)$ for all $f \in C(X)$ and all $g \in C(X)$. Thus, T is an algebraic homomorphism, and hence a lattice homomorphism.

For the converse, assume that $T \in \mathcal{C}$ is a lattice homomorphism. By Theorem 14.18 there exists a continuous function $\xi: Y \to X$ such that $Tf = f \circ \xi$ for each $f \in C(X)$. Now assume that $T = \alpha S + (1 - \alpha)R$ with $S, R \in \mathcal{C}$ and $0 < \alpha < 1$. Clearly, $T', S', R': ca_r(Y) \to ca_r(X)$, the AL-spaces of all regular Borel measures on Y and X, respectively. Also, note that the norm dual of every Markov operator carries regular probability measures to regular probability measures. Next, observe that for each $y \in Y$ and each $f \in C(X)$, we have

$$\langle f, T'\delta_y \rangle = \langle Tf, \delta_y \rangle = (Tf)(y) = f(\xi(y)) = \langle f, \delta_{\xi(y)} \rangle.$$

This shows that $T'\delta_y = \delta_{\xi(y)}$, and consequently

$$\delta_{\xi(y)} = T'\delta_y = \alpha S'\delta_y + (1 - \alpha)R'\delta_y.$$

But clearly every point mass is an extreme point of the convex set of (Borel) regular probability measures on X, so it follows that $T'\delta_y = S'\delta_y = R'\delta_y$ for each $y \in Y$. Therefore,

$$(Tf)(y) = T'\delta_y(f) = S'\delta_y(f) = (Sf)(y) = (Rf)(y).$$

for all $f \in C(X)$ and each $y \in Y$. Hence, $T = S = R$, so T is an extreme point of the convex set $\mathcal{C}$. ∎

Chapter 15

Probability measures

Unless otherwise indicated, in this chapter X is a metrizable topological space, and $\mathcal{P}(X)$ (or simply $\mathcal{P}$) is the set of all probability measures on the Borel sets $\mathcal{B}$ of X. As usual, $C_b(X)$ denotes the Banach lattice of all bounded continuous real functions on X. The reason we focus on probability measures is that the probability measures span the space of all signed measures of bounded variation.

Recall that a **probability measure**, $\mu \colon \mathcal{B} \to [0, 1]$ is a measure with $\mu(X) = 1$. We use the phrase "a probability measure on a topological space X" synonymously with "a probability measure on the Borel σ-algebra $\mathcal{B}_X$." The set $\mathcal{P}(X)$ is endowed with the topology $w^* = \sigma(\mathcal{P}(X)C_b(X))$.

In this chapter we study the topological properties of $\mathcal{P}(X)$. First, we characterize w^*-convergence in $\mathcal{P}(X)$ by means of topological properties of the space X. The space X can be viewed as a subset of $\mathcal{P}(X)$ by identifying each $x \in X$ with the point mass δ_x. This identification is an embedding (Theorem 15.8) and in case X is also separable, each point in X is an extreme point of $\mathcal{P}(X)$ (Theorem 15.9). The space $\mathcal{P}(X)$ inherits many of the properties of X. For instance, for a metrizable topological space X, we prove:

1. X is compact if and only if $\mathcal{P}(X)$ is compact.

2. X is separable if and only if $\mathcal{P}(X)$ is separable.

3. X is Polish if and only if $\mathcal{P}(X)$ is Polish.

4. X is a Borel space if and only if $\mathcal{P}(X)$ is a Borel space.

By the definition of the $w^* = \sigma(\mathcal{P}(X), C_b(X))$ topology, for every bounded continuous real function f on X, the bounded real function on $\mathcal{P}(X)$ defined by $\mu \mapsto \int f \, d\mu$ is w^*-continuous. Moreover, we shall see that bounded semicontinuous functions define on X define bounded semicontinuous functions on $\mathcal{P}(X)$ (Theorem 15.5), and when X is separable, bounded measurable functions on X define bounded measurable functions on $\mathcal{P}(X)$ (Theorem 15.13).

The chapter ends with a discussion of infinite products and the Kolmogorov Extension Theorem 15.26.

15.1   The weak∗ topology on $\mathcal{P}(X)$

Recall that $U_d(X)$ (or simply U_d) denotes the set of all bounded d-uniformly continuous real functions on X. The set U_d contains the constant functions, and by Corollary 3.15 it is pointwise dense in $C_b(X)$. Moreover, U_d is closed under addition, scalar multiplication, pointwise multiplication, and the lattice operations. It is also a uniformly closed (that is, a norm-closed) subset of $C_b(X)$. In other words, U_d is a uniformly closed subalgebra of the algebra $C_b(X)$. If X is also compact, then U_d coincides, of course, with $C_b(X) = C(X)$.

   Our first result shows that U_d is a total set of linear functionals on the probability measures. That is, U_d separates points.

15.1 Theorem   *For two probability measures μ and ν on a metrizable topological space X, the following statements are equivalent.*

1. $\mu = \nu$.

2. $\mu(G) = \nu(G)$ *for all open sets G.*

3. $\mu(F) = \nu(F)$ *for all closed sets F.*

4. $\int f\,d\mu = \int f\,d\nu$ *for all $f \in C_b(X)$.*

5. $\int f\,d\mu = \int f\,d\nu$ *for all $f \in U_d$, where d is any compatible metric.*

6. $\int f\,d\mu = \int f\,d\nu$ *for all $f \in D$, where D is any uniformly dense subset of U_d for some compatible metric d on X.*

Proof: The equivalence of (1), (2) and (3) follows from Corollary 10.11. Also the implications (1) $\implies$ (4) $\implies$ (5) $\implies$ (6) are obviously true. We finish the proof by proving (6) implies (3). So assume that there exists a compatible metric d and a uniformly dense subset D of U_d such that $\int f\,d\mu = \int f\,d\nu$ for all $f \in D$. Now if $f \in U_d$ pick a sequence $\{f_n\} \subset D$ with $\|f_n - f\|_\infty \to 0$. Clearly, $\|f_n\|_\infty < M < \infty$ for all n and some $M > 0$. So by the Lebesgue Dominated Convergence Theorem 11.21, we get

$$\int f\,d\mu = \lim_{n\to\infty} \int f_n\,d\mu = \lim_{n\to\infty} \int f_n\,d\nu = \int f\,d\nu$$

for all $f \in U_d$.

   Finally, let F be a closed subset of X. By Corollary 3.14 there exists a sequence $\{f_n\}$ in U_d such that $f_n(x) \downarrow \chi_F(x)$ for all $x \in X$. Therefore, using the Lebesgue Dominated Convergence Theorem 11.21 once more, we see that

$$\mu(F) = \int \chi_F\,d\mu = \lim_{n\to\infty} \int f_n\,d\mu = \lim_{n\to\infty} \int f_n\,d\nu = \int \chi_F\,d\nu = \nu(F),$$

and the proof is finished.   ∎

The preceding theorem shows that every $\mu \in \mathcal{P}(X)$ gives rise to a unique (linear) mapping $f \mapsto \langle f, \mu \rangle = \int f \, d\mu$ from $C_b(X)$ into $\mathbb{R}$. This means that $\mathcal{P}(X)$ can be identified with a convex subset of $\mathbb{R}^{C_b(X)}$. Similarly, $\mathcal{P}(X)$ can be identified with a convex subset of $\mathbb{R}^{U_d}$, where d is any compatible metric on X, and also with a convex subset of $\mathbb{R}^D$ for any dense subset D of U_d. Under these identifications, $\mathcal{P}(X)$ inherits the product topologies of $\mathbb{R}^{C_b(X)}$, $\mathbb{R}^{U_d}$, and $\mathbb{R}^D$, which are denoted $\sigma(\mathcal{P}, C_b)$, $\sigma(\mathcal{P}, U_d)$, and $\sigma(\mathcal{P}, D)$, respectively. The topology $\sigma(\mathcal{P}, C_b)$ also goes by the names of the weak* topology, the **weak topology**, or possibly the **topology of convergence in distribution**. Unless otherwise specified, $\mathcal{P}(X)$ is always endowed with the $\sigma(\mathcal{P}(X), C_b(X))$-topology, which we simply call the w^*-topology.[1] Remember that a net $\{\mu_\alpha\}$ in $\mathcal{P}(X)$ satisfies $\mu_\alpha \xrightarrow{w^*} \mu$ if and only if $\int f \, d\mu_\alpha \to \int f \, d\mu$ for each $f \in C_b(X)$. It is important to know that the above three topologies are the same.

15.2 Theorem   *Let d be a compatible metric on X, and let D be a uniformly dense subset of U_d. Then $\sigma(\mathcal{P}, C_b) = \sigma(\mathcal{P}, U_d) = \sigma(\mathcal{P}, D)$.*

Proof: It suffices to prove that $\mu_\alpha \xrightarrow{\sigma(\mathcal{P}, C_b)} \mu$ if and only if $\mu_\alpha \xrightarrow{\sigma(\mathcal{P}, U_d)} \mu$. One direction is easy; $\mu_\alpha \xrightarrow{\sigma(\mathcal{P}, C_b)} \mu$ if and only if $\{\mu_\alpha\}$ converges pointwise on $C_b(X)$, so it converges pointwise on $U_d \subset C_b(X)$.

Suppose that $\mu_\alpha \xrightarrow{\sigma(\mathcal{P}, U_d)} \mu$. Let $f \in C_b(X)$ and let $\{g_n\}$ and $\{h_n\}$ be sequences of bounded d-uniformly continuous functions with $g_n(x) \uparrow f(x)$ and $h_n(x) \downarrow f(x)$ for each $x \in X$ (see Corollary 3.15). Fixing n, we have

$$\int g_n \, d\mu_\alpha \leqslant \int f \, d\mu_\alpha \leqslant \int h_n \, d\mu_\alpha$$

for each α. Taking limits with respect to α, we obtain

$$\int g_n \, d\mu \leqslant \liminf_\alpha \int f \, d\mu_\alpha \leqslant \limsup_\alpha \int f \, d\mu_\alpha \leqslant \int h_n \, d\mu.$$

If we take limits with respect to n, the Dominated Convergence Theorem 11.21 implies both $\int g_n \, d\mu \uparrow \int f \, d\mu$ and $\int h_n \, d\mu \downarrow \int f \, d\mu$. Hence $\int f \, d\mu_\alpha \to \int f \, d\mu$. That is, $\mu_\alpha \xrightarrow{\sigma(\mathcal{P}, C_b)} \mu$.   ∎

This result should be compared carefully to Corollary 5.94. If X is not compact, so that $U_d(X) \neq C_b(X)$, Corollary 5.94 says that $\sigma(ca, C_b)$ is strictly finer

---

[1] More precisely, if $ca(X)$ denotes (as usual) the AL-space of all signed measures on $\mathcal{B}_X$ of bounded variation, then

$$\langle C_b(X), ca(X) \rangle, \quad \langle U_d, ca(X) \rangle, \quad \text{and} \quad \langle D, ca(X) \rangle$$

are all dual pairs under the duality $\langle f, \mu \rangle = \int f \, d\mu$ and $\mathcal{P}(X)$ is a $\sigma(ca(X), C_b(X))$-closed convex subset of $ca(X)$. So $\sigma(\mathcal{P}(X), C_b(X))$ is the relativization of $\sigma(ca(X), C_b(X))$ to $\mathcal{P}(X)$ and this justifies the name "w^*-topology." Note well that metrizability is important here. Example 14.13 shows that $C(X)$ need not separate the points of $ca(X)$ when X is compact and Hausdorff, but not metrizable. In this case, $\langle C(X), ca(X) \rangle$ is not a dual pair.

than $\sigma(ca, U_d)$, where ca is the vector space of all (countably additive) signed Borel measures of bounded variation on X. Nevertheless, U_d and C_b induce the same topology on $\mathcal{P}$.

We are now prepared to characterize weak$*$ convergence in $\mathcal{P}(X)$.

15.3 Theorem   *For a net $\{\mu_\alpha\}$ in $\mathcal{P}(X)$ and some $\mu \in \mathcal{P}(X)$ the following statements are equivalent.*

1. $\mu_\alpha \xrightarrow{w^*} \mu$.

2. $\int f \, d\mu_\alpha \to \int f \, d\mu$ *for all $f \in C_b(X)$.*

3. $\int f \, d\mu_\alpha \to \int f \, d\mu$ *for all $f \in U_d$, where d is any compatible metric.*

4. $\int f \, d\mu_\alpha \to \int f \, d\mu$ *for all $f \in D$, where D is any uniformly dense subset of U_d for some compatible metric d.*

5. $\limsup_\alpha \mu_\alpha(F) \leqslant \mu(F)$ *for each closed set F.*

6. $\liminf_\alpha \mu_\alpha(G) \geqslant \mu(G)$ *for each open set G.*

7. $\mu_\alpha(B) \to \mu(B)$ *for each Borel set B with $\mu(\partial B) = 0$.*

Proof: The equivalence of (1) and (2) is a restatement of the definition of the weak$*$ topology. The equivalence of (2), (3), and (4) follows immediately from Theorem 15.2. Also, it is obvious that (5) and (6) are equivalent.

Next, we prove that (3) implies (5). So assume that for a compatible metric d on X, we have $\int f \, d\mu_\alpha \to \int f \, d\mu$ for each $f \in U_d$. Also, let F be a fixed closed set. By Corollary 3.14, there exists a sequence $\{f_n\} \subset U_d$ satisfying $f_n(x) \downarrow \chi_F(x)$ for each $x \in X$. From the inequality $f_n \geqslant \chi_F$, we get $\int f_n \, d\mu_\alpha \geqslant \int \chi_F \, d\mu_\alpha = \mu_\alpha(F)$ for each α, so for each n,

$$\int f_n \, d\mu = \lim_\alpha \int f_n \, d\mu_\alpha \geqslant \limsup_\alpha \mu_\alpha(F).$$

Now apply the Lebesgue Dominated Convergence Theorem 11.21 to get

$$\mu(F) = \int \chi_F \, d\mu = \lim_{n \to \infty} \int f_n \, d\mu \geqslant \limsup_\alpha \mu_\alpha(F).$$

Next we establish that (5) implies (2). So assume (5) and let $f \in C_b(X)$. It suffices to prove $\limsup_\alpha \int f \, d\mu_\alpha \leqslant \int f \, d\mu$. Indeed, if this is done, then by applying the inequality to $-f$, we get

$$\liminf_\alpha \int f \, d\mu_\alpha = -\limsup_\alpha \int (-f) \, d\mu_\alpha \geqslant \int f \, d\mu,$$

so that $\int f \, d\mu_\alpha \to \int f \, d\mu$.

Since f is bounded, there exists some $M > 0$ satisfying $-M < f(x) < M$ for all $x \in X$. Replacing f by $\frac{f+M}{2M}$, we can assume without loss of generality that $0 < f(x) < 1$ for all $x \in X$.

Fix a natural number n, and let $A_i = \{x \in X : \frac{i-1}{n} \leqslant f(x) < \frac{i}{n}\}$ for $i = 1, \ldots, n$. Clearly, $\bigcup_{i=1}^{n} A_i = X$, and the step function $\varphi = \sum_{i=1}^{n} \frac{i}{n} \chi_{A_i}$ satisfies $\|f - \varphi\|_{\infty} \leqslant \frac{1}{n}$. Next, note that if $F_i = \{x \in X : f(x) \geqslant i/n\}$ for each $i = 0, 1, \ldots, n$, then each F_i is closed and $A_i = F_{i-1} \setminus F_i$. In addition, for each $\nu \in \mathcal{P}(X)$ we have

$$\int \varphi \, d\nu = \frac{1}{n} \sum_{i=1}^{n} i \left[\nu(F_{i-1}) - \nu(F_i) \right] = \frac{1}{n} \sum_{i=0}^{n} \nu(F_i).$$

Consequently, from the inequality

$$\int f \, d\mu_\alpha = \int (f - \varphi) \, d\mu_\alpha + \int \varphi \, d\mu_\alpha \leqslant \frac{1}{n} + \frac{1}{n} \sum_{i=0}^{n} \mu_\alpha(F_i)$$

and the hypothesis, it follows that

$$\limsup_{\alpha} \int f \, d\mu_\alpha \leqslant \frac{1}{n} + \frac{1}{n} \limsup_{\alpha} \left[\sum_{i=0}^{n} \mu_\alpha(F_i) \right]$$

$$\leqslant \frac{1}{n} + \frac{1}{n} \sum_{i=0}^{n} \limsup_{\alpha} \mu_\alpha(F_i)$$

$$\leqslant \frac{1}{n} + \frac{1}{n} \sum_{i=0}^{n} \mu(F_i) = \frac{1}{n} + \int \varphi \, d\mu$$

$$= \frac{1}{n} + \int (\varphi - f) \, d\mu + \int f \, d\mu \leqslant \frac{2}{n} + \int f \, d\mu.$$

Since n is arbitrary, we infer that $\limsup_\alpha \int f \, d\mu_\alpha \leqslant \int f \, d\mu$.

Thus, statements (1) through (6) are equivalent. Next, we establish that (5) and (6) imply (7). To this end, let B be a Borel set satisfying $\mu(\partial B) = 0$. From $B^\circ \subset B \subset \overline{B} = B^\circ \cup \partial B$, we get $\mu(B^\circ) = \mu(B) = \mu(\overline{B})$. So

$$\liminf_{\alpha} \mu_\alpha(B) \geqslant \liminf_{\alpha} \mu_\alpha(B^\circ) \geqslant \mu(B^\circ) = \mu(\overline{B})$$

$$\geqslant \limsup_{\alpha} \mu_\alpha(\overline{B}) \geqslant \limsup_{\alpha} \mu_\alpha(B),$$

which implies that $\mu_\alpha(B) \to \mu(B)$.

Finally, to complete the proof, we show that (7) implies (5). To this end, let F be a closed set and let d be a compatible metric on X. For each $\varepsilon > 0$, let $F_\varepsilon = \{x \in X : d(x, F) \leqslant \varepsilon\}$. Clearly, each F_ε is closed, and satisfies $F \subset F_\varepsilon$ and $\partial F_\varepsilon \subset \{x \in X : d(x, F) = \varepsilon\}$. It follows that $\partial F_{\varepsilon_1} \cap \partial F_{\varepsilon_2} = \varnothing$ if $\varepsilon_1 \neq \varepsilon_2$, so there are at most countably many F_ε with $\mu(\partial F_\varepsilon) > 0$. In particular, there exists a sequence $\{\varepsilon_n\}$ with $\varepsilon_n \downarrow 0$ and $\mu(\partial F_{\varepsilon_n}) = 0$ for each n (and, of course, $F_{\varepsilon_n} \downarrow F$). Using our hypothesis, we see that

$$\limsup_{\alpha} \mu_\alpha(F) \leqslant \limsup_{\alpha} \mu_\alpha(F_{\varepsilon_n}) = \mu(F_{\varepsilon_n})$$

for each n, from which we get $\limsup_\alpha \mu_\alpha(F) \leqslant \lim_{n\to\infty} \mu(F_{\varepsilon_n}) = \mu(F)$. ∎

We next present some applications of Theorem 15.3. Let Z be a Borel subset of a metrizable space X. Then $\mathcal{P}(Z)$ can be considered to be a subset of $\mathcal{P}(X)$ by extending every $\mu \in \mathcal{P}(Z)$ to a Borel probability measure on X by letting $\mu(X \setminus Z) = 0$. That is,

$$\mathcal{P}(Z) = \{\mu \in \mathcal{P}(X) : \mu(X \setminus Z) = 0\}.$$

Alternatively, we can consider every $\mu \in \mathcal{P}(Z)$ to be defined on $\mathcal{B}_X$ via the formula

$$\mu(B) = \mu(B \cap Z), \quad B \in \mathcal{B}_X.$$

It turns out that the weak* topology on $\mathcal{P}(X)$ induces the weak* topology on $\mathcal{P}(Z)$.

15.4 Lemma *If Z is a Borel subset of a metrizable space X, then the weak* topology on $\mathcal{P}(Z)$ is the relativization of the weak* topology on $\mathcal{P}(X)$ to $\mathcal{P}(Z)$.*

Proof: Let us denote the weak* topologies on $\mathcal{P}(Z)$ and $\mathcal{P}(X)$ by w_Z^* and w_X^*, respectively. Let $\{\mu_\alpha\}$ be a net in $\mathcal{P}(Z)$ and let $\mu \in \mathcal{P}(Z)$.

Assume first that $\mu_\alpha \xrightarrow{w_Z^*} \mu$ and let $f \in C_b(X)$. Clearly $f|_Z \in C_b(Z)$, so

$$\int_X f \, d\mu_\alpha = \int_Z f|_Z \, d\mu_\alpha \to \int_Z f|_Z \, d\,mu = \int_X f \, d\mu.$$

This shows that $\mu_\alpha \xrightarrow{w_X^*} \mu$.

Next suppose that $\mu_\alpha \xrightarrow{w_X^*} \mu$ and let G be an open subset of Z. Pick an open subset V of X such that $G = V \cap Z$. Then, from condition (6) of Theorem 15.3, we see that

$$\liminf_{\alpha} \mu_\alpha(G) = \liminf_{\alpha} \mu_\alpha(V \cap Z) = \liminf_{\alpha} \mu_\alpha(V)$$
$$\geqslant \mu(V) = \mu(V \cap Z) = \mu(G).$$

Using condition (6) of Theorem 15.3 once more, we see that $\mu_\alpha \xrightarrow{w_Z^*} \mu$. Thus, w_X^* induces w_Z^* on $\mathcal{P}(Z)$. ∎

Another consequence of Theorem 15.3 is that bounded semicontinuous functions on X define semicontinuous functions on $\mathcal{P}(X)$.

15.5 Theorem *If a bounded real-valued function f defined on the metric space (X, d) is lower semicontinuous (respectively upper semicontinuous), then the mapping $\mu \mapsto \int f\, d\mu$, from $\mathcal{P}(X)$ to $\mathbb{R}$, is lower semicontinuous (respectively upper semicontinuous).*

Proof: We prove the result only for lower semicontinuous functions f. So let $\mu_\alpha \xrightarrow{w^*} \mu$ in $\mathcal{P}(X)$. By Theorem 3.13 there exists a sequence $\{f_n\}$ in U_d such that $f_n(x) \uparrow f(x)$ for all $x \in X$. From

$$\int f_n\, d\mu_\alpha \leqslant \int f\, d\mu_\alpha \quad \text{and} \quad \int f_n\, d\mu_\alpha \xrightarrow{\alpha} \int f_n\, d\mu,$$

we see that $\int f_n\, d\mu \leqslant \liminf_\alpha \int f\, d\mu_\alpha$ for each n. Hence,

$$\int f\, d\mu = \lim_{n\to\infty} \int f_n\, d\mu \leqslant \liminf_\alpha \int f\, d\mu_\alpha.$$

By Lemma 2.42 the function $\mu \to \int f\, d\mu$ is lower semicontinuous. ∎

A special case of this result is that indicator functions of open sets define lower semicontinuous functions on $\mathcal{P}(X)$, and indicators of closed sets define upper semicontinuous functions.

15.6 Corollary *If F is a closed subset of a metrizable space X, then for each real number c the set*

$$\{\mu \in \mathcal{P}(X) : \mu(F) \geqslant c\}$$

is a closed subset of $\mathcal{P}(X)$. Similarly, if V is an open set, then the set

$$\{\nu \in \mathcal{P}(X) : \nu(V) > c\}$$

is an open subset of $\mathcal{P}(X)$ for each real number c.

Proof: Let F be a closed subset of X. Then χ_F is upper semicontinuous. So by Theorem 15.5, the map $\mu \mapsto \int \chi_F d\mu = \mu(F)$, from $\mathcal{P}(X)$ to $\mathbb{R}$, is upper semicontinuous. Hence, $\{\mu \in \mathcal{P}(X) : \mu(F) = \int \chi_F d\mu \geqslant c\}$ is closed in $\mathcal{P}(X)$ for each $c \in \mathbb{R}$. The other case follows from the fact that $\mu \mapsto \mu(V) = \int \chi_V d\mu$ is lower semicontinuous. ∎

The following special case of Theorem 6.39 is also worth recalling.

15.7 Corollary *Let X be a metric space and give $\mathcal{P}(X)$ its w^*-topology and $C_b(X)$ the sup norm topology. Then the evaluation $(f, \mu) \mapsto \int f\, d\mu$ is continuous on $C_b(X) \times \mathcal{P}(X)$.*

15.2   Embedding X in $\mathcal{P}(X)$

Recall that if $x \in X$, then the **point mass** δ_x on X is the probability measure defined by $\delta_x(A) = 0$ if $x \notin A$ and $\delta_x(A) = 1$ if $x \in A$. Thus each $x \in X$ gives rise to a probability measure δ_x on the power set of X that is a regular Borel probability measure when restricted to the Borel σ-algebra of X. Integration with respect to δ_x is the same as evaluation at x.

15.8 Theorem   *If X is metrizable, then the mapping $x \mapsto \delta_x$ from X into $\mathcal{P}(X)$ is an embedding. Consequently, X can be topologically identified with a subspace of $\mathcal{P}(X)$. In addition, if X is separable, then X is a closed subset of $\mathcal{P}(X)$.*

Proof: For the first part note that $\delta_{x_\alpha} \xrightarrow{w^*} \delta_x$ if and only if

$$f(x_\alpha) = \int f \, d\delta_{x_\alpha} \to \int f \, d\delta_x = f(x)$$

for each $f \in C_b(X)$, which, by Corollary 2.57, is equivalent to $x_\alpha \to x$ in X.

Now assume that X is a separable metrizable space and let $\delta_{x_\alpha} \xrightarrow{w^*} \mu$ in $\mathcal{P}(X)$. By Theorem 12.14, we know that $\operatorname{supp}\mu$ exists and is nonempty. If $x \in \operatorname{supp}\mu$, then we claim that $x_\alpha \to x$. To see this, let V be an open neighborhood of x. Pick a function $0 \leqslant f \in C_b(X)$ with $f(x) = 1$ and $f(y) = 0$ for all $y \in V^c$ (Lemma 3.20). From $x \in \operatorname{supp}\mu$, it follows that $\int f \, d\mu > 0$, and from $\delta_{x_\alpha} \to \mu$, we infer that $f(x_\alpha) = \int f \, d\delta_{x_\alpha} > 0$ for all $\alpha \geqslant \alpha_0$. The latter shows that $x_\alpha \in V$ for all $\alpha \geqslant \alpha_0$, so $x_\alpha \to x$. To finish the proof note that $\mu = \delta_x$, so X is a closed subset of $\mathcal{P}(X)$.   ∎

In the separable case, the point masses δ_x are the extreme points of the convex subset $\mathcal{P}(X)$ of $ca(X)$.

15.9 Theorem   (Point masses are extreme)   *If X is a separable metrizable topological space, then the set of extreme points of $\mathcal{P}(X)$ is identified with X under the embedding $x \mapsto \delta_x$.*

Proof: Clearly, for any $x \in X$, δ_x is an extreme point of $\mathcal{P}(X)$. Conversely, let μ be an extreme point of $\mathcal{P}(X)$, and suppose that $\operatorname{supp}\mu$ contains two distinct points, x and y. (The support exists by Theorem 12.14 and is clearly nonempty.) Then there are disjoint open neighborhoods V of x and W of y, each having positive μ measure. For any set A satisfying $\mu(A) > 0$, let $\mu(\cdot|A)$ denote the conditional probability measure given A, that is, $\mu(B|A) = \mu(A \cap B)/\mu(A)$.

Since V and W have positive μ measure and $W \subset V^c$, it follows that V^c has positive μ-measure. It is easy to see that $\mu(\cdot|V)$ and $\mu(\cdot|V^c)$ are distinct probability measures in $\mathcal{P}(X)$. But μ is a proper convex combination of $\mu(\cdot|V)$ and $\mu(\cdot|V^c)$, namely $\mu = \mu(V)\mu(\cdot|V) + \mu(V^c)\mu(\cdot|V^c)$, which contradicts the hypothesis that μ is an extreme point of $\mathcal{P}(X)$. Thus the support of μ is a singleton.   ∎

We now consider the convex hull of X as a subset of $\mathcal{P}(X)$. The support of any convex combination of point masses is finite. Conversely, any probability on X that has finite support can be written as a convex combination of point masses corresponding to points in the support. Thus the set $\operatorname{co} X$ regarded as a subset of $\mathcal{P}(X)$ is the set of probabilities with finite support. It turns out this set is dense.

15.10 Density Theorem *If X is metrizable, then the set of probability measures on X with finite support is dense in $\mathcal{P}(X)$. This set may be identified with $\operatorname{co} X$, where X is embedded in $\mathcal{P}(X)$ via $x \mapsto \delta_x$.*

Proof: We need to show that $\mathcal{P}(X) \subset \overline{\operatorname{co}} X$. So suppose that $\mu \notin \overline{\operatorname{co}} X$. Then by Separating Hyperplane Theorem 5.80 there is some $f \in C_b(X)$ that strongly separates μ from $\overline{\operatorname{co}} X$. Since $X \subset \overline{\operatorname{co}} X$, this implies that there exists some $\varepsilon > 0$ with $\int f \, d\mu > \varepsilon + \int f \, d\delta_x = \varepsilon + f(x)$ for every $x \in X$. Since μ is a probability measure, this implies $\int f \, d\mu > \varepsilon + \int f \, d\mu$, which is impossible. Therefore $\mu \in \overline{\operatorname{co}} X$ for every $\mu \in \mathcal{P}(X)$. ∎

When X is identified with a subspace of $\mathcal{P}(X)$, its convex hull need not be a closed set even when X is a compact metric space. Otherwise, by the above theorem, all probability measures would have finite support.

15.3 Properties of $\mathcal{P}(X)$

In this section we discuss additional properties of the weak* topology on $\mathcal{P}(X)$ when X is metrizable. We start with the following result.

15.11 Theorem *A metrizable space X is compact if and only if $\mathcal{P}(X)$ is compact and metrizable.*

Proof: Assume that X is compact and metrizable. Then, by Theorem 9.14, $C(X)$ is a separable Banach lattice. Consequently, by Theorems 6.30 and 6.21, the closed unit ball U' of its norm dual is weak* compact and weak* metrizable. Since $\mathcal{P}(X)$ is a weak*-closed subset of U', we see that $\mathcal{P}(X)$ is compact and metrizable.

Now assume that $\mathcal{P}(X)$ is compact and metrizable. By Theorem 15.8, X is a topological subset of $\mathcal{P}(X)$. As such, X is separable, and hence (by Theorem 15.8 again) closed. Therefore, X is itself compact. ∎

15.12 Theorem *A metrizable space X is separable if and only if $\mathcal{P}(X)$ is separable and metrizable.*

Proof: By Corollary 3.41, there is a compatible metric d so that (X, d) is totally bounded. In particular, the completion $(\hat{X}, \hat{d})$ of (X, d) is a compact metric space.

By Lemma 3.11, there is an isometry $\varphi \colon U_d \to C(\hat{X})$ defined by $\varphi(f) = \hat{f}$, the unique d-uniformly continuous extension of f to $\hat{X}$. By Theorem 9.14 the space $C(\hat{X})$ is separable, so there is a countable dense subset D in U_d. By Theorem 15.2, we see that $\sigma(\mathcal{P}(X), C_b(X))$ and $\sigma(\mathcal{P}(X), D)$ agree on $\mathcal{P}(X)$. Since $\mathbb{R}^D$ is separable and metrizable (Theorem 3.38), we infer that $\mathcal{P}(X)$ is separable and metrizable.

Now assume that $\mathcal{P}(X)$ is a separable metrizable space. By Theorem 15.8, the subset X of $\mathcal{P}(X)$ is likewise separable. ∎

This allows us to prove the following result.

15.13 Theorem   *Let X be a separable metrizable space, and let f be a bounded Borel measurable real function on X. Then the mapping $\mu \mapsto f(\mu) = \int f \, d\mu$, from $\mathcal{P}(X)$ to $\mathbb{R}$, is Borel measurable.*

Proof: This is one of those theorems where it is easier to characterize the set of f with a given property than it is to show that any particular f has the property. So let $\mathcal{F} = \{f \in B_b(X) : \mu \mapsto f(\mu)$ is Borel measurable$\}$, where, as you may recall, $B_b(X)$ is the set of bounded Borel measurable functions on X. Clearly $\mathcal{F}$ is a vector subspace of $B_b(X)$ that includes $C_b(X)$. Moreover by Levi's Theorem, if $f_n \uparrow f$, then $f_n(\mu) \to f(\mu)$ for any $\mu \in \mathcal{P}(X)$. Since X is separable, so is $\mathcal{P}(X)$ by Theorem 15.12, so by Corollary 4.29, f is Borel measurable on $\mathcal{P}(X)$. Similarly if $f_n \downarrow f$. Thus $\mathcal{F}$ is closed under monotone sequential limits, so Theorem 4.33 implies that $\mathcal{F} = B_b(X)$. ∎

15.14 Theorem   *Let $f \colon X \to Y$ be a continuous function between two metrizable spaces. Define $\hat{f} \colon \mathcal{P}(X) \to \mathcal{P}(Y)$ via $\hat{f}(\mu) = \mu \circ f^{-1}$. Then:*

1. *$\hat{f}$ is continuous.*

2. *If X is Polish and f is injective, then $\hat{f}$ is injective.*

3. *If X and Y are Polish spaces and f is injective, then the range of $\hat{f}$ is $\mathcal{P}(f(X))$.*

4. *If X and Y are Polish spaces and f is an embedding, then so is $\hat{f}$, and moreover $\hat{f} \colon \mathcal{P}(X) \to \mathcal{P}(f(X))$ is a homeomorphism.*

5. *If f is surjective and Y is Polish, then $\hat{f}$ is surjective.*

Proof: First note that for each $\mu \in \mathcal{P}(X)$ the formula

$$\hat{f}(\mu)(A) = \mu \circ f^{-1}(A) = \mu(f^{-1}(A))$$

indeed defines a Borel probability measure on Y, that is, $\hat{f}(\mu) \in \mathcal{P}(Y)$.

(1)   To see that $\hat{f}$ is continuous, observe that if h is a bounded continuous real function on Y, then $h \circ f$ is a bounded continuous real function on X. Furthermore, by the Change of Variables Theorem 13.46, for any $\mu \in \mathcal{P}(X)$ we have

$$\int_X (h \circ f)(x)\, d\mu(x) = \int_Y h(y)\, d(\mu \circ f^{-1})(y) = \int_Y h(y)\, d(\hat{f}(\mu))(y).$$

So if $\mu_\alpha \xrightarrow{w^*} \mu$ in $\mathcal{P}(X)$ and $h \in C_b(Y)$, then

$$\int_Y h\, d\hat{f}(\mu_\alpha) = \int_X h \circ f\, d\mu_\alpha \longrightarrow \int_X h \circ f\, d\mu = \int_Y h\, d\hat{f}(\mu).$$

This shows that $\hat{f}(\mu_\alpha) \xrightarrow{w^*} \hat{f}(\mu)$ in $\mathcal{P}(Y)$ proving that $\hat{f}$ is continuous.

(2)   To see that $\hat{f}$ is injective, let $\mu, \nu \in \mathcal{P}(X)$ with $\mu \neq \nu$. Since μ and ν are regular by Theorem 12.7, there exists some compact subset K of X such that $\mu(K) \neq \nu(K)$. Now $f(K)$ is a compact subset of Y and hence a Borel subset. Since f is injective, $K = f^{-1}(f(K))$. Thus $\hat{f}(\mu)(f(K)) = \mu(K) \neq \nu(K) = \hat{f}(\nu)(f(K))$, so $\hat{f}(\mu) \neq \hat{f}(\nu)$, which shows that $\hat{f}$ is injective.

(3)   Since f is continuous, it is Borel measurable, so for any Borel subset B of Y, $f^{-1}(B)$ is a Borel subset of X. Since in addition f is one-to-one, Theorem 12.29 implies that for any Borel subset A of X, $f(A)$ is a Borel subset of Y. In other words, a subset A of X is a Borel set if and only if $f(A)$ is a Borel subset of Y (or equivalently, if $f(A)$ is a Borel subset of the Borel set $f(X)$).

Now notice that if $\mu \in \mathcal{P}(X)$, then $\hat{f}(\mu) = \mu f^{-1} \in \mathcal{P}(f(X))$. On the other hand, if $\nu \in \mathcal{P}(f(X))$, then it is easy to see that the formula $\mu(A) = \nu(f(A))$ defines a Borel measure on X, and moreover $\hat{f}(\mu) = \nu$. In addition, by Lemma 15.4, the weak* topology on $\mathcal{P}(Y)$ induces on $\mathcal{P}(f(X))$ its weak* topology. Therefore, $\hat{f} \colon \mathcal{P}(X) \to \mathcal{P}(f(X))$ is surjective, one-to-one, and continuous.

(4)   To see that $\hat{f} \colon \mathcal{P}(X) \to \mathcal{P}(f(X))$ is a homeomorphism, note that from part (3) we know that the mapping $\hat{f}^{-1} \colon \mathcal{P}(f(X)) \to \mathcal{P}(X)$ is also surjective, one-to-one, and continuous. Moreover, if $\nu \in \mathcal{P}(f(X))$, then for each Borel subset A of X we have

$$\widehat{f^{-1}}(\nu)(A) = \nu(f^{-1})^{-1}(A) = \nu(f(A)) = (\hat{f})^{-1}(\nu)(A).$$

This shows that $(\hat{f})^{-1} = \widehat{f^{-1}}$ proving that $(\hat{f})^{-1}$ is also continuous. Thus $\hat{f}$ is a homeomorphism.   ∎

(5)   The case where f is surjective (but not injective) and Y is Polish is quite subtle. We prove it as Corollary 18.24.

15.15 Theorem   *A metrizable space X is a Polish space if and only if $\mathcal{P}(X)$ is a Polish space.*

Proof: Let X be a Polish space. By the Urysohn Metrization Theorem 3.40, there exists an embedding $\varphi: X \to \mathcal{H}$, where $\mathcal{H} = [0,1]^{\aleph}$ is the Hilbert cube. By Theorem 15.14, φ gives rise to an embedding $\hat{\varphi}: \mathcal{P}(X) \to \mathcal{P}(\mathcal{H})$, defined by

$$\hat{\varphi}(\mu)(B) = \mu(\varphi^{-1}(B)),$$

for $B \in \mathcal{B}_{\mathcal{H}}$, whose range is $\mathcal{P}(\varphi(X))$. By Lemma 15.4, we know that the weak* topology of $\mathcal{P}(\mathcal{H})$ induces on $\mathcal{P}(\varphi(X))$ the weak* topology of $\mathcal{P}(\varphi(X))$. Now invoke Lemma 3.33 to see that $\varphi(X)$ is a $\mathcal{G}_\delta$ in $\mathcal{H}$. Pick a sequence $\{G_n\}$ of open subsets of $\mathcal{H}$ such that $\varphi(X) = \bigcap_{n=1}^{\infty} G_n$. Next, a simple argument shows that

$$\mathcal{P}(\varphi(X)) = \{\mu \in \mathcal{P}(\mathcal{H}) : \mu(\mathcal{H} \setminus \varphi(X)) = 0\} = \bigcap_{n=1}^{\infty} \{\mu \in \mathcal{P}(\mathcal{H}) : \mu(\mathcal{H} \setminus G_n) = 0\},$$

and consequently $\mathcal{P}(\varphi(X)) = \bigcap_{n=1}^{\infty} \bigcap_{k=1}^{\infty} \{\mu \in \mathcal{P}(\mathcal{H}) : \mu(\mathcal{H} \setminus G_n) < \frac{1}{k}\}$.

Since for each n the set $\mathcal{H} \setminus G_n$ is a closed subset of $\mathcal{H}$, it follows (from Corollary 15.6) that each set of the form $\{\mu \in \mathcal{P}(\mathcal{H}) : \mu(\mathcal{H} \setminus G_n) < \frac{1}{k}\}$ is open in $\mathcal{P}(\mathcal{H})$, so $\mathcal{P}(\varphi(X)) = \hat{\varphi}(\mathcal{P}(X))$ is a $\mathcal{G}_\delta$-subset of $\mathcal{P}(\mathcal{H})$. Since $\mathcal{P}(\mathcal{H})$ is compact (Theorem 15.11), it follows from Alexandroff's Lemma 3.34 that $\hat{\varphi}(\mathcal{P}(X))$ (and hence $\mathcal{P}(X)$) is a Polish space.

Now assume that $\mathcal{P}(X)$ is a Polish space. Then X (as a subset of $\mathcal{P}(X)$) is closed. Therefore X is a Polish space in its own right. ∎

Now let X be a topological space. Then each Borel set $E \in \mathcal{B}_X$ defines a function $\theta_E: \mathcal{P}(X) \to [0,1]$ via the formula $\theta_E(\mu) = \mu(E)$. If E is closed (resp. open), then we know from Theorem 15.5 that θ_E is upper (resp. lower) semicontinuous. In particular, when E is either open or closed, the function θ_E is Borel measurable. The next result tells us that θ_E is, in fact, a Borel measurable function for each Borel subset E of X, even without the measurability hypothesis of Theorem 15.13.

15.16 Lemma *If X is a metrizable space, then for each $E \in \mathcal{B}_X$, the function $\theta_E: \mathcal{P}(X) \to [0,1]$, defined by $\theta_E(\mu) = \mu(E)$, is Borel measurable.*

Proof: Let X be a metrizable space (with topology τ) and let

$$\mathcal{A} = \{E \in \mathcal{B}_X : \theta_E \text{ is Borel measurable}\}.$$

Corollary 15.6 implies that $\mathcal{A}$ includes τ, and it is easy to see that $\mathcal{A}$ is a Dynkin-system. Since τ is closed under finite intersections, Dynkin's Lemma 4.11 implies that $\mathcal{B}_X = \sigma(\tau) = \mathcal{A}$. ∎

15.17 Definition *A **Borel space** is any metrizable space that is homeomorphic to a Borel subset of a Polish space.*

Clearly, every Borel subset of a Borel space is itself a Borel space. Every Borel space is a separable metrizable space, but not every separable metrizable space is a Borel space. This is despite the fact that the completion of a separable metric space is a Polish space. It is just that a metric space may not be a Borel subset of its completion. For instance, if we take a nonanalytic subset A of the Baire space $\mathcal{N}$ (see Example 12.33), then A (as a subset of the Polish space $\mathcal{N}$) is separable metrizable space. However, A is not a Borel subset of its completion.

15.18 Theorem   *A metrizable space X is a Borel space if and only if $\mathcal{P}(X)$ is a Borel space.*

Proof: Let X be a Borel space. We can assume without loss of generality that X is a Borel subset of a Polish space Y. Now notice that

$$\mathcal{P}(X) = \{\mu \in \mathcal{P}(Y) : \mu(X) = 1\} = \theta_X^{-1}(\{1\}),$$

where $\theta_X \colon \mathcal{P}(Y) \to [0, 1]$ is defined by $\theta_X(\mu) = \mu(X)$. By Lemma 15.16, the set $\theta_X^{-1}(\{1\}) = \mathcal{P}(X)$ is a Borel subset of $\mathcal{P}(Y)$ as X is Borel measurable. Since (by Theorem 15.15) $\mathcal{P}(Y)$ is a Polish space, we infer that $\mathcal{P}(X)$ is a Borel space.

Now assume that $\mathcal{P}(X)$ is a Borel space. Then X (as a subset of $\mathcal{P}(X)$) is a separable metrizable space. Thus X is a closed subset of $\mathcal{P}(X)$ and hence a Borel space in its own right.   ∎

15.4   The many faces of $\mathcal{P}(X)$

Recall that a **face** of a convex set C is a convex extreme subset of C. We now characterize the closed faces of $\mathcal{P}(X)$.

Let X be a separable metrizable space and consider a closed subset F of X. Then F with the induced topology from X is also a separable metrizable space, and clearly

$$\mathcal{P}(F) = \{\mu \in \mathcal{P}(X) : \operatorname{supp}\mu \subset F\}.$$

When X is also compact, the sets $\mathcal{P}(F)$ are precisely the closed faces of $\mathcal{P}(X)$.

15.19 Theorem   *For a separable metric space X we have the following.*

1. *If F is a closed subset of X, then $\mathcal{P}(F)$ is a closed face of $\mathcal{P}(X)$.*

2. *If X is also compact, then every closed face of $\mathcal{P}(X)$ is of the form $\mathcal{P}(F)$ for a unique closed subset F of X.*

Proof: (1)   We first note that the set of probability measures on X with support included in F is closed in $\mathcal{P}(X)$. Indeed, if $\mu_\alpha \xrightarrow{w^*} \mu$ in $\mathcal{P}(X)$ with $\{\mu_\alpha\} \subset \mathcal{P}(F)$, put $G = F^c$ and then use Theorem 15.3 (6) to get $0 = \liminf_\alpha \mu_\alpha(G) \geq \mu(G)$, or $\mu(G) = 0$. That is, $\operatorname{supp}\mu \subset F$. The set $\mathcal{P}(F)$ is also clearly convex.

Suppose now that supp μ lies in F and that μ is a strict convex combination of v and γ. That is, $\mu = \lambda v + (1 - \lambda)\gamma$, where $0 < \lambda < 1$ and $v, \gamma \in \mathcal{P}(X)$. Then $\mu(A) = 0$ if and only if $v(A) = \gamma(A) = 0$ and $\mu(A) = 1$ if and only if $v(A) = \gamma(A) = 1$. Thus supp $v \subset$ supp $\mu \subset F$, and likewise supp $\gamma \subset$ supp $\mu \subset F$. This shows that the set of probability measures with support in F is an extreme set of $\mathcal{P}(X)$.

(2)   Now suppose X is also compact, and let Φ be a nonempty closed face of $\mathcal{P}(X)$. Then Φ is compact, since $\mathcal{P}(X)$ is compact, so by Corollary 7.66, Φ has extreme points. These extreme points must also be extreme points of $\mathcal{P}(X)$. Therefore they are point masses by Theorem 15.9. So consider the nonempty closed set $F = \{x : \delta_x \in \Phi\}$. Then by the Krein–Milman Theorem 7.68, Φ is the closed convex hull of $\{\delta_x : x \in F\}$. On the other hand, F is also compact, so by Theorem 15.11, $\mathcal{P}(F)$ is compact, and so is the closed convex hull of its extreme points $\{\delta_x : x \in F\}$. Consequently, we have $\Phi = \overline{\mathrm{co}}\{\delta_x : x \in F\} = \mathcal{P}(F)$. ∎

15.5   Compactness in $\mathcal{P}(X)$

The purpose of this section is to characterize the relatively compact subsets of $\mathcal{P}(X)$ when X is a separable metrizable space. We start with a definition.

15.20 Definition   *A family $\mathcal{F}$ of finite Borel measures on X is **tight** if for each $\varepsilon > 0$ there exists a compact set K satisfying $\mu(K) > \mu(X) - \varepsilon$ for each $\mu \in \mathcal{F}$.*

Tight families in $\mathcal{P}(X)$ are relatively compact.

15.21 Lemma   *If X is a separable metrizable space, then every tight family of measures of $\mathcal{P}(X)$ is a relatively compact set.*

Proof: Let X be a separable metrizable space. By the Urysohn Metrization Theorem 3.40, we can identify X with a "topological" subset of the Hilbert cube $\mathcal{H}$.

Now let $\mathcal{F}$ be a tight family in $\mathcal{P}(X)$ and let $\{\mu_n\}$ be a sequence in $\mathcal{F}$. Since $\mathcal{P}(X)$ is metrizable (Theorem 15.12), we must show that $\{\mu_n\}$ has a convergent subsequence in $\mathcal{P}(X)$. To this end, for each m pick some compact set K_m of X (and hence of $\mathcal{H}$) with $\mu_n(K_m) > 1 - \frac{1}{m}$ for each n. Put $E = \bigcup_{m=1}^{\infty} K_m \subset X$ and note that E is a Borel subset of X and $\mathcal{H}$; and, in fact, it is an $\mathcal{F}_\sigma$-set in both spaces. From $1 \geqslant \mu_n(E) \geqslant \mu_n(K_m) > 1 - \frac{1}{m}$ for each m, we see that $\mu_n(E) = 1$ for each n. So if $\hat{\mu}_n : \mathcal{B}_{\mathcal{H}} \to [0, 1]$ is defined by $\hat{\mu}_n(B) = \mu_n(B \cap E)$ for each $B \in \mathcal{B}_{\mathcal{H}}$, then $\hat{\mu}_n \in \mathcal{P}(\mathcal{H})$. Since $\mathcal{P}(\mathcal{H})$ is a compact metrizable space, we can assume (by passing to a subsequence if necessary) that $\hat{\mu}_n \xrightarrow{w^*} \mu$ in $\mathcal{P}(\mathcal{H})$. From

$$\mu(E) \geqslant \mu(K_m) \geqslant \limsup_{n \to \infty} \hat{\mu}_n(K_m) = \limsup_{n \to \infty} \mu_n(K_m) \geqslant 1 - \frac{1}{m}$$

for each m, we get $\mu(E) = 1$. Thus $\{\mu_n\} \subset \mathcal{P}(E)$ and $\mu \in \mathcal{P}(E)$.

Since E is a Borel subset of $\mathcal{P}(\mathcal{H})$, it follows from Lemma 15.4 that $\mu_n \xrightarrow{w^*} \mu$ in $\mathcal{P}(E)$. Since E is also a Borel subset of X, it follows (from Lemma 15.4 again) that $\mu_n \xrightarrow{w^*} \mu$ in $\mathcal{P}(X)$. Thus $\mathcal{F}$ is a relatively compact subset of $\mathcal{P}(X)$. ∎

We now come to the characterization of the relatively compact subsets of probability measures on Polish spaces.

15.22 Theorem   *If X is a Polish space, then a nonempty subset of $\mathcal{P}(X)$ is relatively compact if and only if it is tight.*

Proof: The "if" part is Lemma 15.21. Next, we prove the "only if" part. To this end, let $\mathcal{F}$ be a relatively compact subset of $\mathcal{P}(X)$ and let $0 < \varepsilon < 1$.

Fix a compatible metric d under which X is a Polish space, and let $\{x_1, x_2, \ldots\}$ be countable dense subset of X. For each i and n define the sets

$$C_i^n = \{x \in X : d(x, x_i) \leqslant 1/n\} \quad \text{and} \quad B_i^n = \{x \in X : d(x, x_i) < 1/n\}.$$

Clearly, each C_i^n is closed each B_i^n is open and $B_i^n \subset C_i^n$. We claim that for each fixed n and each $0 < \delta < 1$ there exists some m such that $\mu(\bigcup_{i=1}^m B_i^n) > \delta$ for all $\mu \in \mathcal{F}$. If this is not the case, then for each m there exists some $\mu_m \in \mathcal{F}$ satisfying $\mu_m(\bigcup_{i=1}^m B_i^n) \leqslant \delta$. Since $\mathcal{F}$ is relatively compact, by passing to a subsequence and relabeling, we can assume that $\mu_m \xrightarrow{w^*} \mu$ in $\mathcal{P}(X)$. Note that for $m \geqslant r$,

$$\mu_m\Big(\bigcup_{i=1}^r B_i^n\Big) \leqslant \mu_m\Big(\bigcup_{i=1}^m B_i^n\Big) \leqslant \delta,$$

so from Theorem 15.3, we see that

$$\mu\Big(\bigcup_{i=1}^r B_i^n\Big) \leqslant \liminf_{m\to\infty} \mu_m\Big(\bigcup_{i=1}^r B_i^n\Big) \leqslant \delta < 1 = \mu(X)$$

for each r. However, this contradicts $\bigcup_{i=1}^r B_i^n \uparrow_r \bigcup_{i=1}^\infty B_i^n = X$.

Next, for each n pick some integer k_n satisfying

$$\nu\Big(X \setminus \bigcup_{i=1}^{k_n} C_i^n\Big) \leqslant \nu\Big(X \setminus \bigcup_{i=1}^{k_n} B_i^n\Big) < \varepsilon/2^n$$

for each $\nu \in \mathcal{F}$ and let $C = \bigcap_{n=1}^\infty \bigcup_{i=1}^{k_n} C_i^n$. Now repeat the arguments of the last part of the proof of Theorem 12.7 to conclude that the compact set C satisfies $\nu(C) > 1 - \varepsilon$ for each $\nu \in \mathcal{F}$. ∎

15.6   The Kolmogorov Extension Theorem

Let $\{(X_t, \Sigma_t) : t \in T\}$ be a family of measurable spaces. The index set T is allowed to be infinite, and is generally interpreted as a set of time periods or dates. For any nonempty subset H of T define

$$X_H = \prod_{t \in H} X_t \quad \text{and} \quad X_{-H} = \prod_{t \notin H} X_t.$$

We may thus write $X_T = X_H \times X_{-H}$. To ease notation write X_{-t} for $X_{-\{t\}}$. When $H \subset G \subset T$, let P_{GH} denote the natural projection of X_G on X_H.

We now define the infinite product σ-algebra Σ_T on X_T. For each finite subset F of T, let Σ_F denote the product σ-algebra on X_F. That is, $\Sigma_F = \otimes_{t \in F} \Sigma_t$. Call a set an **$F$-cylinder** if it is of the form $A \times X_{-F}$, where F is a finite subset of T and A belongs to Σ_F. It is easy to verify that the collection of all F-cylinders, where F is a finite subset of T, is an algebra of subsets of X_T. We call the σ-algebra that it generates the **infinite product σ-algebra** $\Sigma_T = \bigotimes_{t \in T} \Sigma_t$ on X_T. It is the smallest σ-algebra on X_T for which each projection on X_t is measurable.

Now for each finite subset F of T, let μ_F be a probability measure (called a finite dimensional distribution) on (X_F, Σ_F). The family $\{\mu_F\}$ is **Kolmogorov consistent** if for all finite subsets F and G of T satisfying $F \subset G \subset T$ the projection $P_{GF} \colon X_G \to X_F$ satisfies $\mu_G \circ P_{GF}^{-1} = \mu_F$, that is, $\mu(P_{GF}^{-1}(B)) = \mu_F(B)$ for each $B \in S_F$. (Note that each such projection P_{GF} is (Σ_G, Σ_F) measurable.)

We say that a Kolmogorov consistent family $\{\mu_F\}$ has a **Kolmogorov extension**, if there is a probability μ on the infinite product X_T with its product σ-algebra that extends each finite dimensional distribution μ_F in the sense that

$$\mu \circ P_{TF}^{-1} = \mu_F.$$

A. N. Kolmogorov [213] proves the existence and uniqueness of such an extension for the case where each $X_t = \mathbb{R}$.

In order to prove the existence of a Kolmogorov extension, we follow in the footsteps of S. Bochner [54], and introduce a more abstract problem. Consider a set X and an increasing net $\{\Sigma_\alpha\}_{\alpha \in I}$ of σ-algebras on X. That is, $\alpha \geqslant \beta$ implies $\Sigma_\alpha \supset \Sigma_\beta$.

For each $\alpha \in I$ let μ_α be a probability measure on Σ_α. We say the net $\{(\Sigma_\alpha, \mu_\alpha)\}_{\alpha \in I}$ is **Kolmogorov consistent** if

$$\Sigma_\alpha \supset \Sigma_\beta \quad \text{implies} \quad \mu_\alpha|_{\Sigma_\beta} = \mu_\beta,$$

where $\mu_\alpha|_{\Sigma_\beta}$ is the restriction of μ_α to the σ-subalgebra Σ_β of Σ_α.

Let $\mathcal{A} = \bigcup_{\alpha \in I} \Sigma_\alpha$. A **Kolmogorov extension** of the net $\{\mu_\alpha\}$ is a probability measure μ on $\sigma(\mathcal{A})$ satisfying $\mu_\alpha = \mu|_{\Sigma_\alpha}$ for every $\alpha \in I$.

Recall that a family $\mathcal{C}$ of subsets of X is a **compact class** if every sequence $\{C_n\}$ in $\mathcal{C}$ with the finite intersection property has a nonempty intersection.

15.23 Theorem (Bochner) *Let $\{(\Sigma_\alpha, \mu_\alpha)\}_{\alpha \in I}$ be a Kolmogorov consistent net. Suppose that there is a compact class $\mathcal{C}$ of subsets of X having the property that for each $\alpha \in I$, and each $E \in \Sigma_\alpha$,*

$$\mu_\alpha(E) = \sup\{\mu_\alpha(C) : C \subset E \text{ and } C \in \mathcal{C} \cap \Sigma_\alpha\}.$$

Then there is a unique Kolmogorov extension to the σ-algebra $\sigma(\bigcup_{\alpha \in I} \Sigma_\alpha)$.

Proof: Define μ on $\mathcal{A} = \bigcup_{\alpha \in I} \Sigma_\alpha$ by $\mu(E) = \mu_\alpha(E)$ for $E \in \Sigma_\alpha$. Kolmogorov consistency guarantees that this is well defined. Now note that μ is nonnegative and $\mu(X) = \mu_\alpha(X) = 1$ for all α. In addition, μ is finitely additive, for if a finite collection of sets belongs to $\mathcal{A}$, then since $\{\Sigma_\alpha\}$ is an increasing net of σ-algebras, there is some α for which every member of the collection belongs to Σ_α. Consequently their union belongs to Σ_α and hence to $\mathcal{A}$. This also proves that $\mathcal{A}$ is an algebra. The finite additivity of μ is then guaranteed by that of each μ_α. Now Theorem 10.13 implies that μ is countably additive on $\mathcal{A}$. We may thus use the Carathéodory Extension Theorem 10.23 to extend it uniquely to $\sigma(\mathcal{A})$. ∎

To apply this theorem to our original problem, identify each Σ_F with the collection $\hat{\Sigma}_F$ of F-cylinder sets in X_T. That is,

$$\hat{\Sigma}_F = \{A \times X_{-F} : A \in \Sigma_F\}.$$

Observe that by definition, the product σ-algebra $\bigotimes_{t \in T} \Sigma_t$ is the σ-algebra generated by $\{\hat{\Sigma}_F : F$ is a finite subset of $T\}$. Define $\hat{\mu}_F$ by $\hat{\mu}_F(A \times X_{-F}) = \mu_F(A)$. Regard the family of finite subsets of T as a net directed upward by inclusion, and (after some minor arguments) we are back in the abstract framework, and the two notions of Kolmogorov consistency coincide.

The real work of proving a standard version of the extension theorem is verifying the existence of a compact class with the desired property. Here are two lemmas that are useful in this regard.

15.24 Lemma *If $\mathcal{C}$ is a compact class, then the smallest family including $\mathcal{C}$ and closed under finite unions and countable intersections is also a compact class.*

Proof: Exercise, or see J. Neveu [261, Lemma I.6.1, p. 26]. ∎

15.25 Lemma *Let μ be a finitely additive probability set function on a semiring $\mathcal{S}$ of subsets of X that contains X. Let $\mathcal{C}$ be a compact subclass of $\mathcal{S}$, and suppose*

$$\mu(E) = \sup\{\mu(C) : C \subset E \text{ and } C \in \mathcal{C}\}$$

for every $E \in \mathcal{S}$. Then μ is countably additive on $\mathcal{S}$.

Proof: By Lemma 15.24 the class $\mathcal{C}_u$ of finite unions of members of $\mathcal{C}$ is a compact class. Further, it is included in the algebra $\mathcal{A}$ generated by $\mathcal{S}$. Extend μ to the set function μ' on $\mathcal{A}$ by $\mu'(\bigcup_{n=1}^m E_n) = \sum_{n=1}^m \mu(E_n)$ whenever $E_1, \ldots, E_m$ are pairwise disjoint sets in $\mathcal{S}$. (The algebra generated by $\mathcal{S}$ consists precisely of such sets.) We leave it as an exercise to show that this is well defined and additive on $\mathcal{A}$, and that μ' and $\mathcal{C}_u$ satisfy the hypotheses of Theorem 10.13. ∎

We can now prove the following version of the Kolmogorov Extension Theorem, taken from J. Neveu [261, p. 82].

15.26 Generalized Kolmogorov Extension Theorem *Let $\{(X_t, \Sigma_t) : t \in T\}$ be a family of measurable spaces, and for each finite subset F of T let μ_F be a probability measure on $X_F = \prod_{t \in F} X_t$ with its product σ-algebra Σ_F. Assume the family $\{\mu_F\}$ is Kolmogorov consistent. Also, suppose that for each t there is a compact class $\mathcal{C}_t \subset \Sigma_t$ satisfying*

$$\mu_t(A) = \sup\{\mu_t(C) : C \subset A \text{ and } C \in \mathcal{C}_t\}$$

for each $A \in \Sigma_t$. Then there is a unique probability measure on the infinite product σ-algebra $\Sigma_T = \bigotimes_{t \in T} \Sigma_t$ that extends each μ_F.

Proof: Clearly the collection of cylinders $\hat{E}_t = E_t \times X_{-t}$, as t ranges over T and E_t ranges over Σ_t is enough to generate the product σ-algebra. Define $\mathcal{C}^0$ to be the collection of cylinders of the form $\hat{C}_t = C_t \times X_{-t}$ in Σ_T, where C_t belongs to $\mathcal{C}_t$. We claim that $\mathcal{C}^0$ is a compact class. For every intersection $\bigcap_{n=1}^\infty \hat{C}_{t_n} = \bigcap_{n=1}^\infty C_{t_n} \times X_{-t_n}$ is actually a product set $\prod_{t \in T} E_t$. (But only countably many E_t differ from X_t.) The only way a product set can be empty is if some factor, say E_s, is empty. But $\mathcal{C}_s$ is a compact class, so there is some N for which $\bigcap_{n=1}^N \hat{C}_{t_n} = \varnothing$. This shows that $\mathcal{C}^0$ is a compact class. Consequently the collection $\mathcal{C}$ closed under finite unions and countable intersections generated by $\mathcal{C}^0$ is also a compact class.

Now let $E = \prod_{t \in F} E_t \times X_{-F}$ be a cylinder, where F is a finite subset of T having n elements. Let $\varepsilon > 0$, and for each $t \in F$ choose $C_t \in \mathcal{C}_t$ so that $C_t \subset E_t$, and $\mu_t(E_t) < \mu_t(C_t) + \frac{\varepsilon}{n}$. Then $C = \prod_{t \in F} C_t \times X_{-F} = \bigcap_{t \in F} C_t \times X_{-t}$ belongs to $\mathcal{C}$, and $E \setminus C \subset \bigcup_{t \in F} \{(E_t \setminus C_t) \times X_{-t}\}$. Since μ is finitely additive and nonnegative, it is finitely subadditive, so

$$\mu(E) - \mu(C) \leqslant \sum_{t \in F} [\mu_t(E_t) - \mu(C_t)] < \varepsilon.$$

This shows that $\mu(E) = \sup\{\mu(C) : C \subset E \text{ and } C \in \mathcal{C}\}$, and the conclusion now follows from Lemma 15.25. ∎

This theorem gives rise to a number of corollaries based on the regularity of Borel measures on topological spaces.

15.27 Corollary *Let $\{X_t : t \in T\}$ be a family of Polish spaces equipped with their Borel σ-algebras. For each finite subset F of T let μ_F be a Borel probability measure on $X_F = \prod_{t \in F} X_t$ with its product (Borel) σ-algebra Σ_F. Assume the distributions $\{\mu_F\}$ are Kolmogorov consistent. Then there is a unique probability measure on the infinite product σ-algebra $\Sigma_T = \bigotimes_{t \in T} \Sigma_t$ that extends each μ_F.*

Proof: By Theorem 12.7 the hypotheses of Theorem 15.26 are satisfied for the class $\mathcal{C}_t$ of compact subsets of X_t. ∎

15.28 Corollary *The conclusion of Corollary 15.27 remains true if each X_t is Hausdorff, and each μ_F is tight.*

Proof: In a Hausdorff space, every compact set is closed, so it follows that the class $\mathcal{C}_t$ of compact subsets of X_t is a compact class, and hence satisfies the hypotheses of Theorem 15.26. ∎

Chapter 16

Spaces of sequences

Among the most important and simplest normed and Banach spaces are the sequence spaces—vector subspaces of the vector space $\mathbb{R}^{\mathbb{N}}$ of all real sequences. In this chapter, we introduce the classical sequence spaces; φ, the space of sequences with only finitely many nonzero terms; c_0, the space of sequences converging to zero; c, the space of all convergent sequences; ℓ_∞, the space of bounded sequences; and ℓ_p $(0 < p < \infty)$, the space of p-absolutely summable sequences. We discuss each of these sequence spaces and investigate its topological and lattice structures, including its duals.

We start with the "universal" sequence space $\mathbb{R}^{\mathbb{N}}$ as a Fréchet lattice. Its topological dual and its order dual coincide, and is φ (Theorem 16.3). The norm dual of φ (with the sup norm) is ℓ_1 (Theorem 16.1). The space c_0 is an AM-space under the sup norm, and its norm dual is also ℓ_1 (Theorem 16.7). The space c is an AM-space with unit. Remarkably the spaces c and c_0 are linearly homeomorphic (Theorem 16.12). The norm dual of c is a little complicated (Theorem 16.14).

The basic properties of the ℓ_p-spaces are discussed with special emphasis on ℓ_1, ℓ_∞, and the symmetric Riesz pair $\langle \ell_\infty, \ell_1 \rangle$. For $1 \leqslant p < \infty$, the norm dual of ℓ_p is ℓ_q where $\frac{1}{p} + \frac{1}{q} = 1$ (Theorem 16.20). The norm dual of ℓ_∞ is $ba(\mathbb{N})$ (Corollary 14.11). We discuss at length the structure of the unit ball in $ba(\mathbb{N})$ in terms of ultrafilters in section 16.8. In particular, we show that ultrafilters on $\mathbb{N}$ correspond to zero-one valued probability charges on $\mathbb{N}$, and that the free ultrafilters are the weak*-limit points of the point-mass probability charges.

The sequence spaces can be thought of as the "building blocks" of Banach spaces and Banach lattices. Whether they are *embeddable* in a Banach space or a Banach lattice reflect the topological and order structure of the space. We investigate the embeddings of c_0, ℓ_1, and ℓ_∞ into Banach spaces and Banach lattices.

The mapping from a sequence to its limit is a linear functional on the vector space c of convergent sequences. A Banach–Mazur limit is a linear functional on the space ℓ_∞ of all bounded sequences that is an extension the limit functional. The extension is required to be bounded between the lim inf and lim sup. We establish the existence of such Banach–Mazur limits (Theorem 16.47), and use them to prove the existence of invariant measures under *flows* on compact metrizable spaces (Theorem 16.48).

We close the chapter with a short discussion of vector-valued sequence spaces, and discuss the generalization of the ℓ_p-spaces from real-valued sequences to vector-valued sequences. In particular, we show that the norm dual of the ℓ_p-sum is the ℓ_q-sum of the sequence of norm duals (Theorem 16.49).

16.1 The basic sequence spaces

Recall that $\mathbb{N}$ denotes the set of natural numbers $\{1, 2, \ldots\}$. Then $\mathbb{R}^{\mathbb{N}}$ is the vector space of all real sequences, that is, real-valued functions on $\mathbb{N}$. Since $\mathbb{N}$ is naturally a separable metric space under the discrete metric, we can choose to think of sequences as continuous functions on $\mathbb{N}$. A **sequence space** is simply any vector subspace of $\mathbb{R}^{\mathbb{N}}$.

As usual, we may write $x = (x_1, x_2, \ldots)$ to denote an element of $\mathbb{R}^{\mathbb{N}}$. If x is a convergent sequence, then we denote its limit by x_∞. Given a sequence x we define the **n-tail** of x by

$$x^{(n)} = (0, \ldots, 0, x_{n+1}, x_{x+2}, \ldots)$$

and the **n-head** by

$$^{(n)}x = (x_1, \ldots, x_n, 0, 0, \ldots).$$

There are some special sequences to which we have occasion to refer, and we assign them special symbols. The constant sequence whose terms are all unity is denoted e, that is, $e = (1, 1, \ldots)$. The kth unit coordinate vector is the sequence whose k^{th}-term is one and every other term is zero, denoted e_k. In a finite dimensional space the unit coordinate vectors form a basis for the space. This is not true in $\mathbb{R}^{\mathbb{N}}$, because any linear combination of unit coordinate vectors is a sequence with only finitely many nonzero terms.

The vector space $\mathbb{R}^{\mathbb{N}}$ is partially ordered by the pointwise ordering, $x \geqslant y$ in $\mathbb{R}^{\mathbb{N}}$ if $x_n \geqslant y_n$ for each n. You should check that $\mathbb{R}^{\mathbb{N}}$ is an order complete Riesz space. Its lattice operations are given pointwise:

$$x \vee y = (\max\{x_1, y_1\}, \max\{x_2, y_2\}, \ldots)$$

and

$$x \wedge y = (\min\{x_1, y_1\}, \min\{x_2, y_2\}, \ldots).$$

For any pair of sequences $x, y \in \mathbb{R}^{\mathbb{N}}$, we define the **dot product** $\langle x, y \rangle$ by

$$\langle x, y \rangle = \sum_{n=1}^{\infty} x_n y_n,$$

provided that the series is convergent in $\mathbb{R}$. We may sometimes write the dot product as $x \cdot y$.

We say that a sequence space Y **represents** the topological dual of the topological sequence space (X, τ) if the pair $\langle X, Y \rangle$ is a dual pair under the duality $\langle x, y \rangle = \sum_{n=1}^{\infty} x_n y_n$ and τ is a consistent topology on X. Often we say that "Y is the dual of X" rather than "Y represents the dual of X."

We are now ready to collect a few important sequence spaces and recite their basic properties. If $0 < p < \infty$, then the ℓ_p-**norm** of a vector $x \in \mathbb{R}^{\mathbb{N}}$ is defined by

$$\|x\|_p = \left(\sum_{n=1}^{\infty} |x_n|^p \right)^{1/p}.$$

The $\| \cdot \|_\infty$-**norm** of a vector $x \in \mathbb{R}^{\mathbb{N}}$ is defined by

$$\|x\|_\infty = \sup_{n \in \mathbb{N}} |x_n|.$$

For $1 \leqslant p \leqslant \infty$ and any vectors $x, y \in \mathbb{R}^{\mathbb{N}}$, we have

- $\|x\|_p \geqslant 0$.

- $\|\lambda x\|_p = |\lambda| \, \|x\|_p$, for $\lambda \in \mathbb{R}$.

- $\|x + y\|_p \leqslant \|x\|_p + \|y\|_p$.

We investigate the following special sequence spaces.

1. $\varphi = \{x \in \mathbb{R}^{\mathbb{N}} : x_i = 0 \text{ except for finitely many indexes } i\}$.

2. $c_0 = \{x \in \mathbb{R}^{\mathbb{N}} : \lim_{n \to \infty} x_n = 0\}$.

3. $c = \{x \in \mathbb{R}^{\mathbb{N}} : x_\infty = \lim_{n \to \infty} x_n \text{ exists in } \mathbb{R}\}$.

4. $\ell_p = \{x \in \mathbb{R}^{\mathbb{N}} : \|x\|_p < \infty\}$, where $0 < p \leqslant \infty$.

In other words, φ is the collection of all continuous real functions on $\mathbb{N}$ with compact support, and ℓ_∞ consists of all bounded continuous functions on $\mathbb{N}$. For $0 < p < \infty$, we have the following Riesz subspace inclusions:

$$\varphi \subset \ell_p \subset c_0 \subset c \subset \ell_\infty \subset \mathbb{R}^{\mathbb{N}}.$$

We discuss these spaces separately in the next several sections.

16.2 The sequence spaces $\mathbb{R}^{\mathbb{N}}$ and φ

The sequence space φ (sequences with finite support) is an ideal in $\mathbb{R}^{\mathbb{N}}$. As such, it is an order complete Riesz space in its own right. In addition, the sup norm $\| \cdot \|_\infty$ on φ is an M-norm. That is, $\|x \vee y\|_\infty = \max\{\|x\|_\infty, \|y\|_\infty\}$ for $x, y \geqslant 0$. The basic properties of φ are given in the next result, whose proof is left as an exercise.

16.1 Theorem   *The sequence space φ of sequences with finite support has the following properties.*

1. *φ is an order complete Riesz space.*

2. *The sup norm on φ is an order continuous M-norm.*

3. *φ is not norm complete (with the sup norm).*

4. *The sup norm completion of φ is the Banach lattice c_0.*

5. *The norm dual of φ equipped with the sup norm is the sequence space ℓ_1. In particular, both $\langle \varphi, \ell_1 \rangle$ and $\langle \ell_1, \varphi \rangle$ are symmetric Riesz pairs.*

Our next goal is to establish that φ represents the dual of $\mathbb{R}^{\mathbb{N}}$. We already know that $\mathbb{R}^{\mathbb{N}}$ equipped with the product topology is a completely metrizable locally convex-solid Riesz space. In particular, $\mathbb{R}^{\mathbb{N}}$ is a Fréchet lattice. It turns out that the product topology is the only Hausdorff locally solid topology on $\mathbb{R}^{\mathbb{N}}$.

16.2 Theorem   *The product topology on the sequence space $\mathbb{R}^{\mathbb{N}}$, which is an order continuous completely metrizable locally convex-solid topology, is the only Hausdorff locally solid topology on $\mathbb{R}^{\mathbb{N}}$.*

Proof: Let τ be the product topology on $\mathbb{R}^{\mathbb{N}}$ and let τ_1 be another locally solid Hausdorff topology on $\mathbb{R}^{\mathbb{N}}$. By Theorem 9.6, we know that the identity operator $I: (\mathbb{R}^{\mathbb{N}}, \tau) \to (\mathbb{R}^{\mathbb{N}}, \tau_1)$ is continuous. To see that its inverse $I: (\mathbb{R}^{\mathbb{N}}, \tau_1) \to (\mathbb{R}^{\mathbb{N}}, \tau)$ is also continuous, let $x_\alpha = (x_\alpha^1, x_\alpha^2, \ldots) \xrightarrow{\ \tau_1\ } 0$ in $\mathbb{R}^{\mathbb{N}}$. The inequality $|x_\alpha^i e_i| \leq |x_\alpha|$ and the solidness of τ_1 imply $x_\alpha^i e_i \xrightarrow{\ \tau_1\ } 0$ for each i. Now the Hausdorffness of τ_1 implies $x_\alpha^i \xrightarrow{\ \alpha\ } 0$ in $\mathbb{R}$ for each i (why?), so $x_\alpha \xrightarrow{\ \tau\ } 0$.[1] Therefore, $\tau_1 = \tau$.   ∎

Unless otherwise stated, the sequence space $\mathbb{R}^{\mathbb{N}}$ is equipped with the product topology. The dual of $\mathbb{R}^{\mathbb{N}}$ is the sequence space φ.

16.3 Theorem   *The topological dual of the Fréchet lattice $\mathbb{R}^{\mathbb{N}}$ (which coincides with the order dual) is the Riesz space φ.*

Proof: Clearly, every $y \in \varphi$ defines, via the formula

$$f_y(x) = \langle x, y \rangle = \sum_{n=1}^{\infty} x_n y_n,$$

a continuous linear functional f_y on $\mathbb{R}^{\mathbb{N}}$.

For the converse, assume that $f: \mathbb{R}^{\mathbb{N}} \to \mathbb{R}$ is a continuous linear functional. The continuity of f at zero guarantees that there exist some $\delta > 0$ and some k such

---

[1] Actually, the arguments of this part show that for any set X the topology of pointwise convergence is the weakest locally solid Hausdorff topology on $\mathbb{R}^X$.

that $x \in \mathbb{R}^{\mathbb{N}}$ and $|x_i| < \delta$ for $i = 1, \ldots, k$ imply $|f(x)| < 1$. So for each $x \in \mathbb{R}^{\mathbb{N}}$ we have $n|f(x^{(k)})| = |f(nx^{(k)})| < 1$ for each n, and hence $f(x^{(k)}) = 0$. Now let $y = (f(e_1), \ldots, f(e_k), 0, 0, \ldots) \in \varphi$ and note that

$$f(x) = f((x_1, x_2, \ldots, x_k, 0, 0, \ldots)) = \sum_{i=1}^{k} x_i f(e_i) = \langle x, y \rangle$$

for each $x \in \mathbb{R}^{\mathbb{N}}$. That is, $f = f_y$ and, of course, y is uniquely determined.

Since $y \geqslant 0$ if and only if $f_y \geqslant 0$, we infer that $y \mapsto f_y$ is a lattice isomorphism from φ onto $(\mathbb{R}^{\mathbb{N}})'$. That is, $\varphi = (\mathbb{R}^{\mathbb{N}})'$. The parenthetical remark follows immediately from Theorem 9.11. ∎

16.4 Corollary *Both Riesz pairs $\langle \mathbb{R}^{\mathbb{N}}, \varphi \rangle$ and $\langle \varphi, \mathbb{R}^{\mathbb{N}} \rangle$ are symmetric.*

The Riesz space $\mathbb{R}^{\mathbb{N}}$ can also be viewed as an appropriate $L_0(\mu)$ space, the Riesz space of all equivalence μ-measurable real functions on some finite measure space. To see this, fix any summable sequence $\varepsilon = (\varepsilon_1, \varepsilon_2, \ldots)$ of strictly positive real numbers, and consider the measure μ_ε induced on $\mathbb{N}$ by ε via the formula

$$\mu_\varepsilon(A) = \sum_{n \in A} \varepsilon_n.$$

Then $\mathbb{R}^{\mathbb{N}} = L_0(\mu_\varepsilon)$. In particular, from Theorem 13.41 and Corollary 9.9, it follows that the topology of convergence in measure coincides with the topology of pointwise convergence. (Can you prove this conclusion directly?) Thus, we have the following characterization of $\mathbb{R}^{\mathbb{N}}$.

16.5 Theorem *If $\varepsilon = (\varepsilon_1, \varepsilon_2, \ldots)$ is a summable sequence of strictly positive real numbers and μ_ε is its induced measure on $\mathbb{N}$, then $\mathbb{R}^{\mathbb{N}} = L_0(\mu_\varepsilon)$.*

16.3 The sequence space c_0

It is easy to see that the sequence space c_0, also called the space of **null sequences**, is an ideal in $\mathbb{R}^{\mathbb{N}}$ (so c_0 is an order complete Riesz space). However, c_0 does not have order units. Indeed, if $u = (u_1, u_2, \ldots) \in c_0$ satisfies $u_i > 0$ for each i, then put $x = (\sqrt{u_1} \sqrt{u_2}, \ldots)$ and note that there is no $\lambda > 0$ satisfying $x \leqslant \lambda u$.

Unless otherwise stated, c_0 is equipped with the sup norm,

$$\|x\|_\infty = \sup_{n \in \mathbb{N}} |x_n| = \max_{n \in \mathbb{N}} |x_n|.$$

Then c_0 is an AM-space, that is, $\|x \vee y\|_\infty = \max\{\|x\|_\infty, \|y\|_\infty\}$ for each $0 \leqslant x, y \in c_0$.

16.6 Theorem *The sequence space c_0 is an AM-space with order continuous norm.*

Proof: It is easy to check that c_0 under the sup norm is complete, that is, a Banach space. Since the sup norm is clearly a lattice norm, it follows that c_0 is a Banach lattice.

To see that c_0 is also an AM-space, let $0 \leqslant x, y \in c_0$. Assume that $x_i \leqslant x_k$ and $y_i \leqslant y_m$ for all i; we can suppose that $x_k \leqslant y_m$. Then

$$(x \vee y)_i = \max\{x_i, y_i\} \leqslant \max\{x_k, y_m\} = y_m = \max\{x_m, y_m\}$$

implies $\|x \vee y\|_\infty = y_m = \max\{x_k, y_m\} = \max\{\|x\|_\infty, \|y\|_\infty\}$.

For the order continuity of the norm, let $x_\alpha = (x_1^\alpha, x_2^\alpha, \ldots) \downarrow 0$; we can assume $x_\alpha \leqslant u \in c_0$ for each α. Note that $x_i^\alpha \downarrow_\alpha 0$ for each fixed i. Now let $\varepsilon > 0$. Pick some k such that $u_n < \varepsilon$ for each $n > k$ and then select some α_0 such that $x_i^\alpha < \varepsilon$ for each $\alpha \geqslant \alpha_0$ and all $1 \leqslant i \leqslant k$. Hence, $\|x_\alpha\|_\infty < \varepsilon$ for each $\alpha \geqslant \alpha_0$, which means that $\|x_\alpha\|_\infty \downarrow 0$. ∎

The next result characterizes the first and second norm duals of c_0 and its easy proof is left as an exercise.

16.7 Theorem   *The norm dual of c_0 is ℓ_1 under the duality*

$$\langle x, y \rangle = \sum_{n=1}^{\infty} x_n y_n, \quad x \in c_0, y \in \ell_1.$$

The norm double dual of c_0 is ℓ_∞ under the duality

$$\langle x, y \rangle = \sum_{n=1}^{\infty} x_n y_n, \quad x \in \ell_1, y \in \ell_\infty.$$

16.8 Corollary   *The Banach lattice c_0 is not reflexive.*

16.9 Corollary   *Both Riesz pairs $\langle c_0, \ell_1 \rangle$ and $\langle \ell_1, c_0 \rangle$ are symmetric.*

The sequence space c_0 provides an example of a non-reflexive Banach lattice E for which E and E' both have order continuous norms.

A **Schauder basis** (or simply a basis) for a Banach space X is a sequence $\{u_n\}$ such that for each $x \in X$ whose span is dense, that is, there exists a unique sequence of scalars $\{\lambda_n\}$ such that $\lim_{n \to \infty} \|x - \sum_{i=1}^{n} \lambda_i u_i\| = 0$. In this case we write $x = \sum_{n=1}^{\infty} \lambda_n u_n$, and say that the series is norm convergent. Every Banach space with a Schauder basis is automatically separable. (The set of all finite linear combinations of basis vectors with rational coefficients is countable and norm dense.) A basis $\{u_n\}$ in a Banach lattice is **positive** if a vector $x = \sum_{n=1}^{\infty} \lambda_n u_n$ is positive ($x \geqslant 0$) if and only if $\lambda_n \geqslant 0$ for each n.

16.10 Theorem *The sequence $\{e_n\}$ of basic unit vectors is a positive basis for c_0. (Consequently, c_0 is separable.) Moreover, for each $x \in c_0$ we have*

$$x = \sum_{n=1}^{\infty} x_n e_n.$$

Proof: If $x = (x_1, x_2, \ldots) \in c_0$, then $\left\| x - \sum_{i=1}^{n} x_i e_i \right\|_\infty = \max_{i>n} |x_i| \xrightarrow{n \to \infty} 0$, and the conclusion follows. ∎

16.4 The sequence space c

The space c of convergent sequences is not an ideal in $\mathbb{R}^{\mathbb{N}}$., However, with the sup norm it is an AM-space with unit. One such unit is the constant sequence $e = (1, 1, 1, \ldots)$. The sup norm is not order continuous on c: For instance, the tails of e satisfy $e^{(n)} \downarrow 0$ while $\|e^{(n)}\|_\infty = 1$ for each n. Also, c is not order complete: If $x_n = (\underbrace{0, 1, 0, 1, \ldots, 0, 1}_{2n}\,0, 0, \ldots)$, then $x_n \uparrow \leqslant e$, but $\sup\{x_n\}$ does not exist in c.[2]

Again, unless otherwise stated, c is equipped with the sup norm.

If we consider $\mathbb{N}$ as a locally compact Hausdorff space with the discrete topology, then it is easy to see that $c = C(\mathbb{N}_\infty)$, where $\mathbb{N}_\infty$ is the one-point compactification of $\mathbb{N}$. In this setting, note that

$$c_0 = \{f \in C(\mathbb{N}_\infty) : f(\infty) = 0\}.$$

We summarize the above discussion in the following theorem.

16.11 Theorem *The sequence space c with the sup norm is an AM-space with unit e. The sup norm is not order continuous and c fails to be order complete.*

The Banach lattice c cannot be lattice isomorphic to c_0. (Any lattice isomorphism preserves order units and order continuity of the norm.) Surprisingly, c and c_0 are linearly homeomorphic as Banach spaces.

16.12 Theorem *The Banach spaces c and c_0 are linearly homeomorphic.*

Proof: Consider the mapping $T : c \to c_0$ defined by

$$T(x_1, x_2, \ldots) = (x_\infty, x_1 - x_\infty, x_2 - x_\infty, \ldots),$$

where recall $x_\infty = \lim_{n \to \infty} x_n$. An easy verification shows that T is linear, one-to-one, and surjective. Now if $x = (x_1, x_2, \ldots) \in c$, then note that

$$|x_i - x_\infty| \leqslant |x_i| + |x_\infty| \leqslant 2\|x\|_\infty \quad \text{and} \quad |x_i| \leqslant |x_i - x_\infty| + |x_\infty| \leqslant 2\|Tx\|_\infty.$$

---

[2] If K is a Hausdorff compact topological space, then in order for the Riesz space $C(K)$ to be order complete it is necessary and sufficient that K be extremally disconnected; see [273, p. 16]

Consequently, it follows that

$$\tfrac{1}{2}\|x\|_\infty \leqslant \|Tx\|_\infty \leqslant 2\|x\|_\infty$$

for each $x \in c$, which shows that T is a linear homeomorphism. From

$$\|T(1,-1,-1,\ldots)\|_\infty = \|(-1,2,0,0,\ldots)\|_\infty = 2$$

and

$$\|T(1,\tfrac{1}{2},\tfrac{1}{2},\ldots)\|_\infty = \|(\tfrac{1}{2},\tfrac{1}{2},0,0,\ldots)\|_\infty = \tfrac{1}{2},$$

we see that the above norm bounds are also exact. ∎

16.13 Theorem *The sequence $\{e, e_1, e_2, \ldots\}$ is a Schauder basis for the AM-space c (and hence c is separable).*

Proof: This follows immediately from the observation that

$$x = x_\infty e + \sum_{n=1}^{\infty}(x_n - x_\infty)e_n$$

for each $x = (x_1, x_2, \ldots) \in c$. (Note that this is not a positive basis.) ∎

We next describe the norm dual of c. Recall that the **direct sum** $X \oplus Y$ of two Banach spaces is also a Banach space under the norm $\|x \oplus y\| = \|x\| + \|y\|$. [3]

16.14 Theorem *The norm dual of the AM-space c can be identified with the AL-space $\ell_1 \oplus \mathbb{R}$ via the lattice isometry $T: \ell_1 \oplus \mathbb{R} \to c'$ defined by*

$$T_{x\oplus r}(y) = ry_\infty + \sum_{n=1}^{\infty} x_n y_n,$$

where $x \oplus r$ belongs to $\ell_1 \oplus \mathbb{R}$, and y belongs to c.

Proof: Clearly, T is linear and one-to-one. We claim that T is also surjective. To see this, let $0 \leqslant f \in c'$. Then, for each $y \in c$ we have

$$f(y) = f\Big(y_\infty e + \sum_{n=1}^{\infty}(y_n - y_\infty)e_n\Big) = y_\infty f(e) + \sum_{n=1}^{\infty}(y_n - y_\infty)f(e_n)$$

$$= \Big[f(e) - \sum_{n=1}^{\infty} f(e_n)\Big]y_\infty + \sum_{n=1}^{\infty} f(e_n)y_n$$

$$= T_{x\oplus r}(y),$$

---

[3] Other commonly used equivalent norms on the direct sum $X \oplus Y$ are $\|x \oplus y\| = \|x\| \vee \|y\|$ and $\|x \oplus y\| = (\|x\|^2 + \|y\|^2)^{\frac{1}{2}}$.

where $r = f(e) - \sum_{n=1}^{\infty} f(e_n)$ and $x = (f(e_1), f(e_2), \ldots)$. To see that $x \in \ell_1$, argue as follows: From the inequality $0 \leqslant \sum_{n=1}^{k} e_n \leqslant e$, we see that $0 \leqslant \sum_{n=1}^{k} f(e_n) \leqslant f(e)$ for each k, so $\sum_{n=1}^{\infty} f(e_n) \leqslant f(e) < \infty$.

Next, note that $T_{x \oplus r} \geqslant 0$ if and only if both $x \geqslant 0$ and $r \geqslant 0$. The latter (in view of Theorem 9.17) implies that T is a lattice isomorphism. Finally, from

$$
\begin{aligned}
\left\| T_{x \oplus r} \right\|_\infty &= \left\| |T_{x \oplus r}| \right\|_\infty = \left\| T_{|x| \oplus |r|} \right\|_\infty \\
&= T_{|x| \oplus |r|}(e) = |r| + \sum_{n=1}^{\infty} |x_n| \\
&= \| x \oplus r \|,
\end{aligned}
$$

we infer that T is, in fact, a lattice isometry. ∎

Note also that the duality of the Riesz pair $\langle c, \ell_1 \oplus \mathbb{R} \rangle$ satisfies

$$
\langle y, x \oplus r \rangle = r y_\infty + \sum_{n=1}^{\infty} x_n y_n.
$$

16.5 The ℓ_p-spaces

Recall that for $0 < p < \infty$, the ℓ_p-norm of a vector $x \in \mathbb{R}^{\mathbb{N}}$ is given by

$$
\|x\|_p = \Big(\sum_{n=1}^{\infty} |x_n|^p \Big)^{1/p}.
$$

The space ℓ_p is defined by

$$
\ell_p = \{ x = (x_1, x_2, \ldots) : \|x\|_p < \infty \}.
$$

16.15 Lemma *We have:*

1. *If $0 < p < q$, then ℓ_p is a Riesz subspace of ℓ_q, and*

2. *If $x \in \ell_r$ for some $0 < r < \infty$, then $\lim_{p \to \infty} \|x\|_p = \|x\|_\infty$.*

Proof: (1) If $y = (y_1, y_2, \ldots) \in \ell_p$, then $y \in c_0$. So $|y_n| < 1$ for all sufficiently large n, and hence $|y_n|^q \leqslant |y_n|^p$ for all sufficiently large n too. This implies $y \in \ell_q$.

(2) Let $x \in \ell_r$ where $0 < r < \infty$; we can assume that $s = \|x\|_\infty > 0$. Since $\lim_{n \to \infty} x_n = 0$, there exists some k such that $|x_k| = \|x\|_\infty$. This implies $\|x\|_\infty \leqslant \|x\|_p$ for each $p > 0$. Next, pick some natural number m such that $\sum_{n=m}^{\infty} |\frac{x_n}{s}|^r < 1$.

Hence, for $p > r$, we have

$$\|x\|_p = \Big(\sum_{n=1}^{\infty} |x_n|^p\Big)^{\frac{1}{p}} = s\Big(\sum_{n=1}^{m-1} |\tfrac{x_n}{s}|^p + \sum_{n=m}^{\infty} |\tfrac{x_n}{s}|^p\Big)^{\frac{1}{p}}$$

$$\leqslant s\Big(\sum_{n=1}^{m-1} 1^p + \sum_{n=m}^{\infty} |\tfrac{x_n}{s}|^r\Big)^{\frac{1}{p}} \leqslant s[(m-1) + 1]^{\frac{1}{p}}$$

$$= sm^{\frac{1}{p}}.$$

So we have shown that there exists some positive integer m such that

$$\|x\|_{\infty} \leqslant \|x\|_p \leqslant \|x\|_{\infty} m^{\frac{1}{p}}$$

for each $p > r$. Since $\lim_{p\to\infty} m^{\frac{1}{p}} = 1$, we have $\lim_{p\to\infty} \|x\|_p = \|x\|_{\infty}$. ∎

Part (2) of the preceding lemma justifies using the symbol $\|\cdot\|_{\infty}$ for the sup norm. If μ denotes the counting measure on $\mathbb{N}$,

$$\mu(A) = \begin{cases} n & \text{if } A \text{ has } n \text{ elements,} \\ \infty & \text{if } A \text{ is infinite,} \end{cases}$$

then $\ell_p = L_p(\mu)$. Therefore, the properties of the L_p-spaces discussed in Chapter 13 apply to ℓ_p-spaces as well. We mention a few of them below.

Recall that p and q are **conjugate** if $1 \leqslant p, q \leqslant \infty$ and $\frac{1}{p} + \frac{1}{q} = 1$, where by convention $\frac{1}{\infty} = 0$. The following inequality—known as Hölder's inequality—is fundamental for studying the ℓ_p-spaces. Its proof is a special case of Theorem 13.2

16.16 Hölder's Inequality *Let p and q be conjugate exponents. If $x \in \ell_p$ and $y \in \ell_q$, then the series $\langle x, y \rangle = \sum_{n=1}^{\infty} x_n y_n$ converges absolutely and*

$$|\langle x, y \rangle| \leqslant \|x\|_p \cdot \|y\|_q.$$

For $1 \leqslant p < \infty$, the ℓ_p-spaces are Banach lattices with order continuous norms. (See Theorems 13.5 and 13.7.)

16.17 Theorem *For each $1 \leqslant p \leqslant \infty$, the function $\|\cdot\|_p$ is indeed a norm on ℓ_p, and each ℓ_p-space with this norm and the pointwise ordering is a Banach lattice. Moreover, for $1 \leqslant p < \infty$ the ℓ_p-norm is order continuous.*

For $0 < p < 1$, the ℓ_p-space in no longer a Banach lattice. In this case, the "ℓ_p-norm" fails to satisfy the triangle inequality. However, the formula

$$d(x, y) = \sum_{n=1}^{\infty} |x_n - y_n|^p$$

defines a metric on ℓ_p. Under this metric the ℓ_p-space is complete. In addition, the metric d generates a locally solid topology; see Theorem 13.30. Summarizing, we have the following result.

16.18 Theorem   *For $0 < p < 1$, the formula*

$$d(x, y) = \sum_{n=1}^{\infty} |x_n - y_n|^p$$

defines a complete metric on ℓ_p. The topology generated by d is locally solid and order continuous. That is, for $0 < p < 1$, each ℓ_p is a Fréchet lattice, but not a Banach lattice.

For $1 \leqslant p < \infty$ the ℓ_p-spaces are separable.

16.19 Theorem   *For each $1 \leqslant p < \infty$ the sequence $\{e_n\}$ of basic unit vectors is a positive basis. In particular, for $1 \leqslant p < \infty$ the Banach lattice ℓ_p is separable.*

Proof: If $x = (x_1, x_2, \ldots) \in \ell_p$ and $1 \leqslant p < \infty$, then note that

$$\left\| x - \sum_{i=1}^{n} x_n e_n \right\|_p = \left(\sum_{i=n+1}^{\infty} |x_i|^p \right)^{\frac{1}{p}} \xrightarrow[n \to \infty]{} 0.$$

This means that $x = \sum_{n=1}^{\infty} x_n e_n$, and the conclusion follows.   ∎

A fundamental difference between sequences in ℓ_∞ and sequences in the other ℓ_p-spaces concerns the behavior of their tails. It is easy to see that for $1 \leqslant p < \infty$, if $x \in \ell_p$, then

$$\lim_{n \to \infty} \|x^{(n)}\|_p = 0.$$

This fails to be true in ℓ_∞. For instance, $\|e^{(n)}\|_\infty = 1$ for all n, where $e^{(n)}$ is the tail of the sequence $e = (1, 1, \ldots)$.

From Theorem 13.26, we know that if $1 < p < \infty$, then $\ell_p' = \ell_q$. We present another proof of this basic result next.

16.20 Theorem   *Let $1 \leqslant p < \infty$ and let q be its conjugate exponent. Then the mapping $y \mapsto f_y$, from ℓ_q to ℓ_p', defined by*

$$f_y(x) = \langle x, y \rangle = \sum_{n=1}^{\infty} x_n y_n, \ x \in \ell_p,$$

is a surjective lattice isometry, so under this identification, $\ell_p' = \ell_q$.

Proof: We consider only the case $1 < p < \infty$ and leave the similar arguments for $p = 1$ as an exercise. From Hölder's inequality, we see that f_y defines indeed a linear functional on ℓ_p and $\|f_y\| \leqslant \|y\|_q$ for each $y \in \ell_q$. Clearly, $y \mapsto f_y$ is a linear one-to-one mapping. We verify below the rest of the properties needed to make $y \mapsto f_y$ a lattice isometry.

- *The mapping $y \mapsto f_y$ is a surjective isometry.*

Let $f \in \ell'_p$ and put $y_n = f(e_n)$. We now show that $y = (y_1, y_2, \ldots) \in \ell_q$, $f_y = f$, and $\|f_y\| = \|y\|_q$.

To this end, define $\lambda_n = y_n |y_n|^{q-2}$ if $y_n \neq 0$ and $\lambda_n = 0$ if $y_n = 0$ and note that $|\lambda_n|^p = |y_n|^q = \lambda_n y_n$. Hence,

$$
\sum_{i=1}^{n} |y_i|^q = \sum_{i=1}^{n} \lambda_i y_i = \sum_{i=1}^{n} \lambda_i f(e_i) = f\left(\sum_{i=1}^{n} \lambda_i e_i\right)
$$

$$
\leqslant \|f\| \cdot \left\|\sum_{i=1}^{n} \lambda_i e_i\right\|_p = \|f\| \cdot \left(\sum_{i=1}^{n} |\lambda_i|^p\right)^{\frac{1}{p}}
$$

$$
= \|f\| \cdot \left(\sum_{i=1}^{n} |y_i|^q\right)^{\frac{1}{p}},
$$

for each n, so $\left(\sum_{i=1}^{n} |y_i|^q\right)^{\frac{1}{q}} \leqslant \|f\| < \infty$ for each n. The latter implies $y \in \ell_q$ and that $\|y\|_q \leqslant \|f\|$. Moreover, note that

$$
f_y(x) = \sum_{n=1}^{\infty} x_n y_n = \sum_{n=1}^{\infty} x_n f(e_n) = f\left(\sum_{n=1}^{\infty} x_n e_n\right) = f(x)
$$

for all $x \in \ell_p$. Now note that $\|f_y\| = \|y\|_q$.

- *The mapping $y \mapsto f_y$ is a lattice isomorphism.*

This follows immediately from Theorem 9.17 by observing that $y \geqslant 0$ if and only if $f_y \geqslant 0$. The proof of the theorem is now complete. ∎

16.21 Corollary   *For $1 < p < \infty$, ℓ_p is a reflexive Banach lattice.*

Finally, we close the section with a useful characterization of the norm totally bounded subsets of the ℓ_p-spaces.

16.22 Theorem   *Let A be a nonempty norm bounded subset of some ℓ_p-space, where $1 \leqslant p < \infty$, and let*

$$
s_n(A) = \sup\left\{\sum_{i=n}^{\infty} |a_i|^p : a = (a_1, a_2, \ldots) \in A\right\}, \; n = 1, 2, \ldots .
$$

Then A is norm totally bounded if and only if $\lim_{n \to \infty} s_n(A) = 0$.

Proof: We prove the "if" part and leave the "only if" as an exercise. So assume $\lim_{n \to \infty} s_n(A) = 0$ and let $\varepsilon > 0$. Pick some k such that $s_n(A) < \varepsilon$ for all $n > k$. Then the set

$$
B = [-s_1, s_1] \times [-s_2, s_2] \times \cdots \times [-s_k, s_k] \times \{0\} \times \{0\} \cdots
$$

is compact and satisfies $A \subset B + \varepsilon U$, where U is the closed unit ball of ℓ_p. This implies that A is a norm totally bounded set. ∎

16.6 ℓ_1 and the symmetric Riesz pair $\langle \ell_\infty, \ell_1 \rangle$

We discuss here a few more interesting properties of the Banach lattice ℓ_1. Recall that the sequence $\{e_n\}$ is a positive Schauder basis for ℓ_1 (so ℓ_1 is separable).

16.23 Definition   *A Banach space X has the **Schur property** if $x_n \xrightarrow{w} 0$ implies $\|x_n\| \to 0$ (or equivalently, if the collections of weakly compact and norm compact subsets of X coincide).*

The parenthetical part of the above definition is, of course, an immediate consequence of the Eberlein–Šmulian Theorem 6.34. The space ℓ_1 has the Schur property.

16.24 Theorem (Banach)   *The Banach lattice ℓ_1 has the Schur property, that is, $x_n \xrightarrow{w} 0$ implies $\|x_n\|_1 \to 0$. In particular, every weakly compact (that is, $\sigma(\ell_1, \ell_\infty)$-compact) subset of ℓ_1 is norm compact.*

Proof: See [12, Theorem 13.1, p. 200].   ∎

The proof of the next theorem is left as an exercise.

16.25 Theorem   *The Banach lattice ℓ_1 has order continuous norm—so the Riesz pairs $\langle \ell_1, \ell_\infty \rangle$ and $\langle \ell_\infty, \ell_1 \rangle$ are both symmetric.*

Although the sup norm on ℓ_∞ is not order continuous, the Mackey topology $\tau(\ell_\infty, \ell_1)$ is order continuous. This is an important result.

16.26 Theorem   *The Mackey topology $\tau(\ell_\infty, \ell_1)$ is an order continuous locally convex-solid topology on ℓ_∞.*

Proof: This is a special case of Theorem 9.36. The order continuity of $\tau(\ell_\infty, \ell_1)$ follows immediately from Theorem 8.60.   ∎

And now we come to an important consequence of Theorem 16.26. Although the tail sequence of an element of ℓ_∞ need not converge in norm to zero, it always converges to zero with respect to the Mackey topology $\tau(\ell_\infty, \ell_1)$.

16.27 Corollary   *If $x \in \ell_\infty$, then $x^{(n)} \xrightarrow{\tau(\ell_\infty, \ell_1)} 0$.*

Proof: Let $x \in \ell_\infty$. Then $|x|^{(n)} \downarrow 0$ and since $\tau(\ell_\infty, \ell_1)$ is an order continuous topology, it follows that $|x|^{(n)} \xrightarrow{\tau(\ell_\infty, \ell_1)} 0$. The local solidness of $\tau(\ell_\infty, \ell_1)$ guarantees $x^{(n)} \xrightarrow{\tau(\ell_\infty, \ell_1)} 0$.[4]   ∎

---

[4] A more direct proof goes as follows: We must show that the sequence $\{x^{(n)}\}$ converges to zero uniformly on every circled, convex and $\sigma(\ell_1, \ell_\infty)$-compact subset of ℓ_1. To this end, let A be such a

16.7   The sequence space ℓ_∞

The Riesz space ℓ_∞ of all bounded sequences is an ideal in $\mathbb{R}^{\mathbb{N}}$. With the sup norm it is an AM-space with unit e, the constant sequence one. Again, unless otherwise stated, ℓ_∞ is equipped with the sup norm. Unlike the other ℓ_p-spaces, ℓ_∞ is not separable.

16.28 Theorem   *The AM-space ℓ_∞ is not separable.*

Proof: The proof uses the classical Cantor diagonal process. Let $\{x^1, x^2, \ldots\}$, where $x^i = (x_1^i, x_2^i, \ldots)$, be a countable subset of ℓ_∞. For each n, let

$$x_n = \begin{cases} 0 & \text{if } |x_n^n| \geq 1, \\ 2 & \text{if } |x_n^n| < 1. \end{cases}$$

Then $x = (x_1, x_2, \ldots) \in \ell_\infty$ and $\|x - x^n\|_\infty \geq |x_n - x_n^n| \geq 1$ for each n. This shows that no countable subset of ℓ_∞ can be norm dense, so ℓ_∞ is a non-separable Banach lattice. ∎

Now let us introduce some notation. As usual, $ba(2^{\mathbb{N}})$ is the AL-space of all signed charges of bounded variation on the σ-algebra $2^{\mathbb{N}}$ of all subsets of $\mathbb{N}$, $ca(2^{\mathbb{N}})$ is the band of all σ-additive signed measures in $ba(2^{\mathbb{N}})$, and $pa(2^{\mathbb{N}})$ is the band of all purely finitely additive signed measures of $ba(2^{\mathbb{N}})$. For brevity, we denote $ba(2^{\mathbb{N}})$ by $ba(\mathbb{N})$ or simply by ba. That is, $ba = ba(\mathbb{N}) = ba(2^{\mathbb{N}})$. Similarly, $ca = ca(\mathbb{N}) = ca(2^{\mathbb{N}})$ and $pa = pa(\mathbb{N}) = pa(2^{\mathbb{N}})$.

If we consider $\mathbb{N}$ equipped with the discrete topology, then $\ell_\infty = C_b(\mathbb{N})$. As such, it follows from Corollary 14.11 that

$$\ell_\infty' = ba(\mathbb{N}) = ca(\mathbb{N}) \oplus pa(\mathbb{N}).$$

16.29 Lemma   *A signed charge in $ba(\mathbb{N})$ is purely finitely additive if and only if it vanishes on the finite subsets of $\mathbb{N}$.*

Proof: Let $\mu \in ba = ba(\mathbb{N})$. Assume first that μ vanishes on the finite subsets of $\mathbb{N}$. From

$$|\mu|(A) = \sup\{\mu(B) - \mu(A \setminus B) : B \subset A\},$$

---

subset.
   By Theorem 16.24, A is a norm compact subset of ℓ_1. This means (in view of Theorem 16.22) that $s_n(A) \to 0$, where $s_n(A) = \sup\{\sum_{i=n}^{\infty} |a_i| : a = (a_1, a_2, \ldots) \in A\}$. Now note that for each $a \in A$, we have

$$\left| \langle x^{(n)}, a \rangle \right| = \left| \sum_{i=n+1}^{\infty} x_i a_i \right| \leq \|x\|_\infty \left(\sum_{i=n+1}^{\infty} |a_i| \right) \leq \|x\|_\infty \cdot s_{n+1}(A).$$

This implies that the tail sequence $\{x^{(n)}\}$ converges uniformly to zero on A.

we see that $|\mu|$ vanishes on the finite subsets of $\mathbb{N}$. Now let $0 \leqslant \nu \in ca$. Put $A_n = \{1, \ldots, n\}$ and $B_n = \{n+1, n+2, \ldots\}$ and note that $B_n \downarrow \varnothing$ implies $\nu(B_n) \downarrow 0$. Now notice that

$$|\mu| \wedge \nu(\mathbb{N}) \leqslant |\mu|(A_n) + \nu(B_n) = \nu(B_n) \downarrow 0$$

implies $|\mu| \wedge \nu(\mathbb{N}) = 0$. Thus, $\mu \perp \nu$ for each $\nu \in ca$, so $\mu \in pa$.

For the converse, assume that $\mu \in pa$. Also, suppose by way of contradiction that $|\mu|(\{k\}) > 0$ for some k. From $\mu \perp \delta_k$, where $\delta_k \in ca$ is the point mass supported at k, we see that

$$0 = |\mu| \wedge \delta_k(\{k\}) = \min\{|\mu|(\{k\}), \delta_k(\{k\})\} = \min\{|\mu|(\{k\}), 1\} > 0,$$

which is a contradiction. Hence, μ vanishes on every singleton and consequently on every finite set. ∎

Since $\ell_1' = \ell_\infty$ and ℓ_1 is an AL-space, it follows (from Theorem 9.34) that ℓ_1 is a band in ℓ_∞' that coincides with the order continuous dual of ℓ_∞. That is, $(\ell_\infty)_n^\sim = \ell_1$. In addition, we have

$$\ell_\infty' = \ell_1 \oplus \ell_1^d,$$

where each $y \in \ell_1$ gives rise to a continuous linear functional on ℓ_∞ via the formula

$$\langle x, y \rangle = \sum_{n=1}^\infty x_n y_n, \quad x \in \ell_\infty.$$

The disjoint complement ℓ_1^d of ℓ_1 consists of all singular functionals. In fact, it consists (up to scalar multiples) of all extensions of the "limit functional" to ℓ_∞.[5]

16.30 Lemma *A linear functional $\theta \in \ell_\infty'$ belongs to ℓ_1^d if and only if there exists some constant k such that $\theta(x) = k \lim_{n\to\infty} x_n$ for each $x \in c$ (and, of course, $k = \theta(e)$). In particular, $\ell_1^d \neq \{0\}$.*

Proof: Assume first that $\theta \in \ell_1^d$. We claim that $\theta(e_n) = 0$ for each n. Indeed, noting that $0 \leqslant x \leqslant e_n$ if and only if there exists some $0 \leqslant \lambda \leqslant 1$ with $x = \lambda e_n$, it follows from

$$0 = |\theta| \wedge e_n(e_n) = \inf\{|\theta|(x) + e_n(e_n - x) : 0 \leqslant x \leqslant e_n\}$$
$$= \inf\{\lambda|\theta|(e_n) + 1 - \lambda : 0 \leqslant \lambda \leqslant 1\}$$

that $|\theta|(e_n) = 0$. Thus, $\theta(e_n) = 0$ for each n, so $\theta(x) = 0$ for each $x \in \varphi$ and $\theta(x) = \theta(x^{(n)})$ for each $x \in \ell_\infty$ and each n. Since $x \in c_0$ implies $\|x^{(n)}\|_\infty \to 0$,

---

[5] The **limit functional** is the positive linear functional $\mathrm{Lim} \colon c \to \mathbb{R}$ defined by $\mathrm{Lim}(x) = \lim_{n\to\infty} x_n$.

we see that $\theta(x) = 0$ for all $x \in c_0$. Therefore, since for each $x \in c$ we have $x - x_\infty e \in c$, it follows that $\theta(x) - \theta(e) \lim_{n\to\infty} x_n = \theta(x - x_\infty e) = 0$. That is,

$$\theta(x) = \theta(e) \lim_{n\to\infty} x_n.$$

Next, assume that θ satisfies $\theta(x) = k \lim_{n\to\infty} x_n$ for each $x \in c$ and let $x \geq 0$ belong to ℓ_1. From

$$
\begin{aligned}
0 \leq |\theta| \wedge x(e) &= \inf\{|\theta|(y) + x(y) : 0 \leq y \leq e\} \\
&\leq |\theta|\big(^{(n)}e\big) + x(e^{(n)}) \\
&= \sup\{|\theta(z)| : |z| \leq {}^{(n)}e\} + \sum_{i=n+1}^{\infty} x_i = \sum_{i=n+1}^{\infty} x_i \xrightarrow[n\to\infty]{} 0,
\end{aligned}
$$

we infer that $|\theta| \wedge x(e) = 0$. Therefore, $|\theta| \wedge x = 0$ for all $x \in \ell_1$. That is, $\theta \in \ell_1^d$.

To see that $\ell_1^d \neq \{0\}$, note that if $\mathrm{Lim} \colon c \to \mathbb{R}$ is the limit functional, then from Theorem 8.32 it follows that Lim has a positive linear extension f to all of ℓ_∞. By the above $0 < f \in \ell_1^d$.   ∎

It turns out that ℓ_1 can be identified with ca and ℓ_1^d with pa. The discussion below clarifies the situation. We establish first that $ca = \ell_1$.

It is not difficult to see that the mapping $x \mapsto \mu_x$, from ℓ_1 to ca, defined by

$$\mu_x(A) = \sum_{n\in A} x_n$$

is a lattice isometry from ℓ_1 onto ca. Indeed, note first that the mapping $x \mapsto \mu_x$ is one-to-one, linear, and surjective. (If $\mu \in ca$, then let $x = (\mu(\{1\}), \mu(\{2\}), \ldots) \in \ell_1$ and note that $\mu_x = \mu$.) Also, $x \geq 0$ if and only if $\mu_x \geq 0$ guarantees that $x \mapsto \mu_x$ is a lattice isomorphism (Theorem 9.17). Moreover, we have

$$\|\mu_x\| = \big\||\mu_x|\big\| = |\mu_x|(\mathbb{N}) = \sum_{n=1}^{\infty} |\mu_x|(\{n\}) = \sum_{n=1}^{\infty} |x_n| = \|x\|_1.$$

Therefore, $x \mapsto \mu_x$ is a lattice isometry from ℓ_1 onto ca. Thus, under this lattice isometry ℓ_1 and ca can be identified.

If A is any subset of $\mathbb{N}$, then χ_A (the characteristic function of A) can be viewed as the sequence that takes the value one at every $n \in A$ and zero at every $n \notin A$. Next, we identify ℓ_1^d and pa. To do this, we employ the following two properties:

1. *A positive linear functional θ on ℓ_∞ is identically equal to zero if and only if $\theta(e) = 0$; and*

2. *For every $\theta \in \ell_\infty'$, we have*

$$\theta^+(e) = \sup\{\theta(\chi_A) : A \subset \mathbb{N}\}, \ \ \theta^-(e) = \sup\{-\theta(\chi_A) : A \subset \mathbb{N}\}, \text{ and}$$

$$|\theta|(e) = \sup\{\theta(\chi_A) - \theta(\chi_{A^c}) : A \subset \mathbb{N}\}.$$

Property (2) is a special case of a general result due to Y. A. Abramovich; see [12, Theorem 3.16, p. 38]. However, in this case, property (2) can be proven easily. We indicate here how to prove the formula for θ^+: Let $s = \sup\{\theta(\chi_A) : A \subset \mathbb{N}\}$. Now $\theta^+(e) = \sup\{\theta(x) : 0 \leqslant x \leqslant e\}$, so $s \leqslant \theta^+(e)$. Now fix $0 \leqslant x \leqslant e$; we can assume $\theta(x) \leqslant \theta(e - x)$. Put $A = \{n \in \mathbb{N} : x_n > 0\}$ and note that $0 \leqslant x \leqslant \chi_A$. Now notice that either $\theta(x) \leqslant \theta(\chi_A) \leqslant s$ or $\theta(x) > \theta(\chi_A)$. In the latter case, we have $\theta(x) \leqslant \theta(e - x) \leqslant \theta(e - \chi_A) = \theta(\chi_{A^c}) \leqslant s$. Therefore, $\theta^+(e) \leqslant s$ and hence $\theta^+(e) = s$.

We now define a mapping $\theta \mapsto \mu_\theta$, from ℓ_1^d to pa, via the formula

$$\mu_\theta(A) = \theta(\chi_A).$$

From Lemmas 16.29 and 16.30, $\mu_\theta \in pa$ for each $\theta \in \ell_1^d$. Clearly, $\theta \mapsto \mu_\theta$ is linear. Next, we verify the properties needed to make $\theta \mapsto \mu_\theta$ a surjective lattice isometry.

- *The mapping $\theta \mapsto \mu_\theta$ is one-to-one.*

If $\mu_\theta = 0$, then from (2) it follows that $\theta^+(e) = \theta^-(e) = 0$. Therefore by (1), $\theta^+ = \theta^- = 0$, so $\theta = \theta^+ - \theta^- = 0$.

- *We have $\theta \geqslant 0$ if and only if $\mu_\theta \geqslant 0$. That is, $\theta \mapsto \mu_\theta$ is a lattice isomorphism.*

If $\theta \geqslant 0$, then clearly $\mu_\theta \geqslant 0$. Now assume $\mu_\theta \geqslant 0$. Then, by (2)

$$0 \leqslant \theta^-(e) = \sup\{-\theta(\chi_A) : A \subset \mathbb{N}\} = \sup\{-\mu_\theta(A) : A \subset \mathbb{N}\} \leqslant 0,$$

so $\theta^-(e) = 0$. Hence, $\theta^- = 0$, and consequently $\theta = \theta^+ - \theta^- = \theta^+ \geqslant 0$.

- *The mapping $\theta \mapsto \mu_\theta$ is surjective.*

Let $0 \leqslant \mu \in pa$. Define the positive linear functional $\theta \colon \ell_\infty \to \mathbb{R}$ by

$$\theta(x) = \int x \, d\mu.$$

If $0 \leqslant u \in \ell_1$, put $A_n = \{1, \ldots, n\}$, $B_n = \{n+1, n+2, \ldots\}$, and note that

$$0 \leqslant \theta \wedge u(e) \leqslant \theta(\chi_{A_n}) + u(\chi_{B_n}) = u(\chi_{B_n}) = \sum_{i=n+1}^{\infty} u_i \xrightarrow[n\to\infty]{} 0.$$

Hence, $\theta \wedge u(e) = 0$, so $\theta \wedge u = 0$. That is, $\theta \in \ell_1^d$.

Now notice that $\theta(\chi_A) = \int \chi_A \, d\mu = \mu(A)$ for each $A \subset \mathbb{N}$, that is, $\mu = \mu_\theta$, and from this it easily follows that $\theta \mapsto \mu_\theta$ is surjective.

- *We have $\|\theta\| = \|\mu_\theta\|$ for each $\theta \in \ell_1^d$.*

Using (2) once more, we see that

$$\begin{aligned}
\|\theta\| = \left\|\|\theta\|\right\| &= |\theta|(e) = \sup\{\theta(\chi_A) - \theta(\chi_{A^c}) : A \subset \mathbb{N}\} \\
&= \sup\{\mu_\theta(A) - \mu_\theta(A^c) : A \subset \mathbb{N}\} = |\mu_\theta|(\mathbb{N}) \\
&= \|\mu_\theta\|.
\end{aligned}$$

These results show that $\theta \mapsto \mu_\theta$ is a lattice isometry from ℓ_1^d onto pa. In sum, we have established the following important theorem.

16.31 Theorem   *The norm dual of the AM-space ℓ_∞ is given by*

$$\ell_\infty' = \ell_1 \oplus \ell_1^d = ca \oplus pa,$$

with the following identifications:

1. *The AL-spaces ℓ_1 and ca are identified via the lattice isometry $x \mapsto \mu_x$ defined by $\mu_x(A) = \sum_{n \in A} x_n$; and*

2. *The AL-spaces ℓ_1^d and pa are identified via the lattice isometry $\theta \mapsto \mu_\theta$ defined by $\mu_\theta(A) = \theta(\chi_A)$.*

Moreover, we have $(\ell_\infty)_n^\sim = \ell_1 = ca$ and $(\ell_\infty)_s^\sim = \ell_1^d = pa$.

To put it another way: *Every countably additive finite signed measure on $\mathbb{N}$ corresponds to exactly one sequence belonging to ℓ_1, and every purely additive finite signed charge corresponds to exactly one extension of a scalar multiple of the limit functional on c.*

16.32 Theorem   *Neither ℓ_1 nor ℓ_∞ is a reflexive Banach lattice.*

Proof: From $\ell_1' = \ell_\infty$ and $\ell_\infty' = \ell_1 \oplus \ell_1^d$, it follows that ℓ_1 cannot be reflexive. By Theorem 6.25, ℓ_∞ cannot be reflexive either.   ∎

We now state an important result on convergence of sequences in ba due to R. S. Phillips [279].

16.33 Phillips' Lemma   *Let $\{\mu_n\} \subset ba$ satisfy $\mu_n \xrightarrow{\sigma(ba,\ell_\infty)} 0$. Then $\{\mu_n\}$ converges to zero uniformly on the closed unit ball of c_0. That is,*

$$\sup\left\{\left|\int x \, d\mu_n\right| : x \in c_0 \text{ and } \|x\|_\infty \leqslant 1\right\} \xrightarrow[n \to \infty]{} 0.$$

Proof: See [12, Theorem 14.19, p. 233].   ∎

Finally, we close the section with a class of AM-spaces that are lattice isometric to ℓ_∞. Let us say that a sequence $u = (u_1, u_2, \ldots)$ of real numbers is **strictly positive** if $u_n > 0$ for each n.

16.34 Theorem *If $u = (u_1, u_2, \ldots)$ is a strictly positive sequence in $\mathbb{R}^\mathbb{N}$, then the ideal A_u generated by u in $\mathbb{R}^\mathbb{N}$ (equipped with its $\|\cdot\|_\infty$-norm) is lattice isometric to ℓ_∞.*

Proof: Let $0 \leqslant u \in \mathbb{R}^\mathbb{N}$ be strictly positive. Then the mapping $\theta\colon A_u \to \ell_\infty$ via

$$\theta(x_1, x_2, \ldots) = \left(\frac{x_1}{u_1}, \frac{x_2}{u_2}, \ldots\right).$$

is a surjective lattice isometry. (Why?)   ∎

This theorem is used implicitly in growth theory in economics. Suppose output can grow at a fixed rate each period. Then the sequence of outputs is unbounded. Nevertheless, Theorem 16.34 guarantees that the space of outputs behaves just like ℓ_∞.

16.8 More on $\ell'_\infty = ba(\mathbb{N})$

Since the norm dual of ℓ_∞ is $ba(\mathbb{N})$, the Alaoglu Compactness Theorem 6.21 asserts that the unit ball of $ba(\mathbb{N})$ is $\sigma(ba(\mathbb{N}), \ell_\infty)$-compact. Now the sequence $\{e_n\}$ of unit coordinate vectors lies in the unit ball of $\ell_1 \subset ba(\mathbb{N})$ (identifying e_n with the charge that puts mass one on $\{n\}$). Consequently, by Theorem 2.31 there is a convergent subnet of $\{e_n\}$. The question is, what are the limit points? Chasing down the answer provides a good test of our understanding of nets, weak topologies, and integration.

We start our quest by looking at a "construction" of purely finitely additive probability charges on $\mathbb{N}$. (We put quotation marks around the term construction because we use Zorn's Lemma, a nonconstructive proposition, in our argument.) Recall that a filter on $\mathbb{N}$ is a collection $\mathcal{F}$ of subsets of $\mathbb{N}$ satisfying:

     i. $\varnothing \notin \mathcal{F}$ and $\mathbb{N} \in \mathcal{F}$.

     ii. If $A \in \mathcal{F}$ and $A \subset B$, then $B \in \mathcal{F}$.

     iii. If $A, B \in \mathcal{F}$, then $A \cap B \in \mathcal{F}$.

By Lemma 2.21, a maximal filter, or ultrafilter, on $\mathbb{N}$, also satisfies:

     iv. $A \notin \mathcal{F}$ if and only if $A^c \in \mathcal{F}$.

Every ultrafilter $\mathcal{U}$ on $\mathbb{N}$ defines a probability charge $\pi_{\mathcal{U}}$ on $\mathbb{N}$ by

$$\pi_{\mathcal{U}}(A) = \begin{cases} 1 & \text{if } A \in \mathcal{U}, \\ 0 & \text{if } A \notin \mathcal{U}. \end{cases}$$

To see that this set function is additive, suppose $A \cap B = \varnothing$. Then properties (i) and (iii) imply that at most one of A and B belongs to $\mathcal{U}$: Suppose first that one of A and B belongs to $\mathcal{U}$, say $A \in \mathcal{U}$ and $B \notin \mathcal{U}$. Then $A \cup B \supset A$, so $A \cup B \in \mathcal{U}$ and

$$\pi_{\mathcal{U}}(A) + \pi_{\mathcal{U}}(B) = 1 + 0 = 1 = \pi_{\mathcal{U}}(A \cup B).$$

If on the other hand, neither A nor B belongs to $\mathcal{U}$, then by (iv) both A^c and B^c belong to $\mathcal{U}$, so $A^c \cap B^c = (A \cup B)^c \in \mathcal{U}$. Thus, by (iv), $A \cup B \notin \mathcal{U}$, so

$$\pi_{\mathcal{U}}(A) + \pi_{\mathcal{U}}(B) = 0 + 0 = \pi_{\mathcal{U}}(A \cup B).$$

A probability charge $\mu \in ba(\mathbb{N})$ is a **zero-one charge** if for each $A \subset \mathbb{N}$ we have either $\mu(A) = 0$ or $\mu(A) = 1$.

By mimicking the proof of Theorem 15.9, we see that the zero-one charges are the extreme points of the set of probability charges. They are also the charges generated by ultrafilters.

16.35 Lemma   *A charge $\mu \in ba(\mathbb{N})$ is a zero-one-charge if and only if $\mu = \pi_{\mathcal{U}}$ for a unique ultrafilter $\mathcal{U}$ on $\mathbb{N}$. Moreover for an ultrafilter $\mathcal{U}$:*

1. *If $\mathcal{U}$ is free, then $\pi_{\mathcal{U}}$ is purely finitely additive.*

2. *If $\mathcal{U}$ is fixed, then $\pi_{\mathcal{U}}$ is countably additive.*

Proof: Let μ be a zero-one-charge. Put $\mathcal{U} = \{A \subset \mathbb{N} : \mu(A) = 1\}$. Then $\mathcal{U}$ is an ultrafilter (why?) on $\mathbb{N}$ satisfying $\mu = \pi_{\mathcal{U}}$.

(1)   Let A be a finite subset of $\mathbb{N}$. By Lemma 2.22, $A \notin \mathcal{U}$, so $\pi_{\mathcal{U}}(A) = 0$. By Lemma 16.29, $\pi_{\mathcal{U}}$ is a purely finitely additive charge.

(2)   Now let $\mathcal{U}$ be fixed, say $\mathcal{U} = \{A \subset \mathbb{N} : x \in A\}$. Then $\pi_{\mathcal{U}} = \delta_x$, the point mass at x, which is countably additive.   ∎

It turns out that we can think of every ultrafilter on $\mathbb{N}$ as a point mass that lives at one of the points of the Stone–Čech compactification of $\mathbb{N}$ (cf. Theorem 2.86), but that is another story. It is also fairly clear that the ultrafilters define probability charges that are the extreme points of the set of probability charges on $\mathbb{N}$. For now though, we are interested in the free ultrafilters because they are precisely the $\sigma(ba, \ell_\infty)$-limit points of the sequence of unit coordinate vectors $\{e_n\}$.

16.36 Theorem   *The set*

$$\{\pi_{\mathcal{U}} : \mathcal{U} \text{ is a free ultrafilter on } \mathbb{N}\}$$

of zero-one charges consists precisely of all $\sigma(ba, \ell_\infty)$-limit points of the sequence of unit coordinate vectors $\{e_n\}$.

Proof: Recall that a subnet of a sequence $\{x_n\}$ in a set X is a net $\xi\colon D \to X$, where D is a directed set under $\geqslant$, for which there is a function $\lambda\colon D \to \mathbb{N}$ satisfying the following two properties. (i) For every $\alpha \in D$, $\xi_\alpha = x_{\lambda(\alpha)}$. (We write ξ_α rather than $\xi(\alpha)$ to simplify the notation.) (ii) For every n, there is an $\alpha \in D$ such that $\beta \geqslant \alpha$ implies $\lambda(\beta) \geqslant n$.

Now let $\mathcal{U}$ be a free ultrafilter on $\mathbb{N}$. The obvious candidate for our directed set is $\mathcal{U}$ itself. This set is directed by $A \geqslant B$ if $A \subset B$. Define $\lambda\colon \mathcal{U} \to \mathbb{N}$ by

$$\lambda(A) = \min A.$$

(Every nonempty set of natural numbers has a smallest element, so λ is well defined.) We claim that the net $\eta\colon \mathcal{U} \to ba(\mathbb{N})$, defined by $\eta_A = e_{\lambda(A)}$, is a subnet of the sequence $\{e_n\}$ and that

$$\eta_A \xrightarrow{\ \sigma(ba,\ell_\infty)\ } \pi_{\mathcal{U}}.$$

To see that $\{\eta_A : A \in \mathcal{U}\}$ is a subnet of $\{e_n : n \in \mathbb{N}\}$, we need to show that for any n there is an $A \in \mathcal{U}$ such that for every $B \geqslant A$ (that is, for every $B \subset A$) we have $\lambda(B) \geqslant n$. But this is easy: set $A = \{n, n+1, \dots\}$. This set A belongs to $\mathcal{U}$ because its complement is finite, and free ultrafilters contain no finite sets.

To see that $\eta_A \xrightarrow{\ \sigma(ba,\ell_\infty)\ } \pi_{\mathcal{U}}$, note that $\sigma(ba, \ell_\infty)$-convergence requires

$$\langle y, \eta_A \rangle = \int_{\mathbb{N}} y_n \, d\eta_A(n) \xrightarrow{\ A\ } \int_{\mathbb{N}} y_n \, d\pi_{\mathcal{U}}(n) = \langle y, \pi_{\mathcal{U}} \rangle$$

for each $y \in \ell_\infty$.

It is easy to see that $\int_{\mathbb{N}} y_n \, d\eta_A(n) = y_{\lambda(A)}$. Since $y \in \ell_\infty$ (that is, y is a bounded measurable function on $\mathbb{N}$), for every $\varepsilon > 0$ there is a step function $s \in \ell_\infty$, satisfying $|s_n - y_n| < \varepsilon$ for each n. Write $s = \sum_{i=1}^m \alpha_i \chi_{A_i}$, where the A_is partition $\mathbb{N}$. Then $\int s(n) \, d\pi_{\mathcal{U}}(n) = \sum_{i=1}^m \alpha_i \pi_{\mathcal{U}}(A_i)$. Since the A_is partition $\mathbb{N}$, exactly one of them, say A_1 belongs to $\mathcal{U}$. Thus, $\int s \, d\pi_{\mathcal{U}} = \alpha_1$. Now for each $A \geqslant A_1$, that is, $A \subset A_1$, $\lambda(A) \in A_1$, so $s_{\lambda(A)} = \alpha_1$. (Remember that $s = \sum_{i=1}^m \alpha_i \chi_{A_i}$.) This means that for $A \geqslant A_1$,

$$\int s \, d\eta_A = s_{\lambda(A)} = \alpha_1 = \int s \, d\pi_{\mathcal{U}}.$$

Recalling how s was chosen, for all $A \geqslant A_1$ we have

$$\left| \int y \, d\eta_A - \int y \, d\pi_{\mathcal{U}} \right|$$
$$\leqslant \left| \int y \, d\eta_A - \int s \, d\eta_A \right| + \left| \int s \, d\eta_A - \int s \, d\pi_{\mathcal{U}} \right| + \left| \int s \, d\pi_{\mathcal{U}} - \int y \, d\pi_{\mathcal{U}} \right|$$
$$< \varepsilon + 0 + \varepsilon \ = \ 2\varepsilon.$$

Thus, $\int y \, d\eta_A \xrightarrow{\ A\ } \int y \, d\pi_{\mathcal{U}}$, so $\eta_A \xrightarrow{\ \sigma(ba,\ell_\infty)\ } \pi_{\mathcal{U}}$.

For the converse, assume that $\{e_{\lambda_\alpha}\}_{\alpha \in D}$ is a subnet of the net $\{e_n\}$ such that $e_{\lambda_\alpha} \xrightarrow{\;\sigma(ba,\ell_\infty)\;}_\alpha \mu$. Let

$$\mathcal{U} = \left\{ A \subset \mathbb{N} : \int \chi_A \, de_{\lambda_\alpha} \xrightarrow{\;\alpha\;} 1 \right\} = \{ A \subset \mathbb{N} : \lambda_\alpha \in A \text{ eventually} \}.$$

To complete the proof, verify that $\mathcal{U}$ is a free ultrafilter on $\mathbb{N}$ satisfying $\pi_{\mathcal{U}} = \mu$. ∎

16.9   Embedding sequence spaces

We say that a Banach space X includes a **copy** of another Banach space Y (or that Y is **embeddable** in X) if there exists a linear operator $T : Y \to X$ and two positive constants K and M satisfying

$$K\|y\| \leqslant \|Ty\| \leqslant M\|y\|$$

for each $y \in Y$. Thus, $T(Y)$ is a closed subspace of X that is linearly homeomorphic (via the **linear embedding** T) to Y; we can think of the Banach space $T(Y)$ as a copy of Y. A Banach lattice E includes a **lattice copy** of another Banach lattice F (or that F is **lattice embeddable** in E) if F is embedded in E via a lattice isomorphism T. That is, in addition to $T : F \to E$ being a linear embedding, it also satisfies $|Ty| = T|y|$ for all $y \in F$. In this case, $T(F)$ is a closed Riesz subspace of E, which can be viewed as a copy of the Banach lattice F.

It is a remarkable fact that whether certain sequence spaces are embeddable in a Banach lattice characterize its lattice and topological structure. The sequence spaces are, in fact, the building blocks of Banach spaces and Banach lattices. We state a few results of this nature below. For a more detailed account see [12, Section 14] or [96].

16.37 Theorem   *The sequence space c_0 is embeddable in a Banach lattice if and only it is lattice embeddable.*

Proof: See [12, Theorem 14.13, p. 227].   ∎

Recall that a sequence $\{x_n\}$ in a Riesz space is called (pairwise) **disjoint** if $|x_n| \wedge |x_m| = 0$ for each $n \neq m$.

16.38 Theorem   *The Banach lattice c_0 is lattice embeddable in a Banach lattice E if and only if there exists a disjoint sequence in E^+ such that:*

1. *$\{x_n\}$ is not norm convergent to zero; and*

2. *$\{x_n\}$ has a norm bounded sequence of partial sums.*

In this case, the linear operator $T: c_0 \to E$, defined by

$$T(\alpha_1, \alpha_2, \ldots) = \sum_{n=1}^{\infty} \alpha_n x_n,$$

is a lattice embedding.

Proof: See [12, Theorem 14.3, p. 220]. ∎

The Banach lattice c_0 can be embedded in any infinite dimensional AM-space.

16.39 Theorem *Every infinite dimensional AM-space includes a lattice copy of the Banach lattice c_0.*

Proof: Let E be an infinite dimensional AM-space. It is well known that if in an Archimedean Riesz space every subset of nonzero pairwise disjoint elements is finite, then the Riesz space is finite dimensional; see [235, Theorem 26.10, p. 152]. So E includes a disjoint sequence $\{x_n\}$ of nonzero vectors. Replacing each x_n by $\frac{|x_n|}{\|x_n\|}$, we can assume that $x_n \geqslant 0$ and $\|x_n\| = 1$ for each n.

Now note that for scalars $\alpha_1, \ldots, \alpha_n$, from the identity $\left|\sum_{i=1}^{n} \alpha_i x_i\right| = \sum_{i=1}^{n} |\alpha_i| x_i$ (see Theorem 8.12), we get

$$\left\|\sum_{i=1}^{n} \alpha_i x_i\right\| = \left\|\sum_{i=1}^{n} |\alpha_i| x_i\right\| = \max\{|\alpha_i| : i = 1, \ldots, n\}.$$

So if $\alpha = (\alpha_1, \alpha_2, \ldots)$ belongs to c_0, then the series $\sum_{i=1}^{\infty} \alpha_i x_i$ is norm convergent, $\left\|\sum_{i=1}^{\infty} \alpha_i x_i\right\| = \|\alpha\|_\infty$, and $\left|\sum_{i=1}^{\infty} \alpha_i x_i\right| = \sum_{i=1}^{\infty} |\alpha_i| x_i$. Therefore, the mapping $T: c_0 \to E$ defined by

$$T(\alpha_1, \alpha_2, \ldots) = \sum_{i=1}^{\infty} \alpha_i x_i,$$

is a lattice isometry, so c_0 is lattice embedded in E. ∎

Let us illustrate Theorem 16.39 when $E = C[0, 1]$. For each n choose a function $0 \leqslant f_n \in C[0, 1]$ such that $\|f_n\|_\infty = 1$ and $f_n(t) = 0$ for every $t \notin [\frac{1}{n+1}, \frac{1}{n}]$. Then the linear operator $T: c_0 \to C[0, 1]$, defined by

$$T(\alpha_1, \alpha_2, \ldots) = \sum_{n=1}^{\infty} \alpha_n f_n,$$

is a lattice embedding.

Regarding ℓ_1, we have the following embedding result.

16.40 Theorem *The Banach lattice ℓ_1 is lattice embeddable into another Banach lattice E if and only if E^+ includes a norm bounded disjoint sequence that does not converge weakly to zero.*

Proof: See [12, Theorem 14.21, p. 238].   ∎

The Banach lattice ℓ_1 can always be lattice embedded in an infinite dimensional AL-space.

16.41 Theorem   *Every infinite dimensional AL-space includes a lattice copy of the Banach lattice ℓ_1.*

Proof: Let E be an infinite dimensional AL-space. As in the proof of Theorem 16.39, we know that E admits a pairwise disjoint sequence $\{x_n\}$ of positive unit vectors. Again from Theorem 8.12, we have $\left|\sum_{i=1}^{n} \alpha_i x_i\right| = \sum_{i=1}^{n} |\alpha_i| x_i$, so

$$\left\|\sum_{i=1}^{n} \alpha_i x_i\right\| = \sum_{i=1}^{n} |\alpha_i| \cdot \|x_i\| = \sum_{i=1}^{n} |\alpha_i|.$$

Thus the series $\sum_{i=1}^{\infty} \alpha_i x_i$ converges in norm for each $\alpha = (\alpha_1, \alpha_2, \ldots) \in \ell_1$ and that $\left|\sum_{i=1}^{\infty} \alpha_i x_i\right| = \sum_{i=1}^{\infty} |\alpha_i| x_i$. It follows that the linear operator $T: \ell_1 \to E$, defined by $T(\alpha) = \sum_{n=1}^{\infty} \alpha_n x_n$, is a lattice isometry, so E includes a lattice copy of ℓ_1.   ∎

16.42 Theorem   *An order complete Banach lattice includes a lattice copy of ℓ_∞ if and only if it does not have order continuous norm.*

Proof: See [12, Theorem 14.4, p. 220].   ∎

Let us demonstrate that ℓ_∞ is lattice embeddable in $L_\infty[0, 1]$. For simplicity, let us write $L_\infty = L_\infty[0, 1]$ and $L_1 = L_1[0, 1]$. For each n, let $f_n = \chi_{(\frac{1}{n+1}, \frac{1}{n}]}$. Clearly, $\{f_n\}$ is a disjoint sequence in L_∞. Now define the linear operator $T: \ell_\infty \to L_\infty$ by

$$T(\alpha_1, \alpha_2, \ldots) = \sum_{n=1}^{\infty} \alpha_n f_n,$$

where now the series converges pointwise—and not in the sup norm. It is a routine matter to verify that T is a lattice isometry (into), so L_∞ includes a lattice copy of ℓ_∞. Also, note that $T(e) = \mathbf{1}$, the constant function one.

The same arguments can be applied to $L_\infty(A)$ for an arbitrary measurable subset A of $\mathbb{R}$ of positive Lebesgue measure. Indeed, choose first a sequence $\{A_n\}$ of pairwise disjoint measurable subsets each of positive measure such that $\bigcup_{n=1}^{\infty} A_n = A$; this is possible since the Lebesgue measure is nonatomic. Then, as above, define the linear operator $T: \ell_\infty \to L_\infty(A)$ by

$$T(\alpha_1, \alpha_2, \ldots) = \sum_{n=1}^{\infty} \alpha_n \chi_{A_n},$$

where the series converges pointwise, and note that T is a lattice isometry satisfying $T(e) = \chi_A$.

These types of embeddings allow us to study the band of singular functionals on L_∞. The next result tells us that (unlike ℓ_∞) there is an abundance of singular functionals on L_∞.

16.43 Theorem   *The band L_1^d of all singular functionals on L_∞ separates the points of L_∞.*

Proof: Let $0 < f \in L_\infty$. It suffices to show that there is some $\theta > 0$ in L_1^d such that $\theta(f) > 0$. Once this is established, then the proof can be completed as follows. Fix $g \in L_\infty$ with $g \neq 0$. Then, either $g^+ > 0$ or $g^- > 0$; we can assume $g^+ > 0$. Fix $0 < \psi_1 \in L_1^d$ such that $\psi_1(g^+) > 0$. Now viewing g as an order bounded linear functional on L_1^d and arguing as in the proof of Theorem 8.34, we see that

$$0 < \psi_1(g^+) = g^+(\psi_1) = \sup\{\psi(g) : \psi \in L_1^d \text{ and } 0 \leqslant \psi \leqslant \psi_1\}.$$

Hence, $\psi(g) \neq 0$ must hold for some $\psi \in L_1^d$.

To this end, pick some $\varepsilon > 0$ and a measurable set $A \subset [0, 1]$ of positive Lebesgue measure such that $f \geqslant \varepsilon \chi_A$. Note that $L_\infty = L_\infty(A) \oplus L_\infty(A^c)$, where $A^c = [0, 1] \setminus A$. Also, fix a pairwise disjoint sequence $\{A_n\}$ of measurable sets each of positive measure such that $\bigcup_{n=1}^\infty A_n = A$. As before, the linear operator $T : \ell_\infty \to L_\infty(A)$ defined by

$$T(\alpha_1, \alpha_2, \ldots) = \sum_{n=1}^\infty \alpha_n \chi_{A_n},$$

where again the convergence of the series is pointwise, is a lattice isometry (into) satisfying $T(e) = \chi_A$.

Next, pick some $0 < \theta \in \ell_1^d$, and consider it as a positive linear functional on the copy $T(\ell_\infty)$ in $L_\infty(A)$. Since $T(e) = \chi_A$, the Riesz subspace $T(\ell_\infty)$ majorizes $L_\infty(A)$. So θ has a positive linear extension $\hat{\theta}$ to all of $L_\infty(A)$; see Theorem 8.32. By assigning zero values on $L_\infty(A^c)$, we can assume that $0 \leqslant \hat{\theta} \in L'_\infty$.

We claim that $\hat{\theta} \in L_1^d$. To see this, let $0 \leqslant g \in L_1$. Then g acts on $T(\ell_\infty)$ as the ℓ_1-sequence $(\int_{A_1} g(x)\,dx, \int_{A_2} g(x)\,dx, \ldots)$. Consequently, from $g \wedge \theta(e) = 0$, we infer that $g \wedge \hat{\theta}(1) = 0$. (Why?) That is, $\hat{\theta} \perp g$ for each $g \in L_1$, which means that $\hat{\theta} \in L_1^d$. To complete the proof, note that $\hat{\theta}(f) \geqslant \hat{\theta}(\varepsilon \chi_A) = \varepsilon > 0$.   ∎

This brings up the following point. You may have had the impression that all Riesz pairs are pleasant and harmless objects to spend some time with. The Riesz pair $\langle L_\infty, L_1^d \rangle$ should dispel any such notions. By Theorem 9.34, we know that $L'_\infty = L_1 \oplus L_1^d$, where L_1 coincides with the order continuous dual of L_∞ and L_1^d is the band of all singular functionals on L_∞. Consequently, the only order continuous functional on L_∞ that belongs to L_1^d is the zero functional. The Riesz pair $\langle L_\infty, L_1^d \rangle$ fails to possess almost any reasonable properties. For instance, no order interval is weakly compact. It is a freak Riesz pair that should be avoided at any cost! Unless, of course, you are in search of counterexamples.

16.10   Banach–Mazur limits and invariant measures

Banach–Mazur limits are special linear extensions of the notion of limit to sequences that do not converge.

16.44 Definition   *A positive linear functional* $\Lambda \colon \ell_\infty \to \mathbb{R}$ *is a **Banach–Mazur limit** if*

- $\Lambda(e) = 1$, *where* $e = (1, 1, 1, \ldots)$, *and*

- $\Lambda(x_1, x_2, \ldots) = \Lambda(x_2, x_3, \ldots)$ *for each* $(x_1, x_2, \ldots) \in \ell_\infty$.

From $\|\Lambda\| = \Lambda(e)$, we see that every Banach–Mazur limit has norm one.

16.45 Lemma   *If* Λ *is a Banach–Mazur limit, then*

$$\liminf_{n \to \infty} x_n \leqslant \Lambda(x) \leqslant \limsup_{n \to \infty} x_n$$

for each $x = (x_1, x_2, \ldots) \in \ell_\infty$. *In particular,* $\Lambda(x) = \lim_{n \to \infty} x_n$ *for each* $x \in c$ *(so every Banach–Mazur limit is an extension of the limit functional).*

Proof: Let $x = (x_1, x_2, \ldots) \in \ell_\infty$ and put $s = \limsup_{n \to \infty} x_n$. Fix $\varepsilon > 0$. Then there is some k such that $x_n \leqslant s + \varepsilon$ for all $n \geqslant k$. That is, $(x_k, x_{k+1}, x_{k+2}, \ldots) \leqslant (s + \varepsilon)e$. Hence, if Λ is any Banach–Mazur limit, then

$$\Lambda(x) = \Lambda(x_k, x_{k+1}, \ldots) \leqslant (s + \varepsilon)\Lambda(e) = s + \varepsilon.$$

Since $\varepsilon > 0$ is arbitrary, $\Lambda(x) \leqslant s$. Similarly, $\liminf_{n \to \infty} x_n \leqslant \Lambda(x)$.   ∎

For each $x = (x_1, x_2, \ldots) \in \ell_\infty$ fix some $M_x > 0$ satisfying $x_i \leqslant M_x$ for each i and let

$$a_n(x) = \sup_{i \in \mathbb{N}} \frac{1}{n} \sum_{j=0}^{n-1} x_{i+j}.$$

If $n > m$, write $n = km + r$ with $0 \leqslant r < m$ and note that

$$\sum_{j=0}^{n-1} x_{i+j} \leqslant \frac{km}{n} a_m(x) + \frac{r}{n} M_x \leqslant a_m(x) + \frac{m}{n} M_x.$$

Therefore, $a_n(x) \leqslant a_m(x) + \frac{m}{n} M_x$ for each $n > m$, from which it follows that $\limsup_{n \to \infty} a_n(x) \leqslant a_m(x)$ for each m and each x. This implies that $\lim_{n \to \infty} a_n(x)$ exists (in $\mathbb{R}$) for each $x \in \ell_\infty$ and $\lim_{n \to \infty} a_n(x) = \inf_{n \in \mathbb{N}} a_n(x)$. Define

$$\mathcal{L}(x) = \lim_{n \to \infty} a_n(x).$$

Clearly, $\mathcal{L} \colon \ell_\infty \to \mathbb{R}$ is a sublinear mapping satisfying $\mathcal{L}(x) = \lim_{n \to \infty} x_n$ for each $x \in c$. (Why?) Also, notice that $x \leqslant 0$ implies $\mathcal{L}(x) \leqslant 0$.

16.46 Lemma   *A linear functional* $\Lambda\colon \ell_\infty \to \mathbb{R}$ *is a Banach–Mazur limit if and only if* $\Lambda(x) \leqslant \mathcal{L}(x)$ *for each* $x \in \ell_\infty$.

Proof: Let $\Lambda\colon \ell_\infty \to \mathbb{R}$ be a linear functional. Assume first that $\Lambda(x) \leqslant \mathcal{L}(x)$ for each $x \in \ell_\infty$. From $\Lambda(e) \leqslant \mathcal{L}(e) = 1$ and $-\Lambda(e) = \Lambda(-e) \leqslant \mathcal{L}(-e) = -1$, we see that $\Lambda(e) = 1$. Now notice that if $x \geqslant 0$, then from $-\Lambda(x) = \Lambda(-x) \leqslant \mathcal{L}(-x) \leqslant 0$, it follows that $\mathcal{L}(x) \geqslant 0$, that is, Λ is a positive linear functional.

Now let $x = (x_1, x_2, \ldots) \in \ell_\infty$ and put $y = (x_1, x_2, \ldots) - (x_2, x_3, \ldots)$. An easy computation shows that $\mathcal{L}(y) = \mathcal{L}(-y) = 0$, and from this we infer that $\Lambda(y) = 0$. That is, $\Lambda(x_1, x_2, \ldots) = \Lambda(x_2, x_3, \ldots)$. So Λ is a Banach–Mazur limit.

Conversely, assume that Λ is a Banach–Mazur limit and let $x = (x_1, x_2, \ldots)$ belong to ℓ_∞. Summing up the n equations below

$$\Lambda(x) = \Lambda(x_1, x_2, x_3, \ldots)$$
$$\Lambda(x) = \Lambda(x_2, x_3, x_4, \ldots)$$
$$\Lambda(x) = \Lambda(x_3, x_4, x_5, \ldots)$$
$$\vdots$$
$$\Lambda(x) = \Lambda(x_n, x_{n+1}, x_{n+2}, \ldots),$$

we easily obtain

$$\Lambda(x) = \Lambda\Big(\frac{1}{n}\sum_{i=0}^{n-1} x_{1+i}, \frac{1}{n}\sum_{i=0}^{n-1} x_{2+i}, \ldots\Big) \leqslant \Lambda(a_n(x), a_n(x), \ldots)$$
$$= a_n(x)\Lambda(e) = a_n(x)$$

for each n. Letting $n \to \infty$, we get $\Lambda(x) \leqslant \lim_{n\to\infty} a_n(x) = \mathcal{L}(x)$ for $x \in \ell_\infty$.   ∎

We are now able to prove the existence of Banach–Mazur limits.[6]

16.47 Theorem   *Banach–Mazur limits exist.*

Proof: Consider the limit functional $\mathrm{Lim}\colon c \to \mathbb{R}$ defined by

$$\mathrm{Lim}(x) = \lim_{n\to\infty} x_n = x_\infty.$$

Clearly, $\mathrm{Lim}(x) = \mathcal{L}(x)$ for each $x \in c$. By the Hahn–Banach Theorem 8.30, Lim has an extension Λ to all of ℓ_∞, satisfying $\Lambda(x) \leqslant \mathcal{L}(x)$ for each $x \in \ell_\infty$. By Lemma 16.46 any such extension Λ is a Banach–Mazur limit.   ∎

The existence of Banach–Mazur limits can be used to prove the existence of invariant measures. Let X be a topological space and let $\xi\colon X \to X$ be a continuous mapping. Such a mapping is sometimes called a **flow**. A Borel measure μ on X is called ξ-**invariant** if $\mu(B) = \mu(\xi^{-1}(B))$ for each Borel set B.

---

[6] Another way of proving the existence of Banach–Mazur limits is indicated in Example 19.16.

16.48 Theorem *Every continuous function on a compact metrizable topological space has an invariant measure.*

Proof: Let $\xi\colon X \to X$ be a continuous mapping on a compact metrizable topological space. As usual, ξ^n denotes the composition of ξ with itself taken n-times. Let Λ be a Banach limit and fix some $x \in X$.

Define a positive linear functional $\theta\colon C(X) \to \mathbb{R}$ by

$$\theta(f) = \Lambda(f(x), f(\xi(x)), f(\xi^2(x)), \ldots)$$
$$= \Lambda(f(x), f \circ \xi(x), f \circ \xi^2(x), \ldots).$$

By the Riesz Representation Theorem 14.12 there exists a unique regular Borel measure μ on X satisfying $\theta(f) = \int f \, d\mu$ for each f in $C(X)$. From

$$\Lambda(f(x), f \circ \xi(x), f \circ \xi^2(x), \ldots) = \Lambda(f \circ \xi(x), f \circ \xi^2(x), f \circ \xi^3(x), \ldots),$$

we see that $\int f \, d\mu = \int f \circ \xi \, d\mu$ for each $f \in C(X)$. Invoking Theorem 13.46, we see that $\int f \, d\mu = \int f \, d\mu\xi^{-1}$ for each $f \in C(X)$. Since (by Theorem 12.7) $\mu\xi^{-1}$ is a regular Borel measure, the uniqueness property of the Riesz Representation Theorem 14.12 guarantees $\mu = \mu\xi^{-1}$. That is, μ is a ξ-invariant measure.[7] ∎

16.11   Sequences of vector spaces

We can form sequence spaces whose elements have components taken from arbitrary vector spaces. To discuss this general framework, let $\{X_1, X_2, \ldots\}$ be a sequence of (real) vector spaces. The **sum** $X_1 \oplus X_2 \oplus \cdots$ of the sequence $\{X_1, X_2, \ldots\}$ is simply another name for the Cartesian product $X_1 \times X_2 \times \cdots$,[8] that is,

$$X_1 \oplus X_2 \oplus \cdots = X_1 \times X_2 \times \cdots$$
$$= \{x = (x_1, x_2, \ldots) : x_i \in X_i \text{ for all } i\}.$$

With the pointwise operations the sum $X_1 \oplus X_2 \oplus \cdots$ is a vector space. If each X_i is also a Riesz space, then with the pointwise lattice operations $X_1 \oplus X_2 \oplus \cdots$ is also

---

[7] Another elegant proof, due to M. G. Krein, goes like this: Consider the positive operator $T\colon C(X) \to C(X)$ defined by $T(f) = f \circ \xi$. Clearly, $T(\mathbf{1}) = \mathbf{1}$, where $\mathbf{1}$ denotes the constant function one. The adjoint operator $T'\colon ca(X) \to ca(X)$ satisfies

$$\langle f, T'\mu \rangle = \langle Tf, \mu \rangle = \int f \circ \xi \, d\mu; \quad f \in C(X), \ \mu \in ca(X),$$

and is continuous for the w^*-topology. It follows that T' maps $\mathcal{P}(X)$ into $\mathcal{P}(X)$. Since $\mathcal{P}(X)$ is a convex w^*-compact subset of $ca(X)$ (see Theorem 15.11), the Brouwer–Schauder–Tychonoff Fixed Point Theorem 17.56 implies $T'\mu = \mu$ for some $\mu \in \mathcal{P}(X)$. Thus we have $\int f \, d\mu = \int f \circ \xi \, d\mu$ for each $f \in C(X)$ and, as above, $\mu = \mu\xi^{-1}$.

[8] A fair question at this point is, Why do we introduce a new notation for the Cartesian product? The answer is simple: Tradition.

a Riesz space. In case each X_i is a normed space, then we can define sequence spaces that are analogues of the c_0- and ℓ_p-spaces.

- The **c_0-sum** of the sequence $\{X_1, X_2, \ldots\}$ of normed spaces:

$$(X_1 \oplus X_2 \oplus \cdots)_0 = \{x = (x_1, x_2, \ldots) : \lim_{n \to \infty} \|x_n\| = 0\}.$$

- The **ℓ_∞-sum** of the sequence $\{X_1, X_2, \ldots\}$ of normed spaces:

$$(X_1 \oplus X_2 \oplus \cdots)_\infty = \{x = (x_1, x_2, \ldots) : \|x\|_\infty = \sup_{n \in \mathbb{N}} \|x_n\| < \infty\}.$$

- The **ℓ_p-sum** (where $0 < p < \infty$):

$$(X_1 \oplus X_2 \oplus \cdots)_p = \Big\{x = (x_1, x_2, \ldots) : \|x\|_p = \Big(\sum_{n=1}^{\infty} \|x_n\|^p\Big)^{\frac{1}{p}} < \infty\Big\}.$$

If each X_i is a Banach space, then a direct verification shows that the above three spaces are all Banach spaces. They are also Banach lattices if all the X_is are Banach lattices.

Now assume that each X_i is a normed space and let $1 < p, q < \infty$ be conjugate exponents, that is, $\frac{1}{p} + \frac{1}{q} = 1$. If $x' = (x'_1, x'_2, \ldots) \in (X'_1 \oplus X'_2 \oplus \cdots)_q$, then an easy computation shows that the formula

$$\langle x, x' \rangle = \sum_{n=1}^{\infty} \langle x_n, x'_n \rangle, \qquad x = (x_1, x_2, \ldots) \in (X_1 \oplus X_2 \oplus \cdots)_p,$$

defines a continuous linear functional on $(X_1 \oplus X_2 \oplus \cdots)_p$ whose norm equals $\|x'\|_q$. As a matter of fact, we have the following important result.

16.49 Theorem *Let $\{X_1, X_2, \ldots\}$ be a sequence of Banach spaces, and let p, q satisfy $1 < p, q < \infty$ and $\frac{1}{p} + \frac{1}{q} = 1$. Then the norm dual of the ℓ_p-sum is the ℓ_q-sum of the sequence of norm duals. That is,*

$$(X_1 \oplus X_2 \oplus \cdots)'_p = (X'_1 \oplus X'_2 \oplus \cdots)_q,$$

where the equality holds under the duality

$$\langle x, x' \rangle = \sum_{n=1}^{\infty} \langle x_n, x'_n \rangle,$$

$x = (x_1, x_2, \ldots) \in (X_1 \oplus X_2 \oplus \cdots)_p; \ x' = (x'_1, x'_2, \ldots) \in (X'_1 \oplus X'_2 \oplus \cdots)_q.$

Proof: We shall establish the result when $1 < p, q < \infty$ satisfy $\frac{1}{p} + \frac{1}{q} = 1$, and leave the other cases for the reader. To this end, it is enough to show and the mapping $x' \mapsto \varphi_{x'}$ where $\varphi_{x'}(x_1, x_2, \ldots) = \sum_{n=1}^{\infty} x'_n(x_n)$, is an isometry from $(X'_1 \oplus X'_2 \oplus \cdots)_q$ onto $(X_1 \oplus X_2 \oplus \cdots)'_p$.

Fix $x' = (x'_1, x'_2, \ldots) \in (X'_1 \oplus X'_2 \oplus \cdots)_q$. Then for each $x = (x_1, x_2, \ldots)$ in $(X_1 \oplus X_2 \oplus \cdots)_p$ we have $|x'_n(x_n)| \leqslant \|x'_n\| \cdot \|x_n\|$, and so

$$\left| \sum_{n=1}^{k} x'_n(x_n) \right| \leqslant \sum_{n=1}^{k} \|x'_n\| \cdot \|x_n\| \leqslant \left[\sum_{n=1}^{\infty} \|x'_n\|^q \right]^{\frac{1}{q}} \cdot \left[\sum_{n=1}^{\infty} \|x_n\|^p \right]^{\frac{1}{p}} = \|x'\|_q \cdot \|x\|_p$$

holds for all k. Thus, the formula $\varphi_{x'}(x) = \sum_{n=1}^{\infty} x'_n(x_n)$ defines a continuous linear functional on $(X_1 \oplus X_2 \oplus \cdots)_p$ satisfying

$$\left\| \varphi_{x'} \right\| \leqslant \left\| x' \right\|_q \tag{$\star$}$$

for all $x' \in (X'_1 \oplus X'_2 \oplus \cdots)_q$. Clearly, $x' \mapsto \varphi_{x'}$ is a linear operator.

Now let $\varphi \in (X_1 \oplus X_2 \oplus \cdots)'_p$. If $x'_n \colon X_n \to \mathbb{R}$ is defined by

$$x'_n(x_n) = \varphi(0, \ldots, 0, x_n, 0, 0, \ldots),$$

then $x'_n \in X'_n$, and $\varphi(x_1, x_2, \ldots) = \sum_{n=1}^{\infty} x'_n(x_n)$ for all $(x_1, x_2, \ldots) \in (X_1 \oplus X_2 \oplus \cdots)_p$. Fix $0 < \varepsilon < 1$. For each n pick some $y_n \in X_n$ with $\|y_n\| = 1$ and $x'_n(y_n) \geqslant \varepsilon \|x'_n\|$. Put $z_n = \|x'_n\|^{q-1} y_n$, and note that for each k we have

$$\varepsilon \sum_{n=1}^{k} \|x'_n\|^q = \sum_{n=1}^{k} \|x'_n\|^{q-1} \varepsilon \|x'_n\| \leqslant \sum_{n=1}^{k} \|x'_n\|^{q-1} x'_n(y_n)$$

$$= \varphi(z_1, \ldots, z_k, 0, 0, \ldots) \leqslant \|\varphi\| \cdot \left[\sum_{n=1}^{k} \|z_n\|^p \right]^{\frac{1}{p}}$$

$$= \|\varphi\| \cdot \left[\sum_{n=1}^{k} \|x'_n\|^q \right]^{\frac{1}{p}}.$$

Therefore, $\varepsilon (\sum_{n=1}^{k} \|x'_n\|^q)^{\frac{1}{q}} \leqslant \|\varphi\|$ holds for all k, and each $0 < \varepsilon < 1$. This implies that $x' = (x'_1, x'_2, \ldots) \in (X'_1 \oplus X'_2 \oplus \cdots)_q$, $\varphi = \varphi_{x'}$, and $\|x'\|_q \leqslant \|\varphi_{x'}\|$. By $(\star)$ we see that $\|\varphi_{x'}\| = \|x'\|_q$, and thus $x' \mapsto \varphi_{x'}$ is an onto linear isometry. Finally, note that if each X_n is a Banach lattice, then $x' \mapsto \varphi_{x'}$ is an onto lattice isometry. $\blacksquare$

In a similar fashion one can prove:

$$(X_1 \oplus X_2 \oplus \cdots)'_1 = (X'_1 \oplus X'_2 \oplus \cdots)_\infty$$

and

$$(X_1 \oplus X_2 \oplus \cdots)'_0 = (X'_1 \oplus X'_2 \oplus \cdots)_1.$$

Chapter 17

Correspondences

A *correspondence* is a set-valued function. That is, a correspondence assigns to each point in one set a set of points in a possibly different set. As such, it can simply be identified with a subset of the Cartesian product of the two sets. It may seem a bit silly to dedicate two chapters to such a topic, but correspondences arise naturally in many applications. For instance, the budget correspondence in economic theory associates the set of affordable consumption plans to each price-income combination; the excess demand correspondence is a useful tool in studying economic equilibria; and the best-reply correspondence is the key to analyzing noncooperative games. The theory of "differential inclusions" deals with set-valued differential equations and plays an important role in control theory.

The biggest difference between functions and correspondences has to do with the definition of an inverse image. The inverse image of a set A under a function f is the set $\{x : f(x) \in A\}$. For a correspondence φ, there are two reasonable generalizations, the *upper inverse* of A, which is $\{x : \varphi(x) \subset A\}$, and the *lower inverse* of A, namely $\{x : \varphi(x) \cap A \neq \varnothing\}$. When φ is singleton-valued, both definitions reduce to the inverse of A treating φ as a function.

Having two distinct notions of the inverse leads to (at least) two definitions of continuity. As a result, the terminology has not been fully standardized. We adopt the following definitions. A correspondence is *upper hemicontinuous* if the upper inverse of any open set is open, and *lower hemicontinuous* if the lower inverse of any open set is open. The Closed Graph Theorem 17.11 for correspondences states that a closed-valued correspondence into a compact Hausdorff space is upper hemicontinuous if and only if its graph is closed. Upper hemicontinuous correspondences with compact values mimic the properties of continuous functions reasonably well. For instance the image of a compact set under such correspondences is compact (Lemma 17.8), and products preserve the property (Theorem 17.28).

One of the most useful results involving correspondences is the Maximum Theorem 17.31. This theorem gives sufficient conditions for the set of solutions of a parametric constrained maximization problem to be upper hemicontinuous, and for the optimal value function to be continuous. This theorem is the key result in control theory, equilibrium theory, and game theory. The essential requirements are that the constraint correspondence be both upper and lower hemicontinuous in the parameters and that the objective function be continuous.

We also present a useful result (Theorem 17.48) on maximal elements of a possibly intransitive and incomplete binary relation. This result is the dual form of K. Fan's extension (Theorem 17.46) of the KKM Lemma. It is the key to a number of useful fixed point theorems. A *fixed point* of a correspondence φ is a point x satisfying $x \in \varphi(x)$. The noted Kakutani–Fan–Glicksberg Fixed Point Theorem 17.55 asserts that an upper hemicontinuous correspondence with nonempty compact convex values from a compact convex subset of a locally convex Hausdorff space into itself has a fixed point. The Brouwer–Schauder–Tychonoff Fixed Point Theorem 17.56 is a special case of this result. These theorems are the fundamental tools of supply and demand analysis, and of the analysis of noncooperative equilibria in games.

In addition, we present the Michael Selection Theorem 17.66, which asserts that a lower hemicontinuous correspondence with nonempty closed convex values into a Banach space admits a continuous *selector*.

The theory of correspondences was first codified by C. Berge [37]. Many of the results of this chapter may be found in K. C. Border [56] for the special case of Euclidean spaces, and more general results may be found in W. Hildenbrand [158], E. Klein and A. C. Thompson [209], and J. C. Moore [253]. More esoteric works include the monographs by J.-P. Aubin and H. Frankowska [24], C. Castaing and M. Valadier [75], J. E. Jayne and C. A. Rogers [183], and I. Kluvánek and G. Knowles [210]. The book by J.-P. Aubin and A. Cellina [22] is a good reference for the theory of differential inclusions. S. Hu and N. S. Papageorgiou [172] have produced an encyclopedic treatment of the whole area.

17.1 Basic definitions

We start with a formal definition of correspondences and related terms.

17.1 Definition   *A **correspondence** φ from a set X to a set Y assigns to each x in X a subset $\varphi(x)$ of Y. We prefer to think of φ as a "multivalued function" from X to Y rather than as a function from X to the power set 2^Y of Y.*

*The terms **multifunction** or **set-valued function** are sometimes used to mean a correspondence. We write $\varphi \colon X \twoheadrightarrow Y$ to distinguish a correspondence from a function from X to Y.*

Let $\varphi \colon X \twoheadrightarrow Y$ be a correspondence. As with functions, we refer to X as the **domain** of φ, and Y as the **range space** (or **codomain**). The **image** of a set $A \subset X$ under φ is the set

$$\varphi(A) = \bigcup_{x \in A} \varphi(x).$$

The **range** of φ is the image of X. We can identify φ with its **graph** $\mathrm{Gr}\,\varphi$, the subset of $X \times Y$ given by

$$\mathrm{Gr}\,\varphi = \{(x, y) \in X \times Y : y \in \varphi(x)\}.$$

If $\varphi(x)$ is always a singleton, then its graph also defines a function from X to Y. Conversely, every function $f \colon X \to Y$ defines a correspondence whose values are singletons. While f and φ are identical twins, they are not the same object. Nevertheless, we sometimes identify a singleton-valued correspondence with its function counterpart.

Just as functions have inverses, so do correspondences. Indeed, each correspondence $\varphi \colon X \twoheadrightarrow Y$ has two natural inverses:

- the **upper inverse** φ^u (also called the **strong inverse**) of a subset A of Y is defined by
$$\varphi^u(A) = \{x \in X : \varphi(x) \subset A\}.$$

- the **lower inverse** φ^ℓ (also called the **weak inverse**) defined by
$$\varphi^\ell(A) = \{x \in X : \varphi(x) \cap A \neq \varnothing\}.$$

Note that if φ is singleton-valued (that is, if φ is a function), then both $\varphi^u(A)$ and $\varphi^\ell(A)$ coincide with the inverse of A viewing φ as a function.

Here are a few handy identities that we use frequently. You should verify them as an exercise.

$$\varphi^u(A) = X \setminus \varphi^\ell(Y \setminus A) = [\varphi^\ell(A^c)]^c,$$
$$\varphi^\ell(A) = X \setminus \varphi^u(Y \setminus A) = [\varphi^u(A^c)]^c,$$

which relate upper and lower inverses, and the identities

$$\varphi^u\Big(\bigcap_{i \in I} A_i\Big) = \bigcap_{i \in I} \varphi^u(A_i) \qquad \text{and} \qquad \varphi^\ell\Big(\bigcup_{i \in I} A_i\Big) = \bigcup_{i \in I} \varphi^\ell(A_i),$$

relating inverses of unions and intersections. If φ has only nonempty values, then it is clear that $\varphi^u(A) \subset \varphi^\ell(A)$ for all subsets A of Y.

Every correspondence has a natural inverse correspondence. If $\varphi \colon X \twoheadrightarrow Y$ is a correspondence, then the **inverse correspondence** of φ is the correspondence $\varphi^{-1} \colon Y \twoheadrightarrow X$ defined by

$$\varphi^{-1}(y) = \{x \in X : y \in \varphi(x)\} = \varphi^\ell(\{y\}).$$

The set $\varphi^{-1}(y)$ is also called the **lower section** of φ at y.

We frequently wish to denote restrictions on the values of a correspondence. For instance, if Y is a topological space, then we say that the correspondence $\varphi \colon X \twoheadrightarrow Y$ is **closed-valued** or **has closed values** if $\varphi(x)$ is a closed set for each x. The terms "open-valued," "compact-valued," "convex-valued," etc., are defined similarly.

17.2   Continuity of correspondences

Recall that a **neighborhood of a set** A is any set B for which there is an open set V satisfying $A \subset V \subset B$. Any open set V that satisfies $A \subset V$ is called an **open neighborhood** of A.

17.2 Definition   *A correspondence $\varphi \colon X \twoheadrightarrow Y$ between topological spaces is:*

- **upper hemicontinuous** *at the point x if for every neighborhood U of $\varphi(x)$, there is a neighborhood V of x such that $z \in V$ implies $\varphi(z) \subset U$ (equivalently, the upper inverse image $\varphi^{\mathrm{u}}(U)$ is a neighborhood of x in X). As with functions, we say φ is upper hemicontinuous on X, abbreviated **uhc**, if it is upper hemicontinuous at every point of X.*

- **lower hemicontinuous** *at x if for every open set U that meets $\varphi(x)$ (that is, for which $\varphi(x) \cap U \neq \varnothing$) there is a neighborhood V of x such that $z \in V$ implies $\varphi(z) \cap U \neq \varnothing$ (equivalently, the lower inverse image $\varphi^{\ell}(U)$ is a neighborhood of x). As above, φ is lower hemicontinuous on X, abbreviated **lhc**, if it is lower hemicontinuous at each point.*

- **continuous** *at x if it is both upper and lower hemicontinuous at x. It is continuous if it is continuous at each point.*

> *Throughout this chapter, if we assert that a function or a correspondence possesses some continuity property, it is tacitly assumed that its domain and range spaces are topological spaces.*

Some authors require additional properties as part of the definition of upper hemicontinuity. For instance, it is common, but not universal, to require that the correspondence have compact values, or nonempty values.[1] These additional requirements do not seem conceptually related to continuity, so we do not make them part of the definition. A few texts use the term *semicontinuity* in place of hemicontinuity. Several authors use the closed graph property as the definition of upper hemicontinuity, which is fine as long as the range space is compact, see Theorem 17.11. When the range space is not compact, this can lead to misunderstandings. For instance, the composition of upper hemicontinuous correspondences (as we define hemicontinuity) is upper hemicontinuous (Theorem 17.23), but the composition of correspondences with closed graph need not have closed graph (Example 17.24).

---

[1] J. C. Moore [252] identifies five slightly different definitions of upper hemicontinuity in use by economists, and points out some of the differences for compositions, etc. T. Ichiishi [175] and E. Klein and A. C. Thompson [209] also give other notions of continuity.

17.3 Example **(Hemicontinuity)**   The following correspondences illustrate the definitions. Define $\varphi, \psi \colon [0, 1] \twoheadrightarrow [0, 1]$ by

$$\varphi(x) = \begin{cases} \{0\} & \text{if } x < 1, \\ [0, 1] & \text{if } x = 1, \end{cases} \quad \text{and} \quad \psi(x) = \begin{cases} [0, 1] & \text{if } x < 1, \\ \{0\} & \text{if } x = 1. \end{cases}$$

Then φ is upper hemicontinuous everywhere, but it is not lower hemicontinuous at the point 1. On the other hand, ψ is lower hemicontinuous everywhere, but it is not upper hemicontinuous at 1. Finally, the correspondence $\gamma \colon [0, 1] \twoheadrightarrow [0, 1]$ given by $\gamma(x) = [0, x]$ is continuous.   ∎

As with functions, perhaps the simplest continuous correspondences are the constant ones. That is, correspondences $\varphi \colon X \twoheadrightarrow Y$ between topological spaces that satisfy $\varphi(x) = C$ for each $x \in X$, where C is a fixed (possibly empty) subset of Y.

The next two lemmas, which are analogues of Theorem 2.27, characterize upper and lower hemicontinuity.

17.4 Lemma   **(Upper Hemicontinuity)**   *For $\varphi \colon X \twoheadrightarrow Y$, a correspondence between topological spaces, the following statements are equivalent.*

1. *φ is upper hemicontinuous.*

2. *$\varphi^{\mathrm{u}}(V)$ is open for each open subset V of Y.*

3. *$\varphi^{\ell}(F)$ is closed for each closed subset F of Y.*

The proof is left as an exercise. For the equivalence of (2) and (3) use the identity $\varphi^{\ell}(A) = [\varphi^{\mathrm{u}}(A^{\mathrm{c}})]^{\mathrm{c}}$.

17.5 Lemma   **(Lower Hemicontinuity)**   *For a correspondence $\varphi \colon X \twoheadrightarrow Y$ between topological spaces, the following statements are equivalent.*

1. *φ is lower hemicontinuous.*

2. *$\varphi^{\ell}(V)$ is open for each open subset V of Y.*

3. *$\varphi^{\mathrm{u}}(F)$ is closed for each closed subset F of Y.*

In particular, if a correspondence $\varphi \colon X \twoheadrightarrow Y$ is upper hemicontinuous, then $\{x \in X : \varphi(x) = \varnothing\} = \varphi^{\mathrm{u}}(\varnothing)$ is open, and if it is lower hemicontinuous, then $\{x \in X : \varphi(x) = \varnothing\} = [\varphi^{\ell}(Y)]^{\mathrm{c}}$ is closed. You should also verify the following simple fact.

17.6 Lemma   *A singleton-valued correspondence is upper hemicontinuous if and only if it is lower hemicontinuous, in which case it is continuous as a function.*

Recall that a function $f: X \to Y$ between topological spaces is:

- an **open mapping** if $f(V)$ is open in Y for each open set V in X.

- a **closed mapping** if $f(F)$ is closed in Y for each closed set F in X.

Open and closed mappings can be characterized in terms of hemicontinuity of the inverse correspondence.

17.7 Theorem   *Let $f: X \to Y$ be a function between topological spaces and consider the inverse correspondence $f^{-1}: Y \twoheadrightarrow X$ defined by the usual formula $f^{-1}(y) = \{x \in X : f(x) = y\}$. Then:*

1. *f is a closed mapping if and only if f^{-1} is upper hemicontinuous.*

2. *f is an open mapping if and only if f^{-1} is lower hemicontinuous.*

Proof: We verify (1) and leave the corresponding proof of (2) as an exercise. Assume first that f is a closed mapping. Fix $y \in Y$ and choose an open subset V of X such that $f^{-1}(y) \subset V$. Put $W = [f(V^c)]^c$ and note that W is an open neighborhood of y such that for each $z \in W$ we have $f^{-1}(z) \subset V$. (Why?) This shows that f^{-1} is upper hemicontinuous at y.

For the converse, suppose that f^{-1} is upper hemicontinuous. Let F be a closed subset of X and pick $y \in [f(F)]^c$. Then $f^{-1}(y) \subset F^c$. So by the upper hemicontinuity of f^{-1}, there exists a neighborhood V of y such that for all $z \in V$ we have $f^{-1}(z) \subset F^c$. This implies $V \cap f(F) = \varnothing$ (why?) or $V \subset [f(F)]^c$. Therefore, every point of $[f(F)]^c$ is an interior point, so $[f(F)]^c$ is an open set, which implies that $f(F)$ is closed.   ∎

The next lemma is the analogue of Theorem 2.34 for correspondences.

17.8 Lemma (Uhc image of a compact set)   *The image of a compact set under a compact-valued upper hemicontinuous correspondence is compact.*

Proof: Let $\varphi: X \twoheadrightarrow Y$ be upper hemicontinuous with compact values, where X is compact. Let $\{U_\alpha\}$ be an open cover of $\varphi(X)$. Since each $\varphi(x)$ is compact, each can be covered by a finite subcover. Let V_x denote the union of such a finite subcover of $\varphi(x)$. Then $W_x = \varphi^u(V_x)$ is an open neighborhood of x, so $\{W_x : x \in X\}$ is an open cover of X. Since X is compact, there is a finite subcover $\{W_{x_1}, \ldots, W_{x_n}\}$. The original U_αs corresponding to $V_{x_1}, \ldots, V_{x_n}$ provide a finite cover of $\varphi(X)$.   ∎

17.9 Definition   *A correspondence $\varphi: X \twoheadrightarrow Y$ between topological spaces is* **closed**, *or* **has closed graph**, *if its graph*

$$\operatorname{Gr}\varphi = \{(x, y) \in X \times Y : y \in \varphi(x)\}$$

is a closed subset of $X \times Y$.

Closed correspondences are always closed-valued. The converse is false. For instance, the correspondence $\varphi\colon [0,1] \twoheadrightarrow [0,1]$ defined by $\varphi(x) = \{0\}$ if $x > 0$ and $\varphi(0) = \{1\}$ is closed-valued but fails to be closed.

Although a closed-valued correspondence need not be closed, upper hemicontinuous closed-valued correspondences are closed under mild conditions.

17.10 Theorem   *An upper hemicontinuous correspondence $\varphi\colon X \twoheadrightarrow Y$ is closed if either:*

1. *φ is closed-valued and Y is regular, or*

2. *φ is compact-valued and Y is Hausdorff.*

Proof: Let $\varphi\colon X \twoheadrightarrow Y$ be an upper hemicontinuous correspondence between topological spaces. Suppose $(x, y) \notin \operatorname{Gr}\varphi$, that is, $y \notin \varphi(x)$. If $\varphi(x)$ is closed and Y is regular, then there are open neighborhoods V of y and W of $\varphi(x)$ with $V \cap W = \varnothing$. If Y is Hausdorff and $\varphi(x)$ is compact, by Lemma 2.32 there are open neighborhoods V of y and W of $\varphi(x)$ with $V \cap W = \varnothing$. In either case, $U = \varphi^{\mathrm{u}}(W)$ is open, and $U \times V$ is a neighborhood of (x, y) disjoint from $\operatorname{Gr}\varphi$. Therefore $\operatorname{Gr}\varphi$ is closed. ∎

For a correspondence having a compact Hausdorff range, the properties of being closed and being upper hemicontinuous coincide.

17.11 Closed Graph Theorem   *A correspondence with compact Hausdorff range space is closed if and only if it is upper hemicontinuous and closed-valued.*

Proof: Since closed subsets of a compact set are compact, Theorem 17.10 (2) implies that a closed-valued upper hemicontinuous correspondence into a compact Hausdorff space is closed.

For the converse, we will prove the contrapositive, namely that if a correspondence with compact Hausdorff range space is not upper hemicontinuous, then it is not closed. (This may appear to be stronger than what we want since we do not assume closed values. But if one value is not a closed set, the graph cannot be closed.) So let $\varphi\colon X \twoheadrightarrow Y$ where Y is compact and Hausdorff, and assume that φ is not upper hemicontinuous. Then there must be some x and an open set $V \supset \varphi(x)$ such that for every neighborhood U of x, there exist $x_U \in U$ and $y_U \in \varphi(x_U)$ with $y_U \notin V$. Note that $\{y_U\}$ is a net in V^c directed by inclusion. Since Y is compact, there is a convergent subnet, say to $y \in Y$. Then $y \notin \varphi(x) \subset V$ since V^c is closed. Then $\{(x_U, y_U)\} \subset \operatorname{Gr}\varphi$, and a subnet of $\{(x_U, y_U)\}$ converges to $(x, y) \notin \operatorname{Gr}\varphi$, so the correspondence φ is not closed. ∎

Note that Theorem 17.11 generalizes the Closed Graph Theorem 2.58 for functions. Compactness of the range space is required for this conclusion, as discussed in Example 2.59.

Correspondences with open graph have open lower sections and are lower hemicontinuous. The proof is straightforward from the definitions.

17.12 Lemma **(Correspondences with open graph)**   *For correspondences between topological spaces we have the following implications:*

Open Graph $\implies$ *Open Lower Sections* $\implies$ *Lower Hemicontinuity*

As an application of the Closed Graph Theorem, we now present a useful little result, which is a special case of C. J. Himmelberg and F. S. van Vleck [162, Theorem 3]. It has applications to the theory of anonymous games; see A. Mas-Colell [242]. Recall that $\mathcal{P}(Y)$ denotes the space of Borel probability measures on Y endowed with its weak* topology (see Chapter 15).

17.13 Theorem   *Let X be a metrizable space, and let Y be a compact metrizable space. Let $\varphi\colon X \twoheadrightarrow Y$ be upper hemicontinuous with nonempty closed values. Let $\psi\colon X \twoheadrightarrow \mathcal{P}(Y)$ map each x to the set of probability measures with support in $\varphi(x)$,*

$$\psi(x) = \{\mu \in \mathcal{P}(Y) : \mu(\varphi(x)) = 1\} = \mathcal{P}(\varphi(x)).$$

Then ψ is upper hemicontinuous with nonempty compact convex values.

Proof: It is easy to see that ψ has nonempty convex values, and Corollary 15.6 implies that the values of ψ are closed in $\mathcal{P}(Y)$. Since by Theorem 15.11, $\mathcal{P}(Y)$ is compact, the values of ψ are also compact. Now to establish that ψ is upper hemicontinuous, it suffices to show that ψ has closed graph (Theorem 17.11).

So suppose $x_n \to x$ in X and $\mu_n \to \mu$ in $\mathcal{P}(Y)$, where $\mu_n \in \psi(x_n)$ for each n. (We can work with sequences since $X \times \mathcal{P}(Y)$ is metrizable.) We need to prove that $\mu \in \psi(x)$. Now fix m, let d be a compatible metric on Y, and let G_m be the open $\frac{1}{m}$-neighborhood of $\varphi(x)$, that is,

$$G_m = \{y \in Y : d(y, \varphi(x)) < 1/m\}.$$

Since φ is upper hemicontinuous, $\varphi(x_n) \subset G_m$ for large enough n. So $\mu_n \in \psi(x_n)$ implies $\mu_n(G_m) = 1$ for large n. Now let F_m denote the closed $\frac{1}{m}$-neighborhood of $\varphi(x)$, that is, $F_m = \{y \in Y : d(y, \varphi(x)) \leq 1/m\}$. Then for large enough n, we have $\varphi(x_n) \subset G_m \subset F_m$, so $\mu_n(F_m) = 1$. Since F_m is closed and $\mu_n \to \mu$, Theorem 15.3 (5) implies that $\mu(F_m) = 1$. Now let $m \to \infty$ and observe that $F_m \downarrow \varphi(x)$, so $\mu(\varphi(x)) = 1$ (Theorem 10.8). That is, $\mu \in \psi(x)$, which proves that ψ has closed graph. ∎

Let X be a separable metrizable space. By Theorem 12.14 every finite Borel measure μ on X has a support $\operatorname{supp}\mu$. Thus, the support defines a correspondence $\mu \mapsto \operatorname{supp}\mu$ from finite Borel measures to closed subsets of X. This correspondence is not closed. To see this, fix $x, y \in X$ with $x \neq y$ and then let $\mu_n = \frac{1}{n}\delta_x + (1 - \frac{1}{n})\delta_y$. Clearly, $\operatorname{supp}\mu_n = \{x, y\}$ for each n and $\mu_n \xrightarrow{w^*} \delta_y$, where $w^* = \sigma(ca(X), C_b(X))$. In particular, note that $(\mu_n, x) \in \operatorname{Gr}\operatorname{supp}$ for each n and $(\mu_n, x) \to (\delta_y, x)$ in $ca(X) \times X$. Since $x \notin \operatorname{supp}\delta_y = \{y\}$, this shows that the support correspondence does not have closed graph. Consequently, by Theorem 17.10, it is not upper hemicontinuous either. However, it turns out that the support correspondence is lower hemicontinuous.

17.14 Theorem *If X is a separable metrizable space, then the support correspondence $\mu \mapsto \operatorname{supp}\mu$, from $\mathcal{P}(X)$ to X, is lower hemicontinuous.*

Proof: Let U be an open subset of X and let $x \in U \cap \operatorname{supp}\mu$. By Lemma 3.20 there is a nonnegative bounded continuous function f on X with $f(x) = 1$ and $f = 0$ on U^c. Thus $\int f\, d\mu > 0$, since $x \in \operatorname{supp}\mu$, so $V = \{v \in \mathcal{P}(X) : \int f\, dv > 0\}$ is a w^*-neighborhood of μ, and $v \in V$ implies $U \cap \operatorname{supp}v \neq \varnothing$. (Why?) ∎

Up to now we have eschewed treating correspondences as functions from one space into another space of sets. But we cannot resist mentioning the following.

17.15 Theorem *Let $\varphi\colon X \twoheadrightarrow Y$ be a nonempty compact-valued correspondence from a topological space into a metrizable space, and let $\mathcal{K}_Y$ denote the space of nonempty compact subsets of Y endowed with its Hausdorff metric topology. Then the function $f\colon X \to \mathcal{K}_Y$ defined by $f(x) = \varphi(x)$ is continuous if and only if the correspondence φ is continuous.*

Proof: This follows from the characterization of the Hausdorff metric topology given in Theorem 3.91. We leave the execution as an exercise. ∎

There are also topologies on $\mathcal{K}_Y$ that characterize upper hemicontinuity and lower hemicontinuity. See E. Klein and A. C. Thompson [209, Theorems 7.1.4 and 7.1.7, pp. 73–75].

17.3 Hemicontinuity and nets

Just as it is often convenient to describe continuity of functions in terms of nets, it is also convenient to characterize hemicontinuity in terms of nets rather than inverses. This is especially true for compact-valued correspondences.

17.16 Theorem *For a correspondence $\varphi\colon X \twoheadrightarrow Y$ between topological spaces and a point $x \in X$ the following statements are equivalent.*

1. *The correspondence φ is upper hemicontinuous at x and $\varphi(x)$ is compact.*

2. *For every net $\{(x_\alpha, y_\alpha)\}$ in the graph of φ, that is, with $y_\alpha \in \varphi(x_\alpha)$ for each α, if $x_\alpha \to x$, then the net $\{y_\alpha\}$ has a limit point in $\varphi(x)$.*

Proof: (1) $\implies$ (2)   Assume that φ is upper hemicontinuous at $x \in X$, with $\varphi(x)$ compact. Let $x_\alpha \to x$, and let $y_\alpha \in \varphi(x_\alpha)$ for each α. Notice that the upper hemicontinuity of φ at x guarantees that $\varphi(x) \neq \varnothing$.

Now suppose by way of contradiction that $\{y_\alpha\}$ has no limit point in $\varphi(x)$. This implies that for every $y \in \varphi(x)$ there is an open neighborhood V_y of y and an index α_y such that for all $\alpha \geqslant \alpha_y$ we have $y_\alpha \notin V_y$. Since $\varphi(x)$ is compact, it lies in a finite union $V = V_{y_1} \cup \cdots \cup V_{y_n}$. Let α_0 satisfy $\alpha_0 \geqslant \alpha_{y_i}$ for each $i = 1, \ldots, n$. Then for all $\alpha \geqslant \alpha_0$ we have $y_\alpha \notin V$. On the other hand, since φ is upper hemicontinuous at x, for large enough α it must be the case that $y_\alpha \in \varphi(x_\alpha) \subset V$, a contradiction. This establishes the validity of (2).

(2) $\implies$ (1)   If φ is not upper hemicontinuous at x, then there exists an open neighborhood U of $\varphi(x)$ such that for each open neighborhood V of x there exist $x_V \in V$ and $y_V \in \varphi(x_V)$ with $y_V \notin U$. Clearly the net $\{y_V\}$ (directed by inclusion) has no limit point in $\varphi(x)$, so (2) is violated. The contrapositive shows that (2) implies that φ is upper hemicontinuous at x.

Now let $\{y_\alpha\}$ be a net in $\varphi(x)$. Then taking $x_\alpha = x$ for all α we trivially have $x_\alpha \to x$. Thus by (2), the net $\{y_\alpha\}$ has a limit point in $\varphi(x)$. Since the net $\{y_\alpha\}$ is arbitrary, by Theorem 2.31 the set $\varphi(x)$ is compact.   ∎

17.17 Corollary   *A compact-valued correspondence φ between topological spaces is upper hemicontinuous if and only if for every net $\{(x_\alpha, y_\alpha)\}$ in the graph of φ that satisfies $x_\alpha \to x$ for some $x \in X$ the net $\{y_\alpha\}$ has a limit point in $\varphi(x)$.*

Note that the corollary provides an alternate proof of the Closed Graph Theorem 2.58. As another application of Theorem 17.16 we present the following result on subcorrespondences. If $\psi(x) \subset \varphi(x)$ for each x, then we say that ψ is a **subcorrespondence** of φ.

17.18 Corollary   *Let $\varphi, \psi \colon X \twoheadrightarrow Y$ be correspondences between topological spaces such that φ is compact-valued, and ψ is a closed subcorrespondence of φ. If φ is upper hemicontinuous, then ψ is also upper hemicontinuous.*

Proof: Assume that a net $\{(x_\alpha, y_\alpha)\}$ in the graph of ψ satisfies $x_\alpha \to x$. Then, by Theorem 17.16, $\{y_\alpha\}$ has a limit point in $\varphi(x)$. We claim that such a limit point must lie in $\psi(x)$. To see this, let y be a limit point of $\{y_\alpha\}$. By passing to a subnet if necessary, we can assume that $y_\alpha \to y$ in Y. But then from $(x_\alpha, y_\alpha) \in \mathrm{Gr}\,\psi$ for each α, $(x_\alpha, y_\alpha) \to (x, y)$, and the closedness of $\mathrm{Gr}\,\psi$, we infer that $(x, y) \in \mathrm{Gr}\,\psi$. Thus $y \in \psi(x)$, and since ψ is clearly compact-valued, it follows from Corollary 17.17 that ψ is upper hemicontinuous.   ∎

The next theorem characterizes lower hemicontinuity in terms of nets.

17.19 Theorem   *For a correspondence $\varphi \colon X \twoheadrightarrow Y$ between topological spaces the following statements are equivalent.*

1. *The correspondence φ is lower hemicontinuous at a point x.*

2. *If $x_\alpha \to x$, then for each $y \in \varphi(x)$ there exists a subnet $\{x_{\alpha_\lambda}\}_{\lambda \in \Lambda}$ of the net $\{x_\alpha\}$ and a net $\{y_\lambda\}_{\lambda \in \Lambda}$ in Y satisfying $y_\lambda \in \varphi(x_{\alpha_\lambda})$ for each $\lambda \in \Lambda$ and $y_\lambda \to y$.*

Proof: (1) $\implies$ (2)   Assume that φ is lower hemicontinuous at x. Also, let a net $\{x_\alpha\}_{\alpha \in A}$ satisfy $x_\alpha \to x$, and fix $y \in \varphi(x)$. As usual, $\mathcal{N}_x$ and $\mathcal{N}_y$ denote the neighborhood systems of x and y, respectively.

If $U \in \mathcal{N}_y$, then $y \in \varphi(x) \cap U$. So by the lower hemicontinuity of φ at x, the set $\varphi^\ell(U)$ is a neighborhood of x. Thus if $V \in \mathcal{N}_x$, then there exists some $\alpha_{U,V} \in A$ such that $x_\alpha \in V \cap \varphi^\ell(U)$ for each $\alpha \geqslant \alpha_{U,V}$. Now consider the directed set $\Lambda = A \times \mathcal{N}_y \times \mathcal{N}_x$, and for each $\lambda = (\alpha, U, V) \in \Lambda$ fix some index $\alpha_\lambda \in A$ with $\alpha_\lambda \geqslant \alpha$ and $\alpha_\lambda \geqslant \alpha_{U,V}$. Clearly, $\{\alpha_\lambda\}$ is a subnet of the net A of indexes. Moreover, if for each $\lambda = (\alpha, U, V)$, we choose some $y_\lambda \in \varphi(x_{\alpha_\lambda}) \cap U$, then it is clear that $y_\lambda \to y$.

(2) $\implies$ (1)   Assume by way of contradiction that φ is not lower hemicontinuous at x. So there exists an open set U of Y with $\varphi(x) \cap U \neq \varnothing$ such that for any neighborhood V of x there exists some $x_V \in V$ with $\varphi(x_V) \cap U = \varnothing$. Clearly, $x_V \to x$. Now if $y \in \varphi(x) \cap U$, then by passing to a subnet if necessary, we can assume that there exists a net $\{y_V\}$ with $y_V \in \varphi(x_V)$ for each V and $y_V \to y$. So $\{y_V\} \subset U^c$, which implies $y \in U^c$, contrary to $y \in U$. ∎

We close the section by stating the sequential companions of Theorems 17.16 and 17.19. We leave the proofs as exercises.

17.20 Theorem   *Assume that the topological space X is first countable and that Y is metrizable. Then for a correspondence $\varphi \colon X \twoheadrightarrow Y$ and a point $x \in X$ the following statements are equivalent.*

1. *The correspondence φ is upper hemicontinuous at x and $\varphi(x)$ is compact.*

2. *If a sequence $\{(x_n, y_n)\}$ in the graph of φ satisfies $x_n \to x$, then the sequence $\{y_n\}$ has a limit point in $\varphi(x)$.*

17.21 Theorem   *For a correspondence $\varphi \colon X \twoheadrightarrow Y$ between first countable topological spaces the following statements are equivalent.*

1. *The correspondence φ is lower hemicontinuous at a point x.*

2. *If $x_n \to x$, then for each $y \in \varphi(x)$ there exists a subsequence $\{x_{n_k}\}$ of $\{x_n\}$ and elements $y_k \in \varphi(x_{n_k})$ for each k such that $y_k \to y$.*

17.4 Operations on correspondences

The next set of propositions concerns the preservation of hemicontinuity under various set theoretic operations on correspondences. Most of these results may be found in Berge [37, Chapter 6, pp. 109–117].

The first operation we consider is taking closures. Given a correspondence φ between topological spaces, the **closure correspondence** of φ, denoted $\overline{\varphi}$, is defined by $\overline{\varphi}(x) = \overline{\varphi(x)}$.

17.22 Lemma   *For a correspondence $\varphi \colon X \twoheadrightarrow Y$ between topological spaces and a point $x_0 \in X$ we have the following.*

1. *The closure correspondence $\overline{\varphi}$ is lower hemicontinuous at x_0 if and only if φ is.*

2. *If Y is normal and φ is upper hemicontinuous at x_0, then $\overline{\varphi}$ is also upper hemicontinuous at x_0.*

Proof: (1)   This follows from the fact that if U is open, $\overline{\varphi(x_0)}$ meets U if and only if $\varphi(x_0)$ meets U. (Why?)

(2)   Suppose that φ is upper hemicontinuous at x_0 and that $\overline{\varphi(x_0)}$ is included in the open set U. Then $\overline{\varphi(x_0)}$ and U^c are disjoint closed sets. Since Y is normal, there is an open set G with $\overline{\varphi(x_0)} \subset G \subset \overline{G} \subset U$. (Why?) Since φ is upper hemicontinuous at x_0, it follows that $W = \varphi^u(G)$ is a neighborhood of x_0. Furthermore, for any $z \in W$ we have $\overline{\varphi(z)} \subset \overline{G} \subset U$. So $\overline{\varphi}$ is upper hemicontinuous at x_0.   ∎

Given correspondences $\varphi \colon X \twoheadrightarrow Y$ and $\psi \colon Y \twoheadrightarrow Z$, there is a natural **composition** $\psi \circ \varphi \colon X \twoheadrightarrow Z$ defined by

$$(\psi \circ \varphi)(x) = \bigcup_{y \in \varphi(x)} \psi(y).$$

It is an easy set theoretic exercise to verify that

$$(\psi \circ \varphi)^u(A) = \varphi^u(\psi^u(A)) \quad \text{and} \quad (\psi \circ \varphi)^{\ell}(A) = \varphi^{\ell}(\psi^{\ell}(A)).$$

Consequently the next result is obvious.

17.23 Theorem (Composition)   *The composition of upper hemicontinuous correspondences is upper hemicontinuous. The composition of lower hemicontinuous correspondences is lower hemicontinuous.*

The composition of closed correspondences need not be closed.

17.24 Example (Composition of closed maps need not be closed)   Consider $\varphi \colon \mathbb{R}_+ \to \mathbb{R}_+$ defined by $\varphi(x) = \frac{1}{x}$ for $x > 0$ and $\varphi(0) = 0$. Then φ has closed graph, but is not upper hemicontinuous. Define $\psi\colon \mathbb{R}_+ \to \mathbb{R}_+$ by $\psi(y) = \frac{y}{1+y}$. Since ψ is a continuous function, it is upper hemicontinuous as a correspondence and has closed graph. (Indeed ψ is a homeomorphism.) However, the composition $\psi \circ \varphi\colon \mathbb{R}_+ \to \mathbb{R}_+$ does not have closed graph, since $(\psi \circ \varphi)(x) = \frac{1}{1+x}$ for $x > 0$ and $(\psi \circ \varphi)(0) = 0$. ∎

Given a family $\{\varphi_i : i \in I\}$ of correspondences from X to Y, we define the union and intersection of the family pointwise. That is,

- the **union**, $\bigcup_{i \in I} \varphi_i$, maps x to $\bigcup_{i \in I} \varphi_i(x)$.

- the **intersection**, $\bigcap_{i \in I} \varphi_i$, maps x to $\bigcap_{i \in I} \varphi_i(x)$.

Observe that the graph of the union is the union of the graphs, and the graph of the intersection is the intersection of the graphs. That is,

$$\operatorname{Gr} \bigcup_{i \in I} \varphi_i = \bigcup_{i \in I} \operatorname{Gr} \varphi_i \quad \text{and} \quad \operatorname{Gr} \bigcap_{i \in I} \varphi_i = \bigcap_{i \in I} \operatorname{Gr} \varphi_i.$$

17.25 Theorem (Intersections of correspondences) *Intersections of correspondences satisfy the following properties.*

1. *The intersection of a family of closed correspondences is closed.*

2. *The intersection of a closed correspondence and an upper hemicontinuous compact-valued correspondence is upper hemicontinuous.*

3. *If the range space is regular, then the intersection of a family of closed-valued upper hemicontinuous correspondences, at least one of which is also compact-valued, is upper hemicontinuous.*

Proof: (1) Since the intersection of a family of closed sets is closed, the intersection has closed graph.

(2) Let $\varphi, \psi\colon X \twoheadrightarrow Y$, where φ is closed, and ψ is upper hemicontinuous with compact values. Let $W \subset Y$ be open and suppose $(\varphi \cap \psi)(x) \subset W$. We need to find a neighborhood N of x such that $(\varphi \cap \psi)(N) \subset W$.

Observe that $K = \psi(x) \setminus W$ is compact (why?), but possibly empty. If K is empty, then we can take $N = \psi^u(W)$. Otherwise, as $\varphi(x) \cap \psi(x) \subset W$, for $y \in K$, we have $y \notin \varphi(x)$, or $(x, y) \notin \operatorname{Gr} \varphi$. Since the graph of φ is closed, for each $y \in K$ there are open neighborhoods U_y of x and V_y of y such that $U_y \times V_y$ is disjoint from $\operatorname{Gr} \varphi$. Thus there is a finite subset $\{y_1, \ldots, y_n\}$ of K such that $V_{y_1}, \ldots, V_{y_n}$ cover K. Let $V = \bigcup_{i=1}^n V_{y_i}$, $U = \psi^u(W \cup V)$ and $N = U \cap \left(\bigcap_{i=1}^n U_{y_i} \right)$. Observe that $\left[\left(\bigcap_{i=1}^n U_{y_i} \right) \times V \right] \cap \operatorname{Gr} \varphi = \varnothing$. Now from $\psi(x) \subset [\psi(x) \setminus W] \cup W \subset V \cup W$ and the upper hemicontinuity of ψ, we see that N is an open neighborhood of x. Then for each $z \in N$, we have $\psi(z) \subset W \cup V$ and $\varphi(z) \cap V = \varnothing$. Thus, $(\varphi \cap \psi)(N) \subset W$.

(3) Let $\{\varphi_i\}_{i\in I}$ be a family of closed-valued upper hemicontinuous correspondences with one of them, say φ_{i_0}, compact-valued. Put $\varphi = \bigcap_{i\in I}\varphi_i$. By Theorem 17.10, each φ_i is closed, so by part (1), the correspondence φ is itself closed. Now observe that $\varphi = \varphi \cap \varphi_{i_0}$ and then use part (2) above to infer that φ is upper hemicontinuous.   ∎

We point out that the intersection of lower hemicontinuous correspondences need not be lower hemicontinuous.

17.26 Example   (Intersection not lhc)   Define $\varphi, \psi \colon [0,1] \twoheadrightarrow [0,1]$ via $\varphi(x) = \{x\}$ and $\psi(x) = \{1 - x\}$. Both are continuous as functions and so lower hemicontinuous as correspondences. Now $(\varphi \cap \psi)(x)$ is equal to $\varnothing$ for $x \neq \frac{1}{2}$ and $(\varphi \cap \psi)(\frac{1}{2}) = \{\frac{1}{2}\}$. This is not lower hemicontinuous. (Why?)   ∎

17.27 Theorem   (Unions of correspondences)   *Unions of correspondences satisfy the following properties.*

1. *The union of a family of lower hemicontinuous correspondences is lower hemicontinuous.*

2. *The union of a finite family of upper hemicontinuous correspondences is upper hemicontinuous.*

Proof: Hint: $(\bigcup_i \varphi_i)^{\ell}(U) = \bigcup_i \varphi_i^{\ell}(U)$ and $(\bigcup_i \varphi_i)^{u}(U) = \bigcap_i \varphi_i^{u}(U)$.   ∎

The **product** of a family $\{\varphi_i \colon X_i \twoheadrightarrow Y_i\}_{i\in I}$ of correspondences is the correspondence $\prod_i \varphi_i$, from $\prod_i X_i$ to $\prod_i Y_i$, defined naturally by $(\prod_i \varphi_i)(x) = \prod_i \varphi_i(x_i)$ for each $x = \{x_i\}_{i\in I}$.

17.28 Theorem   (Products of Correspondences)   *Products of correspondences satisfy the following properties.*

1. *The product of a family of upper hemicontinuous correspondences with compact values is upper hemicontinuous with compact values.*

2. *The product of a finite family of lower hemicontinuous correspondences is lower hemicontinuous.*

Proof: (1)   Let $\{\varphi_i \colon X_i \twoheadrightarrow Y_i\}_{i\in I}$ be a family of upper hemicontinuous compact-valued correspondences. Pick a point $x = (x_i)_{i\in I} \in \prod_{i\in I} X_i$ and suppose that $\varphi(x) = \prod_{i\in I} \varphi_i(x_i) \subset G$, where G is an open subset of $\prod_{i\in I} Y_i$. By Theorem 2.62 there is a basic open set $\prod_{i\in I} V_i$ in $\prod_{i\in I} Y_i$ satisfying $\prod_{i\in I} \varphi_i(x_i) \subset \prod_{i\in I} V_i \subset G$. Let $W_i = \varphi_i^u(V_i)$ for each i. Since $\prod_{i\in I} V_i$ is a basic open set, $V_i = Y_i$ for all but finitely many i. Thus $W_i = \varphi_i^u(Y_i) = X_i$ for all but finitely many i. Hence, $W = \prod_{i\in I} W_i$ is an open neighborhood of x and satisfies $\varphi(W) \subset G$, which proves that φ is upper hemicontinuous at x.

The fact that the product correspondence φ has compact values follows from the Tychonoff Product Theorem 2.61.

(2) Let $\varphi_i \colon X_i \twoheadrightarrow Y_i$, $i = 1, 2$, be lower hemicontinuous and suppose that $(y_1, y_2) \in [\varphi_1(x_1) \times \varphi_2(x_2)] \cap G$, where G is open in $Y_1 \times Y_2$. Then there is a basic open neighborhood $V_1 \times V_2$ of (y_1, y_2) included in G. Put $W = \varphi_1^\ell(V_1) \times \varphi_2^\ell(V_2)$. Then W is an open neighborhood of (x_1, x_2), and moreover $(z_1, z_2) \in W$ implies $[\varphi_1(z_1) \times \varphi_2(z_2)] \cap G \neq \varnothing$. (Why?)   ∎

17.5   The Maximum Theorem

One of the most useful results about correspondences is the **Maximum Theorem**, which is due to C. Berge [37, pp. 115–116]. It states that the set of solutions to a well behaved constrained maximization problem is upper hemicontinuous in its parameters and that the value function is continuous. It is a consequence of the following two lemmas.

17.29 Lemma   *Let $\varphi \colon X \twoheadrightarrow Y$ be a lower hemicontinuous correspondence between topological spaces, and let the function $f \colon \operatorname{Gr} \varphi \to \mathbb{R}$ be lower semicontinuous. Define the extended real function $m \colon X \to \mathbb{R}^*$ by*

$$m(x) = \sup_{y \in \varphi(x)} f(x, y),$$

where as usual, $\sup \varnothing = -\infty$. Then the function m is lower semicontinuous.

Proof: We need to show that $\{x \in X : m(x) > \alpha\}$ is open for any $\alpha \in \mathbb{R}$. So suppose $m(x_0) > \alpha$. Then $f(x_0, y_0) > \alpha$ for some $y_0 \in \varphi(x_0)$. (In particular, $\varphi(x_0)$ is nonempty.) Since f is lower semicontinuous on the graph $\operatorname{Gr} \varphi$ of φ, the upper contour set $W = \{(x, y) \in \operatorname{Gr} \varphi : f(x, y) > \alpha\}$ is an open neighborhood of (x_0, y_0) in $\operatorname{Gr} \varphi$. Thus there are open neighborhoods U of x_0 and V of y_0 such that $(U \times V) \cap \operatorname{Gr} \varphi \subset W$.

Clearly, $N = U \cap \varphi^\ell(V)$ is a neighborhood of x_0. Now for each x in N, there is some $y \in \varphi(x) \cap V$, so that $(x, y) \in (U \times V) \cap \operatorname{Gr} \varphi \subset W$. Thus $f(x, y) > \alpha$, so $m(x) > \alpha$ for each $x \in N$. So $N \subset \{x \in X : m(x) > \alpha\}$, consequently $\{x \in X : m(x) > \alpha\}$ is an open set. Therefore m is lower semicontinuous.   ∎

17.30 Lemma   *Let $\varphi \colon X \twoheadrightarrow Y$ be an upper hemicontinuous correspondence between topological spaces with nonempty compact values, and let $f \colon \operatorname{Gr} \varphi \to \mathbb{R}$ be upper semicontinuous. Define the function $m \colon X \to \mathbb{R}$ by*

$$m(x) = \max_{y \in \varphi(x)} f(x, y).$$

Then the function m is upper semicontinuous.

Proof: First note that for each fixed $x \in X$ the function $f(x, \cdot): \varphi(x) \to \mathbb{R}$ is well defined and upper semicontinuous. Thus, by Theorem 2.43, the maximum actually exists. We must show that the set $\{x \in X : m(x) < \alpha\}$ is open for any $\alpha \in \mathbb{R}$. So fix $\alpha \in \mathbb{R}$ and $x_0 \in X$ satisfying $m(x_0) < \alpha$ and let $W = \{(x, y) \in \mathrm{Gr}\,\varphi : f(x, y) < \alpha\}$. Clearly, for each $y \in \varphi(x_0)$ we have $(x_0, y) \in W$.

Since f is upper semicontinuous on $\mathrm{Gr}\,\varphi$, the set W is open in $\mathrm{Gr}\,\varphi$. Hence for each $y \in \varphi(x_0)$ there is an open neighborhood U_y of x_0 and an open neighborhood V_y of y such that $(U_y \times V_y) \cap \mathrm{Gr}\,\varphi \subset W$. Clearly the family $\{V_y : y \in \varphi(x_0)\}$ is an open cover of $\varphi(x_0)$. Now pick a finite subcover $\{V_{y_1}, \ldots, V_{y_n}\}$ of $\varphi(x_0)$ and put $U = \bigcap_{i=1}^n U_{y_i}$ and $V = \bigcup_{i=1}^n V_{y_i} \supset \varphi(x_0)$. Then $(U \times V) \cap \mathrm{Gr}\,\varphi \subset W$, and the upper hemicontinuity of φ guarantees that $N = U \cap \varphi^u(V)$ is an open neighborhood of x_0. Now for each $x \in N$, if $y \in \varphi(x)$, then $(x, y) \in (U \times V) \cap \mathrm{Gr}\,\varphi \subset W$, so $f(x, y) < \alpha$. In particular, $m(x) < \alpha$. Thus $N \subset \{x \in X : m(x) < \alpha\}$, so $\{x \in X : m(x) < \alpha\}$ is open, proving that m is an upper semicontinuous function. ∎

We are now ready to prove the very useful Maximum Theorem.

17.31 Berge Maximum Theorem   *Let $\varphi: X \twoheadrightarrow Y$ be a continuous correspondence between topological spaces with nonempty compact values, and suppose $f: \mathrm{Gr}\,\varphi \to \mathbb{R}$ is continuous. Define the "value function" $m: X \to \mathbb{R}$ by*

$$m(x) = \max_{y \in \varphi(x)} f(x, y),$$

and the correspondence $\mu: X \twoheadrightarrow Y$ of maximizers by

$$\mu(x) = \{y \in \varphi(x) : f(x, y) = m(x)\}.$$

Then:

1. *The value function m is continuous.*

2. *The "argmax" correspondence μ has nonempty compact values.*

3. *If either f has a continuous extension to all of $X \times Y$ or Y is Hausdorff, then the "argmax" correspondence μ is upper hemicontinuous.*

Proof: From Lemmas 17.29 and 17.30, m is continuous, and Theorem 2.43 implies that μ has compact values.

Now assume that f has a continuous extension (which we denote by f again) to all of $X \times Y$. Then it is easy to see that $\psi(x) = \{y \in Y : f(x, y) = m(x)\}$ has closed graph, and that $\mu = \varphi \cap \psi$. Thus by Theorem 17.25 (2), μ is upper hemicontinuous.

Finally, consider the case where Y is a Hausdorff topological space. In view of Corollary 17.18, it suffices to show that μ has closed graph. To this end, let a net $\{(x_\alpha, y_\alpha)\}$ in $\mathrm{Gr}\,\mu$ satisfy $(x_\alpha, y_\alpha) \to (x, y)$ in $X \times Y$. That is, $x_\alpha \to x$, $y_\alpha \to y$, and $y_\alpha \in \mu(x_\alpha)$ for each α. We claim that $y \in \varphi(x)$. If $y \notin \varphi(x)$, then there exist (by Lemma 2.32) disjoint open sets V and W such that $y \in V$ and $\varphi(x) \subset W$. Now by the upper hemicontinuity of φ at x we have $y_\alpha \in \varphi(x_\alpha) \subset W$ for all eventually large α and in view of $y_\alpha \to y$ we see that $y_\alpha \in V$ for all eventually large α too, contradicting $V \cap W = \varnothing$. Hence, $y \in \varphi(x)$ or $(x, y) \in \mathrm{Gr}\, \varphi$. Finally, from $m(x) = \lim_\alpha m(x_\alpha) = \lim_\alpha f(x_\alpha, y_\alpha) = f(x, y)$ we infer that $y \in \mu(x)$ or $(x, y) \in \mathrm{Gr}\, \mu$, proving that μ is a closed correspondence. ∎

17.6   Vector-valued correspondences

When the range space of a correspondence is a vector space, then there are additional natural operations on correspondences. If $\varphi, \psi \colon X \twoheadrightarrow Y$, where Y is now a vector space, then we define:

- The **sum** correspondence $\varphi + \psi$ of φ and ψ by

$$(\varphi + \psi)(x) = \varphi(x) + \psi(x) = \{y + z : y \in \varphi(x) \text{ and } z \in \psi(x)\}.$$

- The **convex hull** correspondence $\mathrm{co}\,\varphi$ of φ by $(\mathrm{co}\,\varphi)(x) = \mathrm{co}\,\varphi(x)$.

- If Y is a topological vector space, the **closed convex hull** correspondence $\overline{\mathrm{co}}\,\varphi$ of φ by $\overline{\mathrm{co}}\,\varphi(x) = \overline{\mathrm{co}}\,\varphi(x)$.

- If α and β are scalars, the **linear combination** correspondence $\alpha\varphi + \beta\psi$ of φ and ψ by $(\alpha\varphi + \beta\psi)(x) = \alpha\varphi(x) + \beta\psi(x)$.

When Y is a topological vector space, the notions of hemicontinuity can certainly be applied, and it is natural to ask which of these properties are inherited by the sum and convex hull correspondences.

17.32 Theorem   (Linear combinations of correspondences)   *For two correspondences $\varphi, \psi \colon X \twoheadrightarrow Y$ from a topological space into a topological vector space and any two nonzero scalars α and β we have the following:*

1. *If φ is closed-valued and ψ is compact-valued, then $\alpha\varphi + \beta\psi$ is closed-valued.*

2. *If φ and ψ are compact-valued, then the linear combination $\alpha\varphi + \beta\psi$ is compact-valued.*

3. *If φ and ψ are compact-valued and upper hemicontinuous at a point, then $\alpha\varphi + \beta\psi$ is upper hemicontinuous at that point.*

4. *If φ and ψ are lower hemicontinuous at a point, then $\alpha\varphi + \beta\psi$ is also lower hemicontinuous at that point.*

Proof: The validity of (1) and (2) follows immediately from Lemma 5.3.

(3) Let φ and ψ be upper hemicontinuous at the point x_0, and suppose that $\alpha\varphi(x_0) + \beta\psi(x_0) \subset G$, where G is an open subset of Y. By Theorem 5.9, there is a neighborhood V of zero such that $\alpha\varphi(x_0) + \beta\psi(x_0) + V \subset G$. Pick an open neighborhood W of zero such that $W + W \subset V$. Since $\varphi(x_0) \subset \varphi(x_0) + \frac{1}{\alpha}W$ and $\varphi(x_0) + \frac{1}{\alpha}W$ is open, the upper hemicontinuity of φ at x_0 guarantees the existence of a neighborhood N_1 of x_0 such that $x \in N_1$ implies $\varphi(x) \subset \varphi(x_0) + \frac{1}{\alpha}W$ or $\alpha\varphi(x) \subset \alpha\varphi(x_0) + W$. Similarly, there exists a neighborhood N_2 of x_0 satisfying $\beta\psi(x) \subset \beta\psi(x_0) + W$ for all $x \in N_2$. Consequently, if we let $N = N_1 \cap N_2$, then N is a neighborhood of x_0 and for each $x \in N$ we have

$$(\alpha\varphi + \beta\psi)(x) = \alpha\varphi(x) + \beta\psi(x) \subset \alpha\varphi(x_0) + W + \beta\psi(x_0) + W$$
$$\subset \alpha\varphi(x_0) + \beta\psi(x_0) + V \subset G.$$

This shows that $\alpha\varphi + \beta\psi$ is upper hemicontinuous at x_0.

(4)   Suppose $[\alpha\varphi(x_0) + \beta\psi(x_0)] \cap U \neq \varnothing$, where U is open. Then there are $y \in \alpha\varphi(x_0)$ and $z \in \beta\psi(x_0)$ with $y + z \in U$. Pick an open neighborhood V of zero such that $y + V + z + V \subset U$. Since $\frac{y}{\alpha} \in \varphi(x_0) \cap (\frac{y}{\alpha} + \frac{1}{\alpha}V)$ and φ is lower hemicontinuous at x_0, there exists a neighborhood N_1 of x_0 such that $\varphi(x) \cap (\frac{y}{\alpha} + \frac{1}{\alpha}V) \neq \varnothing$ for all $x \in N_1$. Similarly, there exists a neighborhood N_2 of x_0 such that $\psi(x) \cap (\frac{z}{\beta} + \frac{1}{\beta}V) \neq \varnothing$ for all $x \in N_2$. Now if $x \in N_1 \cap N_2$, then pick $v_1, v_2 \in V$ with $x_1 = \frac{y}{\alpha} + \frac{1}{\alpha}v_1 \in \varphi(x)$ and $x_2 = \frac{z}{\beta} + \frac{1}{\beta}v_2 \in \psi(x)$. Consequently, the vector $x_3 = \alpha x_1 + \beta x_2 \in \alpha\varphi(x) + \beta\psi(x)$ satisfies

$$x_3 = \alpha x_1 + \beta x_2 = y + v_1 + z + v_2 \in y + V + z + V \subset U.$$

Thus, $[\alpha\varphi(x) + \beta\psi(x)] \cap U \neq \varnothing$ holds for all $x \in N_1 \cap N_2$, and this shows that $\alpha\varphi + \beta\psi$ is lower hemicontinuous at x_0. ∎

The following is an immediate consequence of part (3) of the preceding.

17.33 Corollary   *Let Y be a topological vector space, let X be a subset of Y, and let $\varphi \colon X \twoheadrightarrow Y$ be an upper hemicontinuous correspondence with compact values. Then for every pair of scalars α and β and every vector $y \in Y$ the correspondence $x \mapsto \alpha x + y + \beta\varphi(x)$ is also upper hemicontinuous.*

In particular, under these hypotheses, the correspondences $x \mapsto y + \varphi(x)$ and $x \mapsto x + \varphi(x)$ are both upper hemicontinuous.

The next example shows that the assumption of compact values for both correspondences in part (3) of Theorem 17.32 cannot be dropped.

17.34 Example (Sum not uhc)   Let $F = \{(x, y) \in \mathbb{R}^2 : x > 0,\ y \geqslant \frac{1}{x}\}$. Define $\varphi, \psi \colon \mathbb{R}^2 \twoheadrightarrow \mathbb{R}^2$ by $\varphi(z) = \{z\}$ and $\psi(z) = F$ for each $z \in \mathbb{R}^2$. Then φ and ψ are upper hemicontinuous, φ has compact values, ψ has closed values, and $\varphi + \psi$ has closed values. Now consider the open set $G = \{(x, y) \in \mathbb{R}^2 : x, y > 0\}$. Note that $[\varphi + \psi]^u(G) = \{z \in \mathbb{R}^2 : z \geqslant 0\}$, which is not open. Thus, $\varphi + \psi$ is not upper hemicontinuous. You can verify that at every point $\varphi + \psi$ fails to be upper hemicontinuous.   ∎

The rest of the discussion in this sections deals with continuity properties of the convex hull correspondence of a given correspondence. We start with upper hemicontinuity.

17.35 Theorem *Assume that X is a topological space and that Y is a locally convex space. For a correspondence $\varphi \colon X \twoheadrightarrow Y$ that is upper hemicontinuous at some point $x_0 \in X$ we have the following:*

1. *If the set $\overline{\mathrm{co}}\,\varphi(x_0)$ is compact, then the closed convex hull correspondence $\overline{\mathrm{co}}\,\varphi \colon X \twoheadrightarrow Y$ (and hence also the convex hull correspondence $\mathrm{co}\,\varphi$) is upper hemicontinuous at x_0.*

2. *If $Y = \mathbb{R}^n$ and $\varphi(x_0)$ is compact, then the convex hull correspondence $\mathrm{co}\,\varphi \colon X \twoheadrightarrow \mathbb{R}^n$ is upper hemicontinuous at x_0.*

Proof: (1) Assume that $\overline{\mathrm{co}}\,\varphi(x_0) \subset U$, where U is an open subset of the locally convex space Y. Since $\overline{\mathrm{co}}\,\varphi(x_0)$ is compact, it follows from Theorem 5.9 that there exists an open convex neighborhood V of zero satisfying $\overline{\mathrm{co}}\,\varphi(x_0) + V \subset U$. Pick another open convex neighborhood of zero W such that $W + W \subset V$. According to part (4) of Lemma 5.3, the set $\overline{\mathrm{co}}\,\varphi(x_0) + \overline{W}$ is a closed and convex subset of Y. Moreover, from $\overline{W} \subset V$, it follows that $\overline{\mathrm{co}}\,\varphi(x_0) + \overline{W} \subset \overline{\mathrm{co}}\,\varphi(x_0) + V \subset U$.

From $\varphi(x_0) \subset \varphi(x_0) + W$ and the upper hemicontinuity of φ at x_0, there is a neighborhood N of x_0 such that $\varphi(x) \subset \varphi(x_0) + W \subset \overline{\mathrm{co}}\,\varphi(x_0) + \overline{W}$ for each $x \in N$. So if $x \in N$, then $\overline{\mathrm{co}}\,\varphi(x) \subset \overline{\mathrm{co}}\,\varphi(x_0) + \overline{W} \subset U$, which proves that $\overline{\mathrm{co}}\,\varphi$ is upper hemicontinuous at x_0.

(2) This follows immediately from part (1) by observing that (according to Corollary 5.33) the convex hull of a compact subset of $\mathbb{R}^n$ is likewise a compact subset of $\mathbb{R}^n$.   ∎

The convex hull correspondence of a lower hemicontinuous correspondence is always lower hemicontinuous.

17.36 Theorem *If a correspondence from a topological space to a locally convex space is lower hemicontinuous at some point, then its convex hull correspondence is also lower hemicontinuous at that point.*

Proof: We use the characterization of lower hemicontinuity in terms of nets given in Theorem 17.19 (2). Let $\varphi \colon X \twoheadrightarrow Y$ be a correspondence from a topological space to a locally convex space that is lower hemicontinuous at the point x_0. Assume that a net $\{x_\alpha\}$ in X satisfies $x_\alpha \to x_0$ and let $y \in \mathrm{co}\,\varphi(x_0)$. Write

$y = \sum_{i=1}^{k} \lambda_i y_i$, where $\lambda_i > 0$ and $y_i \in \varphi(x_0)$ for each $i = 1, \ldots, k$ and $\sum_{i=1}^{k} \lambda_i = 1$. Since φ is lower hemicontinuous at x_0, it easily follows from Theorem 17.19 that there exists a subnet $\{x_{\alpha_\lambda}\}_{\lambda \in \Lambda}$ of the net $\{x_\alpha\}$ and nets $\{y_\lambda^i\}_{\lambda \in \Lambda}$ $(i = 1, \ldots, k)$ in Y such that for each $i = 1, \ldots, k$ we have $y_\lambda^i \in \varphi(x_{\alpha_\lambda})$ for each $\lambda \in \Lambda$ and $y_\lambda^i \xrightarrow{\tau} y_i$ in Y.

Now for each $\lambda \in \Lambda$ put $y_\lambda = \sum_{i=1}^{k} \lambda_i y_\lambda^i$. Clearly, the net $\{y_\lambda\}_{\lambda \in \Lambda}$ in Y satisfies $y_\lambda \in \mathrm{co}\, \varphi(x_{\alpha_\lambda})$ for each $\lambda \in \Lambda$ and $y_\lambda \to y$. Thus $\mathrm{co}\, \varphi$ is lower hemicontinuous at x_0 by Theorem 17.19 (2). ∎

The convex hull of the correspondence formed by a finite number of continuous functions is continuous.

17.37 Theorem   *Let $f_i \colon X \to Y$ $(i = 1, \ldots, k)$ be functions between topological spaces. Consider the correspondence $\varphi \colon X \twoheadrightarrow Y$ via $\varphi(x) = \{f_1(x), \ldots, f_k(x)\}$. If each function f_i is continuous at some point x_0, then:*

1. *The correspondence φ is continuous at x_0.*

2. *If Y is a locally convex tvs, then the convex hull correspondence $\mathrm{co}\, \varphi$ is also continuous at x_0.*

Proof: (1)   We establish first that the correspondence φ is upper hemicontinuous. To this end, assume $\varphi(x_0) = \{f_1(x_0), \ldots, f_k(x_0)\} \subset U$, where U is an open subset of Y. Then $V = \bigcap_{i=1}^{k} f_i^{-1}(U)$ is a neighborhood of x_0 such that $x \in V$ implies $\varphi(x) \subset U$.

For the lower hemicontinuity of φ, suppose $\varphi(x_0) \cap U \neq \varnothing$ for some open subset U. If $f_i(x_0) \in U$, then $V = f_i^{-1}(U)$ is a neighborhood of x_0 and $x \in V$ implies $\varphi(x) \cap U \neq \varnothing$.

(2)   The upper hemicontinuity of φ follows immediately from part (1) of Theorem 17.35 and the lower hemicontinuity from Theorem 17.36. ∎

The converses of both statements are not true. Consider the two functions f, the indicator function of the rational numbers, and g, the indicator function of the irrational numbers. Define $\varphi \colon \mathbb{R} \twoheadrightarrow \mathbb{R}$ by $\varphi(x) = \{f(x), g(x)\} = \{0, 1\}$. It is continuous, but neither f nor g is continuous anywhere. The convex hull correspondence is also a constant correspondence, and so continuous.

17.7   Demicontinuous correspondences

For correspondences with values in a topological vector space, there is yet another natural notion of continuity.

17.38 Definition   *A correspondence $\varphi \colon X \twoheadrightarrow Y$ from a topological space to a topological vector space is:*

- **upper demicontinuous** *if the upper inverse of every open half space in Y is open in X; and*

- **lower demicontinuous** *if the lower inverse of every open half space in Y is open in X.*

It is obvious that an upper hemicontinuous correspondence is upper demicontinuous and a lower hemicontinuous correspondence is lower demicontinuous. However, an upper demicontinuous correspondence need not be upper hemicontinuous, nor need a lower demicontinuous correspondence be lower hemicontinuous.

17.39 Example (Demicontinuity vs. hemicontinuity) To see the claim above, let $X = Y$ be an infinite dimensional normed space and let τ denote the norm topology on X. We know (Theorem 6.26) that the weak topology $\sigma(X, X')$ is strictly coarser than τ. This implies that the identity operator $I \colon (X, \sigma(X, X')) \to (X, \tau)$ is not continuous. Consequently, I viewed as a correspondence from $(X, \sigma(X, X'))$ to (X, τ) is neither upper nor lower hemicontinuous. On the other hand, the operator $I \colon (X, \sigma(X, X')) \to (X, \sigma(X, X'))$ is continuous, and hence I viewed as a correspondence from $(X, \sigma(X, X'))$ to itself is both upper and lower hemicontinuous. This easily implies that I viewed as a correspondence from $(X, \sigma(X, X'))$ to (X, τ) is both upper and lower demicontinuous. So the function $I \colon (X, \sigma(X, X')) \to (X, \tau)$ viewed as a correspondence is upper and lower demicontinuous and is neither upper nor lower hemicontinuous. ∎

17.40 Lemma *A compact-valued correspondence $\varphi \colon X \twoheadrightarrow Y$ from a topological space to a topological vector space is upper demicontinuous if and only if $\overline{\mathrm{co}}\,\varphi$ is upper demicontinuous.*

Proof: Let $H = \{y \in Y : \langle y, y' \rangle < \alpha\}$ be an open half space, where $y' \in Y'$. Since y' is linear and continuous, for each $x \in X$ with $\varphi(x) \neq \varnothing$ we have

$$\max\{\langle y, y' \rangle : y \in \varphi(x)\} = \max\{\langle y, y' \rangle : y \in \overline{\mathrm{co}}\,\varphi(x)\}.$$

(Why?) This implies $\varphi(x) \subset H$ if and only if $\overline{\mathrm{co}}\,\varphi(x) \subset H$. Thus φ is upper demicontinuous if and only if $\overline{\mathrm{co}}\,\varphi$ is. ∎

For a correspondence $\varphi \colon X \to Y$, from a topological space to a locally convex Hausdorff space, with nonempty weakly compact convex values, upper demicontinuity is equivalent to upper hemicontinuity with respect to the weak topology. The real function $x \mapsto h_{\varphi(x)}(y')$, where

$$h_{\varphi(x)}(y') = \max\{\langle y, y' \rangle : y \in \varphi(x)\}$$

is the support function of $\varphi(x)$ evaluated at y', is called the **support mapping** of
φ at y'.[2]

17.41 Theorem   *Let Y be a locally convex Hausdorff space, let X be a topo-
logical space, and let $\varphi\colon X \twoheadrightarrow Y$ have nonempty $\sigma(Y, Y')$-compact convex values.
The following statements are equivalent.*

1. *The correspondence $\varphi\colon X \twoheadrightarrow (Y, \sigma(Y, Y'))$ is upper hemicontinuous.*

2. *The correspondence φ is upper demicontinuous.*

3. *For each $y' \in Y'$ the support mapping $x \mapsto h_{\varphi(x)}(y')$ is upper semicontinuous.*

Proof: (1) $\implies$ (2)   This is immediate since every open half space is a weakly
open set.

(2) $\implies$ (3)   Observe that $\{x \in X : h_{\varphi(x)}(y') < \alpha\}$ is simply $\varphi^{\mathrm{u}}([y' < \alpha])$,
which is open by upper demicontinuity.

(3) $\implies$ (1)   Suppose $x \mapsto h_{\varphi(x)}(y')$ is upper semicontinuous at the point x_0
for each $y' \in Y'$. The first thing to observe is that without loss of generality, we
can assume $0 \in \varphi(x_0)$. (Pick any y_0 in $\varphi(x_0)$. We may replace φ by $\varphi - y_0$ without
changing any continuity properties.)

Now let G be a $\sigma(Y, Y')$-open set including $\varphi(x_0)$. We need to show that there
exists a neighborhood N of x_0 such that $x \in N$ implies $\varphi(x) \subset G$. Observe that
since $\varphi(x_0)$ is $\sigma(Y, Y')$-compact, by Theorems 5.6 and 5.9 there is a $\sigma(Y, Y')$-closed
circled convex neighborhood V of zero satisfying $\varphi(x_0) + V + V \subset G$. We may take
V to be the absolute polar of a finite subset of Y'. Also, since $\varphi(x_0)$ is $\sigma(Y, Y')$-
compact, it has a finite subset $\{y_1, \ldots, y_m\}$ satisfying $\varphi(x_0) \subset \{0, y_1, \ldots, y_m\} + V$.
Let $K = \mathrm{co}\{0, y_1, \ldots, y_m\}$ and put $C = K + V + V = K + 2V$. Since $\varphi(x_0)$ is also
convex, we obtain

$$\varphi(x_0) + V \subset C \subset \varphi(x_0) + V + V \subset G.$$

Now consider the one-sided polar of C,

$$C^\odot = \{y' \in Y' : \langle y, y' \rangle \leqslant 1 \text{ for each } y \in C\}.$$

Since $0 \in K$, Lemma 7.82 implies that $C^\odot$ is the convex hull of a finite subset
$\{y'_1, \ldots, y'_n\}$ of $(2V)^\odot = (2V)^\circ$. By the Bipolar Theorem 5.103 for one-sided polars, $C^{\odot\odot} = C$. That is,

$$\begin{aligned} C &= \{y \in Y : \langle y, y' \rangle \leqslant 1 \text{ for all } y' \in C^\odot\} \\ &= \{y \in Y : \langle y, y'_j \rangle \leqslant 1 \text{ for all } j = 1, \ldots, n\}. \end{aligned}$$

---

[2] Recall that in a dual pair $\langle X, X' \rangle$, the support functional h_K of a nonempty $\sigma(X, X')$-compact
convex set K is a $\tau(X', X)$-continuous sublinear function on X'. (This is Theorem 7.52 with X and X'
interchanged.)

Now let

$$N = \{x \in X : h_{\varphi(x)}(y_j') < 1 \text{ for each } j = 1, \ldots, n\}$$
$$= \{x \in X : \langle y, y_j' \rangle < 1 \text{ for all } y \in \varphi(x) \text{ and each } j = 1, \ldots, n\}.$$

Since $\varphi(x_0) + V \subset C$ and V is circled, it is easy to see that $x_0 \in N$. (Why?) Since each $x \mapsto h_{\varphi(x)}(y_j')$ is upper semicontinuous, N is a neighborhood of x_0. By the above, for each $x \in N$ we have $\varphi(x) \subset C \subset G$, as desired. ∎

See Theorems II-21 and II-24 of Castaing and Valadier [75, pp. 53–55] for additional theorems regarding lower hemicontinuity and continuity in terms of support mappings.

Since a finite dimensional vector space admits a unique Hausdorff topology (Theorem 5.21), we have the following consequence of the preceding result.

17.42 Corollary *A correspondence from a topological space to a finite dimensional Hausdorff topological vector space with nonempty compact convex values is upper demicontinuous if and only if it is upper hemicontinuous.*

17.8 Knaster–Kuratowski–Mazurkiewicz mappings

This section presents a theorem that provides the foundation for many of the most useful fixed point theorems. It is due to B. Knaster, K. Kuratowski, and S. Mazurkiewicz [212].

17.43 KKM Lemma *Let $\{x_1, \ldots, x_m\}$ be a finite subset of some Euclidean space $\mathbb{R}^n$, and let $\{F_1, \ldots, F_m\}$ be a family of closed subsets of $\mathbb{R}^n$ such that for every subset A of indexes $\{1, \ldots, m\}$, we have*

$$\operatorname{co}\{x_i : i \in A\} \subset \bigcup_{i \in A} F_i.$$

Then the intersection $F_1 \cap \cdots \cap F_m \cap \operatorname{co}\{x_1, \ldots, x_m\}$ is compact and nonempty.

Compactness is obvious. The proof of nonemptiness is based on a combinatorial result, Sperner's Lemma [318], whose proof is elementary but longer than we care to address here; but see [56, Chapters 4–5, pp. 23–27] for a treatment more or less from scratch.

This lemma suggests the following definition.

17.44 Definition *Let X be a vector space and let A be a nonempty subset of X. A correspondence $\psi : A \twoheadrightarrow X$ is a **KKM correspondence** or **KKM map** on A provided*

$$\operatorname{co}\{x_1, \ldots, x_n\} \subset \bigcup_{i=1}^{n} \psi(x_i)$$

for every finite subset $\{x_1, \ldots, x_n\}$ of A.

Notice that every KKM correspondence $\psi\colon A \twoheadrightarrow X$ satisfies $x \in \psi(x)$ for each $x \in A$. In particular, every KKM correspondence is nonempty-valued. Two simple but useful properties of KKM correspondences are included in the next lemma.

17.45 Lemma   *For a KKM correspondence $\psi\colon A \twoheadrightarrow X$ we have:*

1. *The restriction of ψ to any subset of A is also a KKM correspondence. That is, if B is a subset of A, then the correspondence $\psi|_B\colon B \twoheadrightarrow X$ (defined by $\psi|_B(x) = \psi(x)$ for each $x \in B$) is also KKM.*

2. *If C is a convex subset of A, then the correspondence $\theta\colon C \twoheadrightarrow X$, defined by $\theta(x) = C \cap \psi(x)$, is likewise a KKM correspondence.*

Proof: The validity of (1) is obvious. To see that (2) is true, assume that $x_1, \ldots, x_n$ belong to C. The convexity of C implies $\mathrm{co}\{x_1, \ldots, x_n\} \subset C$ and the fact that ψ is KKM yields $\mathrm{co}\{x_1, \ldots, x_n\} \subset \bigcup_{i=1}^n \psi(x_i)$. Consequently, we have

$$\mathrm{co}\{x_1, \ldots, x_n\} = C \cap \mathrm{co}\{x_1, \ldots, x_n\}$$
$$\subset C \cap \left[\bigcup_{i=1}^n \psi(x_i)\right] = \bigcup_{i=1}^n C \cap \psi(x_i) = \bigcup_{i=1}^n \theta(x_i).$$

This shows that θ is a KKM correspondence (whose values also lie in C).   ∎

The KKM Lemma is an existence theorem, as it asserts the existence of a point lying in each of $F_1, \ldots, F_m$. While the original KKM Lemma is restricted to finite dimensional spaces, there is a straightforward and useful infinite dimensional generalization due to K. Fan [118]. Further generalizations may be found in S. Park [269] and the references therein.

17.46 Theorem (Fan) *Let A be a subset of a Hausdorff topological vector space X and let $\psi\colon A \twoheadrightarrow X$ be a KKM correspondence. If ψ is closed-valued and $\psi(x)$ is compact for some x, then $\bigcap_{x \in A} \psi(x)$ is compact and nonempty.*

Proof: Since one of the values $\psi(x)$ is compact, by Theorem 2.31 it suffices to show that the collection $\{\psi(x) : x \in A\}$ of closed sets has the finite intersection property. So let $\{x_1, \ldots, x_n\}$ be a finite subset of A. Let M be the finite dimensional subspace of X spanned by $\{x_1, \ldots, x_n\}$. Then $F_i = \psi(x_i) \cap \mathrm{co}\{x_1, \ldots, x_n\}$, $i = 1, \ldots, n$, are closed subsets of M and satisfy the hypothesis of the finite dimensional KKM Lemma 17.43. Thus $\bigcap_{i=1}^n \psi(x_i) \supset \bigcap_{i=1}^n F_i \neq \varnothing$. Therefore the family $\{\psi(x) : x \in A\}$ has the finite intersection property.   ∎

Let $\varphi\colon X \twoheadrightarrow Y$ be a correspondence. Recall that the inverse correspondence of φ is the correspondence $\varphi^{-1}\colon Y \twoheadrightarrow X$ defined by $\varphi^{-1}(y) = \{x \in X : y \in \varphi(x)\}$. The **inverse complement correspondence** of φ is the correspondence $\psi\colon X \twoheadrightarrow X$ defined by $\psi(x) = X \setminus \varphi^{-1}(x)$.

17.47 Lemma   *Let A be a subset of a vector space X, let $\varphi\colon A \twoheadrightarrow X$ be a correspondence (whose values may be empty), and let $\psi\colon X \twoheadrightarrow X$ be the inverse complement correspondence of φ defined by $\psi(x) = X \setminus \varphi^{-1}(x)$. Then the following statements are equivalent:*

1. $x \notin \operatorname{co}\varphi(x)$ *for all* $x \in A$.

2. ψ *is a KKM correspondence.*

Proof: (1) $\implies$ (2)   Assume $x \notin \operatorname{co}\varphi(x)$ for all $x \in A$ and let $x_1, \ldots, x_n$ belong to A. We must show that $\operatorname{co}\{x_1, \ldots, x_n\} \subset \bigcup_{i=1}^{n} \psi(x_i)$.

To this end, let $y \in \operatorname{co}\{x_1, \ldots, x_n\}$ and assume by way of contradiction that $y \notin \bigcup_{i=1}^{n} \psi(x_i)$. The identity

$$X \setminus \bigcup_{i=1}^{n} \psi(x_i) = \bigcap_{i=1}^{n} [X \setminus \psi(x_i)] = \bigcap_{i=1}^{n} \varphi^{-1}(x_i)$$

yields $y \in A$ and $x_i \in \varphi(y)$ for all i. This implies $\operatorname{co}\{x_1, \ldots, x_n\} \subset \operatorname{co}\varphi(y)$. Consequently, $y \in \operatorname{co}\varphi(y)$, a contradiction. Thus, $y \in \bigcup_{i=1}^{n} \psi(x_i)$ and hence $\operatorname{co}\{x_1, \ldots, x_n\} \subset \bigcup_{i=1}^{n} \psi(x_i)$, as desired.

(2) $\implies$ (1)   Let ψ be a KKM correspondence, and let x belong to A. Assume by way of contradiction that $x \in \operatorname{co}\varphi(x)$, that is, there is a convex combination $x = \sum_{i=1}^{n} \lambda_i y_i$ with $y_i \in \varphi(x)$ for each i. Then $x \in \varphi^{-1}(y_i)$ or $x \notin \psi(y_i)$ for each i. By hypothesis, $x \in \operatorname{co}\{y_1, \ldots, y_n\} \subset \bigcup_{i=1}^{n} \psi(y_i)$, which means that $x \in \psi(y_i)$ for some i, a contradiction. Thus, $x \notin \operatorname{co}\varphi(x)$ for each x. ∎

We often find it more useful and intuitive to recast arguments involving KKM correspondences in terms of binary relations. This framework was introduced in economics by K. J. Arrow [21] and H. Sonnenschein [317] as a model of preference relations. Consider a binary relation $>$ on the set X that is not necessarily either transitive or complete. (We interpret the statement $x > y$ to mean that "x is preferred to y.") Define the **upper section** of $>$ at x to be $\varphi(x) = \{y \in X : y > x\}$ and the **lower section** to be $\varphi^{-1}(x) = \{y \in X : x > y\}$. We can identify $>$ with the (possibly empty-valued) correspondence $\varphi\colon X \twoheadrightarrow X$.

The economic content of convexity of preferences is captured by the requirement that $x \notin \operatorname{co}\varphi(x)$. Note that this convexity requirement also implies that $>$ is irreflexive. An element x is **$>$-maximal** in $A \subset X$ if $\varphi(x) \cap A = \varnothing$. Let ψ denote the inverse complement correspondence, defined by $\psi(x) = X \setminus \varphi^{-1}(x)$. If $>$ is asymmetric, that is, if $y > x$ implies $x \not> y$, then $\psi(x) \supset \varphi(x)$. Also note that if $>$ is defined only on a subset A of X, that is, if $x > y$ implies $x, y \in A$, then $\psi(x)$ includes $X \setminus A$.

The next result presents sufficient conditions for a possibly nontransitive relation to possess maximal elements.

17.48 Theorem   (Existence of ≻-maximal elements)   *Let K be a nonempty compact convex subset of a Hausdorff topological vector space X and let the binary relation ≻ on K satisfy:*

1. *For each x in K we have $x \notin \text{co}\{y \in K : y \succ x\}$.*

2. *For each x in K the lower section $\{y \in K : x \succ y\}$ is open in K.*

Then the set of ≻-maximal elements in K is nonempty and compact.

Proof: Define the correspondence $\varphi \colon K \twoheadrightarrow X$ by $\varphi(x) = \{y \in K : y \succ x\}$. Assumption (1) and Lemma 17.47 imply that the inverse complement correspondence $\psi(x) = X \setminus \varphi^{-1}(x) = \{y \in X : y \not\succ x\}$ is a KKM map. Since K is convex, Lemma 17.45 (2) implies that the correspondence $\theta \colon K \twoheadrightarrow X$ defined by

$$\theta(x) = K \cap \psi(x) = K \cap [X \setminus \varphi^{-1}(x)] = K \setminus \varphi^{-1}(x)$$

is also a KKM map with values in K. Assumption (2) guarantees that $\theta(x)$ is a compact subset of K for each $x \in K$. Thus Fan's generalization of the KKM Lemma (Theorem 17.46) guarantees that $\bigcap_{z \in K} \theta(z)$ is nonempty and compact.

But $x \in \theta(z)$ if and only if $x \in K$ and $z \not\succ x$, so $\bigcap_{z \in K} \theta(z)$ is just the set of ≻-maximal elements of K.   ∎

As an application of this result we present a fundamental existence theorem. The finite dimensional version is due to P. Hartman and G. Stampacchia [153]. Related results may be found in F. E. Browder [68].

17.49 Theorem   (Browder–Hartman–Stampacchia)   *Let K be a nonempty compact convex subset of a locally convex Hausdorff space X, and let $p \colon K \to X'$. Suppose the mapping $(x, y) \mapsto \langle y, p(x) \rangle$ is jointly continuous on $K \times K$. Then there exists some $x_0 \in K$ satisfying*

$$\langle x_0, p(x_0) \rangle \leqslant \langle y, p(x_0) \rangle \text{ for all } y \in K.$$

Proof: Define the binary relation ≻ on K by $a \succ b$ if $\langle a, p(b) \rangle < \langle b, p(b) \rangle$. Clearly ≻ is irreflexive. Also note that each $x \in K$ does not belong to the convex set

$$\{y \in K : y \succ x\} = \{y \in K : \langle y, p(x) \rangle < \langle x, p(x) \rangle\}.$$

Furthermore, from the continuity of the mappings

$$(x, y) \mapsto (y, x - y) \mapsto \langle x - y, p(y) \rangle,$$

it follows that the mapping $(x, y) \mapsto \langle x - y, p(y) \rangle$ is jointly continuous on $K \times K$. Therefore, for each $x \in K$ the set

$$\{y \in K : x \succ y\} = \{y \in K : \langle x - y, p(y) \rangle < 0\}$$

is open in K. Thus, by Theorem 17.48, there is a ≻-maximal x_0 in K. That is, for such an x_0, we have $\langle x_0, p(x_0) \rangle \leqslant \langle y, p(x_0) \rangle$ for all $y \in K$.   ∎

17.9 Fixed point theorems

We can use the preceding results on KKM correspondences to prove a number of well-known fixed point existence theorems. These techniques were pioneered by F. E. Browder [70] and K. Fan [118, 119]. An excellent survey of related results is in A. Granas [140], and a more detailed treatment may be found in the definitive work of A. Granas and J. Dugundji [141]. The monographs by D. R. Smart [313] and K. C. Border [56] are accessible treatments of fixed point theorems with applications.

17.50 Definition *Let A be a subset of a set X. A **fixed point** of a function $f : A \to X$ is a point x in A satisfying $f(x) = x$. A **fixed point** of a correspondence $\varphi : A \twoheadrightarrow X$ is a point x in A satisfying $x \in \varphi(x)$.*

We now note two trivial results on the set of fixed points of a correspondence.

17.51 Lemma *If A is a subset of a topological space X and a correspondence $\varphi : A \twoheadrightarrow X$ has closed graph, then the set of fixed points of φ is closed in A.*

Hint of proof: A point x is a fixed point of φ if and only if $(x, x) \in \mathrm{Gr}\, \varphi$. ∎

17.52 Corollary *Let A be a subset of a topological space X and assume that $\varphi : A \twoheadrightarrow X$ is an upper hemicontinuous correspondence. If either*

1. *φ is closed-valued and Y is regular, or*

2. *φ is compact-valued and Y is Hausdorff,*

then the set of fixed points of the correspondence φ is closed.

Proof: Use Theorem 17.10 to show that the correspondence has closed graph. ∎

17.53 Definition *Let A be a subset of a vector space X. A correspondence $\varphi : A \twoheadrightarrow X$ is **inward pointing** (resp. **outward pointing**) if for each $x \in A$ there exists some $y \in \varphi(x)$ and $\lambda > 0$ (resp. $\lambda < 0$) satisfying $x + \lambda(y - x) \in A$.*

Figure 17.1 depicts an inward pointing correspondence. We point out that if φ has nonempty values and maps A into itself, then φ must be inward pointing: To see this, choose any $y \in \varphi(x)$, and notice that $x+1(y - x) = y \in A$. Also, the identity map is both inward and outward pointing, as $x = x + 1(x - x) = x - 1(x - x)$.

The following fundamental fixed point theorem is due to B. R. Halpern [150] and B. R. Halpern and G. M. Bergman [151]. The proof is based on F. E. Browder [69] and K. Fan [119]. As we shall see, many useful fixed point theorems are easy corollaries.

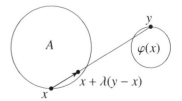

Figure 17.1. φ points inward at x.

17.54 Theorem (Halpern–Bergman) *Let K be a nonempty compact convex subset of a locally convex Hausdorff space X, and let $\varphi \colon K \twoheadrightarrow X$ be an inward pointing upper demicontinuous correspondence with nonempty closed convex values. Then φ has a fixed point.*

Proof: Suppose by way of contradiction that $x \notin \varphi(x)$ for each $x \in K$. Then, by Corollary 5.80, for each $x \in K$ there is a continuous linear functional q_x strongly separating x from $\varphi(x)$. That is, for each $x \in K$, there is a continuous linear functional q_x and a real number α_x such that $q_x(y) < \alpha_x$ for each $y \in \varphi(x)$ and $q_x(x) > \alpha_x$. Since φ is upper demicontinuous, the set

$$U_x = \varphi^{\mathrm{u}}(\{y \in K : q_x(y) < \alpha_x\}) \cap \{y \in K : q_x(y) > \alpha_x\}$$

is an open neighborhood of x in K, and clearly the collection $\{U_x : x \in K\}$ covers K. Since K is compact, there is a finite subcover $\{U_{x_1}, \ldots, U_{x_n}\}$. By Lemma 2.92 there exist continuous functions $f_i \colon K \to [0, 1]$, $i = 1, \ldots, n$, satisfying $f_i = 0$ on $U_{x_i}^c$ for each i, and $\sum_{i=1}^n f_i(x) = 1$ for each $x \in K$. Thus if $f_i(x) > 0$, we have $x \in U_{x_i}$. So $f_i(x) > 0$ implies $q_{x_i}(y) < \alpha_{x_i}$ for each $y \in \varphi(x)$, and $q_{x_i}(x) > \alpha_{x_i}$.

Define $p \colon K \to X'$ by $p(x) = \sum_{i=1}^n f_i(x) q_{x_i}$. Since each q_{x_i} is a continuous linear functional in X, observe that the function

$$(x, y) \mapsto \langle y, p(x) \rangle = \sum_{i=1}^n f_i(x) q_{x_i}(y)$$

is jointly continuous in x and y. Thus by Theorem 17.49 there is some $x_0 \in K$ satisfying

$$\langle y, p(x_0) \rangle \geqslant \langle x_0, p(x_0) \rangle \tag{1}$$

for all $y \in K$. Furthermore, for $x \in K$ and $y \in \varphi(x)$, it follows that

$$\langle x, p(x) \rangle = \sum_{i=1}^n f_i(x) q_{x_i}(x) > \sum_{i=1}^n f_i(x) \alpha_{x_i} > \sum_{i=1}^n f_i(x) q_{x_i}(y) = \langle y, p(x) \rangle. \tag{2}$$

Since φ points inward, there exist $y_0 \in \varphi(x_0)$ and $\lambda > 0$ such that $x_0 + \lambda(y_0 - x_0) \in K$. Now letting $y = x_0 + \lambda(y_0 - x_0)$ in (1), we get

$$\langle y_0, p(x_0) \rangle \geqslant \langle x_0, p(x_0) \rangle,$$

which contradicts (2). Hence, $x \in \varphi(x)$ for some $x \in K$.   ∎

The above proof (with suitable modifications) also shows that we could replace inward with outward pointing in the hypotheses. We can now establish several classical fixed point theorems.

17.55 Corollary (Kakutani–Fan–Glicksberg)   *Let K be a nonempty compact convex subset of a locally convex Hausdorff space, and let the correspondence $\varphi\colon K \twoheadrightarrow K$ have closed graph and nonempty convex values. Then the set of fixed points of φ is compact and nonempty.*

Proof: Let K be a nonempty compact convex subset of a locally convex Hausdorff space X, and let $\varphi\colon K \twoheadrightarrow K$ satisfy the properties. Recall that for compact Hausdorff range spaces, closedness is equivalent to upper hemicontinuity (Closed Graph Theorem 17.11). This implies that φ is upper hemicontinuous so φ as a correspondence from K to X is also upper hemicontinuous (and also has closed graph). In particular, it follows that $\varphi\colon K \twoheadrightarrow X$ is upper demicontinuous. From Lemma 17.51, the set of fixed points of $\varphi\colon K \twoheadrightarrow X$ is closed (and so compact). To see that fixed points exist, observe that since φ maps K into itself, it is inward pointing. Now apply Theorem 17.54.   ∎

The next fixed point theorem is immediate from the fact that continuous functions define upper hemicontinuous correspondences, but is stated separately for historical reasons.

17.56 Corollary (Brouwer–Schauder–Tychonoff)   *Let K be a nonempty compact convex subset of a locally convex Hausdorff space, and let $f\colon K \to K$ be a continuous function. Then the set of fixed points of f is compact and nonempty.*

These fixed point results were first proven for compact convex subsets of finite dimensional spaces. The finite dimensional version of Corollary 17.56, the fixed point theorem for continuous functions, was proven by L. E. J. Brouwer [67], and so is known as the **Brouwer Fixed Point Theorem**. J. Schauder [297, 296, 298] extended the result to cover weakly compact convex subsets of Banach spaces, and A. Tychonoff [333] generalized the theorem to locally convex spaces. The finite dimensional version of Corollary 17.55, a fixed point theorem for upper hemicontinuous convex-valued correspondences, is the **Kakutani Fixed Point Theorem**, due to S. Kakutani [191]. H. F. Bohnenblust and S. Karlin [55] generalized it to Banach spaces, and I. L. Glicksberg [139] and K. Fan [117] generalized the result to locally convex spaces.

There is another, older technology for proving fixed point theorems based on finite dimensional results, which historically were proven first. The idea is that a compact set in an infinite dimensional space, is in a sense, almost finite dimensional. By choosing the finite dimensional approximation cleverly, a finite dimensional fixed point theorem is used to find a fixed point of an approximation of the original correspondence. Taking limits yields a fixed point of the original infinite dimensional correspondence. As an example of this technique we prove the following generalization of Corollary 17.55, which is due to H. F. Bohnenblust and S. Karlin [55].

17.57 Theorem (Bohnenblust–Karlin) *Let C be a nonempty closed convex sub-set of a locally convex Hausdorff space, and let $\varphi\colon C \twoheadrightarrow C$ be a correspondence with closed graph and nonempty convex values. If the range of φ is relatively compact (or, equivalently, if it is included in a compact set), then the set of fixed points of φ is nonempty and compact.*

Proof: Let C be a closed subset of locally convex Hausdorff space (X, τ), and that $\varphi\colon C \twoheadrightarrow C$ is nonempty and convex-valued correspondence with a closed graph. Let $\mathcal{V}$ denote the collection of all closed, convex and symmetric τ-neighborhoods of zero. As usual, $\mathcal{V}$ is directed by inclusion.

Since $\overline{\varphi(C)}$ is τ-compact, for any $V \in \mathcal{V}$ there is a finite subset F_V of $\varphi(C)$ such that $\overline{\varphi}(C) \subset \bigcup_{x \in F_V}(x + V)$. Let C_V be the convex hull of F_V. Then C_V is a nonempty convex compact subset of the finite dimensional vector subspace of X generated by F_V. Now consider the convex-valued correspondence $\varphi_V\colon C_V \twoheadrightarrow C_V$ defined by $\varphi_V(x) = [\varphi(x) + V] \cap C_V$. We claim that each φ_V has closed graph and nonempty values.

To see that φ_V has closed graph, let $\{(x_\alpha, y_\alpha)\}$ be a net in the graph of φ_V satisfying $(x_\alpha, y_\alpha) \xrightarrow{\tau} (x, y)$. We must show that $y \in \varphi_V(x)$.

To see this, for each α pick $z_\alpha \in \varphi(x_\alpha)$ and $v_\alpha \in V$ such that $y_\alpha = z_\alpha + v_\alpha$. Since $\{z_\alpha\} \subset \overline{\varphi(C)}$, by passing to a subnet if necessary, we can assume that $z_\alpha \xrightarrow{\tau} z$ in X. Since C is τ-closed, $z \in C$, and since the graph of φ is closed, we get $z \in \varphi(x)$. Moreover, since V is τ-closed and $v_\alpha = y_\alpha - z_\alpha \xrightarrow{\tau} y - z$ we have $v = y - z \in V$. Thus $y = z + v \in [\varphi(x) + V] \cap C_V = \varphi_V(x)$, as desired.

To see that φ_V has nonempty values, fix $x \in C_V$. Pick $z \in \varphi(x)$ and then $u \in F_V (\subset C_V)$ and $v \in V$ such that $z = u + v$. Since the neighborhood V is symmetric, we have $-v \in V$ and so $u = z - v \in [\varphi(x) + V] \cap C_V = \varphi_V(x)$. This shows that $\varphi_V(x) \neq \varnothing$.

By the Kakutani Fixed Point Theorem applied to C_V, for each $V \in \mathcal{V}$ there is a fixed point $x_V \in \varphi_V(x_V)$. Write $x_V = z_V + u_V$ where $z_V \in \varphi(x_V)$ and $u_V \in V$. Clearly, the net $u_V \xrightarrow{\tau} 0$. In addition, the net $\{z_V\}_{V \in \mathcal{V}}$ lies in the compact set $\overline{\varphi(C)}$ and so it has a τ-convergent subnet. We can assume without loss of generality that $z_V \xrightarrow{\tau} z$. Now C is closed so $z \in C$. Since (x_V, z_V) belongs to the graph of φ for each $V \in \mathcal{V}$ and $(x_V, z_V) \to (z, z)$, it follows once more from the closedness of the graph of φ that (z, z) lies in the graph of φ. In other words, z is a fixed point of φ.

Lemma 17.51 guarantees that the set of fixed points of φ is compact. ∎

As yet another application of Theorem 17.49, we present the following gener-alization due to K. Fan [119] of a "coincidence" theorem of F. E. Browder [69].

17.58 Coincidence Theorem　*Let K be a nonempty compact convex subset of a locally convex Hausdorff space X. Let $\varphi, \psi\colon K \twoheadrightarrow X$ be upper demicontinuous correspondences with nonempty closed convex values. Assume that for each x in K at least one of $\varphi(x)$ or $\psi(x)$ is compact.*

Suppose that for each x in K, there exist $u \in \varphi(x)$, $v \in \psi(x)$, and a real number $\lambda > 0$ such that $x + \lambda(u-v) \in K$. Then there exists x in K satisfying $\varphi(x) \cap \psi(x) \neq \varnothing$.

Proof: Assume by way of contradiction that $\varphi(x)$ and $\psi(x)$ are disjoint for every x in K. In that case, by the Strong Separating Hyperplane Theorem 5.79, for each x we can strongly separate the two sets. So as in the proof of Theorem 17.54, we can use a partition of unity to construct a mapping $p \colon K \to X'$ such that $(x, y) \mapsto \langle y, p(x) \rangle$ is jointly continuous and

$$\langle u, p(x) \rangle < \langle v, p(x) \rangle \tag{3}$$

for every $u \in \varphi(x)$ and every $v \in \psi(x)$. By Theorem 17.49, there exists some $x_0 \in K$ with

$$\langle y, p(x_0) \rangle \geqslant \langle x_0, p(x_0) \rangle \tag{4}$$

for all $y \in K$. By hypothesis there exist $u_0 \in \varphi(x_0)$, $v_0 \in \psi(x_0)$, and $\lambda > 0$ satisfying $x_0 + \lambda(u_0 - v_0) \in K$. Putting $y = x_0 + \lambda(u_0 - v_0)$ in (4), we see that $\lambda \langle u_0, p(x_0) \rangle \geqslant \lambda \langle v_0, p(x_0) \rangle$ or

$$\langle u_0, p(x_0) \rangle \geqslant \langle v_0, p(x_0) \rangle,$$

which contradicts (3). Hence, $\varphi(x) \cap \psi(x) \neq \varnothing$ for some $x \in K$. ∎

The next result can be viewed as saying that an inward pointing set-valued vector field must vanish somewhere on a compact convex set.

17.59 Corollary *Let K be a nonempty compact convex subset of a locally convex Hausdorff space X. Let $\varphi \colon K \twoheadrightarrow X$ be upper demicontinuous with nonempty closed convex values.*

Suppose that for each x in K there exist $y \in \varphi(x)$ and a real number $\lambda > 0$ such that $x + \lambda y \in K$. Then there exists x in K satisfying $0 \in \varphi(x)$.

Proof: Let $\psi \colon K \twoheadrightarrow X$ be the correspondence defined by $\psi(x) = \{0\}$. Now apply the Coincidence Theorem 17.58. ∎

17.10 Contraction correspondences

We now turn our attention to fixed points of contraction correspondences.

17.60 Definition *Let (X, d) be a metric space with induced Hausdorff metric h_d. A correspondence $\varphi \colon X \twoheadrightarrow X$ is a **contraction correspondence** on X if it has nonempty closed d-bounded values and there is a constant $0 < c < 1$ satisfying*

$$h_d(\varphi(x), \varphi(y)) \leqslant c\, d(x, y)$$

*for all $x, y \in X$. The constant c is called a **modulus of contraction** for the correspondence φ.*

The next fixed point theorem, due to S. B. Nadler [259], generalizes the Banach Fixed Point Theorem 3.48 to contraction correspondences.

17.61 Nadler Fixed Point Theorem *Every contraction correspondence on a complete metric space has a fixed point.*

Proof: The argument is similar to that of the proof of the Contraction Mapping Theorem 3.48. Let c be a modulus of contraction for φ. Pick any point $x_0 \in X$, and pick $x_1 \in \varphi(x_0)$, then pick $x_2 \in \varphi(x_1)$ to satisfy $d(x_1, x_2) \leqslant h_d(\varphi(x_0), \varphi(x_1)) + c$. Recursively construct a sequence $\{x_n\}$ satisfying

$$x_{n+1} \in \varphi(x_n) \quad \text{and} \quad d(x_n, x_{n+1}) \leqslant h_d(\varphi(x_{n-1}), \varphi(x_n)) + c^n.$$

For each $n \geqslant 1$, a little arithmetic yields

$$
\begin{aligned}
d(x_n, x_{n+1}) &\leqslant h_d(\varphi(x_{n-1}), \varphi(x_n)) + c^n \\
&\leqslant cd(x_{n-1}, x_n) + c^n \\
&\leqslant c\big[h_d(\varphi(x_{n-2}), \varphi(x_{n-1})) + c^{n-1}\big] + c^n \\
&\leqslant c^2 d(x_{n-2}, x_{n-1}) + 2c^n \\
&\;\;\vdots \\
&\leqslant c^n d(x_0, x_1) + nc^n.
\end{aligned}
$$

Consequently,

$$
\begin{aligned}
d(x_n, x_{n+m}) &\leqslant \sum_{k=n}^{n+m-1} d(x_k, x_{k+1}) \\
&\leqslant \sum_{k=n}^{n+m-1} \big[c^k d(x_0, x_1) + kc^k\big] \\
&\leqslant \Big[\sum_{k=n}^{\infty} c^k\Big] d(x_0, x_1) + \sum_{k=n}^{\infty} kc^k.
\end{aligned}
$$

Since both series $\sum_{k=1}^{\infty} c^k$ and $\sum_{k=1}^{\infty} kc^k$ converge, we see that the sequence $\{x_n\}$ is a d-Cauchy sequence. Since X is d-complete, there exists some $x \in X$ satisfying $d(x_n, x) \to 0$. To complete the proof, we show that $x \in \varphi(x)$.

To this end, notice that for each n we have

$$0 \leqslant d(x_{n+1}, \varphi(x)) \leqslant h_d(\varphi(x_n), \varphi(x)) \leqslant cd(x_n, x).$$

Letting $n \to \infty$, we get $d(x, \varphi(x)) = 0$, so $x \in \overline{\varphi(x)} = \varphi(x)$. ∎

For recent generalizations of this sort of result, see P. Diamond [93].

17.11 Continuous selectors

A selector from a relation $R \subset X \times Y$ is a subset S of Y such that for every $x \in X$ there exists a unique $y_x \in S$ satisfying $(x, y_x) \in R$. Viewing relations as correspondences, we see that only nonempty-valued correspondences can admit selectors. On the other hand, the Axiom of Choice guarantees that nonempty-valued correspondences always admit selectors—but they may have no additional useful properties.

17.62 Definition   *A **selector** or **selection** from a correspondence φ mapping X to Y is a function $f : X \to Y$ that satisfies $f(x) \in \varphi(x)$ for each $x \in X$. If X and Y are topological spaces, then we say that f is a **continuous selector** if f is a selector and is continuous.*

The next theorem is due to F. E. Browder [70], and while it is one of the more straightforward selection theorems, it is also very useful.

17.63 Browder Selection Theorem   *A correspondence having nonempty convex values and open lower sections from a paracompact topological space to a Hausdorff topological vector space admits a continuous selector.*

Proof: Let $\varphi : X \twoheadrightarrow Y$ satisfy the hypotheses, and let F denote the image of X under φ. Since φ has nonempty values, the family $\{\varphi^{-1}(y) : y \in F\}$ of lower sections is an open cover of X. (If $x \in X$, then $y \in \varphi(x)$ for some $y \in F$, so $x \in \varphi^{-1}(y)$.) From the paracompactness of X, it follows that there exists a locally finite continuous partition of unity $\{g_y\}_{y \in F}$ such that $g_y(x) = 0$ for each $x \notin \varphi^{-1}(y)$. In particular, notice that $g_y(x) > 0$ implies $y \in \varphi(x)$. Now for each $x \in X$ let $f(x) = \sum_{y \in F} g_y(x) y$ and note that the local finiteness of $\{g_y\}_{y \in F}$ in conjunction with the convexity of $\varphi(x)$ guarantees that $f(x) \in \varphi(x)$ for all $x \in X$. Now it remains to observe that the formula $f(x)$ defines a function $f : X \to F \subset Y$ that is continuous (why?) and hence it is a continuous selector from φ. ∎

Here is an example of a correspondence from the unit interval into itself that satisfies the hypotheses of Browder's Theorem. It has nonempty closed convex values, and convex open lower sections, but its graph is not closed, nor open, nor convex.

Browder's Selection Theorem applies to any topological vector space but requires the some-what strong assumption of open lower sections. E. Michael [249] proved a series of theorems on the existence of continuous selectors that assume the weaker condition of lower hemicontinuity of the correspondences, but require in addition that the range spaces be Fréchet

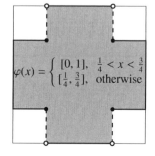

$$\varphi(x) = \begin{cases} [0,1], & \frac{1}{4} < x < \frac{3}{4} \\ [\frac{1}{4}, \frac{3}{4}], & \text{otherwise} \end{cases}$$

spaces (with the metric topology). We present only the half of one of these theorems that guarantees the existence of continuous selectors. But first we need two simple lemmas.

17.64 Lemma *Let X be paracompact, let Y be a locally convex space, and let $\psi\colon X \twoheadrightarrow Y$ be lower hemicontinuous with nonempty convex values. If V is an open convex circled neighborhood of zero, then there exists a continuous function $f\colon X \to Y$ satisfying $f(x) \in \psi(x) + V$ for each $x \in X$.*

Proof: For each $x \in X$ choose $y_x \in \psi(x)$ and note that $\{\psi^\ell(y_x + V) : x \in X\}$ is an open cover of X. Let $\{f_x : x \in X\}$ be a locally finite continuous partition of unity subordinated to this cover. Then $f_x(z) > 0$ implies $z \in \psi^\ell(y_x + V)$, or equivalently, $y_x \in \psi(z) + V$. Since $\psi(z)$ and V are convex, so is $\psi(z) + V$. Consequently, the convex combination $f(z) = \sum_{x \in X} f_x(z)y_x$ belongs to $\psi(z) + V$. Now notice that f is a continuous function with the desired properties. ∎

17.65 Lemma *Let $\varphi\colon X \twoheadrightarrow Y$ be a lower hemicontinuous correspondence into a topological vector space. If $f\colon X \to Y$ is continuous and U is an open neighborhood of zero, then the correspondence $\psi\colon X \twoheadrightarrow Y$ defined by*

$$\psi(x) = \varphi(x) \cap [f(x) + U]$$

is also lower hemicontinuous.

Proof: Let G be an open set in Y and suppose $y_0 \in \varphi(x_0) \cap [f(x_0) + U] \cap G$. In particular, y_0 belongs to the open set $[f(x_0) + U] \cap G$. Thus there is an open symmetric neighborhood V of zero such that

$$y_0 + V + V \subset [f(x_0) + U] \cap G.$$

In particular, note that $G \supset y_0 + V$.

Now let $W = f^{-1}(f(x_0) + V)$. The continuity of f guarantees that W is an open neighborhood of x_0. Next, we claim that

$$f(x) + U \supset y_0 + V$$

for each $x \in W$. To see this, suppose $v \in V$ and $x \in W$. Then $z = f(x_0) - f(x) \in V$. Since $y_0 + V + V \subset f(x_0) + U$, there exists $u \in U$ satisfying $y_0 + v + z = f(x_0) + u$. Rewriting, we get

$$y_0 + v = f(x_0) - z + u = f(x) + u \in f(x) + U.$$

Since $y_0 \in \varphi(x_0) \cap (y_0 + V)$, we see that $x_0 \in \varphi^\ell(y_0 + V)$. Now consider the neighborhood $N = W \cap \varphi^\ell(y_0 + V)$ of x_0. If $x \in N$, then

$$\begin{aligned}
\psi(x) \cap G &= \varphi(x) \cap [f(x) + U] \cap G \\
&\supset \varphi(x) \cap [f(x) + U] \cap (y_0 + V) \\
&= \varphi(x) \cap (y_0 + V) \neq \varnothing,
\end{aligned}$$

so $\psi(x) \cap G \neq \varnothing$. Thus ψ is lower hemicontinuous. ∎

Recall that a **Fréchet space** is a completely metrizable locally convex space. The next theorem is part of Michael's Selection Theorem [249, Theorem 3.2″]. The idea behind the proof is to use Lemma 17.64 to construct a sequence of approximate continuous selectors. The sequence is cleverly designed so that it is uniformly Cauchy. Since Fréchet spaces are complete, it converges to a continuous function, which turns out to be the desired selector.

17.66 Michael Selection Theorem   *A lower hemicontinuous correspondence from a paracompact space into a Fréchet space with nonempty closed convex values admits a continuous selector.*

Proof: Let X be paracompact, let Y be a Fréchet space, and let $\varphi \colon X \twoheadrightarrow Y$ be a lower hemicontinuous correspondence with nonempty closed convex values. Let $\{V_0, V_1, V_2, \ldots\}$ be a base at zero for Y consisting of open convex circled neighborhoods of zero such that $V_{n+1} + V_{n+1} + V_{n+1} \subset V_n$ for each $n = 0, 1, 2, \ldots$. We claim that there exist a selector $f_0 \colon X \to Y$ from the correspondence φ and a sequence of continuous functions $f_1, f_2, \ldots$ from X to Y satisfying

1. $f_n(x) \in \varphi(x) + V_n$, and

2. $f_n(x) \in f_{n-1}(x) + V_{n-1}$

for each $n = 1, 2, \ldots$ and each $x \in X$.

To establish the existence of such a sequence we proceed inductively. For $n = 1$, there exists by Lemma 17.64 a continuous function $f_1 \colon X \to Y$ satisfying $f_1(x) \in \varphi(x) + V_1 \subset \varphi(x) + V_0$ for each $x \in X$. In particular, there exists a (not necessarily continuous) selector $f_0 \colon X \to Y$ from the correspondence φ satisfying $f_1(x) \in f_0(x) + V_0$.

Now, for the inductive step, assume that $f_0 \colon X \to Y$ is a selector from φ and the continuous functions $f_1, \ldots, f_k$ from X to Y have been chosen to satisfy (1) and (2) for each $n = 1, \ldots, k$. Consider the correspondence $\psi \colon X \twoheadrightarrow Y$ defined by

$$\psi(x) = \varphi(x) \cap [f_k(x) + V_k].$$

From (1), it easily follows that $\psi(x) \neq \varnothing$ for each x, and by Lemma 17.65, ψ is lower hemicontinuous. Now, by Lemma 17.64, there exists a continuous function $f_{k+1} \colon X \to Y$ satisfying $f_{k+1}(x) \in \psi(x) + V_{k+1}$ for each x. Therefore, for each $x \in X$ we have:

a. $f_{k+1}(x) \in \varphi(x) + V_{k+1}$, and

b. $f_{k+1}(x) \in f_k(x) + V_{k+1} + V_{k+1} \subset f_k(x) + V_k$,

and the induction argument is complete.

Next, notice that it follows from (2) that $f_{n+1}(x) - f_n(x) \in V_n$ for each $x \in X$ and all $n = 1, 2, \ldots$. This implies

$$f_{n+p}(x) - f_n(x) = \sum_{k=n}^{n+p-1} [f_{k+1}(x) - f_k(x)] \in V_{n+p-1} + V_{n+p} + \cdots + V_n \subset V_{n-1}$$

for each $x \in X$ and all $n = 1, 2, \ldots$. This shows that $\{f_n(x)\}$ is a Cauchy sequence in Y for each $x \in X$. Since Y is a Fréchet space, $f(x) = \lim f_n(x)$ exists in Y for each $x \in X$. Letting $p \to \infty$ and taking into consideration that $f_{n+p}(x) - f_n(x) \subset V_{n-1}$, we see that $f(x) - f_n(x) \subset \overline{V}_{n-1} \subset V_n$ for each $x \in X$ and all $n \in \mathbb{N}$. To finish the proof we shall show that the function $f : X \to Y$ is a continuous selector from the correspondence φ.

From (1), it follows immediately that f is a selector from φ. For the continuity of f fix $x_0 \in X$ and let $n \in \mathbb{N}$. Since the function $f_{n+3} : X \to Y$ is continuous at x_0, there exists a neighborhood N of x_0 such that $f_{n+3}(x) - f_{n+3}(x_0) \in V_{n+1}$ for all $x \in N$. But then for each $x \in N$ we have

$$f(x) - f(x_0) = [f(x) - f_{n+3}(x)] + [f_{n+3}(x) - f_{n+3}(x_0)] + [f_{n+3}(x_0) - f(x_0)]$$
$$\in V_{n+3} + V_{n+1} + V_{n+3} \subset V_{n+1} + V_{n+1} + V_{n+1} \subset V_n.$$

Thus, $f(x) \in f(x_0) + V_n$ for all $x \in N$, proving that f is continuous at x_0 and so continuous everywhere. $\blacksquare$

As an application, we give E. Michael's [249] generalization of a result due to R. G. Bartle and L. M. Graves [33].

17.67 Corollary (Existence of right inverses) *A surjective continuous linear operator between Fréchet spaces has a (not necessarily linear) continuous right inverse.*

Proof: Let $T : X \to Y$ be an onto continuous linear operator between Fréchet spaces. Then the inverse correspondence $\varphi : Y \twoheadrightarrow X$, defined by $\varphi(y) = T^{-1}(y)$, assumes nonempty, closed, and convex values. Since T is an open mapping (Theorem 5.18), it follows from Theorem 17.7 that φ is lower hemicontinuous. Therefore, by Theorem 17.66, φ admits a continuous selector. But any selector $S : Y \to X$ from φ satisfies $T(S(y)) = y$ for each $y \in Y$. $\blacksquare$

Chapter 18

Measurable correspondences

Throughout this chapter S denotes a measurable space and X is a topological space (usually metrizable or even Polish). We let Σ denote the σ-algebra of measurable subsets of S, and equip X with its Borel σ-algebra $\mathcal{B}_X$. A special case is where S is a topological space and Σ is its Borel σ-algebra. Of primary interest is whether a correspondence $\varphi \colon S \twoheadrightarrow X$ admits a selector that is measurable. Ideally we want a notion of measurability for correspondences so that any measurable correspondence has a measurable selector. Unfortunately this is not straightforward.

An obvious approach is to define measurability in terms of the lower inverse images of Borel sets.[1] It turns out to be extremely restrictive to require the lower inverse image of every Borel subset of X to be measurable, as we demonstrate in Example 18.11. Thus we look at definitions that require either the lower inverse image of closed sets to be measurable or the lower inverse image of open sets to be measurable. For functions it makes no difference, since $f^{-1}(A^c) = [f^{-1}(A)]^c$. This is not true for either the upper or lower inverse of a correspondence, and the two approaches lead to different notions of measurability, unless the correspondence has compact values; see Theorem 18.10. Call a correspondence *measurable* if the lower inverse of every closed set is measurable, and *weakly measurable* if the lower inverse of every open set is measurable. This choice of definitions turns out to lead to some nice results; see for instance, Theorem 18.13.

A weaker notion of measurability for a correspondence is that its graph be a measurable set. By Theorem 12.28 a function between Polish spaces is Borel measurable if and only if its graph is a Borel set. This equivalence fails to be true for correspondences. A weakly measurable correspondence has measurable graph (Theorem 18.6), but a correspondence with measurable graph need not be weakly measurable. There are two ways around this problem. One is to use a larger σ-algebra on S than the Borel σ-algebra. Indeed, the σ-algebra of universally measurable sets seems to be the appropriate one; see Theorem 18.21. If we want to avoid topological restrictions on S, we can assume that Σ is complete for some measure μ; for this approach, see the excellent treatment by E. Klein and A. C. Thompson [209]. Yet another natural notion of measurability for closed-

---

[1] The use of lower inverses rather than upper inverses is insignificant. Every definition in terms of lower inverses has a corresponding definition in terms of upper inverses.

valued correspondences arises from treating them as functions into the space $\mathcal{F}$ of nonempty closed sets.

One of the most important results concerning measurable correspondences is the Kuratowski–Ryll-Nardzewski Measurable Selection Theorem 18.13, which asserts that a weakly measurable correspondence with nonempty closed values into a Polish space has a measurable selector. This is applied to prove Filippov's Implicit Function Theorem 18.17 and the Measurable Maximum Theorem 18.19. The Measurable Maximum Theorem is a useful result that gives conditions for the set of solutions of a parametric constrained maximization problem to be measurable as well as for the optimal value function to be measurable.

We also prove a fundamental result (Theorem 18.31) relating the measurability of a correspondence having compact convex values to the measurability of its support functionals. Measurable correspondences can be integrated. The integral is defined to be the set of integrals of selectors from the correspondence. We consider the integration of compact convex-valued correspondences and present the fundamental Theorem 18.37 of V. Strassen.

18.1   Measurability notions

We start with a few natural, but not equivalent, notions of measurability.

18.1 Definition   *Let (S, Σ) be a measurable space and X a topological space. We say that a correspondence $\varphi \colon S \twoheadrightarrow X$ is:*

- **weakly measurable,** *if $\varphi^{\ell}(G) \in \Sigma$ for each open subset G of X.*

- **measurable,** *if $\varphi^{\ell}(F) \in \Sigma$ for each closed subset F of X.*

- **Borel measurable,** *if $\varphi^{\ell}(B) \in \Sigma$ for each Borel subset B of X.*

There is nothing special about using the lower inverse rather than the upper inverse in these definitions. For instance, a correspondence φ is weakly measurable if and only if $\varphi^{u}(F)$ belongs to Σ for every closed set F, since $\varphi^{u}(F) = [\varphi^{\ell}(F^{c})]^{c}$. Note well that weak measurability has nothing to do with weak topologies. Obviously, measurability and weak measurability are weaker conditions than Borel measurability. Also note that we do not require that φ have nonempty values, but observe that if φ is either measurable or weakly measurable, then the set $\{s \in S \; : \; \varphi(s) \neq \varnothing\} = \varphi^{\ell}(X)$ is measurable. Thus requiring nonempty values would not affect the measurability of a correspondence.

If φ is singleton-valued, that is, if it defines a function, then measurability, weak measurability, and Borel measurability of φ all coincide with Borel measurability of φ as a function. The main difference between functions and correspondences in terms of inverse images is that taking the inverse images under a function commutes with complementation, union, and intersection. This is not true for either the upper or lower inverse of a correspondence. Consequently the relationship between weak measurability and measurability is not immediate. For metric spaces though, the situation is clear.

18.2 Lemma *For a correspondence $\varphi \colon (S, \Sigma) \twoheadrightarrow X$ from a measurable space into a metrizable space we have the following.*

1. *If φ is measurable, then it is also weakly measurable.*

2. *If φ is compact-valued and weakly measurable, then it is measurable.*

Proof: (1) Let G be an open subset of X. By Corollary 3.19 the open set G is an $\mathcal{F}_\sigma$. So we can write $G = \bigcup_{n=1}^\infty F_n$ with each F_n closed. Then

$$\varphi^\ell(G) = \varphi^\ell\Big(\bigcup_{n=1}^\infty F_n\Big) = \bigcup_{n=1}^\infty \varphi^\ell(F_n),$$

which belongs to Σ, since φ is measurable.

(2) Fix a compatible metric d for X and let F be a closed subset of X. If F is empty, then $\varphi^\ell(F) = \varnothing$, which belongs to Σ. So assume F is nonempty. For each n put $G_n = \{x \in X : d(x, F) > \frac{1}{n}\}$, and let $F_n = \overline{G_n}$. Then $F^c = \bigcup_{n=1}^\infty G_n = \bigcup_{n=1}^\infty F_n$.

Now suppose $\varphi(s) \subset F^c$. Since $G_n \subset G_{n+1}$ for each n and $\varphi(s)$ is compact, there is some n such that $\varphi(s) \subset G_n \subset F_n$. This shows that $\varphi^u(F^c) = \bigcup_{n=1}^\infty \varphi^u(F_n)$. Consequently,

$$\varphi^\ell(F) = \big[\varphi^u(F^c)\big]^c = \Big[\bigcup_{n=1}^\infty \varphi^u(F_n)\Big]^c = \bigcap_{n=1}^\infty \varphi^\ell(F_n^c).$$

Since φ is weakly measurable, $\varphi^\ell(F_n^c) \in \Sigma$ for each n. Therefore $\varphi^\ell(F) \in \Sigma$, so φ is measurable. ∎

We mention that if the range of φ lies in some subset Y of X, then φ is weakly measurable as a correspondence into Y endowed with its relative topology if and only if is weakly measurable as a correspondence into X. The proof of this is just a matter of definitions. The next lemma points out that taking the closure preserves weak measurability.

18.3 Lemma *A correspondence $\varphi \colon S \twoheadrightarrow X$ from a measurable space into a topological space is weakly measurable if and only if its closure correspondence $\overline{\varphi}$ is weakly measurable.*

Proof: If G is an open subset of X, then note that $\varphi(s) \cap G \neq \varnothing$ if and only if $\overline{\varphi}(s) \cap G \neq \varnothing$, and the conclusion follows. ∎

The next result describes weak measurability properties of countable unions, intersections, and products of correspondences.

18.4 Lemma (Countable operations and measurability)   *For a sequence $\{\varphi_n\}$ of correspondences from a measurable space (S, Σ) into a topological space X we have the following.*

1. *The union correspondence $\varphi \colon S \twoheadrightarrow X$, defined by $\varphi(s) = \bigcup_{n=1}^{\infty} \varphi_n(s)$, is:*

   a. *weakly measurable, if each φ_n is weakly measurable,*

   b. *measurable, if each φ_n is measurable, and*

   c. *Borel measurable, if each φ_n is Borel measurable.*

2. *If X is a separable metrizable space and each φ_n is weakly measurable, then the product correspondence $\psi \colon S \twoheadrightarrow X^{\mathbb{N}}$, defined by $\psi(s) = \prod_{n=1}^{\infty} \varphi_n(s)$, is weakly measurable.*

3. *If X is a separable metrizable space, each φ_n is weakly measurable with closed values, and for each s there is some k such that $\varphi_k(s)$ is compact, then the intersection correspondence $\theta \colon S \twoheadrightarrow X$, defined by $\theta(s) = \bigcap_{n=1}^{\infty} \varphi_n(s)$, is measurable (and hence weakly measurable).*

Proof: (1)   This follows from the identity $\left(\bigcup_{n=1}^{\infty} \varphi_n \right)^{\ell}(A) = \bigcup_{n=1}^{\infty} \varphi_n^{\ell}(A)$.

(2)   Since X is a separable metrizable space, by Theorem 3.38, $X^{\mathbb{N}}$ is a separable metrizable space. Let G be open in $X^{\mathbb{N}}$. Then G can be written as a countable union $\bigcup_{k=1}^{\infty} U_k$ of basic open sets of the form $U_k = \prod_{n=1}^{\infty} U_{k,n}$, where for each k the set $U_{k,n}$ is open for all n and $U_{k,n} = X$ for all but finitely many n. Thus

$$\psi^{\ell}(G) = \psi^{\ell}\left(\bigcup_{k=1}^{\infty} U_k \right) = \bigcup_{k=1}^{\infty} \psi^{\ell}(U_k) = \bigcup_{k=1}^{\infty} \bigcap_{n=1}^{\infty} \varphi_n^{\ell}(U_{k,n}),$$

which belongs to Σ. Thus ψ is weakly measurable.

(3)   First assume that each φ_n has compact values. Then by part (2), the product correspondence $\prod_{n=1}^{\infty} \varphi_n \colon S \twoheadrightarrow X^{\mathbb{N}}$ is weakly measurable, and since it has compact values, Lemma 18.2 (2) implies it is measurable. Now observe that for a closed set $F \subset X$, we have

$$\theta^{\ell}(F) = \left\{ s \in S : \left[\bigcap_{n=1}^{\infty} \varphi_n(s) \right] \cap F \neq \varnothing \right\}$$

$$= \left\{ s \in S : \left[\prod_{n=1}^{\infty} \varphi_n(s) \right] \cap F^{\mathbb{N}} \cap D \neq \varnothing \right\}$$

$$= \left[\prod_{n=1}^{\infty} \varphi_n \right]^{\ell} (F^{\mathbb{N}} \cap D),$$

where D is the diagonal of $X^{\mathbb{N}}$, that is, $D = \{(x)_{n \in \mathbb{N}} : x \in X\}$. Now observe that $F^{\mathbb{N}} \cap D$ is closed in $X^{\mathbb{N}}$. Thus, the measurability of $\prod_{n=1}^{\infty} \varphi_n$ implies $\theta^{\ell}(F) \in \Sigma$ for each closed subset F of X.

Now drop the assumption that every $\varphi_n(s)$ is compact. By Corollary 3.41, the space X has a metrizable compactification $\hat{X}$. For each n define $\hat{\varphi}_n \colon S \twoheadrightarrow \hat{X}$ by $\hat{\varphi}_n(s) = \overline{\varphi_n(s)}$, the closure of $\varphi_n(s)$ in $\hat{X}$. By Lemma 18.3 each $\hat{\varphi}_n$ is weakly measurable. But $\hat{\varphi}_n$ is compact-valued, and so measurable by Lemma 18.2.

The preceding conclusion implies that the correspondence $\hat{\theta} \colon S \twoheadrightarrow \hat{X}$ defined by $\hat{\theta}(s) = \bigcap_{n=1}^{\infty} \hat{\varphi}_n(s)$ is measurable. But if $\varphi_k(s)$ is already compact for some k, then $\hat{\varphi}_k(s) = \varphi_k(s) \subset X$ for that k. So $\hat{\theta}(s) \subset X$ for each $s \in S$. Moreover, since each $\varphi_n \colon S \twoheadrightarrow X$ is closed-valued, it is easy to see that $\hat{\varphi}_n(s) \cap X = \varphi_n(s)$ for each s. This implies

$$\hat{\theta}(s) = \left[\bigcap_{n=1}^{\infty} \hat{\varphi}_n(s) \right] \cap X = \bigcap_{n=1}^{\infty} [\hat{\varphi}_n(s) \cap X] = \bigcap_{n=1}^{\infty} \varphi_n(s) = \theta(s).$$

Therefore θ is a measurable correspondence from S to $\hat{X}$.

Now let F be a closed subset of X. If $\overline{F}$ is the closure of F in $\hat{X}$, then $\overline{F} \cap X = F$, so

$$\theta^{\ell}(F) = \{ s \in S : \theta(s) \cap F \neq \varnothing \} = \{ s \in S : \hat{\theta}(s) \cap \overline{F} \neq \varnothing \} = (\hat{\theta})^{\ell}(\overline{F}) \in \Sigma.$$

This shows that $\theta \colon S \twoheadrightarrow X$ is indeed measurable. ∎

We can employ Carathéodory functions to characterize the measurability of correspondences in terms of the measurability of the distance functions associated with correspondences. If $\varphi \colon (S, \Sigma) \twoheadrightarrow (X, d)$ is a nonempty-valued correspondence from a measurable space into a metric space, then the **distance function** associated with φ is the function $\delta \colon S \times X \to \mathbb{R}$ defined by

$$\delta(s, x) = d(x, \varphi(s)).$$

For each $x \in X$ let δ_x denote the function $\delta_x \colon S \to \mathbb{R}$ defined via the formula $\delta_x(s) = \delta(s, x) = d(x, \varphi(s))$. Recall that a function $f \colon S \times X \to Y$, where (S, Σ) is a measurable space and X and Y are topological spaces, is a Carathéodory function if it is continuous in x and measurable in s. Also recall that Carathéodory functions are jointly measurable under mild conditions; see Lemma 4.51.

18.5 Theorem (Weak measurability and distance functions) *A nonempty-valued correspondence mapping a measurable space into a separable metrizable space is weakly measurable if and only if its associated distance function is a Carathéodory function.*

Proof: Let $\varphi \colon (S, \Sigma) \twoheadrightarrow (X, d)$ be a nonempty-valued correspondence from a measurable space into a separable metric space. Since (X, d) is separable, every open subset of X is the union of a countable family of open d-balls. Since $\varphi^{\ell}(\bigcup_{i \in I} A_i) = \bigcup_{i \in I} \varphi^{\ell}(A_i)$, we see that φ is weakly measurable if and only if $\varphi^{\ell}(B_{\varepsilon}(x))$ belongs to Σ for each $x \in X$ and each $\varepsilon > 0$. But

$$\varphi^{\ell}(B_{\varepsilon}(x)) = \{ s \in S : d(x, \varphi(s)) < \varepsilon \} = \delta_x^{-1}((-\infty, \varepsilon)).$$

Therefore φ is weakly measurable if and only if $\delta_x = \delta(\cdot, x)$ is a measurable function for each $x \in X$. Since $\delta(s, x) = d(x, \varphi(s))$ is automatically continuous in x for each s (Theorem 3.16), this occurs if and only if δ is a Carathéodory function.   ∎

The closure of a weakly measurable correspondence into a metrizable separable space has a measurable graph.

18.6 Theorem   (Weak measurability versus measurable graph)   *Consider a nonempty-valued correspondence $\varphi \colon (S, \Sigma) \twoheadrightarrow X$ from a measurable space into a separable metrizable space. If φ is weakly measurable, then its closure correspondence $\overline{\varphi}$ has measurable graph, that is, $\mathrm{Gr}\,\overline{\varphi} \in \Sigma \otimes \mathcal{B}_X$.*

Proof: Let d be a compatible metric for X. By Theorem 18.5 the distance function $\delta \colon S \times X \to \mathbb{R}$ of φ is a Carathéodory function, and hence jointly measurable by Lemma 4.51. Thus $\mathrm{Gr}\,\overline{\varphi} = \delta^{-1}(\{0\})$ belongs to $\Sigma \otimes \mathcal{B}_X$.   ∎

We can use Carathéodory functions to identify a large class of weakly measurable correspondences.

18.7 Lemma   *Suppose $f \colon S \times X \to Y$ is a Carathéodory function, where (S, Σ) is a measurable space, X is a separable metrizable space, and Y a topological space. For each subset G of Y define the correspondence $\varphi_G \colon S \twoheadrightarrow X$ by*

$$\varphi_G(s) = \{x \in X : f(s, x) \in G\}.$$

If G is open, then φ_G is a measurable correspondence.

Proof: If $G = \varnothing$, then $\varphi_G(s) = \varnothing$ for each $s \in S$ and hence φ_G is measurable. So we can assume that G is a nonempty subset of Y. Let F be a closed subset of X, and fix a countable dense subset $\{x_1, x_2, \ldots\}$ of F. Now

$$\begin{aligned}
\varphi_G^{\ell}(F) &= \{s \in S : \varphi_G(s) \cap F \neq \varnothing\} \\
&= \{s \in S : f(s, x) \in G \text{ for some } x \in F\} \\
&= \{s \in S : f(s, x_n) \in G \text{ for some } n\} \\
&= \bigcup_{n=1}^{\infty} \{s \in S : f(s, x_n) \in G\},
\end{aligned}$$

where the third equality follows from the openness of G and the continuity of f in x. Since the function f is measurable in s for each $x \in X$, each of the sets $\{s \in S : f(s, x_n) \in G\}$ belongs to Σ, so φ_G is measurable.   ∎

18.8 Corollary   *Suppose $f \colon S \times X \to \mathbb{R}$ is a Carathéodory function, where (S, Σ) is a measurable space and X is a separable metrizable space. Define the correspondence $\varphi \colon S \twoheadrightarrow X$ by*

$$\varphi(s) = \{x \in X : f(s, x) = 0\}.$$

Then the correspondence φ:

1. *has measurable graph, and*

2. *if X is compact, φ is also measurable.*

Proof: Define $\varphi_n \colon S \twoheadrightarrow X$ by $\varphi_n(s) = \{x \in X : |f(s, x)| < 1/n\}$. Then by Lemma 18.7, each correspondence φ_n is measurable (and so weakly measurable). Now $\varphi(s) \subset \overline{\varphi}_n(s) \subset \{x \in X : |f(s, x)| \leqslant \frac{1}{n}\}$, so $\varphi(s) = \bigcap_{n=1}^{\infty} \overline{\varphi}_n(s)$.

(1) Since each φ_n is weakly measurable, Theorem 18.6 implies that each $\overline{\varphi}_n$ has measurable graph. But $\operatorname{Gr}\varphi = \bigcap_{n=1}^{\infty} \operatorname{Gr}\overline{\varphi}_n$, so φ too has measurable graph.

(2) If X is compact, then each correspondence $\overline{\varphi}_n$ has compact values and so, by Lemma 18.4 (3), the intersection $\varphi = \bigcap_{n=1}^{\infty} \overline{\varphi}_n$ is measurable. ∎

18.2 Compact-valued correspondences as functions

It is possible to treat nonempty compact-valued correspondences from a measurable space S into a metrizable space X as functions from S into the space $\mathcal{K}$ of nonempty compact subsets of X. Theorem 3.91 assures us that the Hausdorff metric topology on $\mathcal{K}$ is the same for all compatible metrics on X. Furthermore, this topology is generated by the collection of all sets of the form $G^{\mathrm{u}} = \{K \in \mathcal{K} : K \subset G\}$ and $G^{\ell} = \{K \in \mathcal{K} : K \cap G \neq \varnothing\}$ as G ranges over the open subsets of X. In this section we consider functions from a measurable space into $\mathcal{K}$ that are Borel measurable. This requires that in order to study their properties, we need first a description of the Borel σ-algebra of $\mathcal{K}$.

The next theorem characterizes the Borel σ-algebra of $\mathcal{K}$ when X is separable. It is attributed to L. E. Dubins and D. S. Ornstein by G. Debreu [85, p. 355]. But see also E. G. Effros [114] and K. Kuratowski [218, Section 43, pp. 70–72, vol. 2].

18.9 Theorem (Borel σ-algebra of $\mathcal{K}$)   *Let $\mathcal{K}$ be the space of nonempty compact subsets of the separable metrizable space X, equipped with its Hausdorff metric topology. Then the Borel σ-algebra $\mathcal{B}_{\mathcal{K}}$ of $\mathcal{K}$ is generated by the family $\{G^{\mathrm{u}} : G \text{ is open}\}$, and is also generated by the family $\{G^{\ell} : G \text{ is open}\}$.*

Proof: Let $\mathcal{U} = \{G^{\mathrm{u}} : G \text{ is open}\}$ and $\mathcal{L} = \{G^{\ell} : G \text{ is open}\}$. Since $\mathcal{U}$ and $\mathcal{L}$ consist of open sets, $\sigma(\mathcal{U} \cup \mathcal{L}) \subset \mathcal{B}_{\mathcal{K}}$. Theorem 3.91 ensures that $\sigma(\mathcal{U} \cup \mathcal{L})$ includes a basis for $\mathcal{K}$. By Corollary 3.90, since X is separable, so is $\mathcal{K}$. Therefore, every open set is a countable union of basic open sets from $\sigma(\mathcal{U} \cup \mathcal{L})$. Thus, $\mathcal{B}_{\mathcal{K}} = \sigma(\mathcal{U} \cup \mathcal{L})$.

Now let G be an open subset of X different than X. Write $G = \bigcup_{n=1}^{\infty} F_n$, where each F_n is closed. (Recall that every open set in a metric space is an $\mathcal{F}_{\sigma}$.) Then

$$G^{\ell} = \bigcup_{n=1}^{\infty} F_n^{\ell} = \bigcup_{n=1}^{\infty} [(F_n^{\mathrm{c}})^{\mathrm{u}}]^{\mathrm{c}}.$$

This shows that $\sigma(\mathcal{L}) \subset \sigma(\mathcal{U})$.

To show the reverse inclusion, let d be a compatible metric for X. Then we have $G^c = \bigcap_{n=1}^{\infty} N_{\frac{1}{n}}(G^c)$, where $N_{\frac{1}{n}}(G^c) = \{x \in X : d(x, G^c) < 1/n\}$. Clearly, each $N_{\frac{1}{n}}(G^c)$ is an open set for each n. Also, $(G^c)^{\ell} \subset \bigcap_{n=1}^{\infty}[N_{1/n}(G^c)]^{\ell}$. Now suppose that $K \in \bigcap_{n=1}^{\infty}[N_{\frac{1}{n}}(G^c)]^{\ell}$. That is, $K \cap N_{\frac{1}{n}}(G^c) \neq \varnothing$ for each n. Let $x_n \in K \cap N_{\frac{1}{n}}(G^c)$ for each n. If $x \in K$ is a limit point of the sequence $\{x_n\}$, then from $d(x_n, G^c) < 1/n$, it follows that $d(x, G^c) = 0$, or $x \in \overline{G^c} = G^c$. So $x \in K \cap G^c$, which implies $K \in (G^c)^{\ell}$. Therefore $(G^c)^{\ell} = \bigcap_{n=1}^{\infty}[N_{1/n}(G^c)]^{\ell}$. Now observe that

$$G^{\mathrm{u}} = [(G^c)^{\ell}]^{\mathrm{c}} = \left[\bigcap_{n=1}^{\infty}[N_{1/n}(G^c)]^{\ell}\right]^{\mathrm{c}} \in \sigma(\mathcal{L}).$$

Thus, $\sigma(\mathcal{U}) \subset \sigma(\mathcal{L})$, so $\sigma(\mathcal{U}) = \sigma(\mathcal{L}) = \sigma(\mathcal{U} \cup \mathcal{L}) = \mathcal{B}_{\mathcal{K}}$.   ∎

We can use the preceding theorem to show the equivalence of measurability and weak measurability for compact-valued correspondences, when the range space is separable and metrizable (cf. C. Castaing and M. Valadier [75, Theorem III.2, p. 62]).

18.10 Theorem   *Let (S, Σ) be a measurable space, let X be a separable metrizable space, and let $\varphi \colon S \twoheadrightarrow X$ be a correspondence with nonempty compact values. Then the following statements are equivalent.*

1. *The correspondence φ is weakly measurable.*

2. *The correspondence φ is measurable.*

3. *The function $f \colon S \to \mathcal{K}$, defined by $f(s) = \varphi(s)$, is Borel measurable. That is, $f^{-1}(B) \in \Sigma$ for every Borel set $B \in \mathcal{B}_{\mathcal{K}}$.*

Proof: Start by observing that for each subset G of X we have

$$\varphi^{\ell}(G) = \{s \in S : \varphi(s) \cap G \neq \varnothing\} = f^{-1}(G^{\ell}). \quad (\star)$$

(1) $\iff$ (2)   This follows immediately from Lemma 18.2.

(3) $\implies$ (1)   By $(\star)$, if G is open, then Theorem 18.9 guarantees that G^{ℓ} is a Borel set, so $\varphi^{\ell}(G) \in \Sigma$.

(1) $\implies$ (3)   By $(\star)$, if φ is weakly measurable, then $f^{-1}(G^{\ell}) \in \Sigma$ for every open set G. By Theorem 18.9, the family $\{G^{\ell} : G \text{ is open}\}$ generates $\mathcal{B}_{\mathcal{K}}$. Therefore, by Corollary 4.24, f is Borel measurable.   ∎

So far we have been silent on the seemingly natural notion of Borel measurability for correspondences. We are now in a position to show that the requirement of Borel measurability can be unreasonably strong. The two examples below will clarify the situation.

18.11 Example   (A non-Borel measurable correspondence)   Let $S = \mathcal{K}$ denote the space of nonempty compact subsets of the unit interval $X = [0, 1]$ endowed with the Borel σ-algebra from its Hausdorff metric topology. Consider the correspondence $\varphi \colon \mathcal{K} \twoheadrightarrow [0, 1]$ defined by $\varphi(K) = K$. Surely this is as nice a correspondence as one could wish for, since it corresponds to the identity function on $\mathcal{K}$. Consequently by Theorem 18.10, it is measurable according to our Definition 18.1. But φ is *not* Borel measurable according to Definition 18.1 again! The reason is, as we show in Example 18.12 below, that $\varphi^\ell(\mathcal{I}) = \{K \in \mathcal{K} : K \cap \mathcal{I} \neq \varnothing\} = \{K \in \mathcal{K} : K \subset \mathbb{Q}\}^c$ is not a Borel subset of $\mathcal{K}$, where $\mathbb{Q}$ is the Borel set of rationals in $[0, 1]$ and $\mathcal{I}$ is the Borel set of irrationals. (Recall that the set $\mathbb{Q}$ of rationals, being countable, is a Borel subset of $[0, 1]$, and therefore its complement $\mathcal{I}$ is likewise a Borel subset of $[0, 1]$.) ∎

18.12 Example   (A non-Borel subset of $\mathcal{K}([0, 1])$)   Let $\mathcal{K}$ denote the space of closed (hence compact) subsets of $I = [0, 1]$, and endow $\mathcal{K}$ with its Hausdorff metric topology and resulting Borel σ-algebra. Let $\mathbb{Q}$ denote the set of rationals in I. We claim that $\mathbb{Q}^u = \{K \in \mathcal{K} : K \subset \mathbb{Q}\}$ is *not* a Borel subset of $\mathcal{K}$.

To prove this we use the following roundabout approach. Suppose by way of contradiction that $\mathbb{Q}^u$ is a Borel subset of $\mathcal{K}$. Then for any nonempty closed-valued measurable correspondence $\psi \colon I \twoheadrightarrow I$, Theorem 18.10 implies that $\psi^u(\mathbb{Q})$ is a Borel subset of I. (To see this, identify ψ with the Borel measurable function $f \colon I \to \mathcal{K}$, where $f(x) = \psi(x)$. Then $\psi^u(\mathbb{Q}) = \{x \in I : \psi(x) \subset \mathbb{Q}\}$ is just $f^{-1}(\mathbb{Q}^u)$, which is a Borel subset of I.) We now proceed to exhibit a nonempty closed-valued measurable ψ for which $\psi^u(\mathbb{Q})$ is not a Borel set. This contradiction shows that $\mathbb{Q}^u$ is not a Borel subset of $\mathcal{K}$. The following construction is taken from K. Kuratowski [218, Corollary 3, p. 72, vol. 2].

Recall that a set is analytic if it is the continuous image of $\mathcal{N} = \mathbb{N}^{\mathbb{N}}$. Start with an analytic subset A of I that is not a Borel set; see Example 12.33. It follows that A^c is not a Borel set either. Let $f \colon \mathcal{N} \to A$ be a continuous function from $\mathcal{N}$ onto A. By Theorem 3.68 we may identify $\mathcal{N}$ with the set $\mathcal{I}$ of irrationals in I (so that $\mathcal{N} \times A \subset I^2$). Let F be the closure of the graph $\mathrm{Gr}\, f = \{(x, f(x)) : x \in \mathcal{N}\}$ of f in I^2. It is easy to see that

$$x \in \mathcal{N} \text{ and } (x, y) \in F \implies y \in A. \tag{$\star$}$$

For if $\{(x_n, y_n)\}$ is a sequence in $\mathrm{Gr}\, f$ such that $(x_n, y_n) \to (x, y)$, then $y = f(x)$ by continuity, and $f(x) \in A$ by hypothesis.

Now define $\psi \colon I \twoheadrightarrow I$ by $\psi(x) = \{y \in I : (y, x) \in F\}$. Clearly, ψ has nonempty closed values. Moreover, we claim that $A^c = \psi^u(\mathbb{Q})$. To see this, suppose $x \notin A$. Then $(\star)$ implies $\psi(x) \subset \mathbb{Q}$. On the other hand if $x \in A$, then there is some $z \in \mathcal{N}$ with $f(z) = x$. But then $z \in \psi(x)$, so $\psi(x) \not\subset \mathbb{Q}$.

Now the graph of ψ is closed, since $\mathrm{Gr}\, \psi = \{(x, y) : (y, x) \in F\}$. Since the unit interval is compact, ψ is measurable by Theorem 18.20 below. Thus we have found a measurable correspondence ψ such that $\psi^u(\mathbb{Q})$ is the non-Borel set A^c. ∎

18.3   Measurable selectors

A **measurable selector** from a correspondence $\varphi \colon S \twoheadrightarrow X$ between measurable spaces is just what you think it should be, namely, a measurable function $f \colon S \to X$ satisfying $f(s) \in \varphi(s)$ for each $s \in S$. Thus, only nonempty-valued correspondences can possibly admit a measurable (or any other type of) selector. We now present the main selection theorem for measurable correspondences, which is a special case of a result due to K. Kuratowski and C. Ryll-Nardzewski [220].

18.13   Kuratowski–Ryll-Nardzewski Selection Theorem   *A weakly measurable correspondence with nonempty closed values from a measurable space into a Polish space admits a measurable selector.*

Proof: We use a technique similar to the proof of Michael's Selection Theorem 17.66. Let $\varphi \colon (S, \Sigma) \twoheadrightarrow X$ be a weakly measurable nonempty closed-valued correspondence from a measurable space into a Polish space. Fix a countable dense subset $D = \{x_1, x_2, \ldots\}$ of X. Let d be a bounded compatible metric for X satisfying $\operatorname{diam} X < 1$, and for each k and n, put $B_{k,n} = \{x \in X : d(x_k, x) < \frac{1}{2^n}\}$, the open ball of radius $\frac{1}{2^n}$ at x_k.

We shall inductively construct a sequence $\{f_0, f_1, \ldots\}$ of measurable functions (from S into X) with values in D satisfying

1. $d(f_n(s), \varphi(s)) < \frac{1}{2^n}$, and

2. $d(f_n(s), f_{n+1}(s)) < \frac{1}{2^n}$,

for each $s \in S$ and all $n \geqslant 0$. As in the proof of Michael's Theorem, this sequence is uniformly Cauchy, and so converges uniformly to a function $f \colon S \to X$. The closedness of $\varphi(s)$ and condition (1) imply that $f(s) \in \varphi(s)$ for each s. By Lemma 4.29 this f is also measurable.

Start by defining $f_0 \colon S \to X$ by $f_0(s) = x_1$ for all s in S, and note that since $\operatorname{diam} X < 1$, it follows that f_0 satisfies condition (1).

For the inductive step, suppose that f_n is measurable and satisfies condition (1). Then for each s, there is some $x \in \varphi(s)$ satisfying $d(x, f_n(s)) < \frac{1}{2^n}$. Since D is dense, there is an x_k so close to x that $d(x_k, x) + d(x, f_n(s)) < \frac{1}{2^n}$ and $d(x_k, \varphi(s)) < \frac{1}{2^{n+1}}$. See Figure 18.1. This implies $s \in A_k = \varphi^{\ell}(B_{k,n+1}) \cap f_n^{-1}(B_{k,n})$. Let $k_n(s)$ denote the smallest k such that $s \in A_k$, and set $f_{n+1}(s) = x_{k_n(s)}$. By construction, f_{n+1} satisfies

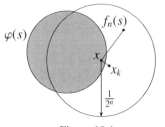

Figure 18.1.

$$d(f_{n+1}(s), \varphi(s)) < \tfrac{1}{2^{n+1}} \text{ and } d(f_n(s), f_{n+1}(s)) < \tfrac{1}{2^n},$$

so we need only verify that f_{n+1} is measurable.

Since φ is weakly measurable and f_n is measurable, each A_k belongs to Σ. For any Borel subset E of X, note that

$$f_{n+1}^{-1}(E) = \{s \in S : x_{k_n(s)} \in E\} = \bigcup_{x_k \in E} f_{n+1}^{-1}(\{x_k\}).$$

But by construction,

$$f_{n+1}^{-1}(\{x_k\}) = \{s \in S : k_n(s) = k\} = A_k \setminus \bigcup_{m=1}^{k-1} A_m \in \Sigma,$$

so $f_{n+1}^{-1}(E) \in \Sigma$. Hence, f_{n+1} is measurable, as desired. ∎

The following corollary to the Kuratowski–Ryll-Nardzewski Theorem is due to C. Castaing [73].

18.14 Corollary (Castaing) *For a correspondence $\varphi: (S, \Sigma) \twoheadrightarrow X$ with non-empty closed values from a measurable space into a topological space we have the following.*

1. *If there exists a sequence $\{f_n\}$ of measurable selectors from φ satisfying $\varphi(s) = \overline{\{f_1(s), f_2(s), \ldots\}}$ for each s, then φ is weakly measurable.*

2. *If X is a Polish space and φ is weakly measurable, then there exists a sequence $\{f_n\}$ of measurable selectors from the correspondence φ satisfying $\varphi(s) = \overline{\{f_1(s), f_2(s), \ldots\}}$ for each s.*

Proof: (1) If G is an open subset of X, then note that

$$
\begin{aligned}
\varphi^{\ell}(G) &= \{s \in S : \varphi(s) \cap G \neq \varnothing\} \\
&= \{s \in S : \overline{\{f_1(s), f_2(s), \ldots\}} \cap G \neq \varnothing\} \\
&= \{s \in S : \{f_1(s), f_2(s), \ldots\} \cap G \neq \varnothing\} \\
&= \{s \in S : f_n(s) \in G \text{ for some } n\} \\
&= \{s \in S : s \in f_n^{-1}(G) \text{ for some } n\} = \bigcup_{n=1}^{\infty} f_n^{-1}(G) \in \Sigma.
\end{aligned}
$$

(2) Assume that X is Polish and that φ is weakly measurable. Fix a countable base $\{U_1, U_2, \ldots\}$ for X, and define $\varphi_n: S \twoheadrightarrow X$ by

$$\varphi_n(s) = \begin{cases} \overline{\varphi(s) \cap U_n} & \text{if } \varphi(s) \cap U_n \neq \varnothing \\ \varphi(s) & \text{otherwise.} \end{cases}$$

From the identity

$$\varphi_n^{\ell}(G) = \varphi^{\ell}(U_n \cap G) \cup [\varphi^{\ell}(G) \cap [\varphi^{\ell}(U_n)]^c]$$

and the weak measurability of φ, we see that each φ_n is weakly measurable and nonempty valued. Hence, each closure correspondence $\overline{\varphi_n} \colon S \twoheadrightarrow X$ is weakly measurable with nonempty closed values and satisfies $\overline{\varphi_n}(s) \subset \varphi(s)$ for each $s \in S$. So by Theorem 18.13, there is a measurable selector f_n from $\overline{\varphi_n}$. The sequence $\{f_n\}$ satisfies the desired properties. ∎

The following immediate consequence of the preceding result is also of some interest.

18.15 Corollary *A correspondence $\varphi \colon (S, \Sigma) \twoheadrightarrow X$ with nonempty compact values from a measurable space into a separable metrizable space is weakly measurable if and only if there exists a sequence $\{f_n\}$ of measurable selectors from φ satisfying $\varphi(s) = \overline{\{f_1(s), f_2(s), \ldots\}}$ for each s.*

In particular, every weakly measurable correspondence with nonempty compact values from a measurable space into a separable metrizable space admits a measurable selector.

Proof: Take a compatible metric d for X and let $\hat{X}$ denote the completion of the metric space (X, d). Now note that φ considered as a correspondence from S to $\hat{X}$ is weakly measurable and has nonempty compact values. By part (2) of Corollary 18.14 there exists a sequence $\{f_n\}$ of measurable selectors from φ satisfying $\varphi(s) = \overline{\{f_1(s), f_2(s), \ldots\}}$ for each s. Clearly, each f_n carries S into X and is measurable as a function from S to X if and only if it is measurable as a function from S to $\hat{X}$. ∎

The next result slightly generalizes a theorem that V. Strassen [323] attributes to K. Jacobs, for the case where X is a compact metric space.

18.16 Jacobs' Selection Theorem *Let X be a locally compact separable metrizable space, and let $\mathcal{F}$ denote the compact metrizable space of all nonempty closed subsets of X endowed with the topology of closed convergence. Then there is a Borel measurable function $f \colon \mathcal{F} \to X$ satisfying $f(F) \in F$ for each nonempty closed set F.*

Proof: By Corollary 3.95 we know that $\mathcal{F}$ is indeed a compact metrizable space, so equip it with its Borel σ-algebra. Define the correspondence $\varphi \colon \mathcal{F} \twoheadrightarrow X$ by $\varphi(F) = F$ and note that φ is weakly measurable: For any open set $G \subset X$ we have $\varphi^{\ell}(G) = \{F \in \mathcal{F} : F \cap G \neq \varnothing\}$, which is open by Lemma 3.92, and hence a Borel set. Now to complete the proof invoke Corollary 18.15. ∎

We can use the Kuratowski–Ryll-Nardzewski Theorem 18.13 to prove the following selection theorem, known as **Filippov's Implicit Function Theorem** after A. F. Filippov [124]. This version is based on C. J. Himmelberg [160, Theorem 7.1].

18.17 Filippov's Implicit Function Theorem *Let* (S, Σ) *be a measurable space and let X and Y be separable metrizable spaces. Suppose that* $f: S \times X \to Y$ *is a Carathéodory function and that* $\varphi: S \twoheadrightarrow X$ *is weakly measurable with nonempty compact values. Assume also that* $\pi: S \to Y$ *is a measurable selector from the range of f on* φ *in the sense that g is measurable and for each s there exists* $x \in \varphi(s)$ *with* $\pi(s) = f(s, x)$.

Then the correspondence $\gamma: S \twoheadrightarrow X$, *defined by*

$$\gamma(s) = \{x \in \varphi(s) : f(s, x) = \pi(s)\},$$

is measurable and admits a measurable selector. That is, in addition to γ *being measurable, there exists a measurable function* $\xi: S \to X$ *with* $\xi(s) \in \varphi(s)$ *and* $\pi(s) = f(s, \xi(s))$ *for each* $s \in S$.

Proof: Fix a compatible metric d on Y and consider the continuous function $g: Y \times Y \to \mathbb{R}$ defined by $g(y_1, y_2) = d(y_1, y_2)$. Also, consider the function $h: S \times X \to \mathbb{R}$ defined by

$$h(s, x) = g(f(s, x), \pi(s)) = d(f(s, x), \pi(s)).$$

Since f is continuous in x, it follows immediately that h is continuous in x. Now since the functions $(s, x) \mapsto f(s, x)$ and $(s, x) \mapsto \pi(s)$ (both from $S \times X$ to Y) are clearly Carathéodory functions, it follows from Theorem 4.53 that h is jointly measurable. In particular, h is measurable in s and so h is itself a Carathéodory function.

Next for each n define the correspondence $\psi_n: S \twoheadrightarrow X$ by

$$\psi_n(s) = \{x \in X : d(f(s, x), \pi(s)) < \tfrac{1}{n}\} = \{x \in X : h(s, x) \in (-\infty, \tfrac{1}{n})\}.$$

By Lemma 18.7 each ψ_n is measurable. Thus, by Lemmas 18.2 and 18.3, the correspondence $\overline{\psi_n}$ is weakly measurable. Observe that

$$\gamma(s) = \varphi(s) \cap \overline{\psi_1(s)} \cap \overline{\psi_2(s)} \cap \cdots,$$

and that $\{\varphi, \overline{\psi_1}, \overline{\psi_2}, \ldots\}$ satisfies the hypotheses of Lemma 18.4 (3). Therefore γ is measurable (and hence weakly measurable), and has compact values. By hypothesis, γ has also nonempty values. By Corollary 18.15, the correspondence γ has measurable selectors, any of which will do for ξ. ∎

As another application we offer a stochastic version of Taylor's Theorem.

18.18 Stochastic Taylor's Theorem *Let* $h: [a, b] \to \mathbb{R}$ *be a function possessing a continuous* n^{th}-*order derivative on* $[a, b]$. *Fix* $c \in [a, b]$ *and let r be a random variable on the probability space* (S, Σ, P) *such that* $c + r(s)$ *belongs to* $[a, b]$

for all s. Then there is a measurable function ξ such that $\xi(s)$ lies in the closed interval with endpoints 0 and $r(s)$ for each $s \in S$, and

$$h(c + r(s)) = h(c) + \sum_{k=1}^{n-1} \frac{1}{k!} h^{(k)}(c) \, r^k(s) + \frac{1}{n!} h^{(n)}(c + \xi(s)) r^n(s).$$

Proof: Taylor's classical theorem[2] asserts that for each s there is such a $\xi(s)$, the trick in this case is to show that we can chose it in a measurable fashion. To this end, define the correspondence $\varphi \colon S \twoheadrightarrow \mathbb{R}$ by

$$\varphi(s) = \begin{cases} [0, r(s)] & \text{if } r(s) > 0, \\ [r(s), 0] & \text{if } r(s) < 0, \\ \{0\} & \text{if } r(s) = 0. \end{cases}$$

If we let $A = r^{-1}((0, \infty))$, $B = r^{-1}((-\infty, 0))$, and $C = r^{-1}(\{0\})$, then the sets A, B, and C belong to Σ. It is easy to see that the distance function associated with the correspondence φ is given by

$$\delta(s, x) = [(-x)^+ + [x - r(s)]^+]\chi_A(s) + [x^+ + [r(s) - x]^+]\chi_B(s) + |x|\chi_C(s),$$

which is clearly a Carathéodory function. It follows from Theorem 18.5 that φ is weakly measurable, and it clearly has compact values.

Now consider the functions

$$\pi(s) = h(c + r(s)) - h(c) - \sum_{k=1}^{n-1} \frac{1}{k!} h^{(k)}(c) r^k(s),$$

and

$$g(s, x) = \frac{1}{n!} h^{(n)}(c + x) r^n(s).$$

Since h has a continuous n^{th}-order derivative on $[a, b]$, it follows that the function $g \colon S \times [a-c, b-c] \to \mathbb{R}$ is a Carathéodory function. Moreover, by Taylor's classical theorem, the function $\pi \colon S \to \mathbb{R}$ is a measurable selector from the range of g on φ. By Filippov's Theorem 18.17 there is a measurable function $\xi \colon S \to \mathbb{R}$ satisfying $\xi(s) \in \varphi(s)$ and $g(s, \xi(s)) = \pi(s)$ for all $s \in S$, and we are done. ∎

---

[2] Taylor's theorem (see, for instance, [18, p. 113]) states: *If a function $f \colon [a, b] \to \mathbb{R}$ has a finite n^{th} order derivative everywhere on the open interval (a, b) and a continuous $(n-1)^{\text{th}}$ order derivative on the closed interval $[a, b]$, then for each $c \in [a, b]$ and every $x \in [a, b]$ with $x \neq c$ there exists some point x_1 interior to the interval joining x and c such that*

$$f(x) = f(c) + \sum_{k=1}^{n-1} \frac{f^{(k)}(c)}{k!}(x - c)^k + \frac{f^{(n)}(x_1)}{n!}(x - c)^n.$$

We can use Castaing's Corollary 18.14 to prove a measurable version of the Maximum Theorem. It has numerous applications in the fields of statistics, econometrics, control theory, dynamic programming, game theory, and mathematical economics. See, for instance, L. E. Dubins and L. J. Savage [100, Lemma 6, p. 38], R. I. Jennrich [186] or M. Stinchcombe and H. White [320]. M. A. Khan and A. Rustichini [200] provide an example involving a nonseparable range space in which this result fails.

18.19 Measurable Maximum Theorem *Let X be a separable metrizable space and (S, Σ) a measurable space. Let $\varphi \colon S \twoheadrightarrow X$ be a weakly measurable correspondence with nonempty compact values, and suppose $f \colon S \times X \to \mathbb{R}$ is a Carathéodory function. Define the value function $m \colon S \to \mathbb{R}$ by*

$$m(s) = \max_{x \in \varphi(s)} f(s, x),$$

and the correspondence $\mu \colon S \twoheadrightarrow X$ of maximizers by

$$\mu(s) = \{x \in \varphi(s) : f(s, x) = m(s)\}.$$

Then:

1. *The value function m is measurable.*

2. *The argmax correspondence μ has nonempty and compact values.*

3. *The argmax correspondence μ is measurable and admits a measurable selector.*

Proof: Clearly, m is well defined and μ has nonempty and compact values. By Corollary 18.15 there is a sequence $\{g_n\}$ of measurable selectors from φ satisfying $\varphi(s) = \{g_1(s), g_2(s), \ldots\}$ for each $s \in S$. Define the function $h_n \colon S \to S \times X$ by letting $h_n(s) = (s, g_n(s))$. Then h_n is $(\Sigma, \Sigma \otimes \mathcal{B}_X)$-measurable for each n. Since f is a Carathéodory function it is $\Sigma \otimes \mathcal{B}_X$-measurable (Lemma 4.51), so $f \circ h_n$ is Σ-measurable for each n. But

$$m(s) = \sup_{n \in \mathbb{N}} f(s, g_n(s)) = \sup_{n \in \mathbb{N}} f(h_n(s)) = \sup_{n \in \mathbb{N}} [f \circ h_n](s),$$

so $m \colon S \to \mathbb{R}$ is Σ-measurable.

Now note that $\mu(s) = \{x \in \varphi(s) : f(s, x) = m(s)\}$ for each s, so by Filippov's Theorem 18.17, the correspondence μ is measurable, and admits a measurable selector $\xi \colon S \to X$ so that $\xi(s) \in \mu(s)$ for each $s \in S$. ∎

18.4   Correspondences with measurable graph

As mentioned earlier, Theorem 12.28 asserts that a function between Polish spaces
is Borel measurable if and only if its graph is a Borel set. It is thus natural to investigate the measurability of the graph of a correspondence. Theorem 18.6 shows
that a (weakly) measurable correspondence with closed values has a measurable
graph. Theorem 18.20 is a partial converse, since it gives conditions under which
a correspondence with closed (hence Borel) graph is measurable. But in general,
measurability of the graph is a weaker condition than even weak measurability. Indeed, we know from Example 12.35 that there is a closed correspondence from the
Baire space $\mathcal{N}$ into itself that has no measurable selector, so by the Kuratowski–
Ryll-Nardzewski Theorem 18.13, it cannot be weakly measurable. (Can you find
a more direct argument?) It turns out that the reason for this is that the Baire space
fails to be σ-compact (Theorem 3.69).

18.20 Theorem (Correspondences with closed graph)   *A closed correspon-*
dence between σ-compact Hausdorff spaces is measurable.

Proof: Let $\varphi\colon S \twoheadrightarrow X$ be a closed correspondence between two σ-compact Hausdorff spaces. Write $S = \bigcup_{n=1}^{\infty} C_n$ and $X = \bigcup_{m=1}^{\infty} K_m$, where each C_n and K_m is
compact. Note that for each subset F of X, we have

$$\varphi^{\ell}(F) = \bigcup_{m=1}^{\infty} \bigcup_{n=1}^{\infty} \pi_S\big([C_n \times (F \cap K_m)] \cap \operatorname{Gr}\varphi\big),$$

where π_S is the projection of $S \times X$ onto S. When F is closed, since $\operatorname{Gr}\varphi$ is
also closed, $D_{n,m} = [C_n \times (F \cap K_m)] \cap \operatorname{Gr}\varphi$ is compact for each n, m. Since π_S is
continuous, $\pi_S(D_{n,m})$ is compact and so a Borel set. Therefore φ is measurable. ∎

The reason the proof of Theorem 18.20 works is that the projection of a compact set is compact and hence a Borel set. In general, the projection of a Borel
set is not a Borel set, but it is analytic. If we are willing to put a larger σ-algebra
than the Borel σ-algebra on the domain, we can get a very nice result. Recall that
an analytic set is universally measurable (Theorem 12.41), that is, μ-measurable
for any Borel probability measure μ, and that the collection of universally measurable sets is a σ-algebra. Thus the σ-algebra Σ_A generated by the analytic sets is
included in the σ-algebra of universally measurable sets, but is generally smaller.
We shall say that a set is **analytically measurable** if it belongs to Σ_A. A correspondence between two Polish spaces is **analytically measurable** if the lower
inverse image of every Borel set belongs to Σ_A.

18.21 Theorem   *Let S be a Polish space and let $\mathcal{A}$ denote the σ-algebra of*
all analytically measurable subsets of S. Let X be a Polish space with Borel
σ-algebra $\mathcal{B}$, and let $\varphi\colon S \twoheadrightarrow X$ be a nonempty closed-valued correspondence. If
the graph of φ is analytic, then φ is analytically measurable, that is, $\varphi^{\ell}(B) \in \mathcal{A}$
for every $B \in \mathcal{B}$.

Proof: Suppose that $\operatorname{Gr}\varphi$ is analytic, and let $B \in \mathcal{B}$. Then $S \times B$ is Borel, so $(S \times B) \cap \operatorname{Gr}\varphi$ is analytic (Theorem 12.25). Now observe that

$$\varphi^\ell(B) = \pi_S((S \times B) \cap \operatorname{Gr}\varphi),$$

so $\varphi^\ell(B)$ is analytic (Theorem 12.24). ∎

We can now prove a selection theorem for correspondences with analytic graph. It is due to J. von Neumann and V. A. Jankov.[3]

18.22 Jankov–von Neumann Selection Theorem   *Let S and X be Polish spaces, and let $\varphi\colon S \twoheadrightarrow X$ be a nonempty closed-valued correspondence with analytic graph. Then φ admits an analytically measurable selector.*

Proof: By Theorem 18.21 the correspondence $\varphi\colon (S, \Sigma_A) \twoheadrightarrow X$ is Borel measurable, and so weakly measurable. Therefore by the Kuratowski–Ryll-Nardzewski Theorem 18.13, it has a Σ_A-measurable selector. ∎

18.23 Corollary   *Every surjective Borel function $f\colon X \to Y$ between Polish spaces admits an **analytically measurable inverse** in the sense that there exists a function $g\colon Y \to X$ such that:*

1. $f(g(y)) = y$ *for any y in Y,*

2. $f^{-1}(g^{-1}(A)) = A$ *for any subset A of X, and*

3. $g^{-1}(B)$ *is analytically measurable in Y for each Borel subset B of X.*

Proof: Start by observing that, according to Theorem 4.60, we can assume that f is a continuous function. By Theorem 12.28, the correspondence $\psi\colon Y \twoheadrightarrow X$, defined by $\psi(y) = f^{-1}(y)$, has Borel (and therefore analytic) graph, since f is Borel measurable. In addition, ψ has nonempty and closed values. Since Y is Polish, by the Jankov–von Neumann Theorem 18.22 there is a selection g from ψ that is analytically measurable. Clearly g has the desired properties. ∎

As an application, we present the following piece of Theorem 15.14.

18.24 Corollary   *Let $f\colon X \to Y$ be a surjective Borel function between two Polish spaces. Then the function $\hat{f}\colon \mathcal{P}(X) \to \mathcal{P}(Y)$, defined by $\hat{f}(\mu) = \mu \circ f^{-1}$, is also surjective.*

---

[3] von Neumann [337] is a summary of papers written over the period 1929–1935. Lemma V in section 15 (pp. 448–451) proves what is essentially Corollary 18.23 for the case where S is Polish and $X = \mathbb{R}$. Jankov [180] proves a related result for the case $S = X = \mathbb{R}$, but attributes the original result to Lusin [230].

Proof: Let v be a Borel probability measure on Y, and let Σ_v denote the collection of v-measurable sets. By Corollary 18.23, the function f has an analytically measurable, and hence $(\Sigma_v, \mathcal{B}_X)$-measurable, inverse g.

Define the Borel measure μ on X by

$$\mu = v \circ g^{-1}.$$

We claim that $\hat{f}(\mu) = v$. But first we verify that μ is a Borel measure: For any Borel subset B of X, the inverse image $g^{-1}(B)$ belongs to Σ_v, so $v(g^{-1}(B))$ is defined, and it is easy to see that μ is a probability measure. Now $\hat{f}(\mu)$ is the Borel probability on Y defined by

$$\hat{f}(\mu)(A) = \mu(f^{-1}(A)) = v(g^{-1}(f^{-1}(A))) = v(A)$$

whenever A is a Borel subset of Y. In other words, $\hat{f}(\mu) = v$. Since v is arbitrary, $\hat{f}$ is surjective. ∎

A more general version of Theorem 18.21 is true. The general result does not place any topological structure on S, but it does require the existence of a σ-finite measure μ on Σ under which Σ is complete, that is, $\Sigma = \Sigma_\mu$, the σ-algebra of all μ-measurable subsets of S. The proof relies on the following general **Projection Theorem**—whose proof is quite deep and involves new constructions (Σ-analytic sets). In the interest of brevity, we do not present a proof. Klein and Thompson [209, Theorem 12.3.4, p. 147] do an excellent job of presenting this theorem and its applications to measurable selectors from correspondences with measurable graph.

18.25 Projection Theorem   *Let X be a separable metrizable space and let (S, Σ, μ) be a σ-finite measure space. If a set A belongs to $\Sigma_\mu \otimes \mathcal{B}_X$, then the projection $\pi_S(A)$ of A on S belongs to Σ_μ.*

The Projection Theorem allows us to prove the following selection theorem due to R. J. Aumann [28].

18.26 Theorem   (Aumann)   *Let X be a Polish space and let (S, Σ, μ) be a complete finite measure space. Let $\varphi\colon S \twoheadrightarrow X$ have $\Sigma \otimes \mathcal{B}_X$-measurable graph and nonempty values. Then φ admits a measurable selector.*

18.27 Corollary   *Let X be a Polish space and let (S, Σ, μ) be a finite measure space. Let $\varphi\colon S \twoheadrightarrow X$ have $\Sigma \otimes \mathcal{B}_X$-measurable graph and nonempty values. Then there exists a Σ-measurable function $f\colon S \to X$ satisfying $f(s) \in \varphi(s)$ μ-a.e.*

18.5 Correspondences with compact convex values

Recall that a sublinear function is a convex positively homogeneous function on a vector space. Let $\langle X, X' \rangle$ be a dual pair. Recall that by Theorem 7.52, there is a one-to-one correspondence between Mackey-continuous sublinear functions on X and nonempty weak∗-compact convex subsets of X' in the following sense. If $h \colon X \to \mathbb{R}$ is a Mackey-continuous sublinear function, then the set K_h of linear functionals dominated by h, that is, the set

$$K_h = \{x' \in X' : \langle x, x' \rangle \leqslant h(x) \text{ for all } x \in X\},$$

is a nonempty $\sigma(X', X)$-compact convex subset of X'. On the other hand, for every nonempty $\sigma(X', X)$-compact convex subset K of X', the support functional $h_K \colon X \to \mathbb{R}$ of K, defined by

$$h_K(x) = \max\{\langle x, x' \rangle : x' \in K\},$$

is a Mackey-continuous sublinear function. Moreover Theorem 7.52 asserts that

$$K_{h_K} = K \quad \text{and} \quad h_{K_h} = h.$$

Thus there is a one-to-one mapping from the space $\mathcal{C}$ of nonempty $w\ast$-compact convex subsets of X' onto the space of Mackey-continuous sublinear functions on X. We know from Lemma 7.54 and Theorem 7.56 and the remarks following it that this mapping preserves addition, multiplication by positive scalars, and the order properties of these spaces. We now turn to functions from a measurable space S into these spaces. Clearly every family $\{h_s : s \in S\}$ of Mackey-continuous sublinear functions defines a function from S into $\mathcal{C}$ via $s \mapsto K_{h_s}$ (or equivalently, the correspondence $s \mapsto K_{h_s}$ into X'). On the other hand, if $\varphi \colon S \twoheadrightarrow X'$ is a correspondence with nonempty, convex and w^*-compact values, then we get a family $\{h_{\varphi(s)}\}$ of Mackey-continuous sublinear functionals indexed by S.

18.28 Definition *Let (S, Σ) be a measurable space and let X be a topological vector space. A function $C \colon S \times X \to \mathbb{R}$ is a **linear** (resp. **sublinear**) **Carathéodory function** if $C(\cdot, x)$ is Σ-measurable for every x in X and $C(s, \cdot)$ is a continuous linear (resp. sublinear) function on X for every s in S. We often write $C_s(x)$ for $C(s, x)$, and say that $s \mapsto C_s$ is a linear (resp. sublinear) Carathéodory function.*

Note that a linear Carathéodory function is simply a (Σ, w^*)-measurable function in the sense of Definition 11.48. Now recall that the Mackey topology on a normed space is just the norm topology (Theorem 6.23). When X is a normed space, and $h \colon X \to \mathbb{R}$ is sublinear we define the (extended) **norm** $\|h\|$ by

$$\|h\| = \sup\{|h(x)| : \|x\| \leqslant 1\}.$$

This is the obvious generalization of the definition of the operator norm to sublinear functions. We say that h is a **bounded** sublinear function if it is bounded on the unit ball of X, that is, if its norm is finite. By Lemma 5.51, h is bounded if and only if it is norm continuous. Clearly,

$$|h(x)| \leqslant \|h\| \cdot \|x\|$$

for each $x \in X$.

18.29 Lemma   *Let X be a separable Banach space, let (S, Σ, μ) be a probability space, and let $C \colon S \times X \to \mathbb{R}$ be a sublinear Carathéodory function. Then the real function $s \mapsto \|C_s\|$ is Σ-measurable.*

Moreover, if $\int_S \|C_s\| \, d\mu(s) < \infty$, then the integral

$$h(x) = \int_S C(s, x) \, d\mu(s)$$

defines a bounded sublinear function on X.

Proof: Since X is separable, there is a countable dense subset $\{x_1, x_2, \ldots\}$ of the unit ball, so $\|C_s\| = \sup\{|C(s, x_n)| : n = 1, 2, \ldots\}$. Thus $s \mapsto \|C_s\|$ is the pointwise supremum of the countable collection $\{|C(\cdot, x_n)|\}$ of Σ-measurable functions, and so Σ-measurable itself (Theorem 4.27).

From the inequalities $|C(s, x)| \leqslant \|C_s\| \cdot \|x\|$ and $\int_S \|C_s\| \, d\mu(s) < \infty$, it follows that the measurable real function $s \mapsto C(s, x)$ is Lebesgue integrable for any x. Thus, h is a well defined function. It is easy to see that h is also sublinear. To see that h is bounded, note that

$$|h(x)| \leqslant \int_S |C(s, x)| \, d\mu(s) \leqslant \left[\int_S \|C_s\| \, d\mu(s) \right] \|x\|,$$

so h is bounded on the unit ball of X and $\|h\| \leqslant \int_S \|C_s\| \, d\mu(s)$. ∎

Sublinear Carathéodory functions define a new measurability notion for correspondences from S into X'.

18.30 Definition   *Let (S, Σ) be a measurable space and let $\langle X, X' \rangle$ be a dual pair. Let $\varphi \colon S \twoheadrightarrow X'$ be a correspondence with nonempty weak*-compact convex values. We say that φ is **scalarly measurable** if the function $C \colon S \times X \to \mathbb{R}$, defined by*

$$C(s, x) = h_{\varphi(s)}(x) = \max\{\langle x, x' \rangle : x' \in \varphi(s)\},$$

is a sublinear Carathéodory function. An analogous definition applies to correspondences $\varphi \colon S \twoheadrightarrow X$.

Clearly any sublinear Carathéodory function C defines a scalarly measurable correspondence φ via $\varphi(s) = \{x' \in X' : x' \leq C_s\}$. Just as Theorem 17.41 characterizes weak upper hemicontinuity in terms of support functionals, support functionals can be also used to characterize measurability for compact convex-valued correspondences. The proof of the next result is based on ideas of C. Castaing and M. Valadier [75, Theorem III.15, p. 70].

18.31 Theorem *Let X be a separable Banach space, let (S, Σ) be a measurable space, and let $\varphi : S \twoheadrightarrow X'$ be a correspondence with nonempty weak*-compact convex values. Then the following statements are equivalent.*

1. *The correspondence φ is scalarly measurable.*

2. *The correspondence φ is measurable, where X' is endowed with the Borel σ-algebra from its weak* topology.*

Proof: (1) $\implies$ (2) Suppose $\varphi \colon S \twoheadrightarrow X'$ is scalarly measurable. That is, assume that the real function $s \mapsto h_{\varphi(s)}(x)$ is measurable for each $x \in X$. Start by fixing a countable dense subset $\{x_1, x_2, \ldots\}$ of the closed unit ball U of X. We shall denote the closed unit ball of X' by U'.

For each k let $S_k = \varphi^u(kU') = \{s \in S : \varphi(s) \subset kU'\}$. Now for nonempty closed convex sets C and K we have $C \subset K$ if and only if $h_C(x) \leq h_K(x)$ for each $x \in U$, so

$$\varphi^u(kU') = \{s \in S : h_{\varphi(s)}(x) \leq h_{kU'}(x) \text{ for each } x \in U\}.$$

Since every w^*-compact subset of X' is norm-bounded (and so belongs to some kU'), we conclude that $S = \bigcup_{k=1}^{\infty} S_k$. Furthermore, each S_k belongs to Σ. This follows immediately from the equality $S_k = \bigcap_{m=1}^{\infty} \{s \in S : h_{\varphi(s)}(x_m) \leq h_{kU'}(x_m)\}$. (We use here the fact that if K is a nonempty, w^*-compact and convex subset of X', then its support functional h_K is norm-continuous; see Theorem 7.52.)

Now let $\varphi_k \colon S_k \twoheadrightarrow X'$ be the restriction of φ to S_k. Then $\varphi(s) = \varphi_k(s)$ for each $s \in S_k$, so each φ_k is scalarly measurable on S_k. In addition, notice that for each subset A of X' we have $\varphi^\ell(A) = \bigcup_{k=1}^{\infty} \varphi_k^\ell(A \cap kU')$. This shows that in order to establish the measurability of $\varphi \colon S \twoheadrightarrow X'$, it suffices to show that each $\varphi_k \colon S_k \twoheadrightarrow kU'$ is measurable. This means that we can assume without loss of generality that the range of φ lies in U', that is, $\varphi \colon S \twoheadrightarrow U'$. So assume in addition that the range of φ lies in U'.

Recall that $d(x', y') = \sum_{m=1}^{\infty} \frac{1}{2^m} |x'(x_m) - y'(x_m)|$ is a metric for the w^*-topology on U'; see the proof of Theorem 6.30. By Alaoglu's Theorem 6.21, (U', d) is a compact metric space. Let ρ_d denote the Hausdorff metric on the set $\mathcal{K}$ of all nonempty closed (and hence compact) subsets of (U', d). By Theorem 3.85 (3), $(\mathcal{K}, \rho_d)$ is also a compact metric space. Thus Theorem 18.10 implies that it suffices to prove that the function $f \colon S \to (\mathcal{K}, \rho_d)$ defined by $f(s) = \varphi(s)$ is Borel measurable.

Now the set $\mathcal{C}$ of all convex nonempty w^*-compact subsets of U' is a closed subset of $(\mathcal{K}, \rho_d)$ (Theorem 7.60) and f takes values in $\mathcal{C}$, so by Lemma 4.20 it suffices to show that the function $f \colon S \twoheadrightarrow (\mathcal{C}, \rho_d)$ is Borel measurable. With this in mind we now examine the compact Hausdorff metric topology on $\mathcal{C}$.

For each m, the semimetric $d_m(x', y') = \left| x'(x_m) - y'(x_m) \right|$ on U' induces a Hausdorff semimetric ρ_m on $\mathcal{C}$ by the usual formula

$$\rho_m(A, B) = \max\left\{ \sup_{a' \in A} d_m(a', B), \sup_{b' \in B} d_m(b', B) \right\}.$$

Clearly, $\rho_m(A, B) \leqslant 2$ for all $A, B \in \mathcal{C}$ and each m. We claim that the metric

$$\rho(A, B) = \sum_{m=1}^{\infty} \tfrac{1}{2^m} \rho_m(A, B)$$

induces the Hausdorff metric topology on $\mathcal{C}$. (You should verify that ρ is indeed a metric on $\mathcal{C}$.) That is, the identity mapping from $(\mathcal{C}, \rho_d)$ to $(\mathcal{C}, \rho)$ is a homeomorphism. By Theorem 2.36 it suffices to prove that $\rho_d(A_n, A) \to 0$ in $\mathcal{C}$ implies $\rho(A_n, A) \to 0$. To see this, suppose $\rho_d(A_n, A) \to 0$ and let $\varepsilon > 0$. From $\rho_d(A, B) = \inf\{\delta > 0 : A \subset N_\delta(B) \text{ and } B \subset N_\delta(A)\}$ (see Lemma 3.71), for each m it follows (how?) that

$$A \subset \{x' \in U' : d_m(x', A_n) < 2^m \varepsilon\} \quad \text{and} \quad A_n \subset \{x' \in U' : d_m(x', A) < 2^m \varepsilon\}$$

for all sufficiently large n. Thus $\rho_d(A_n, A) \to 0$ implies $\rho_m(A_n, A) \to 0$, which in turn implies $\rho(A_n, A) \to 0$. Therefore ρ generates the Hausdorff metric topology on $\mathcal{C}$.

Remember, we want to prove that the function $f \colon S \to (\mathcal{C}, \rho_d)$ is Borel measurable. This is equivalent to the Borel measurability of $f \colon S \to (\mathcal{C}, \rho)$. So by Lemma 4.30, f is measurable if and only if for each $K \in \mathcal{C}$ the distance function $s \mapsto \rho(K, \varphi(s))$ is measurable. Since $\rho = \sum_{m=1}^{\infty} \tfrac{1}{2^m} \rho_m$ it suffices to show that for each m and each $K \in \mathcal{C}$, the real function $s \mapsto \rho_m(K, \varphi(s))$ is measurable.

But $\rho_m(A, B) = \sup\{|h_A(x) - h_B(x)| : x \in L_m\}$ on $\mathcal{C}$, where L_m is the line segment joining x_m and $-x_m$. To see this, let U'_m denote the "closed unit ball" for the semimetric d_m. That is, $U'_m = \{x' \in X' : |x'(x_m)| \leqslant 1\}$, the polar of the singleton set $\{x_m\}$. From the Bipolar Theorem 5.103, the polar of U'_m is L_m, and Lemma 7.58 now applies.

If you have been patient enough to follow us so far, you recall that we are trying proving that $s \mapsto \rho_m(K, \varphi(s)) = \sup\{|h_K(x) - h_{\varphi(s)}(x)| : x \in L_m\}$ is measurable. But L_m is separable, so there is a countable dense subset $\{y_1, y_2, \ldots\}$ of L_m. It follows that

$$\rho_m(K, \varphi(s)) = \sup\{|h_K(y_\ell) - h_{\varphi(s)}(y_\ell)| : \ell = 1, 2, \ldots\}.$$

Now the scalar measurability of φ implies that $s \mapsto |h_K(y_\ell) - h_{\varphi(s)}(y_\ell)|$ is measurable for each ℓ. Since the pointwise supremum of a family of measurable functions is measurable, we are finished at last.

(2) $\implies$ (1) Suppose φ is (Σ, w^*)-measurable and let x belong to X. We must show that the real function $s \mapsto h_{\varphi(s)}(x)$ is measurable. Start by letting B denote the open unit ball of X', that is, $B = \{x' \in X' : \|x'\| < 1\}$. From the identity $\varphi^u(kB) = [\varphi^\ell((kB)^c)]^c$ and the (Σ, w^*)-measurability of φ, we see that each $S_k = \varphi^u(kB)$ belongs to Σ. Since $S_k \uparrow S$, in order to establish the measurability of $s \mapsto h_{\varphi(s)}(x)$, it suffices to show that $s \mapsto h_{\varphi(s)}(x)$ is measurable on each S_k.

To this end, consider the compact metric space (kU', d), where d is the metric defined in the preceding part. Then φ restricted to S_k has its range in kU', so $\varphi \colon S_k \twoheadrightarrow (kU', d)$ is a (Σ, w^*)-measurable correspondence (where Σ is now restricted to S_k). By Corollary 18.14, there exists a sequence $\{f_n\}$ of w^*-measurable selectors from φ such that $\varphi(s) = \{f_1(s), f_2(s), \ldots\}$ for each $s \in S_k$. Then the real function $s \mapsto \langle x, f_n(s) \rangle$, from S_k to $\mathbb{R}$, is measurable for each n, since it is the composition of the w^*-continuous function x on kU' with the w^*-measurable function $f_n \colon S_k \to kU'$. But $h_{\varphi(s)}(x) = \sup_n \langle x, f_n(s) \rangle$, so φ is scalarly measurable on S_k, and the proof is finished. ∎

The next result is taken from C. Castaing and M. Valadier [75, Theorem III.15, p. 70]. It is closely related to Theorem 18.31, and we state it without proof.

18.32 Theorem *Let X be a separable metrizable locally convex space. Suppose (S, Σ) is a measurable space, and let $\varphi \colon S \twoheadrightarrow X$ be a correspondence with $\sigma(X, X')$-compact convex values. Then φ is measurable if and only if it is scalarly measurable.*[4]

Given a correspondence $\varphi \colon S \twoheadrightarrow X'$ having nonempty convex w^*-compact values, the definition of the support functional implies that a function $f \colon S \to X'$ is a selector for φ if and only if $f_s(x) \leqslant h_{\varphi(s)}(x)$ for every $x \in X$. The next measurable selection theorem for scalarly measurable correspondences is due to V. Strassen [324].

18.33 Theorem (Scalarly measurable selectors) *Let X be a separable Banach space and let (S, Σ) a measurable space. Then every scalarly measurable correspondence $\varphi \colon S \twoheadrightarrow X'$ with nonempty weak*-compact convex values admits a (Σ, w^*)-measurable selector. Or, equivalently, there exists a (Σ, w^*)-measurable function $s \mapsto x'_s$ from S to X' satisfying the inequality $x'_s \leqslant h_{\varphi(s)}$ for all $s \in S$.*

Proof: Let $\varphi \colon S \twoheadrightarrow X'$ be a scalarly measurable correspondence with nonempty weak*-compact convex values. By Theorem 18.31 this is equivalent to saying that φ is a (Σ, w^*)-measurable correspondence. For each n define the correspondence $\varphi_n \colon S \twoheadrightarrow X'$ by $\varphi_n(s) = \varphi(s) \cap nU'$, where U' denotes the closed unit ball of X', and let $S_n = \{s \in S : \varphi_n(s) \neq \varnothing\} = \varphi^\ell(nU')$. Clearly, $S_n \in \Sigma$ and $S_n \subset S_{n+1}$ for each n and $\bigcup_{n=1}^\infty S_n = S$.

---

[4] That is, φ is measurable if and only if for each $x' \in X'$ the support mapping $s \mapsto h_{\varphi(s)}(x')$ is a measurable real function.

Now notice that each $\varphi_n \colon S_n \twoheadrightarrow nU'$ is a measurable correspondence. Since nU' is a compact metrizable space, the Kuratowski–Ryll-Nardzewski Selection Theorem 18.13 implies that each correspondence $\varphi_n \colon S_n \twoheadrightarrow nU'$ admits a measurable selector, say f_n. If we consider the sets $A_1 = S_1$ and $A_{n+1} = S_{n+1} \setminus S_n$, then the function $f \colon S \to X'$ defined pointwise via the formula $f = \sum_{n=1}^{\infty} f_n \chi_{A_n}$ is a measurable selector from φ.   ∎

Note that the selector given by Theorem 18.33 is a linear Carathéodory function. For general results on **Carathéodory selectors** (not necessarily linear) from correspondences see, for instance, T. Kim, K. Prikry, and N. C. Yannelis [203, 204], and their references. Another related result is C. J. Himmelberg [160, Theorem 9.3], which in light of Theorem 18.31 implies that under the hypotheses of Theorem 18.33 the extreme point correspondence $s \mapsto \mathcal{E}(\varphi(s))$ has measurable graph.

We now present a different kind of Carathéodory selection theorem, which does not assume convex-valuedness. Let X be a topological space and μ be a finite Borel measure on X. A nonempty set D of functions in $L_1(\mu, \mathbb{R}^n)$ is **decomposable** if $f, g \in D$ implies that for every Borel subset B of X, the function h defined by $h(x) = f(x)$ for $x \in B$ and $h(x) = g(x)$ for $x \notin B$, that is, $h = f\chi_B + g\chi_{B^c}$, also belongs to D. Note that a decomposable set need not be convex. For instance, if $X = [0, 1]$ and $\mathcal{B}$ is the σ-algebra of Borel subsets of $[0, 1]$, then $\{\chi_A : A \in \mathcal{B}\}$ is decomposable, but not convex. More generally the set of measurable selectors from a correspondence is decomposable. To indicate some of the selection results that have been obtained for decomposable-valued correspondences, we present without proof the following selection theorem of A. Fryszkowski [130, 131].

18.34 Fryszkowski's L_1-Selection Theorem   *Let Z be a locally compact Polish space, and let X be a compact metrizable space. Let μ be a finite Borel measure on X. Then every lower hemicontinuous correspondence $\varphi \colon Z \twoheadrightarrow L_1(\mu, \mathbb{R}^n)$ with decomposable values admits a continuous selector.*

18.6   Integration of correspondences

We have already discussed integration of functions. We now turn our attention to the integration of correspondences. The standard definition of the integral of a correspondence is due to R. J. Aumann [26]. The **Aumann integral** of a measurable correspondence $\varphi \colon S \twoheadrightarrow X$ is the set of integrals of measurable selectors from the correspondence. It is denoted $\int_S \varphi(s) \, d\mu(s)$, or more simply $\int \varphi \, d\mu$. The correspondence is **integrable** if its integral is nonempty. Clearly in order for this definition to make sense, X must be a vector space (perhaps $\mathbb{R}$ or $\mathbb{R}^n$) so that we may apply one of our definitions of integrability: Bochner integrability, Pettis integrability, Gelfand integrability, etc. (For $\mathbb{R}^n$ all these definitions agree.)

There are other notions of integrability for correspondences. In particular, G. Debreu [85] proposed embedding the compact convex subsets of X in a normed space Y. A correspondence from S to X with compact convex values can be identified with a function from S to Y. Debreu proposed a modification of the Bochner integral of this function as the integral of the correspondence. This is very close to our approach, which is based on Theorem 18.35 below. It turns out that Debreu's integral coincides with the Aumann integral for correspondences with compact convex values. (See [209, Chapter 17] for a clear discussion of the relation.) The advantage of the Aumann definition is that it does not require convex values.

We do not attempt an exposition of the general theory of integration of correspondences, such as the Radon–Nikodym Theorem or Fatou's Lemma for correspondences. For that we refer you to Klein and Thompson [209] for the case where X is a Banach space. More esoteric results are available in the excellent monographs by Castaing and Valadier [75], Kluvánek and Knowles [210], Aubin and Cellina [22], and Aubin and Frankowska [24]. We also recommend the surveys by N. C. Yannelis [343, 344, 345].

Instead, we confine our attention to the case of scalarly measurable correspondences taking values in the dual of a separable Banach space. The reason for this choice is that this peculiar case is not adequately addressed elsewhere and we actually have a good use for it in Chapter 19. We start with V. Strassen's [324, Theorem 1] result on integrals of sublinear Carathéodory functions.

18.35 Strassen's Sublinearity Theorem   *Let (S, Σ, μ) be a probability space, and let X be a separable Banach space. Let $C \colon S \times X \to \mathbb{R}$ be a sublinear Carathéodory function satisfying*

$$\int_S \|C_s\| \, d\mu(s) < \infty.$$

If $h \colon X \to \mathbb{R}$ is the sublinear function defined by the integral

$$h(x) = \int_S C(s, x) \, d\mu(s),$$

then for an arbitrary $x' \in X'$ the following statements are equivalent.

1. *The linear functional x' is dominated by h, i.e., $x'(x) \leqslant h(x)$ for all $x \in X$.*

2. *There is a linear Carathéodory function $s \mapsto x'_s$ with $x'_s(x) \leqslant C_s(x)$ for every s and x, whose Gelfand integral is x'. That is, the function $s \mapsto x'_s(x)$ is Σ-measurable for each x, C_s dominates x'_s for every s, and*

$$x'(x) = \int_S \langle x, x'_s \rangle \, d\mu(s).$$

Proof: The implication (2) $\implies$ (1) is straightforward. What requires proof is the implication (1) $\implies$ (2). The proof is quite involved and is presented in steps. So suppose that the sublinear function $h: X \to \mathbb{R}$ defined by $h(x) = \int_S C(s, x) \, d\mu(s)$ dominates a continuous linear functional x'. We start with a representation result.

Let L denote the collection of μ-equivalence classes of Σ-simple X-valued functions. That is, an X-valued function $f: S \to X$ belongs to L if and only if there exist vectors $a_1, \ldots, a_n$ in X and pairwise disjoint sets $A_1, \ldots, A_n$ in Σ satisfying $S = \bigcup_{i=1}^n A_i$ and $f = \sum_{i=1}^n a_i \chi_{A_i}$. As usual, under the almost everywhere pointwise algebraic operations, L is a vector space that is also a normed space under the essential sup norm $\|f\| = \text{ess sup}_s \|f(s)\|$. We typically write f_s instead of $f(s)$. Note that every $f \in L$ is weakly Σ-measurable, that is, $s \mapsto x'(f_s)$ is Σ-measurable for every $x' \in X'$. We can isometrically embed X in L as a vector subspace under the mapping $x \mapsto \hat{x}$ defined by $\hat{x}(s) = x$ for all s. Let $\hat{X}$ denote the image of X under this mapping. It is the vector space of μ-almost constant functions. Observe that h satisfies

$$h(x) = \int_S C_s(x) \, d\mu(s) = \int_S C_s(\hat{x}(s)) \, d\mu(s).$$

This suggests that we extend h from $\hat{X}$ to L via the formula

$$\hat{h}(f) = \int_S C_s(f_s) \, d\mu(s).$$

To see that $s \mapsto C_s(f_s)$ is Σ-measurable, note that if $f = \sum_{i=1}^n a_i \chi_{A_i} \in L$, then

$$C_s(f_s) = C(s, f(s)) = \sum_{i=1}^n C(s, a_i) \chi_{A_i}(s).$$

Since $|C_s(f_s)| \leq \|C_s\| \cdot \|f_s\|$, we see that $|\hat{h}(f)| \leq \|h\| \cdot \|f\|$, so $\hat{h}$ is bounded on L. It is also obvious that $\hat{h}$ is sublinear. Since h dominates x' on X, treating x' as a continuous linear functional on $\hat{X}$, we see that $\hat{h}$ dominates x' on $\hat{X}$. Therefore, by the Hahn–Banach Extension Theorem 5.53, we can extend x' to a linear functional ℓ on L satisfying $\ell \leq \hat{h}$. Since $\hat{h}$ is bounded, it follows that $\ell: L \to \mathbb{R}$ is a norm-continuous linear functional. We claim that the linear functional ℓ can be represented by a weak* measurable function as follows.

> *Claim: There exists a set $A \in \Sigma$ with $\mu(A^c) = 0$ and a (Σ, w^*)-measurable function $s \mapsto g_s^*$ from A into X' satisfying $g_s^* \leq C_s$ for all $s \in A$ and $\ell(f) = \int_A \langle f_s, g_s^* \rangle \, d\mu(s)$ for all $f \in L$.*

We shall present the proof of this claim momentarily, but before we forget where we are headed, note that we are nearly finished. It may be that the pesky set A^c from the Claim above is nonempty, but of measure zero. To eliminate this set we need to replace x'_s on A by some $\tilde{x}'_s$ satisfying $\tilde{x}'_s \leqslant h_s$. The trick is to do it in a Σ-measurable fashion. For this we use Theorem 18.33, which asserts that there is a (Σ, w^*)-measurable function $s \mapsto x^*_s$ from S to X' satisfying $x^*_s \leqslant C_s$ for all $s \in S$. (This really means all, not almost all, s.) Then it easily follows that the mapping

$$s \mapsto x'_s = \begin{cases} g^*_s & \text{if } s \in A, \\ x^*_s & \text{if } s \in A^c, \end{cases}$$

satisfies the desired properties.

We now prove the Claim. Fix $x \in X$ and define the set function $\nu_x \colon \Sigma \to \mathbb{R}$ by $\nu_x(B) = \ell(x\chi_B)$. The inequality

$$|\nu_x(B)| = |\ell(x\chi_B)| \leqslant \|\ell\| \cdot \|x\chi_B\| \leqslant \|\ell\| \cdot \|x\| < \infty$$

shows that ν_x is a signed charge of bounded variation. As a matter of fact, ν_x is a signed measure. To see this, let $\{B_n\} \subset \Sigma$ satisfy $B_n \downarrow \varnothing$. Then from

$$\nu_x(B_n) = \ell(x\chi_{B_n}) \leqslant \hat{h}(x\chi_{B_n}) = \int_S C(s, x\chi_{B_n}) \, d\mu(s)$$

$$= \int_S C(s, x)\chi_{B_n}(s) \, d\mu(s) \leqslant \left[\int_S \|C_s\|\chi_{B_n}(s) \, d\mu(s) \right] \cdot \|x\|,$$

it follows that $|\nu_x(B_n)| \leqslant [\int_S \|C_s\|\chi_{B_n}(s) \, d\mu(s)]\|x\|$. Since $\|C_s\|\chi_{B_n}(s) \downarrow 0$ for each s, this implies $\lim_{n\to\infty} \nu_x(B_n) = 0$. This shows that ν_x is a signed measure of bounded variation.

Furthermore, if $\mu(B) = 0$, then $\nu_x(B) = 0$, since $x\chi_B = 0$ in L. So by the Radon–Nikodym Theorem 13.18, there is a Σ-measurable Radon–Nikodym derivative g_x of ν_x with respect to μ. That is, $g_x \colon S \to \mathbb{R}$ is Σ-measurable and satisfies

$$\int_B g_x(s) \, d\mu(s) = \nu_x(B) = \ell(x\chi_B) \leqslant \hat{h}(x\chi_B) = \int_B C_s(x) \, d\mu(s) \qquad (\star)$$

for every $B \in \Sigma_\mu$. Now do this construction for each x in X, and define the function $g \colon S \times X \to \mathbb{R}$ by $g(s, x) = g_x(s)$. By construction $g(\cdot, x)$ is Σ-measurable for each $x \in X$. Since ℓ is linear, for each $\alpha, \beta \in \mathbb{R}$ and $x, y \in X$ we have

$$\int_B g(s, \alpha x + \beta y) \, d\mu(s) = \alpha \int_B g(s, x) \, d\mu(s) + \beta \int_B g(s, y) \, d\mu(s) \qquad (\star\star)$$

for every $B \in \Sigma$. We now show that g is "almost" a linear Carathéodory function.

Fix a countable dense subset $D = \{x_1, x_2, \ldots\}$ of X and let G be the countable subset of X consisting of all rational linear combinations from D. This set is dense in X and is closed under rational linear combinations. For each $y \in G$ let $E_y = \{s \in S : g(s, y) > C_s(y)\} \in \Sigma$. From $(\star)$, we see that $\mu(E_y) = 0$. Therefore, if

$E = \bigcup_{y \in G} E_y$, then $E \in \Sigma$ and $\mu(E) = 0$. In particular, for each $y \in G$ and $s \notin E$ we have $g(s, y) \leqslant C_s(y)$.

Next, notice that if $\alpha, \beta \in \mathbb{R}$ and $x, y \in X$ are fixed, then it follows from ($\star\star$) that $g(s, \alpha x + \beta y) = \alpha g(s, x) + \beta g(s, y)$ for almost all s. In particular, for $y, z \in G$ and rational numbers r, q the set

$$C(y, z, r, q) = \{s \in S : g(s, ry + qz) \neq rg(s, y) + qg(s, z)\}$$

belongs to Σ and satisfies $\mu(C(y, z, r, q)) = 0$. Consequently, if

$$C = \bigcup_{y \in G} \bigcup_{z \in G} \bigcup_{r \in \mathbb{Q}} \bigcup_{q \in \mathbb{Q}} C(y, z, r, q),$$

then $C \in \Sigma$ and $\mu(C) = 0$.

Now put $A = S \setminus (E \cup C) \in \Sigma$ and note that $\mu(A^c) = 0$. Moreover, for each $s \in A$ and $x \in G$, we have $g(s, x) \leqslant C_s(x)$ and

$$g(s, ry + qz) = rg(s, y) + qg(s, z) \tag{$\dagger$}$$

for all $y, z \in G$ and $r, q \in \mathbb{Q}$.

Fix $s \in A$. From $g(s, y) \leqslant C_s(y)$ for each $y \in G$, we see that $g(s, \cdot)$ is uniformly continuous on G. Since G is dense in X, $g(s, \cdot)$ has a continuous extension to all of X, which we denote $g^*(s, \cdot)$; see Lemma 3.11. Now ($\dagger$) guarantees that the extension $g^*(s, \cdot)$ is a continuous linear functional for each $s \in A$ that satisfies $g^*(s, x) \leqslant C_s(x)$ for all $x \in X$ and each $s \in A$. To see that each $g^*(\cdot, x)$ is measurable, fix $x \in X$ and pick a sequence $\{x_n\}$ in G such that $x_n \to x$. Then $g(s, x_n) \to g^*(s, x)$ for each $s \in A$, which in view of the measurability of the functions $g(\cdot, x_n)$ shows that $g^*(\cdot, x) \colon A \to \mathbb{R}$ is measurable for each $x \in X$.

Next, fix $x \in X$ and $B \in \Sigma$. Pick a sequence $\{x_n\} \subset G$ such that $x_n \to x$ in X. Also, choose some $M > 0$ such that $\|x_n\| \leqslant M$ for each n. Then for each $s \in A$, we have

$$g_{x_n}(s) \leqslant C_s(x_n) \leqslant |C_s(x_n)| \leqslant \|C_s\| \cdot \|x_n\| \leqslant M\|C_s\|,$$

and similarly

$$-g_{x_n}(s) = g_{-x_n}(s) \leqslant M\|C_s\|.$$

Together these give the bound, $|g_{x_n}(s)| \leqslant M\|C_s\|$ for all $s \in A$ and all n. Therefore, from $\ell(x_n \chi_B) = \int_B g_{x_n}(s)\,d\mu(s) = \int_{A \cap B} g_{x_n}(s)\,d\mu(s)$ and the Lebesgue Dominated Convergence Theorem, we see that

$$\ell(x \chi_B) = \lim_{n \to \infty} \int_{A \cap B} g_{x_n}(s)\,d\mu(s) = \int_{A \cap B} g_x^*(s)\,d\mu(s)$$
$$= \int_B g_x^*(s)\,d\mu(s) = \int_S g^*(s, x \chi_B(s))\,d\mu(s).$$

This easily implies $\ell(f) = \int_S g^*(s, f_s)\,d\mu(s)$ for each $f \in L$. In particular, we have $x'(x) = \ell(\hat{x}) = \int_S g_x^*(s)\,d\mu(s)$ for every $x \in X$. That is, x' is the Gelfand integral of $s \mapsto g_s^*$ over S. ∎

We can rewrite this result as a theorem on the integration of correspondences. Let $\varphi \colon S \twoheadrightarrow X'$ be a correspondence. A **Gelfand μ-integrable selector** from φ is any Gelfand integrable (Σ_μ, w^*)-measurable function $f \colon S \to X'$ satisfying $f(s) \in \varphi(s)$ for each $s \in S$. Let $\mathcal{G}_\mu(\varphi)$ denote the collection of all Gelfand μ-integrable selectors from φ. Similarly, a **Gelfand Σ-integrable selector** from φ is any Gelfand integrable (Σ, w^*)-measurable function $f \colon S \to X'$ that satisfies $f(s) \in \varphi(s)$ for each $s \in S$. The collection of all Gelfand Σ-integrable selectors of φ is denoted $\mathcal{G}_\Sigma(\varphi)$. Clearly, $\mathcal{G}_\Sigma(\varphi) \subset \mathcal{G}_\mu(\varphi)$.

18.36 Definition   *Let (S, Σ) be a measurable space, let X be a Banach space, and let $\varphi \colon S \twoheadrightarrow X'$ be a correspondence.*

- *The **Gelfand μ-integral** $\int \varphi \, d\mu$ of φ is the set of Gelfand integrals of all Gelfand μ-integrable selectors from φ. That is, $\int \varphi \, d\mu$ is the subset of X' given by*

$$\int \varphi \, d\mu = \Big\{ \int_S f(s) \, d\mu(s) : f \in \mathcal{G}_\mu(\varphi) \Big\}.$$

- *The **Gelfand Σ-integral** of φ is the set of Gelfand integrals of all Gelfand Σ-integrable selectors from φ. That is, the Gelfand Σ-integral is the set of linear functionals*

$$\Big\{ \int_S f(s) \, d\mu(s) : f \in \mathcal{G}_\Sigma(\varphi) \Big\}.$$

Since $\mathcal{G}_\Sigma(\varphi) \subset \mathcal{G}_\mu(\varphi)$, it is harder to find a Gelfand Σ-integrable selector than it is to find a Gelfand μ-integrable selector when μ is not a complete measure. The next consequence of Strassen's Theorem presents conditions that guarantee the existence of Gelfand Σ-integrable selectors and characterizes the integral of the correspondence.

18.37 Corollary   *Let X be a separable Banach space and let (S, Σ, μ) be a probability space. Assume further that $\varphi \colon S \twoheadrightarrow X'$ is a scalarly measurable correspondence with nonempty w^*-compact convex values.*

If $\int_S \|h_{\varphi(s)}\| \, d\mu(s) < \infty$, then the Gelfand Σ-integral of the correspondence φ is a nonempty w^*-compact convex subset of X'.

Proof: The function $C \colon S \times X \to \mathbb{R}$ defined by $C(s, x) = h_{\varphi(s)}(x)$ is (by hypothesis) measurable for each fixed x and (by Theorem 7.52) norm-continuous for each fixed s. Moreover, by Lemma 18.29, the integral $h(x) = \int_S C(s, x) \, d\mu(s)$ is a norm-continuous sublinear function on X. Also, let $\mathcal{J}_\varphi = \{ \int_S f(s) \, d\mu(s) : f \in \mathcal{G}_\Sigma(\varphi) \}$, the Gelfand Σ-integral of φ. We now show that

$$\mathcal{J}_\varphi = K_h = \{ x' \in X' : x'(x) \leqslant h(x) \text{ for each } x \in X \}.$$

Once this is established, the conclusion follows from Theorem 7.52.

To establish the claim, assume first that $x' \in K_h$. Then, by Strassen's Theorem, there exists a linear Carathéodory function $s \mapsto x'_s$ satisfying $x'_s \leqslant C_s$ for each s and having Gelfand integral x'. Since $x'_s(x) \leqslant C_s(x) = h_{\varphi(s)}(x)$ for $s \in S$ and $x \in X$, it follows from Theorem 7.52 again that $x'_s \in \varphi(s)$ for each $s \in S$. In other words, $s \mapsto x'_s$ is a Gelfand Σ-integrable selector from φ whose integral is x'. Hence $x' \in \mathfrak{I}_\varphi$, so $K_h \subset \mathfrak{I}_\varphi$.

Next let $x' \in \mathfrak{I}_\varphi$. This means that there exists a Gelfand Σ-integrable selector $f \colon S \to X'$ from φ satisfying $\int_S f_s \, d\mu(s) = x'$. Since $f_s \in \varphi(s)$ for each $s \in S$, it follows that $f_s(x) \leqslant h_{\varphi(s)}(x) = C(s, x)$ for all $s \in S$ and each $x \in X$. Consequently,

$$x'(x) = \int_S f_s(x) \, d\mu(s) \leqslant \int_S C(s, x) \, d\mu(s) = h(x)$$

for each $x \in X$, which shows that $x' \in K_h$. That is, $x' \in \mathfrak{I}_\varphi \subset K_h$. Therefore, $\mathfrak{I}_\varphi = K_h$, and the proof is finished. ∎

There are slightly weaker versions of Strassen's Theorem that can be proven more easily. If we assume in addition, as in [246, Theorem XI.51, p. 244] or [176, Theorem 6.11, p. 100], that $s \mapsto \|C_s\|$ is μ-essentially bounded, then we can draw the stronger conclusion that $\hat{h}$ is well defined and continuous on $L_1(\mu, X)$, the space of Bochner integrable functions. Then standard results on $L_\infty(\mu, X')$ as the dual of $L_1(\mu, X)$ imply the Claim in the proof. The construction in the proof of Theorem 18.31 can be bypassed if we are willing either to work with μ-measurable functions rather than Σ-measurable functions, or to assume that μ is complete ($\Sigma = \Sigma_\mu$); we only have to patch a set of measure zero, which can be done without affecting μ-measurability. A. and C. Ionescu Tulcea [176, Theorem 7.11, p. 100] provide a version without the separability assumption on X. This generalization is not free—it involves an added hypothesis that we do not wish to discuss here, and only yields μ-measurability. Applications of this result to statistics may be found in P.-A. Meyer [246], T. Kamae, U. Krengel, and G. L. O'Brien [192], and of course V. Strassen [324]. Applications to expected utility theory may be found in K. C. Border [57, 59].

Chapter 19

Markov transitions

A Markov system is a stochastic process for which the state of the system at any future time depends only on the present state. Such processes are called *Markov processes*. In the language of conditional expectation of random variables, a Markov process is a family $\{X_t\}$ of random variables (indexed by *time*) with the property that for any measurable function f, any t, and any $h > 0$, $E(f(X_{t+h})|X_s, s \leqslant t) = E(f(X_{t+h})|X_t)$. This defines a family of *transition functions* relating the distribution of the process at time t to the probability distribution of the process at time $t + h$. The process is *stationary* if such transition functions do not depend on t. Markov processes are generally considered to belong to the realm of probability theory, but some useful results can be derived by purely analytic methods. The main idea is to abstract from the random variables and look at the transition function as a mapping from states to probability measures on the set of states.

The traditional approach uses *stochastic kernels*, which are real-valued functions defined on the Cartesian product of a state space S and a σ-algebra $\mathcal{A}$ of subsets of a possibly different space X, that define probability measures on X for each state s. That is, a kernel k maps $S \times \mathcal{A}$ into $\mathbb{R}$. Presumably this approach is adopted because real-valued functions are not intimidating. We believe that in many applications it is conceptually more natural to think in terms of what we call *Markov transitions*, which are functions from S into $\mathcal{P}(X)$. For instance, in stochastic dynamic programming problems we are given a set S of states and a set A of actions. The probability of tomorrow's state is determined jointly by today's state and action. We are led naturally to consider mappings from $S \times A$ to $\mathcal{P}(S)$, the space of probabilities on the state space S. There are other, independent, reasons to study such mappings. For instance, we may have a model of commodity differentiation, as in A. Mas-Colell [241], in which allocations are mappings from a set S of *traders* to measures on a space of *attributes*. Or we may have a game-theoretic framework in mind, where we are interested in mappings from a space of *players* to *mixed strategies*. One advantage of thinking in terms of mappings from S into $\mathcal{P}(X)$ is that it is easier to generalize to *transition correspondences*, or set-valued transitions. The tradeoff is that it is more work to deal with probability measures than real numbers.

We confine our attention to the reasonably well behaved case where S and X are separable metric spaces. Since most of us are addicted to working with countably additive probabilities, this potentially creates some technical difficulties. Namely, the topological dual of the space of bounded continuous functions on a general separable metric space contains purely finitely additive probabilities. To avoid these, we work with the σ-order continuous dual, which is the space of (countably additive) Borel measures. In the important special case of compact metric spaces, the space of (countably additive) measures is also the norm dual.

Each transition function P gives rise to a *transition operator* P that maps real functions on X to functions on S. The value of Pf at s is the expected value of f next period given that today's state is s. This association is reversible: Given a transition operator on functions, we can recover a transition function that generates it (Theorem 19.10). The truly abstract approach to Markov systems studies only these operators. Indeed a *Markov operator* is a positive operator between AM-spaces with units that maps the unit of the domain onto the unit in the range. Section 19.1 presents the most basic results on this class of operators and their study resumes in Section 19.9.

To tie the operator theoretic treatment to our more concrete model with transition functions, we show that a transition P is Borel measurable as a function from the metric space S into the metric space $\mathcal{P}(X)$ if and only if its associated transition operator P carries $B_b(X)$ into $B_b(S)$ (Theorem 19.7). A transition P is continuous if and only if it has the *Feller property*, that is, if and only if its transition operator P carries $C_b(X)$ into $C_b(S)$ (Theorem 19.14). The adjoint P' of the operator P (either norm adjoint or σ-order continuous adjoint, as appropriate) maps probability measures on S to probability measures on X (Theorem 19.9). The adjoint P' is continuous if and only if P is continuous (Theorem 19.14).

When $S = X$ is a compact metrizable space, then P' has fixed points in $\mathcal{P}(S)$ (Theorem 19.18). These are called *invariant measures*. (Even if S is not compact, P' has fixed points in the space of charges (Theorem 19.4). The charge may be purely finitely additive though; see Examples 19.16 and 19.17.) Given an invariant measure μ, a function f is μ-invariant if $f = Pf$ μ-a.e. The set of invariant measures is compact and convex. An invariant measure μ is *ergodic* if the only μ-invariant functions are constant μ-a.e. The ergodic measures comprise the set of extreme points of the set of invariant measures (Theorem 19.25).

In many applications (see, e.g., L. E. Blume [49], D. Nachman [258] and D. Duffie, J. Geanakoplos, A. Mas-Colell, and A. McLennan [105]) it is natural and useful to consider set-valued transition functions, or transition correspondences. We use Strassen's Sublinearity Theorem 18.35 to prove the basic result on the existence of selectors (that is, transition functions) that have ergodic measures (Theorem 19.31).

We also prove the nice result that continuous transitions (with full support pointwise) correspond in a natural way to random functions. That is, given a continuous Markov transition P from S to $X = [0, 1]$, there is a Borel probability measure μ on the metrizable space $C(S, X)$ such that P_s is the distribution of $f(s)$ when f is drawn at random from $C(S, X)$ according to μ. We prove this for the important case $X = [0, 1]$ in Theorems 19.32 and 19.33.

19.1 Markov and stochastic operators

The abstract theory of Markov and stochastic operators is extremely useful in studying stochastic systems. For the time being we ask you to take our word for this. Abstract Markov operators act between AM-spaces with units, and abstract stochastic operators act between AL-spaces.

19.1 Definition   *Let E and F be AM-spaces with units e and u, respectively. A* **Markov operator** *is a positive operator $T \colon E \to F$ that carries the unit of E onto the unit of F. That is, $T e = u$.*

A **stochastic operator** *is a positive operator $T \colon E \to F$ between AL-spaces satisfying $\|Tx\| = \|x\|$ for each $x \in E^+$.*

The representation theorem 9.32 for AM-spaces with units guarantees that we can represent every Markov operator as a positive operator $T \colon C(Q) \to C(K)$, where Q and K are compact Hausdorff spaces, satisfying $T\mathbf{1}_Q = \mathbf{1}_K$.

Markov and stochastic operators on finite dimensional spaces can be represented by certain types of matrices. So consider the Euclidean spaces $\mathbb{R}^n$ equipped with the usual pointwise algebraic and lattice operations. The Riesz space $\mathbb{R}^n$ with the sup norm

$$\|x\|_\infty = \max\{|x_1|, \ldots, |x_n|\}$$

is an AM-space with unit the vector $e = (1, 1, \ldots, 1)$. Similarly, the Riesz space $\mathbb{R}^n$ equipped with the L_1-norm

$$\|x\|_1 = \sum_{i=1}^{n} |x_i|$$

is an AL-space.

A positive operator $A \colon \mathbb{R}^n \to \mathbb{R}^m$ can be represented by a nonnegative $m \times n$ matrix

$$A = \begin{pmatrix} a_{11} & \cdots & a_{1n} \\ \vdots & \ddots & \vdots \\ a_{m1} & \cdots & a_{mn} \end{pmatrix}.$$

(A matrix A is nonnegative if $a_{ij} \geqslant 0$ for all i and j.) A moment's thought should convince you that A is a Markov operator if and only if each row sum is one. That is, A is a Markov matrix if and only if $\sum_{j=1}^{n} a_{ij} = 1$ for each $i = 1, \ldots, m$. On the other hand, A defines a stochastic operator if and only if each column sums to one. That is, if and only if $\sum_{i=1}^{m} a_{ij} = 1$ for each $j = 1, \ldots, n$.

Recall that if $T : X \to Y$ is a continuous operator between Banach spaces, then its (norm) adjoint $T' : Y' \to X'$ is defined via the duality identity

$$\langle x, T'y' \rangle = \langle Tx, y' \rangle,$$

for each $x \in X$ and $y' \in Y'$. If T is a positive operator, then clearly its adjoint T' is a positive operator too. Also recall that Theorem 9.6 asserts that every positive operator between Banach lattices is (norm) continuous. Thus Markov and stochastic operators are norm-continuous. By Theorem 6.17, they must also be weakly continuous.

It turns out that Markov and stochastic operators are dual to each other.

19.2 Theorem (Markov and stochastic operators) *Let $T : E \to F$ be a positive operator between two AM-spaces with units e and u, respectively. Then T is a Markov operator if and only if its norm adjoint $T' : F' \to E'$ is a stochastic operator.*

Similarly, if $T : E \to F$ is a positive operator between AL-spaces, then T is a stochastic operator if and only if its norm adjoint $T' : F' \to E'$ is a Markov operator.

Proof: It suffices to show the "only if" part of each claim. (Why?) To this end, assume that T is a Markov operator. If $0 \leqslant x' \in F'$, then by Lemma 9.29, we have

$$\|T'x'\| = T'x'(e) = x'(Te) = x'(u) = \|x'\|.$$

Now assume that T is stochastic, and let e' and u' denote the units in E' and F' generated by the norms on E and F, respectively; see Theorem 9.30. Then, for each $x \in E^+$, we have

$$T'u'(x) = u'(Tx) = \|Tx\| = \|x\| = e'(x),$$

which shows that $T'u' = e'$. That is, T' is a Markov operator. ∎

We continue by introducing the concept of invariant linear functionals for positive operators.

19.3 Definition   *Let $T : E \to E$ be a positive operator on a Banach lattice. A* **T-invariant functional** *is a positive unit vector in E' that is a fixed point of T'. In other words, $x' \in E'$ is T-invariant if $x' \geqslant 0$, $\|x'\| = 1$, and $T'x' = x'$.*

The set of invariant functionals of a Markov operator has a particularly pleasing structure.

19.4 Theorem (Invariant functionals) *Every Markov operator has a nonempty, convex, and w^*-compact set of invariant functionals.*

Proof: Let $T: E \to E$ be a Markov operator on an AM-space with unit e. Consider the set

$$\Delta = \{x' \in E'_+ : \|x'\| = x'(e) = 1\},$$

and note that Δ is a nonempty, convex, and w^*-compact subset of the closed unit ball in E'. Let F denote the set of fixed points of T' in Δ, that is,

$$F = \{x' \in \Delta : T'x' = x'\}.$$

Clearly, F is convex and w^*-compact. To see that F is nonempty, start by observing that the adjoint operator $T': E' \to E'$ is w^*-continuous. Indeed, if $x'_\alpha \xrightarrow{w^*} 0$ in E', then $\langle x, T'x'_\alpha \rangle = \langle Tx, x'_\alpha \rangle \to 0$ for each $x \in X$, which shows that $T'x'_\alpha \xrightarrow{w^*} 0$. Now (by Theorem 19.2) T' leaves Δ invariant. That is, $T(\Delta) \subset \Delta$, so from the Brouwer–Schauder–Tychonoff Fixed Point Theorem 17.56 we know that T' has fixed points in Δ. ∎

If S is a compact metrizable space, then the norm dual of $C(S)$ is precisely $ca(S) = ca_r(S)$, the space of regular (countably additive) Borel measures on S, and the positive linear functionals on $C(S)$ of norm one are the Borel probability measures on S. The next result characterizes the invariant measures for Markov operators acting on a $C(S)$-space when S is compact and metrizable. It is enough to check the action of the measure on nonnegative continuous functions.

19.5 Lemma *Let $T: C(S) \to C(S)$ be a Markov operator, where S is compact and metrizable. Then a probability measure $\mu \in \mathcal{P}(S)$ is T-invariant (that is, $T'\mu = \mu$) if and only if*

$$\int_S f \, d\mu = \int_S Tf \, d\mu$$

for each $0 \leqslant f \in C(S)$.

Proof: If $\mu \in \mathcal{P}(S)$ satisfies $T'\mu = \mu$ and $f \in C(S)$, then

$$\int_S f \, d\mu = \langle f, \mu \rangle = \langle f, T'\mu \rangle = \langle Tf, \mu \rangle = \int_S Tf \, d\mu.$$

For the converse, note that if $\int_S f \, d\mu = \int_S Tf \, d\mu$ for each $0 \leqslant f \in C(S)$, then $\langle f, \mu \rangle = \langle Tf, \mu \rangle = \langle f, T'\mu \rangle$ for each $f \in C(S)$. Thus $\mu = T'\mu$. ∎

19.2 Markov transitions and kernels

In this section we relate the fairly abstract notion of a Markov operator to the reasonably concrete case of Markov transitions, which are measurable mappings from a separable metrizable space to the space of probabilities on a possibly different separable metrizable space.

Throughout this section, S and X are separable metrizable spaces.

As such, each comes equipped with a collection of Borel sets $\mathcal{B}$, a family of
bounded continuous functions C_b, and a collection of bounded Borel measurable
functions B_b. Recall that each of the spaces B_b and C_b is an AM-space with unit
(the constant function $\mathbf{1}$).[1] (See Section 14.1 and Theorem 9.32.) As usual, the
space $\mathcal{P}$ of Borel probability measures is endowed with its $\sigma(\mathcal{P}, C_b)$-topology. If
X is also a Borel space, that is, a Borel subset of a separable completely metriz-
able space, then every countably additive probability measure is actually regular.
(This follows from Lemma 4.20 and Theorem 12.7.) Recall (Theorem 15.18)
that $\mathcal{P}$ is a separable metrizable convex set in the dual $ba = ba_n$ of C_b in its
$\sigma(\mathcal{P}, C_b)$-topology. As a separable metrizable space, $\mathcal{P}$ also comes equipped with
its Borel σ-algebra $\mathcal{B}_{\mathcal{P}}$. Remember that ba and ca are AL-spaces (Theorems 10.53
and 10.56).

Let $P: S \to \mathcal{P}(X)$ be a (not necessarily measurable) function. To simplify
notation, we write P_s instead of $P(s)$. The mapping P gives rise to an operator
$P: B_b(X) \to \mathbb{R}^S$ via the formula

$$(Pf)(s) = (f \circ P)(s) = \langle f, P_s \rangle = \int_X f(x)\, dP_s(x)$$

for each $f \in B_b(X)$ and each $s \in S$. (The notation $f \circ P$ is justified because f
defines a real function on $\mathcal{P}(X)$, namely the function $\mu \mapsto \int f\, d\mu$.) We call P the
transition operator induced by P.

19.6 Lemma   *Let S and X be separable metrizable spaces. Any transition
operator $P: B_b(X) \to \mathbb{R}^S$ is a σ-order continuous positive operator satisfying
$P\mathbf{1}_X = \mathbf{1}_S$.*

Proof: Let $P: S \to \mathcal{P}(X)$ and let $P: B_b(X) \to \mathbb{R}^S$ be its transition operator. That
is, $(Pf)(s) = \int_X f(x)\, dP_s(x)$ for each $f \in B_b(X)$ and each $s \in S$. Clearly, P is
a positive operator. Moreover, $P\mathbf{1}_X(s) = \langle \mathbf{1}_X, P_s \rangle = 1$ for each $s \in S$ so that
$P\mathbf{1}_X = \mathbf{1}_S$.

To see that P is σ-order continuous, suppose $f_n \downarrow 0$ in $B_b(X)$. Then, by
Lemma 14.6, we have $f_n(x) \downarrow 0$ for each $x \in X$, so by the Dominated Convergence
Theorem 11.21 we have $(Pf_n)(s) = \int_X f_n(x)\, dP_s(x) \downarrow 0$ for each $s \in S$. This shows
that $Pf_n \downarrow 0$ in $\mathbb{R}^S$, so P is σ-order continuous. ∎

The next theorem characterizes Borel functions from S to $\mathcal{P}(X)$.

---

[1] Theorem 9.32 asserts that each of these spaces is lattice isometric to $C(K)$ for a compact metric
space K. For $C_b(X)$, where X is separable and metrizable, we can use the Stone–Čech compactification
βX for K (Theorem 2.79). That is, $C_b(X)$ is lattice isometric to $C(\beta X)$. However, the Stone–Čech
compactification is too unwieldy to use directly.

19.7 Theorem (Measurable transitions) *Let S and X be separable metrizable spaces. For a function P: S $\to$ $\mathcal{P}(X)$ the following statements are equivalent.*

1. *P is Borel measurable.*

2. $f \in B_b(X) \implies \boldsymbol{P}f \in B_b(S)$.

3. *The function $s \mapsto P_s(A)$ (that is, $\boldsymbol{P}\chi_A$) is Borel measurable for each Borel subset A of X.*

4. $f \in C_b(X) \implies \boldsymbol{P}f \in B_b(S)$.

Proof: (1) $\implies$ (2) Let P be a Borel measurable function from S into $\mathcal{P}(X)$. Let $\mathcal{F} = \{f \in B_b(X) : \boldsymbol{P}f \in B_b(S)\}$. We claim that:

a. $\mathcal{F}$ is a vector subspace of $B_b(X)$,

b. $C_b(X) \subset \mathcal{F}$, and

c. If $\{f_n\} \subset \mathcal{F}$, $f_n \downarrow f$ pointwise and $f \in B_b(X)$, then $f \in \mathcal{F}$.

Property (a) follows from the linearity of the integral. Now if f is a bounded continuous real function on X (that is, $f \in C_b(X)$), then (by the definition of the $\sigma(\mathcal{P}, C_b)$ topology) f defines a bounded continuous real function on $\mathcal{P}(X)$. Therefore the function $\boldsymbol{P}f = f \circ P$ is bounded and measurable (Lemma 4.22), so (b) is true. To see (c), suppose the sequence $\{f_n\}$ in $\mathcal{F}$ satisfies $f_n(x) \downarrow f(x)$ for each $x \in X$ and some $f \in B_b(X)$. By the Dominated Convergence Theorem 11.21 we have

$$\boldsymbol{P}f_n(s) = \int_X f_n(x) \, dP_s(x) \downarrow \int_X f(x) \, dP_s(x) = \boldsymbol{P}f(s).$$

But $B_b(S)$ is closed under bounded pointwise limits, so if $\boldsymbol{P}f_n \in B_b(S)$ for each n, then $\boldsymbol{P}f \in B_b(S)$.

Given (a), (b), and (c), Theorem 4.33 implies that $\mathcal{F} = B_b(X)$.

(2) $\implies$ (3) This follows from $P_s(A) = \boldsymbol{P}\chi_A(s)$.

(3) $\implies$ (4) Note that the relation $P_s(A) = \int_X \chi_A(x) \, dP_s(x)$ implies that the real function $s \mapsto \boldsymbol{P}g(s) = \int_X g(x) \, dP_s(x)$ is Borel measurable for each $\mathcal{B}_X$-simple function g. Since each bounded continuous real function on X is a uniform limit of $\mathcal{B}_X$-simple functions (see Theorem 13.9), a straightforward limiting argument implies that the function $\boldsymbol{P}g$ is Borel measurable for each $g \in C_b(X)$.

(4) $\implies$ (1) Now assume that $\boldsymbol{P}(C_b(X)) \subset B_b(S)$. Since $\mathcal{P}(X)$ is separable and metrizable in its $\sigma(\mathcal{P}(X), C_b(X))$-topology (Theorem 15.12), we see that $\mathcal{P}(X)$ is second countable. Thus every open set in $\mathcal{P}(X)$ is a countable union of finite intersections of open sets from the subbase

$$\mathcal{C} = \{f^{-1}(G) : G \text{ open subset of } \mathbb{R} \text{ and } f \in C_b(X)\},$$

where we write $f^{-1}(G) = \{\mu \in \mathcal{P}(X) : \int_X f \, d\mu \in G\}$ treating $f \in C_b(X)$ as a real function on $\mathcal{P}(X)$. In particular, $\sigma(\mathcal{C})$ (the σ-algebra generated by $\mathcal{C}$) coincides with the Borel σ-algebra on $\mathcal{P}(X)$. By hypothesis Pf is Borel measurable for each $f \in C_b(X)$, so from

$$P^{-1}(f^{-1}(G)) = (f \circ P)^{-1}(G) = (Pf)^{-1}(G),$$

it follows that $P^{-1}(f^{-1}(G))$ is a Borel subset of S for each open subset G of $\mathbb{R}$ and each $f \in C_b(X)$. By Corollary 4.24, P is Borel measurable.   ∎

The preceding theorem leads us to the formal definition of a Markov transition.

19.8 Definition   *A **Markov transition** (or simply a **transition**) from S to X is a Borel measurable function $P\colon S \to \mathcal{P}(X)$, where as usual $\mathcal{P}(X)$ is equipped with its $\sigma(\mathcal{P}, C_b)$-topology. The transition operator P of a Markov transition is called a **Markov transition operator**.*

*A Markov transition P is **deterministic** if for each $s \in S$, the measure P_s is a point mass δ_x for some $x \in X$.*

In other words, a function $P\colon S \to \mathcal{P}(X)$ is a Markov transition if it satisfies any one of the equivalent statements of Theorem 19.7.

The major properties of Markov transition operators follow from our results on abstract Markov operators and their adjoints. We present the implications of these general theorems for our concrete case in the next theorem.

19.9 Theorem (**Markov transition summary**)   *Let S and X be separable metrizable spaces, and let $P\colon S \to \mathcal{P}(X)$ be a Markov transition with Markov transition operator $P\colon B_b(X) \to B_b(S)$. Then:*

1. *P is a σ-order continuous Markov operator.*

2. *The norm adjoint $P'\colon ba(S) \to ba(X)$ is a stochastic operator that carries countably additive measures to countably additive measures, that is, $P'(ca(S)) \subset ca(X)$, and satisfies*

$$P'\mu(A) = \int_S P_s(A) \, d\mu(s)$$

   for each $\mu \in ba(S)$ and all $A \in \mathcal{B}_X$. In other words, $P'\mu$ is the Gelfand integral of P with respect to μ. Moreover, the restriction

   $$P'\colon (ca(S), \sigma(ca(S), B_b(S))) \to (ca(X), \sigma(ca(X), B_b(X)))$$

   is continuous and carries $\mathcal{P}(S)$ into $\mathcal{P}(X)$.

3. *The transition operator $P\colon B_b(X) \to B_b(S)$ is continuous with respect to the $\sigma(B_b(X), ca(X))$ and $\sigma(B_b(S), ca(S))$ weak topologies.*

4. *The Markov transition P that generates $\boldsymbol{P}$ is unique, and satisfies*

$$P_s = \boldsymbol{P}'\delta_s$$

for each $s \in S$.

Proof: (1) The proof is identical to that of Lemma 19.6 with $B_b(S)$ replacing $\mathbb{R}^S$ everywhere.

(2) Part (1) and Theorem 19.2 show that $\boldsymbol{P}' \colon ba(S) \to ba(X)$ is a stochastic operator. Also, the σ-order continuity of $\boldsymbol{P}$ coupled with Theorems 8.39 (1) and 14.5 shows that $\boldsymbol{P}'(ca(S)) \subset ca(X)$. For the formula defining $\boldsymbol{P}'$ note that

$$\boldsymbol{P}'\mu(A) = \langle \chi_A, \boldsymbol{P}'\mu \rangle = \langle \boldsymbol{P}\chi_A, \mu \rangle = \int_S \boldsymbol{P}\chi_A(s)\,d\mu(s) = \int_S P_s(A)\,d\mu(s).$$

Theorem 6.43 thus implies $\sigma(ca(S), B_b(S))$-$\sigma(ca(X), B_b(X))$ continuity. Clearly $\boldsymbol{P}'$ maps probability measures to probability measures.

(3) This conclusion is an easy consequence of part (2) and Theorem 6.43.

(4) Notice that for each $f \in B_b(X)$ and all $s \in S$ we have

$$\langle f, P_s \rangle = \boldsymbol{P}f(s) = \langle \boldsymbol{P}f, \delta_s \rangle = \langle f, \boldsymbol{P}'\delta_s \rangle.$$

Since P_s and $\boldsymbol{P}'\delta_s$ are both probability measures on X, it follows from Theorem 15.1 that $P_s = \boldsymbol{P}'\delta_s$ for each $s \in S$, and the proof is finished. ∎

Note that while the adjoint $\boldsymbol{P}'$ is $\sigma(ca(S), B_b(S))$-$\sigma(ca(X), B_b(X))$ continuous, it may not be $\sigma(ca(S), C_b(S))$-$\sigma(ca(X), C_b(X))$ continuous. As usual, the $\sigma(ca, B_b)$-topology is much stronger than the $\sigma(ca, C_b)$-topology. In particular, if $x_n \to x$, then the point masses satisfy $\delta_{x_n} \to \delta_x$ in the $\sigma(ca, C_b)$-topology, but not generally in the $\sigma(ca, B_b)$-topology. (Consider the Borel measurable function $\chi_{\{x\}}$. If $x_n \neq x$, then $\int \chi_{\{x\}}\,d\delta_{x_n} = 0$, but $\int \chi_{\{x\}}\,d\delta_x = 1$.) Section 19.3 gives the conditions under which $\boldsymbol{P}'$ is continuous in the usual $\sigma(ca, C_b)$-topologies.

The next result asserts that Markov transition operators are the only σ-order continuous Markov operators from $B_b(X)$ to $B_b(S)$. That is, every σ-order continuous Markov operator from $B_b(X)$ into $B_b(S)$ is induced by some measurable transition function.

19.10 Theorem *A positive operator $T \colon B_b(X) \to B_b(S)$ is a Markov transition operator if and only if T is a σ-order continuous Markov operator.*

Proof: The "only if" part is just part (1) of Theorem 19.9. For the converse, assume that $T \colon B_b(X) \to B_b(S)$ is a σ-order continuous Markov operator. The σ-order continuity of T implies that the norm-adjoint operator T' maps $ca(S)$ into $ca(X)$ (see Theorem 8.39). Now define the mapping $P \colon S \to \mathcal{P}(X)$ via the formula

$$P_s = T'\delta_s,$$

where δ_s is the point mass supported at s. (Note that by Theorem 19.2 the function P indeed maps S into $\mathcal{P}(X)$.) In addition, note that for each $f \in B_b(X)$ and all $s \in S$ the transition operator P corresponding to P satisfies

$$Pf(s) = \langle f, P_s \rangle = \langle f, T'\delta_s \rangle = \langle Tf, \delta_s \rangle = Tf(s).$$

That is, $Pf = Tf \in B_b(S)$. In particular, $P(B_b(X)) \subset B_b(S)$. By Theorem 19.7, P is also Borel measurable, and the proof is finished.   ∎

A more popular way of presenting Markov transitions is as Markov kernels.

19.11 Definition   *Let (S, Σ) and $(X, \mathcal{A})$ be measurable spaces. A **Markov kernel** (or a **stochastic kernel**) is a mapping $k: S \times \mathcal{A} \to [0, 1]$ satisfying the following two properties.*

1. *For each $s \in S$, the set function $k(s, \cdot): \mathcal{A} \to [0, 1]$ is a probability measure.*

2. *For each $A \in \mathcal{A}$, the mapping $k(\cdot, A): S \to [0, 1]$ is Σ-measurable.*

Clearly Markov transitions can be used to define Markov kernels. The proof of this is left as an exercise:

19.12 Theorem (Kernels and transitions I) *If S and X are separable metrizable spaces, then for a mapping $P: S \to \mathcal{P}(X)$ the following statements are equivalent.*

1. *P is a Markov transition, that is, P is Borel measurable.*

2. *The function $k: S \times \mathcal{B}_X \to [0, 1]$, defined by $k(s, A) = P_s(A)$, is a Markov kernel.*

The converse is also true in the context of separable metrizable spaces. While it may seem less restrictive to place only a measurable and no topological structure on S and X, but the most useful theorems depend on a topological structure. Again the proof of the next result is straightforward and is left as an exercise.

19.13 Theorem (Kernels and transitions II) *If S and X are separable metrizable spaces, then for a mapping $k: S \times \mathcal{B}_X \to [0, 1]$ the following statements are equivalent.*

1. *The function k is a Markov kernel.*

2. *The function $P: S \to \mathcal{P}(X)$, defined by $s \mapsto P_s(\cdot) = k(s, \cdot)$, is a Markov transition.*

Another common way to create Markov kernels is from a measurable function $\xi: (S, \Sigma) \to (X, \mathcal{B}_X)$ into a topological space via the formula

$$k(s, A) = (\chi_A \circ \xi)(s) = \chi_A(\xi(s)) = \chi_{\xi^{-1}(A)}(s).$$

This is equivalent to the Markov transition $P: S \to \mathcal{P}(X)$ via $s \mapsto \delta_{\xi(s)}$.

19.3   Continuous Markov transitions

We saw in the last section that the measurability of a transition function can be characterized in terms of the range of its transition operator. The same goes for continuity. A Markov transition P from S to X possesses the **Feller Property** if Pf is bounded and continuous whenever f is bounded and continuous. That is, if $P(C_b(X)) \subset C_b(S)$. It turns out that the Feller Property characterizes continuous Markov transitions.

19.14 Theorem   (Continuous Markov transitions)   *For a Markov transition* $P \colon S \to \mathcal{P}(X)$ *the following statements are equivalent.*

1. *P is continuous.*

2. *The transition operator* $\boldsymbol{P} \colon B_b(X) \to B_b(S)$ *of P carries $C_b(X)$ to $C_b(S)$, that is, $\boldsymbol{P}$ has the Feller property.*

3. *The adjoint mapping* $\boldsymbol{P}' \colon \mathcal{P}(S) \to \mathcal{P}(X)$ *is continuous.*

Proof: (1) $\implies$ (2)   By definition, every function $f \in C_b(X)$ defines a continuous real function on $\mathcal{P}(X)$ via $f(\mu) = \langle f, \mu \rangle = \int_X f(x)\, d\mu(x)$. Under this interpretation, if P is continuous, then $f \circ P = \boldsymbol{P}f$ is a bounded continuous real function on S for each $f \in C_b(X)$.

(2) $\implies$ (3)   Let $\mu_\alpha \xrightarrow{w^*} \mu$ in $\mathcal{P}(S)$ and fix $f \in C_b(X)$. Since $\boldsymbol{P}f \in C_b(S)$, it follows that

$$\langle f, \boldsymbol{P}'\mu_\alpha \rangle = \langle \boldsymbol{P}f, \mu_\alpha \rangle \xrightarrow{w^*} \langle \boldsymbol{P}f, \mu \rangle = \langle f, \boldsymbol{P}'\mu \rangle .$$

This shows $\boldsymbol{P}'\mu_\alpha \xrightarrow{w^*} \boldsymbol{P}'\mu$, so $\boldsymbol{P}' \colon \mathcal{P}(S) \to \mathcal{P}(X)$ is continuous.

(3) $\implies$ (1)   Assume that $s_n \to s$ in S and let $f \in C_b(X)$. Now $s_n \to s$ in S implies $\delta_{s_n} \to \delta_s$ in $\mathcal{P}(S)$, so by the continuity of $\boldsymbol{P}' \colon \mathcal{P}(S) \to \mathcal{P}(X)$ we have $\boldsymbol{P}'\delta_{s_n} \to \boldsymbol{P}'\delta_s$ in $\mathcal{P}(X)$. Therefore

$$\langle f, P_{s_n} \rangle = \langle f, \boldsymbol{P}'\delta_{s_n} \rangle \longrightarrow \langle f, \boldsymbol{P}'\delta_s \rangle = \langle f, P_s \rangle$$

for each $f \in C_b(X)$. This shows $P_{s_n} \xrightarrow{w^*} P_s$, so that P is continuous.   ∎

19.4   Invariant measures

We now turn our attention to the case where X and S coincide, that is, $X = S$. Unless otherwise stated, P denotes a Markov transition process from S to $\mathcal{P}(S)$.

19.15 Definition   *Let* $P \colon S \to \mathcal{P}(S)$ *be a Markov transition. A probability measure μ on S is* **P-invariant** *if μ is a fixed point of $\boldsymbol{P}'$. That is, if $\boldsymbol{P}'\mu = \mu$.*

We emphasize that a P-invariant measure is a *countably additive* set function and not merely a finitely additive charge. Also, note that the set of all P-invariant measures is a (possible empty) convex set.

If S denotes the space of possible states of a stochastic system, then the interpretation of a μ-invariant measure is this. For any set $A \in \mathcal{B}$, if the current state s_t is chosen according to the measure μ, the probability that s_t lies in A is just $\mu(A) = \langle \chi_A, \mu \rangle$. If μ is invariant, then the probability that next period's state s_{t+1} lies in A is

$$\int_S P_s(A)\, d\mu(s) = \langle P\chi_A, \mu \rangle = \langle \chi_A, P'\mu \rangle = \langle \chi_A, \mu \rangle = \mu(A),$$

which is exactly the same. At any time in the future, the state will be unconditionally distributed according to μ. (Conditional on the state s, next period's state is distributed according to P_s.)

Although by Theorem 19.4 we know that for an arbitrary Markov transition $P \colon S \to \mathcal{P}(S)$ the convex set

$$\{0 \leqslant \mu \in ba(S) : \mu(S) = 1 \text{ and } P'\mu = \mu\}$$

is nonempty and $\sigma(ba(S), B_b(S))$-compact, the Markov transition P need not admit invariant measures.

19.16 Example (No invariant measure I) Let $S = \mathbb{N}$ with the discrete topology and consider the (continuous and deterministic) Markov transition $P \colon S \to \mathcal{P}(S)$ defined by $P_s = \delta_{s+1}$. We claim that P has no invariant measures.

We start by showing that the positive unit fixed points of P' are precisely the charges that represent Banach–Mazur limits on ℓ_∞. To see this, let $\Lambda \colon \ell_\infty \to \mathbb{R}$ be a Banach–Mazur limit and let μ be the charge representing Λ; see Section 16.10. That is, $\Lambda(f) = \int f\, d\mu$ for each $f \in \ell_\infty = C_b(\mathbb{N})$ and $\mu(A) = \Lambda(\chi_A)$ for each subset A of $\mathbb{N}$. Note that for each $f = (f_1, f_2, \ldots) \in \ell_\infty$ we have

$$\langle f, P'\mu \rangle = \langle Pf, \mu \rangle = \int_S f(s+1)\, d\mu(s)$$
$$= \Lambda(f_2, f_3, \ldots) = \Lambda(f_1, f_2, \ldots) = \langle f, \mu \rangle.$$

Since f is arbitrary, this shows that $P'\mu = \mu$. That is, every Banach–Mazur limit is a fixed point of P'.

Now let $\mu \in ba(\mathbb{N})$ be a charge satisfying $\mu(\mathbb{N}) = 1$ and $P'\mu = \mu$. Defining $\Lambda(f) = \langle f, \mu \rangle = \int f\, d\mu$, then $\Lambda(1, 1, 1, \ldots) = 1$ and as above

$$\Lambda(f_2, f_3, \ldots) = \langle Pf, \mu \rangle = \langle f, P'\mu \rangle = \langle f, \mu \rangle = \Lambda(f_1, f_2, \ldots)$$

for all $f = (f_1, f_2, \ldots) \in \ell_\infty$, so Λ is a Banach–Mazur limit. [2]

---

[2] This characterization of the Banach–Mazur limits coupled with Theorem 19.4 presents an alternate way of proving the existence of Banach–Mazur limits.

Now if μ represents a Banach–Mazur limit, then by Lemma 16.29 the measure μ is purely finitely additive. That is, μ cannot be a measure. ∎

The preceding example used a continuous transition on a noncompact metrizable space. The next example uses a discontinuous transition on a compact metrizable space.

19.17 Example (No invariant measure II) Let $S = \mathbb{N} \cup \{\infty\}$ be the one-point compactification of $\mathbb{N}$. By Corollary 3.45, this is a compact metrizable space—in fact, S is homeomorphic to the compact metric space $\{1, \frac{1}{2}, \frac{1}{3}, \ldots\} \cup \{0\}$. Consider the (deterministic, but discontinuous) Markov transition $P \colon S \to \mathcal{P}(S)$ defined by $P_s = \delta_{s+1}$ for $s \in \mathbb{N}$ and $P_\infty = \delta_1$. Now use the arguments of the preceding example to verify that P has indeed no invariant measures. ∎

When S is compact and metrizable and P is continuous, then invariant measures always exist.

19.18 Theorem *If P is a continuous Markov transition on a compact metrizable topological space S, then the set of P-invariant measures is a nonempty, convex, and w^*-compact subset of $ca(S)$.*

Proof: Since S is compact and metrizable, Corollary 14.15 implies that the norm dual of $C(S)$ coincides with $ca(S)$. Since P restricted to $C(S)$ is a Markov operator, the conclusion follows from Theorem 19.4.[3] ∎

19.19 Definition *Let P be a Markov transition on S and let $\mu \in \mathcal{P}(S)$. A function $f \in B_b(S)$ is $\boldsymbol{\mu}$-invariant if $Pf = f$ μ-almost everywhere. Let $\mathcal{J}_\mu^*$ denote the set of μ-invariant functions in $B_b(S)$. That is,*

$$\mathcal{J}_\mu^* = \{f \in B_b(S) : Pf = f \ \mu\text{-a.e.}\}.$$

A Borel subset A of S is a $\boldsymbol{\mu}$-invariant set if χ_A is μ-invariant. The set of all μ-invariant sets is denoted $\mathcal{J}_\mu$. That is,

$$\mathcal{J}_\mu = \{A \in \mathcal{B}_S : P\chi_A = \chi_A \ \mu\text{-a.e.}\}.$$

In other words, a Borel set A is μ-invariant if and only if $P_s(A) = \chi_A(s)$ for μ-almost all s (or if and only if $P_s(A) = 1$ for μ-almost every $s \in A$ and $P_s(A) = 0$ for μ-almost every $s \in A^c$). For invariant measures, we have the following nice result on the structure of the collections of invariant functions and sets.

---

[3] Another way to prove this is to note that by Theorem 19.14, the $\sigma(ca, C_b)$-continuous mapping P' maps the $\sigma(ca, C_b)$-compact convex set into itself. Therefore by the Brouwer–Schauder–Tychonoff Fixed Point Theorem 17.56, it has a nonempty compact set of fixed points. Convexity of the set of fixed points is easy.

19.20 Lemma   *For a P-invariant measure μ:*

1. *The set $\mathfrak{I}^*_\mu$ of all μ-invariant functions is a Riesz subspace of $B_b(S)$ that contains the constant functions and is closed under monotone pointwise limits in $B_b(S)$.*

2. *The set $\mathfrak{I}_\mu$ of μ-invariant sets is a σ-subalgebra of the Borel σ-algebra $\mathcal{B}_S$.*

Proof: (1)   Clearly, the linearity of P implies that $\mathfrak{I}^*_\mu$ is a vector space. Since each P_s is a probability measure, it follows that $\mathfrak{I}^*_\mu$ contains all constant functions. The σ-order continuity of the operator $P\colon B_b(S) \to B_b(S)$ guarantees that $\mathfrak{I}^*_\mu$ is closed under monotone limits in $B_b(S)$. To see that $\mathfrak{I}^*_\mu$ is closed under the lattice operations, it suffices to show that it is closed under absolute values. So let $f \in \mathfrak{I}^*_\mu$, that is, $Pf = f$ μ-a.e. Then $|f| = |Pf| \leqslant P|f|$ μ-a.e. or $P|f| - |f| \geqslant 0$ μ-a.e. But since μ is P-invariant,

$$\langle |Pf|, \mu \rangle \leqslant \langle P|f|, \mu \rangle = \langle |f|, P'\mu \rangle = \langle |f|, \mu \rangle = \langle |Pf|, \mu \rangle.$$

So $\int (P|f| - |f|)\, d\mu = 0$, and Theorem 11.16(3) implies that $|f| = P|f|$ μ-a.e.

(2)   To see that $\mathfrak{I}_\mu$ is a σ-algebra, apply part (1) to the indicator functions, and use the facts that $\chi_{A^c} = \mathbf{1} - \chi_A, \chi_{A \cap B} = \chi_A \wedge \chi_B$, and $\chi_{\cap^n_{i=1} A_i} \downarrow \chi_{\cap^\infty_{i=1} A_i}$.   ∎

Recall that if μ is a probability measure on S and $\mu(A) > 0$ for some Borel subset A of S, then the **conditional probability** of μ given A, $\mu(\cdot|A)$, defined by $\mu(B|A) = \frac{\mu(B \cap A)}{\mu(A)}$, is a well defined probability measure. Another important property of invariant sets is that conditioning on an invariant set produces another invariant measure.

19.21 Lemma   *Let $P\colon S \to \mathcal{P}(S)$ be a Markov transition. If a probability measure μ on S is P-invariant and a Borel set A is μ-invariant with $\mu(A) > 0$, then the conditional probability measure $\nu = \mu(\cdot|A)$ is likewise P-invariant.*

Proof: The proof is an exercise in applying the definitions. Assume that some μ in $\mathcal{P}(S)$ is P-invariant and let $A \in \mathcal{B}$ be μ-invariant with $\mu(A) > 0$. Let $\nu = \mu(\cdot|A)$. Then for any $g \in B_b(S)$ we have

$$\langle g, \nu \rangle = \int g\, d\mu(\cdot|A) = \frac{1}{\mu(A)} \int g\chi_A\, d\mu = \frac{1}{\mu(A)} \langle g\chi_A, \mu \rangle. \qquad (\star)$$

Now fix $f \in C_b(S)$. Since A is μ-invariant, we have $P\chi_A = \chi_A$ and $P\chi_{A^c} = \chi_{A^c}$ μ-a.e. That is, $P_s(A) = 1$ and $P_s(A^c) = 0$ for μ-almost every $s \in A$ and $P_s(A) = 0$ and $P_s(A^c) = 1$ for μ-almost every $s \in A^c$. Consequently,

$$\int_A \Big[\int_{A^c} f(x)\, dP_s(x)\Big]\, d\mu(s) = \int_{A^c} \Big[\int_A f(x)\, dP_s(x)\Big]\, d\mu(s) = 0. \qquad (\star\star)$$

Now using $(\star\star)$, we get

$$
\begin{aligned}
\langle f\chi_A, \mu \rangle &= \langle f\chi_A, \boldsymbol{P}'\mu \rangle = \langle \boldsymbol{P}(f\chi_A), \mu \rangle = \int_S \Big[\int_A f(x)\, dP_s(x) \Big]\, d\mu(s) \\
&= \int_A \Big[\int_A f(x)\, dP_s(x) \Big]\, d\mu(s) + \int_{A^c} \Big[\int_A f(x)\, dP_s(x) \Big]\, d\mu(s) \\
&= \int_A \Big[\int_A f(x)\, dP_s(x) \Big]\, d\mu(s) \\
&= \int_A \Big[\int_A f(x)\, dP_s(x) \Big]\, d\mu(s) + \int_A \Big[\int_{A^c} f(x)\, dP_s(x) \Big]\, d\mu(s) \\
&= \int_A \Big[\int_S f(x)\, dP_s(x) \Big]\, d\mu(s) \\
&= \langle (\boldsymbol{P}f)\chi_A, \mu \rangle.
\end{aligned}
$$

This combined with $(\star)$ yields $\langle f, \nu \rangle = \langle \boldsymbol{P}f, \nu \rangle = \langle f, \boldsymbol{P}'\nu \rangle$ for any f belonging to $C_b(S)$. So by Theorem 15.1, we have $\boldsymbol{P}'\nu = \nu$. ∎

In Section 16.10, we defined a Borel measure μ on a topological space X to be invariant with respect to a continuous function $\xi\colon X \to X$ if $\mu(A) = \mu(\xi^{-1}(A))$ for each Borel set A. By employing the existence of Banach–Mazur limits, we showed that every continuous function on a compact metrizable space admits invariant measures (Theorem 16.48). We now show that those results are a special case of the results of this section. To see this, fix a continuous function $\xi\colon S \to S$ and define the continuous deterministic Markov transition $P\colon S \to \mathcal{P}(S)$ by $P_s = \delta_{\xi(s)}$. As noted in the preceding section, $\boldsymbol{P}f = f \circ \xi$ for each $f \in B_b(S)$. We claim that a probability measure μ on S is P-invariant (that is, $\boldsymbol{P}'\mu = \mu$) if and only if μ is ξ-invariant (that is, $\mu(A) = \mu(\xi^{-1}(A))$ for each $A \in \mathcal{B}_S$).

Indeed, if $\boldsymbol{P}'\mu = \mu$, then for each $A \in \mathcal{B}_S$ we have

$$
\begin{aligned}
\mu(A) &= \langle \chi_A, \mu \rangle = \langle \chi_A, \boldsymbol{P}'\mu \rangle = \langle \boldsymbol{P}\chi_A, \mu \rangle \\
&= \int \chi_A \circ \xi\, d\mu = \int \chi_{\xi^{-1}(A)}\, d\mu = \mu(\xi^{-1}(A)).
\end{aligned}
$$

That is, μ is ξ-invariant.

On the other hand, if $\mu(A) = \mu(\xi^{-1}(A))$ for each $A \in \mathcal{B}_S$, then by reversing the preceding arguments we see that

$$
\mu(A) = \langle \chi_A, \mu \rangle = \langle \chi_A \circ \xi, \mu \rangle = \langle \boldsymbol{P}\chi_A, \mu \rangle = \langle \chi_A, \boldsymbol{P}'\mu \rangle = \boldsymbol{P}'\mu(A)
$$

for each $A \in \mathcal{B}_S$. That is, $\boldsymbol{P}'\mu = \mu$, so μ is a fixed point of $\boldsymbol{P}'$.

19.5   Ergodic measures

In a sense, P-invariant measures reproduce themselves in that if μ is invariant, then $P'\mu = \mu$. This is not very satisfying however. For instance consider the deterministic transition P on $\{0, 1\}$ defined by $P_0 = \delta_0$ and $P_1 = \delta_1$. That is, once the system is in state s it stays there. Now note that every probability on $\{0, 1\}$ is P-invariant. To see this, identify a probability measure μ on $\{0, 1\}$ with a number p in $[0, 1]$, where $\mu(\{0\}) = p$ and $\mu(\{1\}) = 1 - p$. Then $P'\mu = p\delta_0 + (1 - p)\delta_1 = \mu$. However, μ is not a very good indicator of the distribution of future states of the system, unless μ is zero or one. Observe however, that δ_0 and δ_1 are invariant measures, which really do reproduce themselves, and they also happen to be the extreme points of the set of invariant measures. This is no accident, as we shall soon see.

19.22 Lemma   *For a continuous Markov transition P on a metrizable space S and a P-invariant measure $\mu \in \mathcal{P}(S)$ the following statements are equivalent.*

1. *If $A \in \mathcal{B}$ is μ-invariant, then either $\mu(A) = 0$ or $\mu(A) = 1$.*

2. *If $f \in C_b(S)$ is μ-invariant, then f is constant μ-almost everywhere.*

Proof: (1) $\implies$ (2)   Suppose that $f \in C_b(S)$ satisfies $Pf = f$ μ-a.e. For each $\alpha \in \mathbb{R}$ define the upper contour set $U_f^\alpha = \{s \in S : f(s) > \alpha\}$. By Lemma 19.20, the function $1 \wedge n(f - \alpha)^+$ is μ-invariant for each n, and it is easy to see that $1 \wedge n(f - \alpha)^+(s) \uparrow \chi_{U_f^\alpha}(s)$ for each $s \in S$. Thus, by Lemma 19.20 again, U_f^α is μ-invariant. So by our hypothesis, $\mu(U_f^\alpha)$ is either 0 or 1 for any α. Letting $c = \inf\{\alpha : \mu(U_f^\alpha) = 0\}$, we see that $f = c$ μ-a.e. (Why?)
  (2) $\implies$ (1)   Let $A \in \mathcal{B}$ be μ-invariant. This means that χ_A is μ-invariant, so χ_A is constant μ-a.e. It follows that $\chi_A = 0$ μ-a.e. or $\chi_A = 1$ μ-a.e., which shows that either $\mu(A) = 0$ or $\mu(A) = 1$.   ∎

The preceding lemma motivates the following definition.

19.23 Definition   *Let P be a Markov transition on a metrizable space S and let μ be a P-invariant measure. Then μ is **P-ergodic**, or simply **ergodic**, if it satisfies either of the equivalent statements of Lemma 19.22. That is, μ is ergodic if for every μ-invariant set A, either $\mu(A) = 1$ or $\mu(A) = 0$.*
  *If μ is an ergodic measure and A is μ-invariant with $\mu(A) = 1$, then A is called an **ergodic set**.*

Two distinct ergodic measures cannot be mutually absolutely continuous.

19.24 Lemma   *Let S be a metrizable space, let $P: S \to \mathcal{P}(S)$ be a Markov transition, and let $\mu, \nu \in \mathcal{P}(S)$ be P-invariant. If μ is P-ergodic and ν is absolutely continuous with respect to μ, then $\nu = \mu$.*

Proof: Let $0 \leqslant f = \frac{d\nu}{d\mu}$ be a Radon–Nikodym derivative. That is,

$$\nu(A) = \int_A f \, d\mu = \langle f\chi_A, \mu \rangle$$

for each $A \in \mathcal{B}_S$. Note that the measure ν is equal to the measure μ if and only if $f = \mathbf{1}$ μ-almost everywhere. Let $A = \{s \in S : f(s) \geqslant 1\}$. It suffices to show that the indicator function χ_A is μ-invariant, for since μ is ergodic, χ_A is constant μ-almost everywhere. This fact coupled with $\int f \, d\mu = 1$ implies $f = \mathbf{1}$ μ-almost everywhere.

Let $\gamma = \nu - \mu$, and note that for each $g \in B_b(S) \subset L_\infty(|\gamma|)$, we have

$$\langle g, \gamma \rangle = \int g \, d\gamma = \int g(f - \mathbf{1}) \, d\mu.$$

Also, from $0 \leqslant \chi_A \leqslant \mathbf{1}$ and the positivity of P, we get $0 \leqslant P\chi_A \leqslant P\mathbf{1} = \mathbf{1}$. Now observe that $\chi_{A^c} P\chi_A(f - \mathbf{1}) \leqslant 0$, so

$$\langle \chi_{A^c} P\chi_A, \gamma \rangle = \int \chi_{A^c} P\chi_A(f - \mathbf{1}) \, d\mu \leqslant 0.$$

From $P'\gamma = \gamma$, it follows that $\langle \chi_A, \gamma \rangle = \langle P\chi_A, \gamma \rangle$. So noting that $f - \mathbf{1} \geqslant 0$ on A, we see that $\chi_A(\mathbf{1} - P\chi_A)(f - \mathbf{1}) \geqslant 0$. Hence,

$$\langle \chi_{A^c} P\chi_A, \gamma \rangle = \langle (\mathbf{1} - \chi_A)P\chi_A, \gamma \rangle = \langle P\chi_A - \chi_A P\chi_A, \gamma \rangle$$
$$= \langle \chi_A - \chi_A P\chi_A, \gamma \rangle = \int \chi_A(\mathbf{1} - P\chi_A)(f - \mathbf{1}) \, d\mu \geqslant 0.$$

Therefore,

$$\int_{A^c} P\chi_A(\mathbf{1} - f) \, d\mu = \langle \chi_{A^c} P\chi_A, \gamma \rangle = 0,$$

which (in view of $\mathbf{1} - f < 0$ on A^c and $P\chi_A \geqslant 0$) implies

$$P\chi_A = 0 \; \mu\text{-a.e. on } A^c. \tag{$\star$}$$

Using $(\star)$, we obtain

$$\int_A f \, d\mu = \nu(A) = \langle \chi_A, \nu \rangle = \langle \chi_A, P'\nu \rangle$$
$$= \langle P\chi_A, \nu \rangle = \int P\chi_A f \, d\mu$$
$$= \int_A P\chi_A f \, d\mu + \int_{A^c} P\chi_A f \, d\mu = \int_A P\chi_A f \, d\mu,$$

or $\int_A(\mathbf{1} - P\chi_A)f \, d\mu = 0$. Since $f \geqslant 1$ on A and $\mathbf{1} - P\chi_A \geqslant 0$, this shows that $P\chi_A = \mathbf{1}$ μ-a.e. on A. Combining this with $(\star)$ we see that $P\chi_A = \chi_A$ μ-a.e. Thus, χ_A is μ-invariant, and the proof is finished. ∎

We can now characterize the set of ergodic measures as the extreme points of the set of invariant measures.

19.25 Theorem   *For an arbitrary Markov transition P on a metrizable space S, the P-ergodic measures are precisely the extreme points of the (possibly empty) convex set of P-invariant measures.*

Proof: Suppose first that μ is P-invariant, but not P-ergodic. That is, there is a μ-invariant set A with $0 < \mu(A) < 1$. Then A^c is invariant too, and we have the identity $\mu = \mu(A)\mu(\cdot|A) + \mu(A^c)\mu(\cdot|A^c)$. Lemma 19.21 then shows that μ is not an extreme point of the convex set of P-invariant measures. This implies that every extreme point is P-ergodic.

Now suppose that μ is P-ergodic and $\mu = \lambda\nu + (1 - \lambda)\gamma$, where $0 < \lambda < 1$ and ν and γ are P-invariant. Then $\nu \ll \mu$ and $\gamma \ll \mu$, so according to Lemma 19.24 we have $\mu = \nu = \gamma$. Thus μ is an extreme point.   ∎

Note that the above theorem does not guarantee the existence of any ergodic measures, even if invariant measures exist, since there may be no extreme points. In the case that S is compact and metrizable and P has the Feller property, the set of P-invariant measures is a nonempty w^*-compact and convex subset of the dual space $C'(S) = ca(S)$ (Theorem 19.18), so by the Krein–Milman Theorem 7.68, there are extreme points. Thus we have the following.

19.26 Corollary   *Every continuous Markov transition P on a compact metrizable space S admits P-ergodic measures.*

19.6   Markov transition correspondences

The notion of a set-valued Markov transition, or Markov transition correspondence, was used by L. E. Blume [49] to study equilibria of stochastic economies. The theory has been extended in various ways by D. Nachman [258] and D. Duffie, J. Geanakoplos, A. Mas-Colell, and A. McLennan [105]. Once again:

Throughout this section S and X are metrizable spaces.

19.27 Definition   *A **Markov transition correspondence** from S to X is a closed-valued measurable correspondence from S into $\mathcal{P}(X)$.*

Recall that a correspondence between topological spaces is measurable if the lower inverse of every closed set is a Borel set (Definition 18.1). Every Markov transition correspondence $\pi\colon S \twoheadrightarrow \mathcal{P}(X)$ and each $f \in B_b(X)$ together define a correspondence $\pi f\colon S \twoheadrightarrow \mathbb{R}$ via

$$\pi f(s) = \left\{ \int_X f(x)\, d\nu(x) : \nu \in \pi(s) \right\}.$$

When f is also continuous we have the following result.

19.28 Lemma   *If $\pi\colon S \twoheadrightarrow \mathcal{P}(X)$ is measurable (that is, a Markov transition correspondence) and f belongs to $C_b(X)$, then the correspondence $\pi f\colon S \twoheadrightarrow \mathbb{R}$ is measurable.*

Proof: Let F be a closed subset of $\mathbb{R}$, and write $f(\mu)$ for $\int_X f(x)\,d\mu(x)$. By the definition of the w^*-topology, f is a continuous function from $\mathcal{P}(X)$ to $\mathbb{R}$. Now observe that

$$(\pi f)^\ell(F) = \{s \in S \ : \ \exists\, \nu \in \pi(s) \text{ with } f(\nu) \in F\} = \pi^\ell(f^{-1}(F)).$$

Since f is continuous, $f^{-1}(F)$ is closed, so the measurability of π implies that $\pi^\ell(f^{-1}(F))$ is a Borel subset of S. Thus πf is measurable.   ∎

There is also a "dual correspondence" $\pi'\colon \mathcal{P}(S) \twoheadrightarrow \mathcal{P}(X)$ defined by

$$\pi'\mu = \int \pi\,d\mu = \Big\{ \int_S P_s\,d\mu(s) : P \text{ is a measurable selector from } \pi\Big\},$$

where the integral is a Gelfand vector integral. That is, $\pi'\mu$ is the integral of the correspondence π with respect to μ; see Definition 18.36 and Theorem 11.52. The Kuratowski–Ryll-Nardzewski Selection Theorem 18.13 (and Lemma 18.2) guarantees that every Markov transition correspondence into a Polish space has measurable selectors, so in that case π' has nonempty values.

19.29 Lemma   *If X is Polish and $\pi\colon S \twoheadrightarrow \mathcal{P}(X)$ is measurable with nonempty compact convex values, then $\pi'\colon \mathcal{P}(S) \twoheadrightarrow \mathcal{P}(X)$ is measurable with nonempty compact convex values.*

Proof: Let h_s denote the support function of $\pi(s)$, that is, $h_s\colon C_b(X) \to \mathbb{R}$ is defined via the formula

$$h_s(f) = \max_{\nu \in \pi(s)} \int_X f(x)\,d\nu(x),$$

and for each $\mu \in \mathcal{P}(S)$ define $h_\mu\colon C_b(X) \to \mathbb{R}$ via

$$h_\mu(f) = \int_S h_s(f)\,d\mu(s).$$

Theorem 18.31 implies that $h_s(f)$ is measurable for each $f \in C_b(X)$, so by Theorem 15.13 the mapping $\mu \mapsto h_\mu(f)$ is measurable for each $f \in C_b(X)$. By Strassen's Theorem 18.35, h_μ is the support function of $\pi'\mu$. So by Theorem 18.31 again, π' is measurable.   ∎

Following L. E. Blume [49], we say that a Markov transition correspondence π has the **multivalued Feller property** if πf is upper hemicontinuous whenever f is a bounded continuous real function on X. The next theorem shows that this is equivalent to the upper hemicontinuity of π.

19.30 Theorem   *Let X be a Polish space and let $\pi \colon S \twoheadrightarrow \mathcal{P}(X)$ be measurable with nonempty compact convex values. Then the following statements are equivalent.*

1. *The correspondence π is upper hemicontinuous.*

2. *The correspondence π has the multivalued Feller property. That is, for each $f \in C_b(X)$ the correspondence πf is upper hemicontinuous.*

3. *The correspondence π' is upper hemicontinuous.*

Proof: (1) $\implies$ (2)   Let $U \subset \mathbb{R}$ be open and let f belong to $C_b(X)$. By the definition of the weak∗ topology, the function $\nu \mapsto f(\nu) = \int_X f(x)\,d\nu(x)$, from $\mathcal{P}(X) \to \mathbb{R}$, is continuous, so $f^{-1}(U)$ is open in $\mathcal{P}(X)$. Now observe that

$$
\begin{aligned}
(\pi f)^{\mathrm{u}}(U) &= \left\{ s \in S : \pi f(s) \subset U \right\} \\
&= \left\{ s \in S : \int_X f(x)\,d\nu(x) \in U \text{ for all } \nu \in \pi(s) \right\} \\
&= \left\{ s \in S : \pi(s) \subset f^{-1}(U) \right\} = \pi^{\mathrm{u}}(f^{-1}(U)),
\end{aligned}
$$

which is an open set since π is upper hemicontinuous. Therefore the correspondence πf is upper hemicontinuous.

(2) $\implies$ (1)   Let f belong to $C_b(X)$. Then the correspondence $\pi f \colon S \twoheadrightarrow \mathbb{R}$ has nonempty, compact, and convex values. Moreover, since the correspondence $\pi f \colon S \twoheadrightarrow \mathbb{R}$ is upper hemicontinuous, for every $\beta \in \mathbb{R}' = \mathbb{R}$, the support mapping $s \mapsto \max\{\beta\alpha : \alpha \in \pi f(s)\}$ is upper semicontinuous (Theorem 17.41). In particular, letting $\beta = 1$, the mapping

$$
s \mapsto \max \pi f(s) = \max\{\langle f, \nu \rangle : \nu \in \pi(s)\}
$$

is upper semicontinuous. Since this is true for each $f \in C_b(X)$, the support mapping for π is upper semicontinuous at each $f \in C_b(X)$, so Theorem 17.41 implies that π is upper hemicontinuous.

(1) $\implies$ (3)   Let h_s denote the support function of $\pi(s)$, that is,

$$
h_s(f) = \max_{\nu \in \pi(s)} \int f\,d\nu
$$

for $f \in C_b(X)$. Then by the upper semicontinuous half of the Berge Maximum Theorem 17.30, $s \mapsto h_s(f)$ is upper semicontinuous for each $f \in C_b(X)$. Now define $h_\mu(f) = \int_S h_s(f)\,d\mu(s)$. (This is finite since the integrand is assumed to be bounded.) By Theorem 15.5, $\mu \mapsto h_\mu(f)$ is upper semicontinuous for each f. By Strassen's Theorem 18.35, $h_{\pi'\mu}(f) = \int_S h_s(f)\,d\mu(s)$. Therefore by Theorem 17.41 the mapping $\mu \mapsto \pi'\mu$ is upper hemicontinuous.

(3) $\implies$ (1)   Recall that $s \mapsto \delta_s$ is a homeomorphism and observe that $\pi'\delta_s = \pi(s)$.   ∎

We now consider the case that $S = X$ is compact, and present the central result of D. Duffie, J. Geanakoplos, A. Mas-Colell, and A. McLennan [105].

19.31 Theorem   *Let S be a compact metric space and let $\pi\colon S \twoheadrightarrow \mathcal{P}(S)$ be upper hemicontinuous with nonempty compact convex values. Then there exists a measurable selector P from the correspondence π that has a P-ergodic measure μ in $\mathcal{P}(S)$.*

Proof: Since π is upper hemicontinuous, $\pi'\colon \mathcal{P}(S) \twoheadrightarrow \mathcal{P}(S)$ is upper hemicontinuous by Theorem 19.30. Since $\mathcal{P}(S)$ is compact and convex, the Kakutani–Fan–Glicksberg Fixed Point Theorem 17.55 implies that π' has a nonempty compact set of fixed points. It is easy to verify that the set of fixed points is convex. (Why?) Consequently, the set of fixed points has extreme points. By Corollary 18.37, for every extreme fixed point μ of π', there is a selector P from π with $\mu = \boldsymbol{P}'\mu$. Clearly μ is also an extreme fixed point of $\boldsymbol{P}'$, so by Theorem 19.25 this measure μ is P-ergodic.   ∎

19.7   Random functions

There is another characterization of continuous Markov transitions. A continuous Markov transition from S to X corresponds to choosing a *continuous* function from $C(S, X)$ at random and evaluating it at the state s. For the remainder of this section S denotes a compact metrizable space and X a metrizable space.

Recall that for each $s \in S$ the **evaluation functional** at s is the function $e_s\colon C(S, X) \to X$ defined by $e_s(f) = f(s)$ for each $f \in C(S, X)$. Clearly each evaluation functional is continuous, and therefore Borel measurable.

19.32 Theorem   (Measures on $C(S, X)$ induce Markov transitions)   *Let S be a compact metrizable space and let X be a metrizable topological space. Then every probability measure μ on the metrizable space $C(S, X)$ induces a continuous Markov transition $P\colon S \to \mathcal{P}(X)$ via the formula $P_s = \mu e_s^{-1}$. That is,*

$$P_s(B) = \mu(e_s^{-1}(B)) = \mu(\{f \in C(S, X) : f(s) \in B\})$$

for each Borel subset B of X.

Moreover, the transition operator $\boldsymbol{P}\colon B_b(X) \to B_b(S)$ generated by the continuous transition P is given by

$$\boldsymbol{P}f(s) = \int_{C(S,X)} (f \circ e_s)(g)\, d\mu(g) = \int_{C(S,X)} f(g(s))\, d\mu(g)$$

for all $f \in B_b(X)$ and each $s \in S$.

Proof: We must establish that $P\colon S \to \mathcal{P}(X)$ is continuous. To this end, assume $s_n \to s$ in S and let $f \in C_b(X)$. Fix some $M > 0$ such that $|f(x)| \leqslant M$ for each $x \in X$ and note that

$$|(f \circ e_{s_n})(g)| = |f(g(s_n))| \leqslant M$$

for each $g \in C(S, X)$. Therefore $f \circ e_{s_n} \in C_b(C(S, X))$ for all n. Since

$$(f \circ e_{s_n})(g) = f(g(s_n)) \to f(g(s)) = (f \circ e_s)(g)$$

for each $g \in C(S, X)$, it follows from the Lebesgue Dominated Convergence Theorem 11.21 that

$$\int_{C(S,X)} (f \circ e_{s_n})(g) \, d\mu(g) \longrightarrow \int_{C(S,X)} (f \circ e_s)(g) \, d\mu(g). \qquad (\star)$$

Now, from Theorem 13.46 we know that

$$\langle f, P_{s_n} \rangle = \int_X f(x) \, d\mu e_{s_n}^{-1}(x) = \int_{C(S,X)} (f \circ e_{s_n})(g) \, d\mu(g)$$

and

$$\langle f, P_s \rangle = \int_X f(x) \, d\mu e_s^{-1}(x) = \int_{C(S,X)} (f \circ e_s)(g) \, d\mu(g).$$

This combined with $(\star)$ shows that $\langle f, P_{s_n} \rangle \to \langle f, P_s \rangle$ for each $f \in C_b(X)$. That is, $P_{s_n} \xrightarrow{w^*} P_s$, so P is a continuous Markov transition. The formula for $Pf(s)$ follows immediately from Theorem 13.46. $\blacksquare$

We now prove a partial converse of the previous theorem for the special case $X = [0, 1]$. For this result we require that each P_s have full support on X. The result is due to R. M. Blumenthal and H. H. Corson [51].

19.33 Theorem (Markov transitions generate random functions) *Let S be a compact metrizable space, and let $X = [0, 1]$. If $P\colon S \to \mathcal{P}(X)$ is a continuous Markov transition satisfying* $\operatorname{supp} P_s = X$ *for each s, then there exists a (regular) Borel probability measure μ on $C(S, X)$ satisfying $P_s = \mu e_s^{-1}$ for each $s \in S$. In other words,*

$$P_s(B) = \mu(e_s^{-1}(B)) = \mu(\{f \in C(S, X) : f(s) \in B\})$$

for each Borel subset B of X and every $s \in S$.

Proof: Fix a continuous Markov transition $P\colon S \to \mathcal{P}(X)$ with $\operatorname{supp} P_s = X$ for each $s \in S$. Let Ω denote the unit interval (a different copy of the unit interval from X) considered a probability space equipped with its Borel sets and Lebesgue measure λ. For each s, let F_s be the cumulative distribution function of P_s, that is, $F_s(x) = P_s([0, x])$. Clearly $F_s(0) = P_s(\{0\})$ and $F_s(1) = 1$. Furthermore, F_s is nondecreasing and right continuous. Define $\Phi: \Omega \times S \to [0, 1] = X$ by

$$\Phi(\omega, s) = \inf\{x \in X : F_s(x) \geqslant \omega\}.$$

For convenience, let $\Phi_s(\cdot) = \Phi(\cdot, s)$ and $\Phi_\omega(\cdot) = \Phi(\omega, \cdot)$. You may recognize Φ_s as the standard construction of a random variable on $\Omega = [0, 1]$ having the cumulative distribution function F_s. When F_s is strictly increasing and maps onto $[0, 1]$, then Φ_s is just F_s^{-1}. More generally, flat spots in F_s correspond to jumps in Φ_s and vice-versa. See Figure 19.1.

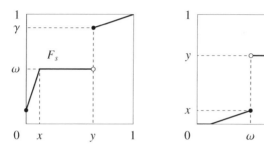

Figure 19.1. Construction of Φ_s from F_s.

This construction has the following properties.

- *For each s, the function $\Phi_s: \Omega \to [0, 1]$ is nondecreasing, and hence Borel measurable.*[4]

Let $\omega < \omega'$. Then $\{x \in X : F_s(x) \geqslant \omega'\} \subset \{x \in X : F_s(x) \geqslant \omega\}$, so

$$\Phi_s(\omega) = \inf\{x \in X : F_s(x) \geqslant \omega\} \leqslant \inf\{x \in X : F_s(x) \geqslant \omega'\} = \Phi_s(\omega').$$

- *For each $\omega \in \Omega$, the function $\Phi_\omega: S \to X$ is continuous.*

To see this let $s_n \to s$, put

$$x_0 = \Phi_\omega(s) = \Phi(\omega, s) = \inf\{x \in X : F_s(x) \geqslant \omega\},$$

and fix $0 < \varepsilon < \min\{x_0, 1 - x_0\}$. Assume that $0 < x_0 < 1$. (The cases $x_0 = 0$ and $x_0 = 1$ are left as an exercise.)

Since F_s is nondecreasing, it has at most countably many points of discontinuity (Theorem 1.3), so there exist points x_1 and x_2 of continuity of F_s satisfying

$$x_0 - \varepsilon < x_1 < x_0 < x_2 < x_0 + \varepsilon.$$

---

[4] If $f: [0, 1] \to \mathbb{R}$ is a nondecreasing function and I is a subinterval of $[0, 1]$, then $f^{-1}(I)$ is also a subinterval (in the generalized sense that includes the empty set and the singletons). This implies that f is Borel measurable.

Now the right continuity of F_s implies $F_s(x_0) \geqslant \omega$, and clearly $x_1 < x_0$ implies $F_s(x_1) < \omega$. Since P_s is assumed to have full support on X, it follows that F_s is strictly increasing. (Why?) Thus we can choose $\eta > 0$ such that

$$F_s(x_1) + \eta < \omega \leqslant F_s(x_0) < F(x_2) - \eta.$$

Now note that if x is a point of continuity F_s, then $P_s(\{x\}) = 0$. (Why?)

Since P is a continuous transition and $s_n \to s$, Theorem 15.3 (7) implies that $F_{s_n}(x) \to F_s(x)$ whenever x is a point of continuity F_s. Thus for n large enough, $|F_{s_n}(x_1) - F_s(x_1)| < \eta$ and $|F_{s_n}(x_2) - F_s(x_2)| < \eta$. So for sufficiently large n we have

$$F_{s_n}(x_1) < \omega < F_{s_n}(x_2).$$

The first inequality implies $\Phi(\omega, s_n) \geqslant x_1 > x_0 - \varepsilon$ and the second inequality yields $\Phi(\omega, s_n) \leqslant x_2 < x_0 + \varepsilon$. Thus $|\Phi_\omega(s_n) - x_0| < 2\varepsilon$, so $\Phi_\omega(s_n) \to x_0 = \Phi_\omega(s)$.

In particular, these two properties imply that Φ is a Carathéodory function. Next we note that Φ_s has cumulative distribution function F_s.

- *For any $x \in [0, 1]$, the probability that $\Phi_s \leqslant x$ is just $F_s(x)$. That is,*

$$\lambda(\{\omega \in \Omega : \Phi_s(\omega) \leqslant x\}) = F_s(x).$$

To see this, note that since F_s is right continuous,

$$x \geqslant \Phi_s(\omega) \iff F_s(x) \geqslant \omega,$$

so $\{\omega \in \Omega : \Phi_s(\omega) \leqslant x\} = [0, F_s(x)]$, which has length $F_s(x)$. That is, Φ_s has cumulative distribution F_s. Consequently by Theorem 10.48, for any Borel subset B of X,

$$\lambda(\{\omega \in \Omega : \Phi_s(\omega) \in B\}) = P_s(B).$$

Now fix $s \in S$ and consider the mappings

$$\Omega \xrightarrow{\hat{\Phi}} C(S, X) \xrightarrow{e_s} X$$

defined by $\hat{\Phi}_\omega(s) = \Phi(\omega, s)$ and $e_s(f) = f(s)$. By Theorem 4.55, the mapping $\hat{\Phi}$ is Borel measurable, so Lebesgue measure λ on Ω induces the probability measure $\mu = \lambda\hat{\Phi}^{-1}$ on $C(S, X)$. Similarly, since e_s is continuous, the probability measure μ induces the probability measure μe_s^{-1} on X. We claim that μ is the Borel measure on $C(S, X)$ we want.

Indeed, if B is a Borel subset of X, then using the Change of Variables Theorem 13.46 consecutively, we get

$$\mu e_s^{-1}(B) = \int_B \chi_B(x) \, d\mu e_s^{-1}(x) = \int_{C(S,X)} (\chi_B \circ e_s)(g) \, d\mu(g)$$

$$= \int_{C(S,X)} (\chi_B \circ e_s)(g) \, d\lambda \hat{\Phi}^{-1}(g) = \int_\Omega [\chi_B \circ e_s \circ \hat{\Phi}](\omega) \, d\lambda(\omega)$$

$$= \int_\Omega \chi_B(\Phi(\omega, s)) \, d\lambda(\omega) = \lambda(\{\omega \in \Omega : \Phi(\omega, s) \in B\})$$

$$= P_s(B),$$

and the proof is finished.   ∎

Note well that the measure on $C(S, X)$ need not be unique. Here is a simple example. Let $S = X = \{0, 1\}$. Then $C(S, X)$ has the four elements $(0, 0)$, $(0, 1)$, $(1, 0)$, $(1, 1)$, where each ordered pair defines the function f with values $(f(0), f(1))$. Let $P: S \to \mathcal{P}(X)$ be defined by $P_s(\{0\}) = P_s(\{1\}) = \frac{1}{2}$, $s = 0, 1$. Define $\mu \in \mathcal{P}(C(S, X))$ by $\mu(\{f\}) = \frac{1}{4}$ for each f, and $\nu \in \mathcal{P}(C(S, X))$ by $\nu(\{(0, 1)\}) = \nu(\{(1, 0)\}) = \frac{1}{2}$. Then $\mu \neq \nu$, but $\mu e_0^{-1} = \nu e_0^{-1} = P_0$ and $\mu e_1^{-1} = \nu e_1^{-1} = P_1$.

R. M. Blumenthal and H. H. Corson [51] prove Theorem 19.33 for the more general case where the space X is connected and locally connected. A topological space is **connected** if the only clopen (simultaneously closed and open) subsets of X are $\varnothing$ and X. A subset of X that is a connected space with its relative topology is called a **connected set**. A topological space is **locally connected** if the neighborhood system of each point has a base consisting of connected sets. Their proof reduces the general case to the case $X = [0, 1]$ by means of a subtle topological argument that is beyond our scope here. They also prove that the connectedness conditions are essential for the result. We mention that there is a version of this result that makes no continuity assumptions on the transition, but does not yield the continuity of the random functions. For details, see Y. Kifer [202, pp. 7–10].

19.8   Dilations

In this section we discuss the relationship between what economists call stochastic dominance relations and Markov transitions. The discussion here follows the treatment by P.-A. Meyer [246, Section XI.3, pp. 239–246].

Throughout this section, we assume:

1. X is a compact metrizable space.

2. D is a convex cone in $C(X)$. That is, if $x, y \in D$ and $\alpha \geqslant 0$, then αx and $x + y$ belong to D. Also,

    a. D contains the nonnegative constant functions.

    b. D is closed under $\wedge$, that is, $f, g \in D$ implies $f \wedge g \in D$.

Recall that $C(X)$ is a separable Banach lattice under the sup norm and pointwise ordering. Its norm dual is $ca(X)$ (Corollary 14.15) with norm $\|\mu\| = |\mu|(X)$. As usual, we may write $\mu(f)$ for $\int f \, d\mu$. We also let $B(X)$ denote the Banach lattice of all bounded real functions on X equipped with the sup norm. Note that $C(X)$ is a majorizing Banach sublattice of $B(X)$.

19.34 Definition　　*Given such a cone D in $C(X)$, define the binary relation $\geqslant_D$ on $ca(X)$ by*

$$\mu \geqslant_D \nu \quad if \quad \int f \, d\mu \leqslant \int f \, d\nu \ for \ every \ f \in D.$$

When D is understood we simply write $\geqslant$ instead of $\geqslant_D$.

The binary relation $\geqslant$ is clearly transitive. If D separates points either of $ca(X)$ or of X, then $\geqslant$ is antisymmetric, that is, $\mu \geqslant \nu$ and $\nu \geqslant \mu$ imply $\mu = \nu$, so $\geqslant$ is a partial order.[5] Note that the relation remains the same if we replace the cone D by its w^*-closure, so we could have required D to be closed, but there is no point in doing so.

Observe that $\{\mu \in ca_+(X) : \mu \geqslant \nu\}$ and $\{\mu \in ca_+(X) : \nu \geqslant \mu\}$ are w^*-closed and convex for each ν. If $\mu \geqslant \nu$, the French call μ a **balayage** of ν. Thus for each $\nu \in ca_+(X)$ the set $\beta(\nu) = \{\mu \in ca_+(X) : \mu \geqslant \nu\}$ of nonnegative balayages of ν is convex and nonempty (it contains ν).

Binary relations of this form appear in economics in the guise of stochastic dominance relations. When X is a real interval, and D is the cone of nondecreasing continuous functions on X, then by definition ν **first degree stochastically dominates** μ if $\mu \geqslant \nu$. (Once again, mathematicians' predilection for minimizing rather than maximizing leads to a definition of $\geqslant$ that seems backwards to economists.) **Second degree stochastic dominance** is defined to be the partial order generated by the cone of nondecreasing concave functions. If D is the cone of continuous concave functions on X, then $\mu \geqslant \nu$ if and only if μ is **riskier** than ν in the sense of M. Rothschild and J. E. Stiglitz [289]. Statisticians will recognize this ordering from D. Blackwell [47]. Other stochastic dominance relations correspond to other cones; see K. C. Border [57] and A. Müller and D. Stoyan [255].

---

[5] Assume that D separates the points of X and that $\int f \, d\mu = \int f \, d\nu$ for each $f \in D$. Then the vector subspace $V = D - D$ of $C(X)$ contains the constants and separates the points of X and $\int f \, d\mu = \int f \, d\nu$ for each $f \in V$. Moreover, from the lattice identity

$$(f_1 - f_2) \wedge (g_1 - g_2) = (f_1 + g_2) \wedge (g_1 + f_2) - (f_2 + g_2),$$

we see that V is a Riesz subspace of $C(X)$. Consequently by the Stone–Weierstrass Theorem 9.12, the subspace V is dense in $C(X)$, so $\int f \, d\mu = \int f \, d\nu$ for each $f \in C(X)$. This implies $\mu = \nu$.

We now prove that there is an intimate connection between Markov transitions and dominance relations induced by cones. We start out with the following definition.

19.35 Definition   *A Markov transition $P\colon X \to \mathcal{P}(X)$ is a **D-dilation** if $P_x \succcurlyeq_D \delta_x$ for every $x \in X$, where δ_x is the point mass at x.*

A dilation is sometimes called a **dispersion** or **dilatation**. We mention that this definition generalizes the definition of dilation used in Choquet theory, which takes D to be the cone of continuous concave functions.[6] The results from that theory inspired the main result of this section, Theorem 19.40 below, which may be found in P.-A. Meyer [246]. The theorem states that for probability measures, $\mu \succcurlyeq_D \nu$ if and only if there is a D-dilation P with $\mu = P'\nu$.

Unfortunately we need some preliminary results before we can prove this theorem. We start with a digression to prove a monotone convergence result for nets in $C(X)$ when X is compact and metrizable.

19.36 Lemma (Monotone Convergence Lemma)   *Let X be a compact metrizable space and let $\{f_\alpha\} \subset C(X)$ be a decreasing net that is bounded from below in $C(X)$. If $f_\alpha(x) \downarrow f(x)$ for each $x \in X$, then the function $f\colon X \to \mathbb{R}$ is upper semicontinuous (hence Borel measurable) and*

$$\int f_\alpha \, d\mu \downarrow \int f \, d\mu$$

for each $\mu \in ca_+(X)$.

Proof: Let $\{f_\alpha\}$ and $f\colon X \to \mathbb{R}$ satisfy the hypotheses of the lemma. Lemma 2.42 implies that the function f is upper semicontinuous, and so Borel measurable. Since f is bounded, it is integrable with respect to every Borel measure.

Let $\mu \in ca_+(X)$ and suppose $\int f_\alpha \, d\mu \downarrow \ell \geqslant \int f \, d\mu$. Let $g \in C(X)$ satisfy $g \geqslant f$. Letting $g_\alpha = g \vee f_\alpha$, we have $g_\alpha(x) \downarrow g(x)$ for each $x \in X$. By Dini's Theorem 2.66, the net $\{g_\alpha\}$ converges uniformly to g on X, so $\int g \, d\mu = \lim_\alpha \int g_\alpha \, d\mu$ by the norm continuity of μ. Since $g_\alpha \geqslant f_\alpha$ for each α, we see that $\int g \, d\mu \geqslant \ell$ whenever $g \in C(X)$ satisfies $g \geqslant f$.

Since f is upper semicontinuous, Theorem 3.13 implies that there is some sequence $\{g_n\} \subset C(X)$ satisfying $g_n(x) \downarrow f(x)$ for each $x \in X$. But then the Lebesgue Dominated Convergence Theorem 11.21 implies $\int g_n \, d\mu \downarrow \int f \, d\mu$. On the other hand, by the conclusion above, $\int g_n \, d\mu \geqslant \ell$ for each n, so $\int f \, d\mu \geqslant \ell$. In other words, $\int f_\alpha \, d\mu \downarrow \int f \, d\mu$, as claimed.   ∎

---

[6] See R. R. Phelps [277] for an excellent account of Choquet theory, which uses vector integrals to significantly generalize the Krein–Milman Theorem.

N. Bourbaki [60, Chapter IV, §1, Theorem 1, pp. 107–108] proves a more general version of this theorem, in which X is only assumed to be locally compact (not necessarily metrizable) and $C(X)$ is replaced by $C_c(X)$, the space of continuous functions with compact support.

The next definition may seem to be leading us astray, so please bear with us.

19.37 Definition   *Given $f \in C(X)$ define the D-envelope $\hat{f}$ of f by*

$$\hat{f}(x) = \inf\{g(x) : g \in D \text{ and } g \geqslant f\}.$$

Since D contains all the positive constants and each f in $C(X)$ is bounded, there is always a constant function g belonging to D satisfying $g \geqslant f$. Thus $\hat{f}$ is always finite.

As an example, if D is the cone of nondecreasing continuous functions on $[0, 1]$, then $\hat{f}(x) = \sup\{f(t) : t \leqslant x\}$. To see this, note that $\hat{f}$ so defined is nondecreasing and satisfies $\hat{f} \geqslant f$. It is also continuous. (Why?) On the other hand, if g is nondecreasing and $g \geqslant f$, then $g \geqslant \hat{f}$. (Why?)

19.38 Lemma   *For each $f \in C(X)$, the D-envelope $\hat{f}$ of f is bounded and upper semicontinuous. Furthermore, the mapping $f \mapsto \hat{f}$ from $C(X)$ into $B(X)$ is monotone and sublinear.*

Proof: Since D contains the nonnegative constants, there is always $g \in D$ with $g \geqslant \hat{f} \geqslant f$, so $\hat{f}$ is bounded. By Lemma 2.41, $\hat{f}$ is upper semicontinuous.

Clearly $f \geqslant g$ implies $\hat{f} \geqslant \hat{g}$, so $f \mapsto \hat{f}$ is monotone. To see that it is subadditive, fix $f, g \in C(X)$ and let $h_1, h_2 \in D$ satisfy $f \leqslant h_1$, $g \leqslant h_2$. Since D is a convex cone, $h = h_1 + h_2 \in D$, and $f + g \leqslant h$, so $\widehat{f + g} \leqslant h_1 + h_2$, which implies $\widehat{f + g} \leqslant \hat{f} + \hat{g}$. Since D is a cone, it is easy to verify that $\widehat{\alpha f} = \alpha \hat{f}$, so $f \mapsto \hat{f}$ is positively homogeneous.   ∎

For each $f \in C(X)$ the set $D_f = \{g \in D : g \geqslant f\}$ is nonempty (every positive constant that majorizes f belongs to D_f) and is closed under finite infima. Clearly $\hat{f}(x) = \inf\{g(x) : g \in D_f\}$. So for each $v \in ca_+(X)$ the Monotone Convergence Lemma 19.36 implies

$$v(\hat{f}) = \int \hat{f} \, dv = \inf\{v(g) : g \in D_f\}.$$

That is, every cone D satisfying our properties and every measure $v \in ca_+(X)$ define a mapping $p_v : C(X) \to \mathbb{R}$ via

$$p_v(f) = v(\hat{f}) = \int \hat{f} \, dv = \inf\{v(g) : g \in D_f\}.$$

It turns out that each p_v is a norm-continuous sublinear mapping.

19.39 Theorem   *For each $v \in ca_+(X)$:*

1. *The function p_v is norm-continuous and sublinear.*

2. *Its set of balayages satisfies*

$$\beta(v) = \{\mu \in ca(X) : \mu(f) \leqslant p_v(f) \text{ for every } f \in C(X)\}$$

In particular, $\beta(v)$ is nonempty, w^*-compact, and convex, with support functional p_v.

Proof: (1)   The sublinearity of $f \mapsto \hat{f}$ and the nonnegativity of v easily imply that p_v is sublinear.

To see that p_v is norm-continuous, it suffices to show that p_v is continuous at zero. To this end, let $\varepsilon > 0$ and assume that $f \in C(X)$ satisfies $\|f\|_\infty < \varepsilon$, that is, $-\varepsilon < f(x) < \varepsilon$ for each $x \in X$. Since the constant function ε belongs to D, we see that $-\varepsilon < f(x) \leqslant \hat{f}(x) \leqslant \varepsilon$ for each $x \in X$. This implies $|p_v(f)| = |v(\hat{f})| \leqslant \varepsilon v(X)$, which shows that p_v is continuous at zero.

(2)   Fix $f \in C(X)$, and consider $g \in D$ with $g \geqslant f$. For each μ in $\beta(v)$ we have $v(g) \geqslant \mu(g)$. By the argument above, $v(\hat{f}) \geqslant \mu(\hat{f}) \geqslant \mu(f)$. Therefore $\mu(f) \leqslant p_v(f)$, so $\beta(v)$ is included in the set of linear functionals dominated by p_v.

For the reverse inclusion, suppose $\mu \leqslant p_v$. If $f \leqslant 0$, then $\hat{f} \leqslant 0$ too, so $\mu(f) \leqslant p_v(f) \leqslant 0$. This implies that $\mu \in ca_+(X)$. Now if $f \in D$, then $f = \hat{f}$, so $\mu(f) \leqslant p_v(f) = v(f)$. That is, $\mu \geqslant v$, and we are done.

Note that $\mu \geqslant v$ implies $0 \leqslant \|\mu\| = \mu(\mathbf{1}) \leqslant v(\mathbf{1})$, so $\beta(v)$ is a norm-bounded subset of the norm dual $ca(X)$ of $C(X)$. Further, $\beta(v)$ is w^*-closed, so by Alaoglu's Theorem 6.21, it is w^*-compact, and plainly it is convex. Since the Mackey topology on a normed space is just the norm topology, p_v is a Mackey-continuous sublinear functional on $C(X)$, so Theorem 7.52 implies that it is the support functional of $\beta(v)$.   ∎

It follows from this result that for nonnegative measures μ and v, we have $\mu \geqslant v$ if and only if $\mu(\hat{f}) \leqslant v(\hat{f})$ for every $f \in C(X)$. We finally have all the machinery needed to prove the main result of this section.

19.40 Theorem   (Dilations)   *Let D be a convex cone in $C(X)$ satisfying the hypotheses of this section. Then for two measures $\mu, v \in \mathcal{P}(X)$ the following statements are equivalent.*

1. *$\mu \geqslant_D v$.*

2. *There is a D-dilation P with $\mu = \mathbf{P}'v$.*

Proof: (1) $\implies$ (2)   Suppose $\mu, v \in \mathcal{P}(X)$ satisfy $\mu \geqslant v$. Define the correspondence $\varphi: X \twoheadrightarrow ca(X)$ via

$$\varphi(x) = \beta(\delta_x) = \{\gamma \in ca_+(X) : \gamma \geqslant \delta_x\}.$$

Observe that a Markov transition is a D-dilation if and only if it is a measurable selection from φ.

Clearly φ has nonempty w^*-compact convex values. It is also scalarly measurable. To see this, note that Theorem 19.39 implies that the support functional h_x of $\varphi(x) = \beta(\delta_x)$ is p_{δ_x}. That is, $h_x(f) = p_{\delta_x}(f) = \hat{f}(x)$, which is upper semicontinuous in x, and therefore Borel measurable for each f. Thus $(x, f) \mapsto h_x(f)$ is a sublinear Carathéodory function from $X \times C(X)$ into $\mathbb{R}$, since $h_x(f) = p_{\delta_x}(f)$ is continuous in f by Theorem 19.39.

Since $\mu \geqslant \nu$, Theorem 19.39 implies $\mu \leqslant p_\nu$. But

$$p_\nu(f) = \int \hat{f} \, d\nu = \int p_{\delta_x}(f) \, d\nu.$$

Recalling that $C(X)$ is a separable Banach space, Strassen's Sublinearity Theorem 18.35 applies: There is a Borel measurable function $g: X \to ca(X)$ satisfying $g_x \leqslant p_{\delta_x}$ for each x and $\mu = \int g_x \, d\nu$. Since μ and ν are probability measures, $1 = \mu(\mathbf{1}) = \int g_x(\mathbf{1}) \, d\nu$. Since $g_x \geqslant \delta_x$ for each x, we have $g_x(\mathbf{1}) \leqslant \delta_x(\mathbf{1}) = 1$, and $g_x \in ca_+(X)$. This implies $g_x(\mathbf{1}) = 1$ except perhaps for $x \in A$ where A is a Borel subset of X with $\nu(A) = 0$. Define the Markov transition $P: X \to \mathcal{P}(X)$ by $P_x = g_x$ for $x \in A^c$ and $P_x = \delta_x$ for $x \in A$. Then P is Borel measurable (why?) and $\mu = \int P_x \, d\nu = \mathbf{P}'\nu$.

(2) $\implies$ (1)   Suppose $\mu = \mathbf{P}'\nu$, where P is a D-dilation and μ, ν belong to $\mathcal{P}(X)$. For any $f \in D$, we have

$$\mu(f) = \langle f, \mathbf{P}'\nu \rangle = \int P_x(f) \, d\nu(x) \leqslant \int \delta_x(f) \, d\nu(x) = \int f(x) \, d\nu(x) = \nu(f).$$

In other words, $\mu \geqslant_D \nu$.   ∎

19.9   More on Markov operators

Recall that a Markov operator from $C_b(X)$ into $C_b(S)$ is a positive operator that maps the constant function $\mathbf{1}_X$ onto $\mathbf{1}_S$. Lemma 19.6 and Theorem 19.14 show that restricting the transition operator of a continuous transition function to $C_b(X)$ defines a Markov operator from $C_b(X)$ into $C_b(S)$. Is there a converse result? That is, suppose we are handed a Markov operator from $C_b(X)$ into $C_b(S)$, do we know that it is the restriction of the transition operator for some continuous transition function?

Recall that for compact metrizable spaces S and X, the Riesz Representation Theorem 14.12 asserts that the norm duals of $C(X)$ and $C(S)$ are $ca(X)$ and $ca(S)$, respectively. Therefore, when $T: C(X) \to C(S)$ is a continuous operator, its norm adjoint T' maps the Banach lattice $ca(S)$ into the Banach lattice $ca(X)$ via the duality identity

$$\langle Tf, \mu \rangle = \langle f, T'\mu \rangle, \quad f \in C(X), \ \mu \in ca(S).$$

19.41 Theorem   *A positive operator* $T: C(X) \to C(S)$, *where S and X are compact and metrizable, is a Markov operator if and only if there exists a unique continuous Markov transition* $P: S \to \mathcal{P}(X)$ *satisfying* $\boldsymbol{P}f = Tf$ *for all* $f \in C(X)$.

Moreover, if T is a Markov operator, the unique continuous Markov transition $P: S \to \mathcal{P}(X)$ *is given by* $P_s = T'\delta_s$ *for each* $s \in S$.

Proof: Assume that $T: C(X) \to C(S)$ is a Markov operator. Then its norm adjoint satisfies $T': ca(S) \to ca(X)$. Now define $P: S \to \mathcal{P}(X)$ via the formula $P_s = T'\delta_s$ for each $s \in S$. We claim first that P is continuous. To see this, let $s_n \to s$ in S. Then for each $f \in C(X)$ we have

$$\langle f, P_{s_n} \rangle = \langle f, T'\delta_{s_n} \rangle = \langle Tf, \delta_{s_n} \rangle = Tf(s_n) \to Tf(s) = \langle f, P_s \rangle,$$

which shows that P is continuous.

To complete the proof note that if $f \in C(X)$ and $s \in S$, then

$$\boldsymbol{P}f(s) = \langle f, P_s \rangle = \langle f, T'\delta_s \rangle = \langle Tf, \delta_s \rangle = Tf(s),$$

so that $Tf = \boldsymbol{P}f$ for each $f \in C(X)$.   ∎

Suppose that $\xi: S \to X$ is a continuous function. Then ξ gives rise to a continuous deterministic Markov transition $P: S \to \mathcal{P}(X)$ via the formula

$$P_s = \delta_{\xi(s)}.$$

Clearly, the transition operator $\boldsymbol{P}$ determined by P satisfies

$$\boldsymbol{P}f(s) = \langle f, P_s \rangle = f(\xi(s)) = (f \circ \xi)(s).$$

In other words, for every continuous function $\xi: S \to X$, the composition operator $f \mapsto f \circ \xi$, from $C(X)$ to $C(S)$, is in fact a Markov operator. It is customary to identify the transition P with the continuous function ξ.

With these observations in mind, an easy application of Theorem 14.23 yields the following result.

19.42 Theorem (Deterministic transition operators)   *Let S and X be compact metrizable spaces, and let* $\boldsymbol{P}: C(X) \to C(S)$ *be a Markov operator. Then the following are equivalent.*

1. $\boldsymbol{P}$ *is a composition operator. That is, there exists a continuous function* $\xi: S \to X$ *such that* $\boldsymbol{P}f = f \circ \xi$ *for each* $f \in C(X)$.

2. *The transition P is deterministic. That is, there is a continuous function* $\xi: S \to X$ *such that* $P_s = \delta_{\xi(s)}$ *for each* $s \in S$.

3. $\boldsymbol{P}$ *is multiplicative. That is,* $\boldsymbol{P}(fg) = \boldsymbol{P}f \cdot \boldsymbol{P}g$ *for all* $f, g \in C(X)$.

We close this section with another extension result for Markov operators when S is compact and metrizable. Not only can Markov operators be extended from $C(S)$ to $B_b(S)$, they can be uniquely extended all the way to $L_\infty(\mu)$ for any invariant measure μ.

19.43 Theorem (Extension to L_∞) *Let $T \colon C(S) \to C(S)$ be a Markov operator, where S is metrizable and compact. If $\mu \in \mathcal{P}(S)$ is a T-invariant measure (that is, $T'\mu = \mu$), then there exists a unique σ-order continuous Markov operator $\hat{T} \colon L_\infty(\mu) \to L_\infty(\mu)$ satisfying $\hat{T}(f) = Tf$ for each $f \in C(S)$.*

Proof: Theorem 19.41 guarantees that T has a unique positive σ-order linear extension $P \colon B_b(S) \to B_b(S)$. Since P is σ-order continuous, it follows from part (2) of Theorem 19.9 that $P'(ca(S)) \subset ca(S)$.

Now let $\mu \in \mathcal{P}(S)$ be a T-invariant measure. If $f \in C(S)$, then

$$\langle f, P'\mu \rangle = \langle Pf, \mu \rangle = \langle Tf, \mu \rangle = \langle f, T'\mu \rangle = \langle f, \mu \rangle.$$

By Theorem 15.1, $P'\mu = \mu$. Consequently, if $f, g \in B_b(S)$ satisfy $f = g$ μ-a.e., then from

$$0 \leqslant \langle |Pf - Pg|, \mu \rangle \leqslant \langle P|f - g|, \mu \rangle = \langle |f - g|, P'\mu \rangle = \langle |f - g|, \mu \rangle = 0,$$

it follows that $Pf = Pg$ μ-a.e. So the mapping $\hat{T} \colon L_\infty(\mu) \to L_\infty(\mu)$, defined by

$$\hat{T}f = Pf,$$

where Pf now denotes the μ-equivalence class of Pf, is a well defined σ-order continuous Markov operator satisfying $\hat{T}f = Tf$ for each $f \in C(S)$.

For uniqueness, assume that another σ-order continuous Markov operator $R \colon L_\infty(\mu) \to L_\infty(\mu)$ satisfies $Rf = \hat{T}f$ for each $f \in C(S)$. Put

$$\mathcal{F} = \{A \in \mathcal{B}_S : R(\chi_A) = \hat{T}(\chi_A)\}.$$

An easy verification shows that $\mathcal{F}$ is a σ-algebra. Also, by using Corollary 3.14, it is easy to see that $F \in \mathcal{F}$ for each closed subset F of S. So $\mathcal{F} = \mathcal{B}_S$. An approximation argument now shows that $Rf = \hat{T}f$ for each $f \in L_\infty(\mu)$, and the proof is finished. ∎

19.10   A note on dynamical systems

There is another framework for defining ergodic measures, which we adumbrate here. A **(measure theoretic) dynamical system** is a quadruple (Ω, Σ, μ, T), where (Ω, Σ, μ) is a probability space and T is a measurable function from Ω into itself. The interpretation usually assigned to these objects is that the initial state ω_0 is distributed randomly according to μ and subsequent states evolve according to the difference equation $\omega_{t+1} = T(\omega_t)$. For an excellent discussion of the application of this model of dynamical systems see D. S. Ornstein [267]. The transformation $T: \Omega \to \Omega$ generates an operator $\boldsymbol{T}: B_b(\Sigma) \to B_b(\Sigma)$ via the formula $\boldsymbol{T}f = f \circ T$. It is easy to see that its adjoint $\boldsymbol{T}': ba(\Sigma) \to ba(\Sigma)$ is given by $\boldsymbol{T}'\mu = \mu T^{-1}$. In this context, T is **measure-preserving**, or μ is $\boldsymbol{T}$-**invariant**, if $\mu = \boldsymbol{T}'\mu$. An **invariant set** is a measurable set A satisfying $A = T^{-1}(A)$. The system is $\boldsymbol{T}$-**ergodic** if it is T-invariant and each invariant set has μ-measure zero or one.

Given a dynamical system (Ω, Σ, μ, T), define the deterministic Markov transition $P: \Omega \to \mathcal{P}(\Omega)$ by $P_\omega = \delta_{T(\omega)}$, the point mass at $T(\omega)$. Then for any $f \in B_b(\Sigma)$,

$$\boldsymbol{P}f(x) = \langle f, P_\omega \rangle = f(T(\omega)) = \boldsymbol{T}f(\omega),$$

so $\boldsymbol{P}f = \boldsymbol{T}f$, and hence $\boldsymbol{P}'\mu = \boldsymbol{T}'\mu$. Thus the system is T-invariant if and only if μ is P-invariant, and ergodic if and only if μ is P-ergodic.

The Markov transition defined this way from a dynamical system is special in that it is deterministic. That is, next period's state is not random given today's state. Clearly, for a given set of states, the notion of Markov transition is more general than a dynamical system. To be fair, given a Markov transition on a set S of states, there is a dynamical system on a larger state space, namely $S^{\mathbb{N}}$, which captures the stochastic nature of the dynamics; again see D. S. Ornstein [267].

Chapter 20

Ergodicity

Ergodic theory can be described as the discipline that studies the *long run average* behavior of *dynamical systems*. There is a set S of possible *states* of the system, and the evolution of the system is usually modeled as a function $T : S \rightarrow S$. If the system is in state s at time t, then Ts is the state of the system at time $t+1$. The sequence $\{s, Ts, T^2 s, \ldots\}$ is called the *orbit* of the state s.

There are several approaches to the mapping $T : S \rightarrow S$, depending on the structure of the state space S and the topological properties of the mapping T. In this chapter, we discuss briefly two approaches. In the first approach S is a probability measure space and T is a measure-preserving transformation. In the second case S is a Banach space and T is a continuous linear operator.

A real function $f : S \rightarrow \mathbb{R}$ (subject to some measurability or continuity requirement) is usually interpreted as some sort of measurement of the system. If a phenomenon follows the evolutionary orbit $\{s, Ts, T^2 s, \ldots\}$, then the sequence of real numbers $\{f(s), f(Ts), f(T^2 s), \ldots\}$ represents the values of the measurements of some quantity during the evolution of the phenomenon. The average of these measurements during the first n periods is given by

$$A_n(f) = \frac{1}{n} \sum_{i=0}^{n-1} f(T^i s).$$

As mentioned above, the concern of ergodic theory is the long run behavior of the sequence of time averages $\{A_n(f)\}$, especially the convergence of these averages. Results on the convergence of the sequence of averages are known as *ergodic theorems*. A limit of the sequence $\{A_n(f)\}$ can be interpreted as an "equilibrium" value of the measurement f.

There are several ergodic theorems in the literature, and our goal here is to describe a few of them that you may find useful. We have no intention of entering into the delicate details of ergodic theory at this time. There are many detailed and extensive treatments of the theory from various points of view and of varying degrees of obscurity. W. Parry [270] and K. E. Petersen [274] offer quite readable treatments of basic ergodic theorems. U. Krengel [217] and P. Walters [339] are highly operator theoretic in nature. R. Mañé [239] studies differentiable structures and ergodic theory. A. Lasota and M. C. Mackey [223] apply ergodic theory to "chaotic" systems. Y. Kifer [202] studies ergodic theory in terms of random sequences of functions. The monograph by D. S. Ornstein [267] addresses the question of isomorphism of dynamical systems using the concepts of coding and entropy. C. A. Futia [132] discusses the use of ergodic theory in economic theory.

20.1   Measure-preserving transformations and ergodicity

In this section (Ω, Σ, μ), or simply Ω, denotes a probability space. We start with the definition of a measure-preserving transformation.

20.1 Definition   *A transformation $\xi \colon \Omega \to \Omega$ is **μ-measure-preserving** (or simply **measure-preserving**) if it is measurable and*

$$\mu(A) = \mu(\xi^{-1}(A))$$

for each $A \in \Sigma$. In other words, ξ is measure-preserving if the measure $\mu\xi^{-1}$ induced by ξ on Σ coincides with μ.

Continuous μ-measure-preserving transformations are precisely those whose composition operators leave μ invariant.

20.2 Theorem   *Let X be a compact metrizable space, and let μ be a Borel probability measure on X. For a continuous function $\xi \colon X \to X$, the following statements are equivalent.*

1. *The transformation ξ is μ-measure-preserving.*

2. *The measure μ is T_ξ-invariant, where $T_\xi \colon C(X) \to C(X)$ is the composition operator defined as usual by $T_\xi(f) = f \circ \xi$.*

Proof: By the Change of Variables Theorem 13.46, for each $f \in C(X)$ we have

$$\langle f, T_\xi'\mu \rangle = \langle T_\xi f, \mu \rangle = \langle f \circ \xi, \mu \rangle = \langle f, \mu\xi^{-1} \rangle.$$

Since every probability measure on X is regular, we infer that $T_\xi'\mu = \mu\xi^{-1}$. Consequently, $T_\xi'\mu = \mu$ if and only if $\mu = \mu\xi^{-1}$.   ∎

For a function $f \colon X \to X$ we employ the following standard notation and terminology.

•   The **iterates** of f are defined inductively by $f^0(x) = x$ and $f^{n+1}(x) = f(f^n(x))$ for each $x \in X$.

•   If A is a subset of X, then we let $f^0(A) = A$ and

$$f^{-(n+1)}(A) = f^{-1}(f^{-n}(A)) = \{x \in X : f^{n+1}(x) \in A\}.$$

In other words, a point belongs to $f^{-n}(A)$ if and only if its nth iterate lies in A.

- The *f*-orbit of a point $x \in X$ is the set

$$\mathcal{O}_f(x) = \{f^n(x) : n = 0, 1, 2, \ldots\}.$$

20.3 Poincaré's Recurrence Theorem *Let $\xi \colon \Omega \to \Omega$ be a measure-preserving transformation and let $E \in \Sigma$. Then the ξ-iterates of almost each point in E visit E infinitely often. That is, there exists a set $F \in \Sigma$ satisfying $F \subset E$, $\mu(F) = \mu(E)$, and for each $x \in F$ and each n there is some $k > n$ such that $\xi^k(x) \in E$.*

Proof: Put $G = \bigcap_{n=0}^{\infty} \bigcup_{k=n}^{\infty} \xi^{-k}(E)$ and let $F = E \cap G$. Clearly, each point of F visits E infinitely often. So it suffices to show that $\mu(F) = \mu(E)$.

To see this, start by observing that $\mu(\xi^{-k}(A)) = \mu(A)$ for each $A \in \Sigma$ and each $k = 0, 1, 2, \ldots$. Next, let $G_n = \bigcup_{k=n}^{\infty} \xi^{-k}(E)$ for $n = 0, 1, 2, \ldots$. Then $\xi^{-1}(G_n) = G_{n+1}$ for each n, so (since ξ is measure-preserving) $\mu(G_n) = \mu(G_{n+1})$ for each n. From $G_n \downarrow G$, we see that $\mu(G) = \mu(G_n)$ for each n. Now from $E \subset G_0$, $G \subset G_0$ and $\mu(G) = \mu(G_0)$, we get

$$\mu(F) = \mu(E \cap G) = \mu(E \cap G_0) = \mu(E),$$

and the proof is finished. ∎

In what follows, we employ the following standard terminology. If $A, B \in \Sigma$, then the notation $A = B$ μ-a.e. means simply that

$$\mu(A \triangle B) = \mu((A \setminus B) \cup (B \setminus A)) = 0.$$

If for a measure-preserving transformation $\xi \colon \Omega \to \Omega$ there exists a set $E \in \Sigma$ with $0 < \mu(E) < 1$ and $\xi^{-1}(E) = E$, then the study of ξ can be reduced to the study of the measure-preserving transformations $\xi \colon E \to E$ and $\xi \colon E^c \to E^c$. Irreducible measure-preserving transformations, those that cannot be reduced in this way, are called ergodic transformations.

20.4 Definition *A measure-preserving transformation $\xi \colon \Omega \to \Omega$ is μ-ergodic (or simply ergodic) if $E \in \Sigma$ and $\xi^{-1}(E) = E$ μ-a.e. imply either $\mu(E) = 0$ or $\mu(E) = 1$.*

The elementary characterizations of ergodic transformations are given in the next result.

20.5 Theorem *For a measure-preserving transformation $\xi \colon \Omega \to \Omega$ the following statements are equivalent.*

1. *The mapping ξ is ergodic. That is, for $E \in \Sigma$, if $\xi^{-1}(E) = E$ μ-a.e., then either $\mu(E) = 0$ or $\mu(E) = 1$.*

2. *If $E \in \Sigma$ satisfies $\xi^{-1}(E) = E$, then either $\mu(E) = 0$ or $\mu(E) = 1$.*

3. *For each $E, F \in \Sigma$ satisfying $\mu(E) > 0$ and $\mu(F) > 0$, there exists some $n \geqslant 1$ such that $\mu(\xi^{-n}(E) \cap F) > 0$.*

4. *If $f \in L_1(\mu)$ satisfies $f \circ \xi = f$ μ-a.e., then f is constant μ-a.e.*

Proof: (1) $\Longrightarrow$ (2)   Obvious.

(2) $\Longrightarrow$ (3)   Assume that $E, F \in \Sigma$ satisfy $\mu(E) > 0$ and $\mu(F) > 0$. For each n let $G_n = \bigcup_{k=n}^{\infty} \xi^{-k}(E)$. Clearly, $G_{n+1} \subset G_n$ and $\xi^{-1}(G_n) = G_{n+1}$ for each n. Since ξ is measure-preserving, we see that $\mu(G_n) = \mu(G_{n+1})$ for each n.

Now let $G = \bigcap_{n=1}^{\infty} G_n$. Then $\xi^{-1}(G) = G$, so from our hypothesis either $\mu(G) = 1$ or $\mu(G) = 0$. From $G_n \downarrow G$ and $\mu(G_n) = \mu(G_1)$ for each n, we infer that $\mu(G_1) = \mu(G)$, so either $\mu(G_1) = 1$ or $\mu(G_1) = 0$.

But $\xi^{-1}(E) \subset G_1$, which implies $\mu(G_1) \geqslant \mu(\xi^{-1}(E)) = \mu(E) > 0$. So $\mu(G_1) = 1$. But then $0 < \mu(F) = \mu(G_1 \cap F) \leqslant \sum_{n=1}^{\infty} \mu(\xi^{-n}(E) \cap F))$ implies $\mu(\xi^{-n}(E) \cap F) > 0$ for some $n \geqslant 1$.

(3) $\Longrightarrow$ (4)   Assume that $f \in L_1(\mu)$ satisfies $f \circ \xi = f$ μ-a.e. Consider the μ-null set $A = \{\omega \in \Omega : f(\xi(\omega)) \neq f(\omega)\}$ and let $B = \bigcup_{n=0}^{\infty} \xi^{-n}(A)$. Clearly, $\mu(B) = 0$ and $\xi(\Omega \setminus B) \subset \Omega \setminus B$. So replacing Ω by $\Omega \setminus B$ (if necessary), we can assume that $f(\xi(\omega)) = f(\omega)$ for each $\omega \in \Omega$.

Now suppose by way of contradiction that f is not μ-a.e. constant. Then by Theorem 10.36 there exists some constant c such that the two disjoint measurable sets

$$E = \{\omega \in \Omega : f(\omega) < c\} \quad \text{and} \quad F = \{\omega \in \Omega : f(\omega) > c\}$$

satisfy $\mu(E) > 0$ and $\mu(F) > 0$. Also, note that if $\omega \in E$, then $f(\xi(\omega)) = f(\omega) < c$, so $\xi(\omega) \in E$. That is, $E \subset \xi^{-1}(E)$. Therefore by induction $E \subset \xi^{-n}(E)$ for all n. Since ξ is measure-preserving, we have $E = \xi^{-n}(E)$ μ-a.e. for each n, from which (taking into account $E \cap F = \varnothing$) it follows that $\mu(\xi^{-n}(E) \cap F) = 0$ for each n, contradicting our hypothesis. This proves that f is a μ-a.e. constant function.

(4) $\Longrightarrow$ (1)   Let $E \in \Sigma$ satisfy $\xi^{-1}(E) = E$ μ-a.e. Then an easy argument shows that $\chi_E \circ \xi = \chi_E$ μ-a.e. Therefore, by our hypothesis, $\chi_E =$ constant μ-a.e. Since χ_E takes only the values 0 or 1, we either have $\chi_E = 0$ μ-a.e. or $\chi_E = 1$ μ-a.e. This means that either $\mu(E) = 0$ or $\mu(E) = 1$, so that ξ is an ergodic transformation.   ∎

The orbits of ergodic transformations are almost always dense.

20.6 Theorem   *Let X be a second countable space, let μ be a Borel probability measure on X with $\operatorname{supp}\mu = X$, and let $\xi \colon X \to X$ be μ-ergodic. Then almost every point of X has a dense orbit. That is,*

$$\mu(\{x \in X : \overline{\mathcal{O}_\xi(x)} = X\}) = 1.$$

Proof: Let $\{V_1, V_2, \ldots\}$ be a countable base for X. An orbit $\mathcal{O}_\xi(x) = \{\xi^k(x) : k \geqslant 0\}$ is dense if and only if x belongs to $G = \bigcap_{n=1}^{\infty} \bigcup_{k=0}^{\infty} \xi^{-k}(V_n)$. Next, notice that

$\xi^{-1}(G) \subset G$. Since ξ is measure-preserving, $\xi^{-1}(G) = G$ μ-a.e. The ergodicity of ξ guarantees that $\mu(G) = 0$ or $\mu(G) = 1$.

From $\xi^{-1}(\bigcup_{k=0}^{\infty} \xi^{-k}(V_n)) \subset \bigcup_{k=0}^{\infty} \xi^{-k}(V_n)$ and the fact that ξ is ergodic, we see that $\mu(\bigcup_{k=0}^{\infty} \xi^{-k}(V_n))$ equals 0 or 1. Since $V_n \subset \bigcup_{k=0}^{\infty} \xi^{-k}(V_n)$ and supp $\mu = X$, we obtain $\mu(\bigcup_{k=0}^{\infty} \xi^{-k}(V_n)) = 1$ for each n. This implies (why?) $\mu(G) = 1$. ∎

20.2 Birkhoff's Ergodic Theorem

In this section $\Omega = (\Omega, \Sigma, \mu)$ is again a probability space. Let $\xi: \Omega \to \Omega$ be a measurable transformation and let $f: \Omega \to \mathbb{R}$ be a Σ-measurable function. The nth-**average** $A_n(f, \omega)$ of f (with respect to ξ) at some point $\omega \in \Omega$ is defined by

$$A_n(f, \omega) = \frac{1}{n} \sum_{i=0}^{n-1} f(\xi^i(\omega)).$$

The importance of the average $A_n(f, \omega)$ can be seen by considering the case $f = \chi_E$ for some $E \in \Sigma$. In this case,

$$f(\xi^i(\omega)) = \chi_E(\xi^i(\omega)) = \chi_{\xi^{-i}(E)}(\omega) = \begin{cases} 1 & \text{if } \xi^i(\omega) \in E, \\ 0 & \text{if } \xi^i(\omega) \notin E. \end{cases}$$

So $A_n(\chi_E, \omega)$ is the proportion of elements in $\{\omega, \xi(\omega), \ldots, \xi^{n-1}(\omega)\}$ (the first part of the ξ-orbit of ω) that lie in E. *What is the behavior of the sequence of averages as n goes to infinity?* G. D. Birkhoff [44] shows that if ξ is measure-preserving, then the averages of any L_1-function converge pointwise almost everywhere. Here is this remarkable result.

20.7 Birkhoff's Ergodic Theorem *If $\xi: \Omega \to \Omega$ is a measure-preserving transformation on a probability space,[1] and if $f \in L_1(\mu)$, then for μ-almost all $\omega \in \Omega$ the sequence $\{A_n(f, \omega)\}$ of averages converges in $\mathbb{R}$. Further, if*

$$A_n(f, \omega) = \frac{1}{n} \sum_{i=0}^{n-1} f(\xi^i(\omega)) \xrightarrow[n \to \infty]{} f^*(\omega) \ \mu\text{-a.e.},$$

then $f^* \in L_1(\mu)$, $f^* \circ \xi = f^*$ μ-a.e., and

$$\int_\Omega f^* \, d\mu = \int_\Omega f \, d\mu.$$

Proof: See [146, pp. 18–21]. ∎

---

[1] The theorem remains true even when Ω is a σ-finite measure space.

20.8 Corollary (G. D. Birkhoff) *If $\xi\colon \Omega \to \Omega$ is an ergodic transformation on a probability space, then for each $f \in L_1(\mu)$ we have*

$$A_n(f, \omega) = \frac{1}{n} \sum_{i=0}^{n-1} f(\xi^i(\omega)) \xrightarrow[n\to\infty]{} \int_\Omega f\,d\mu$$

for μ-almost all $\omega \in \Omega$.

Proof: Let $f \in L_1(\mu)$. Birkhoff's Theorem 20.7 implies that $A_n(f, \omega) \to f^*(\omega)$ for some $f^* \in L_1(\mu)$ and μ-almost all ω. In addition, $f^* \circ \xi = f^*$ μ-a.e. Since ξ is ergodic, it follows from Theorem 20.5 that $f^* = c$ (a constant) μ-a.e. Now note that $c = \int_\Omega f^*\,d\mu = \int_\Omega f\,d\mu$. ∎

20.9 Corollary   *A measure-preserving transformation $\xi\colon \Omega \to \Omega$ on a probability space is ergodic if and only if for any pair of sets $A, B \in \Sigma$ we have*

$$\frac{1}{n} \sum_{i=0}^{n-1} \mu(\xi^{-i}(A) \cap B) \xrightarrow[n\to\infty]{} \mu(A)\mu(B).$$

Proof: Assume first that ξ is ergodic, and let $A, B \in \Sigma$. Letting $f = \chi_A$ in Corollary 20.8, we get $\frac{1}{n} \sum_{i=0}^{n-1} \chi_{\xi^{-i}(A)}(\omega) \xrightarrow[n\to\infty]{} \mu(A)$ for μ-almost all ω. Hence, $\frac{1}{n} \sum_{i=0}^{n-1} \chi_{\xi^{-i}(A)}(\omega)\chi_B(\omega) \xrightarrow[n\to\infty]{} \mu(A)\chi_B(\omega)$ for μ-almost all ω. Now an easy application of the Lebesgue Dominated Convergence Theorem 11.21 implies

$$\frac{1}{n} \sum_{i=0}^{n-1} \mu(\xi^{-i}(A) \cap B) \xrightarrow[n\to\infty]{} \mu(A)\mu(B).$$

For the converse, assume the condition is satisfied, and let $\xi^{-1}(E) = E$ μ-a.e. for some $E \in \Sigma$. It follows that $\xi^{-i}(E) = E$ μ-a.e. for each i. Taking $A = B = E$, we get $\mu(E) = \frac{1}{n} \sum_{i=0}^{n} \mu(\xi^{-i}(E) \cap E) \xrightarrow[n\to\infty]{} [\mu(E)]^2$. That is, $\mu(E) = [\mu(E)]^2$, so either $\mu(E) = 0$ or $\mu(E) = 1$. ∎

We now introduce two useful notions of "mixing" for measure-preserving transformations.

20.10 Definition   *A measure-preserving transformation $\xi\colon \Omega \to \Omega$ on a probability space is:*

- **strongly-mixing**, *if for each $A, B \in \Sigma$ we have*

$$\lim_{n\to\infty} \mu(\xi^{-n}(A) \cap B) = \mu(A)\mu(B).$$

- **weakly-mixing**, *if for each $A, B \in \Sigma$ we have*

$$\lim_{n\to\infty} \frac{1}{n} \sum_{i=0}^{n-1} \left| \mu(\xi^{-i}(A) \cap B) - \mu(A)\mu(B) \right| = 0.$$

20.11 Theorem *For measure-preserving transformations we have the following implications:*

$$\text{Strongly-Mixing} \implies \text{Weakly-Mixing} \implies \text{Ergodic}$$

Proof: Let $\xi\colon \Omega \to \Omega$ be a transformation. The fact that "strongly-mixing" implies "weakly-mixing" is an easy consequence of the following property of real sequences:

If a sequence $\{a_n\} \subset \mathbb{R}$ satisfies $a_n \to a$, then $\frac{1}{n}\sum_{i=0}^{n-1} a_i \to a$.

If ξ is weakly mixing, then from Corollary 20.9 we see that ξ is ergodic. ∎

For a detailed account of mixing properties, we refer you to U. Krengel [217].

20.3 Ergodic operators

In this section, unless otherwise stated, X is a Banach space and $L(X)$ denotes the Banach space $L(X, X)$ of bounded linear operators from X into X, with the operator norm. The **identity operator** $I\colon X \to X$, defined by $I(x) = x$ belongs to $L(x)$. Consider an operator $T\colon X \to X$ belonging to $L(X)$. With such an operator T, we associate its sequence $\{A_n\}$ of operator averages, defined by

$$A_n = \frac{I + T + T^2 + \cdots + T^{n-1}}{n} = \frac{1}{n}\sum_{i=0}^{n-1} T^i.$$

The symbol M_n is also used to denote A_n, and occasionally it is called the nth **Cesàro mean** of T. It is easy to see that A_n is a bounded linear operator on X for each n. If, in addition, X is a Banach lattice and T is a positive operator, then each A_n is also a positive operator. The next lemma presents some elementary identities for the averaging operators.

20.12 Lemma *For a linear operator $T\colon X \to X$ and natural numbers n and k we have:*

1. $\frac{1}{n}T^n = \frac{n+1}{n}A_{n+1} - A_n$;

2. $A_n T = T A_n = \frac{n+1}{n}A_{n+1} - \frac{1}{n}I$; and

3. $A_n T^k - A_n = \frac{1}{n}\sum_{i=0}^{k-1} T^{i+n} - \frac{1}{n}\sum_{i=0}^{k-1} T^i$.

Proof: We prove the first identity only. Note that

$$A_n = \frac{1}{n}\sum_{i=0}^{n-1} T^i = \frac{n+1}{n}\Big[\frac{1}{n+1}\sum_{i=0}^{n} T^i\Big] - \frac{1}{n}T^n = \frac{n+1}{n}A_{n+1} - \frac{1}{n}T^n,$$

from which the desired identity follows. ∎

20.13 Definition   *An operator $T \in L(X)$ is called:*

•   ***uniformly ergodic***, *if the sequence $\{A_n\}$ of operator averages is convergent in the sup norm on $L(X)$.*

•   ***strongly*** (or ***mean***) ***ergodic***, *if for each $x \in X$, the sequence $\{A_n x\}$ of averages is norm-convergent.*

•   ***weakly ergodic***, *if for each $x \in X$, the sequence $\{A_n x\}$ is weakly convergent.*

Clearly,

Uniform Ergodicity $\implies$ Strong Ergodicity $\implies$ Weak Ergodicity.

If $T \in L(X)$ is weakly ergodic, then we introduce the operator $P_T : X \to X$ via

$$P_T x = w\text{-}\lim_{n \to \infty} A_n x.$$

Here are some basic properties of the operator P_T.

20.14 Theorem   *If $T \in L(X)$ is weakly ergodic, then P_T is a continuous projection whose range is the fixed space of T. Moreover, we have*

$$P_T T = T P_T = P_T.$$

Proof: It follows immediately from Corollary 6.18 that P_T is a bounded operator. To see that P_T is a projection, notice that if x belongs to $\mathcal{F}_T = \{y \in X : Ty = y\}$ (the fixed space of T), then $A_n x = x$ for each n, so $P_T x = x$. That is, P_T acts as the identity operator on $\mathcal{F}_T$.

Now let $x \in X$. Then $A_n x \xrightarrow{w} P_T x$, so $T A_n(x) \xrightarrow{w} T(P_T x)$. On the other hand, from Lemma 20.12, we have

$$T A_n(x) = \frac{n+1}{n} A_{n+1}(x) - \frac{x}{n} \xrightarrow[n \to \infty]{w} P_T x,$$

so $T(P_T x) = P_T x$, that is, $P_T x \in \mathcal{F}_T$. The above show that P_T is a projection onto $\mathcal{F}_T$ and that $T P_T = P_T T = P_T$.   ∎

Next, we discuss uniformly ergodic operators. To do this, we need to introduce the spectral radius of an operator.

20.15 Lemma   *If $T : X \to X$ is a continuous operator on a normed space, then the limit $\lim_{n \to \infty} \|T^n\|^{1/n}$ always exists in $\mathbb{R}$.*

Proof: The proof relies on the fact that if $X \xrightarrow{R} Y \xrightarrow{S} Z$ are continuous operators between normed spaces, then $\|SR\| \leq \|S\| \cdot \|R\|$. We claim that the limit coincides with the real number $r = \inf_n \|T^n\|^{\frac{1}{n}} \geq 0$.

To see this, start by observing that $r \leqslant \|T^n\|^{\frac{1}{n}}$ implies

$$r \leqslant \liminf_{n \to \infty} \|T^n\|^{\frac{1}{n}}. \qquad (\star)$$

Now let $\varepsilon > 0$, and fix some k satisfying $\|T^k\|^{\frac{1}{k}} < r + \varepsilon$. For each n, write $n = m_n k + p_n$ with $0 \leqslant p_n < k$. Clearly, $\lim_{n \to \infty} \frac{p_n}{n} = 0$ and $\lim_{n \to \infty} \frac{m_n}{n} = \frac{1}{k}$. From the inequalities

$$\left\|T^n\right\|^{\frac{1}{n}} \leqslant \left\|T^{m_n k}\right\|^{\frac{1}{n}} \cdot \left\|T^{p_n}\right\|^{\frac{1}{n}} \leqslant \left\|T^k\right\|^{\frac{m_n}{n}} \cdot \|T\|^{\frac{p_n}{n}},$$

we get that $\limsup_{n \to \infty} \|T^n\|^{\frac{1}{n}} \leqslant \|T^k\|^{\frac{1}{k}} < r + \varepsilon$. Since $\varepsilon > 0$ is arbitrary, $\limsup_{n \to \infty} \|T^n\|^{\frac{1}{n}} \leqslant r$. From $(\star)$, we see that $r = \lim_{n \to \infty} \|T^n\|^{\frac{1}{n}}$. ∎

20.16 Definition *If $T \colon X \to X$ is a continuous operator on a normed space, then the limit*

$$r(T) = \lim_{n \to \infty} \left\|T^n\right\|^{1/n}$$

*is called the **spectral radius** of T.*[2]

Uniform ergodicity depends on the spectral radius of the operators.

20.17 Theorem *For a continuous operator $T \colon X \to X$ on a normed space:*

1. *If $r(T) < 1$, then T is uniformly ergodic and $P_T = 0$.*

2. *If $r(T) > 1$, then T cannot be uniformly ergodic.*

3. *If $r(T) = 1$, then T may or may not be uniformly ergodic.*

Proof: (1) In this case, there exist some $0 < \alpha < 1$ and some n_0 such that $\|T^n\|^{\frac{1}{n}} \leqslant \alpha$ for each $n \geqslant n_0$. That is, $\|T^n\| \leqslant \alpha^n$ for each $n \geqslant n_0$. This easily implies that $s = \sum_{i=0}^{\infty} \|T^i\| < \infty$. So $\|A_n\| \leqslant \frac{1}{n} \sum_{i=0}^{n-1} \|T^i\| \leqslant \frac{s}{n}$ implies $\|A_n\| \to 0$.

(2) If $r(T) > 1$, then there exist some $\alpha > 1$ and some n_0 such that $\|T^n\| \geqslant \alpha^n$ for each $n \geqslant n_0$. If $\{A_n\}$ is uniformly ergodic, then from Lemma 20.12, we see that $\left\|\frac{T_n}{n}\right\| \to 0$, contradicting $\left\|\frac{T_n}{n}\right\| \geqslant \frac{\alpha^n}{n} \to \infty$.

(3) The identity operator on any Banach space satisfies $r(I) = 1$ and is uniformly ergodic. On the other hand, the operator $T \colon C[0,1] \to C[0,1]$ defined by $Tx(t) = tx(t)$ satisfies $r(T) = 1$ and fails to be weakly ergodic. (Why?) ∎

---

[2] If $X = \mathbb{R}^n$ and T is a matrix with eigenvalues $\{\lambda_1, \lambda_2, \ldots, \lambda_n\}$, then it is well known that

$$r(T) = \max\{|\lambda_i| : i = 1, 2, \ldots, n\}.$$

In general, if X is a complex Banach space and $\sigma(T) = \{\lambda \in \mathbb{C} : \lambda I - T \text{ is not invertible}\}$ (the **spectrum** of T), then it is well known that $\sigma(T)$ is a nonempty compact set and

$$r(T) = \max\{|\lambda| : \lambda \in \sigma(T)\}.$$

For details, see, for example, A. E. Taylor and D. C. Lay [330, Chapter V] or Y. A. Abramovich and C. D. Aliprantis [1, Chapter 6].

A continuous operator $T : X \to X$ on a Banach space is:

- **power bounded**, if there is some M with $\|T^n\| \leqslant M$ for each n.

- **Cesàro bounded**, if there is some M with $\|A_n\| \leqslant M$ for each n.

Clearly, every power bounded operator is Cesàro bounded but the converse is not true. (Can you exhibit an example?) Moreover, if an operator T is power bounded, then $r(T) \leqslant 1$.

20.18 Lemma *If an operator $T : X \to X$ on a normed space is Cesàro bounded, then for a vector $x \in X$ satisfying $\frac{1}{n}\|T^n x\| \to 0$ the following statements are equivalent.*

1. *The sequence $\{A_n x\}$ of averages is norm-convergent.*

2. *The sequence $\{A_n x\}$ of averages is weakly convergent.*

3. *The sequence $\{A_n x\}$ of averages has a weak accumulation point.*

4. *The closed convex hull of the T-orbit of x contains a (unique) fixed point of T. That is, $\overline{co}\, \mathcal{O}_T(x) \cap \mathcal{F}_T \neq \varnothing$.*

Proof: The implications $(1) \implies (2) \implies (3)$ are obvious. For the rest, start by observing that $\frac{1}{n}\|T^{i+n} x\| \xrightarrow[n \to \infty]{} 0$ for each i. Hence, from the third identity of Lemma 20.12, we see that

$$\|A_n T^k x - A_n x\| \xrightarrow[n \to \infty]{} 0 \tag{$\dagger$}$$

for each fixed k.

$(3) \implies (4)$ Let y be a weak accumulation point of the sequence $\{A_n x\}$, and let $\mathcal{O}_T(x) = \{x, Tx, T^2 x, \ldots\}$ denote the orbit of x under T. From $\{A_n x\} \subset co\, \mathcal{O}_T(x)$, we see that y belongs to the weak closure of $co\, \mathcal{O}_T(x)$. By Theorem 5.98, y belongs to the norm closure of $co\, \mathcal{O}_T(x)$. Next, we show that $Ty = y$.

To this end, let $x' \in X'$ satisfy $\|x'\| = 1$ and let $\varepsilon > 0$. From $(\dagger)$, we see that there exists some n_0 such that $\|A_n Tx - A_n x\| < \varepsilon$ for all $n \geqslant n_0$. Now since y is a weak accumulation point of $\{A_n x\}$, there exists some integer $m > n_0$ such that

$$|x'(y - A_m x)| < \varepsilon \quad \text{and} \quad |T'x'(A_m x - y)| = |x'(A_m Tx - Ty)| < \varepsilon.$$

Therefore,

$$|x'(y - Ty)| \leqslant |x'(y - A_m x)| + |x'(A_m x - A_m Tx)| + |x'(A_m Tx - Ty)| < 3\varepsilon.$$

Since $\varepsilon > 0$ is arbitrary, we get $x'(y - Ty) = 0$ for each $x' \in X'$, and from this it follows that $Ty = y$.

(4) $\implies$ (1)   Fix some $M > 0$ satisfying $\|A_n\| < M$ for each n and let $\varepsilon > 0$. Also, let $y \in \mathcal{F}_T \cap \overline{co}\, \mathcal{O}_T(x)$. Then there exists some operator S in $co\{I, T, T^2, \ldots\}$ with $\|y - Sx\| < \varepsilon$. Since S is a convex combination of the powers T^k, it follows from (†) that there exists some n_0 such that $\|A_n Sx - A_n x\| < \varepsilon$ for each $n \geqslant n_0$. Now note that

$$\|A_n x - y\| = \|A_n x - A_n y\| \leqslant \|A_n(y - Sx)\| + \|A_n Sx - A_n x\| \leqslant (1 + M)\varepsilon$$

for all $n \geqslant n_0$. This shows that $\{A_n(x)\}$ is norm-convergent to y. (Clearly, y as the limit of the sequence $\{A_n x\}$ is uniquely determined. That is, $\mathcal{F}_T \cap \overline{co}\, \mathcal{O}_T(x)$ is a singleton.) ∎

We are now ready to state the major characterizations of strongly ergodic operators.

20.19 Theorem   *For a power bounded operator $T : X \to X$ on a normed space, the following statements are equivalent.*

1. *The operator T is strongly ergodic.*

2. *The operator T is weakly ergodic.*

3. *For each x, the sequence $\{A_n x\}$ of averages has a weak accumulation point.*

4. *For each x, we have $\overline{co}\, \mathcal{O}_T(x) \cap \mathcal{F}_T \neq \varnothing$.*

Proof: The proof follows from Lemma 20.18 by noting that every power bounded operator is Cesàro bounded and satisfies $\lim_{n \to \infty} \frac{1}{n}\|T^n x\| = 0$ for each x. ∎

20.20 Corollary   *Every power bounded operator on a reflexive Banach space is strongly ergodic.*

Proof: This follows from statement (3) of Theorem 20.19 and the fact that in a reflexive Banach space every norm-bounded sequence has a weak accumulation point. ∎

Finally, we present an ergodic theorem regarding positive contraction operators on L_1-spaces. Recall (Definition 9.44) that a contraction operator is a continuous operator of norm at most unity.

20.21 Lemma   *If $T : X \to X$ is a Cesàro bounded operator on a Banach space, then T is strongly ergodic if and only if $\{A_n x\}$ is norm-convergent for each x in a (norm) dense subset of X.*

Proof: Assume that $\|A_n\| < M$ for each n and that the set

$$\{x \in X : \{A_n x\} \text{ is norm-convergent}\}$$

is norm-dense. Fix $y \in X$ and let $\varepsilon > 0$. Choose some $x \in X$ such that $\|x - y\| < \varepsilon$ and $\{A_n x\}$ is norm-convergent. So there exists some n_0 such that $\|A_n x - A_m x\| < \varepsilon$ for all $n, m \geqslant n_0$. Now if $n, m \geqslant n_0$, then

$$\|A_n y - A_m y\| \leqslant \|A_n(y - x)\| + \|A_n x - A_m x\| + \|A_m(x - y)\| < (2M + 1)\varepsilon.$$

This shows that $\{A_n y\}$ is a norm-Cauchy sequence, and hence a norm-convergent sequence for each $y \in X$. ∎

20.22 Theorem   *Let μ be a finite measure and let $T : L_1(\mu) \to L_1(\mu)$ be a positive contraction operator. If $T\mathbf{1} \leqslant \mathbf{1}$ (where $\mathbf{1}$ denotes the constant function one), then T is strongly ergodic.*

Proof: Assume that T satisfies the stated properties. Clearly, T is a power (and hence a Cesàro) bounded operator. In particular, we have $\frac{1}{n}\|T^n f\|_1 \xrightarrow[n \to \infty]{} 0$ for each $f \in L_1(\mu)$.

The positivity of T, coupled with $T\mathbf{1} \leqslant \mathbf{1}$, implies that A_n maps the order interval $[-\mathbf{1}, \mathbf{1}]$ into itself. Since $L_1(\mu)$ has order continuous norm, we know (by Theorem 9.22) that $[-\mathbf{1}, \mathbf{1}]$ is weakly compact. This means that $\{A_n f\}$ has a weak accumulation point for each $f \in [-\mathbf{1}, \mathbf{1}]$. It follows that $\{A_n f\}$ has a weak accumulation point for each $f \in L_\infty(\mu)$. By Lemma 20.18, $\{A_n f\}$ is norm-convergent for every f in $L_\infty(\mu)$. Since $L_\infty(\mu)$ is $\|\cdot\|_1$-dense in $L_1(\mu)$, it follows from Lemma 20.21 that $\{A_n f\}$ is norm-convergent for each $f \in L_1(\mu)$. That is, T is a strongly ergodic operator. ∎

20.23 Corollary   *If $\xi : \Omega \to \Omega$ is a measure-preserving transformation on a probability measure space (Ω, Σ, P), then the corresponding composition operator $T_\xi : L_1(P) \to L_1(P)$, defined by $T_\xi(f) = f \circ \xi$, is strongly ergodic.*

Proof: Note that T_ξ is positive, satisfies $T\mathbf{1} = \mathbf{1}$, and (by the Change of Variable Theorem 13.46) T is a lattice isometry. By Theorem 20.22, the operator T is strongly ergodic. ∎

References

1 Y. A. Abramovich and C. D. Aliprantis. 2002. *An invitation to operator theory*, volume 50. Graduate Studies in Mathematics. Providence, RI: American Mathematical Society.

2 Y. A. Abramovich, C. D. Aliprantis, and I. A. Polyrakis. 1994. Lattice-subspaces and positive projections. *Proceedings of the Royal Irish Academy* 94A:237–253.

3 Y. A. Abramovich and A. W. Wickstead. 1993. Remarkable classes of unital AM-spaces. *Journal of Mathematical Analysis and Applications* 180:398–411.

4 D. Adams. 1979. *The hitchhiker's guide to the galaxy*. New York: Pocket Books.

5 L. Alaoglu. 1940. Weak compactness of normed linear spaces. *Annals of Mathematics* 41:252–267.

6 A. D. Alexandroff. 1939. Almost everywhere existence of the second differential of a convex function and surfaces connected with it. *Leningrad State University Annals, Mathematics Series* 6:3–35. In Russian.

7 C. D. Aliprantis. 1996. *Problems in equilibrium theory*. Heidelberg & New York: Springer–Verlag.

8 C. D. Aliprantis, K. C. Border, and W. A. J. Luxemburg, eds. 1991. *Positive operators, Riesz spaces, and economics*. Studies in Economic Theory, 2. Berlin: Springer–Verlag.

9 C. D. Aliprantis, D. J. Brown, and O. Burkinshaw. 1990. *Existence and optimality of competitive equilibria*. New York: Springer–Verlag.

10 ———. 1990. Valuation and optimality in the overlapping generations model. *International Economic Review* 31:275–288.

11 C. D. Aliprantis and O. Burkinshaw. 1980. Minimal topologies and L_p-spaces. *Illinois Journal of Mathematics* 24:164–172.

12 ———. 1985. *Positive operators*. Pure and Applied Mathematics, 119. New York: Academic Press.

13 ———. 1998. *Principles of real analysis*, 3d. ed. San Diego: Academic Press.

14 ———. 1999. *Problems in real analysis*, 2d. ed. San Diego: Academic Press. Date of publication, 1998. Copyright 1999.

15 ———. 2003. *Locally solid Riesz spaces with applications to economics*. Mathematical Surveys and Monographs, 105. Providence: American Mathematical Society.

16 C. D. Aliprantis, R. Tourky, and N. C. Yannelis. 2000. Cone conditions in general equilibrium theory. *Journal of Economic Theory* 92:96–121.

17 T. M. Apostol. 1969. *Calculus*, 2d. ed., volume 2. Waltham, Massachusetts: Blaisdell.

18 ———. 1974. *Mathematical analysis*, 2d. ed. Reading, Massachusetts: Addison Wesley.

19 R. Arens. 1947. Duality in linear spaces. *Duke Mathematical Journal* 14:787–794.

20 T. E. Armstrong and K. Prikry. 1981. Liapounoff's theorem for nonatomic, finitely-additive, bounded, finite-dimensional, vector-valued measures. *Transactions of the American Mathematical Society* 266:499–514.

21 K. J. Arrow. 1969. Tullock and an existence theorem. *Public Choice* 6:105–111.

22 J.-P. Aubin and A. Cellina. 1984. *Differential inclusions*. Berlin: Springer–Verlag.

23 J.-P. Aubin and I. Ekeland. 1984. *Applied nonlinear analysis*. Pure and Applied Mathematics: A Wiley-Interscience Series of Texts, Monographs, and Tracts. New York: John Wiley and Sons.

24 J.-P. Aubin and H. Frankowska. 1990. *Set-valued analysis*. Boston: Birkhäuser.

25 R. J. Aumann. 1961. Borel structures for function spaces. *Illinois Journal of Mathematics* 5:614–630.

26 ———. 1965. Integrals of set-valued functions. *Journal of Mathematical Analysis and Applications* 12:1–12.

27 ———. 1966. Existence of competitive equilibria in markets with a continuum of traders. *Econometrica* 34:1–17.

28 ———. 1969. Measurable utility and the measurable choice problem. In *La Décision*, pages 15–26, Paris. Colloque Internationaux du C. N. R. S.

29 K. Back. 1988. Structure of consumption sets and existence of equilibria in infinite-dimensional spaces. *Journal of Mathematical Economics* 17:89–99.

30 S. Banach. 1923. Sur le probléme de la mesure. *Fundamenta Mathematicae* 4:7–33.

31 S. Banach and K. Kuratowski. 1929. Sur une généralization du probléme de la mesure. *Fundamenta Mathematicae* 14:127–131.

32 S. Banach and A. Tarski. 1924. Sur la décomposition des ensembles de points en parties respectivement congruentes. *Fundamenta Mathematicae* 6:244–277.

33 R. G. Bartle and L. M. Graves. 1952. Mappings between function spaces. *Transactions of the American Mathematical Society* 72:400–413.

34 G. A. Beer. 1991. A Polish topology for the closed subsets of a Polish space. *Proceedings of the American Mathematical Society* 113:1123–1133.

35 ———. 1993. *Topologies on closed and closed convex sets*. Mathematics and Its Applications, 268. Dordrecht: Kluwer Academic Publishers.

36 E. T. Bell. 1965. *Men of mathematics*. New York: Simon and Schuster.

37 C. Berge. 1963. *Topological spaces*. New York: Macmillan. English translation by E. M. Patterson of *Espaces topologiques et fonctions multivoques*, published by Dunod, Paris, 1959.

38 M. Berliant. 1985. An equilibrium existence result for an economy with land. *Journal of Mathematical Economics* 14:53–56.

39 D. P. Bertsekas and S. E. Shreve. 1978. *Stochastic optimal control: The discrete time case*. Mathematics in Science and Engineering, 139. New York: Academic Press.

40 T. F. Bewley. 1972. Existence of equilibria in economies with infinitely many commodities. *Journal of Economic Theory* 4:514–540.

41 K. P. S. Bhaskara Rao and M. Bhaskara Rao. 1983. *Theory of charges*. Pure and Applied Mathematics. London: Academic Press.

42 P. Billingsley. 1968. *Convergence of probability measures*. Wiley Series in Probability and Mathematical Statistics. New York: Wiley.

43 ———. 1979. *Probability and measure*. Wiley Series in Probability and Mathematical Statistics. New York: Wiley.

44 G. D. Birkhoff. 1931. Proof of the ergodic theorem. *Proceedings of the National Academy of Sciences, U.S.A.* 17:656–660.

45 E. Bishop and K. DeLeeuw. 1959. The representation of linear functionals by measures on sets of extreme points. *Annales de l'Institut Fourier (Grenoble)* 9:305–331.

46 E. Bishop and R. R. Phelps. 1963. The support functionals of a convex set. In *Convexity*, Proceedings of Symposia in Pure Mathematics, 7, pages 27–35. Providence, RI: American Mathematical Society.

47 D. Blackwell. 1953. Equivalent comparison of experiments. *Annals of Mathematical Statistics* 24:265–272.

48 ———. 1965. Discounted dynamic programming. *Annals of Mathematical Statistics* 36:226–235.

49 L. E. Blume. 1982. New techniques for the study of stochastic equilibrium processes. *Journal of Mathematical Economics* 9:61–70.

50 R. M. Blumenthal and H. H. Corson. 1970. On continuous collections of measures. *Annales de l'Institut Fourier (Grenoble)* 20:193–199.

51 ———. 1972. On continuous collections of measures. In Le Cam et al. [224], pages 33–40.

52 R. M. Blumenthal and R. K. Getoor. 1968. *Markov processes and potential theory.* Pure and Applied Mathematics, 29. New York: Academic Press.

53 S. Bochner. 1933. Integration von Funktionen deren Werte die Elemente eines Vectorräumes sind. *Fundamenta Mathematicae* 20:262–276.

54 ———. 1960. *Harmonic analysis and the theory of probability.* Berkeley: University of California Press. Second printing.

55 H. F. Bohnenblust and S. Karlin. 1950. On a theorem of Ville. In H. W. Kuhn and A. W. Tucker, eds., *Contributions to the Theory of Games, I,* Annals of Mathematics Studies, 24, pages 155–160. Princeton: Princeton University Press.

56 K. C. Border. 1985. *Fixed point theorems with applications to economics and game theory.* New York: Cambridge University Press.

57 ———. 1991. Functional analytic tools for expected utility theory. In Aliprantis et al. [8], pages 69–88.

58 ———. 1991. Implementation of reduced form auctions: A geometric approach. *Econometrica* 59:1175–1187.

59 ———. 1992. Revealed preference, stochastic dominance, and the expected utility hypothesis. *Journal of Economic Theory* 56:20–42.

60 N. Bourbaki. 1965. *Éléments de mathématique, livre viii: Intégration: Chapitres 1–4,* 2d. revised and augmented ed. Actualités Scientifiques et Industrielles, 1175. Paris: Hermann.

61 ———. 1966. *General topology: Part 2.* Elements of Mathematics, 3. Reading, Massachusetts: Addison–Wesley. Translation of *Éléments de mathématique: Topologie Générale,* published in French by Hermann, Paris.

62 ———. 1968. *Theory of sets.* Elements of Mathematics, 1. Reading, Massachusetts: Addison–Wesley. Translation of *Éléments de mathématique: Théorie des ensembles,* published in French by Hermann, Paris, 1968.

63 ———. 1987. *Topological vector spaces: Chapters 1–5.* Elements of Mathematics. Berlin: Springer–Verlag. Translated by H. G. Eggleston and S. Madan. Originally published in French as *Espaces vectoriels topologiques* by Masson, Paris, 1981.

64 A. Brøndsted and R. T. Rockafellar. 1965. On the subdifferentiability of convex functions. *Proceedings of the American Mathematical Society* 16:605–611.

65 J. K. Brooks. 1980. On a theorem of Dieudonné. *Advances in Mathematics* 36:165–168.

66 J. K. Brooks and R. V. Chacon. 1980. Continuity and compactness of measures. *Advances in Mathematics* 37:16–26.

67 L. E. J. Brouwer. 1912. Über Abbildung von Mannigfaltikeiten. *Mathematische Annalen* 71:97–115.

68 F. E. Browder. 1965. Nonlinear monotone operators and convex sets in Banach spaces. *Bulletin of the American Mathematical Society* 71:271–310.

69 ———. 1967. A new generalization of the Schauder fixed point theorem. *Mathematische Annalen* 174:285–290.

70 ———. 1968. The fixed point theory of multi-valued mappings in topological vector spaces. *Mathematische Annalen* 177:283–301.

71 S. L. Brumelle and R. G. Vickson. 1975. A unified approach to stochastic dominance. In W. T. Ziemba and R. G. Vickson, eds., *Stochastic optimization methods in finance,* pages 101–113. New York: Academic Press.

72 C. Carathéodory. 1918. *Vorlesungen über reelle Funktionen,* 1st. ed. Berlin: Leibzig. 2nd. edition, New York: Chelsea, 1948.

73 C. Castaing. 1967. Sur les multi-applications mesurables. *Revue Francaise d'Information et de Recherche Opèrationnaelle* 1:91–126.

74 ———. 1967. Sur une nouvelle extension du thèoréme de Ljapunov. *Comptes Rendus Hebdomadaire des Séances de l'Académie des Sciences, Série A (Paris)* 264:333–336.

75 C. Castaing and M. Valadier. 1977. *Convex analysis and measurable multifunctions.* Lecture Notes in Mathematics, 580. Berlin: Springer–Verlag.

76 G. Choquet. 1969. *Lectures on analysis.* Reading, Massachusetts: Benjamin. 3 vols.

77 P. J. Cohen. 1966. *Set theory and the continuum hypothesis.* Mathematics Lecture Note Series. Reading, Massachusetts: Benjamin.

78 P. Cousot and R. Cousot. 1979. Constructive versions of Tarski's fixed point theorem. *Pacific Journal of Mathematics* 81:43–57.

79 D. J. Daley. 1968. Stochastically monotone Markov chains. *Zeitschrift für Wahrscheinlichkeitstheorie und Verwandte Gebiete* 10:305–317.

80 P. J. Daniell. 1917. A general form of integral. *Annals of Mathematics* 19:279–294.

81 A. C. Davies. 1955. A characterization of complete lattices. *Pacific Journal of Mathematics* 5:311–319.

82 R. O. Davies and J. Dravecký. 1973. On the measurability of functions of two variables. *Matematický Časopis Slovenskej Akadémie Vied* 23:351–372.

83 M. M. Day. 1940. The spaces L^p with $0 < p < 1$. *Bulletin of the American Mathematical Society* 46:816–823.

84 G. Debreu. 1954. Valuation equilibrium and Pareto optimum. *Proceedings of the National Academy of Sciences, U.S.A.* 40:588–592.

85 ———. 1967. Integration of correspondences. In L. M. Le Cam, J. Neyman, and E. L. Scott, eds., *Proceedings of the Fifth Berkeley Symposium on Mathematical Statistics and Probability II, Part I,* pages 351–372. Berkeley and Los Angeles: University of California Press.

86 C. Dellacherie. 1972. *Capacités et processus stochastiques.* Ergebnisse der Mathematik und ihrer Grenzgebiete, 67. Berlin, Heidelberg, and New York: Springer–Verlag.

87 ———. 1978. Quelques exemples familiers, en probabilités, d'ensembles analytiques, non boréliens. In *Séminaires de Probabiltés, Université de Strasbourg,* Lecture Notes in Mathematics, 649, pages 746–756. Berlin: Springer–Verlag.

88 C. Dellacherie and P.-A. Meyer. 1978. *Probabilities and potential.* Mathematics Studies, 29. Amsterdam: North Holland. Translated from the French by the authors.

89 ———. 1982. *Probabilities and potential B: Theory of martingales.* Mathematics Studies, 72. Amsterdam: North Holland. Translated and prepared by J. P. Wilson.

90 E. V. Denardo. 1967. Contraction mappings in the theory underlying dynamic programming. *SIAM Review* 9:165–177.

91 K. J. Devlin. 1992. *Sets, functions and logic,* 2d. ed. London: Chapman & Hall.

92 ———. 1993. *The joy of sets: Fundamentals of contemporary set theory,* 2d. ed. Undergraduate Texts in Mathematics. New York: Springer–Verlag.

93 P. Diamond. 1989. Fixed points of iterates of multivalued mappings. *Journal of Mathematical Analysis and Applications* 143:252–258.

94 K. Diemling. 1985. *Nonlinear functional analysis.* Berlin: Springer–Verlag.

95 J. Diestel. 1977. Remarks on weak compactness in $L_1(\mu, X)$. *Glasgow Mathematics Journal* 18:87–91.

96 J. Diestel and J. J. Uhl, Jr. 1977. *Vector measures.* Mathematical Surveys, 15. Providence: American Mathematical Society.

97 J. Dieudonné. 1969. *Foundations of modern analysis.* Pure and Applied Mathematics, 10-I. New York: Academic Press. Volume 1 of Treatise on Analysis.

98 J. L. Doob. 1953. *Stochastic processes.* New York: Wiley.

99 ———. 1994. *Measure theory.* Graduate Texts in Mathematics, 143. New York: Springer–Verlag.

100 L. E. Dubins and L. J. Savage. 1976. *Inequalities for stochastic processes (How to gamble if you must).* New York: Dover. Reprint of *How to gamble if you must: Inequalities for stochastic processes* published by McGraw–Hill, New York, 1965. With a bibliographic supplement.

101 L. E. Dubins and E. H. Spanier. 1961. How to cut a cake fairly. *American Mathematical Monthly* 68:1–17.

102 R. M. Dudley. 1966. Convergence of Baire measures. *Studia Mathematica* 27:151–268.

103 ———. 1972. A counterexample on measurable processes. In Le Cam et al. [224], pages 57–18.

104 ———. 1989. *Real analysis and probability*. Pacific Grove, California: Wadsworth & Brooks/Cole.

105 D. Duffie, J. Geanakoplos, A. Mas-Colell, and A. McLennan. 1994. Stationary Markov equilibria. *Econometrica* 62:745–781.

106 J. Dugundji. 1966. *Topology*. Boston: Allyn and Bacon.

107 J. Dugundji and A. Granas. 1978. KKM-maps and variational inequalities. *Annali della Scuola Normale Superiore de Pisa, Serie IV* 5:679–682.

108 ———. 1982. *Fixed point theory*, volume 1. Monografie Matematyczne, 61. Warsaw: Polish Scientific Publishers.

109 N. Dunford. 1937. Integration of vector-valued functions. *Bulletin of the American Mathematical Society* 3:24. Abstract.

110 N. Dunford and J. T. Schwartz. 1957. *Linear operators: Part I*. New York: Interscience.

111 E. B. Dynkin. 1965. *Markov processes II*. Grundlehren der mathematischen Wissenschaften in Einzeldarstellugen mit besonderer Berucksichtigung der Anwendungsgebiete, 122. Berlin: Springer–Verlag.

112 W. F. Eberlein. 1947. Weak compactness in Banach spaces, I. *Proceedings of the National Academy of Sciences, U.S.A.* 33:51–53.

113 F. Echenique. 2005. A short and constructive proof of Tarski's fixed-point theorem. *International Journal of Game Theory* 33:215–218.

114 E. G. Effros. 1965. Convergence of closed sets in a topological space. *Proceedings of the American Mathematical Society* 86:929–931.

115 I. Ekeland and R. Temam. 1976. *Convex analysis and variational problems*. Studies in Mathematics and its Applications, 1. Amsterdam: North Holland.

116 I. Ekeland and T. Turnbull. 1983. *Infinite-dimensional optimization and convexity*. Chicago lectures in mathematics. Chicago: University of Chicago Press.

117 K. Fan. 1952. Fixed-point and minimax theorems in locally convex topological spaces. *Proceedings of the National Academy of Sciences, U.S.A.* 38:121–126.

118 ———. 1961. A generalization of Tychonoff's fixed point theorem. *Mathematische Annalen* 142:305–310.

119 ———. 1969. Extensions of two fixed point theorems of F. E. Browder. *Mathematische Zeitschrift* 112:234–240.

120 K. Fan, I. L. Glicksberg, and A. J. Hoffman. 1957. Systems of inequalities involving convex functions. *Proceedings of the American Mathematical Society* 8:617–622.

121 J. Farkas. 1902. Über die Theorie der einfachen Ungleichungen. *Journal für Reine und Angewandte Mathematik* 124:1–24.

122 J. M. G. Fell. 1962. A Hausdorff topology for the closed subsets of a locally compact non-Hausdorff space. *Proceedings of the American Mathematical Society* 13:472–476.

123 W. Fenchel. 1953. Convex cones, sets, and functions. Lecture notes, Princeton University, Department of Mathematics. From notes taken by D. W. Blackett, Spring 1951.

124 A. F. Filippov. 1962. On certain questions in the theory of optimal control. *Journal of SIAM Series A: Control* 1:76–84. English translation of *Vestnik Moskovskogo Universiteta. Serija I Matematika, Mehanika* 2:(1959), pp. 25–32.

125 M. Florenzano and C. Le Van. 2001. *Finite dimensional convexity and optimization*. Studies in Economic Theory, 13. New York and Heidelberg: Springer–Verlag.

126 M. Foreman and F. Wehrung. 1991. The Hahn–Banach theorem implies the existence of a non-Lebesgue measurable set. *Fundamenta Mathematicae* 138:13–19.

127 M. Frantz. 1991. On Sierpiński's nonmeasurable set. *Fundamenta Mathematicae* 139:17–22.

128 D. H. Fremlin. 1974. *Topological Riesz spaces and measure theory.* Cambridge: Cambridge University Press.

129 B. Fristedt and L. Gray. 1997. *A modern approach to probability theory.* Boston: Birkhäuser.

130 A. Fryszkowski. 1983. Continuous selections for a class of nonconvex multivalued maps. *Studia Mathematica* 75:163–174.

131 ———. 1990. Continuous selections of Aumann integrals. *Journal of Mathematical Analysis and Applications* 145:431–446.

132 C. A. Futia. 1981. Rational expectations in stationary linear models. *Econometrica* 49:171–192.

133 D. Gale. 1960. *Theory of linear economic models.* New York: McGraw-Hill.

134 B. R. Gelbaum and J. M. H. Olmsted. 1990. *Theorems and counterexamples in mathematics.* Heidelberg & New York: Springer–Verlag.

135 I. M. Gelfand. 1936. Sur un lemma de la théorie des espaces linéaires. *Communications de la Societé Mathematique de Kharkoff et de l'Institut des Sciences Mathematiques et Méchaniques de l'Université de Kharkoff (4)* 13:35–40.

136 J. R. Giles. 1982. *Convex analysis with application in differentiation of convex functions.* Research Notes in Mathematics, 58. Boston: Pitman Advanced Publishing Program.

137 C. Gilles. 1989. Charges as equilibrium prices and asset bubbles. *Journal of Mathematical Economics* 18:155–167.

138 L. Gillman and M. Jerison. 1976. *Rings of continuous functions.* Graduate Texts in Mathematics, 43. New York: Springer–Verlag. Reprint of the edition published in the University Series in Higher Mathematics by Van Nostrand, 1960.

139 I. L. Glicksberg. 1952. A further generalization of the Kakutani fixed point theorem, with applications to Nash equilibrium points. *Proceedings of the American Mathematical Society* 3:170–174.

140 A. Granas. 1981. KKM-maps and their applications to nonlinear problems. In R. D. Mauldin, ed., *The Scottish Book: Mathematics from the Scottish Cafe,* pages 45–61. Boston: Birkhäuser.

141 A. Granas and J. Dugundji. 2003. *Fixed point theory.* Springer Monographs in Mathematics. New York: Springer–Verlag.

142 J.-M. Grandmont and W. Hildenbrand. 1974. Stochastic processes of temporary equilibria. *Journal of Mathematical Economics* 1:247–277.

143 A. Grothendieck. 1973. *Topological vector spaces.* New York & London: Gordon and Breach.

144 H. Halkin. 1965. A generalization of LaSalle's bang-bang principle. *SIAM Journal on Control and Optimization* 2:199–202.

145 P. R. Halmos. 1948. The range of a vector measure. *Bulletin of the American Mathematical Society* 54:416–421.

146 ———. 1956. *Lectures on ergodic theory.* New York: Chelsea.

147 ———. 1974. *Finite dimensional vector spaces.* New York: Springer–Verlag. Reprint of the edition published by Van Nostrand, 1958.

148 ———. 1974. *Measure theory.* Graduate Texts in Mathematics, 18. New York: Springer–Verlag. Reprint of the edition published by Van Nostrand, 1950.

149 ———. 1974. *Naive set theory.* New York: Springer–Verlag. Reprint of the edition published by Litton Educational Printing, 1960.

150 B. R. Halpern. 1968. A general fixed point theorem. In *Proceedings of the Symposium on Nonlinear Functional Analysis.* Providence: American Mathematical Society.

151 B. R. Halpern and G. M. Bergman. 1968. A fixed point theorem for inward and outward maps. *Transactions of the American Mathematical Society* 130:353–358.

152 J. D. Halpern. 1964. The independence of the axiom of choice from the Boolean prime ideal theorem. *Fundamenta Mathematicae* 55:55–64.

153 P. Hartman and G. Stampacchia. 1966. On some non-linear elliptic differential-functional equations. *Acta Mathematica* 115:271–310.

154 F. Hausdorff. 1914. *Grundzuge der Mengenlehre.* Leipzig. Translation published as *Set Theory* by Chelsea, New York, 1962.

155 ———. 1991. *Set theory*, 4th. corrected english ed. New York: Chelsea. Translated by John R. Aumann from the 1937 German third edition.

156 R. Henstock. 1991. *The general theory of integration.* Oxford Mathematical Monographs. Oxford: Clarendon Press.

157 I. N. Herstein. 1964. *Topics in algebra.* Lexington, Massachusetts: Xerox College Publishing.

158 W. Hildenbrand. 1974. *Core and equilibria of a large economy.* Princeton: Princeton University Press.

159 T. Hill. 1983. Determining a fair border. *American Mathematical Monthly* 90:438–442.

160 C. J. Himmelberg. 1975. Measurable relations. *Fundamenta Mathematicae* 87:53–72.

161 C. J. Himmelberg and F. S. van Vleck. 1973. Extreme points of multifunctions. *Indiana University Mathematics Journal* 22:719–729.

162 ———. 1975. Multifunctions with values in a space of probability measures. *Journal of Mathematical Analysis and Applications* 50:108–112.

163 J.-B. Hiriart-Urruty and C. Lemaréchal. 1993. *Convex analysis and minimization algorithms I.* Grundlehren der mathematischen Wissenschaften, 305. Berlin: Springer–Verlag.

164 ———. 1993. *Convex analysis and minimization algorithms II.* Grundlehren der mathematischen Wissenschaften, 306. Berlin: Springer–Verlag.

165 ———. 2001. *Fundamentals of convex analysis.* Grundlehren Text Editions. Berlin: Springer–Verlag.

166 R. B. Holmes. 1975. *Geometric functional analysis and its applications.* Graduate Texts in Mathematics, 24. Berlin: Springer–Verlag.

167 L. Hörmander. 1954. Sur la fonction d'appui des ensembles convexes dans une espace localement convexe. *Arkiv för Matematik* 3:181–186.

168 J. Horváth. 1966. *Topological vector spaces and distributions*, volume 1. Reading, Mass.: Addison Wesley.

169 R. F. Hoskins. 1990. *Standard and nonstandard analysis: Fundamental theory, techniques, and applications.* New York: Horwood.

170 P. Howard and J. E. Rubin. 1998. *Consequences of the axiom of choice.* Mathematical Surveys and Monographs, 59. Providence, RI: American Mathematical Society.

171 R. Howard. 1998. Alexandrov's theorem on the second derivatives of convex functions via Rademacher's theorem on the first derivatives of Lipschitz functions. On-line lecture note, Department of Mathematics, University of South Carolina, Columbia, South Carolina. URL: http://www.math.sc.edu/~howard/Notes/alex.pdf.

172 S. Hu and N. S. Papageorgiou. 1997. *Handbook of multivalued analysis, volume I: Theory.* Mathematics and its Applications, 419. Dordrecht, Boston & London: Kluwer Academic Publishers.

173 A. E. Hurd and P. A. Loeb. 1985. *An introduction to nonstandard real analysis.* New York: Academic Press.

174 T. Husain. 1965. *The open mapping and closed graph theorems in topological vector spaces.* Oxford: Oxford University Press.

175 T. Ichiishi. 1983. *Game theory for economic analysis.* New York: Academic Press.

176 A. Ionescu Tulcea and C. Ionescu Tulcea. 1969. *Topics in the theory of lifting.* Ergebnisse der Mathematik und ihrer Grenzgebiete, 48. New York: Springer–Verlag.

177 K. Jacobs. 1978. *Measure and integral.* Probability and Mathematical Statistics. New York: Academic Press.

178 R. C. James. 1964. Weakly compact sets. *Transactions of the American Mathematical Society* 113:129–140.

179 G. J. O. Jameson. 1970. *Ordered linear spaces.* Lecture Notes in Mathematics, 141. Heidelberg and New York: Springer–Verlag.

180 V. A. Jankov. 1941. Sur l'uniformisation des ensembles *A*. *Comptes Rendus (Doklady) de l'Académie des Sciences de l'URSS* 39:591–592. In French, translated from the Russian.

181 H. Jarchow. 1981. *Locally convex spaces.* Mathematical Textbooks. Stuttgart: B. G. Teubner.

182 J. E. Jayne and C. A. Rogers. 1977. The extremal structure of convex sets. *Journal of Functional Analysis* 26:251–288.

183 ———. 2002. *Selectors*. Princeton: Princeton University Press.

184 T. Jech. 1973. *The axiom of choice*. Amsterdam: North Holland.

185 ———. 1978. *Set theory*. New York: Academic Press.

186 R. I. Jennrich. 1969. Asymptotic properties of non-linear least squares estimators. *Annals of Mathematical Statistics* 40:633–643.

187 L. E. Jones. 1983. Existence of equilibria with infinitely many consumers and infinitely many commodities: A theorem based on models of commodity differentiation. *Journal of Mathematical Economics* 12:119–138.

188 ———. 1984. A competitive model of commodity differentiation. *Econometrica* 52:507–530.

189 ———. 1987. The efficiency of monopolistically competitive equilibria in large economies: Commodity differentiation with gross substitutes. *Journal of Economic Theory* 41:356–391.

190 ———. 1987. Existence of equilibria with infinitely many commodities: Banach lattices reconsidered. *Journal of Mathematical Economics* 16:89–104.

191 S. Kakutani. 1941. A generalization of Brouwer's fixed point theorem. *Duke Mathematical Journal* 8:457–459.

192 T. Kamae, U. Krengel, and G. L. O'Brien. 1977. Stochastic inequalities on partially ordered spaces. *Annals of Probability* 5:899–912.

193 L. V. Kantorovich. 1937. On the moment problem for a finite interval. *Doklady Akademii Nauk SSSR* 14:531–537. In Russian.

194 L. V. Kantorovich and G. P. Akilov. 1964. *Functional analysis in normed spaces*. International series of monographs in pure and applied mathematics, 46. Oxford: Pergamon Press. Translated from the Russian by D. E. Brown.

195 D. W. Katzner. 1970. *Static demand theory*. London: Macmillan.

196 A. S. Kechris. 1995. *Classical descriptive set theory*. Graduate Texts in Mathematics, 156. New York: Springer–Verlag.

197 J. L. Kelley. 1950. The Tychonoff product theorem implies the axiom of choice. *Fundamenta Mathematicae* 37:75–76.

198 ———. 1955. *General topology*. New York: Van Nostrand.

199 J. L. Kelley, I. Namioka, et al. 1963. *Linear topological spaces*. Graduate Texts in Mathematics, 36. Berlin: Springer–Verlag. Reprint of the edition published in the University Series in Higher Mathematics by Van Nostrand, 1963.

200 M. A. Khan and A. Rustichini. 1991. Some unpleasant objects in a non-separable Hilbert space. In Aliprantis et al. [8], pages 179–187.

201 M. A. Khan and N. C. Yannelis, eds. 1991. *Equilibrium theory in infinite dimensional spaces*. Studies in Economic Theory, 1. Berlin: Springer–Verlag.

202 Y. Kifer. 1986. *Ergodic theory of random transformations*. Boston: Birkhäuser.

203 T. Kim, K. Prikry, and N. C. Yannelis. 1987. Carathéodory-type selections and random fixed point theorems. *Journal of Mathematical Analysis and Applications* 122:393–407.

204 ———. 1988. On a Carathéodory-type selection theorem. *Journal of Mathematical Analysis and Applications* 135:664–670.

205 V. Klee. 1948. The support property of a convex set. *Duke Mathematical Journal* 15:767–772.

206 ———. 1951. Convex sets in linear spaces. *Duke Mathematical Journal* 18:443–466.

207 ———. 1956. Strict separation of convex sets. *Proceedings of the American Mathematical Society* 7:735–737.

208 ———. 1963. On a question of Bishop and Phelps. *American Journal of Mathematics* 85:95–98.

209 E. Klein and A. C. Thompson. 1984. *Theory of correspondences: Including applications to mathematical economics*. Canadian Mathematical Society Series of Monographs and Advanced Texts. New York: John Wiley and Sons.

210 I. Kluvánek and G. Knowles. 1976. *Vector measures and control systems.* Mathematics Studies, 20. Amsterdam/New York: North-Holland/American Elsevier.

211 B. Knaster. 1928. Une théorème sur les fonctions d'ensembles. *Annales de la Societé Polonaise de Mathématique* 6:133–134.

212 B. Knaster, K. Kuratowski, and S. Mazurkiewicz. 1929. Ein Beweis des Fixpunktsatzes für n-dimensionale simplexe. *Fundamenta Mathematicae* 14:132–137.

213 A. N. Kolmogorov. 1956. *Foundations of the theory of probability.* New York: Chelsea. Translated from the 1933 German edition by N. Morrison.

214 G. Köthe. 1969. *Topological vector spaces.* Grundlehren der mathematischen Wissenschaften, 159. Berlin: Springer–Verlag. Translated from the German by D. J. H. Garling.

215 M. G. Krein and D. Milman. 1940. On extreme points of regular convex sets. *Studia Mathematica* 9:133–138.

216 M. G. Krein and V. L. Šmulian. 1940. On regularly convex sets in the space conjugate to a Banach space. *Annals of Mathematics* 41:556–583.

217 U. Krengel. 1985. *Ergodic theorems.* de Gruyter Studies in Mathematics, 6. Berlin: Walter de Gruyter.

218 K. Kuratowski. 1966–68. *Topology.* New York: Academic Press. 2 vols.

219 ———. 1972. *Introduction to set theory and topology,* revised 2d. English ed. International Series of Monographs in Pure and Applied Mathematics, 101. Warsaw: Pergamon Press.

220 K. Kuratowski and C. Ryll-Nardzewski. 1965. A general theorem on selectors. *Bulletin de l'Académie Polonaise des Sciences; Serie des Sciences Mathématiques, Astronomiques et Physiques* 13:397–403.

221 E. Landau. 1960. *Foundations of analysis.* New York: Chelsea. Translation of *Grundlagen der Analysis,* published in 1930.

222 J. P. LaSalle. 1960. The time optimal control problem. In *Contributions to the Theory of Nonlinear Oscillations,* volume 5, pages 1–24. Princeton, New Jersey: Princeton University Press.

223 A. Lasota and M. C. Mackey. 1994. *Chaos, fractals, and noise,* 2d. ed. New York: Springer–Verlag. Second edition of *Probabilistic Properties of Deterministic Systems,* published by Cambridge University Press, 1985.

224 L. M. Le Cam, J. Neyman, and E. L. Scott, eds. 1972. *Proceedings of the sixth Berkeley symposium on mathematical statistics and probability,* volume 2. Berkeley and Los Angeles: University of California Press.

225 H. Lebesgue. 1902. Intégrale, longueur, aire. *Annali di Matematica Pura ed Applicata. Serie 3* 7:231–359.

226 N. Levinson. 1966. Minimax, Liapunov, and 'bang-bang'. *Journal of Differential Equations* 2:218–241.

227 J. Lindenstrauss. 1966. A short proof of Liapounoff's convexity theorem. *Journal of Mathematics and Mechanics* 15:971–972.

228 P. A. Loeb and E. Talvila. 2004. Lusin's theorem and Bochner integration. *Scientiae Mathematicae Japonicae* 60:113–120.

229 V. I. Lomonosov. 2000. A counterexample to the Bishop–Phelps theorem in complex spaces. *Israel Journal of Mathematics* 115:25–28.

230 N. Lusin. 1930. *Mathematica, IV* 54.

231 ———. 1930. *Leçons sur les ensembles analytiques: et leurs applications.* Paris: Gauthier–Villars. In French.

232 W. A. J. Luxemburg. 1962. Two applications of the method of construction by ultrapowers in analysis. *Bulletin of the American Mathematical Society* 68:416–419.

233 ———. 1991. Integration with respect to finitely additive measures. In Aliprantis et al. [8], pages 109–150.

234 W. A. J. Luxemburg and A. C. Zaanen. 1963. Notes on Banach function spaces, VII. *Koninklijke Nederlandse Akademie van Wetenschappen. Proceedings. Series A* 66:669–681.

235 ———. 1971. *Riesz spaces I*. Amsterdam: North Holland.

236 A. A. Lyapunov. 1940. Sur les fonctions vecteurs complètement additives. *Izvestija Akademija Nauk SSR. Seria Matematičeskaja*. 4:465–478. In Russian.

237 G. W. Mackey. 1946. On convex topological linear spaces. *Transactions of the American Mathematical Society* 60:519–537.

238 S. MacLane and G. Birkhoff. 1993. *Algebra*, 3d. ed. New York: Chelsea.

239 R. Mañé. 1987. *Ergodic theory and differentiable dynamics*. Ergebnisse der Mathematik und ihrer Grenzgebiete, 3.8. Berlin: Springer–Verlag. Translated by Silvio Levy.

240 A. Mas-Colell. 1974. Continuous and smooth consumers: Approximation theorems. *Journal of Economic Theory* 8:305–336.

241 ———. 1975. A model of equilibrium with differentiated commodities. *Journal of Mathematical Economics* 2:263–295.

242 ———. 1984. On a theorem of Schmeidler. *Journal of Mathematical Economics* 13:210–206.

243 ———. 1986. The price equilibrium existence problem in topological vector lattices. *Econometrica* 54:1039–1054.

244 A. Mas-Colell, M. D. Whinston, and J. R. Green. 1995. *Microeconomic theory*. Oxford: Oxford University Press.

245 G. Mehta and E. Tarafdar. 1987. Infinite-dimensional Gale–Nikaidô–Debreu theorem and a fixed-point theorem of Tarafdar. *Journal of Economic Theory* 41:333–339.

246 P.-A. Meyer. 1966. *Probability and potentials*. Waltham, Massachusetts: Blaisdell.

247 P. Meyer-Nieberg. 1991. *Banach lattices*. Berlin: Springer–Verlag.

248 E. Michael. 1951. Topologies on spaces of subsets. *Transactions of the American Mathematical Society* 71:152–182.

249 ———. 1956. Continuous selections I. *Annals of Mathematics* 63:361–382.

250 A. W. Miller. 1995. *Descriptive set theory and forcing: How to prove theorems about Borel sets the hard way*. Lecture Notes in Logic, 4. Berlin: Springer–Verlag.

251 G. H. Moore. 1982. *Zermelo's axiom of choice: Its origins, development, and influence*. Studies in the History of Mathematics and Physical Sciences, 8. New York: Springer–Verlag.

252 J. C. Moore. 1968. A note on point-set mappings. In J. P. Quirk and A. M. Zarley, eds., *Papers in Quantitative Economics, 1*, pages 129–137. Lawrence, Kansas: University of Kansas Press.

253 ———. 1999. *Mathematical methods for economic theory*. Studies in Economic Theory. New York: Springer–Verlag.

254 B. S. Mordukhovich. 2006. *Variational analysis and generalized differentiation*. Grundlehren der mathematischen Wissenschaften, 330–331. Berlin: Springer–Verlag. Two volumes.

255 A. Müller and D. Stoyan. 2002. *Comparison methods for stochastic models and risks*. Wiley Series in Probability and Statistics. Chichester, England: John Wiley & Sons, Ltd.

256 J. R. Munkres. 1975. *Topology: A first course*. Englewood Cliffs, New Jersey: Prentice–Hall.

257 L. Nachbin. 1976. *Topology and order*. New York: Krieger. Reprint of 1965 edition published by van Nostrand.

258 D. C. Nachman. 1988. Stochastic equilibria. *Journal of Mathematical Economics* 17:69–75.

259 S. B. Nadler. 1969. Multivalued contraction mappings. *Pacific Journal of Mathematics* 30:475–488.

260 L. W. Neustadt. 1963. The existence of optimal control in the absence of convexity. *Journal of Mathematical Analysis and Applications* 7:110–117.

261 J. Neveu. 1965. *Mathematical foundations of the calculus of probability*. Holden–Day Series in Probability and Statistics. San Francisco: Holden–Day.

262 H. Nikaidô. 1968. *Convex structures and economic theory*. Mathematics in Science and Engineering. New York: Academic Press.

263 C. D. Olds. 1963. *Continued fractions*. New York: Random House.

264 C. Olech. 1966. Extremal solutions of a control system. *Journal of Differential Equations* 2:74–101.

265 ———. 1967. Lexicographical order, range of integrals and "bang-bang" principle. In A. V. Balakrishnan and L. W. Neustadt, eds., *Mathematical Theory of Control*, pages 35–45. New York: Academic Press. Proceedings of a conference held at the University of Southern California, Los Angeles, January 30-February 1, 1967.

266 ———. 1974. The characterization of the weak∗ closure of certain sets of integrable functions. *SIAM Journal on Control* 12:311–318.

267 D. S. Ornstein. 1974. *Ergodic theory, randomness, and dynamical systems*. Yale Mathematical Monographs, 5. New Haven: Yale University Press.

268 F. H. Page, Jr. 1987. The existence of optimal contracts in the principal agent problem. *Journal of Mathematical Economics* 16:157–167.

269 S. Park. 1994. A unified approach to generalizations of the KKM-type theorems related to acyclic maps. *Numerical Functional Analysis and Optimization* 15:105–119.

270 W. Parry. 1981. *Topics in ergodic theory*. Cambridge Tracts in Mathematics, 75. Cambridge: Cambridge University Press.

271 K. R. Parthasarathy. 1967. *Probability measures on metric spaces*. Probability and Mathematical Statistics. New York: Academic Press.

272 J. E. Pečarić, F. Proschan, and Y. L. Tong. 1992. *Convex functions, partial orderings, and statistical applications*. Mathematics in Science and Engineering, 187. New York: Academic Press.

273 A. L. Peressinni. 1967. *Ordered topological vector spaces*. New York & London: Harper & Row.

274 K. E. Petersen. 1983. *Ergodic theory*. Cambridge Studies in Advanced Mathematics, 2. Cambridge and New York: Cambridge University Press.

275 B. J. Pettis. 1938. On integration in vector spaces. *Transactions of the American Mathematical Society* 44:277–304.

276 J. Pfanzagl and W. Pierlo. 1966. *Compact systems of sets*. Lecture Notes in Mathematics, 16. Heidelberg: Springer–Verlag.

277 R. R. Phelps. 1966. *Lectures on Choquet's theorem*. Van Nostrand Mathematical Studies, 7. New York: Van Nostrand.

278 ———. 1993. *Convex functions, monotone operators and differentiability*, 2d. ed. Lecture Notes in Mathematics, 1364. Berlin: Springer–Verlag.

279 R. S. Phillips. 1940. On linear transformations. *Transactions of the American Mathematical Society* 48:516–541.

280 D. Pollard. 1984. *Convergence of stochastic processes*. Springer Series in Statistics. Berlin: Springer–Verlag.

281 I. A. Polyrakis. 1994. Lattice-subspaces of $C[0, 1]$ and positive bases. *Journal of Mathematical Analysis and Applications* 184:1–18.

282 ———. 1996. Finite-dimensional lattice-subspaces of $C(\Omega)$ and curves of $\mathbb{R}^n$. *Transactions of the American Mathematical Society* 348:2793–2810.

283 M. K. Richter. 1966. Revealed preference theory. *Econometrica* 34:635–645.

284 A. W. Roberts and D. E. Varberg. 1973. *Convex functions*. New York: Academic Press.

285 ———. 1974. Another proof that convex functions are locally Lipschitz. *American Mathematical Monthly* 81:1014–1016.

286 J. W. Roberts. 1977. A compact convex set with no extreme points. *Studia Mathematica* 60:255–266.

287 A. P. Robertson and W. J. Robertson. 1973. *Topological vector spaces*, 2d. ed. Cambridge Tracts in Mathematics, 53. London: Cambridge University Press.

288 R. T. Rockafellar. 1970. *Convex analysis*. Princeton Mathematical Series, 28. Princeton, NJ: Princeton University Press.

289 M. Rothschild and J. E. Stiglitz. 1970. Increasing risk I: A definition. *Journal of Economic Theory* 2:225–243.

290 H. L. Royden. 1988. *Real analysis*, 3d. ed. New York: Macmillan.

291 W. Rudin. 1966. *Real and complex analysis*. New York: McGraw Hill.

678   References

292 ———. 1976. *Principles of mathematical analysis*, 3d. ed. International Series in Pure and Applied Mathematics. New York: McGraw Hill.

293 H. H. Schaefer. 1971. *Topological vector spaces*. Graduate Texts in Mathematics, 3. New York: Springer–Verlag. 3rd. corrected printing.

294 ———. 1974. *Banach lattices and positive operators*. Berlin: Springer–Verlag.

295 M. Schäl. 1975. On dynamic programming: Compactness of the space of policies. *Stochastic Processes and Their Applications* 3:345–364.

296 J. Schauder. 1927. Bemerkung zu meiner "Zur Theorie stetiger Abbildung in Funktionalräumen". *Mathematische Zeitschrift* 26:417–431.

297 ———. 1927. Zur Theorie stetiger Abbildung in Funktionalräumen. *Mathematische Zeitschrift* 26:47–65.

298 ———. 1930. Der Finxpunksatz in Funktionalräumen. *Studia Mathematica* 2:171–180.

299 D. Schmeidler. 1972. On set correspondences into uniformly convex Banach spaces. *Proceedings of the American Mathematical Society* 34:97–101.

300 ———. 1986. Integral representation without additivity. *Proceedings of the American Mathematical Society* 97:255–261.

301 Z. Semadeni. 1971. *Banach spaces of continuous functions*, volume 1. Monografie Matematyczne, 55. Warsaw: Polish Scientific Publishers.

302 L. S. Shapley and R. Vohra. 1991. On Kakutani's fixed point theorem, the K–K–M–S theorem and the core of a balanced game. *Economic Theory* 1:107–116.

303 A. Shen and N. K. Vereshchagin. 2002. *Basic set theory*, volume 17. Student Mathematical Library. Providence RI: American Mathematical Society.

304 K. Sherstyuk. 1998. How to gerrymander, a formal analysis. *Public Choice* 95:27–49.

305 W. Sierpiński. 1920. Sur un problème concernant les ensembles mesurables superficiellement. *Fundamenta Mathematicae* 1:112–115.

306 ———. 1928. Un théorème général sur les familles d'ensembles. *Fundamenta Mathematicae* 12:206–210.

307 R. C. Sine. 1968. Geometric theory of a single Markov operator. *Pacific Journal of Mathematics* 27:155–166.

308 ———, ed. 1983. *Fixed points and nonexpansive mappings*. Contemporary Mathematics, 18. Providence, Rhode Island: American Mathematical Society.

309 I. Singer. 1997. *Abstract convex analysis*. Canadian Mathematical Society Series of Monographs and Advanced Texts. New York: John Wiley and Sons.

310 M. Sion. 1958. On general minimax theorems. *Pacific Journal of Mathematics* 8:171–176.

311 ———. 1960. On analytic sets in topological spaces. *Transactions of the American Mathematical Society* 96:341–354.

312 ———. 1960. On uniformization of sets in topological spaces. *Transactions of the American Mathematical Society* 96:237–245.

313 D. R. Smart. 1974. *Fixed point theorems*. Cambridge Tracts in Mathematics, 66. Cambridge: Cambridge University Press.

314 R. E. Smithson. 1971. Fixed points of order preserving multifunctions. *Proceedings of the American Mathematical Society* 28:304–310.

315 V. L. Šmulian. 1940. Über lineare topologische räume. *Matematičii Sbornik. (N. S.)* 7:425–448.

316 R. M. Solovay. 1970. A model of set theory in which every set is Lebesgue measurable. *Annals of Mathematics* 92:1–56.

317 H. F. Sonnenschein. 1971. Demand theory without transitive preferences, with applications to the theory of competitive equilibrium. In J. S. Chipman, L. Hurwicz, M. K. Richter, and H. F. Sonnenschein, eds., *Preferences, Utility, and Demand: A Minnesota Symposium*, chapter 10, pages 215–223. New York: Harcourt, Brace, Jovanovich.

318 E. Sperner. 1928. Neuer beweis für die invarianz der dimensionszahl und des gebietes. *Abhandlungen aus dem Mathematischen Seminar der Hamburgische Universitat* 6:265–272.

319 S. M. Srivastava. 1998. *A course on Borel sets*. Graduate Texts in Mathematics, 180. New York: Springer–Verlag.

320 M. B. Stinchcombe and H. White. 1992. Some measurability results for extrema of random functions over random sets. *Review of Economic Studies* 59:495–512.

321 J. Stoer and C. Witzgall. 1970. *Convexity and optimization in finite dimensions I.* Die Grundlehren der mathematischen Wissenschaften in Einzeldarstellungen, 163. Berlin: Springer–Verlag.

322 N. Stokey, R. E. Lucas, Jr., and E. C. Prescott. 1989. *Recursive methods in economic dynamics.* Cambridge, Mass.: Harvard University Press.

323 V. Strassen. 1964. Meßfehler und Information. *Zeitschrift für Wahrscheinlichkeitstheorie und Verwandte Gebiete* 2:273–305.

324 ———. 1965. The existence of probability measures with given marginals. *Annals of Mathematical Statistics* 36:423–439.

325 S. Straszewicz. 1935. Über exponierte Punkte abgeschlossener Punktmengen. *Fundamenta Mathematicae* 24:139–143.

326 K. D. Stroyan and W. A. J. Luxemburg. 1976. *Introduction to the theory of infinitesimals.* Pure and Applied Mathematics, 72. New York: Academic Press.

327 E. Szpilrajn. 1930. Sur l'extension de l'ordre partiel. *Fundamenta Mathematicae* 16:386–389.

328 E. Tarafdar. 1977. On nonlinear variational inequalities. *Proceedings of the American Mathematical Society* 67:95–98.

329 A. Tarski. 1955. A lattice-theoretical fixpoint theorem and its applications. *Pacific Journal of Mathematics* 5:285–309.

330 A. E. Taylor and D. C. Lay. 1980. *Introduction to functional analysis*, 2d. ed. New York: John Wiley and Sons.

331 D. M. Topkis. 1998. *Supermodularity and complementarity.* Princeton: Princeton University Press.

332 J. W. Tukey. 1942. Some notes on the separation of convex sets. *Portugaliae Mathematicae* 3:95–102.

333 A. Tychonoff. 1935. Ein Fixpunktsatz. *Mathematische Annalen* 111:767–776.

334 J. J. Uhl, Jr. 1969. The range of a vector-valued measure. *Proceedings of the American Mathematical Society* 23:158–163.

335 K. Vind. 1973. A third remark on the core of an atomless economy. *Econometrica* 40:585–586.

336 X. Vives. 1990. Nash equilibrium with strategic complementarities. *Journal of Mathematical Economics* 19:305–321.

337 J. von Neumann. 1949. On rings of operators: Reduction theory. *Annals of Mathematics* 50:401–485.

338 R. C. Walker. 1974. *The Stone–Čech compactification.* Ergebnisse der Mathematik und ihrer Grenzgebiete, 83. Berlin: Springer–Verlag.

339 P. Walters. 1982. *An introduction to ergodic theory.* Graduate Texts in Mathematics, 79. New York: Springer–Verlag. Previously published as *Ergodic Theory: Introductory Lectures* by Springer–Verlag.

340 A. Wilansky. 1967. Between T_1 and T_2. *American Mathematical Monthly* 74:261–266.

341 ———. 1978. *Modern methods in topological vector spaces.* New York: McGraw Hill.

342 S. Willard. 1970. *General topology.* Reading, Massachusetts: Addison Wesley.

343 N. C. Yannelis. 1988. Fatou's lemma in infinite dimensional spaces. *Proceedings of the American Mathematical Society* 102:303–310.

344 ———. 1991. Integration of Banach-valued correspondences. In Khan and Yannelis [201], pages 2–35.

345 ———. 1991. Set-valued functions of two variables in economic theory. In Khan and Yannelis [201], pages 36–72.

346 K. Yosida and E. Hewitt. 1952. Finitely additive measures. *Transactions of the American Mathematical Society* 72:46–66.

347 A. C. Zaanen. 1983. *Riesz spaces II*. Amsterdam: North Holland.

348 L. Zhou. 1994. The set of Nash equilibria of a supermodular game is a complete lattice. *Games and Economic Behavior* 7:295–300.

349 G. M. Ziegler. 1995. *Lectures on polytopes*. Graduate Texts in Mathematics, 152. New York: Springer–Verlag.

350 M. Zorn. 1935. A remark on method in transfinite algebra. *Bulletin of the American Mathematical Society* 41:667–670.

Index

Miscellaneous Notation

$\mathbf{1}$, constant function one, 54

$A^\odot$, one-sided polar, 215

A^O, absolute polar, 215

$A^\bullet$, absolute polar with respect to the algebraic dual, 221

$A^\perp$, orthogonal complement of A, 250

$A^{\perp\perp}$, the double orthogonal complement of A, 250

$\langle x, x' \rangle$, duality function, 211

∂A, the boundary of A, 27

$\geqslant_D$, order induced by cone D, 646

$[f = \alpha]$, etc. $(= \{x : f(x) = \alpha\})$, 197

$\|\cdot\|_\infty$, ∞-norm, 462, 527

$\|\cdot\|_p$, p-norm, 461, 527

$\|\cdot\|$, norm, 167

$\oplus$, direct sum, 210

$\otimes$, product of σ-algebras, 148

Y^X, set of functions from X to Y, 5

A^ℓ, $= \{B : A \cap B \neq \varnothing\}$, 119

φ^ℓ, lower inverse of φ, 557

order theoretic notation

$\quad |x|$, absolute value of x, 313

$\quad A \uparrow$, $A \downarrow$, etc., 316

$\quad x_\alpha \downarrow$, $x_\alpha \uparrow$, etc., 315

$\quad x \gg 0$, strictly positive vector, 341

$\quad [x, y]$, order interval, 8

$\quad \vee$, maximum or supremum, 8, 9, 312

$\quad \wedge$, minimum or infimum, 8, 9, 312

$\quad x^-$, negative part of x, 313

$\quad E^+$, E_+, positive cone of E, 313

$\quad x^+$, positive part of x, 313

set theoretic notation

$\quad 2^X$, power set of X, 3

$\quad \varnothing$, the empty set, 2

$\quad A^c$, complement of A, 2

$\quad A \setminus B$, complement of B in A, 2

$\quad \triangle$, symmetric difference operator, 3

$\quad \chi_A$, indicator function of A, 5

$\quad A \subset B$, subset notation, 2

$\quad A \subsetneq B$, proper subset notation, 3

$\quad A \supset B$, superset notation, 2

$\quad \cap$, set intersection, 3

$\quad \cup$, set union, 3

$\quad A_n \uparrow A$, $A_n \downarrow A$ monotone sequences of sets, 136

$\quad [x]$, equivalence class of x, 7

$\quad f|_A$, restriction of f to A, 4

$\quad A \times B$, Cartesian product, 3

$\quad A^u$, $= \{B : B \subset A\}$, 119

$\quad \varphi^u$, upper inverse of φ, 557

A

$\mathcal{A}$ operation, 106

$\mathcal{A}_\mathbb{R}$, 145

$\mathcal{A}_X$, algebra generated by open subsets of X, 434

a.e. (= almost everywhere), 387–389, 468

$\quad$ equality of sets, 657

a.s. (= almost surely), 387

Abramovich, Y. A., 365, 445, 541, 663n, 667

absolute topologies, 338, 341

absolutely continuity of measures, 401

absorbing set, 168

$\quad$ with empty interior, 201

accumulation point, 27

Adams, D., ix, 667

adjoint, 243–244

affine function, 256

$\quad$ and convex function, 257

affine subspace, 277

Akilov, G. P., 373n, 674

AL-space, 357–361

Alaoglu, L., 218, 235, 667

Alaoglu Compactness Theorem, 218, 235

Alexandroff, A. D., 88, 275, 667

Alexandroff one-point compactification, 57

Alexandroff's Lemma on complete metrizability, 88

Alexandroff's Theorem on convex functions, 275

algebra, see also σ-algebra

$\quad$ of functions, 352

$\quad$ of sets, 129–134

algebraic dual of a vector space, 195

Aliprantis, C. D., xxii, 1, 252, 269, 275, 311, 332, 365, 372, 379, 434, 663n, 667

AM-space, 357
   as $C(K)$, 359
   has the Dunford–Pettis Property, 361
   with unit, 358
analytic measurability, 606
analytic set, 446, 599
annihilate, annihilator, 199, 219
antiderivative, 422
Apostol, T. M., xxii, 667
Archimedean Riesz space, 316, 318
Archimedes, 403
Arens, R., 221, 667
argmax correspondence, 570, 605
Armstrong, T. E., 478, 667
Arrow, K. J., 579, 667
atom, of a measure, 395, 476
Aubin, J.-P., 164, 556, 615, 667, 668
Aumann, R. J., 156, 614, 668
Aumann integral, 614
Axiom of Choice, 13

B

$B_b(X)$, bounded Borel measurable functions,
   144, 490–491, 626
$B_\varepsilon(x)$, ε-ball at x, 23
$B(X)$, bounded real functions on X, 74, 331, 646
   as a Banach lattice, 348
ba, charges of bounded variation, 314, 337, 396
   on $\mathbb{N}$, 538, 543–546
ba_n, normal signed charges, 440
ba_r, regular signed charges, 440
Back, K., 290, 668
$\mathcal{B}aire$, $\mathcal{B}aire^*$, 158
Baire category (first, second), 93
Baire Category Theorem, 94
Baire measure, 434
Baire property, of a space, 93
Baire sets, 158
   compact, 160
   of a product, 161
   open, 159
   vs. Borel sets, 160
Baire σ-algebra, see Baire sets
Baire space, $\mathcal{N} = \mathbb{N}^{\mathbb{N}}$, the, 101
balanced set, 168
balayage, 646
ball, 23, 71
   unit, 228
Banach, S., 14, 373, 373n, 668
Banach Fixed Point Theorem, 95, 586
Banach lattice, 348
   AL-space, 357
   AM-space, 357

   dual of, 350
   equivalent norms on, 352
   examples of, 348
   order continuity of norm, 355
   order dual, 352
   reflexive, 356
   strictly positive functionals on, 357
   vs. Riesz space, 349
   weak compactness of boxes, 356
   with strictly monotone norm, 364
Banach–Mazur limit, 550
Banach space, 228, see also Banach lattice
   continuity of evaluation duality, 243
   double dual, 231
   Dunford–Pettis Property, 361
   examples of, 228
   finite dimensional, 361
   norm dual of, 230
   reflexive, 237, 361
   separable, 240
   weak compactness in, 240
Banach–Tarski Paradox, 14
band (in a vector lattice), 324
   principal, 324
   projection, 325
Bartle, R. G., 590, 668
base, for a topology, 25
   for a uniformity, 109
basis (Hamel), 15
   positive, 530
   vs. Schauder, 530
Bauer Maximum Principle, 298, 300
Beer, G. A., 109, 117, 668
Bell, E. T., 1n, 668
Beppo Levi Theorem (Levi's Theorem), 413
Berge, C., 556, 566, 569, 668
Berge Maximum Theorem, 570
Bergman, G. M., 581, 672
Berliant, M., 253, 371, 668
Bertsekas, D. P., 106, 135n, 433, 668
βX, Stone–Čech compactification of X, 59
Bewley, T. F., 165, 668
Bhaskara Rao, K. P. S., 372, 668
Bhaskara Rao, M., 372, 668
bijection, 5
Billingsley, P., 372, 373n, 434, 435n, 668
binary relation, see relation
bipolar, 215
Bipolar Theorem, 217
Birkhoff, G., 132n, 676, ($\ne$ G. D. Birkhoff)
Birkhoff, G. D., 659, 668, ($\ne$ G. Birkhoff)
Birkhoff's Ergodic Theorem, 659
Bishop, E., 252, 281, 295, 668

Bishop–Phelps Theorem, 284
 may fail in complex Banach space, 284
Blackwell, D., 97, 646, 668
Blackwell's Theorem, 97
Blume, L. E., 622, 638, 639, 668
Blumenthal, R. M., 135n, 642, 645, 669
Bochner, S., 520, 669
Bochner integral, 426
Bohnenblust, H. F., 359, 583, 669
Bohnenblust–Karlin Fixed Point Theorem, 584
Border, K. C., 253, 556, 581, 667, 669
Borel charge, measure, 393, 434, 435
Borel function, 139
Borel measurable correspondence, 592
Borel sets, 137
 of a product, 149
 vs. Baire sets, 160
Borel σ-algebra, 137, see also Borel sets
Borel space, 516
boundary, 27
bounded operator, 229
bounded set, 71, 234
 τ-bounded, 186, 206, 214
Bourbaki, N., 4n, 45n, 215n, 648, 669
box, in an order vector space, 315
Brøndsted, A., 252, 265, 287, 669
Brøndsted–Rockafellar Theorem, 287
Brooks, J. K., 379, 669
Brouwer, L. E. J., 583, 669
Brouwer Fixed Point Theorem, 583
Brouwer–Schauder–Tychonoff Fixed Point
 Theorem, 583
Browder, F. E., 580, 581, 584, 587, 669
Browder Selection Theorem, 587
Browder–Fan Coincidence Theorem, 584
Brown, D. J., 311, 667
Brumelle, S. L., 253, 669
Burkinshaw, O., xxii, 1, 311, 332, 372, 379, 434,
 667
BV_0, 366
 and ca, 402
 BV_0^ℓ, BV_0^r, 369

C

$C(X)$, continuous real functions on X, 49
 as a Riesz space, 337
 separability of, 353
$C[0, 1]$, continuous real functions on $[0, 1]$, 49
 as a normed Riesz space, 348
 dual pair with $ca[0, 1]$, 211
 has no σ-order continuous functional, 329
 has order unit, 325
 has strictly positive linear functional, 326

 has the countable sup property, 326
 is Archimedean, not order complete, 317
 is not an ideal in $\mathbb{R}^{[0,1]}$, 321
 Riesz subspace of $B[0, 1]$, 321
C_b, bounded continuous functions, 49
 dual space of, 495
 positive functionals on, 491
C_c, continuous functions with compact support,
 49
 and Baire sets, 158
 dual space of, 497
 positive functionals on (Radon measures),
 496
C^∞-function, 466
C^* ($= C_b$), 49n
$C(X, Y)$, continuous functions from X to Y, 123
 Borel σ-algebra of, 154
 completeness of, 124
 equivalent metrics, 124
 is not compact, 126
 measurability and, 155
 metrizability of, 123
 separability of, 125
$\mathfrak{c}$, cardinality of the continuum, 12
c, space of convergent sequences, 527, 531–533
c_0, space of sequences converging to zero, 527,
 529–531
c_0-sum of Banach spaces, 553
ca, signed measures of bounded variation, 399
 on $[0, 1]$, 329
ca_n, normal signed measures, 440
ca_r, regular signed measures, 440
Cantor, G., 10, 11, 98
Cantor–Bernstein Theorem, 10
Cantor Diagonal Theorem, 12
Cantor Intersection Theorem, 75
Cantor set, 98
capacity, 457
 "nice", 457
 Choquet Capacity Theorem, 459
Carathéodory, C., 382, 404, 469, 669
Carathéodory Convexity Theorem, 184
Carathéodory Extension Theorem, 382
Carathéodory function, 153, 595
 and $C(X, Y)$, 155
 linear, 609, 617
 sublinear, 609
cardinality, 10
 measurable cardinal, 372
carrier of a measure, 442
Cartesian product, 3
Castaing, C., 215n, 292, 556, 577, 598, 601,
 611, 613, 615, 669, 670
category, Baire (first, second), 93

Cauchy, A.-L., 403
Cauchy net, filter in tvs, 174
Cauchy sequence, 73
Cauchy–Schwarz inequality, 246
Cellina, A., 556, 615, 667
Cephalonia, 150
Cesàro bounded operator, 664
Cesàro mean, 661
Chacon, R. V., 379, 669
chain, 7, 15
Change of Variables Theorem, 484, 486
characteristic (= indicator) function, 5
charge, 374
    absolute continuity of, 401
    Borel, 434
    bounded variation, 396
    inner regular, 435
    normal, 435
    outer regular, 434
    purely finitely additive, 400
    regular Borel, 435
    set function, 374
    signed, 314, 374
    tight, 435
    Yosida–Hewitt decomposition, 400
    zero-one, 544
Choquet, G., 179n, 190, 215n, 434, 670
Choquet Capacity Theorem, 459
Choquet theory, 647
circled set, 168
clopen set, 23
closed convergence of sets, 121
closed convex circled hull, 183
closed convex hull, 183
    compactness of, 185, 241
    of extreme points, 297
closed function, 41
Closed Graph Theorem, 51, 177, 561
closed limit of sets, 114
closed mapping, 560
closed set, 23
    relatively, 25
closed unit ball, 228, see also unit ball
    is w^*-compact, 235
    metrizability of, 239, 240
    weak compactness of, 237
    weak interior of, 238
closure, 26
    point of, 27
closure correspondence, 566
cluster point, 27
coanalytic set, 446
codimension, 220
codomain, 4

cofinite set, 33
Cohen, P. J., 373, 670
Coincidence Theorem, 584
commutative diagram, 5
compact class of sets, 378, 520
compact space, 38
    is normal, 46
compact subsets of Euclidean space, 87
compactification, 56
    Alexandroff one-point, 57
        metrizability of, 92, 93
    of Ω_0, 62
    of a separable metrizable space, 91
    Stone–Čech, 59
compactness of metric spaces, 86
complement (of a vector subspace), 210, 245
    orthogonal, 250
complement, set theoretic, 2
complete lattice, 8
complete measure, 387
complete metric space, 73
complete topology, tvs, 174
completely metrizable tvs, 174
completely regular topological space, 45
    is uniformizable, 109
    Stone–Čech compactification of, 59
completion
    Dedekind, 319
    of a measure, 387
    of a metric space, 84
    of a normed space, 232
    of a tvs, 175
component (vector lattice), 468n
    order continuous, singular, 329
composition of relations, 5
    of correspondences, 566
    of functions, 5
composition operator, 502, 651
concave function, 187, see also convex function
    extended-valued, 254
    proper, 254
    strictly, 187
conditional probability, 634
cone, 179, 190, 209, 213, 312–313
    and separating hyperplane, 199
    convex, 179
    generated by a finite set, 179
    generated by a set, 179
    nonnegative, 313
    open, 268
    pointed, 179, 312
    polar, 215n
    positive, 313
    vertex of, 268

conjugate exponents, 463, 534
connected space, 645
continuity, 36
    characterizations of, 36
    joint, 51
    norm vs. weak, 233
    of a binary relation, 44
    of a correspondence, 558
    of a limit, 54
    of a limit function, 233
    of a limit operator, 233
    of convex functions, 279
    of correspondences, 558
    of positive operators, 350
    of the evaluation, 241–243, 361, 511
    uniform, *see* uniformly continuous function
continuum, 12
Continuum Hypothesis, 372
contraction, 95
    contraction correspondence, 585
    contraction operator, 364
    modulus of, 95
Contraction Mapping Theorem, 95, 97, 586
convergence, 30
    in a metric space, 72
    in measure, 479
    in probability, 479
    of a filter, 34
    order, 322
        in L_p spaces, 323
    pointwise, 53, 212
    uniform, 54
convex circled hull, 183
convex combination, 181
convex cone, 179, 209, 213
convex function, 187, 271
    continuity of, 188, 279
    extended-valued, 254
    Lipschitz continuity of, 189
    proper, 254
    strictly, 187
        continuity of, 190
convex hull, 182
    Carathéodory Convexity Theorem, 184
    circled, 183
    closed, 183
    compactness of, 183, 185, 241
    Krein–Milman Theorem, 297
    of a correspondence, 571
    of a finite set (= polytope), 184, 301
convex set, 168, 181
    dimension of, 277
    extreme point of, 294
    face of, 294

space of convex sets, 292
correspondence, 4, 556
    analytically measurable, 606
    Berge Maximum Theorem, 570
    closed, or closed graph, 560
    continuity, 558
        and Hausdorff metric, 563
        demicontinuity, 575
        hemicontinuity, 558
        lower hemicontinuity vs. open sections,
            562
        upper hemicontinuity vs. closed graph,
            561
        upper hemicontinuity vs. upper
            demicontinuity, 576
    contraction, 585
    convex hull of, 571
        upper hemicontinuity, 573
    domain of, 556
    fixed point of, 581
    graph of, 4
    image under, 556
        of a compact set, 560
    integration, 614, 619
        Gelfand, 619
    inverse, 578
        lower, 557
        strong, 557
        upper, 557
        weak, 557
    inward pointing, 581
    KKM, 577
        and binary relations, 579
    Markov, 638
    measurability, 592
        Borel, 592, 599
        scalar, 610
        weak, 592
    open graph, 562
    open sections, 562
    operations on, 566–568, 571
    outward pointing, 581
    range of, 557
    singleton-valued, as a function, 559
    support correspondence for measures, 563
Corson, H. H., 642, 645, 669
countable set, 10
countable sup property, 326
    and strictly positive functional, 326
countably additive set function, 374
counting measure, 375, 534
Cousot, P., 670
Cousot, R., 670
cover of a set, 38

cylinder set, 520

D

Daley, D. J., 670
Daniell, P. J., 404, 410n, 670
Davies, A. C., 18, 670
Davies, R. O., 153, 670
Day, M. M., 670
de Morgan's laws, 3
Debreu, G., 165, 597, 615, 670
decomposable set of functions, 614
Dedekind complete space, 316
Dedekind completion, 319
DeLeeuw, K., 295, 668
Dellacherie, C., 433, 670
δ_x, point mass at x, 443, 512
$\delta^*(x \mid C)$, support functional, 288n
demicontinuity, 575
Denardo, E. V., 95, 670
dense set, 28
density of a measure, 470
derivative
    directional, 266
    Fréchet, 274
    Gâteaux, 267, 268
    Gâteaux vs. Fréchet, 273
    of a convex function, 272
    of a correspondence, 275
    Radon–Nikodym, 470
derived set, 27
Devlin, K. J., 3, 4n, 13, 670
Diagonal Theorem, 12
diagonal, of a Cartesian product, 3
diameter of a set, 71
Diamond, P., 586, 670
Diemling, K., 670
Diestel, J., 423, 427, 432, 478, 670
Dieudonné, J., xxii, 670
Dieudonné's Theorem, 379
difference operator, 394
differential (= Fréchet derivative), 274
dilation, 645, 647, 649
Dini's Theorem, 54
Dirac measure, 443n
direct sum, 210
    in Hilbert space, 250
    of Banach spaces, 532
directed set, direction, 29
    in a Riesz space, 316
    product, 29
directional derivative, 266
discrete metric, 24
discrete topology, 23

disjoint vectors, in a Riesz space, 320, 546
disjoint complement of a subspace, 324
dispersion (= dilation), 647
distance, 70
distance function, 80
    of a correspondence, 595
dom f, effective domain of f, 254
domain, 4
Dominated Convergence Theorem, 415, 427
domination, 9, 312
Doob, J. L., 372, 670
dot product, 526
double dual, 224, 231
double order dual, 331
Dravecký, J., 153, 670
dual, see also dual space
    algebraic, 195
    double, 231
    double order, 331
    of a normed Riesz space, 350
    order, 327
    order continuous, 329
    σ-order continuous, 329
    topological, 195
dual norm, 230
dual pair, 211
    Riesz pair, 340
dual space, see also dual
    of selected Banach lattices, 499
    of c, 532
    of c_0, 530
    of ℓ_∞, 542
    of ℓ_p, 535
    of $L_1(\mu)$, 473
    of $L_p(\mu)$, 473
    of $X \times \mathbb{R}$, 256
    of φ, 528
    of $\mathbb{R}^\mathbb{N}$, 528
    representation of
        by dot product, 527
    separates points, 208
duality, 9, 211
Dubins, L. E., 477, 597, 605, 670, 671
Dubins–Spanier Theorem, 478
Dudley, R. M., 143, 158n, 372, 373n, 435, 671
Duffie, D., 622, 638, 641, 671
Dugundji, J., 22, 66, 581, 671, 672
Dunford, N., 163, 191n, 258n, 379, 428, 432n, 498, 671
Dunford integral, 431
Dunford–Pettis Property, 361
dynamical system, 652
Dynkin, E. B., 135n, 671
Dynkin system, 135

Dynkin's Lemma, 136

E

$E^\sim$, order dual of E, 327
$E_c^\sim$, σ-order continuous dual of E, 329
$E_n^\sim$, order continuous dual of E, 329
$E^{\sim\sim}$, double order dual of E, 331
E_x, principal ideal generated by x, 322
$\mathcal{E}(C)$, extreme points of C, 294
e, constant unit sequence, 526
e_k, kth unit coordinate vector, 526
Eberlein, W. F., 241, 671
Eberlein–Šmulian Theorem, 241
Echenique, F., 18, 671
effective domain of a convex function, 254
Effros, E. G., 597, 671
Egoroff's Theorem, 389
eigenvalue, eigenvector, eigenspace, 244, 663
Ekeland, I., 164, 254, 667, 671
embedding, 38, 166
    in space of functions, 84
    isometric, 84
    lattice, 546
    linear, 546
entourage, in a uniform space, 109
envelope, by a cone of functions, 648
envelope, convex (concave), 256
    and support functionals, 291–292
epi f, = epigraph, 8
epigraph, 8, 52, 187, 254
ε-dense set, 85
equivalence relation, 7
equivalent measures, 471
equivalent metrics, 71
equivalent norms, 178
μ-equivalent sets, 468
ergodic dynamical system, 653
ergodic measure, 636
ergodic set, 636
Ergodic Theorem, 659
ergodic transformation, 657
ess sup, essential supremum norm, 462
Euclidean metric, 24
Euclidean norm, 177
Euclidean topology, 24, 177
Eudoxus, 403
evaluation, 9, 53, 154, 641
    joint continuity of, 241–243, 361, 511
example
    analytic set that is not Borel, 453
    Archimedean property, 316
    Archimedean vs. order complete, 317
    Baire sets vs. Borel sets, 160

Bishop–Phelps Theorem vs. James'
    Theorem, 285
boundary point is not a support point, 259
$C(X, Y)$ not compact, 126
Carathéodory extension not unique, 386
closed convex hull not compact, 185
closed convex set with no support point, 260
closed correspondence with no Borel
    selector, 454
closed graph function that is discontinuous
    everywhere, 51
closed set whose projection is not Borel, 453
compact set not closed, 40
compact space not separable, 41
composition of closed maps not closed, 566
continuous functions on one-point
    compactification, 58
convex hull not compact, 185
demicontinuity need not imply
    hemicontinuity, 575
discontinuous positive operator, 351
disjoint unions vs. unions, 130
dual pairs, 211
Hausdorff metric, 112, 118
hemicontinuous correspondences, 559
inseparable disjoint closed convex sets, 203
internal vs. interior points in a convex set, 200
intersection of correspondences, 568
kinds of separation, 199
limit of a sequence of sets, 115
limit of measurable functions not measurable,
    143
locally convex-solid Riesz spaces, 337
locally solid but not locally convex topology,
    337
measure with no support, 442
measure with uncountably many extensions,
    386
net without compact tails, 42
no strictly positive functional, 326
non-Borel measurable correspondence, 599
non-Borel subset of $\mathcal{K}([0, 1])$, 599
non-regular Borel measure, 439
nonunique representation by Borel measures,
    497
normed Riesz spaces, 348
Ω is compact, 41
order continuity of the norm, 355
order vs. norm convergence, 323
ordering with no utility, 11
pointwise convergence, 54
Radon–Nikodym derivative does not exist,
    471
Riesz duals, 329

Riesz spaces, 313
$\mathbb{R}^N$ is not normable, 207
separable vs. second countable, 28
sequences cannot describe closure, 30
sum of closed sets, 168
sum of closed sets is dense, 203
sum of correspondences, 572
support points of the positive cone, 285
topological continuity on boxes, 340
topologies, 23
transition with no invariant measure, 632, 633
exponential topology, 119
exponents, conjugate, 463, 534
exposed point of convex set, 305
extended real numbers, 2
    extended real function, 8
    topology of, 57
extension, 4
    from an ideal, 341
    minimal, of a positive functional, 342
    norm preserving, 231
    of a measure, 382
    of a preorder, 15
    of positive functionals, 330
    of Riesz space-valued operators, 330
extremally disconnected space, 63, 363, 531n
extreme point, 294
extreme ray, 294
extreme subset, 294

F

$\mathcal{F}$, space of nonempty closed sets, 113
    $\mathcal{F}_d$, d-bounded members of $\mathcal{F}$, 113
F_A, flat generated by A, 277
∂f, subdifferential of a convex function, 264
$\mathcal{F}_\sigma$-set, 26, 81, 139
face of a convex set, 294, 517
family, = set, 2
Fan, K., 578, 581, 583, 584, 671
Fan Coincidence Theorem, 584
Fan Fixed Point Theorem, 583
Farkas, J., 671
Farkas' Lemma, 209
Fatou's Lemma, 414
Feinstein, J., 445
Fell, J. M. G., 121, 671
Fell topology, 121
Feller Property, 631
    multivalued, 639
Fenchel, W., 179n, 253, 274, 671
field, of sets, 129
Filippov, A. F., 602, 671
Filippov's Implicit Function Theorem, 603

filter, 32
    base, 34
    Cauchy, in tvs, 174
    finer, 33
    generated by a base, 34
    generated by a net, 35
    on $\mathbb{N}$, 543
    relation to net, 35
    section, 35
    subfilter, 33
    ultrafilter, 33
    $\mathcal{Z}$-, 65
finite dimensional space, 177–181, 361
    and weak topology, 237
    convex functions on, 271–275
finite dimensional subspace, 178, 220
    separating hyperplanes, 275–280
    supporting hyperplanes, 268–271
finite intersection property, 39
finite set, 10
finitely additive set function, 374
first category, 93
first countable space, 27, 42
fixed point, 4, 95
    of a contraction correspondence, 586
    of a correspondence, 581
    of a function, 581
    of a monotone function, 16, 17
fixed space, of an operator, 244
flat, 277
Florenzano, M., 254, 303, 671
Foreman, M., 14, 671
Frankowska, H., 556, 615, 668
Frantz, M., 151, 671
Fréchet derivative, 274
Fréchet lattice, 349–352, 528, 535
    order dual, 352, 528
Fréchet space, 205, 206, 589, 590
    not normable, 207
Fremlin, D. H., 404, 672
Fristedt, B., 135n, 672
Fryszkowski, A., 614, 672
Fryszkowski's L_1-Selection Theorem, 614
Fubini's Theorem, 418
function, 4–9
    affine, 256
    Bochner integrable, 426
    Borel measurable, 139
    bounded sublinear, 610
    C^∞, 466
    Carathéodory, 153, 595, 609, 617
    closed, 41
    concave, see concave function
    continuous, 36

convex, *see* convex function
countably additive, 374
differentiable, 273
distance, 80
explicitly quasiconcave, 300
finitely additive, 374
Gâteaux differentiable, 267, 268
gauge, 191
Gelfand integrable, 429
graph of, 4
homogeneous, 330
increasing, 8
indicator, 5, 54
integrable, 407
    Dunford, 432
    Gelfand, 429
    Lebesgue, 410
    over a set, 412
    Pettis, 432
    Riemann, 420
lattice, 334
left continuous, 393
linear, 166
measurable, 139, 384, 408, 416
Minkowski, 191
monotone, 8, 330
nondecreasing, 8
of bounded variation, 366
open, 41
p-integrable, 462
positively homogeneous, 190
quasiconvex, quasiconcave, 299
right continuous, 393
semicontinuous, 43
set-valued, *see* correspondence
simple, 144
space of continuous real functions, 49
space of functions with compact support, 49
space of real functions, 49
step, 404
strongly measurable, 424
subadditive, 190, 330
sublinear, 190, 330
uniformly continuous, 76, 175
upper, 408
weakly measurable, 431, 432
weak* measurable, 428
with compact support, 49
function space, 9
functional, 166, *see also* function
bilinear, 211
dominated by a seminorm, 209
invariant, 624
limit, 539n

positive on an ideal, 341
singular, 329
support, 288, 293
tangent, 258n
Fundamental Theorem of Calculus, 422
Futia, C. A., 656, 672

G

$\mathcal{G}_\delta$-set, 26, 81, 88, 92, 107, 139, 295
Gale, D., 179n, 253, 303, 672
Gâteaux derivative, 267, 268
gauge, 191
Geanakoplos, J., 622, 638, 641, 671
Gelbaum, B. R., 151, 672
Gelfand, I. M., 429, 672
Gelfand integral, 429, 628
Getoor, R. K., 135n, 669
Giles, J. R., 253, 672
Gilles, C., 311, 672
Gillman, L., 50, 65, 672
Glicksberg, I. L., 583, 671, 672
Glicksberg Fixed Point Theorem, 583
Granas, A., 581, 671, 672
Grandmont, J.-M., 672
graph
    Closed Graph Theorem, 51
    of a correspondence, 4, 557
    of a function, 4, 51, 177
Graves, L. M., 590, 668
Gray, L., 135n, 672
greatest element, 8
Green, J. R., 676
Grothendieck, A., 240, 361, 672
Grothendieck's Theorem, 240

H

Hahn's Theorem, 142
Hahn–Banach Extension Theorem, 195
    Riesz space version, 330
half space, 197
Halkin, H., 672
Halmos, P. R., xxii, 10, 13, 14, 137n, 158n, 160, 372, 434, 439, 476, 672
Halpern, B. R., 581, 672
Halpern, J. D., 14, 672
Hamel basis, 15
Hartman, P., 580, 672
Hartman–Stampacchia Theorem, 580
Hausdorff, F., 114n, 115, 373n, 672

Hausdorff metric, 110
  and closed convex sets, 293
  and correspondences, 563
  on convex sets, 293
  semimetric, 110
  topology, 113
Hausdorff topological space, 27
Heine–Borel Theorem, 87
hemicompact space, 58
hemicontinuity (upper and lower), 558, see
  correspondence
Henstock, R., 422, 673
Herstein, I. N., 132n, 673
Hessian matrix, 274
Hewitt, E., 361n, 399, 400, 680
Hilbert cube, 90
Hilbert space, 203, 247
Hildenbrand, W., 114n, 371, 556, 672, 673
Hill, T., 673
Himmelberg, C. J., 562, 602, 614, 673
Hiriart-Urruty, J.-B., 253, 275, 673
Hoffman, A. J., 671
Hölder's Inequality, 463, 534
Holmes, R. B., 163, 673
homeomorphism, 38
  lattice, 355
  linear, 166, 354
  sufficient condition for, 41
homomorphism, algebraic, 501
Hörmander, L., 673
Horváth, J., 66n, 163, 175, 673
Hoskins, R. F., 2n, 673
Howard, P., 14, 673
Howard, R., 275, 673
Hu, S., 556, 673
hull
  closed convex, 183
  closed convex circled, 183
  convex, 182
  convex circled, 183
  solid, 334
hull-kernel topology, 64
Hurd, A. E., 2n, 673
Husain, T., 175, 673
hyperplane, 197
  supporting, 258
  vertical, 256
hypograph, 9, 52, 187, 254

I

J, irrationals in (0, 1), 106
$\| \cdot \|_\infty$, ∞-norm, 462, 527
$I(x)$, initial segment generated by x, 18

Ichiishi, T., 558n, 673
ideal, 18, 321
  generated by a set, 322
  positive functional on, 341
  principal, 322
  vs. Riesz subspace, 321
image, 4
  under a correspondence, 556
Implicit Function Theorem, 603
increasing function, 8
indicator function, 5, 54
  convex, 255
indiscrete topology, 23
inequality
  Cauchy-Schwarz, 246
  Hölder's, 463, 534
  Jensen's, 417
  linear, 300
  linear system of, 303
  Minkowski's, 464
  triangle, 23, 227
infimum, 8, 312, 315
infinity, defined, 10
initial topology, 48
injection, 5
inner product space, 246
inner regular charge, measure, 435
integrable correspondence, 614
integrable function, 407
  over a set, 412
integral, 404, 407
  change of variables, 484
  Dunford, 431
  extended, 416
  finitely additive, 407
  Gelfand, 429, 628
  iterated, 418
  Lebesgue, 409
  Lebesgue–Stieltjes, 416
  lower, 406
  of correspondence, 614, 619
  of measurable function, 416
  of step function, 404
  Pettis, 431
  Riemann, 420
  σ-order continuous functional, 328
  upper (wrt a charge), 406
  weak*, 429
interior, 26
  relative, of a convex set, 278
Interlacing Lemma, 20
internal point, 199
Intr($A - a$), interior relative to an affine
  subspace, 278

invariant function, functional, 624, 633
invariant measure, 483n, 551, 631, 653, 656
invariant set, 633, 653
inverse
  of a correspondence, 557
  right, 590
inverse correspondence, 557, 578
inward pointing correspondence, 581
Ionescu Tulcea, A., 620, 673
Ionescu Tulcea, C., 620, 673
isolated point, 28
isometry, 76
  embedding, 84
  lattice, 355
  linear, 355
isomorphism, 317
  lattice, 317, 354
  Riesz, 317
isotone (= monotone), 8
iterated integral, 418
iterates of a function, 656

J

Jacobian matrix, determinant, 485
Jacobs, K., 404, 602, 673
Jacobs' Selection Theorem, 602
Jain, N., 22
James, R. C., 241, 673
James' Theorem, 241
Jameson, G. J. O., 281, 673
Jankov, V. A., 607, 673
Jankov–von Neumann Selection Theorem, 607
Jarchow, H., 163, 673
Jayne, J. E., 295, 556, 674
Jech, T., 12–14, 373n, 674
Jennrich, R. I., 605, 674
Jensen's Inequality, 417
Jerison, M., 50, 65, 672
joint continuity, 51
joint measurability, 151, 153
Jones, L. E., 311, 371, 674

K

$\mathcal{K}$, space of nonempty compact sets, 113
  Borel σ-algebra of, 597
Kakutani, S., 359, 360, 583, 674
Kakutani Fixed Point Theorem, 583
Kakutani–Fan–Glicksberg Fixed Point Theorem, 583
Kamae, T., 620, 674
Kantorovich, L. V., 331, 373n, 674
Kantorovich–Banach space, 361
Kaplan, S., 338

Karlin, S., 583, 669
Katzner, D. W., 253, 674
KB-space, 361
Kechris, A. S., 434, 674
Kelley, J. L., 14, 22, 45n, 163, 674
kernel, 212
  stochastic, 630
kernel of a linear topology, 171
Khan, M. A., 605, 674
Kifer, Y., 645, 656, 674
Kim, T., 614, 674
KKM correspondence, 577
  KKM Lemma, 578
Klee, V., 252, 260, 268, 277, 674, (aka V. L. Klee, Jr.)
Klein, E., 556, 558n, 563, 591, 615, 674
Kluvánek, I., 556, 615, 675
Knaster, B., 1, 16, 577, 675
Knaster–Kuratowski–Mazurkiewicz (KKM) Lemma, 577
Knaster–Tarski Fixed Point Theorem, 16
Knowles, G., 556, 615, 675
Kolmogorov, A. N., 520, 675
Kolmogorov Extension Theorem, 521–523
Köthe, G., 215n, 675
Krein, M. G., 241, 297, 359, 552n, 675
Krein, S., 359
Krein–Milman Theorem, 297
Krein–Šmulian Theorem, 241
Krengel, U., 620, 655, 661, 674, 675
Kronecker, L., 1
Kuratowski, K., 22, 45n, 69, 137n, 139, 373, 434, 577, 597, 599, 600, 668, 675
Kuratowski–Ryll-Nardzewski Selection Theorem, 600

L

$L_0(\mu)$, μ-measurable functions, 462
  as a Riesz space, 337
ℓ_1, absolutely summable sequences, 329, 537–542
ℓ_∞, bounded sequences, 329, 537–546
  ℓ_∞-sum of Banach spaces, 553
  Mackey topology on, 537
$L_\infty(\mu)$, essentially bounded measurable functions, 462
$L(\mu)$, μ-step functions, 408
$L_p(\mu)$, p integrable functions wrt μ, 330, 462
  L_p-norm, 462
ℓ_p, p-summable sequences, 527, 533–546
  ℓ_p-norm, 527, 533
  ℓ_p-sum of Banach spaces, 553
$L(X) = L(X, X)$, 661

$L(X, Y)$, bounded linear operators from X into Y, 230
Li, topological lim inf, 114
Ls, topological lim sup, 114
L-norm, L-space, 357
λ-system, 135
Landau, E., 1, 675
LaSalle, J. P., 675
Lasota, A., 655, 675
lattice, 8, 312
    complete, 8
    embedding, 546
    Fréchet, 349
    function, 334
    homeomorphism, 355
    homomorphism, 354
    isometry, 355
        between σ-algebras, 469
    isomorphism, 317, 354
    operations, 312, 334
    subspace, 365
    vector, 313
lattice norm, 348, 530
lattice seminorm, 336
Lay, D. C., 163, 663n, 679
Le Cam, L. M., 675
Le Van, C., 254, 303, 671
least element, 8
Lebesgue, H., 404, 410n, 675
Lebesgue Decomposition Theorem, 325, 401
Lebesgue Dominated Convergence Theorem, 415
Lebesgue integral, 409, 410
Lebesgue measurable set, 390
Lebesgue measure, 390, 394
    n-dimensional, 391
Lebesgue number, 85
Lebesgue–Stieltjes integral, 416
left continuity, 393
Leibniz, G., 403
Lemaréchal, C., 253, 275, 673
Levi's Theorem, 413
Levinson, N., 675
lexicographic order, 11, 103
lhc (= lower hemicontinuous), 558
Liapounoff, see Lyapunov
Lim, 539n
lim inf
    of a net, 32
    of a sequence of sets, 114
lim sup
    of a net, 32
    of a sequence of sets, 114

limit, 30
    Banach–Mazur, 550
    inferior, 32, 323
    of sets, 114
    point, 27
    superior, 32, 323
limit functional, 539n
limit point, 31, 34
Lindelöf space, 46
Lindenstrauss, J., 476, 675
line, 179
line segment, 168
linear embedding, 546
linear function, 166
linear functional, 166, 175, 195
    Carathéodory, 609
    exposing a point, 305
    order bounded, 325
    order continuous, 328
    positive, 325
    separating, 198
    σ-order continuous, 328
    strictly positive, 325
    strongly exposing a point, 306
    supporting, 258
linear homeomorphism, 166, 354
linear isometry, 355
linear operator, 166
    continuity of, 243
    space of, 230
linear order, 6, 7
linear space (subspace), 166
linear topological space, 166
Lipschitz continuity, 76
locally compact space, 55
    one-point compactification of
        metrizability, 92
    separable metrizable, 93
locally connected topological space, 645
locally convex space, 169, 205
    normable, 206
    semi-reflexive, 224
locally convex-solid space, 336
locally finite collection of sets, 65
locally solid Riesz space, 334
Loeb, P. A., 2n, 438, 673, 675
Lomonosov, V. I., 284, 675
lower hemicontinuous correspondence, 558
lower integral, 406
lower inverse, 557
lower Riemann integral, 420
lower section, 557
lower semicontinuous function, 43
lower sum, 419

Lucas, Jr., R. E., 69, 679
Lusin, N., 434, 675
Lusin scheme, 106
Lusin's Separation Theorem, 448
Lusin's Theorem, 438
Luxemburg, W. A. J., 2n, 14, 64n, 311, 339, 372,
    667, 675, 679
Lyapunov, A. A., 475, 676
Lyapunov Convexity Theorem, 476

M

M-norm, M-space, 357
Mackey, G. W., 221, 676
Mackey, M. C., 655, 675
Mackey topology, 223
    vs. norm, 236
Mackey–Arens Theorem, 222
MacLane, S., 132n, 676
majorize
    a subspace, 330
Mañé, R., 655, 676
mapping, 4
Markov kernel, 630
Markov operator, 502, 623
Markov transition, 628
Markov transition correspondence, 638
Mas-Colell, A., 70, 311, 320, 371, 562, 621,
    622, 638, 641, 671, 676
matrix
    Hessian, 274
    Jacobian, 485
    Markov, 623
    positive semidefinite, 275n
maximal element, 8
    existence of, 44, 580
maximum
    existence, 40
Maximum Theorem, 569, 570
    measurable, 605
Mazurkiewicz, S., 577, 675
McLennan, A., 622, 638, 641, 671
meager set (= first category), 93
mean ergodic operator, 662
measurable cardinal, 372
measurable correspondence, 592
measurable function, 139, 384, 408, 416
    jointly, 153
    jointly measurable, 151
    separately measurable, 151
    weak*, 609
Measurable Maximum Theorem, 605
measurable rectangle, 148
Measurable Selection Theorem, 600, 613

measurable selector, 600
    nonexistence, 454
measurable set, 380, 384
measurable space, 139
measurable transformation, 483
measure, 374
    absolute continuity of, 401
    and nondecreasing function, 393
    atomless, 395, 476
    Baire, 434
    Borel, 393, 434
    bounded variation, 396
    complete, completion, 387
    continuity of, 376
    counting, 375, 534
    Dirac, 443n
    finite, 376, 382
        vs. totally finite, 382n
    induced by a transformation, 483
    inner regular, 435
    invariant, 483n, 551
    Lebesgue, 390, 394
    n-dimensional Lebesgue, 391
    nonatomic, 395, 476
    and nondecreasing function, 393
    normal, 435
    outer regular, 434
    probability, 387, 505
    product, 392
    purely atomic, 395
    Radon, 496
    regular Borel, 435
    separable, 469
    set function, 374
    σ-finite, 382
    signed, 374
    support of, 441
        as correspondence, 563
    tight, 435
    total variation of, 396
    totally finite, 376
measure space, 387
measure-preserving transformation, 653, 656
    strongly, weakly mixing, 660
meet (a set), 3
Mehta, G., 676
mesh (of a partition), 419
metric, 23, 70, *see also* metric space
    bounded, 71
    consistent with a topology, 71
    discrete, 24
    equivalent, 71
    Euclidean, 24
    Hausdorff, 110, 563

Lipschitz continuity of, 76
  topology induced by, 23
  translation invariant, 172
  tree, on $\mathfrak{N}$, 103
  ultrametric, 102
  uniform, 74
metric projection, 248
metric space, 23, 70, *see also* metric
  compactness of, 86
  complete, 73, 74
  completion of, 84
  is paracompact, 81
  Polish, 74
  separable = second countable, 73
  sequentially compact, 85
  total boundedness vs. completeness, 87
  totally bounded, 85
metric uniformity, 109
metrizability
  of compact spaces, 353
  of locally convex spaces, 206
  of regular spaces, 91
  of topological vector spaces, 172
  Urysohn Metrization Theorem, 91
  w^*, 239
  weak, 240
metrizable space, 24, 71
Meyer, P.-A., 620, 645, 647, 670, 676
Meyer-Nieberg, P., 311, 676
Michael, E., 587, 589, 590, 676
Michael Selection Theorem, 589
Miller, A. W., 106, 676
Milman, D., 297, 675
minimal element, 8
minimal extension of a positive functional, 342
Minkowski functional, 191
Minkowski's Inequality, 464
modulus of contraction, 95, 585
monotone class, 136
Monotone Class Lemma, 137
Monotone Convergence Theorem, 413, 647
monotone function, 8, 16, 330
Moore, G. H., 676
Moore, J. C., 556, 558n, 676
Mordukhovich, B. S., 676
Müller, A., 676
multifunction, *see* correspondence
multiplicative operator, 501, 651
Munkres, J. R., 22, 676
Müller, A., 646

N

$\mathbb{N}$, natural numbers, 1

$\mathfrak{N}$, $= \mathbb{N}^{\mathbb{N}}$, the Baire space, 101
$\mathbb{N}^{<\mathbb{N}}$, the finite sequences in $\mathbb{N}$, 101
Nachbin, L., 331, 676
Nachman, D. C., 622, 638, 676
Nadler, S. B., 586, 676
Nadler Fixed Point Theorem, 586
Nakano, H., 344
Nakano's Theorem, 344
Namioka, I., 163, 674
natural numbers, 1
neighborhood, 26
  of a set, 80, 110, 558
  weak, 304
neighborhood base, 27
  at zero of a tvs, 169
  at zero of locally convex space, 205
net, 30
  Cauchy, in tvs, 174
  convergence of a net of integrals, 415
  generated by a filter, 35
  relation to filter, 35
  section of, 35
  tail of, 35
Neustadt, L. W., 676
Neveu, J., 378, 434, 521, 522, 676
Newton, I., 403
Neyman, J., 675
Nikaidô, H., 254, 302, 676
nonatomic measure, 476
nondecreasing function, 8
nonexpansive operator, 364n
nonlinear operator, 176
nonnegative cone, 313
norm, 167, 191, *see also* seminorm
  dual, 230
  essential sup, 462
  Euclidean, 177
  induced by an inner product, 246
  $\| \cdot \|_\infty$, 462, 527
  is uniformly continuous, 228
  L_p, 462
  ℓ_p, 527
  L-, 357
  lattice, 336, 348, 530
  M-, 357
  of a sublinear function, 609
  of operator, 229
  order continuous, 355
  σ-order continuous, 355
  strictly monotone, 364
  sup, 529
norm bounded set, 228
norm dual, 230
norm function of a vector-valued function, 423

norm topology, 167, 227
normable locally convex space, 206
normal charge, measure, 435
normal integral, 328
normal topological space, 45
normed space, 227, *see also* Banach space
  examples of, 228
  normed Riesz space, 348
  examples, 348
nowhere dense set, 28, 93
nucleus of a Suslin scheme, 106
null sequence, 529
null set, 380

O

O'Brien, G. L., 620, 674
Ogasawara, T., 329
Olds, C. D., 106, 676
Olech, C., 676
Olmsted, J. M. H., 151, 672
Ω, space of ordinals, 19
  is compact, 41
  is not metrizable, 41
  is not separable, 41
  measure with no support on, 442
  non-regular Borel measure on, 439
  topology on, 41
Ω_0, space of countable ordinals, 19
  Baire sets of, 497
  compactification of, 62
  continuous functions on, 61
  is not σ-compact, 440
ω_0, first infinite ordinal, 19
ω_1, first uncountable ordinal, 19
one-point compactification, 57
  and extension of continuous functions, 58
  metrizability of, 92
  of Ω_0, 62
  of $\mathbb{R}$, 57
one-to-one correspondence, 5
open cone, 268
open cover, 38
open function, 41
open mapping, 176, 560
Open Mapping Theorem, 176
open set, 23
operator, 166
  bounded, 229
  Cesàro bounded, 664
  composition, 502, 651
  contraction, 364
  ergodic, 662
  idempotent, 245

linear, 175, 176
  continuity of, 243
  Markov, 502, 623
  multiplicative, 501, 651
  nonexpansive, 364n
  order adjoint, 333
  order bounded, 333
  positive, 332
  power bounded, 664
  projection, 245
  stochastic, 623
  transition, 626, 628
  unbounded, 229
operator norm, 229
orbit, 657
order
  related terms, 6
  lexicographic, 11, 103
  partial, 312
  pointwise, 9, 314, 316, 366
  related terms, 7–8
  well, 18
order adjoint of positive operator, 333
order bounded functional, 325
order bounded operator, 333
order bounded set, 315
order closed set, 324
order complete space, 316
order continuous component, 329
order continuous dual, 329
order continuous function, 322, 328
order continuous norm, 355
order continuous operator, 332
order continuous topology, 344
order convergence, 322
order dual, 327
  double, 331
order interval, 8, 315
order topology, 41
order unit, 322, 358, 529
  vs. weak unit, 325
ordered vector space, 313, *see also* Riesz space
ordering, *see* order
ordinal, *see also* Ω, Ω_0
ordinals, 19–20
Ornstein, D. S., 597, 653, 656, 677
orthogonal complement, 250
orthogonal projection, 248
orthogonal vectors, in a Riesz space, 320
outer measure, 379
  generated by a measure, 383
outer regular charge, measure, 434
outward pointing correspondence, 581

P

$\| \cdot \|_p$, p-norm, 461
p-integrable function, 462
p.p. (= a.e.), 387
Page, Jr., F. H., 70, 677
Papageorgiou, N. S., 556, 673
paracompact space, 65
   metric spaces are paracompact, 81
parallelogram law, 247
Park, S., 578, 677
Parry, W., 655, 677
Parthasarathy, K. R., 434, 435n, 446, 677
partial function, 4
partial order, 7, 312
partition, 7, 365, 396, 419
partition of unity, 66
Pečarić, J. E., 677
Peressinni, A. L., 677
perfect set, 28
perfectly normal space, 45, 81
Petersen, K. E., 655, 677
Pettis, B. J., 432n, 677
Pettis integral, 431
Pfanzagl, J., 378, 677
Phelps, R. R., 163, 164, 252, 253, 281, 287, 647n, 668, 677
φ, space of sequences with finite support, 527–529
   continuity on boxes of, 340
Phillips, R. S., 542, 677
Phillips' Lemma, 542
π-system, 135
π-λ Lemma, 136
Pierlo, W., 378, 677
Poincaré's Recurrence Theorem, 657
point
   accumulation, 27
   at infinity, 56
   boundary, 27
   of closure, 27
   cluster, 27
   exposed, 305
   extreme, 294
   fixed, 4
   internal, 199
   isolated, 28
   limit, 27
   of support, 258
point mass, 443, 512
pointed convex cone, 312
pointwise bounded set, 232
pointwise convergence, 53, 212

pointwise operations on functions, 9
   on correspondences, 566–569
pointwise ordering, 9, 314, 316, 366
polar
   absolute, 215, 215n
   one-sided, 215
Polish space, 74
Pollard, D., 434, 677
polyhedron, 197, 301
Polyrakis, I. A., 365, 667, 677
polytope, 184, 301
poset, 7
positive basis, 530
positive cone, vector, 313
   strictly positive, 341
positive linear functional, 325
   and countable sup property, 326
   on an ideal, 341
positive operator, 332
   is continuous, 350
   order continuous, 332
   σ-order continuous, 332
positive semidefinite matrix, 275n
positively homogeneous function, 190, 330
power bounded operator, 664
Prescott, E. C., 69, 679
Prikry, K., 478, 614, 667, 674
Prime Ideal Theorem, 14
probability measure, 387, 505
   space of, 562
probability space, 387
Problem of Measure, 372
product measure, 392
product semiring, 133, 148
   vs. product σ-algebra, 148
product σ-algebra, 148
   infinite, 520
product topology, 50
projection, 50, 100n, 245
   lattice properties of range, 364
   metric, in a Hilbert space, 248
   orthogonal, 248
   strictly positive, 364
projection band, 325
Projection Theorem, 608
proper convex (concave) function, 254
properness conditions, 269
Proschan, F., 677
purely finitely additive charge, 400

Q

$\mathbb{Q}$, rational numbers, 2
quasi-interior point, 341

quasiconvex (quasiconcave) function, 299
  explicitly, 300
quotient space, mapping, 7

R

$\mathbb{R}$, real numbers, 2
$\mathbb{R}^*$, extended real numbers, 2
  topology of, 57
$\mathbb{R}^N$, real sequences, 260, 526–529
  is not normable, 207
$\mathbb{R}^X$, real functions on X, 49
  as a Riesz space, 337
radial set, 168
Radon measure, 496
Radon–Nikodym derivative, 470
Radon–Nikodym Theorem, 470, 471
range, 4
  of a correspondence, 557
ray, 179
  extreme, 294
real function, 8
  space of all, 49
real numbers, extended, $\mathbb{R}^*$
  topology of, 57
rectangle, 148
reflexive Banach space, 232
regular Borel charge, measure, 435
regular topological space, 45
relation, 4
  continuous, 44
  kinds of, 6–8
  semicontinuous, 44
relative topology, 25
relatively closed set, 25
relatively compact set, 38
relatively open set, 25
retract, 100
ri(A), relative interior of a convex set A, 278
Richter, M. K., 15, 677
Riemann, G. F. B., 403
Riemann integral, 420
Riemann sum, 420
Riesz, F., 325, 327, 461, 473
Riesz Decomposition Property, 319
Riesz homomorphism, 354
Riesz isomorphism, 317, 354
Riesz pair, 340
  symmetric, 342
Riesz Representation Theorem, 491, 495–499
Riesz seminorm, 336
Riesz space, 313
  but not a Riesz subspace, 364
  locally convex-solid, 336

locally solid, 334
  normed, 348
Riesz subspace, vs. ideal, 321
Riesz–Fischer Theorem, 464
Riesz–Markov Theorem, 496
right continuity, 393
ring, of sets, 131
  σ-ring, 131
Roberts, A. W., 189, 253, 677
Roberts, J. W., 298, 677
Robertson, A. P., 163, 677
Robertson, W. J., 163, 677
Rockafellar, R. T., 179n, 252, 253, 265, 271,
  274, 287, 669, 677
Rogers, C. A., 295, 556, 674
Rothschild, M., 646, 677
Royden, H. L., 26n, 158n, 372, 434, 497, 677
Rubin, J. E., 14, 673
Rudin, W., xxii, 372, 434, 439, 677
Russell, B., 12
Russell's Paradox, 12, 13
Rustichini, A., 605, 674
Ryll-Nardzewski, C., 600, 675

S

$\mathfrak{S}$-topology, 221
saturated family of seminorms, 205
Savage, L. J., 605, 670
scalar measurability, 610
Schaefer, H. H., 163, 174, 215n, 311, 332, 364,
  678
Schäl, M., 678
Schauder, J., 583, 678
Schauder basis, 530
Schauder Fixed Point Theorem, 583
Schauder–Tychonoff Fixed Point Theorem, 583
Schmeidler, D., 678
Schröder–Bernstein Theorem, 10
Schur property, 238, 537
Schwartz, J. T., 163, 191n, 258n, 379, 428, 498,
  671
Scott, E. L., 675
second adjoint, 244
second category, 93
second countable space, 25
second dual, 231
  includes space, 231, 236
section
  of a correspondence, 557
  of a set, 150
section filter, 35
segment, 168

selector, from a correspondence, 454, 587
    Carathéodory, 614
    continuous, 587, 589, 614
    Gelfand integrable, 619
    measurable, 600, 602, 603, 607
    scalarly measurable, 613
Semadeni, Z., 498, 678
semi-reflexive locally convex space, 224
semicontinuity, 43, see also hemicontinuity
semimetric, 23, 70
    topology induced by, 23
seminorm, 191
    absolute, 336
    dominating a functional, 209
    lattice, 336
    monotone, 336
    Riesz, 336
semiring, 132
    product, 133, 148
separable measure, 469
separable space, 28
separately measurable function, 151
separating hyperplane, 198
    proper, 198
    strict, 198
    strong, 198
Separating Hyperplane Theorem
    and cones, 271
    finite dimensional, 276, 279
    interior, 202
    nontopological, 200
    points and closed sets, 208
    strict, 277
    strong, 207
separation, of points
    by a family of functions, 48
    by continuous functions, 44
    by open sets, 27, 44
    completely regular space, 45
    normal space, 45
    regular space, 45
    T_0-,...,T_4-spaces, 46
sequence, 4, 29
    Cauchy, 73
    finite, 101
    head, tail, 526
    null, 529
    strictly positive, 543
sequence space, 526
sequentially compact set, 39
    is compact in a metric space, 86
set, see also space
    basic terminology, 2–12
    absorbing, 168

analytic, 446, 599
analytically measurable, 606
balanced, 168
bounded, 71, 234
Cantor, 98
circled, 168
clopen, 23
closed, 23
coanalytic, 446
compact, 38
connected, 645
convex, 168, 181
countable, 10
dense, 28
derived, 27
ε-dense, 85
ergodic, 636
extreme, 294
$\mathcal{F}_\sigma$, 139
finite, 10
$\mathcal{G}_\delta$, 26, 139, 295
Lebesgue measurable, 390
measurable, 380
nowhere dense, 28, 93
null, 380
open, 23
order bounded, 315
order closed, 324
partially ordered, 312
perfect, 28
pointwise bounded, 232
radial, 168
relatively compact, 38, 240
sequentially compact, 39
Sierpiński, 151
solid, 321, 334
solution, 300
star-shaped, 168
symmetric, 168
topologically bounded, 186, 206, 214
totally bounded, 85
unbounded, 71
uncountable, 10
universal, 452
universally measurable, 456
with measurable sections, 151
$\mathcal{Z}$-, 65
set function
    additive, 374
    countably additive, 374
    finitely additive, 374
    monotone, 374
    σ-additive, 374
    σ-subadditive, 374

subadditive, 374
Shapley, L. S., 678
Shen, A., 13, 678
Sherstyuk, K., 678
Shreve, S. E., 106, 135n, 434, 668
Sierpiński, W., 135n, 151, 678
Sierpiński class, 135
Sierpiński set, 151
σ-algebra, 129
    Baire, 158
    Borel, 137
    induced by a function, 147
    product, 148
        infinite, 520
$|\sigma|(E, A)$, $|\sigma|(E', A)$ absolute weak, weak$*$
    topology, 338
$|\sigma|(E', A)$, absolute weak$*$ topology, 338
$\sigma(\mathcal{C})$, 130
σ-field (= σ-algebra), 129
Σ_μ, 380
σ-ring, 131
$\sigma(X, \mathcal{F})$, weak topology, 48
σ-compact space, 58
σ-order continuous dual, 329
σ-order continuous functional, 328
σ-order continuous norm, 355
σ-order continuous operator, 332
σ-order continuous topology, 344
signed charge, 314, 374
signed measure, 374
    absolute continuity of, 401
simple function, 144
    standard representation, 144
    vector-valued, 422
Sine, R. C., 678
Singer, I., 678
singular component, functional, 329
Sion, M., 678
Smart, D. R., 581, 678
Smithson, R. E., 18, 678
Šmulian, V. L., 241, 675, 678
solid hull, 334
solid set, 321, 334
Solovay, R. M., 14, 678
solution set, 300, 303
Sonnenschein, H. F., 579, 678
Souslin, *see* Suslin
space, *see also* Banach space, topological space
    AL-, 357
    AM-, 357
    Baire, 93, 101
    Banach, 228
    Borel, 516
    compact, 38
    completely regular, 45
    connected, 645
    extremally disconnected, 363
    function, 9
    Hilbert, 247
    inner product, 246
    KB-, 361
    L-, 357
    linear, 166
    locally connected, 645
    locally convex, 205
    M-, 357
    measurable, 139
    metric, 70
    metrizable, 71
    normal, 45
    normed, 227
    of convex sets, 292
    of ultrafilters, 64
    perfect, 28
    perfectly normal, 81
    Polish, 74
    reflexive, 232
    regular, 45
    Riesz, 313
    semimetric, 70
    sequentially compact, 39
    Tychonoff, 46
    uniform, 108
    vector, 166
span, 166
Spanier, E. H., 477, 671
spectral radius, spectrum, 663
Sperner, E., 577, 678
Sperner's Lemma, 577
Srivastava, S. M., 434, 679
Stampacchia, G., 580, 672
standard representation of simple function, 144
star-shaped set, 168
step function, 404, 408
    vector-valued, 423
Stiglitz, J. E., 646, 677
Stinchcombe, M. B., 434, 605, 679
stochastic dominance, 645, 646
stochastic kernel, 630
stochastic operator, 623
Stoer, J., 254, 303, 679
Stokey, N., 69, 679
Stone–Čech compactification, 59
    of $(0, 1]$, 62
    of Ω_0, 62
Stone–Weierstrass Theorem, 352
Stoyan, D., 646, 676
Strassen, V., 592, 602, 613, 615, 620, 679

Strassen's Sublinearity Theorem, 615
Straszewicz, S., 308, 679
strictly monotone norm, 364
strictly positive linear functional, 325
    and countable sup property, 326
strictly positive sequence, 543
strictly positive vector, 341
strong inverse (= upper inverse), 557
strong topology, 223
strongly ergodic operator, 662
strongly measurable vector function, 424
strongly mixing transformation, 660
Stroyan, K. D., 2n, 679
subadditive function, 190, 330
subbase, 25
subdifferential of a convex function, 264
    and directional derivatives, 268
    and Fréchet derivative, 274
    and Gâteaux derivative, 268
    is compact and convex, 265
    nonemptiness of, 265, 272
subfilter, 33
subgradient of a convex function, 264–268
    existence of, 265
    in one dimension, 272
subgraph, see hypograph
sublattice, 8
sublinear function, 190, 330
    bounded, 610
    Carathéodory, 609
subnet, 31
subsequence, 5
subspace
    affine, 277
    dense, 219
    lattice, 365
    linear, 166
    Riesz, 319
    vector, 166
sum
    direct, 325
    of Banach spaces, 552
    c_0, 553
    ℓ_∞, 553
    ℓ_p, 553
    of sets, 3
    of vector spaces, 552
sup norm, 529
supergradient, superdifferential of a concave
    function, 264
supp, support of a measure, 441
support
    of a function, 49
    of a measure, 441

support functional, 288, 293
    and Hausdorff metric, 293
    as defined by Dunford and Schwartz, 191
    of a correspondence, 576
support point, 258
supporting half space, hyperplane, linear
    functional, 258
Supporting Hyperplane Theorem, 259
    finite dimensional, 279
supremum, 8, 312, 315
surjection, 5
surrounding, in a uniform space, 109
Suslin $\mathcal{A}$ operation, 106
    and analytic sets, 447
Suslin scheme, 106
symmetric Riesz pair, 342
symmetric set, 168
Szpilrajn, E., 1, 15, 679
Szpilrajn Extension Theorem, 15

T

T_0-,...,T_4-spaces, 46
tail of a net, 35
Talvila, E., 438, 675
tangent functional, 258n
Tarafdar, E., 676, 679
Tarski, A., 1, 14, 16, 17, 373n, 668, 679
Tarski Fixed Point Theorem, 16, 17
τ_C, topology of closed convergence, 121
$|\tau|(E, E')$, absolute Mackey topology, 341
τ_h, Hausdorff metric topology, 113
τ_V, the Vietoris, or exponential, topology, 119
Taylor, A. E., 163, 663n, 679
Taylor's theorem, 603
Temam, R., 164, 254, 671
Theorem of the Alternative, 204, 208, 209, 212
Thompson, A. C., 556, 558n, 563, 591, 615, 674
Tietze Extension Theorem, 45
tight charge, measure, 435
tight family of measures, 518
Tonelli's Theorem, 419
Tong, Y. L., 677
Topkis, D. M., 8, 8n, 679
topological dual, 195
topological lim inf, lim sup, of a sequence of
    sets, 114
topological space, 23, see also topology, metric
    space
    Borel, 516
    compact, 38
    completely metrizable, 74
    completely regular, 45
    connected, 645

extremally disconnected, 63, 531n
first countable, 27, 42
Hausdorff, 27
hemicompact, 58
locally compact, 55
locally connected, 645
metrizable, 24
normal, 45
paracompact, 65
perfect, 28
perfectly normal, 45
Polish, 74
regular, 45
second countable, 25
separable, 28
separable vs. second countable, 28, 73
separated (= Hausdorff), 27
sequentially compact, 39, 85
σ-compact, 58
subspace of, 25
$T_0,\ldots,T_4$,etc, 46
Tychonoff, 46
topological vector space, 166, *see also* Banach
    space, locally convex space, normed
    space
complete, 174
topologically bounded set, 186, 206, 214
topology, 23, *see also* topological space
    $\mathfrak{S}$-, 221
    absolute Mackey, 341
    absolute weak, weak*, 338
    base for, 25
    comparison of -s, 24
    compatible (= consistent) with a dual pair,
        214
    Euclidean, 24, 177
    exponential, 119
    Fell, 121
    generated by a semimetric, 71
    generated by a uniformity, 109
    generated by seminorms, 205
    hull-kernel, 64
    indiscrete, 23
    initial, 48
    locally convex-solid, 336
    locally solid, 334
        not locally convex, 337
    Mackey, 223
    metric, 23, 71
    norm, 167, 227
    of closed convergence on sets, 121
    of convergence in distribution, 507
    of convergence in measure, 337
    of pointwise convergence, 53, 212

    of uniform convergence, 124, 221
    order, 41
    order continuous, 344
    product, 50
    relative, 25
    semimetric, 23
    σ-order continuous, 344
    strong, 223, 341
    subbase for, 25
    Vietoris, 119
    weak, 47–50, 53, 211
    weak*, 212
    Wijsman, 117
total family of functions, 48, 211
total order, 7
total variation
    norm, 368
    of a function, 366
    of a measure, 396
totally bounded metric, set, 85
Tourky, R., 252, 269, 667
transfinite induction, 82n
transition operator, 626, 628
transition, Markov, 628
    deterministic, 651
    measurability of, 627
transitive closure of a relation, 6
translation invariant metric, topology, 167, 172
tree metric on $\mathbb{N}$, 103
triangle inequality, 23, 70, 227
tribe, = σ-algebra, 129
trichotomy law, 6
trivial topology, 23
Tukey, J. W., 203, 679
Turnbull, T., 164, 254, 671
tvs (= topological vector space), 166
Tychonoff, A., 583, 679
Tychonoff Fixed Point Theorem, 583
Tychonoff Product Theorem, 52
Tychonoff space, 46

U

$U(f, x, \varepsilon)$, base for weak topology, 48
uhc (= upper hemicontinuous), 558
Uhl, Jr., J. J., 423, 427, 432, 478, 670, 679
ultrafilter, 33, *see also* filter
    on $\mathbb{N}$, 544
ultrafilter space, 64
Ultrafilter Theorem, 33
ultrametric, 102
unbounded operator, 229
unbounded set, 71
uncountable set, 10

Uniform Boundedness Principle, 232
uniform continuity, *see* uniformly continuous
     function
uniform convergence, 54, 124
uniform limit of continuous functions, 54
uniform metric, 74
uniform space, 108
uniformity, 108
   metric, 109
   metrizable, 109
   separating, 109
uniformizable topology, 109
uniformly continuous function, 76, 109, 175
   on compact metric space, 87
uniformly ergodic operator, 662
unit, 322, 358, 529
   order, 322, 358
   vs. weak unit, 325
   weak order, 325
unit ball, 228
   is w^*-compact, 235
   metrizability of, 239, 240
   weak compactness of, 237
   weak interior of, 238
unit coordinate vector, 526
universal set (or correspondence), 452
universally measurable set, 456
upper demicontinuous correspondence, 575
upper function, 408
upper hemicontinuous correspondence, 558
upper integral (wrt a charge), 406
upper inverse, 557
upper Riemann integral, 420
upper semicontinuous function, 43
upper sum, 419
Urysohn Metrization Theorem, 91
Urysohn's Lemma, 45
utility function, 11

V

Valadier, M., 215n, 292, 556, 577, 598, 611,
     613, 615, 670
van Vleck, F. S., 562, 673
Varberg, D. E., 189, 253, 677
variation
   of a function, 366
   of a measure, 396
vector lattice, 313
vector measure, range of, 476
vector space, *see also* Banach space, locally
     convex space, normed space, Riesz
     space, topological vector space
   ordered, 313
   subspace, 166

vector space (subspace), 166
Vereshchagin, N. K., 13, 678
Vickson, R. G., 253, 669
Vietoris topology, 119
Vind, K., 679
Vitali's non-measurable set, 390
Vitali–Hahn–Saks Theorem, 379
Vives, X., 18, 679
Vohra, R., 678
von Neumann, J., 607, 679

W

Walker, R. C., 65, 679
Walters, P., 655, 679
weak compactness
   in a Banach space, 240, 241
   sequential, 241
weak inverse (= lower inverse), 557
weak neighborhood, 304
weak topology, 47–50, 53, 211
   for measures, 507
   on a subspace, 49
   on finite dimensional space, 237
weak unit, 325
   vs. order unit, 325
weakly ergodic operator, 662
weakly measurable correspondence, 592
weakly-mixing transformation, 660
weak* integral, 429
weak* measurable function, 609
weak* topology, 212
Wehrung, F., 14, 671
Weierstrass' Theorem, 40
well ordered set, 18
Well Ordering Principle, 18
   implies the Axiom of Choice, 20
Whinston, M. D., 676
White, H., 434, 605, 679
Wickstead, A. W., 445, 667
Wijsman topology, 117
Wilansky, A., 46, 163, 679
Willard, S., 22, 66, 108, 109, 679
Witzgall, C., 254, 303, 679

X

X', topological dual of X, 195
X^*, algebraic dual of X, 195
X_∞, one-point compactification of X, 56

Y

Yannelis, N. C., 252, 269, 614, 615, 667, 674,
     679

Yosida, K., 361n, 399, 400, 680
Yosida–Hewitt decomposition, 361, 400

Z

$\mathbb{Z}$, integers, 2
$\mathcal{Z}$-filter, $\mathcal{Z}$-set, 65
Zaanen, A. C., 64n, 311, 332, 675, 680
Zermelo–Frankel (ZF) set theory, 13
zero-one charge, 544
Zhou, L., 680
Ziegler, G. M., 253, 254, 303, 680
Zorn, M., 15, 680
Zorn's Lemma, 15

This edition was typeset directly from a pdf file produced by the authors on a Macintosh PowerBook G4 using pdftex, written by Hán Thế Thánh, and based on Donald Knuth's typesetting program TEX. We started with Leslie Lamport's LATEX 2_ε Document Preparation System and the American Mathematical Society's amsmath macro package (based on work by Michael Spivak) as the basis for the formatting, which we then customized.

The main text is set in Times Roman and the mathematics is set using Young Ryu's tx fonts. Some of the mathematics uses Hermann Zapf's Euler fonts that were designed for the $\mathcal{AMS}$.

The figures were created by the authors, mostly via hand-written PostScript and a few were by produced using Adobe Illustrator. The PostScript was converted to pdf by the ghostscript program.